What our readers say about REA's Problem Solvers®...

- ***"...the best tools for learning..."***
- ***"...superb..."***
- ***"...taught me more than I imagined..."***

"Your *Problem Solver*® books are the best tools for learning I have ever encountered."

~Instructor, Batavia, Illinois

"I own a large library of your *Problem Solver*® books, which I find to be extremely useful. I use them for reference not only for my own homework and research, but I also tutor undergraduate students and your books help me to do a 'quick refresher' of concepts in certain topics."

~Student, Rochester, New York

"Thank you for the superb work you have done in publishing the *Problem Solvers*®. These books are the best review books on the market."

~Student, New Orleans, Louisiana

"I found your *Problem Solvers*® to be very helpful. I have nine of your books and intend to purchase more."

~Student, Gulfport, Mississippi

"I love your *Problem Solvers*®. The volumes I have, have already taught me more than I imagined."

~Instructor, Atlanta, Georgia

We couldn't have said it any better!

Research & Education Association
Making the world smarter

Visit us online at www.rea.com.
Your comments welcome at info@rea.com.

PROBLEM SOLVERS®

Calculus

Staff of Research & Education Association

Dr. H. Weisbecker, Chief Editor

Research & Education Association
61 Ethel Road West
Piscataway, New Jersey 08854
E-mail: info@rea.com

THE CALCULUS PROBLEM SOLVER®

Published 2010

Printed in the United States of America

Library of Congress Control Number 2006932329

ISBN-13: 978-0-87891-505-7
ISBN-10: 0-87891-505-2

L09

Let REA's Problem Solvers® work for you

REA's ***Problem Solvers*** are for anyone—from student to seasoned professional—who seeks a thorough, practical resource. Each ***Problem Solver*** offers hundreds of problems and clear step-by-step solutions not found in any other publication.

Perfect for self-paced study or teacher-directed instruction, from elementary to advanced academic levels, the ***Problem Solvers*** can guide you toward mastery of your subject.

Whether you are preparing for a test, are seeking to solve a specific problem, or simply need an authoritative resource that will pick up where your textbook left off, the ***Problem Solvers*** are your best and most trustworthy solution.

Since the problems and solutions in each ***Problem Solver*** increase in complexity and subject depth, these references are found on the shelves of anyone who requires help, wants to raise the academic bar, needs to verify findings, or seeks a challenge.

For many, ***Problem Solvers*** are homework helpers. For others, they're great research partners. What will ***Problem Solvers*** do for you?

- Save countless hours of frustration groping for answers
- Provide a broad range of material
- Offer problems in order of capability and difficulty
- Simplify otherwise complex concepts
- Allow for quick lookup of problem types in the index
- Be a valuable reference for as long as you are learning

Each ***Problem Solver*** book was created to be a reference for life, spanning a subject's entire breadth with solutions that will be invaluable as you climb the ladder of success in your education or career.

—Staff of Research & Education Association

How to Use This Book

The genius of the ***Problem Solvers*** lies in their simplicity. The problems and solutions are presented in a straightforward manner, the organization of the book is presented so that the subject coverage will easily line up with your coursework, and the writing is clear and instructive.

Each chapter opens with an explanation of principles, problem-solving techniques, and strategies to help you master entire groups of problems for each topic.

The chapters also present progressively more difficult material. Starting with the fundamentals, a chapter builds toward more advanced problems and solutions—just the way one learns a subject in the classroom. The range of problems takes into account critical nuances to help you master the material in a systematic fashion.

Inside, you will find varied methods of presenting problems, as well as different solution methods, all of which take you through a solution in a step-by-step, point-by-point manner.

There are no shortcuts in ***Problem Solvers***. You are given no-nonsense information that you can trust and grow with, presented in its simplest form for easy reading and quick comprehension.

As you can see on the facing page, the key features of this book are:

- Clearly labeled chapters
- Solutions presented in a way that will equip you to distinguish key problem types and solve them on your own more efficiently
- Problems numbered and conveniently indexed by the problem number, not by page number

Get smarter....Let ***Problem Solvers*** go to your head!

Anatomy of a Problem Solver®

Chapter Subject

CHAPTER 4

CONTINUITY

Problem Topic

A function, f(x) is continuous at a point, a, if it satisfies three separate conditions: (1) f(a) is defined; (2) $\lim_{x \to a} f(x)$ exists; and (3) $\lim_{x \to a} f(x) = f(a)$. If any one or more of these conditions fail, the function is said to be discontinuous.

Problems generally encountered in elementary calculus are found to have continuous functions.

Problem Number

• PROBLEM 50

Determine the domain of the function:

$$y = \frac{2x}{(x - 2)(x + 1)} .$$

The Problem

Solution:

$$y = \frac{2x}{(x - 2)(x + 1)}$$

The Full Solution to the Problem

...d only if the denominator is not equal to

...e denominator is equal to zero when ...2)(x + 1) = 0, or, equivalently, when (x − 2) =0 o... (x + 1) = 0. (x − 2) = 0 when x = 2, and (x ... = 0 when x = − 1.

...herefore, the domain of the function is all values of x where x ≠ − 1 and x ≠ 2.

• PROBLEM 51

Determine the domain of the function:

$$y = \sqrt{\frac{x}{2 - x}} .$$

Solution: Since we are restricted to the real number system, there are two cases to be considered:

Case 1. When the denominator equals zero, the fraction is undefined. This occurs when x = 2.

Case 2. Since we are restricted to real values of y, it is necessary that

27

CONTENTS

UNITS CONVERSION FACTORS

This section includes a particularly useful and comprehensive table to aid students and teachers in converting between systems of units.

The problems and their solutions in this book use SI (International System) as well as English units. Both of these units are in extensive use throughout the world, and therefore students should develop a good facility to work with both sets of units until a single standard of units has been found acceptable internationally.

In working out or solving a problem in one system of units or the other, essentially only the numbers change. Also, the conversion from one unit system to another is easily achieved through the use of conversion factors that are given in the subsequent table. Accordingly, the units are one of the least important aspects of a problem. For these reasons, a student should not be concerned mainly with which units are used in any particular problem. Instead, a student should obtain from that problem and its solution an understanding of the underlying principles and solution techniques that are illustrated there.

To convert	To	Multiply by	For the reverse, multiply by
acres	square feet	4.356×10^4	2.296×10^{-5}
acres	square meters	4047	2.471×10^{-4}
ampere-hours	coulombs	3600	2.778×10^{-4}
ampere-turns	gilberts	1.257	0.7958
ampere-turns per cm.	ampere-turns per inch	2.54	0.3937
angstrom units	inches	3.937×10^{-9}	2.54×10^8
angstrom units	meters	10^{-10}	10^{10}
atmospheres	feet of water	33.90	0.02950
atmospheres	inch of mercury at 0°C	29.92	3.342×10^{-2}
atmospheres	kilogram per square meter	1.033×10^4	9.678×10^{-5}
atmospheres	millimeter of mercury at 0°C	760	1.316×10^{-3}
atmospheres	pascals	1.0133×10^5	0.9869×10^{-5}
atmospheres	pounds per square inch	14.70	0.06804
bars	atmospheres	9.870×10^{-7}	1.0133
bars	dynes per square cm.	10^6	10^{-6}
bars	pascals	10^5	10^{-5}
bars	pounds per square inch	14.504	6.8947×10^{-2}
Btu	ergs	1.0548×10^{10}	9.486×10^{-11}
Btu	foot-pounds	778.3	1.285×10^{-3}
Btu	joules	1054.8	9.480×10^{-4}
Btu	kilogram-calories	0.252	3.969
calories, gram	Btu	3.968×10^{-3}	252
calories, gram	foot-pounds	3.087	0.324
calories, gram	joules	4.185	0.2389
Celsius	Fahrenheit	$(°C \times 9/5) + 32 = °F$	$(°F - 32) \times 5/9 = °C$

To convert	To	Multiply	For the reverse, multiply by
Celsius	kelvin	°C + 273.1 = K	K – 273.1 = °C
centimeters	angstrom units	1×10^{8}	1×10^{-8}
centimeters	feet	0.03281	30.479
centistokes	square meters per second	1×10^{-6}	1×10^{6}
circular mils	square centimeters	5.067×10^{-6}	1.973×10^{5}
circular mils	square mils	0.7854	1.273
cubic feet	gallons (liquid U.S.)	7.481	0.1337
cubic feet	liters	28.32	3.531×10^{-2}
cubic inches	cubic centimeters	16.39	6.102×10^{-2}
cubic inches	cubic feet	5.787×10^{-4}	1728
cubic inches	cubic meters	1.639×10^{-5}	6.102×10^{4}
cubic inches	gallons (liquid U.S.)	4.329×10^{-3}	231
cubic meters	cubic feet	35.31	2.832×10^{-2}
cubic meters	cubic yards	1.308	0.7646
curies	coulombs per minute	1.1×10^{12}	0.91×10^{-12}
cycles per second	hertz	1	1
degrees (angle)	mils	17.45	5.73×10^{-2}
degrees (angle)	radians	1.745×10^{-2}	57.3
dynes	pounds	2.248×10^{-6}	4.448×10^{5}
electron volts	joules	1.602×10^{-19}	0.624×10^{18}
ergs	foot-pounds	7.376×10^{-8}	1.356×10^{7}
ergs	joules	10^{-7}	10^{7}
ergs per second	watts	10^{-7}	10^{7}
ergs per square cm.	watts per square cm.	10^{-3}	10^{3}
Fahrenheit	kelvin	(°F + 459.67)/1.8	1.8K – 459.67
Fahrenheit	Rankine	°F + 459.67 = °R	°R – 459.67 = °F
faradays	ampere-hours	26.8	3.731×10^{-2}
feet	centimeters	30.48	3.281×10^{-2}
feet	meters	0.3048	3.281
feet	mils	1.2×10^{4}	8.333×10^{-5}
fermis	meters	10^{-15}	10^{15}
foot candles	lux	10.764	0.0929
foot lamberts	candelas per square meter	3.4263	0.2918
foot-pounds	gram-centimeters	1.383×10^{4}	1.235×10^{-5}
foot-pounds	horsepower-hours	5.05×10^{-7}	1.98×10^{6}
foot-pounds	kilogram-meters	0.1383	7.233
foot-pounds	kilowatt-hours	3.766×10^{-7}	2.655×10^{6}
foot-pounds	ounce-inches	192	5.208×10^{-3}
gallons (liquid U.S.)	cubic meters	3.785×10^{-3}	264.2
gallons (liquid U.S.)	gallons (liquid British Imperial)	0.8327	1.201
gammas	teslas	10^{-9}	10^{9}
gausses	lines per square cm.	1.0	1.0
gausses	lines per square inch	6.452	0.155
gausses	teslas	10^{-4}	10^{4}
gausses	webers per square inch	6.452×10^{-8}	1.55×10^{7}
gilberts	amperes	0.7958	1.257
grads	radians	1.571×10^{-2}	63.65
grains	grams	0.06480	15.432
grains	pounds	$^{1}/_{7000}$	7000
grams	dynes	980.7	1.02×10^{-3}
grams	grains	15.43	6.481×10^{-2}

To convert	To	Multiply	For the reverse, multiply by
grams	ounces (avdp)	3.527×10^{-2}	28.35
grams	poundals	7.093×10^{-2}	14.1
hectares	acres	2.471	0.4047
horsepower	Btu per minute	42.418	2.357×10^{-2}
horsepower	foot-pounds per minute	3.3×10^{4}	3.03×10^{-5}
horsepower	foot-pounds per second	550	1.182×10^{-3}
horsepower	horsepower (metric)	1.014	0.9863
horsepower	kilowatts	0.746	1.341
inches	centimeters	2.54	0.3937
inches	feet	8.333×10^{-2}	12
inches	meters	2.54×10^{-2}	39.37
inches	miles	1.578×10^{-5}	6.336×10^{4}
inches	mils	10^{3}	10^{-3}
inches	yards	2.778×10^{-2}	36
joules	foot-pounds	0.7376	1.356
joules	watt-hours	2.778×10^{-4}	3600
kilograms	tons (long)	9.842×10^{-4}	1016
kilograms	tons (short)	1.102×10^{-3}	907.2
kilograms	pounds (avdp)	2.205	0.4536
kilometers	feet	3281	3.408×10^{-4}
kilometers	inches	3.937×10^{4}	2.54×10^{-5}
kilometers per hour	feet per minute	54.68	1.829×10^{-2}
kilowatt-hours	Btu	3413	2.93×10^{-4}
kilowatt-hours	foot-pounds	2.655×10^{6}	3.766×10^{-7}
kilowatt-hours	horsepower-hours	1.341	0.7457
kilowatt-hours	joules	3.6×10^{6}	2.778×10^{-7}
knots	feet per second	1.688	0.5925
knots	miles per hour	1.1508	0.869
lamberts	candles per square cm.	0.3183	3.142
lamberts	candles per square inch	2.054	0.4869
liters	cubic centimeters	10^{3}	10^{-3}
liters	cubic inches	61.02	1.639×10^{-2}
liters	gallons (liquid U.S.)	0.2642	3.785
liters	pints (liquid U.S.)	2.113	0.4732
lumens per square foot	foot-candles	1	1
lumens per square meter	foot-candles	0.0929	10.764
lux	foot-candles	0.0929	10.764
maxwells	kilolines	10^{-3}	10^{3}
maxwells	webers	10^{-8}	10^{8}
meters	feet	3.28	30.48×10^{-2}
meters	inches	39.37	2.54×10^{-2}
meters	miles	6.214×10^{-4}	1609.35
meters	yards	1.094	0.9144
miles (nautical)	feet	6076.1	1.646×10^{-4}
miles (nautical)	meters	1852	5.4×10^{-4}
miles (statute)	feet	5280	1.894×10^{-4}
miles (statute)	kilometers	1.609	0.6214
miles (statute)	miles (nautical)	0.869	1.1508
miles per hour	feet per second	1.467	0.6818
miles per hour	knots	0.8684	1.152
millimeters	microns	10^{3}	10^{-3}

To convert	To	Multiply	For the reverse, multiply by
mils	meters	2.54×10^{-5}	3.94×10^{4}
mils	minutes	3.438	0.2909
minutes (angle)	degrees	1.666×10^{-2}	60
minutes (angle)	radians	2.909×10^{-4}	3484
newtons	dynes	10^{5}	10^{-5}
newtons	kilograms	0.1020	9.807
newtons per sq. meter	pascals	1	1
newtons	pounds (avdp)	0.2248	4.448
oersteds	amperes per meter	7.9577×10	1.257×10^{-2}
ounces (fluid)	quarts	3.125×10^{-2}	32
ounces (avdp)	pounds	6.25×10^{-2}	16
pints	quarts (liquid U.S.)	0.50	2
poundals	dynes	1.383×10^{4}	7.233×10^{-5}
poundals	pounds (avdp)	3.108×10^{-2}	32.17
pounds	grams	453.6	2.205×10^{-3}
pounds (force)	newtons	4.4482	0.2288
pounds per square inch	dynes per square cm.	6.8946×10^{4}	1.450×10^{-5}
pounds per square inch	pascals	6.895×10^{3}	1.45×10^{-4}
quarts (U.S. liquid)	cubic centimeters	946.4	1.057×10^{-3}
radians	mils	10^{3}	10^{-3}
radians	minutes of arc	3.438×10^{3}	2.909×10^{-4}
radians	seconds of arc	2.06265×10^{5}	4.848×10^{-6}
revolutions per minute	radians per second	0.1047	9.549
roentgens	coulombs per kilogram	2.58×10^{-4}	3.876×10^{3}
slugs	kilograms	1.459	0.6854
slugs	pounds (avdp)	32.174	3.108×10^{-2}
square feet	square centimeters	929.034	1.076×10^{-3}
square feet	square inches	144	6.944×10^{-3}
square feet	square miles	3.587×10^{-8}	27.88×10^{6}
square inches	square centimeters	6.452	0.155
square kilometers	square miles	0.3861	2.59
stokes	square meter per second	10^{-4}	10^{-4}
tons (metric)	kilograms	10^{3}	10^{-3}
tons (short)	pounds	2000	5×10^{-4}
torrs	newtons per square meter	133.32	7.5×10^{-3}
watts	Btu per hour	3.413	0.293
watts	foot-pounds per minute	44.26	2.26×10^{-2}
watts	horsepower	1.341×10^{-3}	746
watt-seconds	joules	1	1
webers	maxwells	10^{8}	10^{-8}
webers per square meter	gausses	10^{4}	10^{-4}

CHAPTER 1

INEQUALITIES

If we write the equation: $x = 3$, we have an equation that applies to only one value, namely, the value where $x = 3$. But if we now write the inequality: $x > 3$, we have a relation that applies not only to one value, but to an entire region of values, i.e., the region between 3 and $+\infty$. If, on the other hand, we write the equation: $x < 3$, this holds for all values of x in the region $-\infty$ to 3. If we consider the symbol: $|\,|$, defined as the absolute value symbol, such that:

$$|x| = x \text{ if } x \geq 0$$
$$= -x \text{ if } x < 0$$

then the inequality

$$|x| > 3,$$

holds in the regions:

$$-\infty \text{ to } -3, \text{ and } +3 \text{ to } \infty .$$

Inequalities arise generally when considering ranges of values. Once an equality has been written, the solution may be arrived at by applying the basic algebraic operations which hold for inequalities. These differ only slightly from the algebraic operations applicable to equations.

• PROBLEM 1

Solve for x in:

$$-7 - 3x < 5x + 29.$$

Solution: Subtracting 5x from both sides,

$$-7 - 8x < 29.$$

If we now multiply by -1, the direction of the inequality is reversed because of the negative sign.

$$7 + 8x > -29.$$

Subtracting 7 from both sides yields $8x > -36$, and dividing by 8 gives the solution:

$$x > -\frac{9}{2}$$

Accordingly, the solution consists of all values of x in the interval $\left(-\frac{9}{2},\infty\right)$.

• PROBLEM 2

For what set of real numbers of x will $\sqrt{1 - 2x}$ be a real number?

Solution: Recall that the square root of a non-negative number will always be a real number.

$$1 - 2x \geq 0$$

$$2x \leq 1$$

$$x \leq \frac{1}{2}$$

This will give us the set $(-\infty, 1/2)$, which is the region between $-\infty$ and $+1/2$.

• PROBLEM 3

Solve for x when

$$\frac{3}{x} < 5. \qquad x \neq 0.$$

Solution: There may be an immediate inclination to multiply both sides by x. However, since we don't know in advance whether x is positive or negative, we must consider two cases: (1) x is positive, and (2) x is negative.

Case 1. Assume $x > 0$. Then multiplying by x preserves the direction of the inequality, and we get

$$3 < 5x.$$

Dividing by 5, we find that $x > \frac{3}{5}$.

This means that we must find all numbers which satisfy both of the inequalities

$$x > 0 \quad \text{and} \quad x > \frac{3}{5}.$$

Since, any number greater than $\frac{3}{5}$ is also positive, the solution in Case 1 consists of all x in the interval $\left(\frac{3}{5}, \infty\right)$.

Case 2. Let $x < 0$. Multiplying by x reverses the direction of the inequality because of the negative value of x. Therefore,

$$3 > 5x,$$

and $\frac{3}{5} > x$. We now seek all numbers x, such that both of the inequalities:

$$x < 0, \quad \text{and} \quad x < \frac{3}{5},$$

hold. The solution in Case 2 is the collection of all x in the interval $(-\infty, 0)$. A way of combining the answers in the two cases is to state that the solution consists of all numbers x not in the closed interval $\left[0, \frac{3}{5}\right]$.

• PROBLEM 4

Solve for $x (x \neq -2)$ when

$$\frac{2x - 3}{x + 2} < \frac{1}{3}.$$

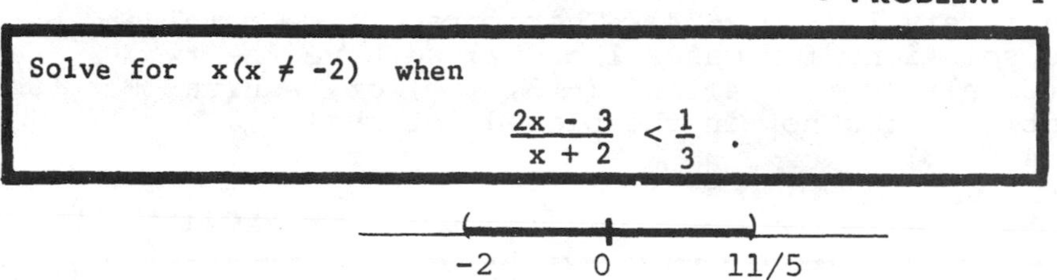

Solution: Here we consider two cases, according to whether $x + 2$ is positive or negative.

Case 1: $x + 2 > 0$. We multiply both sides of the given inequality by $3(x + 2)$, which is positive, and obtain:

$$6x - 9 < x + 2.$$

Adding $9 - x$ to both sides, we have

$$5x < 11, \text{ from which } x < \frac{11}{5}.$$

Since we have already assumed that $x + 2 > 0$, and since we must have $x < \frac{11}{5}$, we see that x must be larger than -2 and smaller than $\frac{11}{5}$. That is, the solution consists of all x in the interval $\left(-2, \frac{11}{5}\right)$.

Case 2: $x + 2 < 0$. Again multiplying by $3(x + 2)$ but now reversing the inequality because we are multiplying by a negative value,

$$6x - 9 > x + 2,$$

or $5x > 11$ and $x > \frac{11}{5}$. In this case, x must be less than -2 and greater than $\frac{11}{5}$, which is impossible. Combining the cases, we get as the solution all numbers in $\left(-2, \frac{11}{5}\right)$.

• PROBLEM 5

Find all the real numbers satisfying the inequality

$$(x + 3)(x + 4) > 0.$$

Solution: The inequality will be satisfied when both factors have the same sign, that is, if $x + 3 > 0$ and $x + 4 > 0$, or if $x + 3 < 0$ and $x + 4 < 0$. Let us consider the two cases.

Case 1: $x + 3 > 0$ and $x + 4 > 0$. That is,

$x > -3$ and $x > -4$

Thus, both inequalities hold if $x > -3$, which is the interval $(-3, +\infty)$.

Case 2: $x + 3 < 0$ and $x + 4 < 0$. That is,

$x < -3$ and $x < -4$

Both inequalities hold if $x < -4$, which is the interval $(-\infty, -4)$. Therefore, if we combine the solutions for Cases 1 and 2, we have the two intervals $(-\infty, -4)$ and $(-3, +\infty)$ or, equivalently, all x not in the closed interval $[-4, -3]$.

• PROBLEM 6

Solve the inequality: $x^2 > x$.

Solution: Subtracting x from both sides of the given inequality, we obtain:

$$x^2 - x > 0.$$

Factoring the left member, we have:

$$x(x - 1) > 0.$$

For x to satisfy this requirement it is necessary and sufficient that x and $x - 1$ be of the same sign. Thus, either

(i) $x > 0$ and $x - 1 > 0$ or

(ii) $x < 0$ and $x - 1 < 0$

Case (i) is satisfied if and only if $x > 1$, whereas case (ii) is satisfied if and only if $x < 0$. Thus, the solution set is the union of the intervals $(-\infty, 0)$ and $(1, \infty)$. Expressed in another way, the solution consists of all numbers of x, which are outside of the closed interval $[0, 1]$.

The same results may also be obtained by dividing the given inequality by x and considering the two cases of x positive and negative.

• PROBLEM 7

Solve the inequality: $2x^2 + 5x < 12$.

Solution: Subtracting 12 from both sides, we have:

$$2x^2 + 5x - 12 < 0,$$

and factoring,

$$(x + 4)(2x - 3) < 0.$$

Now, for a product to be negative, it is

necessary that the factors be of opposite sign. Thus two cases must be considered.

(i) $\quad x + 4 > 0$, and $2x - 3 < 0$; and

(ii) $\quad x + 4 < 0$, and $2x - 3 > 0$.

Condition (i) reduces to $-4 < x < \frac{3}{2}$.

Condition (ii) is impossible since there are no real numbers x satisfying

$$x < -4 \text{ and } x > \frac{3}{2}$$

simultaneously. Hence the solution is the open interval $\left(-4, \frac{3}{2}\right)$.

• PROBLEM 8

Draw the graph of the equation: $x + y = 5$, and describe the inequalities that apply to the regions of the plane which this line separates.

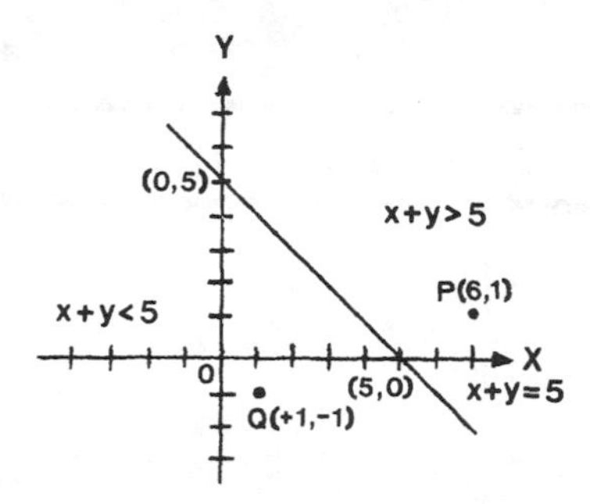

Solution: The graph of this equation is a straight line which divides the plane into two regions. Selecting the point P(6, 1), we see that at P

$$6 + 1 > 5.$$

We conclude that $x + y > 5$ in the entire region on the same side of the line as P. Similarly, at Q(1, - 1), we have $1 + (-1) < 5$, and so $x + y < 5$ in the region containing Q. The portion of the plane on one side of any line is called a half-plane.

• PROBLEM 9

Given the equation: $x^2 - y^2 = 4$, describe what inequalities hold in the regions of the plane which this curve separates.

Solution: The curve divides the plane into three regions; R_1, R_2, and R_3. The region R_1 is the one containing the point A, the region R_2 is the one containing B, and R_3 the one containing C. At point A (4, 1) we have $4^2 - 1^2 > 4$, and

therefore,

$$x^2 - y^2 > 4 \quad \text{in } R_1.$$

Since $1^2 - 0^2 < 4$,

$$x^2 - y^2 < 4 \quad \text{in } R_2.$$

Similarly, $(-4)^2 - 1^2 > 4$, and

$$x^2 - y^2 > 4 \quad \text{in } R_3.$$

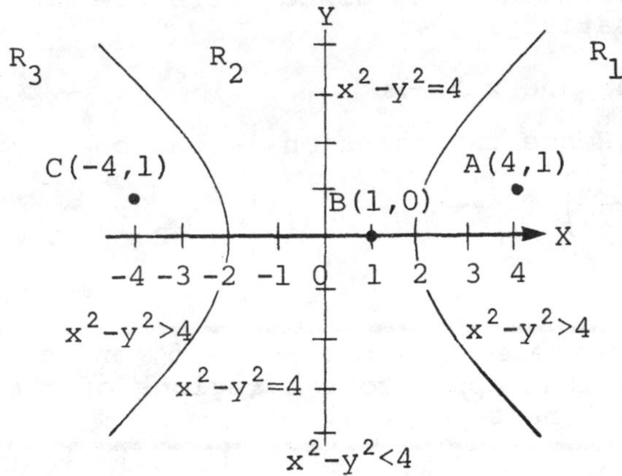

• PROBLEM 10

Graph the inequality: $y < -\frac{1}{2}x + 3$.

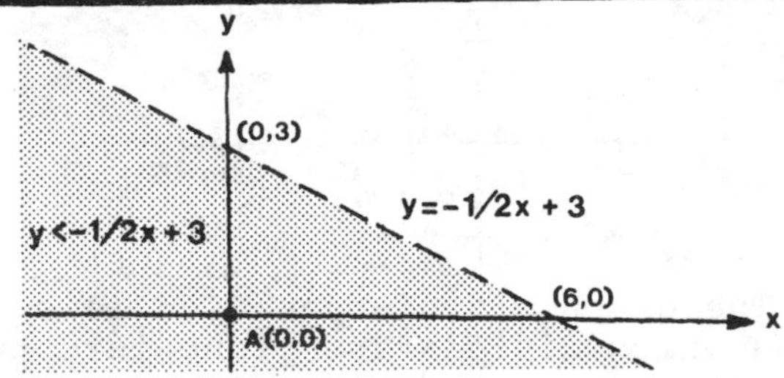

Solution: The graph is represented by shading the appropriate half-plane. The plot of: $y = -\frac{1}{2}x + 3$, is shown by a broken line to indicate that it is not part of the graph. To determine which half-plane is represented by the inequality, we test a point in a half-plane. If we test point A with co-ordinates (0,0), we have:

$$0 < -\tfrac{1}{2} \cdot 0 + 3.$$
$$0 < 3$$

Therefore the shaded portion is the correct half-plane.

• PROBLEM 11

Graph the inequality: $y \geq x - 1$.

Solution: The graph is shown by the shaded area and includes the line $y = x - 1$, which is drawn as a solid line. To verify this, we test a point on the line and in the half-plane. Taking point A with co-ordinates (0,0) and point B with co-ordinates (1,0), we have:

$$1) \quad 0 \geq 0 - 1$$

$$0 \geq -1$$

$$2) \quad 0 \geq 1 - 1$$

$$0 \geq 0$$

The union of a half plane with the line bounding it is called a closed half plane.

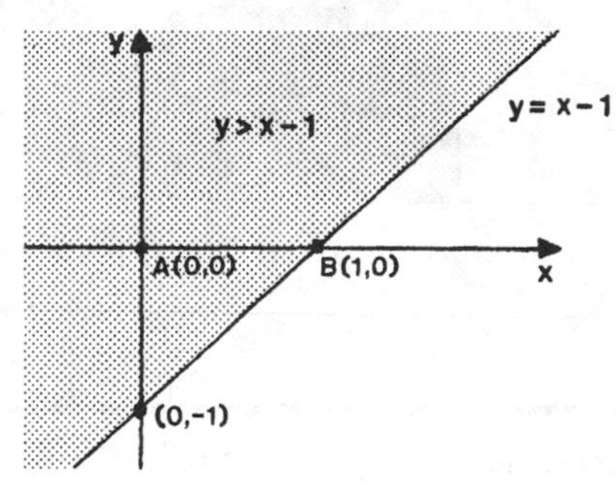

• **PROBLEM 12**

Graph the inequalities:

$$y \leq x + 1 \qquad \text{and} \qquad y \geq 2x + 1.$$

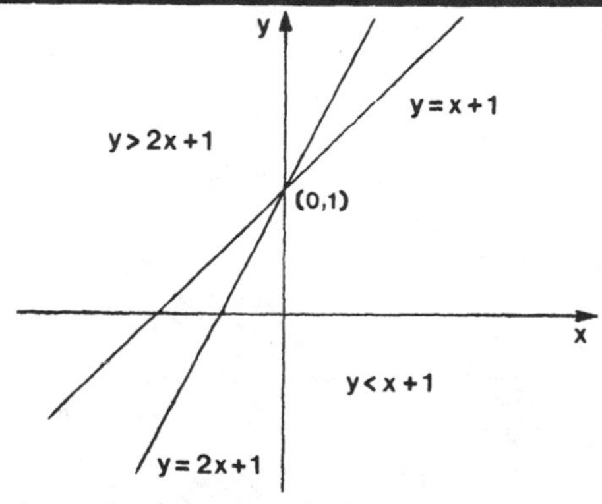

Solution: Each inequality is represented by a shaded half plane. The solid lines of intersection indicate the border lines of the planes. To verify, just test a point in the half-plane.

• **PROBLEM 13**

Graph the following inequalities and indicate the region of their intersection:

$$x \geq 1,\ y \geq 0,\ x + y \leq 4,\ 4x + 3y \leq 14.$$

Solution: The solution is the shaded region, having the quadrilateral as its boundary. The indicated vertices are the intersection points of the inequalities.

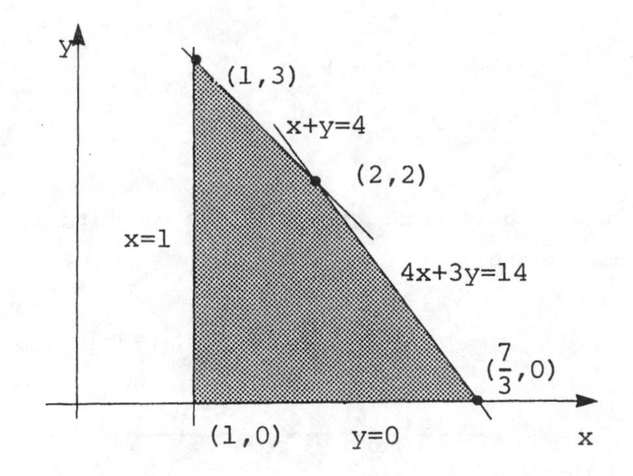

• PROBLEM 14

Graph the following inequalities:

$$(x + 1)^2 + y^2 < 1,$$

and

$$(x + 1)^2 + y^2 > 1.$$

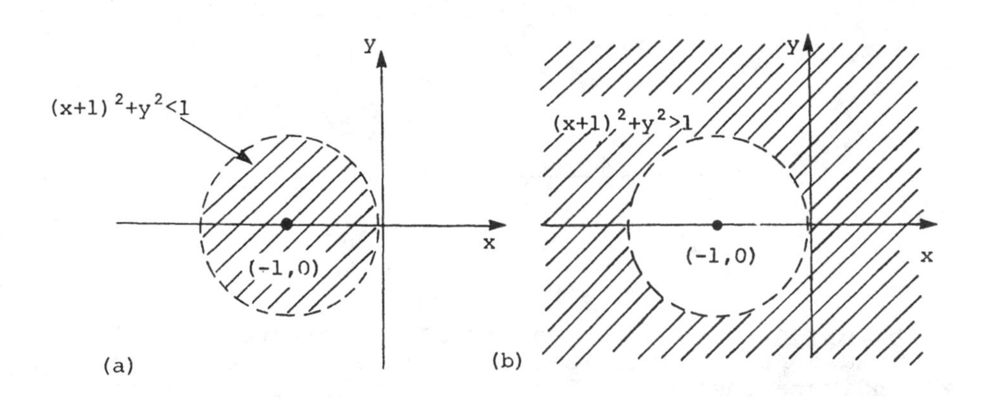

Solution: The graph of

$$(x + 1)^2 + y^2 = 1$$

is a circle. By testing a point inside the circle and a point outside the circle, we will determine which inequality holds.

fig. (a) is the graph of:

$(x + 1)^2 + (y^2) < 1,$ and

fig. (b) is the graph of :

$(x + 1)^2 + (y^2) > 1.$

CHAPTER 2

ABSOLUTE VALUES

When dealing with equations or inequalities involving absolute values, the first consideration should be to replace the given relationship with two relationships: one relationship expresses the value within the "absolute brackets" as a positive quantity, and the second relationship expresses the bracketed values as a negative quantity.

Once these two relationships are written, the basic algebraic operations may be applied to obtain the desired solution. Thus, the equation:

$$|x| = 2, \quad .$$

may be replaced by two equations, namely:

$$x = 2, \quad \text{and}$$
$$-x = 2,$$

Hence, the single equation, $|x| = 2$, is equivalent to two equations.

If more than one absolute value is involved, the situation is a little more complex, but subject to the same kind of analysis. Thus, if we have:

$$|x| = |x^2 - 1| \text{ , it is equivalent to:}$$
$$x = x^2 - 1 ,$$
$$-x = x^2 - 1 ,$$
$$x = -x^2 + 1 ,$$

and,
$$-x = -x^2 + 1 .$$

Note, however, that the first equation is identical to the fourth, and the second is identical to the third, a condition that often arises.

• PROBLEM 15

Find the range of possible values of x when given that $|x| < 7$.

Solution: $-7 < x < 7$.

• PROBLEM 16

Solve for x when $|5 - 3x| = -2$.

Solution: This problem has no solution, since the absolute value can never be negative and we need not proceed further.

• PROBLEM 17

Solve for x when $|5x + 4| = -3$.

Solution: In examining the given equation, it is seen that the absolute value of a number is set equal to a negative value. By definition of an absolute number, however, the number cannot be negative. Therefore the given equation has no solution.

• PROBLEM 18

Solve for x when $|x - 7| = 3$.

Solution: This equation, according to the definition of absolute value, expresses the conditions that $x - 7$ must be 3 or -3, since in either case the absolute value is 3. If $x - 7 = 3$, we have $x = 10$; and if $x - 7 = -3$, we have $x = 4$. We see that there are two values of x which solve the equation.

• PROBLEM 19

Solve for x when $|3x + 2| = 5$.

Solution: First we write expressions which replace the absolute symbols in forms of equations that can be manipulated algebraically. Thus this equation will be satisfied if either

$$3x + 2 = 5 \quad \text{or} \quad 3x + 2 = -5.$$

Considering each equation separately, we find

$$x = 1 \quad \text{and} \quad x = -\frac{7}{3}.$$

Accordingly, the given equation has two solutions.

• PROBLEM 20

Given that $|x - 1| < 5$, write an expression defining the region of possible values of x.

Solution: $|x - 1| < 5$.

From which $-5 < x - 1 < 5$.

Solving for x, we have: $-4 < x < 6$.

• PROBLEM 21

Find the range of possible values of x when $|x - 5| < 4$.

Solution: We rewrite the given equation in a form which can be manipulated algebraically.

$$-4 < x - 5 < 4.$$

Adding 5 to each member of the preceding inequality, we obtain:

$$1 < x < 9$$

Because each step is reversible, we can conclude that

$$|x - 5| < 4 \text{ if and only if } 1 < x < 9 .$$

Therefore, the interval solution of the given inequality is (1,9), as illustrated.

• PROBLEM 22

Solve the inequality $|3 - 2x| < 1$

Solution: $|3 - 2x| < 1$ can be represented as

$$- 1 < 3 - 2x < 1$$

By subtracting 3 from all terms, we have

$$- 4 < - 2x < - 2$$

By multiplying all terms by $-\frac{1}{2}$ and remembering to reverse the signs of the inequalities, we have

$$2 > x > 1$$

or all values of x in the open interval (1, 2).

• PROBLEM 23

Solve for x in $|2 - 5x| < 3.$

Solution: Rewriting the given equation in the equivalent form which can be manipulated algebraically,

$$-3 < 2 - 5x < 3.$$

Subtracting 2 from each side yields:

$$-5 < -5x < 1.$$

Dividing by -5 reverses each of the inequalities because of the negative value of 5, and we obtain:

$$1 > x > - \frac{1}{5} .$$

The solution consists of those x in the open interval $\left(- \frac{1}{5}, 1\right)$.

• PROBLEM 24

Solve: $|3x - 4| \leq 7.$

Solution: We write the inequality in the equivalent form

$$-7 \leq 3x - 4 \leq 7.$$

Adding 4 to each term,

$$-3 \leq 3x \leq 11.$$

Dividing each term by 3,

$$-1 \le x \le \frac{11}{3}.$$

The solution consists of all numbers x in the closed interval $\left[-1, \frac{11}{3}\right]$.

• PROBLEM 25

Find all real numbers satisfying the inequality $|3x + 2| > 5$.

Solution: Writing the given equation into the form which can be dealt with algebraically, we note that the given inequality expressed with an absolute number is equal to the following inequalities:

$$3x + 2 > 5 \quad \text{or} \quad 3x + 2 < -5.$$

Considering, now, the first inequality, we have:

$$3x + 2 > 5,$$

or

$$x > 1.$$

Therefore, the interval $(1, +\infty)$ is a solution.

From the second inequality, we have:

$$3x + 2 < -5,$$

or

$$x < -\frac{7}{3}.$$

Hence, the interval $\left(-\infty, -\frac{7}{3}\right)$ is a solution.

The solution of the given inequality consists of the two intervals $\left(-\infty, -\frac{7}{3}\right)$ and $(1, +\infty)$ or, expressed in another way, all x not in the closed interval $\left[-\frac{7}{3}, 1\right]$.

• PROBLEM 26

Graph the following two inequalities and show where the two graphs coincide:

$$2 \le x < 3 \qquad \text{and} \qquad |y - 2| < \frac{1}{2}.$$

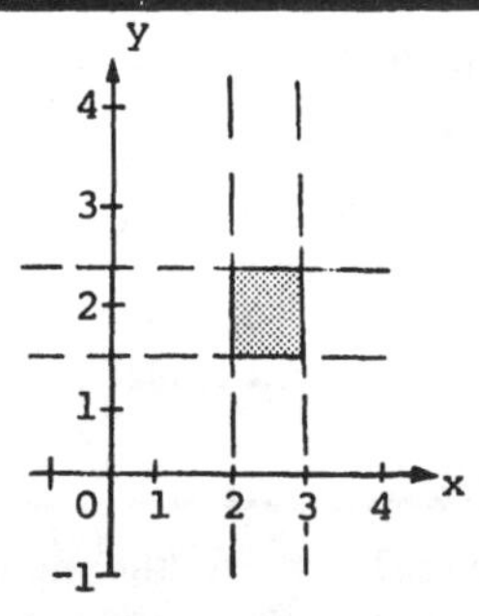

Solution: The first inequality consists of an infinite strip between the lines x = 2 and x = 3. Note that the points on the line x = 2 are included, but the points on x = 3 are not.

For the second inequality, two cases must

be considered, depending on whether $(y - 2)$ is positive or negative. This may be expressed as:

$$-\frac{1}{2} < y - 2 < \frac{1}{2}.$$

Adding 2 to each term to simplify, gives:

$$\frac{3}{2} < y < \frac{5}{2}.$$

This is an infinite strip which intersects the first one in a rectangle, with vertices at the points $\left(2, \frac{3}{2}\right)$, $\left(3, \frac{3}{2}\right)$, $\left(3, \frac{5}{2}\right)$, and $\left(2, \frac{5}{2}\right)$. The result is actually a square.

Therefore, all points inside this square satisfy both inequalities. However, in addition, the points on the boundary of the square along the line $x = 2$ (left boundary), except for the corners, also satisfy both inequalities.

• PROBLEM 27

Solve

$$\left|\frac{2x - 5}{x - 6}\right| < 3.$$

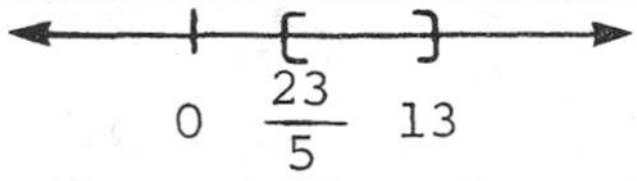

Solution: Writing the equivalent form, for ease in algebraic manipulation:

$$-3 < \frac{2x - 5}{x - 6} < 3.$$

Before we multiply by $x - 6$, we must distinguish two cases, depending on whether $x - 6$ is positive or negative.

Case 1: $x - 6 > 0$. In this case, multiplication by $(x - 6)$ preserves the direction of the inequalities, and we have:

$$-3(x - 6) < 2x - 5 < 3(x - 6).$$

Now the left inequality states that:

$$-3x + 18 < 2x - 5,$$

from which

$$\frac{23}{5} < x.$$

The right inequality states that:

$$2x - 5 < 3x - 18,$$

from which

$$13 < x.$$

We see that in Case 1 we must have: $x - 6 > 0$, $\frac{23}{5} < x$, and $13 < x$. If the third inequality holds, then the other two hold as a consequence. Hence the solution in Case 1 consists of all x in the interval $(13,\infty)$.

Case 2: $x - 6 < 0$. The inequalities reverse when we multiply by

$(x - 6)$ because of the negative value. We then obtain:

$$-3(x - 6) > 2x - 5 > 3(x - 6).$$

The two inequalities now state that: $\frac{23}{5} > x$, and $13 > x$. The three inequalities:

$$x - 6 < 0, \frac{23}{5} > x, \text{ and } 13 > x,$$

all hold if $x < \frac{23}{5}$. In Case 2, the solution consists of all numbers in the interval $\left(-\infty, \frac{23}{5}\right)$. We could also describe the solution by saying that it consists of all numbers not in the closed interval $\left[\frac{23}{5}, 13\right]$.

• PROBLEM 28

Find the possible values of x when $\left|\frac{3 - 2x}{2 + x}\right| \leq 4$.

Solution: The given relationship is first written in the form which can be manipulated algebraically.

$$-4 \leq \frac{3 - 2x}{2 + x} \leq 4$$

If we multiply all members by $2 + x$, we must consider two cases, depending upon whether $2 + x$ is positive or negative.

Case 1: If $2 + x > 0$, then $x > -2$ and $-4(2 + x) \leq 3 - 2x \leq 4(2 + x)$, or

$$-8 - 4x \leq 3 - 2x \leq 4(2 + x).$$

Therefore, if $x > -2$, then $-8 - 4x \leq 3 - 2x$ and $3 - 2x \leq 8 + 4x$. We now proceed to solve these two inequalities. The first inequality is:

$$-8 - 4x \leq 3 - 2x .$$

Adding $2x + 8$ to both sides gives:

$$-2x \leq 11 .$$

Dividing both sides by -2 and reversing the inequality sign because of the negative sign, we obtain

$$x \geq -\frac{11}{2}$$

The second inequality is:

$$3 - 2x \leq 8 + 4x .$$

Adding $-4x - 3$ to both sides gives:

$$-6x \leq 5 .$$

Dividing both sides by -6 and reversing the inequality sign due to the negative value of 6, we obtain:

$$x \geq -\frac{5}{6} .$$

Therefore, if $x > -2$, then the original inequality holds if and only if

$$x \geq -\frac{11}{2} \text{ and } x \geq -\frac{5}{6} .$$

Because all three inequalities: $x > -2$, $x \geq -\frac{11}{2}$, and $x \geq -\frac{5}{6}$, must be satisfied by the same value of x, we have $x \geq -\frac{5}{6}$, or the interval $\left(-\frac{5}{6}, +\infty\right)$.

Case 2: If $2 + x < 0$, then $x < -2$, and $-4(2 + x) \geq 3 - 2x \geq 4(2 + x)$, or

$$-8 - 4x \geq 3 - 2x \geq 8 + 4x .$$

Considering the left inequality, we have:

$$-8 - 4x \geq 3 - 2x$$

$$-2x \geq 11$$

$$x \leq -\frac{11}{2} .$$

From the right inequality we have:

$$3 - 2x \geq 8 + 4x$$

$$-6x \geq 5$$

$$x \leq -\frac{5}{6}$$

Therefore, if $x < -2$, the original inequality holds if and only if $x \leq -\frac{11}{2}$ and $x \leq -\frac{5}{6}$.

Because all three inequalities must be satisfied by the same value of x, we have $x \leq -\frac{11}{2}$, or the interval $\left(-\infty, -\frac{11}{2}\right)$.

Combining the solutions of Case 1 and Case 2, we have as the solution the two intervals $\left(-\infty, -\frac{11}{2}\right)$ and $\left(-\frac{5}{6}, +\infty\right)$ or, more simply, all x not in the interval $\left(-\frac{11}{2}, -\frac{5}{6}\right)$.

• PROBLEM 29

Solve for x in $|2x - 6| = |4 - 5x|$.

<u>Solution:</u> There are four possibilities here. $2x - 6$ and $4 - 5x$ can be either positive or negative. Therefore,

$$2x - 6 = 4 - 5x \qquad (1)$$

$$-(2x - 6) = 4 - 5x \qquad (2)$$

and

$$2x - 6 = -(4 - 5x) \qquad (3)$$

$$-(2x - 6) = -(4 - 5x) \qquad (4)$$

Equations (2) and (3) result in the same solution, as do equations (1) and (4). Therefore it is necessary to solve only for equations (1) and (2). This gives:

$$x = \frac{10}{7}, -\frac{2}{3} .$$

• PROBLEM 30

Solve for x when $|2x - 1| = |4x + 3|$.

<u>Solution:</u> Replacing the absolute sysmbols with equations that can be handled algebraically according to the conditions implied by the given equation, we have:

$$2x - 1 = 4x + 3 \quad \text{or} \quad 2x - 1 = -(4x + 3) .$$

Solving the first equation, we have $x = -2$; solving the second, we obtain $x = -\frac{1}{3}$, thus giving us two solutions to the original equation. (We could also write: $-(2x - 1) = -(4x + 3)$, but this is equivalent to the first of the equations above.)

• PROBLEM 31

Find a positive number M such that

$$|x^3 - 2x^2 + 3x - 4| \leq M$$

for all values of x in the interval $[-3,2]$.

Solution: The expression may be written in the form:

$$|x^3 - 2x^2 + 3x - 4| \leq |x^3| + |2x^2| + |3x| + |4|,$$

and, from the rules of absolute value of products, we obtain:

$$|x^3| + |2x^2| + |3x| + |4| \leq |x|^3 + 2|x|^2 + 3|x| + 4.$$

Now, since $|x|$ can never be larger than 3, we have:

$$|x|^3 + 2|x|^2 + 3|x| + 4 \leq 27 + 2 \cdot 9 + 3 \cdot 3 + 4 = 58.$$

Therefore, the positive number M we seek is 58.

• PROBLEM 32

Find a number M such that $\left|\frac{x + 2}{x - 2}\right| \leq M$, for x in the interval $\left[\frac{1}{2}, \frac{3}{2}\right]$.

Solution: From the rules of absolute values, we obtain

$$\left|\frac{x + 2}{x - 2}\right| = \frac{|x + 2|}{|x - 2|} .$$

If we can estimate the smallest possible value of the denominator and the largest possible value of the numerator, then we will have estimated the largest possible value of the entire expression. For the numerator we have

$$|x + 2| \leq |x| + |2| \leq \frac{3}{2} + 2 = \frac{7}{2} ,$$

with the substitution of the largest limit for x in the numerator.

The minimum value of the denominator, however, is not obtained when x is at the lower limit of $\frac{1}{2}$.

For the denominator we note that the smallest value occurs when x is as close as possible to 2. This occurs when $x = \frac{3}{2}$, and therefore

$$|x - 2| \geq \frac{1}{2} ,$$

if x is in the interval $\left[\frac{1}{2}, \frac{3}{2}\right]$. Consequently,

$$\left|\frac{x+2}{x-2}\right| \le \frac{\frac{7}{2}}{\frac{1}{2}} = 7.$$

• PROBLEM 33

What is the largest possible value of

$$\left|\frac{x^2+2}{x+3}\right|$$

if x is restricted to the interval $[-4,4]$?

Solution: The largest possible value for the expression results when the numerator is largest and the denominator is smallest.

The numerator is largest when the upper limit of $x = 4$ is substituted, as follows:

$$|x^2 + 2| \le |x|^2 + 2 \le 4^2 + 2 = 18.$$

We must now find a smallest value for $x + 3$ if x is in $[-4,4]$. We see that the expression is not defined for $x = -3$, since then the denominator would be zero, and division by zero is always excluded. Furthermore, if x is a number near -3, the denominator is near zero, the numerator has a value near 11, and the quotient is a "large" number, i.e., it approaches ∞. Hence, in this problem, there is no largest value of the given expression in $[-4,4]$.

• PROBLEM 34

Find a number M if $\left|\frac{x+2}{x} - 5\right| \le M$ and if, x is restricted to the interval $(1,4)$.

Solution: The problem is best approached by first writing the given function in the form of a ratio, and then examining the numerator and denominator to obtain the largest value for the ratio.

$$\frac{x+2}{x} - 5 = \frac{-4x+2}{x}.$$

Therefore,

$$\left|\frac{x+2}{x} - 5\right| = \frac{|-4x+2|}{|x|} \le \frac{|4||x| + |2|}{|x|}.$$

Since $|x| < 4$, we have $4|x| + 2 < 18$, while the denominator, $|x|$, can never be smaller than 1. We obtain

$$\left|\frac{x+2}{x} - 5\right| < 18.$$

CHAPTER 3

LIMITS

The limit of a function at a point is the value that the function has as the point is approached. The simplest approach to determine $\lim_{x \to a} f(x)$ is to substitute $x = a$ in the function. If this gives a determinate result, the problem is solved. If the result is indeterminate, we should try to manipulate the function algebraically to obtain a form in which determinate results might be obtained. In this respect, one approach which is often applicable when the function is expressed as a ratio, is to divide numerator and denominator by the term with the highest power. When dealing with trigonometric functions, use can be made of the following expressions for this purpose:

$$\sin x = \tan x = x, \quad \text{for small } x$$

$$\cos x = 1 - \frac{x^2}{2}, \quad \text{for small angles,}$$

and similar approximations for angles in the vicinity of $\pi/2$.

Whatever approach is used, the procedure is not always successful, because sometimes the limit simply does not exist. This is true whenever the value of the function is different, depending on the direction of approach of the point. For example, lim tan x does not exist at $x = \pi/2$, since $\tan x$ is equal to $+\infty$ as it is approached from below, while it is $-\infty$ when approached from above.

If a function has a limit, it can generally be found. It is also useful to plot the function, since, from the plot, it can often be determined whether it has a limit or not.

• PROBLEM 35

Find $\lim_{x \to 2} f(x) = 2x + 1$

Solution: As $x \to 2$, $f(x) \to 5$. Therefore,

$$\lim_{x \to 2} (2x + 1) = 5$$

● **PROBLEM 36**

Find $\lim_{x \to 3} f(x) = \frac{x^2 - 9}{x + 1}$

Solution:

$$\lim_{x \to 3} \frac{x^2 - 9}{x + 1} = \frac{0}{4} = 0$$

● **PROBLEM 37**

Find $\lim_{x \to 2} \left(\frac{x^2 - 5}{x + 3} \right)$

Solution: For the numerator only

$$\lim_{x \to 2} (x^2 - 5) = 4 - 5 = -1$$

For the denominator only

$$\lim_{x \to 2} (x + 3) = 5$$

Therefore,

$$\lim_{x \to 2} \left(\frac{x^2 - 5}{x + 3} \right) = -\frac{1}{5}$$

● **PROBLEM 38**

Find $\lim_{x \to 2} \frac{4x - 3}{x^2 + 2x + 5}$

Solution: Using the method of direct substitution, we obtain 5 for the numerator and 13 for the denominator. Since the ratio 5/13 is determinate, the required limit is 5/13.

● **PROBLEM 39**

Find $\lim_{x \to 3} f(x) = \frac{1}{(x - 3)^2}$, $\quad x \neq 3$.

Solution: Sketching the graph of this function about $x = 3$, we see that it increases without bound as x tends to 3.

Using the method of simple substitution, we find that $\lim_{x \to 3} f(x) = \infty$. There is no limit.

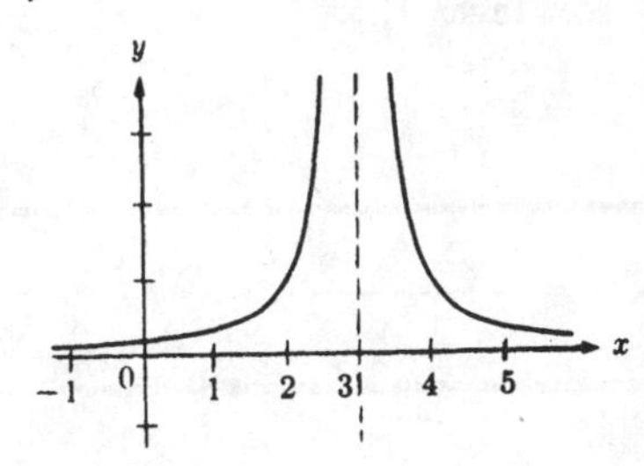

• PROBLEM 40

Find $\lim_{x \to 3} \dfrac{5x}{6 - 2x}$.

<u>Solution:</u>

$$\lim_{x \to 3} 5x = 15$$

and

$$\lim_{x \to 3} [6 - 2x] = 0$$

Therefore,

$$\lim_{x \to 3} \frac{5x}{6 - 2x} = \frac{15}{0} = \infty .$$

The function has no limit.

• PROBLEM 41

Find:

(a) $\lim_{x \to 3^+} \dfrac{x^2 + x + 2}{x^2 - 2x - 3}$

(b) $\lim_{x \to 3^-} \dfrac{x^2 + x + 2}{x^2 - 2x - 3}$

(c) $\lim_{x \to 3} \dfrac{x^2 + x + 2}{x^2 - 2x - 3}$

<u>Solution:</u>

(a) $\lim_{x \to 3^+} \dfrac{x^2 + x + 2}{x^2 - 2x - 3} = \lim_{x \to 3^+} \dfrac{x^2 + x + 2}{(x - 3)(x + 1)}$

The limit of the numerator is 14. To find the limit of the denominator,

$$\lim_{x \to 3^+} (x - 3)(x + 1) = \lim_{x \to 3^+} (x - 3) \cdot \lim_{x \to 3^+} (x + 1)$$

$$= 0 \cdot 4 = 0$$

Thus, the limit of the denominator is 0, as the denominator is approaching 0 through positive values. Consequently,

$$\lim_{x\to 3^+} \frac{x^2 + x + 2}{x^2 - 2x - 3} = +\infty$$

(b) $$\lim_{x\to 3^-} \frac{x^2 + x + 2}{x^2 - 2x - 3} = \lim_{x\to 3^-} \frac{x^2 + x + 2}{(x - 3)(x + 1)}$$

As in part (a), the limit of the numerator is 14 here also. To find the limit of the denominator,

$$\lim_{x\to 3^-} (x - 3)(x + 1) = \lim_{x\to 3^-} (x - 3) \cdot \lim_{x\to 3^-} (x + 1)$$

$$= 0 \cdot 4 = 0$$

In this case, the limit of the denominator is again zero, but since the denominator is approaching zero through negative values,

$$\lim_{x\to 3^-} \frac{x^2 + x + 2}{x^2 - 2x - 3} = -\infty$$

(c) $$\lim_{x\to 3} \frac{x^2 + x + 2}{x^2 - 2x - 3} = |\infty|$$

• PROBLEM 42

Find:

$$\operatorname*{limit}_{x\to 0} \left(\operatorname*{limit}_{y\to 0} \frac{ax^2 + bxy + cy^2}{dx^2 + exy + fy^2} \right) - \operatorname*{limit}_{y\to 0} \left(\operatorname*{limit}_{x\to 0} \frac{ax^2 + bxy + cy^2}{dx^2 + exy + fy^2} \right)$$

$ad \neq 0$, $cf \neq 0$.

Solution: This problem must be evaluated in 5 separate stages. First each limit within the brackets is evaluated, then this result is evaluated with respect to the outer limits, and finally, the two resulting limits are subtracted.

The first limit is

$$\operatorname*{limit}_{x\to 0} \left(\operatorname*{limit}_{y\to 0} \frac{ax^2 + bxy + cy^2}{dx^2 + exy + fy^2} \right).$$

1) What is in the brackets is considered first.

$$\operatorname*{limit}_{y\to 0} \frac{ax^2 + bxy + cy^2}{dx^2 + exy + fy^2}$$

$$= \frac{ax^2 + b \cdot x \cdot 0 + c \cdot 0}{dx^2 + e \cdot x \cdot 0 + f \cdot 0}$$

$$= \frac{ax^2 + 0 + 0}{dx^2 + 0 + 0} = \frac{ax^2}{dx^2} = \frac{a}{d}.$$

2) Replacing this answer in the brackets, the problem now becomes:

$$\lim_{x\to 0} \left(\frac{a}{d}\right).$$

The $\lim_{x\to 0} \left(\frac{a}{d}\right) = \frac{a}{d}$, because the fraction $\frac{a}{d}$ is independent of x, and as x goes to zero, the value remains $\frac{a}{d}$. The fraction $\frac{a}{d}$ is defined only if $d \neq 0$. Since $ad \neq 0$, neither a nor d can equal zero. Therefore $d \neq 0$ and $\frac{a}{d}$ is defined.

3) The second part,

$$\lim_{y\to 0} \left(\lim_{x\to 0} \frac{ax^2 + bxy + cy^2}{dx^2 + exy + fy^2}\right),$$

is evaluated in the same manner. The value of the inner limit is obtained first. As $x \to 0$, the value of each factor is considered.

$$\lim_{x\to 0} \frac{ax^2 + bxy + cy^2}{dx^2 + exy + fy^2}$$

$$= \frac{a \cdot 0 + b \cdot 0 \cdot y + cy^2}{d \cdot 0 + e \cdot 0 \cdot y + fy^2}$$

$$= \frac{0 + 0 + cy^2}{0 + 0 + fy^2} = \frac{cy^2}{fy^2} = \frac{c}{f}.$$

4) Replacing this solution in the brackets, the problem now is:

$$\lim_{y\to 0} \left(\frac{c}{f}\right).$$

The $\lim_{y\to 0} \left(\frac{c}{f}\right) = \frac{c}{f}$, since the value $\frac{c}{f}$ is independent of y. Once again, $\frac{c}{f}$ is defined only if $f \neq 0$. Following the same reasoning as before, since $cf \neq 0$ neither $c = 0$ nor $f = 0$, which means that $f \neq 0$ and $\frac{c}{f}$ is defined.

5) The problem calls for subtracting the two limits, therefore:

$$\lim_{x\to 0} \left(\lim_{y\to 0} \frac{ax^2 + bxy + cy^2}{dx^2 + exy + fy^2}\right)$$

$$- \lim_{y \to 0} \left(\lim_{x \to 0} \frac{ax^2 + bxy + cy^2}{dx^2 + exy + fy^2} \right) = \frac{a}{d} - \frac{c}{f} .$$

• **PROBLEM 43**

Find $\lim_{x \to 0} (x\sqrt{x - 3})$.

Solution: In checking the function by simple substitution, we see that:

$$x\sqrt{x - 3} = 0$$

if $$x = 0.$$

However, this function does not have real values for values of x less than 3. Therefore, since x cannot approach 0, f(x) does not approach 0 and the limit does not exist. This example illustrates that we cannot properly find

$$\lim_{x \to a} f(x)$$

by finding f(a), even though they are equal in many cases. We must consider values of x near a, but not equal to a.

• **PROBLEM 44**

Find $\lim_{\Theta \to \frac{\pi}{2}} [\tan \Theta]$.

Solution: As θ approaches $\frac{\pi}{2}$ from values of θ smaller than $\frac{\pi}{2}$, tan θ increases x towards $+\infty$ without limit. But when θ approaches $\frac{\pi}{2}$ from larger values, then tan θ decreases x towards $-\infty$ without limit. Therefore, $\lim_{\theta \to \frac{\pi}{2}} [\tan \theta] = |\infty|$

• **PROBLEM 45**

Find $\lim_{(x,y) \to (0,0)} \frac{x^2 - y^2}{x^2 + y^2}$.

Solution: We first follow the procedure used for a function of one variable. Thus,

$$\lim_{y \to 0} \frac{x^2 - y^2}{x^2 + y^2} = \frac{x^2}{x^2} .$$

As $x \to 0$, $x^2/x^2 \to 1$. This might make it appear that the required limit is 1. However, let us consider the reverse situation where

$$\lim_{x \to 0} \frac{x^2 - y^2}{x^2 + y^2} = - \frac{y^2}{y^2}.$$

As $y \to 0$, $-y^2/y^2 \to -1$. This is an entirely different result from the preceding one. Consequently, no limit exists for this problem.

• PROBLEM 46

Given that f is the function defined by:

$$f(x) = \begin{cases} x - 3 & \text{if } x \neq 4 \\ 5 & \text{if } x = 4, \end{cases}$$

Find $\lim_{x \to 4} f(x)$.

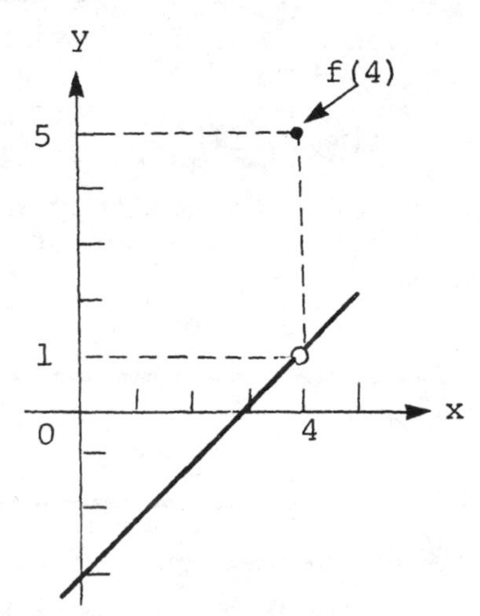

Solution: When plotting f(x) to obtain a visual representation, it is seen that $f(x) = x - 3$ is a straight line which has a break or discontinuity at the point $x = 4$. At $x = 4$, the value of f(x) is given as 5, and not as 1 - the value that f(x) would assume if the line were continuous. However, when evaluating $\lim_{x \to 4} f(x)$, we are considering values of x close to 4 but not equal to 4. Thus, we have

$$\lim_{x \to 4} f(x) = \lim_{x \to 4} (x - 3) = 1.$$

In this example,

$\lim_{x \to 4} f(x) = 1$ but $f(4) = 5$; therefore,

$$\lim_{x \to 4} f(x) \neq f(4).$$

• PROBLEM 47

Find $\lim_{x \to 0} g(x)$ when this function is defined as

$$g(x) = \begin{cases} |x| & \text{if } x \neq 0 \\ 2 & \text{if } x = 0. \end{cases}$$

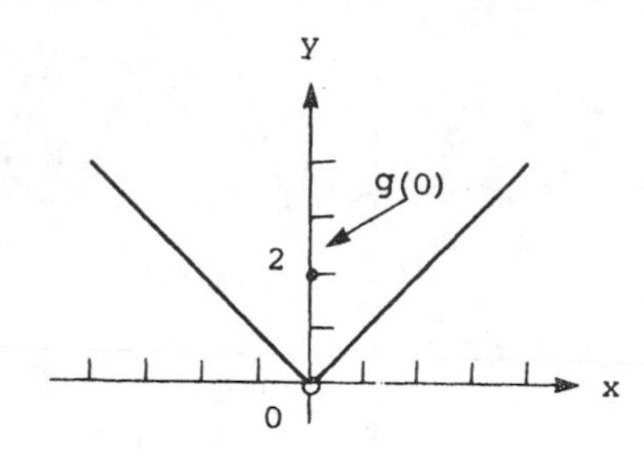

Solution: In thinking about the approach to be taken, it is helpful to sketch the function. It is to be noted that the function is discontinuous at x = 0. Now,

$$\lim_{x \to 0^-} g(x) = \lim_{x \to 0^-} (-x) = 0 \qquad \text{and}$$

$$\lim_{x \to 0^+} g(x) = \lim_{x \to 0^+} x = 0.$$

Therefore, $\lim_{x \to 0} g(x)$ exists and is equal to 0. Note that g(0) = 2, but this has no effect on $\lim_{x \to 0} g(x)$, since, by definition of limits, we are considering values of x close to zero but not equal to zero.

• PROBLEM 48

Let h be defined by:

$$h(x) = \begin{cases} 4 - x^2 & \text{if } x < 1 \\ 2 + x^2 & \text{if } 1 < x. \end{cases}$$

Find each of the following limits if they exist:

$$\lim_{x \to 1^-} h(x), \ \lim_{x \to 1^+} h(x), \ \lim_{x \to 1} h(x).$$

Solution: It is desirable to sketch the given function to aid in visualizing the problem. Now,

$$\lim_{x \to 1^-} h(x) = \lim_{x \to 1^-} (4 - x^2) = 3$$

$$\lim_{x \to 1^+} h(x) = \lim_{x \to 1^+} (2 + x^2) = 3$$

Therefore, $\lim_{x \to 1} h(x)$ exists and is equal to 3. Note that $h(1) = 3$. This holds because the function is continuous.

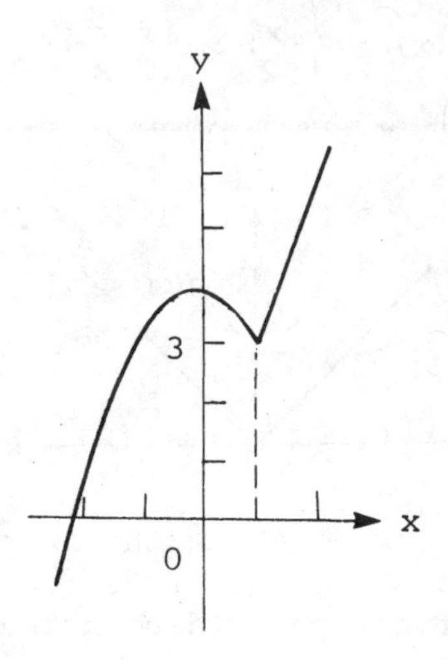

• PROBLEM 49

Let f(x) and g(x) be defined by:

$$f(x) = \begin{cases} x^2 + 2x, & x \leq 1, \\ 2x, & x > 1, \end{cases}$$

$$g(x) = \begin{cases} 2x^3, & x \leq 1, \\ 3, & x > 1. \end{cases}$$

Find $\lim_{x \to 1} [f(x) \cdot g(x)]$ if it exists.

Solution: Neither f(x) nor g(x) have limits as $x \to 1$, but one-sided limits exist for both functions. It is possible that the product of two functions may have a limit, even though the two functions do not have limits individually.

$$\lim_{x \to 1^-} f(x) = 3, \quad \lim_{x \to 1^+} f(x) = 2,$$

$$\lim_{x \to 1^-} g(x) = 2, \quad \lim_{x \to 1^+} g(x) = 3.$$

Therefore,

$$\lim_{x \to 1^-} [f(x) \cdot g(x)] = 6$$

and

$$\lim_{x \to 1^+} [f(x) \cdot g(x)] = 6.$$

Consequently,

$$\lim_{x \to 1} [f(x) \cdot g(x)] = 6.$$

CHAPTER 4

CONTINUITY

A function, $f(x)$ is continuous at a point, a, if it satisfies three separate conditions: (1) $f(a)$ is defined; (2) $\lim_{x \to a} f(x)$ exists; and (3) $\lim_{x \to a} f(x) = f(a)$. If any one or more of these conditions fail, the function is said to be discontinuous.

Problems generally encountered in elementary calculus are found to have continuous functions.

• PROBLEM 50

Determine the domain of the function:

$$y = \frac{2x}{(x-2)(x+1)} .$$

Solution:

$$y = \frac{2x}{(x-2)(x+1)}$$

is defined only if the denominator is not equal to zero.

The denominator is equal to zero when $(x - 2)(x + 1) = 0$, or, equivalently, when $(x - 2) = 0$ or $(x + 1) = 0$. $(x - 2) = 0$ when $x = 2$, and $(x + 1) = 0$ when $x = -1$.

Therefore, the domain of the function is all values of x where $x \neq -1$ and $x \neq 2$.

• PROBLEM 51

Determine the domain of the function:

$$y = \sqrt{\frac{x}{2-x}} .$$

Solution: Since we are restricted to the real number system, there are two cases to be considered:

Case 1. When the denominator equals zero, the fraction is undefined. This occurs when $x = 2$.

Case 2. Since we are restricted to real values of y, it is necessary that

$$\frac{x}{2-x} \geq 0,$$

because, if

$\frac{x}{2-x} < 0$, the square root gives

an imaginary number. Therefore,

$$\frac{x}{2-x} < 0$$

only if $x < 0$, and $x > 2$.

Combining the conditions in cases 1 and 2, the domain of x is the interval

$$0 \leq x < 2.$$

• PROBLEM 52

Investigate the continuity of the expression:

$$y = \frac{x^2 - 9}{x - 3} \quad \text{at} \quad x = 2.$$

Solution: For a function to be continuous at a point, in this case, 2, it must satisfy three conditions: (1) $f(2)$ is defined, (2) $\lim_{x \to 2} f(x)$ exists, (3) $\lim_{x \to 2} f(x) = f(2)$.

For $x = 2$, $y = f(x) = f(2) = \frac{(2)^2 - 9}{2 - 3} = 5$.

Also,

$$\lim_{x \to 2} \frac{x^2 - 9}{x - 3} = 5 .$$

Therefore, the function is continuous at $x = 2$.

• PROBLEM 53

Determine the continuity of the following function and sketch its graph:

$$M(x) = \frac{|x|}{x}(x^2 - 1).$$

Solution: For $x > 0$, $|x| = x$ and for $x < 0$, $|x| = -x$. Therefore for $x > 0$

$$M(x) = \frac{x}{x}(x^2 - 1) = x^2 - 1 \qquad --- (1)$$

for $x < 0$

$$M(x) = \frac{-x}{x}(x^2 - 1) = -(x^2 - 1) = 1 - x^2 \quad ---(2)$$

for x = 0

$$M(x)\big|_{x=0} = \frac{x}{x}(x^2 - 1)\big|_{x=0} = \frac{0}{0}, \text{ undefined} \quad ---(3)$$

To test the continuity of the function at x = 0, we might test whether

$$\lim_{x\to 0^+} M(x) = \lim_{x\to 0^-} M(x) = M(0).$$

It turns out that the former = - 1, the latter = + 1.

However, from (3), we know that M(0) is not even defined, therefore there is no need to calculate the limits, M(x) is not continuous at x = 0.

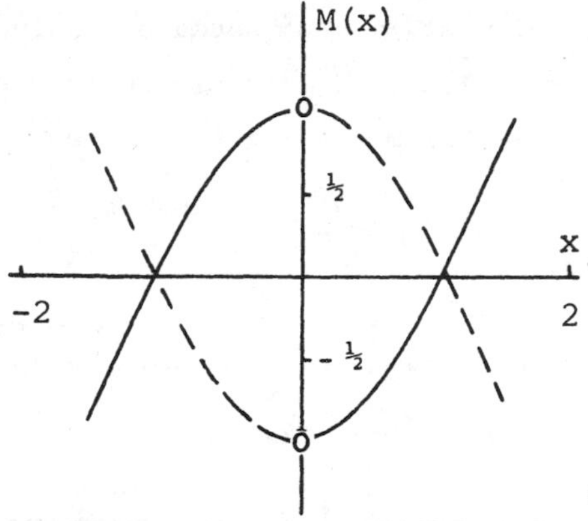

• **PROBLEM 54**

> Determine whether the function $y = \frac{1}{x - 2}$ is continuous at (a) x = 0, (b) x = 1, and (c) x = 2.

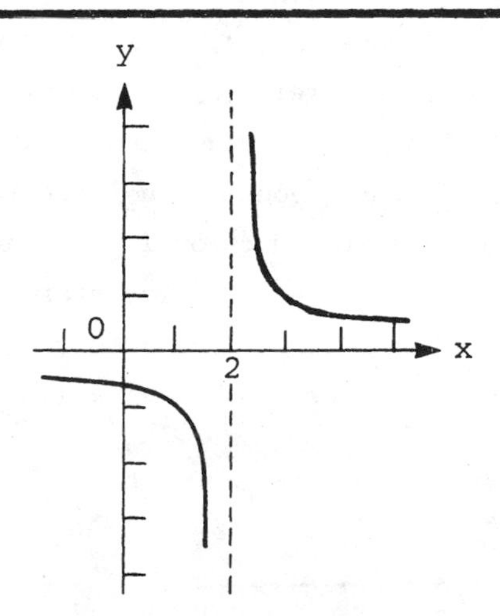

Solution: (a) $\lim_{x\to 0} \frac{1}{x-2} = \lim_{x\to 0} \frac{1}{0-2} = -\frac{1}{2}$

$$f(0) = -\frac{1}{2}.$$

Therefore,

$$\lim_{x \to 0} \frac{1}{x - 2} = f(0) = -\frac{1}{2},$$

and the function is continuous at $x = 0$, because the limit exists at $x = 0$ and $= f(0)$.

(b) $$\lim_{x \to 1} \frac{1}{x - 2} = -1$$

$$f(1) = -1.$$

The function is continuous at $x = 1$, because the limit exists at $x = 1$ and $= f(1)$.

(c) $$\lim_{x \to 2} \frac{1}{x - 2} = \lim_{x \to 2} \frac{1}{0} = \pm \infty .$$

The limit does not exist at $x = 2$ because the function is not defined at this point. I.e., if we approach from the left the function approaches $-\infty$, from the right, it goes to $+\infty$.

• PROBLEM 55

If $h(x) = \sqrt{4 - x^2}$, prove that $h(x)$ is continuous in the closed interval $[-2,2]$.

Solution: To prove continuity we employ the following definition: A function defined in the closed interval $[a,b]$ is said to be continuous in $[a,b]$ if and only if it is continuous in the open interval (a,b), as well as continuous from the right at a and continuous from the left at b. The function h is continuous in the open interval $(-2,2)$. We must now show that the function is continuous from the right at -2 and from the left at 2. Therefore, we must show that $f(-2)$ is defined and $\lim_{x \to -2^+} f(x)$ exists and that these are equal. Also, we must show that $f(2) = \lim_{x \to 2^-} f(x)$. We have:

$$\lim_{x \to -2^+} \sqrt{4 - x^2} = 0 = h(-2),$$

and

$$\lim_{x \to 2^-} \sqrt{4 - x^2} = 0 = h(2).$$

Thus, h is continuous in the closed interval $[-2,2]$.

• PROBLEM 56

Investigate the continuity of the function:

$$g(x) = \begin{cases} \dfrac{1}{x-2} & \text{if } x \neq 2 \\ 3 & \text{if } x = 2 \end{cases}$$

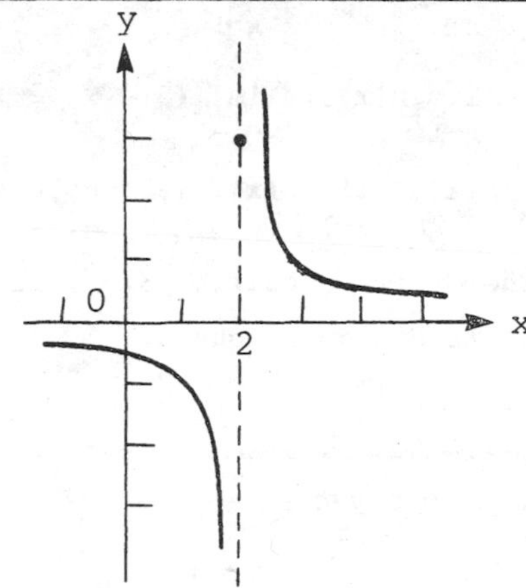

Solution:

For a function to be continuous at a point x_0, it must satisfy three conditions: (1) $f(x_0)$ is defined, (2) $\lim_{x \to x_0} f(x)$ exists, (3) $\lim_{x \to x_0} f(x) = f(x_0)$. At $x = 2$, $g(2) = 3$; therefore, condition (1) is satisfied. $\lim_{x \to 2^-} g(x) = -\infty$, and $\lim_{x \to 2^+} g(x) = +\infty$; therefore, condition (2) is not satisfied.

Thus, g is discontinuous at $x = 2$.

• PROBLEM 57

Investigate the continuity of the function:

$$h(x) = \begin{cases} 3 + x & \text{if } x \leq 1 \\ 3 - x & \text{if } 1 < x \end{cases}.$$

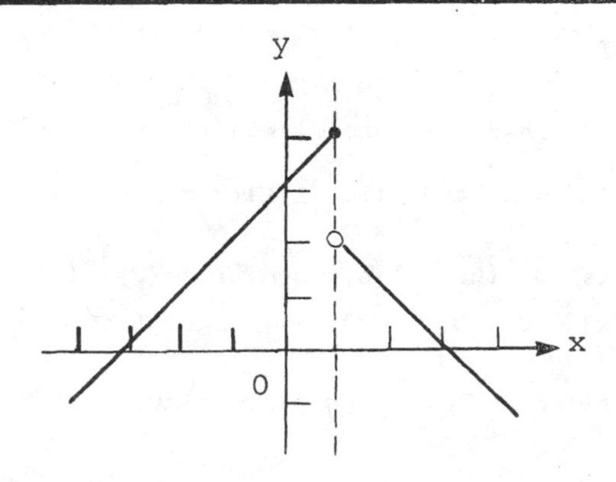

Solution: Because there is a break in the graph at the point $x = 1$, we investigate the three conditions for continuity at the point $x = 1$. The three conditions are: (1) $f(x_0)$ is defined, (2) $\lim_{x \to x_0} f(x)$ exists, (3) $\lim_{x \to x_0} f(x) = f(x_0)$. At $x = 1$, $h(1) = 4$; therefore, condition (1) is satisfied.

$$\lim_{x \to 1^-} h(x) = \lim_{x \to 1^-} (3 + x) = 4$$

$$\lim_{x \to 1^+} h(x) = \lim_{x \to 1^+} (3 - x) = 2.$$

Because $\lim_{x \to 1^-} h(x) \neq \lim_{x \to 1^+} h(x)$, we conclude that $\lim_{x \to 1} h(x)$ does not exist. Therefore, condition (2) fails to hold at 1.

Hence, h is discontinuous at 1.

• PROBLEM 58

Investigate continuity of:

$$F(x) = \begin{cases} |x - 3| & \text{if } x \neq 3 \\ 2 & \text{if } x = 3 \end{cases} .$$

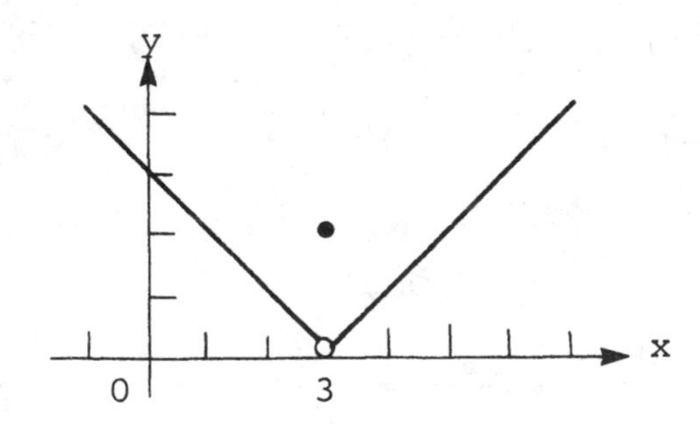

Solution:

We investigate the three conditions for continuity at the point $x = 3$. The three conditions are: (1) $f(x_0)$ is defined, (2) $\lim_{x \to x_0} f(x)$ exists, (3) $\lim_{x \to x_0} f(x) = f(x_0)$. At $x = 3$ we have

$F(3) = 2$; therefore, condition (1) is satisfied.

$\lim_{x \to 3^-} F(x) = 0$ and $\lim_{x \to 3^+} F(x) = 0$. Therefore, $\lim_{x \to 3} F(x)$ exists and is 0; therefore, condition (2) is satisfied.

$\lim_{x \to 3} F(x) = 0$ but $F(3) = 2$. Therefore, condition (3) is not satisfied. F is thus discontinuous at 3.

CHAPTER 5

DERIVATIVE Δ– METHOD

The Δ-method for finding derivatives is based on the following reasoning: Take any curve in the (x, y) - plane, defined analytically by a continuous function $y = f(x)$, and subdivide it into many little sections. Each section of the curve can then be approximated by a straight line. Along this line, the slope represents the rate of change of one variable with respect to the other. If one variable is y, the other x, then the rate of change of y with respect to x along the straight line is the slope. If we now consider a point (x, y) on the curve and a second nearby point $(x + \Delta x, y + \Delta y)$ also on the curve, then the change in x between them is Δx, the change in y is Δy, and their ratio is $\frac{\Delta y}{\Delta x}$. We said that this was an approximation, hence the smaller we make the section, the closer the approximation. If we allow Δx to approach zero, we have an infinite number of sections, and the approximation approaches an exact result. Hence we can define the derivative, $\frac{dy}{dx}$, as the value that the delta-ratio approaches as we let x go to zero. Mathematically, all this is expressed as follows:

$$\frac{dy}{dx} = \lim_{\Delta x \to 0} \frac{f(x+\Delta x) - f(x)}{\Delta x},$$

which is a general definition of the derivative.

• PROBLEM 59

Find the slope of each of the following curves at the given point, using the Δ-method.

(a) $y = 3x^2 - 2x + 4$ at (1,5)

(b) $y = x^3 - 3x + 5$ at (-2,3).

Solution: The slope of a given curve at a specified point is the derivative, in this case $\frac{\Delta y}{\Delta x}$, evaluated at that point.

(a) From the Δ-method we know that:

$$\frac{\Delta y}{\Delta x} = \frac{f(x + \Delta x) - f(x)}{\Delta x}.$$

For the curve $y = 3x^2 - 2x + 4$, we find:

$$\frac{\Delta y}{\Delta x} = \frac{3(x + \Delta x)^2 - 2(x + \Delta x) + 4 - (3x^2 - 2x + 4)}{\Delta x}$$

$$= \frac{3x^2 + 6x\Delta x + 3(\Delta x)^2 - 2x - 2\Delta x + 4 - 3x^2 + 2x - 4}{\Delta x}$$

$$= \frac{6x\Delta x + 3(\Delta x)^2 - 2\Delta x}{\Delta x}$$

$$= 6x + 3\Delta x - 2.$$

$$\lim_{\Delta x \to 0} \frac{\Delta y}{\Delta x} = \lim_{\Delta x \to 0} 6x + 3\Delta x - 2 = 6x - 2.$$

At (1,5) $\frac{\Delta y}{\Delta x} = 4$ is the required slope.

(b) Again using the Δ-method, $\frac{\Delta y}{\Delta x}$ for the curve: $y = x^3 - 3x + 5$, can be found as follows:

$$\frac{\Delta y}{\Delta x} = \frac{f(x + \Delta x) - f(x)}{\Delta x}.$$

$$\frac{\Delta y}{\Delta x} = \frac{(x + \Delta x)^3 - 3(x + \Delta x) + 5 - (x^3 - 3x + 5)}{\Delta x}$$

$$= \frac{x^3 + 3x^2\Delta x + 3x(\Delta x)^2 + (\Delta x)^3 - 3x - 3\Delta x + 5 - x^3 + 3x - 5}{\Delta x}$$

$$= \frac{3x^2\Delta x + 3x(\Delta x)^2 + (\Delta x)^3 - 3\Delta x}{\Delta x}$$

$$= 3x^2 + 3x\Delta x + (\Delta x)^2 - 3.$$

$$\lim_{\Delta x \to 0} \frac{\Delta y}{\Delta x} = \lim_{\Delta x \to 0} 3x^2 + 3x\Delta x + (\Delta x)^2 - 3 = 3x^2 - 3.$$

At (-2,3), $\frac{\Delta y}{\Delta x} = 9$ is the required slope.

• PROBLEM 60

Find the derivative function of $f(x) = 1/\sqrt{2x - 1}$, using the Δ-method.

Solution: By definition,

$$\frac{dy}{dx} = f'(x) = \lim_{\Delta x \to 0} \frac{f(x + \Delta x) - f(x)}{\Delta x}.$$

Since

$$f(x) = \frac{1}{\sqrt{2x - 1}},$$

$$f(x + \Delta x) = \frac{1}{\sqrt{2(x + \Delta x) - 1}} .$$

Substituting, we have:

$$f'(x) = \lim_{\Delta x \to 0} \frac{(1/\sqrt{2(x + \Delta x) - 1}) - (1/\sqrt{2x - 1})}{\Delta x} .$$

Substituting 0 for Δx gives a 0 in the denominator, which is meaningless. We therefore rearrange terms until we obtain a form in which substitution is possible. Now,

$$\frac{1}{\sqrt{2x + 2\Delta x - 1}} - \frac{1}{\sqrt{2x - 1}}$$

$$= \frac{\sqrt{2x - 1} - \sqrt{2x + 2\Delta x - 1}}{\sqrt{2x + 2\Delta x - 1}\ \sqrt{2x - 1}} .$$

Since multiplication by 1 does not change a term, let us multiply

$$\frac{\sqrt{2x - 1} - \sqrt{2x + 2\Delta x - 1}}{\sqrt{2x + 2\Delta x - 1}\ \sqrt{2x - 1}}$$

by

$$\frac{\sqrt{2x - 1} + \sqrt{2x + 2\Delta x - 1}}{\sqrt{2x - 1} + \sqrt{2x + 2\Delta x - 1}} ,$$

which equals 1.

Doing this, we have:

$$\frac{(\sqrt{2x-1}-\sqrt{2x+2\Delta x-1})(\sqrt{2x-1}+\sqrt{2x+2\Delta x-1})}{\sqrt{2x+2\Delta x-1}\ \sqrt{2x-1}(\sqrt{2x-1}+\sqrt{2x+2\Delta x-1})}$$

$$= \frac{(2x - 1) - (2x + 2\Delta x - 1)}{(2x-1)\sqrt{2x+2\Delta x-1}+(2x+2\Delta x-1)\sqrt{2x-1}} ,$$

by multiplying out.

We now have:

$$\lim_{\Delta x \to 0} \frac{-2\Delta x}{\left[(2x-1)\sqrt{2x+2\Delta x-1}+(2x+2\Delta x-1)\sqrt{2x-1}\right](\Delta x)}$$

$$= \lim_{\Delta x \to 0} \frac{-2}{(2x-1)\sqrt{2x+2\Delta x-1}+(2x+2\Delta x-1)\sqrt{2x-1}} .$$

Now we substitute 0 for Δx and obtain:

$$\frac{-2}{(2x - 1)\ \sqrt{2x - 1} + (2x - 1)\ \sqrt{2x - 1}}$$

$$= \frac{-2}{(2x - 1)^{\frac{3}{2}} + (2x - 1)^{\frac{3}{2}}} = \frac{-2}{2(2x - 1)^{\frac{3}{2}}}$$

$$= -\frac{1}{(2x-1)^{\frac{3}{2}}} = f'(x).$$

• PROBLEM 61

Find the instantaneous rate of change of the function:

$$y = \frac{2x}{x+1},$$

for any value of x and for x = 2.

Solution: The instantaneous rate of change of a function is defined as,

$$\lim_{\Delta x \to 0} \frac{\Delta y}{\Delta x} = \lim_{\Delta x \to 0} \frac{f(x+\Delta x) - f(x)}{\Delta x}.$$

Therefore,

$$\Delta y = f(x+\Delta x) - f(x).$$

In this case,

$$f(x) = \frac{2x}{x+1}, \text{ therefore,}$$

$$f(x+\Delta x) = \frac{2(x+\Delta x)}{x+\Delta x+1}.$$

Substituting, we have:

$$\Delta y = \frac{2x + 2 \cdot \Delta x}{x+\Delta x+1} - \frac{2x}{x+1}$$

$$= \frac{(2x+2\cdot\Delta x)(x+1) - 2x(x+\Delta x+1)}{(x+\Delta x+1)(x+1)}$$

$$= \frac{2x^2+2x\cdot\Delta x+2x+2\cdot\Delta x-2x^2-2x\cdot\Delta x-2x}{(x+\Delta x+1)(x+1)}$$

$$= \frac{2\cdot\Delta x}{(x+\Delta x+1)(x+1)}.$$

$$\frac{\Delta y}{\Delta x} = \frac{2\cdot\Delta x}{(x+\Delta x+1)(x+1)(\Delta x)}$$

$$= \frac{2}{(x+\Delta x+1)(x+1)}$$

Now,

$$\lim_{\Delta x \to 0} \frac{\Delta y}{\Delta x} = \lim_{\Delta x \to 0} \frac{2}{(x+\Delta x+1)(x+1)}.$$

Substituting 0 for Δx we have,

$$\lim_{\Delta x \to 0} \frac{\Delta y}{\Delta x} = \frac{2}{(x+1)^2},$$

the instantaneous rate of change for any value of x.

For x = 2, we have,

$$\frac{2}{(x+1)^2} = \frac{2}{(2+1)^2} = \frac{2}{9} .$$

• PROBLEM 62

Compute the average rate of change of $y = f(x) = x^2 - 2$ between x = 3 and x = 4.

Solution: Average rate of change is defined as:

$\frac{\Delta y}{\Delta x}$ with $\Delta y = f(x + \Delta x) - f(x)$.

Given: $x = 3$, $\Delta x = 4 - 3 = 1$,

$y = f(x) = f(3) = 3^2 - 2 = 7$

For $x = 4$,

$y + \Delta y = f(x + \Delta x) = 4^2 - 2 = 14$

$$\Delta y = f(x + \Delta x) - f(x) = f(4) - f(3)$$

$$= (4^2 - 2) - (3^2 - 2) = 14 - 7 = 7$$

$\frac{\Delta y}{\Delta x} = \frac{7}{1} = 7$, the average rate of change.

• PROBLEM 63

Find the average rate of change, by the Δ process, for:

$$y = 1/x.$$

Solution:

$$y = f(x) = \frac{1}{x}$$

The average rate of change is defined to be

$\frac{\Delta y}{\Delta x}$ with $\Delta y = f(x + \Delta x) - f(x)$.

Since

$$f(x) = \frac{1}{x}, \quad f(x + \Delta x) = \frac{1}{x + \Delta x},$$

and

$$\Delta y = \frac{1}{x + \Delta x} - \frac{1}{x} = \frac{x - (x + \Delta x)}{x(x + \Delta x)}$$

$$= \frac{-\Delta x}{x(x + \Delta x)} .$$

Now,

$$\frac{\Delta y}{\Delta x} = \frac{-\ \Delta x}{x\ (x + \Delta x)\ \Delta x} = -\ \frac{1}{x\ (x + \Delta x)} .$$

Therefore, the average rate of change is

$$\frac{-\ 1}{x\ (x + \Delta x)} .$$

• PROBLEM 64

Compute the increment Δy of $y = 3x + 7$ between $x = -\ 1$ and $x = 1$.

Solution: By definition,

$$\Delta y = f(x + \Delta x) - f(x)$$

Given: $x = -\ 1, \quad \Delta x = 1 - (-\ 1) = 2$

$$\Delta y = f(x + \Delta x) - f(x)$$

$$= f(-\ 1 + 2) - f(-\ 1)$$

$$= \big(3(1) + 7\big) - \big(3(-\ 1) + 7\big) = 10 - 4 = 6.$$

Therefore, the increment,

$$\Delta y = 6.$$

• PROBLEM 65

Find the rate of change of y with respect to x at the point $x = 5$, if

$$2y = x^2 + 3x - 1.$$

Solution: Rate of change is defined as

$$\lim_{\Delta x \to 0} \frac{\Delta y}{\Delta x}, \quad \text{with}$$

$$\Delta y = f(x + \Delta x) - f(x).$$

We have:

$$2\Delta y = (x + \Delta x)^2 + 3(x + \Delta x) - 1 - (x^2 + 3x - 1)$$

$$= x^2 + 2x \cdot \Delta x + (\Delta x)^2 + 3x + 3\Delta x - 1 - x^2 - 3x + 1$$

$$= 2x \cdot \Delta x + (\Delta x)^2 + 3\Delta x.$$

Dividing by Δx,

$$\frac{2\ \Delta y}{\Delta x} = \frac{2x \cdot \Delta x}{\Delta x} + \frac{(\Delta x)^2}{\Delta x} + \frac{3\ \Delta x}{\Delta x}$$

$$= 2x + \Delta x + 3$$

and

$$\frac{\Delta y}{\Delta x} = x + \frac{\Delta x}{2} + \frac{3}{2}.$$

Now,

$$\lim_{\Delta x \to 0} \frac{\Delta y}{\Delta x} = \lim_{\Delta x \to 0} x + \frac{\Delta x}{2} + \frac{3}{2} = x + \frac{3}{2}.$$

For $x = 5$,

$$\lim_{\Delta x \to 0} \frac{\Delta y}{\Delta x} = 5 + \frac{3}{2} = 6\frac{1}{2}$$

This means that the instantaneous rate of change of the function represented by the curve at the point $x = 5$ is $6\ 1/2$.

The function, it is seen, changes $6\frac{1}{2}$ times as fast as the independent variable x at $x = 5$.

The slope of the tangent at $x = 5$ is $6\frac{1}{2}$.

• PROBLEM 66

Find the derivative of the function: $y = 2x^2 + 3x$, by the delta process.

Solution: By definition,

$$y'(x) = \lim_{\Delta x \to 0} \frac{f(x+\Delta x) - f(x)}{\Delta x}.$$

Since

$$f(x) = 2x^2 + 3x,$$

$$f(x+\Delta x) = 2(x+\Delta x)^2 + 3(x+\Delta x).$$

Substituting, we obtain:

$$y'(x) = f'(x) = \lim_{\Delta x \to 0} \frac{2(x+\Delta x)^2 + 3(x+\Delta x) - (2x^2+3x)}{\Delta x}.$$

Simplifying, we have:

$$f'(x) = \lim_{\Delta x \to 0} \frac{4x\Delta x + 2(\Delta x)^2 + 3\Delta x}{\Delta x}$$

$$= \lim_{\Delta x \to 0} 4x + 2\Delta x + 3.$$

Now, as $\Delta x \to 0$ the term: $2\Delta x$, drops out and we have

$$f'(x) = 4x + 3.$$

• PROBLEM 67

Using the delta method, find $f'(x)$ for the function:

$$f(x) = x^2 + \frac{1}{x},\ x \neq 0.$$

Solution: By definition,

$$f'(x) = \frac{dy}{dx} = \lim_{\Delta x \to 0} \frac{f(x+\Delta x)-f(x)}{\Delta x}$$

$$f(x+\Delta x) = (x+\Delta x)^2 + \frac{1}{x+\Delta x}$$

$$= x^2 + 2x\Delta x + (\Delta x)^2 + \frac{1}{x+\Delta x}.$$

$$f(x) = x^2 + \frac{1}{x}.$$

Subtracting the two expressions,

$$f(x+\Delta x) - f(x) = 2x\Delta x + (\Delta x)^2 + \frac{1}{x+\Delta x} - \frac{1}{x}$$

$$= 2x\Delta x + (\Delta x)^2 + \frac{x-(x+\Delta x)}{x(x+\Delta x)}$$

$$= \Delta x\left(2x + \Delta x - \frac{1}{x(x+\Delta x)}\right).$$

Dividing by Δx, we have:

$$\frac{f(x+\Delta x) - f(x)}{\Delta x} = 2x + \Delta x - \frac{1}{x(x+\Delta x)}.$$

Since

$$f'(x) = \lim_{\Delta x \to 0} \frac{f(x+\Delta x) - f(x)}{\Delta x},$$

we have,

$$\lim_{\Delta x \to 0} \left(2x + \Delta x - \frac{1}{x(x+\Delta x)}\right)$$

$$= 2x + 0 - \frac{1}{x(x+0)}$$

$$= 2x - \frac{1}{x^2}.$$

• PROBLEM 68

Find the derivative of (a):

$$y = 1/x$$

and (b):

$$y = x^2 + \frac{1}{x+1},$$

by the Δ-process.

Solution: By definition,

$$f'(x) = \frac{dy}{dx} = \lim_{\Delta x \to 0} \frac{f(x+\Delta x) - f(x)}{\Delta x}.$$

Since

$$f(x) = \frac{1}{x},$$

$$f(x+\Delta x) = \frac{1}{(x+\Delta x)}.$$

Substituting, we have:

$$f'(x) = \lim_{\Delta x \to 0} \frac{\frac{1}{x+\Delta x} - \frac{1}{x}}{\Delta x}$$

$$= \lim_{\Delta x \to 0} \frac{x - (x+\Delta x)}{(x+\Delta x)(x)} \cdot \frac{1}{\Delta x}$$

$$= \lim_{\Delta x \to 0} \frac{-\Delta x}{(x+\Delta x)(x)(\Delta x)}$$

$$= \lim_{\Delta x \to 0} \frac{-1}{(x+\Delta x)(x)}$$

$$= \frac{-1}{x^2}.$$

Therefore, the derivative $= -\frac{1}{x^2}$.

(b) We write:

$$f'(x) = \lim_{\Delta x \to 0} \frac{f(x+\Delta x) - f(x)}{\Delta x}.$$

We are given:

$$f(x) = x^2 + \frac{1}{x+1};$$

therefore

$$f(x+\Delta x) = (x+\Delta x)^2 + \frac{1}{x + \Delta x + 1}.$$

By substitution,

$$f'(x) = \lim_{\Delta x \to 0} \frac{(x+\Delta x)^2 + \frac{1}{x + \Delta x + 1} - \left(x^2 + \frac{1}{x+1}\right)}{\Delta x}$$

$$= \lim_{\Delta x \to 0} \left(x^2+2x\cdot\Delta x+(\Delta x)^2+\frac{1}{x+\Delta x+1} - x^2 - \frac{1}{x+1}\right)\cdot\frac{1}{\Delta x}.$$

The problem is most easily handled algebraically if the fractions are combined and the other terms are handled separately. We have:

$$f'(x) = \lim_{\Delta x \to 0} \left(2x\Delta x + (\Delta x)^2 + \frac{1}{x+\Delta x+1} - \frac{1}{x+1}\right) \cdot \frac{1}{\Delta x}$$

$$= \lim_{\Delta x \to 0} \left(2x\Delta x + (\Delta x)^2 + \frac{(x+1) - (x+\Delta x+1)}{(x+\Delta x+1)(x+1)}\right) \cdot \frac{1}{\Delta x}$$

$$= \lim_{\Delta x \to 0} \left(2x\Delta x + (\Delta x)^2 - \frac{\Delta x}{(x+\Delta x+1)(x+1)}\right) \cdot \frac{1}{\Delta x}$$

$$= \lim_{\Delta x \to 0} 2x + \Delta x - \frac{1}{(x+\Delta x+1)(x+1)}.$$

Therefore,

$$f'(x) = 2x - \frac{1}{(x+1)^2}.$$

• PROBLEM 69

Find the derivative of f(x) = 2/(3x+1), using the Δ-method.

Solution: By definition,

$$\frac{dy}{dx} = f'(x) = \lim_{\Delta x \to 0} \frac{f(x+\Delta x) - f(x)}{\Delta x}.$$

Since

$$f(x) = \frac{2}{3x+1},$$

$$f(x+\Delta x) = \frac{2}{3(x+\Delta x)+1}.$$

Substituting, we obtain:

$$f'(x) = \lim_{\Delta x \to 0} \frac{2/[3(x+\Delta x)+1] - 2/(3x+1)}{\Delta x}$$

At this point we cannot substitute 0 for Δx because we would obtain a 0 in the denominator. Multiplying out and obtaining a common denominator we have:

$$\lim_{\Delta x \to 0} \frac{6x + 2 - 6x - 6\Delta x - 2}{(3x+3\Delta x+1)(3x+1)\Delta x}$$

$$= \frac{-6\Delta x}{(3x+3\Delta x+1)(3x+1)(\Delta x)}$$

$$= \lim_{\Delta x \to 0} \frac{-6}{(3x+3\Delta x+1)(3x+1)}$$

We can now substitute 0 for Δx, and we obtain:

$$-\frac{6}{(3x+1)^2} = f'(x).$$

• PROBLEM 70

Find dy/dx of the function $y = \sqrt{x}$ by the Δ-process.

Solution: By definition,

$$f'(x) = \lim_{\Delta x \to 0} \frac{f(x+\Delta x) - f(x)}{\Delta x}.$$

Since

$$f(x) = \sqrt{x},$$

$$f(x+\Delta x) = \sqrt{x+\Delta x}.$$

Substituting, we obtain:

$$f'(x) = \frac{dy}{dx} = \lim_{\Delta x \to 0} \frac{\sqrt{x+\Delta x} - \sqrt{x}}{\Delta x}.$$

We now rationalize the numerator to obtain a form in which Δx can meaningfully approach its limit, 0. Doing this, we obtain:

$$f'(x) = \lim_{\Delta x \to 0} \frac{\sqrt{x+\Delta x} - \sqrt{x}}{\Delta x} \cdot \frac{(\sqrt{x+\Delta x} + \sqrt{x})}{(\sqrt{x+\Delta x} + \sqrt{x})}$$

$$= \lim_{\Delta x \to 0} \frac{x + \Delta x - x}{\Delta x(\sqrt{x+\Delta x} + \sqrt{x})}$$

$$= \lim_{\Delta x \to 0} \frac{1}{\sqrt{x+\Delta x} + \sqrt{x}}.$$

Now, as

$$\Delta x \to 0, \quad \frac{1}{\sqrt{x+\Delta x} + \sqrt{x}} \to \frac{1}{\sqrt{x} + \sqrt{x}}.$$

Therefore,

$$f'(x) = \frac{1}{2\sqrt{x}}.$$

• PROBLEM 71

Find the derivative of:

$$f(x) = \sqrt{2x - 1},$$

using the Δ-method.

Solution: By definition,

$$\frac{dy}{dx} = f'(x) = \lim_{\Delta x \to 0} \frac{f(x+\Delta x) - f(x)}{\Delta x}.$$

Since

$$f(x) = \sqrt{2x - 1},$$

$$f(x+\Delta x) = \sqrt{2(x+\Delta x) - 1}.$$

Substituting, we obtain:

$$f'(x) = \lim_{\Delta x \to 0} \frac{\sqrt{2(x+\Delta x) - 1} - \sqrt{2x - 1}}{\Delta x}.$$

Substituting 0 for Δx gives a 0 in the denominator. Therefore, we rearrange the numerator. Since multiplication by 1 does not change a term, we multiply

$$(\sqrt{2x + 2\Delta x - 1} - \sqrt{2x - 1})$$

by 1, writing 1 as:

$$\frac{\sqrt{2x + 2\Delta x - 1} + \sqrt{2x - 1}}{\sqrt{2x + 2\Delta x - 1} + \sqrt{2x - 1}}.$$

We have:

$$\frac{1}{\Delta x} \cdot \left(\sqrt{2x + 2\Delta x - 1} - \sqrt{2x - 1}\right)$$

$$= \frac{[\sqrt{2x+2\Delta x-1} - \sqrt{2x-1}][\sqrt{2x+2\Delta x-1} + \sqrt{2x-1}]}{\sqrt{2x+2\Delta x-1} + \sqrt{2x-1}} \cdot \frac{1}{\Delta x}.$$

Multiplying out the numerator, we obtain:

$$\frac{(2x+2\Delta x-1) - (2x-1)}{\sqrt{2x+2\Delta x-1} + \sqrt{2x-1}} \cdot \frac{1}{\Delta x}$$

$$= \frac{2\Delta x}{(\sqrt{2x+2\Delta x-1} + \sqrt{2x-1})(\Delta x)}$$

$$f'(x) = \lim_{\Delta x \to 0} \frac{2}{\sqrt{2x+2\Delta x-1} + \sqrt{2x-1}}.$$

We can now substitute for Δx, and we obtain:

$$\frac{2}{2\sqrt{2x-1}} = \frac{1}{\sqrt{2x-1}} = f'(x).$$

• PROBLEM 72

Given $f(x) = x^{2/3}$, find $f'(x)$ using the Δ-method.

Solution: By definition,

$$\frac{dy}{dx} = f'(x) = \lim_{\Delta x \to 0} \frac{f(x+\Delta x) - f(x)}{\Delta x}$$

$$f(x) = x^{2/3}.$$

Therefore,

$$f(x+\Delta x) = (x+\Delta x)^{2/3}.$$

Substituting, we have:

$$\lim_{\Delta x \to 0} \frac{(x+\Delta x)^{2/3} - x^{2/3}}{\Delta x}.$$

We wish to rewrite this in a form in which Δx can approach the limit, 0. Direct substitution at this point would leave a 0 in the denominator. Since multiplication by 1 does not change the value, we multiply

$$\frac{(x+\Delta x)^{2/3} - x^{2/3}}{\Delta x}$$

by

$$\frac{(x+\Delta x)^{4/3} + (x+\Delta x)^{2/3}x^{2/3} + x^{4/3}}{(x+\Delta x)^{4/3} + (x+\Delta x)^{2/3}x^{2/3} + x^{4/3}},$$

which is equal to 1. Doing this we have,

$$f'(x) = \lim_{\Delta x \to 0}$$

$$\frac{\left[(x+\Delta x)^{2/3} - x^{2/3}\right]\left[(x+\Delta x)^{4/3} + (x+\Delta x)^{2/3}x^{2/3} + x^{4/3}\right]}{\Delta x\left[(x+\Delta x)^{4/3} + (x+\Delta x)^{2/3}x^{2/3} + x^{4/3}\right]}$$

$$= \lim_{\Delta x \to 0} \frac{(x+\Delta x)^2 - x^2}{\Delta x[x+\Delta x)^{4/3} + (x+\Delta x)^{2/3}x^{2/3} + x^{4/3}]}$$

$$= \lim_{\Delta x \to 0} \frac{x^2 + 2x(\Delta x) + (\Delta x)^2 - x^2}{\Delta x[(x+\Delta x)^{4/3} + (x+\Delta x)^{2/3}x^{2/3} + x^{4/3}]}$$

$$= \lim_{\Delta x \to 0} \frac{2x(\Delta x) + (\Delta x)^2}{\Delta x[(x+\Delta x)^{4/3} + (x+\Delta x)^{2/3}x^{2/3} + x^{4/3}]}$$

$$= \lim_{\Delta x \to 0} \frac{2x + \Delta x}{(x+\Delta x)^{4/3} + (x+\Delta x)^{2/3}x^{2/3} + x^{4/3}}.$$

Now, substituting 0 for Δx, we obtain:

$$\frac{2x}{x^{4/3} + x^{2/3}x^{2/3} + x^{4/3}}$$

$$= \frac{2x}{3x^{4/3}} = \frac{2x}{3x^{3/3}x^{1/3}}$$

$$= \frac{2}{3x^{1/3}}.$$

CHAPTER 6

DIFFERENTIATION OF ALGEBRAIC FUNCTIONS

Speaking most generally, the derivative is the RATE OF CHANGE of one variable with respect to another. If y is a function of x, then, as x changes, y changes. The derivative tells us the rate at which y is changing with respect to x. To employ a simple example, we can consider the relation:

$$y = 2x,$$

where it can be seen that y changes twice as fast as x. Hence, the derivative of y with respect to x is 2. If the curve is other than a straight line, then the process of finding its derivative is also somewhat more complex. Moreover, the value of the derivative changes from point to point, along the curve a fact which is not true for the straight line.

To find the derivative, one method, the Δ-method, is always applicable, and yields a result, provided the derivative exists. However, the Δ-method is lengthy, and involved, and is generally not required, except for practice.

When the Δ-method is not specifically called for, it is best to apply the basic derivative relationships that have been developed in calculus. These should be memorized and when applied properly, it is usually possible to obtain easily the derivative of a function, regardless of its complexity or the number of terms involved in the function. This, however, is not true when attempting to integrate a function, as discussed in subsequent chapters.

Algebraic functions are generally differentiated by the use of a few general rules. Among these, the most prominent one is the power rule. This rule states that if

$$y = x^n,$$

then,

$$\frac{dy}{dx} = nx^{n-1},$$

i.e., the derivative has a power one less than the function, and is multiplied by the original exponent. The rule holds for all exponents, positive, negative, fractional, etc. Hence, we have, for:

$$y = x^3 , \quad \frac{dy}{dx} = 3x^2 ,$$

$$y = \frac{1}{x} , \quad \frac{dy}{dx} = -\frac{1}{x^2}$$

and

$$y = x^{\frac{1}{4}} , \quad \frac{dy}{dx} = \tfrac{1}{4} x^{-\frac{3}{4}} = \frac{1}{4x^{\frac{3}{4}}} .$$

Also of importance is the chain rule, which states that, if $y = f(u)$, and $u = F(x)$, then, $\frac{dy}{dx} = \frac{dy}{du} \cdot \frac{du}{dx}$, a rule that is not restricted to two variables, but holds for any number.

Additionally, we can give here the rules for the derivatives of products and quotients, which are:

if $y = uv$, where u and v are functions of x, then

$$\frac{dy}{dx} = u \frac{dv}{dx} + v \frac{du}{dx} ;$$

and, if $y = \frac{u}{v}$, then

$$\frac{dy}{dx} = \frac{v \frac{du}{dx} - u \frac{dv}{dx}}{v^2} .$$

By use of the formulas given above, singly or in combination, most algebraic functions can be differentiated.

• PROBLEM 73

Find the derivative of: $y = x^{3b}$.

Solution: Applying the theorem for $d(u^n)$,

$$\frac{dy}{dx} = 3b \cdot x^{3b-1} .$$

• PROBLEM 74

Find the derivative of: $y = (2x^3 - 5x^2 + 4)^5$.

Solution: $D_x = \frac{d}{dx}$. This problem can be solved by simply applying the theorem for $d(u^n)$. However, to illustrate the use of the chain rule, make the following substitutions:

$$y = u^5 \qquad \text{where } u = 2x^3 - 5x^2 + 4$$

Therefore, from the chain rule,

$$D_x y = D_u y \cdot D_x u = 5u^4 (6x^2 - 10x)$$

$$= 5(2x^3 - 5x^2 + 4)^4 (6x^2 - 10x).$$

• PROBLEM 75

Find the derivative of: $y = (x^2 + 2)^3$.

Solution: Method 1. We may expand the cube and write:

$$\frac{dy}{dx} = \frac{d}{dx}[(x^2 + 2)^3] = \frac{d}{dx}(x^6 + 6x^4 + 12x^2 + 8)$$

$$= 6x^5 + 24x^3 + 24x.$$

Method 2. Let $u = x^2 + 2$, then $y = (x^2 + 2)^3 = u^3$; Using the chain rule we have:

$$\frac{dy}{dx} = \frac{dy}{du} \cdot \frac{du}{dx} = \frac{d(u^3)}{du} \cdot \frac{d(x^2 + 2)}{dx} = 3u^2(2x)$$

$$= 3(x^2 + 2)^2 \cdot (2x) = 3(x^4 + 4x^2 + 4) \cdot (2x)$$

$$= 6x^5 + 24x^3 + 24x.$$

• PROBLEM 76

Find the derivative of $y = (x^3 - 4)^5$ with respect to x.

Solution: Applying the theorem for $d(u^n)$,

$$\frac{dy}{dx} = 5(x^3 - 4)^4 \cdot \frac{d}{dx}(x^3 - 4)$$

$$= 5(x^3 - 4)^4(3x^2) = 15x^2(x^3 - 4)^4.$$

The same result is, of course, obtained when $(x^3 - 4)^5$ is multiplied out, and the derivative is taken term by term of the resultant expansion.

• PROBLEM 77

Find $D_x(2x^2 + 3x)^{10}$.

Solution: $D_x = \frac{d}{dx}$. Using the theorem for $d(u^n)$,

i.e. $d(u^n) = nu^{n-1}du$,

$$D_x(2x^2 + 3x)^{10} = 10(2x^2 + 3x)^9 D_x(2x^2 + 3x)$$

$$= 10(2x^2 + 3x)^9(4x + 3).$$

• **PROBLEM 78**

Find the derivative of:

$$v = \sqrt[4]{\frac{1}{y^7}}.$$

<u>Solution:</u> Rewrite the expression to replace the radical symbol with an exponent, and apply the theorem for $d(u^n)$.

$$v = y^{-\frac{7}{4}}$$

$$\frac{dv}{dy} = -\frac{7}{4} \cdot y^{-\frac{7}{4}-1} = -\frac{7}{4} \cdot y^{-\frac{11}{4}}$$

$$= -\frac{7}{4}\sqrt[4]{\frac{1}{y^{11}}}.$$

• **PROBLEM 79**

Find $f'(x)$ from $f(x) = 2\sqrt[5]{x^3}$.

<u>Solution:</u> $f(x) = 2x^{3/5}$ and, by the rule for rational exponents,

$$f'(x) = 2\left(\frac{3}{5}\right)x^{\frac{3}{5}-1} = \frac{6}{5}\,x^{-\frac{2}{5}}.$$

• **PROBLEM 80**

Find the derivative of:

$$y = \frac{ax^n - 2}{b}.$$

<u>Solution:</u>

$$y = \frac{a}{b}\,x^n - \frac{2}{b}$$

Applying the theorem for $d(u^n)$,

$$\frac{dy}{dx} = \frac{a}{b} \cdot n \cdot x^{n-1} - 0,$$

or

$$\frac{dy}{dx} = \frac{na}{b} \cdot x^{n-1}.$$

The same result may be obtained by writing

$$y = \frac{1}{b}(ax^n-2),$$

and solving for $dy/dx = \frac{1}{b}(nax^{n-1})$.

• PROBLEM 81

Find the derivative of:

$$y = \sqrt[a]{\frac{1}{x^c}} .$$

Solution: Rewrite the expression in the form to which the theorem for $d(u^n)$ may be applied.

$$y = x^{-\frac{c}{a}}$$

$$\frac{dy}{dx} = -\frac{c}{a} \cdot x^{-\frac{c}{a} - 1} = -\frac{c}{a} x^{-\frac{(c + a)}{a}} .$$

• PROBLEM 82

Find the derivative of: $h(x) = \sqrt{2x^3 - 4x + 5}$.

Solution: Rewrite the expression in the form, to which it is possible to apply the theorem for $d(u^n)$.

$$h(x) = (2x^3 - 4x + 5)^{1/2}.$$

Then

$$h'(x) = \frac{1}{2}(2x^3 - 4x + 5)^{-1/2}(6x^2 - 4),$$

and

$$h' = \frac{dh}{dx} = \frac{3x^2 - 2}{\sqrt{2x^3 - 4x + 5}} .$$

• PROBLEM 83

Find the derivative of:

$$f(x) = \sqrt{x^2 + 1} .$$

Solution:

$$f(x) = (x^2 + 1)^{\frac{1}{2}},$$

and

$$f'(x) = \frac{1}{2}(x^2 + 1)^{-\frac{1}{2}}(2x),$$

or

$$f'(x) = \frac{2x}{2\sqrt{x^2 + 1}} = \frac{x}{\sqrt{x^2 + 1}} .$$

• PROBLEM 84

Given $y = \sqrt{3x^2 + 4}$, find dy/dx.

Solution: Using the chain rule method in combination with substitution, let

$$u = 3x^2 + 4, \quad \text{then } y = \sqrt{u};$$

and

$$\frac{dy}{dx} = \frac{dy}{du} \cdot \frac{du}{dx} = \frac{1}{2\sqrt{u}} \cdot 6x$$

$$= \frac{6x}{2\sqrt{3x^2 + 4}} = \frac{3x\sqrt{3x^2 + 4}}{3x^2 + 4}.$$

The same result may be obtained by another method:

Square both sides, and take the derivative of each side of the resulting equation with respect to x:

$$\frac{d(y^2)}{dx} = \frac{d(3x^2 + 4)}{dx} = 6x.$$

But

$$\frac{d(y^2)}{dx} = \frac{d(y^2)}{dy} \cdot \frac{dy}{dx} = 2y\frac{dy}{dx}.$$

Hence,

$$2y\frac{dy}{dx} = 6x, \text{ or } \frac{dy}{dx} = \frac{3x}{y} = \frac{3x}{\sqrt{3x^2 + 4}}$$

$$= \frac{3x\sqrt{3x^2 + 4}}{3x^2 + 4}.$$

• PROBLEM 85

Find the derivative of $p = \sqrt[3]{4 - t}$.

Solution: Replacing the cube root with a fractional exponent, the given equation may be written as

$$p = (4 - t)^{1/3}$$

The equation is now in the form, to which we can apply the theorem for $d(u^n)$. Therefore,

$$\frac{dp}{dt} = \frac{1}{3}(4 - t)^{-2/3} \cdot \frac{d}{dt}(4 - t)$$

$$= \frac{1}{3}(4 - t)^{-2/3}(-1) = -\frac{1}{3}(4 - t)^{-2/3}.$$

• PROBLEM 86

If $f(t) = \sqrt[3]{t^3 + 3t + 1}$, find $f'(t)$.

Solution:

$$f'(t) = \frac{df}{dt}.$$

Replace the radical symbol by an exponent, and apply the theorem for $d(u^n)$.

$$f(t) = (t^3 + 3t + 1)^{1/3},$$

$$f'(t) = \frac{1}{3}(t^3 + 3t + 1)^{-2/3}(3t^2 + 3)$$

$$= \frac{t^2 + 1}{(t^3 + 3t + 1)^{2/3}}.$$

• PROBLEM 87

Find the derivative of:

$$f(x) = \frac{1}{4x^3 + 5x^2 - 7x + 8}.$$

Solution: This function is best differentiated by rewriting it so that the theorem for

$d(u^n)$ i.e.: $d(u^n) = nu^{n-1}du$, may be applied.

$$f(x) = (4x^3 + 5x^2 - 7x + 8)^{-1}$$

$$f'(x) = -1(4x^3 + 5x^2 - 7x + 8)^{-2}(12x^2 + 10x - 7)$$

$$= \frac{-12x^2 - 10x + 7}{(4x^3 + 5x^2 - 7x + 8)^2}.$$

• PROBLEM 88

Given $y = x^3 - 2\sqrt{3x^2 + 4}$, find dy/dx.

Solution:

$$\frac{dy}{dx} = \frac{d(x^3)}{dx} - 2\frac{d}{dx}\sqrt{3x^2 + 4}$$

$$= 3x^2 - \frac{6x\sqrt{3x^2 + 4}}{3x^2 + 4},$$

where

$$\frac{d}{dx}\sqrt{3x^2 + 4} = \frac{d}{dx}(3x^2 + 4)^{\frac{1}{2}}$$

$$= d(u^n)$$

$$= 6x(3x^2 + 4)^{-\frac{1}{2}}.$$

• PROBLEM 89

Find the derivative of:

$$f(x) = 3 + \frac{5}{\sqrt{x}} + 2\sqrt{x} - \frac{1}{x\sqrt{x}} .$$

Solution: Rewriting the function f(x) in terms of expressions with exponents, so that the basic theorem for $d(u^n)$ may be applied,

$$f(x) = 3 + 5x^{-1/2} + 2x^{1/2} - x^{-3/2}.$$

$$f'(x) = -\frac{5}{2}x^{-3/2} + x^{-1/2} + \frac{3}{2}x^{-5/2}$$

$$= \frac{-5}{2x\sqrt{x}} + \frac{1}{\sqrt{x}} + \frac{3}{2x^2\sqrt{x}} .$$

• PROBLEM 90

Find the derivative of y^n with respect to y^7.

Solution: Using the definition for the derivative, and recognizing that $\frac{du}{dv} = \frac{\frac{du}{dy}}{\frac{dv}{dy}}$,

$$\frac{d}{d(y^7)}(y^n) = \frac{ny^{n-1}}{7y^{7-1}} = \frac{n}{7} \cdot y^{n-1-6} = \frac{n}{7} \cdot y^{n-7}.$$

• PROBLEM 91

Find the sixth derivative of $y = x^6$.

Solution:

First derivative $= 6x^{6-1} = 6x^5$

Second derivative $= 5 \cdot 6x^{5-1} = 30x^4$

Third derivative $= 4 \cdot 30x^{4-1} = 120x^3$

Fourth derivative $= 3 \cdot 120x^{3-1} = 360x^2$

Fifth derivative $= 2 \cdot 360x^{2-1} = 720x^1 = 720x$

Sixth derivative $= 1 \cdot 720x^{1-1} = 720x^0 = 720$

The seventh derivative is seen to be zero, and therefore the function $y = x^6$ has seven derivatives.

• PROBLEM 92

Find the derivative of:
$y = (x^2 - 3x)(4x + 5)$.

Solution: Using the theorem: $d(uv) = udv + vdu$,

$$\frac{dy}{dx} = (x^2 - 3x) \cdot \frac{d}{dx}(4x + 5) + (4x + 5) \cdot \frac{d}{dx}(x^2 - 3x).$$

$$\frac{dy}{dx} = (x^2 - 3x) \cdot 4 + (4x + 5)(2x^{2-1} - 3)$$

$$\frac{dy}{dx} = 4(x^2 - 3x) + (4x + 5)(2x - 3)$$

$$= 4x^2 - 12x + 8x^2 - 2x - 15$$

$$= 12x^2 - 14x - 15.$$

• PROBLEM 93

Find the derivative of:
$f(x) = (x^2 + 1)(1 - 3x)$.

Solution: Using the theorem for differentiating a product of terms, i.e.: $d(uv) = udv + vdu$

$$f'(x) = (x^2 + 1)(-3) + (1 - 3x)2x = -9x^2 + 2x - 3.$$

• PROBLEM 94

Find the derivative of:
$y = (x^2 + 1)^3(x^3 - 1)^2$.

Solution: We could, of course, expand everything here and write y as a polynomial in x, but this would make the solution laborious. The preferred method is to note that the equation is a product of two expressions to which the theorem for d(uv) i.e.: $d(uv) = vdu + udv$, may be applied.

$$\frac{dy}{dx} = (x^2 + 1)^3 \frac{d}{dx}(x^3 - 1)^2 + (x^3 - 1)^2 \frac{d}{dx}(x^2 + 1)^3.$$

But

$$\frac{d}{dx}(x^3 - 1)^2 = 2(x^3 - 1)\frac{d}{dx}(x^3 - 1)$$

$$= 2(x^3 - 1) \cdot 3x^2 = 6x^2(x^3 - 1),$$

and

$$\frac{d}{dx}(x^2 + 1)^3 = 3(x^2 + 1)^2\frac{d}{dx}(x^2 + 1)$$

$$= 3(x^2 + 1)^2 \cdot 2x = 6x(x^2 + 1)^2.$$

We substitute these into the earlier equation and have

$$\frac{dy}{dx} = (x^2 + 1)^3 6x^2(x^3 - 1) + (x^3 - 1)^2 6x(x^2 + 1)^2$$

$$= 6x(x^2 + 1)^2(x^3 - 1)(2x^3 + x - 1).$$

• PROBLEM 95

If $y = (x^3 - 2)\sqrt{3x^2 + 4}$, find dy/dx.

Solution:

$$\frac{dy}{dx} = \sqrt{3x^2 + 4}\,\frac{d}{dx}(x^3 - 2) + (x^3 - 2)\frac{d}{dx}(\sqrt{3x^2 + 4})$$

since $\frac{d}{dx}(\sqrt{3x^2 + 4}) = \frac{d}{dx}(3x^2 + 4)^{\frac{1}{2}}$

$$= 3x(3x^2 + 4)^{-\frac{1}{2}}$$

$$= \frac{3x}{\sqrt{3x^2 + 4}} \cdot \frac{\sqrt{3x^2 + 4}}{\sqrt{3x^2 + 4}} = \frac{3x\sqrt{3x^2 + 4}}{3x^2 + 4}$$

Then

$$\frac{dy}{dx} = \sqrt{3x^2 + 4} \cdot 3x^2 + (x^3 - 2)\frac{3x\sqrt{3x^2 + 4}}{3x^2 + 4}$$

$$= \sqrt{3x^2 + 4}\left[3x^2 + (x^3 - 2) \cdot \frac{3x}{3x^2 + 4}\right]$$

$$= \sqrt{3x^2 + 4} \cdot \frac{12x^4 + 12x^2 - 6x}{3x^2 + 4}.$$

• PROBLEM 96

Given: $f(x) = (x + 1)^3(2x - 1)^{4/3}$. Find $f'(x)$.

Solution: We use the rule for differentiating a product, by setting $u(x) = (x + 1)^3$ and

$v(x) = (2x - 1)^{4/3}$, so that $f(x) = u(x)v(x)$. Then

$$f'(x) = \left[(x + 1)^3\right] v'(x) + (2x - 1)^{4/3} u'(x).$$

Since

$$u'(x) = 3(x + 1)^2 \cdot 1, \text{ and}$$

$$v'(x) = \frac{4}{3}(2x - 1)^{1/3} \cdot 2,$$

we have

$$f'(x) = \frac{8}{3}(x + 1)^3 (2x - 1)^{1/3} + 3(2x - 1)^{4/3}(x + 1)^2$$

$$= (x + 1)^2 (2x - 1)^{1/3} \left[\frac{8}{3}(x + 1) + 3(2x - 1) \right]$$

$$= \frac{1}{3}(x + 1)^2 (2x - 1)^{1/3} (26x - 1).$$

• PROBLEM 97

Find the derivative of:

$$x = \frac{y}{a} \sqrt{2by - y^2}$$

using substitution and the chain rule.

Solution:

$$\frac{dx}{dy} = \frac{y}{a} \cdot \frac{d}{dy}(2by - y^2)^{\frac{1}{2}} + (2by - y^2)^{\frac{1}{2}} \cdot \frac{d}{dy}\left(\frac{y}{a}\right)$$

$$= \frac{y}{a} \cdot \frac{d}{dy} (2by - y^2)^{\frac{1}{2}} + \frac{1}{a} \cdot (2by - y^2)^{\frac{1}{2}}$$

Let $w = (2by - y^2)$

Then $\frac{dw}{dy} = (2b - 2y)$

also let $s = (2by - y^2)^{\frac{1}{2}}$

Then $s = w^{\frac{1}{2}}$

and $\frac{ds}{dw} = \frac{1}{2} w^{-\frac{1}{2}} = \frac{1}{2w^{\frac{1}{2}}} = \frac{1}{2(2by - y^2)^{\frac{1}{2}}}$.

$$\frac{dx}{dy} = \frac{y}{a} \frac{ds}{dy} + \frac{s}{a}.$$

Therefore,

$$\frac{ds}{dy} = \frac{ds}{dw} \cdot \frac{dw}{dy} = \frac{1}{2(2by - y^2)^{\frac{1}{2}}} \cdot (2b - 2y)$$

$$= \frac{(b - y)}{(2by - y^2)^{\frac{1}{2}}} \cdot$$

Finally,

$$\frac{dx}{dy} = \frac{y}{a} \cdot \frac{(b - y)}{(2by - y^2)^{\frac{1}{2}}} + \frac{1}{a}(2by - y^2)^{\frac{1}{2}},$$

or

$$\frac{dx}{dy} = \frac{y(b - y) + (2by - y^2)}{a(2by - y^2)^{\frac{1}{2}}}$$

$$= \frac{by - y^2 + 2by - y^2}{a(2by - y^2)^{\frac{1}{2}}} = \frac{3by - 2y^2}{a(2by - y^2)^{\frac{1}{2}}}$$

The method used for this problem is not the best. A shorter solution is obtained by rewriting the expression in the form of:

$$x = \frac{y}{a}(2by - y^2)^{\frac{1}{2}}$$

and applying the theorems for $d(uv)$ and $d(u^n)$ in succession.

• PROBLEM 98

Find the derivative of:

$$y = (x + 1)(x^2 + 2)(x - 9).$$

Solution: Using the theorem: $d(uv) = udv + vdu$,

$$\frac{dy}{dx} = D_x y = (x - 9)D_x\left[(x + 1)(x^2 + 2)\right]$$

$$+ (x + 1)(x^2 + 2)D_x(x - 9)$$

$$= (x - 9)\left[(x^2 + 2)D_x(x + 1)\right.$$

$$\left.+ (x + 1)D_x(x^2 + 2)\right]$$

$$+ (x + 1)(x^2 + 2)D_x(x - 9)$$

$$= (x - 9)(x^2 + 2)D_x(x + 1)$$

$$+ (x + 1)(x - 9)D_x(x^2 + 2)$$

$$+ (x + 1)(x^2 + 2)D_x(x - 9)$$

$$= (x - 9)(x^2 + 2)(1) + (x + 1)(x - 9)$$

$$(2x) + (x + 1)(x^2 + 2)(1)$$

$$= 4x^3 - 24x^2 - 14x - 16.$$

• **PROBLEM 99**

Find the derivative of:

$$y = \frac{x^2 + 1}{x^2 - 1}, \; x^2 \neq 1.$$

<u>Solution:</u> We apply the theorem for $d\left(\frac{u}{v}\right)$ i.e.: $\frac{vdu - udv}{v^2}$, to obtain:

$$\frac{dy}{dx} = \frac{(x^2 - 1)\cdot 2x - (x^2 + 1)\cdot 2x}{(x^2 - 1)^2}$$

$$= \frac{-4x}{(x^2 - 1)^2}.$$

• **PROBLEM 100**

Find the derivative of:

$$f(x) = \frac{x^2 + 1}{1 - 3x}.$$

<u>Solution:</u> Using the theorem for differentiating a quotient of terms, i.e.:

$$\frac{d}{dx}\left(\frac{u}{v}\right) = \frac{v\frac{du}{dx} - u\frac{dv}{dx}}{v^2},$$

$$f'(x) = \frac{(1 - 3x)(2x) - (x^2 + 1)(-3)}{(1 - 3x)^2}$$

$$= \frac{-3x^2 + 2x + 3}{(1 - 3x)^2}.$$

• **PROBLEM 101**

Find

$$D_m\left(\frac{m^2}{1 + m^{1/2}}\right).$$

<u>Solution:</u>

$$D_m = \frac{d}{dm}.$$

Using the theorem for $d\left(\frac{u}{v}\right)$, i.e.: $\frac{udv - vdu}{v^2}$,

$$D_m\left[\frac{m^2}{1+m^{1/2}}\right] = \frac{\left(1+m^{1/2}\right)D_m m^2 - m^2 D_m\left(1+m^{1/2}\right)}{(1+m^{1/2})^2}$$

$$= \frac{\left(1+m^{1/2}\right)(2m) - m^2\left(\frac{1}{2}m^{-1/2}\right)}{\left(1+m^{1/2}\right)^2}.$$

$$= \frac{4m+3m^{3/2}}{2\left(1+m^{1/2}\right)^2}.$$

• PROBLEM 102

Find the derivative of:

$$f(s) = \frac{s}{\sqrt{s^2-1}}.$$

Solution: Write:

$$f(s) = \frac{s}{(s^2-1)^{1/2}}.$$

Using the theorem for the derivative of a quotient:

$$f'(s) = \frac{(s^2-1)^{1/2}\cdot 1 - s\,\frac{1}{2}(s^2-1)^{-1/2}(2s)}{s^2-1}.$$

This expression may be simplified in several ways. One easy method is to multiply through both the numerator and the denominator by

$(s^2-1)^{1/2}$. This gives

$$f'(s) = \frac{(s^2-1)^1 - s^2(s^2-1)^0}{(s^2-1)^{3/2}}.$$

Since $(s^2-1)^0 = 1$, we obtain:

$$f'(s) = \frac{-1}{(s^2-1)^{3/2}}.$$

• PROBLEM 103

Find the derivative of:

$$g(x) = \frac{x^3}{\sqrt[3]{3x^2-1}}.$$

Solution: First rewrite the expression to change the radical symbol to an exponent. This allows the application of the theorem for $d(u^n)$.

$$g(x) = x^3(3x^2 - 1)^{-1/3}.$$

Now, using the theorems for d(uv) and d(u^n) in succession,

$$g'(x) = 3x^2(3x^2 - 1)^{-1/3}$$

$$- \frac{1}{3}(3x^2 - 1)^{-4/3}(6x)(x^3)$$

$$= x^2(3x^2 - 1)^{-4/3}\left[3(3x^2 - 1) - 2x^2\right]$$

$$= \frac{x^2(7x^2 - 3)}{(3x^2 - 1)^{4/3}} .$$

• PROBLEM 104

Find the derivative of: $x = \dfrac{1}{\sqrt{1 + y^2}}$.

Solution: The method of solving this problem by the combination of the chain rule and substitution is illustrated, here. However, the shortest solution is obtained by rewriting the expression as:

$$x = (1 + y^2)^{-\frac{1}{2}},$$

and then applying the theorem for d(u^n).

Let $u = 1 + y^2$.

Then, $\dfrac{du}{dy} = 2y$, and $x = u^{-\frac{1}{2}}$.

Now,

$$\frac{dx}{du} = - \frac{1}{2} u^{-\frac{3}{2}} = - \frac{1}{2}(1 + y^2)^{-\frac{3}{2}} .$$

Therefore,

$$\frac{dx}{dy} = \frac{dx}{du} \cdot \frac{du}{dy}$$

$$= - \frac{1}{2}(1 + y^2)^{-\frac{3}{2}} \cdot 2y = - \frac{y}{\sqrt{(1 + y^2)^3}} .$$

• PROBLEM 105

Find $\dfrac{dx}{dy}$ when $y = \sqrt{\dfrac{5}{x} - 1}$.

Use the chain rule method.

Solution: Substitute

$$u = \left(\frac{5}{x} - 1\right) .$$

Then,

$$\frac{du}{dx} = - \frac{5}{x^2} ,$$

and $\quad y = u^{\frac{1}{2}}$.

Now,

$$\frac{dy}{du} = \frac{1}{2} u^{-\frac{1}{2}} = \frac{1}{2\sqrt{\frac{5}{x} - 1}} .$$

Then,

$$\frac{dy}{dx} = \frac{dy}{du} \cdot \frac{du}{dx} = \frac{1}{2\sqrt{\frac{5}{x} - 1}} \cdot - \frac{5}{x^2}$$

$$= - \frac{5}{2x^2\sqrt{\frac{5}{x} - 1}} .$$

Since dx/dy is the reciprocal of dy/dx,

$$\frac{dx}{dy} = - \frac{2x^2\sqrt{\frac{5}{x} - 1}}{5} .$$

A more direct solution is obtained if first we solve for x as a function of y, and dx/dy is derived directly from this function. Thus, by squaring, we obtain:

$$y^2 = \frac{5}{x} - 1 ,$$

from which

$$x = \frac{5}{y^2 + 1} .$$

Using the theorem for $d\left(\frac{u}{v}\right)$,

$$\frac{dx}{dy} = - \frac{5 \cdot 2y}{(y^2 + 1)^2} = - \frac{10y}{(y^2 + 1)^2}$$

$$= - \frac{10\left(\frac{5}{x} - 1\right)^{\frac{1}{2}}}{\left(\frac{5}{x} - 1 + 1\right)^2}$$

$$= - \frac{10\left(\frac{5}{x} - 1\right)^{\frac{1}{2}}}{\frac{25}{x^2}} = - \frac{10x^2\left(\frac{5}{x} - 1\right)^{\frac{1}{2}}}{25}$$

$$= - \frac{2x^2\sqrt{\frac{5}{x} - 1}}{5} .$$

• **PROBLEM 106**

Find the derivative, $d\Theta/dy$, from:

$$y = \frac{1}{\sqrt[4]{\Theta + 6}} .$$

Solution: Rewriting the function to solve for Θ expressed in terms of y,

$$y = (\Theta + 6)^{-\frac{1}{4}}, \quad y^{-4} = \Theta + 6, \quad \Theta = y^{-4} - 6$$

$$\Theta = \frac{1}{y^4} - 6.$$

$$\frac{d\Theta}{dy} = - 4y^{-5} = - 4(\Theta + 6)^{\frac{5}{4}} = - 4 \sqrt[4]{(\Theta + 6)^5} .$$

The result could also have been obtained by finding $dy/d\Theta$, and then simply taking the reciprocal.

• **PROBLEM 107**

Find the derivative of the function: $y = \frac{10^x}{x+2}$.

Solution: To find the derivative of the given function we use the quotient rule and also the formula: $\frac{d}{dx}a^u = a^u \frac{du}{dx} \ln a$. We find:

$$\frac{dy}{dx} = \frac{(x+2)(10^x \ln 10) - (10^x)(1)}{(x+2)^2}$$

$$= \frac{(x \cdot 10^x \ln 10) + (2 \cdot 10^x \ln 10) - (10^x)}{(x+2)^2}$$

$$= \frac{10^x(x \ln 10 + 2 \ln 10 - 1)}{(x+2)^2} .$$

Find D^2y and D^3y, if $y = \dfrac{x}{1 + x^2}$.

Solution:

$$D^2y = \frac{d^2y}{dx^2}, \quad D^3y = \frac{d^3y}{dx^3}, \quad Dy = \frac{dy}{dx}.$$

It is necessary to obtain the first derivative before the second and third derivatives may be found in successive operations of differentiation.

Applying the theorem for $d\left(\dfrac{u}{v}\right)$,

$$Dy = \frac{(1 + x^2) - x \cdot 2x}{(1 + x^2)^2} = \frac{1 - x^2}{(1 + x^2)^2}.$$

Now using $d\left(\dfrac{u}{v}\right)$ again,

$$D^2y = D\left(\frac{1 - x^2}{(1 + x^2)^2}\right)$$

$$= \frac{(1 + x^2)^2 \cdot (-2x) - (1 - x^2) \cdot 2(1 + x^2) \cdot 2x}{(1 + x^2)^4}$$

$$= \frac{(1 + x^2)(-6x + 2x^3)}{(1 + x^2)^4}$$

$$= \frac{2x^3 - 6x}{(1 + x^2)^3}.$$

Similarly, applying $d\left(\dfrac{u}{v}\right)$ once more,

$$D^3y = D\left(\frac{2x^3 - 6x}{(1 + x^2)^3}\right)$$

$$= \frac{(1 + x^2)^3(6x^2 - 6) - (2x^3 - 6x) \cdot 3(1 + x^2)^2 \cdot 2x}{(1 + x^2)^6}$$

$$= \frac{(1 + x^2)^2 \cdot (-6x^4 + 36x^2 - 6)}{(1 + x^2)^6}$$

$$= \frac{-6x^4 + 36x^2 - 6}{(1 + x^2)^4}.$$

• **PROBLEM 109**

Find the derivative of:

$$y = x^2 + 1/x^2, \quad x \neq 0.$$

Solution: The equation may be written in the form of

$$y = x^2 + x^{-2}.$$

Then applying the theorem for $d(u^n)$,

$$\frac{dy}{dx} = 2x^{2-1}\frac{dx}{dx} + (-2)x^{-2-1}\frac{dx}{dx}$$

$$= 2x - 2x^{-3}.$$

• **PROBLEM 110**

Find the derivative of the function:

$$s = \frac{9t^3}{3t-2} - (t-3)(3t-1).$$

Solution: To find the derivative of the given function we must apply the quotient rule to the first term and the product rule to the second term. Doing this, we obtain:

$$\frac{ds}{dt} = \frac{(3t-2)(27t^2)-(9t^3)(3)}{(3t-2)^2}$$

$$- \left[(t-3)(3) + (3t-1)(1)\right].$$

Multiplying out and simplifying, we have:

$$\frac{ds}{dt} = \frac{54t^3 - 54t^2}{(3t-2)^2} - (6t - 10).$$

Using $(3t-2)^2$ as a common denominator, we write:

$$\frac{ds}{dt} = \frac{54t^3 - 54t^2 - \left[(6t-10)(3t-2)^2\right]}{(3t-2)^2}$$

$$= \frac{108t^2 - 144t + 40}{(3t-2)^2}.$$

• **PROBLEM 111**

Find the derivative of:

$$f(x) = \frac{(x^2 + 1)\sqrt{x^2 - 1}}{3x + 2}.$$

Solution: Using the theorems for d(uv) and $d\left(\frac{u}{v}\right)$,

$$f'(x) = \frac{(3x+2)\left[(x^2+1)\frac{2x}{2\sqrt{x^2-1}} + \sqrt{x^2-1}\,2x\right]}{(3x+2)^2}$$

$$\frac{-(x^2+1)\sqrt{x^2-1}\,(3)}{(3x+2)^2}$$

$$= \frac{(3x+2)\left[x(x^2+1) + 2x(x^2-1)\right]}{(3x+2)^2\sqrt{x^2-1}}$$

$$\frac{-(x^2+1)(x^2-1)(3)}{(3x+2)^2\sqrt{x^2-1}}$$

$$= \frac{(3x+2)(3x^3-x) - 3x^4 + 3}{(3x+2)^2\sqrt{x^2-1}}$$

$$= \frac{6x^4 + 6x^3 - 3x^2 - 2x + 3}{(3x+2)^2\sqrt{x^2-1}}.$$

A simpler solution results when logarithms are taken of both sides of the equation, as this reduces products and quotients to operations of additions and subtractions.

• PROBLEM 112

Find the derivative of:

$$f(x) = \left(\frac{2x+1}{3x-1}\right)^4.$$

Solution: Applying first the theorem for $d(u^n)$ and then the theorem for $d\left(\frac{u}{v}\right)$,

$$f'(x) = 4\left(\frac{2x+1}{3x-1}\right)^3 \frac{(3x-1)(2) - (2x+1)(3)}{(3x-1)^2}$$

$$= \frac{4(2x+1)^3(-5)}{(3x-1)^5}$$

$$= -\frac{20(2x+1)^3}{(3x-1)^5}.$$

• PROBLEM 113

If

$$y = \frac{\sqrt{x+1} - \sqrt{x}}{\sqrt{x+1} + \sqrt{x}}$$

find dy/dx.

Solution: First reduce y to its simplest form:

$$y = \frac{\sqrt{x+1} - \sqrt{x}}{\sqrt{x+1} + \sqrt{x}} \cdot \frac{\sqrt{x+1} - \sqrt{x}}{\sqrt{x+1} - \sqrt{x}}$$

$$= \frac{2x + 1 - 2\sqrt{x^2 + x}}{(x+1) - x} = 2x + 1 - 2\sqrt{x^2 + x}$$

Then

$$\frac{dy}{dx} = \frac{d}{dx}(2x + 1) - 2\frac{d}{dx}\sqrt{x^2 + x}$$

$$= 2 - 2\frac{d\sqrt{u}}{du}\frac{du}{dx},$$

where $u = x^2 + x$; note that $\sqrt{u} = u^{\frac{1}{2}}$ and

$$\frac{d}{du}(u^{\frac{1}{2}}) = \frac{1}{2}u^{-\frac{1}{2}}$$

$$\frac{dy}{dx} = 2 - 2\frac{1}{2\sqrt{u}}\frac{du}{dx} = 2 - \frac{1}{\sqrt{x^2 + x}}(2x + 1).$$

This example may be done also by first applying the rule for the derivative of a fraction. The work is simpler, in this example, however, if the given expression is first simplified.

• PROBLEM 114

Find the derivative of:

$$y = \sqrt{\frac{x^2 + 1}{x^2 - 1}}.$$

Solution: Rewrite y in the form of u^n, so that the theorem for $d(u^n)$ may be applied.

$$y = \left(\frac{x^2 + 1}{x^2 - 1}\right)^{1/2}.$$

Then we have

$$\frac{dy}{dx} = \frac{1}{2}\left(\frac{x^2 + 1}{x^2 - 1}\right)^{-1/2} D_x\left(\frac{x^2 + 1}{x^2 - 1}\right).$$

Using the theorem for $d\left(\frac{u}{v}\right)$,

$$\frac{dy}{dx} = \frac{1}{2}\left(\frac{x^2 - 1}{x^2 + 1}\right)^{1/2} \frac{(x^2 - 1)(2x) - (x^2 + 1)(2x)}{(x^2 - 1)^2}$$

$$= \frac{-2x}{(x^2 + 1)^{1/2}(x^2 - 1)^{3/2}}.$$

● **PROBLEM 115**

Find the derivative of:

$$y = \sqrt{1 - \frac{1}{x^2 + 1}}.$$

Solution:

$$D_x = \frac{d}{dx}, \quad D_v = \frac{d}{dv}, \quad D_u = \frac{d}{du}.$$

To illustrate the use of the chain rule, make the following substitutions:

$$y = u^{1/2}$$

$$u = 1 - \frac{1}{v}$$

$$v = x^2 + 1.$$

Then, $D_x y = (D_u y)(D_v u)(D_x v)$, where $D_u y = \frac{1}{2} u^{-1/2}$, $D_v u = v^{-2}$, and $D_x v = 2x$. Hence,

$$D_x y = \frac{1}{2} u^{-1/2} \frac{1}{v^2}(2x)$$

$$= \frac{x}{(x^2 + 1)^2}\left(1 - \frac{1}{x^2 + 1}\right)^{-1/2}$$

$$= \frac{x}{(x^2 + 1)^2} \frac{\sqrt{x^2 + 1}}{|x|}.$$

● **PROBLEM 116**

Find dy/dx from;

$$y = \sqrt{\frac{(x - 1)(x - 2)}{(x - 3)(x - 4)}}$$

Solution: When the equation is a combination of products and quotients of terms, the simplest method is to apply logarithms to reduce the products and quotients to operations of additions and subtractions. Therefore, taking natural logarithms of both sides,

$$\ln y = \frac{1}{2}[\ln (x - 1) + \ln (x - 2) - \ln (x - 3) - \ln(x - 4)].$$

Differentiating both sides with respect to x,

$$\frac{1}{y}\frac{dy}{dx} = \frac{1}{2}\left[\frac{1}{x - 1} + \frac{1}{x - 2} - \frac{1}{x - 3} - \frac{1}{x - 4}\right].$$

Combining fractions over a common denominator,

$$\frac{1}{y}\frac{dy}{dx} = -\frac{2x^2 - 10x + 11}{(x - 1)(x - 2)(x - 3)(x - 4)},$$

and

$$\frac{dy}{dx} = -\frac{2x^2 - 10x + 11}{(x - 1)^{\frac{1}{2}}(x - 2)^{\frac{1}{2}}(x - 3)^{\frac{3}{2}}(x - 4)^{\frac{3}{2}}}.$$

• PROBLEM 117

Find the derivative of:

$$f(x) = \frac{(x - 1)^2\sqrt{2x + 3}}{(x + 4)^3(2x - 1)^4}.$$

Solution: An interesting approach to differentiating such a combination of terms is to try using logarithms, so that the product and quotient of terms are simplified to additions and subtractions, respectively.

$$\ln|f(x)| = 2\ln|x - 1| + \frac{1}{2}\ln|2x + 3| - 3\ln|x + 4| - 4\ln|2x - 1|.$$

Differentiating and using the relation that $(d/dx) \ln |f(x)| = \left(f'(x)/f(x)\right)$,

$$\frac{f'(x)}{f(x)} = \frac{2}{(x - 1)} + \frac{1}{2x + 3} - \frac{3}{x + 4} - \frac{8}{2x - 1}.$$

Therefore,

$$f'(x) = \left(\frac{2}{x-1} + \frac{1}{2x+3} - \frac{3}{x+4} - \frac{8}{2x-1}\right) \left(\frac{(x-1)^2\sqrt{2x+3}}{(x+4)^3(2x-1)^4}\right).$$

• **PROBLEM** 118

Find the derivative of the function:

$$y = \left[\sqrt{x-3} + \frac{3}{\sqrt{x-3}}\right]^4.$$

Solution: To find the derivative of the given function we must apply the chain rule. We find:

$$\frac{dy}{dx} = 4\left[(x-3)^{\frac{1}{2}} + 3(x-3)^{-\frac{1}{2}}\right]^3 \cdot \left[\frac{1}{2}(x-3)^{-\frac{1}{2}} - \frac{1}{2}(3)(x-3)^{-\frac{3}{2}}\right].$$

Obtaining common denominators and simplifying, we obtain:

$$\frac{dy}{dx} = 4\left[\frac{x}{(x-3)^{\frac{1}{2}}}\right]^3 \cdot \left[\frac{x-6}{2(x-3)^{\frac{3}{2}}}\right]$$

$$= \frac{4x^3}{(x-3)^{\frac{3}{2}}} \cdot \frac{x-6}{2(x-3)^{\frac{3}{2}}}$$

$$= \frac{4x^3(x-6)}{2(x-3)^3}.$$

• **PROBLEM** 119

Find the derivative of the function:

$$f(x) = \sqrt{x + \sqrt{x + \sqrt{x}}}\ .$$

Solution: Rewriting the problem, f(x) becomes

$$\left(x + \left(x + (x^{\frac{1}{2}})\right)^{\frac{1}{2}}\right)^{\frac{1}{2}}$$

This example requires the chain rule.

$$\frac{d\left(x + \left(x + (x^{\frac{1}{2}})\right)^{\frac{1}{2}}\right)^{\frac{1}{2}}}{dx} = \frac{1}{2}\left(x + \left(x + (x^{\frac{1}{2}})\right)^{\frac{1}{2}}\right)^{-\frac{1}{2}}$$

$$\cdot \frac{d\left(x + \left(x + (x^{\frac{1}{2}})\right)^{\frac{1}{2}}\right)}{dx}$$

$$= \frac{1}{2}\left(x + \left(x + (x^{\frac{1}{2}})\right)^{\frac{1}{2}}\right)^{-\frac{1}{2}} \left[1 + \frac{1}{2}\left(x + (x^{\frac{1}{2}})\right)^{-\frac{1}{2}} \cdot \frac{d\left(x + (x^{\frac{1}{2}})\right)}{dx}\right]$$

$$= \frac{1}{2}\left(x + \left(x + (x^{\frac{1}{2}})\right)^{\frac{1}{2}}\right)^{-\frac{1}{2}} \left[1 + \frac{1}{2}\left(x + (x^{\frac{1}{2}})\right)^{-\frac{1}{2}} \left(1 + \frac{1}{2}x^{-\frac{1}{2}}\right)\right]$$

$$= \left(\frac{1}{2\sqrt{x + \sqrt{x + \sqrt{x}}}}\right)\left[1 + \frac{1}{2\sqrt{x + \sqrt{x}}}\left(1 + \frac{1}{2\sqrt{x}}\right)\right].$$

• PROBLEM 120

If $y = u^2$, and $x = (u - 1)/(u + 1)$, find dy/dx.

Solution: We can readily obtain dy/du and dx/du from the parametric equations. If we then apply the chain rule, we obtain:

$$\frac{dy}{dx} = \frac{dy/du}{dx/du}.$$

Consequently,

$$\frac{dy}{du} = 2u$$

$$\frac{dx}{du} = \frac{2}{(u + 1)^2},$$

and

$$\frac{dy}{dx} = \frac{2u}{2/(u + 1)^2} = u(u + 1)^2.$$

• PROBLEM 121

If $y = t^2 + 2$ and $x = 3t + 4$, find dy/dx.

Solution: Method 1. We may solve the equation: $x = 3t + 4$, for t and substitute this value of t in the first equation:

$$y = \left(\frac{x - 4}{3}\right)^2 + 2 = \frac{x^2}{9} - \frac{8}{9}x + \frac{34}{9},$$

$$\frac{dy}{dx} = \frac{2}{9}x - \frac{8}{9} = \frac{2}{9}(3t + 4) - \frac{8}{9} = \frac{2t}{3}.$$

Method 2. Using the chain rule we have:

$$\frac{dy}{dx} = \frac{dy}{dt} \cdot \frac{dt}{dx} \qquad \text{or}$$

$$\frac{dy}{dx} = \frac{dy}{dt} \div \frac{dx}{dt} = \frac{d(t^2 + 2)}{dt} \div \frac{d(3t + 4)}{dt}$$
$$= 2t \div 3 = \frac{2}{3}t.$$

• PROBLEM 122

If $y = (x^2 + 2x + 1)^3$, while simultaneously $x = 3t^2 + 2t - 1$, find dy/dt.

Solution: The difficult procedure would be to express y in terms of t, y(t), and then to differentiate. A far easier solution may be obtained by noting that

$$\frac{dy}{dt} = \frac{dy}{dx} \cdot \frac{dx}{dt}$$

since

$$\frac{dy}{dx} = 3(x^2 + 2x + 1)^2(2x + 2),$$

and

$$\frac{dx}{dt} = 6t + 2,$$

then

$$\frac{dy}{dt} = 6(x + 1)(x^2 + 2x + 1)^2 \cdot 2(3t + 1)$$

$$=12(x+1)^5(3t+1).$$

• PROBLEM 123

Find dy/dt from

$$y = x^3 - 3x^2 + 5x - 4,$$

where $x = t^2 + t.$

Solution: From these equations, we find

$$\frac{dy}{dx} = 3x^2 - 6x + 5$$

$$= 3(t^2 + t)^2 - 6(t^2 + t) + 5,$$

$$\frac{dx}{dt} = 2t + 1.$$

Since

$$\frac{dy}{dt} = \frac{dy}{dx}\frac{dx}{dt} \quad \text{from the chain rule,}$$

$$\frac{dy}{dt} = \left[3(t^2 + t)^2 - 6(t^2 + t) + 5\right](2t + 1).$$

We can also first substitute the value of x in terms of t into the equation for y. We then have:

$$y = (t^2 + t)^3 - 3(t^2 + t)^2 + 5(t^2 + t) - 4.$$

When we differentiate this with respect to t, we obtain:

$$\frac{dy}{dt} = 3(t^2 + t)^2(2t + 1)$$

$$- 6(t^2 + t)(2t + 1) + 5(2t + 1)$$

$$= \left[3(t^2 + t)^2 - 6(t^2 + t) + 5\right](2t + 1),$$

which agrees with the previous answer.

The first method using the chain rule, however, often results in the simpler solution when dealing with problems involving parametric equations.

• PROBLEM 124

If $y = a\sqrt{x} + \frac{b}{\sqrt{x}}$, show that $4x^2D_x^2y + 4xD_xy - y = 0.$

Solution: We first rewrite the given function as

$$y = ax^{\frac{1}{2}} + bx^{-\frac{1}{2}}.$$

$$D_x y = \frac{dy}{dx} = \frac{1}{2}ax^{-\frac{1}{2}} - \frac{1}{2}bx^{-\frac{3}{2}}.$$

$$D_x^2 y = \frac{d^2y}{dx^2} = -\frac{1}{4}ax^{-\frac{3}{2}} + \frac{3}{4}bx^{-\frac{5}{2}}.$$

Substituting, we have:

$$4x^2D_x^2y + 4xD_xy - y = 4x^2\left[-\frac{a}{4x^{\frac{3}{2}}} + \frac{3b}{4x^{\frac{5}{2}}}\right] + 4x\left[\frac{a}{2x^{\frac{1}{2}}} - \frac{b}{2x^{\frac{3}{2}}}\right]$$

$$- \left[ax^{\frac{1}{2}} + \frac{b}{x^{\frac{1}{2}}}\right]$$

$$= -\frac{4ax^2}{4x^{\frac{3}{2}}} + \frac{12bx^2}{4x^{\frac{5}{2}}} + \frac{4ax}{2x^{\frac{1}{2}}} - \frac{4bx}{2x^{\frac{3}{2}}} - ax^{\frac{1}{2}}$$

$$- \frac{b}{x^{\frac{1}{2}}}$$

$$= -ax^{\frac{1}{2}} + \frac{3b}{x^{\frac{1}{2}}} + 2ax^{\frac{1}{2}} - \frac{2b}{x^{\frac{1}{2}}} - ax^{\frac{1}{2}} - \frac{b}{x^{\frac{1}{2}}}$$

$$= 2ax^{\frac{1}{2}} - 2ax^{\frac{1}{2}} + \frac{3b}{x^{\frac{1}{2}}} - \frac{3b}{x^{\frac{1}{2}}} = 0.$$

• **PROBLEM 125**

Find the general expression for the nth derivative of:

$$f(x) = \frac{1}{3x + 2}.$$

Solution:

$$f(x) = \frac{1}{3x + 2} = (3x + 2)^{-1}$$

First derivative:

$$f'(x) = -1(3x + 2)^{-2}(3) = -3(3x + 2)^{-2}.$$

Second derivative:

$$f''(x) = 6(3x + 2)^{-3} (3) = 18(3x + 2)^{-3}.$$

Third derivative:

$$f'''(x) = - 54(3x + 2)^{-4} (3) = - 162(3x + 2)^{-4}.$$

Fourth derivative:

$$f''''(x) = 648(3x + 2)^{-5} (3) = 1944(3x + 2)^{-5}.$$

To express the nth derivative, a pattern that the sequence of derivatives follows must be found.

1) First, it is noted that the derivatives are alternately negative and positive. The odd order of differentiation results in a negative value, and the even order of differentiation, in a positive value. This property can be expressed by $(- 1)^n$ for the nth derivative.

2) The coefficients of (3x + 2)(ignoring the sign) are 3, 18, 162, 1944 ...

First derivative:

$$n = 1: \quad 3 = 3^1 \cdot 1 = 3^n \cdot n\,!$$

Second derivative:

$$n = 2: 18 = 3^2 \; (2!) = 3^n \cdot n\,!$$

Third derivative:

$$n = 3: \quad 162 = 3^3(3!) = 3^n \cdot n!$$

Fourth derivative:

$$n = 4: 1944 = 3^4(4!) = 3^n \cdot n!$$

3) The power of (3x + 2) is considered.

First derivative:

$$n = 1: \quad (3x + 2)^{-2} = \frac{1}{(3x + 2)^2} = \frac{1}{(3x + 2)^{n+1}}$$

Second derivative:

$$n = 2: \quad (3x + 2)^{-3} = \frac{1}{(3x + 2)^3} = \frac{1}{(3x + 2)^{n+1}}$$

Third derivative:

$$n = 3: \quad (3x + 2)^{-4} = \frac{1}{(3x + 2)^4} = \frac{1}{(3x + 2)^{n+1}}$$

Fourth derivative:

$$n = 4: \quad (3x + 2)^{-5} = \frac{1}{(3x + 2)^5} = \frac{1}{(3x + 2)^{n+1}}$$

Combining these results, the nth derivative is:

$$f^n(x) = (-1)^n \frac{3^n \cdot n!}{(3x+2)^{n+1}} .$$

CHAPTER 7

DIFFERENTIATION OF TRIGONOMETRIC FUNCTIONS

Once such elementary relationships as:

$$\frac{d}{dx}(\sin x) = \cos x$$

$$\frac{d}{dx}(\cos x) = -\sin x$$

and

$$\frac{d}{dx}(\tan x) = \sec^2 x\,,$$

are understood, any given trigonometric function can be differentiated by applying these in combination with the basic rules that are applicable to algebraic functions, such as the power rule, chain rule, etc. Where the trigonometric function given is in terms of the cot, csc, sec, etc., it may be useful to change these functions into sines and cosines first, because considerable simplification may result from such a procedure. Also, trigonometric identities may be used to simplify the function before or after differentiation.

The basic trigonometric derivatives should be memorized, and when applied in combination as described above, any trigonometric function, regardless of complexity, can usually be differentiated easily.

• PROBLEM 126

Find the derivative of: $y = \sin(x + b)$.

<u>Solution:</u> Let $u = (x + b)$

Then, $y = \sin u$, and

$$\frac{dy}{dx} = \frac{dy}{du} \cdot \frac{du}{dx}$$

Since $\frac{du}{dx} = 1$, and

$$\frac{dy}{du} = \cos u = \cos(x + b),$$

$$\frac{dy}{dx} = \frac{dy}{du} \cdot \frac{du}{dx} = \cos(x + b) \cdot 1 = \cos(x + b)$$

This illustration shows how the solution may be obtained in a systematic manner. After the student has obtained some facility in applying the differentiating theorems, many of the intermediate steps above can be omitted.

• PROBLEM 127

Find the derivative of:

$$y = \sin ax^2.$$

Solution: Applying the theorem for the derivative of the sine of a function,

$$\frac{dy}{dx} = \cos ax^2 \cdot \frac{d}{dx}(ax^2)$$

$$= 2\,ax \cos ax^2.$$

• PROBLEM 128

Find the derivative of:

$G(x) = \sin \sqrt{\ln x}$, for $x > 1$.

Solution: Let $u = \sqrt{\ln x}$

Then $G(x)$ is in the form of $\sin u$ and $G'(x) = d(\sin u) = \cos u \, d\, u$.

Now,

$$du = \frac{1}{2\sqrt{\ln x}} \cdot \frac{1}{x}.$$

Substituting for u and du in $G'(x)$,

$$G'(x) = (\cos \sqrt{\ln x}) \frac{1}{2x\sqrt{\ln x}}.$$

• PROBLEM 129

Find the derivative of: $y = \tan 3\Theta$.

Solution: Let $u = 3\Theta$.

Then, $y = \tan u$, and

$$\frac{dy}{d\Theta} = \frac{dy}{du} \cdot \frac{du}{d\Theta}$$

$$\frac{du}{d\Theta} = 3,$$

and $\frac{dy}{du} = \sec^2 u$.

Therefore,

$$\frac{dy}{d\Theta} = \frac{dy}{du} \cdot \frac{du}{d\Theta} = \sec^2 u \cdot 3 = 3 \sec^2 (3\Theta).$$

• PROBLEM 130

Differentiate:

$$y = \tan\sqrt{1 - x}.$$

Solution: Rewriting the function:

$$y = \tan\sqrt{1 - x}, \qquad \text{as}$$

$$y = \tan(1 - x)^{\frac{1}{2}}, \text{ and,}$$

applying the theorem for the derivative of the tangent,

$$\frac{dy}{dx} = D_x\left(\tan((1 - x)^{\frac{1}{2}}\right).$$

$$\frac{dy}{dx} = \sec^2(1 - x)^{\frac{1}{2}} \cdot D_x(1 - x)^{\frac{1}{2}}.$$

By further differentiation,

$$\frac{dy}{dx} = \left(\sec^2(1 - x)^{\frac{1}{2}}\right)\left(\frac{-1}{2}(1 - x)^{-\frac{1}{2}}\right),$$

or rewriting in terms of square roots,

$$\frac{dy}{dx} = -\frac{\sec^2\sqrt{1 - x}}{2\sqrt{1 - x}}.$$

• PROBLEM 131

Find the derivative of:

$$f(x) = 2\tan\frac{1}{2}x - x.$$

Solution:

$$f'(x) = 2\sec^2\frac{1}{2}x\,\frac{d\left(\frac{1}{2}x\right)}{dx} - 1.$$

From the identity: $\sec^2 x - 1 = \tan^2 x$,

$$f'(x) = \tan^2\frac{1}{2}x.$$

• PROBLEM 132

Find the derivative of:

$$y = \tanh(1 - x^2).$$

Solution: Applying the theorem for $\frac{d(\tanh u)}{dx}$, we have:

$$D_x y = \operatorname{sech}^2(1 - x^2)D_x(1 - x^2),$$

and, substituting for $D_x(1 - x^2)$,

$$D_x\, y = -\ 2x\ \text{sech}^2(1 - x^2).$$

• PROBLEM 133

Find the derivative of the expression:

$$y = \sin^2 x.$$

Solution: Applying the theorem for $d(u^n)$ and $d(\sin u)$,

$$\frac{dy}{dx} = 2(\sin x)\ \frac{d\ \sin x}{dx} = 2\ \sin x\ \cos x = \sin 2x.$$

• PROBLEM 134

Differentiate: $y = \cos^3 x$.

Solution: Applying the theorem for finding the derivative of a power of a function $d(u^n)$, the equation may also be written:

$$y = (\cos x)^3,$$

and

$$\frac{dy}{dx} = 3(\cos x)^2\ \frac{d}{dx}\left(\cos x\right)$$

$$= 3\ \cos^2 x(-\ \sin x)$$

$$= -\ 3\ \sin x\ \cos^2 x.$$

• PROBLEM 135

Find the derivative of: $y = \sec x$.

Solution: The attempt is made here to derive the desired result from basic sine and cosine functions.

$$y = \sec x$$

$$= \frac{1}{\cos x}$$

Applying the theorem for the derivative of $u^n (n = -\ 1)$,

$$y' = \frac{-\ (-\ \sin x)}{\cos^2 x}$$

$$= \frac{\sin x}{\cos^2 x} = \frac{\sin x}{\cos x}\ \frac{1}{\cos x}$$

Simplifying, we have:

$y' = \tan x \sec x.$

• PROBLEM 136

Find the derivative of: $y = \tan^2 x$.

Solution: $y = \tan^2 x.$

We apply the theorem:

$$d(u^n) = nu^{n-1},$$

$$y' = \frac{dy}{dx} = 2 \tan x \, D_x \tan x$$

$$= 2 \tan x \frac{d}{dx}(\tan x)$$

Applying the theorem for $\frac{d(\tan x)}{dx}$, we have:

$$y' = 2 \tan x \sec^2 x.$$

• PROBLEM 137

Find the derivative of the function: $y = \sec^2 x$.

Solution: Applying the theorem for finding the derivative of a power function, $d(u^n)$,

$$\frac{dy}{dx} = 2 \sec x \frac{d}{dx}(\sec x).$$

Substituting,

$$\frac{dy}{dx} = 2 \sec x(\sec x \tan x) = 2 \sec^2 x \tan x.$$

• PROBLEM 138

Find the derivative of: $y = \cos^2 (2x)$.

Solution: By applying the theorem for $d(u^n)$

$$\frac{dy}{dx} = \left(2 \cos 2x\right)\frac{d}{dx}\left(\cos 2x\right).$$

$$\frac{d}{dx}(\cos 2x) = - 2 \sin 2x.$$

Hence, by substitution,

$$\frac{dy}{dx} = 2 \cos(2x)(- \sin 2x)(2)$$

$$= - 4 \sin 2x \cos 2x = - 2 \sin 4x.$$

• PROBLEM 139

Find the derivative of: $y = \tan^2 4x$.

Solution: By applying the theorem for $d(u^n)$,

$$\frac{dy}{dx} = 2 \tan 4x \frac{d}{dx} (\tan 4x).$$

Substituting for $\frac{d}{dx}(\tan 4x)$,

$$\frac{dy}{dx} = 2 \tan 4x(\sec^2 4x)(4)$$

$$= 8 \tan 4x \sec^2 4x.$$

• PROBLEM 140

Find the derivative of:

$$f(x) = \sec^4 3x.$$

Solution: Applying the theorem for $d(u^n)$, we have:

$$f'(x) = \frac{df(x)}{dx} = 4 \sec^3 3x \cdot D_x(\sec 3x)$$

$$= 4 \sec^3 3x \frac{d}{dx}(\sec 3x).$$

Now, applying the theorem for $d(\sec u)$,

$$f'(x) = 4 \sec^3 3x(\sec 3x \tan 3x)(3)$$

$$= 12 \sec^4 3x \tan 3x.$$

• PROBLEM 141

Find the derivative of $y = \cot^4 2x$.

Solution: Applying the theorem for $d(u^n)$,

$$\frac{dy}{dx} = 4 \cot^3 2x \frac{d}{dx}\left(\cot 2x\right).$$

Substituting,

$$\frac{dy}{dx} = 4(\cot^3 2x)(-\csc^2 2x)(2)$$

$$= -8 \cot^3 2x \csc^2 2x.$$

• PROBLEM 142

Find the derivative of the function:

$$y = 2 \sin^3 (2x^4).$$

Solution: Applying the theorem for $\frac{d(u^n)}{dx}$

$$\frac{dy}{dx} = 2(3) \sin^2 (2x^4) \frac{d(\sin 2x^4)}{dx}.$$

By applying the theorem for $\frac{d(\sin u)}{dx}$,

$$\frac{dy}{dx} = 6 \sin^2 (2x^4) \cos (2x^4) \frac{d(2x^4)}{dx}$$

$$= 48x^3 \sin^2 (2x^4) \cos (2x^4).$$

• PROBLEM 143

Find dy/dx if $y = (1/3) \sin 2x - \tan^2 3x$.

Solution: By rewriting the function as the sum of two functions:

$$f(x) = (1/3) \sin 2x, \quad \text{and} \quad g(x) = \tan^2 3x,$$

then $y = f(x) - g(x)$.

Applying the theorem for $\frac{d(\sin u)}{dx}$ in $f(x)$ and the theorem for $d(u^n)$ in $g(x)$,

$$\frac{dy}{dx} = 1/3(\cos 2x)(2) - 2 \tan 3x \, D_x(\tan 3x).$$

By applying the theorem for $\frac{d(\tan u)}{dx}$ and substituting,

$$\frac{dy}{dx} = \frac{1}{3} (\cos 2x)(2) - 2(\tan 3x)(\sec^2 3x)(3).$$

Simplifying,

$$\frac{dy}{dx} = \frac{2}{3} \cos 2x - 6 \tan 3x \sec^2 3x.$$

• PROBLEM 144

Find the derivative of:

$$y = (1 + \cos 3x^2)^4.$$

Solution: Applying the theorem for $d(u^n)$,

$$D_x y = 4(1 + \cos 3x^2)^3 D_x(1 + \cos 3x^2).$$

$$D_x y = 4(1 + \cos 3x^2)^3(- \sin 3x^2)(6x)$$

$$= - 24x \sin 3x^2(1 + \cos 3x^2)^3.$$

• PROBLEM 145

Find the derivative of:

$$y = (\tan 2x + \sec 2x)^3.$$

Solution: By applying the theorem for the derivative of a power function, $d(u^n)$,

$$\frac{dy}{dx} = 3(\tan 2x + \sec 2x)^2 D_x(\tan 2x + \sec 2x).$$

By substitution of:

$$D_x(\tan 2x + \sec 2x) = 2\sec^2 2x + 2\sec 2x \tan 2x.$$

we have:

$$\frac{dy}{dx} = 3(\tan 2x + \sec 2x)^2 \left[2\sec^2 2x + 2\sec 2x \tan 2x\right]$$

$$= 6\sec 2x(\tan 2x + \sec 2x)^3.$$

• PROBLEM 146

Find the derivative of the function:
$x = 2\sqrt{4\sin y - 6\cos y}.$

Solution: The given function can be rewritten as: $x = 2(4\sin y - 6\cos y)^{1/2}$. Applying the chain rule, we obtain:

$$\frac{dx}{dy} = 2\cdot\frac{1}{2}(4\sin y - 6\cos y)^{-\frac{1}{2}}\cdot(4\cos y + 6\sin y)$$

$$= \frac{1}{(4\sin y - 6\cos y)^{\frac{1}{2}}}\cdot\left(4\cos y + 6\sin y\right)$$

$$= \frac{4\cos y + 6\sin y}{\sqrt{4\sin y - 6\cos y}}.$$

$$\frac{dy}{dx} = \frac{1}{\frac{dx}{dy}} = \frac{\sqrt{4\sin y - 6\cos y}}{4\cos y + 6\sin y} = \frac{x}{4(2\cos y + 3\sin y)}.$$

• PROBLEM 147

Find the derivative of:

$$y = \sqrt{1 + 3\tan^2\Theta}.$$

Solution: Let $u = 3\tan^2\Theta$. Then $y = \sqrt{1+u}$, and

$$\frac{dy}{d\Theta} = \frac{dy}{du}\cdot\frac{du}{d\Theta}.$$

$$\frac{du}{d\Theta} = 3\cdot 2\tan\Theta\cdot\frac{d}{d\Theta}(\tan\Theta) = 6\tan\Theta\sec^2\Theta.$$

Since $y = (1+u)^{\frac{1}{2}}$, $\frac{dy}{du} = \frac{1}{2\sqrt{1+u}}$

Substituting,

$$\frac{dy}{d\Theta} = \frac{dy}{du}\cdot\frac{du}{d\Theta} = \frac{1}{2\sqrt{1+u}}\cdot 6\tan\Theta\sec^2\Theta$$

$$= \frac{6 \tan \Theta \sec^2 \Theta}{2\sqrt{1 + 3\tan^2 \Theta}}.$$

• PROBLEM 148

Find the derivative of:

$$f(x) = \cot x \csc x.$$

Solution: By the theorem for the derivative of the product of functions, $d(uv) = udv + vdu$,

$$f'(x) = \cot x \cdot D_x(\csc x) + \csc x \cdot D_x(\cot x).$$

Applying the theorems for

$$\frac{d(\csc x)}{dx} \quad \text{and} \quad \frac{d(\cot x)}{dx},$$

$$f'(x) = \cot x(-\csc x \cot x) + \csc x(-\csc^2 x)$$

$$= -\csc x \cot^2 x - \csc^3 x.$$

• PROBLEM 149

Find the derivative of the function:
$y = \sin(\cos x) + \sin x \cos x.$

Solution: To find the derivative of the given function, we use the formulas: $\frac{d}{dx} \sin u = \cos u \frac{du}{dx}$, and $\frac{d}{dx} \cos u = -\sin u \frac{du}{dx}$, and apply the product rule to the last term. We find:

$$\frac{dy}{dx} = \cos(\cos x)(-\sin x)$$

$$+ \left[(\sin x)(-\sin x) + (\cos x)(\cos x)\right]$$

$$= -\sin x \cos(\cos x) - \sin^2 x + \cos^2 x$$

$$= \cos^2 x - \sin^2 x - \sin x \cos(\cos x).$$

We now make use of the identity, $\cos^2 x - \sin^2 x = \cos 2x$, obtaining:

$$\frac{dy}{dx} = \cos 2x - \sin x \cos(\cos x).$$

• PROBLEM 150

Find the derivative of: $y = x \csc^3 2x.$

Solution: This is the product of two functions. Hence applying the theorem for finding the derivative of the product of two functions $d(uv)$,

$$\frac{dy}{dx} = x D_x \csc^3 2x + \csc^3 2x D_x x.$$

Applying the theorem for the derivative of a power of a function $D(u^n)$,

$$D_x \csc^3 2x = 3 \csc^2 2x \, D_x \csc 2x.$$

By substitution,

$$\frac{dy}{dx} = x(3 \csc^2 2x)(- \csc 2x \cot 2x)(2) + \csc^3 2x$$

$$= \csc^3 2x(- 6x \cot 2x + 1).$$

• PROBLEM 151

Find the derivative of the expression:

$$y = \sin 2x \cos x^2.$$

Solution: Using the product rule, $d(uv) = udv + vdu$, and the derivatives of the sine and cosine functions,

$$\frac{dy}{dx} = \sin 2x \frac{d}{dx}\left(\cos x^2\right) + \cos x^2 \left(\frac{d}{dx} \sin 2x\right).$$

Applying the theorems for

$$\frac{d}{dx}\left(\cos u\right) \quad \text{and} \quad \frac{d}{dx}\left(\sin v\right)$$

and substituting, we have:

$$\frac{dy}{dx} = \sin 2x(- \sin x^2)(2x) + \cos x^2 (\cos 2x)(2)$$

$$= - 2x \sin 2x \sin x^2 + 2 \cos 2x \cos x^2.$$

• PROBLEM 152

Differentiate: $y = \sin nx \sin^n x.$

Solution: Applying the theorem for finding the derivative of the product of two functions, $d(uv)$,

$$\frac{dy}{dx} = \sin nx \frac{d}{dx}\left(\sin x\right)^n + \sin^n x \frac{d}{dx}\left(\sin nx\right)$$

$$= \sin nx \cdot n (\sin x)^{n-1} \frac{d}{dx}\left(\sin x\right)$$

$$+ \sin^n x \cos nx \frac{d}{dx}\left(nx\right)$$

$$= n \sin nx \cdot \sin^{n-1} x \cos x$$

$$+ n \sin^n x \cos nx$$

$$= n \sin^{n-1} x(\sin nx \cos x + \cos nx \sin x)$$

$$= n \sin^{n-1} x \sin(n + 1)x.$$

• PROBLEM 153

Find the derivative of: $y = \tan\Theta\sqrt{3\sec\Theta}$.

Solution: Rewriting $y = \tan\Theta\sqrt{3\sec\Theta}$ as

$$y = \tan\Theta(3\sec\Theta)^{1/2},$$

and applying the theorems for d(uv) and d(u^n),

$$\frac{dy}{d\Theta} = \tan\Theta \cdot \frac{d}{d\Theta}\left(3\sec\Theta\right)^{\frac{1}{2}} + (3\sec\Theta)^{\frac{1}{2}} \cdot \frac{d}{d\Theta} \cdot \tan\Theta.$$

Since

$$\frac{d}{d\Theta}(3\sec\Theta)^{\frac{1}{2}} = \frac{1}{2}(3\sec\Theta)^{-\frac{1}{2}}(3\sec\Theta\tan\Theta),$$

$$\frac{dy}{d\Theta} = \frac{3\sec\Theta \cdot \tan\Theta \cdot \tan\Theta}{2\sqrt{3\sec\Theta}} + \sqrt{3\sec\Theta} \cdot \sec^2\Theta$$

$$\frac{dy}{d\Theta} = \frac{3\tan^2\Theta \cdot \sec\Theta}{2\sqrt{3\sec\Theta}} + \sqrt{3\sec\Theta} \cdot \sec^2\Theta$$

$$= \frac{3\tan^2\Theta \cdot \sec\Theta + 2 \cdot 3\sec\Theta \cdot \sec^2\Theta}{2\sqrt{3\sec\Theta}}$$

$$= \frac{3\tan^2\Theta\sec\Theta + 6\sec^3\Theta}{2\sqrt{3\sec\Theta}}$$

$$= \frac{3\sec\Theta(\tan^2\Theta + 2\sec^2\Theta)}{2\sqrt{3\sec\Theta}}$$

$$= \frac{(3\sec\Theta)^{\frac{1}{2}}(\tan^2\Theta + 2\sec^2\Theta)}{2}.$$

Now $\tan^2\Theta = \sec^2\Theta - 1$

Therefore,

$$\frac{dy}{d\Theta} = \frac{(3\sec\Theta)^{\frac{1}{2}}(\sec^2\Theta - 1 + 2\sec^2\Theta)}{2}$$

$$= \frac{\sqrt{3\sec\Theta}\,(3\sec^2\Theta - 1)}{2}.$$

• PROBLEM 154

If $r = \left[\sec^4(\Theta/2) - \cos^2(\Theta/2)\right]\cot(\Theta/2)$, find $dr/d\Theta$.

Solution: This function is of the form u · v.

Hence, applying the theorem for

$$\frac{d(uv)}{dx},$$

$$\frac{dr}{d\Theta} = \left(\sec^4 \frac{\Theta}{2} - \cos^2 \frac{\Theta}{2}\right) D_x \cot \left(\frac{\Theta}{2}\right)$$

$$+ \cot \left(\frac{\Theta}{2}\right) \left[D_x\left(\sec^4 \frac{\Theta}{2}\right) - D_x\left(\cos^2 \frac{\Theta}{2}\right)\right].$$

Applying the theorems for $D_x(\cot u)$ and $D_x(u^n)$,

$$\frac{dr}{d\Theta} = \left(\sec^4 \frac{\Theta}{2} - \cos^2 \frac{\Theta}{2}\right) \left(- \csc^2 \frac{\Theta}{2}\right)\left(\frac{1}{2}\right) +$$

$$\cot\left(\frac{\Theta}{2}\right)\left[4 \sec^3 \frac{\Theta}{2} D_x(\sec u)\right.$$

$$\left. + 2 \cos \frac{\Theta}{2} D_x(\cos u)\right].$$

Now, substituting for $D_x(\sec u)$ and $D_x(\cos u)$,

$$\frac{dr}{d\Theta} = \left(\sec^3 \frac{\Theta}{2} - \cos^2 \frac{\Theta}{2}\right)\left[\left(- \cos^2 \frac{\Theta}{2}\right)\left(\frac{1}{2}\right)\right]$$

$$+ \cot\frac{\Theta}{2}\left[\left(4 \sec^3 \frac{\Theta}{2} \sec \frac{\Theta}{2} \tan \frac{\Theta}{2}\right) \left(\frac{1}{2}\right)\right.$$

$$\left. - 2 \cos \frac{\Theta}{2} \left(- \sin \frac{\Theta}{2}\right) \left(\frac{1}{2}\right)\right].$$

By multiplying the trigonometric functions together and simplifying,

$$\frac{dr}{d\Theta} = \frac{1}{2} \cot^2 \frac{\Theta}{2} - \frac{1}{2} \sec^4 \frac{\Theta}{2} \csc^2 \frac{\Theta}{2}$$

$$+ 2 \sec^4 \frac{\Theta}{2} + \cos^2 \frac{\Theta}{2}.$$

• PROBLEM 155

Find the derivative of:

$$f(x) = x^2 \cos x(1 + x^4)^{-7}.$$

Solution: When a combination of products are given, it is often best to use logarithms to reduce the multiplications to steps of addition. We take the logarithm of the absolute value of f(x).

Hence,

$$g(x) = \ln|f(x)| = \ln x^2 + \ln |\cos x + \ln (1 + x^4)^{-7}$$

$$= 2 \ln |x| + \ln |\cos x| - 7 \ln (1 + x^4).$$

Applying the theorem for the derivative of a logarithm,

$$g'(x) = \frac{f'(x)}{f(x)} = \frac{2}{x} - \frac{\sin x}{\cos x} - \frac{28x^3}{1 + x^4}.$$

To find f'(x), we multiply by f(x) to obtain:

$$f'(x) = \frac{2x \cos x}{(1 + x^4)^7} - \frac{x^2 \sin x}{(1 + x^4)^7} - \frac{28x^5 \cos x}{(1 + x^4)^8}.$$

• PROBLEM 156

Find the derivative of:

$$f(x) = \frac{\sin x}{1 - 2 \cos x}.$$

Solution: Applying the theorem for $d\left(\frac{u}{v}\right)$,

$$f'(x) = \frac{(1 - 2 \cos x) D_x (\sin x) - \sin x \cdot D_x (1 - 2 \cos x)}{(1 - 2 \cos x)^2}$$

Substituting for $\frac{d\ (\sin x)}{dx}$ and $\frac{d\ (1 - 2 \cos x)}{dx}$,

$$= \frac{(1 - 2 \cos x)(\cos x) - \sin x(2 \sin x)}{(1 - 2 \cos x)^2}.$$

Simplifying the result,

$$f'(x) = \frac{\cos x - 2(\cos^2 x + \sin^2 x)}{(1 - 2 \cos x)^2}$$

$$= \frac{\cos x - 2}{(1 - 2 \cos x)^2}.$$

• PROBLEM 157

Find the derivative of the function:

$$y = \frac{2 - 3 \tan 4x}{\sec 4x}.$$

Solution: We apply the quotient rule, obtaining:

$$\frac{dy}{dx} =$$

$$\frac{\left[(\sec 4x)(-3 \sec^2 4x)(4)\right] - \left[(2 - 3 \tan 4x)(\sec 4x \tan 4x)(4)\right]}{\sec^2 4x}.$$

Multiplying out and factoring sec 4x from the numerator we have,

$$\frac{dy}{dx} = \frac{\sec 4x(-12 \sec^2 4x - 8 \tan 4x + 12 \tan^2 4x)}{\sec^2 4x}$$

$$= \frac{-12 \sec^2 4x}{\sec 4x} - \frac{8 \tan 4x}{\sec 4x} + \frac{12 \tan^2 4x}{\sec 4x}$$

$$= -12 \sec 4x - 8 \tan 4x \cos 4x + 12 \tan^2 4x \cos 4x$$

$$= -12 \sec 4x - 8 \sin 4x + 12 \sin^2 4x \sec 4x$$

$$= -8 \sin 4x - 12 \sec 4x(1 - \sin^2 4x).$$

Making use of the identity, $\cos^2 x = 1 - \sin^2 x$, we obtain,

$$\frac{dy}{dx} = -8 \sin 4x - 12 \sec 4x \cos^2 4x.$$

Using the fact that $\sec x = \frac{1}{\cos x}$, we have:

$$\frac{dy}{dx} = -8 \sin 4x - 12 \cdot \frac{1}{\cos 4x} \cdot \cos^2 4x$$

$$= -8 \sin 4x - 12 \cos 4x.$$

• PROBLEM 158

Find the derivative of the function: $r = \frac{\sqrt{1 - \cos 2\theta}}{\tan \theta}$.

Solution: We can use the quotient rule, which states:

$$\frac{d}{dx}\left(\frac{u}{v}\right) = \frac{v\frac{du}{dx} - u\frac{dv}{dx}}{v^2},$$

to find the derivative of this function. We obtain,

$$r = \frac{\sqrt{1 - \cos 2\theta}}{\tan \theta} = \frac{(1 - \cos 2\theta)^{\frac{1}{2}}}{\tan \theta}.$$

$$\frac{dr}{d\theta} = \left[\tan \theta \left[\frac{1}{2}(1 - \cos 2\theta)^{-\frac{1}{2}}(\sin 2\theta)(2)\right] - (1 - \cos 2\theta)^{\frac{1}{2}} \sec^2 \theta\right] / \tan^2 \theta$$

$$= \frac{\tan \theta \sin 2\theta - (1 - \cos 2\theta) \sec^2 \theta}{\tan^2 \theta \sqrt{1 - \cos 2\theta}}.$$

Find $D_x^2 y$ of: $x^2 + \cos 2y = 3$.

Solution: To find $D_x^2 y = \frac{d^2 y}{dx^2}$ of the given function we first determine $D_x y = \frac{dy}{dx}$, using implicit differentiation. We find:

$$2x + (-\sin 2y)(2)D_x y = 0$$

$$2x = 2 \sin 2y \, D_x y$$

$$D_x y = \frac{2x}{2 \sin 2y} = \frac{x}{\sin 2y}.$$

To obtain $D_x^2 y$, we differentiate $D_x y$ using the quotient rule.

$$D_x^2 y = \frac{(\sin 2y)(1) - (x)(\cos 2y)(2)D_x y}{\sin^2 2y}.$$

Substituting for $D_x y$, we obtain:

$$D_x^2 y = \frac{\sin 2y - 2x \cdot \cos 2y\left(\frac{x}{\sin 2y}\right)}{\sin^2 2y} =$$

$$\frac{\sin 2y}{\sin^2 2y} - \frac{2x^2 \cos 2y}{\sin^3 2y} =$$

$$\frac{1}{\sin 2y} - \left(2x^2 \cdot \frac{\cos 2y}{\sin 2y} \cdot \frac{1}{\sin^2 2y} \right) =$$

$$\csc 2y - 2x^2 \cot 2y \csc^2 2y =$$

$$\csc 2y(1 - 2x^2 \cot 2y \csc 2y).$$

Find $D_x^2 y$ of $2y^2 - \cos y^2 = x$.

Solution: We first find $D_x y = \frac{dy}{dx}$ by using implicit differentiation. We obtain:

$$4y\ D_xy - (-\sin y^2 \cdot 2y)D_xy = 1.$$

$$4y\ D_xy + 2y \sin y^2\ D_xy = 1.$$

$$D_xy(4y + 2y \sin y^2) = 1.$$

$$D_xy = \frac{1}{4y + 2y \sin y^2}.$$

Using the quotient rule, we find:

$$D_x^2y = \frac{d^2y}{dx^2} =$$

$$\frac{(4y + 2y \sin y^2)(0) - (1)\left[4D_xy + (2y \cos y^2\ 2y\ D_xy + \sin y^2\ 2\ D_xy)\right]}{(4y + 2y \sin y^2)^2}$$

$$= -\frac{4D_xy + 4y^2 \cos y^2\ D_xy + 2 \sin y^2\ D_xy}{(4y + 2y \sin y^2)^2}$$

$$= -\frac{2D_xy(2 + 2y^2 \cos y^2 + \sin y^2)}{(4y + 2y \sin y^2)^2}.$$

Substituting for D_xy:

$$D_x^2y = -\frac{\frac{2}{4y + 2y \sin y^2}\left(2 + 2y^2 \cos y^2 + y^2\right)}{(4y + 2y \sin y^2)^2}$$

$$= -\frac{2}{(4y + 2y \sin y^2)^3}\left(2 + 2y^2 \cos y^2 + \sin y^2\right)$$

$$= -\frac{2}{[2y(2 + \sin y^2)]^3}\left(2 + 2y^2 \cos y^2 + \sin y^2\right)$$

$$= -\frac{2 + 2y^2 \cos y^2 + \sin y^2}{4y^3(2 + \sin y^2)^3}.$$

CHAPTER 8

DIFFERENTIATION OF INVERSE TRIGONOMETRIC FUNCTIONS

The basic derivatives for arc sin x, arc cos x, arc tan x should be memorized and then applied in the manner described for direct trigonometric functions.

Inverse trigonometric functions may be also handled sometimes by inverting the expressions, and applying the rules for the direct trigonometric functions. The procedure is best explained with a simple example. If we have $y = \text{arc}\sin x$, then, $x = \sin y$. We can differentiate this expression now, treating, in this case, x as the dependent variable and y as the independent one.

$$x = \sin y.$$

$$\frac{dx}{dy} = \cos y.$$

But $\frac{dy}{dx} = 1 \Big/ \frac{dx}{dy}$. Hence, $\frac{dy}{dx} = \left(\frac{dy}{dx} = \frac{1}{\cos y}\right)$. This is the answer, but we wish to have it in terms of x, not y. From the identity: $\sin^2\theta + \cos^2\theta = 1$, we find that:

$$\cos\theta = \sqrt{1 - \sin^2\theta}.$$

But $x = \sin y$. Hence $\cos y = \sqrt{1 - x^2}$. Thus, the final answer is:

$$\left(\frac{dy}{dx} = \frac{1}{\sqrt{1 - x^2}}\right).$$

• PROBLEM 161

Find the derivative of $y = \text{arc}\sin 4x$.

Solution: We use the formula for differentiation of the $\sin^{-1}$ or arc sin function, which states:

$$\frac{d}{dx}\sin^{-1} u = \frac{1}{\sqrt{1-u^2}}\;\frac{du}{dx}.$$

Hence,

$$\frac{dy}{dx} = \frac{1}{\sqrt{1 - 16x^2}}\left(4\right) = \frac{4}{\sqrt{1 - 16x^2}}.$$

• PROBLEM 162

Given: $y = \sin^{-1} x^2$, find $\frac{dy}{dx}$.

Solution: In order to differentiate this function, we apply the formula for differentiation of the $\sin^{-1}$ or arc sin function.

$$\frac{d}{dx} \sin^{-1} u = \frac{1}{\sqrt{1 - u^2}} \frac{du}{dx} .$$

Hence,

$$\frac{dy}{dx} = \frac{1}{\sqrt{1 - (x^2)^2}} \left(2x\right) = \frac{2x}{\sqrt{1 - x^4}} .$$

• PROBLEM 163

Differentiate the following: $y = \text{arc} \tan ax^2$, where a is a constant.

Solution: In order to solve this problem we use the formula for differentiation of arc tan or $\tan^{-1}$ functions,

$$\frac{d}{dx} \tan^{-1} u = \frac{1}{1 + u^2} \frac{du}{dx} .$$

Therefore,

$$\frac{dy}{dx} = \frac{\frac{d}{dx}\left(ax^2\right)}{1 + \left(ax^2\right)^2}$$

$$= \frac{2ax}{1 + a^2x^4} .$$

• PROBLEM 164

What is the slope of: $y = 4x$ arc tan $2x$ at the point: $x = \frac{1}{2}$?

Solution: To find the slope of a function we find its derivative. The given function is the product of two functions of x: $4x$ and arc tan $2x$. We therefore apply the product rule in differentiating the given function. Hence,

$$\frac{dy}{dx} = 4x\left[\frac{1}{1 + (2x)^2} \cdot \frac{d}{dx}(2x)\right] + \text{arc} \tan 2x \cdot \frac{d}{dx}(4x)$$

$$= \frac{8x}{1 + 4x^2} + 4 \text{ arc} \tan 2x .$$

Letting $x = \frac{1}{2}$, we have

$$\frac{dy}{dx} = \frac{4}{2} + 4 \text{ arc} \tan 1 = 2 + 4\left(\frac{\pi}{4}\right) = 5.14$$

• PROBLEM 165

Find the derivative of the expression: $y = \text{arc}\cos(1 - 2x)$.

Solution: In order to find the derivative of this function, we use the formula for differentiation of the $\cos^{-1}$, inverse, or arc cos function, which states:

$$\frac{d}{dx}\cos^{-1}u = \frac{-1}{\sqrt{1 - u^2}}\frac{du}{dx},$$

hence,

$$\frac{dy}{dx} = \frac{-1}{\sqrt{1 - (1 - 2x)^2}}\left(-2\right) = \frac{2}{\sqrt{1 - 1 + 4x - 4x^2}}$$

$$= \frac{2}{\sqrt{4x - 4x^2}} = \pm\frac{1}{\sqrt{x - x^2}}.$$

• PROBLEM 166

$y = \text{arc}\sin\left(3x - 4x^3\right)$. Find $\frac{dy}{dx}$.

Solution: We use the formula for the differentiation of the $\sin^{-1}$ or arc sin function:

$$\frac{d}{dx}\sin^{-1}u = \frac{1}{\sqrt{1 - u^2}}\frac{du}{dx}.$$

$$\frac{dy}{dx} = \frac{\frac{d}{dx}\left(3x - 4x^3\right)}{\sqrt{1 - \left(3x - 4x^3\right)^2}}$$

$$= \frac{3 - 12x^2}{\sqrt{1 - 9x^2 + 24x^4 - 16x^6}} = \frac{3(1 - 4x^2)}{\sqrt{(1 - x^2)(16x^4 - 8x^2 + 1)}}$$

$$= \frac{3(1 - 4x^2)}{\sqrt{(4x^2 - 1)^2(1 - x^2)}} = \pm\frac{3}{\sqrt{1 - x^2}} .$$

• PROBLEM 167

Find the derivative of the function:
$y = \sqrt{\arcsin 2x}$.

Solution: To solve this problem we use the rules that state:

$$\frac{d}{dx}(u^n) = nu^{n-1}\frac{du}{dx},$$

and

$$\frac{d}{dx}\sin^{-1} u = \frac{1}{\sqrt{1 - u^2}}\frac{du}{dx}.$$

We obtain:

$$\frac{d}{dx}(\arcsin 2x)^{\frac{1}{2}} = \frac{1}{2}(\arcsin 2x)^{-\frac{1}{2}}\left[\frac{1}{\sqrt{1 - (2x)^2}}\right](2)$$

$$= \left(\frac{1}{\sqrt{\arcsin 2x}}\right)\left(\frac{1}{\sqrt{1 - 4x^2}}\right)$$

$$= \frac{1}{\sqrt{\arcsin 2x - 4x^2 \arcsin 2x}}.$$

• **PROBLEM 168**

Find the derivative of the function:
$y = 2 \arcsin \sqrt{1 - 2x}$.

Solution: We solve this problem using the formula for the derivative of arcsin u:

$$\frac{d}{dx} \operatorname{Sin}^{-1} u = \frac{1}{\sqrt{1 - u^2}} \frac{du}{dx},$$

and the formula:

$$\frac{d}{dx} u^n = nu^{n-1} \frac{du}{dx}.$$

Hence, we have:

$$\frac{d}{dx}\left(2 \arcsin \sqrt{1-2x}\right) = 2\left[\frac{1}{\sqrt{1 - (\sqrt{1-2x})^2}}\right]\left[\frac{1}{2}(1-2x)^{-\frac{1}{2}}\right](-2)$$

$$= \left(\frac{1}{\sqrt{1 - 1 + 2x}}\right)\left(\frac{1}{\sqrt{1 - 2x}}\right)(-2)$$

$$= \frac{-2}{\sqrt{2x - 4x^2}}.$$

• **PROBLEM 169**

Find the derivative of the function:
$y = 4 \arctan \sqrt{x^2 - 1}$.

Solution: To solve this problem we use the formula for the derivative of the arctangent function:

$$\frac{d}{dx} \tan^{-1} u = \frac{1}{1 + u^2} \frac{du}{dx},$$

and the formula:

$$\frac{d}{dx} u^n = nu^{n-1} \frac{du}{dx}.$$

Therefore we find:

$$\frac{d}{dx}\left(4 \arctan \sqrt{x^2-1}\right) = 4\left[\frac{1}{1 + (\sqrt{x^2-1})^2}\right]\left[\frac{1}{2}(x^2-1)^{-\frac{1}{2}}\right](2x)$$

$$= 4\left(\frac{1}{1 + x^2 - 1}\right)\left[\frac{1}{2}\left(\frac{1}{\sqrt{x^2-1}}\right)\right](2x)$$

$$= \frac{4x}{x^2\sqrt{x^2-1}}$$

$$= \frac{4}{x\sqrt{x^2-1}}.$$

• PROBLEM 170

Given: $y = \text{arc tan } \frac{3}{x}$. Find $\frac{dy}{dx}$.

Solution: In this example, we use the formula:

$$\frac{d(\text{arc tan } u)}{dx} = \frac{1}{1+u^2} \frac{du}{dx}.$$

For

$$y = \text{arc tan } \frac{3}{x}, \quad u = \frac{3}{x}, \quad \text{and} \quad du = \frac{-3}{x^2}.$$

Therefore,

$$\frac{dy}{dx} = \frac{1\left(\frac{-3}{x^2}\right)}{1+\left(\frac{3}{x}\right)^2} = \frac{\frac{-3}{x^2}}{\frac{x^2+9}{x^2}} = \frac{-3}{x^2+9}.$$

• PROBLEM 171

Find the derivative of the function: $\theta = \arccos \frac{r-1}{3r}$.

Solution: Using the rule that

$$\frac{d}{dr}(\arccos u) = \frac{-1}{\sqrt{1-u^2}} \frac{du}{dr},$$

and the quotient rule that states:

$$\frac{d}{dr}\left(\frac{u}{v}\right) = \frac{v\frac{du}{dr} - u\frac{dv}{dr}}{v^2},$$

we have:

$$\frac{d}{dr}\left(\arccos \frac{r-1}{3r}\right) = \left[\frac{-1}{\sqrt{1-\left(\frac{r-1}{3r}\right)^2}}\right]\left[\frac{3r(1)-(r-1)(3)}{(3r)^2}\right]$$

$$= \left(\frac{-1}{\sqrt{1-\frac{(r^2-2r+1)}{9r^2}}}\right)\left(\frac{3r-3r+3}{9r^2}\right)$$

$$= \left(\frac{-1}{\sqrt{\frac{8r^2+2r-1}{9r^2}}}\right)\left(\frac{3}{9r^2}\right)$$

$$= \frac{-1}{r\sqrt{8r^2+2r-1}}$$

• PROBLEM 172

> Find the derivative of the function:
> $y = \arctan \frac{x - a}{1 + ax}$, a = a constant.

Solution: We solve this problem by using the formula for the derivative of the arctangent:

$$\frac{d}{dx} \tan^{-1} u = \frac{1}{1 + u^2} \frac{du}{dx},$$

and the quotient rule:

$$\frac{d}{dx}\left(\frac{u}{v}\right) = \frac{v\frac{du}{dx} - u\frac{dv}{dx}}{v^2}.$$

$$\frac{d}{dx}\left(\arctan \frac{x-a}{1+ax}\right) = \left[\frac{1}{1 + \left(\frac{x-a}{1+ax}\right)^2}\right]\left[\frac{(1+ax)(1) - (x-a)(a)}{(1 + ax)^2}\right]$$

$$= \left(\frac{1}{1 + \frac{x^2-2ax+a^2}{1+2ax+a^2x^2}}\right)\left(\frac{1+ax-ax+a^2}{1+2ax+a^2x^2}\right)$$

$$= \frac{1 + a^2}{1+2ax+a^2x^2+x^2-2ax+a^2}$$

$$= \frac{1 + a^2}{1+a^2x^2+x^2+a^2}$$

$$= \frac{1 + a^2}{(1+a^2)(1+x^2)}$$

$$= \frac{1}{1+x^2}.$$

• PROBLEM 173

Differentiate: $y = \text{arc sec}\,\frac{x^2+1}{x^2-1}$.

Solution: In order to differentiate this function, we use the formula for the differentiation of the $\sec^{-1}$ or arcsec function,

$$\frac{d}{dx}\sec^{-1}u = \frac{-1}{u\sqrt{u^2-1}}\frac{du}{dx},$$

and the quotient rule, which states:

$$\frac{d}{dx}\frac{u}{v} = \frac{v\frac{du}{dx} - u\frac{dv}{dx}}{v^2}.$$

Therefore,

$$\frac{dy}{dx} = \frac{\frac{d}{dx}\left(\frac{x^2+1}{x^2-1}\right)}{\frac{x^2+1}{x^2-1}\sqrt{\left(\frac{x^2+1}{x^2-1}\right)^2-1}}$$

$$= \frac{\frac{(x^2-1)2x-(x^2+1)2x}{(x^2-1)^2}}{\frac{x^2+1}{x^2-1}\cdot\frac{2x}{x^2-1}} = -\frac{2}{x^2+1}.$$

• PROBLEM 174

Find the derivative of the function:
$y = \arccos\left(\frac{\tan 2x}{2x}\right)$.

Solution: We use the chain rule and the quotient rule to obtain this derivative. If we have a function $y = f(u)$, where u is a function of x, the chain rule states that

$$\frac{dy}{dx} = \frac{dy}{du}\frac{du}{dx}.$$

The quotient rule states:

$$\frac{d}{dx}\left(\frac{u}{v}\right) = \frac{v\frac{du}{dx} - u\frac{dv}{dx}}{v^2}.$$

From the chain rule,

$$y = \arccos\left(\frac{\tan 2x}{2x}\right),\ u = \left(\frac{\tan 2x}{2x}\right).$$

$$\frac{dy}{du} = \frac{-1}{\sqrt{1 - \left(\frac{\tan 2x}{2x}\right)^2}}.$$

Using the quotient rule, we find:

$$\frac{du}{dx} = \frac{2x(\sec^2 2x)(2) - \tan 2x(2)}{4x^2}.$$

Therefore,

$$\frac{dy}{dx} = \left(\frac{-1}{\sqrt{1 - \frac{\tan^2 2x}{4x^2}}}\right)\left(\frac{4x \sec^2 2x - 2 \tan 2x}{4x^2}\right)$$

$$= \frac{2 \tan 2x - 4x \sec^2 2x}{4x^2 \frac{\sqrt{4x^2 - \tan^2 2x}}{2x}}$$

$$= \frac{\tan 2x - 2x \sec^2 2x}{x\sqrt{4x^2 - \tan^2 2x}}.$$

• PROBLEM 175

Find the derivative of the function:
$y = \cos(2 \arctan 3x)$.

Solution: We can use the chain rule to find this derivative, since it is a function of a function. If $y = f(u)$ and $u = g(x)$ then $y = f\left[g(x)\right]$, and

$$\frac{dy}{dx} = \frac{dy}{du}\frac{du}{dx}.$$

$$y = \cos(2 \arctan 3x), \; u = (2 \arctan 3x).$$

$$\frac{dy}{du} = -\sin(2 \arctan 3x).$$

$$\frac{du}{dx} = 2\left(\frac{1}{1 + (3x)^2}\right)\left(3\right).$$ Therefore

$$\frac{dy}{dx} = -\sin(2 \arctan 3x)(2)\left(\frac{1}{1 + 9x^2}\right)(3)$$

$$= \frac{-6 \sin(2 \arctan 3x)}{1 + 9x^2}.$$

• **PROBLEM 176**

Find the derivative of the function:
$y = \arccos\left[\tan(\arcsin x)\right]$.

Solution: To take the derivative of a function of a function, we use the chain rule, which states that, if u is a function of x, $\frac{dy}{dx} = \frac{dy}{du}\frac{du}{dx}$.

$$\frac{d}{dx}\left(\arccos\left[\tan(\arcsin x)\right]\right) =$$

$$\frac{-1}{\sqrt{1 - \tan^2(\arcsin x)}}\left[\sec^2(\arcsin x)\right]\left(\frac{1}{\sqrt{1 - x^2}}\right)$$

$$= \frac{-\sec^2(\arcsin x)}{\sqrt{1 - \tan^2(\arcsin x)}\sqrt{1 - x^2}}.$$

• **PROBLEM 177**

Find the derivative of the function:
$y = \sqrt{x^2 - a^2} - a \arccos \frac{a}{x}$, with a = constant.

Solution: To solve this problem we use the power rule:

$$\frac{d}{dx} u^n = nu^{n-1} \frac{du}{dx},$$

the formula for the derivative of the arccosine:

$$\frac{d}{dx} \cos^{-1} u = \frac{-1}{\sqrt{1 - u^2}} \frac{du}{dx},$$

and the quotient rule:

$$\frac{d}{dx}\left(\frac{u}{v}\right) = \frac{v\frac{du}{dx} - u\frac{dv}{dx}}{v^2}.$$

$$\frac{d}{dx}\left(\sqrt{x^2 - a^2} - a \arccos \frac{a}{x}\right)$$

$$= \frac{1}{2}\left(x^2 - a^2\right)^{-\frac{1}{2}}(2x) - a\left(\frac{-1}{\sqrt{1 - (\frac{a}{x})^2}}\right)\left(\frac{(x)(0) - (a)(1)}{x^2}\right)$$

$$= \frac{2x}{2\sqrt{x^2 - a^2}} - \left(\frac{-a}{\sqrt{1 - \frac{a^2}{x^2}}}\right)\left(\frac{-a}{x^2}\right)$$

$$= \frac{x}{\sqrt{x^2 - a^2}} - \frac{a^2}{x^2\sqrt{1 - \frac{a^2}{x^2}}}$$

$$= \frac{x}{\sqrt{x^2 - a^2}} - \frac{a^2}{x\sqrt{x^2 - a^2}}$$

$$= \frac{x^2 - a^2}{x\sqrt{x^2 - a^2}}$$

$$= \frac{\sqrt{x^2 - a^2}}{x}.$$

• PROBLEM 178

Find the derivative of the expression: $y = x \text{ arc} \tan^2 2x$.

Solution: In order to differentiate this function, we make use of the following: the product rule, $\frac{d}{dx} uv = u \frac{dv}{dx} + v \frac{dy}{dx}$; the formula for differentiation of the $\tan^{-1}$ or arc tan function, which states:

$$\frac{d}{dx} \tan^{-1} u = \frac{1}{1 + u^2} \frac{du}{dx};$$

and the formula: $\frac{d}{dx} u^n = nu^{n-1} \frac{du}{dx}$. Thus we obtain:

$$\frac{dy}{dx} = x(2)(\arctan 2x)\left(\frac{1}{1 + 4x^2}\right)(2) + \arctan^2 2x$$

$$= \arctan 2x\left(\frac{4x}{1 + 4x^2} + \arctan 2x\right).$$

• PROBLEM 179

Given: $y = x^3 \cot^{-1} \frac{1}{3}x$, find $\frac{dy}{dx}$.

Solution: In order to differentiate this function, we make use of both the product rule; $\frac{d}{dx} uv = \frac{dv}{dx} + v \frac{du}{dx}$, and the formula for differentiation of $\cot^{-1}$ or arccot function, which states:

$$\frac{d}{dx} \cot^{-1} u = \frac{-1}{1 + u^2} \frac{du}{dx}.$$

We find:

$$\frac{dy}{dx} = 3x^2 \cot^{-1} \frac{1}{3}x + x^3 \cdot \frac{-1}{1 + \frac{1}{9}x^2} \cdot \frac{1}{3}$$

$$= 3x^2 \cot^{-1} \frac{1}{3}x - \frac{3x^3}{9 + x^2}.$$

• PROBLEM 180

Differentiate: $y = 2 \text{ arc sin } \frac{1}{2}x - \frac{1}{2}x\sqrt{4 - x^2}$.

Solution: The first term of this function, $2 \sin^{-1} \frac{1}{2}x$, is differentiated using the formula for differentiation of the arc sin or $\sin^{-1}$ function,

$$\frac{d}{dx} \sin^{-1} u = \frac{1}{\sqrt{1-u^2}} \frac{du}{dx} .$$

The second term, $-\frac{1}{2}x\sqrt{4 - x^2}$, is differentiated using both the product rule; $\frac{d}{dx} uv = u\frac{dv}{dx} + v\frac{du}{dx}$, and the formula: $\frac{d}{dx} u^n = nu^{n-1} \frac{du}{dx}$.

Hence,

$$\frac{dy}{dx} = 2\cdot \frac{1}{\sqrt{1 - \frac{1}{4}x^2}} \cdot \frac{1}{2} - \left[\frac{1}{2}x \cdot \frac{1}{2}\left(4 - x^2\right)^{-\frac{1}{2}}(-2x) + \frac{1}{2}\left(4 - x^2\right)^{\frac{1}{2}}\right]$$

$$= \frac{2}{\sqrt{4 - x^2}} + \frac{x^2}{2\sqrt{4 - x^2}} - \frac{4 - x^2}{2\sqrt{4 - x^2}} = \frac{x^2}{\sqrt{4 - x^2}} .$$

• PROBLEM 181

Find the derivative of the function: $y = x \text{ arcsin } 2x + \frac{1}{2}\sqrt{1 - 4x^2}$.

Solution: The first term of this function, $x \text{ arcsin } 2x$, is differentiated using the product rule; $\frac{d}{dx} uv = u\frac{dv}{dx} + v\frac{du}{dx}$, and the formula for differentiating the $\sin^{-1}$ or arcsin function,

$$\frac{d}{dx} \sin^{-1} u = \frac{1}{\sqrt{1 - u^2}} \frac{du}{dx} .$$

The second term, $\frac{1}{2}\sqrt{1 - 4x^2}$, is differentiated using the formula $\frac{d}{dx} u^n = nu^{n-1} \frac{du}{dx}$. Therefore,

$$\frac{dy}{dx} = x\left(\frac{2}{\sqrt{1 - 4x^2}}\right) + \text{arcsin } 2x + \frac{1}{2}\cdot\frac{1}{2}\left(1 - 4x^2\right)^{-\frac{1}{2}}(-8x)$$

$$= \frac{2x}{\sqrt{1 - 4x^2}} + \text{arcsin } 2x - \frac{2x}{\sqrt{1- 4x^2}} = \text{arcsin } 2x.$$

• PROBLEM 182

Find the derivative of the expression: $y = (x^2 + 1)$ arc tan $x - x$.

Solution: The formula for differentiation of the arc tan or $\tan^{-1}$ function states:

$$\frac{d}{dx} \tan^{-1} u = \frac{1}{1+u^2} \frac{du}{dx}.$$

Using this formula, as well as the product rule:

$$\frac{d}{dx} uv = u\frac{dv}{dx} + v\frac{du}{dx},$$

we find:

$$\frac{dy}{dx} = (x^2+1)\ \frac{1}{1+x^2} + (\text{arc tan } x)(2x) - 1 = 2x \text{ arctan } x.$$

• PROBLEM 183

Find the derivative of the function:
$y = (2x^2 - 1)\arccos x - x\sqrt{1-x^2}$.

Solution: Using the product rule: $\frac{d}{dx}(uv) = u\frac{dv}{dx} + v\frac{du}{dx}$, and the formula for the derivative of the arccosine function,

$$\frac{d}{dx} \cos^{-1} u = \frac{-1}{\sqrt{1-u^2}} \frac{du}{dx}, \qquad \text{we have:}$$

$$\frac{d}{dx}\left[(2x^2-1)\arccos x - x\sqrt{1-x^2}\right]$$

$$= (2x^2-1)\ \frac{-1}{\sqrt{1-x^2}}(1) \quad + 4x(\arccos x)$$

$$-\left[x\left(\frac{1}{2}\right)\left(1-x^2\right)^{-\frac{1}{2}}\ (-2x) + \sqrt{1-x^2}\right]$$

$$= \frac{1-2x^2}{\sqrt{1-x^2}} + 4x \arccos x + \frac{x^2}{\sqrt{1-x^2}} - \sqrt{1-x^2}$$

$$= \frac{1-2x^2}{\sqrt{1-x^2}} + 4x \arccos x + \frac{x^2}{\sqrt{1-x^2}} - \frac{1-x^2}{\sqrt{1-x^2}}$$

$$= 4x \arccos x.$$

Given: $y = (x - a)\sqrt{2ax - x^2} + a^2 \text{arc sin } \dfrac{x - a}{a}$.

Find $\dfrac{dy}{dx}$, where a is a constant.

Solution:

$$y = (x - a)\sqrt{2ax - x^2} + a^2 \text{arc sin } \frac{x - a}{a} .$$

To obtain $\dfrac{dy}{dx}$, the chain rule for the derivative of a product is used for the first term, and the rule;

$$\frac{d(\text{arc sin } u)}{dx} = \frac{1}{\sqrt{1 - u^2}} \frac{du}{dx} ,$$

is used for the second term.

We first find:

$$\frac{d\left[(x - a)(2ax - x^2)^{1/2}\right]}{dx}$$

$$= (x - a)\frac{1}{2}\left(2ax - x^2\right)^{-1/2}(2a - 2x) + (2ax - x^2)^{1/2}(1)$$

$$= \frac{(x - a)(a - x)}{\sqrt{2ax - x^2}} + \sqrt{2ax - x^2}$$

$$= \frac{(x - a)(a - x) + (2ax - x^2)}{\sqrt{2ax - x^2}}$$

$$= \frac{- x^2 + 2ax - a^2 + 2ax - x^2}{\sqrt{2ax - x^2}}$$

$$= \frac{2(- x^2 + 2ax) - a^2}{\sqrt{2ax - x^2}} .$$

Next, we find

$$\frac{d\left[a^2 \text{arc sin } \dfrac{x - a}{a}\right]}{dx} = a^2 \left(\frac{1 \left(\dfrac{1}{a}\right)}{\sqrt{1 - \left(\dfrac{x - a}{a}\right)^2}}\right)$$

$$= \frac{a}{\sqrt{1 - \left(\dfrac{x^2 - 2ax + a^2}{a^2}\right)}}$$

$$= \frac{a}{\sqrt{\frac{-x^2 + 2ax}{a^2}}} = \frac{a^2}{\sqrt{-x^2 + 2ax}}$$

Combining the results:

$$\frac{dy}{dx} = \frac{2(-x^2 + 2ax) - a^2}{\sqrt{2ax - x^2}} + \frac{a^2}{\sqrt{2ax - x^2}}$$

$$= \frac{2(-x^2 + 2ax)}{\sqrt{2ax - x^2}} = 2\sqrt{2ax - x^2}\ .$$

• PROBLEM 185

> Find a point on the cycloid: $x = a \arccos \frac{a - y}{a} - \sqrt{2ay - y^2}$, at which the slope is equal to 1. Consider a to be a constant.

Solution: To solve this problem, we solve for $\frac{dx}{dy}$, the slope in terms of y. Then, setting this value for the slope equal to one, we solve for the values of y, and then find the corresponding x. We use the derivative of the arccosine function: $\frac{d}{dx}\cos^{-1} u = \frac{-1}{\sqrt{1 - u^2}}\frac{du}{dx}$, the quotient rule:

$\frac{d}{dx}\left(\frac{u}{v}\right) = \frac{v\frac{du}{dx} - u\frac{dv}{dx}}{v^2}$, and the power rule:

$\frac{d}{dx} u^n = nu^{n-1}\frac{du}{dx}$.

$$x = a \arccos \frac{a - y}{a} - \sqrt{2ay - y^2}.$$

$$\frac{dx}{dy} = \left[a \frac{-1}{\sqrt{1 - \left(\frac{a-y}{a}\right)^2}}\left(\frac{a(-1) - (a-y)(0)}{a^2}\right)\right] -$$

$$\frac{1}{2}\left(\frac{1}{\sqrt{2ay - y^2}}\right)(2a - 2y)$$

$$= \frac{1}{\sqrt{\frac{2ay - y^2}{a^2}}} - \frac{a - y}{\sqrt{2ay - y^2}}$$

$$= \frac{a}{\sqrt{2ay - y^2}} - \frac{a - y}{\sqrt{2ay - y^2}}$$

$$= \frac{y}{\sqrt{2ay - y^2}}, \text{ the slope of the cycloid.}$$

To find whether any points exist where this slope is equal to one, we set this value equal to one and solve for y. Then, we substitute the value for y into the equation for the cycloid, and solve for x. We find,

$$\frac{y}{\sqrt{2ay - y^2}} = 1$$

$$y = \sqrt{2ay - y^2}$$

$$2y^2 - 2ay = 0$$

$$2y(y - a) = 0$$

$$2y = 0 \text{ or } y - a = 0$$

$$y = 0 \qquad y = a.$$

Substituting these values for y into the given equation, we can calculate x.

When $y = 0$,

$$x = a \arccos\left(\frac{a - 0}{a}\right) - \sqrt{0 - 0}$$

$$= a \arccos 1$$

$$= a(0).$$

$$x = 0.$$

When $y = a$,

$$x = a \arccos\left(\frac{a - a}{a}\right) - \sqrt{2a^2 - a^2}$$

$$= a \arccos 0 - a.$$

$$x = a\left(\frac{\pi}{2}\right) - a = \frac{\pi a}{2} - a.$$

Therefore, the slope of the cycloid is one at the points $(0,0)$ and $\left(\frac{\pi a}{2} - a, a\right)$.

CHAPTER 9

DIFFERENTIATION OF EXPONENTIAL AND LOGARITHMIC FUNCTIONS

The function that has the simplest derivative is the function, e^x, where e is the base of the natural logarithms. Its derivative is also e^x, as are any succeeding derivatives. However, if, instead of e^x, we have e^u, where u is a function of x, then $\frac{d}{dx}\left(e^u\right) = e^u \frac{du}{dx}$, so that it is necessary to differentiate u. If, instead of e^x, we wish to differentiate a^x, where a is some other constant, then $\frac{dy}{dx} = a^x \ln a$, i.e., we have not simply the function itself, but the function multiplied by the natural logarithm of a.

Consider the function: $y = \ln x$. Then we can write:

$$x = e^y.$$

Differentiating,

$$\frac{dx}{dx} = e^y \frac{dy}{dx}.$$

$$1 = e^y \frac{dy}{dx}.$$

$$\frac{dy}{dx} = \frac{1}{e^y} = \frac{1}{x}.$$

Hence we have the widely-used derivative of a logarithmic function.

If the given function is of the exponential type but somewhat more complex, it is often advantageous to begin by taking the logarithm of both sides. For example, to differentiate $y = x^x$, we can write:

$$\ln y = x \ln x$$

$$\frac{1}{y}\frac{dy}{dx} = x\frac{(1)}{(x)} + \ln x$$

$$= 1 + \ln x$$

$$\frac{dy}{dx} = y(1 + \ln x)$$

$$= (1 + \ln x)x^x.$$

• PROBLEM 186

Differentiate the expression: $y = a^{3x^2}$.

Solution: To differentiate, we use the formula,

$$\frac{d}{dx} a^u = a^u \frac{du}{dx} \ln a,$$

letting $u = 3x^2$. **Then** $\frac{du}{dx} = 6x$. Applying the formula, we obtain:

$$\frac{dy}{dx} = a^{3x^2} \cdot 6x \cdot \ln a = 6\ x \ln a \cdot a^{3x^2}.$$

• PROBLEM 187

Find the derivative of $y = 2^{4x}$.

Solution: To find $\frac{dy}{dx}$, we use the differentiation formula:

$$\frac{d}{dx} a^u = a^u \frac{du}{dx} \ln a,$$

letting $a = 2$, $u = 4x$. **Then** $\frac{du}{dx} = 4$. Applying the formula, we obtain:

$$\frac{dy}{dx} = 2^{4x} \cdot 4 \cdot \ln 2 = 2^{2+4x} \ln 2.$$

• PROBLEM 188

Given $y = 3^{x^2}$, find $D_x y$.

Solution: To find $D_x y$, we apply the differentiation formula:

$$\frac{d}{dx} a^u = a^u \frac{du}{dx} \ln a.$$

In this problem, $a = 3$ and $u = x^2$. Differentiating, we obtain:

$$D_x y = 3^{x^2} \cdot 2x \cdot \ln 3$$

$$= 2(\ln 3)x3^{x^2}.$$

• PROBLEM 189

Find the derivative of $y = e^x$.

Solution: To find $\frac{dy}{dx}$, we use the formula,

$$\frac{d}{dx} e^u = e^u \frac{du}{dx},$$

letting $u = x$. Then $\frac{du}{dx} = 1$. Applying this formula, we obtain:

$$\frac{dy}{dx} = e^x(1) = e^x.$$

• **PROBLEM 190**

If $y = e^{1/x^2}$, find $D_x y$.

Solution: To find $D_x y = \frac{dy}{dx}$, we use the differentiation formula:

$$\frac{d}{dx} e^u = e^u \frac{du}{dx}, \quad \text{with } u = \frac{1}{x^2}. \text{ We}$$

obtain:

$$D_x y = e^{\frac{1}{x^2}} \cdot \left(- \frac{2x}{x^4} \right)$$

$$= e^{1/x^2} \left(- \frac{2}{x^3} \right) = - \frac{2e^{1/x^2}}{x^3}.$$

• **PROBLEM 191**

Find the derivative of:

$$y = \left(e^{1/x} \right)^2.$$

Solution: We can first rewrite the function as:

$$y = e^{\frac{2}{x}}.$$

Now we use the formula:

$$\frac{d}{dx} e^u = e^u \frac{du}{dx},$$

letting $u = \frac{2}{x}$. Then,

$$\frac{du}{dx} = \frac{(x)(0) - (2)(1)}{x^2} = - \frac{2}{x^2}.$$

Applying the formula, we obtain:

$$\frac{dy}{dx} = e^{\frac{2}{x}} \cdot - \frac{2}{x^2}$$

$$= - \frac{2e^{\frac{2}{x}}}{x^2}.$$

• PROBLEM 192

If $y = e^{\frac{x^2}{4}}$, what is $\frac{dy}{dx}$?

Solution: To find $\frac{dy}{dx}$, we use the differentiation formula:

$$\frac{d}{dx} e^u = e^u \frac{du}{dx}, \quad \text{letting } u = \frac{x^2}{4}.$$

Then

$$\frac{du}{dx} = \frac{2x}{4} = \frac{x}{2}.$$

Applying the formula we obtain:

$$\frac{dy}{dx} = e^{\frac{x^2}{4}} \cdot \frac{x}{2} = \frac{x}{2} e^{\frac{x^2}{4}}.$$

• PROBLEM 193

Differentiate: $y = be^{c^2 + x^2}$.

Solution: To differentiate, we use the formula:

$$\frac{d}{dx} e^u = e^u \frac{du}{dx},$$

letting $u = c^2 + x^2$. Then, $\frac{du}{dx} = 2x$. Applying the formula,we obtain:

$$\frac{dy}{dx} = b \cdot e^{c^2 + x^2} \cdot 2x = 2bxe^{c^2 + x^2}.$$

• PROBLEM 194

If $y = e^{\sqrt{x^3 + b}}$, what is $\frac{dy}{dx}$?

Solution: To find $\frac{dy}{dx}$, we use the differentiation formula:

$$\frac{d}{dx} e^u = e^u \frac{du}{dx},$$

letting $u = (x^3 + b)^{\frac{1}{2}}$. Then

$$\frac{du}{dx} = \frac{1}{2}(x^3 + b)^{-\frac{1}{2}} \cdot 3x^2 = \frac{3x^2}{2\sqrt{x^3 + b}}.$$

Applying the formula, we obtain:

$$\frac{dy}{dx} = e^{\sqrt{x^3 + b}} \cdot \frac{3x^2}{2\sqrt{x^3 + b}} .$$

• PROBLEM 195

Find the derivative of the expression:

$$y = xe^{\tan x} .$$

Solution: In this example, we make use of the product rule, $d(uv) = vdu + udv$. We find:

$$\frac{dy}{dx} = x\left(e^{\tan x}\right) \sec^2 x + e^{\tan x}$$

$$= e^{\tan x} (x \sec^2 x + 1).$$

• PROBLEM 196

Find the derivative of: $y = \log 4x$.

Solution: To find $\frac{dy}{dx}$, we use the differentiation formula:

$$\frac{d}{dx} \log_a u = \frac{1}{u} \frac{du}{dx} \log_a e,$$

letting $u = 4x$, with the a understood to equal 10. (We recall that when the base 10 is used it need not be written.) Applying the formula, we obtain:

$$\frac{dy}{dx} = \frac{1}{4x} \cdot 4 \cdot \log e$$

$$= \frac{1}{x} \log e . \quad (\log e = 0.434).$$

• PROBLEM 197

Differentiate: $y = \log \frac{2x}{1 + x^2}$.

Solution: To find $\frac{dy}{dx}$, we use the formula,

$$\frac{d}{dx} \log_a u = \frac{1}{u} \frac{du}{dx} \log_a e,$$

letting $u = \frac{2x}{1 + x^2}$, and the a is understood to equal 10. (We recall that when base 10 is used, it need not be written.) Applying the formula, we obtain:

$$\frac{dy}{dx} = \frac{1 + x^2}{2x} \cdot \frac{(1 + x^2)(2) - (2x)(2x)}{(1 + x^2)^2} \cdot \log e.$$

Simplifying, we obtain,

$$\frac{dy}{dx} = \left(\log e\right)\frac{1 - x^2}{x(1 + x^2)} .$$

• PROBLEM 198

Given the expression:

$$y = \log_{10} \frac{x + 1}{x^2 + 1},$$

find D_x y.

Solution: To find D_x y we can apply the differentiation formula:

$$\frac{d}{dx} \log_a u = \frac{1}{u} \frac{du}{dx} \log_a e,$$

with $a = 10$, $u = \frac{x + 1}{x^2 + 1}$.

Applying this, we obtain:

$$D_x y = \frac{1}{\frac{x + 1}{x^2 + 1}} \cdot \frac{(x^2 + 1)(1) - (x + 1)(2x)}{(x^2 + 1)^2} \cdot \log_{10} e$$

$$= \frac{\log_{10} e(1 - 2x - x^2)}{(x + 1)(x^2 + 1)} .$$

• PROBLEM 199

Find the derivative of the expression:

$$y = \ln 3x^4.$$

Solution: To find $\frac{dy}{dx}$, we use the differentiation formula:

$$\frac{d}{dx} \ln u = \frac{1}{u} \frac{du}{dx},$$

letting $u = 3x^4$, $\frac{du}{dx} = 12x^3$.

Applying the formula, we obtain:

$$\frac{dy}{dx} = \frac{1}{3x^4} (12x^3) = \frac{4}{x} .$$

• PROBLEM 200

> If $y = \ln(b + x^4)$, what is $\frac{dy}{dx}$?

Solution: To find $\frac{dy}{dx}$, we use the differentiation formula,

$$\frac{d}{dx} \ln u = \frac{1}{u} \frac{du}{dx},$$

with

$$u = (b + x^4), \qquad \frac{du}{dx} = 4x^3.$$

Applying the formula,we obtain:

$$\frac{dy}{dx} = \frac{1}{(b + x^4)} \cdot 4x^3$$

$$= \frac{4x^3}{b + x^4}.$$

• PROBLEM 201

> Differentiate the expression:
>
> $$y = \ln(x^2 + a).$$

Solution: To find $\frac{dy}{dx}$, we use the formula:

$$\frac{d}{dx} \ln u = \frac{1}{u} \frac{du}{dx},$$

letting $u = (x^2 + a)$. $\frac{du}{dx} = 2x.$

Applying the formula, we obtain:

$$\frac{dy}{dx} = \frac{1}{x^2 + a} \cdot 2x = \frac{2x}{x^2 + a}.$$

• PROBLEM 202

> Find the derivative of $y = \ln(1 - 2x)^3$.

Solution: It is best to rewrite the equation as:

$$y = 3 \ln(1 - 2x).$$

Then we apply the formula:

$$\frac{d}{dx} \ln u = \frac{1}{u} \frac{du}{dx},$$

letting $u = (1 - 2x)$. Then $\frac{du}{dx} = -2$. We obtain:

$$\frac{dy}{dx} = 3\left(\frac{1}{1-2x}\right)(-2) = \frac{-6}{1-2x}.$$

• PROBLEM 203

Differentiate the expression:

$$y = \ln\sqrt{1 - x^2}.$$

Solution: We may write this in a form free from radicals, as follows:

$$y = \frac{1}{2}\ln(1 - x^2).$$

To differentiate, we use the formula:

$$\frac{d}{dx}\ln u = \frac{1}{u}\frac{du}{dx},$$

letting $u = (1 - x^2)$. Then $\frac{du}{dx} = -2x$.

Applying the formula, we have:

$$\frac{dy}{dx} = \frac{1}{2}\left[\frac{1}{(1-x^2)}(-2x)\right]$$

$$= \frac{1}{2}\cdot\frac{-2x}{1-x^2} = \frac{x}{x^2-1}.$$

• PROBLEM 204

Find the derivative of: $y = \ln\tan x$.

Solution: To find $\frac{dy}{dx}$, we use the differentiation formula:

$$\frac{d}{dx}\ln u = \frac{1}{u}\frac{du}{dx},$$

letting $u = \tan x$, $\frac{du}{dx} = \sec^2 u$.

Applying the formula, we obtain:

$$\frac{dy}{dx} = \frac{1}{\tan x}\sec^2 x = \frac{\cos x}{\sin x}\frac{1}{\cos^2 x}$$

$$= \frac{1}{\sin x}\frac{1}{\cos x} = \csc x \sec x.$$

• PROBLEM 205

Given the expression: $y = \ln \sinh x$, find $D_x y$.

Solution: To find

$$D_x y = \frac{dy}{dx},$$

we use the differentiation formula:

$$\frac{d}{dx} \ln u = \frac{1}{u} \frac{du}{dx},$$

with $u = \sinh x$. Applying this formula, we have:

$$D_x y = \frac{1}{\sinh x} \cdot \cosh x = \frac{\cosh x}{\sinh x}$$

$$= \coth x.$$

• PROBLEM 206

If $y = \dfrac{1}{\ln x}$, what is $\dfrac{dy}{dx}$?

Solution: To find $\frac{dy}{dx}$, we use the quotient rule, obtaining:

$$\frac{dy}{dx} = \frac{\ln x\ (0) - 1\left(\frac{1}{x}\right)}{(\ln x)^2}$$

$$= \frac{-\frac{1}{x}}{(\ln x)^2}$$

$$= -\frac{1}{x\ (\ln x)^2}.$$

• PROBLEM 207

Find the derivative of the expression:

$$y = \ln[(x - 1)/(x + 1)].$$

Solution: To find $\frac{dy}{dx}$, we use the differentiation formula:

$$\frac{d}{dx} \ln u = \frac{1}{u} \frac{du}{dx},$$

letting $u = \dfrac{x - 1}{x + 1}$.

Then

$$\frac{du}{dx} = \frac{(x+1)(1) - (x-1)(1)}{(x+1)^2} = \frac{2}{(x+1)^2}.$$

Applying the formula we obtain:

$$\frac{dy}{dx} = \frac{x+1}{x-1} \cdot \frac{2}{(x+1)^2} = \frac{2}{x^2 - 1}.$$

• PROBLEM 208

> Find the derivative of the function:
> $y = \ln^2(2x - 4)$.

Solution: To find the derivative of the given function we use the chain rule. We find:

$$\frac{dy}{dx} = 2\ln(2x-4) \cdot \frac{1}{2x-4} \cdot 2$$

$$= \frac{2\ln(2x-4) \cdot 2}{2(x-2)}$$

$$= \frac{2\ln(2x-4)}{x-2}.$$

• PROBLEM 209

> If $f(x) = \ln(x^2 + \ln x)$, compute $f'(x)$.

Solution: We use the differentiation formula,

$$\frac{d}{dx}\ln u = \frac{1}{u}\frac{du}{dx},$$

with $u = x^2 + \ln x$. We obtain:

$$f'(x) = \frac{1}{x^2 + \ln x}\left(2x + \frac{1}{x}\right)$$

$$= \frac{2x + (1/x)}{x^2 + \ln x} = \frac{2x^2 + 1}{x^3 + x\ln x}.$$

• PROBLEM 210

> Find the derivative of the expression:
> $$y = \ln\cos e^{2x}.$$

Solution: In this example we make use of the chain rule for differentiation. We find:

$$\frac{dy}{dx} = \frac{1}{\cos e^{2x}}\left(-\sin e^{2x}\right)\left(e^{2x}\right)\left(2\right)$$

$$= -2e^{2x} \tan e^{2x}.$$

• PROBLEM 211

Differentiate:

$$y = \ln \sqrt{\frac{1 + x^2}{1 - x^2}} .$$

Solution: Simplifying,

$$y = \frac{1}{2} \left[\ln (1 + x^2) - \ln (1 - x^2)\right].$$

Then,

$$\frac{dy}{dx} = \frac{1}{2} \left(\frac{\frac{d}{dx}(1 + x^2)}{1 + x^2} - \frac{\frac{d}{dx}(1 - x^2)}{1 - x^2} \right)$$

$$= \frac{1}{2} \left(\frac{2x}{1 + x^2} - \frac{-2x}{1 - x^2} \right)$$

$$= \frac{1}{2} \left(\frac{2x(1 - x^2) + 2x(1 + x^2)}{(1 + x^2)(1 - x^2)} \right)$$

$$= \frac{1}{2} \left(\frac{4x}{1 - x^4} \right) = \frac{2x}{1 - x^4} .$$

• PROBLEM 212

Find the derivative of: $y = x^x$.

Solution: We begin by taking the logarithm of both sides of the equation. This gives:

$$\ln y = \ln x^x , \quad \text{or,}$$

$$\ln y = x \ln x.$$

Now we differentiate implicity with respect to x. We obtain:

$$\frac{1}{y} \frac{dy}{dx} = x \frac{1}{x} + \ln x(1) = 1 + \ln x$$

To solve for dy/dx, we multiply both sides of this equation by y.

$$\frac{dy}{dx} = (1 + \ln x)y$$

But, from the given equation $y = x^x$. Therefore,

$$\frac{dy}{dx} = (1 + \ln x)x^x .$$

● PROBLEM 213

Differentiate the expression: $y = x^{e^x}$.

Solution: We can simplify this by taking the logarithm of both sides of the equation.
We obtain:

$$\ln y = e^x \ln x .$$

Now we differentiate implicitly with respect to x, using the product rule on the right side of the equation. We have:

$$\frac{1}{y} \frac{dy}{dx} = e^x \cdot \frac{1}{x} + \ln x \cdot e^x .$$

Solving for $\frac{dy}{dx}$, we find

$$\frac{dy}{dx} = y \left(e^x \cdot \frac{1}{x} + e^x \ln x \right) .$$

Substituting the value for y, we have:

$$\frac{dy}{dx} = x^{e^x} \left(e^x \cdot \frac{1}{x} + e^x \ln x \right)$$

$$= e^x x^{e^x} \left(\frac{1}{x} + \ln x \right).$$

● PROBLEM 214

$$y = (x + 4)^2 \sqrt{x - 3} .$$

Find $\frac{dy}{dx}$, using logarithms.

Solution: Taking the logarithm of both sides of the equation, we obtain:

$$\ln y = 2 \ln (x + 4) + \frac{1}{2} \ln (x - 3).$$

Now, we differentiate implicitly with respect to x, and solve for

$$\frac{dy}{dx} .$$

Differentiating, we obtain:

$$\frac{1}{y} \frac{dy}{dx} = \left(2 \cdot \frac{1}{(x + 4)} \right) + \left(\frac{1}{2} \cdot \frac{1}{(x - 3)} \right)$$

$$= \frac{2}{(x + 4)} + \frac{1}{2(x - 3)} .$$

Solving for $\frac{dy}{dx}$, we obtain:

$$\frac{dy}{dx} = y \left[\frac{2}{(x + 4)} + \frac{1}{2(x - 3)} \right] .$$

Substituting the original value of y, we find:

$$\frac{dy}{dx} = (x + 4)^2 \sqrt{x - 3} \left[\frac{2}{x + 4} + \frac{1}{2(x - 3)} \right]$$

$$= (x + 4)^2 \sqrt{x - 3} \left[\frac{4x - 12 + x + 4}{2(x + 4)(x - 3)} \right]$$

$$= (x + 4)^2 \sqrt{x - 3} \left[\frac{5x - 8}{2(x + 4)(x - 3)} \right]$$

$$= \frac{(x + 4)(5x - 8)}{2(x - 3)^{\frac{1}{2}}}$$

• PROBLEM 215

$y = (x^2 + 4)^4 (x^3 - 3)^{\frac{3}{4}}$.

Find $\frac{dy}{dx}$, using logarithms.

Solution: We begin by taking the logarithm of both sides of the equation. We have:

$$\ln y = 4 \ln (x^2 + 4) + \frac{3}{4} \ln (x^3 - 3).$$

Now we differentiate implicitly with respect to x, then solve for $\frac{dy}{dx}$. Differentiating, we obtain:

$$\frac{1}{y} \frac{dy}{dx} = \left[4 \left(\frac{1}{x^2 + 4} \right) \cdot 2x \right] + \left[\frac{3}{4} \left(\frac{1}{x^3 - 3} \right) \cdot 3x^2 \right].$$

$$\frac{1}{y} \frac{dy}{dx} = \frac{8x}{x^2 + 4} + \frac{9x^2}{4(x^3 - 3)} .$$

Solving for $\frac{dy}{dx}$, we obtain:

$$\frac{dy}{dx} = y\left[\frac{8x}{x^2 + 4} + \frac{9x^2}{4(x^3 - 3)}\right].$$

Substituting the original value for y, we obtain:

$$\frac{dy}{dx} = (x^2 + 4)^4 (x^3 - 3)^{\frac{3}{4}} \left[\frac{8x}{x^2 + 4} + \frac{9x^2}{4(x^3 - 3)}\right].$$

• PROBLEM 216

If $y = \left(\frac{1}{b^{2x}}\right)^{2bx}$, **find** $\frac{dy}{dx}$.

Solution: We take the logarithm of both sides of this equation. This gives:

$$\ln y = 2bx \cdot \ln \frac{1}{b^{2x}}.$$

Now, we examine the term,

$$\ln \frac{1}{b^{2x}},$$

and we see that this can be rewritten as:

$$\ln \frac{1}{b^{2x}} = \ln 1 - \ln b^{2x},$$

but $\ln 1 = 0$, therefore, $\ln \frac{1}{b^{2x}} = -2x \ln b$.

Substituting into the equation, we have:

$$\ln y = 2bx(-2x \ln b)$$

$$\ln y = -4bx^2 \ln b.$$

We now differentiate implicitly with respect to x, and we obtain:

$$\frac{1}{y}\frac{dy}{dx} = -8bx \ln b$$

$$\frac{dy}{dx} = y(-8bx \ln b).$$

We replace y by its original value,

$$\left(\frac{1}{b^{2x}}\right)^{2bx}$$

and obtain:

$$\frac{dy}{dx} = \left(\frac{1}{b^{2x}}\right)^{2bx} \cdot (-\ 8bx \ln b)$$

$$= -\ 8x \cdot b \cdot \frac{1}{b^{4bx^2}} \ln b$$

$$= -\ 8x(b)^{(1-4bx^2)} \ln b.$$

• PROBLEM 217

Differentiate the expression:

$$y = (4x^2 - 7)^{2+\sqrt{x^2 - 5}} .$$

Solution: Taking the natural logarithm of both sides,

$$\ln y = (2 + \sqrt{x^2 - 5}) \ln (4x^2 - 7).$$

Differentiating both sides with respect to x, using the product rule we have:

$$\frac{1}{y}\frac{dy}{dx} = \left[(2 + \sqrt{x^2 - 5}) \cdot \frac{1}{4x^2 - 7} \cdot 8x\right]$$

$$+ \left[\ln (4x^2 - 7) \cdot \frac{1}{2} (x^2 - 5)^{-\frac{1}{2}} \cdot 2x\right]$$

$$= (2 + \sqrt{x^2 - 5}) \frac{8x}{4x^2 - 7}$$

$$+ \ln (4x^2 - 7) \cdot \frac{x}{\sqrt{x^2 - 5}} .$$

Solving for $\frac{dy}{dx}$ we have:

$$\frac{dy}{dx} = y \left[\frac{8x\left(2 + \sqrt{x^2 - 5}\right)}{4x^2 - 7} + \frac{x\left(\ln (4x^2 - 7)\right)}{\sqrt{x^2 - 5}}\right]$$

$$= x\left(4x^2 - 7\right)^{2+\sqrt{x^2-5}} \left[\frac{8\left(2 + \sqrt{x^2 - 5}\right)}{4x^2 - 7} + \frac{\ln\,(4x^2 - 7)}{\sqrt{x^2 - 5}}\right].$$

• PROBLEM 218

Given the function:

$$\ln\,(x + y) = \tan^{-1}\left(\frac{x}{y}\right),$$

find $D_x\, y$.

Solution: To find

$$D_x\, y = \frac{dy}{dx},$$

we differentiate implicitly on both sides of the given equation with respect to x, and then solve for $D_x\, y$. Doing this, we obtain:

$$\frac{1}{x + y}\left(1 + D_x\, y\right) = \frac{1}{1 + \frac{x^2}{y^2}} \cdot \frac{y - x\,D_x\, y}{y^2},$$

or

$$\frac{1 + D_x\, y}{x + y} = \frac{y - x\,D_x\, y}{y^2 + x^2}.$$

Cross-multiplying, we have:

$$y^2 + x^2 + (y^2 + x^2)D_x\, y = xy + y^2 - (x^2 + xy)D_x\, y.$$

Finally, combining coefficients of $D_x\, y$ and solving, we find:

$$D_x\, y = \frac{xy - x^2}{2x^2 + xy + y^2}.$$

• PROBLEM 219

In the disintegration of a radioactive element it is found by experiment that the amount present at any time after the beginning of the process can be expressed as a negative exponential function of the time. Show that the rate of disintegration at any instant is proportional to the amount of radium present at that time.

Solution: Let A be the amount remaining after t seconds. Then, to express this as a negative exponential of the time we have:

$$A = ke^{-at},$$

and a and k are constants. The rate of disintegration (rate of change of A) is

$$\frac{dA}{dt}.$$

We use the formula:

$$\frac{d}{dt} e^{u} = e^{u} \frac{du}{dt},$$

letting $u = -at$. Then $\frac{du}{dt} = -a$. Applying the formula, we have:

$$\frac{dA}{dt} = -kae^{-at},$$

or

$$-\frac{dA}{dt} = kae^{-at},$$

the negative rate showing that A, the amount present, decreases.

But e^{-at} is A. Therefore the value of the negative rate is:

$$\frac{dA}{dt} = kaA,$$

which shows that it is proportional to A, the amount present, the constant of proportionality being ka.

CHAPTER 10

DIFFERENTIATION OF HYPERBOLIC FUNCTIONS

In differentiating hyperbolic functions, simplification is often obtained by expressing these functions in terms of e^u. Thus,

$$\sinh x = \frac{e^x - e^{-x}}{2},$$

$$\frac{d}{dx}(\sinh x) = \frac{1}{2}\left(e^x + e^{-x}\right)$$

$$= \cosh x.$$

Consequently, $\cosh x = \dfrac{e^x + e^{-x}}{2}$. Now,

$$\frac{d}{dx}(\cosh x) = \frac{1}{2}\left(e^x - e^{-x}\right)$$

$$= \sinh x.$$

For $\tanh x$, we make use of $\dfrac{\sinh x}{\cosh x} = \tanh x$ from which,

$$\frac{d}{dx}(\tanh x) = \frac{\cosh x(\cosh x) - \sinh x(\sinh x)}{\cosh^2 x}$$

$$= \frac{1}{\cosh^2 x} = \operatorname{sech}^2 x.$$

The remaining hyperbolic functions can be differentiated in a similar manner.

• PROBLEM 220

From the definitions: $\sinh x = \dfrac{e^x - e^{-x}}{2}$,

$\cosh x = \dfrac{e^x + e^{-x}}{2}$, $\tanh x = \dfrac{e^x - e^{-x}}{e^x + e^{-x}}$, and

$\operatorname{sech} x = \dfrac{2}{e^x + e^{-x}}$, show the following:

(a) $D_x(\sinh x) = \cosh x$

(b) $D_x(\cosh x) = \sinh x$

(c) $D_x(\tanh x) = \operatorname{sech}^2 x$.

Solution: (a) Since $\sinh x = \dfrac{e^x - e^{-x}}{2}$, $D_x(\sinh x) =$ $D_x\left(\dfrac{e^x - e^{-x}}{2}\right) = \dfrac{2(e^x + e^{-x}) - 0}{4} = \dfrac{e^x + e^{-x}}{2}$. But

$\frac{e^x + e^{-x}}{2} = \cosh x$. Therefore $D_x(\sinh x) = \cosh x$.

(b) Since $\cosh x = \frac{e^x + e^{-x}}{2}$, $D_x(\cosh x) =$

$D_x\left(\frac{e^x + e^{-x}}{2}\right) = \frac{2(e^x - e^{-x}) - 0}{4} = \frac{e^x - e^{-x}}{2}$. But

$\frac{e^x - e^{-x}}{2} = \sinh x$. Therefore $D_x(\cosh x) = \sinh x$.

(c) Since $\tanh x = \frac{e^x - e^{-x}}{e^x + e^{-x}}$, $D_x(\tanh x) =$

$$D_x\left(\frac{e^x - e^{-x}}{e^x + e^{-x}}\right) = \frac{(e^x + e^{-x})(e^x + e^{-x}) - (e^x - e^{-x})(e^x - e^{-x})}{(e^x + e^{-x})^2}$$

$$= \frac{e^{2x} + 2 + e^{-2x} - (e^{2x} - 2 + e^{-2x})}{(e^x + e^{-x})^2}$$

$$= \frac{4}{(e^x + e^{-x})^2} = \left(\frac{2}{e^x + e^{-x}}\right)^2.$$

But $\frac{2}{e^x + e^{-x}} = \frac{1}{\cosh x} = \text{sech } x$. Therefore,

$D_x(\tanh x) = \text{sech}^2\, x$.

• PROBLEM 221

From the fact that $\sinh^{-1} y = \ln\left(y + \sqrt{1 + y^2}\right)$, $\cosh^{-1} y = \pm \ln\left(y + \sqrt{y^2 - 1}\right)$, and $\tanh^{-1} y = \frac{1}{2} \ln \frac{1 + y}{1 - y}$, show that

(a) $D_y(\sinh^{-1} y) = \frac{1}{\sqrt{1 + y^2}}$

(b) $D_y(\cosh^{-1} y) = \frac{\pm 1}{\sqrt{y^2 - 1}}$

(c) $D_y(\tanh^{-1} y) = \frac{1}{1 - y^2}$.

<u>Solution:</u> (a) Since $\sinh^{-1} y = \ln\left(y + \sqrt{1 + y^2}\right)$,

$D_y(\sinh^{-1} y) = D_y\left[\ln\left(y + \sqrt{1 + y^2}\right)\right] =$

$$\frac{1}{y + \sqrt{1 + y^2}}\left[1 + \frac{1}{2}(1 + y^2)^{-\frac{1}{2}}\, 2y\right] =$$

$$\frac{1}{y+\sqrt{1+y^2}}\left[1+\frac{2y}{2\sqrt{1+y^2}}\right]=\frac{1}{y+\sqrt{1+y^2}}\left[1+\frac{y}{\sqrt{1+y^2}}\right]$$

$$=\frac{1}{y+\sqrt{1+y^2}}\left[\frac{\sqrt{1+y^2}+y}{\sqrt{1+y^2}}\right]$$

$$=\frac{(\sqrt{1+y^2}+y)}{\sqrt{1+y^2}(y+\sqrt{1+y^2})}=\frac{1}{\sqrt{1+y^2}}.$$

Therefore, we have shown that

$$D_y(\sinh^{-1} y)=\frac{1}{\sqrt{1+y^2}}.$$

(b) Since $\cosh^{-1} y = \pm\ln(y+\sqrt{y^2-1})$,

$$D_y(\cosh^{-1} y) = D_y\left[\pm\ln(y+\sqrt{y^2-1})\right]$$

$$=\frac{1}{y+\sqrt{y^2-1}}\left[1+\frac{1}{2}(y^2-1)^{-\frac{1}{2}}\cdot 2y\right]$$

$$=\frac{1}{y+\sqrt{y^2-1}}\left[1+\frac{y}{\sqrt{y^2-1}}\right]=\frac{1}{y+\sqrt{y^2-1}}\left[\frac{\sqrt{y^2-1}+y}{\sqrt{y^2-1}}\right]$$

$$=\frac{(\sqrt{y^2-1}+y)}{(y+\sqrt{y^2-1})(\sqrt{y^2-1})}=\frac{1}{\sqrt{y^2-1}}.$$

Therefore, we have shown that $D_y(\cosh^{-1} y) = \dfrac{1}{\sqrt{y^2-1}}$.

(c) Since $\tanh^{-1} y = \frac{1}{2}\ln\frac{1+y}{1-y}$,

$$D_y(\tanh^{-1} y) = D_y\left[\frac{1}{2}\ln\frac{1+y}{1-y}\right]$$

$$=\frac{1}{2}\cdot\frac{1-y}{1+y}\left[\frac{(1-y)(1)-(1+y)(-1)}{(1-y)^2}\right]$$

$$=\frac{(1-y)(1-y+1+y)}{2(1+y)(1-y)^2}=\frac{(1-y)(2)}{2(1+y)(1-y)^2}$$

$$=\frac{1}{(1+y)(1-y)}=\frac{1}{1-y^2}.$$

Therefore, we have shown that

$$D_y(\tanh^{-1} y)=\frac{1}{1-y^2}.$$

If $y = \sinh x = \frac{1}{2}(e^x - e^{-x})$, the inverse function is written $x = \sinh^{-1} y$. Similar notations are employed for the inverses of the remaining hyperbolic functions. Show that:

(a) $\sinh^{-1} y = \ln\left(y + \sqrt{1 + y^2}\right)$

(b) $\cosh^{-1} y = \pm\ln\left(y + \sqrt{y^2 - 1}\right)$

(c) $\tanh^{-1} y = \frac{1}{2} \ln \frac{1 + y}{1 - y}$.

Solution: (a) To show that $\sinh^{-1} y = \ln\left(y + \sqrt{1+y^2}\right)$, we solve for x in the equation:

$$y = \frac{e^x - e^{-x}}{2}.$$

We have:

$$2y = e^x - e^{-x}$$

$$2y = e^x - \frac{1}{e^x}$$

$$2ye^x = e^{2x} - 1$$

$$e^{2x} - 2ye^x - 1 = 0$$

To solve for e^x, we use the quadratic formula, with $a = 1$, $b = -2y$ and $c = -1$, obtaining:

$$e^x = \frac{2y \pm \sqrt{4y^2 + 4}}{2} = y + \sqrt{1 + y^2}.$$

Therefore, $x = \ln\left(y + \sqrt{1 + y^2}\right) = \sinh^{-1} y$.

(b) If $y = \cosh x = \frac{e^x + e^{-x}}{2}$, solving for x gives $\cosh^{-1} y$. We have:

$$y = \frac{e^x + e^{-x}}{2}$$

$$2y = e^x + e^{-x}$$

$$2y = e^x + \frac{1}{e^x}$$

$$2ye^x = e^{2x} + 1$$

$$e^{2x} - 2ye^x + 1 = 0.$$

We use the quadratic formula to solve for e^x, letting $a = 1$, $b = -2y$ and $c = 1$, obtaining:

$$e^x = \frac{2y \pm \sqrt{4y^2 - 4}}{2} = y \pm \sqrt{y^2 - 1}.$$

Therefore,

$$x = \ln\left(y \pm \sqrt{y^2 - 1}\right) = \cosh^{-1} y.$$

(c) If $y = \tanh x = \dfrac{e^x - e^{-x}}{e^x + e^{-x}}$,

solving for x gives $\tanh^{-1} y$. We have

$$y = \frac{e^x - e^{-x}}{e^x + e^{-x}}$$

$$y\left(e^x + \frac{1}{e^x}\right) = e^x - \frac{1}{e^x}$$

$$ye^x + \frac{y}{e^x} = e^x - \frac{1}{e^x}$$

$$ye^{2x} + y = e^{2x} - 1$$

$$ye^{2x} - e^{2x} = -y - 1$$

$$e^{2x}(y - 1) = -y - 1$$

$$e^{2x} = \frac{-y - 1}{y - 1}$$

$$2x = \ln\left(\frac{-y - 1}{y - 1}\right).$$

Then

$$x = \frac{1}{2} \ln \frac{1 + y}{1 - y} = \tanh^{-1} y.$$

CHAPTER 11

IMPLICIT DIFFERENTIATION

An implicit function of, for example, y and x is a function in which one of the variables (y) is not directly expressed in terms of the other variable (x). This may be because it is either impossible, difficult, or not immediately obvious how the expression may be solved for y in terms of x. Instead, the variables may be related in such a way that separation is not a practical approach. When this is the case, differentiation can still be performed in a straightforward manner. We simply differentiate the expression as it stands, applying all the normal rules, such as the product rule, quotient rule, power rule, etc., except that we must also remember one additional rule: the chain rule. This states that $\frac{du}{dx} \times \frac{dx}{dt} = \frac{du}{dt}$. Applied to this particular case, the rule calls for following each y-term obtained in the differentiation by the factor: $\frac{dy}{dx}$, while each x-term is followed by the factor: $\frac{dx}{dx} = 1$, which can be omitted. This leaves a result consisting of some terms containing the factor: $\frac{dy}{dx}$, and some terms which do not contain this factor. If we move all terms not containing the factor to the right, all the terms containing the factor to the left, we can solve for the derivative $\frac{dy}{dx}$ by dividing both sides of the equation by the factor of $\frac{dy}{dx}$.

• PROBLEM 223

If $x^2 + y^2 = 16$,

find $\frac{dy}{dx}$ as an implicit function of x and y.

Solution: Since y is a function of x, we differentiate the equation implicitly in terms of x and y. We have:

$$2x + 2y \cdot \frac{dy}{dx} = 0 \quad \text{or} \quad 2y\frac{dy}{dx} = -2x.$$

$$\frac{dy}{dx} = -\frac{x}{y}.$$

• PROBLEM 224

Find $\frac{dy}{dx}$ for the expression: $2x^4 - 3x^2y^2 + y^4 = 0.$

Solution: The equation: $2x^4 - 3x^2y^2 + y^4 = 0$, could be solved for y and then differentiated to obtain $\frac{dy}{dx}$, but an easier method is to differentiate implicitly and then solve for $\frac{dy}{dx}$.

Hence, from $2x^4 - 3x^2y^2 + y^4 = 0$ we obtain:

$$8x^3 - 6x^2y\frac{dy}{dx} - 6xy^2 + 4y^3\frac{dy}{dx} = 0 .$$

Solving for $\frac{dy}{dx}$,

$$4y^3\frac{dy}{dx} - 6x^2y\frac{dy}{dx} = 6xy^2 - 8x^3 .$$

$$\left(4y^3 - 6x^2y\right)\frac{dy}{dx} = 6xy^2 - 8x^3 .$$

$$\frac{dy}{dx} = \frac{6xy^2 - 8x^3}{4y^3 - 6x^2y}$$

$$= \frac{3xy^2 - 4x^3}{2y^3 - 3x^2y} .$$

• PROBLEM 225

Find dy/dx for a 5 cm. circle both explicitly and implicitly.

Solution: The equation of a circle with radius = 5 and center at the origin is:

$$x^2 + y^2 = 25$$

To use the explicit method, we solve for y in terms of x

$$y = \sqrt{25 - x^2}$$

$$y = f(x) = \sqrt{25 - x^2}$$

$$= (25 - x^2)^{1/2}$$

We use the theorem for $d(u^n)$

$$\frac{dy}{dx} = \frac{1}{2}(25 - x^2)^{-1/2}(-2x) = \frac{-x}{\sqrt{25 - x^2}} .$$

But $\sqrt{25 - x^2} = y.$

Therefore, by substitution,

$$\frac{dy}{dx} = -\frac{x}{y} .$$

Using the implicit method,

$$f(x,y) = x^2 + y^2 = 25.$$

Differentiating each term with respect to x,

$$\frac{df}{dx}(x,y) = 2x + 2y \frac{dy}{dx} = 0$$

$$2y \frac{dy}{dx} = -2x.$$

$$\frac{dy}{dx} = \frac{-2x}{2y} + -\frac{x}{y},$$

which, in this case, is seen to be simpler.

• PROBLEM 226

Find $\frac{dy}{dx}$ for the equation for the hyperbola:

$$b^2x^2 - a^2y^2 = a^2b^2,$$

implicitly.

Solution: We differentiate implicitly with respect to x, remembering that a^2 and b^2 are constants of the hyperbola, and therefore are treated as such when differentiating.

$$b^2 \cdot 2x - a^2 \cdot 2y \cdot \frac{dy}{dx},$$

or

$$2b^2x = 2a^2y \frac{dy}{dx}.$$

Solving for $\frac{dy}{dx}$,

$$\frac{dy}{dx} = \frac{2b^2x}{2a^2y} = \frac{b^2x}{a^2y}.$$

• PROBLEM 227

Find $D_xy = \frac{dy}{dx}$, when

$$(x + y)^2 - (x - y)^2 = x^4 + y^4.$$

Solution: Differentiating implicitly with respect to x, using the theorem for $d(u^n)$, we obtain:

$$2(x + y)(1 + D_xy) - 2(x - y)(1 - D_xy)$$

$$= 4x^3 + 4y^3 \cdot D_xy.$$

We perform the required multiplication and factor out common terms.

$$2x + 2y + (2x + 2y)D_xy - 2x + 2y + (2x - 2y)D_xy$$

$$= 4x^3 + 4y^3D_xy.$$

Now we add the terms in the previous

equation and obtain:

$$4y + 4xD_xy = 4x^3 + 4y^3D_xy.$$

Then we factor D_xy from the appropriate terms to obtain:

$$D_xy(4x - 4y^3) = 4x^3 - 4y$$

$$D_xy = \frac{x^3 - y}{x - y^3}.$$

• PROBLEM 228

Find dy/dx of the rectangular hyperbola, $xy = 1$, by the explicit and implicit methods.

Solution: Using the explicit method, we solve for y in terms of x.

$$xy = 1$$

$$y = f(x) = \frac{1}{x} = x^{-1}.$$

Now we use the theorem for $d(u^n)$ and obtain:

$$\frac{dy}{dx} = -1x^{-2} = -x^{-2}.$$

Using the implicit method,

$$xy = 1.$$

By the product rule, we differentiate with respect to x,

$$x\frac{dy}{dx} + y\frac{dx}{dx} = 0$$

$$x\frac{dy}{dx} + y(1) = 0$$

$$x\frac{dy}{dx} = -y, \quad \text{and} \quad \frac{dy}{dx} = -\frac{y}{x}.$$

Substitute $y = 1/x$:

$$\frac{dy}{dx} = -\frac{1/x}{x} = -\frac{1}{x^2} = -x^{-2}.$$

• PROBLEM 229

Differentiate the expression: $3xy^2$.

Solution: This is the product of x by y^2 with the constant coefficient 3; therefore

$$d(3xy^2) = 3d(x \cdot y^2)$$

and

$$d(x \cdot y^2) = x \cdot d(y^2) + y^2 \cdot d(x)$$

$$= x \cdot 2y\ dy + y^2 \cdot dx = 2xy\ dy + y^2\ dx.$$

Therefore,

$$d(3xy^2) = 3(2xy\ dy + y^2\ dx) = 3y(2x\ dy + y\ dx).$$

• PROBLEM 230

Find $D_x y = \frac{dy}{dx}$ from

$$x^2 + xy + y^2 - 3 = 0.$$

Solution: We use implicit differentiation with respect to x, applying the product rule to the second term.

$$2x + (y + x\ D_x y) + 2y\ D_x y = 0.$$

Collecting terms:

$$(x + 2y)\ D_x y = -\ (2x + y),$$

and, solving,

$$D_x y = -\frac{2x + y}{x + 2y}, \quad x \neq -\ 2y.$$

• PROBLEM 231

Find $y' = \frac{dy}{dx}$ for

$$xy^2 + y^2 - 2x = 5.$$

Solution: We apply the theorem for $d(u^n)$, and the product rule to the first term.

$$x(2yy') + y^2(1) + 2yy' - 2 = 0$$

$$y'(2xy + 2y) = 2 - y^2.$$

$$y' = \frac{2 - y^2}{2xy + 2y}.$$

• PROBLEM 232

Find $D_x y = \frac{dy}{dx}$ for

$$y^4 - xy^3 + x^2 - 7 = 0.$$

Solution: Let $y = f(x)$

Then, substituting into the given equation, we have:

$$\left[f(x)\right]^4 - x\left[f(x)\right]^3 + x^2 - 7 = 0.$$

Differentiating, and applying the product rule:

$$4[f(x)]^3 D_x f(x) - x3[f(x)]^2 D_x f(x)$$

$$-[f(x)]^3 + 2x - 0 = 0$$

Factoring out $D_x f(x)$:

$$D_x f(x)\{4[f(x)]^3 - 3x[f(x)]^2\} = [f(x)]^3 - 2x$$

$$D_x f(x) = \frac{[f(x)]^3 - 2x}{4[f(x)]^3 - 3x[f(x)]^2} .$$

Substituting back $f(x) = y$,

$$D_x y = \frac{y^3 - 2x}{4y^3 - 3xy^2} .$$

• PROBLEM 233

Given: $y^3 + x^2y^4 + x^3 = 1$, find dy/dx.

Solution: To find $\frac{dy}{dx}$ **we use implicit** differentiation. We differentiate each term with respect to x, that is:

$$\frac{d}{dx}(y^3) + \frac{d}{dx}(x^2y^4) + \frac{d}{dx}(x^3) = \frac{d}{dx}(1)$$

Note that the second term must be handled as a product of two functions of x, since x^2 is a function of x, and y^4 is an implied function of x. Using the product rule, the derivative of the second term is

$$\frac{d}{dx}(x^2y^4) = x^2\left|4y^3\frac{dy}{dx}\right| + y^4[2x]$$

$$= 4x^2y^3\frac{dy}{dx} + 2xy^4$$

Completing the differentiation of the other terms with respect to x, we have

$$3y^2\frac{dy}{dx} + \left|4x^2y^3\frac{dy}{dx} + 2xy^4\right| + 3x^2 = 0$$

Solving for dy/dx,

$$\frac{dy}{dx} = -\frac{3x^2 + 2xy^4}{3y^2 + 4x^2y^3} .$$

• **PROBLEM 234**

Find y' in terms of x and y, using implicit differentiation, where

$$y' = \frac{dy}{dx},$$

in the expression:

$$y^3 + 3xy + x^3 - 5 = 0.$$

Solution: The derivative of y^3 is $3y^2y'$. The term 3xy must be treated as a product. The derivative of 3xy is $3xy' + 3y$. The derivative of x^3 is $3x^2$. The derivative of - 5 is 0. Therefore,

$$3y^2y' + 3xy' + 3y + 3x^2 = 0.$$

We can now solve for y':

$$y' = -\frac{y + x^2}{y^2 + x}.$$

• **PROBLEM 235**

Solve for dy/dx in

$$y^5 + 3x^2y^3 - 7x^6 - 8 = 0.$$

Solution: Let $y' = dy/dx$. In this problem we have no choice as to whether to use the explicit or the implicit method, since it is impossible to solve for one variable in terms of the other. We obtain

$$5y^4y' + 3x^2 \cdot 3y^2y' + 3y^3 \cdot 2x - 42x^5 = 0,$$

from which we can solve for y'.

$$y' = \frac{42x^5 - 6xy^3}{5y^4 + 9x^2y^2}.$$

• **PROBLEM 236**

Given the expression:

$$x \cos y + y \cos x = 1,$$

find $$D_xy = \frac{dy}{dx}.$$

Solution: In differentiating implicitly with respect to x, the product rule must be used on both terms of the equation.

$$1 \cdot \cos y + x(-\sin y)\, D_xy + D_xy(\cos x) + y(-\sin x) = 0.$$

We factor out $D_x y$.

$$D_x y(\cos x - x \sin y) = y \sin x - \cos y$$

$$D_x y = \frac{y \sin x - \cos y}{\cos x - x \sin y}.$$

• **PROBLEM 237**

Differentiate: $\frac{u^2}{y}$.

Solution: This is a quotient. Therefore, using the quotient rule,

$$d\left(\frac{u^2}{y}\right) = \frac{y \cdot d(u^2) - u^2 \cdot d(y)}{y^2}$$

$$= \frac{y \cdot 2u\,du - u^2\,dy}{y^2}$$

$$= \frac{2uy\,du - u^2\,dy}{y^2}$$

$$d\left(\frac{u^2}{y}\right) = \frac{u}{y^2}(2y\,du - u\,dy).$$

• **PROBLEM 238**

Differentiate: $(x + y)/(x - y)$.

Solution: This is the quotient of $(x + y)$ divided by $(x - y)$. Therefore, using the quotient rule,

$$d\left[\frac{(x + y)}{(x - y)}\right]$$

$$= \frac{(x - y) \cdot d(x + y) - (x + y) \cdot d(x - y)}{(x - y)^2}$$

$$= \frac{(x - y)(dx + dy) - (x + y)(dx - dy)}{(x - y)^2}.$$

Now, by multiplying out the terms,

$$[(x - y)(dx + dy)] - [(x + y)(dx - dy)]$$

$$= [x\,dx + x\,dy - y\,dx - y\,dy] - [x\,dx - x\,dy + y\,dx - y\,dy]$$

$$= x\,dx + x\,dy - y\,dx - y\,dy - x\,dx + x\,dy - y\,dx + y\,dy$$

$$= 2x\,dy - 2y\,dx = 2(x\,dy - y\,dx).$$

This completes only the numerator. The complete expression is:

$$d\left(\frac{x+y}{x-y}\right) = \frac{2(x\,dy - y\,dx)}{(x-y)^2}.$$

• PROBLEM 239

Find $D_x{}^2y = \frac{d^2y}{dx^2}$ by implicit differentiation, when

$$4x^2 + 9y^2 = 36.$$

Solution: Differentiating implicitly with respect to x, we obtain:

$$8x + 18y \cdot D_xy = 0,$$

from which:

$$D_xy = -\frac{4x}{9y}$$

To find $D_x{}^2y$, we differentiate D_xy above, by applying the quotient rule

$$D_x{}^2y = \frac{9y(-4) - (-4x)(9\,D_xy)}{81y^2}$$

Now we simplify the previous equation and substitute:

$$D_xy = \left(-\frac{4x}{9y}\right).$$

$$D_x{}^2y = \frac{-36y + (36x)\frac{-4x}{9y}}{81y^2} = \frac{-36y^2 - 16x^2}{81y^3}.$$

• PROBLEM 240

Find $y'' = \frac{d^2y}{dx^2}$ for the expression: $xy^3 = 1$.

Solution: To find the second derivative, y'', we must first find the first derivative and then differentiate that to obtain the second derivative. We could solve for y and then differentiate to obtain y', but an alternative is implicit differentiation.

$$xy^3 = 1.$$

Differentiating implicityly,

$$3xy^2 \cdot y' + y^3 = 0 .$$

$$3xy^2 \cdot y' = - y^3 .$$

$$y' = \frac{-y^3}{3xy^2}$$

$$= - \frac{y}{3x} .$$

Now we take the derivative of y' to find y''.

$$y'' = - \frac{1}{3}\left[\frac{x \cdot y' - y}{x^2}\right] .$$

Substituting $y' = - \frac{y}{3x}$ in the expression for y'' and simplifying

$$y'' = - \frac{1}{3}\left[\frac{x(\frac{-y}{3x}) - y}{x^2}\right]$$

$$= - \frac{1}{3}\left[\frac{- \frac{y}{3} - y}{x^2}\right]$$

$$= - \frac{1}{3}\left[\frac{- \frac{4}{3} y}{x^2}\right]$$

$$= \frac{4y}{9x^2} .$$

• PROBLEM 241

Find D_x^2y of the function: $x^{\frac{2}{3}} + y^{\frac{2}{3}} = a^{\frac{2}{3}}$.

Solution:

We first find $D_xy = \frac{dy}{dx}$, using implicit differentiation. We obtain:

$$\frac{2}{3}x^{-\frac{1}{3}} + \frac{2}{3}y^{-\frac{1}{3}}D_xy = 0.$$

Solving for D_xy ,

$$D_xy = \frac{-x^{-\frac{1}{3}}}{y^{-\frac{1}{3}}} = -\frac{y^{\frac{1}{3}}}{x^{\frac{1}{3}}}.$$

To find D_x^2y we differentiate D_xy implicitly, obtaining:

$$D_x^2 y = \frac{-x^{\frac{1}{3}}\left(\frac{1}{3}y^{-\frac{2}{3}}D_x y\right) - \left[y^{\frac{1}{3}}\left(-\frac{1}{3}x^{-\frac{2}{3}}\right)\right]}{\left(-x^{\frac{1}{3}}\right)^2}$$

$$= \frac{\frac{-x^{\frac{1}{3}}}{3y^{\frac{2}{3}}}D_x y + \frac{y^{\frac{1}{3}}}{3x^{\frac{2}{3}}}}{x^{\frac{2}{3}}} .$$

We now substitute for $D_x y$, obtaining:

$$D_x^2 y = \frac{\frac{-x^{\frac{1}{3}}}{3y^{\frac{2}{3}}} \cdot \frac{-y^{\frac{1}{3}}}{x^{\frac{1}{3}}} + \frac{y^{\frac{1}{3}}}{3x^{\frac{2}{3}}}}{x^{\frac{2}{3}}} = \frac{1}{3x^{\frac{2}{3}}y^{\frac{1}{3}}} + \frac{y^{\frac{1}{3}}}{3x^{\frac{4}{3}}} .$$

Combining these fractions over a common denominator gives:

$$D_x^2 y = \frac{x^{\frac{2}{3}} + y^{\frac{2}{3}}}{3x^{\frac{4}{3}}y^{\frac{1}{3}}} .$$

Since

$$x^{\frac{2}{3}} + y^{\frac{2}{3}} = a^{\frac{2}{3}},$$

we write:

$$D_x^2 y = \frac{a^{\frac{2}{3}}}{3x^{\frac{4}{3}}y^{\frac{1}{3}}} .$$

• PROBLEM 242

Find the slope of the tangent line to the ellipse $4x^2 + 9y^2 = 40$ at the point (1,2).

Solution: The slope of the line tangent to the curve $4x^2 + 9y^2 = 40$ is the slope of the curve and can be found by taking the derivative, $\frac{dy}{dx}$, of the function and evaluating it at the point (1,2). We could solve the equation for y and then find y'. However it is easier to find y' by implicit differentiation.

If
$$4x^2 + 9y^2 = 40 ,$$
then
$$8x + 18y(y') = 0 .$$

$$18y(y') = -8x\ .$$

$$y' = \frac{-8x}{18y} = \frac{-4x}{9y}\ .$$

At the point (1,2), $x = 1$ and $y = 2$. Therefore, substituting these points into $y' = \frac{-4x}{9y}$, we obtain:

$$y' = \frac{-4(1)}{9(2)}$$

$$= -\frac{2}{9}\ .$$

The slope is $-\frac{2}{9}$.

• PROBLEM 243

Prove that the curves:

$$5y - 2x + y^3 - x^2y = 0,$$

and

$$2y + 5x + x^4 - x^3y^2 = 0,$$

intersect at right angles at the origin.

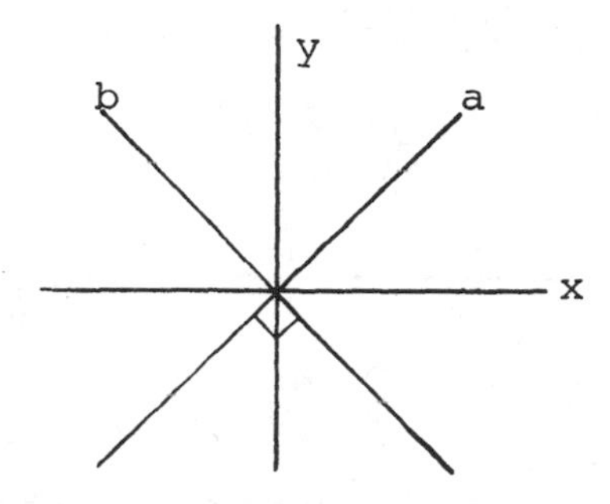

Solution: If any two curves intersect at right angles at a given point, then the product of the slopes equals -1. For instance, consider the two lines, a and b. Assume that the slope of $a = 1$ and the slope of $b = -1$. The two lines intersect at a right angle, because $m_a \cdot m_b = -1$.

Since the slope is equivalent to

$$\frac{dy}{dx},$$

to show that the given curves intersect perpendicularly at the specific point (0,0), the equations must be differentiated implicitly to find $\frac{dy}{dx}$.

$$5y - 2x + y^3 - x^2y = 0$$

Differentiating:

$$5 \frac{dy}{dx} - 2 + 3y^2 \frac{dy}{dx} - 2xy - x^2 \frac{dy}{dx} = 0.$$

Solving for $\frac{dy}{dx}$ for the first curve,

$$\frac{dy}{dx}(5 + 3y^2 - x^2) = 2 + 2xy .$$

$$\frac{dy}{dx} = \frac{2 + 2xy}{5 + 3y^2 - x^2} .$$

At (0,0):

$$\frac{dy}{dx} = \frac{2 + 2 \cdot 0 \cdot 0}{5 + 3 \cdot 0 \cdot 0} = \frac{2}{5} .$$

For the second curve,

$$2y + 5x + x^4 - x^3y^2 = 0$$

Differentiating:

$$2 \frac{dy}{dx} + 5 + 4x^3 - 3x^2y^2 - 2x^3y \frac{dy}{dx} = 0$$

Solving for $\frac{dy}{dx}$,

$$\frac{dy}{dx}(2 - 2x^3y) = 3x^2y^2 - 4x^3 - 5.$$

$$\frac{dy}{dx} = \frac{3x^2y^2 - 4x^3 - 5}{2 - 2x^3y} .$$

At (0,0):

$$\frac{dy}{dx} = \frac{3 \cdot 0 \cdot 0 - 4 \cdot 0 - 5}{2 - 2 \cdot 0 \cdot 0} = -\frac{5}{2} .$$

The product of their slopes at (0,0)

$$= \left(\frac{2}{5}\right)\left(-\frac{5}{2}\right) = -1.$$

Therefore they intersect at right angles at (0,0).

CHAPTER 12

PARAMETRIC EQUATIONS

Often the equation of a curve is not given as $y = f(x)$, but, instead, both x and y are given as functions of a third variable, such as t. When this is the case, the equations are called parametric equations and are interrelated by the third variable, t. Thus, $y = f(x)$ may be obtained from $y(t)$ and $x(t)$ by eliminating t between them.

When a function is given in terms of parametric equations, the problem is generally best solved by performing operations such as differentiation separately on each of the parametric equations and then combining the results. For example,

$$\frac{dy}{dx} = \frac{dy}{dt}\frac{dt}{dx} = \frac{dy}{dt}\frac{1}{\frac{dx}{dt}} .$$

Thus, even though the equations are given in parametric form, the derivatives can be obtained from each equation separately without undue difficulty.

For the second derivatives, the chain rule can also be applied, so that we have:

$$\frac{d^2y}{dx^2} = \frac{d}{dx}\left(\frac{dy}{dx}\right) = \frac{d}{dt}\left(\frac{dy}{dx}\right)\frac{dt}{dx}$$

$$= \frac{\frac{d}{dt}\left(\frac{dy}{dx}\right)}{\frac{dx}{dt}} .$$

• PROBLEM 244

Find d^2y/dx^2 if $x = t - t^2$ and $y = t - t^3$.

Solution: We see here that we can readily obtain $\frac{dx}{dt}$ and $\frac{dy}{dt}$ from the parametric equations. To obtain $\frac{dy}{dx}$, therefore, we apply the chain rule.

$$y' = \frac{dy}{dx} = \frac{dy/dt}{dx/dt} = \frac{1 - 3t^2}{1 - 2t} ,$$

for which

$$\frac{d^2y}{dx^2} = y'' = \frac{dy'}{dx} = \frac{dy'/dt}{dx/dt} = \frac{\frac{d}{dt}\left[\frac{1 - 3t^2}{1 - 2t}\right]}{(1 - 2t)}$$

Using the theorem for $d\left(\frac{u}{v}\right)$,

$$= \frac{(1 - 2t) \cdot (-6t) - (1 - 3t^2) \cdot (-2)}{(1 - 2t)^3}$$

$$= \frac{2 - 6t + 6t^2}{(1 - 2t)^3}.$$

y" cannot be directly obtained from y' without using the chain rule, by merely applying the derivative of a quotient, since y' is a function of t rather than of x.

• PROBLEM 245

Find $D_x^2 y$, given: $x = 2 \sec \theta - 3$, $y = 4 \tan \theta + 2$.

Solution: The two given relations are parametric equations, with parameter θ. For parametric equations, $\frac{dy}{dx}$ is given by:

$$\frac{dy}{dx} = \frac{\frac{dy}{d\theta}}{\frac{dx}{d\theta}},$$

and $\frac{d^2y}{dx^2}$ is given by:

$$\frac{d^2y}{dx^2} = \frac{d}{d\theta}\left(\frac{dy}{dx}\right) \cdot \frac{d\theta}{dx}.$$

To find $D_x^2 y$ we determine $\frac{dy}{d\theta}$, $\frac{dx}{d\theta}$, and $\frac{d\theta}{dx}$, and substitute. We find:

$$\frac{dy}{d\theta} = 4 \sec^2 \theta,$$

and

$$\frac{dx}{d\theta} = 2 \sec \theta \tan \theta.$$

By substitution,

$$\frac{dy}{dx} = \frac{4 \sec^2 \theta}{2 \sec \theta \tan \theta}.$$

$$\frac{d^2y}{dx^2} = \frac{d}{d\theta}\left(\frac{4 \sec^2 \theta}{2 \sec \theta \tan \theta}\right) \cdot \frac{d\theta}{dx},$$

but $\frac{d\theta}{dx} = \frac{1}{2 \sec \theta \tan \theta}$. Using the quotient rule for differentiation and substituting for $\frac{d\theta}{dx}$ we obtain:

$$\frac{d^2y}{dx^2} = (2\sec\theta\tan\theta)(8\sec\theta\sec\theta\tan\theta)$$

$$- \frac{[(4\sec^2\theta)(2\sec\theta\sec^2\theta + 2\tan\theta\sec\theta\tan\theta)]}{4\sec^2\theta\tan^2\theta}$$

$$\cdot \frac{1}{2\sec\theta\tan\theta}.$$

Multiplying out and factoring $8\sec^3\theta$ from the numerator, we obtain:

$$\frac{d^2y}{dx^2} = \frac{8\sec^3\theta(2\tan^2\theta - \sec^2\theta - \tan^2\theta)}{8\sec^3\theta\tan^3\theta}$$

$$= \frac{2\tan^2\theta}{\tan^3\theta} - \frac{\text{see}^2\theta}{\tan^3\theta} - \frac{\tan^2\theta}{\tan^3\theta}$$

$$= 2\cdot\frac{1}{\tan\theta} - \frac{\cos\theta}{\sin^3\theta} - \frac{1}{\tan\theta}.$$

But, $\frac{\cos\theta}{\sin^3\theta}$ can be rewritten as $\frac{\cos\theta}{\sin\theta}\cdot\frac{1}{\sin^2\theta}$, or, $\cot\theta\cdot\csc^2\theta$. Therefore,

$$\frac{d^2y}{dx^2} = 2\cot\theta - \cot\theta - \cot\theta\csc^2\theta$$

$$= \cot\theta - \cot\theta\csc^2\theta.$$

Now we use the identity: $\csc^2\theta = 1 + \cot^2\theta$, and, substituting, we obtain:

$$\frac{d^2y}{dx^2} = \cot\theta - \cot\theta(1 + \cot^2\theta)$$

$$= \cot\theta - \cot\theta - \cot^3\theta$$

$$= -\cot^3\theta.$$

Therefore,

$$D_x^2 y = -\cot^3\theta.$$

• PROBLEM 246

Obtain $D_x{}^2y$ and $D_x{}^3y$ if

$$x = u^2 + 1, \qquad y = 2u - 1.$$

Solution:

$$D_x^2 y = \frac{d^2 y}{dx^2}, \quad D_x^3 y = \frac{d^3 y}{dx^3}, \quad D_x y = \frac{dy}{dx}.$$

It is necessary to obtain the first derivative $D_x y$ before obtaining $D_x^2 y$ and $D_x^3 y$.

Since parametric equations are involved, we use the chain rule to derive $D_x y$ as

$$\frac{dy}{dx} = \frac{dy/du}{dx/du}$$

$$D_x y = 2/2u = 1/u = \phi(u).$$

For the second derivative,

$$D_x^2 y = D_x y' = \frac{-1/u^2}{2u} = -1/2u^3.$$

It is important to note that $D_x^2 y$ is not equal to $\phi'(u)$. The latter would be $D_u(D_x y)$ instead of $D_x(D_x y)$. Each differentiation will involve a division by $D(u^2 + 1)$, or $2u$. Thus

$$D_x^2 y = \frac{dy'}{dx} = \frac{dy'/du}{dx/du}.$$

$$D_x^3 y = D_x(D_x^2 y) = \frac{D_u(-1/2u^3)}{D_u(u^2 + 1)} = \frac{3/2u^4}{2u} = \frac{3}{4u^5}.$$

• PROBLEM 247

Find dz/ds when

$$z = \frac{4y^3}{\sqrt[4]{y - 3}},$$

$$y = u^4 + \frac{u^2}{3},$$

$$u = \frac{1}{8\sqrt{s}}.$$

Solution: When the expression to be differentiated is in parametric form, it is best to apply the chain rule. This form is illustrated here in combination with substitutions, but the operations can be carried out directly without first resorting to substitutions, after some experience has been gained through repeated practice.

$$\frac{dz}{dy} = \frac{(y - 3)^{\frac{1}{4}} \cdot 12y^2 - 4y^3 \cdot \frac{d}{dy}(y - 3)^{\frac{1}{4}}}{(y - 3)^{\frac{1}{2}}}$$

In differentiating $\frac{d}{dy}(y - 3)^{\frac{1}{4}}$, let $w = y - 3$

Then $\frac{dw}{dy} = 1$,

and

$$\frac{d}{dy}(y - 3)^{\frac{1}{4}} = \frac{d}{dy}(w^{\frac{1}{4}}) = \frac{1}{4}w^{-\frac{3}{4}} \cdot \frac{dw}{dy}$$

$$= \frac{1}{4(y - 3)^{\frac{3}{4}}} .$$

Then

$$\frac{dz}{dy} = 12y^2(y - 3)^{-\frac{1}{4}} - \frac{y^3}{(y - 3)^{\frac{5}{4}}}$$

$$= \frac{12y^2(y - 3) - y^3}{(y - 3)^{\frac{5}{4}}} .$$

$$\frac{dz}{dy} = \frac{12y^3 - 36y^2 - y^3}{(y - 3)^{\frac{5}{4}}} = \frac{11y^3 - 36y^2}{(y - 3)^{\frac{5}{4}}} .$$

Since

$$\frac{dy}{du} = 4u^3 + \frac{2}{3}u = \frac{12u^3 + 2u}{3} ,$$

and

$$\frac{du}{ds} = \frac{1}{8} \cdot -\frac{1}{2}s^{-\frac{3}{2}} = -\frac{1}{16s^{\frac{3}{2}}} ,$$

Then,

$$\frac{dz}{ds} = \frac{dz}{dy} \cdot \frac{dy}{du} \cdot \frac{du}{ds}$$

from the chain rule.

$$\frac{dz}{ds} = \frac{11y^3 - 36y^2}{(y - 3)^{\frac{5}{4}}} \cdot \left(\frac{12u^3 + 2u}{3}\right) \cdot -\frac{1}{16s^{\frac{3}{2}}}$$

$$= -\frac{(11y^3 - 36y^2)(12u^3 + 2u)}{48s^{\frac{3}{2}}(y - 3)^{\frac{5}{4}}} .$$

The values of y and u must be substituted in this expression in order to get the value in terms of s.

Find the point where the slope of the curve described by the parametric equations:
$x = 2t^2 - 1$, $y = 3t^3 + t$, is a minimum.

Solution: The slope m is given, in the parametric case, by the formula:

$$m = \frac{dy}{dx} = \frac{\frac{dy}{dt}}{\frac{dx}{dt}} .$$

We find:

$$\frac{dx}{dt} = 4t$$

$$\frac{dy}{dt} = 9t^2 + 1.$$

Therefore,

$$m = \frac{dy}{dx} = \frac{\frac{dy}{dt}}{\frac{dx}{dt}} = \frac{9t^2+1}{4t} .$$

It is the slope that is to be minimized. Therefore, we must find

$$m' = \frac{dm}{dt} = \frac{d\left(\frac{dy}{dx}\right)}{dt}$$

and set this equal to zero in order to obtain the critical points for slope. We find:

$$m' = \frac{(4t)(18t) - (9t^2+1)(4)}{(4t)^2} = \frac{9t^2 - 1}{4t^2} = 0.$$

Therefore the numerator, $9t^2 - 1$, equals 0, $t^2 = \frac{1}{9}$, and $t = \pm\frac{1}{3}$, the critical values. Let us also consider $t = 0$ as a critical value because m' becomes infinite as t approaches 0. We must now test to determine which value of t makes the slope a minimum. We will do this by examining m' at different intervals. We find:

$$\text{when } t < -\frac{1}{3}, \quad m' > 0, +$$

$$0 > t > -\frac{1}{3}, \quad m' < 0, -$$

$$\frac{1}{3} > t > 0, \quad m' < 0, -$$

$$t > \frac{1}{3}, \quad m' > 0, +.$$

Since m' changes sign from - to + at $t = \frac{1}{3}$ we can conclude that this value of t makes the slope a minimum. To find the point where the slope is a minimum in terms of x and y, we substitute $t = \frac{1}{3}$, into the two original parametric equations and obtain the point $\left(-\frac{7}{9}, \frac{4}{9}\right)$.

CHAPTER 13

INDETERMINATE FORMS

Whenever we encounter one of the expressions: $\frac{0}{0}$, $0 \cdot \infty$, 1^{∞}, $\frac{\infty}{\infty}$, $\infty - \infty$, 0^{0}, ∞^{0}, we have an indeterminate form. I.e., we cannot assign a definite value to such an expression without further investigation. Perhaps the result is finite, perhaps it is not. We often cannot predict in advance.

However, the Calculus provides an opportunity for evaluating such expressions. It is L'Hospital's rule. The rule states: If $\frac{f(x)}{g(x)} = \frac{0}{0}$ or $\frac{\infty}{\infty}$, then we can differentiate the numerator separately and the denominator separately, and arrive at an expression that has the same limit as the original expression. Thus,

$$\lim \frac{f'(x)}{g'(x)} = \lim \frac{f(x)}{g(x)} .$$

Ifit turns out that, after the application of L'Hospital's rule, the expression is still indeterminate, the rule can be applied a second time, third time, etc.

It must be remembered that the rule can only be applied to the two indeterminate forms: $\frac{0}{0}$ and $\frac{\infty}{\infty}$. Any other form must first be reduced to one of these forms before the rule is applied.

• PROBLEM 249

The function $f(x) = \dfrac{2x^2 - x - 3}{x + 1}$ is defined for all values of x except $x = -1$, since at $x = -1$ both numerator and denominator vanish. Does

$$\lim_{x \to -1} f(x)$$

exist, and, if so, what is its value?

Solution: As a visual aid, let us sketch f(x). This appears to be a straight line with a "hole" at the point (-1, -5). From the diagram we see that

$$\lim_{x \to -1} f(x) = -5.$$

However, a more systematic method of obtaining limits, without relying on pictorial representation and intuition is desirable. By means of factoring, we can write f(x) in the form

$$f(x) = \frac{(2x - 3)(x + 1)}{x + 1}.$$

Now if $x \neq -1$, we are allowed to divide both the numerator and denominator by $(x + 1)$. Then

$$f(x) = 2x - 3, \qquad \text{if } x \neq -1.$$

This function tends to -5 as x tends to -1, since simple substitution now works. Therefore,

$$\lim_{x \to -1} f(x) = -5.$$

Note that we never substituted the value $x = -1$ in the original expression.

In the above solution, it was possible to factor the numerator and then proceed with direct substitution to obtain a definite answer. However, factoring is not always possible, and when the function f(x) takes the form 0/0, as in this case, the preferable approach is to apply L'Hospital's rule as follows:

$$\lim_{x \to -1} \frac{2x^2 - x - 3}{x + 1} = \lim_{x \to -1} \frac{4x - 1}{1} = -5$$

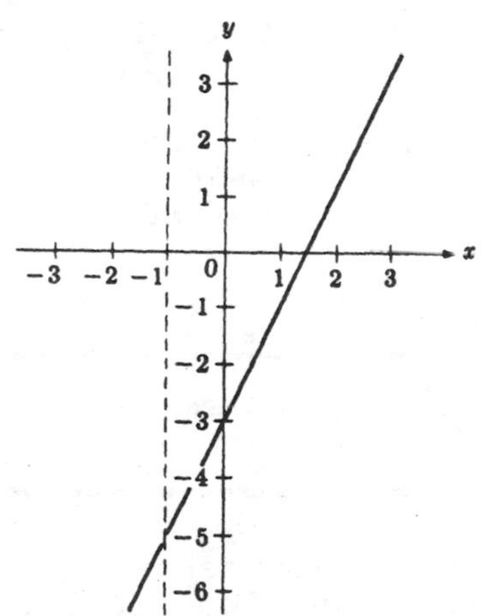

• PROBLEM 250

Evaluate $\lim_{x \to 2} \frac{(2x^2 - 4x)}{x - 2}$.

Solution: The function takes the form 0/0, and therefore we can apply L'Hospital's rule to obtain:

$$\lim_{x \to 2} \frac{4x - 4}{1} = 4$$

We can also solve the problem in a different way by noting that the numerator can be factored.

$$\lim_{x \to 2} \frac{2x^2 - 4x}{x - 2} = \lim_{x \to 2} \frac{2x(x - 2)}{x - 2}$$

$$= \lim_{x \to 2} 2x$$

$$= 4$$

• PROBLEM 251

Find $\lim_{x \to 3} \frac{x^2 - x - 6}{x - 3}$.

Solution: This limit may be found by writing

$$\frac{x^2 - x - 6}{x - 3} = \frac{(x + 2)(x - 3)}{x - 3} = x + 2.$$

Hence $\lim_{x \to 3} (x + 2) = 5.$

Since $\frac{0}{0}$ (indeterminate) is obtained by substitution in the original function, the limit may also be obtained by L'Hospital's rule by differentiating separately numerator and denominator. Thus,

$$\lim_{x \to 3} \frac{x^2 - x - 6}{x - 3} = \lim_{x \to 3} \frac{2x - 1}{1} = 5.$$

The application of L'Hospital's rule is the more systematic approach and should generally be tried first, if another method is not immediately apparent.

• PROBLEM 252

Evaluate $\lim_{x \to 4} \frac{x^2 - 16}{x - 4}$.

Solution: The function takes the form 0/0 and therefore we can apply L'Hospital's rule.

$$\lim_{x \to 4} \frac{x^2 - 16}{x - 4} = \lim_{x \to 4} \frac{2x}{1} = 8$$

It should be noted that in this problem **it does** not help to divide numerator and denominator by x^2, since x does not approach ∞. We can, however, factor the numerator and proceed as follows:

$$\lim_{x \to 4} \frac{x^2 - 16}{x - 4} = \lim_{x \to 4} \frac{(x + 4)(x - 4)}{(x - 4)}$$

$$= \lim_{x \to 4} (x + 4)$$

$$= 8$$

We are not dividing by zero here, although the direct substitution of 4 in $(x - 4)$ results in zero. This is because, in considering limits, we are concerned with values of the function close to $x = 4$, but not actually at $x = 4$. Therefore we are not dividing by zero. Another way to look at this is to remember that we divide first, then allow x to approach its limit.

• PROBLEM 253

Evaluate: $\lim_{x \to 1} \dfrac{x - 1}{\sqrt{x^2 + 3} - 2}$

Solution:

$$\lim_{x \to 1} \frac{x - 1}{\sqrt{x^2 + 3} - 2} = \frac{1 - 1}{\sqrt{1 + 3} - 2} = \frac{0}{0}$$

This is an indeterminate form, therefore L'Hospital's rule is applicable. L'Hospital's rule states that

$$\lim_{x \to a} \frac{f(x)}{g(x)} = \lim_{x \to a} \frac{f'(x)}{g'(x)},$$

under the assumption that $f(x)$, $g(x)$ are continuously differentiable in the interval about the point in question, and where $g'(x)$ does not change sign.

Therefore,

$$\lim_{x \to 1} \frac{x - 1}{\sqrt{x^2 + 3} - 2} = \lim_{x \to 1} \frac{1}{\dfrac{x}{\sqrt{x^2 + 3}}}$$

$$= \lim_{x \to 1} \frac{\sqrt{x^2 + 3}}{x} = 2.$$

• PROBLEM 254

Find the limit, if it exists, of:

$$\lim_{r \to 1} \frac{1 - r^3}{2 - \sqrt{r^2 + 3}}.$$

Solution: As r approaches 1, $r \to 1$, $r^3 \to 1^3$, or $r^3 \to 1$; and $r^2 \to 1^2$, or $r^2 \to 1$.

Therefore,

$$\underset{r \to 1}{\text{limit}} \frac{1 - r^3}{2 - \sqrt{r^3 + 3}} = \frac{1 - 1}{2 - \sqrt{4}} = \frac{0}{2 - \sqrt{4}}$$

Two cases must be considered.

I) Case when $\sqrt{4} = -2$:

$$\lim_{r \to 1} \frac{1 - r^3}{2 - \sqrt{r^3 + 3}} = \frac{0}{2 - \sqrt{4}} = \frac{0}{2 - (-2)}$$

$$= \frac{0}{4} = 0$$

II) Case when $\sqrt{4} = 2$:

$$\lim_{r \to 1} \frac{1 - r^3}{2 - \sqrt{r^2 + 3}} = \frac{0}{2 - \sqrt{4}} = \frac{0}{2 - 2} = \frac{0}{0},$$

which is in indeterminate form, indicating that L'Hospital's rule is applicable. Using this rule,

$\left[\lim_{x \to a} \frac{f(x)}{g(x)} = \lim_{x \to a} \frac{f'(x)}{g'(x)} \right.$ where $f(x)$ and $g(x)$ are continuously differentiable in the interval about a and where $g'(x)$ does not change sign $\left. \right]$,

$$\lim_{r \to 1} \frac{1 - r^3}{2 - \sqrt{r^2 + 3}} = \lim_{r \to 1} \frac{-3r^2}{\left(\frac{-r}{\sqrt{r^2 + 3}} \right)}$$

$$= \lim_{r \to 1} 3r \left(\sqrt{r^2 + 3} \right) = 6.$$

• PROBLEM 255

Show that $\lim_{x \to 1} \frac{x^3 - 3x + 2}{x^3 - x^2 - x + 1} = \frac{3}{2}$.

<u>Solution:</u> Let $f(x) = x^3 - 3x + 2$, $F(x) = x^3 - x^2 - x + 1$. Then $f(1) = 0$, and $F(1) = 0$. Therefore, the function takes the indeterminate form 0/0, and L'Hospital's rule may be applied directly.

$$\lim_{x \to 1} \frac{f(x)}{F(x)} = \lim_{x \to 1} \frac{f'(x)}{F'(x)} = \lim_{x \to 1} \frac{3x^2 - 3}{3x^2 - 2x - 1} = \frac{0}{0}.$$

The function is still indeterminate, but we may apply again L'Hospital's rule.

$$\lim_{x \to 1} \frac{f(x)}{F(x)} = \lim_{x \to 1} \frac{f''(x)}{F''(x)} = \lim_{x \to 1} \frac{6x}{6x - 2} = \frac{3}{2}.$$

Note that the method of dividing numerator and denominator by the highest power in x **does** not help here, since x does not approach ∞ in the given problem.

• PROBLEM 256

Evaluate $\lim_{x \to a} \frac{a - x}{\ln \frac{x}{a}}$.

Solution: The function takes the form 0/0, and therefore it is possible to apply L'Hospital's rule. Now

$$\lim_{x \to a} \frac{a - x}{\ln \frac{x}{a}} = \lim_{x \to a} \frac{a - x}{\ln x - \ln a}.$$

Differentiating numerator and denominator gives:

$$\lim_{x \to a} \frac{-1}{\frac{1}{x}} = -a.$$

• PROBLEM 257

Find $\lim_{x \to 1} \frac{1 - x + \ln x}{x^3 - 3x + 2}$.

Solution: Call the numerator f(x) and the denominator g(x), then

$$\lim_{x \to 1} \frac{f(x)}{g(x)} = \frac{0}{0}.$$

Since the function takes the indeterminate form 0/0, we apply L'Hospital's rule.

$$\lim_{x \to 1} \frac{f'(x)}{g'(x)} = \lim_{x \to 1} \frac{-1 + \frac{1}{x}}{3(x^2 - 1)} = \frac{0}{0}.$$

The function again takes the indeterminate form 0/0, and we therefore apply the rule a second time.

$$\lim_{x \to 1} \frac{f''(x)}{g''(x)} = \lim_{x \to 1} \frac{-\frac{1}{x^2}}{6x} = -\frac{1}{6}.$$

• PROBLEM 258

Find: $\lim_{x \to 0} \frac{x}{1 - e^x}$

Solution: Because, for the numerator

$$\lim_{x \to 0} x = 0, \text{ and } \lim_{x \to 0} (1 - e^x) = 0,$$

for the denominator, the function takes the form 0/0, to which L'Hospital's rule may be applied. Therefore,

$$\lim_{x \to 0} \frac{x}{1 - e^x} = \lim_{x \to 0} \frac{1}{-e^x} = \frac{1}{-1} = -1$$

• PROBLEM 259

Find $\lim_{x \to 0^+} \frac{\sin x}{\sqrt{x}}$

Solution: The function is in the form 0/0, and therefore L'Hospital's rule may be applied directly.

$$\lim_{x \to 0^+} \frac{\sin x}{\sqrt{x}} = \lim_{x \to 0^+} \frac{\cos x}{1/(2\sqrt{x})} = \lim_{x \to 0^+} 2\sqrt{x} \cos x = 0.$$

• PROBLEM 260

Find $\lim_{x \to 0} \frac{x \cos x - \sin x}{x}$.

Solution: This function is in the form 0/0. Consequently, applying L'Hospital's rule,

$$\lim_{x \to 0} \frac{x \cos x - \sin x}{x} = \lim_{x \to 0} \frac{-x \sin x}{1}$$

$$= \frac{0}{1} = 0.$$

Another approach is to recognize that, for small angles, sin x = x, cos x = 1. Thus the given expression becomes:

$$\lim_{x \to 0} \frac{x(1) - x}{x} = \lim_{x \to 0} \frac{1 - 1}{1} = 0.$$

• PROBLEM 261

Find $\lim_{x \to 0} \frac{\sin 2x \tan x}{3x}$.

Solution: The function takes the indeterminate form 0/0, and therefore we apply L'Hospital's rule. Differentiating numerator and denominator,

$$\lim_{x \to 0} \frac{\sin 2x \sec^2 x + 2 \cos 2x \tan x}{3} = 0.$$

A simpler approach is to recognize that, for small angles, $\sin \Theta = \tan \Theta = \Theta$, so that the given expression becomes:

$$\lim_{x \to 0} \frac{2x(x)}{3x} = \lim_{x \to 0} \frac{2x}{3} = 0.$$

● PROBLEM 262

Find $\lim_{x \to 0} \frac{\sin 2x + \tan x}{3x}$.

Solution: The function takes the form 0/0 which is indeterminate. We therefore apply L'Hospital's rule.

$$\lim_{x \to 0} \frac{2 \cos 2x + \sec^2 x}{3} = 1.$$

Another approach is to recognize that, for small angles, sin Θ = tan Θ = Θ, so that the given expression can be written as

$$\lim_{x \to 0} \frac{2x + x}{3x} = \frac{3x}{3x} = 1.$$

● PROBLEM 263

Find $\lim_{x \to 0} \frac{\sin 2x + \tan x}{3x^2}$.

Solution: The function takes the form 0/0 which is indeterminate, and therefore we apply L'Hospital's rule by differentiating numerator and denominator:

$$\lim_{x \to 0} \frac{2 \cos 2x + \sec^2 x}{6x} = \frac{3}{0} = \infty .$$

We can also recognize that, for small angles, sin Θ = tan Θ = Θ, therefore the original expression becomes

$$\lim_{x \to 0} \frac{2x + x}{3x^2} = \lim_{x \to 0} \frac{3x}{3x^2} = \lim_{x \to 0} \frac{1}{x} = \infty .$$

● PROBLEM 264

Prove $\lim_{x \to 0} \frac{\sin nx}{x} = n.$

Solution: Let the numerator be

$$f(x) = \sin nx,$$

and let the denominator be

$$F(x) = x.$$

Then $f(0) = 0$, $F(0) = 0$. Therefore, the function takes the form 0/0 and we apply L'Hospital's rule.

$$\lim_{x \to 0} \frac{f(x)}{F(x)} = \lim_{x \to 0} \frac{f'(x)}{F'(x)} = \lim_{x \to 0} \frac{n \cos nx}{1} = n.$$

• PROBLEM 265

Find $\lim_{x\to+\infty} \dfrac{\sin \frac{1}{x}}{\tan^{-1}\left(\frac{1}{x}\right)}$

Solution: Checking the numerator and denominator, it is found that

$$\lim_{x\to+\infty} \sin(1/x) = 0 \text{ and } \lim_{x\to+\infty} \tan^{-1}(1/x) = 0.$$

Therefore, the function takes the indeterminate form 0/0, so that L'Hospital's rule may be applied.

$$\lim_{x\to+\infty} \frac{\sin \frac{1}{x}}{\tan^{-1}\left(\frac{1}{x}\right)} = \lim_{x\to+\infty} \frac{\cos \frac{1}{x} \cdot \left(-\frac{1}{x^2}\right)}{\frac{1}{1+\frac{1}{x^2}} \cdot \left(-\frac{1}{x^2}\right)} =$$

$$\lim_{x\to+\infty} \frac{\cos \frac{1}{x}}{\frac{x^2}{x^2+1}} = \frac{1}{1} = 1$$

A simpler approach is to recognize that, as

$$x \to \infty, \ \frac{1}{x} \to 0,$$

and that, for small angles, $\sin \Theta = \tan \Theta = \Theta$. Thus,

$$\lim_{x\to\infty} \frac{\sin \frac{1}{x}}{\tan^{-1} \frac{1}{x}} = \lim_{\Theta\to 0} \frac{\sin \Theta}{\tan^{-1} \Theta} = \lim_{\Theta\to 0} \frac{\Theta}{\Theta} = 1.$$

• PROBLEM 266

Find $\lim_{x\to 0} (\tan x - \sin x)/x^2$.

Solution: After checking that the function takes an indeterminate form, 0/0, we can apply L'Hospital's rule.

$$\lim_{x\to 0} \frac{\tan x - \sin x}{x^2} = \lim_{x\to 0} \frac{\sec^2 x - \cos x}{2x}$$

$$= \lim_{x \to 0} \frac{2 \sec^2 x \tan x + \sin x}{2} = \frac{0}{2} = 0.$$

A simpler approach is to recognize that, for small angles, tan x = sin x = x. The given expression can be written as

$$\lim_{x \to 0} \frac{x - x}{x^2} = \lim_{x \to 0} \frac{1 - 1}{x} = \frac{0}{x} = 0.$$

• PROBLEM 267

> Find $\lim_{x \to 0} \frac{\tan x - x}{x - \sin x}$.

Solution: The function takes the form of 0/0, and therefore we apply L'Hospital's rule.

$$\lim_{x \to 0} \frac{\tan x - x}{x - \sin x} = \lim_{x \to 0} \frac{\sec^2 x - 1}{1 - \cos x},$$

The function remains in the indeterminate form 0/0, but another application of the rule here does not help, because another indeterminate form is obtained. We therefore proceed to rewrite the expression to avoid this problem. The most obvious step to be taken first is to substitute

$$\sec^2 x = 1/\cos^2 x.$$

$$\frac{\sec^2 x - 1}{1 - \cos x} = \frac{\frac{1}{\cos^2 x} - 1}{1 - \cos x} = \frac{1 - \cos^2 x}{\cos^2 x(1 - \cos x)}$$

$$= \frac{(1 + \cos x)(1 - \cos x)}{\cos^2 x(1 - \cos x)} = \frac{1 + \cos x}{\cos^2 x}.$$

Now $$\lim_{x \to 0} \frac{(1 + \cos x)}{\cos^2 x} = 2.$$

• PROBLEM 268

> Prove: $\lim_{x \to 0} \frac{e^x - e^{-x} - 2x}{x - \sin x} = 2.$

Solution: Let the numerator be:

$$f(x) = e^x - e^{-x} - 2x,$$

and let the denominator be:

$$F(x) = x - \sin x.$$

Then, $f(0) = 0, \quad F(0) = 0.$

Since the function takes the form 0/0, we apply L'Hospital's rule.

$$\lim_{x\to 0} \frac{f(x)}{F(x)} = \lim_{x\to 0} \frac{f'(x)}{F'(x)} = \lim_{x\to 0} \frac{e^x + e^{-x} - 2}{1 - \cos x} = \frac{0}{0}.$$

Since 0/0 is obtained again, we reapply the rule.

$$\lim_{x\to 0} \frac{f(x)}{F(x)} = \lim_{x\to 0} \frac{f''(x)}{F''(x)} = \lim_{x\to 0} \frac{e^x - e^{-x}}{\sin x} = \frac{0}{0}.$$

We apply the rule a third time.

$$\lim_{x\to 0} \frac{f(x)}{F(x)} = \lim_{x\to 0} \frac{f'''(x)}{F'''(x)} = \lim_{x\to 0} \frac{e^x + e^{-x}}{\cos x} = 2.$$

• PROBLEM 269

Find: $\lim_{x\to 0} \frac{10^{2x} - 2 + 10^{-2x}}{10^{2x} - 10^{-2x}}$.

Solution: We first attempt to solve this problem by substituting 0 for x. Doing this, we obtain an indeterminate form of the type $\frac{0}{0}$. To this we can apply L'Hospital's Rule, using the differentiation formula, $\frac{d}{dx} a^u = a^u \frac{du}{dx} \ln a$. We obtain:

$$\lim_{x\to 0} \frac{(10^{2x} \cdot 2 \ln 10) - 0 + (10^{-2x})(-2 \ln 10)}{(10^{2x} \cdot 2 \ln 10) - (10^{-2x})(-2 \ln 10)}$$

$$= \lim_{x\to 0} \frac{10^{2x} \cdot 2 \ln 10 - \frac{2 \ln 10}{10^{2x}}}{10^{2x} \cdot 2 \ln 10 + \frac{2 \ln 10}{10^{2x}}}$$

$$= \lim_{x\to 0} \frac{10^{4x} \cdot 2 \ln 10 - 2 \ln 10}{10^{4x} \cdot 2 \ln 10 + 2 \ln 10}$$

$$= \lim_{x\to 0} \frac{2 \ln 10 (10^{4x} - 1)}{2 \ln 10 (10^{4x} + 1)} = \lim_{x\to 0} \frac{10^{4x} - 1}{10^{4x} + 1} = \frac{1 - 1}{1 + 1}$$

$$= \frac{0}{2} = 0.$$

Therefore,

$$\lim_{x\to 0} \frac{10^{2x} - 2 + 10^{-2x}}{10^{2x} - 10^{-2x}} = 0.$$

Evaluate: $\lim\limits_{x \to a} \dfrac{x^x - a^a}{a^x - x^a}$.

Solution: If we attempt to evaluate this limit as it stands, we obtain:

$$\lim_{x \to a} \frac{x^x - a^a}{a^x - x^a} = \frac{a^a - a^a}{a^a - a^a} = \frac{0}{0}.$$

This is an indeterminate form, and we therefore attempt to use L'Hospital's Rule. L'Hospital's Rule states that if there is an indeterminate ratio of the type $\frac{0}{0}$, we differentiate the numerator and denominator, and then take the limit. Following this procedure, we obtain,

$$\lim_{x \to a} \frac{x^x - a^a}{a^x - x^a} = \lim_{x \to a} \frac{\frac{d}{dx}\left(x^x - a^a\right)}{\frac{d}{dx}\left(a^x - x^a\right)}.$$

We can find $\frac{d}{dx}\left(x^x\right)$ as follows:

$$\text{let} \quad y = x^x.$$

Then,

$$\ln y = x \ln x.$$

Differentiating,

$$\frac{1}{y}\frac{dy}{dx} = \ln x + x\left(\frac{1}{x}\right).$$

$$\frac{dy}{dx} = \left(\ln x + 1\right)y.$$

$$\frac{dy}{dx} = \left(\ln x + 1\right)x^x.$$

We can find $\frac{d}{dx}\left(a^x\right)$ using the formula: $\frac{d}{dx}\, a^u = a^u \frac{du}{dx} \ln a$. Therefore,

$$\lim_{x \to a} \frac{\frac{d}{dx}\left(x^x - a^a\right)}{\frac{d}{dx}\left(a^x - x^a\right)} = \lim_{x \to a} \frac{\left(\ln x + 1\right)x^x - 0}{a^x \ln a - ax^{a-1}}.$$

Now we can evaluate the limit, to obtain:

$$\lim_{x \to a} \frac{x^x - a^a}{a^x - x^a} = \lim_{x \to a} \frac{\left(\ln x + 1\right)x^x - 0}{a^x \ln a - ax^{a-1}}$$

$$= \frac{(\ln a + 1)a^a}{a^a(\ln a) - aa^{a-1}}$$

$$= \frac{(\ln a + 1)a^a}{a^a(\ln a) - a^a}.$$

$$\lim_{x \to a} \frac{x^x - a^a}{a^x - x^a} = \frac{\ln a + 1}{\ln a - 1}.$$

• PROBLEM 271

Let P(3,4) and $Q(x, \sqrt{25 - x^2})$ be two distinct points on the semicircle with equation

$$y = \sqrt{25 - x^2}.$$

Let M(x) designate the slope of the secant line through P and Q. Find

$$\underset{x \to 3}{\text{limit}}\ M(x).$$

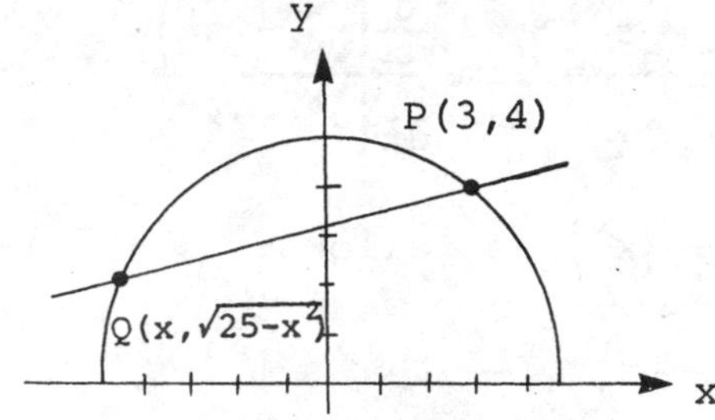

Solution: The slope $M(x) = \Delta y/\Delta x = y_2 - y_1/x_2 - x_1$. Therefore,

$$M(x) = \frac{\sqrt{25 - x^2} - 4}{x - 3}$$

Upon simple substitution of x = 3, the function M(x) takes the form 0/0, and therefore we can apply L'Hospital's rule.

$$\lim_{x \to 3} \frac{(25 - x^2)^{\frac{1}{2}} - 4}{x - 3} =$$

$$\lim_{x \to 3} \frac{\frac{1}{2}(25 - x^2)^{-\frac{1}{2}}(-2x)}{1} = -\frac{3}{4}.$$

• PROBLEM 272

Find $\lim_{x \to \infty} \frac{x^2 + 1}{2x^2 + 3}$.

Solution: As $x \to \infty$ both the numerator and denominator become large without bound. The function, therefore, takes the form ∞/∞. The most direct method of solution here, is to apply L'Hospital's rule, giving:

$$\lim_{x\to\infty} \frac{(x^2 + 1)}{2x^2 + 3} = \lim_{x\to\infty} \frac{2x}{4x} = \frac{1}{2} .$$

However, when the function is a ratio in powers of x, it is also possible to obtain the solution simply by dividing numerator and denominator by the highest power of x. Thus, if we divide numerator and denominator by x^2, the function becomes:

$$\frac{1 + 1/x^2}{2 + 3/x^2} .$$

We see that $1/x^2$ and $3/x^2$ both approach zero as $x \to \infty$. This means that the numerator approaches 1 and the denominator approaches 2. Therefore, the answer of 1/2 is also obtained with the second method.

• PROBLEM 273

Determine $\lim_{x\to\infty} \frac{x^2 - 1}{4x^2 + x}$

Solution: Since the function takes the form ∞/∞, we apply L'Hospital's rule twice, since x^2 is present in both numerator and denominator. It is essential, however, to check that the function is in the required indeterminate form before applying the rule.

As an alternative, we can write

$$\lim_{x\to\infty} \frac{x^2 - 1}{4x^2 + x} = \frac{1 - \frac{1}{x^2}}{4 + \frac{1}{x}}$$

by dividing through by x^2.

As $x \to \infty$, the x-terms vanish, giving $\frac{1}{4}$.

$$\lim_{x\to\infty} \frac{x^2 - 1}{4x^2 + x} = \lim_{x\to\infty} \frac{2x}{8x + 1} = \lim_{x\to\infty} \frac{1}{4} = \frac{1}{4}.$$

• PROBLEM 274

Find $\lim_{x\to\infty} \left| \frac{x^4 + x^3 + x^2 + 2}{3x^4 + x^2 - 1} \right|$.

Solution: Testing the function by simple substitution, we obtain the indeterminate form ∞/∞. Since x approaches ∞, we divide numerator and denominator by x^4, the highest given power of x.

Then

$$\frac{x^4 + x^3 + x^2 + 2}{3x^4 + x^2 - 1} = \frac{1 + \frac{1}{x} + \frac{1}{x^2} + \frac{2}{x^4}}{3 + \frac{1}{x^2} - \frac{1}{x^4}}$$

As x becomes very large, the fractions:

$\frac{1}{x}$, $\frac{1}{x^2}$, $\frac{2}{x^4}$, and $\frac{1}{x^4}$, become very small

and approach 0 as a limit, and the value of the fraction approaches $\frac{1}{3}$ as a limit. Therefore,

$$\lim_{x \to \infty} \left| \frac{x^4 + x^3 + x^2 + 2}{3x^4 + x^2 - 1} \right| = \frac{1}{3}$$

Note that, whereas L'Hospital's rule may be applied here directly, it is necessary to apply this rule four times before arriving at the final answer of $\frac{1}{3}$. This is because powers of x^4 are present. For lower powers of x, this rule may be applied with greater simplicity.

• PROBLEM 275

Evaluate $\lim_{x \to \infty} \frac{3x^3 - x + 7}{x^3 + 4x^2 + x - 3}$.

Solution: The function takes the form ∞/∞, and therefore L'Hospital's rule may be applied. However, the rule would have to be applied several times to arrive at the final answer.

The shorter method is to note that $x \to \infty$, and divide through by the highest power of x.

$$\lim_{x \to \infty} \frac{3x^3 - x + 7}{x^3 + 4x^2 + x - 3} = \lim_{x \to \infty} \frac{3 - \frac{1}{x^2} + \frac{7}{x^3}}{1 + \frac{4}{x} + \frac{1}{x^2} - \frac{3}{x^3}}$$

$$= \frac{3}{1}$$

$$= 3.$$

• PROBLEM 276

Evaluate $\lim_{x \to \infty} \frac{2x^2 + 4x - 7}{x^3 + 3x^2 - 5}$.

Solution: The method of solution used for some

other problems also applies here, even though the highest power of x in the denominator is different from that in the numerator.

$$\lim_{x\to\infty} \frac{2x^2 + 4x - 7}{x^3 + 3x^2 - 5} = \lim_{x\to\infty} \frac{\frac{2}{x} + \frac{4}{x^2} - \frac{7}{x^3}}{1 + \frac{3}{x} - \frac{5}{x^3}}$$

$$= \frac{0}{1}$$

$$= 0.$$

• PROBLEM 277

Find $\lim_{x\to\infty} \frac{x^2 - 2x + 4}{3x^2 + x - 1}$.

Solution: This function takes the indeterminate form ∞/∞. We therefore should try to rewrite the function in the form which has either a limit, or allows the application of L'Hospital's rule. This may be achieved by dividing numerator and denominator by x^2. We have:

$$\frac{1 - \frac{2}{x} + \frac{4}{x^2}}{3 + \frac{1}{x} - \frac{1}{x^2}},$$

As $x \to \infty$, the limits of the new numerator and denominator exist, since $1/x \to 0$ as $x \to \infty$. Hence

$$\lim_{x\to\infty} \frac{x^2 - 2x + 4}{3x^2 + x - 1} = \lim_{x\to\infty} \frac{1 - \frac{2}{x} + \frac{4}{x^2}}{3 + \frac{1}{x} - \frac{1}{x^2}} = \frac{1}{3}.$$

• PROBLEM 278

Find $\lim_{x\to-\infty} \frac{2x^2 - x + 5}{4x^3 - 1}$

Solution: The function is of the indeterminate form ∞/∞. When the given function is a ratio of polynomials, the first approach should be to divide numerator and denominator by the highest power of x appearing in the function. Accordingly,

$$\lim_{x\to-\infty} \frac{2x^2 - x + 5}{4x^3 - 1} = \lim_{x\to-\infty} \frac{2/x - 1/x^2 + 5/x^3}{4 - 1/x^3}$$

$$= \frac{\lim_{x\to-\infty} (2/x - 1/x^2 + 5/x^3)}{\lim_{x\to-\infty} (4 - 1/x^3)} = \frac{0}{4} = 0$$

Find $\lim_{x\to+\infty} \frac{3x - 2}{5x + 4}$,

Solution: Divide each term in the numerator and denominator by x, since some ratios may become zero thereby.

$$\frac{3x - 2}{5x + 4} = \frac{3 - (2/x)}{5 + (4/x)} .$$

For the numerator, we obtain

$$\lim_{x\to+\infty}\left(3 - \frac{2}{x}\right) = \lim_{x\to+\infty} 3 + \lim_{x\to+\infty}\left(- \frac{2}{x}\right),$$

since the limit of a sum is the sum of the limits. Now

$$\lim_{x\to+\infty} 3 = 3 \text{ (using limit of a constant)}$$

and $\lim_{x\to+\infty}\left(- \frac{2}{x}\right) = - 2 \lim_{x\to+\infty} \frac{1}{x} = 0$ (using limit of a product). Therefore

$$\lim_{x\to+\infty}\left(3 - \frac{2}{x}\right) = 3.$$

Similarly,

$$\lim_{x\to+\infty}\left(5 + \frac{4}{x}\right) = 5.$$

Therefore

$$\lim_{x\to+\infty} \frac{3 - (2/x)}{5 + (4/x)} = \frac{\lim_{x\to+\infty}\left[3 - (2/x)\right]}{\lim_{x\to+\infty}\left[5 + (4/x)\right]},$$

since the limit of a quotient is the quotient of the limits. Consequently

$$\lim_{x\to+\infty} \frac{3x - 2}{5x + 4} = \frac{3}{5} .$$

• PROBLEM 280

Find $\lim_{x\to+\infty} \frac{x^2}{x + 1}$

Solution: Since the function takes the indeterminate form ∞/∞, we divide the numerator and the denominator by x^2 which is the highest power of x in the function.

$$\lim_{x\to+\infty} \frac{x^2}{x + 1} = \lim_{x\to+\infty} \frac{1}{1/x + 1/x^2}$$

Evaluating the limit of the denominator, we have

$$\lim_{x\to+\infty}\left(\frac{1}{x}+\frac{1}{x^2}\right)=\lim_{x\to+\infty}\frac{1}{x}+\lim_{x\to+\infty}\frac{1}{x^2}=0+0=0$$

Therefore, the limit of the denominator is 0, and the denominator is approaching 0 through positive values.

The limit of the numerator is 1, and therefore

$$\lim_{x\to+\infty}\frac{x^2}{x+1}=+\infty$$

• PROBLEM 281

Find $\lim_{x\to\infty}\frac{x^2+10}{6x^2+2}$

Solution: As $x\to\infty$, the function takes the form ∞/∞ which is indeterminate. Using the method of dividing numerator and denominator by the largest power in x, we obtain

$$\frac{1+10/x^2}{6+2/x^2}$$

As $x\to\infty$, the fractional terms of both numerator and denominator approach zero. This means that the numerator approaches 1 and the denominator approaches 6. Therefore

$$\lim_{x\to\infty}\frac{x^2+10}{6x^2+2}=\frac{1}{6}$$

Another approach to this problem is to apply L'Hospital's rule directly.

$$\lim_{x\to\infty}\frac{x^2+10}{6x^2+2}=\frac{2x}{12x}=\frac{1}{6}$$

• PROBLEM 282

Find $\lim_{x\to0^+}\frac{\ln x}{\frac{1}{x}}$.

Solution: An examination of the numerator and denominator yields:

$$\lim_{x\to0^+}\ln x=-\infty \quad\text{and}\quad \lim_{x\to0^+}(1/x)=+\infty.$$

The ratio takes the indeterminate form ∞/∞, and therefore L'Hospital's rule may be applied.

$$\lim_{x\to0^+}\frac{\ln x}{\frac{1}{x}}=\lim_{x\to0^+}\frac{\frac{1}{x}}{-\frac{1}{x^2}}=\lim_{x\to0^+}(-x)=0.$$

• PROBLEM 283

Find $\lim_{x\to\infty} \frac{x \ln x}{x + \ln x}$.

Solution: This function has the indeterminate form ∞/∞, and therefore, after differentiating numerator and denominator, in the application of L'Hospital's rule,

$$\lim_{x\to\infty} \frac{x \ln x}{x + \ln x} = \lim_{x\to\infty} \frac{1 + \ln x}{1 + 1/x} = \infty ,$$

since the denominator has the limit 1, while the limit of the numerator is infinite. Thus, this function has no limit.

• PROBLEM 284

Find: $\lim_{x\to+\infty} e^x/\ln x$.

Solution: This function takes the indeterminate form ∞/∞, and therefore it is permissible to apply L'Hospital's rule directly.

$$\lim_{x\to+\infty} \frac{e^x}{\ln x} = \lim_{x\to\infty} \frac{e^x}{1/x} = \lim_{x\to\infty} xe^x = \infty.$$

The function thus has no limit.

• PROBLEM 285

Find: $\lim_{x\to+\infty} \frac{x^2}{e^x}$.

Solution: Checking the numerator and denominator shows that:

$$\lim_{x\to+\infty} x^2 = +\infty \text{ and } \lim_{x\to+\infty} e^x = +\infty.$$

Since the ratio takes the indeterminate form ∞/∞, L'Hospital's rule may be applied.

$$\lim_{x\to+\infty} \frac{x^2}{e^x} = \lim_{x\to+\infty} \frac{2x}{e^x} .$$

Now $\lim_{x\to+\infty} 2x = +\infty$ and $\lim_{x\to+\infty} e^x = +\infty$.

The ratio again takes the form ∞/∞, and therefore we apply L'Hospital's rule a second time to obtain:

$$\lim_{x\to+\infty} \frac{2x}{e^x} = \lim_{x\to+\infty} \frac{2}{e^x} = 0.$$

• PROBLEM 286

Find: $\lim_{x\to+\infty} x^3/e^x$.

Solution: This function takes the indeterminate form ∞/∞, and therefore we can directly apply L'Hospital's rule.

$$\lim_{x\to+\infty} \frac{x^3}{e^x} = \lim_{x\to+\infty} \frac{3x^2}{e^x}.$$

The function remains of the indeterminate form, and therefore we apply the rule a second and a third time to obtain

$$\lim_{x\to\infty} \frac{6x}{e^x} = \lim_{x\to\infty} \frac{6}{e^x} = 0.$$

• PROBLEM 287

Find $\lim_{x\to+\infty} \frac{3x + 4}{\sqrt{2x^2 - 5}}$.

Solution: Since the function is of the form ∞/∞, which is indeterminate, we divide the numerator and the denominator of the fraction by x. (The highest power in the function is x.) In the denominator we let

$$x = \sqrt{x^2}$$

in order to operate under the square root sign. Thus,

$$\lim_{x\to+\infty} \frac{3x + 4}{\sqrt{2x^2 - 5}} = \lim_{x\to+\infty} \frac{3 + 4/x}{\sqrt{\frac{2x^2 - 5}{x^2}}}$$

$$= \frac{\lim_{x\to+\infty} (3 + 4/x)}{\lim_{x\to+\infty} \sqrt{2 - 5/x^2}} = \frac{3}{\sqrt{2}}.$$

• PROBLEM 288

Evaluate $\lim_{x\to+\infty} \frac{\sqrt{x^2 - 1}}{2x + 1}$.

Solution: The function takes an indeterminate form to which L'Hospital's rule is not applicable. The approach to be taken, then, is to manipulate the function into a different form by dividing, for example numerator and denominator by x, the highest power of x in the function

$$\frac{\sqrt{x^2 - 1}}{2x + 1} = \frac{\sqrt{1 - (1/x^2)}}{2 + (1/x)}.$$

Since $$\frac{1}{x^2} = \frac{1}{x} \cdot \frac{1}{x},$$

we have
$$\lim_{x \to +\infty} \frac{1}{x^2} = \lim_{x \to +\infty} \frac{1}{x} \cdot \lim_{x \to +\infty} \frac{1}{x} = 0,$$

using the limit of a product theorem. Then

$$\lim_{x \to +\infty} \left(1 - \frac{1}{x^2}\right) = \lim_{x \to +\infty} 1 - \lim_{x \to +\infty} \frac{1}{x^2} = 1$$

(using limit of a sum). Further,

$$\lim_{x \to +\infty} \sqrt{1 - \frac{1}{x^2}} = 1,$$

and
$$\lim_{x \to +\infty} \left(2 + \frac{1}{x}\right) = 2.$$

Applying the quotient rule,

$$\lim_{x \to +\infty} \frac{\sqrt{x^2 - 1}}{2x + 1} = \frac{1}{2}.$$

• PROBLEM 289

Find: $\lim_{x \to \infty} \frac{x^n}{e^x}$

Solution: The function takes the form ∞/∞ which is indeterminate, but allows the application of L'Hospital's rule. Accordingly,

$$\lim_{x \to \infty} \frac{x^n}{e^x} = \lim_{x \to \infty} \frac{nx^{n-1}}{e^x}.$$

It is seen that ∞/∞ continues to be obtained after successive use of the rule. We note, however, that the nth derivative tends to n! for the numerator. Therefore, since n is assumed finite,

$$\lim_{x \to \infty} \frac{n!}{e^x} = 0.$$

• PROBLEM 290

Find $\lim_{x \to \pi/2} \frac{\sec^2 x}{\sec^2 3x}$.

Solution: Substituting directly in the numerator and denominator, we obtain

$$\lim_{x\to\pi/2} \sec^2 x = +\infty \text{ and } \lim_{x\to\pi/2} \sec^2 3x = +\infty.$$

Since the ratio takes the form ∞/∞, we apply L'Hospital's rule.

$$\lim_{x\to\pi/2} \frac{\sec^2 x}{\sec^2 3x} = \lim_{x\to\pi/2} \frac{2\sec^2 x \tan x}{6\sec^2 3x \tan 3x}$$

Upon direct substitution in the numerator and denominator, it is found that the ratio again takes the form ∞/∞. Furthermore, continued succesive application of L'Hospital's rule results in this same indeterminate form. It is therefore necessary to seek another approach, and this consists of rewriting the ratio in terms of cosines.

$$\lim_{x\to\pi/2} \frac{\sec^2 x}{\sec^2 3x} = \lim_{x\to\pi/2} \frac{\cos^2 3x}{\cos^2 x}$$

Now,

$$\lim_{x\to\pi/2} \cos^2 3x = 0 \text{ and } \lim_{x\to\pi/2} \cos^2 x = 0.$$

The ratio still takes an indeterminate form, but of a different one, 0/0. We therefore try the application of L'Hospital's rule.

$$\lim_{x\to\pi/2} \frac{\cos^2 3x}{\cos^2 x} = \lim_{x\to\pi/2} \frac{-6\cos 3x \sin 3x}{-2\cos x \sin x}$$

$$= \lim_{x\to\pi/2} \frac{3\sin 6x}{\sin 2x}$$

Direct substitution of $\pi/2$ into numerator and denominator again results in 0/0. However, a reapplication of L'Hospital's rule now results in a definite answer. Thus,

$$\lim_{x\to\pi/2} \frac{3\sin 6x}{\sin 2x} = \lim_{x\to\pi/2} \frac{18\cos 6x}{2\cos 2x} = 9$$

A simpler approach is in recognizing that, as

$$\Theta \to \frac{\pi}{2},\ \cos\Theta \to \Theta,\ \sec\Theta \to \frac{1}{\Theta}.$$

Therefore the given expression can be written:

$$\lim_{x\to\pi/2} \frac{\frac{1}{\Theta^2}}{\frac{1}{(3\Theta)^2}} = \lim_{x\to\pi/2} \frac{9\Theta^2}{\Theta^2} = 9.$$

• PROBLEM 291

Find $\lim_{x \to 0} (x \ln x)$.

Solution: Since $\ln x \to -\infty$ when $x \to 0$, the product $x \ln x$ takes the indeterminate form $0 \cdot \infty$ when $x \to 0$. We must, therefore, write the expression in the form of a ratio, so that we can apply L'Hospital's rule. If we write

$$x \ln x = \frac{\ln x}{\frac{1}{x}},$$

this fraction takes the indeterminate form ∞/∞ when $x \to 0$. Applying L'Hospital's rule, therefore,

$$\lim_{x \to 0} (x \ln x) = \lim_{x \to 0} \left[\frac{\ln x}{\frac{1}{x}} \right] = \lim_{x \to 0} \left[-\frac{1}{x} \Big/ \frac{1}{x^2} \right]$$

$$= \lim_{x \to 0} (-x) = 0.$$

• PROBLEM 292

Evaluate $\lim_{x \to 0^+} x \ln \sin x$.

Solution: The approach to be taken here is to write the function as a ratio, to which L'Hospital's rule may be applied. Therefore,

$$x \ln \sin x = \frac{\ln \sin x}{1/x}.$$

This is in a form, to which L'Hospital's rule applies. Hence,

$$\lim_{x \to 0^+} \frac{\ln \sin x}{1/x} = \lim_{x \to 0^+} \frac{\cos x / \sin x}{-1/x^2}$$

$$= \lim_{x \to 0^+} (-x) \frac{x}{\sin x} \cos x = \frac{0}{0}.$$

Since the result is still indeterminate, we apply L'Hospital's rule once more:

$$\lim_{x \to 0^+} x \ln \sin x = \lim_{x \to 0} \frac{2x \cos x - x^2 \sin x}{\cos x}$$

$$= \frac{0}{1} = 0.$$

• PROBLEM 293

Find $\lim_{x \to 0} \sin^{-1} x \csc x$.

Solution: The function takes the form $0 \cdot \infty$. In order to apply L'Hospital's rule, it is necessary to write the function first as a ratio. This may be done by noting that $\csc x = 1/\sin x$, so that $\sin^{-1} x \csc x = \sin^{-1} x/\sin x$. Now

$$\lim_{x \to 0} \frac{\sin^{-1} x}{\sin x}$$

takes the form 0/0, and therefore we may apply L'Hospital's rule as follows:

$$\lim_{x \to 0} \frac{\sin^{-1} x}{\sin x} = \lim_{x \to 0} \frac{\frac{1}{\sqrt{1 - x^2}}}{\cos x} = \frac{1}{1} = 1$$

A simpler approach is to recognize that, for small angles,

$$\sin x = x.$$

Therefore,

$$\lim_{x \to 0} \frac{\sin^{-1} x}{\sin x} = \frac{x}{x} = 1.$$

• PROBLEM 294

Prove $\lim_{x \to \frac{1}{2}\Pi} (\sec 3x \cos 5x) = \frac{5}{3}$.

Solution: Since $\sec \frac{3}{2}\Pi = \infty$ and $\cos \frac{5}{2}\Pi = 0$,

the function takes the indeterminate form $\infty \cdot 0$. We therefore proceed to rewrite the function into a fraction, to which L'Hospital's rule might be applied.

$$\sec 3x \cos 5x = \frac{1}{\cos 3x} \cdot \cos 5x = \frac{\cos 5x}{\cos 3x}.$$

Let $f(x) = \cos 5x$, and $F(x) = \cos 3x$. Then

$$f\left(\frac{1}{2}\Pi\right) = 0, \quad F\left(\frac{1}{2}\Pi\right) = 0.$$

The function now takes the indeterminate form 0/0, and we apply the rule as follows:

$$\lim_{x \to \frac{1}{2}\Pi} \frac{f(x)}{F(x)} = \lim_{x \to \frac{1}{2}\Pi} \frac{f'(x)}{F'(x)} = \lim_{x \to \frac{1}{2}\Pi} \frac{-5 \sin 5x}{-3 \sin 3x} =$$

$$= \frac{5}{3}$$

A simpler approach is in recognizing that, as

$$\Theta \to \frac{\Pi}{2}, \ \cos\Theta \to \Theta, \ \sec\Theta \to \frac{1}{\Theta}.$$

Therefore, the given expression can be written:

$$\lim_{x\to\frac{\Pi}{2}} \left[\frac{1}{3x} \cdot 5x\right] = \frac{5}{3}.$$

• PROBLEM 295

Find $\lim_{x\to 0}$ (csc x - cot x).

Solution: We first write the expression in a form involving sines and cosines, preferably to obtain a ratio to which L'Hospital's rule may be applied.

$$\lim_{x\to 0} (\csc x - \cot x) = \lim_{x\to 0} \left[\frac{1}{\sin x} - \frac{\cos x}{\sin x}\right]$$

$$= \lim_{x\to 0} \frac{1 - \cos x}{\sin x}$$

The expression is now in the form 0/0, so that L'Hospital's rule gives:

$$\lim_{x\to 0} \frac{\sin x}{\cos x} = 0.$$

• PROBLEM 296

Prove $\lim_{x\to\frac{1}{2}\Pi}$ (sec x - tan x) = 0.

Solution: We have $\sec \frac{1}{2}\Pi - \tan \frac{1}{2}\Pi = \infty - \infty$.

Since the function takes an indeterminate form, we attempt to write it as a ratio, to which L'Hospital's rule might be applied. Thus,

$$\sec x - \tan x = \frac{1}{\cos x} - \frac{\sin x}{\cos x} = \frac{1 - \sin x}{\cos x}.$$

Let f(x) = 1 - sin x, F(x) = cos x. Then

$$f\left(\frac{1}{2}\Pi\right) = 0, \ F\left(\frac{1}{2}\Pi\right) = 0.$$

The function is now in the form of 0/0, and therefore we apply the rule

$$\lim_{x \to \frac{1}{2}\pi} \frac{f(x)}{F(x)} = \lim_{x \to \frac{1}{2}\pi} \frac{f'(x)}{F'(x)} = \lim_{x \to \frac{1}{2}\pi} \frac{-\cos x}{-\sin x} = 0.$$

• PROBLEM 297

Find $\lim_{x \to 0} \left(\frac{1}{\sin x} - \frac{1}{x} \right)$.

Solution: Since

$$\frac{1}{\sin x} \to \infty \text{ and } \frac{1}{x} \to \infty \text{ when } x \to 0,$$

the given function takes the indeterminate form $\infty - \infty$. We should, consequently, write the function in the form of a ratio which takes an indeterminate form, so that L'Hospital's rule can be applied. Therefore,

$$\frac{1}{\sin x} - \frac{1}{x} = \frac{x - \sin x}{x \sin x},$$

which becomes 0/0 when $x \to 0$. Applying L'Hospital's rule twice, we obtain

$$\lim_{x \to 0} \left(\frac{1}{\sin x} - \frac{1}{x} \right) = \lim_{x \to 0} \frac{x - \sin x}{x \sin x}$$

$$= \lim_{x \to 0} \frac{1 - \cos x}{x \cos x + \sin x}$$

$$= \lim_{x \to 0} \frac{\sin x}{-x \sin x + \cos x + \cos x} = 0.$$

A simpler approach is to recognize that, for small angles, $\sin x = x$. The original expression therefore becomes

$$\lim_{x \to 0} \left(\frac{1}{x} - \frac{1}{x} \right) = 0.$$

• PROBLEM 298

Find $\lim_{x \to 0} \left(\frac{1}{x^2} - \frac{1}{x^2 \sec x} \right)$.

Solution: The function takes the form $\infty - \infty$, which is indeterminate. We therefore try the approach of rewriting the function into a ratio, to which L'Hospital's rule might be applied, if necessary.

$$\lim_{x\to 0}\left(\frac{1}{x^2}-\frac{1}{x^2 \sec x}\right) = \lim_{x\to 0}\frac{\sec x - 1}{x^2 \sec x}.$$

$\lim_{x\to 0}(\sec x - 1) = 0$, and $\lim_{x\to 0}(x^2 \sec x) = 0$. We can, therefore, apply L'Hospital's rule.

$$\lim_{x\to 0}\frac{\text{sex } x - 1}{x^2 \sec x} = \lim_{x\to 0}\frac{\sec x \tan x}{2x \sec x + x^2 \sec x \tan x}$$

$$= \lim_{x\to 0}\frac{\tan x}{2x + x^2 \tan x}$$

The ratio still takes the form of 0/0, and therefore we apply L'Hospital's rule a second time.

$$\lim_{x\to 0}\frac{\tan x}{2x + x^2 \tan x} = \lim_{x\to 0}\frac{\sec^2 x}{2 + 2x \tan x + x^2 \sec^2 x} =$$

$$= \frac{1}{2}$$

A simpler approach is to write the given expression as

$$\lim_{x\to 0}\frac{1}{x^2}\left(1 - \frac{1}{\sec x}\right)$$

$$= \lim_{x\to 0}\frac{1}{x^2}(1 - \cos x).$$

Recognizing that, for small angles,

$$\cos x = 1 - \frac{x^2}{2},$$

we can substitute, obtaining:

$$\lim_{x\to 0}\frac{1}{x^2}\left(1 - \left[1 - \frac{x^2}{2}\right]\right) = \frac{1}{2}.$$

• PROBLEM 299

Find $\lim_{x\to 1}\left(\frac{1}{\ln x} - \frac{1}{x - 1}\right)$.

Solution: The function takes the indeterminate form $\infty - \infty$, and we therefore proceed to rewrite the function as a fraction, to which L'Hospital's rule is applicable. The first step is to express the function with a common denominator.

$$\frac{1}{\ln x} - \frac{1}{x - 1} = \frac{x - 1 - \ln x}{(\ln x)(x - 1)} .$$

The function is now in the form 0/0, and therefore we can apply the rule.

$$\lim_{x \to 1} \frac{x - 1 - \ln x}{(\ln x)(x - 1)} = \lim_{x \to 1} \frac{1 - (1/x)}{(x - 1)/x + \ln x}$$

The function still takes the form 0/0, and therefore we apply the rule a second time.

$$= \lim_{x \to 1} \frac{1/x^2}{(1/x^2) + (1/x)} = \lim_{x \to 1} \frac{1}{1 + 1} = \frac{1}{2} .$$

• PROBLEM 300

Find: $\lim_{x \to 0} \left(\frac{1}{e^x - 1} - \frac{1}{x} \right)$.

Solution: This function has the indeterminate form$(\infty - \infty)$ for x = 0. We write it in the form:

$$(x - e^x + 1)/x(e^x - 1),$$

which results from combining fractions for the purpose of obtaining a ratio, to which L'Hospital's rule might be applied.

$$\lim_{x \to 0} \frac{x - e^x + 1}{x(e^x - 1)} = \lim_{x \to 0} \frac{1 - e^x}{xe^x + e^x - 1} = \frac{0}{0} ,$$

after applying the rule. Since an indeterminate answer is still obtained, we apply the rule a second time.

$$\lim_{x \to 0} \frac{1 - e^x}{xe^x + e^x - 1} = \lim_{x \to 0} \frac{- e^x}{xe^x + 2e^x} =$$

$$\lim_{x \to 0} \frac{- 1}{0 + 2} = - \frac{1}{2} .$$

• PROBLEM 301

Evaluate: $\lim_{x \to 0} \left[\frac{e^{2x}}{x} - \frac{x + 2}{e^{2x} - 1} \right]$.

Solution: We first attempt to evaluate this problem by simply substituting 0 for x. Doing this, we obtain $\infty - \infty$. We must now rearrange the problem so as to transform this indeterminate form into one of the types $\frac{0}{0}$ or $\frac{\infty}{\infty}$, to which we may apply l'Hospital's

Rule. We can rewrite the given limit by finding a common denominator. We obtain:

$$\lim_{x\to 0}\left[\frac{e^{2x}}{x} - \frac{x+2}{e^{2x}-1}\right] = \lim_{x\to 0}\left[\frac{e^{4x} - e^{2x} - x^2 - 2x}{xe^{2x} - x}\right].$$

Substituting 0 we now obtain $\frac{0}{0}$. Therefore we can apply l'Hospital's Rule, obtaining:

$$\lim_{x\to 0}\left[\frac{4e^{4x} - 2e^{2x} - 2x - 2}{2xe^{2x} + e^{2x} - 1}\right].$$

Again, substitution of 0 gives $\frac{0}{0}$, so we apply the rule again, obtaining:

$$\lim_{x\to 0}\left[\frac{16e^{4x} - 4e^{2x} - 2}{4xe^{2x} + 2e^{2x} + 2e^{2x}}\right] = \lim_{x\to 0}\frac{10}{4} = \frac{5}{2}.$$

Therefore,

$$\lim_{x\to 0}\left(\frac{e^{2x}}{x} - \frac{x+2}{e^{2x}-1}\right) = \frac{5}{2}.$$

• PROBLEM 302

Find: $\lim_{x\to\infty} y = (1+x)^{\frac{1}{x}}$

Solution: The function takes the form ∞^0, which is indeterminate. We proceed to manipulate the function to arrive at a determinate form or one to which L'Hospital's rule may be applied. Since a power in $1/x$ is present, we attempt to use logarithms. Thus,

$$\ln y = \lim_{x\to\infty}\frac{\ln(1+x)}{x}$$

The function now takes the form ∞/∞, so that we can apply L'Hospital's rule.

$$\ln y = \lim_{x\to\infty}\frac{\frac{1}{1+x}}{1} = 0$$

$$y = e^0 = 1.$$

• PROBLEM 303

Find: $\lim_{x\to 1} x^{1/(1-x)}$.

Solution: This function takes the indeterminate form 1^∞. We should, therefore, write the function as a ratio, to which L'Hospital's rule may be applied. This may be done by taking the logarithms of both sides of the equation.

$$y = x^{1/(1-x)}.$$

Then,

$$\ln y = \frac{\ln x}{1 - x}.$$

$$\lim_{x \to 1} (\ln y) = \lim_{x \to 1} \frac{\ln x}{1 - x} = \lim_{x \to 1} \frac{1/x}{-1} = -1.$$

Thus $\lim_{x \to 1} (\ln y) = -1$,

and $\lim_{x \to 1} x^{1/(1-x)} = \frac{1}{e}$.

• PROBLEM 304

Evaluate $\lim_{x \to 0+} (\cos 2x)^{1/x}$.

Solution: The function is of the indeterminate form 1^∞. In view of the exponent involved, we take logs of both sides of the equation:

$$y = (\cos 2x)^{1/x}. \qquad \ln y = (1/x) \ln \cos 2x.$$

$$\lim_{x \to 0^+} \ln y = \lim_{x \to 0^+} \frac{\ln \cos 2x}{x}$$

The expression is now in the form of 0/0, and therefore we apply L'Hospital's rule.

$$\frac{d}{dx} (\ln \cos 2x) = \frac{-2 \sin 2x}{\cos 2x}$$

in the numerator,

$$\frac{dx}{dx} = 1$$

in the denominator. We obtain:

$$\lim_{x \to 0^+} \frac{-2 \sin 2x}{\cos 2x} = 0.$$

Since $\ln y \to 0$ as $x \to 0^+$, then $y \to 1$, so that

$$\lim_{x \to 0^+} (\cos 2x)^{1/x} = 1.$$

• PROBLEM 305

Prove: $\lim_{x \to 0} x^x = 1$.

Solution: The function assumes the indeterminate form 0^0, and we therefore use the approach of rewriting the function as a fraction, to which L'Hospital's rule may be applied. Since a power of x is involved, we make use of logarithms.

Let $$y = x^x.$$

Then, $$\ln y = x \ln x.$$

The function is now in the form of $0\cdot\infty$. To write the function in the form of ∞/∞, so that the rule may be applied, We write:

$$\ln y = \frac{\ln x}{\frac{1}{x}} = \frac{-\infty}{\infty}.$$

Applying the rule,

$$\lim_{x\to 0} \frac{\ln x}{\frac{1}{x}} = \lim_{x\to 0} \frac{\frac{1}{x}}{-\frac{1}{x^2}} = \lim_{x\to 0}\left(-x\right) = 0.$$

Now, since $\lim_{x\to 0}\left(\ln y\right) = 0$, we have:

$$\lim_{x\to 0} y = e^0 = 1.$$

• **PROBLEM 306**

Find $\lim_{x\to 0} (x + 1)^{\cot x}$.

Solution: This takes the form 1^∞. To write the expression in a form, to which L'Hospital's rule may be applied, we take logarithms of both sides of the equation. Let

$$y = (x + 1)^{\cot x};$$

then

$$\ln y = \cot x \cdot \ln (x + 1) = \frac{\ln (x + 1)}{\tan x}.$$

By L'Hospital's rule,

$$\lim_{x\to 0} (\ln y) = \lim_{x\to 0} \frac{\left(\frac{1}{x + 1}\right)}{\sec^2 x} = \lim_{x\to 0} \frac{\cos^2 x}{x + 1} = 1.$$

Since $\ln y \to 1$ as $x \to 0$, we have $y \to e$. The required limit is therefore e.

• **PROBLEM 307**

Find $\lim_{x\to\frac{\pi}{2}} (\tan x)^{\cos x}$.

Solution: This takes the indeterminate form ∞^0. We should now write the function in the form of an indeterminate ratio to which L'Hospital's rule may be applied. This may be achieved by taking logs of both sides of the equation:

$$y = (\tan x)^{\cos x}.$$

Then

$$\ln y = \cos x \cdot \ln \tan x = \frac{\ln \tan x}{\sec x}.$$

Differentiating,

$$\lim_{x\to\frac{\pi}{2}} (\ln y) = \lim_{x\to\frac{\pi}{2}} \frac{\frac{1}{\tan x} \cdot \sec^2 x}{\sec x \tan x}$$

$$= \lim_{x\to\frac{\pi}{2}} \frac{\sec x}{\tan^2 x} = \lim_{x\to\frac{\pi}{2}} \frac{\cos x}{\sin^2 x} = 0,$$

Therefore,

$$\lim_{x\to\frac{\pi}{2}} (y) = e^0 = 1.$$

• PROBLEM 308

Prove $\lim_{x\to 0} (\text{ctn } x)^{\sin x} = 1.$

Solution: The function assumes the indeterminate form ∞^0 when $x = 0$. We therefore proceed to write the function as a fraction, to which L'Hospital's rule might be applied. Since a power in x is present, we try the use of logs for this purpose.
Let

$$y = (\text{ctn } x)^{\sin x}.$$

Then

$$\ln y = \sin x \ln \text{ctn } x = 0 \cdot \infty,$$

But

$$\ln y = \frac{\ln \text{ctn } x}{\csc x} = \frac{\infty}{\infty},$$

We can now apply the rule.

$$\lim_{x\to 0} \frac{\ln \text{ctn } x}{\csc x} = \lim_{x\to 0} \frac{\frac{-\csc^2 x}{\text{ctn } x}}{-\csc x \text{ ctn } x}$$

$$= \lim_{x\to 0} \frac{\sin x}{\cos^2 x} = 0.$$

Since $\lim_{x \to 0} \ln y = 0,$

$$\lim_{x \to 0} y = e^0 = 1.$$

• PROBLEM 309

Prove: $\lim_{x \to 1} (2 - x)^{\tan \frac{\pi x}{2}} = e^{\frac{2}{\pi}}.$

Solution: The function takes the indeterminate form 1^∞. We therefore proceed to write the function as a ratio, to which L'Hospital's rule may be applied. Since a power of x is present, we try the use of logarithms.

Let $$y = (2 - x)^{\tan \frac{\pi x}{2}}.$$

Then,
$$\ln y = \tan \frac{\pi x}{2} \left[\ln (2 - x) \right] = \infty \cdot 0,$$

But
$$\ln v = \frac{\ln (2 - x)}{\operatorname{ctn} \frac{\pi x}{2}} = \frac{0}{0}.$$

The function is now of the indeterminate form 0/0, and we therefore apply the rule.

$$\lim_{x \to 1} \frac{\ln (2 - x)}{\operatorname{ctn} \frac{\pi x}{2}} = \lim_{x \to 1} \frac{-\frac{1}{2 - x}}{-\frac{1}{2} \pi \csc^2 \frac{\pi x}{2}} = \frac{2}{\pi}.$$

Since $\lim_{x \to 1} \ln y = \frac{2}{\pi},$

$$\lim_{x \to 1} y = e^{\frac{2}{\pi}}.$$

• PROBLEM 310

Evaluate: $\lim_{x \to 0} \left(\frac{a^x + b^x}{2} \right)^{\frac{1}{x}}.$

Solution: We first attempt to solve this problem by sutstituting 0 for x. This gives the indeterminate form 1^∞. We now let

$$y = \left(\frac{a^x + b^x}{2} \right)^{\frac{1}{x}}$$

and take lim (ln y). We find:

$$\lim_{x\to 0} (\ln y) = \lim_{x\to 0}\left[\frac{1}{x}\ln\left(\frac{a^x + b^x}{2}\right)\right],$$

and substituting 0 for x gives $\infty \cdot 0$. We try to rearrange the expression so that we obtain an indeterminate form of the type $\frac{0}{0}$ or $\frac{\infty}{\infty}$, to which we can apply l'Hospital's Rule. We write:

$$\lim_{x\to 0}\left[\frac{1}{x}\ln\left(\frac{a^x + b^x}{2}\right)\right] = \lim_{x\to 0}\left[\frac{\ln\left(\frac{a^x + b^x}{2}\right)}{x}\right].$$

Substitution of 0 now gives $\frac{0}{0}$, and, applying l'Hospital's Rule, we obtain:

$$\lim_{x\to 0}\frac{2}{a^x + b^x}\cdot\frac{1}{2}\left(a^x \ln a + b^x \ln b\right) =$$

$$\lim_{x\to 0}\frac{a^x \ln a + b^x \ln b}{a^x + b^x} = \frac{\ln a + \ln b}{2}.$$

Now, since $\lim_{x\to 0} \ln y = \frac{\ln a + \ln b}{2}$,

$$\lim_{x\to 0} y = \lim_{x\to 0}\left(\frac{a^x + b^x}{2}\right)^{\frac{1}{x}} = e^{\frac{\ln a + \ln b}{2}}.$$

Now, let $e^{\frac{\ln a + \ln b}{2}} = Z$. Then, $\frac{\ln a + \ln b}{2} = \ln Z$, and $\frac{\ln a + \ln b}{2}$ can be rewritten as: $\frac{1}{2}\ln ab$. Therefore, we have:

$$\ln Z = \frac{1}{2}\ln ab,$$

or, taking ln of both sides of the equation,

$$Z = (ab)^{\frac{1}{2}} = \sqrt{ab}.$$

Therefore,

$$\lim_{x\to 0}\left(\frac{a^x + b^x}{2}\right)^{\frac{1}{x}} = \sqrt{ab}.$$

• PROBLEM 311

Find $\lim_{x\to 0} x \sin\frac{1}{x}$.

Solution: $\lim_{x \to 0} \sin(1/x)$ does not exist.

The method of rewriting the expression into a form to which L'Hospital's rule may be applied, for example, is here not apparent. The next approach, then, is to write the facts that we know about the expression. Thus, we know

$$|\sin(1/x)| \leqq 1,$$

multiplying both sides by **x**,

$$\left|x \sin \frac{1}{x}\right| \leqq |x|,$$

and since $x \to 0$, $|x \sin(1/x)| \to 0$. Therefore,

$$\lim_{x \to 0} x \sin \frac{1}{x} = 0.$$

CHAPTER 14

TANGENTS AND NORMALS

By definition, a line that is tangent to a curve at a point, must have the same slope as the curve. I.e., if we have y as a function of x, then the slope of the curve at any point is $\frac{dy}{dx}$.

If we now consider a specific point, having the coordinates: (x_1, y_1), then the slope at that point is $m_t = \frac{dy}{dx} \big| x = x_1$, i.e., we first find the equation for the derivative, and then substitute the value: $x = x_1$ in that equation. Since the tangent line has the same slope, the value found for the derivative at the point is also the slope of the tangent line.

On the other hand, a line normal to a curve at a point has to have a slope that is perpendicular to the slope of the tangent line. Since $\tan(\theta + 90°) = -\cot\theta = -\frac{1}{\tan\theta}$, we can write that the slope of a line normal to the curve is,

$$m_n = -\frac{1}{\frac{dy}{dx}\big| x = x_1}$$

Since we know the coordinates of the point (x_1, y_1) as well as the slopes, equations can be written for the tangent and normal lines.

• PROBLEM 312

Find the equations of the tangent line and the normal to the curve: $y = x^2 - x + 3$, at the point (2,5).

Solution: Since the equation of a straight line passing through a given point can be expressed in the form: $y - y_1 = m(x - x_1)$, this is appropriate for finding the equations of the tangent and normal. Here $x_1 = 2$ and $y_1 = 5$. The slope, m, of the tangent line is found by taking the derivative, dy/dx, of the curve: $y = x^2 - x + 3$.

$$dy/dx = 2x - 1.$$

At (2,5), $dy/dx = 2(2) - 1 = 3$, therefore the slope, m, of the tangent line is 3. Substituting x_1, y_1 and m into the equation $y - y_1 = m(x - x_1)$ we obtain:

$$y - 5 = 3(x - 2),$$

as the equation of the tangent line, or

$$3x - y - 1 = 0.$$

Since the slope of the normal is given by: $m' = -1/m$, and since $m = 3$, the slope of the normal is $-1/3$. Substituting $x_1 = 2$, $y_1 = 5$ and the slope of the normal, $m' = -1/3$, into the equation: $y - y_1 = m'(x - x_1)$, we obtain:

$$y - 5 = -\frac{1}{3}(x - 2),$$

or,

$$x + 3y - 17 = 0.$$

This is the equation of the normal.

• PROBLEM 313

Find the equation of the line that is tangent to the curve: $2x^2 + y^2 + 2xy = 5$, at the point $(1,1)$.

Solution: The equation of a straight line passing through a given point $(1,1)$ can be expressed in the form $y - y_1 = m(x - x_1)$. Here $x_1 = 1$ and $y_1 = 1$. We must find m.

The slope, m, of the line tangent to the curve: $2x^2 + y^2 + 2xy = 5$ is found by taking the derivative of the curve, using implicit differentiation.

$$2x^2 + y^2 + 2xy = 5$$

$$4x + 2yy' + 2xy' + 2y = 0$$

$$y'(2y + 2x) = -4x - 2y$$

$$y' = \frac{-4x - 2y}{2x + 2y} = -\frac{2x + y}{x + y}$$

At the point $(1,1)$:

$$m = y' = -\frac{2 + 1}{1 + 1} = -\frac{3}{2}$$

Substituting for x_1, y_1 and m into the equation: $y - y_1 = m(x - x_1)$, we obtain the equation of the tangent line:

$$y - 1 = -\frac{3}{2}(x - 1).$$

• PROBLEM 314

Find the equation of the line that is tangent to the parabola $y^2 = 2px$ at the point (x_1, y_1).

Solution: The equation of a straight line passing through a given point (x_1, y_1) can be expressed in the form $y - y_1 = m(x - x_1)$. To find the slope, m, of the line tangent to the curve: $y^2 = 2px$, we differentiate implicitly and obtain:

$$2yy' = 2p \quad \text{or} \quad y' = p/y;$$

now, $y' = m$, and therefore $m = p/y$. At (x_1, y_1) $m = p/y_1$.

Substituting m into the equation: $y - y_1 = m(x - x_1)$, we obtain:

$$y - y_1 = \frac{p}{y_1}(x - x_1),$$

or,

$$y_1 y - y_1^2 = px - px_1.$$

Using the given equation, $y^2 = 2px$, we find that at (x_1, y_1) $y_1^2 = 2px_1$. Substituting $y_1^2 = 2px_1$ into the equation: $y_1 y - y_1^2 = px - px_1$, we obtain:

$$y_1 y - 2px_1 = px - px_1.$$

$$y_1 y = px + px_1.$$

$$y_1 y = p(x + x_1).$$

This is the required equation of the tangent line.

• PROBLEM 315

Find the equation of the tangent line to the ellipse: $4x^2 + 9y^2 = 40$, at the point $(1,2)$.

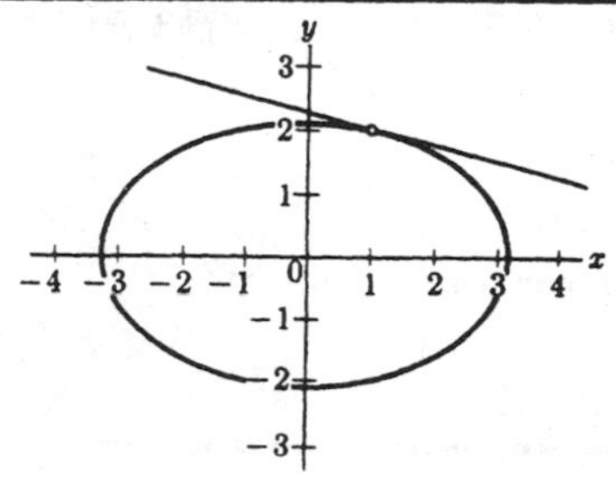

Solution: Since y is not given explicitly, the slope of the tangent to the ellipse at any point is best found by treating it as an implicit function. Differentiating, we have:

$$8x + 18yy' = 0,$$

from which

$$y' = -4x/9y.$$

Evaluating this derivative at the point $(1,2)$, we have

$$y' = -4/18 = -2/9.$$

Thus the slope of the desired tangent line is $-2/9$.

The equation of a straight line at a given point can be expressed in the form $y - y_1 = m(x - x_1)$. Here $x_1 = 1$ and $y_1 = 2$, and the slope $m = -2/9$. Substituting, we obtain:

$$y - 2 = -2/9(x - 1).$$

$$9y - 18 = -2x + 2.$$

$$2x + 9y - 20 = 0, \text{ which}$$

is the equation of the tangent line.

The slope could also have been found by solving the equation of the curve for y, and then differentiating.

• PROBLEM 316

Find the equation of the line that is tangent to the curve

$$e^x + \cos y - 2 = 0 \quad \text{at} \quad (0, \pi/3).$$

Solution: The equation of a straight line passing through a given point can be expressed in the form $y - y_1 = m(x - x_1)$. Here $x_1 = 0$ and $y_1 = \pi/3$. The slope, m, of the line tangent to the curve: $y = e^x + \cos y - 2 = 0$, is found by taking the derivative of the curve, using implicit differentiation.

$$e^x + \cos y - 2 = 0.$$

$$e^x - \sin y y' = 0.$$

$$e^x = \sin y y'.$$

$$y' = \left[\frac{e^x}{\sin y}\right]_{0,\pi/3} = \frac{1}{\frac{1}{2}\sqrt{3}} = \frac{2}{\sqrt{3}}$$

. Since $m = y' = 2/\sqrt{3}$, we substitute this into the equation: $y - y_1 = m(x - x_1)$, and obtain:

$$y - \frac{\pi}{3} = \frac{2x}{\sqrt{3}}.$$

This is the equation of the tangent line.

• PROBLEM 317

Find the equations of the tangent line and the normal line to the curve: $y = 3x^2$, at the point where $x = 3$.

Solution: The equation of a straight line passing through a given point (x_1, y_1) can be expressed in the form: $y - y_1 = m(x - x_1)$. Here $x_1 = 3$. Since $y = 3x^2$, $y_1 = 3(3^2) = 27$, therefore the given point is $(3, 27)$. The slope, m, of the line tangent to the curve: $y = 3x^2$, is found by taking $\frac{dy}{dx}$.

$$\frac{dy}{dx} = 6x.$$

At $(3, 27)$, $6x = 18$. Substituting $x_1 = 3$, $y_1 = 27$, and $m = 18$ into the equation: $y - y_1 = m(x - x_1)$, we obtain: $y - 27 = 18(x - 3)$. $18x - y - 27 = 0$. This is the equation of the tangent line.
The equation of the normal can also be expressed using $y - y_1 = m_1(x - x_1)$ But in this case the slope, m_1, is the negative reciprocal of the slope of the tangent line, or $m_1 = -1/m = -1/18$. Substituting $x_1 = 3$, $y_1 = 27$, and $m_1 = -1/18$, we find: $y - 27 = -1/18(x - 3)$, or, $x + 18y - 489 = 0$. This is the equation of the normal.

• PROBLEM 318

Find the equations of the tangent and the normal to the curve: $y = x^2 + 4x + 2$, at a point where the tangent is also perpendicular to the line $2x - 4y + 5 = 0$.

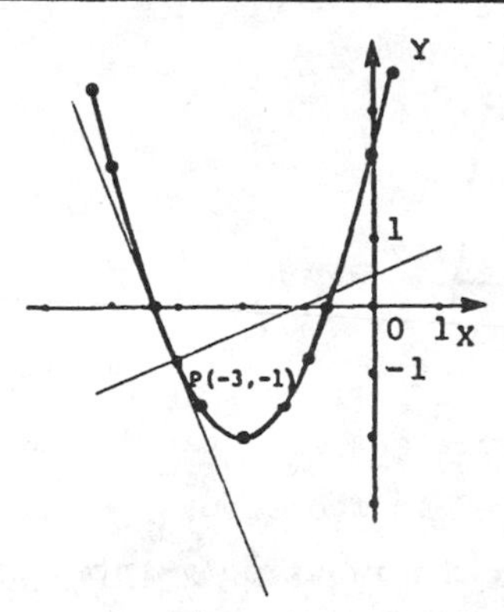

Solution: An appropriate equation of a straight line is $y - y_1 = m(x - x_1)$. For this equation we need a point (x_1, y_1) and the slope, m. Since the tangent is perpendicular to the line $2x - 4y + 5 = 0$ or $y = \frac{1}{2}x + 5/4$, the slope of the tangent is the negative reciprocal of the slope of the line perpendicular to the tangent. Now, the slope of the line perpendicular to the tangent is $\frac{1}{2}$, hence the slope, m, of the tangent is -2. To find the point where the tangent to the curve has a slope of -2, we find dy/dx of the curve and set it equal to -2. For the curve: $y = x^2 + 4x + 2$, $dy/dx = 2x + 4$, and $2x + 4 = -2$. Therefore, $x = -3$. Substituting $x = -3$ back into the equation: $y = x^2 + 4x + 2$, we find $y = -1$. Therefore the point (x_1, y_1) is the point $(-3,1)$. Substituting $x = -3$, $y = -1$, and $m = -2$ into the expression: $y - y_1 = m(x - x_1)$, we find that:

$$y + 2x + 7 = 0, \text{ which}$$

is the equation of the tangent.

Using the relation: $y - y_1 = m_1(x - x_1)$, for the equation of the normal, $x_1 = -3$, $y_1 = -1$ and the slope, m_1, of the normal is the negative reciprocal of the slope of the tangent, or $m_1 = 1/2$. Substituting these values, we obtain:

$$x - 2y + 1 = 0,$$

the equation of the normal.

• PROBLEM 319

Find the y intercept of the line normal to the curve: $y = 3x - x^3$, at $(3,-18)$.

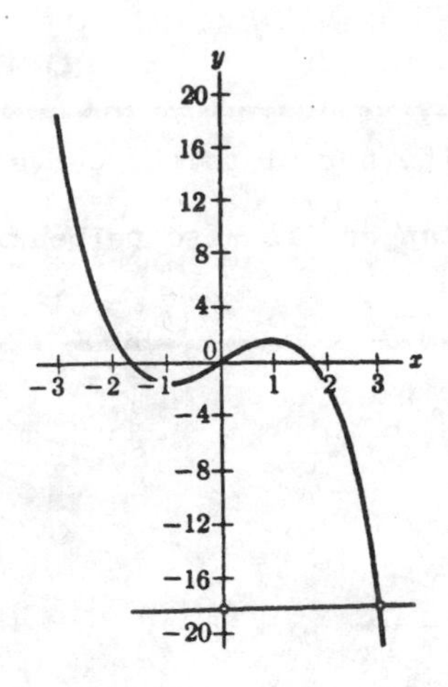

Solution: The equation of a straight line passing through a given point (3,-18) can be expressed in the form: $y - y_1 = m(x - x_1)$. From this equation we can find the required y-intercept by setting $x = 0$ and solving for y.

Here $x_1 = 3$ and $y_1 = -18$. We need to find the slope of the normal in order to write the equation. The slope of the tangent to a curve is given by dy/dx. Since the slope of the normal is the negative reciprocal of the slope of the tangent, we can write:

$$\frac{dy}{dx} = 3 - 3x^2$$

$$\frac{dy}{dx} \text{ at } (3,-18) \text{ is } -24$$

therefore the slope of the normal is $-\left(\frac{1}{-24}\right)$ or $\frac{1}{24}$. We can now write the equation of the normal by substituting $x_1 = 3$, $y_1 = -18$, and $m = \frac{1}{24}$ into $y - y_1 = m(x - x_1)$ to obtain:

$$y + 18 = \frac{1}{24}(x - 3),$$

or

$$24y - x + 435 = 0.$$

To find the y-intercept we set $x = 0$ in the equation: $24y - x + 435 = 0$, and solve for y. In this manner we find the y-intercept to be

$$\frac{-145}{8}.$$

• PROBLEM 320

Using the Δ-method, find the points on the curve: $y = \frac{x}{3} + \frac{3}{x}$, at which the tangent line is horizontal.

Solution: When the slope of a curve equals zero, the curve has a horizontal tangent. We can find the points at which the tangent line is horizontal by calculating the slope, $\frac{dy}{dx}$, setting it equal to zero and solving for x. By the Δ-method,

$$\frac{\Delta y}{\Delta x} = \frac{f(x + \Delta x) - f(x)}{\Delta x}$$

$$\frac{\Delta y}{\Delta x} = \frac{\frac{(x + \Delta x)}{3} + \frac{3}{(x + \Delta x)} - \left(\frac{x}{3} + \frac{3}{x}\right)}{\Delta x}$$

$$= \frac{\frac{x}{3} + \frac{\Delta x}{3} + \frac{3}{x + \Delta x} - \frac{x}{3} - \frac{3}{x}}{\Delta x}$$

$$= \frac{\frac{\Delta x}{3} + \frac{3}{x + \Delta x} - \frac{3}{x}}{\Delta x}.$$

$$\frac{\Delta y}{\Delta x} = \frac{\frac{x^2\Delta x + x(\Delta x)^2 - 9\Delta x}{3(x + \Delta x)x}}{\Delta x}$$

$$= \frac{x^2 + x\Delta x - 9}{3x^2 + 3x\Delta x}.$$

$$\frac{dy}{dx} = \lim_{\Delta x \to 0} \frac{\Delta y}{\Delta x} = \lim_{\Delta x \to 0} \frac{x^2 + x\Delta x - 9}{3x^2 + 3x\Delta x} = \frac{x^2 - 9}{3x^2}.$$

We can set this value, which is the slope, equal to zero and solve for x.

$$\frac{x^2 - 9}{3x^2} = 0$$

$$\frac{(x-3)(x+3)}{x(3x)} = 0$$

$$x = 3, \quad x = -3.$$

(Remember that x cannot be zero, for that would give an infinite slope.)

Substituting these values for x back into $y = \frac{x}{3} + \frac{3}{x}$, we can obtain the y coordinates. We find (3,2) and (-3,-2) to be the required points.

• PROBLEM 321

Show that the curve; $y = x^3 + 3x - 4$, has no horizontal tangents by using the Δ-method.

Solution: When the slope of a curve is equal to zero at a specific point, the curve has a horizontal tangent at that point. We can find whether a horizontal tangent exists by finding the slope of the curve, setting it equal to zero and finding any values for x. The slope of a curve is the derivative, $\frac{dy}{dx}$, of the curve. Hence, by the Δ method we have:

$$\frac{\Delta y}{\Delta x} = \frac{f(x + \Delta x) - f(x)}{\Delta x}$$

$$= \frac{(x + \Delta x)^3 + 3(x + \Delta x) - 4 - (x^3 + 3x - 4)}{\Delta x}$$

$$= \frac{x^3 + 3x^2\Delta x + 3x(\Delta x)^2 + (\Delta x)^3 + 3x + 3\Delta x - 4 - x^3 - 3x + 4}{\Delta x}$$

$$= \frac{3x^2\Delta x + 3x(\Delta x)^2 + (\Delta x)^3 + 3\Delta x}{\Delta x}$$

$= 3x^2 + 3x\Delta x + (\Delta x)^2 + 3$. We have: $\frac{dy}{dx} =$

$$\lim_{\Delta x \to 0} \frac{\Delta y}{\Delta x} = \lim_{\Delta x \to 0} 3x^2 + 3x\Delta x + (\Delta x)^2 + 3 = 3x^2 + 3.$$

We can now set this value, which is the slope, equal to zero and solve for possible values of x.

$$3x^2 + 3 = 0$$

$$x^2 = -1$$

$$x = \pm\sqrt{-1}.$$

The value $\sqrt{-1}$ is an imaginary number, and therefore the curve: $y = x^3 + 3x - 4$, has no horizontal tangents.

• PROBLEM 322

For what values of Θ is the tangent to the cosine curve horizontal?

Solution: The equation of the cosine curve is

$y = \cos \Theta$.

The tangent to the cosine curve is given by the derivative. Differentiating, we have:

$$\frac{dy}{d\Theta} = -\sin \Theta.$$

This is the slope of the tangent line to the curve at any point, given by the value of the angle Θ. For the tangent to be horizontal the slope is zero. Therefore

$-\sin \Theta = 0$, or $\sin \Theta = 0$.

When the sine is zero,

$\Theta = 0^\circ$,

or

$$n\pi, \ n = 0, 1, 2, 3, \ldots\ldots$$

• PROBLEM 323

Find the points where the curve described by the following parametric equations has a zero slope:

$$x = 3t/(1 + t^3), \ y = 3t^2/(1 + t^3).$$

Solution: To find the slope of the curve, we first take the derivative of each of the given equations (dx/dt and dy/dt) and then find dy/dx which is the slope. To find dx/dt and dy/dt, we use the quotient rule for differentiation.

$$\frac{dx}{dt} = \frac{3\left[(1 + t^3) - 3t^3\right]}{(1 + t^3)^2} = \frac{3(1 - 2t^3)}{(1 + t^3)^2},$$

$$\frac{dy}{dt} = \frac{3\left[2t(1 + t^3) - 3t^4\right]}{(1 + t^3)^2} = \frac{3(2t - t^4)}{(1 + t^3)^2}.$$

Then $\frac{dy}{dx} = \frac{dy/dt}{dx/dt} = \frac{2t - t^4}{1 - 2t^3}$. The slope is therefore

$$\frac{2t - t^4}{1 - 2t^3}.$$

To find the points at which the slope of the curve is zero, we set $\frac{2t - t^4}{1 - 2t^3}$ equal to zero and solve for t. When $\frac{2t - t^4}{1 - 2t^3} = 0$, $t = 0$ and $t = 2^{1/3}$. To find the points x and y, we substitute $t = 0$ and $t = 2^{1/3}$ into the given equations for the curve.
For $t = 0$, we find $x = 0$, $y = 0$ and for $t = 2^{1/3}$, we find $x = 2^{1/3}$, $y = 4^{1/3}$. So the points where there is a zero slope are (0,0) and $\left(2^{1/3}, 4^{1/3}\right)$.

• PROBLEM 324

Find the following:

a) the length of the tangent

b) the length of the normal

c) the length of the subtangent

d) the length of the subnormal

for the curve: $y^2 = 2x$, at $x = 1$.

Solution: a) The length of the tangent = TP in the diagram, the distance from the point of tangency to the intersection of the tangent line with the x-axis.

To derive a formula for the length of the tangent we can use trigonometric identities:

TP = length of the tangent

$TP = PC/\sin\theta$

$TP = PC\cdot\operatorname{cosec}\theta$.

But

$PC = y_1$. Therefore,

$TP = y_1\cdot\operatorname{cosec}\theta$.

Now, $\operatorname{cosec}^2\theta = 1 + \cot^2\theta$. Therefore,

$\operatorname{cosec}\theta = \sqrt{1+\cot^2\theta}$. Substituting: $\cot^2\theta = \dfrac{1}{\tan^2\theta}$, and,

multiplying by $\tan^2\theta/\tan^2\theta = 1$, which leaves the expression unchanged, we obtain:

$$\operatorname{cosec}\theta = \sqrt{\left(1+\frac{1}{\tan^2\theta}\right)\left(\frac{\tan^2\theta}{\tan^2\theta}\right)}.$$

Simplifying this, we find:

$$\frac{\sqrt{\left(1+\frac{1}{\tan^2\theta}\right)\tan^2\theta}}{\tan\theta} = \frac{\sqrt{1+\tan^2\theta}}{\tan\theta}.$$

But, $\tan\theta = \dfrac{dy}{dx}$ = slope. Substituting, we find:

$$\operatorname{cosec}\theta = \frac{\sqrt{1+\left(\frac{dy}{dx}\right)^2}}{\frac{dy}{dx}}, \text{ and,}$$

returning to the original expression and substituting, we find:

$$TP = y_1 \operatorname{cosec}\theta = y_1\frac{\sqrt{1+\left(\frac{dy}{dx}\right)^2}}{\frac{dy}{dx}}, \quad \text{which}$$

is the formula for the length of the tangent. Since dy/dx is also the slope, m, this equation may be written as:

$$y_1\frac{\sqrt{1+m^2}}{m}.$$

To find the length of the tangent for the given curve, we find y_1 and dy/dx. Since $x = 1$, substituting into $y^2 = 2x$, $y_1 = \sqrt{2}$.

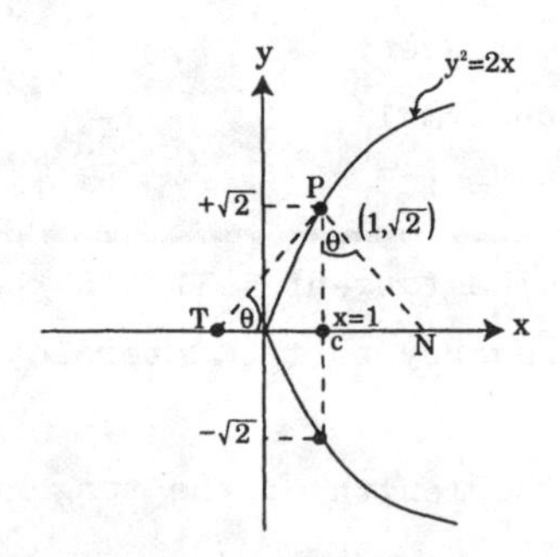

To find dy/dx, we differentiate $y^2 = 2x$ implicitly, and we obtain:

$$2y \frac{dy}{dx} = 2, \text{ or } \frac{dy}{dx} = \frac{1}{y} .$$

When we evaluate dy/dx at $x = 1$ and $y_1 = \sqrt{2}$, we find

$$\frac{dy}{dx} = \frac{1}{y_1} = \frac{1}{\sqrt{2}} .$$

Substituting $y_1 = \sqrt{2}$ and $dy/dx = 1/\sqrt{2}$ into the formula for the length of the tangent, we obtain:

$$\text{length of tangent} = \sqrt{2} \cdot \frac{\sqrt{1 + \left(\frac{1}{\sqrt{2}}\right)^2}}{\frac{1}{\sqrt{2}}}$$

$$= 2\sqrt{1.5} = 2.45.$$

b) The length of the normal = PN in the diagram = the distance from the tangent point to the intersection of the normal line with the x-axis.

To derive a formula for the length of the **normal**, we can use trigonometric identities:

$$PN = \text{length of the normal}$$

$$PN = PC/\cos\theta = PC \sec\theta.$$

But $$PC = y_1 .$$

Substituting, we find:

$$PN = y_1 \sec\theta .$$

$$\text{Sec}\,\theta = \sqrt{1 + \tan^2\theta} .$$

But $$\tan\theta = \frac{dy}{dx} .$$

Hence, $\sec\theta = \sqrt{1 + (dy/dx)^2}$. **Returning** to the original expression, and substituting for $\sec\theta$, we have: $PN = y_1 \sec\theta = y_1\sqrt{1 + (dy/dx)^2}$.

Since dy/dx is also the slope, m, this equation may be written:

$$y_1 \sqrt{1 + m^2} .$$

To find the length of the normal for the given curve, we can again use the values of y_1 and dy/dx found previously:

$$y_1 = \sqrt{2}, \text{ and } \frac{dy}{dx} = \frac{1}{\sqrt{2}} .$$

Substituting these values into the formula for the length of the normal we find:

length of the normal $= \sqrt{2}\sqrt{1 + (1/\sqrt{2})^2}$

$= 1.73$

c) The length of the subtangent = TC in the diagram = the projection of the tangent, TP, onto the x-axis.

To derive a formula for the length of the subtangent, we can use trigonometric identities:

$$TC = \text{length of subtangent}$$

$$= \frac{PC}{\tan\theta} .$$

But $PC = y_1$. **Substituting,** we find:

$$TC = y_1/\tan\theta .$$

But, $\tan\theta = dy/dx$. **Therefore,**

$$TC = y_1/dy/dx .$$

Since dy/dx is also the slope, m, this equation may be written:

$$\frac{y_1}{m}.$$

To find the length of the subtangent for the given curve, we can use the values of y_1 and dy/dx found previously:

$$y_1 = \sqrt{2} \text{ and } dy/dx = 1/\sqrt{2} .$$

Substituting these values into the formula for the length of the subtangent, we find:

$$\text{length of subtangent} = \frac{\sqrt{2}}{\frac{1}{\sqrt{2}}} = 2 .$$

d) The length of the subnormal = CN in the diagram = the projection of the normal, PN onto the x-axis.

To derive a formula for the length of the subnormal we can use trigonometric identities:

$$CN = \text{length of subnormal}$$

$$= PC \cdot \tan\theta.$$

But

$$PC = y_1, \text{ and}$$

$$\tan\theta = \frac{dy}{dx} .$$

Substituting, we find:

$$CN = y_1 \frac{dy}{dx} .$$

Since dy/dx is also the slope, m, the equation may be written as $y_1 m$.

To find the length of the subnormal for the given curve we use the values of y_1 and dy/dx previously obtained: $y_1 = \sqrt{2}$ and $dy/dx = 1/\sqrt{2}$. Substituting these values into the formula just derived for the length of the subnormal we find:

$$\text{length of subnormal} = (\sqrt{2})(1/\sqrt{2}) = 1.$$

• PROBLEM 325

> Find the equations of the tangent and normal and the lengths of the subtangent and subnormal of the ellipse: $x = 3\cos\theta$ and $y = 4\sin\theta$, at the point where $\theta = 30^\circ$.

Solution: The equation of a straight line passing through the point where $\theta = 30^\circ$ can be expressed in the form: $y - y_1 = m(x - x_1)$. To find the slope, m, we first find $dx/d\theta$, then $dy/d\theta$, and finally dy/dx, which is the slope. $dx/d\theta = -3\sin\theta$. $dy/d\theta = 4\cos\theta$. Therefore,

$$\frac{dy}{dx} = \frac{dy/d\theta}{dx/d\theta} = \frac{4\cos\theta}{-3\cdot\sin\theta} = \frac{-4}{3}\cot\theta = \text{slope}, m .$$

When $\theta = 30^\circ$, $m = -\frac{4}{3}\cot 30^\circ = -\frac{4}{3}\cdot 1.7321 = -2.3095$. We substitute $\theta = 30^\circ$ in the given equations to obtain:

$$x_1 = 3\cdot\cos 30^\circ = 3\cdot 0.866 = 2.598$$

$$y_1 = 4\sin 30^\circ = 4\cdot\frac{1}{2} = 2.0$$

$$y - y_1 = m(x - x_1). \quad y - 2 = -2.31(x - 2.6),$$

or, using the standard equation for a straight line:

$$y = 2 + 6 - 2.31x = 8 - 2.31x,$$

the equation of the tangent. Similarly,

$$y - y_1 = -\frac{1}{m}(x - x_1)$$

is the equation for the normal, the slope of the normal line being the negative reciprocal of the slope of the tangent line, or $-\frac{1}{m}$.

Therefore,

$$y - 2 = -\frac{1}{-2.31}(x - 2.6) = \frac{x}{2.31} - \frac{2.6}{2.31}$$

$$= \frac{x}{2.31} - 1.125$$

$$y = 2 - 1.125 + \frac{x}{2.31} = 0.875 + \frac{x}{2.31}$$

$$= \frac{2.02 + x}{2.31}$$

$2.31y = x + 2.02$, the equation of the normal.

Length of subtangent $\frac{y_1}{m} = -\frac{2}{2.31} = -0.87$

Length of subnormal $my_1 = -2.31\cdot 2 = -4.62$.

• PROBLEM 326

Find the coordinates of the point of contact of a straight line drawn through the point: $x = 3$, $y = -2$, and tangent to the curve: $y = x^2 - 3x + 2$.

Solution: In order to find the required points of contact we find the equation of the tangent line passing through the point (3,-2), and set this equation equal to the equation for the curve, solving for x and y.

The equation of a straight line passing through a given point (3,-2) can be expressed in the form $y - y_1 = m(x = x_1)$. Here $x_1 = 3$ and $y_1 = -2$. The slope of this straight line, which is also tangent to the curve: $y = x^2 - 3x + 2$, is found by taking dy/dx. $dy/dx = 2x - 3$, therefore $2x - 3$ is the slope m. Substituting $x_1 = 3$, $y_1 = -2$, and $m = (2x - 3)$ into the equation: $y - y_1 = m(x - x_1)$, we obtain: $y + 2 = (2x - 3)(x - 3)$, or, $y = 2x^2 - 9x + 7$, as the equation of the tangent.

We now set the equation for the tangent equal to the equation for the curve and solve for x.

$$2x^2 - 9x + 7 = x^2 - 3x + 2$$
$$x^2 - 6x + 5 = 0.$$
$$x = 5 \quad \text{and} \quad 1.$$

Now we substitute $x = 5$ and $x = 1$ into the equation for the curve and solve for the value(s) of y. When $x = 5$, $y = 12$ and when $x = 1$, $y = 0$. Therefore the coordinates of the points of contact are (5,12) and (1,0).

• PROBLEM 327

Find the points of contact of the horizontal and vertical tangents to $x = 2 - 3 \sin \theta$, $y = 3 + 2 \cos \theta$.

Solution: The procedure for finding the points of contact of the horizontal tangents to a curve given in parametric form is to find $dy/d\theta$, set $dy/d\theta = 0$, and solve for θ. Then we substitute the value(s) obtained for θ into the given x and y equations and solve for x and y.

$$\frac{dy}{d\theta} = -2 \sin \theta = 0, \quad \text{and} \quad \sin \theta = 0. \quad \theta = 0 \text{ or } \pi$$

Substituting $\theta = 0$ and π into the x and y equations, we obtain:

$$x = 2 - 3 \sin(0) = 2 \qquad x = 2 - 3 \sin(\pi) = 2$$
$$y = 3 + 2 \cos(0) = 5 \qquad y = 3 + 2 \cos(\pi) = 1$$

Therefore the points (2,5) and (2,1) are the points of contact of the horizontal tangents.

The procedure for finding the points of contact of the vertical tangents to a curve given in parametric form is to find $dx/d\theta$, set $dx/d\theta = 0$ and solve for θ. Then we substitute the values obtained for θ into the given x and y equations and solve for x and y.

$$\frac{dx}{d\theta} = -3 \cos \theta = 0, \quad \text{and} \quad \cos \theta = 0, \quad \text{or } \theta = \frac{\pi}{2} \text{ or } \frac{3\pi}{2}$$

Substituting $\theta = \pi/2$ and $3\pi/2$ into the x and y equations we obtain:

$$x = 2 - 3 \sin\left(\frac{\pi}{2}\right) = -1 \qquad x = 2 - 3 \sin\left(\frac{\pi}{2}\right) = 5$$

$$y = 3 + 2 \cos\left(\frac{\pi}{2}\right) = 3 \qquad y = 3 + 2 \cos\left(\frac{3\pi}{2}\right) = 3 .$$

The points (-1,3) and (5,3) are the points of contact of the vertical tangents.

$$x = 2 - 3 \sin\left(\frac{\pi}{2}\right) = -1 \qquad x = 2 - 3 \sin\left(\frac{3\pi}{2}\right) = 5$$

$$y = 3 + 2 \cos\left(\frac{\pi}{2}\right) = 3 \qquad y = 3 + 2 \cos\left(\frac{3\pi}{2}\right) = 3$$

Therefore, the points (2,5) and (2,1) are the points of contact of the horizontal tangents; and (-1,3) and (5,3) are the points of contact of the vertical tangents.

• PROBLEM 328

Find the points of contact of the horizontal and vertical tangents to the cardioid:

$$x = a \cos \theta - \tfrac{1}{2} a \cos 2\theta - \tfrac{1}{2} a,$$

$$y = a \sin \theta - \tfrac{1}{2} a \sin 2\theta.$$

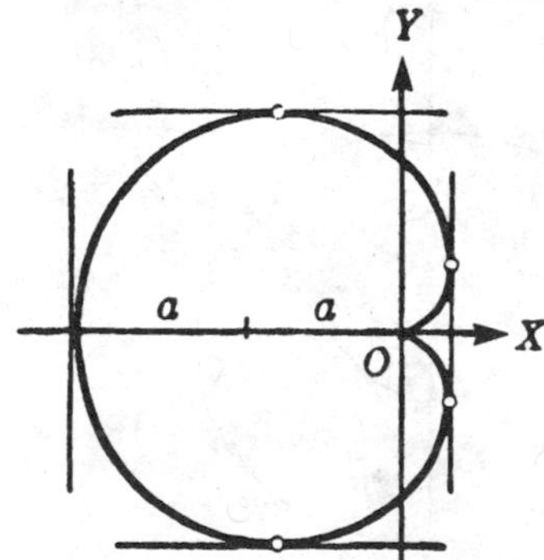

Solution: To find the points of contact of the horizontal tangents we first find:

$$\frac{dy}{d\theta} = a(\cos \theta - \cos 2\theta).$$

We then set $dy/d\theta = 0$ to solve for the value(s) of θ.

$$a(\cos \theta - \cos 2\theta) = 0, \text{ but } \cos 2\theta = 2 \cos^2\theta - 1,$$

by the double angle formula. Substituting $\cos 2\theta = 2\cos^2\theta - 1$ into $a(\cos\theta - \cos 2\theta) = 0$, we obtain:

$$a(\cos\theta - 2\cos^2\theta + 1) = 0,$$

which simplifies to $(-2\cos\theta - 1)(\cos\theta - 1) = 0$. Solving for θ, we obtain: $\theta = 0, 120^\circ, 240^\circ$. (In order to find the corresponding rectangular coordinates, we substitute the values for θ into the given equations and solve for x and y).

To find the points of contact of the vertical tangents we first solve for:

$$\frac{dx}{d\theta} = a(\sin 2\theta - \sin\theta)$$

We then set $dx/d\theta = 0$ to solve for the value(s) of θ. $a(\sin 2\theta - \sin\theta) = 0$, but $\sin 2\theta - 2\sin\theta\cos\theta$, by the double angle formula. Substituting $\sin 2\theta = 2\sin\theta\cos\theta$ into $a(\sin 2\theta - \sin\theta) = 0$, we obtain: $a(2\sin\theta\cos\theta - \sin\theta) = 0$, which simplifies to: $\sin\theta(2\cos\theta - 1) = 0$. Solving for θ, we find $\theta = 0^\circ, 60^\circ, 180^\circ, 300^\circ$. (Again, the rectangular coordinates can be found by substituting the values for θ into the given x and y equations and then solving for x and y).)

• PROBLEM 329

Find the angle between the curves:

$$x^2 - 3y = 3 \qquad (1)$$

and

$$2x^2 + 3y^2 = 30 \qquad (2)$$

at their point of intersection in the second quadrant.

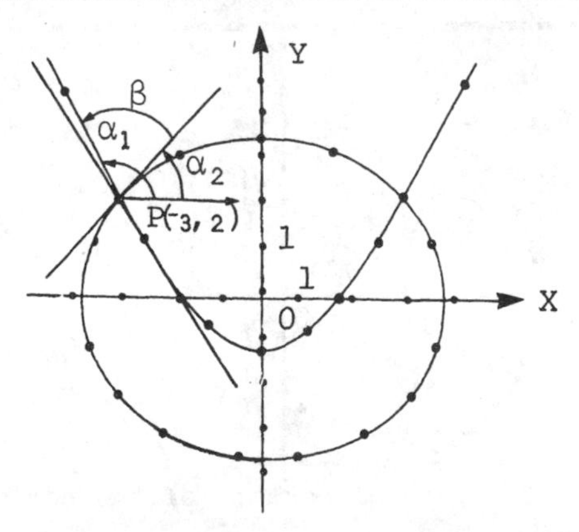

Solution: The angle of intersection of two curves is the angle between the tangents of the two curves at their point of intersection. To find the angle between the two tangents, we first find the point at which they intersect; then the slopes of the tangents at that point. The following formula can be used to obtain the angle:

$$\tan \beta = \tan(\alpha_1 - \alpha_2) = \frac{m_1-m_2}{1+m_1m_2} .$$

This is based on the standard equation for the tangent of the difference of two angles. Here β is the angle of intersection, m_1 is the slope of one tangent, and m_2 is the slope of the other.

The point(s) of intersection can be found by setting the two curves equal to one another and solving for x and y. Two points are obtained, (3,2) and (-3,2), but since the problem requires the point of intersection to be in the second quadrant, only (-3,2) should be used, since x is negative and y is positive in the second quadrant.

The slope of the tangent to the curve is dy/dx. Solving equation (1) for y and differentiating, $dy/dx = 2x/3 = m_1$. Evaluating m_1 at (-3,2), we obtain $m_1 = -2$. Differentiating equation (2) implicitly, we obtain:

$$4x + 6y \frac{dy}{dx} = 0 \text{ , and } \frac{dy}{dx} = -\frac{2x}{3y} = m_2 .$$

Evaluating m_2 at (-3,2), we obtain $m_2 = 1$.

We can now use the above formula to solve for the angle of intersection, β:

$$\tan \beta = \frac{m_1-m_2}{1+m_1m_2}$$

$$\tan \beta = \frac{-2-1}{1+(-2)} = 3$$

$$\beta = 71.5^\circ,$$

the angle of intersection.

• PROBLEM 330

Find the angle of intersection of the circles:

$$x^2 + y^2 - 4x = 1, \qquad \text{(A)}$$
$$x^2 + y^2 - 2y = 9. \qquad \text{(B)}$$

Solution: The angle of intersection of two circles is the angle between the tangents to each circle at the points of intersection. To find the angle between the two tangents, we first obtain the slopes of the two lines and then use the following formula:

$$\tan \theta = \frac{m_1-m_2}{1+m_1m_2} ,$$

where:

m_1 = slope of the tangent to the circle A at (x,y),

and

m_2 = slope of the tangent to the circle B at (x,y).

First we solve for the points of intersection of the two curves by setting the two equations for the circles equal to one another, and solving for x and y. We find the points of intersection to be (3,2) and (1,-2).

To find the slope of the tangents at these points, we find dy/dx for each of the given curves.

From (A), $$m_1 = \frac{dy}{dx} = \frac{2-x}{y},$$

and from (B), $$m_2 = \frac{dy}{dx} = \frac{x}{1-y}.$$

Substituting x = 3, y = 2, we have

$m_1 = -\frac{1}{2}$ = slope of tangent to (A) at (3,2).

$m_2 = -3$ = slope of tangent to (B) at (3,2).

Using the formula: $\tan\theta = \frac{m_1 - m_2}{1 + m_1 m_2}$, and

substituting for m_1 and m_2, we obtain:

$$\tan\theta = \frac{-\frac{1}{2}+3}{1+\frac{3}{2}} = 1 \;; \quad \theta = 45^\circ.$$

This is also the angle of intersection at the point (1,-2), where

$$m_1 = -\tfrac{1}{2};\; m_2 = \frac{1}{3};\; \tan\theta = \frac{-\frac{1}{2} - \frac{1}{3}}{1 - \frac{1}{6}} = -1$$

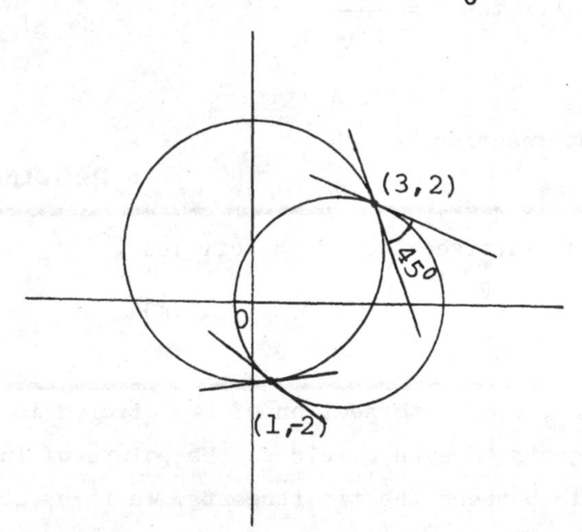

• PROBLEM 331

Find the angles of intersection between the parabola: $y^2 = 6 - 2x$, and the semicubical parabola: $y^2 = 4x^3$.

Solution: We first find the points of intersection. These are found by setting the parabola equal to the semicubical parabola and solving for x. We obtain:

$$y^2 = 6 - 2x.$$

$$y^2 = 4x^3.$$

$$6 - 2x = 4x^3.$$

$$2x^3 + x - 3 = 0.$$

This can be broken up into two factors:

$$(-2x^2 - 2x - 3)(1 - x) = 0.$$

$$(2x^2 + 2x + 3)(1 - x) = 0.$$

$$(2x^2 + 2x + 3) = 0.$$

Also,

$$(1 - x) = 0,$$

$$x = 1.$$

We can solve for x in the factor: $2x^2 + 2x + 3 = 0$, which has the form: $ax^2 + bx + c$, by using the quadratic formula:

$$x = \frac{-b \pm \sqrt{b^2 - 4ac}}{2a}.$$

We obtain:

$$x = \frac{-2 \pm \sqrt{4 - 24}}{4} = \frac{-2 \pm \sqrt{-20}}{4}.$$

These roots contain imaginary numbers, hence the only real root is $x = 1$, for which we can find y by substituting into either equation.

$$y^2 = 6 - 2x$$

$$y^2 = 6 - 2$$

$$y^2 = 4$$

$$y = \pm 2.$$

Therefore the points of intersection are (1,2) and (1,-2). Now that we have the points of intersection, we find the slope of the tangent to each curve at the points of intersection. The slope of the tangent to a curve is the first derivative of the curve. The slope of the parabola: $y^2 = 6 - 2x$, can be found, using implicit differentiation as follows:

$$y^2 = 6 - 2x.$$

$$2y\frac{dy}{dx} = -2.$$

$$\frac{dy}{dx} = \frac{-2}{2y}.$$

$$\frac{dy}{dx} = \frac{-1}{y}.$$

At $(1,2)$, $\frac{dy}{dx} = \frac{-1}{2}$, and at $(1,-2)$, $\frac{dy}{dx} = \frac{1}{2}$. Therefore, the slope, m_1, of $y^2 = 6 - 2x$ is $\frac{1}{2}$ or $\frac{-1}{2}$. We can also find m_2, the slope of the semicubical parabola: $y^2 = 4x^3$, using implicit differentiation.

$$y^2 = 4x^3$$

$$2y\frac{dy}{dx} = 12x^2$$

$$\frac{dy}{dx} = \frac{12x^2}{2y} = \frac{6x^2}{y}.$$

At the point $(1,2)$, $\frac{dy}{dx} = \frac{6}{2} = 3$, and at the point $(1,-2)$, $\frac{dy}{dx} = \frac{6}{-2} = -3$. Therefore the slope of $y^2 = 4x^3$ is 3 or -3. Having the slopes of the two curves at their points of intersection, we can apply the formula:

$$\tan\theta = \frac{m_1 - m_2}{1 + m_1 m_2},$$

to find the angles of intersection at the two points. At the point $(1,2)$, $m_1 = \frac{-1}{2}$ and $m_2 = 3$.

$$\tan\theta = \frac{\frac{-1}{2} - 3}{1 + \left(\frac{-1}{2}\right)(3)}.$$

$$\tan\theta = \frac{\frac{-7}{2}}{\frac{-1}{2}}.$$

$$\tan\theta = 7.$$

$$\theta = \arctan 7.$$

At the point $(1,-2)$, $m_1 = \frac{1}{2}$ and $m_2 = -3$ and the angle of intersection is,

$$\tan\theta = \frac{\frac{1}{2} + 3}{1 + \left(\frac{1}{2}\right)(-3)}.$$

$$\tan\theta = \frac{\frac{7}{2}}{\frac{-1}{2}} .$$

$$\tan\theta = -7.$$

$$\theta = \arctan(-7).$$

• PROBLEM 332

> Show that the parabolas: $x^2 = 8(y+2)$ and $x^2 = -12(y-3)$, intersect at right angles at each point of intersection. Use the Δ-method.

Solution: To determine the angles of intersection of the two parabolas we first solve for the points of intersection and then find the slopes of the tangents to the two curves at these points. From the slopes we can determine the angles of intersection. The points of intersection are found by setting the two curves equal and solving for y and then x. Hence,

$$x^2 = 8(y+2), \text{ and } x^2 = -12(y-3).$$

$$8(y+2) = -12(y-3).$$

$$8y + 16 = -12y + 36.$$

$$20y = 20.$$

$$y = 1.$$

$$x = \pm 2\sqrt{6}.$$

There are two points of intersection: $(2\sqrt{6},1)$ and $(-2\sqrt{6},1)$. We solve for the slopes of the tangents to the two curves at these points. The slope of the tangent to a curve is the derivative, $\frac{dy}{dx}$, of the curve. For the first parabola, $x^2 = 8(y+2)$, the slope, m_1, can be found as follows:

$$\frac{\Delta y}{\Delta x} = \frac{f(x+\Delta x) - f(x)}{\Delta x}$$

$$x^2 = 8(y+2) = 8y + 16.$$

$$y = \frac{x^2 - 16}{8}.$$

$$\frac{\Delta y}{\Delta x} = \frac{\frac{(x+\Delta x)^2 - 16}{8} - \left(\frac{x^2 - 16}{8}\right)}{\Delta x}$$

$$= \frac{\frac{x^2 + 2x\Delta x + (\Delta x)^2 - 16 - x^2 + 16}{8}}{\Delta x}$$

$$= \frac{2x + \Delta x}{8} . \quad \frac{dy}{dx} =$$

$$\lim_{\Delta x \to 0} \frac{\Delta y}{\Delta x} = \lim_{\Delta x \to 0} \frac{2x + \Delta x}{8} = \frac{x}{4} = m_1.$$

At the point $(2\sqrt{6},1)$, $m_1 = \frac{\sqrt{6}}{2}$, and at the point $(-2\sqrt{6},1)$, $m_1 = \frac{-\sqrt{6}}{2}$.

To find m_2, the slope of the second parabola: $x^2 = -12(y-3)$, we follow the same procedure. Hence,

$$\frac{\Delta y}{\Delta x} = \frac{f(x + \Delta x) - f(x)}{\Delta x}$$

$$x^2 = -12(y-3) = -12y + 36.$$

$$y = \frac{x^2 - 36}{-12}.$$

$$\frac{\Delta y}{\Delta x} = \frac{\frac{(x + \Delta x)^2 - 36}{-12} - \frac{x^2 - 36}{-12}}{\Delta x}$$

$$= \frac{\frac{x^2 + 2x\Delta x + (\Delta x)^2 - 36 - x^2 + 36}{-12}}{\Delta x}$$

$$= \frac{2x + \Delta x}{-12} \cdot \frac{dy}{dx} =$$

$$\lim_{\Delta x \to 0} \frac{\Delta y}{\Delta x} = \lim_{\Delta x \to 0} \frac{2x + \Delta x}{-12} = \frac{x}{-6} = m_2.$$

At the point $(2\sqrt{6},1)$, m_2 is $-\frac{\sqrt{6}}{3}$ and at the point $(-2\sqrt{6},1)$, m_2 is $\frac{\sqrt{6}}{3}$. From the values of the slopes of the tangents to the curves at the two points of intersection we can determine the angles of intersection. The angle of intersection of two curves is 90° when the slopes of their tangents are related by:

$$m_1 = \frac{-1}{m_2}.$$

For the point $(2\sqrt{6},1)$, $m_1 = \frac{\sqrt{6}}{2}$ and $m_2 = -\frac{\sqrt{6}}{3}$. Substituting these values,

$$\frac{\sqrt{6}}{2} = -\frac{1}{-\frac{\sqrt{6}}{3}}, \text{ or } \frac{\sqrt{6}}{2} = \frac{3}{\sqrt{6}}.$$

Now, $$\frac{\sqrt{3}\sqrt{2}}{\sqrt{2}\sqrt{2}} = \frac{\sqrt{3}\sqrt{3}}{\sqrt{3}\sqrt{2}}.$$

Finally, $$\frac{\sqrt{3}}{\sqrt{2}} \quad \frac{\sqrt{3}}{\sqrt{2}}.$$

The angle of intersection at the point $(2\sqrt{6},1)$ is therefore 90°. We can check the angle of intersection for the point $(-2\sqrt{6},1)$ in the same manner. For this point, $m_1 = -\frac{\sqrt{6}}{2}$ and $m_2 = \frac{\sqrt{6}}{3}$. Hence,

$$m_1 = \frac{-1}{m_2}$$

$$\frac{-\sqrt{6}}{2} = \frac{-1}{\frac{\sqrt{6}}{3}}$$

$$\frac{-\sqrt{6}}{2} = \frac{3}{\sqrt{6}}$$

$$\frac{-\sqrt{3}\sqrt{2}}{\sqrt{2}\sqrt{2}} = \frac{-\sqrt{3}\sqrt{3}}{\sqrt{3}\sqrt{2}}$$

$$\frac{-\sqrt{3}}{\sqrt{2}} = \frac{-\sqrt{3}}{\sqrt{2}} .$$

The angle of intersection at the point $(-2\sqrt{6},1)$ is also 90°.

• PROBLEM 333

Show that the equation of the tangent to the ellipse: $b^2x^2 + a^2y^2 = a^2b^2$, at any point (x_1,y_1) may be written in the form $b^2x_1x + a^2y_1y = a^2b^2$. Obtain the analogous result for the hyperbola: $b^2x^2 - a^2y^2 = a^2b^2$.

Solution: The equation of the tangent to a curve at a point (x_1,y_1) is given by: $y - y_1 = m(x - x_1)$, where m is the slope. The slope of the tangent to the curve is the first derivative of the curve, $\frac{dy}{dx}$. Using implicit differentiation, we obtain the first derivative of $b^2x^2 + a^2y^2 = a^2b^2$ as follows:

$$b^2x^2 + a^2y^2 = a^2b^2$$

$$2b^2x + 2a^2y\frac{dy}{dx} = 0$$

$$\frac{dy}{dx} = \frac{-2b^2x}{2a^2y} .$$

$$\frac{dy}{dx} = \frac{-b^2x}{a^2y} = m.$$

Now that we have the slope, $m = \frac{-b^2x}{a^2y}$, we can substitute this value into the equation for the tangent.

$$y - y_1 = m(x - x_1).$$

$$y - y_1 = \frac{-b^2x}{a^2y}(x - x_1).$$

$$(a^2y)(y - y_1) = -b^2x^2 + b^2xx_1.$$

$$a^2y^2 - a^2yy_1 = -b^2x^2 + b^2xx_1.$$

$$b^2xx_1 + a^2yy_1 = b^2x^2 + a^2y^2.$$

From the equation of the ellipse, we know that $b^2x^2 + a^2y^2 = a^2b^2$, therefore the equation for the tangent can be written as:

$$b^2xx_1 + a^2yy_1 = a^2b^2.$$

This is the required equation for the tangent to the ellipse.

In a similar manner we can find the equation for the tangent to the hyperbola $b^2x^2 - a^2y^2 = a^2b^2$. We can find the slope, using implicit differentiation, as follows:

$$b^2x^2 - a^2y^2 = a^2b^2$$

$$2b^2x - 2a^2y\frac{dy}{dx} = 0$$

$$\frac{dy}{dx} = \frac{-2b^2x}{-2a^2y}.$$

$$\frac{dy}{dx} = \frac{b^2x}{a^2y} = m,$$

the slope.

Knowing that the slope, $m = \frac{b^2x}{a^2y}$, and the point of tangency is (x_1,y_1) we can use the equation for the tangent as follows:

$$y - y_1 = m(x - x_1).$$

$$y - y_1 = \frac{b^2x}{a^2y}(x - x_1).$$

$$a^2y(y - y_1) = b^2x(x - x_1).$$

$$a^2y^2 - a^2yy_1 = b^2x^2 - b^2xx_1.$$

$$b^2xx_1 - a^2yy_1 = b^2x^2 - a^2y^2.$$

From the equation of the hyperbola, we know that $b^2x^2 - a^2y^2 = a^2b^2$, therefore the equation of the tangent to the hyperbola can be written as:

$$b^2xx_1 - a^2yy_1 = a^2b^2.$$

• PROBLEM 334

> Find the equation of the curve passing through (1,2) which has the property that its slope at any point is equal to twice the abscissa of the point.

Solution: The slope of the curve is found by finding dy/dx of the curve. The problem requires that the slope be twice the abscissa, i.e., dy/dx = 2x. To find the equation of the curve, we integrate:

$$\frac{dy}{dx} = 2x.$$

$$dy = 2xdx.$$

$$y = \int 2xdx$$

$$= x^2 + c.$$

Therefore, $y = x^2 + c$ is the equation.

This particular curve is to pass through (1,2). To find c, we substitute (1,2) into the equation. We obtain:

$$2 = (1)^2 + c, \ 2 = 1 + c, \text{ or } c = 1.$$

Substituting $c = 1$ back into the equation $y = x^2 + c$, we have $y = x^2 + 1$, the required equation.

• PROBLEM 335

> Show that the tangents at the ends of the latus rectum of the parabola $y^2 = 2px$ are perpendicular to each other.

Solution: The latus rectum of this parabola is the vertical line passing through the focal point F and perpendicular to the axis. If the parabola were given by: $y^2 = 4px$, the focal point would be $(p,0)$. This parabola can be written as: $y^2 = 2px = 4\left(\frac{p}{2}\right)x$. Hence, the focal point is $\left(\frac{p}{2},0\right)$.

To find the points where the latus rectum meets the parabola we can substitute the x value, $x = \frac{p}{2}$, into $y^2 = 2px$ and solve for y. We obtain:

$$y^2 = 2p\left(\frac{p}{2}\right)$$

$$y^2 = p^2.$$

$$y = \pm p.$$

Therefore, the two points where the latus rectum meets the parabola are $\left(\frac{p}{2}, p\right)$ above the x-axis and $\left(\frac{p}{2}, -p\right)$ below the x-axis. Two lines are perpendicular if their slopes are negative reciprocals. The slope of a tangent to a curve is the first derivative, $\frac{dy}{dx}$, of the curve.

Using implicit differentiation, we find $\frac{dy}{dx}$ as follows:

$$y^2 = 2px.$$

$$2y\frac{dy}{dx} = 2p.$$

$$\frac{dy}{dx} = \frac{2p}{2y}.$$

$$\frac{dy}{dx} = \frac{p}{y}.$$

At the point $\left(\frac{p}{2}, p\right)$ the slope is found as follows,

$$y^2 = 2px.$$

$$p = \frac{y^2}{2x}.$$

$$\frac{dy}{dx} = \frac{p}{y} = \frac{\frac{y^2}{2x}}{y} = \frac{y}{2x} = \frac{p}{2\left(\frac{p}{2}\right)} = \frac{p}{p} = 1.$$

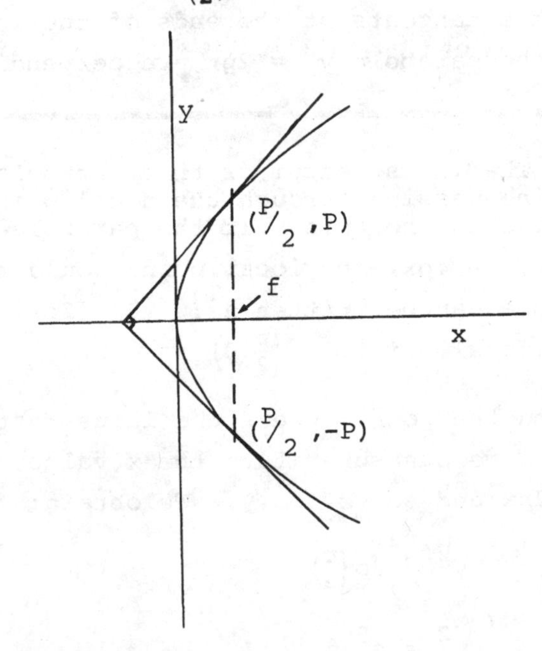

At the point $\left(\frac{p}{2},-p\right)$ the slope is:

$$\frac{dy}{dx} = \frac{p}{y} = \frac{y}{2x} = \frac{-p}{2\left(\frac{p}{2}\right)} = \frac{-p}{p} = -1.$$

Since the slope of the tangent to the curve at the point $\left(\frac{p}{2},p\right)$ is 1, and the slope of the tangent to the curve at the point $\left(\frac{p}{2},-p\right)$ is -1, they are negative reciprocals, and the tangents are therefore perpendicular.

• PROBLEM 336

Obtain the equations of the tangent and the normal to a curve, the equation of which, in polar coordinates, is of the form $r = f(\theta)$.

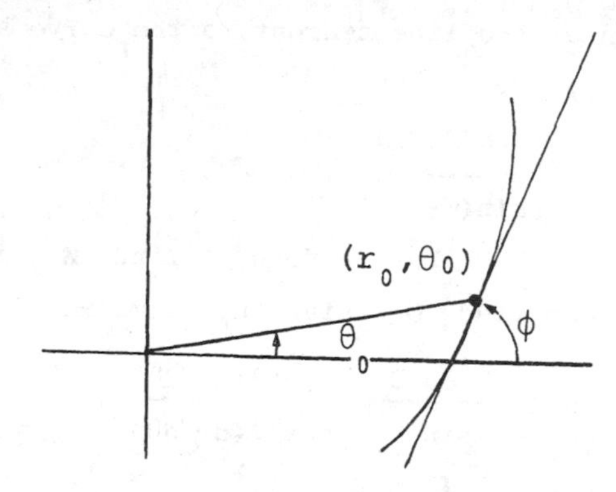

Solution: The equations of the tangent and the normal to a curve at a point, in polar coordinates, are:

$$r = r_0 \frac{\sin(\emptyset-\theta)_0}{\sin(\emptyset-\theta)}, \text{ and } r = r_0 \frac{\cos(\emptyset-\theta)_0}{\cos(\emptyset-\theta)},$$

respectively, where (r_0,θ_0) is the point where the tangent and the normal meet the curve, and $\tan \emptyset$ is the slope of the tangent. In order to solve such equations, we must find $\emptyset$.

Now, $\tan \emptyset = dy/dx$, $x = r \cos \theta$ and $y = r \sin \theta$. Differentiating implicitly, we find:

$$\tan \emptyset = \frac{dy}{dx} = \frac{dy/d\theta}{dx/d\theta} = \frac{r \cos \theta + \sin \theta (dr/d\theta)}{-r \sin \theta + \cos \theta (dr/d\theta)}$$

Given a point (r_0,θ_0), we can substitute the coordinates into the equation and solve for $\emptyset$, and then use $\emptyset$, r_0, and θ_0 in the equations for the tangent and the normal. Thus at the point (r_0,θ_0),

$$\emptyset = \arctan \frac{r_0 \cos \theta_0 + \sin \theta_0 (dr/d\theta)]_{\theta_0}}{-r_0 \sin \theta_0 + \cos \theta_0 (dr/d\theta)]_{\theta_0}} .$$

Find the equation of the tangent to the locus of:

$$r = 1 - \cos\theta, \text{ at } (1, \pi/2).$$

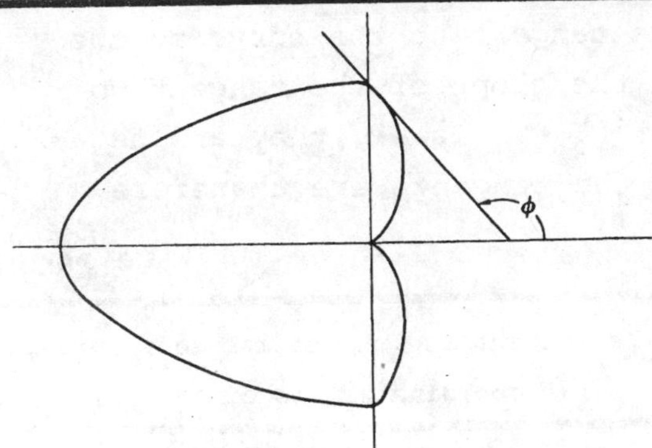

Solution: The equation of the line tangent to the curve at (r_0, θ_0), (here given as $(1, \pi/2)$), is:

$$r = r_0 \frac{\sin(\emptyset - \theta_0)}{\sin(\emptyset - \theta)} .$$

We have $r_0 = 1$ and $\theta_0 = \pi/2$, but we need to find $\emptyset$. $r = 1 - \cos\theta$, and $dr/d\theta = \sin\theta$. We can find $\emptyset$, using the formula:

$$\tan\emptyset = \frac{r\cos\theta + \sin\theta(dr/d\theta)}{-r\sin\theta + \cos\theta(dr/d\theta)} ,$$

and evaluating the formula at $(1, \pi/2)$.

$$\tan\emptyset = \left.\frac{(1 - \cos\theta)\cos\theta + \sin^2\theta}{-(1 - \cos\theta)\sin\theta + \cos\theta\sin\theta}\right]_{1, \pi/2} = -1.$$

$\emptyset = 3\pi/4$. Substituting $r_0 = 1$, $\theta_0 = \pi/2$, and $\emptyset = 3\pi/4$ into:

$$r = r_0 \frac{\sin\emptyset - \theta_0}{\sin\emptyset - \theta} ,$$

we obtain:

$$r = \frac{\sin[(3\pi/4) - (\pi/2)]}{\sin[(3\pi/4) - \theta]} = \frac{\frac{1}{2}\sqrt{2}}{\sin[(3\pi/4) - \theta]} ,$$

the equation for the tangent.

CHAPTER 15

MAXIMUM AND MINIMUM VALUES

If we think of the derivative, $\frac{dy}{dx}$, as the slope of a curve, then it may be seen that there may be certain points where that curve has a zero derivative. The meaning of a zero value of the derivative is that, at the point where this occurs, the tangent line to the curve is horizontal. When a tangent is horizontal, the curve can either be in the process of changing from a rise to a drop, from a drop to a rise, or, finally, it can continue after a pause in the same direction as it has been going.

Mathematically, such points can be found by differentiating the equation for the curve, setting the result equal to zero, and solving for the values of x. These values of x are called the critical values. In accordance with what has been said above, these critical points can be either maxima, minima, or points of inflection.

Once the critical points have been found, there are two ways of determining whether they constitute maxima or minima. At a maximum, the second derivative, i.e., the change in slope, found by differentiating the equation for the first derivative, is negative. At a minimum, the second derivative is positive. At a point of inflection, the second derivative, as the first derivative, is zero. Hence, evaluating the second derivative as well as the first derivative provides identification of the type of critical point.

If we wish to avoid taking the second derivative, we can find whether a point is maximum or minimum by substituting into the given equation a value a little less, and a value a little greater than, the critical value, and noting whether the function has a value there that is greater or less than the value at the critical point.

• PROBLEM 338

Find the maxima and minima of the function $f(x) = x^4$.

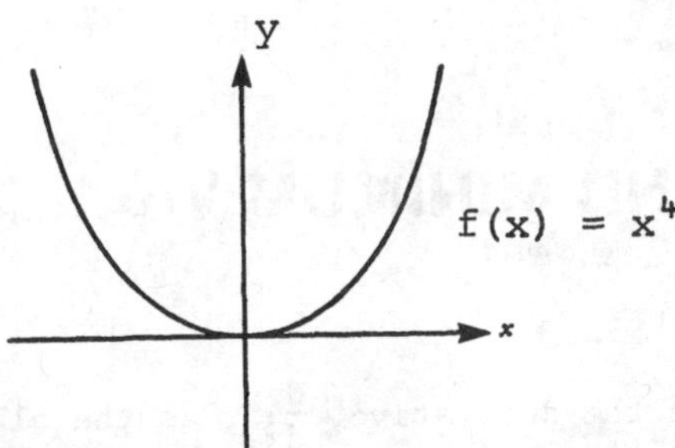

Solution: To determine maxima and minima we find $f'(x)$, set it equal to 0, and solve for x to obtain the critical points. We find: $f'(x) = 4x^3 = 0$, therefore $x = 0$ is the critical value. We must now determine whether $x = 0$ is a maximum or minimum value. In this example the Second Derivative Test fails because $f''(x) = 12x^2$ and $f''(0) = 0$. We must, therefore, use the First Derivative Test. We examine $f'(x)$ when $x < 0$ and when $x > 0$. We find that for $x < 0$, $f'(x)$ is negative, and for $x > 0$, $f'(x)$ is positive. Therefore there is a minimum at (0,0). (See figure).

• PROBLEM 339

Determine the critical points of $f(x) = 3x^4 - 4x^3$ and sketch the graph.

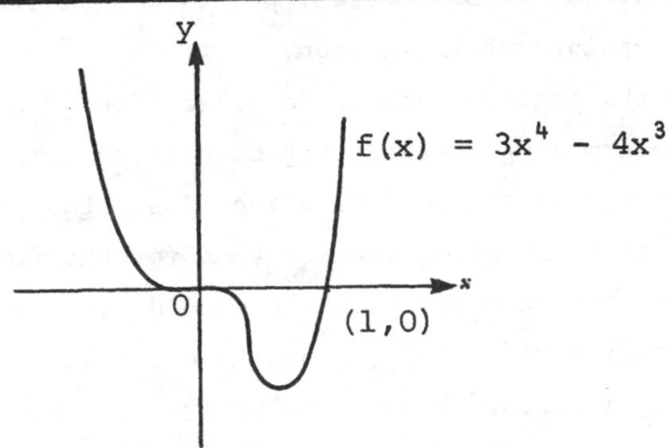

Solution: To determine the critical points we find $f'(x)$, set it equal to 0 and solve for x. These are the abscissas of the critical points. Differentiating, we have $f'(x) = 12x^3 - 12x^2 = 12x^2(x-1)$. Therefore, $x = 0, 1$ are the critical values. We now examine $f'(x)$ when $x < 0$, when $1 > x > 0$, and when $x > 1$ to determine whether there is a maximum, minimum or neither at each critical point. We find that, when $x < 0$, $f'(x)$ is negative. When $1 > x > 0$, $f'(x)$ is also negative, and when $x > 1$, $f'(x)$ is positive. Because $f'(x)$ changes sign from - to + at $x = 1$, this is a minimum. At $x = 0$ there is no change in sign, $f'(x)$ is negative when $x < 0$ and when $0 < x < 1$, therefore this is neither a maximum nor a minimum. Because $f'(0) = 0$, f has a horizontal tangent at $x = 0$, as shown in the figure. Such a point is known as a point of inflection.

Additional insight can be obtained by taking the second derivative:

$$f''(x) = \frac{d}{dx}\left[f'(x)\right] = \frac{d}{dx}(12x^3 - 12x^2)$$
$$= 36x^2 - 24x$$
$$= + \text{ at } x = 1, \text{ a minimum}$$
$$\text{but} = 0 \text{ at } x = 0.$$

A maximum, on the other hand, would yield a negative second derivative.

• PROBLEM 340

Where are the maxima and minima of $y = 2x^2 - 3x$?

Solution: To obtain a maximum or minimum we find $\frac{dy}{dx}$, set it equal to 0, and solve for x, obtaining the critical points. Then we use the Second Derivative Test to determine whether the critical value is a maximum or minimum. Doing this we have:

$$\frac{dy}{dx} = 4x - 3, \quad \frac{d^2y}{dx^2} = 4 = + \text{ (positive)},$$

and indicates a minimum.

Setting $\frac{dy}{dx}$ equal to zero, and solving to locate the minimum,

$$\frac{dy}{dx} = 4x - 3 = 0,$$

therefore

$$x = \frac{3}{4},$$

which we substitute in the original equation and get

$$y = 2\left(\frac{3}{4}\right)^2 - 3 \cdot \frac{3}{4} = \frac{18}{16} - \frac{9}{4} = -\frac{9}{8}.$$

Therefore, the minimum is at $x = \frac{3}{4}$, $y = -\frac{9}{8}$.

• PROBLEM 341

Locate the maxima and minima of $y = 2x^2 - 8x + 6$.

Solution: To obtain the minima and maxima we find

$\frac{dy}{dx}$, set it equal to 0, and solve for x. We find:

$$\frac{dy}{dx} = 4x - 8 = 0.$$

Therefore, x = 2 is the critical point. We now use the Second Derivative Test to determine whether x = 2 is a maximum or a minimum. We find:

$\frac{d^2y}{dx^2} = 4$, (positive). The second derivative is positive, hence x = 2 is a minimum.

Now substitute this back into the original equation to get the corresponding ordinate.

$$y = 2x^2 - 8x + 6 = 2 \cdot 2^2 - 8 \cdot 2 + 6 = 8 - 16 + 6$$
$$= -2.$$

Therefore, the minimum is at x = 2, y = -2.

• PROBLEM 342

Where are the maxima and minima of $y = 3x^3 + 4x + 7$?

Solution: $y = 3x^3 + 4x + 7$, then $\frac{dy}{dx} = 9x^2 + 4 = 0$, $x^2 = -\frac{4}{9}$,

$$x = \pm\sqrt{-\frac{4}{9}} = \pm\frac{2i}{3}.$$

This is an imaginary quantity. Since these are not real roots, in this example y has neither a maximum nor a minimum.

• PROBLEM 343

Find the maxima and minima of $f(x) = 3x^5 - 5x^3$.

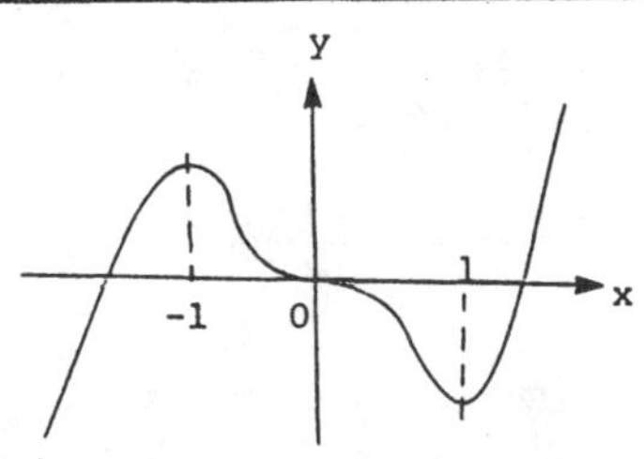

Solution: To determine maxima and minima we find f'(x), set it equal to 0 and solve for x, obtaining the critical points. We find:

$f'(x) = 15x^4 - 15x^2 = 15x^2(x^2-1)$. Therefore, x = 0, ± 1 are the critical points. We must now determine whether the function reaches a maximum, minimum or neither at each of these values. To do this we will use the Second Derivative Test. Computing the second derivative f", we have

$$f''(x) = 60x^3 - 30x = 30x(2x^2-1).$$

$f''(1) = 30 > 0$. Therefore, $x = 1$ is a relative minimum point and $x = -1$ is a relative maximum since $f''(-1) = -30 < 0$. Now, $f''(0) = 0$. Therefore the Second Derivative Test indicates a point which is neither maximum nor minimum. This is known as a point of inflection. For further study of the behavior of f at 0 we must use the first derivative test. We examine $f'(x)$ when $-1 < x < 0$ and when $0 < x < 1$. Let us select a representative value from each interval. We will use $x = -\frac{1}{2}$ and $x = \frac{1}{2}$. For $f'\left(-\frac{1}{2}\right)$ we obtain a negative value, and for $f'\left(\frac{1}{2}\right)$ we again obtain a negative value. Because there is no change in sign we conclude that at $x = 0$ there is neither a maximum nor a minimum, as can also be seen from the graph.

• PROBLEM 344

Find the maxima and minima of the function

$$f(x) = 3x^4 - 4x^3.$$

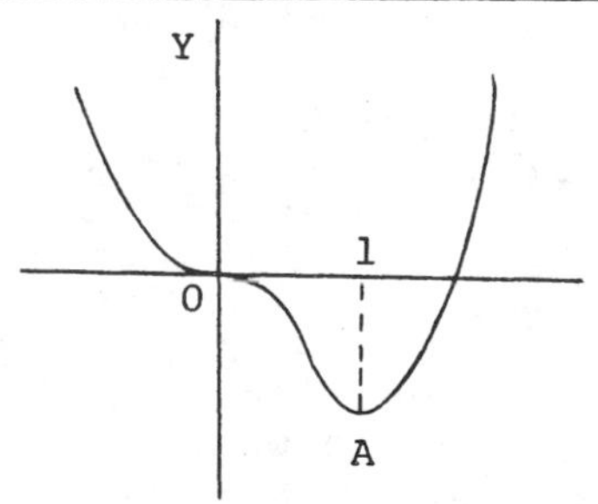

Solution: To find the maxima and minima we find $f'(x)$, set it equal to 0, and solve for x to obtain the critical points. Since

$$f'(x) = 12x^2(x-1),$$

the critical values are $x = 0$ and $x = 1$. We must now determine whether the function reaches a maximum, minimum or neither at each of the critical points. To do this, we examine $f'(x)$ when $x < 0$, $1 > x > 0$, and $x > 1$. If $f'(x)$ changes from + to - at the critical points, we have a maximum; from - to +, we have a minimum, and if there is no change in sign we have neither one at that critical point, but a point of inflection. Let us first examine the critical value $x = 1$. If $x < 1$, then $f'(x) < 0$ and if $x > 1$, then $f'(x) > 0$. Hence, the given function has a minimum value at

$x = 1$. If $x < 0$, we have $f'(x) < 0$, but if $0 < x < 1$, then $f'(x) < 0$. Therefore, since $f'(x)$ does not change sign in passing through $x = 0$, $f(x)$ does not have a maximum nor a minimum at the critical value $x = 0$. See Figure.

• PROBLEM 345

Find the maxima and minima of the function

$$f(x) = 2x^3 - 3x^2 - 12x + 13.$$

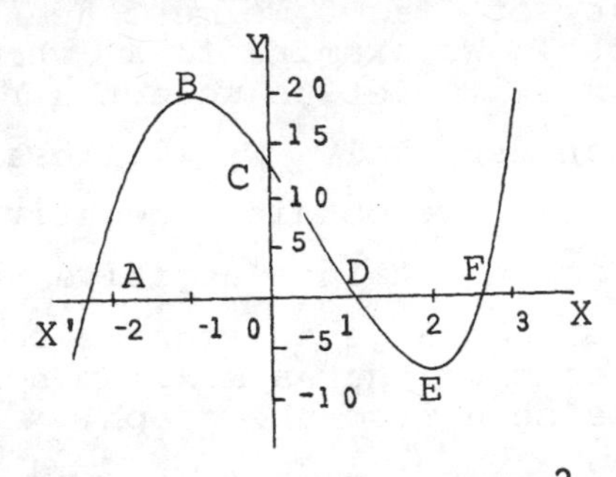

Solution: We first find $f'(x) = 6x^2 - 6x - 12 = 6(x-2)(x+1)$. Putting $f'(x) = 0$ and solving for x gives the critical values $x = 2$ and $x = -1$. We must now determine whether the function reaches a maximum or minimum at each of the critical values. For this problem, the best method for doing this is the Second Derivative Test. We find $f''(x)$ and evaluate it at both critical values, $x = 2$ and $x = -1$. If $f''(x)$ is negative, we have a maximum value, and if it is positive we have a minimum value. We find: $f''(x) = 12x - 6$. Therefore, $f''(2) = 18$ is positive and $f''(-1) = -18$ is negative. Thus, there is a minimum at $x = 2$ and a maximum at $x = -1$. To find the corresponding y values for $x = 2$ and $x = -1$ we substitute back into the original function. We find that, when $x = 2$, $f(x) = y = -7$ and when $x = -1$, $f(x) = y = 20$. Therefore, the function reaches a maximum at $(-1,20)$ and a minimum at $(2,-7)$.

• PROBLEM 346

What is the location of the maxima and minima of

$$y = \sqrt{2+x} + \sqrt{2-x}?$$

Solution: To determine the maxima and minima we find $\frac{dy}{dx}$, set it equal to 0, and solve for x, obtaining the critical values. We find:

$$\frac{dy}{dx} = \frac{1}{2\sqrt{2+x}} - \frac{1}{2\sqrt{2-x}} = 0$$

$$\sqrt{2+x} = \sqrt{2-x}$$

$$2+x = 2-x$$

$$x = 0,$$

the only critical value.

To test whether this is a maximum or minimum we use the Second Derivative Test. We find $\frac{d^2y}{dx^2}$ and evaluate it at x = 0. If it is negative we have a maximum and if positive, a minimum. We find:

$$\frac{d^2y}{dx^2} = \left[\frac{1}{2} \cdot -\frac{1}{2}(2+x)^{-3/2} \cdot 1\right] - \left[\frac{1}{2} \cdot -\frac{1}{2}(2-x)^{-3/2} \cdot (-1)\right]$$

or

$$\frac{d^2y}{dx^2} = -\frac{1}{4}(2+x)^{-3/2} - \frac{1}{4}(2-x)^{-3/2}$$

$$= -\frac{1}{4\sqrt{(2+x)^3}} - \frac{1}{4\sqrt{(2-x)^3}}.$$

For x = 0,

$$\frac{d^2y}{dx^2} = -\frac{1}{4\sqrt{2^3}} - \frac{1}{4\sqrt{2^3}} = -, \text{ (negative)}$$

which indicates a maximum.

Now substitute the value x = 0 into the original equation to determine the y value. Doing this we find, $y = 2\sqrt{2}$. Therefore, maximum is at $x = 0$, $y = 2\sqrt{2}$.

• PROBLEM 347

Examine the function: $(x-1)^2(x+1)^3$ for the points at which it reaches its maximum and minimum values.

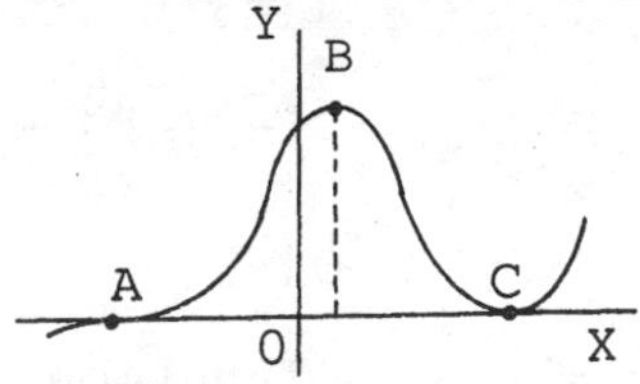

Solution: We have: $f(x) = (x-1)^2(x+1)^3$. To obtain maximum and minimum values we find f'(x), set it equal to 0, and solve for x to find the critical points. We find:

$$f'(x) = 2(x-1)(x+1)^3 + 3(x-1)^2(x+1)^2$$

$$= (x-1)(x+1)^2\left[2(x+1)+3(x-1)\right]$$

$$= (x-1)(x+1)^2(5x-1) = 0.$$

Solving, we obtain $x = 1, -1, \frac{1}{5}$, the critical values.

We must now test each critical point to determine whether it is maximum, minimum or neither. We can do this by the First Derivative Test. Examine $f'(x)$ at values that are, respectively, less than and greater than each of the three critical values. If the sign of $f'(x)$ changes from + to -, the critical value is a maximum; if it changes from - to +, it is a minimum, and if there is no change in sign, the critical value is neither. We rewrite $f'(x)$ as: $f'(x) = 5(x-1)(x+1)^2\left(x-\frac{1}{5}\right)$, to facilitate evaluation.

First, we examine the critical value $x = 1$ (C in figure).

$$\text{When } x < 1,\ f'(x) = 5(-)(+)^2(+) = -.$$

$$\text{When } x > 1,\ f'(x) = 5(+)(+)^2(+) = +.$$

Therefore, when $x = 1$ the function has a minimum value $f(1) = 0$ (the ordinate of C).

Now we examine the critical value $x = \frac{1}{5}$ (B in figure).

$$\text{When } x < \frac{1}{5},\ f'(x) = 5(-)(+)^2(-) = +.$$

$$\text{When } x > \frac{1}{5},\ f'(x) = 5(-)(+)^2(+) = -.$$

Therefore, when $x = \frac{1}{5}$ the function has a maximum value $f\left(\frac{1}{5}\right) = 1.11$ (the ordinate of B).

Finally, we examine the critical value $x = -1$ (A in figure).

$$\text{When } x < -1,\ f'(x) = 5(-)(-)^2(-) = +.$$

$$\text{When } x > -1,\ f'(x) = 5(-)(+)^2(-) = +.$$

Therefore, when $x = -1$ the function has neither a maximum nor a minimum value. This type of critical point is called a point of inflection.

• PROBLEM 348

Find the maxima and minima of: $y = \frac{4}{x^2 - 4}$, and trace the curve.

<u>Solution:</u> To find the maxima and minima we deter-

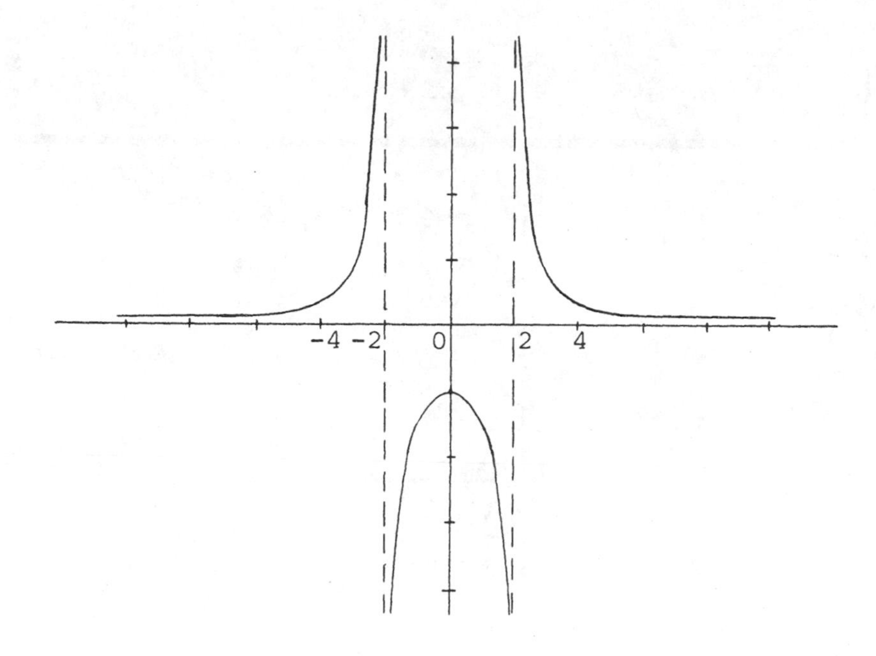

mine $\frac{dy}{dx}$, equate it to 0, and solve for x, obtaining the critical values. We find:

$$\frac{dy}{dx} = \frac{(x^2 - 4)(0) - 4(2x)}{(x^2 - 4)^2} = -\frac{8x}{(x^2 - 4)^2}.$$

$\frac{-8x}{(x^2 - 4)^2} = 0$, or $-8x = 0$, therefore $x = 0$. Substituting $x = 0$, into the original equation, $y = -1$, therefore the critical point is $(0,-1)$. To determine whether a maximum, minimum, or neither occurs at this point, we use the First Derivative Test. We examine $\frac{dy}{dx}$ at a point less than 0, (use -1), and at a point greater than 0 (use 1). If $\frac{dy}{dx}$ changes sign from + to -, a maximum occurs at $x = 0$, from - to +, a minimum occurs, and if there is no change in sign, neither a maximum nor a minimum occurs. We find that at $x = -1$, $\frac{dy}{dx} = \frac{8}{9}$, a positive value, and at $x = 1$, $\frac{dy}{dx} = -\frac{8}{9}$, a negative value. Therefore, at the point $(0,-1)$, a maximum occurs.

Upon further investigation of the curve: $y = \frac{4}{x^2 - 4}$, we observe that for the values $x = 2$, -2, y is undefined or $y = \pm\infty$. Therefore, the graph of the curve has asymptotes at 2 and -2, as shown in the accompanying graph.

• PROBLEM 349

Locate the maxima and minima of

$$y = \frac{x^3}{3} - \frac{5x^2}{2} + 6x + 4.$$

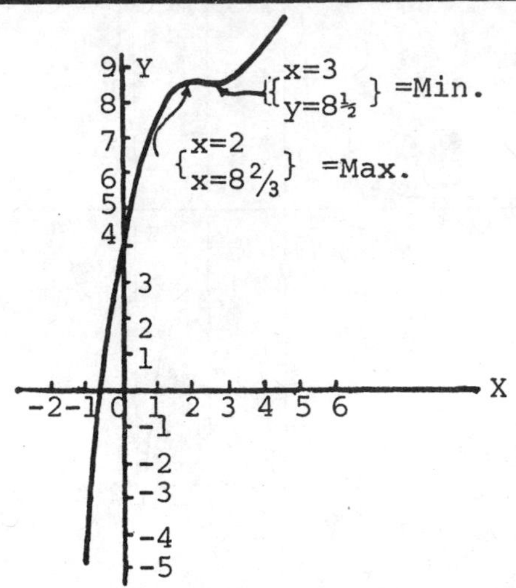

Solution: To find the maxima and minima we find $\frac{dy}{dx}$, set it equal to 0, and solve for x, obtaining the critical points. Doing this we have:

$$\frac{dy}{dx} = x^2 - 5x + 6 = 0, \quad (x-2)(x-3) = 0,$$

therefore,

$$x = 3 \text{ and } 2.$$

We now use the Second Derivative Test to determine whether the critical values are maximum, minimum or neither. We find:

$$\frac{d^2y}{dx^2} = 2x - 5.$$

For $x = 3$,

$$\frac{d^2y}{dx^2} = 2x - 5 = 2 \cdot 3 - 5 = + \text{ (positive)},$$

which indicates a minimum.

For $x = 2$,

$$\frac{d^2y}{dx^2} = 2x - 5 = 2 \cdot 2 - 5 = - \text{ (negative)},$$

which indicates a maximum.

Therefore, we have a minimum at $x = 3$ and a maximum at $x = 2$. We now wish to find the corresponding ordinates. Going back to the original equation, we have:

For $x = 3$,

$$y = \frac{x^3}{3} - \frac{5x^2}{2} + 6x + 4 = \frac{3^3}{3} - \frac{5 \cdot 3^2}{2} + 6 \cdot 3 + 4$$

$$= 9 - \frac{45}{2} + 18 + 4 = 8\frac{1}{2}.$$

For $x = 2$,

$$y = \frac{x^3}{3} - \frac{5x^2}{2} + 6x + 4 = \frac{2^3}{3} - \frac{5 \cdot 2^2}{2} + 6 \cdot 2 + 4$$

$$= \frac{8}{3} - 10 + 12 + 4 = 8\frac{2}{3}.$$

Therefore, minimum is at $x = 3$, $y = 8\frac{1}{2}$, and maximum is at $x = 2$, $y = 8\frac{2}{3}$.

• PROBLEM 350

> In the expression $f(x) = 1/x + x$, determine the relative maxima and minima for f and the intervals in which f is increasing or decreasing. Sketch the graph for f.

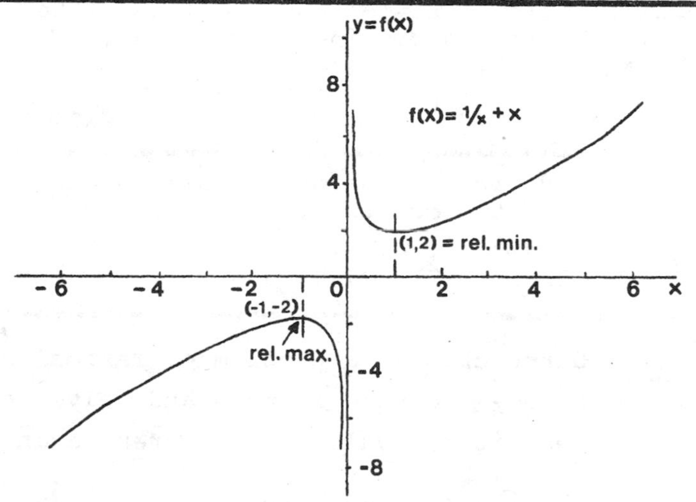

Solution: To determine the relative extreme points we find $f'(x)$, equate it to 0 and solve for x to obtain critical values. Computing the derivative for f, we obtain

$$f'(x) = -\frac{1}{x^2} + 1 = \frac{x^2-1}{x^2} = \frac{(x-1)(x+1)}{x^2} = 0.$$

Clearly, $f'(x) = 0$ when $x = 1$ or $x = -1$. We must now determine whether the function reaches a maximum or a minimum at $x = 1$ and -1, and the intervals in which f increases and those in which it decreases. We know that a function increases where $f'(x) > 0$ and decreases where $f'(x) < 0$, and that a maximum or minimum occurs at the point where $f'(x)$ changes sign. Therefore, let us look at $f'(x)$ in the following intervals;

	$f'(x) = \frac{(x-1)(x+1)}{x^2}$
$x < -1$	+, positive
$0 > x > -1$	-, negative
$1 > x > 0$	-, negative
$x > 1$	+, positive

We conclude that there is a maximum at $x = -1$, because $f'(x)$ changes from + to - at this value; and there is a minimum at $x = 1$, because $f'(x)$ changes from - to +. Also, the function is increasing in $x < -1$ and $x > 1$ and decreasing in $0 > x > -1$ and $1 > x > 0$.

We can also obtain the same result by differentiating $f'(x)$: $f''(x) = \frac{d}{dx}\left(-\frac{1}{x^2} + 1\right) = \frac{2}{x^3}$. This is + at $x = 1$, indicating a minimum. It is - at $x = -1$, hence this is a maximum.

Now substitute $x = -1$ and $x = 1$ into the original function to obtain the y values. We find: $(-1,-2)$ and $(1,2)$. (See graph.)

• PROBLEM 351

Find the x and y positions of the maxima and minima of the graph of the equation.

$$y = 3x^3 - 9x^2 - 27x + 30. \qquad (a)$$

Solution: Since this is a problem of maxima and minima, we find $\frac{dy}{dx}$, equate it to 0 and solve for x to obtain the critical values. Differentiating,

$$\frac{dy}{dx} = 9x^2 - 18x - 27. \qquad (b)$$

Hence, for maximum or minimum:

$$9x^2 - 18x - 27 = 0.$$

Dividing by 9 and factoring, we have $x^2 - 2x - 3 = (x-3)(x+1) = 0$. Therefore,

$$x = 3, \; x = -1 \qquad (c)$$

are the abscissas at which the maximum and minimum ordinates are located. In order to distinguish between them, we must find the second derivative

$f''(x)$. Differentiating equation (b),

$$f''(x) = 18x - 18. \qquad (d)$$

Inserting, in this, the first value of x from (c) we find

$$f''(3) = 18.3 - 18 = +36$$

and since this is positive, a minimum ordinate occurs at the point where x = 3. Using the second value of x from (c) in (d), we get

$$f''(-1) = 18(-1) - 18 = -36$$

and, this being negative, a maximum ordinate occurs at the point where x = -1.

In order to find the length of the greatest and least ordinates we substitute the values of the corresponding abscissas from (c) back in the equation (a), obtaining y. Thus we find at x = -1,

$$y_{max.} = 3(-1)^3 - 9(-1)^2 - 27(-1) + 30 = +45,$$

and at the point where x = 3,

$$y_{min.} = 3(3)^3 - 9(3)^2 - 27(3) + 30 = -51.$$

It is to be noted that, in these results the word greatest does not necessarily mean actually longest, nor does least mean actually shortest. But, what is meant is that mathematically all negative numbers are "less" than positive numbers, or graphically, the ordinate y = -51 is below the ordinate y = +45.

• PROBLEM 352

Determine the maxima and minima of $f(x) = x^3 - x$ in the interval from x = -1 to x = 2.

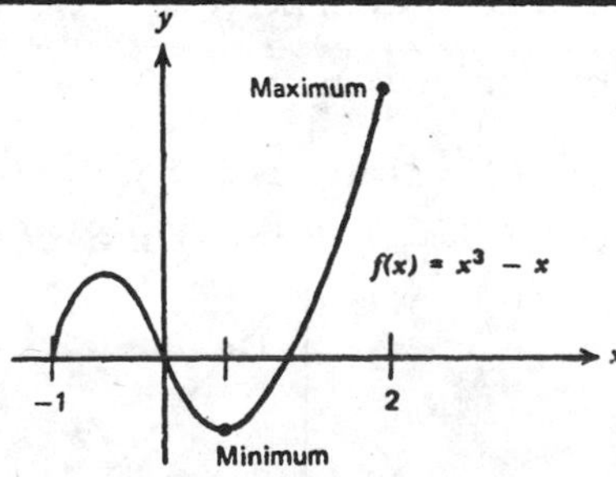

Solution: To determine the extreme points, we find $f'(x)$, equate it to 0, and solve for x to obtain the critical points. We have:

$$f'(x) = 3x^2 - 1 = 0. \quad x^2 = \frac{1}{3}.$$

Therefore, the critical values are $x = \pm \frac{1}{\sqrt{3}}$. Now

$$f\left(\frac{1}{\sqrt{3}}\right) = -\frac{2}{3\sqrt{3}} \quad \text{and} \quad f\left(-\frac{1}{\sqrt{3}}\right) = \frac{2}{3\sqrt{3}}.$$

Evaluating f at the end points of the interval we have $f(-1) = 0$ and $f(2) = 6$. Therefore, $x = 2$, an end point, is the maximum point for f, and $x = 1/\sqrt{3}$ is the minimum point as can be seen in the Figure. The extreme values of f in $[-1,2]$ are 6 and $-2/3\sqrt{3}$. The point $\left(-\frac{1}{\sqrt{3}}, \frac{2}{3\sqrt{3}}\right)$ is not an absolute maximum, but it is a relative maximum.

• PROBLEM 353

Find the maximum, minimum, and inflection points of $y = \sin x + \cos x$, and trace the curve.

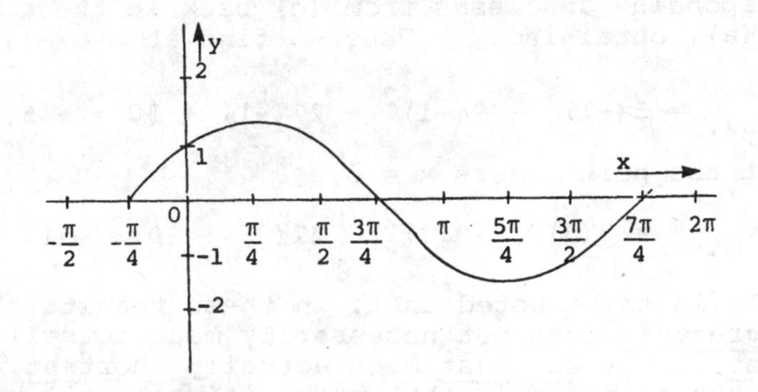

Solution: To find the maximum and minimum points we determine $\frac{dy}{dx}$, equate it to 0, and solve for x, obtaining critical values.

$$\frac{dy}{dx} = \cos x - \sin x.$$

$$\cos x - \sin x = 0,$$

or,

$$\cos x = \sin x.$$

In the 45-45-90 right triangle, $\sin 45° = \frac{\sqrt{2}}{2}$ while $\cos 45° = \frac{\sqrt{2}}{2}$. Therefore $\cos x = \sin x$ when $x = \frac{\pi}{4}$.

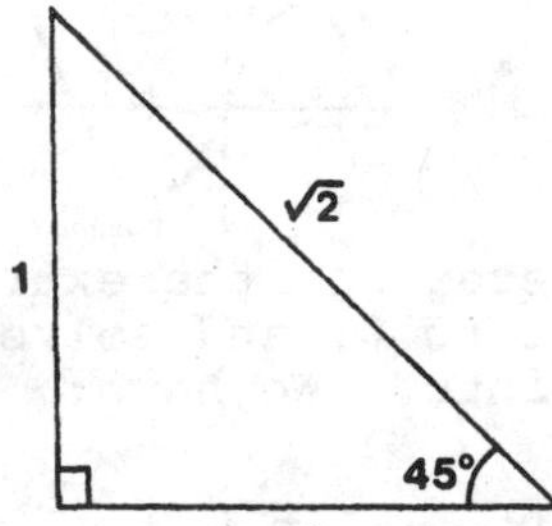

However, in the third quadrant of the coordinate axes, both sin x and cos x are negative, hence

$\sin x = \cos x$ when $x = 225°$ or $\frac{5\pi}{4}$. If we add $2\pi n$, $n = \ldots,-2,-1,0,1,2,\ldots$, to both these values, because the curve has a period of 2π, we find infinitely many critical values. Therefore, the critical values are:

$$\frac{\pi}{4} + 2\pi n, \quad \text{and} \quad \frac{5\pi}{4} + 2\pi n; \quad n = \ldots,-2,-1,0,1,2,\ldots .$$

But we restrict the investigation to $0 \le x \le 2\pi$. We can now use the First Derivative Test to determine whether a maximum, minimum, or neither occurs at each critical value. We must choose values greater and less than each critical value, and examine $\frac{dy}{dx}$ at these points. We construct the follow- table:

		$\frac{dy}{dx} = \cos x - \sin x$	
$x < \frac{\pi}{4}$	$x = 0$	1	(+)
$\frac{5\pi}{4} > x > \frac{\pi}{4}$	$x = \frac{\pi}{2}$	-1	(−)
$x > \frac{5\pi}{4}$	$x = 2\pi$	1	(+)

Now, since $\frac{dy}{dx}$ changes from (+) when $x < \frac{\pi}{4}$ to (−) when $x > \frac{\pi}{4}$, a maximum occurs at $x = \frac{\pi}{4}$. Since $\frac{dy}{dx}$ changes from (−) when $x < \frac{5\pi}{4}$ to (+) when $x > \frac{5\pi}{4}$, a minimum occurs at $x = \frac{5\pi}{4}$.

To find the inflection points we find $\frac{d^2y}{dx^2}$, equate it to 0, and solve for x.

$$\frac{d^2y}{dx^2} = -\sin x - \cos x.$$

$$-\sin x - \cos x = 0$$

$$-\sin x = \cos x.$$

When $x = 45°$ or $\frac{\pi}{4}$, $\sin x = \cos x$.

Since in the second quadrant $\sin x$ is (+) and $\cos x$ is (−), $-\sin x = \cos x$ when $x = 135° = \frac{3\pi}{4}$. Also, in the fourth quadrant, sin is (−) and cos is (+), therefore $-\sin x = \cos x$ when $x = 315° = \frac{7\pi}{4}$. We now test whether each value is indeed an inflection point. We choose values greater and less than each possible inflection point and examine $\frac{d^2y}{dx^2}$. The following table results:

		$\frac{d^2y}{dx^2} = -\sin x - \cos x$	
$x < \frac{3\pi}{4}$	$x = \frac{\pi}{2}$	-1	(-)
$\frac{7\pi}{4} > x > \frac{3\pi}{4}$	$x = \pi$	1	(+)
$x > \frac{7\pi}{4}$	$x = 2\pi$	-1	(-)

Since $\frac{d^2y}{dx^2}$ changes sign from (-) when $x<\frac{3\pi}{4}$ to (+) when $x > \frac{3\pi}{4}$, $x = \frac{3\pi}{4} + 2n\pi$, $n = \ldots,-1,0,1,\ldots$ are inflection points, and since $\frac{d^2y}{dx^2}$ changes from (+) when $x < \frac{7\pi}{4}$ to (-) when $x > \frac{7\pi}{4}$, $x = \frac{7\pi}{4} + 2n\pi$ are also inflection points. See graph.

• PROBLEM 354

Investigate the function $y = (x-a)^{1/3}(2x-a)^{2/3}$ for maxima and minima.

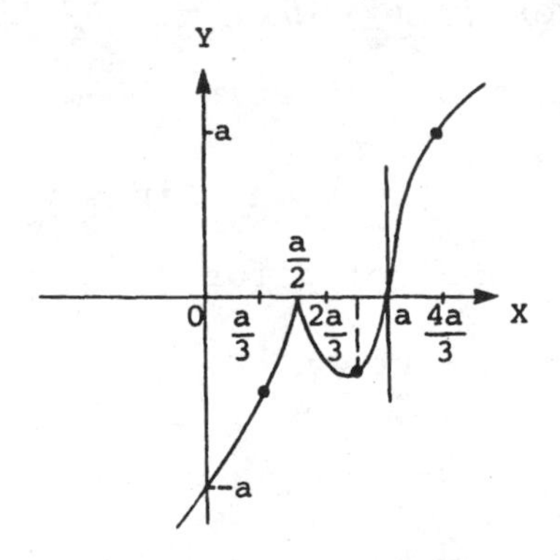

Solution: Differentiating

$$y' = \frac{(2x-a)^{2/3}}{3(x-a)^{2/3}} + \frac{4(x-a)^{1/3}}{3(2x-a)^{1/3}} = \frac{6x - 5a}{3(x-a)^{2/3}(2x-a)^{1/3}}.$$

From $y' = 0$, and $1/y' = 0$, the critical points are

$$x = \frac{a}{2}, \quad x = \frac{5a}{6}, \quad x = a.$$

We must now determine whether each of these critical points is a maximum, minimum or neither. We choose a value of x less than and a value greater than each of the critical values and evaluate y' at these values. If the sign changes from positive to negative, we have a maximum. If it changes from negative to positive, we have a minimum. If the sign does not change there is neither one at that critical value.

Setting $x = \frac{a}{3}$ and $\frac{2a}{3}$ in turn, we have y' positive and negative respectively. Hence $x = \frac{a}{2}$ makes y a maximum.

Test $x = \frac{5a}{6}$, using the values $\frac{2a}{3}$ and $\frac{9a}{10}$. These show y' to be successively negative and positive, so the function has a minimum value at $x = \frac{5a}{6}$.

Apply the test to x = a, with the values $\frac{9a}{10}$ and 2a. These show y' positive in both cases, therefore, at x = a there is neither a maximum or a minimum. We observe that $y' = \frac{a}{0} = \infty$ at x = a, therefore the graph has a vertical tangent at this value. At x = a there is a point of inflection as shown. The maximum point at (a/2,0) is called a cusp.

• PROBLEM 355

Determine the extreme points of $f(x) = 2 - |1-x|$ in the interval between x = 0 and x = 2.

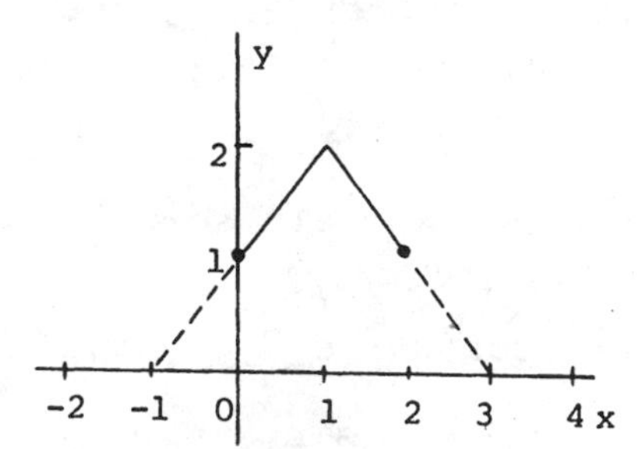

Solution: To determine extreme points we generally find f'(x), equate it to 0 and solve for x, obtaining the critical values. In this example we find that $f'(x) = \pm 1$ at each point x where the derivative exists. Since f' never vanishes, the extreme points must either be end points of the interval or the point x = 1 where f' does not exist. If we compute the values of f at these points we have f(0) = 1, f(1) = 2, and f(2) = 1. Therefore, x = 1 is the maximum point for f and x = 0, 2 are the minimum points. The graph of f is sketched in the figure.

• PROBLEM 356

Determine the extremes of the expression $f(x) = \sqrt{|x|}$ and sketch the graph.

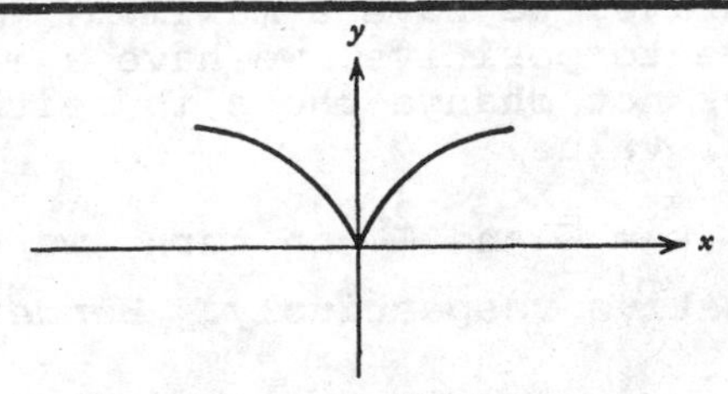

Graph of $f(x) = \sqrt{|x|}$.

Solution: The function f is continuous everywhere. If $x > 0$, then $f'(x) = \frac{1}{2\sqrt{x}}$, and if $x < 0$ then $f'(x) = -\frac{1}{2\sqrt{x}}$. The point $x = 0$ is a critical point since $f'(0)$ does not exist. Since $f'(x) < 0$ if $x < 0$ and $f'(x) > 0$ if $x > 0$, f is decreasing if $x \leq 0$ and increasing if $x \geq 0$. Therefore, f has a relative minimum at $x = 0$. Indeed $x = 0$ is an absolute minimum, and at $x = 0$, $f = 0$. The graph of f is shown in the figure. From the graph we can conclude that f does not have an absolute maximum.

• PROBLEM 357

Find the maxima and minima of $y = e^{-x} \sin x$.

Solution: To find the maxima and minima of the given function, we determine $\frac{dy}{dx}$, equate it to 0, and solve for x to obtain critical values. We find:

$$\frac{dy}{dx} = e^{-x} \cos x + \sin x(-e^{-x})$$

$$= e^{-x} \cos x - e^{-x} \sin x.$$

$$e^{-x} \cos x - e^{-x} \sin x = 0.$$

$$e^{-x}(\cos x - \sin x) = 0.$$

$$\cos x = \sin x.$$

We know that when $x = 45° = \frac{\pi}{4}$, and when $x = 225° = \frac{5\pi}{4}$, $\cos x = \sin x$. Limiting the investigation to $0 \leq x \leq 2\pi$, because an infinite number of critical values can be found, the critical values are $x = \frac{\pi}{4}$ and $x = \frac{5\pi}{4}$.

To find the corresponding y's, we substitute back each critical value into the original equation, obtaining:

for $\quad x = \frac{\pi}{4}, \quad y = e^{-\frac{\pi}{4}} \sin \frac{\pi}{4}$

$$= e^{-\frac{\pi}{4}}\left(\frac{\sqrt{2}}{2}\right)$$

$$= \frac{1}{2}\sqrt{2}e^{-\frac{\pi}{4}}.$$

For $\quad x = \frac{5\pi}{4}, \quad y = e^{-\frac{5\pi}{4}} \sin \frac{5\pi}{4}$

$$= e^{-\frac{5\pi}{4}}\left(-\frac{\sqrt{2}}{2}\right)$$

$$= -\frac{1}{2}\sqrt{2}e^{-\frac{5\pi}{4}}.$$

Therefore, the critical points are

$$\left(\frac{\pi}{4}, \frac{1}{2}\sqrt{2}e^{-\frac{\pi}{4}}\right) \text{ and } \left(\frac{5\pi}{4}, -\frac{1}{2}\sqrt{2}e^{-\frac{5\pi}{4}}\right).$$

Now, we determine whether a maximum, minimun, or neither occurs at each critical point, by using the First Derivative Test.
We examine $\frac{dy}{dx}$ for values greater and less than each critical value. If $\frac{dy}{dx}$ changes from + to -, a maximum occurs at that critical point, if $\frac{dy}{dx}$ changes from - to + a minimum occurs, and if $\frac{dy}{dx}$ does not change sign, neither occurs. The representative points must be chosen from the intervals $x < \frac{\pi}{4}$, $\frac{5\pi}{4} > x > \frac{\pi}{4}$, and $x > \frac{5\pi}{4}$. We choose the values as shown in the following table:

		$\frac{dy}{dx} = e^{-x} \cos x - e^{-x} \sin x$	
$x < \frac{\pi}{4}$	$x = 0$	1	(+)
$x = \frac{\pi}{4}$	$x = \frac{\pi}{4}$	0	(0)
$\frac{5\pi}{4} > x > \frac{\pi}{4}$	$x = \frac{\pi}{2}$	$-e^{\frac{\pi}{2}}$	(-)
$x = \frac{5\pi}{4}$	$x = \frac{5\pi}{4}$	0	(0)
$x > \frac{5\pi}{4}$	$x = 2\pi$	$e^{-2\pi}$	(+)

From the table we conclude that $\left(\frac{\pi}{4}, \frac{1}{2}\sqrt{2}e^{-\frac{\pi}{4}}\right)$ is a maximum point and $\left(\frac{5\pi}{4}, -\frac{1}{2}\sqrt{2}e^{-\frac{5\pi}{4}}\right)$ is a minimum point.

• PROBLEM 358

Find all maximum and minimum points and draw the graph of: $y = e^{2x} + e^{-2x}$.

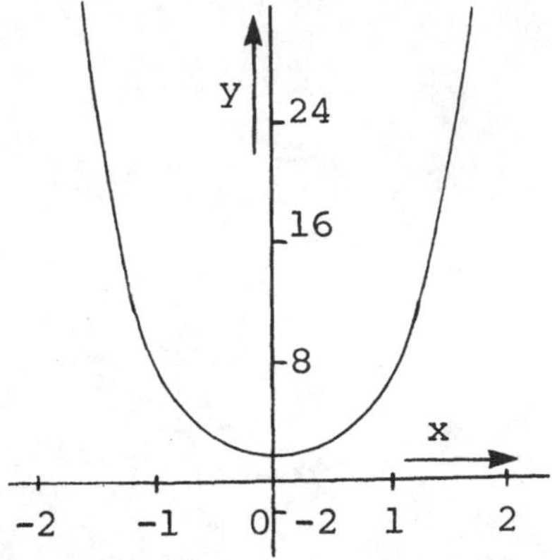

Solution: To find the maximum and minimum points, we find $\frac{dy}{dx}$, equate it to 0 and solve for x, obtaining the critical values. Doing this, we find:

$$\frac{dy}{dx} = e^{2x}(2) + e^{-2x}(-2)$$

$$= 2e^{2x} - 2e^{-2x} = 2(e^{2x} - e^{-2x}).$$

$$2(e^{2x} - e^{-2x}) = 0.$$

$$e^{2x} - e^{-2x} = 0.$$

$$e^{2x} = \frac{1}{e^{2x}}.$$

$$e^{4x} = 1.$$

Taking the natural logarithm of both sides, we have: $4x = \ln 1$, or $4x = 0$. Therefore $x = 0$ is the critical value. We now determine whether a maximum, minimum or neither occurs at this value. We do this by using the First Derivative Test. We choose values greater and less than the critical value, and evaluate $\frac{dy}{dx}$ at these points. If $\frac{dy}{dx}$ changes from + to −, the critical point is a maximum, from − to +, it is a minimum, and if there is no change in sign, neither. We use $-\frac{1}{2}$ and $\frac{1}{2}$. For $x = -\frac{1}{2}$, $\frac{dy}{dx} = \frac{2}{e} - 2e$, a negative value. For $x = \frac{1}{2}$, $\frac{dy}{dx} = 2e - \frac{2}{e}$, a positive value. Since $\frac{dy}{dx}$ changes from − to +, a minimum occurs at the critical point.

Substituting $x = 0$ into the original equation, $y = 2$. Therefore a minimum occurs at $(0,2)$, and there are no maxima. See graph.

• PROBLEM 359

Find the maxima and minima of $y = \frac{\ln x}{x}$.

Solution: To find a maxima and minima we find y', set it equal to 0 and solve for x to obtain critical values. We have:

$$y' = \frac{x\left(\frac{1}{x}\right) - \ln x}{x^2} = \frac{1 - \ln x}{x^2} = 0.$$

If $y' = 0$, then $1 - \ln x = 0$ or $\ln x = 1$, so that $x = e$. To test whether the value $x = e$ is a maximum or minimum we evaluate y' for values less than e and greater than e. If y' changes from positive to negative we have a maximum and if it changes from negative to positive we have a minimum. If $x < e$, then $\ln x < 1$ and y' is positive, and if $x > e$, then $\ln x > 1$ and y' is negative; therefore, y has a maximum at $x = e$.

• PROBLEM 360

Find the maxima and minima of the function $a - b(x-c)^{2/3}$.

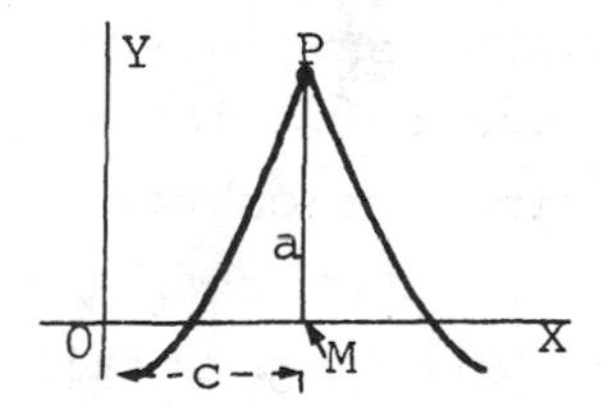

Solution: To obtain maxima and minima we find $f'(x)$, equate it to 0 and solve to obtain critical values.

$$f(x) = a - b(x-c)^{2/3}.$$

$$f'(x) = -\frac{2b}{3(x-c)^{1/3}}.$$

This is nowhere equal to zero, but

$$\frac{1}{f'(x)} = -\frac{3(x-c)^{1/3}}{2b} = 0 \text{ at } x = c.$$

Since $x = c$ is a critical value for which $\frac{1}{f'(x)} = 0$ (and $f'(x) = \infty$), but for which $f(x)$ itself is not infinite, let us test the function for maximum and minimum values when $x = c$, using the First Derivative Test.

When $x < c, f'(x) = +$.

When $x > c, f'(x) = -$.

Since $f'(x)$ changes sign from + to −, we conclude that $x = c$ is a maximum value. Hence, when $x = c = OM$ on the accompanying diagram, the function has a maximum value $f(c) = a = MP$.

• PROBLEM 361

Find the maximum, minimum and inflection points of the function given by: $x = \tan\theta$, $y = \cos^2\theta$, and trace the curve.

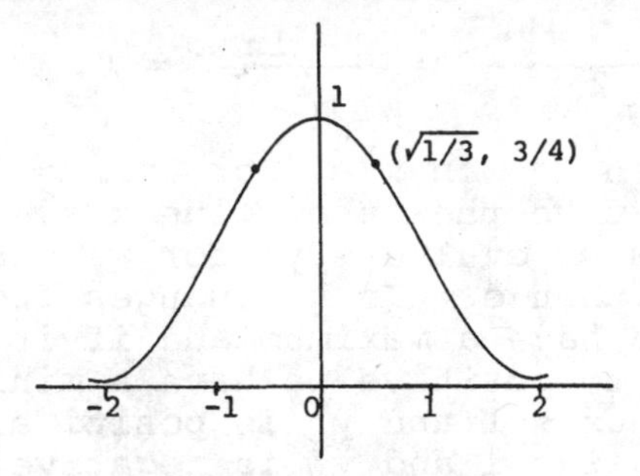

Solution: Since $x = \tan\theta$,

$$y = \cos^2\theta = \frac{1}{\sec^2\theta} = \frac{1}{1 + \tan^2\theta} = \frac{1}{1 + x^2}.$$

To find the maximum and minimum points we solve for the first derivative, $\frac{dy}{dx}$, in this expression, set it equal to zero and solve for x, the critical value. We determine the change in sign of $\frac{dy}{dx}$ when numbers higher and lower than the critical value are substituted. First we obtain $\frac{dy}{dx}$, using the quotient rule.

$$y = \frac{1}{1 + x^2}.$$

$$\frac{dy}{dx} = \frac{-2x}{(1 + x^2)^2}.$$

$$\frac{dy}{dx} = \frac{-2x}{1 + 2x^2 + x^4}.$$

Setting $\frac{dy}{dx}$ equal to zero and solving for x, we obtain:

$$\frac{dy}{dx} = 0.$$

$$\frac{-2x}{1 + 2x^2 + x^4} = 0.$$

$$-2x = 0.$$

$$x = 0,$$

a critical value.

We can now determine the sign change by substituting $x = -1$ and $x = 1$ into $\frac{dy}{dx}$. For $x = -1$,

$$\frac{dy}{dx} = \frac{-2(-1)}{1 + 2(-1)^2 + (-1)^4} = \frac{2}{1 + 2 + 1} = \frac{1}{2},$$

a positive number.

For $x = 1$,

$$\frac{dy}{dx} = \frac{-2(1)}{1 + 2(1)^2 + (1)^4} = \frac{-2}{1 + 2 + 1} = -\frac{1}{2},$$

a negative number.

Since the sign changes from positive to negative, $x = 0$ is a maximum value. When $x = 0$ is substituted into $y = \frac{1}{1 + x^2}$, $y = 1$. Hence, the critical point, a maximum, is $(0,1)$. Since there is no other critical value for x, no minima exist. To find the inflection point, we find $\frac{d^2y}{dx^2}$, set it equal to zero, and solve for x. Using the quotient rule, we obtain:

$$y = \frac{1}{1 + x^2}.$$

$$\frac{dy}{dx} = \frac{1}{(1 + x^2)^2}.$$

$$\frac{d^2y}{dx^2} = \frac{(-2)(1 + x^2)^2 - (-2x)(2)(1 + x^2)(2x)}{(1 + x^2)^4}.$$

$$\frac{d^2y}{dx^2} = \frac{(1 + x^2)\left[-2(1 + x^2) + 8x^2\right]}{(1 + x^2)^4}.$$

$$\frac{d^2y}{dx^2} = \frac{6x^2 - 2}{(1 + x^2)^3}.$$

Setting $\frac{d^2y}{dx^2}$ equal to zero and solving for x, we obtain:

$$\frac{d^2y}{dx^2} = 0.$$

$$\frac{6x^2 - 2}{(1 + x^2)^3} = 0.$$

$$6x^2 - 2 = 0.$$

$$x = \pm\sqrt{\frac{1}{3}}.$$

Substituting this value for x into $y = \dfrac{1}{1 + x^2}$, we obtain:

$$y = \frac{1}{1 + \left(\pm\sqrt{\frac{1}{3}}\right)^2} = \frac{1}{\frac{4}{3}} = \frac{3}{4}.$$

The inflection points are $\left(\pm\sqrt{\frac{1}{3}}, \frac{3}{4}\right)$. Using the maximum point (0,1), the highest point on the curve, the inflection points, $\left(\pm\sqrt{\frac{1}{3}}, \frac{3}{4}\right)$, where the curve changes from concave downward to concave upward, and the facts that $y > 0$ and that there are no minima (the curve never reaches the x-axis) we can draw the graph as shown.

• PROBLEM 362

Show that, in the interval:

$$-\frac{\pi}{2} \leq x \leq \frac{\pi}{2}, \quad \text{the curve } y = \sin^n x,$$

where n is a positive integer greater than unity, has two or three inflection points according to whether n is even or odd.

Solution: To find the inflection points we solve for the second derivative,

$$\frac{d^2y}{dx^2},$$

set it equal to zero and then solve for x. We obtain:

$$y = \sin^n x$$

$$\frac{dy}{dx} = n \sin^{n-1} x \cos x.$$

Using the product rule, we find $\dfrac{d^2y}{dx^2}$ as follows:

$$\frac{d^2y}{dx^2} = (n - 1) n \sin^{n-2} x \cos^2 x + n \sin^{n-1} x (- \sin x)$$

$$\frac{d^2y}{dx^2} = (n - 1)n \sin^{n-2} x \cos^2 x - n \sin^n x = 0$$

$$\sin^{n-2} x \left((n - 1) \cos^2 x - \sin^2 x\right) = 0$$

$$\sin^{n-2} x = 0, \ x = 0,$$

or:

$$(n - 1)\cos^2 x - \sin^2 x = 0$$

$$(n - 1)\cos^2 x = \sin^2 x$$

$$n - 1 = \tan^2 x$$

$$\tan x = \pm \sqrt{n - 1}.$$

From these results it would appear that there are always three points of inflection: $0, \pm \sqrt{n - 1}$. However, an inspection of $\frac{dy}{dx}$ in the region $x < 0$ shows that $\frac{dy}{dx}$ is positive for n even, due to the fact that $\sin^n(- x) = (- \sin x)^n$ is positive for n even. Hence, for that case, $x = 0$ is not a point of inflection but a minimum.

• PROBLEM 363

Show that the curve

$$y = \sin \frac{1}{x}$$

has infinitely many maxima and minima in the interval $0 < x < 1$.

Solution: Let $\frac{1}{x} = u$, then $y = \sin u$. Using the chain rule,

$$\frac{dy}{dx} = \frac{dy}{du} \ \frac{du}{dx} .$$

$$\frac{dy}{dx} = - \frac{\cos \frac{1}{x}}{x^2} .$$

For maxima and minima, the slopes have to vanish. Hence,

$$- \frac{\cos \frac{1}{x}}{x^2} = 0,$$

which implies that $\cos \frac{1}{x} = 0$. This happens only when

$$\frac{1}{x} = \frac{2n - 1}{2} \pi = \pi\left(n - \frac{1}{2}\right)$$

where n is any integer.

As x approaches zero, $\frac{1}{x} \to \infty$, hence $2n - 1 \to \infty$, $n \to \infty$. For $x = 1$, $n = \frac{1}{\pi} + \frac{1}{2}$.

Thus, for the interval $0 < x < 1$ we have

$$\frac{1}{\pi} + \frac{1}{2} < n < \infty.$$

Show that the curve: $y = x^n$, where n is a positive integer, has a minimum point at the origin if n is even but has neither a maximum nor a minimum if n is odd.

Solution: First we show that the origin is indeed a critical point by determining $\frac{dy}{dx}$ for $y = x^n$, equating it to 0, and solving for x. We find:

$$\frac{dy}{dx} = n \cdot x^{n-1}.$$

$$n \cdot x^{n-1} = 0$$

$$x^{n-1} = 0.$$

For $n > 0$, $x^{n-1} = 0$ only if $x = 0$. Substituting $x = 0$ in the original equation, we find the critical point (0,0).

To determine whether this point is a maximum, minimum, or neither we use the First Derivative Test. Choosing $x = -1$ and $x = 1$, we examine $\frac{dy}{dx}$ at each. When $x = -1$, $\frac{dy}{dx} = n(-1)^{n-1}$. If n is even then n-1 is odd, and $(-1)^{n-1}$ is negative, therefore $\frac{dy}{dx}$ has a negative value. If n is odd, n-1 is even, and $(-1)^{n-1}$ is positive, therefore $\frac{dy}{dx}$ = positive value.

Now, when $x = 1$, $\frac{dy}{dx} = n \cdot 1^{n-1}$, a positive value whether n is even or odd. Since $\frac{dy}{dx}$ changes sign from - to + when n is even, a minimum occurs at (0,0) when n is even. Since, when n is odd, $\frac{dy}{dx}$ does not change sign (+ to +), there is neither a maximum nor a minimum at (0,0).

CHAPTER 16

APPLIED PROBLEMS IN MAXIMA AND MINIMA

The first step to be taken in solving these problems is to determine which variable is to be maximized or minimized, i.e., the dependent variable (y). Next, the variable, with respect to which the first variable varies, the independent variable (x) is found. An equation involving y and x should then be written. Any other variables in the problem should then be eliminated by substitution, so that only the one independent variable (x) and one dependent variable (y) remain.

At that point we are ready to proceed, differentiating the dependent variable (y) with respect to the independent variable (x). The derivative is then set equal to zero, to find the critical values. Finally, a determination is made of whether the critical values are maxima or minima.

• PROBLEM 365

Find the point at which the tangent to the curve $y = 2xe^{2x}$ is horizontal. Also determine the region in which the curve is concave downward and that in which the curve is concave upward.

Solution: The tangent to the curve is horizontal at the point where $\frac{dy}{dx} = 0$. Therefore, we determine $\frac{dy}{dx}$, equate it to 0, and solve for x. By the product rule,

$$\frac{dy}{dx} = (2x \cdot 2e^{2x}) + (e^{2x} \cdot 2)$$

$$= 4xe^{2x} + 2e^{2x}$$

$$= 2e^{2x}(2x + 1) = 0.$$

$2e^{2x} = 0$ and $2x + 1 = 0$. $x = -\infty$, which can be rejected as a critical value, and $x = -\frac{1}{2}$. From the original equation, when $x = -\frac{1}{2}$, $y = -\frac{1}{e}$. Therefore, the tangent to the curve: $y = 2xe^{2x}$, is horizontal at the point $\left(-\frac{1}{2}, -\frac{1}{e}\right)$.

To determine the concavity of the curve we find the inflection points by equating $\frac{d^2y}{dx^2}$ to 0 and solving for x, and determine whether $\frac{d^2y}{dx^2}$ is positive or negative in the interval around the inflection points. If $\frac{d^2y}{dx^2} > 0$, the curve is concave upward, and if $\frac{d^2y}{dx^2} < 0$, it is concave downward. We find:

$$\frac{d^2y}{dx^2} = (2e^{2x})(2) + \left[(2x + 1)(4e^{2x})\right]$$
$$= 4e^{2x} + 8xe^{2x} + 4e^{2x}$$
$$= 8e^{2x} + 8xe^{2x}$$
$$= 8e^{2x}(1 + x) = 0.$$

$8e^{2x} = 0$ and $1 + x = 0$. $x = -\infty$, which can be rejected as an inflection point, and $x = -1$. When $x > -1$, $\frac{d^2y}{dx^2} = +$, and when $x < -1$, $\frac{d^2y}{dx^2} = -$. Therefore, the curve: $y = 2xe^{2x}$, is concave upward in the region where $x > -1$, and concave downward where $x < -1$.

• PROBLEM 366

The velocity of a particle in feet per second is given by

$$v(t) = t^2 - 4t + 5, \quad t \geq 0,$$

where t is in seconds. Find the time at which the velocity is a relative maximum or a relative minimum.

Solution: To determine maxima and minima we find $v'(t)$, equate it to 0 and solve for t, obtaining the critical values.

$$v'(t) = \frac{dv}{dt} = 2t - 4 = 2(t-2).$$

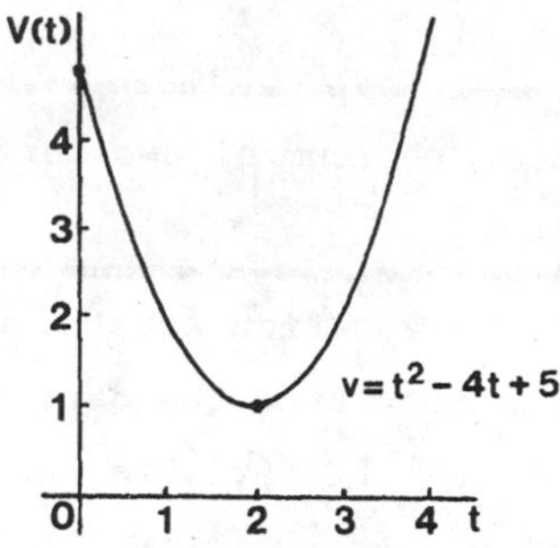

If $v'(t) = 0$, then $t = 2$, the critical value. To determine whether $t = 2$ is a maximum or minimum value we examine $v'(t)$ when $0 \leq t < 2$ and when $t > 2$.

We see that $v'(t) < 0$ for $0 \leq t < 2$ and $v'(t) > 0$ for $t > 2$. It follows that v has a relative minimum at $t = 2$, and $v(2) = 1$ foot per second. Now, we must test the endpoint $t = 0$. We find $v(0) = 5$. If we examine $v'(0)$ we obtain -4, which is a negative value. Then, v is a decreasing function at $t = 0$, therefore in the interval $t \geq 0$, $v(0) = 5$ feet per second is a maximum.

• PROBLEM 367

How can any given number be represented as the sum of two parts so that their product is a maximum?

Solution: Let a = any number
and x = one part.

Then, $a - x$ is the other part, and

$x(a-x) = xa - x^2 = y$, the product of the two parts.

For a maximum,

$$\frac{dy}{dx} = a - 2x = 0.$$

We solve for x, obtaining $x = \frac{a}{2}$. To test whether this value is a maximum we choose a value less than $x = \frac{a}{2}$, say $\frac{a}{4}$, and another greater than $\frac{a}{2}$, we will say $\frac{3a}{4}$. We now evaluate $\frac{dy}{dx}$ at $x = \frac{a}{4}$ and at $x = \frac{3a}{4}$. If there is a change in sign from + to −, then $x = \frac{a}{2}$ is a maximum. We find: $\frac{dy}{dx}$ at $x = \frac{a}{4}$, $= a - 2 \cdot \frac{a}{4} = a - \frac{a}{2} = \frac{a}{2}$, which is +. Now $\frac{dy}{dx}$, at $x = \frac{3a}{4}$, $= a - \frac{6a}{4} = -\frac{a}{2}$ which is negative. Therefore $x = \frac{a}{2}$ is indeed a maximum. Therefore the number must be split equally to obtain the maximum product.

• PROBLEM 368

Consider the number 36 as the sum of two parts, the product of which is to be a maximum.

Solution: We formulate an equation for the product, p, in terms of one variable. We can then take the derivative of p, set it equal to zero and solve for the variable, which is a critical value, and then determine whether it is a maximum. Let x be one part of 36 and 36 - x be the other part. We can write the product, p, as

$$p = x(36 - x) = 36x - x^2 .$$

We differentiate, set the derivative equal to zero and solve for x and 36 - x.

$$p' = \frac{dp}{dx} = 36 - 2x = 0$$

$$-2x = -36$$

$$x = 18, \text{ and } 36 - x = 18.$$

Since $p = 36x - x^2$ is a parabola opening downward, we know that the critical value is at a maximum point. Therefore, the two numbers 18 and 18 are the parts of 36 that give the maximum product, 324.

• PROBLEM 369

Find the number which exceeds its square by the greatest amount.

Solution: We first write a function which expresses the difference between a number and its square. Let the number be x, and let the difference be D. Then $D = x - x^2$. To maximize this equation we find $\frac{dD}{dx}$, set it equal to 0 and solve for x. Doing this we have:

$$\frac{dD}{dx} = 1 - 2x.$$

Setting the derivative equal to zero, we have $1 - 2x = 0$. Solving for x, we obtain the result $x = \frac{1}{2}$. To be certain that this is a maximum value we use the Second Derivative Test. The second derivative gives $\frac{d^2D}{dx^2} = -2$, which states that the second derivative is always negative. This means that whenever the first derivative is zero, it represents a maximum. Therefore the number which exceeds its square by the greatest amount is $\frac{1}{2}$. $\frac{1}{2} - \frac{1}{4} = \frac{1}{4}$. This answer is reasonable, since, for integers: $N^2 > N$, $0 - 0^2 = 0$, and $1 - 1^2 = 0$.

Find two positive numbers, the sum of which is 20, and their product is as large as possible.

Solution: If one of the numbers is x, the other is (20 - x), and their product is

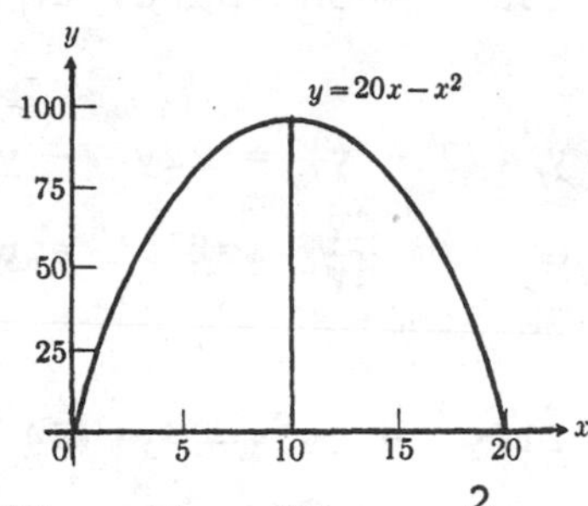

$$y = x(20 - x) = 20x - x^2.$$

Since both numbers are to be positive,

$$x > 0 \quad \text{and} \quad 20 - x > 0,$$

or

$$0 < x < 20.$$

To maximize, we find $\frac{dy}{dx}$ and set it equal to 0. We then solve for x. We find

$$\frac{dy}{dx} = 20 - 2x = 2(10 - x) = 0.$$

Therefore, x = 10.

We must now test x = 10 to be sure it is a maximum. To do this we look at $\frac{dy}{dx}$ when x < 10 and x > 10. If there is a change in sign from + to - then x = 10 is a maximum. We find: $\frac{dy}{dx}$ is,

positive when x < 10,

negative when x > 10,

zero when x = 10.

The value x = 10 is therefore a maximum value. We obtain the same results using the second derivative test.

$$\frac{d^2y}{dx^2} = -2$$

is always negative. Therefore x = 10 is a maximum value. The curve representing $y = 20x - x^2$ is concave downward at every point on it. The graph has an absolute maximum at x = 10. The two numbers are thus x = 10, 20 - x = 10. The product is 100.

• PROBLEM 371

The number 72 is to be represented as the sum of two positive parts, such that the product of one of the parts by the cube of the other is a maximum. It is desired to find the two parts.

Solution: We let y = one part, then (72 - y) = the other. We have:

$$P = y^3(72 - y) = 72y^3 - y^4.$$

To maximize, we find $\frac{dP}{dy}$, set it equal to 0 and solve for y. We find:

$$\frac{dP}{dy} = 216y^2 - 4y^3 = 4y^2(54 - y).$$

Setting $\frac{dP}{dy} = 0$, we obtain y = 0 and y = 54. Clearly, y = 0 is a trivial solution and does not lead to a maximum value for P.

To determine whether y = 54 is a maximum value we can use the Second Derivative Test. We find $\frac{d^2P}{dy^2}$ and evaluate at y = 54. If the value is a negative one, then y = 54 is a maximum. We have:

$$\frac{d^2P}{dy^2} = 432y - 12y^2.$$

At y = 54, $\frac{d^2P}{dy^2} = -11,664$. Hence P is a maximum when y = 54. Thus the two parts are 18 and 54.

• PROBLEM 372

Find two positive numbers, the sum of which is 100, and the square of one number times twice the cube of the other number is to be a maximum.

Solution: We will let x = one number, then (100 - x) = the other. We have: $2(x^2)\left[(100 - x)^3\right] = f(x)$. Since both numbers are to be positive,

$$x \geq 0 \quad \text{and} \quad 100 - x \geq 0$$

or

$$100 \geq x \geq 0.$$

We want to maximize $f(x) = 2x^2(100 - x)^3$, where $0 \leq x \leq 100$.

We do this by finding f'(x), setting it equal to 0 and solving for x. We find:

$$f'(x) = 4x(100 - x)^3 - 6x^2(100 - x)^2$$

$$= 2x(100 - x)^2[2(100 - x) - 3x]$$

$$= 2x(100 - x)^2(200 - 5x) = 0.$$

Solving for x we have:

$$x = 0, \ x = 100, \ x = 40.$$

Therefore, the critical values are 0, 40, and 100. Since $f(0) = 0$, $f(100) = 0$, and $f(40) > 0$, the absolute maximum will be $f(40)$ and the two desired numbers will be 40 and 60.

• PROBLEM 373

A piece of wire 20 inches long is to be cut and made into a rectangular frame. What dimensions should be chosen so that the area of the rectangle enclosed is maximal?

Solution: If we let two sides of the rectangle equal x, then the other two sides combined = $20 - 2x$, therefore each side is $10 - x$. The area of the rectangle = $A(x) = x(10-x) = 10x - x^2$. To maximize A we find $A'(x) = \frac{dA}{dx}$, set it equal to 0, and solve for x. We write:

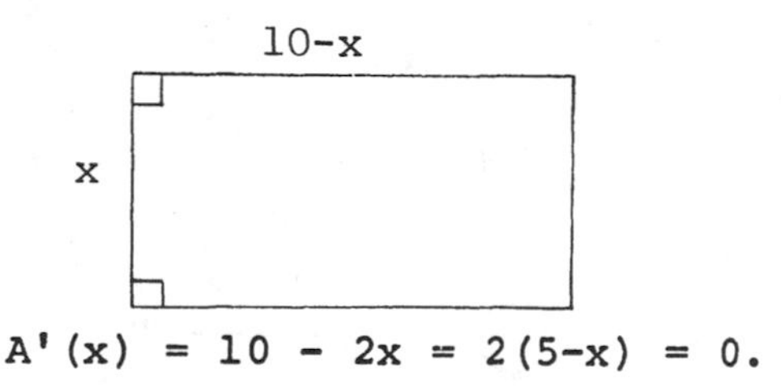

$$A'(x) = 10 - 2x = 2(5-x) = 0.$$

Therefore, $x = 5$. To show that this is a maximum value we choose a value less than 5, say 4, and a value greater than 5, say 6, and show that $A'(x)$ changes sign from + to −.

We find: $A'(4) = 2$ and $A'(6) = -2$, therefore 5 is indeed a maximum value. Letting $x = 5$, we find the dimensions of the rectangle to be 5 and (10-5) or 5. Therefore, the rectangle is a square of area 25.

• PROBLEM 374

What rectangle of maximum area can be inscribed in a circle of radius r?

Solution: Let x = one side of the rectangle. The other side is obtained by the square root of the diagonal squared minus the square of the one side.

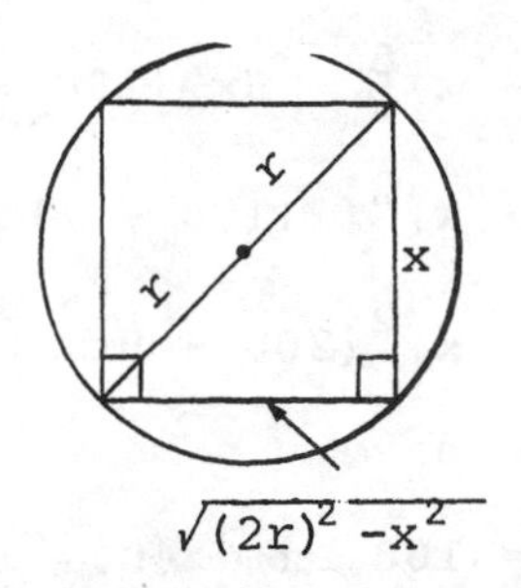

The diagonal is equal to 2r. The other side = $\sqrt{(2r)^2 - x^2} = \sqrt{4r^2 - x^2}$. (See diagram.)

$$A = x\sqrt{4r^2 - x^2} = \text{area of rectangle}$$

To obtain a maximum value we find $A' = \frac{dA}{dx}$ and set it equal to 0. Solving for x we obtain a critical value. We have:

$$\frac{dA}{dx} = x \cdot \frac{d}{dx}(\sqrt{4r^2 - x^2}) + \sqrt{4r^2 - x^2} \cdot 1 = 0$$

for a maximum or minimum.

$$x \cdot \frac{1}{2}(4r^2 - x^2)^{-1/2} \cdot (-2x) + \sqrt{4r^2 - x^2} = 0$$

or
$$-\frac{x^2}{(4r^2 - x^2)^{1/2}} + (4r^2 - x^2)^{1/2} = 0$$

which reduces to

$$\frac{-x^2 + 4r^2 - x^2}{(4r^2 - x^2)^{1/2}} = 0$$

$$\frac{4r^2 - 2x^2}{(4r^2 - x^2)^{1/2}} = 0$$

Now, the denominator, $\sqrt{4r^2 - x^2}$ cannot be 0 because this is the value of a side of the rectangle. Therefore the numerator,

$$4r^2 - 2x^2 = 0$$

and

$$x^2 = 2r^2$$

or

$$x = r\sqrt{2} = \text{one side}$$

and

$$(4r^2 - x^2)^{1/2} = (4r^2 - r^2 \cdot 2)^{1/2} = (2r^2)^{1/2} = r\sqrt{2}$$

$$= \text{other side}$$

The figure is a square.

• **PROBLEM 375**

In a given semi-circle of radius 2, in which a rectangle is to be inscribed, what are the dimensions for maximum area of the rectangle?

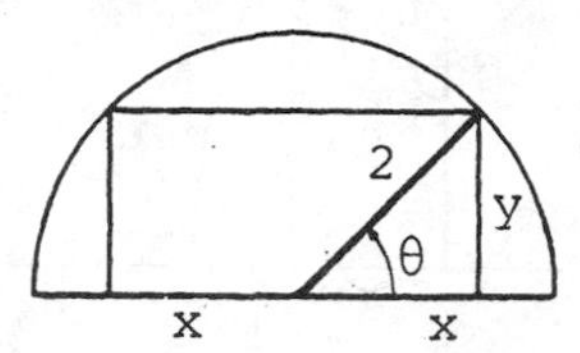

Solution: From the diagram, $A = 2xy$, where $x = 2 \cos \theta$ and $y = 2 \sin \theta$. Thus $A = 8 \sin \theta \cos \theta$. We now use the identity, $\sin \theta \cos \theta = \frac{\sin 2\theta}{2}$, and obtain $4 \sin 2\theta$. We shall find the value of θ corresponding to the maximum area, by finding $\frac{DA}{d\theta}$, setting it equal to 0, and solving for θ. We have:

$$\frac{dA}{d\theta} = 8 \cos 2\theta, \quad 8 \cos 2\theta = 0, \; \cos 2\theta = 0,$$

$$2\theta = \frac{\pi}{2}, \quad \theta = \frac{\pi}{4}.$$

We can now use the Second Derivative Test to test whether the value $\theta = \frac{\pi}{4}$ is a maximum. We find $A'' = -16 \sin 2\theta$. If A'' at $\theta = \frac{\pi}{4}$ is < 0, then the value is a maximum. We have,

$$A'' = -16 \sin 2\theta, \; \theta = \frac{\pi}{4}$$

$$A'' = -16 \sin \frac{\pi}{2} = -16.$$

The value $\theta = \frac{\pi}{4}$ is indeed a maximum. The maximum ares is $A = 4 \sin 2\left(\frac{\pi}{4}\right) = 4$.

• **PROBLEM 376**

A rectangle has two of its vertices on the x axis and the other two above the x axis and on the graph of the parabola: $y = 16 - x^2$. Of all such possible rectangles, find the dimensions of the one of maximum area.

Solution: The parabola is symmetric with respect to the y axis (y is unchanged when x or - x is substituted into the equation). Therefore if one vertex on the parabola has the coordinate (x,y), the corresponding one has (- x, y). Hence, the coordinates of the four vertices are (x, y), (- x, y), (x, 0) and (- x, 0) as indicated on the graph. The area of the rectangle

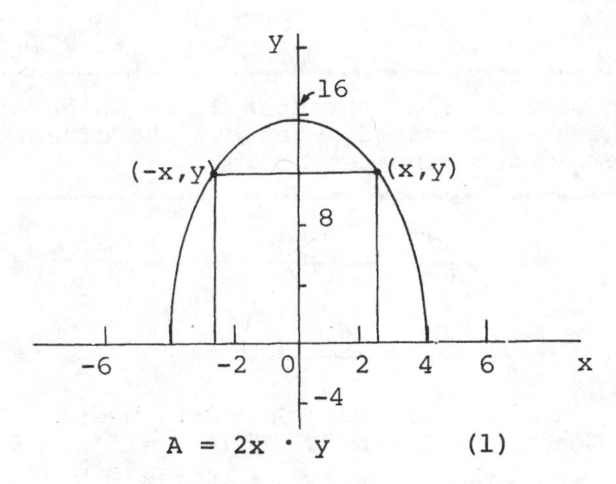

$$A = 2x \cdot y \qquad (1)$$

To obtain values of x or y for maximum A, we first express A in terms of x or y alone, then calculate x or y when either

$$\frac{dA}{dx} = 0, \quad \text{or} \quad \frac{dA}{dy} = 0.$$

Substituting into (1),

$$A = 2x \cdot (16 - x^2) = 2(16x - x^3) \quad . \ . \ . \ (2)$$

Differentiating (2),

$$\frac{dA}{dx} = 2(16 - 3x^2). \qquad (3)$$

When $\dfrac{dA}{dx} = 0, \qquad x = \pm \dfrac{4}{3}\sqrt{3}$

To find which root gives the maximum, we test whether $\dfrac{d^2A}{dx^2} < 0$ for that root.

The second derivative of A can be obtained by differentiating (3).

$$\frac{d^2A}{dx^2} = 2(-6x)$$

Only for

$$x = \frac{4}{3}\sqrt{3}, \text{ do we have: } \frac{d^2A}{dx^2} < 0$$

Substituting this value for x in (2), the dimensions for maximum area are

$$\frac{8}{3}\sqrt{3} \text{ by } 10\frac{2}{3}.$$

• PROBLEM 377

A field of rectangular shape is to be fenced off along the bank of a river. No fence is required on the side lying along the river. If the material for the fence costs $2 per running foot for the two ends and $3 per running foot for the side parallel to the river, find the dimensions of the field of maximum area that can be enclosed with $900 worth of fence.

Solution: Let x = the number of feet of length of an end of the field;

y = the number of feet of length of the side parallel to the river;

A = the number of square feet in the area of the field.

Then, $$A = xy. \quad (1)$$

Since the cost of the material for each end is $2 per running foot and the length of an end is x ft, the total cost for the fence for each end is $2x$ dollars. Similarly, the total cost of the fence for the third side is $3y$ dollars. We have, then,

$$2x + 2x + 3y = 900. \quad (2)$$

To express A in terms of a single variable, we solve Eq. (2) for y in terms of x and substitute this value into Eq. (1), yielding A as a function of x alone, and

$$A(x) = x\left(300 - \frac{4}{3}x\right) \quad (3)$$

$$A(x) = 300x - \frac{4}{3}x^2.$$

We now find $A'(x)$, equate it to 0 and solve for x, obtaining the critical values. We find:

$$A'(x) = 300 - \frac{8}{3}x = 0.$$

Therefore $$x = 112\frac{1}{2}.$$

Substituting back in equation (2) yields

$$y = 150$$

and A is therefore $xy = 16,875$.

Therefore, the largest possible area that can be enclosed for $900 is 16,875 square feet, and this is obtained when the side parallel to the river is 150 ft long and the ends are each $112\frac{1}{2}$ ft long.

• PROBLEM 378

Determine the dimensions of a right circular cone

containing a sphere of radius a, if the volume of the cone is to be a minimum.

Solution: Since the volume V is to be a minimum, it is the function in the problem. Hence the function is

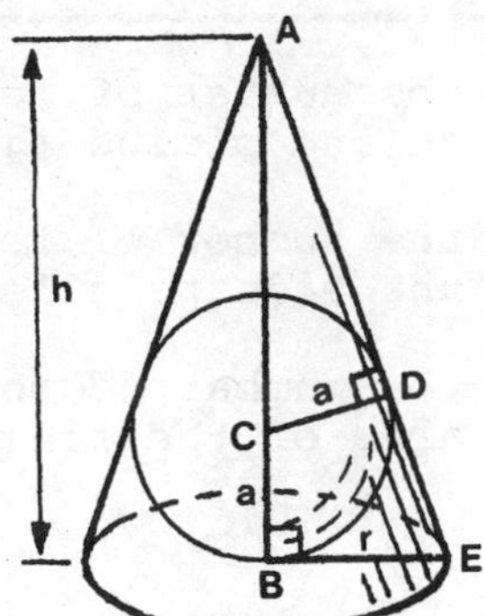

(1) $$V = \frac{1}{3}\pi r^2 h = \text{volume of cone}$$

To express the function in terms of a single variable, we find a relation connecting h and r. From similar triangles,

$$\frac{AE}{BE} = \frac{AC}{DC},$$

or

(2) $$\frac{\sqrt{h^2 + r^2}}{r} = \frac{h - a}{a}.$$

Squaring both sides and solving for r^2,

$$a^2(h^2+r^2) = r^2(h^2-2ah+a^2)$$

(3) $$r^2 = \frac{a^2 h}{h - 2a}.$$

Substituting (3) in (1),

(4) $$V = \frac{\pi a^2 h^2}{3(h - 2a)}.$$

To obtain the dimensions of the cone we must find $\frac{dV}{dh}$, set it equal to 0 and solve for h, obtaining critical values. Doing this we obtain:

(5) $$\frac{dV}{dh} = \frac{\pi a^2 h}{3}\left[\frac{h - 4a}{(h - 2a)^2}\right].$$

Setting the derivative equal to zero,

$$h = 0, \quad h = 4a.$$

Both the function and its derivative become infinite if $h = 2a$ and therefore this value of h cannot be considered. The value $h = 0$ is extraneous since it does not satisfy (2). By applying the critical value $h = 4a$, we observe that $\frac{dV}{dh}$ is negative for $h < 4a$, and positive for $h > 4a$. Hence $h = 4a$ makes V a minimum, and the dimensions of the cone are $h = 4a$, $r = a\sqrt{2}$.

• PROBLEM 379

An open cylindrical can is to have a given volume. Find the dimensions of the can if the least amount of material is to be used in making the can.

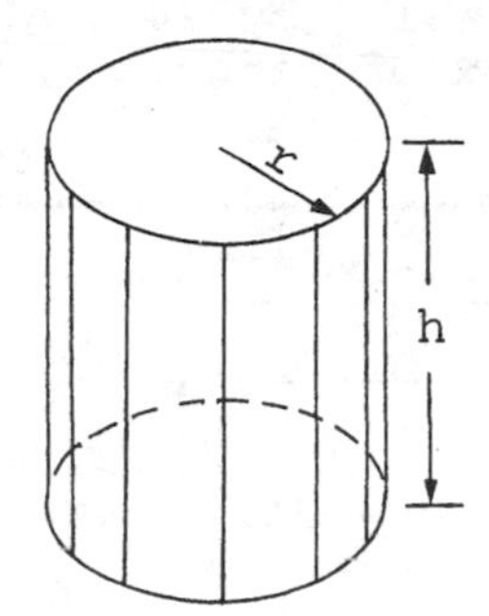

Solution: The conditions to be observed here are that the volume is constant while the surface area or the amount of sheet material, M, is to be a minimum.

Calling the radius r, and the altitude h,

(1) $M = \pi r^2 + 2\pi rh$, (one end is open)

where r and h are connected by the relation,

(2) $\pi r^2 h = V$, a constant.

Here it is easier to eliminate h, hence from (2),

(3) $$h = \frac{V}{\pi r^2}.$$

Substituting (3) in (1), we find M in terms of r, or

(4) $$M = \pi r^2 + \frac{2V}{r}.$$

We wish to minimize M. Find $\frac{dM}{dr}$, equate it to 0 and solve for r, obtaining critical values. We find:

(5) $$\frac{dM}{dr} = 2\pi r - \frac{2V}{r^2}.$$

Using the value of V from (2) we have

(6) $$\frac{dM}{dr} = 2\pi r - \frac{2(\pi r^2 h)}{r^2} = 2\pi r - 2\pi h = 2\pi(r-h).$$

Hence $$\frac{dM}{dr} = 0 \text{ when } r = h.$$

We now use the First Derivative Test to show that the value $r = h$ is a minimum. Examining $\frac{dM}{dr}$ when $r < h$ and when $r > h$ we find that if $r < h$, $\frac{dM}{dr} < 0$, and if $r > h$, $\frac{dM}{dr} > 0$. Since $\frac{dM}{dr}$ changes sign from negative to positive, the critical value $r = h$ is indeed a minimum. Therefore M is a minimum for an open cylinder with fixed volume when $r = h$.

• PROBLEM 380

An open box is to be made by cutting out squares from the corners of a rectangular piece of cardboard and then turning up the sides. If the piece of cardboard is 12" by 24", what are the dimensions of the box of largest volume made in this way?

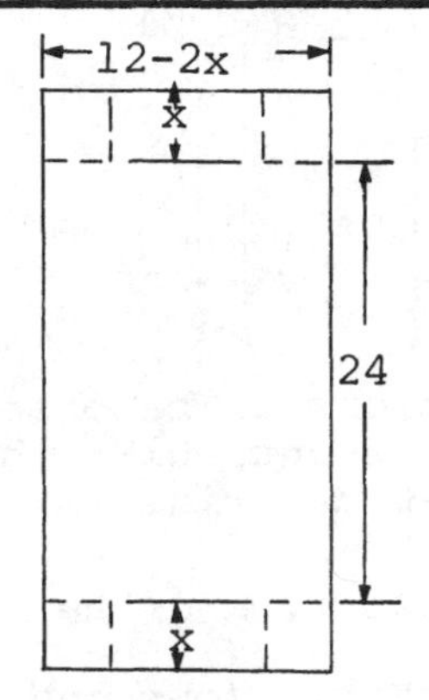

Solution: Assume the squares cut out from the cardboard have the dimension x by x. The base of the open box then has the dimension $(12 - 2x)$ by $(24 - 2x)$. The height of the box is x, therefore, the volume is given by

$$V = x(12 - 2x)(24 - 2x) \quad . \; . \; . \quad (1)$$

$$= 4x^3 - 72x^2 + 288x$$

To find x for the maximum volume, we calculate x when

$$\frac{dV}{dx} = 0.$$

Differentiating (1),

$$\frac{dV}{dx} = 4(3x^2 - 36x + 72) \qquad (2)$$

This equals 0, when

$$x^2 - 12x + 24 = 0$$

$$x = 6 \pm 2\sqrt{3}$$

To determine which value of x gives the maximum volume, we check whether

$$\frac{d^2V}{dx^2} < 0 \text{ for these values.}$$

The second derivative of x can be obtained by differentiating (2).

$$\frac{d^2V}{dx^2} = 24(x - 6) \quad . . . \quad (3)$$

For

$$x = 6 + 2\sqrt{3}, \quad \left.\frac{d^2V}{dx^2}\right|_{6+2\sqrt{3}} = 48\sqrt{3} > 0.$$

For

$$x = 6 - 2\sqrt{3}, \quad \left.\frac{d^2V}{dx^2}\right|_{6-2\sqrt{3}} = - 48\sqrt{3} < 0$$

Therefore, $x = 6 - 2\sqrt{3}$ is the answer.

From (1), the maximum volume now has the dimensions:

$$(6 - 2\sqrt{3}) \times 4\sqrt{3} \times (12 + 4\sqrt{3}) .$$

• PROBLEM 381

Find the relative dimensions of the right circular cone of maximum volume inscribed in a sphere of radius a.

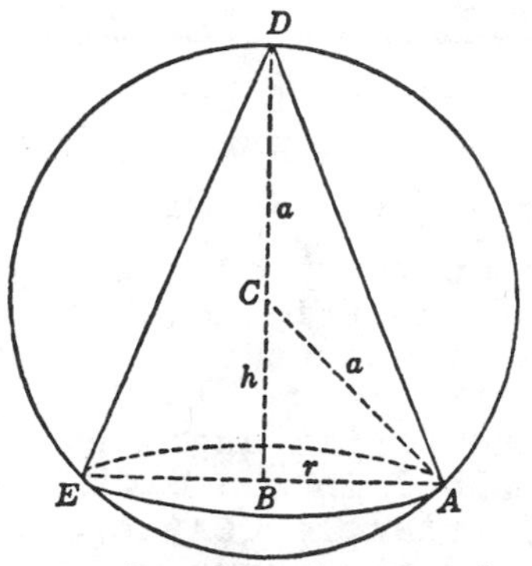

Solution: From the drawing we have

a = radius of sphere
r = radius of cone
(h+a) = height of cone, and

$$h^2 + r^2 = a^2.$$

We also have the volume of the cone =

$$V = \frac{1}{3}\pi r^2 (h+a).$$

Let r be the independent variable and consider h to be a function of r. Differentiating both equations above with respect to r, we obtain

$$2hh' + 2r = 0,$$

$$\frac{1}{3}\pi r^2 h' + \frac{1}{3}\pi h \cdot 2r + \frac{1}{3}\pi a \cdot 2r = 0,$$

where $$h' = \frac{dh}{dr},$$

and where $\frac{dV}{dr}$ has been set equal to 0 to find the maximum volume.

Eliminating h' from these two equations, we find

$$r^2 = 2h(a+h)$$

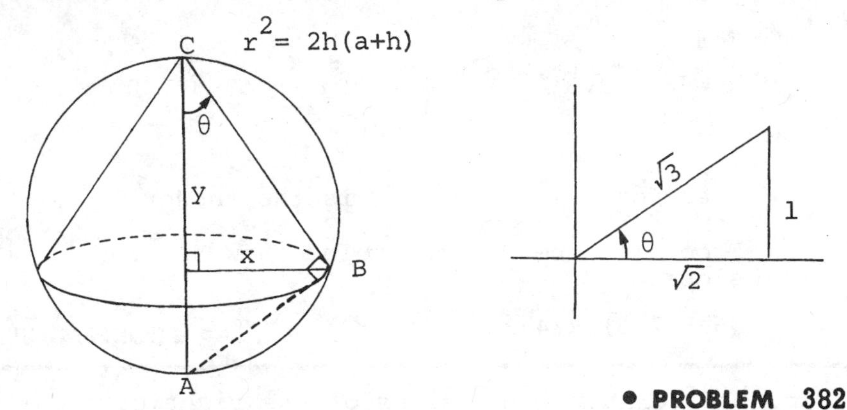

• PROBLEM 382

Find the dimensions of the cone of largest volume that can be inscribed in a sphere of radius 1.

Solution: Let x = radius of cone and y = height of cone, as shown in the accompanying diagram. We have: volume = $V = \frac{1}{3}\pi x^2 y$. We will choose a particular angle in the geometry of the figure and express V in terms of the trigonometric functions of this angle and the constants of the geometry.

Denote half the vertex angle by θ, and recall that an angle inscribed in a semicircle is a right angle. Therefore, triangle CBA is a right triangle, with CA = diameter of the spere = 2. From this configuration we obtain:

$$\overline{CB} = 2\cos\theta,$$

$$x = \overline{CB}\sin\theta = 2\sin\theta\cos\theta,$$

by substitution.

$$y = \overline{CB}\cos\theta = 2\cos^2\theta.$$

Therefore, substituting these values in the equation for V we obtain:

$$V = \frac{\pi}{3}(4\sin^2\theta\cos^2\theta)(2\cos^2\theta).$$

$$V = \frac{8\pi}{3}\sin^2\theta\cos^4\theta.$$

We wish to maximize V. We find $\frac{dV}{d\theta}$, equate it to 0 and solve to obtain the critical values. We find:

$$\frac{dV}{d\theta} = \frac{8\pi}{3}\left[\sin^2\theta\ 4\cos^3\theta(-\sin\theta) + \cos^4\theta\ 2\sin\theta\cos\theta\right].$$

Setting $\frac{dV}{d\theta} = 0$, we obtain,

$$2\sin\theta\cos^3\theta(-2\sin^2\theta + \cos^2\theta) = 0.$$

We see that for the volume to be a maximum,

$$-2\sin^2\theta + \cos^2\theta = 0;\ \cos^2\theta = 2\sin^2\theta,$$

and from this we obtain:

$$\tan^2\theta = \frac{1}{2} \text{ or } \tan\theta = 1/\sqrt{2}.$$

Thus

$$\sin\theta = \frac{1}{\sqrt{3}} \quad \text{and} \quad \cos \quad = \frac{\sqrt{2}}{\sqrt{3}}$$

These yield

$$x = 2\sin\theta\cos\theta$$

$$= 2\left(\frac{1}{\sqrt{3}}\right)\left(\frac{\sqrt{2}}{\sqrt{3}}\right) = \frac{2}{3}\sqrt{2}$$

and

$$y = 2\cos^2\theta, \quad y = 2\left(\frac{2}{3}\right) = \frac{4}{3}.$$

These values of x and y give the maximum volume. It is equal to

$$\frac{1}{3}\pi x^2 y = \frac{\pi}{3}\cdot\frac{8}{9}\cdot\frac{4}{3} = \frac{32\pi}{81}.$$

• PROBLEM 383

For maximum volume, what is the radius of the opening of a conical vessel with a given slant height?

Solution: Let V = volume = $\frac{1}{3}$ height × area of base. The given slant height is **s**. Then, from the diagram we see,

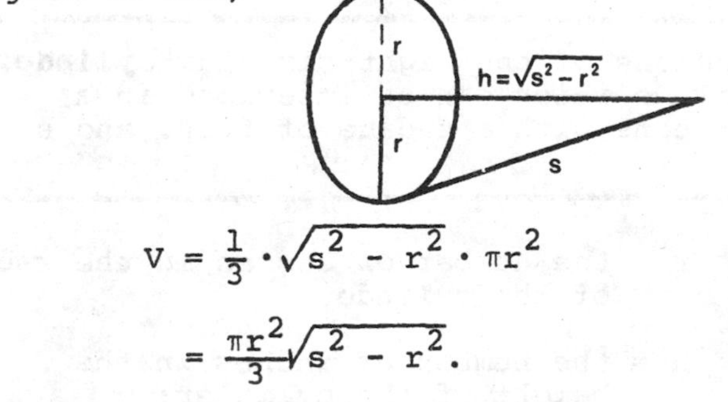

$$V = \frac{1}{3}\cdot\sqrt{s^2 - r^2}\cdot\pi r^2$$

$$= \frac{\pi r^2}{3}\sqrt{s^2 - r^2}.$$

In this problem we wish to maximize V. To do this we find $\frac{dV}{dr}$, equate it to 0 and solve for r, obtaining the critical values. We find:

$$\frac{dV}{dr} = \left[\frac{\pi r^2}{3} \cdot \frac{1}{2}\left(s^2 - r^2\right)^{-1/2} \cdot (-2r)\right]$$

$$+ \left[\left(s^2 - r^2\right)^{1/2} \cdot \frac{2\pi r}{3}\right],$$

by use of the product rule for differentiation. We now simplify and, equating to 0, we obtain:

$$\frac{dV}{dr} = \frac{\pi r^2}{3} \cdot - \frac{r}{(s^2 - r^2)^{1/2}} + \frac{2\pi r}{3} \cdot (s^2 - r^2)^{\frac{1}{2}} = 0$$

$$= - \frac{\pi r^3}{3\left(s^2 - r^2\right)^{1/2}} + \frac{2\pi r}{3} \cdot \left(s^2 - r^2\right)^{1/2} = 0$$

$$= \frac{-\pi r^3 + 2\pi r(s^2 - r^2)}{3\left(s^2 - r^2\right)^{1/2}} = 0 = \frac{2\pi r(s^2 - r^2) - \pi r^3}{3\left(s^2 - r^2\right)^{1/2}}.$$

Now, setting the numerator alone equal to 0, we have:

$$2\pi r(s^2 - r^2) - \pi r^3 = 0$$

$$2\pi r s^2 - 2\pi r^3 - \pi r^3 = 0$$

$$2\pi r s^2 - 3\pi r^3 = 0$$

$$2\pi s^2 \equiv 3\pi r^2$$

$$2s^2 = 3r^2$$

and

$$r = \pm s\sqrt{\frac{2}{3}}.$$

Obviously, the negative value cannot be used. Therefore, the radius is $\sqrt{\frac{2}{3}}s$ for maximum volume.

• PROBLEM 384

Find the dimensions of the right-circular **cylinder** of greatest volume which can be inscribed in a right-circular cone with a radius of 5 in. and a height of 12 in.

Solution: Let r = the number of inches in the radius of the cylinder;

h = the number of inches in the height of the cylinder;

V = the number of cubic inches in the volume of the cylinder.

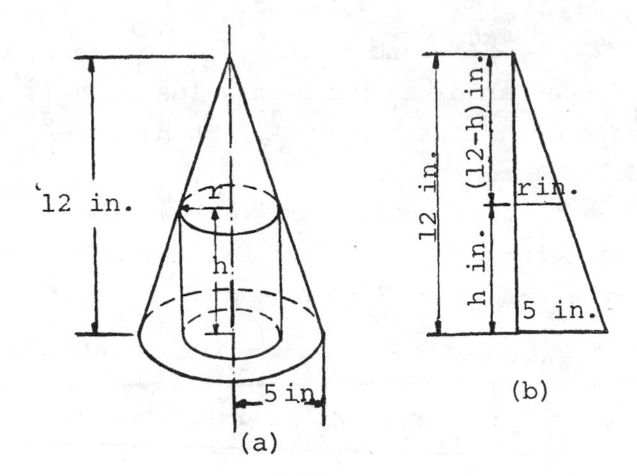

Figure a illustrates the cylinder inscribed in the cone, and Fig. b illustrates a plane section through the axis of the cone.

The following formula expreses V in terms of r and h:

$$V = \pi r^2 h. \tag{1}$$

To express V in terms of a single variable we need another equation involving r and h. From Fig. (b), and using similar triangles, we have

$$\frac{12 - h}{r} = \frac{12}{5}$$

or

$$h = \frac{60 - 12r}{5}. \tag{2}$$

Substituting from Eq. (2) into formula (1), we obtain V as a function of r and write

$$\begin{aligned} V(r) &= \pi r^2 \left(\frac{60 - 12r}{5}\right), \\ &= \frac{12}{5}\pi(5r^2 - r^3) \end{aligned} \tag{3}$$

We now find V'(r), equate it to 0 and solve for r to obtain the critical values. We find:

$$V'(r) = \frac{12}{5}\pi(10r - 3r^2) = 0,$$

$$r(10 - 3r) = 0$$

from which we obtain

$$r = 0 \quad \text{and} \quad r = \frac{10}{3}.$$

The absolute maximum value of V on [0,5] must occur at either 0, $\frac{10}{3}$, or 5. From Eq. (3) we obtain

$V(0) = 0$, $V\left(\frac{10}{3}\right) = \frac{400}{9}\pi$, and $V(5) = 0$. We therefore conclude that the absolute maximum value of V is $\frac{400}{9}\pi$, and this occurs when $r = \frac{10}{3}$. When $r = \frac{10}{3}$, we find from Eq. (2) that $h = 4$.

Thus, the greatest volume of an inscribed cylinder in the given cone is $\frac{400}{9}\pi$ in.^{3}, which occurs when the radius is $\frac{10}{3}$ in. and the height is 4 in.

• PROBLEM 385

A right circular cylinder is to be inscribed in a sphere of radius a. Find the dimensions for maximum volume.

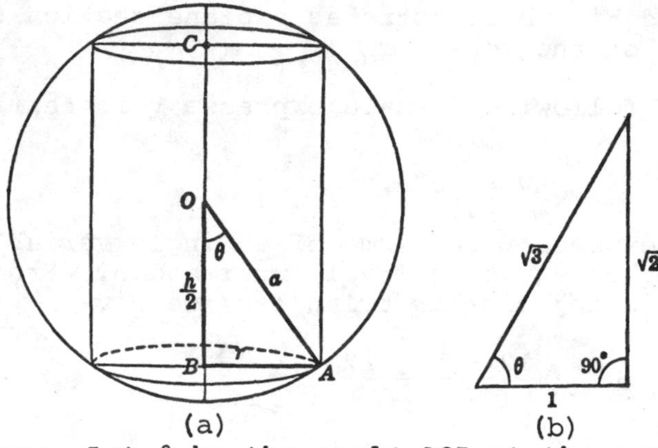

(a) (b)

Solution: Let θ be the angle AOB at the center of the sphere between the half-altitude OB and the radius OA of the sphere. Then, from the right triangle OBA, we have for the radius of base and altitude of the cylinder:

$$r = a \sin \theta, \quad h = 2a \cos \theta.$$

The volume of the cylinder is

$$V = \pi r^2 h = 2\pi a^3 \sin^2 \theta \cos \theta.$$

Using the identity: $\sin^2 \theta = 1 - \cos^2 \theta$, we have:

$$V = 2\pi a^3 (\cos \theta - \cos^3 \theta).$$

In this problem we wish to maximize the volume. We find $\frac{dV}{d\theta} = D_\theta V$, equate it to 0 and solve for θ to obtain the critical values. We find:

$$D_\theta V = -2\pi a^3 \sin \theta (1 - 3 \cos^2 \theta).$$

If $D_\theta V = 0$, we have $\sin \theta = 0$ or $1 - 3 \cos^2 \theta = 0$. If $\sin \theta = 0$, then $\theta = 0$ or π, a trivial solution which obviously does not give a maximum cylinder. But if $1 - 3\cos^2\theta = 0$, then $\cos\theta = \frac{1}{\sqrt{3}}$, which gives a maximum value for V. We now use Fig. (b) to determine $\sin\theta$. We find this to be $\frac{\sqrt{2}}{\sqrt{3}}$. Now, solving for r and h we have:

$$r = a\sin\theta = \frac{1}{3}a\sqrt{6}, \quad h = 2a\cos\theta = \frac{2}{3}a\sqrt{3}.$$

Substituting back in the equation for volume gives:

$$V = \pi r^2 h = \pi\frac{a^2}{9}(6)\cdot\frac{2}{3}a\sqrt{3}$$

$$= \pi a^3\left(\frac{4}{9}\right)\sqrt{3}.$$

• PROBLEM 386

A closed tin can is to be made of a given quantity of material. For maximum volume, what are its dimensions?

Solution: The given quantity of material means that the total surface area (top, bottom, and side) is specified; call it S and let r = radius and h = height of the can. Then

$$S = 2\pi rh + 2\pi r^2.$$

Now the quantity to be maximized is the volume, and

$$V = \pi r^2 h.$$

We wish to eliminate the h in this equation, and obtain an expression using S, which is a constant, and r. We have: $S = 2\pi rh + 2\pi r^2$, therefore,

$$S_r = 2\pi r^2 h + 2\pi r^3.$$

Solving for $\pi r^2 h$, we substitute and obtain:

$$V = \frac{Sr}{2} - \pi r^3.$$

We now find $\frac{dV}{dr}$, equate it to 0 and solve for r to obtain the critical values. Doing this we have:

$$\frac{dV}{dr} = \frac{S}{2} - 3\pi r^2 = 0$$

$$r = \pm\sqrt{\frac{S}{6\pi}},$$

the critical values. But we reject the negative value. We use the Second Derivative Test to determine whether this value, $r = \sqrt{\frac{S}{6\pi}}$, is a maximum or minimum. We find:

$$\frac{d^2V}{dr^2} = -6\pi r,$$

which is negative for all (positive) values of r. Therefore, $r = \sqrt{\frac{S}{6\pi}}$ corresponds to a maximum. We now need to find the value of h. Using the equation, $S = 2\pi rh + 2\pi r^2$, we solve for h, obtaining:

$$h = \frac{S - 2\pi r^2}{2\pi r} = \frac{S - 2\pi\left|\frac{S}{6\pi}\right|}{2\pi\sqrt{\frac{S}{6\pi}}} = \frac{\frac{2}{3}S}{2\pi\sqrt{\frac{S}{6\pi}}}$$

$$= 2 \cdot \frac{S}{6\pi} \cdot \frac{1}{\sqrt{\frac{S}{6\pi}}} = 2r^2 \cdot \frac{1}{r} = 2r.$$

Hence the relative dimensions are h = 2r.

• PROBLEM 387

A sheet of copper which is 20 in. on a side is to be made into a box of maximum volume. Calculate the size of the squares which should be cut out of the corners.

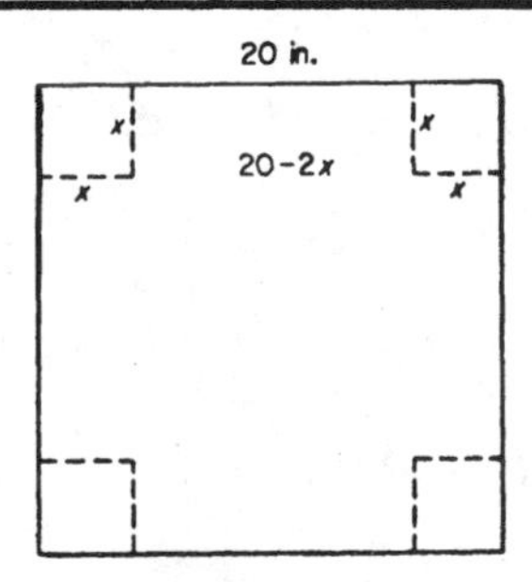

Solution: Let x = height of box = side of square to be cut.

$$V = \text{length} \times \text{width} \times \text{height}$$

$$= (20 - 2x)^2 x$$

$$= 400x - 80x^2 + 4x^3$$

$$\frac{dV}{dx} = 400 - 160x + 12x^2.$$

Minimum or maximum is obtained when $\frac{dv}{dx} = 0$:

$$12x^2 - 160x + 400 = 0.$$

Simplyfing:

$$3x^2 - 40x + 100 = 0.$$

Factoring:

$$(3x - 10)(x - 10) = 0$$

$$x = 10, \frac{10}{3} \text{ in.}$$

The obviously correct answer is $\frac{10}{3}$ in., since the other answer means cutting away all the copper.

If the correct value were not obvious, we would proceed to get the second derivative of the equation:

$$\frac{dv}{dx} = 400 - 160x + 12x^2.$$

Since $x = \frac{10}{3}$,

$$\frac{d^2v}{dx^2} = -160 + 24x = -160 + 24\left(\frac{10}{3}\right) = -80.$$

The quantity $\frac{10}{3}$, when substituted in the second derivative, gives a negative value; therefore, that quantity gives a maximum for the conditions of the problem.

Note that the other answer, $x = 10$, represents the minimum volume - so "minimum," in fact, that the volume is zero.

• PROBLEM 388

If the sum of the volumes of a sphere and a cube is constant, show that the sum of their surface areas is greatest when the diameter of the sphere is equal to the edge of the cube.

Solution: Volume of sphere $= \frac{4}{3}\pi r^3$.

Volume of cube $= s^3$.

Surface area of sphere $= 4\pi r^2$.

Surface area of cube $= 6s^2$.

Total volume: $\frac{4}{3}\pi r^3 + s^3 = K$, where K = a constant.

We wish to maximize $4\pi r^2 + 6s^2$. We let $y = 4\pi r^2 + 6s^2$. But this equation is in two variables, r and s. From the first equation we can solve for r, and substitute this value in the second equation, so as to have an equation in one variable. Since $\frac{4}{3}\pi r^3 + s^3 = K$, $\frac{4}{3}\pi r^3 = K - s^3$.

$$r^3 = \frac{K - s^3}{\frac{4\pi}{3}} = \frac{3(K - s^3)}{4\pi}.$$

Therefore,

$$r = \left[\frac{3(K - s^3)}{4\pi}\right]^{\frac{1}{3}} = \frac{(3)^{\frac{1}{3}}(K - s^3)^{\frac{1}{3}}}{(4\pi)^{\frac{1}{3}}}.$$

By substitution, $y = 4\pi r^2 + 6s^2$ becomes:

$$y = 4\pi\left[\frac{(3)^{\frac{1}{3}}(K - s^3)^{\frac{1}{3}}}{(4\pi)^{\frac{1}{3}}}\right]^2 + 6s^2$$

$$= 4\pi\left[\frac{(3)^{\frac{2}{3}}(K - s^3)^{\frac{2}{3}}}{(4\pi)^{\frac{2}{3}}}\right] + 6s^2$$

$$= \left(4\pi\right)^{\frac{1}{3}}\left(3\right)^{\frac{2}{3}}\left(K - s^3\right)^{\frac{2}{3}} + 6s^2.$$

To maximize this equation, we find $\frac{dy}{ds}$ and equate it to 0.

$$\frac{dy}{ds} = \left(4\pi\right)^{\frac{1}{3}}\left(3\right)^{\frac{2}{3}}\frac{2}{3}\left(K - s^3\right)^{-\frac{1}{3}}\left(-3s^2\right) + 12s$$

$$= \frac{(3)^{\frac{2}{3}}(4\pi)^{\frac{1}{3}}(-6s^2)}{3(K - s^3)^{\frac{1}{3}}} + 12s.$$

Equating this to 0,

$$\frac{(3)^{\frac{2}{3}}(4\pi)^{\frac{1}{3}}(-6s^2) + 12s\left[3(K - s^3)^{\frac{1}{3}}\right]}{3(K - s^3)^{\frac{1}{3}}} = 0.$$

Then,

$$(3)^{\frac{2}{3}}(4\pi)^{\frac{1}{3}}(-6s^2) = -12s\left[3(K - s^3)^{\frac{1}{3}}\right].$$

$$(3)^{\frac{2}{3}}(4\pi)^{\frac{1}{3}}(-6s^2) = -36s(K - s^3)^{\frac{1}{3}}.$$

Cubing both sides:

$$(3)^2(4\pi)(-6)^3s^6 = (-36)^3s^3(K - s^3).$$

$$(36)(\pi)(-216)s^6 = -46656s^3(K - s^3)$$

$$-7776\pi s^6 = -46656s^3(K - s^3)$$

$$K - s^3 = \frac{-7776\pi s^6}{-46656s^3} = \frac{\pi s^6}{6s^3} = \frac{\pi s^3}{6}.$$

We need not solve for s since, by use of the equation: $\frac{4}{3}\pi r^3 + s^3 = K$, we have: $K - s^3 = \frac{4}{3}\pi r^3$. Therefore,

$$\frac{\pi s^3}{6} = \frac{4\pi r^3}{3},$$

or,

$$3\pi s^3 = 24\pi r^3.$$

$$s^3 = 8r^3.$$

Taking the cube root of both sides: $s = 2r$, but $2r$ = diameter. Therefore, we have shown that the sum of the surface areas is greatest when the diameter, 2r, of the sphere is equal to the edge, s, of the cube.

• PROBLEM 389

Find the most economical shape for a box (minimum surface) with a square bottom and vertical **sides**, if it is to hold 4 cu. ft.

Solution: Let x be the length of one side of the base, and let h be the height. Let V be the volume **and A the total area.** Then $V = hx^2 = 4$, and $h = \frac{4}{x^2}$; bottom area is x^2, each side has an area hx. Total area, $A = x^2 + 4hx = x^2 + \frac{16}{x}$;

We wish to find $A' = \frac{dA}{dx}$, equate it to 0 and solve for x, obtaining critical values. We find

$$A' = \frac{dA}{dx} = 2x - \frac{16}{x^2};\ 2x - \frac{16}{x^2} = 0,\ x^3 = 8,\ x = 2.$$

We now test whether $x = 2$ is a maximum or a minimum value.

When $x < 2$, $A' = 2(x^3-8)/x^2$ is negative; when $x > 2$, m is positive; therefore $x = 2$ gives the minimum total area, $A = 12$. Notice that the height is $h = \frac{4}{x^2} = 1$. The correct dimensions are $x = 2$, $h = 1$ (in feet).

A water tank in the form of a right circular cone is to be designed to hold 120 ft^3. What should be the dimensions of the cone in order to have the minimum lateral surface area for the purpose of using a minimum amount of material in the construction of the tank.

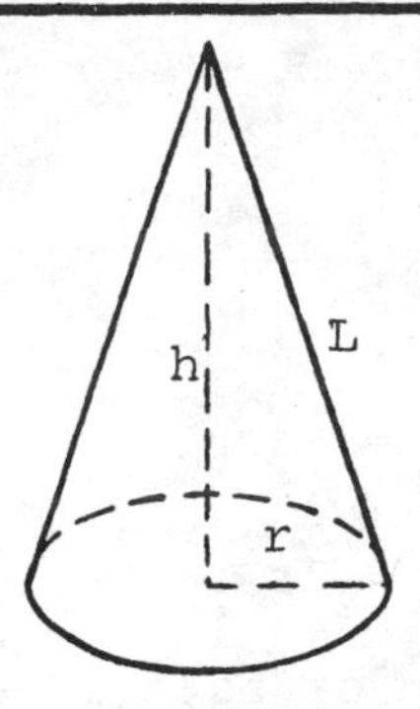

Solution: Assume a cone with radius of base r, altitude h, and slant height L. We denote the volume by V and the lateral surface area by S. Then

$$S = \pi r L,$$

while

$$V = \frac{1}{3}\pi r^2 h = 120.$$

From the figure,

$$r^2 + h^2 = L^2.$$

Regard everything as a function of r. That is, $S = S(r)$, $h = h(r)$, and $L = L(r)$. Taking derivatives,

$$S' = \pi r L' + \pi L,$$

and

$$\frac{1}{3}\pi r^2 h' + \frac{2}{3}\pi r h = 0, \quad 2r + 2hh' = 2LL'.$$

Our aim here is to minimize the surface area, S, therefore equating it to 0; but we first wish to obtain an equation in r and h. We must therefore replace L' in this equation. From the last equation we find, $L' = \frac{r+hh'}{L}$ and from the second equation, $h' = -\frac{2}{r}h$. Substituting this into L' we obtain, $L' = \frac{r^2 - 2h^2}{rL}$. This yields

$$S' = \pi r\left(\frac{r^2 - 2h^2}{rL}\right) + \pi L = 0,$$

and, factoring out π,

$$\frac{r^2 - 2h^2}{L} + L = 0 \quad \text{or} \quad r^2 - 2h^2 + L^2 = 0.$$

Since $L^2 = r^2 + h^2$, by substitution we arrive at the relation

$$h = \pm r\sqrt{2},$$

We use the positive value for h only. From the geometry of the problem we deduce that the positive value must give the minimum area. Since the volume is 120 ft^3, we can now solve for r to get

$$\frac{1}{3}\pi r^3\sqrt{2} = 120, \quad r = \left(\frac{360}{2^{1/2}\pi}\right)^{1/3}.$$

The altitude h is

$$\sqrt{2}\left(\frac{360}{2^{1/2}\pi}\right)^{1/3} = \left(\frac{720}{\pi}\right)^{1/3}.$$

The slant height L is

$$L = \sqrt{\left(\frac{360}{\sqrt{2}\pi}\right)^{2/3} + \left(\frac{720}{\pi}\right)^{2/3}}.$$

• PROBLEM 391

What is the largest rectangle that can be inscribed in a right triangle of sides 5, 12, and 13 inches, if one vertex of the rectangle is on the longest side of the triangle.

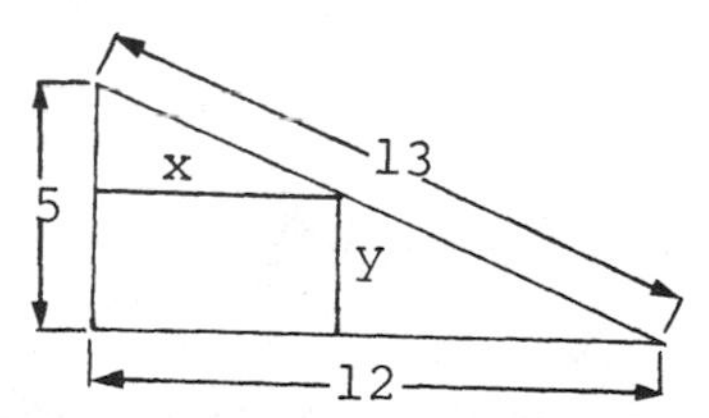

Solution: Let the sides of the rectangle be represented by x and y as shown in the figure. The area, A, = xy. To express A as a function of only one of these variables we must find an equation involving x and y, from which we can substitute for one variable in terms of the other. By using similar triangles, we may write the proportion

$$\frac{y}{5} = \frac{12 - x}{12},$$

from which it follows that

$$y = \frac{5}{12}(12 - x).$$

Thus,

$$A = \frac{5}{12}(12-x)x = \frac{5}{12}(12x-x^2), \quad 0 \leq x \leq 12.$$

We wish to maximize the area. Find $\frac{dA}{dx}$, equate it to 0 and solve, to obtain the critical values. We find:

$$\frac{dA}{dx} = \frac{5}{12}(12-2x), \; 12 - 2x = 0, \; x = 6.$$

This critical value yields a maximum for A because, by the second Derivative Test,

$$\frac{d^2A}{dx^2} = \frac{5}{12}(-2) = -\frac{5}{6} < 0,$$

indicating a maximum. For x = 6, we find

$$y = \frac{5}{12}(12 - 6) = \frac{5}{2} \text{ inches}$$

and

$$A = (6)\left(\frac{5}{2}\right) = 15 \text{ square inches.}$$

It is evident that the end point values x = 0 and x = 12 give the minimum value A = 0. Therefore, the area of the largest rectangle is 15 sq. in.

• PROBLEM 392

A 10-ft display sign is 10 ft. off the ground. Find the distance from the sign of a point on the ground where the subtended angle of the sign is a maximum.

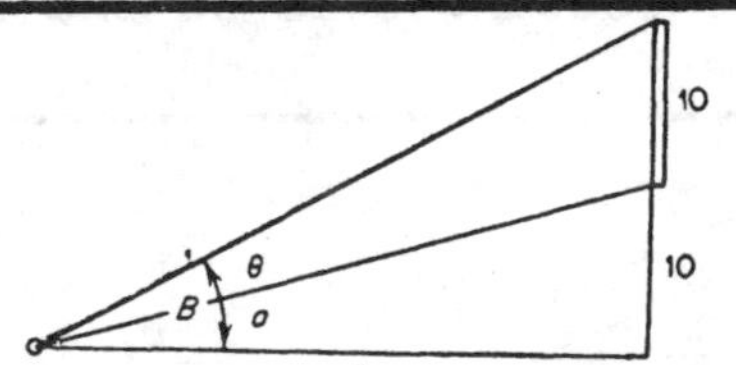

Solution: Let θ be the required angle. Then,

$$\tan B = \frac{20}{x}$$

$$\tan a = \frac{10}{x}$$

$$\tan \theta = \tan(B-a).$$

We make use of the identity

$$\tan(A-B) = \frac{\tan A - \tan B}{1 + \tan A \tan B},$$

and obtain:

$$\frac{\tan B - \tan a}{1 + \tan B \tan a} = \frac{20/x - 10/x}{1 + (20/x)(10/x)} = \frac{10/x}{(x^2+200)/x^2}$$

$$\frac{d}{dx}(\tan \theta) = \frac{d}{dx}\left(\frac{10x}{x^2 + 200}\right)$$

$$\frac{d}{dx}\left(\frac{u}{v}\right) = \frac{v\ du/dx - u\ dv/dx}{v^2}$$

$$= \frac{(x^2 + 200)10 - 10x(2x)}{(x^2 + 200)^2}.$$

For a maximum value of θ (and hence of tan θ),

$$\frac{10x^2 + 200(10) - 20x^2}{(x^2 + 200)^2} = 0$$

$$10x^2 = 200(10)$$

$$x = \underline{+}14.1 \text{ ft.}$$

• PROBLEM 393

A car traveling at 60 mph along a road perpendicular to a railroad track crosses the track at t = 0 hours. A train going 80 mph has 20 miles to traverse before crossing the road. Find the minimum and maximum relative speeds between car and train.

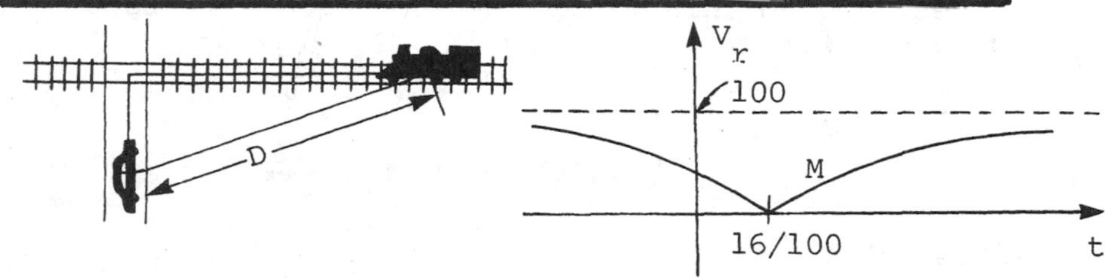

Solution: The procedure for solving this problem is to first find the distance, D(t), between car and train at any time t. Since our aim here is to find the minimum and maximum speeds, we must find $D' = \frac{dD}{dt}$, which gives us v(t), the velocity and $|v(t)|$ the speed, called M(t).

At time t the car will be a distance $|60t|$ miles from the crossing, while the train will be a distance $|80t - 20|$ from the crossing. Hence by the Pythagorean Theorem the distance between car and train is

$$D(t) = \sqrt{(60t)^2 + (80t - 20)^2}.$$

Multiplying out and factoring out 10,000 we have: $\sqrt{10{,}000(t^2 - \frac{32}{100}t + \frac{4}{100})}$. Now, completing the square we have,

$$D(t) = 100\sqrt{(t - \frac{16}{100})^2 + \frac{144}{10{,}000}}.$$

It follows from this expression without the necessity of differentiation that the car and the train are closest at $t = \frac{16}{100}$ hours, and that at that time they are 12 miles apart. Their relative velocity, $v(t)$, at time t is

$$D'(t) = v(t) = \frac{100(t - \frac{16}{100})}{\sqrt{(t - \frac{16}{100})^2 + \frac{144}{10,000}}},$$

while their relative speed $M(t)$ at time t is

$$|v(t)| = M(t) = \frac{|100(t - \frac{16}{100})|}{\sqrt{(t - \frac{16}{100})^2 + \frac{144}{10,000}}}$$

Again, the relative speed is a minimum at $t = \frac{16}{100}$, at which time it is 0. This is not surprising, since before $t = \frac{16}{100}$ car and train are approaching each other, while after $t = \frac{16}{100}$ they are going away from each other.

Now that we have found an equation for speed, $M(t)$ we wish to obtain the maximum and minimum values. To do this we find $M'(t)$, equate it to 0 and solve for t.

Since

$$M'(t) = \begin{cases} -v'(t) & \text{for } t < \frac{16}{100} \\ v'(t) & \text{for } t > \frac{16}{100} \end{cases}$$

we see that

$$M'(t) = \begin{cases} \dfrac{-144}{100[(t - \frac{16}{100})^2 + \frac{144}{10,000}]^{3/2}} & \text{for } t < \frac{16}{100} \\ \dfrac{144}{100[(t - \frac{16}{100})^2 + \frac{144}{10,000}]^{3/2}} & \text{for } t > \frac{16}{100}. \end{cases}$$

Now M does not have a derivative at $t = \frac{16}{100}$. We see therefore that $M'(t)$ is never 0, and fails to exist at only one value, $t = \frac{16}{100}$, at which M takes on a minimum. Thus relative speed never achieves a maximum. However, since $M'(t)$ is negative for $t < \frac{16}{100}$ and positive for $t > \frac{16}{100}$, we see that as we consider the past $\left(t < \frac{16}{100}\right)$ or future $\left(t > \frac{16}{100}\right)$ the relative speed increases in both cases. Examining $M(t)$, we see that for very large or very small t, $M(t)$ approaches 100 mph asymptotically. However, 100 mph relative speed is never actually achieved.

Note that we have here a perfectly "physical" problem in which the derivative fails to exist at the point of interest, namely where the car and train are closest, which is where their relative speed is zero (they are changing from "approaching" each other to "going away" from each other), and in which there is no achieved maximum value (see diagram). The number 100 results from the sum of the two speeds, i.e. 80 + 20.

• PROBLEM 394

> Fuel cost for operating a train is proportional to the square of the speed, and is $50 per hr when the speed is 20 mph. Other charges, such as labor, for example, are $200 per hr. What should be the speed for a 500-mile trip in order to minimize the total cost?

Solution: First we express the cost for fuel per hour. If C represents this cost,

$$C = ks^2,$$

where $\qquad s$ = speed in mph.

We now evaluate k. Since C = 50 when s = 20,

$$50 = 400k,$$

and

$$k = 50/400 = 1/8.$$

The cost for fuel per hour is therefor $\frac{1}{8}s^2$. Since other costs are $200 per hr, the total cost per hour is $\frac{1}{8}s^2 + 200$. The number of hours needed to make a 500-mile trip at s miles per hour is 500/s. Therefore, the total cost for the trip, A dollars is

$$A = \frac{500}{s}\left(\frac{1}{8}s^2 + 200\right) = \frac{125}{2}s + \frac{100{,}000}{s},$$

in which $s > 0$. We wish to minimize A, the total cost for the trip. We find $\frac{dA}{ds}$, equate it to 0, and solve for s to obtain the critical values. We find:

$$\frac{dA}{ds} = \frac{125}{2} - \frac{100{,}000}{s^2}.$$

Setting this quantity equal to 0 we obtain:

$$\frac{125}{2} - \frac{100{,}000}{s^2} = 0,$$

therefore,

$$125s^2 = 200{,}000, \quad s^2 = 1600, \quad \text{and} \quad s = 40.$$

Therefore, 40 is our critical value. We wish to show that this value does indeed represent a minimum. To do this we examine $\frac{dA}{ds}$ when s is less than 40 and when s is greater than 40. We observe that when $s < 40$, $\frac{dA}{ds}$ is negative, and when $s > 40$, $\frac{dA}{ds}$ is positive. We can therefore conclude that the value $s = 40$ is a minimum. The total cost is found by substituting $s = 40$ back in the cost equation:

$$A = \frac{(125)(40)}{2} + \frac{100,000}{40}$$

$$= 2500 + 2500$$

$$= 5000.$$

• PROBLEM 395

A ship sailing between two points separated by a distance ℓ incurs two types of cost: Fuel costs are proportional to the square of the speed of the ship and amount to \$100 when the speed is 10 mph, and there is a cost of \$60 per hour. What is the most economical speed?

Solution: Let x denote the speed in miles per hour.

The total time of the trip is $\frac{\ell}{x}$ hours. The hourly cost of the fuel is given by kx^2, where k is the proportionality constant. Since the hourly cost of the fuel is 100 dollars when $x = 10$, it is seen that $k = 1$.

Let C = total cost of the trip. Then, C = (hourly fuel cost · # of hours) + (\$60 per hr · # of hrs), or,

$$C = x^2 \frac{\ell}{x} + 60\frac{\ell}{x} = \ell\left(x + \frac{60}{x}\right)$$

We wish to minimize C, total cost. We find $\frac{dC}{dx}$, equate it to 0 and solve for x, obtaining critical values. We find:

$$\frac{dC}{dx} = \ell\left(1 - \frac{60}{x^2}\right)$$

Setting $\frac{dC}{dx} = 0$, we obtain $x = 2\sqrt{15}$ = critical value. We must now test to see if the value $x = 2\sqrt{15}$ gives us a minimum. This can be done by the Second Derivative Test. We find: $\frac{d^2C}{dx^2} = \frac{120\ell}{x^3}$. When $x = 2\sqrt{15}$ this becomes $\frac{\ell}{\sqrt{15}} > 0$. Therefore the value is a minimum and the most economical speed is $2\sqrt{15}$ miles per hour.

• PROBLEM 396

Ten thousand pounds of beef in cold storage are worth sixteen cents a pound wholesale. If the price increases steadily one cent per week while the beef loses a hundred pounds a week in weight, and the storage charges are sixty dollars a week, how long should the beef be held before selling for the greatest net value? What is that value?

Solution: Here we must formulate the net value any number of weeks n from the present time and find n such that the total net value at that time shall be a maximum.

With 100 lb. a week shrinkage the total weight in n weeks will be 10,000 - 100n, or 100(100-n). The present price is 16¢ = $\frac{4}{25}$ dollars and with a 1¢ = $\frac{1}{100}$ dollar advance per week the price per pound in n weeks will be $\left(\frac{4}{25}\right) + \left(\frac{1}{100}\right)n = \left(4 + \frac{n}{4}\right)/25$. The gross value will then be the weight times the price per pound, or $100(100-n).\left(4 + \frac{n}{4}\right)/25$. The gross value at the end of n weeks is, therefore,

$$V_g = 1600 + 84n - n^2, \tag{a}$$

In the meantime, however, there is a steady charge of 60 dollars a week which in n weeks amounts to 60n dollars. Subtracting this from the gross value given by equation (a) the net value at the end of n weeks is $V_g - 60n = V$. The net value is therefore

$$V = 1600 + 24n - n^2,$$

and the value of n which makes V a maximum is to be found. We obtain this value by finding $\frac{dV}{dn}$, equating this to 0 and solving for n.

Differentiating, we find:

$$\frac{dV}{dn} = 24 - 2n = 2(12 \cdot n) = 0$$

$$n = 12.$$

That is, the meat should be held twelve weeks in order that the net value shall be the greatest. We can now apply the Second Derivative Test to show that this is indeed a maximum value. We find: $\frac{d^2V}{dn^2} = -2$. Since $\frac{d^2V}{dn^2} < 0$ we conclude that n = 12 is a maximum value. To find V, we substitute 12 in the expression for V:

$$V = 1600 + 24(12) - 144$$

$$= 1600 + 268 - 144$$

$$= \$16,144.$$

What is the best height for a light to be placed over the center of a circle to provide maximum illumination to the circumference? The intensity varies as the sine of the angle at which the rays strike the illuminated surface, divided by the square of the distance from the light? (Assume point-source.)

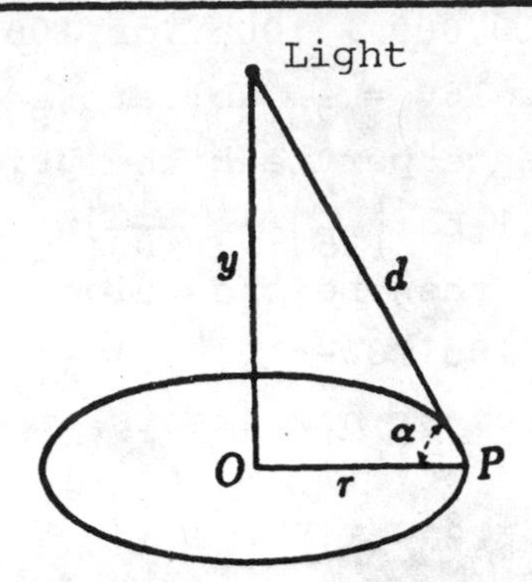

<u>Solution:</u> Given $I = k\frac{\sin\alpha}{d^2}$ = illumination at P, any point on the circumference. Distance,

$$d = \sqrt{y^2 + r^2},$$

by the Pythagorean theorem.

$$\sin\alpha = \frac{y}{\sqrt{y^2 + r^2}}.$$

Then $I = k \cdot \frac{y}{\sqrt{y^2 + r^2}} \cdot \frac{1}{(y^2 + r^2)} = k \cdot \frac{y}{(y^2 + r^2)^{3/2}}.$

We wish to maximize I. To do this, we find $\frac{dI}{dy}$, set it equal to 0 and solve for y. We find:

$$\frac{dI}{dy} = k\left[\frac{(y^2 + r^2)^{3/2} \cdot 1 - y \cdot \frac{3}{2}(y^2 + r^2)^{1/2} \cdot 2y}{(y^2 + r^2)^3}\right]$$

by use of the quotient rule.

$$\frac{dI}{dy} = k\,\frac{(y^2 + r^2)^{3/2} - 3y^2(y^2 + r^2)^{1/2}}{(y^2 + r^2)^3}$$

$$= \frac{k(y^2 + r^2)^{1/2}\left[(y^2 + r^2) - 3y^2\right]}{(y^2 + r^2)^3}$$

$$= \frac{k(y^2 + r^2)^{1/2}(y^2 + r^2 - 3y^2)}{(y^2 + r^2)^3},$$

and simplifying we have:

$$\frac{dI}{dy} = \frac{k(r^2 - 2y^2)}{(y^2 + r^2)^{5/2}}$$

$\frac{dI}{dy} = 0$ only when the numerator = 0. Then

$$r^2 - 2y^2 = 0 \text{ or } y = \frac{r}{2}\sqrt{2} \text{ for a maximum.}$$

We must now test that this is a maximum value. To do this we use the Second Derivative Test. We find:

$$\frac{d^2I}{dy^2} = k\left[\frac{(y^2+r^2)^{5/2}(-4y) - (r^2-2y^2)^{5/2}(y^2+r^2)^{3/2}(2y)}{\left[(y^2+r^2)^{5/2}\right]^2}\right]$$

and, by factoring out $k(y^2+r^2)^{3/2}$ and simplifying, we obtain:

$$\frac{d^2I}{dy^2} = \frac{k(6y^3 - 9yr^2)}{(y^2 + r^2)^{7/2}}.$$

Evaluating $\frac{d^2I}{dy^2}$ at $y = \frac{r}{2}\sqrt{2}$ we obtain a negative value, which according to the Second Derivative Test, indicates a maximum. Therefore, the height to obtain maximum illumination is $\frac{r}{2}\sqrt{2}$.

• PROBLEM 398

Two lamps of intensities a and b, respectively, are d feet apart. If the intensity of illumination at any point due to a given point source is directly proportional to the intensity of the source and inversely proportional to the square of the distance from the source, find the darkest point on the line joining the two sources.

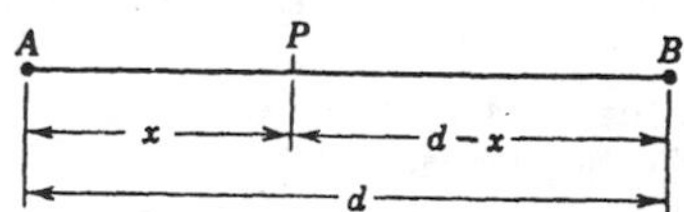

Solution: Considering the lamps, for simplicity, to be point-sources, let A be the source of intensity a, B be the source of intensity b, P be a point on AB, x be the distance from A to P and (d-x) be the distance from B to P, with d, the distance between the two sources. This arrangement can be seen in the accompanying diagram. Then, from the definition of intensity of illumination, at point P, the intensity of illumination due to A is $\frac{ka}{x^2}$, and that due to B is $\frac{kb}{(d-x)^2}$. If I is total illumination of P due to both sources,

$$I = \frac{ka}{x^2} + \frac{kb}{(d-x)^2}.$$

It is clear that at a point very close to either

source the value of I is great and decreases as the point recedes. Thus, if P is near either source, one or the other of the denominators in small, and the corresponding fraction is large. Since I decreases as P moves away from either source, it must reach a minimum somewhere between them. Hence, we expect a minimum, and we solve for its location. To obtain this minimum value we must find $\frac{dI}{dx}$, equate it to 0 and solve for x. Differentiating, we find:

$$\frac{dI}{dx} = k\left(\frac{-2a}{x^3} + \frac{2b}{(d-x)^3}\right) = 0$$

$$k\left(\frac{-2a}{x^3} + \frac{2b}{(d-x)^3}\right) = 0,$$

$$\frac{b}{(d-x)^3} = \frac{a}{x^3}\ .$$

Clearing fractions, $bx^3 = a(d-x)^3.$

Taking the cube roots of both sides,

$$\sqrt[3]{b}x = \sqrt[3]{a}(d-x).$$

Multiplying out and transposing,

$$x(\sqrt[3]{a} + \sqrt[3]{b}) = d\sqrt[3]{a}.$$

Therefore,
$$x = \frac{d\sqrt[3]{a}}{\sqrt[3]{a} + \sqrt[3]{b}},$$

and
$$d - x = d - \frac{d\sqrt[3]{a}}{\sqrt[3]{a} + \sqrt[3]{b}} = \frac{d\sqrt[3]{b}}{\sqrt[3]{a} + \sqrt[3]{b}}.$$

We can now find the ratio of the distances x: (d-x). Doing this we have:

$$\frac{d\sqrt[3]{a}}{\sqrt[3]{a} + \sqrt[3]{b}} : \frac{d\sqrt[3]{b}}{\sqrt[3]{a} + \sqrt[3]{b}} = d\sqrt[3]{a} : d\sqrt[3]{b}.$$

Therefore, the ratio of the distances x: (d-x), for the minimum, is $\sqrt[3]{a} : \sqrt[3]{b}$.

• PROBLEM 399

A cylindrical can is to be made to contain 1 quart. Find the relative dimensions so that the least amount of material is required.

Solution: Let r and h denote the radius of base and the altitude of the cylinder. Then

$$\text{Volume} = \pi r^2 h = 1,$$

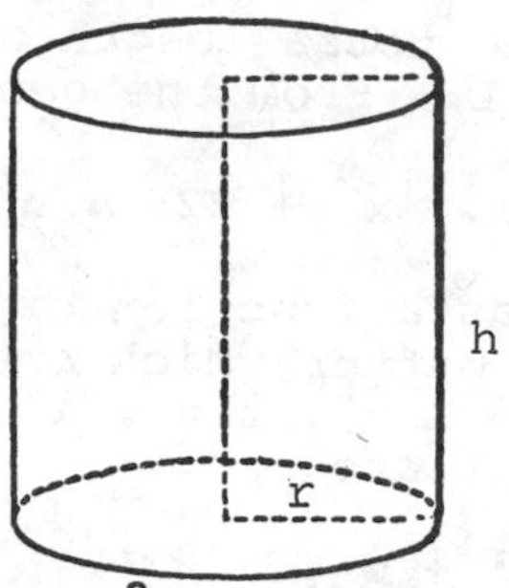

and,

$$2\pi rh + 2\pi r^2 = S \text{ (total surface)}.$$

Let r be regarded as the independent variable and consider h a function of r. Differentiate the above equations with respect to r; denoting $\frac{dh}{dr} = D_r h$ by h', we obtain:

$$\pi r^2 h' + 2\pi rh = 0$$

or

$$rh' + 2h = 0,$$

and

$$2\pi rh' + 2\pi h + 4\pi r.$$

We set this equal to 0 because S is to be a minimum. Doing this we have,

$$2\pi rh' + 2\pi h + 4\pi r = 0$$

or

$$rh' + h + 2r = 0.$$

We now solve:

$$rh' + 2h = 0$$

and,

$$rh' + h + 2r = 0$$

simultaneously for h. We eliminate h' by subtraction, obtaining $h - 2r = 0$ or $h = 2r$. Hence, for minimum total surface S, we must have: altitude = diameter of base.

• PROBLEM 400

Two vertical poles 15 and 20 ft above the ground are to be reinforced by a wire connected to the top of the poles and tied to a stake driven in the ground between the poles. Locate the stake with respect to one of the poles if the poles are 21 ft. apart, and a minimum amount of wire is to be used.

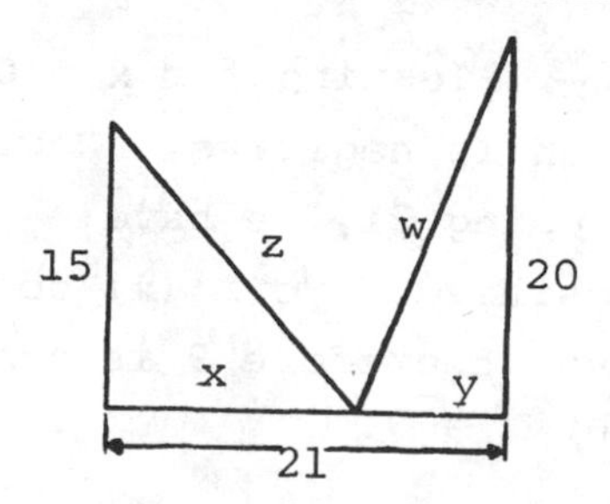

Solution: Let L be the total length of wire. We then wish to minimize L. From the geometry of the figure,

$L = z + w, \quad x + y = 21, \quad x^2 + 225 = z^2, \quad y^2 + 400 = w^2.$

Now we should seek L as a function of one of the variables; it doesn't matter which one we choose. Let it be x. Then $L = L(x)$, $z = z(x)$, $w = w(x)$, and $y = y(x)$. We differentiate,

$$L' = z' + w',$$

and differentiating the above equations implicitly with respect to x we have,

$$1 + y' = 0, \quad 2x = 2zz', \quad 2yy' = 2ww'.$$

Now, $z' = \frac{2x}{2z} = \frac{x}{z}$ and $w' = \frac{2yy'}{2w} = \frac{yy'}{w}$. But $1 + y' = 0$, so $y' = 1$. Therefore, $w' = -\frac{y}{w}$. Substituting these in the relation for L' yields

$$L' = \frac{x}{z} - \frac{y}{w}.$$

To obtain a minimum we set this = 0 and solve. Therefore, $\frac{x}{z} = \frac{y}{w}$. We can rewrite this as $\frac{x}{y} = \frac{z}{w}$ to make computations easier. Since $y = 21 - x$, $z = \sqrt{x^2 + 225}$, and $w = \sqrt{y^2 + 400}$, we obtain

$$\frac{x}{y} = \frac{z}{w}; \quad \frac{x}{21 - x} = \frac{\sqrt{x^2 + 225}}{\sqrt{y^2 + 400}} \quad \text{or}$$

$$\frac{x^2}{(21 - x)^2} = \frac{x^2 + 225}{(21 - x)^2 + 400}.$$

Cross multiplying and simplifying, we obtain: $175(x^2 + 54x - 567) = 0$. Therefore, $x^2 + 54x - 567 = 0$. Factoring we have, $(x-9)(x+63) = 0$. So, $x - 9 = 0$ and $x + 63 = 0$. Now, $x = 9$ and $x = -63$. We can reject the value of -63. This yields $x = 9$. We must now test that this value is a minimum. To do this we show that $L'(x)$ changes from - to +.

We have: $L' = \frac{x}{z} - \frac{y}{w}$. Testing for $x = 0$ we obtain, $L' = 0 - \frac{21}{w}$, which is negative. Now, choos- a value greater than 9, using 21, we have, $L' = \frac{21}{z} - 0$, which is positive. So $L'(x)$ does change from negative to positive, therefore 9 is a minimum value. The minimum is at $x = 9$.

A picture 7 feet in height is hung on a wall with the lower edge 9 feet above the level of the observer's eye. How far from the wall should the observer stand in order to obtain the most favorable view? (I.e., the picture should subtend a maximum angle.)

Solution: Referring to the diagram, we must find the value of x which will make angle θ a maximum. From the figure, we see that

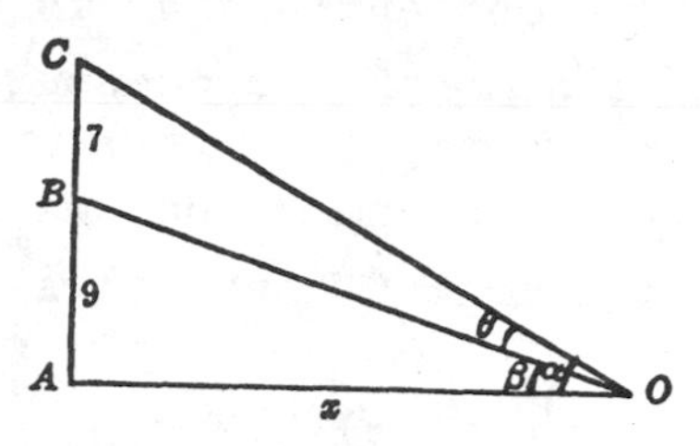

$$\tan \alpha = \frac{16}{x}, \quad \tan \beta = \frac{9}{x},$$

so that

$$\theta = \alpha - \beta = \text{Arc tan}\frac{16}{x} - \text{Arc tan}\frac{9}{x}.$$

To find the maximum value we find $D_x\theta = \frac{d}{dx}$, and, set it equal to 0, and solve to obtain critical values. We find:

$$D_x\theta = \frac{1}{1 + \frac{256}{x^2}}\left(-\frac{16}{x^2}\right) - \frac{1}{1 + \frac{81}{x^2}}\left(-\frac{9}{x^2}\right)$$

$$= -\frac{16}{x^2 + 256} + \frac{9}{x^2 + 81}.$$

Putting $D_x\theta = 0$ for a maximum and solving for x, we have

$$16(x^2+81) = 9(x^2+256), \; 7x^2 = 1008, \; x^2 = 144$$

from which x = 12. We must now test that 12 is a maximum. To do this we choose a value less than 12, say 10 and a value greater than 12, say 15. Now we find $D_x\theta$ at 10 and 15. If the sign changes from + to - then 12 is a maximum value. For x = 10 we have: $-\frac{16}{356} + \frac{9}{181} = D_x\theta$, which is a positive value. For x = 15, $D_x\theta = -\frac{16}{481} + \frac{9}{306}$, which is a negative value. Therefore, x = 12 is indeed a maximum. The observer should therefore stand 12 feet back from the wall.

Whenever a person coughs, the radius of the main air passages to the lungs decreases. We wish to determine at what radius the velocity of the expelled air will be a maximum. Let v denote the function which gives the velocity of air through a particular air passage at a radius r of that passage. Under usual conditions, it has been found that the following relationship prevails:

$$v(r) = cr^2(r_o - r)$$

where c is a positive constant and r_o is the radius of the air passage at atmospheric pressure.

Solution: To find the maximum value, we must find $v'(r) = \frac{dv}{dr}$, set it equal to 0, and solve for r. We write:

$$v(r) = cr_o r^2 - cr^3, \quad v'(r) = 2crr_o - 3cr^2$$

$$cr(2r_o - 3r) = 0.$$

We now have, $cr = 0$ and $2r_o - 3r = 0$. We can reject $cr = 0$ because c = a positive constant and r = a radius, also positive. Their product **cannot** be 0. So, we have:

$$2r_o - 3r = 0$$

$$-3r = -2r_o$$

$$r = \frac{2}{3}r_o.$$

To test if this is a maximum value we use the Second Derivative Test. If $v''(r) < 0$ at $r = \frac{2}{3}r_o$, then this is a maximum value. We find:

$$v''(r) = c(2r_o - 6r)$$

and

$$v''\left(\frac{2}{3}r_o\right) = c\left[2r_o - 6\left(\frac{2}{3}r_o\right)\right]$$

$$= c(-2r_o)$$

$$= -2cr_o,$$

which is less than 0. Therefore this is a maximum value. We find that the maximum velocity is obtained when the radius of tha air passage is two-thirds that of the passage at rest.

In the planning of a restaurant it is estimated that if there are seats for 40 to 80 people, the weekly profit will be $8 per place. However, if the seating capacity goes above 80, the weekly profit on each seat will be decreased by 4 cents times the number of seats above 80. What should be the seating capacity in order to yield the greatest weekly profit?

Solution: Let x = the number of seats in the restaurant. P = the number of dollars of total weekly profit.

The value of P depends upon x, and it is obtained by multiplying x by the number of dollars of profit per seat. When $40 \le x \le 80$, $8 is the profit per seat, and so $P = 8x$. However, when $x > 80$, the number of dollars of profit per place is $[8 - 0.04(x-80)]$, thus giving

$$P = x[8 - 0.04(x-80)] = 11.20x - 0.04x^2.$$

So we have:

$$P(x) = \begin{cases} 8x & \text{if } 40 \le x \le 80 \\ 11.20x - 0.04x^2 & \text{if } 80 \le x \le 280 \end{cases}$$

The upper bound of 280 for x is obtained by noting that $11.20x - 0.04x^2 = 0$ when $x = 280$; and when $x > 280$, $11.20x - 0.04x^2$ is negative.

Even though x is, by definition, an integer, we desire a continuous function that can be differentiated and let x take on all real values in the interval [40, 280]. Note that there is continuity at 80 because both equations:

$$8x = 8 \cdot 80 = 640$$

and $$(11.20x-0.04x^2) = 11.20(80) - 0.04(80)^2 = 640$$

give the same result, from which it follows that the limit: $\lim_{x \to 80} P(x) = 640$, and P is continuous in the closed interval [40, 280].

We wish to maximize P(x). We therefore find P'(x), equate it to 0 and solve for x to obtain critical values. In this case, we must examine P'(x) when $40 < x < 80$, when $x = 80$ and when $80 < x < 280$.

When $40 < x < 80$, $P'(x) = 8$. When $80 < x < 280$, $P'(x) = 11.20 - 0.08x$. P'(80) does not exist. Setting $P'(x) = 0$, we have

$$11.20x - 0.08x = 0$$

$$x = 140.$$

The critical values of P are therefore 80 and 140. Evaluating P(x) at the endpoints of the interval [40, 280] and at the critical numbers, we have P(40) = 320, P(80) = 640, P(140) = 784, and P(280) = 0. The absolute maximum value of P, then, is 784 occurring when x = 140.

The seating capacity should be 140, which gives a total weekly profit of $784.

• PROBLEM 404

If the stiffness of a beam with rectangular cross-section is directly proportional to the width and the cube of the depth, what are the dimensions of the stiffest beam that can be made from a log of circular cross-section of diameter a?

<u>Solution:</u> Let x denote the width and y the depth of the cross-section of the beam, and let a be the diameter of the log. We have

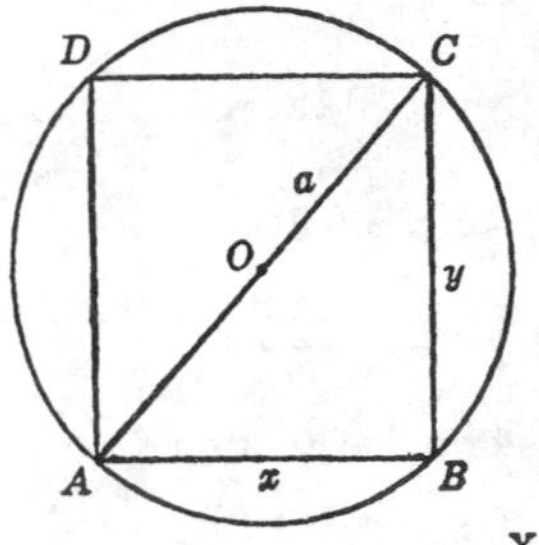

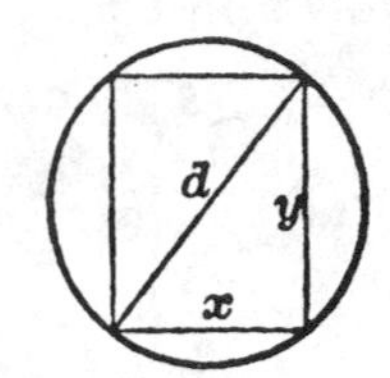

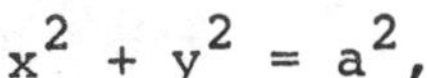

$$x^2 + y^2 = a^2,$$

by the Pythagorean theorem. From the given condition, it can be formulated that

$$S = kxy^3,$$

where S is the stiffness and k and a are constant. We wish to make S a maximum. To do this we differentiate the above equations, set the derivatives equal to 0 and solve them simultaneously for y. Regarding x as the independent variable and y as a function of x, and differentiating the above equations with respect to x, we obtain

$$2x + 2yy' = 0,$$

$$D_xS = ky^3 + 3kxy^2y' = 0.$$

or,

$$[2x + 2yy' = 0]\,3kxy$$

$$\frac{-[ky^3 + 3kxy^2y' = 0]\,2}{2x(3kxy) - ky^3(2) = 0}$$

$$6x^2ky - 2ky^3 = 0$$

$$2ky(3x^2 - y^2) = 0.$$

We can reject $2ky = 0$ because k and y are constants and their product $\neq 0$. We find $y^2 = 3x^2$, or $y = x\sqrt{3}$, which gives the required relative dimensions.

• PROBLEM 405

> If the strength of a beam with rectangular cross-section is directly proportional to the breath and the square of the depth, what are the dimensions of the strongest beam that can be made from a log of circular cross-section of diameter d?

Solution: If x = breadth and y = depth, then the beam will have maximum strength when the function xy^2 is a maximum. From the figure, $y^2 = d^2 - x^2$; hence we should test the function

$$f(x) = x(d^2-x^2) = d^2x - x^3.$$

We now find $f'(x)$, equate it to 0 and solve for x, to obtain the critical values. We find:

$$f'(x) = d^2 - 3x^2 = 0.$$

Therefore, $x = \frac{d}{\sqrt{3}}$ = critical value which gives a maximum. Therefore, if the beam is cut so that

$$\text{Depth} = \sqrt{\frac{2}{3}} \text{ of diameter of log}$$

and $\text{Breath} = \sqrt{\frac{1}{3}}$ of diameter of log.

the beam will have maximum strength.

• PROBLEM 406

> A wall in danger of collapse is to be braced by means of a beam which must pass over a second, lower wall b feet high and located a feet from the first wall. What is the shortest beam that can be used?

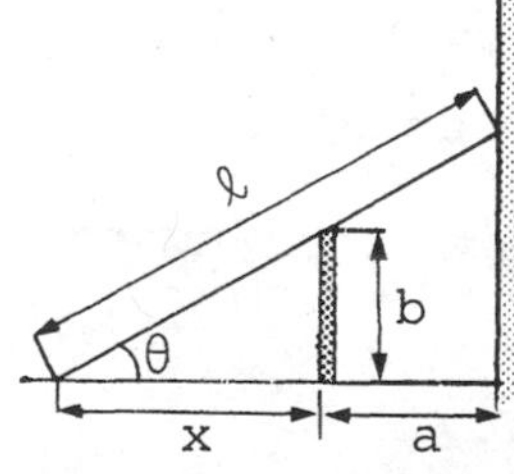

Solution: Let ℓ be the length of the beam and let it make an angle θ with the horizontal. The best approach here is to derive an equation for ℓ, involving θ, from the trigonometric configuration. We first write:

$$\ell = (a + x)\sec\theta \quad \text{and} \quad x = b \operatorname{ctn}\theta.$$

Hence, $\ell = a \sec\theta + x \sec\theta$.

substituting for x, and remembering that $\text{ctn}\,\theta \cdot \sec\theta = \csc\theta$,

$$\ell = a \sec\theta + b \csc\theta.$$

To find maximum and minimum values we find $\frac{d\ell}{d\theta}$, equate it to 0 and solve for θ to obtain critical values. We obtain:

$$\frac{d\ell}{d\theta} = a \sec\theta \cdot \tan\theta - b \csc\theta \cdot \text{ctn}\,\theta = 0.$$

Solving, $\frac{a \sin\theta}{\cos^2\theta} = \frac{b \cos\theta}{\sin^2\theta}$,

or $\frac{\sin^3\theta}{\cos^3\theta} = \frac{b}{a}$, $\tan\theta = \sqrt[3]{\frac{b}{a}}$,

To test whether this is a maximum or a minimum, we find

$$\frac{d^2\ell}{d\theta^2} = a\Big[(\sec\theta)(\sec^2\theta) + (\tan\theta)(\sec\theta\tan\theta)\Big]$$

$$- b\Big[(\cos\theta)(-\csc^2\theta) + (\text{ctn}\,\theta)(-\csc\theta\,\text{ctn}\,\theta)\Big]$$

$$= a \sec\theta(\sec^2\theta + \tan^2\theta)$$

$$+ b \csc\theta(\csc^2\theta + \text{ctn}^2\theta).$$

$\frac{d^2\ell}{d\theta^2}$ is positive, therefore ℓ is a minimum. The length of the beam is $\ell = a \sec\theta + b \csc\theta$. We have obtained; $\tan\theta = \sqrt[3]{\frac{b}{a}}$. We make use of the identities, $\sec\theta = \sqrt{1 + \tan^2\theta}$ and $\csc\theta = \sqrt{1 + \text{ctn}^2\theta}$ to obtain:

$$\ell = a\sqrt{1 + \left(\frac{b}{a}\right)^{2/3}} + b\sqrt{1 + \left(\frac{a}{b}\right)^{2/3}}$$

$$= \left(a^{2/3} + b^{2/3}\right)^{3/2}.$$

• PROBLEM 407

A man in a boat at a location five miles from the nearest point on a straight beach wishes to reach a place five miles along the shore from that point in the shortest possible time. If he can row at a rate of four miles an hour and run six miles an hour, where should he land?

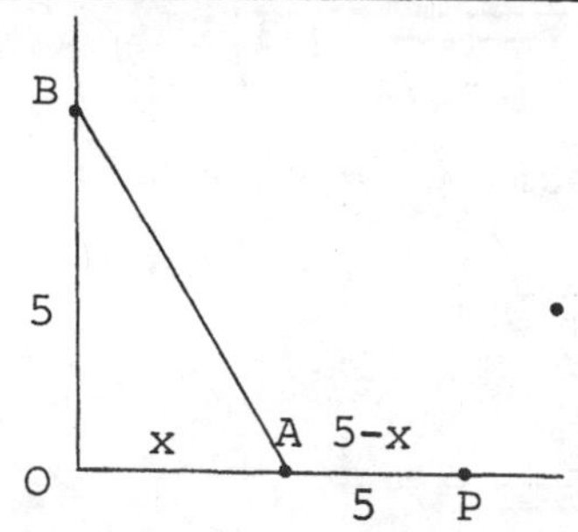

Solution: Let OP be the beach line and B the position of the boat at a distance OB = 5 miles from the nearest point O on the beach; let P be the place which the man wishes to reach, at a distance OP = 5 miles from O. He must then row straight to some point A on the beach at the distance OA = x from O and run the remaining distance AP = 5 - x. We formulate the time taken to row the distance BA and run the distance AP to obtain the total time, and find the value of x which makes this time a minimum.

In the right triangle AOB the rowing distance is:

$$\overline{BA} = \sqrt{\overline{BO}^2 + \overline{OA}^2} = \sqrt{x^2 + 25}$$

and at 4 miles an hour the time it will take him is $t_1 = \overline{BA}/4$. Therefore

$$t_1 = \frac{1}{4}\sqrt{x^2 + 25}. \qquad \text{(a)}$$

After landing he must run the distance $\overline{AP} = 5 - x$ and at 6 miles an hour it will require a time $t_2 = \overline{AP}/6$. Therefore

$$t_2 = \frac{1}{6}(5 - x). \qquad \text{(b)}$$

The total time required will be $t = t_1 + t_2$. Hence, adding (a) and (b),

$$t = \frac{1}{4}\sqrt{x^2 + 25} + \frac{1}{6}(5 - x). \qquad \text{(c)}$$

This is the formula for the time as a function of the distance x, and the value of x, making t a minimum is to be determined.

Differentiating equation (c),

$$\frac{dt}{dx} = \left[\frac{1}{4}\cdot\frac{1}{2}(x^2 + 25)^{-\frac{1}{2}}(2x)\right] + \left[\frac{1}{6}(-1)\right]$$

$$= \frac{2x}{8(x^2 + 25)^{\frac{1}{2}}} - \frac{1}{6}.$$

Therefore,

$$\frac{dt}{dx} = \frac{1}{2}\left(\frac{x}{2\sqrt{x^2 + 25}} - \frac{1}{3}\right).$$

For t to be a minimum, $\frac{dt}{dx} = 0$.

$$\frac{x}{2\sqrt{x^2 + 25}} - \frac{1}{3} = 0.$$

$$\frac{x}{\sqrt{x^2 + 25}} = \frac{2}{3}.$$

Squaring, we have:

$$\frac{x^2}{x^2 + 25} = \frac{4}{9}.$$

Cross-multiplying, $9x^2 = 4x^2 + 100$.

$$5x^2 = 100.$$

Therefore

$$x^2 = 20,$$

and

$$x = \sqrt{20} = \pm 4.47 \text{ miles}.$$

If we use the negative value, the landing place must be 4.47 miles to the left of O in the figure, which will not allow the time to be a minimum. He must therefore land 4.47 miles from O in the direction of P, or a little more than half a mile from his destination.

• PROBLEM 408

A rancher has 100 feet of fencing to make a chicken yard. One side of the yard will be formed by the side of a barn, so no wire will be needed there. What are the dimensions of the yard if it is to be rectangular in shape and contain maximum area?

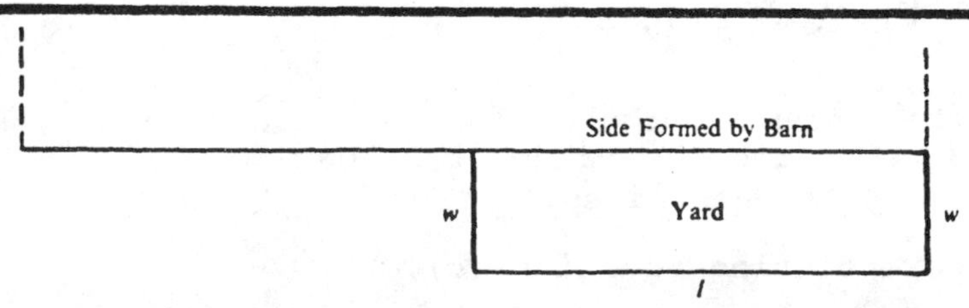

Solution: To find the dimensions of the maximum area, A, we first formulate an equation for A in terms of a single variable, which is one of the dimensions. We then take the first derivative of A, set it equal to zero, and solve for the variable, giving the value for one of the dimensions. We then use this value to calculate the other dimension.

From the diagram it is evident that A can be expressed as $A = wl$, where w is the width and l is the length. The perimeter is 100 feet. Therefore we can write,

$$2w + l = 100.$$

Using these two equations, we can write the equation for A in terms of one variable, w. Hence,

$$2w + l = 100.$$
$$l = 100 - 2w.$$
$$A = wl = w(100 - 2w) = 100w - 2w^2.$$

Taking the derivative of A, setting it equal to zero, and solving for w, we have:

$$A = 100w - 2w^2.$$
$$\frac{dA}{dw} = A' = 100 - 4w = 0.$$
$$4w = 100.$$
$$w = 25.$$

We substitute this value for the width into the equation:

$$2w + l = 100,$$

and find the length, l.

$$w = 25.$$
$$2w + l = 100.$$
$$2(25) + l = 100.$$
$$l = 50.$$

The area = 25(50) = 1250 sq. ft. The yard should be 25 feet wide and 50 feet long in order to have maximum area.

• PROBLEM 409

In an apple orchard there are 30 trees per acre, and the average yield per tree is 400 apples. For each additional tree planted per acre, the average yield per tree is reduced by 10 apples. How many trees per acre will maximize the crop?

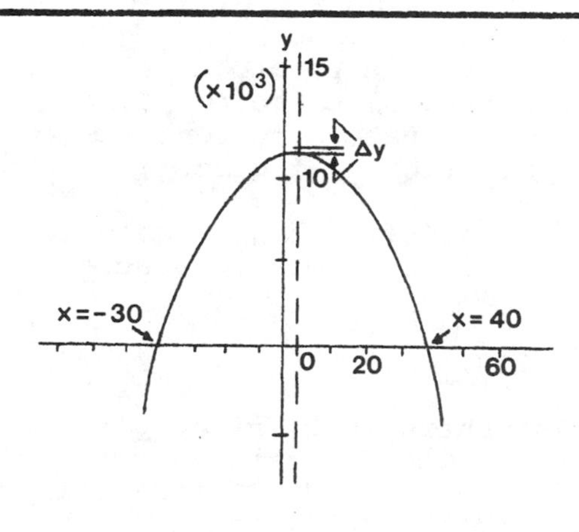

Solution: If x is the number of new trees planted per acre, there are 30 + x trees per acre, having an average yield of 400 - 10x apples per tree. Hence the total yield y of apples per acre is given by the product:

$$y = (30 + x)(400 - 10x)$$

$$= 12{,}000 + 100x - 10x^2$$

Therefore

$$\frac{dy}{dx} = 100 - 20x.$$

For maximum yield we set $dy/dx = 0$, obtaining $x = 5$.

Hence an addition of 5 trees per acre will give the largest crop. There will be 35 trees per acre, each yielding an average of 350 apples.

The yield is (35)(350) = 12,250

This can be shown graphically:

Since $y = 12{,}000 + 100x - 10x^2$

or $x^2 - 10x - 1200 = -y/10.$

This can be written as

$$x^2 - 10x + 25 - 25 - 1200 = -y/10.$$

Where the third term completes the square, and thus we have

$$(x-5)^2 - 1225 = -y/10.$$

For convenience in tracing the curve, we can change coordinates:

$$1225 - X^2 = Y$$

where $X = x-5$ and $Y = y/10$.

We can easily draw the parabola: $1225-X^2 = Y$, shifting the Y-axis five units to the left and multiplying every y graduation by ten. Hence the required graph looks like the figure.

The maximum number of apples is seen to be achieved by planting 5 trees more per acre and is calculated as

$$\Delta y = 250.$$

The graph also shows that, if we planted about 40 trees per acre, we would obtain 70 trees per acre with no yield, and beyond 40 trees our orchard would be spoiled. Furthermore, if we removed 30 trees per acre, as on the left hand side of the graph, the orchard would be empty with no output. Any point beyond this is, of course, meaningless.

• PROBLEM 410

In a certain type of chemical reaction, the weight x of substance formed varied thus with the elapsed time:

(a) $$\frac{x}{a - x} = e^{ka(t-c)} \quad (k > 0),$$

where a, k and c are constants. Find the rate of reaction; also find the value of t for which this rate is a maximum.

Solution: Taking the logarithm of each side of (a), we have

$$\ln x - \ln(a-x) = ka(t-c).$$

In finding the rate of reaction we are looking for the rate of change of the weight, x, with respect to time, t. Therefore, let us find $D_t x = \frac{dx}{dt}$. Differentiating the equation we have:

$$\frac{1}{x} \cdot D_t x + \frac{1}{a - x} \cdot D_t x = ka.$$

We now solve for $D_t x$ and denote this by r. We find:

(b) $$r = D_t x = kx(a - x).$$

To find when this rate r is a maximum, we differentiate (b) with respect to x, equate it to 0, and solve for t. We get

$$D_x r = kx(-1) + k(a-x) = -2kx + ka.$$

If $D_x r = 0$, we have $x = \frac{1}{2}a$. Now, using the Second Derivative Test to determine whether this value is a maximum we find, $D_x^2 r = -2k < 0$, r has a maximum when $x = \frac{1}{2}a$. Putting $x = \frac{1}{2}a$ in equation (a), we find $1 = e^{ka(t-c)}$. Taking the log of each side of the equation we obtain, $0 = ka(t-c)$. Therefore, the maximum rate occurs when $t = c$.

• PROBLEM 411

Find the shortest distance from a given point on the positive y-axis (0,b) to the parabola described by the equation: $x^2 = 4y$.

Solution: The parabola is shown in the figure. The quantity to be minimized is the distance d, where

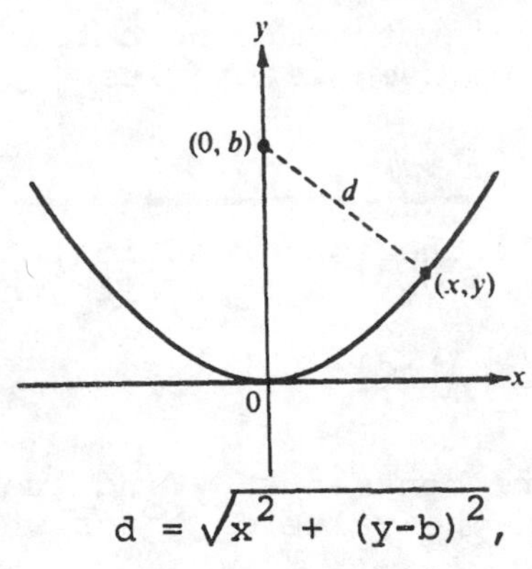

$$d = \sqrt{x^2 + (y-b)^2},$$

subject to the restriction $x^2 = 4y$.

First of all, we observe that the point (x,y) that minimizes d also minimizes d^2. (This observation enables us to avoid differentiation of square roots.)

Therefore the function f to be minimized is given by the formula

$$f(y) = d^2 = 4y + (y-b)^2.$$

To minimize f(y) we find f'(y), equate it to 0 and solve for y to obtain the critical values. We find:

$$f'(y) = 4 + 2(y-b) = 0.$$

Therefore $y = b - 2$. We must examine y when $b < 2$, and when it is ≥ 2. When $b < 2$, this leads to a negative critical point y which is excluded by the restriction $y \geq 0$. In other words, if $b < 2$, the minimum does not occur at a critical point. In fact, when $b < 2$, we see that $f'(y) > 0$ when $y \geq 0$, and hence f is strictly increasing for $y \geq 0$. Therefore the absolute minimum occurs at the endpoint $y = 0$. The corresponding minimum d is $\sqrt{b^2} = |b|$.

If $b \geq 2$, there is a legitimate critical point at $y = b - 2$. Since $f''(y) = 2$ for all y, the <u>absolute minimum of f occurs at this critical point.</u> The minimum d is $\sqrt{4y + (y-b)^2} = \sqrt{4(b-2) + (-2)^2} = \sqrt{4(b-2) + 4} = 2\sqrt{b - 1}$. Thus we have shown that the minimum distance is $|b|$ if $b < 2$ and is $2\sqrt{b - 1}$ if $b \geq 2$.

• PROBLEM 412

At what angle of elevation with the horizontal should a cannon be pointed in order to achieve maximum horizontal range?

<u>Solution:</u> Some simplifying assumptions are made: We neglect wind resistance and curvature of the earth. The muzzle velocity or speed, s, is in-

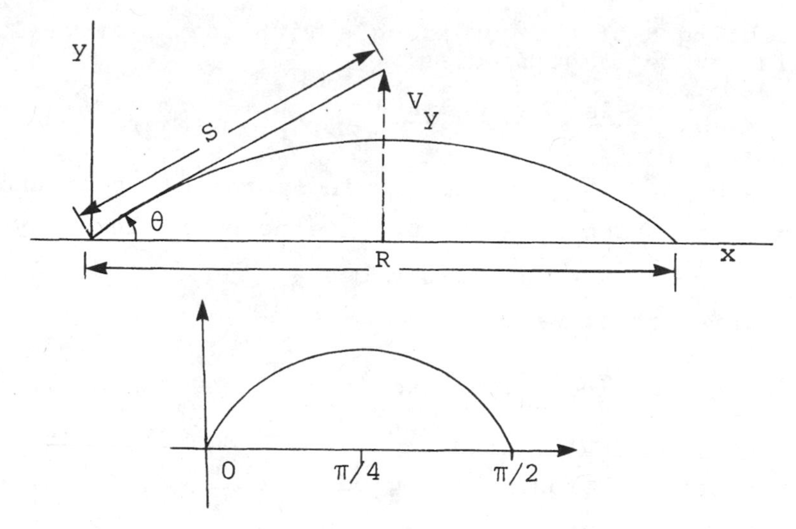

Range versus angle

dependent of the angle with the horizontal. Vertical and horizontal motion can be treated separately. Initially (at time 0) vertical speed is $s \times \sin\theta$. Under the influence of a constant gravitational force of mg (where m is the projectile mass), the height of the object at time t is

$$s \times \sin\theta \times t + \left(\frac{1}{2}\right)gt^2$$

where $g = -32.2$.

Since we assume no forces in the horizontal direction, the projectile travels at constant speed $s \times \cos\theta$. Thus it will have moved $s \times \cos\theta \times t$ after t time units have elapsed.

To determine the range of the gun, we simply ask:

(a) at what time will the projectile have returned to the ground and

(b) how far will the projectile have moved horizontally at <u>this</u> time?

To answer (a) we set

$$s \times \sin\theta \times t + \left(\frac{1}{2}\right)gt^2 = 0.$$

Then, $s \times \sin\theta \times t = -\frac{1}{2}gt^2$ and, $\frac{-2s \times \sin\theta \times t}{g} = t^2$. Dividing both sides by t we find:

(a) the projectile returns to ground at a time $t = \frac{-2s \times \sin\theta}{g}$ after it was fired.

Therefore,

(b) the projectile has moved $s \times \cos\theta \times \left[\frac{-2s \times \sin\theta}{g}\right]$ = units horizontally for a setting θ of the gun's angle with the horizontal. This can be simplified as

$$\frac{-2s^2 \sin\theta \cos\theta}{g}.$$

Since $0 \leq \theta \leq \frac{\pi}{2}$ ($\theta = 0$ is firing parallel to ground; $\theta = \frac{\pi}{2}$ is firing straight up), we need examine only

$$R = -2s^2 \sin\theta \cos\frac{\theta}{g}$$

for this set of values.

We now find $\frac{dR}{d\theta}$, equate it to 0 and solve for θ. We can rewrite R as $\frac{-2s^2}{g}(\sin\theta \cos\theta)$. Differentiating by using the product rule we obtain:

$$R' = \frac{-2s^2}{g}(\sin\theta)(-\sin\theta) + (\cos\theta)(\cos\theta)$$

$$= \frac{-2s^2}{g}(\cos^2\theta - \sin^2\theta).$$

Using the identity: $\sin^2\theta = 1 - \cos^2\theta$, we have,

$$R = \frac{-2s^2}{g}\left[2\cos^2\theta - 1\right].$$

Equating this to 0 we have,

$$2\cos^2\theta - 1 = 0, \text{ and } \cos\theta = \frac{1}{\sqrt{2}}.$$

Therefore $\theta = \frac{\pi}{4}$, the critical value. We must now examine R at the critical value, $\frac{\pi}{4}$, and also at the endpoints, 0 and $\frac{\pi}{2}$. We find, $R(0) = 0$, $R\left(\frac{\pi}{2}\right) = 0$ and $R\left(\frac{\pi}{4}\right) = \frac{-s^2}{g}$, remembering that $g = -32.2$. Therefore, $R\left(\frac{\pi}{4}\right) = \frac{s^2}{32.2}$. It is clear that a setting of $\frac{\pi}{4}$ maximizes the range. This is illustrated by the plot of R vs. θ in the graph.

• PROBLEM 413

Find the point on the parabola $y = x^2$ which is closest to the point (6,3).

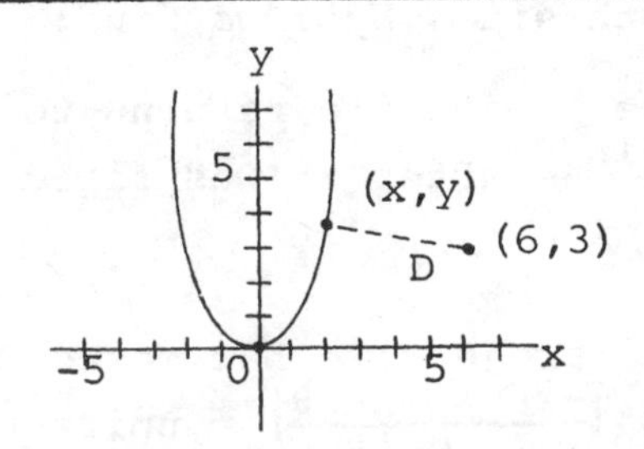

Solution: We write a function for the distance between a general point (x,y) on the parabola and the point (6,3). This relation, well-known from Analytic Geometry, is

$$D = \sqrt{(x - 6)^2 + (y - 3)^2}.$$

To make it easier to take derivatives, we shall square both sides of this expression. Also replace y by x^2 to express D in terms of one variable. Thus,

$$D^2 = (x - 6)^2 + (x^2 - 3)^2 = x^2 - 12x + 36 + x^4 - 6x^2 + 9$$

$$= x^4 - 5x^2 - 12x + 45;$$

$$\frac{d(D^2)}{dx} = 4x^3 - 10x - 12.$$

Setting this equal to 0, and dividing by 2, we have:

$$2x^3 - 5x - 6 = 0.$$

From which x = 2. The required point on the parabola is therefore x = 2, y = 4.

• PROBLEM 414

Find the values of a,b, and c if the parabola: $y = ax^2 + bx + c$, is to be tangent to the line: $y = 4x + 1$, at the point (-1,-3), and is to have a critical point when x = -2.

Solution: If the parabola: $y = ax^2 + bx + c$, is to have a critical point when x = -2, the first derivative, $\frac{dy}{dx}$, = 0 when x = -2. Therefore, we can find b as follows:

$$y = ax^2 + bx + c.$$

$$\frac{dy}{dx} = 2ax + b.$$

$$\frac{dy}{dx} = 0, \text{ at } x = -2.$$

$$2ax + b = 0.$$

$$2a(-2) + b = 0, \quad b = 4a.$$

The parabola: $y = ax^2 + 4ax + c$, and the line: $y = 4x + 1$ are tangent at the point (-1,-3). The slope of a line tangent to a curve at a certain point is the slope of the curve at that point. The slope of the line tangent to the parabola is the first derivative of the parabola. The slope of the line tangent to the parabola is equal to the slope of the parabola, and can be obtained as follows:

$$y = ax^2 + 4ax + c.$$

$$\frac{dy}{dx} = 2ax + 4a.$$

The slope of the given line is found by differentiating: $y = 4x + 1$.

$$\frac{dy}{dx} = 4, \text{ the slope of the line.}$$

At the point (-1,-3), these two slopes are equal.

$$2ax + 4a = 4.$$

$$2a(-1) + 4a = 4.$$

$$-2a + 4a = 4.$$

$$a = 2.$$

$$b = 8.$$

Having a = 2, and the point (-1,-3), find c by substituting these values into the quation for the parabola.

$$y = 2x^2 + 8x + c.$$

$$-3 = 2(-1)^2 + 8(-1) + c.$$

$$-3 = 2 - 8 + c.$$

$$3 = c.$$

With the values: a = 2, b = 8, c = 3, the parabola is:

$$y = 2x^2 + 8x + 3.$$

• **PROBLEM** 415

A corridor of width a makes a right-angle turn. A straight member, (which may be considered to have a thickness of zero) is to be transported around the corner. What is the longest such member that can pass the corner?

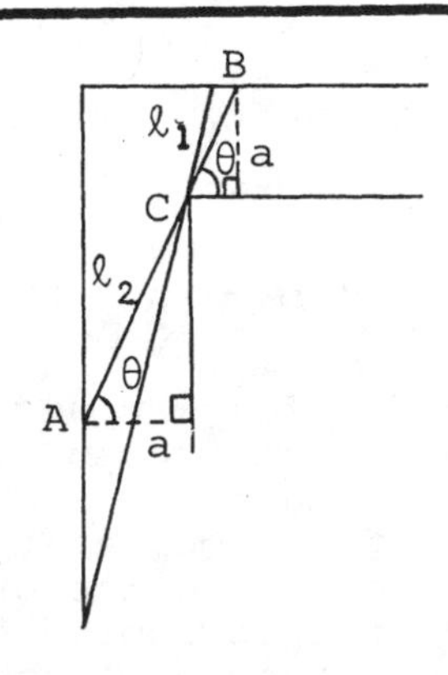

Solution: Referring to the accompanying diagram, we see that many lines can be drawn connecting such points as, for example, A and B and touching corner C. The length of the longest member that will go around the corner is the length of the shortest line ABC. Call the length ℓ and write $\ell = \ell_1 + \ell_2$. Our aim here is to rewrite ℓ, using the trigonometric configuration of the diagram, and then to minimize the equation for ℓ. From the diagram we have: $\ell_1 = a \csc \theta$ and $\ell_2 = a \sec \theta$. Therefore

$$\ell = a \csc \theta + a \sec.\theta.$$

We now find $\frac{d\ell}{d\theta}$, equate it to 0 and solve for θ to obtain the critical values. We find:

$$\ell' = \frac{d\ell}{d\theta} = a(-\csc \theta \cot \theta + \sec \theta \tan \theta) = 0.$$

From this we can write: $\sec \theta \tan \theta = \csc \theta \cot \theta$. Substituting their sin and cos equivalents,

$$\frac{\sin \theta}{\cos^2 \theta} = \frac{\cos \theta}{\sin^2 \theta}.$$

Cross-multiplying and dividing by $\cos^3 \theta$ we have,

$$\tan^3 \theta = 1.$$

$\theta = 45°, 225°$, the critical values. The only one **applicable to this problem is:** $\theta = 45°$.

We must examine this critical value to be certain it gives a minimum ℓ, length of line. It is important to keep in mind that a minimum length of line gives a maximum length of member. To test, we choose a value less than 45° and one greater than 45° and examine ℓ' at these values. If the sign changes from - to +, the critical value is indeed a minimum. We select 30° and 60° as the testing values. We get

$$\ell'(30°) = a\left(-2\sqrt{3} + \frac{2}{3}\right) < 0,$$

$$\ell'(60°) = a\left(2\sqrt{3} - \frac{2}{3}\right) > 0.$$

Therefore, 45° produces a minimum line. Substituting $\csc \theta = \sec \theta = 1$ in the expression for 1, we have: $\ell(45°) = 2a\sqrt{2}$ = Minimum length of line and a Maximum length of member that will go around the corner.

• PROBLEM 416

Lever, beam, and truss problems are usually solved by summation of moments about a selected point, Moments are equal to the product of the force and distance to the point, which causes a torque about the point. All clockwise moments are put on one side of the equation and all counterclockwise moments on the other side. Determine the length of a bar which will give a minimum force F. The bar weighs 3 lb/ft, and the load of 300 lb is located at 2 ft from R.

Solution: Taking moments about R so as to obtain an expression in F only,

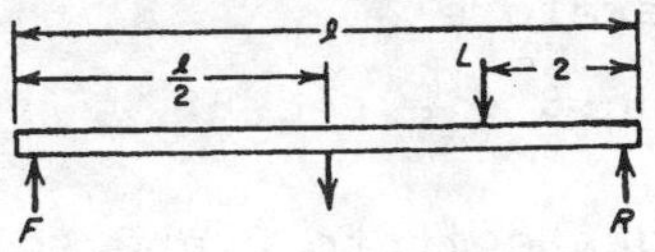

$$F\ell = \frac{\ell}{2}(3\ell) + 300(2) = \frac{3\ell^2}{2} + 600$$

$$F = \frac{3\ell}{2} + \frac{600}{\ell}$$

We wish to minimize F, the force. We find $\frac{dF}{d\ell}$, equate it to 0, and solve for ℓ. We find:

$$\frac{dF}{d\ell} = \frac{3}{2} - \frac{600}{\ell^2} = 0$$

$$3\ell^2 = 1200$$

$$\ell = \pm 20.$$

Rejecting the negative value we have ℓ = 20 ft. Therefore the length of the bar which will give a minimum force is 20 feet.

To determine the force required,

$$F = \frac{3}{2}(20) + \frac{600}{20} = 30 + 30 = 60 \text{ lb.}$$

CHAPTER 17

CURVE TRACING

Any curve can, of course be plotted by "brute force", i.e., substitution of many values for x and obtaining many values for y. However, for the purpose of sketching a curve or determining its trends, several short-cut methods can be applied by using such techniques as:

1) Finding maxima, minima, and points of inflection
2) Noting behavior as x and y go to ∞
3) Substituting in just a few values such as 0, 1, 2.
4) Noting any symmetry
5) Finding what happens at any points of discontinuity
6) Finding asymptotes.

These and other techniques may not be applicable in all cases, but, depending on the particular problem, some can always be used. A few additional points may be required where the behavior of a curve in a particular region is in doubt.

Such techniques used for sketching or tracing curves do provide results as accurate as obtainable from plotting a large number of closely-spaced points. The application of such techniques, however, is usually less time-consuming.

• PROBLEM 417

Determine where the curve $y = x^3$ is rising and where it is falling.

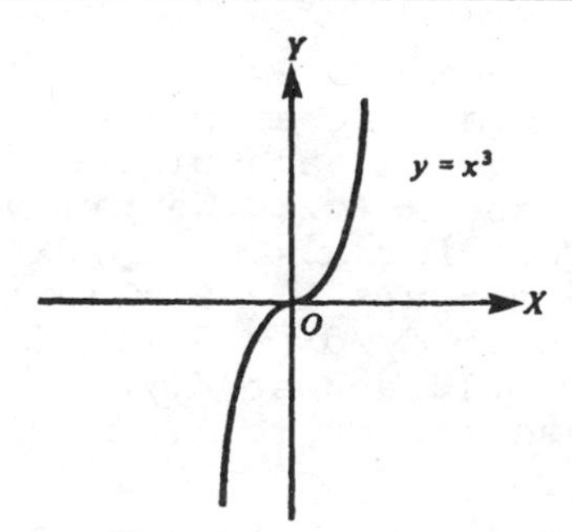

Solution:

$$y = x^3$$

$$y' = \frac{dy}{dx} = 3x^2$$

When $x = 0$, the curve is neither rising nor falling.

When $x \neq 0$, y' is always positive.

Thus, the curve is rising at all points except at the point (0,0).

• PROBLEM 418

Determine the regions in which the function:

$$y = (x-a)^3$$

is increasing and decreasing.

Solution: To determine when a function is increasing or decreasing, we must calculate the derivative.

Since $y = (x-a)^3$, $\frac{dy}{dx} = 3(x-a)^2$.

Note that the term $(x-a)^2$ is always non-negative $\left(\text{if } x \neq a, \frac{dy}{dx} > 0 \text{ and if } x = a, \frac{dy}{dx} = 0\right)$. Therefore the function is increasing for all values of x other than the value $x = a$.

• PROBLEM 419

During which intervals of x does the function

$$y = f(x) = -3 + 12x - 9x^2 + 2x^3$$

increase and decrease?

Solution: First we find the values of x at which the derivative is zero and the function is neither increasing nor decreasing.

$$\frac{dy}{dx} = 12 - 18x + 6x^2 = 6(x-1)(x-2).$$

When $dy/dx = 0$, $x = 1,2$. Now, the sign of the derivative can change only at $x = 1$ and $x = 2$. Hence it has one sign in each of the intervals $x < 1$, $1 < x < 2$, and $x > 2$. We try values of x in each interval. Thus for $x = 0$, $dy/dx = 12$ and hence is positive for $x < 1$. Similarly, $x = 1.5$ makes $dy/dx = -3/2$ or $dy/dx < 0$ for $1 < x < 2$. Also $x = 3$ makes $dy/dx = 12$ and so $dy/dx > 0$ for $x > 2$. This may be represented by the following diagram.

	$x = 1$		$x = 2$	
$-\infty$ ←				→ $+\infty$
$x < 1$		$1 < x < 2$		$x > 2$
dy/dx positive		dy/dx negative		dy/dx positive
y increasing		y decreasing		y increasing

• PROBLEM 420

Is (3x+5)/(x+1) an increasing or decreasing function of the variable x?

Solution: If the first derivative is positive, the function is increasing and if the first derivative is negative, the function is decreasing.

$$y = \frac{3x+5}{x+1} \qquad \text{(Note: y is not defined for } x = -1)$$

$$\frac{dy}{dx} = \frac{(x+1)3 - (3x+5)}{(x+1)^2} = \frac{3x + 3 - 3x - 5}{(x+1)^2} = \frac{-2}{(x+1)^2}$$

Since the denominator is positive for all values of x, dy/dx is always negative. Hence, y is a decreasing function, having no maxima or minima.

• PROBLEM 421

Examine the curve

$$y = (x-2)^{1/3}$$

for inflection points.

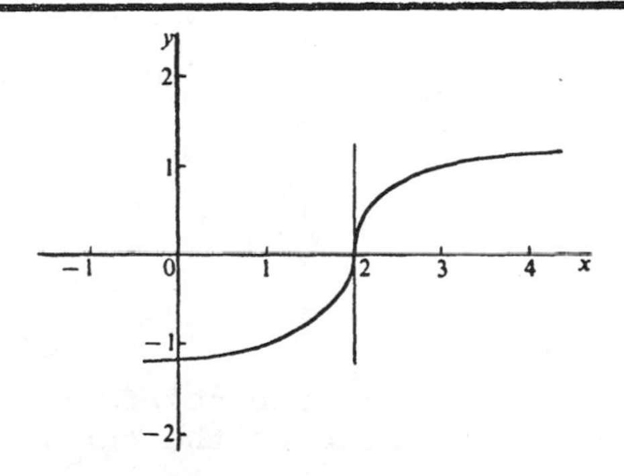

Solution:

$$y = (x-2)^{1/3}$$

$$\frac{dy}{dx} = \frac{1}{3}(x-2)^{-2/3}$$

$$\frac{d^2y}{dx^2} = -\frac{2}{9}(x-2)^{-5/3}$$

There is no point x at which $\frac{d^2y}{dx^2} = 0$. $\frac{d^2y}{dx^2}$ is undefined at x = 2, but the curve is defined at x = 2. Therefore x = 2 is a critical point. By checking points near x = 2 we find:

$$\frac{d^2y}{dx^2} > 0 \text{ for } x < 2.$$

$$\frac{d^2y}{dx^2} < 0 \text{ for } x > 2.$$

Hence x = 2 is an inflection point.

By inspecting $\frac{dy}{dx}$, we note that, as $x \to 2$, $\frac{dy}{dx} \to \infty$ which tells us that the curve has a vertical tangent.

• PROBLEM 422

Determine the concavity and points of inflection for

$$y = x^3 - 3x.$$

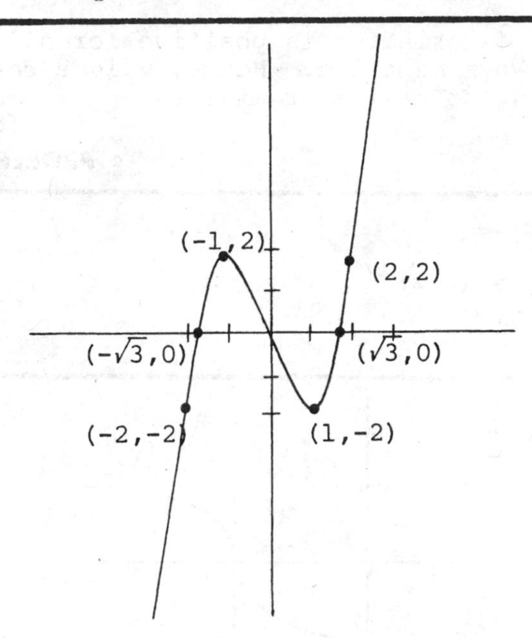

Solution: Concavity and points of inflection are determined by the second derivative of the function

If $\frac{d^2y}{dx^2} > 0$, the curve is concave up.

$\frac{d^2y}{dx^2} = 0$, the curve has an inflection point.

$\frac{d^2y}{dx^2} < 0$, the curve is concave down.

$$y = x^3 - 3x$$

$$\frac{dy}{dx} = 3x^2 - 3$$

$$\frac{d^2y}{dx^2} = 6x$$

By inspecting $\frac{d^2y}{dx^2} = 6x$, for $x > 0$ the curve is concave up, for $x < 0$ the curve is concave down, and $x = 0$ is the inflection point.

• PROBLEM 423

Let $f(x) = x^3 - 6x^2 + 9x$. Determine the inflection points for f and the intervals of concavity of the graph.

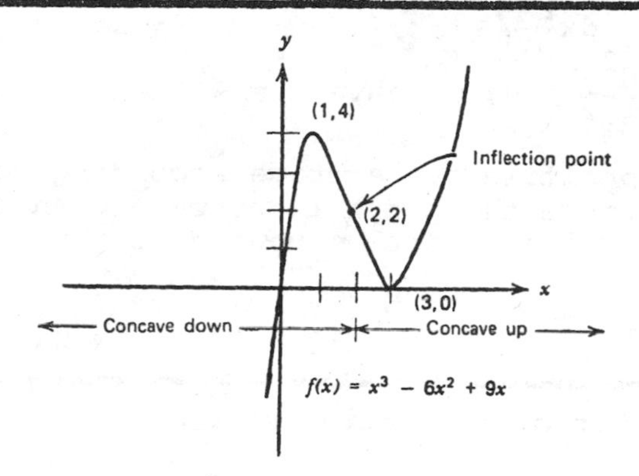

Solution: Since $f(x) = x^3 - 6x^2 + 9x$, by differentiating

$$f'(x) = 3x^2 - 12x + 9$$

and
$$f''(x) = 6x - 12 = 6(x-2).$$

Clearly, $f''(x) > 0$ if $x > 2$, and $f''(x) < 0$ if $x < 2$. Therefore, the graph of f is concave up for $x > 2$ and concave down for $x < 2$. Since the sign of f" changes at $x = 2$, this point is an inflection point for f. To sketch the graph of f we need only determine the intervals on which f is increasing or decreasing.

However, $f'(x) = 3(x^2-4x+3) = 3(x-1)(x-3)$. Therefore, $f'(x) \geq 0$ if $x < 1$ or $x > 3$, and f is increasing on those intervals. If $1 < x < 3$, $f'(x) < 0$ and f is decreasing. The point $x = 1$ is a relative maximum and $x = 3$ is a relative minimum, since $f''(1) < 0$ and $f''(3) > 0$ (applying second derivative test).

• PROBLEM 424

Find the intervals of x for which the curve

$$y = 2x^3 - 9x^2 + 12x - 3$$

is concave downward and concave upward.

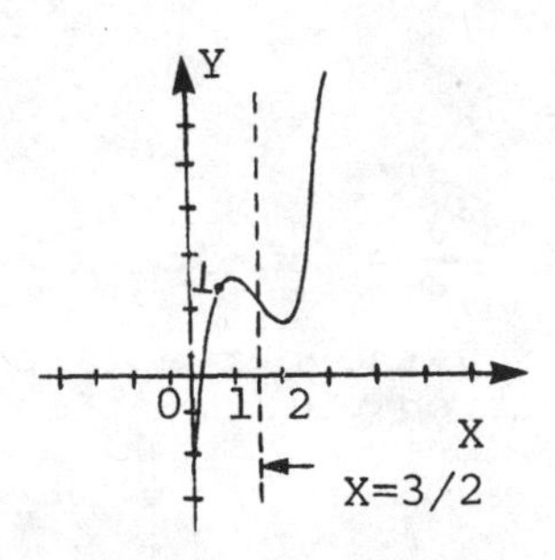

<u>Solution:</u> Differentiating twice,

$$\frac{dy}{dx} = 6(x^2-3x+2)$$

and

$$\frac{d^2y}{dx^2} = 6(2x-3)$$

By setting $\frac{d^2y}{dx^2} = 0$, we have $x = \frac{3}{2}$

y" is positive or negative according to $x > 3/2$ or $< 3/2$. Hence the graph is concave downward to the left of $x = 3/2$ and concave upward to the right of $x = 3/2$.

• PROBLEM 425

Find the point of inflection of:

$$y = 2x^3 - x^2 + 3x - 5.$$

<u>Solution:</u>

$$\frac{dy}{dx} = 6x^2 - 2x + 3$$

$$\frac{d^2y}{dx^2} = 12x - 2, \quad 12x - 2 = 0, \quad \text{and} \quad x = \frac{1}{6}$$

$$= 0.1167$$

Substitute this in the original equation, obtaining

$$y = 2\left(\frac{1}{6}\right)^3 - \left(\frac{1}{6}\right)^2 + 3\left(\frac{1}{6}\right) - 5$$

$$= 2\left(\frac{1}{216}\right) - \frac{1}{36} + \frac{1}{2} - 5.$$

$$y = \frac{1}{108} - \frac{1}{36} + \frac{1}{2} - 5 = \frac{1 - 3 + 54 - 540}{108} = -4.5185.$$

Therefore, the point of inflection is at $x = 1/6$, $y = -4.52$.

Now, for $x = 5/32 = 0.156$, which is a little less than $x = 0.166$ (the point of inflection),

$$\frac{d^2y}{dx^2} = 12x - 2 = 12 \cdot \frac{5}{32} - 2 = -$$

For $x = 3/16 = -.1875$, which is a little greater than $x = 0.1667$,
then

$$\frac{d^2y}{dx^2} = 12x - 2 = 12 \cdot \frac{3}{16} - 2 = +$$

Therefore the criterion for a point of inflection is satisfied.

• PROBLEM 426

> Sketch the graph of
>
> $$f(x) = 6x^5 - 5x^3 = x^3(6x^2-5)$$

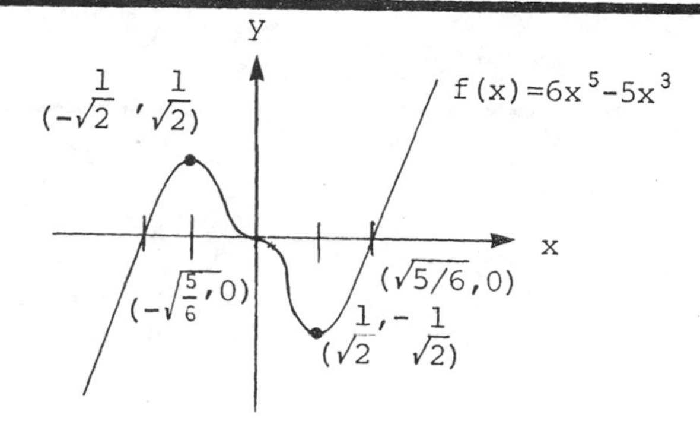

Solution: Note that since $f(x) = -f(-x)$ the function is an odd function.

If $\quad f(x) = 0$ then $x = 0$, $x = \pm\sqrt{\frac{5}{6}}$

Since

$$f(x) = 6x^5 - 5x^3,$$

$$f'(x) = 30x^4 - 15x^2 = 15x^2(2x^2-1)$$

$$f''(x) = 120x^3 - 30x = 30x(4x^2-1)$$

We consider $f'(x) = 15x^2(2x^2-1)$. If $f'(x) = 0$ then $x = 0$, $x = \pm\sqrt{\frac{1}{2}}$. There is no extreme at $x = 0$ since $f'(x)$ does not change sign there. The point $x = \sqrt{\frac{1}{2}}$ is a relative minimum and the point $x = -\sqrt{\frac{1}{2}}$ is a relative maximum.

Now we consider $f''(x) = 30x(4x^2-1)$. If $f''(x) = 0$ then $x = 0$, $x = \pm\frac{1}{2}$. The function f is concave up in the interval $x \geq \frac{1}{2}$ and down in the interval $0 \leq x \leq \frac{1}{2}$. Similarly, the function f is concave up in the interval $-\frac{1}{2} \leq x \leq 0$ and concave down for $x \leq -\frac{1}{2}$.

We make use of the fact that the function is an odd function. We first draw the graph for $x \geq 0$. Then, by folding the graph around the x-axis first and then around the y-axis, we obtain the graph for $x < 0$. By folding the graph around the x-axis, we cause all values of $f(x)$ to change sign. By folding the graph around the y-axis, we cause all values of $x \geq 0$ to become $x \leq 0$.

By the first fold $f(x)$ becomes $-f(x)$

By the second fold $-f(x)$ becomes $-f(-x)$

$f(x)$ becomes $-f(-x)$, indicative of the anti-symmetric property of the graph.

• PROBLEM 427

Draw the graph of

$$y = x^4 - 4x^3.$$

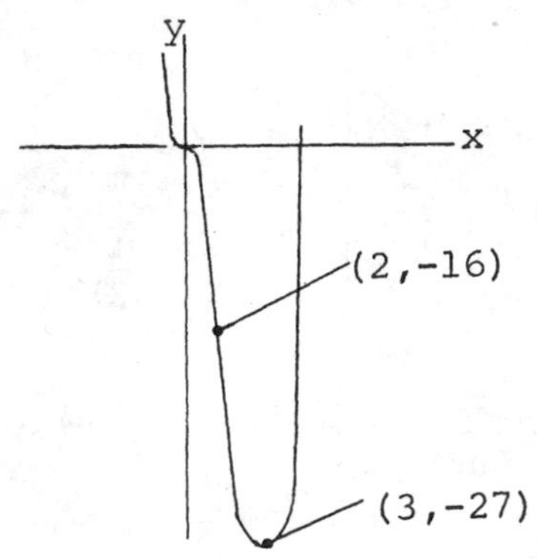

Solution: The graph intercepts the x-axis at $x = 0$ and $x = 4$. For $x < 0$ and $x > 4$, $y > 0$.

To calculate the critical points, we solve for the first derivative and set it equal to zero. We have $y' = 4x^3 - 12x^2$. The critical points are $x = 0$ and $x = 3$. To determine whether they are relative maxima or relative minima, we use the second derivative test.

$y'' = 12x^2 - 24x$. For $x = 3$, $y'' > 0$, which means that it is a relative minimum. For $x = 0$, $y'' = 0$ which tells us nothing. Therefore, we have to see how the first derivative changes in a small region around $x = 0$. The slope does not change sign around $x = 0$, therefore it is neither a relative maximum nor a relative minimum.

We note that $y'' = 0$ at $x = 2$ and $x = 0$.

Therefore these two points are inflection points of the graph.

• PROBLEM 428

Sketch the graph $y = x^3 - 3x^2 - 9x - 3$ after finding maximum and minimum points, and points of inflection.

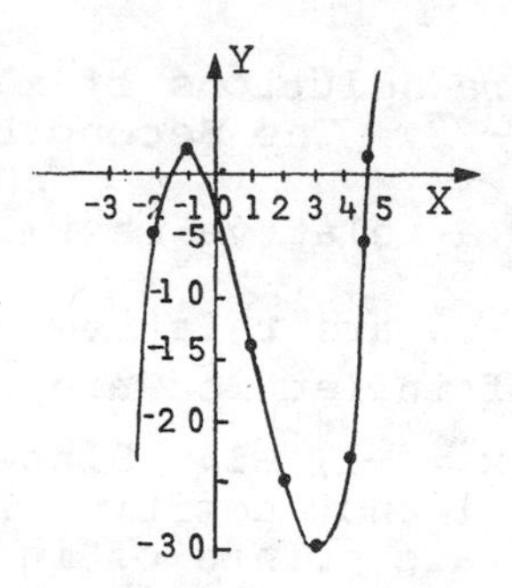

Solution: Differentiating twice,

$$y' = 3(x^2-2x-3)$$
$$= 3(x-3)(x+1),$$

and

$$y'' = 6(x-1).$$

Setting $y' = 0$ to find critical values,

$$x = -1, 3.$$

Setting $y'' = 0$, gives

$$x = 1.$$

If $x = -1$, $y' = 0$, y'' is negative, hence $y = f(-1) = 2$ is a maximum. If $x = 3$, $y' = 0$, y'' is positive, therefore $y = f(3) = -30$ is a minimum.

Since y'' changes sign at $x = 1$, the point $(1,-14)$ is a point of inflection, with a slope of curve at that point of -12. Note that different scales are used for abscissas and ordinates in the graph and this must be taken into consideration in estimating the slope at any point.

• PROBLEM 429

Determine the relative maxima, relative minima, and points of inflection of the function:

$$f(x) = \frac{1}{4}x^4 - \frac{3}{2}x^2.$$

Sketch the graph.

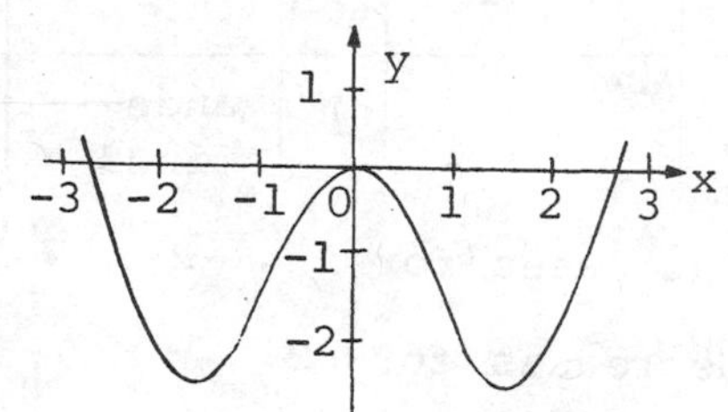

Solution: The derivatives are

$$f'(x) = x^3 - 3x \quad \text{and} \quad f''(x) = 3x^2 - 3.$$

The critical points are solutions of $x^3 - 3x = 0$. We obtain $x = 0, \sqrt{3}, -\sqrt{3}$. The Second Derivative Test tells us that

$$x = 0 \text{ is a relative maximum;}$$

$$x = \sqrt{3}, -\sqrt{3} \text{ are relative minima.}$$

The possible points of inflection are solutions of $3x^2 - 3 = 0$; that is $x = +1, -1$. Since $f''(x)$ is negative for $-1 < x < 1$ and positive for $|x| > 1$, both $x = 1$ and $x = -1$ are points of inflection. We construct the table:

x	-2	$-\sqrt{3}$	-1	0	1	$\sqrt{3}$	2
f	-2	$-\frac{9}{4}$	$-\frac{5}{4}$	0	$-\frac{5}{4}$	$-\frac{9}{4}$	-2
f'	$-$	0	$+$	0	$-$	0	$+$
f''	$+$	$+$	0	$-$	0	$+$	$+$

The graph is symmetrical with respect to the y-axis.

• PROBLEM 430

Determine the relative maxima and minima of the function f, defined by

$$f(x) = x^{5/3} + 5x^{2/3},$$

and determine the intervals in which f is increasing and those in which f is decreasing. Sketch the graph.

Solution: The derivative is:

$$f'(x) = \frac{5}{3}x^{2/3} + \frac{10}{3}x^{-1/3} = \frac{5}{3}x^{-1/3}(x+2).$$

The critical point is $x = -2$. The derivative is not defined at $x = 0$. We construct the following table:

x	-5	-2	-1	0	1
$f(x)$	0	$3\cdot(-2)^{2/3}$	4	0	6
$f'(x)$	$+$	0	$-$	undefined	$+$

We conclude that

$$f \text{ increases for } x < -2;$$

$$f \text{ decreases for } -2 < x < 0;$$

$$f \text{ increases for } x > 0.$$

There is a relative maximum at $x = -2$ and a relative minimum at $x = 0$. Since $f'(x)$ is not defined for $x = 0$, to determine whether this point is a relative maximum or minimum, it is necessary to test values of x which lie a small distance away to the left and to the right of $x = 0$.

If x_1 lies to the left of x and x_2 to the right, and $f(x_1)$ and $f(x_2)$ are both greater than $f(x)$, then x is a relative minimum.

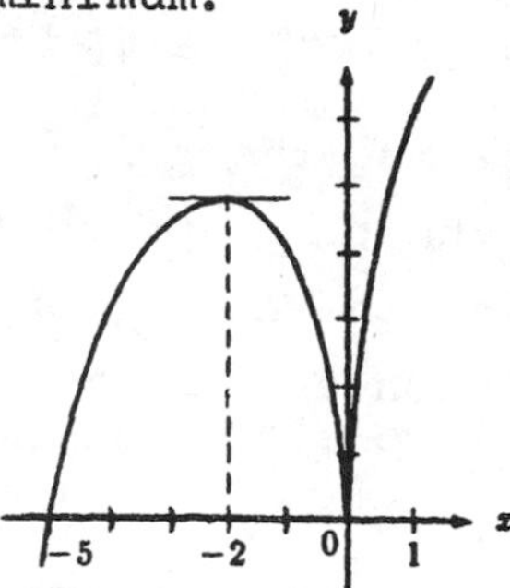

If $f(x_1)$ and $f(x_2)$ are both less than $f(x)$, then x is a relative maximum. In general, it is necessary to carry out this procedure for all values of x where $f'(x)$ is undefined.

• PROBLEM 431

> Sketch the graph of the equation:
>
> $$y = 2x^3 + 3x^2 - 12x.$$

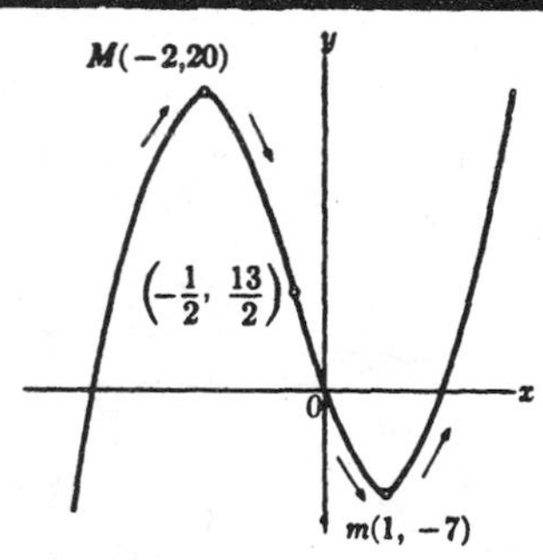

Solution: From the first two derivatives, we have

$$y' = 6x^2 + 6x - 12 = 6(x+2)(x-1)$$

and

$$y'' = 12x + 6 = 6(2x+1).$$

$y' = 0$ when $x = -2$ and $x = 1$. Using these values in the second derivative we find that y'' is negative (-18) for $x = -2$ and y'' is positive (+18) when $x = 1$. When $x = -2$, $y = 20$; and when $x = 1$, $y = -7$. Therefore, (-2,20) is a relative maximum and (1,-7) is a relative minimum.

Next we see that $y' > 0$ if $x < -2$ and $x > 1$.

Also, $y' < 0$ if $-2 < x < 1$. Therefore, y is increasing if $x < -2$ and $x > 1$, and y is decreasing if $-2 < x < 1$.

Now we note that $y'' = 0$ when $x = -\frac{1}{2}$, and that $y'' < 0$ when $x < -\frac{1}{2}$, and $y'' > 0$ when $x > -\frac{1}{2}$. When $x = -\frac{1}{2}$, $y = \frac{13}{2}$. Therefore, there is a point of inflection at $\left(-\frac{1}{2}, \frac{13}{2}\right)$, the curve is concave down if $x < -\frac{1}{2}$ and the curve is concave up if $x > -\frac{1}{2}$

Finally, by locating the points (-2,20), $\left(-\frac{1}{2}, \frac{13}{2}\right)$, and (1,-7), we draw the curve up to (-2,20) and then down to $\left(-\frac{1}{2}, \frac{13}{2}\right)$, with the curve concave down. Continuing down, but concave up, we draw the curve to (1,-7) at which point we start up and continue up.

• PROBLEM 432

Determine where the curve:

$$24y = x^3 - 6x^2 - 36x + 16$$

is concave up and where it is concave down.

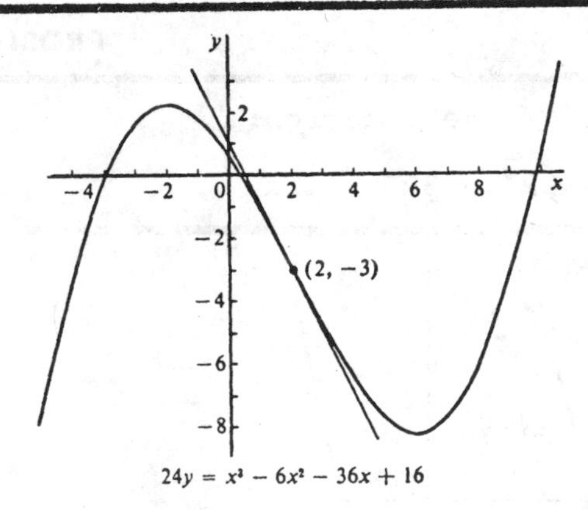

$24y = x^3 - 6x^2 - 36x + 16$

Solution:

$$24y = x^3 - 6x^2 - 36x + 16$$

$$24\frac{dy}{dx} = 3x^2 - 12x - 36$$

$$24\frac{d^2y}{dx^2} = 6x - 12$$

$$\frac{d^2y}{dx^2} = \frac{1}{4}(x-2)$$

$\frac{d^2y}{dx^2} > 0$ for $x > 2$, hence the curve is concave.

$$\frac{d^2y}{dx^2} < 0 \text{ for } x < 2, \text{ therefore the curve}$$

is convex.

• PROBLEM 433

Draw the graph of $f(x) = x^3 - \frac{21}{4}x^2 + 9x - 4$ and indicate the points of relative maxima and minima.

Solution:

$$f(x) = x^3 - \frac{21}{4}x^2 + 9x - 4$$

$$f'(x) = 3x^2 - \frac{21}{2}x + 9 = 3\left(x^2 - \frac{7}{2}x + 3\right)$$

$$f''(x) = 6x - \frac{21}{2}$$

By setting $f'(x) = 0$, the critical points are

$$x = \frac{3}{2} \qquad \text{and} \qquad x = 2$$

Applying the second derivative test, for

$$x = \frac{3}{2},\ f''(x) > 0 \text{ and for}$$

$$x = 2,\ f''(x) < 0.$$

This shows that $x = \frac{3}{2}$ is a relative maximum and $x = 2$ is a relative minimum.

By setting $f''(x) = 0$, we obtain $x = \frac{7}{4}$.

For $x < \frac{7}{4}$, $f''(x) < 0$ (concave down), and for $x > \frac{7}{4}$, $f''(x) > 0$ (concave up).

All the information can be placed into the following table.

x	0	$\frac{3}{2}$	$\frac{7}{4}$	2	3
$f(x)$	-4	$\frac{17}{16}$	$1\frac{1}{32}$	1	$\frac{11}{16}$
$f'(x)$	+	0	-	0	+
$f''(x)$	-	-	0	+	+

For the first and second derivatives, it is only necessary to know the sign of the answer in order to plot the graph.

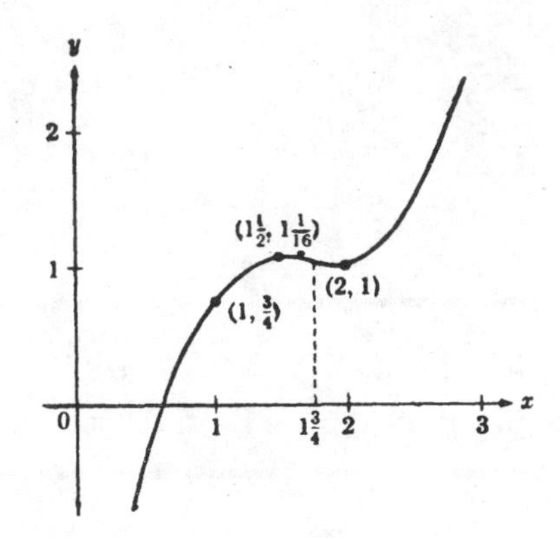

• PROBLEM 434

Graph the function

$$f(x) = \frac{x^2 + x + 7}{\sqrt{2x + 1}}, \quad x > -\frac{1}{2}$$

and find points which are relative maxima and minima.

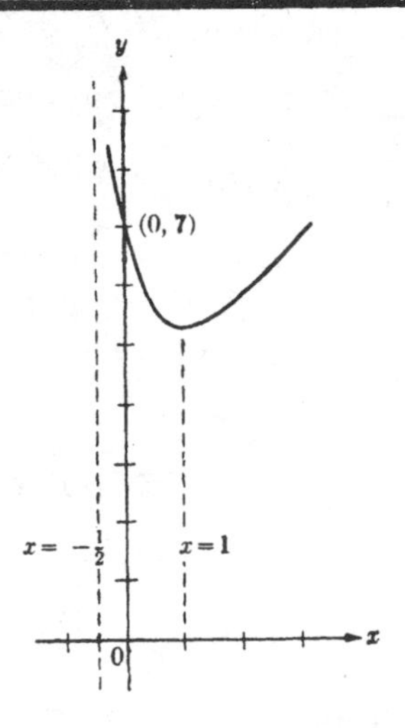

Solution: $f(x) = \dfrac{x^2 + x + 7}{\sqrt{2x + 1}}$

$$f'(x) = \frac{(2x+1)^{1/2}(2x+1) - (x^2+x+7)\frac{1}{2}(2x+1)^{-1/2} \cdot 2}{2x + 1}$$

By multiplying the numerator and denominator by $(2x+1)^{1/2}$, we obtain:

$$f'(x) = \frac{(2x+1)^2 - (x^2+x+7)}{(2x+1)^{3/2}} = \frac{3(x+2)(x-1)}{(2x+1)^{3/2}}$$

We note that $(2x+1)^{1/2}$ never equals zero because we are only concerned with values of $x > \frac{1}{2}$. Upon setting $f'(x) = 0$, we obtain $x = -2$ and $x = 1$. We eliminate the point $x = -2$ because it is not in our domain.

By testing a value in the interval $\left(-\frac{1}{2},1\right)$ and in the interval $(1,\infty)$ we find that:

$$f(x) \text{ decreases for } -\frac{1}{2} < x < 1$$

$$f(x) \text{ increases for } 1 < x < \infty$$

By taking the second derivative and applying it at $x = 1$, $f''(1) > 0$ which indicates that it is a relative minimum.

• PROBLEM 435

> Find the maximum and minimum points, and the points of inflection of the curve
>
> $y = 3 \sin x - 4 \cos x$ in the interval $(0, 2\pi)$

Solution:

$$y = 3 \sin x - 4 \cos x$$

$$\frac{dy}{dx} = 3 \cos x + 4 \sin x$$

$$\frac{d^2y}{dx^2} = -3 \sin x + 4 \cos x$$

To determine maximum and minimum points, we set $\frac{dy}{dx} = 0$. $3 \cos x + 4 \sin x = 0$

$$\text{or} \quad \tan x = -\frac{3}{4}$$

From a table, $x = 2.50$ and $x = 5.64$. By substituting these points into $\frac{d^2y}{dx^2}$, we see that $\frac{d^2y}{dx^2} < 0$ for $x = 2.50$

$$\frac{d^2y}{dx^2} > 0 \text{ for } x = 5.64$$

This tells us that $x = 2.50$ is a maximum and $x = 5.64$ is a minimum.

To find the points of inflection, we set $\frac{d^2y}{dx^2} = 0$.

$$-3 \sin x + 4 \cos x = 0$$

$$\text{or} \quad \tan x = \frac{4}{3}$$

From a table, we find that $x = 0.92$ and $x = 4.06$. By testing points to the left and to the right of $x = 0.92$ and $x = 4.06$, we see that $\frac{d^2y}{dx^2}$ changes sign at these two points. Therefore, they are points of inflection.

• **PROBLEM 436**

Sketch the graph of

$$f(x) = xe^{-x^2}.$$

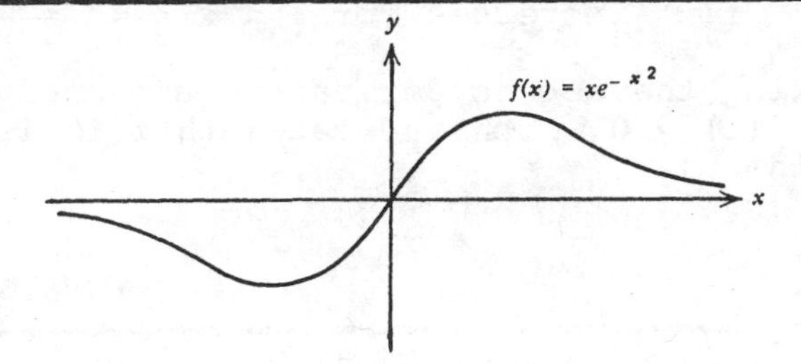

Solution: $f(x) = xe^{-x^2}$

$$f'(x) = e^{-x^2} + (x)\left(e^{-x^2}\right)(-2x)$$

$$= e^{-x^2}\left(1-2x^2\right)$$

$$f''(x) = e^{-x^2}(-4x) + \left(1-2x^2\right)\left(e^{-x^2}\right)(-2x)$$

$$= e^{-x^2}\left(4x^3-6x\right)$$

If $f'(x) = 0$, $x = \frac{1}{\sqrt{2}}$, $x = \frac{-1}{\sqrt{2}}$

Applying the second derivative test, $x = \frac{1}{\sqrt{2}}$ is a maximum and $x = \frac{-1}{\sqrt{2}}$ is a minimum.

If $f''(x) = 0$, $x = 0$, $x = \pm\sqrt{\frac{3}{2}}$.

These are the points of inflection. It is necessary to consider the behavior as $x \to \infty$ and $x \to -\infty$. Therefore

$$\lim_{x\to\infty} x e^{-x^2} = \lim_{x\to\infty} \frac{x}{e^{x^2}} = \lim_{x\to\infty} \frac{1}{2xe^{x^2}} = 0$$

by application of L'Hospital's rule.

Similarly, $\lim_{x\to-\infty} \frac{x}{e^{x^2}} = \lim_{x\to-\infty} \frac{1}{2xe^{x^2}} = 0$

• **PROBLEM 437**

Sketch the graph of

$$f(x) = e^{-x^2} \sin x.$$

Solution: Notice that f(x) is equal to the product of two functions that are known. We sketch the graph on this basis. Since $|\sin x| \leq 1$, it follows that

$$-e^{-x^2} \leq e^{-x^2} \sin x \leq e^{-x^2}$$

This tells us that each point on the sin curve is multiplied by a factor of e^{-x^2}. Therefore, we draw the sin x curve first and then compress the curve by a factor of e^{-x^2}. an exponential expression tending toward 0 as $x \to \infty$.

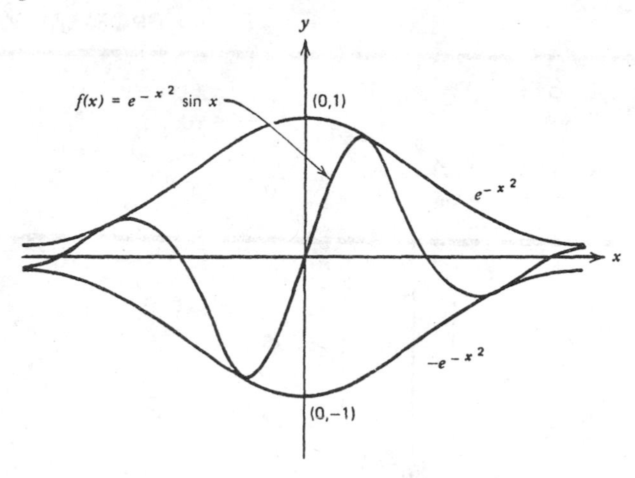

• PROBLEM 438

Draw the graph and discuss the equation:

$$y = \frac{4x^2}{x^2 + 1}.$$

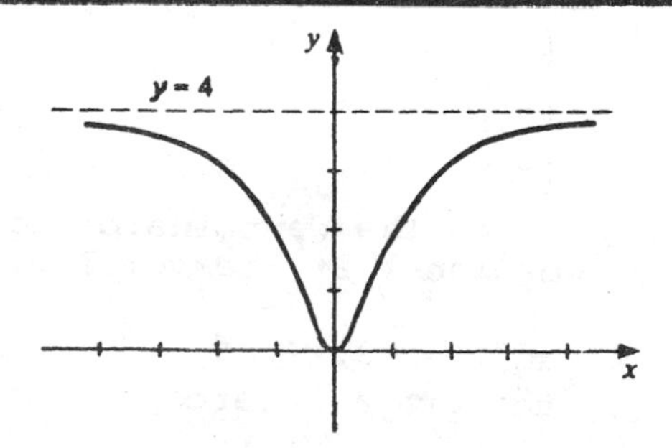

Solution: The graph is symmetric to the y axis, since x occurs to even powers only.

Since $$0 \leq \frac{x^2}{x^2 + 1} < 1,$$

$0 \leq y < 4$ for every number x. Thus the graph lies between the lines $y = 0$ and $y = 4$.

Since $$\lim_{x \to \pm\infty} \frac{4x^2}{x^2 + 1} = \lim_{x \to \pm\infty} \frac{4}{1 + 1/x^2} = 4,$$

the line $y = 4$ is an asymptote of the graph of the given equation. When a few points are plotted, the graph may be drawn as shown.

When $x = 0$, $y = 0$.

$$\frac{dy}{dx} = \frac{(x^2+1)8x - 4x^2(2x)}{(x^2+1)^2} = \frac{8x^3 + 8x - 8x^3}{(x^2+1)^2}$$

$= 0$ when $x = 0$. The curve has a minimum at

$x = 0$.

• PROBLEM 439

Draw the graph of:

$$y = \frac{x^3 + x^2}{x^2 - 4}$$

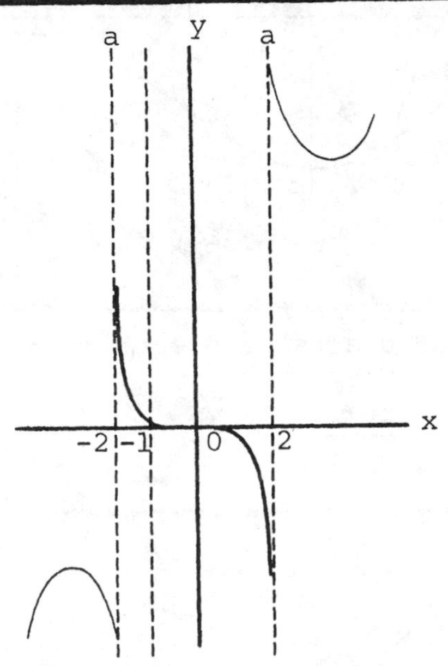

<u>Solution:</u> Upon setting the denominator equal to zero, we find that the vertical asymptotes are

$x = 2$ and $x = -2$.

Setting $y = 0$, the graph intercepts the x-axis at $x = 0$ and $x = -1$.

The function $y = \dfrac{x^3 + x^2}{x^2 - 4}$

can be written as

$$y = \frac{\left(1 + \frac{1}{x}\right)}{\left(\frac{1}{x} - \frac{4}{x^3}\right)}.$$

As $x \to \infty$, $y \to \infty$. This shows that there are no horizontal asymptotes.

To obtain the critical points, we take the first

derivative.

$$y' = \frac{dy}{dx} = \frac{x\left(x^3 - 12x - 8\right)}{\left(x^2 - 4\right)^2}$$

The critical points are $x = 0$ and the solutions of the equation $x^3 - 12x - 8$ (which are, approx., $x = -3.1$, $x = -0.7$ and $x = 3.7$).

Using the second derivative test, we find that $x = -3.1$ and $x = 0$ are relative maxima and $x = -0.7$ and $x = 3.7$ are relative minima.

• PROBLEM 440

Determine the asymptotes and sketch the graph of

$$r(x) = \frac{x^2 - 1}{x + 2}$$

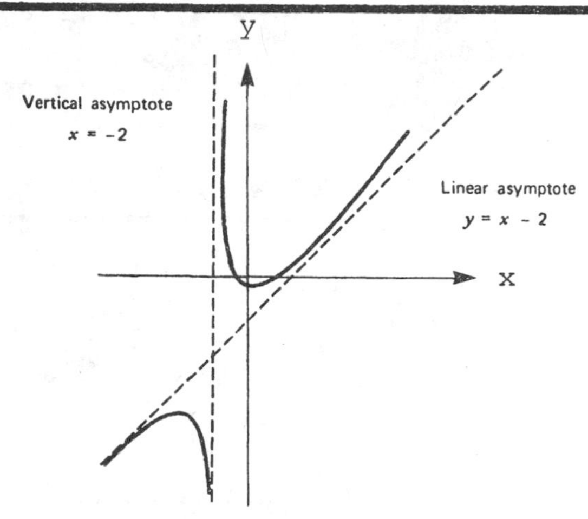

Solution: Setting the denominator equal to zero, $x = -2$ is a vertical asymptote. There is also a linear asymptote (also called oblique) having the form: $mx + b$. To solve for this, we must let $r(x) = mx + b$. Therefore, the equation

$r(x) = \frac{x^2 - 1}{x + 2}$ is written as

$$(mx+b) = \frac{x^2 - 1}{x + 2}$$

$$(mx+b)(x+2) - \left(x^2-1\right) = 0$$

or

$$x^2(m-1) + x(b+2m) + (2b-1) = 0$$

Set the coefficients of the two highest degrees of x equal to zero and we obtain

$$m = 1 \qquad \text{and } b = -2.$$

Therefore $\quad y = x - 2$ is the linear asymptote.

Another way to arrive at this result is to divide $x + 2$ into $x^2 - 1$ by long division. This gives: $r(x) = x - 2 + 3/6 + \ldots$, which, as x increases, approaches $x - 2$.

To obtain the critical points, we calculate the first derivative.

$$r'(x) = \frac{x^2 + 4x + 1}{(x+2)^2}$$

If $r'(x) = 0$, then $x = -2 \pm 2\sqrt{3}$. By using the second derivative test $x = -2 - 2\sqrt{3}$ is shown to be a relative maximum and $x = -2 + 2\sqrt{3}$ is a relative minimum.

• PROBLEM 441

Sketch the graph of $f(x) = x^2/(x^2-1)$.

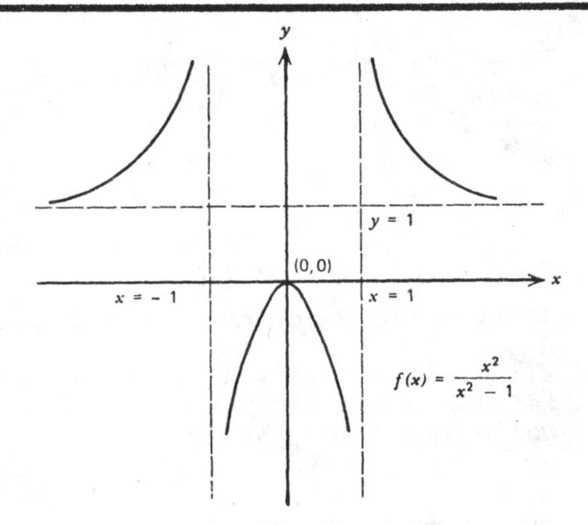

Solution: Note first that f(x) is symmetrical.

Hence, it suffices to sketch the graph for $x > 0$ and then reflect about the y axis. Computing the first derivative we have:

$$f'(x) = \frac{2x(x^2-1) - 2x^3}{(x^2-1)^2} = -\frac{2x}{(x^2-1)^2}.$$

By the first derivative test $x = 0$ is a relative maximum, and $f(0) = 0$. To determine the regions of concavity, we compute the second derivative, obtaining:

$$f''(x) = \frac{-2(x^2-1)^2 + 4x(x^2-1)2x}{(x^2-1)^4} = \frac{6x^2+2}{(x^2-1)^3}.$$

Clearly, for $x > 0$, $f''(x) > 0$ if $x > 1$, and $f''(x) < 0$ if $0 < x < 1$. Therefore, the graph of f is concave up if $x > 1$ and concave down if $0 < x < 1$.

$$f(x) = \frac{x^2}{(x^2-1)}$$

can be written as

$$\frac{1}{1 - \frac{1}{x^2}}.$$

Hence, as $x \to \infty$, $f(x) \to 1$, which shows that $y = 1$ is a horizontal asymptote It only remains to determine the behavior of f near $x = 1$. Now

$$\frac{x^2}{x^2-1} = \frac{x^2}{x+1}\,\frac{1}{x-1} \quad \text{and} \quad \lim_{x\to 1} \frac{x^2}{x+1} = \frac{1}{2}$$

But $\lim_{x\to 1^+}\left(\frac{1}{x-1}\right) = \infty$, while $\lim_{x\ 1^-}\left(\frac{1}{x-1}\right) = -\infty$.

Therefore, $\lim_{x\to 1^+} f(x) = \infty$ and $\lim_{x\to 1^-} f(x) = -\infty$.

Combining these statements we have the sketch as shown. Note we have plotted just one point, the relative maximum (0,0).

• PROBLEM 442

Draw the graph of: $y^2 = x^2(1-x)$

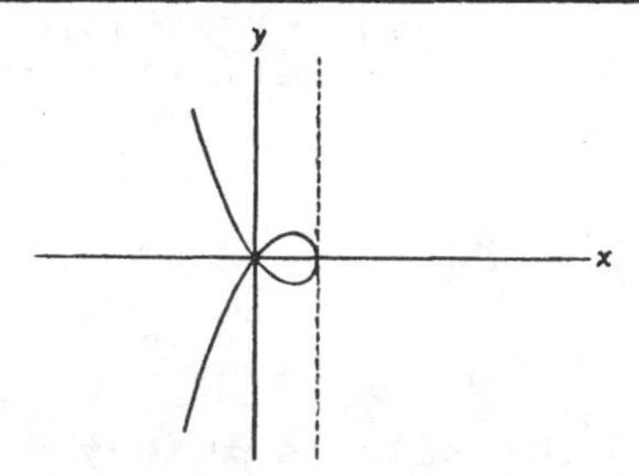

Solution: There are no horizontal or vertical asymptotes.

The graph cuts the x-axis at $x = 0$ and $x = 1$.

For $x > 1$, $Y^2 < 0$. Hence no portion of the graph lies to the right of $x = 1$.

The function $y^2 = x^2(1-x)$ can be written as

$y = \pm\sqrt{x^2(1-x)}$. Considering the positive root only, we can obtain that:

$$y' = \frac{2 - 3x}{2\sqrt{1-x}} = 0$$

This shows that $x = 2/3$ is a critical value, resulting in $y = \sqrt{\frac{4}{27}}$.

By using the second derivative test, this point can be shown to be a relative maximum. When $x = 1$, $y' \to \infty$, indicating a vertical slope.

• PROBLEM 443

Draw the graph of:

$$y^2 = \frac{x^2 - 1}{x^2 + 1}$$

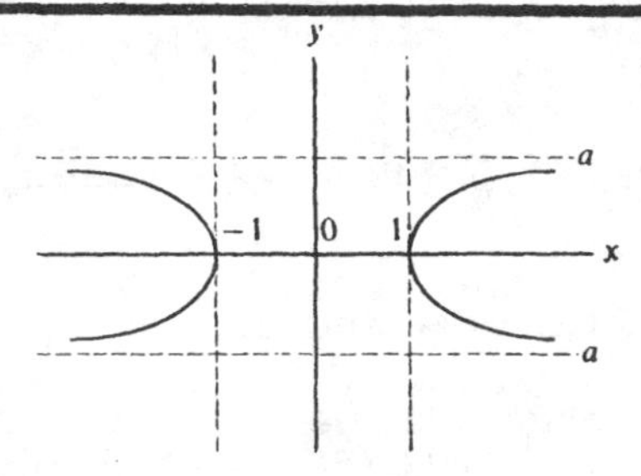

Solution: Since we can replace x with (-x) and y with (-y) without changing the value of the equation, this tells us that the graph is symmetric with the x-axis and y-axis.

To determine vertical asymptotes, we set the denominator equal to zero. There are no real values of x that will satisfy the equation, therefore there are no vertical asymptotes.

Note that $y^2 = \dfrac{x^2 - 1}{x^2 + 1} = \dfrac{1 - \frac{1}{x^2}}{1 + \frac{1}{x^2}}$.

As $x \to \infty$, $y^2 \to 1$, which tells us that $y = 1$ and $y = -1$ are **horizontal asymptotes.**

For values of x in the interval [0,1], y^2 is negative which is impossible. Therefore, the equation is not defined and there is no graph between 0 and 1, and since the graph is symmetric about the y-axis, there is no graph between -1 and 0.

$$y = \sqrt{\frac{x^2 - 1}{x^2 + 1}}$$

$$y' = \frac{dy}{dx} = \frac{1}{2}\left(\frac{x^2-1}{x^2+1}\right)^{-1/2} \left(\frac{(x^2+1)(2x) - (x^2-1)(2x)}{(x^2+1)^2}\right)$$

$$= \frac{2x}{(x^2+1)^2 \sqrt{(x^2-1)/(x^2+1)}}$$

$y' = 0$ only if $x = 0$. Since the graph is not defined in the interval $[-1,1]$, the graph has no critical points.

• PROBLEM 444

Draw the graph of the equation:

$$y^2 = \frac{(x+1)(x+2)}{x}$$

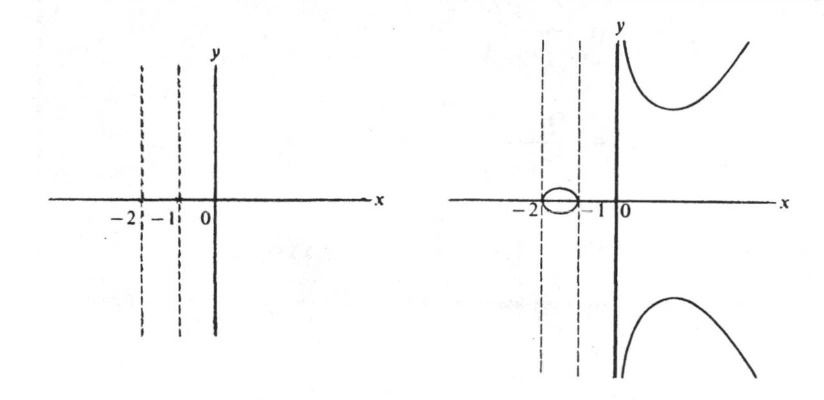

Solution: The vertical asymptote is determined by setting the denominator equal to zero. Therefore $x = 0$ is a vertical asymptote.

The function $y^2 = \frac{(x+1)(x+2)}{x}$ can be written as

$$y^2 = \frac{x^2 + 3x + 2}{x}$$

$$= \frac{1 + \frac{3}{x} + \frac{2}{x^2}}{\frac{1}{x}}$$

As $x \to \infty$, $y^2 \to \infty$. Therefore, there are no horizontal asymptotes. By setting $y^2 = 0$, we obtain the values $x = -2$ and $x = -1$. To the left of $x = -2$, the equation is not defined and there is no portion of the graph because $y^2 < 0$. For x in the interval $(-2,-1)$, $y^2 > 0$. For x in the interval $(-1,0)$, $y^2 < 0$, hence there is no graph. For $x > 0$, $y^2 > 0$.

Since $y^2 = \frac{x^2 + 3x + 2}{x}$,

$$y = \sqrt{\frac{x^2 + 3x + 2}{x}}$$

By considering the positive root only,

$$y' = \frac{x^2 - 2}{2x^2 \sqrt{\frac{x^2 + 3x + 2}{x}}}$$

If $y' = 0$, $x = \pm\sqrt{2}$ which are the critical points.

• PROBLEM 445

A curve is given by the parametric equations

$$x = \frac{2t}{(1+t^2)}$$

$$y = \frac{(1-t^2)}{(1+t^2)}$$

Find the coordinates of the point where $t = 1/2$, the slope at this point and the direction of concavity.

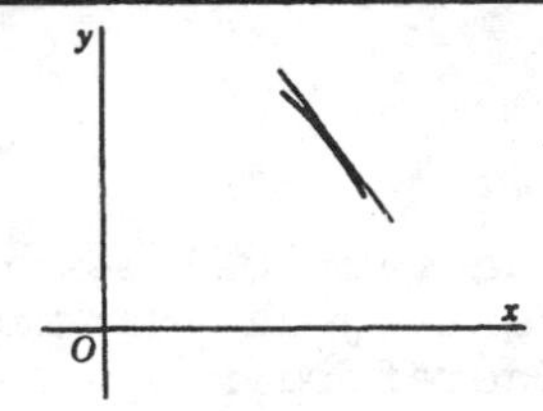

Solution: The coordinates of the point are found by substituting $t = 1/2$ into the parametric equations.

Therefore $x = \frac{4}{5}$, $y = \frac{3}{5}$. Note that

$$\frac{dy}{dx} = \frac{dy/dt}{dx/dt}, \text{ and } \frac{d^2y}{dx^2} = \left(\frac{d}{dt}\left(\frac{dy}{dx}\right)\right) \cdot \frac{dt}{dx},$$

where $\frac{dt}{dx} = \frac{1}{\left(\frac{dx}{dt}\right)}$.

Applying these formulas:

$$\frac{dy}{dx} = \left(\frac{(1+t^2)(-2t) - (1-t^2)(2t)}{(1+t^2)^2}\right) \Big/ \left(\frac{(1+t^2)\cdot 2-2t\cdot 2t}{(1+t^2)^2}\right)$$

$$= -\left(\frac{4t}{(1+t^2)^2}\right) \Big/ \left(\frac{2 - 2t^2}{(1+t^2)^2}\right) = \frac{-2t}{1 - t^2}$$

$$\frac{d^2y}{dx^2} = \left(\frac{d}{dt}\left(\frac{-2t}{1-t^2}\right)\right)\left[\frac{1}{\left(\frac{2-2t^2}{(1+t^2)^2}\right)}\right]$$

$$= \frac{(1-t^2)(-2) + 2t(-2t)}{(1-t^2)^2}\left(\frac{(1+t^2)^2}{2-2t^2}\right)$$

$$= \frac{-2(1+t^2)(1+t^2)^2}{(1-t^2)^2(2-2t^2)}$$

$$= -(1+t^2)^3 \big/ (1-t^2)^3$$

For $t = 1/2$, $dy/dx < 0$ and $\frac{d^2y}{dx^2} < 0$. This shows that the slope is negative, and the curve is concave downward at $t = \frac{1}{2}$.

• PROBLEM 446

Determine the interval of time t in which a particle moves so that its distance s from a fixed point is measured positively, if

$$x = -20 - 24t + 9t^2 - t^3.$$

Solution: Set the derivative equal to zero. Then

$$\frac{ds}{dt} = -24 + 18t - 3t^2 = -3(t-2)(t-4) = 0,$$

and $t = 2, 4$.

Therefore, using $t = 0$, 3, and 5, we obtain the information shown below.

$-\infty$ ←——— $t = 2$ ——— $t = 4$ ———→ $+\infty$

$t < 2$	$2 < t < 4$	$t > 4$
ds/dt negative	ds/dt positive	ds/dt negative
s decreasing	s increasing	s decreasing

Hence the particle moves in the direction in which s is measured positively in the interval $2 < t < 4$.

CURVE TRACING-POLAR COORDINATES

• PROBLEM 447

Sketch the locus of the equation:

$$r = 2 \sin \theta.$$

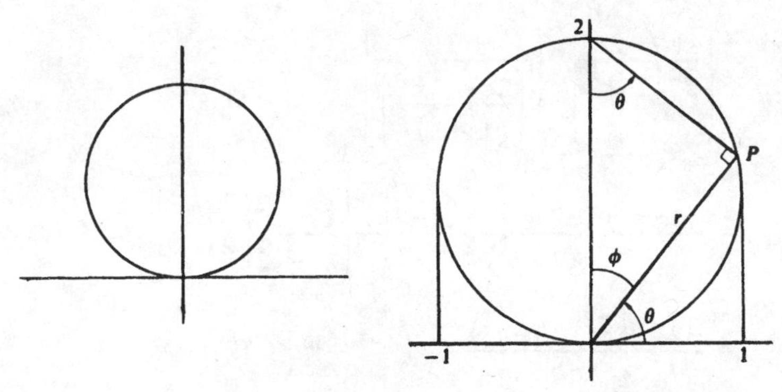

Solution: Since $\sin(\pi + \theta) = \sin\theta$, the locus is symmetric with respect to the y-axis. When $\theta = 0$, $r = 0$. As θ increases from 0 to $\pi/2$, r increases from 0 to 2. On the other hand, as θ decreases from 0 to $-\pi/2$, r decreases from 0 to -2. The general shape is indicated in the upper diagram. That the locus is a circle is seen at once from the lower diagram.

• PROBLEM 448

Draw the locus of the equation:

$$r = 2\cos\theta.$$

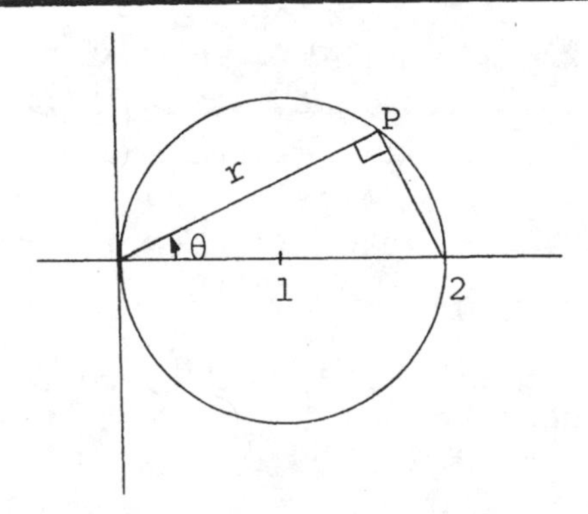

Solution: Since $\cos(-\theta) = \cos\theta$, the locus is symmetric with respect to the x-axis.

When $\theta = 0$, $r = 2$. As θ increases from 0 to $\pi/2$, r decreases from $r = 2$ to 0. As θ increases from $\pi/2$ to π, r decreases from 0 to -2. Since the locus is symmetric with respect to the x-axis, we need not consider any further variation of θ.

• PROBLEM 449

Draw the locus of the equation

$$r = 1 + 2\cos\theta.$$

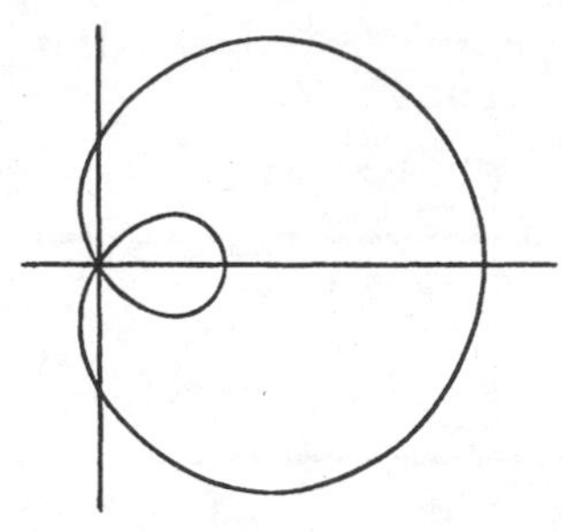

Solution: The locus is symmetric with respect to the x-axis since $\cos(-\theta) = \cos(\theta)$.

When $\theta = 0$, $r = 3$. As θ increases from 0 to $\pi/2$, r decreases from 3 to 1. $r = 0$ when $\theta = 2\pi/3$, thus as θ increases from $\pi/2$ to $2\pi/3$, r decreases from 1 to 0. As θ increases from $2\pi/3$ to π, r decreases from 0 to -1. The general shape is as shown.

• PROBLEM 450

Draw the graph of:

$$r = 3 + 2\sin\theta$$

θ	0	$\frac{1}{6}\pi$	$\frac{1}{3}\pi$	$\frac{1}{2}\pi$	π	$\frac{7}{6}\pi$	$\frac{4}{3}\pi$	$\frac{3}{2}\pi$
r	3	4	$3+\sqrt{3}$	5	3	2	$3-\sqrt{3}$	1

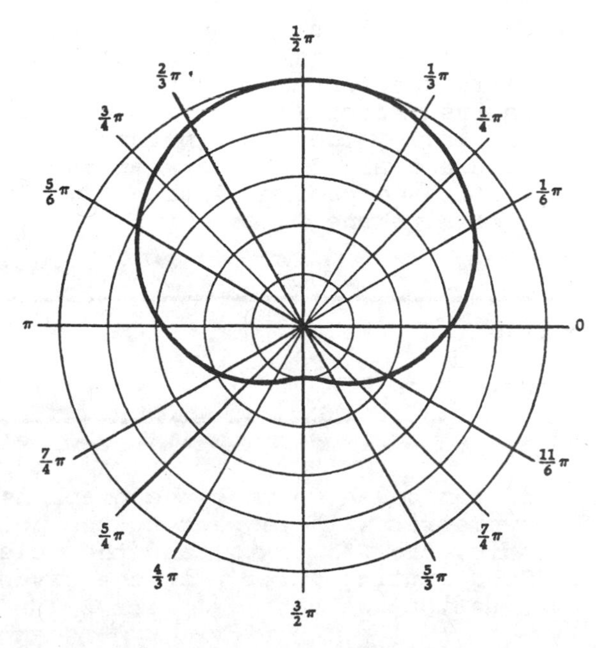

Solution: The graph is symmetric with respect to the $\pi/2$ axis because if (r,θ) is replaced by $(r,\pi-\theta)$, an equivalent equation is obtained.

The table gives the coordinates of some of the points on the graph.

• PROBLEM 451

Sketch the locus of

$$r^2 = \cos\theta.$$

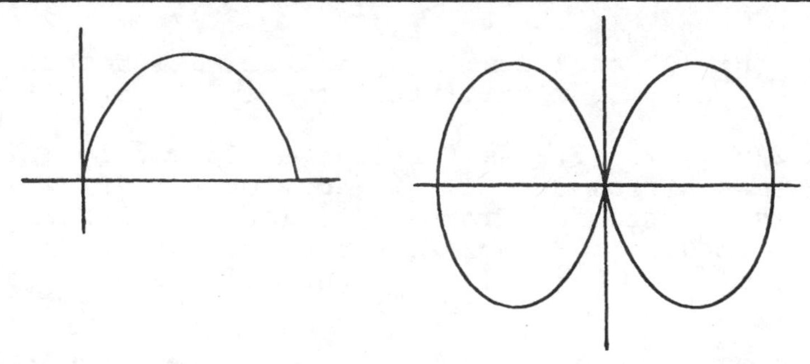

Solution: The locus is symmetric with respect to the x-axis and also with respect to the origin. When $\theta = 0$, $r = \pm 1$. In the first quadrant the locus is seen to be of the general shape shown. When θ is between $\pi/2$ and $3\pi/2$, $\cos\theta < 0$. Hence as the radius vector sweeps from $\pi/2$ to $3\pi/2$ no real locus points are obtained. However, this does not mean that no part of the locus lies in either the second or the third quadrant. The fact that $r = \pm\sqrt{\cos\theta}$ shows that indeed there are locus points in both of these quadrants. The locus points in the third quadrant are obtained when the radius vector sweeps from 0 to $\pi/2$. Similarly, the locus points in the second quadrant are obtained when the radius vector sweeps from 0 to $-\pi/2$. The general shape is as shown.

• PROBLEM 452

Draw the four-leafed rose:

$$r = 4 \cos 2\theta$$

Solution: It can be shown that the graph is symmetric with respect to the polar axis, the $\pi/2$ axis, and the pole. Substituting 0 for r in the given equation, we get

$$\cos 2\theta = 0$$

from which we obtain, for $0 \le \theta < 2\pi$, $\theta = \frac{\pi}{4}, \frac{3\pi}{4}, \frac{7\pi}{4}$, and $\frac{11\pi}{4}$.

The table gives values of r for some values of θ from 0 to $\pi/2$. From these values and the symmetry properties, we draw the graph as shown.

θ	0	$\frac{1}{12}\pi$	$\frac{1}{6}\pi$	$\frac{1}{4}\pi$	$\frac{1}{3}\pi$	$\frac{5}{12}\pi$	$\frac{1}{2}\pi$
r	4	$2\sqrt{3}$	2	0	-2	$-2\sqrt{3}$	-4

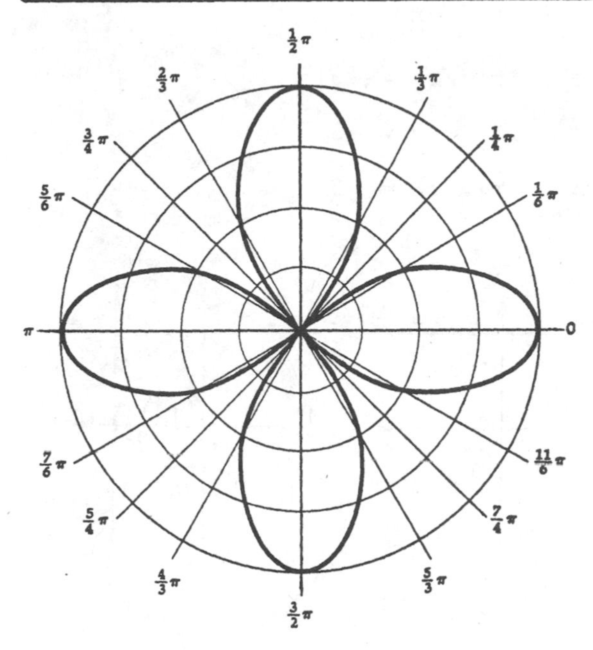

• **PROBLEM** 453

Draw the graph of each of the following equations:

(a) $r \sin\theta = 3$;

(b) $r \cos\theta = 3$;

(c) $r \sin\theta = -3$;

(d) $r \cos\theta = -3$.

Solution: Each graph is a line, as shown in Figures a, b, c, d.

This can be seen by converting the polar coordinates to cartesian coordinates. This is achieved by the following formulas

$$\sin\theta = \frac{y}{r}$$

$$\cos\theta = \frac{x}{r}$$

$$r = \sqrt{x^2 + y^2}$$

For part a, $r \sin\theta = 3$ is converted into

$$\left[\sqrt{x^2 + y^2}\right]\left[\frac{y}{\sqrt{x^2+y^2}}\right] = 3,$$ from which we obtain $y = 3$.

The other graphs are drawn similarly:

For (b): $r(x/r) = 3$

$x = 3$

For (c): $r(y/r) = -3$

$y = -3$

For (d): $r(x/r) = -3$

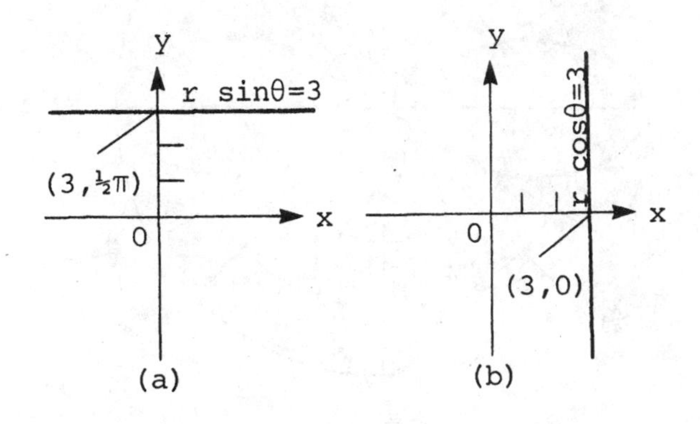

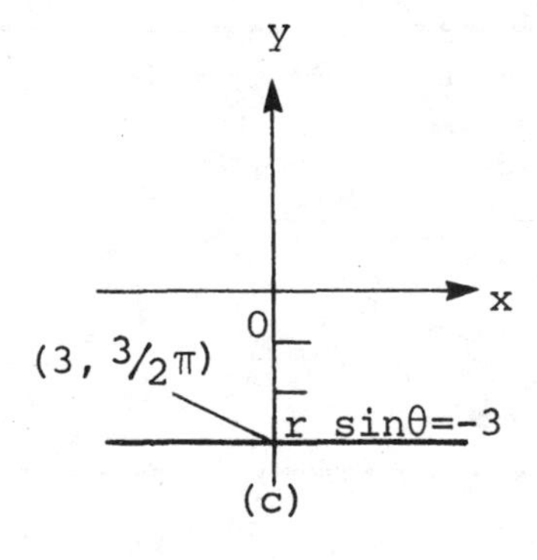

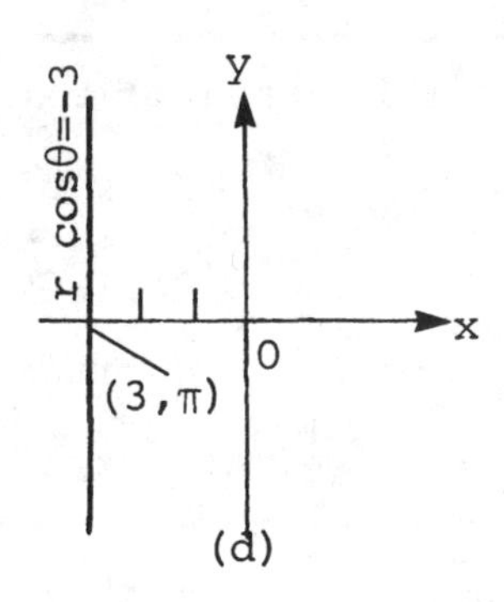

CHAPTER 18

CURVATURE

A useful application of the calculus is for the determination of curvature and radius of curvature. A curve having zero curvature is a straight line. Other curves (not straight lines) have a finite curvature everywhere except at points of discontinuity. The greater the deviation of a curve from a straight line at any point, the greater is its curvature at that point. If we take a small arc of the curve at the point, and we consider that arc to be a part of a circle, then the radius of that circle is the radius of curvature. Taking the straight line again, the only circle, of which it could form a part, would be one of infinite radius. A curve having a constant curvature and a constant radius of curvature is a circle. The more the curvature of a curve, the smaller its radius of curvature. The radius of curvature is defined as the reciprocal of the curvature.

To find the curvature and the radius of curvature, we find the first and second derivatives of the curve, and then substitute in the formula appropriate for the system of coordinates being used.

• PROBLEM 454

Find the point of maximum curvature of the curve $y = e^x$.

Solution. We first solve for the curvature of $y = e^x$, then we find the critical values of the curvature, and determine whether they are maxima or minima . To solve for the curvature, we use the formula:

$$K = \frac{\frac{d^2y}{dx^2}}{\left[1 + \left(\frac{dy}{dx}\right)^2\right]^{\frac{3}{2}}}.$$

We solve for $\frac{dy}{dx}$ and $\frac{d^2y}{dx^2}$ first.

$$y = e^x$$

$$\frac{dy}{dx} = e^x$$

$$\frac{d^2y}{dx^2} = e^x.$$

Substituting these values into the formula for curvature we obtain:

$$K = \frac{e^x}{\left[1 + e^{2x}\right]^{\frac{3}{2}}}.$$

To find the critical values we solve for the first derivative, $\frac{dK}{dx}$, set it equal to zero, and solve for x. Using the quotient rule we obtain $\frac{dK}{dx}$ as follows:

$$K = \frac{e^x}{\left[1 + e^{2x}\right]^{\frac{3}{2}}}.$$

$$\frac{dK}{dx} = \frac{e^x(1+e^{2x})^{\frac{3}{2}} - e^x\,\frac{3}{2}(1+e^{2x})^{\frac{1}{2}}\,e^{2x}(2)}{(1 + e^{2x})^3}.$$

$$\frac{dK}{dx} = \frac{e^x(1+e^{2x})^{\frac{1}{2}}\left[(1+e^{2x}) - 3e^{2x}\right]}{(1 + e^{2x})^3}.$$

Now we equate $\frac{dK}{dx}$ to zero and solve for x, to obtain:

$$\frac{e^x(1+e^{2x})^{\frac{1}{2}}\left[(1+e^{2x}) - 3e^{2x}\right]}{(1 + e^{2x})^3} = 0.$$

$$e^x(1+e^{2x})^{\frac{1}{2}}\left[(1+e^{2x}) - 3e^{2x}\right] = 0.$$

$$e^x = 0.$$

$$(1 + e^{2x})^{\frac{1}{2}} = 0,$$

or,

$$(1 + e^{2x}) - 3e^{2x} = 0.$$

When $e^x = 0$, $x = -\infty$. $(1 + e^{2x})^{\frac{1}{2}} = 0$ is undefined, since, in solving, we obtain $e^{2x} = -1$, $2x \ln e = \ln(-1)$ and $\ln(-1)$ is undefined.

The remaining factor: $(1+e^{2x}) - 3e^{2x} = 0$. We solve for x as follows:

$$(1+e^{2x}) - 3e^{2x} = 0.$$

$$1 - 2e^{2x} = 0.$$

$$2e^{2x} = 1.$$

$$e^{2x} = \frac{1}{2}.$$

$$2x \ln e = \ln 1 = \ln 2$$

$$x = \frac{1}{2}\,\frac{\ln 1 - \ln 2}{\ln e}.$$

$\ln 1 = 0$ and $\ln e = 1$ and $-\frac{1}{2}(\ln 2) = \ln \frac{1}{\sqrt{2}}$. Therefore,

$$x = \frac{1}{2}\left(\frac{0 - \ln 2}{1}\right)$$

$$= -\frac{1}{2} \ln 2$$

$$= \ln 2^{(-\frac{1}{2})}$$

$$= \ln \frac{1}{\sqrt{2}}.$$

Therefore, $x = \ln \frac{1}{\sqrt{2}}$ is a critical value. We can determine whether it is maximum or minimum by substituting values greater and less than $x = \ln \frac{1}{\sqrt{2}}$ into $\frac{dk}{dx}$, and observing the sign change. A value less than $x = \ln \frac{1}{\sqrt{2}}$ is $x = \ln \frac{1}{\sqrt{2}} - \ln e = \ln \frac{1}{e\sqrt{2}}$. Substituting this value for x into $\frac{dk}{dx}$, we obtain,

$$\frac{dk}{dx} = e^{(\ln \frac{1}{e\sqrt{2}})}\left(1 + e^{2(\ln \frac{1}{e\sqrt{2}})}\right)^{\frac{1}{2}}\left(1 - 2e^{2(\ln \frac{1}{e\sqrt{2}})}\right).$$

Since $e^{\ln a} = a$ we can simplify as follows:

$$\frac{dK}{dx} = \frac{1}{e\sqrt{2}}\left(1 + \frac{1}{2e^2}\right)^{\frac{1}{2}}\left(1 - \frac{1}{e^2}\right)$$

$$= \left(\frac{1}{e\sqrt{2}} - \frac{1}{e^3\sqrt{2}}\right)\sqrt{\frac{2e^2 + 1}{2e^2}}.$$

Without further steps, it can be seen that this is a positive value.

As a value greater than $x = \ln \frac{1}{\sqrt{2}}$, we employ $x = \ln \frac{1}{\sqrt{2}} + \ln e = \ln \frac{e}{\sqrt{2}}$. Substituting this value for x into $\frac{dK}{dx}$, we obtain:

$$\frac{dK}{dx} = e^{\left(\ln\frac{e}{\sqrt{2}}\right)}\left(1 + e^{2\left(\ln\frac{e}{\sqrt{2}}\right)}\right)^{\frac{1}{2}}\left(1 - 2e^{2\left(\ln\frac{e}{\sqrt{2}}\right)}\right).$$

Using $e^{\ln a} = a$, we can simplify as follows:

$$\frac{dK}{dx} = \frac{e}{\sqrt{2}}\left(1 + \frac{e^2}{2}\right)^{\frac{1}{2}}(1 - e^2).$$

$$\frac{dK}{dx} = \left(\frac{e}{\sqrt{2}} - \frac{e^3}{\sqrt{2}}\right)\sqrt{1 + \frac{e^2}{2}}.$$

Without further steps it can be seen that this yields a negative number. Since the sign change is from positive to negative, $x = \ln \frac{1}{\sqrt{2}}$ gives a maximum value. Substituting into the equation: $y = e^x$, we obtain:

$$y = e^{\ln\frac{1}{\sqrt{2}}}$$

$$y = \frac{1}{\sqrt{2}}.$$

The point $\left(\ln \frac{1}{\sqrt{2}}, \frac{1}{\sqrt{2}}\right)$ is the point of maximum curvature of the curve $y = e^x$.

• PROBLEM 455

Find the curvature and the radius of curvature of the parabola $y^2 = 2x$, when $x = 2$.

<u>Solution:</u> An appropriate formula for curvature for this problem is:

$$K = \frac{|y''|}{\left[1+(y')^2\right]^{3/2}} \quad .$$

Since the radius of curvature is the reciprocal of the curvature, $R = 1/K$, we can find the radius of curvature from the value we obtain for the curvature. To solve for the curvature we find y' and y'' from the equation: $y^2 = 2x$. Hence,

$$y^2 = 2x$$
$$y = (2x)^{\frac{1}{2}}$$
$$y' = \tfrac{1}{2}(2x)^{-\frac{1}{2}}(2) = (2x)^{-\frac{1}{2}}$$
$$y'' = -\tfrac{1}{2}(2x)^{-3/2}(2) = -(2x)^{-3/2} .$$

When $x = 2$, $y' = \frac{1}{2}$ and $y'' = -\frac{1}{8}$. We can now solve for the curvature, substituting the values found for y' and y'':

$$K = \frac{\left|-\frac{1}{8}\right|}{\left[1+(\frac{1}{2})^2\right]^{3/2}} = \frac{+\frac{1}{8}}{(5/4)^{3/2}} = \frac{1}{\sqrt{125}} = \frac{\sqrt{5}}{25} .$$

$$R = \frac{1}{K} = \sqrt{125} = 5\sqrt{5} .$$

• PROBLEM 456

Find the radius of curvature of the semi-cubical parabola: $3y^2 = x^3$ at the point (3,3).

Solution: An appropriate formula for curvature, K, is:

$$K = \frac{|y''|}{\left[1+(y')^2\right]^{3/2}} .$$

Since curvature equals the reciprocal of the radius of curvature, $K = \frac{1}{R}$, the radius of curvature, R, can be written as: $R = \frac{1}{K}$.
To use this formula, we find y' and y'' from the equation: $3y^2 = x^3$. We could solve for y first and then differentiate, but an easier method is to use implicit differentiation and then solve for y' and y''. Therefore, the first derivative, y', using implicit differentiation is,

$$6yy' = 3x^2, \text{ and } y' = \frac{3x^2}{6y} .$$

The second derivative, y'', using implicit differentiation is,

$$6yy'' + 6(y')^2 = 6x, \text{ and}$$
$$y'' = \frac{x - (y')^2}{y} .$$

To find the radius of curvature at the point (3,3), we can solve for y' and y'' numerically, using $x = 3$ and $y = 3$. Hence,

$$y' = \frac{3(3)^2}{18} = \frac{3}{2}$$
$$y'' = \frac{3 - (3/2)^2}{3} = \frac{1}{4} .$$

Substituting in the formula for the radius of curvature, we obtain,

$$R = \frac{\left[1 + (3/2)^2\right]^{3/2}}{\left|\frac{1}{4}\right|}$$

$$= \frac{(1 + \frac{9}{4})^{3/2}}{\frac{1}{4}} = 4\left(\frac{13}{4}\right)^{3/2} = \frac{13}{2}\sqrt{13} .$$

• PROBLEM 457

Find the curvature of the cissoid $y^2(2a - x) = x^3$ at the point (a,a).

Solution: An appropriate formula for the curvature for this problem is:

$$K = \frac{|y''|}{\left[1+(y')^2\right]^{3/2}} .$$

In order to solve for the curvature we find y' and y'' from the equation: $y^2(2a - x) = x^3$. Differentiating implicitly, we obtain the first derivative as follows:

$$2yy'(2a - x) + y^2(-1) = 3x^2$$

$$y' = \frac{3x^2 + y^2}{2y(2a - x)} .$$

The second derivative, using implicit differentiation is:

$$-2yy' + (2a - x)2yy'' + (2a - x)2y'y' + 2yy'(-1) = 6x$$

$$y'' = \frac{6x + 2yy' - (2a - x)2y'y' + 2yy'}{2y(2a - x)} .$$

At the point (a,a) we can determine y' and y'' by substituting $x = a$ and $y = a$.

$$y' = \frac{3a^2 + a^2}{2a(2a - a)} = 2 .$$

$$y'' = \frac{6a + 2(a)(2) - (2a - a)2(2)(2) + 2(a)(2)}{2a(2a - a)}$$

$$= \frac{6a + 4a - 8a + 4a}{2a^2} = \frac{3}{a} .$$

Now, substituting for y' and y'' in the formula for curvature and solving, we find:

$$K = \frac{\left|\frac{3}{a}\right|}{\left[1+(2)^2\right]^{3/2}}$$

$$= \frac{\frac{3}{a}}{(1+4)^{3/2}} = \frac{3}{5\sqrt{5}\,a} .$$

• PROBLEM 458

Find the curvature of the ellipse: $x = a \cos \varphi$, $y = b \sin \varphi$, at the point where $\varphi = \frac{1}{2}\pi$.

Solution: An appropriate formula for curvature is:

$$K = \frac{|y''|}{\left[1+(y')^2\right]^{3/2}} .$$

To solve this formula for K we find y' and y''. $y' = \frac{dy}{dx}$, and $y'' = \frac{dy'}{dx}$. Therefore, we find dy, dx and dy'. We find:

$$dy = \left(\frac{dy}{d\varphi}\right)d\varphi = b \cos \varphi \, d\varphi , \quad dx = \left(\frac{dx}{d\varphi}\right)d\varphi = -a \sin \varphi \, d\varphi, \text{ and}$$

$$y' = \frac{dy}{dx} = \frac{b \cos \varphi \, d\varphi}{-a \sin \varphi \, d\varphi} = \frac{-b}{a} \cot \varphi .$$

Since $y' = \frac{-b}{a} \cot \varphi$, $dy' = \frac{b}{a} \csc^2 \varphi \, d\varphi$, and

$$y'' = \frac{dy'}{dx} = \frac{b}{a} \cdot \frac{\csc^2 \varphi \, d\varphi}{a -a \sin \varphi \, d\varphi} = -\frac{b}{a^2} \csc^3 \varphi .$$

When $\varphi = \frac{1}{2}\pi$, we have: $y' = 0, y'' = -\frac{b}{a^2}$.

Substituting these values into the formula for K, we obtain:

$$K = \frac{\left|\frac{-b}{a^2}\right|}{\left[1+(0)^2\right]^{3/2}}$$

$$= \frac{b}{a^2} .$$

• PROBLEM 459

Find the radius of curvature of the curve: $x = 2t$, $y = t^2 - 1$, at the point where $t = 1$.

Solution: The radius of curvature is the reciprocal of curvature, $R = \frac{1}{K}$. An appropriate formula for curvature is:

$$K = \frac{|y''|}{\left[1+(y')^2\right]^{3/2}} .$$

Therefore,

$$R = \frac{\left[1+(y')^2\right]^{3/2}}{|y''|} .$$

We find $y' = \frac{dy}{dx}$ and $y'' = \frac{dy'}{dx}$.

$$dy = \left(\frac{dy}{dt}\right)dt = 2t\,dt , \quad dx = \left(\frac{dx}{dt}\right)dt = 2dt, \text{ and}$$

$y' = \frac{dy}{dx} = \frac{2t\,dt}{2dt} = t$. Since $y' = \frac{dy}{dx} = t$, $dy' = dt$, and

$$y'' = \frac{dy'}{dx} = \frac{dt}{2dt} = \frac{1}{2} .$$

At $t = 1$, $y' = 1$ and $y'' = \frac{1}{2}$. Hence, we can substitute these values in the equation for the radius of curvature to obtain:

$$R = \frac{(1+1)^{3/2}}{\frac{1}{2}} = 2 \cdot 2^{3/2} = 4\sqrt{2} .$$

• PROBLEM 460

Find the radius of curvature of the hyperbola $x = a \tan \theta$, $y = a \cot \theta$ at the point (a,a).

Solution: The formula for the radius of curvature is: $R = \frac{1}{K}$, where K is the curvature. Since

$$K = \frac{\frac{d^2y}{dx^2}}{\left[1 + \left(\frac{dy}{dx}\right)^2\right]^{\frac{3}{2}}},$$

the radius of curvature is:

$$R = \frac{1}{K} = \frac{\left[1 + \left(\frac{dy}{dx}\right)^2\right]^{\frac{3}{2}}}{\frac{d^2y}{dx^2}}.$$

In order to solve for the radius of curvature of the hyperbola at the point (a,a) we find $\frac{dy}{dx}$ and $\frac{d^2y}{dx^2}$. We can rewrite $y = a \cot \theta$, using:

$$x = a \tan\theta.$$

$$\tan \theta = \frac{x}{a}.$$

Hence,

$$\cot \theta = \frac{a}{x}$$

$$y = a \cot \theta$$

$$y = \frac{a^2}{x} .$$

We now solve for $\frac{dy}{dx}$.

$$y = \frac{a^2}{x}$$

$$\frac{dy}{dx} = \frac{-a^2}{x^2}$$

At the point (a,a), $\frac{dy}{dx} = \frac{-a^2}{x^2} = \frac{-a^2}{a^2} = -1$. From: $\frac{dy}{dx} = \frac{-a^2}{x^2}$, we obtain:

$$\frac{d^2y}{dx^2} = \frac{2a^2}{x^3}.$$

We use the values: $\frac{dy}{dx} = -1$, and $\frac{d^2y}{dx^2} = \frac{2}{a}$, in the formula for the radius of curvature,

$$R = \frac{\left[1 + (-1)^2\right]^{\frac{3}{2}}}{\frac{2}{a}} = \frac{(2)^{\frac{3}{2}}}{\frac{2}{a}}$$

$$R = \frac{a\sqrt{8}}{2} = a\sqrt{2}.$$

• PROBLEM 461

Show that the curvature of the curve: $y = \sin x$, is numerically equal to unity at every critical point.

Solution: We first solve for the critical points of the curve: $y = \sin x$, then find the curvature of the curve at these points and determine whether or not it is equal to one.

To find the critical points, we first solve for $\frac{dy}{dx}$, set it equal to zero, and solve for x.

$$y = \sin x$$

$$\frac{dy}{dx} = \cos x$$

$$\frac{dy}{dx} = 0$$

$$\cos x = 0$$

$$x = \frac{\pi}{2} \text{ and } \frac{3\pi}{2}.$$

Substituting these values for x into the equation: $y = \sin x$, we obtain,

$$x = \frac{\pi}{2} \qquad\qquad x = \frac{3\pi}{2}$$

$$y = \sin\frac{\pi}{2} = 1 \quad y = -1.$$

The critical points are $\left(\frac{\pi}{2},1\right)$ and $\left(\frac{3\pi}{2},-1\right)$.

We find the curvature at these points. The formula for curvature is:

$$K = \frac{\frac{d^2y}{dx^2}}{\left[1 + \left(\frac{dy}{dx}\right)^2\right]^{\frac{3}{2}}}.$$

We next find $\frac{d^2y}{dx^2}$.

$$\frac{dy}{dx} = \cos x$$

$$\frac{d^2y}{dx^2} = -\sin x.$$

At the point $\left(\frac{\pi}{2},1\right)$, the curvature can be found as follows:

$$K = \frac{-\sin x}{\left[1 + (\cos x)^2\right]^{\frac{3}{2}}} = \frac{-\sin\left(\frac{\pi}{2}\right)}{\left[1 + \left(\cos\frac{\pi}{2}\right)^2\right]^{\frac{3}{2}}}$$

$$K = \frac{-1}{\left[1 + (0)^2\right]^{\frac{3}{2}}}.$$

$K = -1$, which is numerically equal to unity.

At the point $\left(\frac{3\pi}{2},-1\right)$ the curvature is found to be,

$$K = \frac{-\sin x}{\left[1 + (\cos x)^2\right]^{\frac{3}{2}}} = \frac{-\sin\left(\frac{3\pi}{2}\right)}{\left[1 + \left(\cos\frac{3\pi}{2}\right)^2\right]^{\frac{3}{2}}}.$$

$K = 1$ which is numerically equal to unity.

Therefore at both critical points, $\left(\frac{\pi}{2},1\right)$ and $\left(\frac{3\pi}{2},-1\right)$ the curvature of the curve: $y = \sin x$, is numerically equal to unity.

Show that the curvature is zero at: (a) any point of a straight line; (b) an inflection point of a curve.

Solution: (a) We can state the equation of a straight line as y = mx + b. The formula for curvature is:

$$K = \frac{\frac{d^2y}{dx^2}}{\left[1 + \left(\frac{dy}{dx}\right)^2\right]^{\frac{3}{2}}}.$$

In order to solve for the curvature, we find the first derivative, $\frac{dy}{dx}$ of y = mx + b, and the second derivative, $\frac{d^2y}{dx^2}$.

$$y = mx + b$$

$$\frac{dy}{dx} = m$$

$$\frac{d^2y}{dx^2} = 0.$$

Substituting these values into the formula for curvature, we obtain:

$$K = \frac{0}{\left[1 + (m)^2\right]^{\frac{3}{2}}}$$

This proves that, at any point of a straight line, the curvature is zero.

(b) The inflection point of a curve is, by definition, a point at which the second derivative, $\frac{d^2y}{dx^2}$, is equal to zero. Therefore $\frac{d^2y}{dx^2} = 0$ and,

$$K = \frac{0}{\left[1 + \left(\frac{dy}{dx}\right)^2\right]^{\frac{3}{2}}} = 0.$$

This proves that the curvature is zero at an inflection point of a curve.

Show, using a figure, that the center of curvature, corresponding to a point P:(x,y) of a curve, has the coordinates:

$$x - \frac{\frac{dy}{dx}\left[1 + \left(\frac{dy}{dx}\right)^2\right]}{\frac{d^2y}{dx^2}}, \quad y + \frac{1 + \left(\frac{dy}{dx}\right)^2}{\frac{d^2y}{dx^2}}.$$

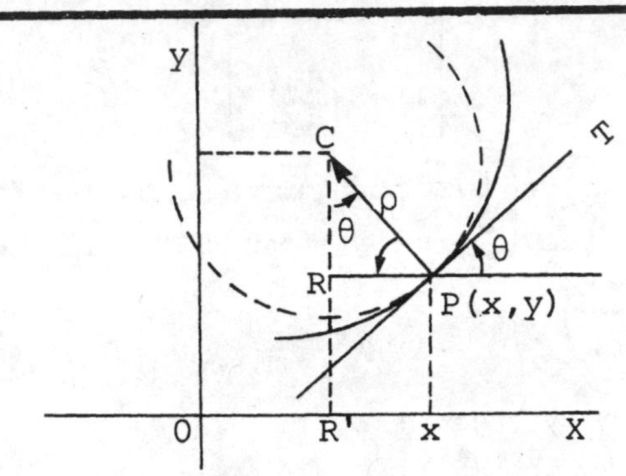

Solution: ρ is the radius of curvature of the curve at point P(x,y) and is given by

$$\rho = \left[1 + \left(\frac{dy}{dx}\right)^2\right]^{\frac{3}{2}} \Bigg/ \left(\frac{d^2y}{dx^2}\right).$$

Since the angle between a radius line and a tangent vector of a circle is 90°, the angle CPT is 90°, because ρ is the radius and the line T is tangent to both the curve and its osculating circle at P(x,y). From this, angle RCP is Θ and

Θ is $\tan^{-1} dy/dx$ at P.

Now,

$$CR = \rho \cos \Theta = \frac{\rho\, dx}{\sqrt{(dx)^2 + (dy)^2}}.$$

Factoring out dx from the radical,

$$CR = \frac{\rho}{\left[1 + \left(\frac{dy}{dx}\right)^2\right]^{\frac{1}{2}}}$$

$$= \left[1 + \left(\frac{dy}{dx}\right)^2\right]^{\frac{3}{2}} \Bigg/ \left(\frac{d^2y}{dx^2}\left[1 + \left(\frac{dy}{dx}\right)^2\right]^{\frac{1}{2}}\right)$$

$$= \left[1 + \left(\frac{dy}{dx}\right)^2\right] \Big/ \; d^2y/dx^2 .$$

But since RR' = y, the y-coordinate of the point C is

$$y + \frac{1 + \left(\frac{dy}{dx}\right)^2}{d^2y/dx^2} .$$

Similarly, using the triangle RCP

$$RP = \rho \sin \Theta = \rho \frac{dy}{\sqrt{(dx)^2 + (dy)^2}} .$$

Factoring out dx from the radical, we have:

$$RP = \frac{dy/dx}{\left[1 + \left(\frac{dy}{dx}\right)^2\right]^{\frac{1}{2}}},$$

and substituting back the value of ρ, we obtain:

$$RP = \frac{\left[1 + \left(\frac{dy}{dx}\right)^2\right]^{\frac{3}{2}} dy/dx}{\frac{d^2y}{dx^2} \left[1 + \left(\frac{dy}{dx}\right)^2\right]^{\frac{1}{2}}}$$

$$= \frac{dy/dx \left[1 + \left(\frac{dy}{dx}\right)^2\right]}{d^2y/dx^2} \equiv R'x.$$

Thus

$$OR' = x - R'x$$

$$= x - \frac{dy}{dx}\left[1 + \left(\frac{dy}{dx}\right)^2\right] \; d^2y/dx^2 .$$

The center of curvature of a curve at point P is located at

$$\left[x - \frac{\frac{dy}{dx}\left[1 + \left(\frac{dy}{dx}\right)^2\right]}{d^2y/dx^2}, \quad y + \frac{1 + (dy/dx)^2}{d^2y/dx^2}\right] .$$

CHAPTER 19

RELATED RATES

Problems classified under "Related Rates" are those in which a relationship is usually given or implied between at least two variables, and the rate (derivative) of one variable is to be found when the rate (derivative) of another variable is given.

The approach that is best taken in these problems, is to write the relationship which involves the variables dealt with in the problem. This relationship may often be derived from the physical and geometrical conditions stated in the problem. Substitute for variables until you obtain an equation involving only the dependent and independent variables. Then take the derivative(s) to enable you to solve for the required rate.

• PROBLEM 464

Find the rate of change of the volume of a sphere with respect to its radius, using the Δ-process.

Solution: The volume of a sphere is:

$$V = \frac{4}{3}\pi r^3.$$

The rate of change is defined as:

$$\lim_{\Delta x \to 0} \frac{\Delta y}{\Delta x} \quad \text{with } \Delta y = f(x + \Delta x) - f(x).$$

In this case we have:

$$\lim_{\Delta r \to 0} \frac{\Delta V}{\Delta r} \quad \text{with } \Delta V = f(r + \Delta r) - f(r).$$

The expression:

$$V = \frac{4}{3}\pi r^3,$$

is f(r). Therefore,

$$f(r + \Delta r) = \frac{4}{3}\pi(r + \Delta r)^3 .$$

By substitution we have:

$$\Delta V = \frac{4}{3}\pi(r+\Delta r)^3 - \frac{4}{3}\pi r^3$$

$$= \frac{4}{3}\pi\left(r^3 + 3r^2\Delta r + 3r(\Delta r)^2 + (\Delta r)^3\right) - \frac{4}{3}\pi r^3$$

$$= \frac{4}{3}\pi\left(3r^2\Delta r + 3r(\Delta r)^2 + (\Delta r)^3\right).$$

Therefore,

$$\Delta V = \frac{4\pi}{3}\left(3r^2\Delta r + 3r(\Delta r)^2 + (\Delta r)^3\right),$$

and

$$\frac{\Delta V}{\Delta r} = \frac{4\pi}{3}\left(3r^2 + 3r\Delta r + (\Delta r)^2\right),$$

$\lim\limits_{\Delta r \to 0} \frac{\Delta V}{\Delta r} = 4\pi r^2$, the rate of change of the volume of the sphere.

• PROBLEM 465

The volume of a sphere is increasing at the constant rate of 24 in.^{3}/sec. Find the radius when its time-rate of change is numerically equal to three times the radius in inches. Use the Δ-method.

Solution: The volume of a sphere is $V = \frac{4}{3}\pi r^3$. We can find $\frac{dv}{dr}$ in terms of r by taking the derivative of this equation. We can then find $\frac{dv}{dt}$ in terms of r and $\frac{dr}{dt}$ because $\frac{dv}{dr}\frac{dr}{dt} = \frac{dv}{dt}$. Given that $\frac{dV}{dt} = 24$, calculate r when $\frac{dr}{dt} = 3r$. Hence, using the Δ-method:

$$\frac{\Delta V}{\Delta r} = \frac{f(r+\Delta r) - f(r)}{\Delta r}.$$

$$\frac{\Delta V}{\Delta r} = \frac{\frac{4}{3}\pi(r+\Delta r)^3 - \frac{4}{3}\pi r^3}{\Delta r}.$$

$$\frac{\Delta V}{\Delta r} = \frac{\frac{4}{3}\pi\left(r^3 + 3r^2\Delta r + 3r(\Delta r)^2 + (\Delta r)^3\right) - \frac{4}{3}\pi r^3}{\Delta r}$$

$$= \frac{\frac{4}{3}\pi r^3 + 4\pi r^2\Delta r + 4\pi r(\Delta r)^2 + \frac{4}{3}\pi(\Delta r)^3 - \frac{4}{3}\pi r^3}{\Delta r}$$

$$= \frac{4\pi r^2\Delta r + 4\pi r(\Delta r)^2 + \frac{4}{3}\pi(\Delta r)^3}{\Delta r}$$

$$= 4\pi r^2 + 4\pi r\Delta r + \frac{4}{3}\pi(\Delta r)^2.$$

$$\frac{dv}{dr} = \lim_{\Delta r \to 0}\frac{\Delta V}{\Delta r} = \lim_{\Delta r \to 0}\ 4\pi r^2 + 4\pi r\Delta r + \frac{4}{3}\pi(\Delta r)^2 = 4\pi r^2$$

$$\frac{dv}{dr} \cdot \frac{dr}{dt} = 4\pi r^2 \frac{dr}{dt}$$

$$\frac{dv}{dt} = 4\pi r^2 \frac{dr}{dt}.$$

We can now substitute $\frac{dv}{dt} = 24$ and $\frac{dr}{dt} = 3r$, and solve for r.

$$24 = 4\pi r^2 (3r)$$

$$\frac{2}{\pi} = r^3$$

$$r = \sqrt[3]{\frac{2}{\pi}}.$$

• PROBLEM 466

A spherical balloon is being inflated. As the radius of the balloon increases, there is a corresponding increase in the volume. Find the rate of increase of the volume when the radius is 3.

Solution: For a sphere,

$$V = \frac{4}{3}\pi r^3.$$

The rate of change is

$$\lim_{\Delta r \to 0} \frac{\Delta V}{\Delta r}, \quad \text{with}$$

$$\Delta V = f(r + \Delta r) - f(r).$$

The expression:

$$V = \frac{4}{3}\pi r^3, \qquad \text{is } f(r). \text{ Therefore,}$$

$$f(r + \Delta r) = \frac{4}{3}\pi (r + \Delta r)^3.$$

Then,

$$\Delta V = \frac{4}{3}\pi (r + \Delta r)^3 - \frac{4}{3}\pi r^3$$

$$= \frac{4}{3}\pi (r^3 + 3r^2\Delta r + 3r\,\Delta r^2 + \Delta r^3) - \frac{4}{3}\pi r^3$$

$$= 4\pi r^2\,\Delta r + 4\pi r\,\Delta r^2 + \frac{4}{3}\pi\Delta r^3$$

$$\frac{\Delta V}{\Delta r} = 4\pi r^2 + 4\pi r\,\Delta r + \frac{4}{3}\pi\Delta r^2$$

Now,

$$\lim_{\Delta r \to 0} \frac{\Delta V}{\Delta r} = \lim_{\Delta r \to 0} 4\pi r^2 + 4\pi r\Delta r + \frac{4}{3}\pi\Delta r^2 = 4\pi r^2.$$

This is the rate of increase of the volume as

the radius increases. When $r = 3$,

$$4\pi r^2 = 36\pi \text{ ft}^3 \text{ per foot of change of radius.}$$

• PROBLEM 467

A circular metal plate is subject to heat expansion, so that its radius increases at the rate of 0.01 in. per second. At what rate is the area on one side increasing when the radius is 2 in.?

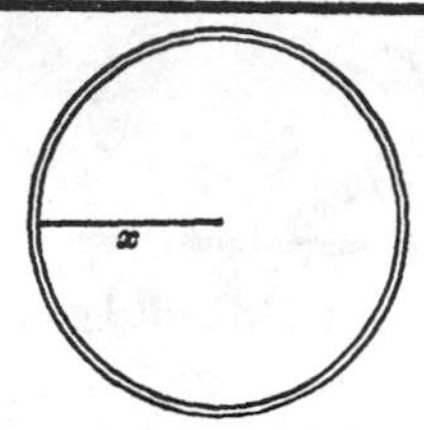

Solution: Let x = radius and y = area of plate.

Then $y = \pi x^2$.

(1) $$\frac{dy}{dt} = 2\pi x \frac{dx}{dt}.$$

When $$x = 2, \quad \frac{dx}{dt} = 0.01,$$

Substituting in (1),

$$\frac{dy}{dt} = 2\pi \times 2 \times 0.01 = 0.04\pi \text{ sq. in. per second.}$$

• PROBLEM 468

A point moves on the parabola $6y = x^2$ in such a way that when $x = 6$ the abscissa is increasing at the rate of 2 ft. per second. At what rate is the ordinate increasing at that instant?

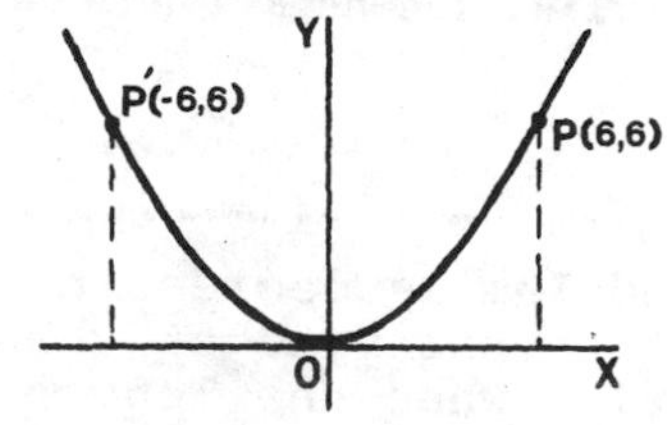

Solution: Since

$$6y = x^2,$$

$$6\frac{dy}{dt} = 2x\frac{dx}{dt}, \text{ or}$$

(1) $$\frac{dy}{dt} = \frac{x}{3} \cdot \frac{dx}{dt}.$$

This means that, at any point on the parabola, the rate of change of ordinate

$$= \left(\frac{x}{3}\right)$$

times the rate of change of abscissa.

When $x = 6$, $\frac{dx}{dt} = 2$ ft. per second.

Thus, substitution gives:

$$\frac{dy}{dt} = \frac{6}{3} \cdot 2$$

$$= 4 \text{ ft/sec.}$$

• **PROBLEM 469**

A painted sphere has a radius of exactly 6 inches. The old paint on the sphere is then ground off and it is repainted, with a net loss in radius of 0.1 inch. How much was the volume changed?

Solution: To find the change in volume, ΔV, we evaluate the volume at the new radius, $r + \Delta r$, and at the original radius, r, and then calculate the difference between the two. The original radius is given to be 6 inches, or $r = 6$. The change in the radius is $\Delta r = 0.1$, so the new radius is $r + \Delta r = 6 + (-0.1) = 5.9$.
The volume of a sphere is $V = \frac{4}{3}\pi r^3$. Δr is negative because the radius was reduced.
The change in volume can now be calculated by inserting the values for the radii into the formula for volume. Hence,

$$\Delta V = \frac{4}{3}\pi(5.9)^3 - \frac{4}{3}\pi 6^3 = \frac{4}{3}(205.379 - 216)\pi$$

$$= -\frac{4}{3}(10.621)\pi$$

$$= \text{(approx.)} - 45 \text{ cubic inches.}$$

The change in volume is likewise negative because there is a reduction.

• **PROBLEM 470**

The length of a pendulum is decreasing at the rate of 0.1 in/sec. What is the rate of change of the period of the pendulum when the length is 16 in., if the relation between the period and length is:

$$T = \Pi\sqrt{L|96},$$

(L = length and T = period)?

Solution: Since we want to find $\frac{dT}{dt}$,

$$T = \Pi\sqrt{L|96} = \Pi\left(\frac{L}{96}\right)^{1/2}.$$

$$\frac{dT}{dt} = \frac{\Pi}{2}\left(\frac{L}{96}\right)^{-1/2} \frac{1}{96}\frac{dL}{dt} = \frac{\Pi \frac{dL}{dt}}{2(96)\sqrt{\frac{L}{96}}} .$$

Substituting the given conditions,

$$L = 16, \frac{dL}{dt} = -0.1$$

$$\frac{dT}{dt} = -\frac{(.1)\,\Pi}{192\sqrt{\frac{16}{96}}} = -0.004 \text{ sec.}$$

• PROBLEM 471

(A) Water flows into a cylindrical tank. Compare the rates of increase of the total volume and the increase in height of the water in the tank, if the radius of the base of the tank is 10 ft. Find the rate of inflow which causes a rise of 2 in. per second, and find the increase in height due to an inflow of 10 cu. ft. per second. (B) Consider the same problem for a conical tank.

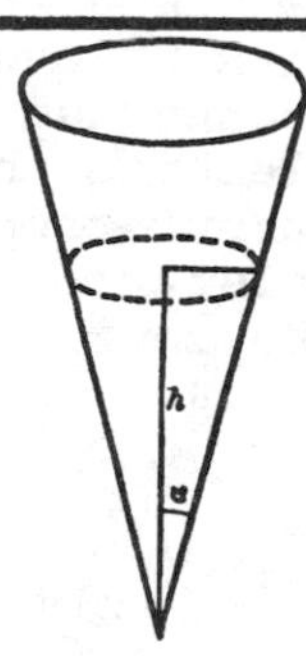

Solution:

(A) The volume V is given in terms of the height h by the formula:

$$V = \pi r^2 h = 100\pi h,$$

hence
$$\frac{dV}{dt} = 100\pi\frac{dh}{dt} ;$$

or, the rate of increase in volume (in cubic feet per second) is 100π times the rate of increase in height (in feet per second).

If $dh/dt = 1/6$ (measured in feet per second), $dV/dt = 100\pi/6 = 52.3$ (cubic feet per second). If $dV/dt = 10$, $dh/dt = 10 \div 100\pi = .031$ (in feet per second).

(B) If the reservoir is conical, we have

$$V = \frac{1}{3}\pi r^2 h = \frac{1}{3}\pi h^3 \tan^2 \alpha,$$

where r is the radius of the water surface, h is the height of the water, and α is the half-angle of the cone; for $r = h \tan \alpha$. In this case

$$\frac{dV}{dt} = \pi h^2 \tan^2 \alpha \frac{dh}{dt} ,$$

which varies with h. If $\alpha = 45°$ ($\tan \alpha = 1$), at a height of 10 ft., a rise 1/6 (feet per second) means an inflow of $\pi h^2 \times (1/6) = 100\pi/6 = 52.3$ (cubic feet per second). At a height of 15 feet, a rise of 1/6 (feet per second) means an inflow of $225\pi/6 = 117.8$ (cubic feet per second). An inflow of 100 cubic feet per second) means a rise in height of $100/\pi h^2$, which varies with the height; at a height of 5 ft., the rate of rise is $4/\pi = 1.28$ (feet/second).

• PROBLEM 472

The light intensity, I, from a carbon filament is found to vary with the applied voltage, v, according to the expression: $I = \sqrt{p\ RV}$, where p is a constant and R is the resistance of the filament. Determine the rate of change of light intensity with respect to voltage for a given filament.

Solution: The resistance R is assumed to be fixed since we are concerned with a particular filament, neglecting the effect of change in filament temperature that alters the resistance. The given equation can be rewritten as

$$I = (p\ Rv)^{1/2}$$

$$\frac{dI}{dv} = \frac{1}{2}(pRv)^{-1/2} \cdot \frac{d}{dv}(pRV) = \frac{1}{2}(pRv)^{-1/2}(pR)$$

$$= \frac{pR}{2\sqrt{pRV}} = \frac{1}{2}\sqrt{\frac{pR}{v}}.$$

• PROBLEM 473

The bottom of a 20-ft ladder is sliding along the ground at a speed of 3 ft/min. Find the velocity of the top at any instant by using the implicit method of differentiation.

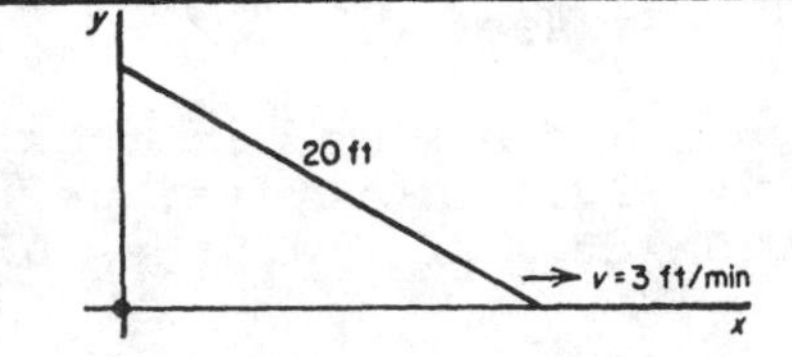

Solution: $x^2 + y^2 = (20)^2$ (Pythagorean theorem)

$$\frac{d(x^2)}{dx} + \frac{d(y^2)}{dx} = \frac{d(20^2)}{dx}$$

$$2x\frac{dx}{dt} + 2y\frac{dy}{dt} = 0 \quad \text{(Implicit derivative with respect to t)}$$

$$v = \frac{dx}{dt} = 3$$

Substituting and transposing,

$$2y\frac{dy}{dt} = -6x$$

$$\frac{dy}{dt} = -\frac{6x}{2y} = -\frac{3x}{y}.$$

The negative sign indicates that the top of the ladder is descending.

• PROBLEM 474

A spherical balloon is being inflated at the rate of 20 cubic feet per minute. At the instant when the radius is 15 feet, at what rate is the surface area increasing?

Solution: Let r = radius

The surface area, $S = 4\pi r^2$

The volume, $V = \frac{4}{3}\pi r^3$

$$\frac{ds}{dt} = 8\pi r \frac{dr}{dt}. \quad \text{Also, } \frac{dv}{dt} = \frac{4}{3} \cdot 3\pi r^2 \frac{dr}{dt}$$

$$= 4\pi r^2 \frac{dr}{dt}.$$

According to the second equation,

$$\frac{dr}{dt} = \frac{1}{4\pi r^2} \frac{dv}{dt}.$$

Substituting in the first equation,

$$\frac{ds}{dt} = \frac{8\pi r}{4\pi r^2} \frac{dv}{dt} = \frac{2}{r} \frac{dv}{dt}. \quad \text{Substituting } \frac{dv}{dt} = 20,$$

r = 15, we have:

$$\frac{ds}{dt} = \frac{2}{15}(20) = \frac{8}{3}$$

• PROBLEM 475

What is the rate of increase of the number of bacteria in a culture described by the formula: $N = 2{,}000\varepsilon^{0.6t}$, where N = number and t = time in hours?

Solution: Given $N = 2{,}000\varepsilon^{0.6t}$, and taking the logarithm of both sides:

$$\ln N = \ln 2{,}000 + 0.6t \ln e,$$

We differentiate to obtain the rate of increase:

$$\frac{1}{N}\frac{dN}{dt} = 0.6$$

$$\frac{dN}{dt} = 0.6N$$

The increase is 0.6, or 60%.

• PROBLEM 476

Along the graph of the equation:

$$y = x^3 - 6x^2 + 3x + 5,$$

both the ordinate y and the slope m change, but generally at different rates. Find the point or points, if any, where the ordinate and slope are momentarily changing at the same rate.

Solution: The rate of change of the ordinate is $\frac{dy}{dx}$ and the rate of change of slope, $\frac{dm}{dx} = \frac{d^2y}{dx^2}$.

When these are the same,

$$\frac{dy}{dx} = \frac{d^2y}{dx^2}.$$

Differentiating the equation,

$$dy = (3x^2 - 12x + 3)dx,$$

is the rate of change of the ordinate.

Hence,

$$\frac{dy}{dx} = m = 3x^2 - 12x + 3.$$

Differentiating this with respect to x to find the rate of change of m,

$$\frac{d^2y}{dx^2} = (6x - 12)$$

$$(3x^2 - 12x + 3) = (6x - 12)$$

$$3x^2 - 12x + 3 = 6x - 12.$$

This is the condition which must be fulfilled for the two rates to be equal. To determine the abscissa at the point or points where this condition holds, this equation must be solved for x. Transposing,

$$x^2 - 6x + 5 = 0.$$

Factoring,

$$(x - 5)(x - 1) = 0.$$

Hence,

$$x - 5 = 0, \quad \text{or } x - 1 = 0,$$

and $x = 5$, $x = 1$

are the abscissas of the points at which the rates of change of the ordinate and the slope are equal.

• PROBLEM 477

> A particle is constrained to move along the parabola: $y = x^2$. (a) At what point on the curve are the abscissa and the ordinate changing at the same rate? (b) Find this rate if the motion is such that at time t we have $x = \sin t$ and $y = \sin^2 t$.

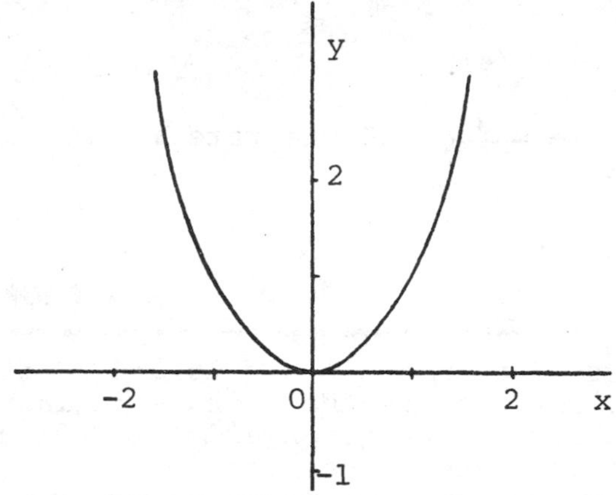

Solution: (a) To find the point at which y and x change at the same rate, we must find an equation that relates $\frac{dy}{dt}$ to $\frac{dx}{dt}$. Since it is given that the particle is constrained to move along

$$y = x^2 \qquad (1),$$

differentiating (1) with respect to t gives

$$\frac{dy}{dt} = \frac{dy}{dx} \cdot \frac{dx}{dt} = 2x \frac{dx}{dt} \qquad (2).$$

If the coordinates are to change at the same rate, then

$$\frac{dy}{dt} = \frac{dx}{dt},$$

and substituting into (2), $2x = 1$; $x = \frac{1}{2}$.

Substituting $x = \frac{1}{2}$ in (1), $y = \frac{1}{4}$.

Therefore, when $x = \frac{1}{2}$, $y = \frac{1}{4}$, the abscissa and the ordinate change at the same rate.

(b) To find this rate, we can find either $\frac{dx}{dt}$ or $\frac{dy}{dt}$, since they are equal at this moment. For the case:

$$x = \sin t \qquad (3)$$

$$y = \sin^2 t \qquad (4),$$

$\frac{dx}{dt}$ can be obtained by differentiating (3).

$$\frac{dx}{dt} = \cos t \qquad (4)$$

Since, at this moment, $x = \frac{1}{2}$, from (3)

$$t = \sin^{-1} \frac{1}{2} = \frac{\Pi}{6}.$$

Substituting into (4),

$\frac{dx}{dt} = \cos \frac{\Pi}{6} = \frac{\sqrt{3}}{2}$, and the rate is

$\frac{\sqrt{3}}{2}$ units/sec.

• PROBLEM 478

> Two sides of a triangle are 15 and 20 ft. long, respectively. How fast is the third side increasing if the angle between the given sides is 60° and is increasing at the rate 2° per sec.?

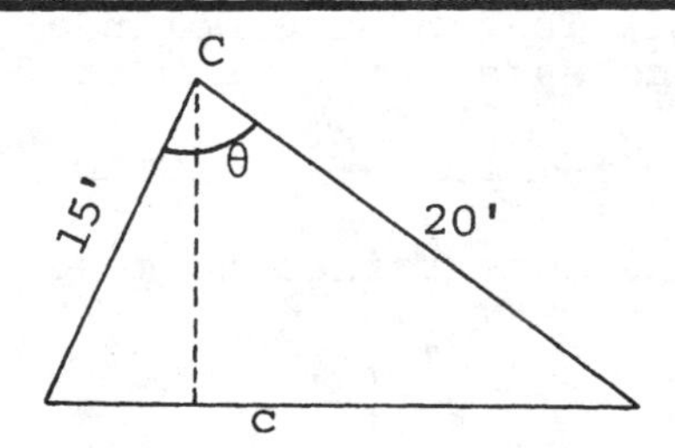

Solution: If Θ is changing at a rate of 2° per sec. then

$$\frac{d\Theta}{dt} = 2° = \frac{\Pi}{90} \text{ radian.}$$

We wish to find the rate of change of side C, i.e., $\frac{dC}{dt}$, when $\Theta = 60°$.

The law of cosines states:

$$C^2 = a^2 + b^2 - 2ab \cos C.$$

This relationship can be used, letting Θ = angle C.

$$C^2 = 15^2 + 20^2 - 2(15)(20)\cos\Theta$$

$$C = \sqrt{625 - 600\cos\Theta} = (625 - 600\cos\Theta)^{1/2}$$

$$\frac{dC}{dt} = \frac{-600(-\sin\Theta)\frac{d\Theta}{dt}}{2\sqrt{625 - 600\cos\Theta}}.$$

Substituting the conditions that $\frac{d\Theta}{dt} = \frac{\Pi}{90}$ and $\Theta = 60°$,

$$\frac{dC}{dt} = \frac{(-600)\left(-\frac{\sqrt{3}}{2}\right)\left(\frac{\Pi}{90}\right)}{2\sqrt{625 - 600(1/2)}} = \frac{\sqrt{3}\,\Pi}{3\sqrt{13}} \text{ ft/sec.}$$

Side C is increasing at a rate of

$$\frac{\sqrt{3}\,\Pi}{3\sqrt{13}} \text{ ft/sec.}$$

• PROBLEM 479

The hands of a tower clock are 4 1/2 ft. and 6 ft. long, respectively. How fast are the ends approaching at four o'clock?

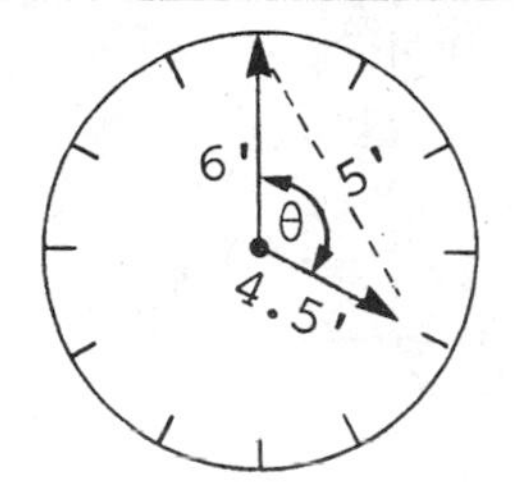

Solution: Let s be the distance between the ends, and θ the angle between them. Then, using the law of cosines,

$$s^2 = 6^2 + (4.5)^2 - 12(4.5)\cos\theta$$

$$= 56.25 - 54\cos\theta.$$

Hence
$$2s\frac{ds}{dt} = 54\sin\theta \cdot \frac{d\theta}{dt},$$

and
$$\frac{ds}{dt} = \frac{27\sin\theta}{s} \cdot \frac{d\theta}{dt}.$$

At four o'clock $\theta = 2\pi/3$,

$$s^2 = 56.25 - 54\cos(2\pi/3)$$

$$= 56.25 + 27 = 83.25.$$

In one full hour the minute hand turns 2π radians and the hour hand turns $2\pi/12 = \pi/6$ rad. Thus we can conclude that θ decreases at the rate of $(2\pi - \pi/6) = 11\pi/6$ radians per hour, and we have

$$\frac{d\theta}{dt} = -\frac{11\pi}{6}\left(\frac{rad}{hr.}\right) \cdot \frac{1}{60}\left(\frac{hr.}{min}\right) = -\frac{11\pi}{360} \text{ rad./min.}$$

Therefore,

$$\frac{ds}{dt} = \frac{27 \sin\theta}{s}\left(\frac{d\theta}{dt}\right)$$

or

$$= \frac{27}{\sqrt{83.25}}\left(\frac{\sqrt{3}}{2}\right)\left(-\frac{11\pi}{360}\right)$$

$$= -.246 \text{ ft./min.}$$

• PROBLEM 480

As shown in the diagram, block A is allowed to move along the x-axis while block B is fastened at a point 10 in. above the origin. If OA is increasing at a constant rate of 2.0 in/sec what is the rate at which block A is moving along the slot when OA = 7 in?

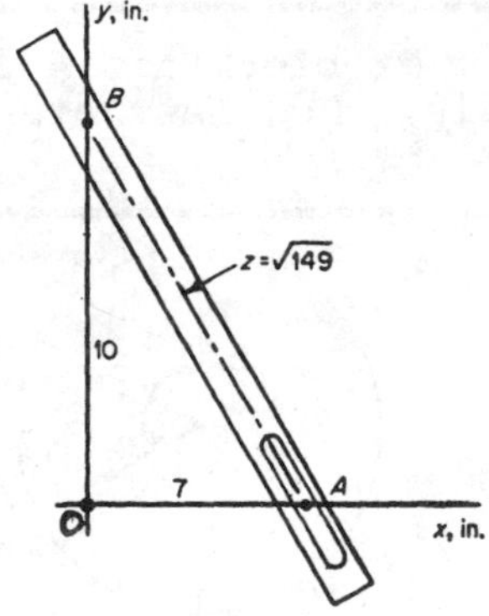

Solution: We let x be the position of block A from the origin 0 at any time t. Since x is increasing at 2.0 in/sec we have:

$$x = 2t.$$

The distance from A to B, by using the Pythagorean theorem for the triangle, is:

$$z^2 = x^2 + y^2 = 4t^2 + (10)^2 ,$$

or

$$z = \sqrt{4t^2 + 100}.$$

We can differentiate this with respect to time to obtain the desired rate of increase of z:

$$\frac{dz}{dt} = \frac{4t}{\sqrt{4t^2 + 100}}$$

We can find the time elapsed for x to reach 7 in, that is:

$$x = 2t = 7, \quad \text{or} \quad t = 7/2 \text{ sec.}$$

Substituting 7/2 for t in the differential equation above, we obtain:

$$\frac{dz}{dt} = \frac{4(7/2)}{\sqrt{4(7/2)^2 + 100}} = \frac{14}{\sqrt{49 + 100}}$$

or

$$\frac{dz}{dt} = \frac{14}{\sqrt{149}} = 1.15 \text{ in/sec, increasing.}$$

• PROBLEM 481

A man approaching a pole 75 feet high is walking at the rate of 3 miles per hour. At what rate is he approaching the top when he is 100 feet away from the pole?

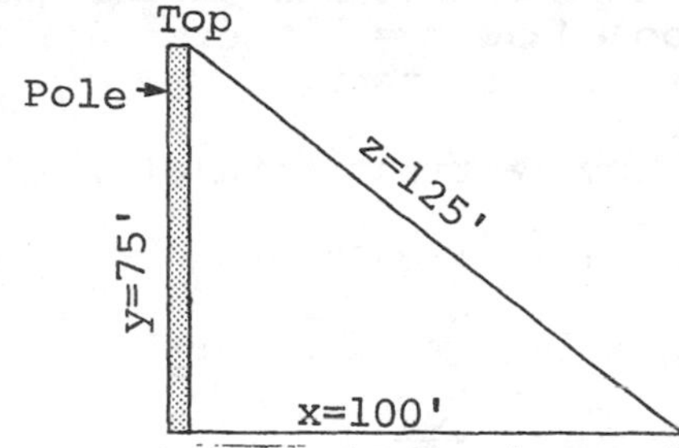

<u>Solution:</u> Let x denote the horizontal distance of the man at some time t. Then the distance between the man and the top of the 75 ft. high pole can be calculated as

$$z^2 = x^2 + (75)^2.$$

Differentiation with respect to time gives:

$$2z\frac{dz}{dt} = 2x\frac{dx}{dt} + 0$$

$$\frac{dz}{dt} = \frac{x}{z}\frac{dx}{dt}$$

At the instant x = 100 ft., we obtain:

$$z = \sqrt{x^2+y^2} = \sqrt{(100)^2+(75)^2} = \sqrt{15.625}$$

$$= 125 \text{ ft.}$$

Since dx/dt = -3m.p.h. = $-3 \cdot 5{,}280$ ft./hr., we substitute in the differential equation:

$$\frac{dz}{dt} = \frac{x}{z}\frac{dx}{dt}$$

$$= -\frac{100}{125} \cdot 3 \cdot 5280 \text{ ft/hr}$$

$$= -2.4 \text{ m.p.h.}$$

The negative sign indicates that z is decreasing.

A man is walking at the rate of 5 mi. per hour toward the foot of a 60 ft. pole. At what rate is he approaching the top when he is 80 ft. from the pole? (Neglect man's height.)

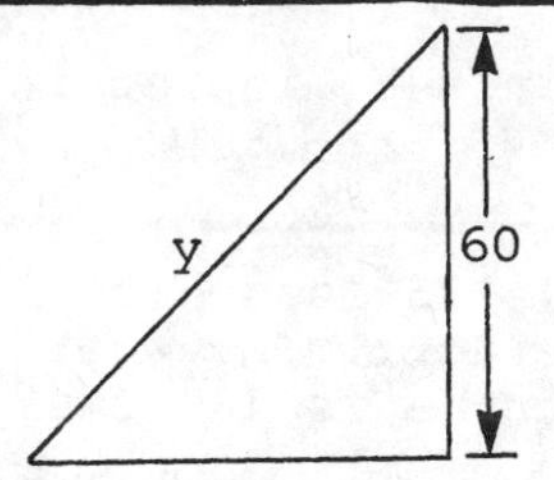

Solution: Let x = distance of the man from the pole and y = his distance from the top at any instant.

Since we have a right triangle,

$$y^2 = x^2 + 3600.$$

Differentiating, we obtain:

$$2y\frac{dy}{dt} = 2x\frac{dx}{dt}, \text{ or}$$

$$\frac{dy}{dt} = \frac{x}{y}\frac{dx}{dt}$$

Substituting $\sqrt{x^2+3600}$ for y, we obtain:

$$\frac{dy}{dt} \quad \frac{x}{\sqrt{x^2+3600}}\frac{dx}{dt}$$

When x = 80,
$$\frac{dx}{dt} = -5 \text{ mi/hr}$$
$$= -5 \cdot 5280 \text{ ft/hr}$$

Substituting these values,

$$\frac{dy}{dt} = \frac{80}{\sqrt{(80)^2+3600}}(-5 \cdot 5280)$$

$$= -\frac{80}{100} \cdot 5 \cdot 5280 \text{ ft/hr}$$

$$= -4 \text{ mi/hr}$$

• PROBLEM 483

Water flows at the rate of 2ft.^{3}/min. into a vessel in the form of an inverted right circular cone of altitude 2ft. and radius of base 1 ft. At what rate is the surface rising when the vessel is one-eighth filled?

Solution: To solve this problem we use the fact

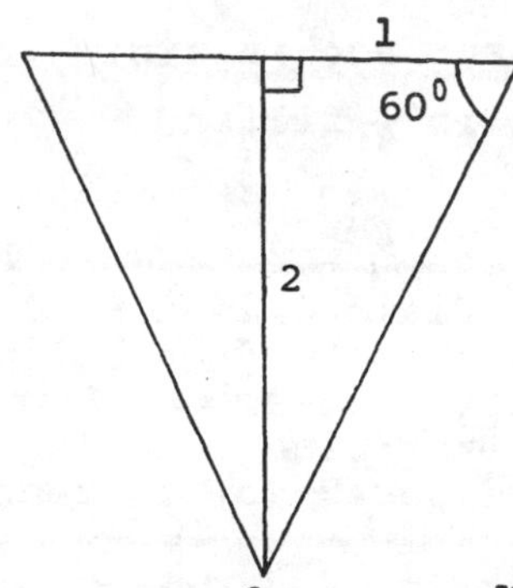

that the rate of flow, 2ft.^{3}/min. = $\frac{dV}{dt}$, the change in volume with respect to time. We wish to find the rate at which the surface is rising, or $\frac{dh}{dt}$, the change in height with respect to time. The volume of a cone is given by, $V = \frac{1}{3}\pi r^2 h$. We wish to find $\frac{dV}{dt}$. To do this, we rewrite the equation for volume in terms of one variable, choose h. Since r = 1 when h = 2, we can write: $r = \frac{h}{2}$. Then, by substitution:

$$V = \frac{1}{3}\pi\left(\frac{h}{2}\right)^2 h = \frac{\pi h^3}{12}.$$

We find,

$$\frac{dV}{dt} = \frac{\pi}{12}(3h^2)\frac{dh}{dt} = \frac{\pi h^2}{4}\frac{dh}{dt}.$$

Solving for $\frac{dh}{dt}$:

$$\frac{dh}{dt} = \frac{\frac{dV}{dt}}{\frac{\pi h^2}{4}}. \qquad (1)$$

We now determine h, at the time when the vessel is $\frac{1}{8}$ filled. The volume of the cone is $\frac{1}{3}\pi r^2 h$, and when r = 1, h = 2, $V = \frac{2}{3}\pi$. This is the volume when the cone is full. Therefore the volume when the cone is $\frac{1}{8}$ filled = $\frac{1}{8}\left(\frac{2}{3}\pi\right) = \frac{\pi}{12}$.

To find h when the vessel is $\frac{1}{8}$ filled, we write:

$$\frac{\pi}{12} = \frac{1}{3}\pi r^2 h = \frac{\pi h^3}{12}.$$

Therefore, $\pi = \pi h^3$, $h^3 = 1$ and h = 1. We now go back to equation (1) and substitute. Since $\frac{dV}{dt}$ = 2ft.^{3}/min. and h = 1 we obtain:

$$\frac{dh}{dt} = \frac{2}{\frac{\pi(1)}{4}} = \frac{8}{\pi}\text{ ft./min.}$$

Therefore, the surface is rising at $\frac{8}{\pi}$ ft./min. when the vessel is $\frac{1}{8}$ filled.

• PROBLEM 484

A baseball diamond is a 90 foot square. A ball is batted along the third-base line at a constant speed of 100 feet per second. How fast is its distance from first-base changing when (a) it is halfway to third-base? (b) it reaches third-base?

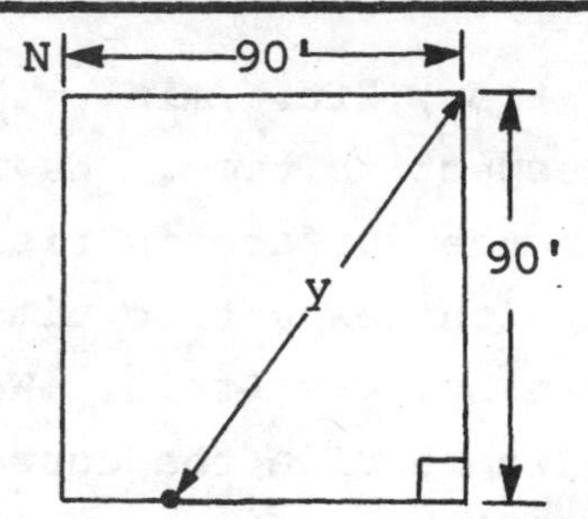

Solution: Let the distance the ball travels along the third-base line be x. The distance between the ball and first base is then given by:

$$y = \sqrt{(90)^2 + x^2}\ . \qquad (1)$$

To find the rate of change of distance of the ball from first base (y), it is necessary to find $\frac{dy}{dt}$. Since equation (1) expresses y as a function of x, the chain rule should be applied to obtain $\frac{dy}{dt}$.

$$\frac{dy}{dt} = \frac{dy}{dx}\,\frac{dx}{dt} = \frac{1}{2}\,\frac{1}{\sqrt{(90)^2 + x^2}}\,2x\,\frac{dx}{dt}$$

$$= \frac{x}{\sqrt{(90)^2 + x^2}}\,\frac{dx}{dt} \quad . \; . \; . \; (2)$$

It is given that the speed at which the ball travels along the third-base line, $\frac{dx}{dt}$ = 100 ft/sec. Substituting this value in (2),

$$\frac{dy}{dt} = \frac{x}{\sqrt{(90)^2 + x^2}} \cdot 100 \text{ ft/sec.} \quad . \; . \; . \; (3)$$

To find $\frac{dy}{dt}$ when the ball travels halfway to the base, we substitute $x = \frac{90}{2} = 45$ ft in (3), and obtain:

$$\left.\frac{dy}{dt}\right|_{x=45} = \frac{45}{\sqrt{(90)^2 + (45)^2}} \cdot 100 = 20\sqrt{5} \text{ ft/sec.}$$

Similarly, for the ball reaching third base, we let x = 90 ft.

$$\left.\frac{dy}{dt}\right|_{x=90} = \frac{90}{\sqrt{(90)^2 + (90)^2}} \quad 100 = 50\sqrt{2} \text{ ft/sec.}$$

• PROBLEM 485

A boy flies a kite, which is 120 ft. above his hand. If the wind carries the kite horizontally at the rate of 30 ft/min, at what rate is the string being pulled out when the length of string out is 150 ft?

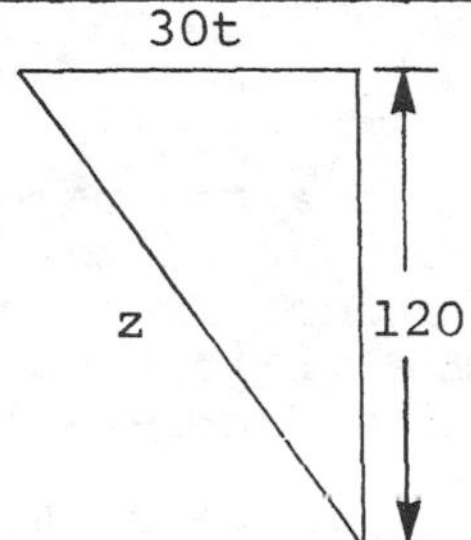

Solution: Formulating the given data, $\frac{dx}{dt} = 30$, $y = 120$, $z = 150$; $\frac{dz}{dt}$ is to be found.

We can construct a triangle with sides proportional to the height of the kite, the length of the string, z, and the horizontal distance the kite has traveled in time t (as shown in the figure).

Using the Pythagorean theorem, we find that:

$$z^2 = (30t)^2 + (120)^2, \quad t = \frac{1}{30}\sqrt{z^2 - (120)^2},$$

or

$$z = \sqrt{(30t)^2 + (120)^2}$$

Differentiation with respect to t gives:

$$2z\frac{dz}{dt} = (30)^2 2t, \qquad \frac{dz}{dt} = \frac{900t}{z}$$

Substituting $\frac{\left(z^2 - \overline{120}^2\right)^{1/2}}{30}$ for t in the above equation and solving for dz/dt, we obtain:

$$\frac{dz}{dt} = \frac{30\sqrt{z^2 - (120)^2}}{z}.$$

At z = 150, the rate of the string being pulled out is:

$$\frac{dz}{dt} = \frac{30\sqrt{(150)^2 - (120)^2}}{150} = \frac{1}{5}\sqrt{8100} = \frac{1}{5}90.$$

$$\frac{dz}{dt} = 18 \text{ ft./min.}$$

• PROBLEM 486

An airplane loses altitude at the rate of 400 mi/hr. What is the rate of decrease of the visible portion of the surface of the earth when the plane is one mile high?

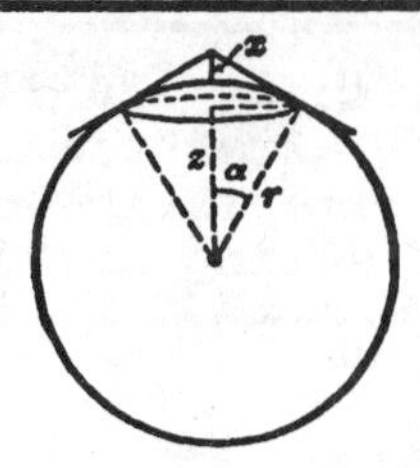

Solution: First call the radius of the earth r and call the height of the plane x; let z be the distance from the center to the base of the visible spherical cap, as shown in the figure. We have

$$\frac{r}{r + x} = \frac{z}{r}, \text{ since each term equals cos a.}$$

Hence,

$$z = \frac{r^2}{r + x}.$$

The area of the spherical cap cut off by a plane z units from the center is, from spherical geometry:

$$A = 2\pi r(r - z)$$

$$= 2\pi r^2 - \frac{2\pi r^3}{r + x}$$

$$\frac{dA}{dt} = \frac{2\pi r^3}{(r + x)^2} \frac{dx}{dt}$$

Putting $r = 4,000$, $x = 1$, $\frac{dx}{dt} = -400$, we can evaluate $\frac{dA}{dt}$ exactly. However, since $x << r$, $r + x$ is approximately equal to r. Thus we can obtain an approximation for $\frac{dA}{dt}$ with little labor.

$$\frac{da}{dt} \approx 2\pi r \frac{dx}{dt} = 2\pi 4,000(-400)$$

$$= -32\pi 10^5 \text{ sq. mi./hr.}$$

$$= -\frac{32 \cdot 22 \cdot 10^5}{7 \cdot 60 \cdot 60} \text{ sq. mi./sec.}$$

$$= -2,800 \text{ sq. mi./sec.}$$

The visible area decreases at the rate of approximately 2,800 sq. mi./sec.

• PROBLEM 487

A road runs at right angles to a wall. A man approaches the wall at 10 feet per minute. There is a lamp on the ground 20 feet from the road and 40 feet from the wall. Find the rate at which the man's shadow is moving along the wall at the instant when the man is 20 feet from the wall.

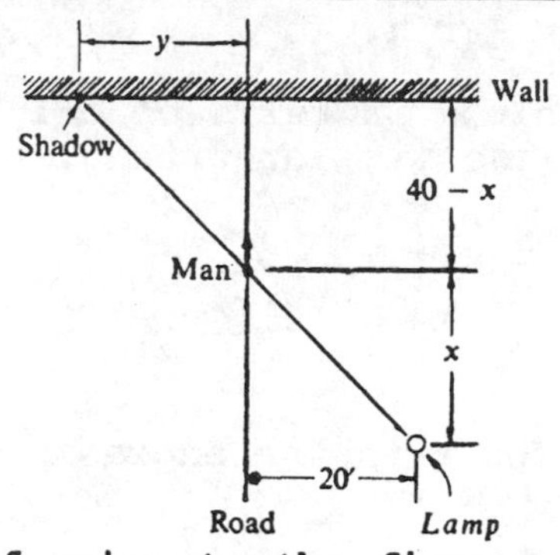

Solution: Referring to the figure, we are given that $dx/dt = 10$. We are seeking the value of dy/dt at the instant when $x = 20$. From the similarity of the relevant triangles,

$$\frac{x}{20} = \frac{40 - x}{y}.$$

$$xy = 800 - 20x.$$

Since x andy are both functions of t, we may differentiate with respect to t.

$$\frac{d(xy)}{dt} = x\frac{dy}{dt} + y\frac{dx}{dt} = -20\frac{dx}{dt}.$$

When $x = 20$, $y = 20$.

$$20\frac{dy}{dt} + 20(10) = -20(10).$$

$$20\frac{dy}{dt} = -400$$

and $dy/dt = -20$. Thus, at the instant under consideration, the shadow is moving at the rate of 20 feet per minute. The negative sign of dy/dt indicates that y is decreasing with t.

• PROBLEM 488

Water runs into the conical tank shown at a constant rate of 2 cubic feet per minute. How fast is the water rising in the tank at any instant?

Solution: At any time t min. after the water starts running, let $h(t)$ be the depth of the water and $r(t)$ the radius of the surface of the water in the tank.

Since the volume of water is dependent on time, we write:

$$v(t) = \frac{1}{3}\pi\left[r(t)\right]^2 h(t)$$

From the given conditions, the volume of water in the tank at any time t, v(t) = 2t. Therefore

$$2t = \frac{1}{3}\pi r^2 h.$$

To express the volume in terms of only one variable, preferably h, we use similar triangles to obtain the expression:

$$\frac{r}{5} = \frac{h}{10}$$

or $\frac{1}{2}h = r$.

Substituting $\frac{1}{2}h$ for r in the expression for volume,

$$v(t) = 2t = \frac{1}{3}\pi\left(\frac{h}{2}\right)^2 h,$$

and hence

$$24t = \pi h^3 \quad \text{or} \quad h = \left(\frac{24t}{\pi}\right)^{1/3}.$$

Now h is expressed as a function of time, from which we can find the instantaneous rate of change of h:

from

$$h = \left(\frac{24}{\pi}\right)^{1/3} \cdot t^{1/3},$$

$$\frac{dh}{dt} = \left(\frac{24}{\pi}\right)^{1/3} \cdot \frac{1}{3}\, t^{-2/3} = \frac{2}{3}\sqrt[3]{\frac{3}{\pi t^2}}$$

This is the rate of rise of the water.

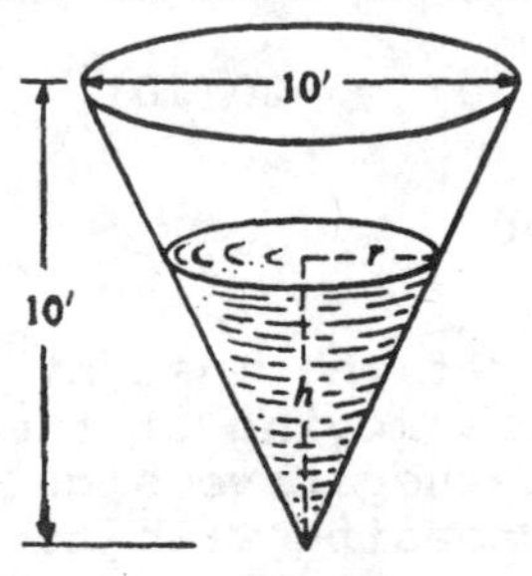

• PROBLEM 489

A lamp hangs 12 ft. directly above a horizontal plane on which a boy 5 ft. tall is walking. How fast is the boy's shadow lengthening when he is walking away from the light at the rate of 168 ft. per minute?

__Solution:__ Let x = distance of the boy from a point

directly under the light and
y = length of boy's shadow on the ground. From the figure,

$$y : y + x = 5 : 12,$$

$$\frac{y}{x+y} = \frac{5}{12}$$

$$12y = 5x + 5y$$

$$7y = 5x$$

or

$$y = \frac{5}{7}x.$$

Differentiating,

$$\frac{dy}{dt} = \frac{5}{7}\frac{dx}{dt};$$

that is, the shadow is lengthening 5/7 as fast as the boy is walking, or 120 ft. per minute.

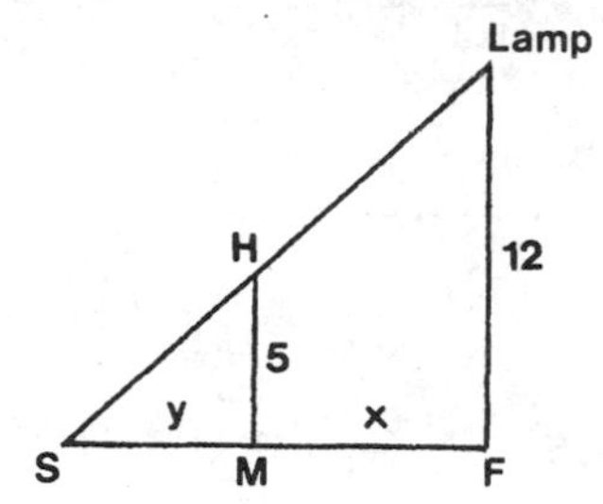

• PROBLEM 490

A ship, s_1, is 41 miles due north of a second ship, s_2. Ship s_1 sails south at the rate of 9 miles per hour, s_2 sails west at 10 miles per hour. (a) How rapidly are they approaching $1\frac{1}{2}$ hours later? (b) How long will they continue to approach?

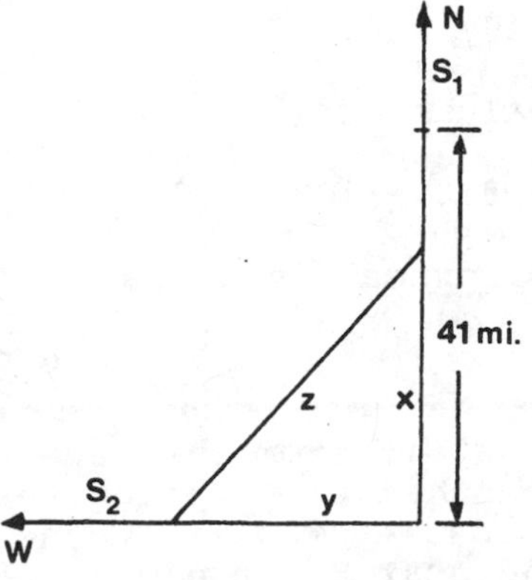

Solution: After t hours of sailing let x be the distance of s_1 from the intersection of the courses, and y the distance of s_2. Let z be the distance between them. Since x is decreasing 9 mi/hr, and y is increasing 10 mi/hr,

$$\frac{dx}{dt} = -9 \text{ mi/hr}, \qquad \frac{dy}{dt} = 10 \text{ mi/hr}.$$

We are asked in (a) to find dz/dt when t=1 1/2; in (b) to find what value of t makes z a minimum.

(a) To find dz/dt, we express z in terms of x and y, the rates of change of which are given. Since

$$(1) \qquad z^2 = x^2 + y^2,$$

and x, y, and therefore z are each functions of t,

$$(2) \qquad x = 41 - 9t, \qquad y = 10t.$$

We differentiate (1) with respect to t. Then

$$2z\frac{dz}{dt} = 2x\frac{dz}{dt} + 2y\frac{dy}{dt},$$

or

$$(3) \qquad \frac{dz}{dt} = \frac{x\frac{dx}{dt} + y\frac{dy}{dt}}{z} = \frac{-9x + 10y}{z}.$$

When t = 1 1/2, from (2) and (1),

$$x = 27.5, \quad y - 15, \quad z = 31.325.$$

Hence $$\frac{dz}{dt} = \frac{-247.5 + 150}{31.325} = -3.11 \text{ mi/hr},$$

that is, in 1 1/2 hours the ships are approaching at the rate of 3.11 mi/hr.

(b) The ships continue to approach until z is a minimum, or until

$$\frac{dz}{dt} = \frac{-9x + 10y}{z} = 0,$$

which is, when

$$-9x + 10y = 0,$$

Substituting values of (2),

$$-369 + 81t + 100t = 0,$$

from which t = approximately 2 hrs.

• PROBLEM 491

Two boats, A and B, start from the same point and move away from each other along paths that are perpendicular. If boat A travels at the rate of 6 miles per hour and boat B travels at 8 miles per hour, at what rate are they separating two hours later?

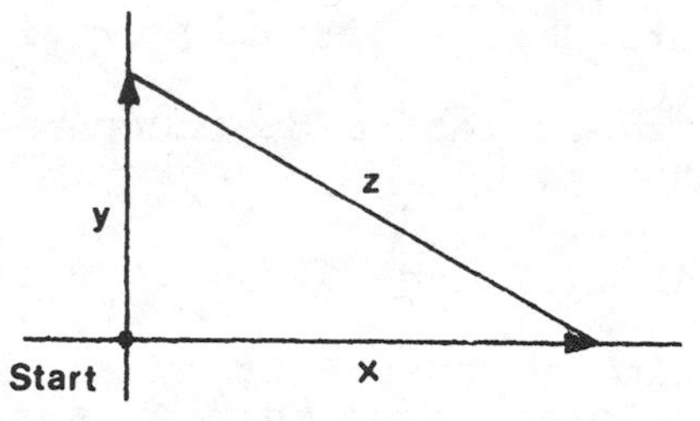

Solution: If y is the distance traveled by A, and x is the distance traveled by B, we can find z, the distance between, from the expression: $z^2 = x^2+y^2$. Taking derivatives of this expression, we find that

$$2z\frac{dz}{dt} = 2x\frac{dx}{dt} + 2y\frac{dy}{dt}, \text{ or}$$

$$\frac{dz}{dt} = \frac{x\frac{dx}{dt} + y\frac{dy}{dt}}{z}$$

Since both boats are moving with constant velocities, the distance traveled by boat A after 2 hours is 6 x 2 = 12 miles, and that of B is 8 x 2 = 16 miles. The distance between them at the specified time is

$$z = \sqrt{(12)^2+(16)^2} = 20$$

Substituting in the differential equation, we can find $\frac{dz}{dt}$:

$$\frac{dz}{dt} = \frac{(12)(6) + (16)(8)}{20} = 10 \text{ mi/hr},$$

the rate of their separation.

• PROBLEM 492

Ship A sails north at 10 miles per hour and ship B, which is 12 miles south and 24 miles west of A, sails east at 18 miles per hour. What is the rate of change of the distance between the ships and how far does ship B travel before the distance between them begins to increase?

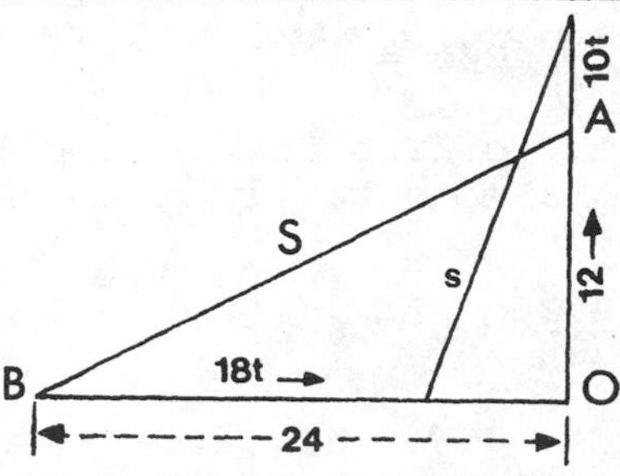

Solution: Let s = the distance between ships in the position after they have traveled for t hr. from their original locations.

At time $t = 0$, the relative distance between the ships is s_o, namely,

$$s_o = \sqrt{(24)^2+(12)^2} = \sqrt{720} = 26.8 \text{ miles.}$$

At some other time t, ship B is at a distance of $24 - 18t$ from origin 0, and ship A is at a distance $12 + 10t$ from 0. Thus the new distance is:

$$s = \sqrt{(24 - 18t)^2 + (12 + 10t)^2}$$

$$\text{or} \quad s = \sqrt{720 - 624t + 424t^2} .$$

For rate of change, $\dfrac{ds}{dt} = \dfrac{1 \cdot (-624 + 848t)}{2\sqrt{720 - 624t + 424t^2}}$

$$= \frac{-312 + 424t}{\sqrt{720 - 624t + 424t^2}} .$$

With the ships at their original position, $t = 0$, and

$$\frac{ds}{dt} = \frac{-312 + 424 \cdot 0}{\sqrt{720 - 624 \cdot 0 + 424 \cdot 0}}$$

$$= -\frac{312}{\sqrt{720}} = -\frac{312}{26.83}$$

$$\text{or} \quad \frac{ds}{dt} = -11.63$$

The distance between the ships is decreasing due to the negative sign of ds/dt, and they are approaching each other at 11.63 m.p.h. when they are at their original positions.

The distance between the ships will stop decreasing when $\dfrac{ds}{dt} = 0$, which means that

$$\frac{ds}{dt} = \frac{-312 + 424t}{\sqrt{720 - 624t + 424t^2}} = 0,$$

$$\text{or} \qquad -312 + 424t = 0.$$

Solving for t, to find the time at which the relative distance is momentarily constant, after which it increases:

$$t = \frac{312}{424} = 0.736 \text{ hr.}$$

In other words, 0.736 hr. after they have left their original positions, the distance between the ships begins to increase. Ship B has meanwhile traveled $0.736 \cdot 18 = 13.25$ miles before the distance begins to increase.

Car A travels east at 30 miles per hour while Car B is traveling north at 22.5 miles per hour. Both cars are approaching a junction O of two roads as indicated in the figure. (a) At what rate are the cars approaching each other at the instant when car A is 300 feet and car B is 400 feet from the junction? (b) What is the rate of change of the speed of one car with respect to the other?

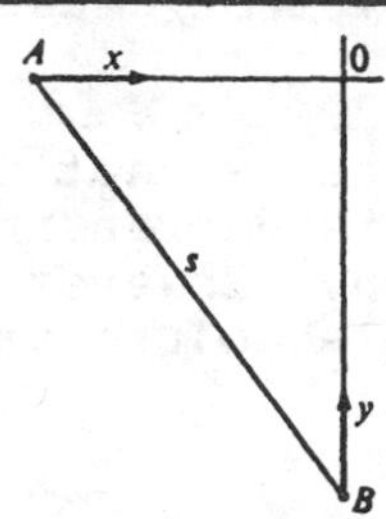

Solution: (a) Let s be the distance between the two cars. Then, ds/dt is the rate at which the cars are approaching each other. By the Pythagorean theorem,

(a) $$s^2 = x^2 + y^2,$$

as shown in the figure. Differentiation with respect to t gives

(b) $$s\frac{ds}{dt} = x\frac{dx}{dt} + y\frac{dy}{dt} \Rightarrow \frac{ds}{dt} = \frac{x\frac{dx}{dt} + y\frac{dy}{dt}}{s}$$

Where $$\frac{dx}{dt} = -30\frac{\text{miles}}{\text{hour}}, \text{ or}$$

$$\frac{dx}{dt} = -30\left(\frac{\text{miles}}{\text{hour}}\right)\frac{5280\,(\text{ft/mile})}{3600\,(\text{sec/hr})} = -44 \text{ ft/sec}$$

$$\frac{dy}{dt} = -22.5\frac{\text{miles}}{\text{hour}}, \quad \text{or } \frac{dy}{dt} = -33 \text{ ft/sec.}$$

The negative signs indicate that the distances are decreasing with time. Using equation (a) for x = 300 and y = 400, s = 500. Hence,

$$\frac{ds}{dt} = \frac{(300)(-44) + (400)(-33)}{500}$$

$$= -52.8 \text{ feet/sec}$$

Thus, s is also decreasing with **time.**

(b) By differentiating equation (b) once more, we can obtain the rate of change of speed of one car with respect to the other.

Differentiation of equation (b) with respect to time gives:

$$\left(\frac{ds}{dt}\right)^2 + s\frac{d^2s}{dt^2} = \left(\frac{dx}{dt}\right)^2 + x\frac{d^2x}{dt^2} + \left(\frac{dy}{dt}\right)^2 + y\frac{d^2y}{dt^2}.$$

But dx/dt and dy/dt are both constant in this problem, so that

$$d^2x/dt^2 = d^2y/dt^2 = 0.$$

Substituting $x = 300$, $y = 400$, $s = 500$, and $ds/dt = -52.8$, we have:

$$(-52.8)^2 + (500)\frac{d^2s}{dt^2}$$
$$= (-44)^2 + (-33)^2$$

and
$$\frac{d^2s}{dt^2} \approx -.47 \text{ foot/sec}^2.$$

The negative sign indicates that the speed of one car relative to the other is decreasing algebraically at the instant in question. However, since the relative speed is negative, the absolute value of the speed must be increasing.

• PROBLEM 494

A boat is being hauled toward a pier at a height of 20 ft. above the water level. The rope is drawn in at the rate of 6 ft./sec. Neglecting sag, how fast is the boat approaching the base of the pier when 25 ft. of rope remain to be pulled in?

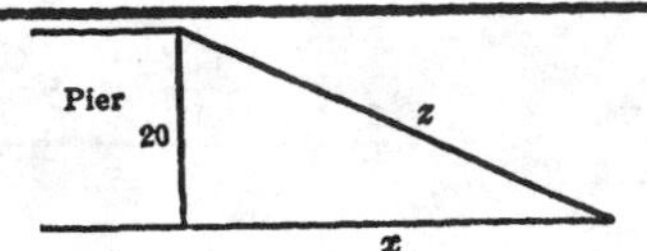

Solution: Formulating the given data, we have:

$$\frac{dz}{dt} = 6, \ z = 25, \text{ and } \frac{dx}{dt} \text{ is to be found.}$$

At any time t we have, from the Pythagorean theorem,

$$20^2 + x^2 = z^2$$

By differentiation, we obtain :

$$x\frac{dx}{dt} = z\frac{dz}{dt}$$

When $z = 25$, $x = \sqrt{25^2 - 20^2} = 15$; therefore

$$15\frac{dx}{dt} = 25\ (-6)$$

$$\frac{dx}{dt} = -10 \text{ ft./sec.}$$

(The boat approaches the base at 10 ft./sec.)

• PROBLEM 495

A rope attached to a boat is being pulled in at a rate of 10ft/sec. If the water is 20 ft. below the level at which the rope is being drawn in, how fast is the boat approaching the wharf when 36 ft of rope are yet to be pulled in?

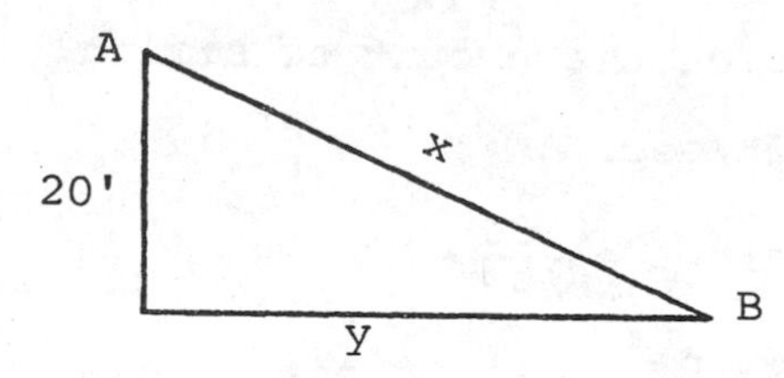

Solution: The length AB denotes the rope, and the position of the boat is at B. Since the rope is being drawn in at a rate of 10ft/sec.

$$\frac{dx}{dt} = 10.$$

To find how fast the boat is being towed in when 36 ft. of rope are left,

$$\frac{dy}{dt} \text{ must be found at } x = 36.$$

From the right triangle, $20^2 + y^2 = x^2$

or $y = \sqrt{x^2 - 400}$. Differentiating with respect to t,

$$\frac{dy}{dt} = \frac{dy}{dx} \cdot \frac{dx}{dt} = \frac{1}{2} (x^2 - 400)^{-1/2} (2x) \frac{dx}{dt}$$

$$= \frac{x \cdot \frac{dx}{dt}}{\sqrt{x^2 - 400}} .$$

Substituting the conditions that:

$$\frac{dx}{dt} = -10, \text{and } x = 36,$$

$$\frac{dy}{dt} = \frac{-360}{\sqrt{896}} = -\frac{45}{\sqrt{14}} .$$

It has now been found that, when there are 36 ft. of rope left, the boat is moving in at the rate of:

$$\frac{45}{\sqrt{14}} \text{ ft/sec.}$$

• PROBLEM 496

A boy flies a kite, which is 100 ft above the ground. If the string is pulled out at a rate of 10 ft/sec, because the wind carries the kite horizontally directly away from the boy, what is the rate of change of the angle of elevation of the kite when the angle is 30°?

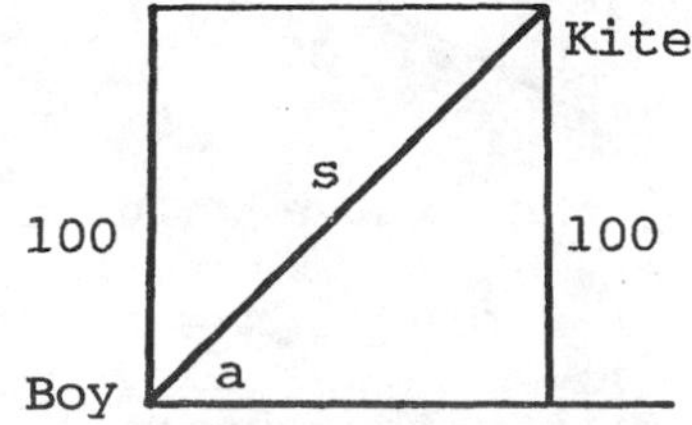

Solution: Let a be the angle of elevation and let

s be the amount of string.

$$\frac{ds}{dt} = 10 \text{ ft/sec.}$$

$$\sin a = \frac{100}{s}$$

We want to find $\frac{da}{dt}$ when $a = 30°$. Since $30° = \frac{\pi}{6}$ radians, when $a = \frac{\pi}{6}$ radians,

$$\sin a = 1/2$$

Therefore, $\frac{100}{s} = \frac{1}{2}$, $s = 200$.

Now, $\sin a = \frac{100}{s}$.

Differentiating with respect to t,

$$\cos a \frac{da}{dt} = -\frac{100}{s^2}\frac{ds}{dt}$$

Substituting the values: $\cos a = \frac{\sqrt{3}}{2}$, $s = 200$, and $ds/dt = 10$,

$$\frac{\sqrt{3}}{2}\frac{da}{dt} = \frac{-100}{200(200)}(10)$$

$$\frac{\sqrt{3}}{2}\frac{da}{dt} = -\frac{1}{40}$$

$$\frac{da}{dt} = -\frac{1}{20\sqrt{3}}$$

$$= \frac{-\sqrt{3}}{60} \text{ radians/sec}$$

• PROBLEM 497

A revolving light located 5 miles from a straight shore line has a constant angular velocity. With what velocity does the light revolve if the spot of light moves along the shore at the rate of 15 miles per minute when the beam makes an angle of 60° with the shore line?

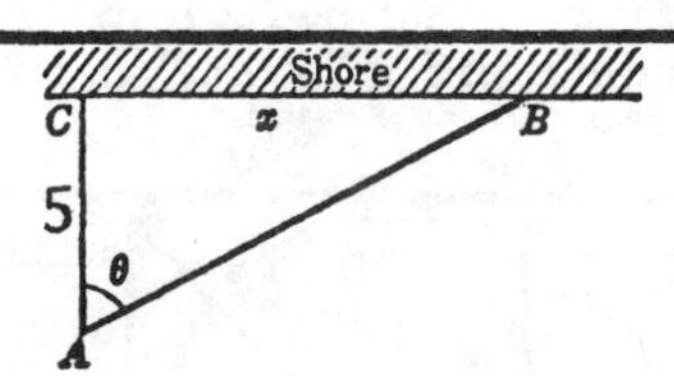

Solution: From the figure we have

$$\tan\theta = x/5 \qquad \text{or} \qquad x = 5\tan\theta.$$

Differentiation with respect to t gives:

$$\frac{dx}{dt} = 5\sec^2\theta\ \frac{d\theta}{dt}.$$

But $\frac{dx}{dt}$ = 15 miles per minute, and when angle ABC = 60°, θ = 30°.

Solving for $d\theta/dt$ and substituting these values, we obtain:

$$\frac{d\theta}{dt} = \frac{\cos^2\theta}{5}\frac{dx}{dt} = 3\cos^2\theta. \quad \text{But } \theta = 30°,$$

$$\cos\theta = \frac{\sqrt{3}}{2}. \quad \text{Therefore } \frac{d\theta}{dt} = 3\left(\frac{3}{4}\right) = 9/4 \text{ radians}$$

per minute, which is the required rate of revolution of the light.

• PROBLEM 498

A man lifts a bucket of cement to a scaffold 30 ft. above his hand by means of a rope passing over a pulley on the scaffold. If he holds his end of the rope at a constant height and walks away from beneath the pulley at 4 ft./sec., how fast is the bucket rising when he is 22 1/2 ft. away?

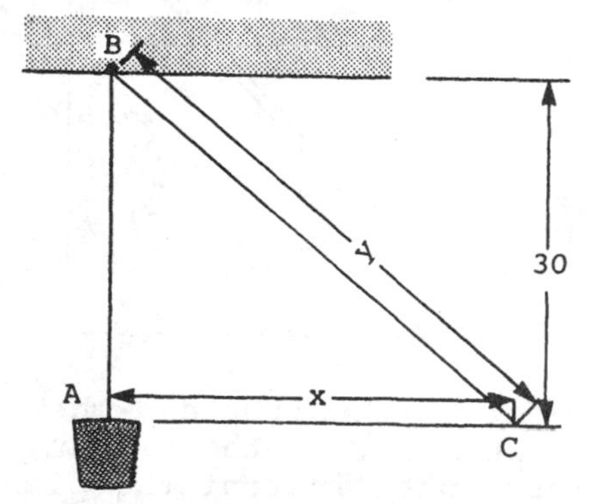

Solution: The bucket begins rising from point A and is being lifted by a rope passing over the pulley at point B, while the man is holding the rope at point C. Determining how fast the bucket is rising is the same as finding how fast the rope is moving, or finding $\frac{dy}{dt}$.

$\frac{dy}{dt}$ is to be calculated

at the point where x = 22 1/2 ft. while

$$\frac{dx}{dt} = 4 \text{ ft./sec.}$$

The right triangle gives:

$$x^2 + 30^2 = y^2$$

$$(x^2 + 900)^{1/2} = y$$

$$\frac{dy}{dt} = \frac{1}{2}(x^2 + 900)^{-1/2}(2x)\frac{dx}{dt}.$$

Substituting the given conditions for $\frac{dx}{dt}$ and x, we have:

$$\frac{dy}{dt} = \frac{(22.5)(4)}{\sqrt{(22.5)^2 + 900}} = \frac{90}{\sqrt{1406}} \ .$$

Therefore, the bucket is rising $\frac{90}{\sqrt{1406}}$ ft/sec. when the man is 22 1/2 ft. away.

• PROBLEM 499

A plane is flying west at 500 ft/sec at an altitude of 4000 ft. The plane is tracked by a searchlight on the ground. If the light is to be kept on the plane, find the change in the angle of the searchlight when the plane is due east of the searchlight at a horizontal distance of 2000 ft.

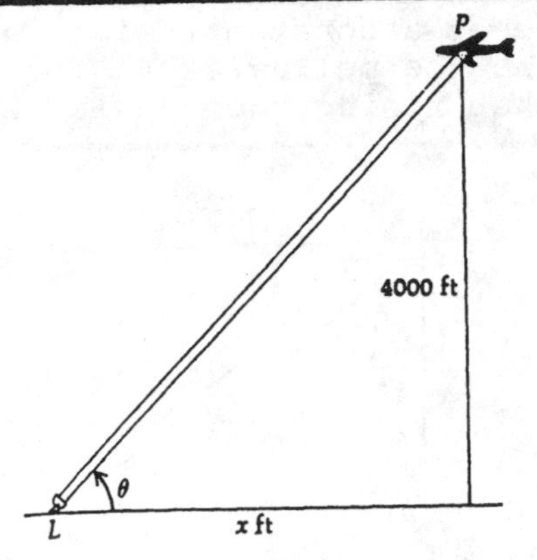

Solution: Refer to the figure. The searchlight is at point L, the origin, and at a particular instant the plane is at point P.

Let x = the number of feet due east in the horizontal distance of the plane from the searchlight at time t sec;

θ = the number of radians in the angle of elevation of the plane at the searchlight at time t sec.

We are given dx/dt = -500, where the negative sign indicates that x is decreasing with time. We wish to find dθ/dt when x = 2000.

$$\tan\theta = \frac{4000}{x}$$

Differentiating both sides of the above equation with respect to t, we obtain

$$\sec^2\theta \frac{d\theta}{dt} = -\frac{4000}{x^2}\frac{dx}{dt}.$$

Substituting dx/dt = -500 in the above, and dividing $\sec^2\theta$ gives

$$\frac{d\theta}{dt} = \frac{2,000,000}{x^2 \sec^2 \theta}$$

When x = 2000, tan θ = 2. Therefore, $\sec^2 \theta = 1 + \tan^2 \theta = 5$. Substituting these values in the above expression for dθ/dt, we have, when x=2000,

$$\frac{d\theta}{dt} = \frac{2,000,000}{4,000,000(5)} = \frac{1}{10}$$

We conclude that, at the given instant, the angle is increasing at the rate of 1/10 rad/sec.

• PROBLEM 500

A balloon is composed of a right circular cone joined at its base to a hemisphere, as shown in the diagram. The diameter of the base is equal to the height of the cone. If the balloon is inflated, at what rate is the volume v changing with respect to the height h.

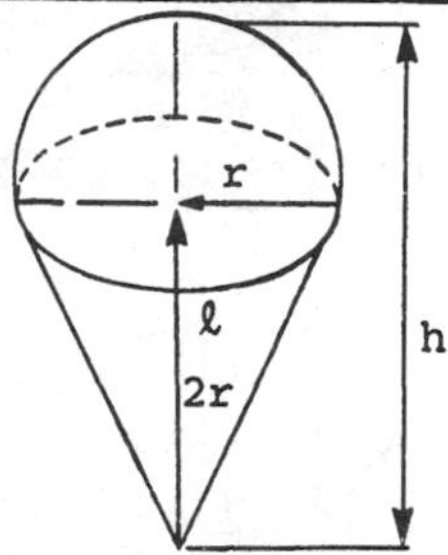

Solution: From the diagram, h = 3r, and we wish to find dv/dh. Hence we must express v in terms of h. The volume of the cone is $\frac{1}{3}\pi r^2 \cdot 2r$, and that of the hemisphere is $\frac{1}{2} \cdot \frac{4}{3}\pi r^3$. Adding these terms,

$$v = \frac{4}{3}\pi r^3 = \frac{4}{3}\pi \left(\frac{h}{3}\right)^3 = \frac{4}{81}\pi h^3.$$

Then $$\frac{dv}{dh} = \frac{4\pi h^2}{27}.$$

Thus, v is changing $4\pi h^2/27$ times as fast as h.

• PROBLEM 501

The lens in an optical tracking device has a focal length 6 inches. It forms the image of a moving object on a small screen. If the object is moving away at a speed of 20 feet per second, how fast must the lens be moving in order to keep the object in focus at the instant when the object is 50 feet away from the lens?

Solution: The simple lens equation is:

$$\frac{1}{u} + \frac{1}{v} = \frac{1}{f},$$

where u is the object distance, v the image distance, and f the focal length, all measured in the same units.

By differentiation with respect to t, we obtain from the lens equation:

$$-\frac{1}{u^2}\frac{du}{dt} - \frac{1}{v^2}\frac{dv}{dt} = 0, \text{ since f is a constant.}$$

or $$\frac{dv}{dt} = -\frac{v^2}{u^2}\frac{du}{dt}$$

With f = 1/2 (foot) and u = 50 (feet), the lens equation becomes

$$\frac{1}{50} + \frac{1}{v} = 2,$$

so that $$v = \frac{50}{99}.$$

Since du/dt = 20, we obtain:

$$\frac{dv}{dt} = -\left(\frac{50}{99}\right)^2 \left(\frac{1}{50}\right)^2 (20)$$

$$= -\frac{20}{9801} \text{ foot per second}$$

$$= -0.024 \text{ (approx). inch per second.}$$

Thus the lens must be moving toward the screen at the rate of approximately 0.024 inch per second.

• PROBLEM 502

A rocket is propelled upward from a launching pad 600 ft. away from an observation station. If the angle of elevation of a tracking instrument in the station is changing at the rate of 0.5 radians per second, what is the vertical speed of the rocket at the instant when the angle is 45°?

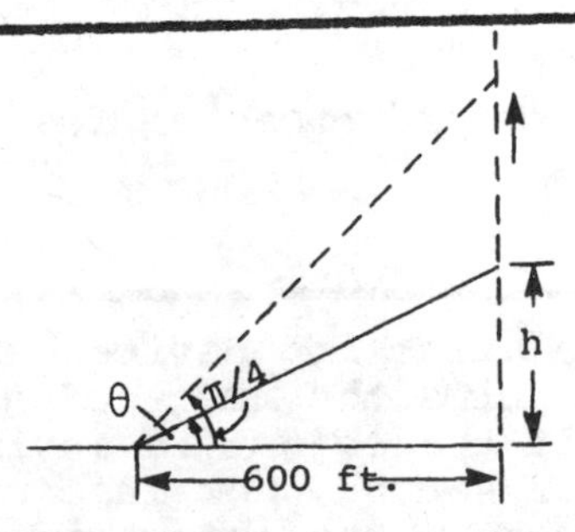

Solution: Let θ be the angle of elevation and h be the height the rocket moved upward after some time t. Then, from the triangle shown in the figure,

$$\frac{h}{600} = \tan\theta$$

or $\qquad h = 600\tan\theta$

Taking derivatives with respect to time, we have

$$\frac{dh}{dt} = 600\sec^2\theta\,\frac{d\theta}{dt},$$

and, for any time t and θ, $0 \le \theta < \pi/2$. From the given information, $\frac{d\theta}{dt} = 1/2$ rad./sec. Therefore, the speed of the rocket is:

$$\frac{dh}{dt} = 600\sec^2\theta\left(\frac{1}{2}\right) = 300\sec^2\theta .$$

At the instant when $\theta = \pi/4$ rad., the speed of the rocket is:

$$\frac{dh}{dt} = 300(\sqrt{2})^2 = 600 \text{ ft./sec.}$$

This problem may also be solved by recognizing that

$$\frac{d\theta}{dt} = \frac{d\theta}{dh}\cdot\frac{dh}{dt}$$

$$\theta = \arctan\frac{h}{600}, \quad \frac{d\theta}{dh} = \frac{1}{600}\left(\frac{1}{1+\frac{h^2}{(600)^2}}\right)$$

Therefore, $\frac{d\theta}{dt} = \frac{1}{600 + \frac{h^2}{600}}\,\frac{dh}{dt}$.

$$\frac{dh}{dt} = \left(600 + \frac{h^2}{600}\right)\frac{d\theta}{dt}$$

At $\theta = \pi/4$, $h = 600$,

and thus

$$\left.\frac{dh}{dt}\right|_{\theta=\pi/4} = \left(600 + \frac{(600)^2}{(600)}\right)\cdot\frac{1}{2} = 600 \text{ ft./sec.}$$

• PROBLEM 503

The theory of relativity states that the apparent mass m_a of an object moving at a velocity v is given by:

$$m_a = \frac{m_o}{\sqrt{1 - \left(\frac{v}{c}\right)^2}}$$

where m_o is the rest mass, v is the velocity of the object, and c is the speed of light. Find the expression for the rate of change of mass with respect to the velocity of the object.

<u>Solution:</u> The expression for the apparent mass

can be rewritten as

$$m_a = m_o\left(1 - \frac{v^2}{c^2}\right)^{-1/2}$$

This is of the form u^n, where $u = 1 - v^2/c^2$ and $n = -1/2$ and, therefore

$$\frac{dm_a}{dv} = -\frac{m_o}{2}\left(1 - \frac{v^2}{c^2}\right)^{-3/2} \cdot \frac{d}{dv}\left(1 - \frac{v^2}{c^2}\right)$$

$$= -\frac{m_o}{2}\left(1 - \frac{v^2}{c^2}\right)^{-3/2}\left(-\frac{2v}{c^2}\right)$$

$$\frac{dm_a}{dv} = \frac{m_o v}{c^2}\left(1 - \frac{v^2}{c^2}\right)^{-3/2} = \frac{m_o v}{c^2\left(1 - \frac{v^2}{c^2}\right)^{3/2}}$$

It can be seen that, as the velocity of the object approaches close to the speed of light, the apparent mass will increase indefinitely. The equation shows that the slope is almost constant at velocities less than 75% of c, because the factor outside the bracket dominates and is linear. For velocities of more than 95% of c, the slope changes very rapidly, and ends up as an asymptotic curve along a line parallel to the y-axis at v = c. For greater velocities, the expression is imaginary.

The main point of the equation is that the "relativistic" mass of any object increases if the object is accelerated sufficiently to attain a velocity close to the speed of light. Remember that, if the speed of light is expressed in meters/sec., i.e., $c \approx 3 \times 10^8$ m/sec., the mass is in kilograms.

• PROBLEM 504

At the start of a race the transverse distance between two cars is 12 feet. The two cars race parallel to each other along straight tracks with constant acceleration. If the acceleration of the first car is 10 feet per second per second, and the acceleration of the second is 8 feet per second per second, determine the rate of change of the distance between the cars 6 seconds after the start of the race.

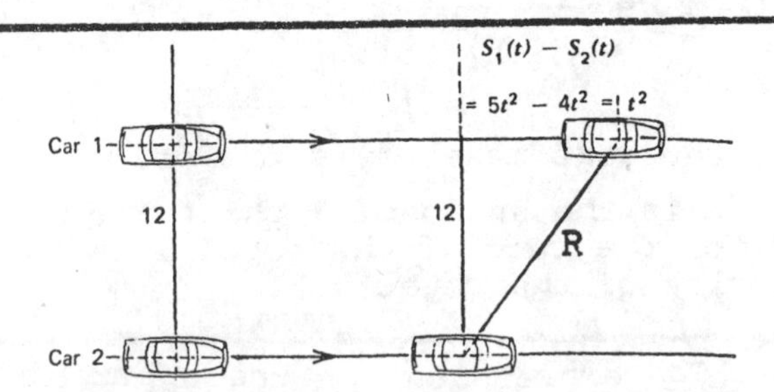

Solution: Let S_1 = distance traveled by the first car, S_2 = distance of the second.

The cars start from the left, where $S_1 = S_2 = t = 0$. Furthermore, the initial velocities are zero. The respective accelerations are:

$$a_1 = \frac{d^2s_1}{dt^2} = 10 \text{ ft/sec}^2 \quad \text{and}$$

$$a_2 = \frac{d^2s_2}{dt^2} = 8 \text{ ft/sec}^2.$$

Thus, the first car has a velocity

$V_1 = \int a_1 dt = 10t + c_1$, but $c_1 = 0$ since the initial velocity = 0

For the second car,

$V_2 = \int a_2 dt = 8\,t + c_2$, but $c_2 = 0$ since the initial velocity = 0

We can integrate further to find the distances they travel in t sec.

Hence,

$$s_1 = \int V_1 dt = \int 10t dt = 5t^2 + c_3, \text{ but}$$

$c_3 = 0$ since $s_1 = 0$ when $t = 0$.

And $s_2 = \int 8t dt = 4t^2 + c_4$, but $c_4 = 0$ since $s_2 = 0$ when $t = 0$.

Therefore, at time t, car 1 is leading and is $S_1 - S_2 = 5t^2 - 4t^2 = t^2$ ahead of car 2.

Letting R be the distance between the cars, and, using the Phythagorean theorem for the triangle shown in the figure,

$$R^2 = (12)^2 + (t^2)^2 \cdot R = \sqrt{144 + t^4}$$

Differentiating with respect to t and applying the chain rule, we obtain

$$\frac{dR}{dt} = \frac{1}{2}\left(\frac{1}{\sqrt{144+t^4}}\right)(4t^3) = \left(\frac{2t^3}{\sqrt{144+t^4}}\right)$$

and when t = 6 sec,

$$\frac{dR}{dt} = \frac{2(6)^3}{\sqrt{144+(6)^4}} = \frac{2(216)}{12\sqrt{1+\frac{64}{12^2}}} = \frac{36}{\sqrt{10}} \text{ ft/sec.}$$

• **PROBLEM 505**

In many cultures the rate of growth of the number of bacteria is proportional to the amount already present. If there are 1000 bacteria present initially in such a culture, and the amount doubles in 12 min, how long will it take before the number of bacteria = 1,000,000?

Solution: Let t = the number of minutes,
A = the number of bacteria present in t min.

The growth can be expressed mathematically as follows:

$$\frac{dA}{dt} = kA$$

To solve the differential equation we separate the variables,

$$\frac{dA}{A} = kdt,$$

and integration gives:

$$\ln A = kt + \ln C.$$

Taking the antilogarithm, we have

$$A = Ce^{kt}$$

When $t = 0$, $A = 1000$. Hence, $C = 1000$, which gives:

$$A = 1000e^{kt}$$

From the condition that $A = 2000$ when $t = 12$, we obtain:

$$e^{12k} = 2. \quad 12k\ln e = \ln 2$$

$$k = \frac{\ln 2}{12} = .05776.$$

Hence, we have:

$$A = 1000e^{0.05776t}$$

Replacing t by T and A by 1,000,000, we obtain:

$$1{,}000{,}000 = 1000e^{0.05776T}$$

$$e^{0.05776T} = 1000$$

$$0.05776T = \ln 1000$$

$$T = \frac{\ln 1000}{0.05776} = 119.6 \approx 120 \text{ min.} = 2 \text{ hrs.}$$

• **PROBLEM 506**

The cost in dollars of selling x items is given by $C = 500 + x + \frac{1}{x}$. When the 50th item is sold, it is noted that the rate of sales is 20 items per hour. What is the rate of change of the cost with respect to time at that moment?

Solution: The rate of sales is the change in the number of items, x, with time, which we can write as $\frac{dx}{dt}$. We are given that this $\frac{dx}{dt} = 20$. We are asked to find the rate of change of the cost with respect to time, or $\frac{dC}{dt}$, at the moment at which $x = 50$.

$C = 500 + x + \frac{1}{x}$. Therefore,

$$\frac{dC}{dt} = \frac{d}{dt}(500) + \frac{d}{dt}(x) + \frac{d}{dt}\left(\frac{1}{x}\right)$$

$$= 0 + \frac{dx}{dt} + \left(-\frac{1}{x^2}\right)\frac{dx}{dt}$$

$$= \frac{dx}{dt} - \frac{1}{x^2}\frac{dx}{dt}$$

$$= \left(1 - \frac{1}{x^2}\right)\frac{dx}{dt}.$$

Substituting 50 for x and 20 for $\frac{dx}{dt}$ gives:

$$\frac{dC}{dt} = \left(1 - \frac{1}{50^2}\right)20$$

$$= \left(1 - \frac{1}{2,500}\right)20$$

$$= \frac{2,499}{2,500}(20)$$

$$= \frac{2,499}{125} \approx \$20/\text{hr}.$$

CHAPTER 20

DIFFERENTIALS

A differential is a change, which the dependent variable (y) experiences when the independent variable (x) undergoes an infinitesimal change. It is also another way of looking at the derivative. If

$$y = f(x),$$

$$\frac{dy}{dx} = f'(x).$$

$$dy = f'(x)dx.$$

This is the differential form, which is seen here to be merely the derivative, with the two sides multiplied by dx.

However, when we have more than two variables, the matter is somewhat more complex. For example, with three variables, x, y and z, the differential of z,

$$dz = \frac{\partial z}{\partial x}\,dx + \frac{\partial z}{\partial y}\,dy .$$

Here the symbol, $\frac{\partial z}{\partial x}$, is the partial derivative of z with respect to x, i.e., the derivative of z with respect to x while considering y as a constant. The partial derivative of z with respect to y is defined similarly, and the concept can be extended to any number of variables.

If we wish to find the change in a dependent variable where the change in the independent variables is small but not infinitesimal, then the same equation can be applied as an approximation, i.e.:

$$\Delta z = \frac{\partial z}{\partial x}\,\Delta x + \frac{\partial z}{\partial y}\,\Delta y ,$$

where the Δ's are small but not infinitesimal changes. This approach of approximating dz by Δz is very useful in the numerical computation of changes in quantities.

• PROBLEM 507

If $y = x^2$, find dy.

Solution: By definition, the differential, $dy = f'(x)dx = \left(\frac{dy}{dx}\right)dx$. We find $f'(x) = 2x$. Therefore,

$$dy = 2x\ dx .$$

• PROBLEM 508

If $y = 3x^4 - 2x^2 + 3x + 5$, find dy.

Solution: The definition for the differential states that

$$dy = \left(\frac{dy}{dx}\right)dx .$$

We find $\frac{dy}{dx} = 12x^3 - 4x + 3$. Therefore $dy = (12x^3 - 4x + 3)dx$.

• PROBLEM 509

Differentiate: $\frac{2}{3} x^{3/2}$.

Solution:

$$d\left(\frac{2}{3} x^{3/2}\right) = \frac{2}{3} \cdot d\left(x^{3/2}\right),$$

$$d\left(x^{3/2}\right) = \frac{3}{2} x^{3/2-1} dx = \frac{3}{2} x^{1/2} dx = \frac{3}{2} \sqrt{x}\, dx .$$

$$d\left(\frac{2}{3}x^{3/2}\right) = \frac{2}{3}\cdot\frac{3}{2} \sqrt{x}\, dx = \sqrt{x}\, dx .$$

• PROBLEM 510

Find the differential of $(x^2 + 2)^3$.

Solution: This is a variable, $(x^2 + 2)$, raised to the power 3.

$$\begin{aligned} d\left[(x^2 + 2)^3\right] &= 3(x^2 + 2)^{3-1} \cdot d(x^2 + 2) \\ &= 3(x^2 + 2)^2 \cdot d(x^2) \\ &= 3(x^2 + 2)^2 \cdot 2x\, dx \\ &= 6x(x^2 + 2)^2\, dx . \end{aligned}$$

• PROBLEM 511

Find $d\left(2x^3 + 3\sqrt{x} - \frac{3}{2} x^2\right)$.

Solution: This is equal to $d(2x^3) + d(3\sqrt{x}) - d\left(\frac{3}{2} x^2\right)$.

$$d(2x^3) = 2d(x^3) = 2(3x^2dx) = 6x^2dx.$$

$$d(3\sqrt{x}) = 3d(\sqrt{x}) = 3\left(\frac{1}{2\sqrt{x}}\, dx\right) = \frac{3}{2}\frac{dx}{\sqrt{x}} .$$

$$d\left(\frac{3}{2} x^2\right) = \frac{3}{2}d(x^2) = \frac{3}{2}(2x\, dx) = 3x\, dx.$$

Therefore the required differential is:

$$6x^2dx + \frac{3}{2}\frac{dx}{\sqrt{x}} - 3xdx = 3\left(2x^2 + \frac{1}{2\sqrt{x}} - x\right)dx .$$

• PROBLEM 512

Find dy if $y = (x - x^2)(2 - 2x - x^2)^{1/2}$.

Solution: By definition, $dy = f'(x)dx = \left(\frac{dy}{dx}\right)dx$. Therefore, we find $f'(x)$. We do this by applying the product rule. We find:

$$f'(x) = \left[(x-x^2)\tfrac{1}{2}(2-2x-x^2)^{-\frac{1}{2}}(-2-2x)\right] + \left[(2-2x-x^2)^{\frac{1}{2}}(1-2x)\right]$$

$$= \left[\frac{-(x-x^2)(1+x)}{(2-2x-x^2)^{\frac{1}{2}}} + (2-2x-x^2)^{\frac{1}{2}}(1-2x) \right]$$

$$= \frac{-(x-x^2)(1+x) + (2-2x-x^2)(1-2x)}{(2-2x-x^2)^{\frac{1}{2}}}$$

$$= (2-7x+3x^2+3x^3)(2-2x-x^2)^{-\frac{1}{2}} .$$

Therefore, $dy = (2-7x + 3x^2 + 3x^3)(2-2x-x^2)^{-\frac{1}{2}} \, dx$.

• PROBLEM 513

If $y = x^{3/2}$, what is the approximate change in y when x changes from 9 to 9.01?

Solution: The differential, dy, is found from:

$$\frac{dy}{dx} = \frac{3}{2} x^{\frac{1}{2}} .$$

$$dy = \frac{3}{2} x^{\frac{1}{2}} \, dx .$$

The numerical value of dy may be found by letting $x = 9.00$ and $dx = 0.01$ in this equation. Thus

$$dy = \frac{3}{2}(9.00)^{\frac{1}{2}}(0.01) = 0.045$$

The exact change in y may be found by subtracting $(9.00)^{3/2}$ from $(9.01)^{3/2}$. However, this is a more laborious task than finding dy. The exact value is 0.0444.

• PROBLEM 514

If $y = f(x) = x^2 + 3$, find Δy when $x = 2$ and $\Delta x = 0.5$.

Solution: We first derive a formula for Δy, enabling us to solve this problem. $\Delta x = x_2 - x_1$ and $\Delta y = y_2 - y_1$. If we let $y = f(x)$, then $y_1 = f(x_1)$, $y_2 = f(x_2)$, and $\Delta y = \Delta f(x) = f(x_2) - f(x_1)$. We can solve for x_2 from the equation: $\Delta x = x_2 - x_1$, to obtain: $x_2 = \Delta x + x_1$. Substituting this value for x_2 into $\Delta y = f(x_2) - f(x_1)$, we obtain:

$$\Delta y = f(\Delta x + x_1) - f(x_1) .$$

Since x_1 can be any point in the domain of f, the subscript is superfluous, and the formula for Δy becomes

$$\Delta y = f(x + \Delta x) - f(x).$$

We can now use this formula to solve the problem. Substituting the values $x = 2$ and $\Delta x = .5$ enables us to find Δy in terms of a difference in two functions of f. We can then use the term $f(x) = x^2 + 3$ to find the numerical value of Δy.

$$\Delta y = f(2 + 0.5) - f(2)$$
$$= f(2.5) - f(2).$$

$f(x) = x^2 + 3$. Hence, $f(2.5) = \left[(2.5)^2 + 3\right]$ and $f(2) = (2^2 + 3)$. Therefore,

$$\Delta y = \left[(2.5)^2 + 3\right] - (2^2 + 3)$$
$$= 6.25 + 3 - 4 - 3$$
$$= 2.25.$$

In this example, a 0.5 change in x results in a 2.25 change in y.

• PROBLEM 515

Find an approximation to the value of $y = x^3 - 3x^2 + 2x - 1$ when $x = 1.998$, by using differentials.

Solution: We can consider the value 1.998 as the result of applying an increment of -0.002 to an original value of 2. Then $x = 2$ and $\Delta x = -0.002 \approx dx$. Now, $\frac{dy}{dx} = 3x^2 - 6x + 2$, and $dy = (3x^2 - 6x + 2)dx$. We substitute $x = 2$ and $dx = -0.002$, obtaining:

$$dy = \left[3\cdot(2)^2 - 6\cdot 2 + 2\right](-0.002) = -0.004,$$

which is approximately the change in y caused by going from $x = 2$ to $x = 1.998$.

We must now find the value of y when $x = 2$ and add this value to dy.

When $x = 2$ in the original equation,

$$y = 2^3 - 3\cdot 2^2 + 2\cdot 2 - 1 = -1.$$

Therefore, y at $x = 1.998 = y + dy = -1 - 0.004 = -1.004$, which is approximately the value of the polynomial for $x = 1.998$. Often such approximations save much labor. However, to check the closeness of this method, we substitute 1.998 in the original function to get the exact value for this point, and obtain: $y + \Delta y = -1.003988008$. This shows that the method of approximations by differentials is very exact.

• PROBLEM 516

What is the increment in y, (Δy) and the differential of y, (dy) for an increment of $\Delta x = 1$ at $x = 2$, if $y = x^2$? For an increment $\Delta x = 0.01$?

Solution: If $y = x^2$, $x = 2$, and $\Delta x = 1$, then

$$y + \Delta y = (x + \Delta x)^2,$$

so that $\Delta y = (x + \Delta x)^2 - y$, or,

$$\Delta y = (x + \Delta x)^2 - x^2 = (2 + 1)^2 - (2)^2 = 9 - 4 = 5.$$

Therefore the increment in y is 5. To obtain the differential of y, we differentiate and find: $dy = 2x\,dx = (2)(2)(1) = 4$.

If $y = x^2$, $x = 2$, and $\Delta x = 0.01$, then

$$\Delta y = (x + \Delta x)^2 - x^2 = (2.01)^2 - 2^2 = 0.0401 \quad \text{, the increment,}$$

and

$$dy = 2x\,dx = (2)(2)(0.01) = 0.0400 \quad \text{, the differential.}$$

In the second case, Δy is very close to dy, but in the first case the difference is very substantial, so that the relationship: $\Delta y \approx dy$, holds only when $\frac{\Delta x}{x}$ is small.

• PROBLEM 517

The side of a square, x, is increased by an amount $dx = \Delta x$. See accompanying figure. Analyze and interpret the increases in the area in terms of the increment of the area and the differential.

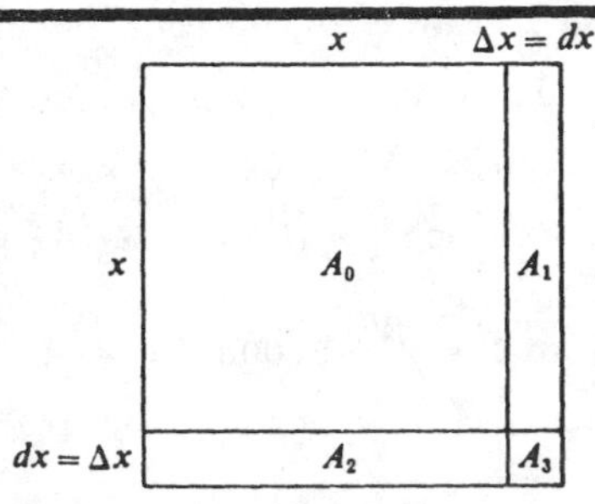

Solution: For the original area, we have $A_0 = x^2$. The enlarged area is:

$$A = (x + \Delta x)^2 = x^2 + 2x\Delta x + (\Delta x)^2 .$$

The term x^2 is, of course, A_0. The term $2x\Delta x$ corresponds to the two areas A_1 and A_2. The term $(\Delta x)^2$ corresponds to A_3. The increment, $A_1 + A_2 + A_3$, is equivalent to $2x\Delta x + (\Delta x)^2$; and the differential of the area, $dA = 2xdx$, corresponds to $A_1 + A_2$.

• PROBLEM 518

A is the area of a square of side x. Find ΔA, dA, and $\Delta A - dA$ for any x and Δx, and for $x = 5$ and $\Delta x = 0.0002$ in.

Solution: The area A, of a square of side x is given by $A = x^2$. Then ΔA is the difference between the areas of a square with side $= (x + \Delta x)$ and one with side $= x$. Therefore,

$$\begin{aligned} \Delta A &= (x + \Delta x)^2 - x^2 \\ &= x^2 + 2x\Delta x + (\Delta x)^2 - x^2 \\ &= 2x\Delta x + (\Delta x)^2 . \end{aligned}$$

We wish to find dA. From $A = x^2$, we have $\frac{dA}{dx} = 2x$, therefore

$dA = 2xdx$.

To obtain $\Delta A - dA$, we use the values we have found for ΔA and dA. Doing this we have:

$$\Delta A - dA = 2x\Delta x + (\Delta x)^2 - 2xdx.$$

If we let $\Delta x = dx$, then,

$$\Delta A - dA = (\Delta x)^2.$$

We now find ΔA, dA and $\Delta A - dA$ for $x = 5$ and $\Delta x = 0.0002$ in. Keeping in mind that $\Delta x = dx$, we find, by substitution:

$$\Delta A = 2 \cdot 5 \cdot 0.0002 + (0.0002)^2 = 0.00200004$$

$$dA = 2 \cdot 5 \cdot 0.0002 = 0.002$$

$$\Delta A - dA = 0.00200004 - 0.002 = 0.00000004$$

Because Δx is so small, dA closely approximates ΔA.

• PROBLEM 519

A circular metal plate is to be increased in size by applying heat. What is the resultant change in area when the diameter is increased from 15.00" to 15.14"?

Solution: The function is: $A = \pi r^2$. We wish to find dA when $r = 7.5$ in. and $dr = 0.07$ in., the difference between the initial radius and the increased radius. We find:

$$dA = 2\pi r \cdot dr,$$

and, by substitution,

$$\begin{aligned} dA &= 15\,\pi(0.07) \\ &= 1.05\,\pi \text{ sq. in. approximately.} \end{aligned}$$

We note in this example that dA may be interpreted as a rectangular strip of length $2\pi r$ and width dr, whereas ΔA is an annular ring of thickness Δr bordering a circle of radius r.

• PROBLEM 520

A new spherical ball bearing has a 3.00 inch radius, r. What is the approximate volume of metal lost after it wears down to $r = 2.98$ inches?

Solution: We can solve this problem by means of differentials. The equation for volume of a sphere is given by:

$$V = \frac{4}{3}\pi r^3 .$$

Then,

$$dV = 4\pi r^2 dr.$$

Since $dV \approx \Delta V$, the exact amount of metal lost, we wish to find the value of dV. To do this, we substitute the values for r and dr into the equation. We have, $r = 3$, the initial radius, and $dr = .02$, the difference between the initial and final radii. By substitution,

$$\begin{aligned} dV &= 4\pi(3^2)(.02) = .72\pi \\ &= 2.26 \text{ cu. in.} \end{aligned}$$

If we assume the values: $r = 3.00"$ and $r = 2.98"$, to be exact, then the exact answer is:

$$\Delta V = \frac{4}{3}\pi\left[3^3 - (2.98)^3\right]$$
$$= (.71521066...)\pi$$
$$= 2.26 \text{ cu. in. to two decimal places.}$$

• PROBLEM 521

Find the approximate volume of a spherical shell of outside diameter 12 inches and thickness $\frac{3}{32}$ inch.

Solution: V = volume of sphere = $\frac{1}{6}\pi x^3$ (x = diameter). The exact volume of the shell = ΔV, which is the difference between the volume of a sphere of diameter = 12 in. and the volume of a sphere of diameter $12'' - \frac{6''}{32}$ or $11\frac{13}{16}$ in. (Two thicknesses must be substracted.)

For an approximate value of ΔV, find dV, since $\Delta V \approx dV$.

$$\frac{dV}{dx} = \frac{\pi x^2}{2} \quad \text{and} \quad dV = \frac{\pi x^2}{2} \cdot dx$$

We wish to find dV, therefore we must substitute for x and dx. We use $x = 12$ and $dx = -\frac{3}{16}$, the difference between 12 and $11\frac{13}{16}$. We have a negative sign because the diameter, x, is decreasing. By substitution,

$$dV = \frac{\pi}{2} \cdot 12^2 \cdot \left(-\frac{3}{16}\right) = -42.4116 \text{ cu. in.} = 42.41 \text{ cu.in.}$$

where the sign is dropped, since it merely means that V decreases as x decreases.

The exact value of ΔV is

$$\frac{1}{6}\pi\left[12^3 - \left(11\frac{13}{16}\right)^3\right] = 41.75237 = 41.75 \text{ cu. in.}$$

To appreciate how close the approximation is, we should keep in mind that the error of .66 is made on a total volume of $\frac{\pi}{6}12^3 \approx 905$ cu. in.

• PROBLEM 522

A balloon has a diameter of 4 ft. at sea level. After rising a certain distance in the atmosphere it swells to 4 feet, 2 inches in diameter. What is (a) the exact, (b) the approximate change in volume of the enclosed gas?

Solution: (a) the volume of a sphere is given by the equation: $V = \frac{4}{3}\pi r^3$, where r is the radius of the sphere. Therefore, at sea level, with r = 2 ft. = 24 inches, $V = \frac{4}{3}\pi(24)^3$ cubic inches, and at the higher altitude it is $\frac{4}{3}(25)^3$ cubic inches. The exact increase in volume is

$$\frac{4}{3}\pi(25)^3 - \frac{4}{3}\pi(24)^3 = \frac{4}{3}\pi\left[(25)^3 - (24)^3\right]$$

$$= \frac{4}{3}\pi(1801)$$

$$= 7540 \text{ cubic inches}$$

(b) The approximate increase in volume can be obtained by finding the differential of the volume:

$$V = \frac{4}{3}\pi r^3 \quad .$$

$$dv = 4\pi r^2 dr \quad .$$

For $r = 24$ inches, $dr = 1$ inch, $dv = 4\pi(24)^2(1) = 7235$ cubic inches. The error of 305 cu. in. is small when we consider the total volume involved = 65,450 cu. in., i.e., it is about 0.5%.

• PROBLEM 523

The expression for the horsepower of an engine is: $P = 0.4nx^2$, where n = number of cylinders and x = bore of cylinders. Determine the power differential added when a four-cylinder Volkswagen has the cylinders rebored from 3.250 in. to 3.265 in.

Solution: $P = 0.4nx^2$.

We wish to find $\frac{dp}{dx}$ because x is changing in this example.

We find:

$$\frac{dP}{dx} = 0.8nx,$$

and

$$dP = 0.8nx\, dx.$$

Here $dx = 0.015$, since x changes from 3.250 to 3.265, $n = 4$, and $x = 3.250$ or $\frac{13}{4}$, the initial value of x.

Substituting these values, we have:

$$dP = 0.8(4)\left(\frac{13}{4}\right)(0.015) = 0.156 \text{ hp.}$$

• PROBLEM 524

A surveying instrument is placed at a point 180 ft. from the base of a building on level ground. The angle of elevation of the top of the building is 30°, as measured by the instrument. What would be the error in the height of the building due to an error of 15' in this measured angle?

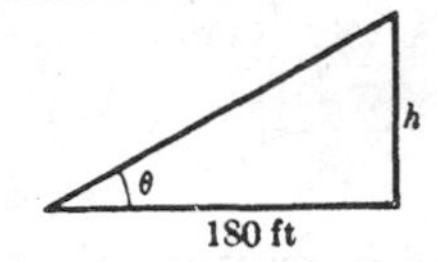

Solution: From the diagram, $h = 180 \tan\theta$. Finding the differential of h, we have

$$dh = 180 \sec^2\theta \, d\theta .$$

Since $\theta = 30°$, $\sec\theta = 2/\sqrt{3}$, or $\sec^2\theta = 4/3$. The possible error in

θ is 15', or 0.25°, which is 0.00436 rad. This is the value we employ for $d\theta$ in the calculation. Thus,

$$dh = (180)\left(\frac{4}{3}\right)(0.00436) = 1.05 \text{ ft.}$$

• PROBLEM 525

Find the approximate maximum error in the area of a circle if the diameter is measured as 10 inches, with a maximum error of 0.01 inch.

Solution: The area of a circle is given as: $A = \frac{1}{4}\pi x^2$, where x = diameter. The approximate error is dA.

$$\frac{dA}{dx} = \frac{1}{2}\pi x \quad \text{or} \quad dA = \frac{\pi x}{2} \cdot dx.$$

We wish to obtain the value of dA. Therefore we substitute for x and dx. We are given $x = 10$ and $dx = .01$. Hence,

$$dA = \frac{\pi}{2} \cdot 10 \cdot 0.01 = 0.157 \text{ sq. in.}$$

The exact maximum error in A is the change (ΔA) in its value when x changes from 10 to 10.01 in., which is determined to be 0.1572 sq. in. Observe that $dA \approx \Delta A$.

• PROBLEM 526

The diameter of a circle is found by measurement to be 5.2 in., with a maximum error of 0.05 in. Find the approximate maximum error in the area when calculated by the formula:

$$A = \frac{1}{4}\pi x^2,$$

where x is the diameter.

Solution: To find the approximate maximum error in the area we use differentials. We find:

$$\frac{dA}{dx} = \tfrac{1}{2}\pi x \quad \text{and,} \quad dA = \tfrac{1}{2}\pi x\, dx.$$

Since $dA \approx \Delta A$, the value we obtain for dA will be the approximate maximum error in area. The given diameter, $x = 5.2$, and $dx = .05$. By substitution, we have:

$$dA = \tfrac{1}{2} \times (3.14) \times (5.2) \times (.05) = 0.41 \text{ sq. in.}$$

Therefore, .41 sq. in. is the approximate maximum error in the area.

The exact maximum error in A is the change (ΔA) in its value when x changes from 5.2 in. to 5.25 in.

• PROBLEM 527

Find the error obtained in computing the volume of a cube if a 1% error is made in measuring the length of an edge.

Solution: The volume, V, of a cube is given as:

$$V = x^3 .$$

Then,

$$dV = 3x^2 dx.$$

We are given that the percentage error made in measuring the length of an edge = 1%. Percentage error is defined to be the relative error × 100. In this case, relative error is $\frac{dx}{x}$, therefore we have, $100\frac{dx}{x} = 1$, or $\frac{dx}{x} = .01$. We modify dV to read

$$dV = 3x^3 \frac{dx}{x}$$

by multiplying and dividing by x. Substituting, we obtain:

$$dV = 3x^3(.01) = .03x^3,$$

which depends, necessarily, upon x. Therefore, the error in volume is $.03x^3$. To obtain the percentage error made in computing the volume we find the relative error, $\frac{dV}{V}$ and multiply by 100. We use $dV = 3x^2 dx$ and $V = x^3$ and find:

$$\frac{dV}{V} = \frac{3x^2}{x^3} dx$$

$$= 3\frac{dx}{x}$$

$$= .03 = 3\%.$$

Hence, a 3% error in the calculation of the volume is obtained if a 1% error is made in the measurement of an edge.

• PROBLEM 528

The surface area of a sphere is to be computed from a measurement of the radius of the sphere. In measuring the radius as 3'', a possible error of ± 0.03'' might have been incurred. Compute the possible percentage error in the surface area due to the possible error in measurement.

Solution: The surface area S is given by $S = 4\pi r^2$, and for $r = 3$, the area is $36\pi \text{ in}^2$. The error is approximated by $dS = S'(r)dr = \frac{dS}{dr} = 8\pi r dr$. The quantity $dr = \pm 0.03$, and therefore the error is: $dS = 8\pi(3)(\pm 0.03) = \pm 0.72\pi \text{ in}^2$. The proportional error is:

$$\frac{dS}{S} = \pm \frac{0.72\pi}{36\pi} = \pm 0.02,$$

and the percentage error is 2%.

• PROBLEM 529

Evaluate, using differentials:
(a) the cube root of 0.124
(b) the cube root of 4100.

Solution: (a) We let $y = x^{\frac{1}{3}}$, with x = the closest known cube to .124. We know that $.5^3 = .125$, therefore let $x = .125$. Then $\Delta x = .124 - .125 = -.001$. This value, $\Delta x \simeq dx$. Differentiating the given expression, we find:

$$dy = \frac{1}{3}x^{-\frac{1}{3}}dx = \frac{dx}{3x^{\frac{1}{3}}}.$$

Substituting for dx and x we obtain:

$$dy = -\frac{.001}{3(.125)^{\frac{1}{3}}} = -\frac{.001}{3(.5)} = -\frac{1}{1500} = -.00067.$$

Now, since $dy \simeq \Delta y$, we write:

$$\sqrt[3]{.124} \simeq y + dy = .5 - .0067 = .49933.$$

(b) We let $y = x^{\frac{1}{3}}$, with x = the closest known cube to 4100. We know that $16^3 = 4096$, therefore let $x = 4096$. Then $\Delta x = 4100 - 4096 = 4$. This value, $\Delta x \simeq dx$. We have:

$$dy = \frac{1}{3}x^{-\frac{1}{3}}dx = \frac{dx}{3x^{\frac{1}{3}}},$$

and, substituting for dx and x, we obtain:

$$dy = \frac{4}{3(4096)^{\frac{1}{3}}} = \frac{4}{3(16)} = \frac{1}{12} = .0833.$$

Now, since $dy \simeq \Delta y$, we write:

$$\sqrt[3]{4100} \simeq y + dy = 16 + .0833 = 16.0833.$$

• PROBLEM 530

Calculate $\tan 46^\circ$, approximately, using differentials, given $\tan 45^\circ = 1$, $\sec 45^\circ = \sqrt{2}$, $1^\circ = 0.01745$ radians.

Solution: Let $y = \tan x$. Then, $dy = \sec^2 x\, dx$. When x changes to $x + dx$, y will change to $y + dy$, approximately. We now substitute into the equation; $x = 45^\circ = \frac{\pi}{4}$, and $dx = 0.0175$, the change in x from 45° to 46°, obtaining: $dy = \sec^2 \frac{\pi}{4} \times (0.0175)$

$$= 2 \times .0175$$
$$= .0350 .$$

Since $y = \tan 45^\circ = 1$, $y + dy = 1.0350 = \tan 46^\circ$, approximately. (Four-place tables give $\tan 46^\circ = 1.0355$).

If $xy + x^3 = y - 3x/y$ evaluate dy for $x = 1$, and $dx = 0.03$.

Solution: The differential of such an implicit function can be written as follows:

$$xdy + ydx + 3x^2dx = dy - 3\frac{ydx - xdy}{y^2} .$$

When $x = 1$, $y = -3$, by substituting for x in the original equation, and $dx = 0.03$. Substituting, we have:

$$dy - 3(0.03) + 3(0.03) = dy - 3\left[(-3)(0.03) - dy\right]/9,$$

or

$$dy = -0.09 \text{ units}.$$

CHAPTER 21

PARTIAL DERIVATIVES

When a function is expressed in terms of several variables rather than in terms of only one variable, the concept of the partial derivative is usually applicable. If, for example, z is a function of x and y, i.e., $z = f(x,y)$, then

$\frac{\partial z}{\partial x}$ is the derivative of z with respect to x, with y treated as a constant, and

$\frac{\partial z}{\partial y}$ is the derivative of z with respect to y, with x treated as a constant. Hence, other than treating the variables one at a time, while the others are considered constant for the moment, there is no fundamental difference between the partial derivative and the total derivative.

In considering the total differential, however, an additional concept is involved. Where, for one variable, we have: $dy = \left(\frac{dy}{dx}\right) dx$, for two variables, the differential must take account of both. Thus,

$$dz = \frac{\partial z}{\partial x} dx + \frac{\partial z}{\partial y} dy .$$

The total differential of z is the sum of the partial derivatives, each multiplied by its proper differential. The concept can be extended from two variables to any number.

• **PROBLEM** 532

If $z = 3x^2 - 2y^2 + 2xy$, find $\frac{\partial z}{\partial x}$ and $\frac{\partial z}{\partial y}$.

Solution: When solving for the partial derivative, $\frac{\partial z}{\partial x}$ we differentiate with respect to x, treating y as a constant. We obtain:

$$\frac{\partial z}{\partial x} = 6x + 2y.$$

Solving for $\frac{\partial z}{\partial y}$ we differentiate with respect to y, treating x as a constant to find:

$$\frac{\partial z}{\partial y} = 2x - 4y.$$

• **PROBLEM 533**

Solve the following equation for its first partial derivatives: $u = x^2 + xy + y^2$.

Solution: When finding the partial derivatives, we first differentiate the equation with respect to x, treating y as a constant. Then we differentiate with respect to y, treating x as a constant.

$$\frac{\partial u}{\partial x} = 2x + y. \quad \frac{\partial u}{\partial y} = x + 2y.$$

• **PROBLEM 534**

Find $\frac{\partial y}{\partial x}$ and $\frac{\partial y}{\partial z}$ when $xy = 5z$.

Solution: Solving for y, we find $y = \frac{5z}{x}$. To find $\frac{\partial y}{\partial x}$ we differentiate with respect to x, treating z as a constant.

$$\frac{\partial y}{\partial x} = -\frac{5z}{x^2}.$$

To find $\frac{\partial y}{\partial z}$ we differentiate with respect to z, treating x as a constant.

$$\frac{\partial y}{\partial z} = \frac{5}{x}.$$

• **PROBLEM 535**

Find Z_{xy} and Z_{yx} from the expression: $z = x^2y + 2xe^{1/y}$ and show that $Z_{xy} = Z_{yx}$.

Solution: The first step in finding second partial derivatives is finding the first partial derivatives.

$$Z_x = 2xy + 2e^{1/y}$$

and

$$Z_y = x^2 - \frac{2xe^{1/y}}{y^2}.$$

To find Z_{xy}, we differentiate Z_y with respect to x.

$$Z_{xy} = 2x - \frac{2e^{1/y}}{y^2}.$$

To find Z_{yx} we differentiate Z_x with respect to y, and obtain:

$$Z_{yx} = 2x - \frac{2e^{1/y}}{y^2}.$$

Therefore $Z_{yx} = Z_{xy}$.

• PROBLEM 536

Find: $\frac{\partial^2 y}{\partial x^2}$, $\frac{\partial^2 y}{\partial t^2}$, $\frac{\partial^2 y}{\partial t \cdot \partial x}$, and $\frac{\partial^2 y}{\partial x \cdot \partial t}$, when $y = e^{xt}$.

Solution: To find $\frac{\partial^2 y}{\partial x^2}$, we first solve for the first partial derivative of $y = e^{xt}$ with respect to x, treating t as a constant. We find $\frac{\partial y}{\partial x} = te^{xt}$. To find $\frac{\partial^2 y}{\partial x^2}$ we differentiate $\frac{\partial y}{\partial x}$ again with respect to x, treating t as a constant. Solving we find $\frac{\partial^2 y}{\partial x^2} = t^2 e^{xt}$.

To find $\frac{\partial^2 y}{\partial t^2}$, we first determine the first partial derivative of $y = e^{xt}$ with respect to t, treating x as a constant. Solving, we find $\frac{\partial y}{\partial t} = xe^{xt}$. To find $\frac{\partial^2 y}{\partial t^2}$, we differentiate $\frac{\partial y}{\partial t}$ again with respect to t, with x considered a constant. Solving, we find $\frac{\partial^2 y}{\partial t^2} = x^2 e^{xt}$.

To find $\frac{\partial^2 y}{\partial t \cdot \partial x}$, we differentiate $\frac{\partial y}{\partial x}$ with respect to t, treating x as a constant. We found $\frac{\partial y}{\partial x}$ to be equal to te^{xt}. Differentiating now with respect to t, we find $\frac{\partial^2 y}{\partial t \cdot \partial x} = e^{xt} + xte^{xt} = e^{xt}(1 + xt)$.

To find $\frac{\partial^2 y}{\partial x \cdot \partial t}$, we differentiate $\frac{\partial y}{\partial t}$ with respect to x considering t a constant. Now, $\frac{\partial y}{\partial t} = xe^{xt}$, therefore $\frac{\partial^2 y}{\partial x \cdot \partial t} = e^{xt} + xte^{xt} = e^{xt}(1 + xt)$.

Note that $\frac{\partial^2 y}{\partial t \cdot \partial x}$ and $\frac{\partial^2 y}{\partial x \cdot \partial t}$ are always equal for a function, y, of two variables, x and t.

• PROBLEM 537

Determine the partial derivatives of:
$A = x^2 + y^2 - 2xy \cos \theta$, where x, y, and θ are

independent variables.

Solution: To find the partial derivative A_x, or $\frac{\partial A}{\partial x}$, θ and y are treated as constants, and A is differentiated with respect to x:

$$\frac{\partial A}{\partial x} = 2x - 2y \cos \theta.$$

The partial derivative A_y, or $\frac{\partial A}{\partial y}$, is found by differentiating A with respect to y, and treating x and θ as constants:

$$\frac{\partial A}{\partial y} = 2y - 2x \cos \theta.$$

Likewise, A_θ or $\frac{\partial A}{\partial \theta}$ is found by differentiating A with respect to θ and treating x and y as constants:

$$\frac{\partial A}{\partial \theta} = 2xy \sin \theta.$$

• PROBLEM 538

If x and y depend parametrically on the two independent variables r and θ, such that $x = e^{2r} \cos \theta$, $y = e^r \sin \theta$, find x_r, x_θ, y_r, y_θ.

Solution: To find the partial derivative x_r, or $\frac{\partial x}{\partial r}$, θ is held constant.

$$\frac{\partial x}{\partial r} = 2e^{2r} \cos \theta.$$

To find x_θ, or $\frac{\partial x}{\partial \theta}$, r is held constant.

$$\frac{\partial x}{\partial \theta} = -e^{2r} \sin \theta.$$

To find y_r, or $\frac{\partial y}{\partial r}$, θ is held constant.

$$\frac{\partial y}{\partial r} = e^r \sin \theta.$$

To find y_θ, or $\frac{\partial y}{\partial \theta}$, r is held constant.

$$\frac{\partial y}{\partial \theta} = e^r \cos \theta.$$

• PROBLEM 539

If $u = \frac{x}{y} + \frac{y}{z} + \frac{z}{x}$, show that $x\frac{\partial u}{\partial x} + y\frac{\partial u}{\partial y} + z\frac{\partial u}{\partial z} = 0$.

Solution :

In order to prove this, we find the partial derivatives: $\frac{\partial u}{\partial x}$, $\frac{\partial u}{\partial y}$ and $\frac{\partial u}{\partial z}$, and then substitute them into the equation: $x\frac{\partial u}{\partial x} + y\frac{\partial u}{\partial y} + z\frac{\partial u}{\partial z}$, to see if this does equal zero.

$$\frac{\partial u}{\partial x} = \frac{1}{y} + z\left(-\frac{1}{x^2}\right) = \frac{1}{y} - \frac{z}{x^2}.$$

$$\frac{\partial u}{\partial y} = x\left(-\frac{1}{y^2}\right) + \frac{1}{z} = -\frac{x}{y^2} + \frac{1}{z},$$

$$\frac{\partial u}{\partial z} = y\left(-\frac{1}{z^2}\right) + \frac{1}{x} = -\frac{y}{z^2} + \frac{1}{x}.$$

Substituting, we find:

$$x\frac{\partial u}{\partial x} + y\frac{\partial u}{\partial y} + z\frac{\partial u}{\partial z} = \frac{x}{y} - \frac{z}{x} - \frac{x}{y} + \frac{y}{z} - \frac{y}{z} + \frac{z}{x} = 0.$$

• **PROBLEM** 540

If f is a differentiable function and a and b are constants, prove that $z = f\left(\frac{1}{2}bx^2 - \frac{1}{3}ay^3\right)$ satisfies the partial differential equation

$$ay^2 \frac{\partial z}{\partial x} + bx \frac{\partial z}{\partial y} = 0.$$

Solution: If we let $u = \frac{1}{2}bx^2 - \frac{1}{3}ay^3$ then $z = f(u)$. By the chain rule we obtain:

$$\frac{\partial z}{\partial x} = \frac{\partial u}{\partial x}\frac{dz}{du}$$

$$\frac{\partial z}{\partial y} = \frac{\partial u}{\partial y}\frac{dz}{du}.$$

Now $\frac{\partial u}{\partial x} = bx$ and $\frac{dz}{du} = f'(u)$, therefore $\frac{\partial z}{\partial x} = (bx)f'(u)$.
$\frac{\partial u}{\partial y} = -ay^2$ and $\frac{dz}{du} = f'(u)$, therefore $\frac{\partial z}{\partial y} = (-ay^2)f'(u)$.
To prove that $z = f\left(\frac{1}{2}bx^2 - \frac{1}{3}ay^3\right)$ satisfies the equation: $ay^2\frac{\partial z}{\partial x} + bx\frac{\partial z}{\partial y} = 0$, we substitute the values we obtained for $\frac{\partial z}{\partial x}$ and $\frac{\partial z}{\partial y}$ into the equation above:

Therefore,

$$ay^2 \frac{\partial z}{\partial x} + bx \frac{\partial z}{\partial y} = ay^2\left[f'(u)(bx)\right] + bx\left[f'(u)(-ay^2)\right] = 0.$$

If $z = f(y + ax) + \phi(y - ax)$, show that

$$\frac{\partial^2 z}{\partial x^2} = a^2 \frac{\partial^2 z}{\partial y^2}.$$

Solution: Let $\alpha = y + ax$, $\beta = y - ax$, so that $z = f(\alpha) + \phi(\beta)$. We now determine $\frac{\partial^2 z}{\partial x^2}$ and $\frac{\partial^2 z}{\partial y^2}$ to show that $\frac{\partial^2 z}{\partial x^2} = a^2 \frac{\partial^2 z}{\partial y^2}$. We find:

$\frac{\partial z}{\partial x} = \frac{\partial f}{\partial \alpha}\frac{\partial \alpha}{\partial x} + \frac{\partial \phi}{\partial \beta}\frac{\partial \beta}{\partial x}$, with $\frac{\partial \alpha}{\partial x} = a$ and $\frac{\partial \beta}{\partial x} = -a$.

Substituting, we have:

$$\frac{\partial z}{\partial x} = a\frac{\partial f}{\partial \alpha} - a\frac{\partial \phi}{\partial \beta}.$$

From this, $\frac{\partial^2 z}{\partial x^2} = a\frac{\partial^2 f}{\partial \alpha^2}\frac{\partial \alpha}{\partial x} - a\frac{\partial^2 \phi}{\partial \beta^2}\frac{\partial \beta}{\partial x}$.

Again substituting for $\frac{\partial \alpha}{\partial x}$ and $\frac{\partial \beta}{\partial x}$, we obtain:

$$\frac{\partial^2 z}{\partial x^2} = a^2\frac{\partial^2 f}{\partial \alpha^2} + a^2\frac{\partial^2 \phi}{\partial \beta^2} = a^2\left[\frac{\partial^2 f}{\partial \alpha^2} + \frac{\partial^2 \phi}{\partial \beta^2}\right].$$

We now determine $\frac{\partial^2 z}{\partial y^2}$. We find:

$$\frac{\partial z}{\partial y} = \frac{\partial f}{\partial \alpha}\frac{\partial \alpha}{\partial y} + \frac{\partial \phi}{\partial \beta}\frac{\partial \beta}{\partial y},$$

with $\frac{\partial \alpha}{\partial y} = 1$ and $\frac{\partial \beta}{\partial y} = 1$. Then, by substitution,

$$\frac{\partial z}{\partial y} = \frac{\partial f}{\partial \alpha} + \frac{\partial \phi}{\partial \beta},$$

and

$$\frac{\partial^2 z}{\partial y^2} = \frac{\partial^2 f}{\partial \alpha^2} + \frac{\partial^2 \phi}{\partial \beta^2}.$$

Substituting for $\frac{\partial^2 z}{\partial x^2}$ and $\frac{\partial^2 z}{\partial y^2}$ we have:

$$a^2\left[\frac{\partial^2 f}{\partial \alpha^2} + \frac{\partial^2 \phi}{\partial \beta^2}\right] = a^2\left[\frac{\partial^2 f}{\partial \alpha^2} + \frac{\partial^2 \phi}{\partial \beta^2}\right].$$

Therefore, we have shown $\frac{\partial^2 z}{\partial x^2} = a^2 \frac{\partial^2 z}{\partial y^2}$.

• PROBLEM 542

Find $f_x = \frac{\partial f}{\partial x}$ and $f_y = \frac{\partial f}{\partial y}$ by the Δ method if

$$f(x,y) = x^2y + y^2.$$

Solution: To find f_x,

$$f(x + \Delta x,y) - f(x,y) = \left[(x + \Delta x)^2 y + y^2\right] - [x^2y + y^2]$$

$$= (x + \Delta x)^2 y + y^2 - x^2y - y^2.$$

We expand and combine terms to obtain:

$$2xy\ \Delta x + y(\Delta x)^2,$$

$$f_x = \lim_{\Delta x \to 0} \frac{f(x + \Delta x,y) - f(x,y)}{\Delta x}$$

We substitute:

$2xy\Delta x + y(\Delta x)^2$ for

$f(x + \Delta x,y) - f(x,y)$, and obtain:

$$f_x = \lim_{\Delta x \to 0} \frac{2xy\Delta x + y(\Delta x)^2}{\Delta x} = \lim_{\Delta x \to 0} \frac{\Delta x(2xy + y\Delta x)}{\Delta x}$$

$$= \lim_{\Delta x \to 0} 2xy + y\Delta x = 2xy.$$

Similarly, to obtain f_y, we have

$$f(x,y + \Delta y) - f(x,y) = x^2(y + \Delta y) + (y + \Delta y)^2 - x^2y - y^2$$

$$= x^2\ \Delta y + 2y\ \Delta y + (\Delta y)^2,$$

$$f_y = \lim_{\Delta y \to 0} \frac{f(x,y + \Delta y) - f(x,y)}{\Delta y} = x^2 + 2y.$$

• PROBLEM 543

Find $z_x = \frac{\partial z}{\partial x}$ if

$$x^2z^2 + y \sin xz = 2.$$

Solution: To find z_x we make use of the theorem which states:

$$z_x = -\frac{F_x(x,\ y,\ z)}{F_z(x,\ y,\ z)} = \frac{\frac{\partial F}{\partial x}}{\frac{\partial F}{\partial z}}.$$

If

$$F(x, y, z) = x^2z^2 + y \sin xz - 2 = 0,$$

then $F_x(x, y, z) = 2xz^2 + yz \cos xz,$

and $F_x(x, y, z) = 2x^2z + xy \cos xz,$

so that,
$$z_x = -\frac{2xz^2 + yz \cos xz}{2x^2z + xy \cos xz}$$

$$= -\frac{z(2xz + y \cos xz)}{x(2xz + y \cos xz)} = -\frac{x}{z}.$$

• PROBLEM 544

If $x^2 + 2y^2 + 3z^2 = 6$, find $\frac{\partial z}{\partial x}$ and $\frac{\partial z}{\partial y}$.

Solution: We can define $F(x,y,z) = x^2 + 2y^2 + 3z^2 - 6.$

$$\frac{\partial z}{\partial x} = -\frac{\frac{\partial F}{\partial x}}{\frac{\partial F}{\partial z}}, \text{ and } \frac{\partial z}{\partial y} = -\frac{\frac{\partial F}{\partial y}}{\frac{\partial F}{\partial z}}.$$

We can now solve for $\frac{\partial F}{\partial x}$, $\frac{\partial F}{\partial y}$ and $\frac{\partial F}{\partial z}$ and substitute them into the equations for $\frac{\partial z}{\partial x}$ and $\frac{\partial z}{\partial y}$.

$$\frac{\partial F}{\partial x} = 2x, \quad \frac{\partial F}{\partial y} = 4y, \quad \frac{\partial F}{\partial z} = 6z;$$

substituting, we find:

$$\frac{\partial z}{\partial x} = -\frac{2x}{6z} = -\frac{x}{3z}, \quad \frac{\partial z}{\partial y} = -\frac{4y}{6z} = -\frac{2y}{3z} \quad (\text{if } z \neq 0).$$

• PROBLEM 545

If $x^2y + xy^2 = 1$, find $\frac{dy}{dx}$.

Solution: If we let $f(x,y) = x^2y + xy^2 - 1 = 0,$
then we can find $\frac{dy}{dx} = \frac{-\frac{\partial f}{\partial x}}{\frac{\partial f}{\partial y}}$. Therefore, solving for $\frac{\partial f}{\partial x}$ and $\frac{\partial f}{\partial y}$, we find $\frac{\partial f}{\partial x} = 2xy + y^2$, $\frac{\partial f}{\partial y} = x^2 + 2xy$, and

$$\frac{dy}{dx} = -\frac{2xy + y^2}{x^2 + 2xy}.$$

Implicit differentiation gives (as a check):

$$2xy + x^2 \frac{dy}{dx} + y^2 + 2xy \frac{dy}{dx} = 0.$$

Solving, $\frac{dy}{dx} = -\frac{2xy + y^2}{x^2 + 2xy}$.

• PROBLEM 546

If u is any function of x and y in which x and y occur only in the combination $x \cdot y$, show that

$$x\frac{\partial u}{\partial x} = y\frac{\partial u}{\partial y}.$$

Solution: Since u is a function of x and y in which x and y only occur in the form $x \cdot y$, we can set $r = x \cdot y$ and, u is a function only of r. Now, since $r = x \cdot y$, we can find $\frac{\partial r}{\partial x}$ and $\frac{\partial r}{\partial y}$. I.e., $\frac{\partial r}{\partial x} = y$ and $\frac{\partial r}{\partial y} = x$. Having found $\frac{\partial r}{\partial x}$ and $\frac{\partial r}{\partial y}$, we can use the following formulas from the chain rule to find $\frac{\partial u}{\partial x}$ and $\frac{\partial u}{\partial y}$:

$$\frac{\partial u}{\partial x} = \frac{du}{dr}\frac{\partial r}{\partial x}, \text{ and } \frac{\partial u}{\partial y} = \frac{du}{dr}\frac{\partial r}{\partial y}.$$

Substituting $\frac{\partial r}{\partial x} = y$ and $\frac{\partial r}{\partial y} = x$, we find:

$$\frac{\partial u}{\partial x} = \frac{du}{dr}(y) \text{ and } \frac{\partial u}{\partial y} = \frac{du}{dr}(x).$$

Multiplying $\frac{\partial u}{\partial x}$ by x and $\frac{\partial u}{\partial y}$ by y we have:

$$x\frac{\partial u}{\partial x} = xy\frac{du}{dr}, \text{ and } y\frac{\partial u}{\partial y} = xy\frac{du}{dr}.$$

Since $\frac{du}{dr}$ is the same for both equations, we find that $x\frac{\partial u}{\partial x} = y\frac{\partial u}{\partial y}$.

• PROBLEM 547

Given the expression: $u = \ln\sqrt{x^2 + y^2}$, where $x = re^s$, and $y = re^{-s}$, find $\frac{\partial u}{\partial r}$ and $\frac{\partial u}{\partial s}$.

Solution: The chain rule states that:

$$\frac{\partial u}{\partial r} = \frac{\partial u}{\partial x}\frac{\partial x}{\partial r} + \frac{\partial u}{\partial y}\frac{\partial y}{\partial r}, \quad (1)$$

and $$\frac{\partial u}{\partial s} = \frac{\partial u}{\partial x}\frac{\partial x}{\partial s} + \frac{\partial u}{\partial y}\frac{\partial y}{\partial s}. \quad (2)$$

In order to solve equation (1), we need to find

$\frac{\partial u}{\partial x}$, $\frac{\partial x}{\partial r}$, $\frac{\partial u}{\partial y}$ and $\frac{\partial y}{\partial r}$:

$$\frac{\partial u}{\partial x} = \frac{x}{x^2+y^2}, \quad \frac{\partial x}{\partial r} = e^s, \quad \frac{\partial u}{\partial y} = \frac{y}{x^2+y^2}, \quad \frac{\partial y}{\partial r} = e^{-s}.$$

Substituting these values into equation (1), we find:

$$\frac{\partial u}{\partial r} = \frac{x}{x^2+y^2}(e^s) + \frac{y}{x^2+y^2}(e^{-s}) = \frac{xe^s+ye^{-s}}{x^2+y^2}.$$

In order to solve equation (2) we need $\frac{\partial u}{\partial x}$, $\frac{\partial x}{\partial s}$, $\frac{\partial u}{\partial y}$ and $\frac{\partial y}{\partial s}$.

$$\frac{\partial u}{\partial x} = \frac{x}{x^2+y^2}, \quad \frac{\partial x}{\partial s} = re^s, \quad \frac{\partial u}{\partial y} = \frac{y}{x^2+y^2}, \quad \frac{\partial y}{\partial s} = -re^{-s}.$$

Substituting these values into equation (2) we find:

$$\frac{\partial u}{\partial s} = \frac{x}{x^2+y^2}(re^s) + \frac{y}{x^2+y^2}(-re^{-s}) = \frac{r(xe^s-ye^{-s})}{x^2+y^2}.$$

• PROBLEM 548

If $w = u^2v + uv^2 + 2u - 3v$;

$u = \sin(x + y + z)$,

$v = \cos(x + 2y - z)$,

find $\frac{\partial w}{\partial x}$ at $x = \frac{\pi}{2}$, $y = \frac{\pi}{4}$, $z = \frac{\pi}{6}$.

Solution: According to the chain rule,

$$\frac{\partial w}{\partial x} = \frac{\partial w}{\partial u}\frac{\partial u}{\partial x} + \frac{\partial w}{\partial v}\frac{\partial v}{\partial x}.$$

At $x = \frac{\pi}{2}$, $y = \frac{\pi}{4}$, $z = \frac{\pi}{6}$,

$$u = \sin\frac{11\pi}{12} \quad \text{and} \quad v = -\cos\frac{\pi}{6}.$$

We need to find $\frac{\partial w}{\partial u}$, $\frac{\partial u}{\partial x}$, $\frac{\partial w}{\partial v}$ and $\frac{\partial v}{\partial x}$ in order to solve the equation for $\frac{\partial w}{\partial x}$.

$$\frac{\partial w}{\partial u} = 2uv + v^2 + 2$$

$$\frac{\partial u}{\partial x} = \cos(x + y + z)$$

$$\frac{\partial w}{\partial v} = u^2 + 2uv - 3$$

$$\frac{\partial v}{\partial x} = -\sin(x + 2y - z).$$

Substituting $x = \frac{\pi}{2}$, $y = \frac{\pi}{4}$, $z = \frac{\pi}{6}$, $u = \sin\frac{11\pi}{12}$, and $v = -\cos\frac{\pi}{6}$ into these equations, we find:

$$\frac{\partial w}{\partial u} = 2 + \cos^2\frac{\pi}{6} - 2\cos\frac{5\pi}{12}\cos\frac{\pi}{6}.$$

(The $\cos\frac{5\pi}{12}$ is obtained from $\sin\frac{11\pi}{12}$ by using the formula: $\sin\left(\frac{\pi}{2} - A\right) = \cos A$, and **the fact that: cos (-A) = cos A.**)

$$\frac{\partial u}{\partial x} = \left(-\sin\frac{5\pi}{12}\right).$$

(The $-\sin\frac{5\pi}{12}$ is obtained from $\cos\frac{11\pi}{12}$ by using the formula $\cos\left(\frac{\pi}{2} - A\right) = \sin A$. Here the sin of a minus angle is minus the sin of the angle.)

$$\frac{\partial w}{\partial v} = \cos^2\frac{5\pi}{12} - 2\cos\frac{5\pi}{12}\cos\frac{\pi}{6} - 3$$

$$\frac{\partial v}{\partial x} = -\sin\frac{\pi}{6}.$$

Substituting these values into the formula for $\frac{\partial w}{\partial x}$, we find:

$$\frac{\partial w}{\partial x} = \left(2 + \cos^2\frac{\pi}{6} - 2\cos\frac{5\pi}{12}\cos\frac{\pi}{6}\right)\left(-\sin\frac{5\pi}{12}\right)$$

$$- \left(\cos^2\frac{5\pi}{12} - 2\cos\frac{5\pi}{12}\cos\frac{\pi}{6} - 3\right)\left(\sin\frac{\pi}{6}.\right)$$

• PROBLEM 549

If $u = x^3 + y^3 - 3xy$, and $x = r^2 + s$, $y = r + s^2$, find $\frac{\partial u}{\partial r}$ and $\frac{\partial u}{\partial s}$.

Solution: We use the chain rule in order to solve for $\frac{\partial u}{\partial r}$ and $\frac{\partial u}{\partial s}$. The chain rule states:

$$\frac{\partial u}{\partial r} = \frac{\partial u}{\partial x}\frac{\partial x}{\partial r} + \frac{\partial u}{\partial y}\frac{\partial y}{\partial r},$$

$$\frac{\partial u}{\partial s} = \frac{\partial u}{\partial x}\frac{\partial x}{\partial s} + \frac{\partial u}{\partial y}\frac{\partial y}{\partial s}.$$

Now, $\frac{\partial u}{\partial x} = 3x^2 - 3y$, $\frac{\partial u}{\partial y} = 3y^2 - 3x$,

$$\frac{\partial x}{\partial r} = 2r,\ \frac{\partial x}{\partial s} = 1,\ \frac{\partial y}{\partial r} = 1,\ \frac{\partial y}{\partial s} = 2s.$$

Substituting these values into the formulas for the chain rule we find:

$$\frac{\partial u}{\partial r} = (3x^2 - 3y)2r + (3y^2 - 3x) = 6x^2r - 6yr + 3y^2 - 3x.$$

$$\frac{\partial u}{\partial s} = (3x^2 - 3y) + (3y^2 - 3x)2s = 3x^2 - 3y + 6y^2s - 6xs.$$

• PROBLEM 550

Find $D_xy = \frac{\partial y}{\partial x}$, if

$$xy + yz^2 + 3 = 0,$$
$$x^2y + yz - z^2 + 1 = 0.$$

Solution: By differentiation with respect to x, we obtain:

$$(x\ D_xy + y) + (2yz\ D_xz + z^2D_xy)$$
$$= y + (x + z^2)\ D_xy + 2yz\ D_xz = 0,$$

and

$$(x^2D_xy + 2xy) + (y\ D_xz + z\ D_xy) - 2z\ D_xz$$
$$= 2xy + (x^2 + z)\ D_xy + (y - 2z)\ D_xz = 0.$$

We can now solve these two equations simultaneously for D_xy as follows:

Multiplying the first equation by $(y - 2z)$ gives:

$$\left[y + (x + z^2)\ D_xy + 2yz\ D_xz\right]\left[y - 2z\right] = 0$$

Multiplying the second equation by $(2yz)$, we obtain:

$$\left[2xy + (x^2 + z)\ D_xy + (y - 2z)\ D_xz\right]\ (2yz) = 0.$$

Multiplying out and subtracting eliminates the D_xz term, and leaves us with:

$$\left[y(y - 2z) - (2xy)(2yz)\right] + \left[(x + z^2)(y - 2z)\right.$$
$$\left. - (2yz)(x^2 + z)\right]\ D_xy = 0.$$

Solving for D_xy we have:

$$\left[xy + yz^2 - 2xz - 2z^3 - (2x^2yz + 2yz^2)\right]\ D_xy$$
$$= -\ (y^2 - 2yz - 4xy^2z)$$

$$(xy + yz^2 - 2xz - 2z^3 - 2x^2yz - 2yz^2)\ D_xy$$
$$= -\ (y^2 - 2yz - 4xy^2z)$$

$$(xy - yz^2 - 2xz - 2z^3 - 2x^2yz)\ D_xy$$
$$= -\ (y^2 - 2yz - 4xy^2z),$$

and,

$$D_x y = -\frac{y^2 - 2yz - 4xy^2 z}{xy - yz^2 - 2xz - 2z^3 - 2x^2 yz}.$$

• PROBLEM 551

A function F(x,y,z) is called homogeneous of order n if, for any quantity p, we have, identically,

$$F(px,py,pz) = p^n F(x,y,z).$$

Differentiate this relation partially with respect to p, set p = 1 in the result, and hence show that

$$x \frac{\partial F}{\partial x} + y \frac{\partial F}{\partial y} + z \frac{\partial F}{\partial z} = nF.$$

This is known as Euler's theorem on homogeneous functions for three variables.

Solution: Let F(px,py,pz) be expressed as F(u,v,w) such that px = u, py = v and pz = w. Then, the total differential over the parameter p is given by

$$\frac{\partial F}{\partial p} = \frac{\partial F}{\partial u} \frac{\partial u}{\partial p} + \frac{\partial F}{\partial v} \frac{\partial u}{\partial p} + \frac{\partial F}{\partial w} \frac{\partial w}{\partial p}$$

where $\frac{\partial u}{\partial p} = x$, $\frac{\partial v}{\partial p} = y$ and $\frac{\partial w}{\partial p} = z$.

If we let p = 1, then

$$\frac{\partial F}{\partial p} = \frac{\partial F}{\partial x} x + \frac{\partial F}{\partial y} y + \frac{\partial F}{\partial z} z.$$

On the other hand, since F(u,v,w) $= p^n$ F(x,y,z)

$$\frac{\partial F}{\partial p} = \frac{\partial}{\partial p}\left[p^n F(x,y,z)\right] = np^{n-1} F(x,y,z) + 0$$

Again, by letting p = 1, $\frac{\partial F}{\partial p} = nF$, and combining both equations

$$\frac{\partial F}{\partial p} = \frac{\partial F}{\partial x} x + \frac{\partial F}{\partial y} y + \frac{\partial F}{\partial z} z = nF.$$

• PROBLEM 552

If z = f(x,y), and x = r cos Θ, y = r sin Θ, show that

$$\left(\frac{\partial z}{\partial x}\right)^2 + \left(\frac{\partial z}{\partial y}\right)^2 = \left(\frac{\partial z}{\partial r}\right)^2 + \frac{1}{r^2}\left(\frac{\partial z}{\partial \Theta}\right)^2.$$

Solution: z = f(x,y)

x = r cos Θ, y = r sin Θ.

Hence, introducing the parameters r and Θ

$$z = f(r \cos \Theta, r \sin \Theta)$$

also expresses the point on the surface z.

The differential dz of z is

$$dz = \frac{\partial z}{\partial x} dx + \frac{\partial z}{\partial y} dy = \frac{\partial f}{\partial x} dx + \frac{\partial f}{\partial y} dy.$$

Using the new variables,

$$\frac{\partial z}{\partial r} = \frac{\partial f}{\partial x} \frac{\partial x}{\partial r} + \frac{\partial f}{\partial y} \frac{\partial y}{\partial r}, \text{ and}$$

$$\frac{\partial z}{\partial \Theta} = \frac{\partial f}{\partial x} \frac{\partial x}{\partial \Theta} + \frac{\partial f}{\partial y} \frac{\partial y}{\partial \Theta}.$$

But, since $x = r \cos \Theta$ and $y = r \sin \Theta$, we have:

$$\frac{\partial x}{\partial r} = \cos \Theta, \ \frac{\partial y}{\partial r} = \sin \Theta, \text{ and } \frac{\partial x}{\partial \Theta} = - r \sin \Theta,$$

$$\frac{\partial y}{\partial \Theta} = r \cos \Theta.$$

Substituting these values in the above equation and squaring yields:

$$\left(\frac{\partial z}{\partial r}\right)^2 = \cos^2 \Theta\left(\frac{\partial f}{\partial x}\right)^2 + \sin^2 \Theta\left(\frac{\partial f}{\partial y}\right)^2 + 2 \sin \Theta \cos \Theta \left(\frac{\partial f}{\partial x}\right)\left(\frac{\partial f}{\partial y}\right), \text{ and}$$

$$\left(\frac{\partial z}{\partial \Theta}\right)^2 = r^2 \sin^2 \Theta \left(\frac{\partial f}{\partial x}\right)^2 + r^2 \cos^2 \Theta\left(\frac{\partial f}{\partial y}\right)^2 - 2r^2 \sin \Theta \cos \Theta \left(\frac{\partial f}{\partial x}\right)\left(\frac{\partial f}{\partial y}\right).$$

To eliminate the angle Θ, we multiply the first equation by r^2, add both equations and upon factoring out like terms and using the trigonometric identity

$$\sin^2 \Theta + \cos^2 \Theta = 1,$$

we obtain

$$r^2 \left(\frac{\partial z}{\partial r}\right)^2 + \left(\frac{\partial z}{\partial \Theta}\right)^2 = r^2 \left(\frac{\partial f}{\partial x}\right)^2 + r^2 \left(\frac{\partial f}{\partial y}\right)^2$$

from which

$$\left(\frac{\partial f}{\partial x}\right)^2 + \left(\frac{\partial f}{\partial y}\right)^2 = \left(\frac{\partial z}{\partial x}\right)^2 + \left(\frac{\partial z}{\partial y}\right)^2$$

$$= \left(\frac{\partial z}{\partial r}\right)^2 + \frac{1}{r^2} \left(\frac{\partial z}{\partial \Theta}\right)^2.$$

Let u and v be functions of x and y such that

$$\frac{\partial u}{\partial x} = \frac{\partial v}{\partial y}, \quad \frac{\partial u}{\partial y} = -\frac{\partial v}{\partial x}.$$

Set $x = r \cos \Theta$, $y = r \sin \Theta$, and show that

$$r\frac{\partial u}{\partial r} = \frac{\partial v}{\partial \Theta}, \quad \frac{\partial u}{\partial \Theta} = -r\frac{\partial v}{\partial r},$$

$$r^2 \frac{\partial^2 u}{\partial r^2} + r\frac{\partial u}{\partial r} + \frac{\partial^2 u}{\partial \Theta^2} = 0.$$

Solution: $u = f_1(x,y)$ and $v = f_2(x,y)$.

Hence the differential du of u is:

$$du = \frac{\partial u}{\partial x} dx + \frac{\partial u}{\partial y} dy.$$

Using r and Θ as parametric variables,

$$\frac{\partial u}{\partial r} = \frac{\partial u}{\partial x}\frac{\partial x}{\partial r} + \frac{\partial u}{\partial y}\frac{\partial y}{\partial r},$$

$$\frac{\partial u}{\partial \Theta} = \frac{\partial u}{\partial x}\frac{\partial x}{\partial \Theta} + \frac{\partial u}{\partial y}\frac{\partial y}{\partial \Theta}.$$

If $x = r \cos \Theta$ and $y = r \sin \Theta$, as given, then

$$\frac{\partial x}{\partial r} = \cos \Theta, \quad \frac{\partial y}{\partial r} = \sin \Theta;$$

$$\frac{\partial x}{\partial \Theta} = -r \sin \Theta, \quad \frac{\partial y}{\partial \Theta} = r \cos \Theta.$$

Thus,

$$\frac{\partial u}{\partial r} = \cos \Theta \frac{\partial u}{\partial x} + \sin \Theta \frac{\partial u}{\partial y}, \text{ and}$$

$$\frac{\partial u}{\partial \Theta} = -r \sin \Theta \frac{\partial u}{\partial x} + r \cos \Theta \frac{\partial u}{\partial y}.$$

Similarly,

$$\frac{\partial v}{\partial r} = \cos \Theta \frac{\partial v}{\partial x} + \sin \Theta \frac{\partial v}{\partial y}, \text{ and}$$

$$\frac{\partial v}{\partial \Theta} = -r \sin \Theta \frac{\partial v}{\partial x} + r \cos \Theta \frac{\partial v}{\partial y}.$$

Using the given identities

$$\frac{\partial u}{\partial x} = \frac{\partial v}{\partial y} \text{ and } \frac{\partial u}{\partial y} = -\frac{\partial v}{\partial x}$$

$$r\frac{\partial u}{\partial r} = r \cos \Theta \frac{\partial u}{\partial x} + r \sin \Theta \frac{\partial u}{\partial y}$$

$$= r \cos \Theta \left(\frac{\partial v}{\partial y}\right) + r \sin \Theta \left(-\frac{\partial v}{\partial x}\right)$$

$$= - r \sin \Theta \frac{\partial v}{\partial x} + r \cos \Theta \frac{\partial v}{\partial y}$$

$$= \frac{\partial v}{\partial \Theta} .$$

Therefore

$$r \frac{\partial u}{\partial r} = \frac{\partial v}{\partial \Theta} .$$

Again

$$- r \frac{\partial v}{\partial r} = - r \cos \Theta \frac{\partial v}{\partial x} - r \sin \Theta \frac{\partial v}{\partial y}$$

$$= - r \cos \Theta \left(- \frac{\partial u}{\partial y}\right) - r \sin \Theta \left(\frac{\partial u}{\partial x}\right)$$

$$= - r \sin \Theta \frac{\partial u}{\partial x} + r \cos \Theta \frac{\partial u}{\partial y} = \frac{\partial u}{\partial \Theta},$$

or $$- r \frac{\partial v}{\partial r} = \frac{\partial u}{\partial \Theta} .$$

Furthermore,

$$\frac{\partial^2 u}{\partial r^2} = \frac{\partial}{\partial r} (\cos \Theta) \frac{\partial u}{\partial x} + \cos \Theta \frac{\partial}{\partial r} \left(\frac{\partial u}{\partial x}\right)$$

$$+ \frac{\partial}{\partial r} (\sin \Theta) \frac{\partial u}{\partial y} + \sin \Theta \frac{\partial}{\partial r}\left(\frac{\partial u}{\partial y}\right)$$

$$= \cos \Theta \frac{\partial^2 u}{\partial r \partial x} + \sin \Theta \frac{\partial^2 u}{\partial r \partial y}$$

or $$r^2 \frac{\partial^2 u}{\partial r^2} = r^2 \cos \Theta \frac{\partial^2 u}{\partial r \partial x} + r^2 \sin \Theta \frac{\partial^2 u}{\partial r \partial y} .$$

Similarly,

$$\frac{\partial^2 u}{\partial \Theta^2} = \frac{\partial}{\partial \Theta} (r) \sin \Theta \frac{\partial u}{\partial x}$$

$$+ r \frac{\partial}{\partial \Theta} (\sin \Theta) \frac{\partial u}{\partial x} + r \sin \Theta \frac{\partial}{\partial \Theta} \left(\frac{\partial u}{\partial x}\right)$$

$$+ \frac{\partial}{\partial \Theta} (r) \cos \Theta \frac{\partial u}{\partial y} + r \frac{\partial}{\partial \Theta} (\cos \Theta) \frac{\partial u}{\partial y}$$

$$+ r \cos \Theta \frac{\partial}{\partial \Theta} \left(\frac{\partial u}{\partial y}\right)$$

$$= r \cos \Theta \frac{\partial u}{\partial x} + r \sin \Theta \frac{\partial^2 u}{\partial \Theta \partial x} - r \sin \Theta \frac{\partial u}{\partial y}$$

$$+ r \cos \Theta \frac{\partial^2 u}{\partial \Theta \partial y} .$$

Hence,

$$r^2 \frac{\partial^2 u}{\partial r^2} + r \frac{\partial u}{\partial r} + \frac{\partial^2 u}{\partial \Theta^2}$$

$$= r^2 \cos \Theta \frac{\partial^2 u}{\partial r \partial x} + r^2 \sin \Theta \frac{\partial^2 u}{\partial r \partial y} .$$

• PROBLEM 554

Find the values of a and b in order that the cone $ax^2 + by^2 + z^2 = 0$ shall cut the ellipsoid $2x^2 + 4y^2 + z^2 = 42$ orthogonally at the point (1, 3, 2).

Solution: Let the scalar function $\delta(x,y,z)$ denote the equation of the cone surface,

$$\delta(x,y,z) = ax^2 + by^2 + z^2 = 0.$$

Similarly, let

$$\psi(x,y,z) = 2x^2 + 4y^2 + z^2.$$

The gradient of these functions at (1, 3,2) are given by

$$\nabla\delta(x,y,z) = 2ax\hat{\imath} + 2by\hat{\jmath} + 2z\hat{k},$$

$$\nabla\delta(1,3,2) = 2a\hat{\imath} + 6b\hat{\jmath} + 4\hat{k}.$$

Also

$$\nabla\psi(x,y,z) = 4x\hat{\imath} + 8y\hat{\jmath} + 2z\hat{k}$$

or $$\nabla\psi(1,3,2) = 4\hat{\imath} + 24\hat{\jmath} + 4\hat{k}.$$

For orthogonality,

$$\nabla\delta\Big|_{(1,3,2)} \cdot \nabla\psi\Big|_{(1,3,2)} = 0,$$

or $8a + 144b + 16 = 0$, that is, $a + 18b + 2 = 0.$

$$a = -2 - 18b.$$

Substituting in the given equation

$$ax^2 + by^2 + z^2 = 0$$

$$(-2 - 18b)x^2 + by^2 + z^2 = 0$$

$$(-2 - 18b)(1) + b(9) + (4) = 0$$

$$-2 - 18b + 9b + 4 = 0$$

$$9b = 2$$

$$b = \frac{2}{9}$$

$$a = -2 - 4 = -6.$$

If $z = f(x,y)$, and x and y are connected by a relation $\phi(x,y) = 0$, show that

$$\frac{dz}{dx} = \frac{\frac{\partial f}{\partial x}\frac{\partial \phi}{\partial y} - \frac{\partial f}{\partial y}\frac{\partial \phi}{\partial x}}{\frac{\partial \phi}{\partial y}}$$

Solution: $z = f(x,y)$.

The total derivative is:

$$dz = \frac{\partial f}{\partial x} dx + \frac{\partial f}{\partial y} dy.$$

Since x and y are the only independent variables in z

$$\frac{dz}{dx} = \frac{\partial f}{\partial x} + \frac{\partial f}{\partial y}\frac{dy}{dx}.$$

From the x - y relationship, let u be a constant expression for the surface such that $u = \phi(x,y)$, and

$$du = \frac{\partial \phi}{\partial x} dx + \frac{\partial \phi}{\partial y} dy = 0$$

Solving for $\frac{dy}{dx}$, we obtain

$$\frac{dy}{dx} = -\left(\frac{\frac{\partial \phi}{\partial x}}{\frac{\partial \phi}{\partial y}}\right).$$

Substituting this in the differential of z:

$$\frac{dz}{dx} = \frac{\partial f}{\partial x} + \frac{\partial f}{\partial y}\left(-\frac{\frac{\partial \phi}{\partial x}}{\frac{\partial \phi}{\partial y}}\right)$$

$$= \frac{\frac{\partial f}{\partial x}\frac{\partial \phi}{\partial y} - \frac{\partial f}{\partial y}\frac{\partial \phi}{\partial x}}{\frac{\partial \phi}{\partial y}}.$$

• PROBLEM 556

Two surfaces, $F(x,y,z) = 0$ and $G(x,y,z) = 0$ have the point (x_1, y_1, z_1) in common. Find the condition that must be satisfied by the six first partial derivatives at the point if the surfaces (a) are tangent; (b) intersect orthogonally.

Solution: The problem suggests the use of the following operations and properties:

(a) for tangency, we consider $F(x,y,z) = 0$,

and let $\vec{r} = x\hat{i} + y\hat{j} + z\hat{k}$ denote a vector from the origin to some point (x,y,z) in the neighborhood of (x_1,y_1,z_1) on the surface $F(x,y,z) = 0$. Let

$\vec{r}_1 = x_1\hat{i} + y_1\hat{j} + z_1\hat{k}$ be a vector from the origin to the point (x_1, y_1, z_1) The difference $\vec{r} - \vec{r}_1$ denotes a vector tangent to the surface of $F(x,y,z) = 0$. We also know that the gradient vector of F(x,y,z) at this point is perpendicular to the surface. Thus

$$\nabla F(x_1,y_1,z_1) \cdot (\vec{r} - \vec{r}_1) = 0.$$

If the two surfaces F(x,y,z) and G(x,y,z) are tangential at (x_1,y_1,z_1) then

$$\nabla G(x_1,y_1,z_1) \cdot (\vec{r} - \vec{r}_1) = 0$$

is satisfied. This means that ∇F and ∇G are parallel at (x_1,y_1,z_1). That can be expressed as:

$$\nabla F(x_1,y_1,z_1) = k\, \nabla G(x_1,y_1,z_1),$$

where k is a constant. This can also be explicitly represented at a point (x_1,y_1,z_1), as:

$$\frac{\partial F}{\partial x} = k \frac{\partial G}{\partial x}, \quad \frac{\partial F}{\partial y} = k \frac{\partial G}{\partial y} \quad \text{and} \quad \frac{\partial F}{\partial z} = k \frac{\partial G}{\partial z}.$$

(b) for orthogonality, $\nabla G(x_1, y_1, z_1)$ must be parallel to $(\vec{r} - \vec{r}_1)$. Hence,

$$\nabla F(x_1, y_1, z_1) \cdot \nabla G(x_1, y_1, z_1) = 0, \text{ or}$$

$$\frac{\partial F}{\partial x}\frac{\partial G}{\partial x} + \frac{\partial F}{\partial y}\frac{\partial G}{\partial y} + \frac{\partial F}{\partial z}\frac{\partial G}{\partial z} = 0.$$

APPLIED PROBLEMS USING PARTIAL DERIVATIVES

• PROBLEM 557

If a + b + c = a constant number where all factors are positive, when is a • b • c a maximum?

Solution: In this problem we are given that a + b + c = S, a constant. We wish to maximize the product, $P = a \cdot b \cdot c$. Eliminating c by use of the equation $c = S - a - b$, we have the function of two variables:

$$P = (a)(b)(S - a - b),$$

$$\text{or} \quad P = abS - a^2b - ab^2,$$

which is to be maximized. We must find the first partial derivatives, $\frac{\partial P}{\partial a}$ and $\frac{\partial P}{\partial b}$, equate them to 0 and solve for a and b. We find:

$$\frac{\partial P}{\partial a} = bS - 2ab - b^2 = 0; \quad \frac{\partial P}{\partial b} = aS - a^2 - 2ab = 0.$$

We now solve for a and b. Eliminating the terms containing S in both equations we have:

$$a^2b = ab^2 \quad \text{or} \quad a = b.$$

From the equation $bS - 2ab - b^2 = 0$, we have $bS = 2ab + b^2$. This can be rewritten as $bS = 3b^2$ or $S = 3b = 3a$, because $a = b$. From $c = S - a - b$, we get $c = 3b - 2b = b$ or $c = 3a - 2a = a$. Therefore, $a \cdot b \cdot c$ is a maximum when $a = b = c$. This applies to any number of factors.

• PROBLEM 558

Let the number 12 be the sum of three positive parts. Find these parts if their product is to be a maximum.

Solution: Let the three parts be x,y, and $12 - x - y$. Their product $= P \Rightarrow P = xy(12 - x - y)$

To maximize the equation we find the first partial derivatives, set them equal to 0, and then solve the two resulting equations simultaneously for x and y. Next, we find the second partial derivatives, and evaluate at the values found for x and y. Let $A = \frac{\partial^2 P}{\partial x^2}$, $B = \frac{\partial^2 P}{\partial y \partial x}$ and $C = \frac{\partial^2 P}{\partial y^2}$. We then find $AC - B^2$. If this is > 0, with $A < 0$ then we have a maximum. Using this procedure we have:

$$\frac{\partial P}{\partial x} = 12y - 2xy - y^2, \quad \frac{\partial P}{\partial y} = 12x - x^2 - 2xy.$$

Set $12y - 2xy - y^2 = 0$ and $12x - x^2 - 2xy = 0$. The simultaneous system

$$2x + y = 12 \quad \text{and} \quad 2y + x = 12$$

has the unique solution $x = y = 4$. We now use the criterion

$$A = \frac{\partial^2 P}{\partial x^2} = -2y, \quad B = \frac{\partial^2 P}{\partial y \partial x} = 12 - 2x - 2y,$$

$$C = \frac{\partial^2 P}{\partial y^2} = -2x.$$

At (4,4),

$$A = -2(4) = -8$$

$$B = 12 - 2(4) - 2(4) = -4$$

$$C = -2(4) = -8$$

Now, $AC - B^2 = (-8)(-8) - (-4)^2$

$= 64 - 16$

$= 48.$

Because $AC - B^2 > 0$ and $A < 0$, P has a maximum at (4,4). Therefore 12 should be divided into 4,4 and 4 to obtain a maximum product.

• PROBLEM 559

Find the dimensions of a covered, rectangular box of given volume with minimum surface area.

Solution: If the dimensions of the box are x, y, z, the volume is $V = xyz$ = constant, and the surface **area is defined** to be $S = 2(xy + yz + zx)$; eliminating z by use of the equation $z = V/xy$, we have the function of two variables:

$$S = 2\left(xy + \frac{V}{x} + \frac{V}{y}\right), \text{ (V constant)},$$

which is to be made a minimum. To obtain minimum values we must find the first partial derivatives of S, $\frac{\partial S}{\partial x}$ and $\frac{\partial S}{\partial y}$, equate them to 0 and solve for x and y. We find

$$\frac{\partial S}{\partial x} = 2\left(y - \frac{V}{x^2}\right), \quad \frac{\partial S}{\partial y} = 2\left(x - \frac{V}{y^2}\right).$$

Now, $x^2y - V = 0$ and, $xy^2 - V = 0$. Eliminating V from these equations we obtain, $x^2y = xy^2$, from which we have, $x = y$. From the above equations we obtain, $x^2y = V$ and $xy^2 = V$, therefore $x^3 = y^3 = V$ and $x = y = \sqrt[3]{V}$. From $z = \frac{V}{xy}$, we get $z = \sqrt[3]{V}$, so that $x = y = z$. Therefore, the box of minimum surface area must be a cube.

• PROBLEM 560

A strip of sheet steel 18 inches wide is to be made into a gutter, the cross-section of which is an isosceles trapezoid. Find the dimensions of the trapezoid if the cross-sectional area is to be a maximum. Let the dimensions be as shown in the figure.

Solution: The area of the trapezoid is the sum of the areas of the two right triangles and the rectangle. Therefore, letting k = area of the trapezoid, we have:

$$k = (2x)\left[(9-x)\sin\theta\right] + \frac{1}{2}\left[(9-x)\sin\theta\right]\left[(9-x)\cos\theta\right]$$

$$+ \frac{1}{2}\left[(9-x)\sin\theta\right]\left[(9-x)\cos\theta\right]$$

$$= (2x)\left[(9-x)\sin\theta\right] + \left[(9-x)\sin\theta\right]\left[(9-x)\cos\theta\right]$$

$$= \left(18x-2x^2\right)\sin\theta + (9-x)^2\sin\theta\cos\theta.$$

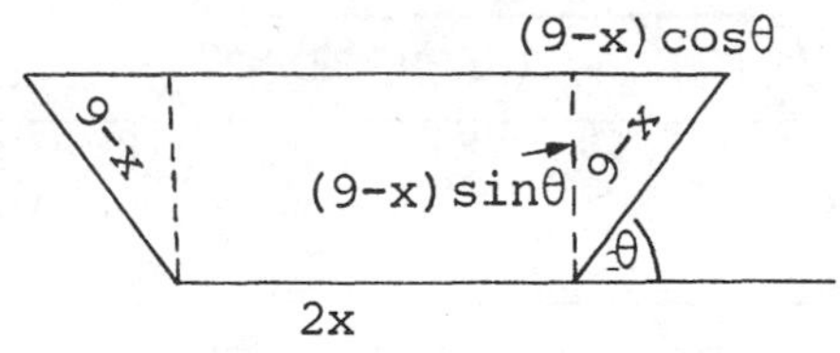

Making use of the identity, $\sin\theta\cos\theta = \frac{\sin 2\theta}{2}$, we write:

$$k = (18x-2x^2)\sin\theta + \frac{(9-x)^2}{2}\sin 2\theta.$$

We wish to maximize k, the area. To do this we find the first partial derivatives of k, $\frac{\partial k}{\partial x}$ and $\frac{\partial k}{\partial \theta}$, equate them to 0 and solve for x and θ. We find:

$$\frac{\partial k}{\partial x} = (18-4x)\sin\theta - (9-x)\sin 2\theta.$$

Making use of the identity: $\sin 2\theta = 2\sin\theta\cos\theta$, we have:

$$\text{(a)} \quad \frac{\partial k}{\partial x} = \sin\theta\left[(18-4x) - (18-2x)\cos\theta\right] = 0.$$

$$\frac{\partial k}{\partial \theta} = (18x-2x^2)\cos\theta + (9-x)^2\cos 2\theta.$$

We now replace $\cos 2\theta$ by use of the identities, $\cos 2\theta = \cos^2\theta - \sin^2\theta$, and $\sin^2\theta = 1 - \cos^2\theta$. This gives us:

$$\text{(b)} \quad \frac{\partial k}{\partial \theta} = (18x-2x^2)\cos\theta + (9-x)^2\left[2\cos^2\theta - 1\right] = 0$$

We now proceed to solve for x and θ. From equation (a) we have: $\sin\theta = 0$, which can be rejected because θ is obviously not 0 or π. This leaves us with,

$$(18-4x) - (18-2x)\cos\theta = 0,$$

$$\text{or} \quad \cos\theta = \frac{9-2x}{9-x}.$$

From equation (b) we obtain:

$$\frac{2\cos^2\theta - 1}{\cos\theta} = -\frac{(18x-2x^2)}{(9-x)^2} = -\frac{2x}{(9-x)}.$$

By substituting the value for $\cos\theta$, $\frac{9-2x}{9-x}$, into this equation and rearranging terms we have,

$$\cos^2\theta = \frac{1}{2}\left[\frac{(-2x)(9-2x)}{(9-x)^2} + 1\right] = \frac{1}{2}\left[\frac{5x^2 - 36x + 81}{(9-x)^2}\right].$$

Therefore,

$$\cos\theta = \frac{\sqrt{\frac{1}{2}(5x^2 - 36x + 81)}}{(9-x)} = \frac{9-2x}{(9-x)}$$

and

$$\sqrt{\frac{1}{2}(5x^2 - 36x + 81)} = 9 - 2x.$$

Squaring both sides and simplifying, gives us

$$3(x^2 - 12x + 27) = 0$$
$$\text{or,}\quad (x-9)(x-3) = 0.$$

We therefore find $x = 3$ and $x = 9$. We now solve for θ. We have, $\cos\theta = \frac{9-2x}{9-x}$. When $x = 3$, $\cos\theta = \frac{1}{2}$, therefore $\theta = \frac{\pi}{3}$. We can reject the value $x = 9$, because in evaluating $\cos\theta$, this value gives us a 0 in the denominator. Therefore, we find $x = 3$, $\theta = \frac{\pi}{3}$. It is now necessary to determine that these values, $x = 3$, $\theta = \frac{\pi}{3}$, are indeed the values that will give us maximal area. To do this we must find $\frac{\partial^2 k}{\partial x \partial x}$, $\frac{\partial^2 k}{\partial\theta\partial x}$ and $\frac{\partial^2 k}{\partial\theta\partial\theta}$, and evaluate each at $x = 3$, $\theta = \frac{\pi}{3}$. We find:

$$\frac{\partial^2 k}{\partial x \partial x} = -4\sin\theta + \sin 2\theta$$

$$\frac{\partial^2 k}{\partial\theta\partial x} = (18 - 4x)\cos\theta - 2(9-x)\cos 2\theta$$

$$\frac{\partial^2 k}{\partial\theta\partial\theta} = -(18x - 2x^2)\sin\theta - 2(9-x)^2\sin 2\theta.$$

Now, we evaluate each second partial derivative at $x = 3$, $\theta = \frac{\pi}{3}$, letting the value for $\frac{\partial^2 k}{\partial x \partial x} = A$, $\frac{\partial^2 k}{\partial\theta\partial x} = B$ and $\frac{\partial^2 k}{\partial\theta\partial\theta} = C$. We find: $\frac{\partial^2 k}{\partial x\partial x} = -3 = A$, $\frac{\partial^2 k}{\partial\theta\partial x} = 9 = B$ and $\frac{\partial^2 k}{\partial\theta\partial\theta} = -54\sqrt{3} = C$. We use the test, if $A < 0$ and $AC - B^2 > 0$, then $(x,\theta) = \left(3,\frac{\pi}{3}\right)$ is a maximum value. We find $A = -3 < 0$ and $AC - B^2 = (-3)(-54\sqrt{3}) - 81 = 162\sqrt{3} - 81 > 0$. Therefore this is indeed a maximum, and the trapezoid must have a base = 2(x) = 2(3) = 6 inches, with sides = (9 - 3) = 6 inches, to yield maximal area.

• PROBLEM 561

Find the equations of the tangent plane and normal line to the surface $x^2 + y^2 = 4z$ (paraboloid) at the point (6,8,25).

Solution: The equation of the tangent plane to a surface $F(x,y,z) = 0$ at a point $P_1(x_1,y_1,z_1)$ is:

$$\frac{\partial F}{\partial x}(x - x_1) + \frac{\partial F}{\partial y}(y - y_1) + \frac{\partial F}{\partial z}(z - z_1) = 0.$$

In this problem, the surface $F(x,y,z) = x^2 + y^2 - 4z$ and the point $P_1(x_1,y_1,z_1)$ is (6,8,25). In order to solve the equation of the tangent plane we need to find $\frac{\partial F}{\partial x}$, $\frac{\partial F}{\partial y}$ and $\frac{\partial F}{\partial z}$.

$$\frac{\partial F}{\partial x} = 2x, \quad \frac{\partial F}{\partial y} = 2y, \quad \frac{\partial F}{\partial z} = -4.$$

At the given point (6,8,25), these derivatives have the numerical values 12, 16, -4. Substituting these values, and $x_1 = 6$, $y_1 = 8$, and $z_1 = 25$, into the formula for the tangent plane, we obtain:

$$12(x - 6) + 16(y - 8) - 4(z - 25) = 0, \text{ or } 3x + 4y - z = 25,$$

as the equation of the tangent plane.

The equations of the normal line to the surface $F(x,y,z) = 0$ at the point $P(x_1,y_1,z_1)$ are:

$$\frac{x - x_1}{\frac{\partial F}{\partial x}} = \frac{y - y_1}{\frac{\partial F}{\partial y}} = \frac{z - z_1}{\frac{\partial F}{\partial z}}.$$

Substituting the values obtained previously, $\frac{\partial F}{\partial x} = 12$, $\frac{\partial F}{\partial y} = 16$, and $\frac{\partial F}{\partial z} = -4$, and $x_1 = 6$, $y_1 = 8$ and $z_1 = 25$, into the formula for the equation of the normal line we find:

$$\frac{x - 6}{12} = \frac{y - 8}{16} = \frac{z - 25}{-4}, \text{ or } \frac{x - 6}{3} = \frac{y - 8}{4} = \frac{z - 25}{-1}$$

as the equation of the normal line.

CHAPTER 22

TOTAL DIFFERENTIALS, TOTAL DERIVATIVES, AND APPLIED PROBLEMS

If w is a function of x, y, and z, then the total differential, dw, is defined as:

$$dw = \frac{\partial w}{\partial x}dx + \frac{\partial w}{\partial y}dy + \frac{\partial w}{\partial z}dz .$$

We can also write the total derivative, with respect to any of the variables we choose, say, x, as follows:

$$\frac{dw}{dx} = \frac{\partial w}{\partial x} + \frac{\partial w}{\partial y}\frac{dy}{dx} + \frac{\partial w}{\partial z}\frac{dz}{dx} .$$

Similar expressions are possible for the other variables.

One of the most useful applications of these concepts is for finding changes in a quantity due to changes in the other quantities. For this purpose, we use Δ's rather than d's to indicate that finite changes, even though small, are involved, rather than infinitesimal changes. Accordingly,

$$\Delta w = \frac{\partial w}{\partial x}\Delta x + \frac{\partial w}{\partial y}\Delta y + \frac{\partial w}{\partial z}\Delta z ,$$

where it is assumed that Δx, Δy, and Δz are known, and Δw is to be found. This formula has wide applications in estimating possible errors in a computation due to errors in individual parameters.

• PROBLEM 562

Find the total differential of the function: $z = x^3y + x^2y^2 - 3xy^3$.

Solution: By definition,

$$dz = \frac{\partial z}{\partial x}dx + \frac{\partial z}{\partial y}dy .$$

$$\frac{\partial z}{\partial x} = 3x^2y + 2xy^2 - 3y^3,$$

and

$$\frac{\partial z}{\partial y} = x^3 + 2x^2y - 9xy^2 .$$

Then,

$$dz = (3x^2y + 2xy^2 - 3y^3)dx + (x^3 + 2x^2y - 9xy^2)dy.$$

• PROBLEM 563

If $z = x^2y + xy^2 + 3x - 2y - 1$,

$x = t^2 + t,$

$y = 2t - 1,$

find $\frac{dz}{dt}$ at $t = 2$.

Solution: From the chain rule we obtain:

$$\frac{dz}{dt} = \frac{\partial z}{\partial x}\frac{dx}{dt} + \frac{\partial z}{\partial y}\frac{dy}{dt}.$$

In order to solve for $\frac{dz}{dt}$, we need to find $\frac{\partial z}{\partial x}$, $\frac{dx}{dt}$, $\frac{\partial z}{\partial y}$ and $\frac{dy}{dt}$.

$$\frac{\partial z}{\partial x} = 2xy + y^2 + 3$$

$$\frac{dx}{dt} = 2t + 1$$

$$\frac{\partial z}{\partial y} = x^2 + 2xy - 2$$

$$\frac{dy}{dt} = 2$$

at $t = 2$, $x = 6$ and $y = 3$, and the above values become:

$$\frac{\partial z}{\partial x} = 48, \quad \frac{dx}{dt} = 5, \quad \frac{\partial z}{\partial y} = 70, \quad \frac{dy}{dt} = 2.$$

Now, substituting these values into the equation for $\frac{dz}{dt}$ we obtain:

$$\frac{dz}{dt} = 48(5) + 70(2) = 380.$$

• PROBLEM 564

Find $\frac{du}{dx}$ from the expression: $u = xy + yz + zx$, when $y = e^x$ and $z = \sin x$.

Solution: In order to find $\frac{du}{dx}$, we use the chain rule:

$$\frac{du}{dx} = \frac{\partial u}{\partial x}\frac{dx}{dx} + \frac{\partial u}{\partial y}\frac{dy}{dx} + \frac{\partial u}{\partial z}\frac{dz}{dx}$$

or $$\frac{du}{dx} = \frac{\partial u}{\partial x} + \frac{\partial u}{\partial y}\frac{dy}{dx} + \frac{\partial u}{\partial z}\frac{dz}{dx}.$$

Now,

$$\frac{\partial u}{\partial x} = y + z, \quad \frac{\partial u}{\partial y} = x + z, \quad \frac{\partial u}{\partial z} = y + x, \quad \frac{dy}{dx} = e^x, \quad \frac{dz}{dx} = \cos x.$$

Substituting these values into the formula for the chain rule, we find:

$$\frac{du}{dx} = y + z + (x + z)e^x + (x + y)\cos x.$$

• **PROBLEM 565**

If

$u = x^2 - xy + y^2$, and if $x = 1 + t^2$, $y = 1 - t^2$,

what is the total derivative of u?

Solution: u is a function of x and y which are each, in turn, functions of t. The total derivative of u with respect to t is given by,

$$\frac{du}{dt} = \frac{\partial u}{\partial x}\frac{dx}{dt} + \frac{\partial u}{\partial y}\frac{dy}{dt}.$$

Note that if u were a function of three variables, $u = F(x, y, z)$, which were all, in turn, functions of t, the total derivative of u with respect to t would be written as:

$$\frac{du}{dt} = \frac{\partial u}{\partial x}\frac{dx}{dt} + \frac{\partial u}{\partial y}\frac{dy}{dt} + \frac{\partial u}{\partial z}\frac{dz}{dt}.$$

To solve this problem, we need to find

$\frac{\partial u}{\partial x}$, $\frac{\partial u}{\partial y}$, $\frac{dx}{dt}$ and $\frac{dy}{dt}$, and then to substitute in the equation for the total derivative, $\frac{du}{dt}$. Hence,

$$\frac{\partial u}{\partial x} = 2x - y \qquad \frac{dx}{dt} = 2t$$

$$\frac{\partial u}{\partial y} = 2y - x \qquad \frac{dy}{dt} = -2t$$

and

$$\frac{du}{dt} = (2x - y)(2t) + (2y - x)(-2t) = 2t(3x - 3y)$$

which, to obtain a relation in terms of t alone, is:

$$2t[3(1 + t^2) - 3(1 - t^2)]$$

$$= 2t(3 + 3t^2 - 3 + 3t^2)$$

$$= 12t^3.$$

• **PROBLEM 566**

Given the expression: $u = x^2 + 2xy + y^2$, where $x = t\cos t$, and $y = t\sin t$, find $\frac{du}{dt}$ by two methods:

(a) using the chain rule; (b) expressing u in terms of t before differentiating.

Solution: (a) The chain rule states that:

$$\frac{du}{dt} = \frac{\partial u}{\partial x}\frac{dx}{dt} + \frac{\partial u}{\partial y}\frac{dy}{dt}.$$

$\frac{\partial u}{\partial x} = 2x + 2y$, $\frac{\partial u}{\partial y} = 2x + 2y$, $\frac{dx}{dt} = \cos t - t \sin t$,

$\frac{dy}{dt} = \sin t + t \cos t$.

Therefore, substituting these values into the equation for $\frac{du}{dt}$, we find:

$$\frac{du}{dt} = (2x+2y)(\cos t - t \sin t) + (2x+2y)(\sin t + t \cos t)$$

$$= 2(x+y)(\cos t - t \sin t + \sin t + t \cos t).$$

Substituting $x = t \cos t$ and $y = t \sin t$, we obtain:

$$2(t \cos t + t \sin t)(\cos t - t \sin t + \sin t + t \cos t).$$

Simplifying, we obtain:

$$2t(\cos^2 t - t \sin t \cos t + \sin t \cos t + t \cos^2 t$$
$$+ \sin t \cos t - t \sin^2 t + \sin^2 t$$
$$+ t \sin t \cos t).$$

Using the identity: $\sin^2 t + \cos^2 t = 1$, we find:

$\frac{du}{dt} = 2t\left[1 + 2 \sin t \cos t + t(\cos^2 t - \sin^2 t)\right]$, and

$2t(1 + \sin 2t + t \cos 2t) = 2t + 2t \sin 2t + 2t^2 \cos 2t$.

(b) Expressing u in terms of t, we obtain:

$$u = (t \cos t)^2 + 2(t \cos t)(t \sin t) + (t \sin t)^2$$

$$= t^2 \cos^2 t + t^2(2 \sin t \cos t) + t^2 \sin^2 t$$

$$= t^2 + t^2 \sin 2t.$$

Now we can differentiate to obtain:

$$\frac{du}{dt} = 2t + 2t \sin 2t + 2t^2 \cos 2t.$$

• PROBLEM 567

If $z = \sin^{-1}\left[(1+x)/(1+y)\right]$ and $x = \sin t$, $y = \cos t$, find the rate of change of z with respect to t when $t = 0$.

Solution: From the chain rule we have:

$$\frac{dz}{dt} = \frac{\partial z}{\partial x}\frac{dx}{dt} + \frac{\partial z}{\partial y}\frac{dy}{dt}.$$

Now,

$$\frac{\partial z}{\partial x} = \frac{1}{\sqrt{1-\left(\frac{1+x}{1+y}\right)^2}}\left(\frac{1}{1+y}\right) \text{ and } \frac{\partial z}{\partial y} = \frac{1}{\sqrt{1-\left(\frac{1+x}{1+y}\right)^2}}\left(-\frac{1+x}{(1+y)^2}\right)$$

and $\frac{dx}{dt} = \cos t$ and $\frac{dy}{dt} = -\sin t$. Substituting these values into the equation for $\frac{dz}{dt}$, we find:

$$\frac{dz}{dt} = \frac{1}{\sqrt{1-\left(\frac{1+x}{1+y}\right)^2}}\left(\frac{1}{1+y}\right)(\cos t)$$

$$+ \frac{1}{\sqrt{1-\left(\frac{1+x}{1+y}\right)^2}}\left(-\frac{1+x}{(1+y)^2}\right)(-\sin t).$$

Now, at $t = 0$, $x = \sin t = 0$, and $y = \cos t = 1$. Substituting $t = 0$, $x = 0$ and $y = 1$ into the equation just obtained for $\frac{dz}{dt}$, we obtain:

$$\frac{dz}{dt} = \frac{1}{\sqrt{1-\left(\frac{1}{1+1}\right)^2}}\left[\frac{1}{1+1}\cdot 1 - \frac{1}{(1+1)^2}\cdot 0\right] = \frac{1}{\sqrt{3}}.$$

• PROBLEM 568

In the Ohm's Law equation, $V = IR$, find dV if R is constant, and also if R is not constant.

Solution: $\frac{dV}{dI} = R$. Therefore: $dV = R\,dI$. If R is not constant, we write, by definition:

$$dV = \frac{\partial V}{\partial I}dI + \frac{\partial V}{\partial R}dR$$

$$= R\,dI + I\,dR .$$

• PROBLEM 569

Find the approximate volume of a thin cylindrical cup of inside diameter 4 inches, inside height 6 inches, and thickness $\frac{1}{16}$ inch.

Solution: This problem can be solved by means of differentials. Let x = diameter, y = height, and V = volume. Then, the volume of the cylinder,

$$V = \frac{\pi x^2}{4}\cdot y .$$

By definition, $dV = \frac{\partial V}{\partial x}dx + \frac{\partial V}{\partial y}dy$. In this example $\frac{\partial V}{\partial x} = \frac{\pi x}{2}y$

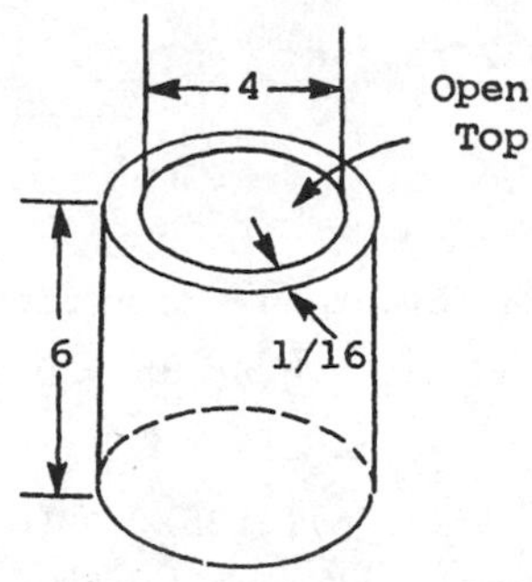

and $\frac{\partial V}{\partial y} = \frac{\pi x^2}{4}$. By substitution, we have,

$$dV = \frac{\pi x}{2} \cdot y \cdot dx + \frac{\pi x^2}{4} \cdot dy$$

We wish to find the value of dV. This is the approximate volume. To obtain dV we must substitute values for x, y, dx and dy. We use x = 4, the inside diameter, and y = 6, inside height.

To determine dx and dy, let us observe the following figure. From this figure we see that $dx = \frac{1}{16} + \frac{1}{16} = \frac{1}{8}$, the difference between the inner and outer diameters, and $dy = \frac{1}{16}$, the difference between inner and outer heights. Two diameter increments are required, but only one thickness increment. Substituting these values, we have:

$$dV = \frac{\pi \cdot 4}{2} \cdot 6 \cdot \frac{1}{8} + \frac{\pi \cdot 4^2}{4} \cdot \frac{1}{16} = \frac{3\pi}{2} + \frac{\pi}{4} = \frac{7\pi}{4} = 5.4978 \text{ cu. in.}$$

Approximate volume = 5.50 cu. in.

$$\text{The exact volume} = \frac{\pi}{4} \cdot \overline{4.125}^2 \cdot 6.0625 - \frac{\pi}{4} \cdot \overline{4}^2 \cdot 6$$

$$= \frac{\pi}{4}(103.16 - 96.00) = 0.7854 \cdot 7.16$$

$$= 5.62 \text{ cu. in.}$$

The error of .12 cu. in. is about 2%.

• PROBLEM 570

A cylindrical piece of steel is initially 8 in. long and has a diameter of 8 in. During heat treating, the length and the diameter each increase by 0.1 in. Find the approximate increase in the volume of the piece during heat treatment.

Solution: We solve this problem by using differentials. The volume of a cylinder is given by

$$v = \pi r^2 h,$$

where r is the radius and h is the height of the cylinder. Writing the total differential of this equation, we have:

$$dv = \frac{\partial v}{\partial r} dr + \frac{\partial v}{\partial h} dh.$$

But

$$\frac{\partial v}{\partial r} = 2\pi rh, \text{ and } \frac{\partial v}{\partial h} = \pi r^2 .$$

Substituting, we have:

$$dv = 2\pi rhdr + \pi r^2 dh$$

Because $dv \approx \Delta v$ we wish to obtain a numerical value for dv. We use the initial values: $r = 4$, $h = 8$, $dh = 0.1$, and $dr = .05$. Substituting these values, we have:

$$dv = 2\pi(4)(8)(0.05) + \pi(4)^2(0.1)$$
$$= 10.05 + 5.03 = 15.08 \text{ in.}^3$$

The exact change in volume can be found by taking the difference between the volume when $r = 4.05$ and $h = 8.10$ and the volume when $r = 4$ and $r = 8$. The exact value is:

$$\Delta v = 15.53 \text{ in.}^3$$

We observe that $dv \approx \Delta v$.

• PROBLEM 571

A closed metal can in the shape of a right-circular cylinder is to have an inside height of 6 in., an inside radius of 2 in., and a thickness of 0.1 in. If the cost of the metal to be used is 10 cents per in.^{3}, use differentials to find the approximate cost of the metal to be used in manufacturing the can.

Solution: The formula for the volume of a right-circular cylinder, where the volume is V in.^{3}, the radius is r in., and the height is h in., is:

$$V = \pi r^2 h .$$

The exact volume of metal in the can is the difference between the volumes of two right-circular cylinders. The smaller of the two cylinders has $r = 2$, $h = 6$. The larger cylinder has $r = 2$, the inside radius, plus .1, the thickness of the cylinder. $h = 6$, the inside height, plus .2, the thickness of the top and bottom of the can.

ΔV would yield the exact volume of metal, but because we only want an approximate value, and since $dV \approx \Delta V$, we find dV instead. We have, by definition,

$$dV = \frac{\partial V}{\partial r} dr + \frac{\partial V}{\partial h} dh,$$

and we find:

$$\frac{\partial V}{\partial r} = 2\pi rh, \text{ and } \frac{\partial V}{\partial h} = \pi r^2 .$$

Now, by substitution, $dV = 2\pi rh\, dr + \pi r^2 dh$. We wish to obtain a numerical value for dV. To do this we must substitute for r,h,dr and dh in the equation. Because $r = 2$, $h = 6$, $dr = 0.1$, and $dh = 0.2$, we have

$$dV = 2\pi(2)(6)(0.1) + \pi(2)^2(0.2)$$
$$= 3.2\pi .$$

Hence, $\Delta V \approx 3.2\pi$, and so there are approximately 3.2π in.^{3} of metal in the can. Since the cost of the metal is 10 cents per in.^{3} and $10 \cdot 3.2\pi = 32\pi \approx 100.53$, the approximate cost of the metal to be used in the manufacture of the can is $1.

• PROBLEM 572

The power delivered to a resistor in an electric circuit is given by the equation: $P = i^2R$. Find the approximate change in power if the current changes from 4.0 to 4.1 amps, and the resistance from 22.0 to 22.4 ohms.

Solution: By definition, $dP = \frac{\partial P}{\partial i} di + \frac{\partial P}{\partial R} dR$.

We find: $\frac{\partial P}{\partial i} = 2iR$ and $\frac{\partial P}{\partial R} = i^2$. Therefore, $dP = (2iR)di + (i^2)dR$.

We are given that i changes from 4.0 to 4.1 amps, Therefore, $di = 0.1$. Also, R changes from 22.0 to 22.4 ohms. Therefore, $dR = 0.4$. Evaluating, we have:

$$\Delta P \approx dP = \left[2(4.0)(22.0)\right](0.1) + (4.0)^2(0.4)$$
$$= 17.6 + 6.4 = 24 \text{ watts.}$$

• PROBLEM 573

An acute angle A of a right triangle is computed from measurement of the sides a and b. If 1% error is assumed in each measurement, estimate the largest possible error in the calculated value of A.

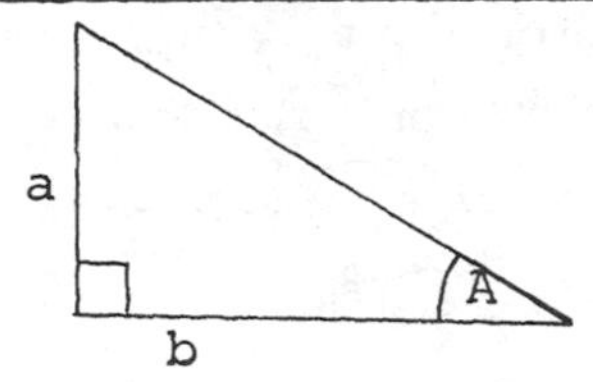

Solution: From the given information we can draw the figure, from which we have; $\tan A = a/b$, $A = \tan^{-1}(a/b)$.

We wish to find dA, which $\approx$ the possible error in the calculated value of A. By definition,

$$dA = \frac{\partial A}{\partial a} da + \frac{\partial A}{\partial b} db .$$

In this example we find:

$$\frac{\partial A}{\partial a} = \frac{1}{1+\left(\frac{a}{b}\right)^2} \cdot \frac{b}{b^2} \quad \text{and} \quad \frac{\partial A}{\partial b} = \frac{1}{1+\left(\frac{a}{b}\right)^2} \cdot \left(-\frac{a}{b^2}\right) .$$

Substituting we have,

$$dA = \frac{b}{b^2+a^2} \cdot da + \left(-\frac{a}{b^2+a^2}\right) \cdot db$$

$$= \frac{bda}{b^2+a^2} - \frac{adb}{b^2+a^2}$$

$$= \frac{bda-adb}{a^2+b^2} .$$

We are given the percentage error in each measurement, a and b, as 1%. Since percentage error = relative error × 100, and the relative errors in this example are $\frac{da}{a}$ and $\frac{db}{b}$, we can write,

$$\frac{da}{a}(100) = \pm 1 \quad \text{and} \quad \frac{db}{b}(100) = \pm 1 ,$$

or

$$\frac{da}{a} = \pm .01 \quad \text{and} \quad \frac{db}{b} = \pm .01 .$$

Then,

$$da = \pm .01 a \quad \text{and} \quad db = \pm .01 b.$$

In order to approximate the <u>largest</u> possible error we let $da = 0.01a$ and $db = -0.01b$. Then, substituting for da and db, we have:

$$dA = \frac{ab}{50(a^2+b^2)} \text{ radians.}$$

• PROBLEM 574

> The radius of a right circular cone is measured as 5 in. with a possible error of 0.02 in., and the altitude as 8 in. with a possible error of 0.025 in. Find the maximum possible percentage error in the volume as computed from these measurements.

<u>Solution:</u> We have the values: $r = 5$ in., $\Delta r \approx dr = \pm 0.02$ in., $h = 8$ in., $\Delta h \approx dh = \pm 0.025$ in. The double sign must be used, since the error may be positive or negative. The function is:

$$V = \frac{\pi}{3} r^2 h.$$

We wish to find dV. By definition, $dV = \frac{\partial V}{\partial r} dr + \frac{\partial V}{\partial h} dh$. In this example, $\frac{\partial V}{\partial r} = \frac{2\pi}{3} rh$ and $\frac{\partial V}{\partial h} = \frac{\pi}{3} r^2$. By substitution we have, $dV = \frac{2\pi}{3} rh\, dr + \frac{\pi}{3} r^2 dh$. Dividing by $V = \frac{\pi}{3} r^2 h$, we obtain:

$$\frac{dV}{V} = 2\frac{dr}{r} + \frac{dh}{h} .$$

Using the positive values for dh and dr, we write:

$$\frac{dV}{V} = .008 + 0.0031 = 0.0111,$$

which is the approximate relative error in V; or

$$100 \frac{dV}{V} = 1.11\% ,$$

which is the approximate percentage error in V.

If negative values are taken for dr and dh, the results are numerically the same as those above. However, if the values of dr and dh differ in sign, the results are numerically less than those given above. Hence only one set of calculations is necessary to determine the maximum possible errors.

• PROBLEM 575

The lengths of two sides of a triangle, which include an angle of 30°, are 27" and 13". In measuring these sides of the triangles, possible errors of 0.1" and 0.05", respectively, may be incurred. Compute the approximate largest possible error in the area of the triangle resulting from the measurements.

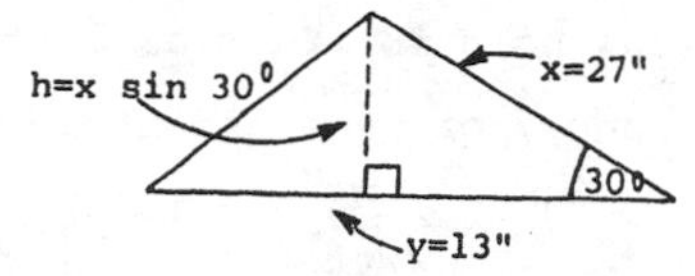

Solution: We can draw the accompanying figure from the given information. The area of a triangle, $A = \frac{1}{2} \times \text{altitude} \times \text{base}$, and for this example, $A = (1/2)xy \sin 30^\circ = (1/4)xy$, since $\sin 30^\circ = 1/2$. An approximate error in A due to errors in x and y is dA.

By definition, $dA = \frac{\partial A}{\partial x} dx + \frac{\partial A}{\partial y} dy$. In this example $\frac{\partial A}{\partial x} = \frac{1}{4}y$ and $\frac{\partial A}{\partial y} = \frac{1}{4}x$. Therefore, by substitution, we have:

$$dA = \frac{1}{4}\left(ydx + xdy\right).$$

We are given: $x = 27$in., $y = 13$ in., $dx = \pm 0.10$ in., and $dy = \pm 0.05$ in. Since the largest possible value for dA is desired, we assume dx and dy to have the same sign, their coefficients having the same sign. Substitution gives: $dA = (1/4)\cdot 13\cdot(1/10) + (1/4)\cdot 27\cdot(1/20) = 53/80 = .663$ sq. in. The actual error is $\Delta A = A_2 - A_1$, where A_2 is computed by using $x = 27.1"$ and $y = 13.05"$, and A_1 by using $x = 27"$, $y = 13"$. $A_1 = 87.75$ sq.in. Such computation being usually laborious, the value of dA is used as an approximation of the error ΔA.

The relative error is defined as $\Delta A/A$, but the approximate relative error dA/A is generally adequate. The percentage error is 100 times the relative error. It is convenient to take the logarithm of a function if the approximate relative error is desired. Thus for this example,

$$\log A = \log x + \log y - \log 4.$$

Then the differential gives

$$\frac{dA}{A} = \frac{dx}{x} + \frac{dy}{y} = \frac{0.10}{27} + \frac{0.05}{13} = 0.0075.$$

This shows that an approximation for the relative error of a function is the sum of the approximations of the relative errors of each factor. This, expressed as a percentage, is (3.4)%.

Going from this particular problem to a general case, and employing a function u, the expressions are readily extended to the cases of three or more variables. Thus, if $u=f(x,y,z)$, we have

$$\Delta u = \frac{\partial u}{\partial x}\cdot\Delta x + \frac{\partial u}{\partial y}\cdot\Delta y + \frac{\partial u}{\partial z}\cdot\Delta z + e_1\cdot\Delta x + e_2\cdot\Delta y + e_3\cdot\Delta z,$$

The differential of u is defined as

$$du = \frac{\partial u}{\partial x}\cdot dx + \frac{\partial u}{\partial y}\cdot dy + \frac{\partial u}{\partial z}\cdot dz\ ,$$

and is used to approximate the value of Δu just as da is used to approximate Δa.

Exactly similar relations hold for four or more independent variables as long as the number of independent variables is finite.

• PROBLEM 576

Using the ideal gas law with k = 10, find the rate at which the temperature is changing at the instant when the volume of the gas is 120 in.^{3} and the gas is under a pressure of 8 lb/in.^{2} if the volume is increasing at the rate of 2 in.^{3}/sec and the pressure is decreasing at the rate of 0.1 lb/in.^{2} per sec.

Solution: Let

t = the time, in seconds, that has elapsed since the volume of gas started to increase.
T = the temperature, in degrees, at t sec.
P = the pressure, in pounds per square inch at t sec.
V = the volume of gas, in cubic inches, at t sec.

The ideal gas law states that PV = kT. Since k is given as 10, PV = 10T, and $T = \frac{PV}{10}$ at time (t). The rate of change of the temperature is the change in temperature (T), over the change in time (t), or $\frac{dT}{dt}$. Using the chain rule, we can write an equation for $\frac{dT}{dt}$:

$$\frac{dT}{dt} = \frac{\partial T}{\partial P}\frac{dP}{dt} + \frac{\partial T}{\partial V}\frac{dV}{dt}.$$

Now, $\frac{\partial T}{\partial P}$ = partial derivative of $T = \frac{PV}{10}$ with respect to P, therefore, $\frac{\partial T}{\partial P} = \frac{V}{10}$. $\frac{dP}{dt}$ = rate of change of the pressure. This is given as decreasing at the rate of .1 lb/in.^{2} per second. Therefore, $\frac{dP}{dt} = -1.$

$\frac{\partial T}{\partial V}$ = partial derivative of $T = \frac{PV}{10}$ with respect to V, therefore, $\frac{\partial T}{\partial V} = \frac{P}{10}$. $\frac{dV}{dt}$ = rate of change of the volume which is given as increasing at 2 in.^{3} per second, so that $\frac{dV}{dt} = 2$.

Substituting these values, as well as the values for pressure and volume at the given instant, P = 8 and V = 120, into the equation for $\frac{dT}{dt}$, we obtain:

$$\frac{dT}{dt} = \frac{\partial T}{\partial P}\frac{dP}{dt} + \frac{\partial T}{\partial V}\frac{dV}{dt}$$

$$= \frac{V}{10}\frac{dP}{dt} + \frac{P}{10}\frac{dV}{dt}$$

$$= \frac{120}{10}(-0.1) + \frac{8}{10}(2)$$

$$= -1.2 + 1.6$$

$$= 0.4.$$

Therefore the temperature is increasing at the rate of 0.4 degrees per second at the given instant.

• PROBLEM 577

What is the rate of change of the volume of a cone if the radius of its base R and its height H vary at the same time?

Solution: We know that the volume of the cone is

$$V = \frac{1}{3}\pi R^2 H.$$

Since V is a function of both R and H, the total differential of V is

$$dV = \frac{\partial V}{\partial R}\cdot dR + \frac{\partial V}{\partial H}\cdot dH$$

or

$$dV = \frac{2\pi RH}{3}\cdot dR + \frac{\pi R^2}{3}\cdot dH$$

That is, the change in volume is

$$\frac{2\pi RH}{3}\cdot(\text{change in radius})$$

$$+ \frac{\pi R^2}{3}\cdot(\text{change in height}).$$

• PROBLEM 578

If the height of a cylinder increases at the rate of 0.1 inch per minute and the radius of the base decreases at the rate of 0.2 inch per minute, how fast is the volume of the cylinder changing when the height is 12 inches and the radius of the base is 8 inches?

<u>Solution:</u> Let V = volume, r = radius, h = height.

$$V = \pi r^2 h, \frac{dV}{dt} = \frac{\partial}{\partial t}\left[\frac{\partial V}{\partial r}dr + \frac{\partial V}{\partial h}dh\right]$$

$$= 2\pi rh\frac{dr}{dt} + \pi r^2\frac{dh}{dt}.$$

But: $r = 8,\ h = 12,\ \frac{dr}{dt} = -0.2,\ \frac{dh}{dt} = 0.1$

Thus,

$$\frac{dV}{dt} = 2\pi(8)(12)(-0.2) + \pi(64)(0.1) = -32\pi.$$

• PROBLEM 579

The radius of a right circular cone decreases at 2 in/min, and the height increases at 3 in/min. Determine the rate of change in its volume when height is 8 in and the radius is 4 in.

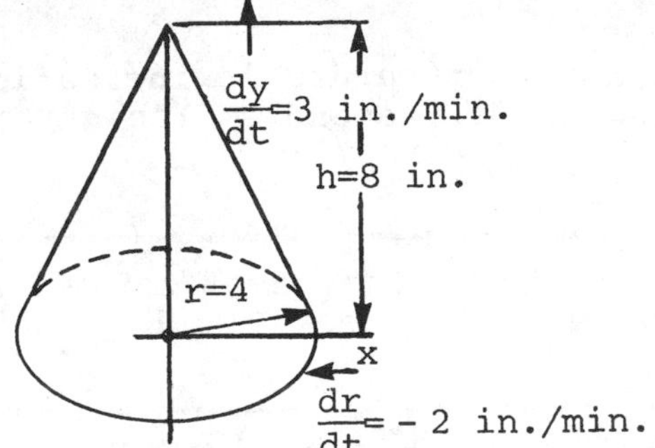

<u>Solution:</u> The volume of the cone is:

$$V = \frac{1}{3}\pi r^2 h,$$

where the radius r and the height h are functions of time. To obtain change of volume with time, we write

$$\frac{dV}{dt} = \frac{d}{dt}\left(\frac{\partial V}{\partial r}dr + \frac{\partial V}{\partial h}dh\right)$$

or

$$\frac{dV}{dt} = \frac{\partial V}{\partial r}\frac{dr}{dt} + \frac{\partial V}{\partial h}\frac{dh}{dt}$$

which gives

$$\frac{dV}{dt} = \frac{2}{3}\pi rh\frac{dr}{dt} + \frac{1}{3}\pi r^2\frac{dh}{dt}.$$

Substituting:

for decreasing r, $\frac{dr}{dt} = -2$ in/min;

and for increasing height, $\frac{dh}{dt} = 3$ in/min, we have,

for dV/dt, at the instant when h = 8 in and r = 4 in:

$$\frac{dV}{dt} = \frac{2}{3}\pi(4)(8)(-2) + \frac{1}{3}\pi(4)^2(3)$$

$$= -\frac{4}{3}\pi 20.$$

$$\approx -84$$

The volume decreases at the rate of about 84 cu in/min.

• PROBLEM 580

What is the rate of change of the hypotenuse u of a variable right triangle with respect to time when x = 10 and y = 7, if x is increasing at the rate of 4 inches per minute and y is increasing at the rate of 6 inches per minute?

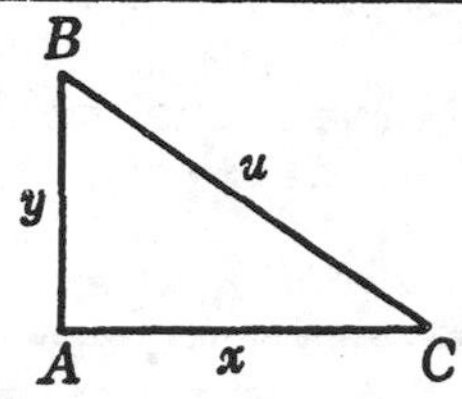

Solution: We formulate the given quantities:

$\frac{dx}{dt} = 4$, and $\frac{dy}{dt} = 6$, where t = time in minutes.

Now: $u^2 = x^2 + y^2$, or $u = \sqrt{x^2 + y^2}$

Then $\frac{\partial u}{\partial x} = \frac{x}{\sqrt{x^2 + y^2}}$ and $\frac{\partial u}{\partial y} = \frac{y}{\sqrt{x^2 + y^2}}$

$$\frac{du}{dt} = \frac{\partial u}{\partial x} \cdot \frac{dx}{dt} + \frac{\partial u}{\partial y} \cdot \frac{dy}{dt} = \frac{x}{\sqrt{x^2 + y^2}} \frac{dx}{dt} + \frac{y}{\sqrt{x^2 + y^2}} \frac{dy}{dt},$$

the total derivative or rate of change of u with respect to t.

At the given point (10,7),

$$\frac{du}{dt} = \frac{10}{\sqrt{(10)^2 + (7)^2}} \cdot 4 + \frac{7}{\sqrt{(10)^2 + (7)^2}} \cdot 6$$

$$= \frac{40}{12.2066} + \frac{42}{12.2066}$$

or $\frac{du}{dt} = 6.72$ in./min.

Therefore, at the instant when x = 10 and y = 7, the hypotenuse is increasing at the rate of 6.72 in./min.

• PROBLEM 581

The radius of a right circular cone is increasing

3 inches per second and its height is decreasing 4 inches per second. How fast is its volume changing when r = 6 inches and h = 12 inches?

Solution: The volume of the cone is $V = (1/3)\pi r^2 h$. Taking partial derivatives,

$$\frac{dV}{dt} = \frac{d}{dt}\left(\frac{\partial V}{\partial r}dr + \frac{\partial V}{\partial h}dh\right) = \frac{\partial V}{\partial r}\cdot\frac{dr}{dt} + \frac{\partial V}{\partial h}\cdot\frac{dh}{dt} = \frac{2}{3}\pi rh\cdot\frac{dr}{dt} + \frac{1}{3}\pi r^2\cdot\frac{dh}{dt}.$$

r = 6", h = 12", dr/dt = 3 in/sec, and dh/dt = -4 in/sec. (The negative sign indicates that the height is decreasing with time.)

Substituting,

$$\frac{dV}{dt} = \frac{2}{3}\pi\cdot 6\cdot 12\cdot 3 + \frac{1}{3}\pi\cdot 36(-4) = 96\pi \text{ cu in/sec.}$$

Since dV/dt is positive, the volume is increasing 96π cu.in/sec at that instant.

• **PROBLEM** 582

The length, width, and depth of a rectangular box are each increasing at the rate of 2 inches per minute. Find the rate at which the volume of the box is increasing at the instant when the length is 6 feet, the width is 4 feet, and the depth is 3 feet.

Solution: Let x denote the depth, w the width, and l the length. The volume of the box

$$v = xwl.$$

Then its rate of increase can be found by partially differentiating the volume expression and, since all are functions of time,

$$\frac{dv}{dt} = \frac{d}{dt}\left[\frac{\partial v}{\partial x}dx + \frac{\partial v}{\partial y}dy + \frac{\partial v}{\partial z}dz\right] = wl\frac{dx}{dt} + xl\frac{dw}{dt} + xw\frac{dl}{dt}$$

We know that $\frac{dx}{dt} = \frac{dw}{dt} = \frac{dl}{dt} = 2$ in/min $= \frac{1}{6}$ ft/min, and we wish to find dv/dt at the instant, at which x = 3 ft, w = 4 ft, l = 6 ft.

Thus

$$\frac{dv}{dt} = \frac{1}{6}(4\cdot 6 + 3\cdot 6 + 3\cdot 4)$$

$$= \frac{1}{6}\cdot 54$$

or

$$\frac{dv}{dt} = 9\ \frac{\text{cubic feet}}{\text{minute}}.$$

This problem can also be solved by defining the dimensions in terms of x alone.

Let x denote the depth as before, then we can assign x + 1 to denote the width, and x + 3 the length. Hence the volume

$$v = x(x + 1)(x + 3) = x^3 + 4x^2 + 3x.$$

$$\frac{dv}{dt} = \left(3x^2 + 8x + 3\right)\frac{dx}{dt}.$$

Substituting 3 for x and 1/6 for dx/dt, we have

$$\frac{dv}{dt} = 54 \cdot \frac{1}{6} = 9 \text{ (ft)}^3/\text{min}.$$

• PROBLEM 583

Two sides and the included angle of a triangle are measured, and the third side is computed. (a) Find a formula for the approximate error in the third side due to errors in the three measurements.
(b) If the two sides and the included angle are 3 ± 0.1 in., 5 ± 0.1 in., and $60° \pm 10'$ respectively, find the maximum possible error in the third side.

Solution: Using the law of cosines, the third side f is given by:

$$f = \left(x^2 + y^2 - 2xy \cos \Theta\right)^{\frac{1}{2}},$$

where x and y are the measured sides, and Θ is the acute angle, which side x makes with side y.

(a) The change in the third side f due to errors in the three measurements is, from

$$f = f(x, y, \Theta),$$

$$df = \frac{\partial f}{\partial x} dx + \frac{\partial f}{\partial y} dy + \frac{\partial f}{\partial \Theta} d\Theta,$$

where

$$\frac{\partial f}{\partial x} = \frac{x - y \cos \Theta}{\sqrt{x^2 + y^2 - 2xy \cos \Theta}},$$

$$\frac{\partial f}{\partial y} = \frac{y - x \cos \Theta}{\sqrt{x^2 + y^2 - 2xy \cos \Theta}},$$

and

$$\frac{\partial f}{\partial \Theta} = \frac{xy \sin \Theta}{\sqrt{x^2 + y^2 - 2xy \cos \Theta}}.$$

Hence,

$$df = \frac{1}{f}\Big[(x - y \cos \Theta)dx + (y - x \cos \Theta)dy + xy \sin \Theta \, d\Theta\Big].$$

(b) $x = 3$ in., $dx = \pm 0.1$ in.; $y = 5$ in., $dy = \pm 0.1$ in., and, for the angle in radians,

$$\Theta = \frac{\Pi}{3}, \quad \text{and} \quad d\Theta = \frac{\pi}{180} \cdot \frac{1}{6}.$$

Given $\cos \pi/3 = \frac{1}{2}$, $\sin \pi/3 = \sqrt{3}/2$

$$f = \left(9 + 25 - 30 \cos \pi/3\right)^{\frac{1}{2}} = \sqrt{19}$$

$$\frac{\partial f}{\partial x} = \left[3 - 5\left(\frac{1}{2}\right)\right] = \frac{1}{2}$$

$$\frac{\partial f}{\partial y} = 5 - \frac{3}{2} = \frac{7}{2},$$

$$\frac{\partial f}{\partial \theta} = 3\ (5) \sin \pi/3 = \frac{15}{2}\sqrt{3}.$$

Thus

$$df = \left[\frac{1}{2}(0.1) + \frac{7}{2}(0.1) + \frac{15}{2}\sqrt{3}\left(\frac{\pi}{180} \cdot \frac{1}{6}\right)\right] \sqrt{19}$$

$$= \frac{0.4377}{\sqrt{19}}$$

$$\approx 0.1.$$

• PROBLEM 584

(a) Find a formula for the approximate error in the quotient, $\frac{x}{y}$, of two numbers, due to small errors in the numbers.

(b) Evaluate $\frac{4.02}{0.597}$ to three significant figures.

Solution: Let $z = \frac{x}{y}$.

The total derivative,

$$dz = \frac{\partial z}{\partial x} dx + \frac{\partial z}{\partial y} dy = \frac{1}{y} dx - \frac{x}{y^2} dy.$$

If dx and dy are the errors, then the contribution to the quotient $\frac{x}{y}$ is expressed as:

$$\Delta z = \frac{1}{y} \Delta x - \frac{x}{y^2} \Delta y.$$

(b) Let $x = 4$ and $x + \Delta x = 4.02$, so that $\Delta x = 0.02$.

Similarly, let $y = 0.6$ and hence $\Delta y = -0.003$, while $y + \Delta y = 0.597$. Thus,

$$dz = \frac{1}{0.6}(0.02) - \frac{4}{(0.6)^2}(-0.003)$$

$$= \frac{0.024}{0.36}$$

Since $z = \frac{x}{y} = \frac{4}{0.6}$,

we obtain:

$$z + dz = \frac{4.02}{0.597} = \frac{4}{0.6} + \frac{0.024}{0.36}$$

$$= 6.67 + 0.0667 = 6.7367.$$

CHAPTER 23

FUNDAMENTAL INTEGRATION

In carrying out integration, the first step is to determine whether the function to be integrated fits one of the basic integration formulas. If it doesn't, the next step is to try to convert the function into a form to which the basic formulas may be applied.

In determining which basic rule is applicable, try the power rule first, since this rule is usually involved most often. This rule states that:

$$\int x^n \, dx = \frac{x^{n+1}}{n+1} + C \, .$$

If the function does not fit this rule directly, try to find some function u, such that the given function may be made to fit the form:

$$\int u^n \, du = \frac{u^{n+1}}{n+1} + C \, .$$

This approach is particularly useful where trigonometric functions are involved. In trying to make functions fit this form, it is permissible to multiply by any constant under the integral sign, while, at the same time, to leave the function unchanged, we have to divide by the same constant outside the integral sign.

Other fundamental formulas can be applied after or in combination with the power rule.

Often one integral can be broken up to give several integrals, each of which may be simpler to integrate than the given one.

When the previous methods fail, integration by parts should be attempted. This is done according to the formula:

$$\int u \, dv = uv - \int v \, du \, ,$$

where u and v are functions of x, and where perhaps the second integral turns out to be simpler than the first.

• PROBLEM 585

Integrate the expression: $\int \frac{2^{\sqrt{x}}}{\sqrt{x}} dx$.

Solution: We can rewrite the given integral as $\int 2^{x^{1/2}} x^{-1/2} dx$. We can now consider the formula, $\int a^u du = \frac{a^u}{\ln a}$. Here we let $u = x^{1/2}$. Then $du = \frac{1}{2}x^{-1/2}dx$, and $a = 2$. Applying the formula, we obtain:

$$\int \frac{2^{\sqrt{x}}}{\sqrt{x}} dx = 2 \int 2^{x^{1/2}} \frac{1}{2}x^{-1/2} dx$$

$$= 2\left[\frac{2^{x^{1/2}}}{\ln 2}\right] + C = \frac{2^1\ 2^{x^{1/2}}}{\ln 2} + C = \frac{2^{1+x^{1/2}}}{\ln 2} + C$$

$$= \frac{2^{1+\sqrt{x}}}{\ln 2} + C.$$

• PROBLEM 586

Integrate the expression: $\int \sqrt{10^{3x}}\ dx$.

Solution: $\int \sqrt{10^{3x}}\ dx = \int 10^{3x/2}\ dx$. We wish to apply the formula $\int a^u du = \frac{a^u}{\ln a}$, where a is a positive constant > 0 and $\neq 1$. In this case we have $a = 10$, and $u = \frac{3}{2}x$. Then $du = \frac{3}{2}dx$; We write:

$$\int 10^{3x/2}\ dx = \frac{2}{3} \int 10^{3x/2}\ \frac{3}{2}dx.$$

Applying the formula for $\int a^u du$, we have:

$$\frac{2}{3} \cdot \frac{10^u}{\ln 10} + C = \frac{2 \cdot 10^{3x/2}}{3 \ln 10} + C.$$

• PROBLEM 587

Integrate the expression: $\int a^x b^x\ dx$.

Solution: In evaluating this integral we rewrite it in a form suitable for application of the formula: $\int k^u du = \frac{k^u}{\ln k}$, with $k = ab$, $u = x$, $du = dx$. Doing this, we obtain:

$$\int a^x \cdot b^x \cdot dx = \int (ab)^x \cdot dx = \frac{(ab)^x}{\ln(ab)} = \frac{a^x \ b^x}{\ln a + \ln b} + C.$$

• PROBLEM 588

$\frac{dy}{dx} = \frac{1}{x + a}$. Find $y = F(x)$.

Solution: $dy = \frac{1}{x + a}\ dx.$

Then, $y = \int \frac{1}{x + a} \cdot dx = \int \frac{dx}{x + a}.$

We can now make use of the formula: $\int \frac{du}{u} = \ln u + C.$ Here $u = x + a$ and $du = dx$. Applying the formula, we obtain:

$$y = \int \frac{dx}{x + a} = \ln(x + a) + C$$

• PROBLEM 589

$\frac{dy}{dx} = (a + bx)^{-1}$. Find $y = F(x)$.

Solution: $\frac{dy}{dx} = (a + bx)^{-1}$ can be rewritten as $dy = (a + bx)^{-1} dx = \frac{dx}{a + bx}$. We can now write:

$$\int dy = \int \frac{dx}{a + bx}$$

or,

$$y = \int \frac{dx}{a + bx}.$$

In integrating this expression, we wish to consider the formula: $\int \frac{du}{u} = \ln u + C$. In this case, we let $u = a + bx$. Then $du = bdx$ (treating a and b as constants). Applying the formula, we obtain:

$$y = \int \frac{dx}{a + bx} = \frac{1}{b} \int \frac{bdx}{a + bx} = \frac{1}{b} \ln(a + bx) + C$$

• PROBLEM 590

Integrate the expression: $\int \frac{2x}{(x^2 + 1)} dx.$

Solution: Use the rule: $\int \frac{du}{u} = \ln |u| + C$, letting $u = x^2 + 1$, making $du = 2xdx$.

$$\int \frac{2xdx}{x^2 + 1} = \ln(x^2 + 1) + C.$$

We may omit the absolute value sign here, since a number, x, when squared, is always positive.

• **PROBLEM 591**

Integrate the expression: $\int \frac{x}{(4 - x^2)} dx$.

Solution: Let $u = 4 - x^2$ from which $du = -2xdx$. Using the rule: $\int \frac{du}{u} = \ln |u| + C$ we obtain:

$$\int \frac{(-2x)}{4 - x^2} dx = \ln |4 - x^2| + C.$$

But, to make this result applicable to the original problem, we require a (-2) in the numerator to obtian the form $\frac{du}{u}$. Because this is a constant, it is permissible to multiply the numerator under the integral sign by -2, as long as we multiply the integral by $\left(-\frac{1}{2}\right)$ outside of the integral sign, in order to leave the resultant value unchanged. Hence,

$$\int \frac{x}{4 - x^2} dx = -\frac{1}{2} \int \frac{(-2x)}{4 - x^2} dx.$$

Now we can use $-\frac{1}{2} \int \frac{du}{u} = -\frac{1}{2} \ln |u| + C$

$$= \frac{1}{2} \ln |4 - x^2| + C.$$

• **PROBLEM 592**

Integrate the expression: $\int \frac{dx}{1 + e^x}$.

Solution: We wish to convert the given integral into the form $\int \frac{du}{u}$. If we multiply $\frac{1}{1+e^x}$ by $\frac{e^{-x}}{e^{-x}}$ (which is equal to 1) we obtain:

$$\frac{e^{-x}(1)}{e^{-x}(1+e^x)} = \frac{e^{-x}}{e^{-x} + e^0} = \frac{e^{-x}}{e^{-x} + 1}.$$

In integrating this, we apply the formula, $\int \frac{du}{u} = \ln |u| + C$, letting $u = e^{-x} + 1$. Then $du = -e^{-x}dx$. We obtain:

$$\int \frac{e^{-x}}{e^{-x} + 1} dx = -\int \frac{-e^{-x}dx}{e^{-x} + 1} = -\ln(1 + e^{-x}) + C.$$

• **PROBLEM 593**

Integrate: $\int \frac{2x}{x + 1} dx$.

Solution: To integrate the given expression we manipulate the integrand to obtain the form $\int \frac{du}{u}$. This can be done as follows:

$$\int \frac{2x}{x+1}dx = 2\int \frac{x}{x+1}dx$$

$$= 2\int \left(\frac{x+1}{x+1} - \frac{1}{x+1}\right)dx$$

$$= 2\int \left(1 - \frac{1}{x+1}\right)dx$$

$$= 2\int dx - 2\int \frac{dx}{x+1}.$$

Now, applying the formula $\int \frac{du}{u} = \ln u$, we obtain:

$$\int \frac{2x}{x+1}dx = 2x - 2\ln(x+1) + C.$$

• **PROBLEM** 594

> $\frac{dy}{dx} = \varepsilon^x$. Find $y = F(x)$.

Solution: $dy = \varepsilon^x \cdot dx$,

then, $y = \int \varepsilon^x \cdot dx.$

We can now apply the formula, $\int e^u du = e^u + C$. Here $u = x$ and $du = dx$, so we obtain:

$$y = \int e^x \cdot dx = e^x + C = \text{the integral.}$$

• **PROBLEM** 595

> If $\frac{dy}{dx} = \varepsilon^{-x}$, find $y = F(x)$.

Solution: $dy = \varepsilon^{-x} \cdot dx$

$$y = \int \varepsilon^{-x} \cdot dx.$$

We now integrate using the formula, $\int e^u du = e^u + C$, letting $u = -x$, $du = -dx$. We obtain:

$$\int e^{-x}dx = -\int e^{-x} - dx = -e^{-x} + C$$

• **PROBLEM** 596

> Integrate the expression: $\int x^2 e^{x^3} dx.$

Solution: To integrate this expression, we wish to consider the formula: $\int e^u du = e^u + C$. In this case we have $u = x^3$. Then $du = 3x^2 dx$. Applying the formula, we obtain:

$$\int x^2 e^{x^3} dx = \frac{1}{3}\int e^{x^3}(3x^2 dx) = \frac{1}{3}e^{x^3} + C.$$

• **PROBLEM** 597

> Integrate: $\int \left(e^{2x} - e^{-2x}\right)^2 dx.$

Solution: To integrate the given expression we expand the integrand, obtaining:

$$\int \left(e^{4x} - 2e^0 + e^{-4x}\right)dx$$

$$= \int \left(e^{4x} - 2 + e^{-4x}\right)dx$$

$$= \int e^{4x}dx - 2\int dx + \int e^{-4x}dx.$$

We now apply the formula $\int e^u du = e^u$, obtaining:

$$\frac{1}{4}e^{4x} - 2x - \frac{1}{4}e^{-4x} + C.$$

• PROBLEM 598

Integrate: $\int (x^2 + 4)^5 \; 2x\,dx.$

Solution: We use the formula: $\int u^n du = \frac{u^{n+1}}{n+1} + C$, letting $u = x^2 + 4$, $du = 2x\,dx$, and $n = 5$. Then,

$$\int (x^2 + 4)^5 \; 2x\,dx = \frac{(x^2 + 4)^6}{6} + C.$$

• PROBLEM 599

Integrate: $\int (x^3 - 7)^8 \; 3x^2dx.$

Solution: We use the formula: $\int u^n du = \frac{u^{n+1}}{n+1} + C$, letting $u = (x^3 - 7)$, $du = 3x^2dx$, and $n = 8$. Therefore,

$$\int (x^3 - 7)^8 \; 3x^2dx = \frac{(x^3 - 7)^9}{9} + C.$$

• PROBLEM 600

Integrate: $\int (x^3 - 2x)^5(3x^2 - 2)dx.$

Solution: Let $u = x^3 - 2x$, $du = (3x^2 - 2)dx$, and $n = 5$. Now we use the formula: $\int u^n du = \frac{u^{n+1}}{n+1} + C$. Therefore,

$$\int (x^3 - 2x)^5(3x^2 - 2)dx = \frac{(x^3 - 2x)^6}{6} + C$$

• PROBLEM 601

Integrate the expression: $\int (x^3 - 3x^2)^5(3x^2 - 6x)dx.$

Solution: We can think of this as being in the form $\int u^n du$, where $u = x^3 - 3x^2$, $n = 5$, and $du = (3x^2 - 6x)dx$. Now we apply the formula for $\int u^n du$.

$$\int (x^3 - 3x^2)^5(3x^2 - 6x)dx = \frac{(x^3 - 3x^2)^6}{6} + C.$$

• PROBLEM 602

$\frac{dy}{dx} = 8.60x^{1.15}$. Find $y = F(x)$.

Solution: $\frac{dy}{dx} = 8.60x^{1.15}$ can be rewritten as $dy = 8.60x^{1.15}dx$. We can now write $\int dy = \int 8.60x^{1.15}$ or,

$$y = 8.60 \int x^{1.15} \cdot dx.$$

We can make use of the formula: $\int u^n du = \frac{u^{n+1}}{n+1}$, letting $u = x$, $du = dx$ and $n = 1.15$. Applying the formula, we obtain:

$$8.60 \cdot \frac{x^{1.15+1}}{1.15 + 1} = \frac{8.60x^{2.15}}{2.15}$$

$$= 4x^{2.15} + C.$$

• PROBLEM 603

Integrate the expression: $\int \sqrt[3]{x^2}\, dx$.

Solution: $\int \sqrt[3]{x^2}\, dx = x^{2/3}dx$.

Since this function is in the form $\int u^n du$, with $u = x$, $du = dx$ and $n = \frac{2}{3}$, we can apply the formula:

$\int u^n du = \frac{u^{n+1}}{n+1} + C$, obtaining:

$$\frac{x^{2/3+1}}{\frac{2}{3} + 1} + C = \frac{3}{5}x^{5/3} + C.$$

• PROBLEM 604

$\frac{dy}{dx} = \sqrt{5 + x}$. Find $y = F(x)$.

Solution: $dy = \sqrt{5 + x} \cdot dx$. Then

$$y = \int \sqrt{5 + x} \cdot dx = \int (5 + x)^{1/2}dx.$$

We can apply the formula for $\int u^n du$, with $u = (5 + x)$, $du = dx$ and $n = \frac{1}{2}$. We obtain:

$$y = \frac{2}{3}(5 + x)^{3/2}.$$

• PROBLEM 605

$\frac{dy}{dx} = (2x^2 + 1)(x^2 - 3x + 2)$. Find $y = F(x)$.

Solution: First we multiply out the two factors. We obtain:

$$\frac{dy}{dx} = 2x^4 - 6x^3 + 5x^2 - 3x + 2,$$

which can be written as,

$$dy = (2x^4 - 6x^3 + 5x^2 - 3x + 2)dx.$$

We can now write

$$\int dy = \int(2x^4 - 6x^3 + 5x^2 - 3x + 2)dx$$

or,

$$y = \int(2x^4 - 6x^3 + 5x^2 - 3x + 2)dx$$

$$= 2\int x^4 \cdot dx - 6\int x^3 \cdot dx + 5\int x^2 \cdot dx - 3\int x \cdot dx + 2\int dx.$$

We can now integrate by applying the formula for $\int u^n du$, obtaining:

$$y = \frac{2x^5}{5} - \frac{3x^4}{2} + \frac{5x^3}{3} - \frac{3x^2}{2} + 2x.$$

• PROBLEM 606

> Evaluate the expression: $\int_0^1 \left(x^{4/3} + 4x^{1/3}\right)dx.$

Solution: In evaluating the given integral we use the formula: $\int u^n du = \frac{u^{n+1}}{n+1}$. We obtain:

$$\int_0^1 \left(x^{4/3}+4x^{1/3}\right)dx = \frac{3}{7}x^{7/3} + 4 \cdot \frac{3}{4}x^{4/3}\Big|_0^1 .$$

Now, evaluating the integral between 0 and 1, we have:

$$\frac{3}{7}\left[\sqrt[3]{1^7} - 0\right] + 3\left[\sqrt[3]{1^4} - 0\right]$$

$$= \frac{3}{7} + 3$$

$$= \frac{24}{7}.$$

• PROBLEM 607

> Evaluate the expression: $\int\left(\frac{1}{x^4} + \frac{1}{\sqrt[4]{x}}\right)dx.$

Solution: $\int\left(\frac{1}{x^4} + \frac{1}{\sqrt[4]{x}}\right)dx = \int\left(x^{-4} + x^{-1/4}\right)dx.$

Both terms in this integral are in the form $\int u^n du$. Applying the formula, we have:

$$\frac{x^{-4+1}}{-4+1} + \frac{x^{-1/4+1}}{-\frac{1}{4}+1} + C$$

$$= \frac{x^{-3}}{-3} + \frac{x^{3/4}}{\frac{3}{4}} + C$$

$$= -\frac{1}{3x^3} + \frac{4}{3}x^{3/4} + C$$

• PROBLEM 608

$\frac{dy}{dx} = (a - bx)^n$. What is $y = F(x)$ when $n = 2$?

Solution: $\frac{dy}{dx} = (a - bx)^n$ can be rewritten as $dy = (a - bx)^n dx$. We can now write: $\int dy = \int (a - bx)^n dx$ or, $y = \int (a - bx)^n dx$. To integrate, we consider the formula:

$\int u^n du = \frac{u^{n+1}}{n+1} + C$, with $u = (a - bx)$ and $du = -bdx$. Applying the formula, we obtain:

$$y = \int (a - bx)^n dx = -\frac{1}{b} \cdot \frac{(a - bx)^{n+1}}{n + 1}$$

$$= -\frac{(a - bx)^{n+1}}{b(n + 1)} + C,$$

the integral in the general form.

For $n = 2$,

$$y = \int (a - bx)^2 \cdot dx = -\frac{(a - bx)^3}{3b} + C.$$

• PROBLEM 609

$\frac{dy}{dx} = (a + bx)^n$. What is $y = F(x)$ when $n = 1$, $n = 2$, and $n = -2$?

Solution: In solving, we first find the integral in the general form and then substitute and find the integral for $n = 1$, $n = 2$ and $n = -2$.

$\frac{dy}{dx} = (a + bx)^n$ can be rewritten as $dy = (a + bx)^n dx$. To find the integral, we write: $\int dy = \int (a + bx)^n dx$ or, $y = \int (a + bx)^n dx$. We now consider the formula: $\int u^n du = \frac{u^{n+1}}{n+1} + C$, with $u = (a + bx)$ and $du = bdx$. Applying the formula, we have:

$$y = \int (a + bx)^n \cdot dx$$

$$= \frac{(a + bx)^{n+1}}{b(n + 1)} + C,$$

the integral in the general form.

For $n = 1$,

$$\int (a + bx)^1 dx = \frac{(a + bx)^{1+1}}{b(1 + 1)} = \frac{(a + bx)^2}{2b} + C.$$

For $n = 2$,

$$\int (a + bx)^2 dx = \frac{(a + bx)^{2+1}}{b(2 + 1)} = \frac{(a + bx)^3}{3b} + C.$$

For $n = -2$,

$$\int \frac{dx}{(a + bx)^2} = \int (a + bx)^{-2} \cdot dx = \frac{(a + bx)^{-2+1}}{b(-2 + 1)}$$

$$= -\frac{1}{b(a + bx)} + C.$$

• PROBLEM 610

Evaluate the expression: $F(t) = \int_0^t x\sqrt{a^2 - x^2}\, dx$, $|t| \leq a$.

<u>Solution:</u> Let $u = a^2 - x^2$, then $du = -2xdx$. We use the formula: $\int u^n du = \frac{u^{n+1}}{n+1} + C$, with $u = a^2 - x^2$, $n = \frac{1}{2}$, $du = -2xdx$. We obtain:

$$\int x\sqrt{a^2 - x^2}\, dx = \int -\tfrac{1}{2}u^{1/2}du = -\tfrac{1}{2}\int u^{1/2}du$$

$$= -\frac{1}{2}\,\frac{u^{3/2}}{\frac{3}{2}} + C$$

$$= -\frac{1}{3}\left(a^2 - x^2\right)^{3/2} + C.$$

Evaluating the integral between t and 0, we have:

$$\int_0^t x\sqrt{a^2 - x^2}\, dx = \left[-\frac{1}{3}\left(a^2 - x^2\right)^{3/2}\right]_0^t$$

$$= -\frac{1}{3}\left(a^2 - t^2\right)^{3/2} - \left[-\frac{1}{3}\left(a^2 - 0^2\right)^{3/2}\right]$$

$$= -\frac{\left(a^2 - t^2\right)^{3/2}}{3} + \frac{a^3}{3}.$$

• PROBLEM 611

Integrate the expression: $\int \sqrt{3x + 4}\, dx$.

Solution: $\int\sqrt{3x + 4}\,dx = \int(3x + 4)^{1/2}dx.$

In integrating this function, we use the formula:

$\int u^n du = \frac{u^{n+1}}{n+1} + C$, with $u = (3x + 4)$, $du = 3dx$, and $n = \frac{1}{2}$. We obtain:

$$\int(3x + 4)^{1/2}dx = \frac{1}{3}\int(3x + 4)^{1/2}3dx$$

$$= \frac{1}{3}\left[\frac{(3x + 4)^{3/2}}{\frac{3}{2}}\right] + C$$

$$= \frac{2}{9}(3x + 4)^{3/2} + C.$$

• PROBLEM 612

Integrate the expression: $\int t(5 + 3t^2)^8 dt.$

Solution: $\int t(5 + 3t^2)^8 dt = \int(5 + 3t^2)^8 t\,dt.$ In integrating this function, we apply the formula:

$\int u^n du = \frac{u^{n+1}}{n+1} + C$, with $u = (5 + 3t^2)$, $du = 6t$, and $n = 8$. We obtain:

$$\int(5 + 3t^2)^8 t\,dt = \frac{1}{6}\int(5 + 3t^2)^8 6t\,dt$$

$$= \frac{1}{6}\cdot\frac{(5 + 3t^2)^9}{9} + C$$

$$= \frac{1}{54}(5 + 3t^2)^9 + C.$$

• PROBLEM 613

Integrate the expression: $\int(x^2 - 2x)^5(x - 1)dx.$

Solution: We can use the formula for $\int u^n du$ if we multiply $(x - 1)$ by the constant, 2. We also have to multiply by $\frac{1}{2}$, so as not to change the value of the integral. Thus, let $u = x^2 - 2x$, and $n = 5$. Then

$$du = (2x - 2)\,dx = 2(x - 1)dx.$$

$$\int(x^2 - 2x)^5(x - 1)dx = \int\frac{1}{2}(x^2 - 2x)^5\cdot 2(x - 1)dx$$

$$= \frac{1}{2}\int(x^2 - 2x)^5\cdot(2x - 2)dx.$$

Now we apply the formula for $\int u^n du$, including the factor of $\frac{1}{2}$.

$$= \frac{1}{2}\left[\frac{(x^2 - 2x)^6}{6} + C\right]$$

• **PROBLEM 614**

Integrate the expression: $\int (x^3 + 9)^2 \cdot x^2 dx.$

Solution: We can let $u = x^3 + 9$ and $n = 2$. Then $du = 3x^2 dx$. The expression is almost in the form $\int u^n du$. The only thing that is missing is the factor 3 in du. This can easily be remedied in the following way: We multiply x^2 by the necessary factor, 3, under the integral sign, while also multiplying by $\frac{1}{3}$, so as not to change the original integral. Then,

$$\int (x^3 + 9)^2 \cdot x^2 dx = \int \frac{1}{3}(x^3 + 9)^2 \cdot 3x^2 dx.$$

We can take the $\frac{1}{3}$ outside of the integral because it is a constant. This gives:

$$\frac{1}{3}\int (x^3 + 9)^2 \cdot 3x^2 dx.$$

Now we can apply the formula.

$$\frac{1}{3} \int u^n du = \frac{1}{3}\left[\frac{u^{n+1}}{n+1} + C\right]$$

$$= \frac{1}{3}\left[\frac{(x^3 + 9)^3}{3} + C\right].$$

• **PROBLEM 615**

Integrate the expression: $\int (x^3 - 3x)^{1/2}(x^2 - 1)dx.$

Solution: Let $u = x^3 - 3x$. Then

$$du = (3x^2 - 3)dx$$

$$= 3(x^2 - 1)dx.$$

Multiply $(x^2 - 1)$ by 3 and the integral by $\frac{1}{3}$. We can now use the formula for $\int u^n du$.

$$\int (x^3 - 3x)^{1/2}(x^2 - 1)dx = \frac{1}{3}\int (x^3 - 3x)^{1/2} 3(x^2 - 1)dx$$

$$= \frac{1}{3}\left[\frac{(x^3 - 3x)^{3/2}}{\frac{3}{2}} + C\right]$$

$$= \frac{2}{9}(x^3 - 3x)^{3/2} + C.$$

• **PROBLEM 616**

Integrate the expression: $\int \sqrt{x^2 + 1}\, x\, dx.$

Solution: $\int\sqrt{x^2 + 1} = \int(x^2 + 1)^{1/2}$.

Let $u = x^2 + 1$ and $n = \frac{1}{2}$. Then $du = 2x\,dx$.
We multiply x in the original integral by 2, and multiply the integral by $\frac{1}{2}$. Now we apply the formula for $\int u^n du$.

$$\int (x^2 + 1)^{1/2} x\,dx = \int\frac{1}{2}(x^2 + 1)^{1/2} 2x\,dx$$

$$= \frac{1}{2}\int(x^2 + 1)^{1/2} 2x\,dx$$

$$= \frac{1}{2}\left[\frac{(x^2 + 1)^{3/2}}{\frac{3}{2}} + C\right]$$

$$= \frac{(x^2 + 1)^{3/2}}{3} + C\,.$$

• **PROBLEM 617**

Integrate the expression: $\int x^2 \sqrt[5]{7 - 4x^3}\,dx$.

Solution: We can rewrite the given integral as $\int (7 - 4x^3)^{1/5} x^2\,dx$. We now wish to consider the formula: $\int u^n du = \frac{u^{n+1}}{n+1} + C$, with $u = (7 - 4x^3)$, $du = -12x^2 dx$ and $n = \frac{1}{5}$. Applying the formula, we obtain:

$$\int x^2 \sqrt[5]{7 - 4x^3}\,dx = -\frac{1}{12}\int (7 - 4x^3)^{1/5}(-12x^2)\,dx$$

$$= -\frac{1}{12}\cdot\frac{(7 - 4x^3)^{6/5}}{\frac{6}{5}} + C$$

$$= -\frac{5}{72}(7 - 4x^3)^{6/5} + C.$$

• **PROBLEM 618**

Evaluate the expression: $\int_0^2 2x^2\sqrt{x^3 + 1}\,dx$.

Solution: We wish to convert the given integral into a form, to which we can apply the formula for $\int u^n du$, with $u = (x^3 + 1)$, $du = 3x^2$ and $n = \frac{1}{2}$. We obtain:

$$\int_0^2 2x^2\sqrt{x^3 + 1}\,dx = \frac{2}{3}\int_0^2 (x^3 + 1)^{1/2}\left(\frac{3}{2}\cdot 2x^2 dx\right).$$

Applying the formula for $\int u^n du$, we obtain:

$$\frac{2}{3}\left[\frac{(x^3+1)^{3/2}}{\frac{3}{2}}\right]_0^2 = \frac{4}{9}(x^3+1)^{3/2}\Big]_0^2.$$

Evaluating between 2 and 0, we have:

$$\frac{4}{9}(8+1)^{3/2} - \frac{4}{9}(0+1)^{3/2}$$

$$= \frac{4}{9}(27 - 1)$$

$$= \frac{104}{9}.$$

• PROBLEM 619

Integrate the expression: $\int \frac{x}{\sqrt{x^2-4}}dx.$

<u>Solution:</u> $\int \frac{x}{\sqrt{x^2-4}}dx = \int (x^2-4)^{-1/2}x\,dx.$

Let $u = (x^2-4)$, then $du = 2x\,dx$, $n = -\frac{1}{2}$. We multiply x by 2 and the integral by $\frac{1}{2}$, and apply the formula for $\int u^n du$.

$$\int (x^2-4)^{-1/2}x\,dx = \frac{1}{2}\int (x^2-4)^{-1/2}2x\,dx$$

$$= \frac{1}{2}\left[\frac{(x^2-4)^{1/2}}{\frac{1}{2}} + C\right]$$

$$= (x^2-4)^{1/2} + C$$

$$= \sqrt{x^2-4} + C.$$

• PROBLEM 620

Evaluate the expression: $\int_2^3 \frac{(x+1)dx}{\sqrt{x^2+2x+3}}.$

<u>Solution:</u> We can rewrite the given integral as:

$$\int_2^3 (x^2+2x+3)^{-1/2}(x+1)dx,$$

and make use of the formula: $\int u^n du = \frac{u^{n+1}}{n+1}.$

Let $u = x^2 + 2x + 3$. Then $du = (2x+2)dx$, and $n = -\frac{1}{2}$. Applying the formula, we obtain:

$$\int_2^3 \frac{(x+1)dx}{\sqrt{x^2+2x+3}} = \frac{1}{2}\int_2^3 (x^2+2x+3)^{-1/2}2(x+1)dx$$

$$= \frac{1}{2}\left[\frac{(x^2+2x+3)^{1/2}}{\frac{1}{2}}\right]_2^3 = \sqrt{x^2+2x+3}\,\Big]_2^3.$$ We

now evaluate the definite integral between 3 and 2, obtaining:

$$\sqrt{3^2 + (2)(3) + 3} - \sqrt{2^2 + (2)(2) + 3} = \sqrt{18} - \sqrt{11}.$$

• PROBLEM 621

> Integrate the expression: $\int \frac{dx}{x^2 + 2x + 1}$.

<u>Solution:</u> $\int \frac{dx}{x^2 + 2x + 1} = \int \frac{dx}{(x + 1)^2} = \int (x + 1)^{-2} dx.$

Now, using the formula: $\int u^n du = \frac{u^{n+1}}{n+1}$, we obtain:

$$-\frac{1}{x+1} + C.$$

• PROBLEM 622

> Integrate the expression: $\int \frac{x\,dx}{\sqrt{1 + x^2}}$.

<u>Solution:</u> We can rewrite the given integral as: $\int (1 + x^2)^{-1/2} x\,dx$ and make use of the formula: $\int u^n du = \frac{u^{n+1}}{n+1}$. Let $u = 1 + x^2$. Then $du = 2x\,dx$, and $n = -\frac{1}{2}$. Applying the formula, we obtain:

$$\int \frac{x\,dx}{\sqrt{1 + x^2}} = \frac{1}{2}\int (1 + x^2)^{-1/2} 2x\,dx = \frac{1}{2}\left[\frac{(1 + x^2)^{1/2}}{\frac{1}{2}}\right] + C$$

$$= (1 + x^2)^{1/2} + C$$

$$= \sqrt{1 + x^2} + C.$$

• PROBLEM 623

> Find the integral of $\int \frac{2x \cdot dx}{\sqrt[3]{5 - 4x^2}}$ and prove.

<u>Solution:</u> The given integral can be rewritten as $\int (5 - 4x^2)^{-1/3} 2x \cdot dx$. We can now make use of the formula: $\int u^n du = \frac{u^{n+1}}{n+1} + C$, letting $u = (5 - 4x^2)$, $du = -8x\,dx$ and $n = -\frac{1}{3}$. **We obtain:**

$$\int \frac{2x\ dx}{\sqrt[3]{5-4x^2}} = -\frac{1}{4}\int (5-4x^2)^{-1/3} \cdot (-8x)dx$$

$$= -\frac{1}{4}\left[\frac{(5-4x^2)^{-1/3+1}}{-\frac{1}{3}+1}\right]$$

$$= -\frac{3}{8}(5-4x^2)^{2/3} + C.$$

To prove the integral, we differentiate it by the chain rule:

$$d\left[-\frac{3}{8}(5-4x^2)^{2/3} + C\right] = -\frac{3}{8}\cdot\frac{2}{3}(5-4x^2)^{2/3-1} \cdot (-8x)dx$$

$$= \frac{2x \cdot dx}{\sqrt[3]{5-4x^2}}$$

• **PROBLEM** 624

Integrate the expression: $p(t) = \int_a^t \frac{1}{\sqrt{x}\left(1-\sqrt{x}\right)^2}dx$, $1 < a < t$.

Solution: We wish to convert the given integral into a form, to which we can apply: $\int u^n du = \frac{u^{n+1}}{n+1} + C$. Let $u = (1-\sqrt{x})$, then $du = -\frac{1}{2}\cdot\frac{1}{\sqrt{x}}dx$. Rewriting the given integral as $-2\int_a^t \left(1-\sqrt{x}\right)^{-2}\left(-\frac{1}{2}\cdot\frac{1}{\sqrt{x}}\right)dx$ gives the desired form, with $n = -2$. Applying the formula, we obtain:

$$\int \frac{dx}{\sqrt{x}(1-\sqrt{x})^2} = \frac{-2\left(1-\sqrt{x}\right)^{-1}}{-1}\Bigg|_a^t = \frac{2}{1-\sqrt{x}}\Bigg]_a^t.$$

Evaluating the integral between t and a, we have:

$$\frac{2}{1-\sqrt{t}} - \frac{2}{1-\sqrt{a}}.$$

• **PROBLEM** 625

Integrate the expression: $\int\sqrt{9x^2 - 16}\ dx$.

Solution: The given integral is in the form:

$\int\sqrt{u^2 - a^2}\,du$, with $u^2 = 9x^2$, $u = 3x$, $du = 3dx$, and $a^2 = 16$. We can therefore apply the formula,

$$\int \sqrt{u^2 - a^2}\,du = \frac{u}{2}\sqrt{u^2 - a^2} - \frac{a^2}{2}\ln\left|u + \sqrt{u^2 - a^2}\right| + C,$$

after multiplying by 3 and by $\frac{1}{3}$. Applying the formula, we obtain:

$$\int\sqrt{9x^2 - 16}\ dx = \frac{1}{3}\int\sqrt{9x^2 - 16}\cdot 3dx$$

$$= \frac{1}{3}\cdot\frac{3x}{2}\sqrt{9x^2 - 16}$$

$$- \frac{1}{3}\cdot\frac{16}{2}\ln\left(3x + \sqrt{9x^2 - 16}\right)$$

$$= \frac{x}{2}\sqrt{9x^2 - 16}$$

$$- \frac{8}{3}\ln\left(3x + \sqrt{9x^2 - 16}\right) + C.$$

• PROBLEM 626

Integrate the expression: $\int\sqrt{16 - 9x^2}\ dx.$

Solution: The given integral is in the form $\int\sqrt{a^2 - u^2}\ du$, with $a^2 = 16$, $a = 4$, $u^2 = 9x^2$, $u = 3x$ and $du = 3dx$. Applying the formula, $\int\sqrt{a^2 - u^2}\ du = \frac{1}{2}u\sqrt{a^2 - u^2} + \frac{1}{2}a^2 \sin^{-1}\frac{u}{a} + C$, after multiplying by 3 and by $\frac{1}{3}$, we obtain:

$$\int\sqrt{16 - 9x^2}\cdot dx = \frac{1}{3}\int\sqrt{16 - 9x^2}\cdot 3\cdot dx$$

$$= \frac{1}{3}\cdot\frac{3x}{2}\sqrt{16 - 9x^2}$$

$$+ \frac{1}{3}\cdot\frac{16}{2}\cdot\sin^{-1}\cdot\frac{3x}{4}$$

$$= \frac{x}{2}\sqrt{16 - 9x^2} + \frac{8}{3}\sin^{-1}\frac{3x}{4} + C.$$

• PROBLEM 627

Integrate the expression: $\int\frac{dx}{\sqrt{x^2 + 6x}}.$

Solution: We wish to convert the given integral into a form that will allow us to apply the formula:

$$\int\frac{du}{\sqrt{u^2 - a^2}} = \cosh^{-1}\frac{u}{a} + C.$$

We first complete the square under the radical to obtain:

$$\int\frac{dx}{\sqrt{x^2 + 6x}} = \int\frac{dx}{\sqrt{x^2 + 6x + 9 - 9}} = \int\frac{dx}{\sqrt{(x+3)^2 - 9}}.$$

The integral is now in the desired form, with

$u = (x + 3)$, $du = dx$, and $a = 3$. Applying the formula, we obtain:

$$\int \frac{dx}{\sqrt{(x+3)^2 - (3)^2}} = \cosh^{-1} \frac{x + 3}{3} + C.$$

• PROBLEM 628

> Integrate the expression: $\int \frac{dx}{\sqrt{9x^2 + 4}}$.

Solution: We wish to convert the given integral into a form, to which we can apply the formula: $\int \frac{du}{\sqrt{u^2 + 1}} = \sin h^{-1} u + C$. We multiply the expression under the radical by $\frac{1}{4}$ to obtain the desired form. Here, $u = \frac{3x}{2}$, $du = \frac{3}{2}dx$. We must also remember to multiply by $\sqrt{\frac{1}{4}} = \frac{1}{2}$ outside the integral sign. Doing this, and using the formula for $\int \frac{du}{\sqrt{u^2 + 1}}$, we obtain:

$$\int \frac{dx}{\sqrt{9x^2 + 4}} = \frac{1}{2}\int \frac{dx}{\sqrt{(3x/2)^2 + 1}}$$

$$= \frac{1}{3}\int \frac{3/2dx}{\sqrt{(3x/2)^2 + 1}}$$

$$= \frac{1}{3}\sinh^{-1} \frac{3x}{2} + C.$$

• PROBLEM 629

> Integrate: $\int \frac{dx}{9x^2 - 4}$.

Solution: To integrate this function, we use the formula: $\int \frac{du}{u^2 - a^2} = \frac{1}{2a}\ln\frac{u - a}{u + a} + C$, letting $u = 3x$, $du = 3\,dx$, $a = 2$. Applying the formula, we obtain:

$$\int \frac{dx}{9x^2 - 4} = \frac{1}{3}\int \frac{3 \cdot dx}{(3x)^2 - (2)^2} = \frac{1}{3} \quad \frac{1}{2 \cdot 2} \ln\frac{3x - 2}{3x + 2}$$

$$= \frac{1}{12}\ln\frac{3x - 2}{3x + 2} + C.$$

• PROBLEM 630

> Integrate the expression: $\int \frac{dx}{4x^2 + 4x - \frac{21}{4}}$.

Solution: We wish to write the denominator of the given integral as the difference between two squares so as to apply the formula: $\int \frac{du}{u^2 - a^2} = \frac{1}{2a}\ln\left(\frac{u-a}{u+a}\right) + C.$

We can do this by completing the square as follows: The denominator,

$$4x^2 + 4x - \frac{21}{4} = 4\left(x^2 + x - \frac{21}{16}\right)$$
$$= 4\left(x^2 + x + \frac{1}{4} - \frac{25}{16}\right)$$
$$= 4\left[\left(x + \frac{1}{2}\right)^2 - \left(\frac{5}{4}\right)^2\right].$$

We now have:

$$\int \frac{dx}{4x^2 + 4x - \frac{21}{4}} = \int \frac{dx}{4\left[\left(x + \frac{1}{2}\right)^2 - \left(\frac{5}{4}\right)^2\right]}$$
$$= \frac{1}{4}\int \frac{dx}{\left(x + \frac{1}{2}\right)^2 - \left(\frac{5}{4}\right)^2}.$$

We can apply the formula, letting $u = \left(x + \frac{1}{2}\right)$, $a = \frac{5}{4}$, $du = dx$, obtaining:

$$\int \frac{dx}{4\left[\left(x + \frac{1}{2}\right)^2 - \left(\frac{5}{4}\right)^2\right]} = \frac{1}{4} \cdot \frac{1}{2 \cdot \frac{5}{4}}\ln\frac{\left(x + \frac{1}{2}\right) - \frac{5}{4}}{\left(x + \frac{1}{2}\right) + \frac{5}{4}}$$
$$= \frac{1}{10}\ln\frac{x - \frac{3}{4}}{x + \frac{7}{4}}$$
$$= \frac{1}{10}\ln\frac{4x - 3}{4x + 7} + C$$

• PROBLEM 631

Integrate the expression: $\int \frac{dx}{9x^2 + 16}.$

Solution: We wish to make use of the formula: $\int \frac{du}{u^2 + a^2} = \frac{1}{a}\tan^{-1}\frac{u}{a} + C.$ In the given integral, $u^2 = 9x^2$ and $a^2 = 16$; then $u = 3x$, $du = 3 \cdot dx$, and $a = 4$. Applying the formula, we obtain:

$$\int \frac{dx}{9x^2 + 16} = \frac{1}{3}\int \frac{3 \cdot dx}{(3x)^2 + (4)^2} = \frac{1}{12}\tan^{-1}\frac{3x}{4} + C.$$

• PROBLEM 632

Integrate the expression: $\frac{dy}{dx} = \frac{1}{x^2 + 6x + 25}.$

Solution: $\frac{dy}{dx} = \frac{1}{x^2 + 6x + 25}$ can be rewritten as

$dy = \frac{dx}{x^2 + 6x + 25}.$ We can now write:

$\int dy = \int \frac{dx}{x^2 + 6x + 25}$ or, $y = \int \frac{dx}{x^2 + 6x + 25}$. We wish to convert the integral into a form, to which we can apply the formula: $\int \frac{du}{u^2 + a^2} = \frac{1}{a}\tan^{-1}\left(\frac{u}{a}\right) + C.$

To do this, we complete the square in the denominator, obtaining:

$$\int dy = \int \frac{dx}{x^2 + 6x + 9 + 16} = \int \frac{dx}{(x + 3)^2 + (4)^2}.$$

We can now apply the formula, letting $u = x + 3$, $du = dx$, and $a = 4$. We obtain:

$$y = \frac{1}{a}\tan^{-1}\frac{u}{a} = \frac{1}{4}\tan^{-1}\frac{x + 3}{4} + C.$$

• PROBLEM 633

Integrate the expression: $\int \frac{dx}{3x^2 - 2x + 5}$.

Solution: We wish to convert the integral into a form, to which we can apply the formula: $\int \frac{du}{a^2 + u^2} = \frac{1}{a}\tan^{-1}\left(\frac{u}{a}\right)$. To do this, we rewrite the denominator as the sum of two squares by completing the square. We first factor a 3, so that we have:

$$\int \frac{dx}{3x^2 - 2x + 5} = \int \frac{dx}{3\left(x^2 - \frac{2}{3}x\right) + 5}.$$

To complete the square of $x^2 - \frac{2}{3}x$ we add $\frac{1}{9}$, and because $\frac{1}{9}$ is multiplied by 3, we actually add $\frac{1}{3}$ to the denominator, and so we also subtract $\frac{1}{3}$ from the denominator. Therefore, we have

$$\int \frac{dx}{3x^2 - 2x + 5} = \int \frac{dx}{3\left(x^2 - \frac{2}{3}x + \frac{1}{9}\right) + 5 - \frac{1}{3}}$$

$$= \int \frac{dx}{3\left(x - \frac{1}{3}\right)^2 + \frac{14}{3}}$$

$$= \frac{1}{3}\int \frac{dx}{\left(x - \frac{1}{3}\right)^2 + \frac{14}{9}}.$$

We now apply the formula for $\int \frac{du}{a^2 + u^2}$, with $a = \frac{\sqrt{14}}{3}$, $u = \left(x - \frac{1}{3}\right)$, $du = dx$. We have:

$$\frac{1}{3} \cdot \frac{1}{\sqrt{14}}\tan^{-1}\left(\frac{x - \frac{1}{3}}{\frac{1}{3}\sqrt{14}}\right) + C$$

$$= \frac{1}{\sqrt{14}}\tan^{-1}\left(\frac{3x - 1}{\sqrt{14}}\right) + C.$$

• **PROBLEM 634**

Integrate: $\int \frac{dx}{3x^2 - 6x + 15}$.

Solution: We convert the integral into a form, to which we can apply the formula:

$\int \frac{du}{a^2 + u^2} = \frac{1}{a}\text{arc tan}\left(\frac{u}{a}\right) + C$, where a = a constant.

To do this, we write the denominator as the sum of squares. By completing the square, we obtain:

$$3x^2 - 6x + 15 = 3\left[(x - 1)^2 + 2^2\right].$$

We now have

$$\int \frac{dx}{3x^2 - 6x + 15} = \int \frac{dx}{3\left[(x-1)^2 + 2^2\right]} = \frac{1}{3}\int \frac{dx}{(x-1)^2 + (2)^2}.$$

This is in the form $\int \frac{du}{a^2 + u^2}$, a = constant. Here a = 2, u = (x - 1), du = dx. Applying the formula, we obtain:

$$\frac{1}{3}\int \frac{dx}{2^2 + (x-1)^2} = \frac{1}{3}\left[\frac{1}{2}\text{arc tan}\ \frac{x - 1}{2} + C\right]$$

$$= \frac{1}{6}\text{arc tan}\ \frac{x - 1}{2} + C$$

• **PROBLEM 635**

Integrate: $\int \frac{dx}{x^2 + 4x + 20}$.

Solution: We wish to convert this integral to the form: $\int \frac{du}{a^2 + u^2}$. To accomplish this, we write the denominator as the sum of two squares, one being a constant. We have:

$$\int \frac{dx}{x^2 + 4x + 20} = \int \frac{dx}{(x^2 + 4x + 4) + 16}$$

$$= \int \frac{dx}{4^2 + (x + 2)^2}.$$

Now we apply the formula:

$$\int \frac{du}{a^2 + u^2} = \frac{1}{a}\ \text{arc tan}\ \left(\frac{u}{a}\right) + C,$$

letting a = 4, u = x + 2, du = dx. We obtain:

$$\int \frac{dx}{x^2 + 4x + 20} = \frac{1}{4}\text{arc tan}\ \frac{x + 2}{4} + C.$$

• **PROBLEM 636**

Integrate: $\int \frac{dx}{\sqrt{4 - 9x^2}}$.

Solution: To integrate this function, we make use of the formula:

$$\int \frac{du}{\sqrt{a^2 - u^2}} = \sin^{-1} \frac{u}{a} + C,$$

with $a = 2$, $u = 3x$, and $du = 3dx$. We have, after multiplying by 3 under the integral sign and by $\frac{1}{3}$ outside, the expression:

$$\int \frac{dx}{\sqrt{4 - 9x^2}} = \frac{1}{3}\int \frac{3dx}{\sqrt{2^2 - (3x)^2}}.$$

Applying the formula, we obtain:

$$\frac{1}{3}\sin^{-1} \frac{3x}{2} + C.$$

• **PROBLEM 637**

Integrate the expression. $\int \frac{dx}{\sqrt{5 - 4x - x^2}}$.

Solution: We try to convert the integral into a form, to which we can apply the formula:

$\int \frac{du}{\sqrt{a^2 - u^2}} = \text{arc sin} \left(\frac{u}{a}\right) + C$. To do this, we write $(5 - 4x - x^2)$ as the difference between two squares, with a = constant. By completing the square we obtain:

$$5 - 4x - x^2 = 3^2 - (x + 2)^2.$$

We now have $\int \frac{dx}{\sqrt{5 - 4x - x^2}} = \int \frac{dx}{\sqrt{3^2 - (x+2)^2}}$. This is in the form $\int \frac{du}{\sqrt{a^2 - u^2}}$, with $a = 3$, $u = x + 2$, $du = dx$.

Applying the formula, we obtain:

$$\int \frac{dx}{\sqrt{5 - 4x - x^2}} = \text{arc sin} \frac{x + 2}{3} + C.$$

• **PROBLEM 638**

Integrate the expression: $\int \frac{1}{\sqrt{21 - 18x - 2x^2}} dx$.

Solution: In integrating this expression, we wish to rewrite it in the form of $\int \frac{du}{\sqrt{a^2 - u^2}}$ and apply the formula, $\int \frac{du}{\sqrt{a^2 - u^2}} = \sin^{-1} \frac{u}{a}$. To do this, we first factor out the (-2) in the denominator and then complete the square. We obtain:

$$\int \frac{dx}{\sqrt{-2\left[\left(x^2+9x+\frac{81}{4}\right) - \frac{21}{2} - \frac{81}{4}\right]}} = \int \frac{dx}{\sqrt{-2\left[\left(x+\frac{9}{2}\right)^2 - \left(\frac{\sqrt{123}}{2}\right)^2\right]}}$$

$$= \int \frac{dx}{\sqrt{2\left[\left(\frac{\sqrt{123}}{2}\right)^2 - \left(x+\frac{9}{2}\right)^2\right]}}$$

$$= \frac{1}{\sqrt{2}} \int \frac{dx}{\sqrt{\left(\frac{\sqrt{123}}{2}\right)^2 - \left(x+\frac{9}{2}\right)^2}}.$$

We can now apply the formula, letting $a = \frac{\sqrt{123}}{2}$, $u = x + \frac{9}{2}$, and $du = dx$, obtaining:

$$\frac{1}{\sqrt{2}} \sin^{-1} \frac{(2x + 9)}{\sqrt{123}} = C.$$

• PROBLEM 639

Integrate the expression: $\frac{dy}{dx} = \frac{1}{\sqrt{\frac{14}{25} + \frac{12x}{5} - 2x^2}}$.

Solution: $\frac{dy}{dx} = \frac{1}{\sqrt{\frac{14}{25} + \frac{12x}{5} - 2x^2}}$ can be rewritten as

$dy = \frac{dx}{\sqrt{\frac{14}{25} + \frac{12x}{5} - 2x^2}}$. We can now write:

$$\int dy = \int \frac{dx}{\sqrt{\frac{14}{25} + \frac{12x}{5} - 2x^2}} \text{ or, } y = \int \frac{dx}{\sqrt{\frac{14}{25} + \frac{12x}{5} - 2x^2}}.$$

We now wish to convert the integral into a form, to which we can apply the formula: $\int \frac{du}{\sqrt{a^2 - u^2}} = \sin^{-1}\left(\frac{u}{a}\right) + C$. This can be obtained by factoring out a 2, and then completing the square in the denominator. We obtain:

$$\frac{1}{\sqrt{2}} \int \frac{dx}{\sqrt{\frac{7}{25} + \frac{9}{25} - \left(x^2 - \frac{6x}{5} + \frac{9}{25}\right)}}$$

$$= \frac{1}{\sqrt{2}} \int \frac{dx}{\sqrt{\left(\frac{4}{5}\right)^2 - \left(x - \frac{3}{5}\right)^2}}.$$

We can now apply the formula, with $a = \frac{4}{5}$, $u = \left(x - \frac{3}{5}\right)$, and $du = dx$.

$$y = \frac{1}{\sqrt{2}} \sin^{-1} \frac{\left(x - \frac{3}{5}\right)}{\frac{4}{5}} = \frac{1}{\sqrt{2}} \sin^{-1} \frac{(5x - 3)}{4} + C.$$

• **PROBLEM 640**

Integrate the expression: $\int \frac{2x - 1}{\sqrt{9x^2 + 16}} \cdot dx.$

Solution: We wish to divide the given integral into two integrals, to which we can apply the formulas:

$\int u^n du = \frac{u^{n+1}}{n+1}$; and: $\int \frac{du}{\sqrt{u^2 + a^2}} = \ln\left(u + \sqrt{u^2 + a^2}\right) + C.$

Doing this, we obtain:

$$\int \frac{2x - 1}{\sqrt{9x^2 + 16}} dx = \int \frac{2x\ dx}{\sqrt{9x^2 + 16}} - \int \frac{dx}{\sqrt{9x^2 + 16}}$$

$$= \int \left(9x^2 + 16\right)^{-1/2} 2x\ dx - \int \frac{dx}{\sqrt{(3x)^2 + (4)^2}}.$$

We now apply the formula for $\int u^n du$ to the first integral with $u = \left(9x^2 + 16\right)$, $du = 18x\ dx$ and $n = -\frac{1}{2}$, remembering to multiply by 9 and by $\frac{1}{9}$.

We apply the formula for $\int \frac{du}{\sqrt{u^2 + a^2}}$ to the second integral, with $u = 3x$, $du = 3dx$ and $a^2 = 16$, obtaining:

$$\int \frac{2x - 1}{\sqrt{9x^2 + 16}} \cdot dx = \frac{2}{9}\sqrt{9x^2 + 16} - \frac{1}{3}\ln\left(3x + \sqrt{9x^2 + 16}\right) + C.$$

• **PROBLEM 641**

Integrate the expression: $\int \frac{(2x + 7)dx}{x^2 + 2x + 5}.$

Solution: We wish to rewrite the given integral in a form, to which we can apply the formula: $\int \frac{du}{u} = \ln u$. Because $d\left(x^2 + 2x + 5\right) = (2x + 2)dx$, we write the numerator as $(2x + 2)dx + 5dx$, and express the original integral as the sum of two integrals. We have:

$$\int \frac{(2x + 7)dx}{x^2 + 2x + 5} = \int \frac{(2x + 2)dx}{x^2 + 2x + 5} + 5 \int \frac{dx}{x^2 + 2x + 5}.$$

We can now apply the formula for $\int \frac{du}{u}$ to the first integral and, writing the denominator of the second integral as the sum of two squares, can apply the formula: $\int \frac{du}{u^2 + a^2} = \frac{1}{a}\tan^{-1} \frac{u}{a} + C.$ We obtain:

$$\ln|x^2 + 2x + 5| + 5\int\frac{dx}{(x + 1)^2 + 4}$$

$$= \ln(x^2 + 2x + 5) + \frac{5}{2}\tan^{-1}\frac{x + 1}{2} + C.$$

• PROBLEM 642

Integrate the expression: $\int\frac{x + 3}{x^2 + 2x + 2}dx.$

Solution: We wish to convert the given integral into a form, to which we can apply the formulas for $\int\frac{du}{u}$ and $\int\frac{du}{a^2 + u^2}$. By completing the square in the denominator of the integrand, we obtain:

$$x^2 + 2x + 2 = (x + 1)^2 + 1,$$

and $\int\frac{(x + 3)dx}{x^2 + 2x + 2} = \int\frac{(x + 3)dx}{(x + 1)^2 + 1}.$

Let $u = (x + 1)^2 + 1$, $du = 2(x + 1)dx = (x + 1)2dx$. To obtain the form $\int\frac{du}{u}$, we rearrange the numerator to obtain $(x + 1)2dx$ from $(x + 3)dx$. Doing this, we have:

$$\int\frac{(x + 3)dx}{(x + 1)^2 + 1} = \frac{1}{2}\int\frac{(x + 1)2dx}{(x + 1)^2 + 1} + \int\frac{2dx}{(x + 1)^2 + 1}.$$

We now integrate the first integral, using the formula: $\int\frac{du}{u} = \ln u + C$, and the second, using: $\int\frac{du}{u^2 + a^2} = \frac{1}{a}\tan^{-1}\frac{u}{a}$, with $a = 1$, $u = (x + 1)$. We have:

$$\frac{1}{2}\ln(x^2 + 2x + 2) + 2\tan^{-1}(x + 1) + C.$$

• PROBLEM 643

Integrate: $\int\frac{x + 2}{x^2 + 16}dx.$

Solution: $\int\frac{x + 2}{x^2 + 16}dx = \int\frac{x\,dx}{x^2 + 16} + \int\frac{2dx}{x^2 + 16}.$

We try to convert $\int\frac{x\,dx}{x^2 + 16}$ into the form $\int\frac{du}{u}$.

If $u = x^2 + 16$, then $du = 2x\,dx$. Therefore we multiply the integral by 2 and also by $\frac{1}{2}$, obtaining: $\frac{1}{2}\int\frac{2x\,dx}{x^2 + 16}$. Now we attempt to convert the second integral, $\int \frac{2dx}{x^2 + 16}$ into the form: $\int \frac{du}{a^2 + u^2}$. Writing the denominator as the sum of squares we have: $x^2 + 16 = x^2 + 4^2$. Let $a = 4$, $u = x$ and $du = dx$. Taking the 2 in the numerator out of the integral sign, we have the form $2\int \frac{du}{a^2 + u^2} = 2\int \frac{dx}{4^2 + x^2}$.

Adding and applying the formulas we obtain:

$$\frac{1}{2}\int \frac{2x\, dx}{x^2 + 16} + 2\int \frac{dx}{x^2 + 16} = \frac{1}{2}\ln(x^2 + 16) + \frac{1}{2} \text{ arc tan } \frac{x}{4} + C.$$

• **PROBLEM 644**

Integrate: $\int \frac{(4x - 3)dx}{x^2 - 6x + 5}$.

Solution: We wish to convert the integral into a form, which will enable us to use the formulas: $\int \frac{du}{u} = \ln u + C$; and: $\int \frac{du}{u^2 - a^2} = \frac{1}{2a}\ln\left|\frac{u - a}{u + a}\right| + C$.

Since, in the denominator, $x^2 - 6x + 5 = x^2 - 6x + 9 - 4 = (x - 3)^2 - 2^2$, we can write:

$$\int \frac{(4x - 3)dx}{x^2 - 6x + 5} = \int \frac{4x - 12 + 9}{x^2 - 6x + 5}dx = \int \frac{2(2x - 6)dx}{x^2 - 6x + 5}$$

$$+ \int \frac{9dx}{x^2 - 6x + 5} = 2\int \frac{(2x - 6)dx}{x^2 - 6x + 5}$$

$$+ 9\int \frac{dx}{(x - 3)^2 - 2^2}.$$

For the first integral we can use the formula for $\int \frac{du}{u}$, and for the second integral we apply the formula for $\int \frac{du}{u^2 - a^2}$, a = constant, here = 2.

In these formulas we obtain:

$$2 \ln(x^2 - 6x + 5) + 9 \quad \frac{1}{4}\ln\frac{x - 3 - 2}{x - 3 + 2} + C$$

$$= 2 \ln(x^2 - 6x + 5) + \frac{9}{4}\ln\frac{x - 5}{x - 1} + C.$$

Or, using an alternative formula for the second integral, $\int \frac{du}{u^2 - a^2} = -\frac{1}{a} \coth^{-1} \frac{u}{a} + C$, we obtain:

$$2 \ln(x^2 - 6x + 5) - \frac{9}{2} \coth^{-1} \frac{x - 3}{2} + C.$$

• **PROBLEM 645**

Integrate the expression: $\frac{dy}{dx} = \frac{x}{\sqrt{4x - x^2}}$.

Solution: $\frac{dy}{dx} = \frac{x}{\sqrt{4x - x^2}}$ can be written as,

$dy = \int \frac{x}{\sqrt{4x - x^2}} dx$. We can now write:

$\int dy = \int \frac{x\,dx}{\sqrt{4x - x^2}}$ or, $y = \int \frac{x \cdot dx}{\sqrt{4x - x^2}}$. We wish to convert the given integral into a form, to which we can apply the formulas: $\int \frac{du}{\sqrt{2au - u^2}} = \sin^{-1} \frac{u - a}{a}$;

and: $\int u^n du = \frac{u^{n+1}}{n+1}$. To do this, we rewrite the numerator, separate the integral into two integrals, and then apply the formulas.

We obtain:

$$\int \frac{x\,dx}{\sqrt{4x - x^2}} = \int \frac{\left[2 - (2 - x)\right]dx}{\sqrt{4x - x^2}}$$

$$= \int \frac{2dx}{\sqrt{4x - x^2}} - \int \frac{(2 - x)dx}{\sqrt{4x - x^2}}$$

$$= 2\int \frac{dx}{\sqrt{4x - x^2}} - \int \left(4x - x^2\right)^{-1/2} (2 - x)dx.$$

We can now apply the formula for $\int \frac{du}{\sqrt{2au - u^2}}$ to the first integral, letting $u = x$, $du = dx$, $a = 2$, and the formula for $\int u^n du$ to the second, with $u = \left(4x - x^2\right)$, $du = (4 - 2x)$, $n = -\frac{1}{2}$, obtaining:

$$y = 2 \sin^{-1} \frac{x - 2}{2} - \frac{1}{2} \left[\frac{\left(4x - x^2\right)^{-1/2+1}}{-\frac{1}{2} + 1} \right]$$

$$= 2 \sin^{-1} \frac{x - 2}{2} - \left(4x - x^2\right)^{1/2} + C.$$

• **PROBLEM 646**

Integrate: $\int \frac{5x - 2}{\sqrt{x^2 + 2x}} dx$.

Solution: We wish to convert the integral into a form enabling use of the formulas:

$\int u^n du = \frac{u^{n+1}}{n+1} + C$: and: $\int \frac{du}{\sqrt{u^2 - a^2}} = \ln \left| u + \sqrt{u^2 - a^2} \right| + C.$

To do this, we rearrange the numerator in the following manner:

$$\int \frac{5x - 2}{\sqrt{x^2 + 2x}}dx = \frac{5}{2}\int \frac{2x - \frac{4}{5}}{\sqrt{x^2 + 2x}}dx$$

$$= \frac{5}{2}\int \frac{(2x + 2)\,dx}{\sqrt{x^2 + 2x}} - \frac{5}{2} \cdot \frac{14}{5}\int \frac{dx}{\sqrt{x^2 + 2x}}.$$

Now, writing the denominator of the second integral as the difference between two squares, we obtain the desired form:

$$\frac{5}{2}\int \left(x^2 + 2x\right)^{-1/2}(2x + 2)\,dx - 7\int \frac{dx}{\sqrt{(x + 1)^2 - 1^2}}.$$

Applying the formulas for $\int u^n du$ and $\int \frac{du}{u^2 - a^2}$, we obtain:

$$5\sqrt{x^2 + 2x} - 7\ln\left(x + 1 + \sqrt{x^2 + 2x)}\right) + C.$$

Or, applying the formula $\int \frac{du}{u^2 - a^2} = -\frac{1}{a}\text{ arc coth }\left(\frac{u}{a}\right) + C$, we obtain:

$$5\sqrt{x^2 + 2x} - 7\text{ arc coth }(x + 1) + C.$$

• **PROBLEM 647**

Integrate: $\int \frac{1 - e^{-2x}}{1 + e^{-4x}}dx.$

Solution: To integrate the given expression we first separate it into two integrals, obtaining:

$$\int \frac{1 - e^{-2x}}{1 + e^{-4x}}dx = \int \frac{1}{1 + e^{-4x}}dx - \int \frac{e^{-2x}}{1 + e^{-4x}}dx.$$

The first integral can be rewritten as

$$\int \frac{1}{1 + e^{-4x}}dx = \int \frac{1}{1 + \frac{1}{e^{4x}}}dx.$$

Now, multiplying both numerator and denominator of the integrand by e^{4x} we have:

$$\int \frac{e^{4x}}{e^{4x} + 1}dx.$$

To integrate this expression, we use the formula: $\int \frac{du}{u} = \ln u$, letting $u = e^{4x} + 1$. Then $du = 4e^{4x}dx$. Applying the formula, we find:

$$\int \frac{1}{1 + e^{-4x}}dx = \frac{1}{4}\ln\left(e^{4x} + 1\right).$$

The second integral can be rewritten as:

$$\int \frac{e^{-2x}dx}{1^2 + \left(e^{-2x}\right)^2},$$

which is nearly in the form:

$$\int \frac{du}{a^2 + u^2},$$

with $a = 1$, $u = e^{-2x}$, $du = -2e^{-2x}dx$. After supplying the necessary constant, we can apply the formula

$$\int \frac{du}{a^2 + u^2} = \frac{1}{a} \tan^{-1} \frac{u}{a},$$

obtaining:

$$\int \frac{e^{-2x}}{1 + e^{-4x}}dx = -\frac{1}{2}\left(\frac{1}{1} \tan^{-1} \frac{e^{-2x}}{1}\right)$$

$$= -\frac{1}{2} \tan^{-1} e^{-2x}.$$

Therefore,

$$\int \frac{1 - e^{-2x}}{1 + e^{-4x}}dx = \frac{1}{4} \ln\left(e^{4x} + 1\right) - \left(-\frac{1}{2} \tan^{-1} e^{-2x} + C\right)$$

$$= \frac{1}{4} \ln\left(e^{4x} + 1\right) + \frac{1}{2} \tan^{-1} e^{-2x} + C.$$

• PROBLEM 648

Integrate: $\int \frac{8x^3dx}{4x^2 + 4x + 5}$.

Solution: We can rewrite the given integral by dividing the denominator of the integrand into the numerator. Doing this, we obtain:

$$\int \frac{8x^3dx}{4x^2 + 4x + 5} = \int \left[2x - 2 - \left(\frac{2x - 10}{4x^2 + 4x + 5}\right)\right]dx,$$

which can be rewritten as three separate integrals. Therefore, we have,

$$2 \int xdx - 2 \int dx - \int \frac{2x - 10}{4x^2 + 4x + 5}dx.$$

We can easily integrate the first two integrals, using the formula $\int u^n du = \frac{u^{n+1}}{n + 1}$ on the first, and we can factor out the constant $\frac{1}{4}$ from the last integral, obtaining:

$$2\left(\frac{x^2}{2}\right) - 2x - \frac{1}{4} \int \frac{2x - 10}{x^2 + x + \frac{5}{4}}dx.$$

Now we rewrite this integral, so that we can apply the formula: $\int \frac{du}{u} = \ln u$. Let $u = x^2 + x + \frac{5}{4}$, then $du = (2x + 1)dx$. We can rewrite the numerator of the integrand as:

$$\frac{(2x + 1) - 11}{x^2 + x + \frac{5}{4}},$$

and, separating this into two integrals, we obtain:

$$x^2 - 2x - \frac{1}{4}\left[\int \frac{2x + 1}{x^2 + x + \frac{5}{4}}dx - 11 \int \frac{dx}{x^2 + x + \frac{5}{4}}\right].$$

Before applying the formula for $\int \frac{du}{u}$ to the first integral, we complete the square in the denominator of the second integral and then apply the formula,

$$\int \frac{du}{a^2 + u^2} = \frac{1}{a} \tan^{-1} \frac{u}{a}.$$

We obtain:

$$x^2 - 2x - \frac{1}{4}\left[\int \frac{2x + 1}{x^2 + x + \frac{5}{4}}dx - 11 \int \frac{dx}{\left(x + \frac{1}{2}\right)^2 + 1}\right].$$

Now we apply the formulas, letting $u = x^2 + x + \frac{5}{4}$ in the first integral. Then $du = (2x + 1)dx$, and, letting $a = 1$, $u = x + \frac{1}{2}$, $du = dx$ in the second integral. Therefore,

$$\int \frac{8x^3 dx}{4x^2 + 4x + 5} = x^2 - 2x - \frac{1}{4} \ln\left(x^2 + x + \frac{5}{4}\right)$$

$$+ \frac{11}{4} \tan^{-1}\left(x + \frac{1}{2}\right) + C.$$

• PROBLEM 649

Evaluate

$$\int_0^8 xy \, dx,$$

subject to the functional relation: $x = t^3$, $y = t^2$.

Solution:

$$y = t^2$$

$$x = t^3, \qquad dx = 3t^2 \, dt$$

and for $0 \le x \le 8$ we have $0 \le t \le 2$.

$$\int_0^8 xy \, dx = \int_0^2 (t^3)(t^2)(3t^2 dt)$$

$$= \int_0^2 3t^7 \, dt = \frac{3}{8} t^8 \Big|_0^2$$

$$= 3\ (2)^5$$

$$= 96.$$

• PROBLEM 650

Evaluate:

$$\int_0^e y\ dx,$$

subject to the functional relation:

$$x = e^y + y - 1.$$

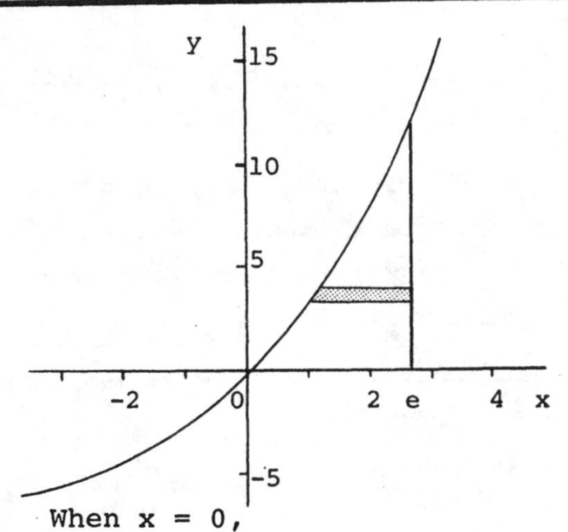

Solution: When $x = 0$,

$$e^y + y - 1 = 0$$

$$e^y + y = 1$$

$$y = 0.$$

When $x = e$, $y = 1$.

As can be seen in the figure, the integral:

$$\int_0^e y\ dx,$$

represents the area under the curve. Since y is difficult to find explicitely, we can achieve the same result by considering an elemental area as shown. This element has the area $(e - x)dy$. Hence the integral:

$$I = \int_0^1 (e - x)dy = \int_0^1 (e - e^y - y + 1)dy,$$

can be evaluated instead of the given one. We have:

$$I = (e + 1)y - e^y - \frac{y^2}{2}\Bigg|_0^1$$

$$= (e + 1) - e - \frac{1}{2} - 0 + 1 + 0$$

$$= \frac{3}{2} .$$

• PROBLEM 651

Find the equation of the curve which passes through the point (0,-4), and for which the slope at any point is equal to 4 sin 2x.

Solution: The slope of the curve is 4 sin 2x. Since the slope of a curve is the derivative, $\frac{dy}{dx}$, of the equation of the curve, we write $\frac{dy}{dx} = 4 \sin 2x$. To find the equation we have to integrate the equation: $\frac{dy}{dx} = 4 \sin 2x$.

$$\int \frac{dy}{dx} = \int 4 \sin 2x \, dx.$$

$$y = \int 4 \sin 2x \, dx.$$

$$y = \frac{1}{2} \int (4 \sin 2x) 2dx.$$

$$y = -2 \cos 2x + C.$$

Since the curve passes through the point (0,-4), we find the term C by substituting the abscissa, 0, for x, the ordinate, -4, for y, and solving for C.

$$y = -2 \cos 2x + C.$$

$$-4 = -2(\cos (0)) + C.$$

$$-4 = -2 + C.$$

$$-2 = C.$$

Therefore the equation of the curve is:

$$y = -2 \cos 2x - 2.$$

$$y = -2(\cos 2x + 1).$$

• PROBLEM 652

Find the equation of the curve which has a horizontal tangent at the point (0,-1), and for which the rate of change, with respect to x, of the slope at any point is equal to $8e^{2x}$.

Solution: The slope of a curve at a point is the derivative of the curve, $\frac{dy}{dx}$. The rate of change of the slope is the derivative of the slope, or the second derivative of the curve, $\frac{d^2y}{dx^2}$. Therefore, $\frac{d^2y}{dx^2} = 8e^{2x}$. To find the equation of the curve we:

1) integrate the rate of change of the slope to obtain the slope, and
2) integrate the slope to obtain the equation of the curve.

Integrating the rate of change we obtain, from:

$$\frac{d^2y}{dx^2} = 8e^{2x},$$

the expression:

$$\int \frac{d^2y}{dx^2} = \frac{dy}{dx} = \int 8e^{2x}dx.$$

$$\frac{dy}{dx} = \frac{1}{2}\int \left(8e^{2x}\right)2dx.$$

$$\frac{dy}{dx} = 4e^{2x} + C_1 = m, \text{ the slope.}$$

Integrating the slope, m, to obtain the equation, we obtain:

$$\int \frac{dy}{dx} = y = \int 4e^{2x} + C_1 dx$$

$$y = \frac{1}{2}\int \left(4e^{2x} + C_1\right)2dx$$

$$y = 2e^{2x} + C_1x + C_2,$$

the equation of the curve.

To find the value of the constant C_1, we use the information given in the problem: that there is a horizontal tangent at the point (0,-1).

A horizontal tangent occurs at a point where the slope is equal to zero. Therefore $\frac{dy}{dx} = 0$ or $4e^{2x} + C_1 = 0$, when $x = 0$ and $y = -1$. Substituting $x = 0$ into the equation: $4e^{2x} + C_1 = 0$, we find C_1 as follows:

$$\frac{dy}{dx} = 4e^{2x} + C_1 = 0$$

$$4e^{2(0)} + C_1 = 0$$

$$4 + C_1 = 0$$

$C_1 = -4$

We now solve for C_2 by substituting $C_1 = -4$ and the point (0,1) into the equation for the curve: $y = 2e^{2x} + C_1x + C_2$. We obtain:

$$y = 2e^{2x} + C_1x + C_2.$$
$$-1 = 2e^{2(0)} - 4(0) + C_2.$$
$$-1 = 2 + C_2.$$
$$-3 = C_2.$$

Substituting C_1 and C_2 into the equation for the curve: $y = 2e^{2x} + C_1x + C_2$, we obtain:

$$y = 2e^{2x} - 4x - 3.$$

• PROBLEM 653

Integrate: $\int x^2\sqrt{1 + x}\, dx.$

Solution: We wish to convert the given integral into a form, which can be integrated by the formula: $\int u^n du = \frac{u^{n+1}}{n+1} + C$. Let $v = \sqrt{1 + x}$; then $v^2 = 1 + x$. Hence, $x = v^2 - 1$, and $dx = 2v\, dv$. Making these substitutions, we have:

$$\int x^2\sqrt{1+x}\, dx = \int (v^2-1)^2\sqrt{1+v^2-1}\; 2v\, dv$$
$$= \int (v^2-1)^2 \cdot v \cdot (2v\, dv) = 2\int (v^6-2v^4+v^2)\,dv$$
$$= 2\int v^6\, dv - 4\int v^4 dv + 2\int v^2\, dv.$$

Applying the formula for $\int u^n du$, we have:

$$\frac{2}{7}v^7 - \frac{4}{5}v^5 + \frac{2}{3}v^3 + C$$
$$= \frac{2}{7}(1+x)^{7/2} - \frac{4}{5}(1+x)^{5/2} + \frac{2}{3}(1+x)^{3/2} + C,$$

by substitution.

• PROBLEM 654

Integrate the expression: $\int x^3\sqrt{1 + 2x^2}\, dx.$

Solution: We wish to convert the given integral into a form, to which we can apply the formula for $\int u^n du$. Let $y^2 = 1 + 2x^2$, so that $2y\, dy = 4x\, dx$. Then,

$$\int x^3\sqrt{1+2x^2}\, dx = \int x^2\sqrt{1+2x^2}\; x\, dx$$

$$= \int \frac{y^2-1}{2} \cdot \sqrt{1+(y^2-1)} \cdot \frac{2y\,dy}{4}$$

$$= \int \left[\frac{1}{2}(y^2-1)\right](y)\left(\frac{1}{2}y\,dy\right)$$

$$= \frac{1}{4}\int\left(y^4-y^2\right)dy.$$

We can now apply the formula for $\int u^n du$. Doing this, we obtain:

$$\frac{1}{4}\left(\frac{1}{5}y^5 - \frac{1}{3}y^3\right) + C = \frac{1}{20}y^5 - \frac{1}{12}y^3 + C$$

$$= \frac{1}{60}y^3(3y^2 - 5) + C$$

$$= \frac{1}{60}(1 + 2x^2)^{3/2}(6x^2 - 2) + C,$$

by substitution.

• PROBLEM 655

Integrate the expression: $\int x^5\sqrt{x^2 + 4}\,dx.$

Solution: We wish to convert the given integral into a form, to which we can apply the formula for $\int u^n du$. We can obtain the desired form by substitution and rearrangement of the integral as follows: Let $z = \sqrt{x^2 + 4}$. Then $z^2 = x^2 + 4$, and $2z\,dz = 2x\,dx$ by differentiating implicitly. Substituting, we have:

$$\int x^5\sqrt{x^2+4}\,dx = \int (x^2)^2\sqrt{x^2+4}\,x\,dx = \int (z^2-4)^2 z(z\,dz)$$

$$= \int (z^6-8z^4+16z^2)dz.$$

Applying the formula: $\int u^n du = \frac{u^{n+1}}{n+1} + C$, we obtain:

$$\frac{1}{7}z^7 - \frac{8}{5}z^5 + \frac{16}{3}z^3 + C = \frac{1}{105}z^3\left[15z^4 - 168z^2 + 560\right] + C.$$

Substituting, we have:

$$\frac{1}{105}(x^2+4)^{3/2}\left[15(x^2+4)^2 - 168(x^2+4) + 560\right] + C,$$

$$= \frac{1}{105}(x^2+4)^{3/2}[15x^4 - 48x^2 + 128] + C,$$

by expansion.

• PROBLEM 656

Integrate the expression: $\int \frac{x\,dx}{\sqrt{x+1} + \sqrt[3]{x+1}}.$

Solution: In evaluating this integral we use substitution to convert the given integral into a form to which we can apply the formula for $\int u^n du$. Since we have the square root and the cube root of the same expression, this integrand is rationalized by setting

$$x + 1 = u^6, \; x = u^6 - 1, \; dx = 6u^5 du,$$

$$\sqrt{x+1} = u^3, \; \sqrt[3]{x+1} = u^2.$$

By substitution, the integral then becomes:

$$\int \frac{(u^6-1)\,6u^5 du}{u^3 + u^2} = \int \frac{6u^3(u^6-1)du}{u+1}.$$

We can now use synthetic division to simplify the integrand, obtaining:

$$\int 6u^3(u^5-u^4+u^3-u^2+u-1)du.$$

Integrating each term using the formula for $\int u^n du$ and then factoring out the common factor, $6u^4$, we have:

$$\frac{6u^9}{9} + \frac{6u^8}{8} + \frac{6u^7}{7} - \frac{6u^6}{6} + \frac{6u^5}{5} - \frac{6u^4}{4} + C$$

$$= 6u^4\left(\frac{u^5}{9} - \frac{u^4}{8} + \frac{u^3}{7} - \frac{u^2}{6} + \frac{u}{5} - \frac{1}{4}\right) + C.$$

Since $u = \sqrt[6]{x+1}$, this is equal to

$$6\sqrt[6]{(x+1)^4}\left[\sqrt[6]{(x+1)^5}/9 - \sqrt[6]{(x+1)^4}/8 + \sqrt[6]{(x+1)^3}/7 - \sqrt[6]{(x+1)^2}/6 + \sqrt[6]{x+1}/5 - 1/4\right] + C,$$

by substitution.

• PROBLEM 657

Integrate the expression: $\displaystyle\int \frac{dx}{x\sqrt{3x^2 + 2x - 1}}$.

Solution: By substituting for x we wish to convert the given integral into a form, to which we can apply the formula, $\displaystyle\int \frac{du}{\sqrt{a^2 - u^2}} = \sin^{-1}\left(\frac{u}{a}\right)$. Substitute $\frac{1}{u}$ for x. Then

$$x = \frac{1}{u}, \quad dx = -\frac{du}{u^2}$$

and
$$\int \frac{dx}{x\sqrt{3x^2 + 2x - 1}} = \int \frac{-\frac{du}{u^2}}{\frac{1}{u}\sqrt{\frac{3}{u^2} + \frac{2}{u} - 1}}$$

$$= -\int \frac{\frac{du}{u^2}}{\frac{1}{u}\sqrt{\frac{1}{u^2}(3 + 2u - u^2)}}$$

$$= -\int \frac{\frac{du}{u^2}}{\frac{1}{u^2}\sqrt{3 + 2u - u^2}}$$

$$= -\int \frac{du}{\sqrt{3 + 2u - u^2}} .$$

We complete the square:

$$\int \frac{dx}{x\sqrt{3x^2 + 2x - 1}} = -\int \frac{du}{\sqrt{4 - (u^2 - 2u + 1)}}$$

$$= -\int \frac{du}{\sqrt{(2)^2 - (u - 1)^2}}.$$

We can now apply the formula for $\int \frac{du}{\sqrt{a^2 - u^2}}$,

obtaining:

$$\int \frac{dx}{x\sqrt{3x^2 + 2x - 1}} = \sin^{-1}\frac{(u-1)}{2} = \sin^{-1}\frac{(1-x)}{2x},$$

by substitution.

• PROBLEM 658

Evaluate the expression: $\int_0^3 x\sqrt{1 + x}\,dx.$

Solution: We wish to convert the given integral into a form, to which we can apply the formula for $\int u^n du$. To evaluate the indefinite integral $\int x\sqrt{1 + x}\,dx$, we let

$$u = \sqrt{1 + x},\ u^2 = 1 + x,\ x = u^2 - 1,\ dx = 2u\,du.$$

Substituting, we have:

$$\int x\sqrt{1 + x}\,dx = \int (u^2 - 1)u(2u\,du)$$

$$= 2\int (u^4 - u^2)du.$$

We can now apply the formula for $\int u^n du$, and we obtain:

$$\frac{2}{5}u^5 - \frac{2}{3}u^3 + C = \frac{2}{5}\left(1 + x\right)^{5/2} - \frac{2}{3}\left(1 + x\right)^{3/2} + C,$$

by substitution. Therefore, the definite integral

$$\int_0^3 x\sqrt{1+x}\,dx = \frac{2}{5}\left(1+x\right)^{5/2} - \frac{2}{3}\left(1+x\right)^{3/2}\Bigg]_0^3$$

$$= \frac{2}{5}\left(4\right)^{5/2} - \frac{2}{3}\left(4\right)^{3/2} - \frac{2}{5}\left(1\right)^{5/2} + \frac{2}{3}\left(1\right)^{3/2}$$

$$= \frac{64}{5} - \frac{16}{3} - \frac{2}{5} + \frac{2}{3}$$

$$= \frac{116}{15}.$$

• **PROBLEM 659**

Evaluate the expression: $\int_1^2 \frac{x}{(1+2x)^3}dx.$

Solution: This integral is difficult because of the expression: $1 + 2x$, in the denominator. Hence we choose our substitution to eliminate this expression. We let

$$u = 1 + 2x, \text{ then } x = \frac{u-1}{2} \text{ and } dx = \frac{1}{2}du.$$

Now

$$u = 3 \text{ when } x = 1,$$

$$u = 5 \text{ when } x = 2,$$

giving us the new limits. Using the substitution, we obtain:

$$\int_1^2 \frac{x}{(1+2x)^3}dx = \int_3^5 \left(\frac{\frac{(u-1)}{2}}{u^3}\right)\left(\frac{1}{2}\right)du$$

$$= \frac{1}{4}\int_3^5 \left(\frac{1}{u^2} - \frac{1}{u^3}\right)du.$$

We can now use the formula for $\int u^n du$ on both terms of the integrand, obtaining:

$$\frac{1}{4}\left[-\frac{1}{u} + \frac{1}{2u^2}\right]_3^5 = \frac{11}{450}.$$

• **PROBLEM 660**

$\frac{dy}{dx} = \frac{2}{\varepsilon^{2x} + \varepsilon^{-2x}}$. Find $y = F(x)$.

Solution: $\frac{dy}{dx} = \frac{2}{\varepsilon^{2x} + \varepsilon^{-2x}}$ can be rewritten as

$dy = \frac{2}{\varepsilon^{2x} + \varepsilon^{-2x}}dx$. To find the integral we write:

$$\int dy = \int \frac{2dx}{\varepsilon^{2x} + \varepsilon^{-2x}}$$

or,

$$y = 2 \int \frac{dx}{\varepsilon^{2x} + \varepsilon^{-2x}}.$$

We wish to get the integral into a form to which we can apply the formula, $\int \frac{du}{a^2 + u^2} = \frac{1}{a} \text{ arc tan } \frac{u}{a} + C.$

Let

$$u = \varepsilon^{2x}.$$

Then

$$\frac{du}{dx} = 2\varepsilon^{2x}$$

$$dx = \frac{du}{2\varepsilon^{2x}}.$$

Substituting these in the original function,

$$y = 2 \int \frac{dx}{\varepsilon^{2x} + \varepsilon^{-2x}} = 2 \int \frac{\frac{du}{2\varepsilon^{2x}}}{\varepsilon^{2x} + \varepsilon^{-2x}}$$

$$= 2 \int \frac{du}{2\varepsilon^{2x}(\varepsilon^{2x} + \varepsilon^{-2x})}$$

or

$$y = \frac{2}{2} \int \frac{du}{u\left(u + \frac{1}{u}\right)} = \int \frac{du}{u^2 + 1}.$$

We can now apply the formula, obtaining:

$$y = \text{arc tan } u = \text{arc tan } \varepsilon^{2x}.$$

CHAPTER 24

TRIGONOMETRIC INTEGRALS

When integrating trigonometric functions, the power rule is often involved. Before applying the fundamental integration formulas, it also may be necessary to simplify the function. For that purpose, the common trigonometric identities are most often applicable as, for example, the half-angle formulas and the double-angle formulas. Again, no general rule can be given for finding the solutions. It takes a combination of experience and trial-and-error to learn what to substitute to arrive at the best solution method.

• PROBLEM 661

Integrate: $\int \sin 3ax \cdot dx$.

Solution: In integrating this expression we use the formula: $\int \sin u \ du = -\cos u$, letting $u = 3ax$, and $du = 3a \cdot dx$. Applying the formula, we obtain:

$$\int \sin 3ax \ dx = \frac{1}{3a} \int \sin 3ax \cdot dx$$

$$= -\frac{1}{3a} \cdot \cos 3ax + C.$$

• PROBLEM 662

Integrate and check: $\int \cos 2x \ dx$.

Solution: Consider the formula: $\int \cos u \ du = \sin u + C$. Let $u = 2x$, then $du = 2dx$. To obtain $\int \cos u \ du$, we multiply dx by 2 under the integral sign, and by $\frac{1}{2}$ outside the sign. We obtain:

$$\frac{1}{2} \int \cos 2x \cdot 2dx.$$

We apply the formula: $\int \cos u \ du = \sin u + C$. Then,

$$\int \cos 2x \ dx = \frac{1}{2} \sin 2x + C.$$

To check our answer, we take the derivative of $\left(\frac{1}{2} \sin 2x + C\right)$. We should obtain $\cos 2x$ because integration is the inverse of differentiation:

$$y = \frac{1}{2} \sin 2x + C,$$

$$y' = \frac{1}{2}\left(\cos 2x\right)(2)$$
$$= \cos 2x.$$

• PROBLEM 663

Integrate: $\int \tan(4x + 1)dx.$

Solution: We wish to make use of the formula:

$\int \tan u \ du = -\ln \cos u$, letting $u = (4x + 1)$, and $du = 4 \ dx$. We obtain:

$$\int \tan(4x + 1) \ dx = \frac{1}{4} \int \tan(4x + 1) \cdot 4 \ dx$$

$$= -\frac{1}{4} \ln \cos(4x + 1) + C.$$

• PROBLEM 664

Integrate: $\int x \sin x^2 \ dx.$

Solution: We attempt to apply the formula for $\int \sin u \ du$. To do this, we let $u = x^2$. Then,

$$du = 2x \ dx.$$

We must introduce a factor of 2 within the integral in order to have the form sin u du, and to multiply by $\frac{1}{2}$ outside the integral. We obtain:

$$\int x \sin x^2 \ dx = \int \sin x^2 \ x \ dx$$

$$= \frac{1}{2} \int \sin x^2 \ 2x \ dx.$$

This is in the form $\frac{1}{2} \int \sin u \ du$,

and we can now apply the formula, $\int \sin u \ du = -\cos u + C$. Thus,

$$\frac{1}{2} \int \sin x^2 \cdot 2x \ dx = -\frac{1}{2} \cos x^2 + C.$$

• PROBLEM 665

Integrate the expression: $\int \cos^2 x \sin x \ dx.$

Solution: In evaluating this integral we use the formula:

$$\int u^n \ du = \frac{u^{n+1}}{n+1} .$$

Let $u = \cos x$, $du = -\sin x \ dx$, and $n = 2$. Applying the formula, we obtain:

$$\int \cos^2 x \sin x \, dx = -\frac{\cos^3 x}{3} + C.$$

• **PROBLEM 666**

Integrate the expression: $\int \frac{\sin \sqrt{x}}{\sqrt{x}} \, dx.$

Solution: We can rewrite the given integral as, $\int \sin x^{1/2} \quad x^{-1/2} \, dx$. This is now in a form, to which we can apply the formula:

$$\int \sin u \, du = -\cos u + C.$$

Let $u = \sqrt{x} = x^{1/2}$. Then $du = \frac{1}{2} x^{-1/2} \, dx$. Therefore,

$$\int \frac{\sin \sqrt{x}}{\sqrt{x}} \, dx = 2\int \sin \sqrt{x} \cdot \frac{1}{2} x^{-1/2} \, dx$$

$$= 2(-\cos \sqrt{x}) + C = -2 \cos \sqrt{x} + C.$$

• **PROBLEM 667**

Find: $\int_0^{\pi} (1 + \sin x) \, dx.$

Solution:

$$\int_0^{\pi} (1 + \sin x) \, dx = \int_0^{\pi} dx + \int_0^{\pi} \sin x \, dx = x - \cos x \Big|_0^{\pi}$$

$$= \pi - \cos \pi - (0 - \cos 0).$$

At this point we recall that $\cos \pi = -1$ and $\cos 0 = 1$. Using these facts, we obtain:

$$\pi + 1 - (0 - 1)$$

$$= \pi + 2.$$

• **PROBLEM 668**

Find: $\int \frac{\sin x}{1 - \cos x} \, dx.$

Solution: The given integral is in the form

$$\int \frac{du}{u}.$$

We let $u = 1 - \cos x$. Then $du = \sin x\, dx$. We can apply the formula,

$$\int \frac{du}{u} = \ln|u| + C, \text{ obtaining}$$

$$\int \frac{\sin x\, dx}{1 - \cos x} = \ln|1 - \cos x| + C.$$

• **PROBLEM** 669

Find: $\int \frac{\cos(\ln x)}{x}\, dx.$

<u>Solution:</u> The given integral can be rewritten as,

$$\int \cos(\ln x) \cdot \frac{1}{x}\, dx.$$

This is in the form,

$$\int \cos u\, du,$$

with $u = \ln x$ and $du = \frac{1}{x}\, dx$. Applying the formula,

$$\int \cos u\, du = \sin u + C, \text{ we obtain:}$$

$$\sin(\ln x) + C.$$

• **PROBLEM** 670

$\frac{dy}{dx} = \varepsilon^x \cdot \cos \varepsilon^x$. Find $y = F(x)$.

<u>Solution:</u> $\frac{dy}{dx} = \varepsilon^x \cdot \cos \varepsilon^x$ can be rewritten as:

$$dy = \varepsilon^x \cdot \cos \varepsilon^x dx.$$

To find the integral, we find $\int dy = \int \varepsilon^x \cdot \cos \varepsilon^x \cdot dx$ or $y = \int \varepsilon^x \cdot \cos \varepsilon^x \cdot dx$. The integral is in the form $\int \cos u\, du$, with $u = \varepsilon^x$, $du = \varepsilon^x dx$. We now apply the formula $\int \cos u\, du = \sin u + C$ obtaining:

$$\int \cos \varepsilon^x \cdot \varepsilon^x dx = \sin \varepsilon^x + C$$

or: $$y = \sin \varepsilon^x + C.$$

• **PROBLEM** 671

What is $\int \varepsilon^{\cos x} \cdot \sin x \cdot dx$?

<u>Solution:</u> In evaluating this integral we will use the formula: $\int \varepsilon^u \cdot du = \varepsilon^u$. Here $u = \cos x$, and $du = -\sin x \cdot dx$. Applying the formula, we obtain:

$$-\int \varepsilon^{\cos x}(-\sin x \cdot dx) = -\varepsilon^{\cos x} + C.$$

• **PROBLEM 672**

Integrate:

$$\int \cos x \, e^{2 \sin x} \, dx.$$

Solution: This problem is best solved by the method of substitution. We let u = 2 sin x. Then du = 2 cos x dx. Substituting, we obtain:

$$\int \cos x \, e^{2 \sin x} \, dx = \frac{1}{2} \int e^{2 \sin x} (2 \cos x \, dx)$$

$$= \frac{1}{2} \int e^u \, du = \frac{1}{2} e^u + C$$

$$= \frac{1}{2} e^{2 \sin x} + C.$$

• **PROBLEM 673**

Integrate: $\int (\cot 2x - \csc 2x)^2 dx.$

Solution: We first multiply out the term $(\cot 2x - \csc 2x)^2$. We obtain:

$$\int (\cot^2 2x - 2 \csc 2x \cot 2x + \csc^2 2x) dx.$$

We can separate this integral into the sum of three integrals and operate on each. We use the formulas: $\int \cot^2 u \, du = -\cot u - u$, $\int \csc u \cot u \, du = -\csc u$, and $\int \csc^2 u \, du = -\cot u$.

$$\int (\cot^2 2x - 2 \csc 2x \cot 2x + \csc^2 2x) dx$$

$$= \int \cot^2 2x \, dx - \int (\csc 2x \cot 2x) 2dx + \int \csc^2 2x \, dx$$

$$= \frac{1}{2} \int (\cot^2 2x) 2dx - \int (\csc 2x \cot 2x) 2dx + \frac{1}{2} \int (\csc^2 2x) 2dx$$

$$= -\frac{1}{2} (\cot 2x + 2x) + \csc 2x - \frac{1}{2} \cot 2x$$

$$= \csc 2x - \cot 2x - x + C.$$

• **PROBLEM 674**

Integrate: $\int \frac{2 \cos x + 3 \sin x}{\sin^3 x} dx.$

Solution: We first separate this integral into the sum of two separate integrals. We can rewrite the integral as:

$$\int \frac{2 \cos x + 3 \sin x}{\sin^3 x} dx = \int \frac{2 \cos x}{\sin^3 x} dx + \int \frac{3 \sin x}{\sin^3 x} dx.$$

This expression can be further simplified as follows:

$$\int \frac{2 \cos x}{\sin^3 x} dx + \int \frac{3 \sin x}{\sin^3 x} dx = \int 2 \cot x \csc^2 x \, dx + \int 3 \csc^2 x \, dx,$$

since $\frac{\cos u}{\sin u} = \cot u$ and $\frac{1}{\sin u} = \csc u$. We can now integrate as follows:

$$\int 2 \cot x \csc^2 x \, dx + \int 3 \csc^2 x \, dx.$$

We integrate the first term using the formula: $\int u^n du = \frac{u^{n+1}}{n+1}$, where $u = \cot x$, and $du = -\csc^2 x \, dx = d(\cot x)$. Hence,

$$-\int 2 \cot x d(\cot x) dx + 3 \int \csc^2 x \, dx$$

$$= -2\left(\frac{\cot^2 x}{2}\right) + 3(-\cot x) + C$$

$$= -\cot^2 x - 3 \cot x + C.$$

• PROBLEM 675

Integrate: $\int \tan^5 x \csc^2 x \, dx.$

Solution: To integrate the given expression, we use the fact that

$$\tan x = \frac{\sin x}{\cos x},$$

and

$$\csc x = \frac{1}{\sin x}.$$

We can therefore rewrite the given integral as:

$$\int \frac{\sin^5 x}{\cos^5 x} \frac{1}{\sin^2 x} dx = \int \frac{\sin^3 x}{\cos^5 x} dx$$

$$= \int \frac{\sin^3 x}{\cos^3 x} \cdot \frac{1}{\cos^2 x} dx$$

$$= \int \tan^3 x \sec^2 x \, dx.$$

The last integral is in the form: $\int u^n du$, with

$u = \tan x$, $n = 3$ and $du = \sec^2 x\, dx$. Applying the formula: $\int u^n\, du = \frac{u^{n+1}}{n+1}$, we obtain:

$$\int \tan^3 x \sec^2 x\, dx = \frac{\tan^4 x}{4} + C.$$

Hence,

$$\int \tan^5 x \csc^2 x\, dx = \frac{\tan^4 x}{4} + C.$$

• PROBLEM 676

$\frac{dy}{dx} = \cos^2 x$. Find the integral.

Solution: $y = \int \cos^2 x \cdot dx.$

In evaluating this integral we make use of the identity:

$$\cos^2 x = \tfrac{1}{2}(\cos 2x + 1),$$

and substituting, we obtain:

$$y = \int \cos^2 x \cdot dx = \tfrac{1}{2}\int (\cos 2x + 1)dx$$
$$= \tfrac{1}{2}\int \cos 2x \cdot dx + \tfrac{1}{2}\int dx.$$

We can now integrate, using the formula: $\int \cos u\, du = \sin u$, on the first integral, letting $u = 2x$ and $du = 2dx$, We obtain:

$$\frac{\sin 2x}{4} + \frac{x}{2} + C.$$

• PROBLEM 677

Integrate: $\int \cos^3 x\, dx.$

Solution: We break up the given integral into two integrals, to which we can apply the formulas:

$\int \cos u\, du = \sin u + C$; and: $\int u^n\, du = \frac{u^{n+1}}{n+1}$.

We obtain:

$$\int \cos^3 x\, dx = \int (\cos^2 x)\cos x\, dx$$
$$= \int (1 - \sin^2 x)\cos x\, dx,$$

by use of the identity $\cos^2 x + \sin^2 x = 1$.

$$\int \cos^3 x\, dx = \int \cos x\, dx - \int \sin^2 x \cos x\, dx.$$

Applying the formulas, we have:

$$\sin x - \tfrac{1}{3}\sin^3 x + C.$$

• **PROBLEM** 678

Find: $\int \cos^4 x \, dx.$

Solution: $\int \cos^4 x \, dx = \int (\cos^2 x)^2 dx = \int \left(\frac{1 + \cos 2x}{2}\right)^2 dx$,

by use of the identity: $\cos^2 x = \frac{1 + \cos 2x}{2}$. Expanding the integrand, we obtain: $\frac{1}{4}\int (1 + 2\cos 2x + \cos^2 2x)dx$,

$$= \frac{1}{4}\int dx + \frac{1}{2}\int \cos 2x \, dx + \frac{1}{4}\int \cos^2 2x \, dx :$$

We can now rewrite the third integral by using the above identity, and applying the formula: $\int \cos u \, du = \sin u$, to the second and third integrals. We obtain:

$$\frac{1}{4} x + \frac{1}{4} \sin 2x + \frac{1}{4}\int \frac{1 + \cos 4x}{2} dx$$

$$= \frac{1}{4} x + \frac{1}{4} \sin 2x + \frac{1}{8} x + \frac{1}{32} \sin 4x + C$$

$$= \frac{3}{8} x + \frac{1}{4} \sin 2x + \frac{1}{32} \sin 4x + C .$$

• **PROBLEM** 679

Integrate: $\int \cos^4 5y \, dy.$

Solution: We use the identity:

$$\cos^2 5y = \frac{1 + \cos 10y}{2}.$$

Then,

$$\int \cos^4 5y \, dy = \int \left(\frac{1 + \cos 10y}{2}\right)^2 dy$$

$$= \frac{1}{4} \int (1 + 2 \cos 10y + \cos^2 10y)dy.$$

We again make use of the above identity, obtaining:

$$\frac{1}{4} \int \left[1 + 2 \cos 10y + \left(\frac{1 + \cos 20y}{2}\right)\right] dy$$

$$= \frac{1}{4} \int (1 + 2 \cos 10y + \frac{1}{2} + \frac{1}{2} \cos 20y)dy$$

$$= \frac{1}{4} \int \left(\frac{3}{2} + 2 \cos 10y + \frac{1}{2} \cos 20y\right)dy.$$

Applying the formula for $\int \cos u \, du$ we obtain: $\int \cos^4 5y \, dy$

$$= \frac{3}{8}y + \frac{1}{20} \sin 10y + \frac{1}{160} \sin 20y + C.$$

• **PROBLEM** 680

Find: $\int \sin^2 x \, dx$.

Solution: To evaluate this integral we make use of the identity:

$$\sin^2 x = \frac{1 - \cos 2x}{2}.$$

Substituting, we obtain:

$$\int \sin^2 x \, dx = \int \frac{1 - \cos 2x}{2} dx = \tfrac{1}{2}\left[\int dx - \int \cos 2x \, dx\right].$$

We can now apply the formula: $\int \cos u \, du = \sin u$, to the second integral letting $u = 2x$ and $du = 2dx$. Doing this, we have:

$$\tfrac{1}{2}x - \tfrac{1}{4}\sin 2x + C.$$

• PROBLEM 681

Integrate: $\int \sin^4 x \, dx$.

Solution: $\int \sin^4 x \, dx = \int (\sin^2 x)^2 \, dx$.

We can make use of the identity: $\sin^2 x = \dfrac{1-\cos 2x}{2}$,

and, substituting, we obtain:

$$\int \left(\frac{1 - \cos 2x}{2}\right)^2 dx .$$

Expanding and placing the constant outside the integral, gives:

$$\tfrac{1}{4}\int \left(1 - 2\cos 2x + \cos^2 2x\right) dx$$

Now we use the identity: $\cos^2 x = \dfrac{1 + \cos 2x}{2}$, obtaining:

$$\tfrac{1}{4}\int \left(1 - 2\cos 2x + \frac{1 + \cos 4x}{2}\right) dx =$$

$\tfrac{1}{4}\left[\int dx - \int \cos 2x \cdot 2dx + \tfrac{1}{2}(\int dx + \int \cos 4x \cdot dx)\right]$. We can use the formula: $\int \cos u \, du = \sin u$ on the second and last integrals, obtaining:

$$\tfrac{1}{4}\left[x - \sin 2x + \tfrac{1}{2}(x + \tfrac{1}{4}\sin 4x)\right]$$

$$= \tfrac{1}{4}\left[x - \sin 2x + \frac{x}{2} + \frac{1}{8}\sin 4x\right]$$

$$= \tfrac{1}{4}\left(\frac{3}{2}x - \sin 2x + \frac{1}{8}\sin 4x\right) + C = \frac{3}{8}x - \tfrac{1}{4}\sin 2x + \frac{1}{32}\sin 4x + C.$$

• PROBLEM 682

Integrate: $\int \sin^5 x \, dx$.

Solution: $\int \sin^5 x \, dx = \int (\sin^2 x)^2 \sin x \, dx$

$$= \int (1 - \cos^2 x)^2 \sin x \, dx,$$

by use of the identity, $\sin^2 x = 1 - \cos^2 x$. Expanding the integrand we have:

$$\int (1 - 2\cos^2 x + \cos^4 x) \sin x \, dx.$$

Therefore,

$$\int \sin^5 x \, dx = \int \sin x \, dx - 2\int \cos^2 x \sin x \, dx + \int \cos^4 x (\sin x \, dx).$$

To evaluate the first integral we use the formula: $\int \sin u \, du = - \cos u$. To evaluate the second and third integrals we let $u = \cos x$ and $du = - \sin x \, dx$, using the formula for $\int u^n \, du$. We obtain:

$$\int \sin^5 x \, dx = - \cos x + \frac{2}{3} \cos^3 x - \frac{1}{5} \cos^5 x + C.$$

• PROBLEM 683

Integrate: $\int \sin^2 x \cos^3 x \, dx$.

Solution: $\int \sin^2 x \cos^3 x \, dx = \int \sin^2 x \cos^2 x (\cos x \, dx)$.

We can make use of the identity, $\cos^2 x = 1 - \sin^2 x$, and, substituting, we obtain: $\int \sin^2 x (1 - \sin^2 x)(\cos x \, dx) =$

$$\int (\sin^2 x - \sin^4 x)(\cos x \, dx).$$

Breaking this up into two integrals, we have:

$$\int \sin^2 x (\cos x \, dx) - \int \sin^4 x (\cos x \, dx).$$

Now we apply the formula:

$$\int u^n \, du = \frac{u^{n+1}}{n+1}$$

to both integrals, letting $u = \sin x$ and $du = \cos x \, dx$. We obtain:

$$\frac{1}{3} \sin^3 x - \frac{1}{5} \sin^5 x + C.$$

• PROBLEM 684

Find: $\int \sin^3 x \cos^4 x \, dx$.

Solution: $\int \sin^3 x \cos^4 x \, dx = \int \sin^2 x \cos^4 x (\sin x \, dx)$.

We can make use of the identity; $\sin^2 x = 1 - \cos^2 x$. Substituting, we have:

$$\int (1 - \cos^2 x)\cos^4 x (\sin x \, dx) = \int (\cos^4 x - \cos^6 x)(\sin x \, dx)$$

$$= \int \cos^4 x \sin x \, dx - \int \cos^6 x \sin x \, dx.$$

We integrate the expressions by using the formula, $\int u^n \, du = \frac{u^{n+1}}{n+1}$, letting $u = \cos x$ and $du = -\sin x \, dx$. $n = 4$ and 6. Applying this formula, we obtain: $-\frac{1}{5}\cos^5 x + \frac{1}{7}\cos^7 x + C.$

• PROBLEM 685

Integrate $\int \sin^4 x \cos^5 x\, dx$

Solution: This is an integral of the form

$\int \sin^m x \cos^n x$ where m is even and n is odd.

To evaluate this type of integral let

$$\cos^n x = (\cos^{n-1} x)(\cos x) = (1 - \sin^2 x)^{\frac{n-1}{2}} \cos x.$$

Thus rewrite the integral as $\int \sin^4 x \cos^2 x \cos^2 x (\cos x\, dx)$ (1)

Then by substitution of the identity $\cos^2 x = 1 - \sin^2 x$

(1) becomes $\int \sin^4 x (1 - \sin^2 x)(1 - \sin^2 x)(\cos x\, dx)$

$$= \int (\sin^4 x - \sin^6 x)(1 - \sin^2 x)(\cos x\, dx)$$

$$= \int (\sin^4 x - \sin^6 x - \sin^6 x + \sin^8 x)(\cos x\, dx)$$

$$= \int \sin^4 x \cos x\, dx - 2\int \sin^6 x \cos x\, dx + \int \sin^8 x \cos x\, dx$$

Now to integrate these expressions, use $\int u^n\, du = \frac{u^{n+1}}{n+1}$

To do this, let $u = \sin x$ so that $du = \cos x\, dx$. Hence, with $n = 4, 6,$ and 8 (for the first, second, and third integrals, respectively) application of this formula yields

$$\frac{\sin^5 x}{5} - \frac{2}{7}\sin^7 x + \frac{\sin^9 x}{9} + c$$

• PROBLEM 686

Integrate the expression: $\int \sin^4 \theta \cdot \cos^4 \theta \cdot d\theta.$

Solution: In integrating this expression, we attempt to rewrite it in a form, to which we can apply some known integration formulas. We first make use of the identity, $\sin 2\theta = 2 \sin \theta \cdot \cos \theta;$

then $(\sin 2\theta)^4 = 16 \sin^4 \theta \cdot \cos^4 \theta.$

By substitution we have:

$$\int \sin^4 \theta \cdot \cos^4 \theta \cdot d\theta = \frac{\sin^4 2\theta}{16} \cdot d\theta.$$

Now we use the identity, $\sin^2 2\theta = \frac{1 - \cos 4\theta}{2} =$

$\frac{1}{2} - \frac{1}{2}\cos 4\theta = \frac{1}{2}\left(1 - \cos 4\theta\right).$ Then,

$$(\sin^2 2\theta)^2 = \sin^4 2\theta = \left[\frac{1}{2}\left(1 - \cos 4\theta\right)\right]^2 = \frac{1}{4}\left(1 - \cos 4\theta\right)^2.$$

Substituting, we have:

$$\frac{1}{16}\int \sin^4 2\theta \cdot d\theta = \frac{1}{64}\int (1-\cos 4\theta)^2 \cdot d\theta$$

$$= \frac{1}{64}\int (1-2\cos 4\theta + \cos^2 4\theta)d\theta$$

$$= \frac{1}{64}\left(\int d\theta - 2\int \cos 4\theta \cdot d\theta + \int \cos^2 4\theta \cdot d\theta\right).$$

We now use the formula: $\int \cos u\, du = \sin u$, on the second integral, and use the identity:

$$\cos^2 4\theta = \frac{1}{2} + \frac{1}{2}\cos 8\theta,$$

on the third. We obtain:

$$\frac{1}{16}\int \sin^4 2\theta \cdot d\theta = \frac{1}{64}\left[\theta - 2\cdot\frac{1}{4}\sin 4\theta + \frac{1}{2}\int (1+\cos 8\theta)d\theta\right]$$

$$= \frac{1}{64}\left(\theta - \frac{1}{2}\sin 4\theta + \frac{1}{2}\int d\theta + \frac{1}{2}\int \cos 8\theta \cdot d\theta\right).$$

Using the formula for $\int \cos u\, du$ again, we obtain:

$$\frac{1}{64}\left(\theta - \frac{\sin 4\theta}{2} + \frac{\theta}{2} + \frac{1}{2}\cdot\frac{1}{8}\cdot \sin 8\theta\right)$$

$$\int \sin^4 \theta \cdot \cos^4 \theta \cdot d\theta = \frac{3\theta}{128} - \frac{\sin 4\theta}{128} + \frac{\sin 8\theta}{1024} + C.$$

• **PROBLEM** 687

Find: $\int \sin^2 x \cos^4 x\, dx$.

Solution: The given integral can be rewritten by making use of the identities: $\sin^2 x = \frac{1-\cos 2x}{2}$, and $\cos^2 x = \frac{1+\cos 2x}{2}$.

We write:

$$\int \sin^2 x \cos^4 x\, dx = \int \left(\frac{1-\cos 2x}{2}\right)\left(\frac{1+\cos 2x}{2}\right)^2 dx.$$

Expanding the integrand, we have:

$$\int \frac{1+\cos 2x - \cos^2 2x - \cos^3 2x}{8}\, dx$$

$$= \frac{1}{8}\int dx + \frac{1}{8}\int \cos 2x\, dx$$

$$- \frac{1}{8}\int \cos^2 2x\, dx - \frac{1}{8}\int \cos^3 2x\, dx.$$

We can now integrate, using the formula: $\int \cos u\, du = \sin u$, on the second integral, and letting $u = 2x$ and $du = 2dx$. The third and last integrals can be rewritten, using the identities:

$$\cos^2 x = \frac{1+\cos 2x}{2},$$

and $\cos^2 x = 1 - \sin^2 x$. Doing this, we obtain:

$$\frac{1}{8}x + \frac{1}{16}\sin 2x - \frac{1}{8}\int \frac{1 + \cos 4x}{2}dx$$

$$-\frac{1}{8}\int(1 - \sin^2 2x)\cos 2x\,dx = \frac{x}{8} + \frac{\sin 2x}{16} - \frac{1}{16}\int dx$$

$$-\frac{1}{16}\int \cos 4x\,dx - \frac{1}{8}\int \cos 2x\,dx + \frac{1}{8}\int \sin^2 2x \cos 2x\,dx.$$

We can now integrate, using the formula for $\int \cos u\,du$ on the second and third integrals and the formula for $\int u^n\,du$ on the last, with $u = \sin 2x$, $du = 2\cos 2x\,dx$, and $n = 2$. Applying these formulas we obtain:

$$\frac{x}{16} + \frac{\sin 2x}{16} - \frac{\sin 4x}{64} - \frac{\sin 2x}{16} + \frac{\sin^3 2x}{48} + C$$

$$= \frac{x}{16} + \frac{\sin^3 2x}{48} - \frac{\sin 4x}{64} + C.$$

• PROBLEM 688

Integrate the expression: $\int \tan x\,dx$.

Solution: The given integral can be rewritten as, $\int \frac{\sin x}{\cos x}dx$. With the integral in this form we can use the formula

$$\int \frac{du}{u} = \ln|u| + C,$$

letting $u = \cos x$ and $du = -\sin x\,dx$. Applying the formula, we obtain: $\int \tan x\,dx = -\ln|\cos x| + C$.

• PROBLEM 689

Evaluate the expression: $\int_0^{\pi/2} \tan^2 \frac{x}{2}dx$.

Solution: Using the identity: $\tan^2 x = \sec^2 x - 1$, we have:

$$\int_0^{\pi/2} \tan^2 \frac{x}{2}dx = \int_0^{\pi/2}\left(\sec^2 \frac{x}{2} - 1\right)dx = \int_0^{\pi/2} \sec^2 \frac{x}{2}dx$$

$$-\int_0^{\pi/2} dx = 2\int_0^{\pi/2} \sec^2 \frac{x}{2} \cdot \frac{1}{2}dx + \int_0^{\pi/2} dx.$$

Applying the formula: $\int \sec^2 u\,du = \tan u + C$, and evaluating between $\frac{\pi}{2}$ and 0, we obtain:

$$\left[2\tan\frac{x}{2} - x\right]_0^{\pi/2} = 2\tan\frac{\pi}{4} - \frac{\pi}{2} = 2 - \frac{\pi}{2}.$$

• PROBLEM 690

Integrate: $\int \tan^3 x\,dx$.

Solution: The given integral can be rewritten as, $\int \tan x \cdot \tan^2 x$. We can now make use of the identity:

$$\tan^2 x = \sec^2 x - 1.$$

Substituting into the given integral, we obtain:

$$\int \tan x(\sec^2 x - 1)dx = \int(\tan x \sec^2 x - \tan x)dx$$

$$= \int \tan x(\sec^2 x\, dx) - \int \tan x\, dx .$$

We can integrate the first integral by use of the formula for $\int u^n du$, with $u = \tan x$, $du = \sec^2 x\, dx$, and $n = 1$; and the second integral by use of the formula; $\int \tan u = \ln \cos u$, obtaining:

$$\tfrac{1}{2} \tan^2 x + \ln \cos x + C .$$

• PROBLEM 691

Integrate the expression: $\int \tan^5 \theta \cdot d\theta$.

Solution: We wish to rewrite the given integral as the sum of three integrals, to which we can apply the formulas, $\int u^n du = \frac{u^{n+1}}{n+1} + C$ and $\int \tan u\, du = \ln \sec u + C$.

$$\int \tan^5 \theta \cdot d\theta = \int \tan^4 \theta \cdot \tan \theta \cdot d\theta.$$

We can make use of the identity, $\tan^2 \theta = \sec^2 \theta - 1$, obtaining:

$$\int (\sec^2 \theta - 1)^2 \tan \theta \cdot d\theta$$

$$= \int (\sec^4 \theta - 2 \sec^2 \theta + 1)\tan \theta \cdot d\theta$$

$$= \int \sec^3 \theta \sec \theta \tan \theta \cdot d\theta$$

$$- 2 \int \sec \theta \sec \theta \tan \theta\, d\theta + \int \tan \theta\, d\theta.$$

We can now apply the formula for $\int u^n du$ to the first integral, with $u = \sec \theta$, $du = \sec \theta \tan \theta\, d\theta$, $n = 3$, and also to the second integral with $u = \sec \theta$, $du = \sec \theta \tan \theta\, d\theta$, $n = 1$. For the third integral we use the formula for $\int \tan u\, du$, with $u = \theta$, $du = d\theta$. Doing this, we obtain:

$$\int \tan^5 \theta \cdot d\theta = \frac{\sec^4 \theta}{4} - \sec^2 \theta + \ln(\sec \theta) + C.$$

• PROBLEM 692

Integrate: $\int \tan^5 3x\, dx$.

Solution: In this problem, we make use of the identity: $\tan^2 3x = \sec^2 3x - 1$, and proceed as follows:

$$\int \tan^5 3x\, dx = \int \tan^3 3x \tan^2 3x\, dx$$

$$= \int \tan^3 3x(\sec^2 3x - 1)dx.$$

We wish to convert the integral into the form: $\int u^n du$ with $u = \tan 3x$, $n = 3$, $du = \sec^2 3x \cdot 3dx$. From the above integral we obtain:

$$\frac{1}{3} \int \tan^3 3x\ (\sec^2 3x)(3dx) - \int \tan^3 3x\, dx$$

$$= \frac{1}{3} \int \tan^3 3x\ (\sec^2 3x)(3dx) - \int \tan 3x(\sec^2 3x-1)dx.$$

We now apply the formula: $\int u^n du = \frac{u^{n+1}}{n+1} + C$, to the first integral, and write the second integral as the sum of two integrals in the form $\int u^n du$ and $\int \tan u\, du = \ln |\sec u| + C$, respectively. We have:

$$\frac{1}{12} \tan^4 3x - \frac{1}{3} \int \tan 3x(\sec^2 3x)(3dx) + \int \tan 3x\, dx.$$

Applying the formulas we obtain: $\int \tan^5 3x$

$$= \frac{1}{12} \tan^4 3x - \frac{1}{6} \tan^2 3x + \frac{1}{3} \ln |\sec 3x| + C.$$

• PROBLEM 693

Integrate: $\int \cot^3 2\theta \cdot d\theta$.

Solution: We wish to rewrite the given integral so that we can apply the formulas: $\int u^n\, du = \frac{u^{n+1}}{n+1} + C$, and $\int \cot u\, du = \ln(\sin u) + C$.

$$\int \cot^3 2\theta \cdot d\theta = \cot^2 2\theta \cdot \cot 2\theta \cdot d\theta.$$

We can now make use of the identity, $\cot^2 \theta = \csc^2 \theta - 1$. Substituting, we obtain:

$$\int (\csc^2 2\theta - 1)(\cot 2\theta)d\theta$$

$$= \int (\csc^2 2\theta \cdot \cot 2\theta \cdot d\theta - \int \cot 2\theta \cdot d\theta$$

$$= \int \csc 2\theta \cdot \csc 2\theta \cdot \cot 2\theta \cdot d\theta - \int \cot 2\theta \cdot d\theta.$$

We multiply by -2 under the integral sign and by $-\frac{1}{2}$ outside in the first term, by $+2$ and $+\frac{1}{2}$ in the second. We can now apply the formulas for $\int u^n du$, with $u = \csc 2\theta$, $n = 1$, $du = -2 \csc 2\theta \cot 2\theta \cdot d\theta$ and for $\int \cot u\, du$, with $u = 2\theta$, $du = 2d\theta$. Doing this we obtain:

$$\int \cot^3 2\theta \cdot d\theta = -\frac{1}{4}\csc^2 2\theta - \frac{1}{2}\ln(\sin 2\theta) + C.$$

• PROBLEM 694

Integrate: $\int \cot^4 3x\, dx.$

Solution: To integrate this expression we make use of the identity: $\cot^2 x = \csc^2 x - 1$. The given integral can be rewritten as:

$$\int (\cot^2 3x)(\cot^2 3x)dx = \int \cot^2 3x(\csc^2 3x - 1)dx$$

$$= \int (\cot^2 3x \csc^2 3x - \cot^2 3x)dx$$

$$= \int \cot^2 3x \csc^2 3x\, dx - \int \cot^2 3x\, dx.$$

We now integrate the first term using the formula for $\int u^n du$, letting $u = \cot 3x$, $du = -3\csc^2 3x\, dx$ and $n = 2$. We can rewrite the second integral by again using the above identity. This gives us:

$$-\frac{1}{3}\int \cot^2 3x(-3\csc^2 3x)dx - \int(\csc^2 3x - 1)dx$$

$$= -\frac{1}{3}\left(\frac{\cot^3 3x}{3}\right) - \int \csc^2 3x\, dx - \int dx.$$

We can now integrate, using the formula: $\int \csc^2 u\, du = -\cot u$, letting $u = 3x$, $du = 3dx$. We obtain:

$$-\frac{1}{9}\cot^3 3x + \frac{1}{3}\cot 3x + x + C.$$

• PROBLEM 695

Integrate the expression: $\int \csc^6 x\, dx.$

Solution: The given integral can be rewritten as:

$$\int (\csc^4 x)\csc^2 x\, dx.$$

We now make use of the identity: $\csc^2 x = \cot^2 x + 1$, obtaining:

$$\int (\cot^2 x + 1)^2 \csc^2 x\, dx.$$

Expanding the integrand, we have:

$$\int (\cot^4 x + 2\cot^2 x + 1)\csc^2 x\, dx$$

$$= \int \cot^4 x \csc^2 x\, dx + 2\int \cot^2 x \csc^2 x\, dx + \int \csc^2 x\, dx.$$

To integrate the first two terms we use the formula for $\int u^n du$, letting $u = \cot x$, and $du = -\csc^2 x\, dx$. To integrate the last term we use the formula: $\int \csc^2 u\, du = -\cot u$. Applying these formulas, we obtain:

$$-\frac{1}{5}\cot^5 x - \frac{2}{3}\cot^3 x - \cot x + C.$$

• PROBLEM 696

Integrate the expression: $\int \sec^4 2\theta \cdot d\theta$.

Solution: We wish to rewrite the given integral as two integrals in the form $\int u^n du$ and $\int \sec^2 u\, du$.

$$\int \sec^4 2\theta \cdot d\theta = \int \sec^2 2\theta \cdot \sec^2 2\theta \cdot d\theta.$$

Making use of the identity: $\sec^2 \theta = \tan^2 \theta + 1$, we have:

$$\int (\tan^2 2\theta + 1) \cdot (\sec^2 2\theta) \cdot d\theta$$

$$= \int (\tan^2 2\theta \cdot \sec^2 2\theta) d\theta + \int \sec^2 2\theta \cdot d\theta.$$

We multiply by 2 under the integral sign and by $\frac{1}{2}$ outside. We now have the first integral in the form $\int u^n du$, with $u = \tan 2\theta$, $n = 2$, $du = 2\sec^2 2\theta d\theta$, and the second integral in the form $\int \sec^2 u\, du$, with $u = 2\theta$, $du = 2d\theta$. Applying the formulas, we obtain:

$$\int \sec^4 2\theta \cdot d\theta = \frac{1}{6}\tan^3 2\theta + \frac{1}{2}\tan 2\theta + C.$$

• PROBLEM 697

Integrate the expression: $\int \tan^3 x \sec^4 x\, dx$.

Solution: The given expression can be written as:

$$\int \tan^3 x \sec^2 x \sec^2 x\, dx.$$

Making use of the identity: $\sec^2 x = 1 + \tan^2 x$, we have

$$\int \tan^3 x \sec^2 x (1 + \tan^2 x)\, dx$$

$$= \int \tan^3 x \sec^2 x\, dx + \int \tan^5 x \sec^2 x\, dx.$$

Letting $\tan x = u$, $du = \sec^2 x\, dx$, the two integrals can be written in the form:

$$\int u^n du = \frac{u^{n+1}}{n+1} + C,$$ where $n = 3$ in the first integral, $n = 5$ in the second. We have:

$$\frac{\tan^4 x}{4} + \frac{\tan^6 x}{6} + C.$$

• PROBLEM 698

Integrate: $\int \tan^5 x \sec^4 x\, dx.$

<u>Solution:</u> The given integral can be rewritten as;

$$\int \tan^5 x(\sec^2 x)\sec^2 x\, dx.$$

Making use of the identity; $\sec^2 x = \tan^2 x + 1$, we have:

$$\int \tan^5 x(\tan^2 x + 1) \sec^2 x\, dx$$

$$= \int (\tan^7 x + \tan^5 x) \sec^2 x\, dx$$

$$= \int \tan^7 x \sec^2 x\, dx + \int \tan^5 x \sec^2 x\, dx.$$

We can now integrate, using the formula:

$$\int u^n\, du = \frac{u^{n+1}}{n+1},$$

letting $u = \tan x$, $du = \sec^2 x\, dx$. Applying this formula, we obtain:

$$\frac{1}{8} \tan^8 x + \frac{1}{6} \tan^6 x + C.$$

• PROBLEM 699

Find: $\int \tan^5 x \sec^7 x\, dx.$

<u>Solution:</u>

$$\int \tan^5 x \sec^7 x\, dx = \int \tan^4 x \sec^6 x \sec x \tan x\, dx.$$

Making use of the identity; $\tan^2 x = \sec^2 x - 1$, we substitute and obtain: $\int (\sec^2 x - 1)^2 \sec^6 x(\sec x \tan x\, dx).$

Expanding the integrand, we have:

$$\int (\sec^2 x - 1)^2 \sec^6 x(\sec x \tan x\, dx)$$

$$= \int (\sec^{10} x)\sec x \tan x\, dx - 2\int (\sec^8 x)\sec x \tan x\, dx$$

$$+ \int (\sec^6 x)\sec x \tan x\, dx$$

The three expressions can now be integrated, using the formula:

$$\int u^n\, du = \frac{u^{n+1}}{n+1},$$

letting $u = \sec x$, $du = \sec x \tan x\, dx$. Applying the formula, we obtain:

$$\int \tan^5 x \sec^7 x\, dx = \frac{1}{11}\sec^{11} x - \frac{2}{9}\sec^9 x + \frac{1}{7}\sec^7 x + C.$$

• PROBLEM 700

Integrate: $\int \tan^7 \theta \sec^5 \theta \cdot d\theta.$

Solution: In integrating this expression we break it up into several integrals, to which we can apply the formula: $\int u^n \, du = \frac{u^{n+1}}{n+1} + C.$

$$\int \tan^7 \theta \cdot \sec^5 \theta \cdot d\theta = \int \tan^6 \theta \cdot \sec^4 \theta \cdot \sec \theta \cdot \tan \theta \cdot d\theta.$$

We now use the identity: $\tan^2 \theta = \sec^2 \theta - 1$, and let $\tan^6 \theta = (\tan^2 \theta)^3 = (\sec^2 \theta - 1)^3$, obtaining:

$$\int (\sec^2 \theta - 1)^3 \sec^4 \theta \cdot \sec \theta \cdot \tan \theta \cdot d\theta.$$

We can now **expand**, and, using the distributive law, we obtain:

$$\int (\sec^6 \theta - 3 \sec^4 \theta + 3 \sec^2 \theta - 1) \sec^4 \theta \cdot \sec \theta \cdot \tan \theta \cdot d\theta$$

$$= \int (\sec^{10} \theta - 3 \sec^8 \theta + 3 \sec^6 \theta - \sec^4 \theta) \sec \theta \tan \theta \cdot d\theta$$

$$= \int \sec^{10} \theta \cdot \sec \theta \cdot \tan \theta \cdot d\theta - 3 \int \sec^8 \theta \cdot \sec \theta \cdot \tan \theta \cdot d\theta + 3 \int \sec^6 \theta \cdot \sec \theta \cdot \tan \theta \cdot d\theta - \int \sec^4 \theta \cdot \sec \theta \cdot \tan \theta \cdot d\theta.$$

The **four** integrals are in the form: $\int u^n \, du$, with $u = \sec \theta$ and $du = \sec \theta \tan \theta \, d\theta$. Applying the formula, we obtain:

$$\frac{\sec^{11} \theta}{11} - \frac{3 \sec^9 \theta}{9} + \frac{3}{7} \sec^7 \theta - \frac{\sec^5 \theta}{5}.$$

• PROBLEM 701

Integrate: $\int \text{cosec}^4 \, 2\theta \cdot d\theta.$

Solution: We wish to rewrite the given integral as two integrals, the first in the form $\int u^n \, du$ and the second: $\int \text{cosec}^2 u \, du$. $\int \text{cosec}^4 \, 2\theta \, d\theta$ can be re-written as: $\int (\cot^2 2\theta + 1)(\text{cosec}^2 \, 2\theta) d\theta$, by making use of the identity: $\text{cosec}^2 \theta = \cot^2 \theta + 1$. Using the distributive law, we obtain: $\int \cot^2 2\theta \cdot \text{cosec}^2 \, 2\theta \cdot d\theta + \int \text{cosec}^2 \, 2\theta \cdot d\theta$. We multiply by 2 under the integral sign and by $\frac{1}{2}$ outside. The first integral is now in the form: $\int u^n \, du$, with $u = \cot 2\theta$, $n = 2$ and $du = -2 \, \text{cosec}^2 \, 2\theta \, d\theta$; and the second is in the form: $\int \operatorname{cosec}^2 u\, du$, with $u = 2\theta$, $du = 2d\theta$. Applying the formulas, we obtain:

$$\int \operatorname{cosec}^4 2\theta \cdot d\theta = -\frac{1}{6} \cot^3 2\theta - \frac{1}{2} \cot 2\theta.$$

• PROBLEM 702

Integrate the expression: $\int \cot^8 \theta \cdot \operatorname{cosec}^6 \theta \cdot d\theta$.

Solution: We wish to convert the given integral into a form, to which we can apply the formula:

$\int u^n\, du = \dfrac{u^{n+1}}{n+1} + C$, with $u = \cot \theta$ and $du = -\operatorname{cosec}^2 \theta$.

We can rewrite $\int \cot^8 \theta \cdot \operatorname{cosec}^6 \theta \cdot d\theta$ as:

$\int \cot^8 \theta \cdot \operatorname{cosec}^4 \theta \cdot \operatorname{cosec}^2 \theta \cdot d\theta$. We now make use of the identity: $\operatorname{cosec}^2 \theta = \cot^2 \theta + 1$, and we have:

$$\int \cot^8 \theta (\cot^2 \theta + 1)^2 \operatorname{cosec}^2 \theta \cdot d\theta$$

$$= \int \cot^8 \theta (\cot^4 \theta + 2 \cot^2 \theta + 1) \operatorname{cosec}^2 \theta \cdot d\theta$$

$$= \int \cot^{12} \theta \operatorname{cosec}^2 \theta \cdot d\theta + 2 \int \cot^{10} \theta \cdot \operatorname{cosec}^2 \theta \cdot d\theta + \int \cot^8 \theta \cdot \operatorname{cosec}^2 \theta\, d\theta,$$

by the distributive law. We now apply the formula and obtain:

$$\cot^8 \theta \cdot \operatorname{cosec}^6 \theta \cdot d\theta = \frac{\cot^{13} \theta}{13} - \frac{2}{11}\cot^{11} \theta - \frac{\cot^9 \theta}{9} + C.$$

• PROBLEM 703

Integrate: $\int \dfrac{\cos^2 x}{1 - \sin x} dx$.

Solution: We simplify this integral by using the identity: $\cos^2 u + \sin^2 u = 1$, or, $\cos^2 u = 1 - \sin^2 u$. We rewrite the integral as:

$$\int \frac{\cos^2 x}{1 - \sin x} dx = \int \frac{1 - \sin^2 x}{1 - \sin x} dx.$$

We can further simplify by factoring $(1 - \sin^2 x)$ into $(1 - \sin x)(1 + \sin x)$. We can again rewrite the integral as:

$$\int \frac{(1 - \sin x)(1 + \sin x)}{(1 - \sin x)} dx = \int (1 + \sin x) dx.$$

We now integrate:

$$\int 1 + \sin x \, dx = \int 1 \, dx + \int \sin x \, dx$$

$$\int \frac{\cos^2 x}{1 - \sin x} dx = x + \cos x + C.$$

• PROBLEM 704

Integrate: $\int \sin^3 2x \cos^3 2x \, dx.$

Solution: To solve this problem we rewrite the integral in a more integrable form. Using the identity: $\cos^2 \theta = 1 - \sin^2 \theta$, we can rewrite the integral as follows:

$$\int \sin^3 2x \cos^3 2x \, dx$$

$$= \int \sin^3 2x \cos 2x (\cos^2 2x) dx$$

$$= \int \sin^3 2x \cos 2x (1 - \sin^2 2x) dx.$$

We now simplify this expression to obtain:

$$\int \sin^3 2x \cos 2x (1 - \sin^2 2x) dx$$

$$= \int (\sin^3 2x \cos 2x - \sin^5 2x \cos 2x) dx$$

$$= \int \sin^3 2x \cos 2x \, dx - \int \sin^5 2x \cos 2x \, dx.$$

We can integrate each of these expressions, using the formula:

$$\int u^n \, du = \frac{u^{n+1}}{n+1} + C.$$

$$\int \sin^3 2x \cos 2x \, dx - \int \sin^5 2x \cos 2x \, dx$$

$$= \int (\sin 2x)^3 \cos 2x \, dx - \int (\sin 2x)^5 \cos 2x \, dx$$

$$= \frac{1}{2} \int (\sin 2x)^3 (\cos 2x) 2dx - \frac{1}{2} \int (\sin 2x)^5 (\cos 2x) 2dx$$

$$= \frac{1}{2} \frac{(\sin 2x)^4}{4} - \frac{1}{2} \frac{(\sin 2x)^6}{6}$$

$$= \frac{\sin^4 2x}{8} - \frac{\sin^6 2x}{12} + C.$$

• PROBLEM 705

Integrate: $\int \sin^3 2x \cos^4 x \, dx.$

Solution: To make the function more easily integrable, we rewrite the integral, using the identity: $\sin 2x = 2 \sin x \cos x$, and then the identity: $\sin^2 \theta = 1 - \cos^2 \theta$. We obtain:

$$\int \sin^3 2x \cos^4 x\,dx = \int (\sin 2x)^3 \cos^4 x\,dx$$

$$= \int (2\sin x \cos x)^3 \cos^4 x\,dx = \int 8\sin^3 x \cos^7 x\,dx$$

$$= 8\int \cos^7 x(\sin^2 x)\sin x\,dx$$

$$= 8\int \cos^7 x(1 - \cos^2 x)\sin x\,dx$$

$$= 8\int (\cos^7 x \sin x - \cos^9 x \sin x)dx.$$

We can now write this as the sum of two integrals and integrate, using the formula $\int u^n\,du = \frac{u^{n+1}}{n+1} + C$. We proceed as follows:

$$8\int (\cos^7 x \sin x - \cos^9 x \sin x)dx$$

$$= -8\int \cos^7 x(-\sin x)dx - (-8)\int \cos^9 x(-\sin x)dx$$

$$= -8\frac{(\cos^8 x)}{8} + 8\frac{(\cos^{10} x)}{10}$$

$$= -\cos^8 x + \frac{4}{5}\cos^{10} x + C.$$

• PROBLEM 706

Integrate: $\int \frac{\cos^3 x}{\sqrt{\sin x}}dx.$

Solution: To integrate the given expression, we make use of the identity: $\cos^2 x = 1 - \sin^2 x$. We rewrite the integral as

$$\int \frac{(1 - \sin^2 x)\cos x}{\sin^{\frac{1}{2}} x}dx = \int \left[\frac{\cos x - \sin^2 x \cos x}{\sin^{\frac{1}{2}} x}\right]dx$$

$$= \int \left[\frac{\cos x}{\sin^{\frac{1}{2}} x} - \frac{\sin^2 x \cos x}{\sin^{\frac{1}{2}} x}\right]dx.$$

Rewriting this as two integrals we obtain:

$$\int \sin^{-\frac{1}{2}} x \cos x\,dx - \int \sin^{\frac{3}{2}} x \cos x\,dx.$$

We can now apply the formula $\int u^n\,du = \frac{u^{n+1}}{n+1}$, to each integral. In the first integral we let $u = \sin x$, $du = \cos x\,dx$, $n = -\frac{1}{2}$, and in the second integral, $u = \sin x$, $du = \cos x\,dx$, $n = \frac{3}{2}$. Applying the formula, we obtain:

$$\int \frac{\cos^3 x}{\sqrt{\sin x}} dx = \frac{\sin^{\frac{1}{2}} x}{\frac{1}{2}} - \frac{\sin^{\frac{5}{2}} x}{\frac{5}{2}}$$

$$= 2 \sin^{\frac{1}{2}} x - \frac{2}{5} \sin^{\frac{5}{2}} x$$

$$= \frac{2}{5}\sqrt{\sin x}\left(5 - \sin^2 x\right).$$

• PROBLEM 707

Integrate: $\int \frac{\sin 2x}{1 + \sin^2 x} dx.$

Solution: To integrate this expression, we use the formula: $\int \frac{du}{u} = \ln u$. We can rewrite the denominator, using the identity:

$$\sin^2 A = \frac{1 - \cos 2A}{2},$$

which makes it easier to see whether the numerator is the derivative of the denominator. We rewrite the denominator as $1 + \left(\frac{1 - \cos 2x}{2}\right)$. The derivative of $1 + \left(\frac{1 - \cos 2x}{2}\right)$ is $\sin 2x$. Therefore,

$$\int \frac{\sin 2x}{1 + \sin^2 x} dx = \int \frac{\sin 2x}{1 + \frac{1 - \cos 2x}{2}} dx = \int \frac{du}{u}$$

$$= \ln(1 + \sin^2 x) + C.$$

• PROBLEM 708

Integrate: (a) $\int \sin 5x \cos 3x \, dx$ and

(b) $\int \cos 3x \cos 2x \, dx$.

Solution: We make use of the identity: $\sin A \cos B = \frac{1}{2}\left[\sin(A+B) + \sin(A-B)\right]$. We have:

(a) $\int \sin 5x \cos 3x \, dx = \frac{1}{2}\int\left[\sin(5x+3x) + \sin(5x-3x)\right]dx$

$= \frac{1}{2}\int \sin 8x \, dx + \frac{1}{2}\int \sin 2x \, dx$.

We can now integrate both expressions using the formula, $\int \sin u \, du = -\cos u$. We obtain:

$$-\frac{1}{16} \cos 8x - \frac{1}{4} \cos 2x + C ;$$

We use the identity: $\cos A \cos B = \frac{1}{2}\left[\cos(A+B) + \cos(A-B)\right]$. Applying this, we obtain:

(b) $\int \cos 3x \cos 2x \, dx = \frac{1}{2}\int\left[\cos(3x+2x) + \cos(3x-2x)\right]dx$

$$= \frac{1}{2}\int \cos 5x \, dx + \frac{1}{2}\int \cos x \, dx = \frac{1}{10} \sin 5x + \frac{1}{2} \sin x + C,$$

by use of the formula: $\int \cos u \, du = \sin u + C.$

• PROBLEM 709

Find: $\int \sin 3x \cos 2x \, dx.$

Solution: To evaluate the given integral we use the formula

$$\sin a \cos b = \frac{1}{2} \sin(a - b) + \frac{1}{2} \sin(a + b).$$

In this case $a = 3x$ and $b = 2x$. Applying the formula, we obtain:

$$\int \sin 3x \cos 2x \, dx = \int\left[\frac{1}{2} \sin x + \frac{1}{2} \sin 5x\right] dx$$

$$= \frac{1}{2}\int \sin x \, dx + \frac{1}{2}\int \sin 5x \, dx.$$

We can evaluate both integrals by using the formula: $\int \sin u \, du = - \cos u$, obtaining:

$$- \frac{1}{2} \cos x - \frac{1}{10} \cos 5x + C.$$

• PROBLEM 710

Integrate: $\int \sin 3x \cos 6x \, dx.$

Solution: To make this function more easily integrable, we use the identity:

$$\sin A \cos B = \frac{1}{2}\left[\sin(A + B) + \sin(A - B)\right].$$

We obtain:

$$\sin 3x \cos 6x = \frac{1}{2}\left[\sin(9x) + \sin(-3x)\right].$$

Now we can integrate by separating the integral into the sum of two separate integrals. We proceed as follows:

$$\int \sin 3x \cos 6x \, dx = \int \frac{1}{2}\left[\sin(9x) + \sin(-3x)\right]dx$$

$$= \int \frac{1}{2} \sin 9x \, dx + \int \frac{1}{2} \sin(-3x)dx$$

$$= \frac{1}{2}\left(\frac{1}{9}\right) \int \sin 9x(9dx)$$

$$+ \frac{1}{2}\left(-\frac{1}{3}\right) \int \sin(-3x)(-3dx)$$

$$= \frac{1}{18}\left(-\cos 9x\right) - \frac{1}{6}\left[-\cos(-3x)\right]$$

$$= \frac{1}{6} \cos(-3x) - \frac{1}{18} \cos 9x.$$

• PROBLEM 711

Integrate: $\int \frac{2 + \sin x + 2 \cos x}{1 + \cos x} dx.$

Solution: We can rewrite this integral as the sum of two integrals which are easier to manipulate. We can write:

$$\int \frac{2 + \sin x + 2 \cos x}{1 + \cos x} dx = \int \frac{2 + 2 \cos x}{1 + \cos x} dx - \int \frac{-\sin x}{1 + \cos x} dx$$

$$= 2 \int \frac{1 + \cos x}{1 + \cos x} dx - \int \frac{-\sin x}{1 + \cos x} dx$$

$$= 2 \int dx - \int \frac{-\sin x}{1 + \cos x} dx.$$

If we now let $(1 + \cos x) = u$, $du = -\sin x\, dx$, and we can apply the rule: $\int \frac{1}{u} du = \ln u$. This gives:

$$\int \frac{2 + \sin x + 2 \cos x}{1 + \cos x} dx = 2x - \ln(1 + \cos x) + C.$$

• PROBLEM 712

Integrate: $\int \frac{\cos^2 x}{\sin^5 x} dx.$

Solution: We can simplify this problem by using the identities: $\sin^2 \theta + \cos^2 \theta = 1$, and $\frac{1}{\sin \theta} = \csc \theta$, as follows:

$$\int \frac{\cos^2 x}{\sin^5 x} dx = \int \frac{1 - \sin^2 x}{\sin^5 x} dx$$

$$= \int \frac{1}{\sin^5 x} dx - \int \frac{1}{\sin^3 x} dx$$

$$= \int \csc^5 x\, dx - \int \csc^3 x\, dx.$$

We use the following formula to perform each integration:

$$\int \csc^n u\, du = \frac{-\cot u \csc^{(n-2)} u}{n - 1} + \frac{n - 2}{n - 1} \int \csc^{(n-2)} u\, du.$$

We first integrate the first term using this formula, as follows:

$$\int \csc^5 x\, dx = \frac{-\cot x \csc^3 x}{4} + \frac{3}{4} \int \csc^3 x\, dx.$$

We use the formula again for the second term.

$$\frac{3}{4} \int \csc^3 x\, dx = \frac{3}{4}\left(\frac{-\cot x \csc x}{2} + \frac{1}{2} \int \csc x\, dx\right).$$

We now use the formula:

$$\int \csc x \, dx = \ln|\csc x - \cot x|,$$

to complete the integration. We can write the total result for the integral: $\int \csc^5 x \, dx$, as follows:

$$\csc^5 x \, dx = \frac{-\cot x \csc^3 x}{4} + \frac{3}{4}\left(\frac{-\cot x \csc x}{2} + \frac{1}{2}\ln|\csc x - \cot x|\right)$$

$$= \frac{-\cot x \csc^3 x}{4} - \frac{3 \cot x \csc x}{8} + \frac{3}{8}\ln|\csc x - \cot x|.$$

We now use the formula for $\int \csc^n u \, du$ to solve the second integral: $\int \csc^3 x \, dx$, as follows:

$$\int \csc^3 x \, dx = \frac{-\cot x \csc x}{2} + \frac{1}{2}\int \csc x \, dx.$$

Using the formula: $\int \csc x \, dx = \ln|\csc x - \cot x|$, we obtain:

$$\int \csc^3 x \, dx = \frac{-\cot x \csc x}{2} + \frac{1}{2}\ln|\csc x - \cot x|.$$

The final result for $\int \frac{\cos^2 x}{\sin^5 x} dx$ is:

$$\int \frac{\cos^2 x}{\sin^5 x} dx = \int \csc^5 x \, dx - \int \csc^3 x \, dx$$

$$= \frac{-\cot x \csc^3 x}{4} - \frac{3 \cot x \csc x}{8} + \frac{3}{8}\ln|\csc x - \cot x| - \left(\frac{-\cot x \csc x}{2} + \frac{1}{2}\ln|\csc x - \cot x|\right)$$

$$= \frac{-\cot x \csc^3 x}{4} + \frac{\cot x \csc x}{8} - \frac{\ln|\csc x - \cot x|}{8}.$$

• PROBLEM 713

Integrate: $\int \frac{dx}{1 - \cos x}$.

Solution: To integrate the given expression, we multiply the integrand by $\frac{1 + \cos x}{1 + \cos x}$, which is equivalent to multiplying by 1, hence not changing the integral. We obtain:

$$\int \frac{dx}{1 - \cos x} = \int \frac{(1 + \cos x)dx}{1 - \cos^2 x}.$$

Now, using the identity: $\sin^2 x = 1 - \cos^2 x$, we write:

$$\int \frac{(1 + \cos x)dx}{\sin^2 x},$$

which can be rewritten as two separate integrals. We have:

$$\int \frac{1}{\sin^2 x}dx + \int \frac{\cos x}{\sin^2 x}dx$$

$$= \int \frac{1}{\sin^2 x}dx + \int \frac{\cos x}{\sin x} \cdot \frac{1}{\sin x}dx.$$

Now, using the fact that $\frac{1}{\sin x} = \csc x$ and $\frac{\cos x}{\sin x} = \cot x$, and substituting we obtain:

$$\int \frac{dx}{1 - \cos x} = \int \csc^2 x\, dx + \int \cot x \csc x\, dx.$$

We now apply the formulas:

$$\int \csc^2 x\, dx = -\cot x,$$

and $$\int \cot x \csc x\, dx = -\csc x,$$

obtaining:

$$\int \frac{dx}{1 - \cos x} = -\cot x + (-\csc x)$$

$$= -\cot x - \csc x + C.$$

• PROBLEM 714

Integrate: $\int \frac{dx}{1 - \sin x}.$

Solution: We apply some algebra to the integrand and use a few identities in order to make it more easily integrable. If we multiply the integrand by $\left(\frac{1 + \sin x}{1 + \sin x}\right) = 1$, we obtain the following:

$$\int \frac{dx}{1 - \sin x}\left(\frac{1 + \sin x}{1 + \sin x}\right) = \int \frac{1 + \sin x}{1 - \sin^2 x}dx.$$

We use the following identities to rewrite the integral:

$$\sin^2 \theta + \cos^2 \theta = 1, \quad \frac{1}{\cos \theta} = \sec \theta, \quad \frac{\sin \theta}{\cos \theta} = \tan \theta.$$

Using these identities, we find:

$$\int \frac{1+\sin x}{1-\sin^2 x}dx = \int \frac{1+\sin x}{\cos^2 x}dx$$

$$= \int \left[\frac{1}{\cos^2 x} + \frac{\sin x}{\cos^2 x}\right]dx = \int \left[\sec^2 x + \left(\frac{\sin x}{\cos x}\right)\left(\frac{1}{\cos x}\right)\right]dx$$

$$= \int (\sec^2 x + \tan x \sec x)dx.$$

$$= \int \sec^2 x\, dx + \int \tan x \sec x\, dx.$$

We now integrate, using the formulas:

$\int \sec^2 u\, du = \tan u$, and $\int \sec u \tan u\, du = \sec u$.
We obtain:

$$\int \sec^2 x\, dx + \int \tan x \sec x\, dx = \tan x + \sec x + C.$$

• PROBLEM 715

Integrate: $\int \frac{\sin\theta\cos\theta}{2-\cos\theta}d\theta.$

Solution: To simplify integration, we perform some algebraic operations. First we add

$$\frac{2\sin\theta}{2-\cos\theta} - \frac{2\sin\theta}{2-\cos\theta},$$

which is in effect adding zero, to the integrand, and then rearrange terms as follows,

$$\int \frac{\sin\theta\cos\theta}{2-\cos\theta}d\theta = \int\left(\frac{2\sin\theta}{2-\cos\theta} - \frac{2\sin\theta}{2-\cos\theta} + \frac{\sin\theta\cos\theta}{2-\cos\theta}\right)d\theta$$

$$= \int\left(\frac{\sin\theta\cos\theta - 2\sin\theta}{2-\cos\theta} + \frac{2\sin\theta}{2-\cos\theta}\right)d\theta$$

$$= \int\left(\frac{-\sin\theta(2-\cos\theta)}{2-\cos\theta} + \frac{2\sin\theta}{2-\cos\theta}\right)d\theta$$

$$= \int\left(\frac{-\sin\theta(2-\cos\theta)}{2-\cos\theta} + \frac{2\sin\theta}{2-\cos\theta}\right)d\theta$$

$$= \int\left(-\sin\theta + \frac{2\sin\theta}{2-\cos\theta}\right)d\theta.$$

We can now separate this integral into the sum of two integrals:

$$\int\left(-\sin\theta + \frac{2\sin\theta}{2-\cos\theta}\right)d\theta = -\int \sin\theta\, d\theta + \int \frac{2\sin\theta}{2-\cos\theta}d\theta$$

$$= -(-\cos\theta) + 2\ln(2-\cos\theta) + C$$

$$= \cos\theta + 2\ln(2-\cos\theta) + C.$$

• PROBLEM 716

Integrate: $\int \operatorname{sech}^4 x\, dx.$

Solution:

$$\int \operatorname{sech}^4 x\, dx = \int \operatorname{sech}^2 x\, (\operatorname{sech}^2 x\, dx)$$

Making use of the identity:

$$\operatorname{sech}^2 x = 1 - \tanh^2 x,$$

and substituting, we obtain:

$$\int (1 - \tanh^2 x)(\operatorname{sech}^2 x\, dx)$$

$$= \int \operatorname{sech}^2 x\, dx - \int \tanh^2 x\, (\operatorname{sech}^2 x\, dx).$$

We can now perform the first integration, using the formula:

$\int \operatorname{sech}^2 u\, du = \tanh u$; and the second,

using the formula for

$\int u^n\, du$, letting $u = \tanh x$,

$du = \operatorname{sech}^2 x\, dx$, and $n = 2$. Applying the formulas, we obtain:

$$\tanh x - \frac{1}{3} \tanh^3 x + C.$$

• PROBLEM 717

Integrate: $\int \sinh^3 x \cosh^2 x\, dx$

Solution:

$$\int \sinh^3 x \cosh^2 x\, dx = \int \sinh^2 x \cosh^2 x\, (\sinh x\, dx),$$

We can now make use of the identity,

$\sinh^2 x = \cosh^2 x - 1$, obtaining:

$$\int (\cosh^2 x - 1) \cosh^2 x\, (\sinh x\, dx)$$

$$= \int (\cosh^4 x - \cosh^2 x) \sinh x\, dx$$

$$= \int \cosh^4 x\, (\sinh x\, dx) - \int \cosh^2 x\, (\sinh x\, dx).$$

We can now integrate, using the formula:

$\int u^n\, du = \frac{u^{n+1}}{n+1}$, letting $u = \cosh x$,

$du = \sinh x\, dx$. We let $n = 4$ in the first integral and 2 in the second. Applying the formula, we obtain:

$$\frac{1}{5} \cosh^5 x - \frac{1}{3} \cosh^3 x + C.$$

Evaluate the expression: $\int_0^1 f(y)dy$, where $f(y) = \int_0^y y^2 \sin xy\, dx$.

Solution: In this problem we must first integrate the expression which defines f(y), and then substitute this value into the original integral. In treating the integral defining f(y), it is necessary to keep in mind that the variable of integration is x, and that y enters this integral in the role of a constant. We use the formula: $\int \sin u\, du = -\cos u$, with $u = xy$ and $du = ydx$, obtaining:

$$f(y) = y^2 \int_0^y \sin xy\, dx = \frac{1}{y} \cdot y^2 \int_0^y \sin xy \cdot ydx$$

$$= \Big[-y \cos xy\Big]_{x=0}^{x=y} = [-y \cos y \cdot y\,] - [-y \cos 0 \cdot y]$$

$$= -y \cos y^2 + y(1)$$

$$= -y \cos y^2 + y.$$

Substituting this result in the first integral, we find:

$$\int_0^1 f(y)dy = \int_0^1 (-y \cos y^2 + y)dy$$

$$= -\frac{1}{2} \int_0^1 (\cos y^2)\, 2ydy + \int_0^1 ydy.$$

Using the formula: $\int \cos u\, du = \sin u$, on the first integral, we obtain:

$$\left[-\frac{1}{2} \sin y^2 + \frac{1}{2}y^2\right]_0^1 = \left[-\frac{1}{2} \sin 1^2 + \frac{1}{2}(1^2)\right]$$

$$- \left[-\frac{1}{2} \sin 0^2 + \frac{1}{2}(0^2)\right]$$

$$= \frac{1}{2}(1 - \sin 1).$$

CHAPTER 25

INTEGRATION BY PARTIAL FRACTIONS

When a function to be integrated is given in the form of a ratio in which the denominator can be factored, the best approach to use often, is to break up the single given ratio into a number of simpler ratios which may be integrated easier.

In this approach, each factor of the denominator of the given ratio, becomes the denominator of a separate fraction, so that the resulting number of separate fractions is equal to the number of factors of the given ratio. The numerators of the separate fractions are then solved from a set of simultaneous equations which impose the condition that the sum of the separate fractions, is equal to the value of the given

• **PROBLEM** 719

Integrate: $\int \frac{2 - x}{x^2 + x} dx.$

Solution: To integrate this expression, we use partial fractions. We first find the factors of the denominator. They are: (x) and (x + 1). Now we find two numbers, A and B, such that

$$\frac{2 - x}{x^2 + x} = \frac{A}{x} + \frac{B}{x + 1}.$$

Multiplying both sides of this equation by the common denominator, $x(x + 1)$, we obtain: $2 - x = A(x + 1) + B(x)$. When $x(x + 1) = 0$, $x = 0$ or $x = -1$. When $x = 0$, $2 - 0 = A(0 + 1) + B(0)$, and $A = 2$. When $x = -1$, $2 - (-1) = A(0) + B(-1)$ and $B = -3$. We can now write the integral as follows:

$$\int \frac{2 - x}{x^2 + x} dx = \int \left[\frac{2}{x} - \frac{3}{x + 1}\right] dx$$

$$= 2 \int \frac{1}{x} dx - 3 \int \frac{1}{x + 1} dx$$

$$= 2 \ln x - 3 \ln(x + 1) + C.$$

• **PROBLEM** 720

Integrate the expression:

$$\int \frac{3x + 11}{(x + 2)(x + 3)} dx.$$

Solution: To integrate the given function we use the method of partial fractions. We write:

$$\frac{3x + 11}{(x + 2)(x + 3)} = \frac{A}{x + 2} + \frac{B}{x + 3}$$

$$= \frac{A(x + 3) + B(x + 2)}{(x + 2)(x + 3)}$$

Since these two fractions are merely different ways of writing the same quantity and their denominators are identical, their numerators must be equal. We have:

$$A(x + 3) + B(x + 2) = 3x + 11.$$

We now solve for A and B. We write:

$$Ax + 3A + Bx + 2B = 3x + 11$$

$$(A + B)x + (3A + 2B) = 3x + 11$$

We now equate coefficients of like powers of x, and we have:

$$A + B = 3$$

$$3A + 2B = 11$$

Solving for A and B by substitution, we find that $A = 5$, $B = -2$. Therefore:

$$\frac{3x + 11}{(x + 2)(x + 3)} = \frac{5}{x + 2} - \frac{2}{x + 3},$$

and

$$\int \frac{3x + 11}{(x + 2)(x + 3)} dx = \int \frac{5}{x + 2} dx - \int \frac{2}{x + 3} dx$$

$$= 5 \int \frac{dx}{x + 2} - 2 \int \frac{dx}{x + 3}.$$

We now integrate, using the formula:

$\int \frac{du}{u} = \ln |u| + C$, letting $u = (x + 2)$ in the first integral and $(x + 3)$ in the second, and $du = dx$. We obtain:

$$5 \ln |x + 2| - 2 \ln |x + 3| + C.$$

• PROBLEM 721

Integrate:

$$\int \frac{1}{x^2 - a^2} dx.$$

Solution: To integrate the given function we use the method of partial fractions. We write:

$$\frac{1}{x^2 - a^2} = \frac{1}{(x - a)(x + a)} = \frac{A}{x - a} + \frac{B}{x + a}$$

$$= \frac{A(x + a) + B(x - a)}{(x - a)(x + a)}$$

We have:

$$\frac{1}{(x - a)(x + a)} = \frac{A(x + a) + B(x - a)}{(x - a)(x + a)} .$$

Since the denominators are the same, we can set the numerators equal. Therefore:

$$1 = A(x + a) + B(x - a)$$

$$1 = Ax + Aa + Bx - Ba$$

$$1 = (A + B)x + (Aa - Ba)$$

We now equate coefficients of like powers of x, and we have:

$$A + B = 0$$
$$Aa - Ba = 1.$$

Solving for A and B by substitution, we obtain:

$$A = \frac{1}{2a} , \quad B = -\frac{1}{2a} .$$

Therefore,

$$\int \frac{1}{x^2 - a^2} dx = \frac{1}{2a} \int \frac{1}{x - a} dx$$

$$- \frac{1}{2a} \int \frac{1}{x + a} dx.$$

We now integrate, using the formula;

$$\int \frac{du}{u} = \ln|u|$$

obtaining:

$$\frac{1}{2a} \ln |x - a| - \frac{1}{2a} \ln |x + a| + C$$

$$= \frac{1}{2a} \left(\ln |x - a| - \ln |x + a| \right).$$

Therefore,

$$\int \frac{1}{x^2 - a^2} dx = \frac{1}{2a} \ln \left| \frac{x - a}{x + a} \right| + C.$$

• PROBLEM 722

Integrate: $\int \frac{dx}{x^2 + 5x + 6}$.

Solution: In performing the integration we use the method of partial fractions. We can write $\int \frac{dx}{x^2+5x+6}$ as $\int \frac{1(dx)}{(x+2)(x+3)} = \frac{A}{(x+2)} + \frac{B}{(x+3)}$. $A(x+3) + B(x+2) = 1$. Solving for A and B, we obtain:

$$(A+B)x + 3A + 2B = 1$$
$$A + B = 0, \quad B = -A$$

$$3A + 2B = 1, \quad 3A - 2A = 1, \quad A = 1, \quad B = -1.$$

We therefore have:

$$\int \frac{dx}{x^2 + 5x + 6} = \int \frac{dx}{x + 2} - \int \frac{dx}{x + 3}.$$

Applying the formula: $\int \frac{du}{u} = \ln \ u + c$, we obtain:

$$\ln(x-2) - \ln(x+3) + C = \ln\frac{x + 2}{x + 3} + c.$$

• PROBLEM 723

Integrate: $\int \frac{x^3 + 6x^2 + 3x + 6}{x^3 + 2x^2} dx$

Solution: To integrate this expression, we use partial fractions. We first find the factors of the denominator. The denominator, $x^3 + 2x^2$, can be factored as $x^2(x + 2)$. We then have three terms: $\frac{A}{x}$, $\frac{B}{x^2}$, and $\frac{C}{(x + 2)}$. Since the denominator is of the same power as the numerator, we can divide the numerator by the denominator, and write the expression as:

$$1 + \frac{4x^2 + 3x + 6}{x^3 + 2x^2}.$$

We now find three numbers: A, B, and C, such that:

$$1 + \frac{4x^2 + 3x + 6}{x^3 + 2x^2} = \frac{A}{x} + \frac{B}{x^2} + \frac{C}{x + 2}.$$

Considering the 1 as a constant, we can ignore it momentarily and solve for the numerical values of A, B and C by multiplying both sides of the equation by $x^3 + 2x^2$. We find,

$$4x^2 + 3x + 6 = Ax(x+2) + B(x+2) + Cx^2.$$

We can combine like powers of x to find

$$4x^2 + 3x + 6 = Ax^2 + 2Ax + Bx + 2B + Cx^2.$$

$$4x^2 + 3x + 6 = (A+C)x^2 + (2A+B)x + 2B.$$

Now we equate like powers of x to find A, B, and C. We find,

$$A + C = 4.$$
$$2A + B = 3.$$
$$2B = 6.$$

Therefore, $B = 3,\ A = 0,\ C = 4.$

We substitute these values, and integrate:

$$\int \frac{x^3 + 6x^2 + 3x + 6}{x^3 + 2x^2}dx = \int\left(1 + \frac{0}{x} + \frac{3}{x^2} + \frac{4}{x + 2}\right)dx$$

$$= x - \frac{3}{x} + 4 \ln(x + 2) + C.$$

• PROBLEM 724

Integrate the expression:

$$\int \frac{2x^2 + 5x - 1}{x^3 + x^2 - 2x} dx.$$

Solution: To integrate the given expression we use the method of partial fractions. Since

$$x^3 + x^2 - 2x = x(x - 1)(x + 2),$$

the denominator is a product of distinct linear factors, and we try to find A_1, A_2, and A_3 such that

$$\frac{2x^2 + 5x - 1}{x^3 + x^2 - 2x} = \frac{A_1}{x} + \frac{A_2}{x - 1} + \frac{A_3}{x + 2}$$

$$= \frac{A_1(x - 1)(x + 2) + A_2(x)(x + 2) + A_3(x)(x - 1)}{x^3 + x^2 - 2x}$$

Setting the two numerators equal, we obtain:

$$2x^2 + 5x - 1 = A_1(x - 1)(x + 2) + A_2x(x + 2)$$
$$+ A_3x(x - 1).$$

Multiplying out and collecting like powers of x yields:

$$(A_1 + A_2 + A_3)x^2 + (A_1 + 2A_2 - A_3)x - 2A_1$$
$$= 2x^2 + 5x - 1.$$

Equating coefficients of like powers of x, we have the equations:

$$A_1 + A_2 + A_3 = 2$$
$$A_1 + 2A_2 - A_3 = 5$$
$$- 2A_1 = - 1.$$

Solving these equations we find that

$$A_1 = \frac{1}{2}, \quad A_2 = 2 \quad \text{and} \quad A_3 = -\frac{1}{2}.$$

Therefore, we have:

$$\int \frac{2x^2 + 5x - 1}{x^3 + x^2 - 2x} dx = \frac{1}{2}\int \frac{dx}{x} + 2\int \frac{dx}{x - 1}$$
$$- \frac{1}{2}\int \frac{dx}{x + 2}.$$

We now integrate, using the formula:

$$\int \frac{du}{u} = \ln |u|,$$

obtaining: $\displaystyle\int \frac{2x^2 + 5x - 1}{x^3 + x^2 - 2x} dx$

$$= \frac{1}{2} \ln |x| + 2 \ln |x - 1| - \frac{1}{2} \ln |x + 2| + C.$$

• PROBLEM 725

Integrate:

$$\int \frac{3x + 6}{x^3 + 2x^2 - 3x} \, dx$$

by the method of partial fractions.

Solution: Factoring the denominator, we can write:

$$\frac{3x + 6}{x^3 + 2x^2 - 3x} = \frac{3x + 6}{x(x - 1)(x + 3)}$$

$$= \frac{A}{x} + \frac{B}{(x - 1)} + \frac{C}{(x + 3)}$$

$$= \frac{A(x - 1)(x + 3) + B(x)(x + 3) + C(x)(x - 1)}{x(x - 1)(x + 3)} .$$

Since the denominators of the fractions are equal, the numerators must also be the same, and we write:

$$3x + 6 = A(x - 1)(x + 3) + Bx(x + 3) + Cx(x - 1)$$

$$= A(x^2 + 2x - 3) + Bx^2 + 3Bx + Cx^2 - Cx$$

$$= x^2(A + B + C) + x(2A + 3B - C) - 3A .$$

Equating coefficients of like powers of x, we have:

$$A + B + C = 0$$

$$2A + 3B - C = 3$$

$$- 3A = 6,$$

therefore $A = - 2.$

Solving for B and C by substitution, we find that $B = \frac{9}{4}$ and $C = - \frac{1}{4}$. Therefore,

$$\frac{3x + 6}{x^3 + 2x^2 - 3x} = - \frac{2}{x} + \frac{9}{4(x - 1)} - \frac{1}{4(x + 3)}$$

We now write:

$$\int \frac{(3x + 6)\,dx}{x^3 + 2x^2 - 3x} = -2 \int \frac{dx}{x} + \frac{9}{4} \int \frac{dx}{x - 1}$$

$$- \frac{1}{4} \int \frac{dx}{x + 3} .$$

We now integrate, using the formula:

$$\int \frac{du}{u} = \ln u + C_1 ,$$

obtaining: $\displaystyle\int \frac{(3x+6)\,dx}{x^3+2x-3x}$

$$= 2 \ln x + \frac{9}{4} \ln(x - 1) - \frac{1}{4} \ln(x + 3) + C_1 .$$

• PROBLEM 726

Integrate:

$$\int \frac{3x^2 + 2x - 2}{x^3 - 1}\,dx.$$

Solution: To integrate the given expression, we use the method of partial fractions.

The denominator can be factored into the product

$$x^3 - 1 = (x - 1)(x^2 + x + 1),$$

and we write:

$$\frac{3x^2 + 2x - 2}{x^3 - 1} = \frac{A}{x - 1} + \frac{Bx + C}{x^2 + x + 1}$$

$$= \frac{A(x^2 + x + 1) + (Bx + C)(x - 1)}{(x - 1)(x^2 + x + 1)} .$$

Setting the numerators of the above fractions equal, we have:

$$3x^2 + 2x - 2 = A(x^2 + x + 1) + (Bx + C)(x - 1)$$

Now we multiply out and collect like powers of x. We obtain:

$$3x^2 + 2x - 2 = (A + B)x^2 + (A - B + C)x + (A - C).$$

Equating coefficients of like powers of x, we obtain:

$$A + B = 3$$

$$A - B + C = 2$$

$$A - C = - 2.$$

Solving for A, B, and C, we find A = 1, B = 2, and C = 3. Therefore we have:

$$\int \frac{3x^2 + 2x - 2}{x^3 - 1} dx = \int \frac{dx}{x - 1} + \int \frac{2x + 3}{x^2 + x + 1} dx.$$

We now integrate using the formula;

$$\int \frac{du}{u} = \ln |u|,$$

on the first integral. We rewrite the second integral as:

$$\int \frac{2x + 3}{x^2 + x + 1} dx = \int \frac{2x + 1}{x^2 + x + 1} dx$$

$$+ \int \frac{2}{x^2 + x + 1} dx,$$

using the formula for $\int \frac{du}{u}$ on the first part.

In treating the second part of this integral we wish to consider the formula,

$$\int \frac{du}{u^2 + a^2} = \frac{1}{a} \arctan \frac{u}{a}.$$

We complete the square in the denominator to obtain the appropriate form. This gives:

$$\ln(x^2 + x + 1) + 2 \int \frac{dx}{\left(x + \frac{1}{2}\right)^2 + \frac{3}{4}}.$$

If we let

$$u = x + \frac{1}{2} \quad \text{and} \quad a = \sqrt{\frac{3}{4}},$$

and apply the formula, we obtain:

$$2\int \frac{dx}{\left(x+\frac{1}{2}\right)^2+\left(\frac{\sqrt{3}}{2}\right)^2} = \frac{1}{\frac{\sqrt{3}}{2}} \arctan \frac{x+\frac{1}{2}}{\frac{\sqrt{3}}{2}}$$

$$= \frac{4\sqrt{3}}{3} \arctan \frac{2x+1}{\sqrt{3}} .$$

Therefore, we have:

$$\int \frac{3x^2+2x-2}{x^3-1}\,dx = \ln|x-1| + \ln(x^2+x+1)$$

$$+ \frac{4}{3}\sqrt{3} \arctan \frac{2x+1}{\sqrt{3}} + C_1.$$

• PROBLEM 727

Find the integral:

$$\int \frac{x^3+5x^2+2x-4}{(x^4-1)} \cdot dx.$$

Solution: To integrate the given expression, we use the method of partial fractions. The denominator $(x^4 - 1)$ can be factored into:

$$(x-1)(x+1)(x^2+1).$$

Then

$$\frac{x^3+5x^2+2x-4}{(x^4-1)} = \frac{A}{x-1} + \frac{B}{x+1} + \frac{Cx+D}{x^2+1}.$$

When the denominator contains a term in x^2, an x term must appear in the numerator in addition to a constant.

Now,

$$\frac{x^3+5x^2+2x-4}{(x^4-1)}$$

$$= \frac{A(x+1)(x^2+1)+B(x-1)(x^2+1)+(Cx+D)(x-1)(x+1)}{(x-1)(x+1)(x^2+1)}$$

Since the denominators are the same, the numerators are also equal and we have:

$$x^3 + 5x^2 + 2x - 4 = A(x+1)(x^2+1)$$

$$+ B(x-1)(x^2+1) + (Cx+D)(x-1)(x+1) =$$

$$(Ax^3 + Ax^2 + Ax + A + Bx^3 - Bx^2 + Bx - B + Cx^3$$

$$+ Dx^2 - Cx - D) = (A+B+C)x^3 + (A-B+D)x^2$$

$$+ (A+B-C)x + (A-B-D), \quad \text{collecting terms.}$$

We have:

$$x^3 + 5x^2 + 2x - 4 = (A+B+C)x^3$$

$$+ (A-B+D)x^2 + (A+B-C)x + (A-B-D).$$

Equating coefficients of like powers of x, we obtain:

$$A + B + C = 1$$

$$A - B + D = 5$$

$$A + B - C = 2$$

$$A - B - D = -4.$$

Solving these equations simultaneously, we obtain:

$$A = 1, \quad B = \frac{1}{2}, \quad C = -\frac{1}{2} \quad \text{and } D = 4\frac{1}{2}.$$

Then

$$\frac{x^3 + 5x^2 + 2x - 4}{(x^4 - 1)} = \frac{1}{(x-1)} + \frac{1}{2(x+1)}$$

$$+ \frac{-\frac{1}{2} \cdot x + 4\frac{1}{2}}{x^2 + 1}$$

$$= \frac{1}{x-1} + \frac{1}{2(x+1)} + \frac{9-x}{2(x^2+1)}$$

$$= \frac{1}{x - 1} + \frac{1}{2(x + 1)} + \frac{9}{2}\left(\frac{1}{x^2 + 1}\right)$$

$$- \frac{1}{2}\left(\frac{x}{x^2 + 1}\right).$$

Then

$$\int \frac{x^3 + 5x^2 + 2x - 4}{(x^4 - 1)} \cdot dx = \int \frac{dx}{x - 1}$$

$$+ \frac{1}{2}\int \frac{dx}{x + 1} + \frac{9}{2}\int \frac{dx}{(x^2 + 1)}$$

$$- \frac{1}{2}\int \frac{x \cdot dx}{(x^2 + 1)}.$$

We can now integrate, using the formula;

$$\int \frac{du}{u} = \ln u + C,$$

for the first, second and last integrals, and the formula;

$$\int \frac{du}{a^2 + u^2} = \frac{1}{a}\tan^{-1}\frac{u}{a},$$

for the third integral. We obtain:

$$\int \frac{x^3 + 5x^2 + 2x - 4}{(x^4 - 1)} \cdot dx = \ln(x - 1)$$

$$+ \frac{1}{2}\ln(x + 1) + \frac{9}{2}\tan^{-1} x - \frac{1}{4}\ln(x^2 + 1)$$

$$+ C_1.$$

• PROBLEM 728

Integrate the expression:

$$\int \frac{x^2 + 4x}{(x - 2)^2 (x^2 + 4)} \cdot dx.$$

Solution: To integrate the given expression, we use the method of partial fractions. We write:

$$\frac{x^2 + 4x}{(x - 2)^2 (x^2 + 4)} = \frac{A}{(x - 2)} + \frac{B}{(x - 2)^2} + \frac{Cx + D}{(x^2 + 4)}$$

$$= \frac{A(x-2)(x^2+4) + B(x^2+4) + (Cx+D)(x-2)^2}{(x - 2)^2 (x^2 + 4)},$$

where the rules for square and repeated factors have been followed. Equating the numerators (denominators are equal) we have:

$$x^2 + 4x = A(x - 2)(x^2 + 4) + B(x^2 + 4) + (Cx + D)(x - 2)^2.$$

Multiplying out and collecting like powers of x we obtain:

$$(A + C)x^3 + (-2A + B - 4C + D)x^2 + (4A + 4C - 4D)x + (-8A + 4B + 4D) = x^2 + 4x.$$

Now we equate the coefficients of like powers of x, obtaining the equations:

$$A + C = 0$$

$$-2A + B - 4C + D = 1$$

$$4A + 4C - 4D = 4$$

$$-8A + 4B + 4D = 0.$$

Solving these equations, we find the values of A, B, C and D to be: $A = \frac{1}{4}$, $B = \frac{3}{2}$, $C = -\frac{1}{4}$ and $D = -1$.

By substitution:

$$\frac{x^2 + 4x}{(x - 2)^2 (x^2 + 4)} = \frac{\frac{1}{4}}{(x - 2)} + \frac{\frac{3}{2}}{(x - 2)^2}$$

$$- \frac{\frac{1}{4} x - 1}{(x^2 + 4)} .$$

We can now write:

$$\int \frac{x^2 + 4x}{(x - 2)^2 (x^2 + 4)} dx = \int \frac{\frac{1}{4}}{(x - 2)} dx$$

$$+ \int \frac{\frac{3}{2}}{(x - 2)^2} dx - \int \frac{\frac{1}{4} x - 1}{(x^2 + 4)} dx.$$

Extracting constants and breaking up the last integral, we have:

$$\int \frac{x^2 + 4x}{(x - 2)^2 (x^2 + 4)} dx = \frac{1}{4} \int \frac{dx}{(x - 2)}$$

$$+ \frac{3}{2} \int \frac{dx}{(x - 2)^2} - \frac{1}{4} \int \frac{xdx}{(x^2 + 4)}$$

$$- \int \frac{dx}{(x^2 + 4)} .$$

Each of these integrals is easily found by one of the known integration formulas. We make use of the formula:

$\int \frac{du}{u} = \ln |u|$, on the first and third

integrals. The second integral can be rewritten as

$\frac{3}{2} \int (x - 2)^{-2}\, dx$, and we can then apply the formula for $\int u^n\,du$. Finally, we use

$$\int \frac{du}{u^2 + a^2} = \frac{1}{a} \tan^{-1} \frac{u}{a}$$

on the last integral. Applying these formulas, we obtain:

$$\int \frac{x^2 + 4x}{(x - 2)^2 (x^2 + 4)} dx = \frac{1}{4} \ln |x - 2|$$

$$- \frac{3}{2(x - 2)} - \frac{1}{8} \ln (x^2 + 4)$$

$$- \frac{1}{2} \tan^{-1} \frac{x}{2} + C_1.$$

• PROBLEM 729

Integrate the function:

$$\int \frac{x^2 + 2x + 3}{(x - 1)(x + 1)^2} dx.$$

Solution: To integrate the given function we use the method of partial fractions. We try to find A_1, A_2, A_3. Since one factor occurs twice, we use the rule for denominators composed of repeated linear factors. We have:

$$\frac{x^2 + 2x + 3}{(x - 1)(x + 1)^2} = \frac{A_1}{x - 1} + \frac{A_2}{x + 1} + \frac{A_3}{(x + 1)^2}.$$

We can now write: $\frac{x^2 + 2x + 3}{(x - 1)(x + 1)^2}$

$$= \frac{A_1(x + 1)^2 + A_2(x - 1)(x + 1) + A_3(x - 1)}{(x - 1)(x + 1)^2}.$$

Since the denominators of the above fractions are the same, we can set the numerators equal, obtaining:

$$x^2 + 2x + 3 = A_1(x + 1)^2 + A_2(x - 1)(x + 1) + A_3(x - 1).$$

Multiplying out and collecting like powers of x gives:

$$(A_1 + A_2)x^2 + (2A_1 + A_3)x + (A_1 - A_2 - A_3) = x^2 + 2x + 3.$$

We can now equate the coefficients of like powers of x, obtaining:

$$A_1 + A_2 = 1, \quad 2A_1 + A_3 = 2 \quad \text{and } A_1 - A_2 - A_3 = 3.$$

Solving for A_1, A_2 and A_3 we find that

$$A_1 = \frac{3}{2}, \quad A_2 = -\frac{1}{2} \quad \text{and} \quad A_3 = -1.$$

We now have:

$$\int \frac{x^2 + 2x + 3}{(x - 1)(x + 1)^2}\,dx = \frac{3}{2}\int \frac{dx}{x - 1} - \frac{1}{2}\int \frac{dx}{x + 1} - \int \frac{dx}{(x + 1)^2},$$

by substituting for A_1, A_2, A_3. We can integrate, using the formula:

$\int \frac{du}{u} = \ln |u|$, on the first two integrals,

and $\int u^n\,du$ on the last. Applying these formulas, we obtain:

$$\int \frac{x^2 + 2x + 3}{(x - 1)(x + 1)^2}\,dx = \frac{3}{2}\ln |x - 1|$$

$$- \frac{1}{2} \ln |x + 1| - \left(\frac{(x + 1)^{-1}}{- 1} \right) + C$$

$$= \frac{3}{2} \ln |x - 1| - \frac{1}{2} \ln |x + 1| + \frac{1}{x + 1} + C.$$

• PROBLEM 730

Integrate the expression:

$$\int \frac{x^4 - x^3 + 2x^2 - x + 2}{(x - 1)(x^2 + 2)^2} dx.$$

Solution: To integrate the given expression, we use the method of partial fractions. We write

$$\frac{x^4 - x^3 + 2x^2 - x + 2}{(x - 1)(x^2 + 2)^2} = \frac{A}{x - 1}$$

$$+ \frac{Bx + C}{x^2 + 2} + \frac{Dx + E}{(x^2 + 2)^2}$$

$$= \frac{A(x^2+2)^2+(Bx+C)(x-1)(x^2+2)+(Dx+E)(x-1)}{(x - 1)(x^2 + 2)^2}.$$

We can now equate the numerators of the first and last fractions, obtaining:

$$x^4 - x^3 + 2x^2 - x + 2 = A(x^2 + 2)^2 + (Bx + C)$$

$$(x - 1)(x^2 + 2) + (Dx + E)(x - 1).$$

Multiplying out and collecting like powers of x,we obtain:

$$(A + B)x^4 + (C - B)x^3 + (4A - C + 2B + D)x^2$$

$$+ (2C - 2B + E - D)x + (4A - 2C - E)$$

$$= x^4 - x^3 + 2x^2 - x + 2.$$

Now equate the coefficients of like powers of x. Doing this, we have:

$$A + B = 1, \quad C - B = -1,$$

$$4A - C + 2B + D = 2,$$

$$2C - 2B + E - D = -1,$$

$$4A - 2C - E = 2.$$

Solving for A, B, C, D, and E, we find that

$$A = \frac{1}{3}, \quad B = \frac{2}{3}, \quad C = -\frac{1}{3}, \quad D = -1, \quad E = 0.$$

Substituting these values, we have:

$$\int \frac{x^4 - x^3 + 2x^2 - x + 2}{(x - 1)(x^2 + 2)^2} dx = \frac{1}{3} \int \frac{dx}{x - 1} + \int \frac{\frac{2}{3} x - \frac{1}{3}}{x^2 + 2} dx - \int \frac{x\, dx}{(x^2 + 2)^2} .$$

The second integral can be rewritten as:

$$\frac{1}{3} \int \frac{2x - 1}{x^2 + 2} dx,$$

and then broken up into two integrals, giving:

$$\frac{1}{3} \int \frac{dx}{x - 1} + \frac{1}{3} \int \frac{2x\, dx}{x^2 + 2} - \frac{1}{3} \int \frac{dx}{x^2 + 2} - \int \frac{x\, dx}{(x^2 + 2)^2} .$$

We can now integrate, using $\int \frac{du}{u}$ for the first and second integrals. For the third integral we use the formula;

$$\int \frac{du}{u^2 + a^2} = \frac{1}{a} \arctan \frac{u}{a} ,$$

letting $u = x$, and $a = \sqrt{2}$. Finally, the last integral is in a form, to which we can apply the formula for $\int u^n\, du$. Applying these formulas, we obtain:

$$\int \frac{x^4 - x^3 + 2x^2 - x + 2}{(x-1)(x^2+2)^2}\,dx = \frac{1}{3}\ln|x-1|$$

$$+ \frac{1}{3}\ln(x^2+2) - \frac{\sqrt{2}}{6}\arctan\frac{x}{\sqrt{2}}$$

$$+ \frac{1}{2}\,\frac{1}{x^2+2} + C_1.$$

• PROBLEM 731

Integrate the expression:

$$\int \frac{18 + 11x - x^2}{(x-1)(x+1)(x^2+3x+3)}\,dx$$

Solution: To integrate the given expression, we use the method of partial fractions. Recognizing the presence of a quadratic factor, we use the rule for that case and write:

$$\frac{18 + 11x - x^2}{(x-1)(x+1)(x^2+3x+3)} = \frac{A}{x-1} + \frac{B}{x+1}$$

$$+ \frac{Cx + D}{x^2 + 3x + 3}$$

$$= \frac{A(x+1)(x^2+3x+3)+B(x-1)(x^2+3x+3)+(Cx+D)(x-1)(x+1)}{(x-1)(x+1)(x^2+3x+3)}.$$

Because the denominators are the same, the two numerators must be equal. We now write:

$$A(x+1)(x^2+3x+3) + B(x-1)(x^2+3x+3)$$

$$+ (Cx+D)(x-1)(x+1)$$

$$= 18 + 11x - x^2.$$

To evaluate A, B, C, and D we multiply out and collect like powers of x. We obtain:

$$(A+B+C)x^3 + (4A+2B+D)x^2 + (6A-C)x$$
$$+ (3A-3B-D) = 18 + 11x - x^2.$$

The two polynomials that are the coefficients of like powers of x must be equal. Therefore,

$$A + B + C = 0.$$

$$4A + 2B + D = -1,$$

$$6A - C = 11,$$

$$3A - 3B - D = 18.$$

We now solve these equations simultaneously for A, B, C, and D. Adding the first and third, and the second and fourth,

$$7A + B = 11,$$

$$7A - B = 17.$$

From these two equations we find that $B = -3$ and $A = 2$. Substituting these values above, we obtain $C = 1$ and $D = -3$. Therefore,

$$\frac{18 + 11x - x^2}{(x-1)(x+1)(x^2+3x+3)}$$

$$= \frac{2}{x-1} - \frac{3}{x+1} + \frac{x-3}{x^2+3x+3},$$

and

$$\int \frac{18 + 11x - x^2}{(x-1)(x+1)(x^2+3x+3)}\,dx$$

$$= \int \frac{2}{x-1}\,dx - \int \frac{3}{x+1}\,dx + \int \frac{x-3}{x^2+3x+3}\,dx$$

$$= 2\int \frac{dx}{(x-1)} - 3\int \frac{dx}{(x+1)}$$

$$+ \int \frac{x-3}{x^2+3x+3}\,dx.$$

We can now integrate. The first and second integrals can be performed by using the formula:

$$\int \frac{du}{u} \ln |u|.$$

We wish to divide the third integral into two integrals, with the first in a form, to which we can apply the formula for

$$\int \frac{du}{u}.$$

To do this, we must have (2x + 3) in the numerator. Rearranging this integral we obtain:

$$\int \frac{x - 3}{x^2 + 3x + 3} dx = \frac{1}{2}\left(\int \frac{\left[2(x - 3) + 9\right] - 9}{x^2 + 3x + 3} dx\right)$$

$$= \frac{1}{2}\int \frac{2x + 3}{x^2 + 3x + 3} dx - \frac{1}{2}\left(9\int \frac{dx}{x^2 + 3x + 3}\right)$$

We can now apply the formula for $\int \frac{du}{u}$

to the first integral, and, after completing the square in the second, we have;

$$-\frac{1}{2}\left(9\int \frac{dx}{\left(x + \frac{3}{2}\right)^2 + \left(\frac{\sqrt{3}}{2}\right)^2}\right).$$

We can now use the formula:

$$\int \frac{du}{u^2 + a^2} = \frac{1}{a} \tan^{-1} \frac{u}{a},$$

on this integral. Applying all the above formulas, we obtain:

$$\int \frac{18 + 11x - x^2}{(x - 1)(x + 1)(x^2 + 3x + 3)} dx$$

$$= 2 \ln |x - 1| - 3 \ln |x + 1| + \frac{1}{2} \ln |x^2 + 3x + 3|$$

$$- \frac{9}{\sqrt{3}} \tan^{-1} \frac{2x + 3}{\sqrt{3}} + C_1.$$

Integrate the expression:

$$\int \frac{x^3 + 5x^2 + 2x - 4}{x(x^2 + 4)^2} dx.$$

Solution: To integrate the given expression we use the method of partial fractions. We write:

$$\frac{x^3 + 5x^2 + 2x - 4}{x(x^2 + 4)^2} = \frac{A}{x} + \frac{Bx + C}{(x^2 + 4)} + \frac{Dx + E}{(x^2 + 4)^2}$$

$$= \frac{A(x^2+4)^2+(Bx+C)(x^2+4)(x)+(Dx+E)(x)}{x(x^2 + 4)^2}.$$

Setting the numerators equal, we obtain:

$$x^3 + 5x^2 + 2x - 4 = A(x^2 + 4)^2 + (Bx + C)(x^2 + 4)(x) + (Dx + E)(x).$$

Multiplying out and collecting like powers of x, we obtain:

$$(A + B)x^4 + Cx^3 + (8A + 4B + D)x^2 + (4C + E)x + 16A = x^3 + 5x^2 + 2x - 4.$$

Now we equate the coefficients of like powers of x. This gives the following equations:

$$A + B = 0$$
$$C = 1$$
$$8A + 4B + D = 5$$
$$4C + E = 2$$
$$16A = -4$$

Solving these equations, we find:

$$A = -\frac{1}{4}, \quad B = \frac{1}{4}, \quad C = 1, \quad D = 6, \quad E = -2.$$

Substituting, we have:

$$\frac{x^3 + 5x^2 + 2x - 4}{x(x^2 + 4)^2} = \frac{-\frac{1}{4}}{x} + \frac{\frac{1}{4}x + 1}{(x^2 + 4)} + \frac{6x - 2}{(x^2 + 4)^2} .$$

We can now write;

$$\int \frac{x^3 + 5x^2 + 2x - 4}{x(x^2 + 4)^2} dx = \int \frac{-\frac{1}{4}}{x} dx + \int \frac{\frac{1}{4}x + 1}{(x^2 + 4)} dx + \int \frac{6x - 2}{(x^2 + 4)^2} dx .$$

We rewrite the integrals by extracting constants and breaking up the second and third integrals. This gives:

$$-\frac{1}{4}\int \frac{dx}{x} + \frac{1}{4}\int \frac{x\,dx}{x^2 + 4} + \int \frac{dx}{x^2 + 4} + 6\int \frac{x\,dx}{(x^2 + 4)^2} - 2\int \frac{dx}{(x^2 + 4)^2} .$$

We can integrate the first and second integrals by the formula;

$$\int \frac{du}{u} = \ln |u| ,$$

the third by using the formula:

$$\int \frac{du}{u^2 + a^2} = \frac{1}{a} \tan^{-1} \frac{u}{a} .$$

Rewriting the fourth integral as

$$6 \int (x^2 + 4)^{-2}\, x\, dx ,$$

we can apply the formula,

$$\int u^n\, du = \frac{u^{n+1}}{n + 1} .$$

To treat the last integral we use the formula:

$$\int \frac{dx}{(a + bx^2)^2} = \frac{x}{2a(a + bx^2)} + \frac{1}{2a} \int \frac{dx}{a + bx^2} .$$

The integral we are treating is:

$$- 2 \int \frac{dx}{(x^2 + 4)^2} ,$$

with a = 4, b = 1. Applying the above formula, we have:

$$- 2 \int \frac{dx}{(x^2 + 4)^2}$$

$$= - 2 \left(\frac{x}{8(4 + x^2)} + \frac{1}{8} \int \frac{dx}{4 + x^2} \right) .$$

To integrate this expression, we employ another formula:

$$\int \frac{dx}{a + bx^2} = \frac{1}{\sqrt{ab}} \tan^{-1} \frac{x \sqrt{ab}}{a} ,$$

with a = 4, b = 1.

Applying this, we have:

$$- 2 \int \frac{dx}{(x^2 + 4)^2} = - 2 \left[\frac{x}{8(4 + x^2)} + \frac{1}{8} \left(\frac{1}{\sqrt{4}} \tan^{-1} \frac{x \sqrt{4}}{4} \right) \right]$$

$$= - \frac{x}{4(4 + x^2)} - \frac{1}{8} \tan^{-1} \frac{x}{2} .$$

Applying all the above formulas, we obtain:

$$\int \frac{x^3 + 5x^2 + 2x - 4}{x(x^2 + 4)^2} dx = - \frac{1}{4} \ln |x|$$

$$+ \frac{1}{8} \ln (x^2 + 4) + \frac{1}{2} \tan^{-1} \frac{x}{2} - \frac{3}{(x^2 + 4)} -$$

$$\frac{x}{4(4 + x^2)} - \frac{1}{8} \tan^{-1} \frac{x}{2} + C_1.$$

• PROBLEM 733

Evaluate:

$$\int_{\frac{3}{2}}^{3} xy \, dx,$$

subject to the functional relation: $xy^2 + x = 3$.

Solution: $xy^2 + x = 3$

Solving for x, we have:

$$x = \frac{3}{y^2 + 1},$$

from which

$$dx = \frac{-6y}{(y^2 + 1)^2} dy.$$

Hence,

$$\int xy \, dx = \int \left(\frac{3}{y^2 + 1}\right) y \left(\frac{-6y}{(y^2 + 1)^2}\right) dy$$

$$= -18 \int \frac{y^2}{(y^2 + 1)^3} dy.$$

To integrate, we need to express the integrand in partial fractions form:

$$\frac{y^2}{(y^2 + 1)^3} = \frac{2y A_1 + A_2}{y^2 + 1} + \frac{2y A_3 + A_4}{(y^2 + 1)^2} + \frac{2y A_5 + A_6}{(y^2 + 1)^3}$$

$$= \frac{(2yA_1 + A_2)(y^2+1)^2 + (2yA_3 + A_4)(y^2+1) + 2yA_5 + A_6}{(y^2 + 1)^3}$$

Canceling out the denominator, and expanding, yields:

$$y^2 = 2A_1y^5 + 4A_1y^3 + 2A_1y + A_2y^4 + 2A_2y^2 + A_2$$
$$+ 2A_3y^3 + 2A_3y + A_4y^2 + A_4 + 2A_5y + A_6$$

Equating like coefficients,

y^5 yields: $2A_1 = 0.$

y^4 : $A_2 = 0.$

y^3 : $4A_1 + 2A_3 = 0.$

y^2 : $2A_2 + A_4 = 1.$

y : $2A_1 + 2A_3 + 2A_5 = 0.$

1 : $A_2 + A_4 + A_6 = 0.$

We have:

$A_1 = A_3 = A_5 = 0$, $A_2 = 0$, $A_4 = 1$, and $A_6 = -1.$

Thus,

$$-18 \int \frac{y^2}{(y^2+1)^3}\,dy$$

$$= 18 \int \frac{dy}{(y^2+1)^3} - 18 \int \frac{dy}{(y^2+1)^2}.$$

Using the method of power reduction,

$$\int \frac{dy}{(y^2+1)^3} = \frac{y}{4(y^2+1)^2} + \frac{3}{4}\int \frac{dy}{(y^2+1)^2},$$

and

$$\int \frac{dy}{(y^2+1)^2} = \frac{y}{2(y^2+1)} + \frac{1}{2}\int \frac{dy}{y^2+1},$$

where

$$\int \frac{dy}{y^2+1} = \tan^{-1} y.$$

Therefore,

$$-18\int\frac{y^2}{(y^2+1)^3}\,dy = 18\left[\frac{y}{4(y^2+1)^2}\right.$$

$$+\frac{3}{4}\left[\frac{y}{2(y^2+1)}+\frac{1}{2}\tan^{-1}y\right]$$

$$\left.-\left[\frac{y}{2(y^2+1)}+\frac{1}{2}\tan^{-1}y\right]\right]$$

$$= 18\left[\frac{y}{4(y^2+1)^2}-\frac{y}{8(y^2+1)}-\frac{\tan^{-1}y}{8}\right]$$

To determine the limits, we use the given equation once more. $xy^2 + x = 3$, and, for $x = \frac{3}{2}$, we obtain $y = \pm 1$, and, for $x = 3$, $y = 0$. Since the curve: $x = \frac{1}{(y^2+1)}$, is non-analytic for $x = 0$, symmetric and asymptotic to the line $x = 0$, we write:

$$18\left[\frac{y}{4(y^2+1)}-\frac{y}{8(y^2+1)}-\frac{\tan^{-}y}{8}\right]_{-1}^{1}$$

$$= 18\left[\frac{1}{8}-\frac{1}{16}-\frac{\frac{\pi}{4}}{8}+\frac{1}{8}-\frac{1}{16}-\frac{\frac{\pi}{4}}{8}\right]$$

$$= 18\left[\frac{1}{8}-\frac{\pi}{16}\right] = 9\left[\frac{1}{4}-\frac{\pi}{8}\right].$$

CHAPTER 26

TRIGONOMETRIC SUBSTITUTIONS

When faced with the problem of integrating functions that contain radicals and/or ordinary powers of the variable, a good technique to try is trigonometric substitution. However, trigonometric substitution should not be attempted until the more fundamental operations have been tried first or it can be immediately recognized from experience that trigonometric substitution is applicable.

There are no set rules that prescribe which trigonometric substitution to use. The selection of the particular trigonometric function is best based on a combination of experience and trial-and-error.

• PROBLEM 734

Integrate the expression: $\displaystyle\int \frac{dx}{x^2\sqrt{4 - x^2}}$.

Solution: Let $x = 2 \sin \theta$. Then $dx = 2 \cos \theta \, d\theta$, so that

$$\int \frac{dx}{x^2\sqrt{4 - x^2}} = \int \frac{2 \cos \theta \, d\theta}{4 \sin^2\theta \sqrt{4 - 4 \sin^2 \theta}}$$

$$= \int \frac{2 \cos \theta \, d\theta}{4 \sin^2 \theta \sqrt{4\left(1 - \sin^2 \theta\right)}}.$$

Using the identity: $\cos^2 \theta = 1 - \sin^2 \theta$, and the fact that $\dfrac{1}{\sin^2 \theta} = \csc^2 \theta$, we obtain:

$$\int \frac{2 \cos \theta \, d\theta}{4 \sin^2 \theta (2 \cos \theta)} = \frac{1}{4} \int \csc^2 \theta \, d\theta.$$

Now applying the formula $\int \csc^2 u \, du = -\cot u + c$, we have:

$$-\frac{1}{4}\cot \theta + C.$$

• PROBLEM 735

Integrate the expression: $\int \frac{du}{\sqrt{u^2 - a^2}}$.

Solution: Let $u = a \sec \theta$, then $du = a \sec \theta \tan \theta \, d\theta$. We have:

$$\int \frac{du}{\sqrt{u^2 - a^2}} = \int \frac{a \sec \theta \tan \theta \, d\theta}{\sqrt{a^2 \sec^2 \theta - a^2}}$$

$$= \int \frac{a \sec \theta \tan \theta \, d\theta}{\sqrt{a^2 (\sec^2 \theta - 1)}}$$

$$= \int \sec \theta \, d\theta,$$

by use of the identity $\tan^2 \theta = \sec^2 \theta - 1$. We now apply the formula: $\int \sec u \, du = \ln(\sec u + \tan u) + C$, obtaining:

$$\int \sec \theta \, d\theta = \ln\left[\sec \theta + \tan \theta\right] + C$$

$$= \ln\left[u + \sqrt{u^2 - a^2}\right] + C .$$

• PROBLEM 736

Integrate: $\int \frac{\sqrt{x^2 - 1}}{x} dx$.

Solution: To integrate the given function we use trigonometric substitutions. Let $x = \sec \theta$. Then by substitution, $\sqrt{x^2 - 1} = \sqrt{\sec^2\theta - 1}$. Making use of the identity, $\tan^2\theta = \sec^2\theta - 1$, we can write: $\sqrt{x^2 - 1} = \tan \theta$. To obtain dx we use $x = \sec \theta$. Then, $\frac{dx}{d\theta} = \sec \theta \tan \theta$ and $dx = \sec \theta \tan \theta \, d\theta$. By substitution we have:

$$\int \frac{\sqrt{x^2 - 1}}{x} dx = \int \frac{\tan \theta \cdot \sec \theta \tan \theta \, d\theta}{\sec \theta} = \int \tan^2\theta \, d\theta$$

$$= \int (\sec^2\theta - 1) d\theta,$$

by again using the above identity. We now integrate, using the formula: $\int \sec^2 u \, du = \tan u$, and we obtain:

$$\tan \theta - \theta + C = \sqrt{x^2 - 1} - \text{Arc} \sec x + C,$$

by substitution.

• PROBLEM 737

Find: $\int \sqrt{a^2 - x^2} \, dx$.

Solution: Let $x = a \sin\theta$. substituting this into the integrand, we find that

$$\sqrt{a^2 - x^2} = \sqrt{a^2 - a^2 \sin^2\theta} = \sqrt{a^2(1-\sin^2\theta)}.$$

We now use the identity, $\cos^2\theta = 1 - \sin^2\theta$, obtaining:

$$\sqrt{a^2 - x^2} = a \cos\theta.$$

To find dx we use $x = a \sin\theta$. We have: $\frac{dx}{d\theta} = a \cos\theta$, therefore $dx = a \cos\theta \, d\theta$. Substituting these values into the integral, we obtain: $\int\sqrt{a^2 - x^2}dx = \int a \cos\theta \cdot a \cos\theta \, d\theta = a^2\int \cos^2\theta \, d\theta$.

Making use of the identity, $\cos^2\theta = \frac{1}{2}(1 + \cos 2\theta)$, and substituting, we obtain:

$$\frac{1}{2} a^2\int(1 + \cos 2\theta)d\theta = \frac{1}{2} a^2\int d\theta + \cos^2\theta \, d\theta.$$

Integrating, we have:

$$\frac{1}{2} a^2(\theta + \frac{1}{2} \sin 2\theta) + C.$$

We now use the formula; $\frac{\sin 2\theta}{2} = \sin\theta \cos\theta$ and, substituting, we have: $\frac{1}{2} a^2(\theta + \sin\theta \cos\theta) + C.$

Now we substitute back the original values for θ, $\sin\theta$ and $\cos\theta$, obtaining:

$$\frac{1}{2} a^2\left(\text{Arc} \sin\frac{x}{a} + \frac{x}{a} \cdot \frac{\sqrt{a^2 - x^2}}{a}\right) + C$$

$$= \frac{1}{2} a^2 \text{ Arc} \sin\frac{x}{a} + \frac{1}{2}x\sqrt{a^2 - x^2} + C.$$

• PROBLEM 738

Integrate the expression: $\int \frac{dx}{x^2\sqrt{x^2 + 4}}$

Solution: To integrate the given expression we use trigonometric substitutions. Let $x = 2 \tan\theta$. Substituting this in the integrand, we find that $\sqrt{x^2 + 4} = \sqrt{4 \tan^2\theta + 4} = \sqrt{4(\tan^2\theta + 1)}$. We can now use the identity, $\sec^2\theta = \tan^2\theta + 1$, obtaining: $\sqrt{x^2 + 4} = 2 \sec\theta$. To obtain the value of dx we use: $x = 2 \tan\theta$. Then $\frac{dx}{d\theta} = 2 \sec^2\theta$ and $dx = 2 \sec^2\theta \, d\theta$. Substituting these values into the integral, we have:

$$\int \frac{dx}{x^2\sqrt{x^2 + 4}} = \int \frac{2 \sec^2\theta \, d\theta}{4 \tan^2\theta \cdot 2 \sec\theta} = \frac{1}{4} \int \frac{\sec\theta \, d\theta}{\tan^2\theta}.$$

Making use of the facts that $\sec\theta = \frac{1}{\cos\theta}$ and $\tan^2\theta = \frac{\sin^2\theta}{\cos^2\theta}$, we can write:

$$\frac{1}{4}\int \frac{\cos\theta\, d\theta}{\sin^2\theta} = \frac{1}{4}\int\left(\frac{\cos\theta}{\sin\theta}\cdot\frac{1}{\sin\theta}\right)d\theta = \frac{1}{4}\int \cot\theta \csc\theta\, d\theta.$$

We can now integrate, using the formula, $\int \cot u \csc u\, du = -\csc u$, obtaining:

$$-\frac{1}{4}\csc\theta + C = -\frac{1}{4}\left(\frac{\sec\theta}{\tan\theta}\right) + C = -\frac{1}{4}\frac{\sqrt{x^2+4}}{x} + C,$$

by substitution.

• **PROBLEM 739**

> Integrate the expression: $\int \frac{x^2\, dx}{\sqrt{a^2 - x^2}}$.

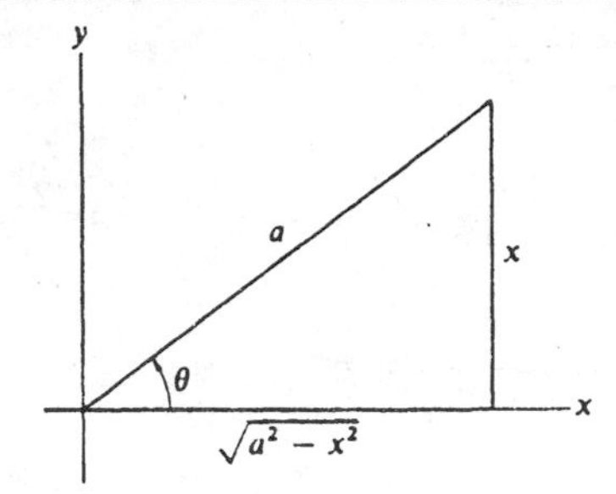

Solution: Let $x = a \sin\theta$. $dx = a\cos\theta\, d\theta$. By substitution and the use of the identity: $\cos^2\theta = 1 - \sin^2\theta$, we obtain:

$$a^2\int \frac{\sin^2\theta\, a\cos\theta\, d\theta}{a\cos\theta} = a^2\int \sin^2\theta\, d\theta = a^2\int \frac{1-\cos 2\theta}{2}d\theta,$$

using the identity $\sin^2\theta = \frac{1-\cos 2\theta}{2}$. Now we evaluate the integral, using: $\int \cos u\, du = \sin u + c$, and then the identity: $\sin 2\theta = 2\sin\theta\cos\theta$, from which $\frac{\sin 2\theta}{2} = \sin\theta\cos\theta$. Doing this, we obtain:

$$\frac{a^2}{2}\left(\theta - \frac{1}{2}\sin 2\theta\right) + c = \frac{a^2}{2}\left(\theta - \sin\theta\cos\theta\right) + c.$$

Hence, substituting values from the shown right triangle we obtain:

$$\int \frac{x^2\, dx}{\sqrt{a^2 - x^2}} = \frac{a^2}{2}\left(\text{arc}\sin\frac{x}{a} - \frac{x}{a^2}\sqrt{a^2 - x^2}\right) + c.$$

• **PROBLEM 740**

> Integrate the expression: $\int \frac{x^3\, dx}{\sqrt{x^2 - a^2}}$.

Solution: Let $x = a\sec\theta$. Then $dx = a\sec\theta\tan\theta\, d\theta$. Substituting for x and dx in the given integral, we obtain:

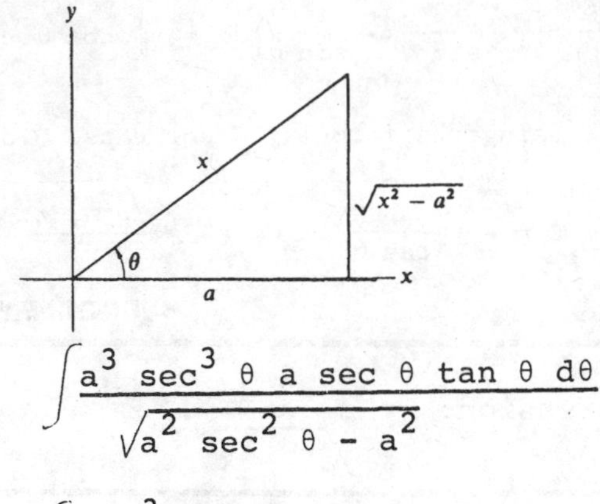

$$\int \frac{a^3 \sec^3 \theta \; a \sec \theta \tan \theta \; d\theta}{\sqrt{a^2 \sec^2 \theta - a^2}}$$

$$= a^3 \int \frac{\sec^3 \theta \; a \sec \theta \tan \theta \; d\theta}{\sqrt{a^2\left(\sec^2 \theta - 1\right)}}$$

We now use the identity: $\sec^2 \theta - 1 = \tan^2 \theta$, and obtain:

$$a^3 \int \frac{\sec^3 \theta \; a \sec \theta \tan \theta \; d\theta}{a \tan \theta} = a^3 \int \sec^4 \theta \; d\theta$$

$$= a^3 \int \left(\tan^2 \theta + 1\right) \sec^2 \theta \; d\theta .$$

By use of the above identity, we have:

$$a^3 \int \tan^2 \theta \sec^2 \theta \; d\theta + a^3 \int \sec^2 \theta \; d\theta .$$

The first integral is now in the form $\int u^n \, du$, with $u = \tan \theta$, $n = 2$, $du = \sec^2 \theta \, d\theta$, and the second is in the form: $\int \sec^2 u \, du = \tan u + c$. Applying the formulas, we obtain:

$$\frac{a^3}{3} \tan^3 \theta + a^3 \tan \theta + c .$$

Hence by substituting the values for $\tan \theta$ from the right triangle, we obtain:

$$\int \frac{x^3 dx}{\sqrt{x^2 - a^2}} = a^3 \left[\frac{1}{3} \frac{\left(\sqrt{x^2 - a^2}\right)^3}{a^3} + \frac{\sqrt{x^2 - a^2}}{a}\right]$$

$$= a^3 \left[\frac{1}{3} \frac{\left(x^2 - a^2\right)^{3/2}}{a^3} + \frac{\sqrt{x^2 - a^2}}{a}\right]$$

$$= \frac{\left(x^2 - a^2\right)^{3/2}}{3} + a^2 \sqrt{x^2 - a^2} .$$

• PROBLEM 741

Integrate the expression: $\int \frac{x \, dx}{\sqrt{4x - x^2}}$.

Solution: To integrate the given expression we rewrite the denominator and then use trigonometric substitutions. We have:

$$\int \frac{x\,dx}{\sqrt{4x - x^2}} = \int \frac{x\,dx}{\sqrt{2^2 - (x-2)^2}} .$$

We now let $x - 2 = 2 \sin \theta$, then $x = 2 + 2 \sin \theta$, $dx = 2 \cos \theta \, d\theta$ and $\sqrt{4x - x^2} = \sqrt{2^2 - (x-2)^2} = \sqrt{4 - (2 \sin \theta)^2} = \sqrt{4 - 4 \sin^2\theta}$ $= \sqrt{4(1 - \sin^2\theta)}$. Making use of the identity: $\cos^2\theta = 1 - \sin^2\theta$, we obtain: $\sqrt{4x - x^2} = 2 \cos \theta$. Substituting, we obtain:

$$\int \frac{x\,dx}{\sqrt{4x - x^2}} = \int \frac{2 + 2 \sin \theta)\cdot 2 \cos \theta \, d\theta}{2 \cos \theta} = \int (2 + 2 \sin \theta) d\theta$$

$$= 2\int d\theta + 2\int \sin \theta \, d\theta .$$

We now integrate, using the formula $\int \sin u \, du = - \cos u$, obtaining:

$$2\theta - 2 \cos \theta + C = 2 \text{ Arc} \sin \frac{x - 2}{2} - \sqrt{4x - x^2} + C,$$

by substitution.

• PROBLEM 742

Integrate the expression: $\displaystyle\int \frac{x\,dx}{\sqrt{ax - x^2}}$, $a > 0$.

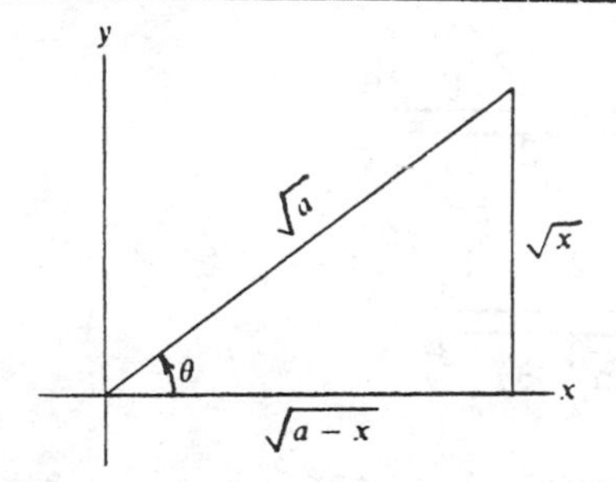

Solution: Let $x = a \sin^2 \theta$. Then $dx = 2a \sin \theta \cos \theta \, d\theta$. Substituting for values of x and dx, and using the identity: $\cos^2 \theta = 1 - \sin^2 \theta$, we obtain:

$$a \int \frac{\sin^2 \theta \; 2a \sin \theta \cos \theta \, d\theta}{a \sin \theta \cos \theta} = 2a \int \sin^2 \theta \, d\theta$$

$$= \frac{2a}{2} \int (1 - \cos 2\theta) d\theta,$$

by use of the identity: $\sin^2 \theta = \frac{1 - \cos 2\theta}{2}$. Evaluating the integral, using the formula: $\int \cos u \, du = \sin u + c$, we obtain:

$$a\left[\theta - \tfrac{1}{2}\sin 2\theta\right] + c = a\left[\theta - \sin \theta \cos \theta\right] + c,$$

by use of the identity: $\sin 2\theta = 2 \sin\theta \cos\theta$,
from which $\frac{\sin 2\theta}{2} = \sin\theta\cos\theta$.

Hence, by substituting values from the right triangle shown into the result, we obtain:

$$\int \frac{x\,dx}{\sqrt{ax - x^2}} = a\left[\arc\sin\sqrt{\frac{x}{a}} - \frac{\sqrt{ax - x^2}}{a}\right] + c.$$

• PROBLEM 743

Integrate the expression: $\int \frac{x^3 dx}{\left(a^2 + x^2\right)^{3/2}}$.

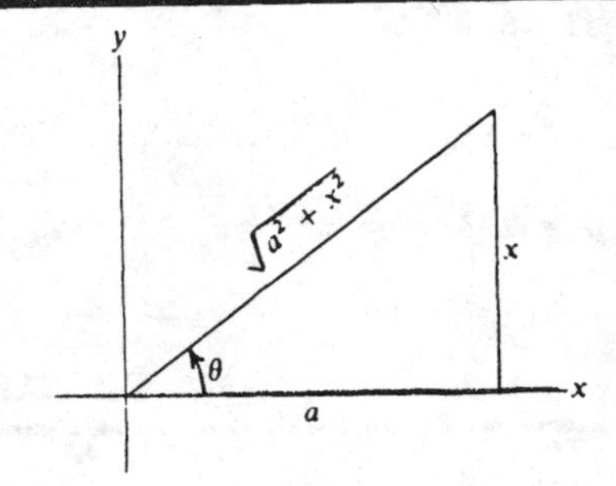

Solution: Let $x = a\tan\theta$. Then $dx = a\sec^2\theta\, d\theta$.
Substituting for x and dx in the integral we obtain:

$$\int \frac{a^3 \tan^3\theta\; a\sec^2\theta\, d\theta}{\sqrt{\left(a^2 + a^2\tan^2\theta\right)^3}}$$

$$= a^3 \int \frac{\tan^3\theta\; a\sec^2\theta\, d\theta}{\sqrt{\left[a^2\left(1 + \tan^2\theta\right)\right]^3}}$$

Using the identity: $\sec^2\theta = \tan^2\theta + 1$, we obtain:

$$a^3 \int \frac{\tan^3\theta\; a\sec^2\theta\, d\theta}{a^3\sec^3\theta} = a\int \frac{\tan^3\theta}{\sec\theta}d\theta.$$

Using the facts that $\tan\theta = \frac{\sin\theta}{\cos\theta}$ and $\sec\theta = \frac{1}{\cos\theta}$, the integral becomes:

$$a\int \frac{\sin^3\theta}{\cos^2\theta}d\theta = a\int \frac{\sin^2\theta}{\cos^2\theta}\sin\theta\, d\theta$$

$$= a\int \frac{1 - \cos^2\theta}{\cos^2\theta}\sin\theta\, d\theta.$$

By use of the identity $\sin^2\theta + \cos^2\theta = 1$,

$$a\int \frac{\sin^3\theta}{\cos^2\theta} = a \int (\cos^{-2}\theta - 1)\sin\theta \, d\theta$$

$$= a \int \cos^{-2}\theta \sin\theta \, d\theta - a \int \sin\theta \, d\theta$$

$$= -a \int \cos^{-2}\theta (-\sin\theta) d\theta - a \int \sin\theta \, d\theta.$$

Using the formulas: $\int u^n du = \frac{u^{n+1}}{n+1} + c$ and $\int \sin u \, du = -\cos u + c$, and the fact that $\cos^{-1}\theta = \frac{1}{\cos\theta} = \sec\theta$ we obtain:

$$a \cos^{-1}\theta + a \cos\theta + c = a(\cos\theta + \sec\theta) + c.$$

Hence according to the above right triangle:

$$\int \frac{x^3 \, dx}{(a^2 + x^2)^{3/2}} = a\left(\frac{a}{\sqrt{a^2 + x^2}} + \frac{\sqrt{a^2 + x^2}}{a}\right),$$

by substituting values for cos θ and sec θ into the integral.

• PROBLEM 744

Integrate: $\int \frac{dx}{2 - \sqrt{3x}}$.

Solution: In order to simplify this integral, we use the substitution method of integration. If we let $\sqrt{3x} = 2\sin^2\theta$, we can find dx as follows,

$$\sqrt{3x} = 2\sin^2\theta$$

$$3x = 4\sin^4\theta$$

$$x = \frac{4}{3}\sin^4\theta$$

$$dx = \frac{16}{3}\sin^3\theta\cos\theta.$$

We now rewrite the integral as follows:

$$\int \frac{dx}{2 - \sqrt{3x}} = \int \frac{\frac{16}{3}\sin^3\theta\cos\theta}{2 - 2\sin^2\theta} d\theta.$$

Using the identity: $\cos^2\theta + \sin^2\theta = 1$, we rewrite and simplify the integral as follows:

$$\int \frac{\frac{16}{3}\sin^3\theta\cos\theta}{2 - 2\sin^2\theta} d\theta = \int \frac{\frac{16}{3}\sin^3\theta\cos\theta}{2(1 - \sin^2\theta)} d\theta$$

$$= \int \frac{8}{3} \frac{\sin^3\theta \cos\theta \, d\theta}{\cos^2\theta} \qquad = \frac{8}{3}\int \frac{\sin^3\theta}{\cos\theta} d\theta.$$

Now we can use the identity: $\cos^2\theta + \sin^2\theta = 1$ again to rewrite the integral. We obtain:

$$\frac{8}{3}\int \frac{\sin^3\theta}{\cos\theta} d\theta = \frac{8}{3}\int \frac{\sin\theta(1 - \cos^2\theta)}{\cos\theta} d\theta$$

$$= \frac{8}{3}\int \frac{\sin\theta - \sin\theta\cos^2\theta}{\cos\theta} d\theta.$$

We separate this integral into the sum of two separate integrals and integrate, using the formulas: $\int \frac{du}{u} = \ln|u| + C$, and $\int u^n \, du = \frac{u^{n+1}}{n+1} + C$

$$\frac{8}{3}\int \frac{\sin\theta - \sin\theta\cos^2\theta}{\cos\theta} d\theta = \frac{8}{3}\int \frac{\sin\theta}{\cos\theta} d\theta - \frac{8}{3}\int \frac{\sin\theta\cos^2\theta}{\cos\theta} d\theta$$

$$= \frac{8}{3}\int \frac{\sin\theta}{\cos\theta} d\theta - \frac{8}{3}\int \sin\theta\cos\theta \, d\theta$$

$$= -\frac{8}{3}\int \frac{-\sin\theta}{\cos\theta} d\theta - \frac{8}{3}\int \sin\theta\cos\theta \, d\theta$$

$$= -\frac{8}{3}\ln|\cos\theta| - \frac{8}{3}\frac{\sin^2\theta}{2} + C$$

$$= -\frac{8}{3}\ln(\cos\theta) - \frac{4}{3}\sin^2\theta + C.$$

We substitute back for $\cos\theta$ and $\sin^2\theta$ in terms of x. We let $\sqrt{3x} = 2\sin^2\theta$, therefore $\sin^2\theta = \frac{\sqrt{3x}}{2}$. We can find $\cos\theta$ by using the identity: $\cos^2\theta + \sin^2\theta = 1$.

$$\cos^2\theta + \sin^2\theta = 1.$$

$$\sin^2\theta = \frac{\sqrt{3x}}{2}.$$

$$\cos^2\theta \; \frac{\sqrt{3x}}{2} = 1.$$

$$\cos^2\theta = 1 - \frac{\sqrt{3x}}{2}.$$

$$\cos\theta = \sqrt{1 - \frac{\sqrt{3x}}{2}}.$$

We can now write:

$$-\frac{8}{3}\ln(\cos\theta) - \frac{4}{3}\sin^2\theta + C = -\frac{8}{3}\ln\sqrt{1 - \frac{\sqrt{3x}}{2}}$$

$$- \frac{4}{3}\frac{\sqrt{3x}}{2} + C.$$

$$= -\frac{8}{3}\ln\left(1 - \frac{\sqrt{3x}}{2}\right)^{1/2}$$

$$- \frac{2}{3}\sqrt{3x} + C$$

$$= -\frac{8}{3}\left(\frac{1}{2}\right)\ln\left(1 - \frac{\sqrt{3x}}{2}\right)$$

$$- \frac{2}{3}\sqrt{3x} + C$$

$$= -\frac{4}{3}\ln\left(\frac{2 - \sqrt{3x}}{2}\right) - \frac{2}{3}\sqrt{3x} + C.$$

• PROBLEM 745

Integrate: $\displaystyle\int \frac{dx}{\left(x^2 + 4\right)^2}$.

Solution: To integrate the given function we use trigonometric substitutions. Let $x = 2\tan\theta$, then $x^2 + 4 = (2\tan\theta)^2 + 4 = 4\tan^2\theta + 4 = 4(1 + \tan^2\theta) = 4\sec^2\theta$, by use of the identity: $\sec^2\theta = 1 + \tan^2\theta$. $(x^2 + 4)^2 = 16\sec^4\theta$, $dx = 2\sec^2\theta\, d\theta$.

Substituting, we obtain:

$$\int \frac{dx}{(x^2 + 4)^2} = \int \frac{2\sec^2\theta\, d\theta}{16\sec^4\theta} = \frac{1}{8}\int \frac{1}{\sec^2\theta\, d\theta} = \frac{1}{8}\int \cos^2\theta\, d\theta.$$

We can now use the formula, $\cos^2\theta = \frac{1}{2}\left(1 + \cos 2\theta\right)$, obtaining:

$$\frac{1}{16}\int (1 + \cos 2\theta)d\theta = \frac{1}{16}\left(\int d\theta + \int \cos 2\theta\, d\theta\right).$$

We now integrate, using the formula: $\int \cos u\, du = \sin u$. This gives:

$$\frac{1}{16}\left(\theta + \frac{1}{2}\sin 2\theta\right) + C = \frac{1}{16}\left(\theta + \sin\theta\cos\theta\right) + C,$$

by use of the formula: $\frac{\sin 2\theta}{2} = \sin\theta\cos\theta$.

We can rewrite $\frac{1}{16}\left(\theta + \sin\theta\cos\theta\right) + C$ as $\frac{1}{16}\left(\theta + \frac{\tan\theta}{\sec\theta}\cdot\frac{1}{\sec\theta}\right) + C$ and, by substitution, we obtain:

$$\frac{1}{16}\text{Arc tan}\,\frac{1}{2}x + \frac{1}{16}\,\frac{x}{\sqrt{x^2 + 4}}\cdot\frac{2}{\sqrt{x^2 + 4}} + C$$

$$= \frac{1}{16}\text{Arc tan}\,\frac{1}{2}x + \frac{1}{8}\,\frac{x}{x^2 + 4} + C.$$

Integrate: $\int y\sqrt{2y - y^2}\,dy$.

Solution: To integrate the given expression, we use the method of trigonometric substitutions. Let $y = 1 + \sin\theta$. Then, $dy = \cos\theta\,d\theta$. Substituting, we obtain:

$$\int y\sqrt{2y-y^2}\,dy = \int (1+\sin\theta)\sqrt{2(1+\sin\theta)-(1+\sin\theta)^2}\cos\theta\,d\theta.$$

$$= \int (1+\sin\theta)\sqrt{2+2\sin\theta-1-2\sin\theta-\sin^2\theta}\cos\theta\,d\theta.$$

$$= \int (1+\sin\theta)\sqrt{1-\sin^2\theta}\cos\theta\,d\theta.$$

Making use of the identity: $\cos^2\theta = 1 - \sin^2\theta$,

$$\int y\sqrt{2y-y^2}\,dy = \int (1+\sin\theta)(\cos\theta)(\cos\theta)d\theta$$

$$= \int (\cos^2\theta + \cos^2\theta\sin\theta)d\theta$$

$$= \int \cos^2\theta\,d\theta + \int \cos^2\theta\sin\theta\,d\theta.$$

To integrate the first term, we use the formula:

$$\int \cos^2 u\,du = \frac{1}{2}u + \frac{1}{4}\sin 2u,$$

letting $u = \theta$, $du = d\theta$. To integrate the second term, we use the formula for $\int u^n\,du$, letting $u = \cos\theta$, $du = -\sin\theta\,d\theta$ and $n = 2$. Applying these formulas we obtain:

$$\frac{1}{2}\theta + \frac{1}{4}\sin 2\theta - \frac{\cos^3\theta}{3}.$$

We now substitute for θ, $\sin 2\theta$ and $\cos\theta$. Since $y = 1 + \sin\theta$, $\sin\theta = y - 1$, therefore, $\theta = \arcsin(y-1)$, and, since $\cos\theta = \sqrt{1 - \sin^2\theta}$, $\cos\theta = \sqrt{1-(y-1)^2} = \sqrt{2y - y^2}$. We can rewrite $\sin 2\theta$ by using the formula: $\sin 2\theta = 2\sin\theta\cos\theta$, and, substituting the values found for $\sin\theta$ and $\cos\theta$, we obtain:

$$\frac{1}{2}\arcsin(y-1) + \frac{1}{4}\left[2\,(y-1)\sqrt{2y-y^2}\right] - \frac{1}{3}\left(2y-y^2\right)^{\frac{3}{2}}$$

$$= \frac{1}{2}\arcsin(y-1) + \frac{1}{2}(y-1)\sqrt{2y-y^2} - \frac{1}{3}\left(2y-y^2\right)^{\frac{3}{2}}.$$

Factoring out from the last two terms the common factor, $\frac{1}{6}\sqrt{2y-y^2}$, we have:

$$\frac{1}{2}\arcsin(y-1) + \frac{1}{6}\sqrt{2y-y^2}\left[3(y-1) - 2\left(2y-y^2\right)\right]$$

$$= \frac{1}{2}\arcsin(y-1) + \frac{1}{6}\sqrt{2y-y^2}\left(3y - 3 - 4y + 2y^2\right)$$

$$= \frac{1}{2}\arcsin(y-1) + \frac{1}{6}\sqrt{2y-y^2}\,(2y^2 - y - 3).$$

• PROBLEM 747

Integrate: $\int y^2\sqrt{a^2 - y^2}\, dy.$

Solution: To integrate the given expression we use the method of trigonometric substitutions. We let $y^2 = a^2 \sin^2 \theta$. Then, $y = a \sin \theta$, and $dy = a \cos \theta\, d\theta$. Now, by substitution:

$$\int y^2\sqrt{a^2-y^2}\, dy = \int a^2 \sin^2 \theta\sqrt{a^2-(a^2 \sin^2 \theta)}\; a \cos \theta\, d\theta$$

$$= \int a^2 \sin^2 \theta\sqrt{a^2(1-\sin^2 \theta)}\; a \cos \theta\, d\theta.$$

By use of the identity: $\cos^2 \theta = 1 - \sin^2 \theta$,

$$\int y^2\sqrt{a^2-y^2}\, dy = \int a^2 \sin^2 \theta\sqrt{a^2 \cos^2 \theta}\; a \cos \theta\, d\theta$$

$$= \int a^2 \sin^2 \theta\; a \cos \theta\; a \cos \theta\, d\theta$$

$$= \int a^4 \sin^2 \theta \cos^2 \theta\, d\theta.$$

We use the formulas: $\sin^2 \theta = \frac{1 - \cos 2\theta}{2}$, and $\cos^2 \theta = \frac{1 + \cos 2\theta}{2}$, and substitute. Then,

$$\int a^4 \sin^2 \theta \cos^2 \theta\, d\theta = a^4 \int \frac{1 - \cos 2\theta}{2} \cdot \frac{1 + \cos 2\theta}{2}\, d\theta.$$

$$= a^4 \int \frac{1 - \cos^2 2\theta}{4}\, d\theta = \frac{a^4}{4}\int \left(1 - \cos^2 2\theta\right) d\theta$$

$$= \frac{a^4}{4}\left(\int d\theta - \int \cos^2 2\theta\, d\theta\right).$$

To integrate the second term, we use the formula:

$$\int \cos^2 u\, du = \frac{1}{2} u + \frac{1}{4} \sin 2u,$$

letting $u = 2\theta$, $du = 2d\theta$. Then we obtain:

$$\frac{a^4}{4}\theta - \frac{a^4}{4} \cdot \frac{1}{2}\left(\frac{1}{2}\, 2\theta + \frac{1}{4} \sin 4\theta\right)$$

$$= \frac{a^4}{4}\theta - \frac{a^4}{8}\,\theta - \frac{a^4}{32} \sin 4\theta$$

$$= \frac{a^4}{8}\theta - \frac{a^4}{32}\sin 4\theta.$$

Now we substitute the original values. Since $y = a \sin\theta$, $\sin\theta = \frac{y}{a}$, and $\theta = \arcsin\frac{y}{a}$. Therefore, the first term becomes:

$$\frac{1}{8}a^4 \arcsin\frac{y}{a}.$$

We manipulate the second term before substituting. Since $\sin 4\theta = \sin(2\theta + 2\theta)$, we use the formula $\sin(A + B) = \sin A \cos B + \cos A \sin B$. Therefore,

$$\sin(2\theta+2\theta) = \sin 2\theta \cos 2\theta + \cos 2\theta \sin 2\theta$$

$$= 2 \sin 2\theta \cos 2\theta.$$

Using the formulas: $\sin 2\theta = 2 \sin\theta \cos\theta$, and $\cos 2\theta = \cos^2\theta - \sin^2\theta$, and substituting, we obtain:

$$2(2 \sin\theta \cos\theta)(\cos^2\theta - \sin^2\theta)$$

$$= 4 \sin\theta \cos\theta(\cos^2\theta - \sin^2\theta)$$

$$= 4(\sin\theta \cos^3\theta - \sin^3\theta \cos\theta).$$

Since $y = a \sin\theta$, we can substitute $\frac{y}{a}$ for $\sin\theta$. Since $\cos\theta = \sqrt{1 - \sin^2\theta}$, we have:

$$\cos\theta = \sqrt{1 - \frac{y^2}{a^2}} = \sqrt{\frac{a^2 - y^2}{a^2}} = \frac{\sqrt{a^2 - y^2}}{a}.$$

Substituting, we obtain:

$$4\left[\frac{y}{a}\ \frac{(a^2 - y^2)^{\frac{3}{2}}}{a^3} - \frac{y^3}{a^3}\ \frac{\sqrt{a^2 - y^2}}{a}\right]$$

$$= \frac{4}{a^4}\left[y(a^2 - y^2)^{\frac{3}{2}} - y^3\sqrt{a^2 - y^2}\right] = \sin 4\theta.$$

But the complete second term is $-\frac{a^4}{32}\sin 4\theta$, therefore we have:

$$-\frac{a^4}{32}\ \frac{4}{a^4}\left[y(a^2 - y^2)^{\frac{3}{2}} - y^3\sqrt{a^2 - y^2}\right]$$

$$= -\frac{1}{8}y(a^2 - y^2)^{\frac{3}{2}} + \frac{1}{8}y^3\sqrt{a^2 - y^2}.$$

Factoring out a common term, we obtain:

$$-\frac{1}{8}y\sqrt{a^2 - y^2}\left[\left(a^2 - y^2\right)^{\frac{3}{2}} - y^2\right] = -\frac{1}{8}y\sqrt{a^2 - y^2}\left(a^2 - 2y^2\right) .$$

Now, combining the first and second terms we find:

$$\int y^2\sqrt{a^2 - y^2}\, dy = \frac{1}{8}a^4 \arcsin \frac{y}{a} - \frac{1}{8}y\sqrt{a^2 - y^2}\left(a^2 - 2y^2\right) + C.$$

• PROBLEM 748

Integrate: $\int\sqrt{\frac{x}{1 - x}}dx.$

Solution: To simplify this problem, we use integration by substitution. If we let $x = \sin^2 \theta$, then $dx = 2 \sin \theta \cos \theta$, and we rewrite the integral as:

$$\int\sqrt{\frac{x}{1 - x}}dx = \int\sqrt{\frac{\sin^2 \theta}{1 - \sin^2 \theta}}\left(2 \sin \theta \cos \theta\right)d\theta.$$

Using the identity: $\sin^2 \theta + \cos^2 \theta = 1$, we write the integral as:

$$\int\sqrt{\frac{\sin^2 \theta}{1 - \sin^2 \theta}}\left(2 \sin \theta \cos \theta\right)d\theta$$

$$= \int\sqrt{\frac{\sin^2 \theta}{\cos^2 \theta}}\left(2 \sin \theta \cos \theta\right)d\theta$$

$$= \int\frac{\sin \theta}{\cos \theta}\left(2 \sin \theta \cos \theta\right)d\theta$$

$$= \int 2 \sin^2 \theta \, d\theta.$$

We use the formula:

$$\int \sin^2 u \, du = \frac{1}{2}u - \frac{1}{4} \sin 2u,$$

to integrate as follows:

$$\int 2 \sin^2 \theta \, d\theta = 2 \int \sin^2 \theta \, d\theta$$

$$= 2\left(\frac{1}{2}\theta - \frac{1}{4} \sin 2\theta\right) + C$$

$$= \theta - \frac{1}{2} \sin 2\theta + C.$$

We can rewrite this result, using the identity: $\sin 2A = 2 \sin A \cos A$, to obtain:

$$\theta - \frac{1}{2} \sin 2\theta + C = \theta - \sin \theta \cos \theta + C.$$

We substitute the x terms back into the result. We previously let $\sin^2\theta = x$, therefore $\sin\theta = \sqrt{x}$, and $\theta = \sin^{-1}\sqrt{x}$. We can find $\cos\theta$, using the identity: $\sin^2\theta + \cos^2\theta = 1$, as follows:

$$\sin^2\theta = x.$$

$$\sin^2\theta + \cos^2\theta = 1.$$

$$x + \cos^2\theta = 1.$$

$$\cos^2\theta = 1 - x.$$

$$\cos\theta = \sqrt{1 - x}.$$

Substituting these values into the integrated function, we obtain:

$$\theta - \sin\theta\cos\theta + C = \sin^{-1}(\sqrt{x}) - \sqrt{x}\sqrt{1 - x} + C$$

$$= \sin^{-1}(\sqrt{x}) - \sqrt{x - x^2} + C.$$

• PROBLEM 749

Integrate: $\displaystyle\int \frac{x^2\,dx}{\sqrt{x^2 + 1}}.$

Solution: To solve this problem we substitute for x, which simplifies the problem. We let $\tan\theta = x$, and substitute this into the integrand to obtain the following:

$$\int \frac{x^2\,dx}{\sqrt{x^2 + 1}} = \int \frac{\tan^2\theta\,d(\tan\theta)}{\sqrt{\tan^2\theta + 1}}.$$

We use the formula:

$$\frac{d}{dx}(\tan u) = \sec^2 u\,\frac{du}{dx},$$

and the identity: $\sec^2\theta - \tan^2\theta = 1$. We have:

$$\int \frac{\tan^2\theta\sec^2\theta\,d\theta}{\sqrt{\tan^2\theta + 1}} = \int \frac{\tan^2\theta\sec^2\theta\,d\theta}{\sqrt{\tan^2\theta + 1}}$$

$$= \int \frac{(\sec^2\theta - 1)(\sec^2\theta)\,d\theta}{\sqrt{\sec^2\theta}} = \int \frac{\sec^4\theta - \sec^2\theta}{\sec\theta}\,d\theta$$

$$= \int \sec^3\theta - \sec\theta\,d\theta = \int \sec^3\theta\,d\theta - \int \sec\theta\,d\theta.$$

We can now integrate. For $\int \sec^3\theta\,d\theta$, we apply the formula:

$$\int \sec^n u\, du = \frac{\tan u \sec^{n-2} u}{n-1} + \frac{n-2}{n-1} \int \sec^{n-2} u\, du,$$

and for $\int \sec\theta\, d\theta$ we use the formula:

$$\int \sec u\, du = \ln|\sec u + \tan u|.$$

We obtain:

$$\int \sec^3 \theta\, d\theta - \int \sec\theta\, d\theta = \frac{\tan\theta \sec\theta}{2} + \frac{1}{2} \int \sec\theta\, d\theta$$

$$- \int \sec\theta\, d\theta = \frac{\tan\theta \sec\theta}{2}$$

$$+ \frac{\ln|\sec\theta + \tan\theta|}{2} - \ln|\sec\theta + \tan\theta|$$

$$= \frac{\tan\theta \sec\theta}{2} - \frac{\ln|\sec\theta + \tan\theta|}{2} + C.$$

We now substitute x back in the equation. We let $\tan\theta = x$, hence we can find $\sec\theta$ by knowing that $\sec^2\theta = 1 + \tan^2\theta$. $\text{Sec}^2\theta = 1 + x^2$ and $\sec\theta = \sqrt{1 + x^2}$. Substituting $\tan\theta = x$ and $\sec\theta = \sqrt{1 + x^2}$ back into the equation, we obtain:

$$\frac{\tan\theta \sec\theta}{2} - \frac{\ln|\sec\theta + \tan\theta|}{2} + C$$

$$= \frac{x\sqrt{1 + x^2}}{2} - \frac{\ln\left(x + \sqrt{1 + x^2}\right)}{2} + C.$$

(Note: the absolute value sign can be dropped since the value is positive.)

• **PROBLEM** 750

Integrate: $\displaystyle\int \frac{t\, dt}{1 - \sqrt{t}}.$

Solution: This problem is simplified if we use integration by substitution. If we let $\sqrt{t} = \sin^2\theta$, then $t = \sin^4\theta$, and $dt = 4 \sin^3\theta \cos\theta\, d\theta$. We rewrite the integral as:

$$\int \frac{t\, dt}{1 - \sqrt{t}} = \int \frac{\sin^4\theta \left(4 \sin^3\theta \cos\theta\right)}{1 - \sin^2\theta} d\theta.$$

Using the identity: $\sin^2\theta + \cos^2\theta = 1$, we can rewrite this as:

$$\int \frac{\sin^4\theta \left(4 \sin^3\theta \cos\theta\right)}{1 - \sin^2\theta} d\theta = \int \frac{4 \sin^7\theta \cos\theta}{\cos^2\theta} d\theta$$

$$= 4\int \frac{\sin^7\theta}{\cos\theta}d\theta.$$

Again using the identity: $\sin^2\theta + \cos^2\theta = 1$, we rewrite the integral as,

$$4\int \frac{\sin^7\theta}{\cos\theta}d\theta = 4\int \frac{\sin^5\theta(\sin^2\theta)}{\cos\theta}d\theta$$

$$= 4\int \frac{\sin^5\theta(1-\cos^2\theta)}{\cos\theta}d\theta$$

$$= 4\int \frac{\sin^5\theta - \sin^5\theta\cos^2\theta}{\cos\theta}d\theta.$$

We write this integral as the sum of two separate integrals:

$$4\int \frac{\sin^5\theta - \sin^5\theta\cos^2\theta}{\cos\theta}d\theta = 4\int \frac{\sin^5\theta}{\cos\theta}d\theta$$

$$- 4\int \sin^5\theta\cos\theta\, d\theta.$$

The second integral, $(-4)\int \sin^5\theta\cos\theta\, d\theta$, can be integrated using the formula: $\int u^n\, du = \frac{u^{n+1}}{n+1}$. But the first integral , $4\int \frac{\sin^5\theta}{\cos\theta}d\theta$, must be further rearranged before it can be integrated. We again use the identity: $\sin^2\theta + \cos^2\theta = 1$, as follows:

$$4\int \frac{\sin^5\theta}{\cos\theta}d\theta - 4\int \sin^5\theta\cos\theta\, d\theta$$

$$= 4\int \frac{\sin^3\theta(\sin^2\theta)}{\cos\theta}d\theta - 4\int \sin^5\theta\cos\theta\, d\theta$$

$$= 4\int \frac{\sin^3\theta(1-\cos^2\theta)}{\cos\theta}d\theta - 4\int \sin^5\theta\cos\theta\, d\theta$$

$$= 4\int \frac{\sin^3\theta - \sin^3\theta\cos^2\theta}{\cos\theta}d\theta - 4\int \sin^5\theta\cos\theta\, d\theta.$$

We can split the first integral into the sum of two integrals to obtain:

$$4\int \frac{\sin^3\theta - \sin^3\theta\cos^2\theta}{\cos\theta}d\theta - 4\int \sin^5\theta\cos\theta\, d\theta$$

$$= 4\int \frac{\sin^3\theta}{\cos\theta}d\theta - 4\int \frac{\sin^3\theta\cos^2\theta}{\cos\theta}d\theta$$

$$- 4\int \sin^5\theta\cos\theta\, d\theta$$

$$= 4 \int \frac{\sin^3 \theta}{\cos \theta} d\theta - 4 \int \sin^3 \theta \cos \theta \, d\theta$$

$$- 4 \int \sin^5 \theta \cos \theta \, d\theta.$$

The second term can be integrated in the same manner used for the third. But the first term, $4 \int \frac{\sin^3 \theta}{\cos \theta} d\theta$, has to be rearranged. We again use the identity: $\sin^2 \theta + \cos^2 \theta = 1$, to obtain:

$$4 \int \frac{\sin^3 \theta}{\cos \theta} d\theta - 4 \int \sin^3 \theta \cos \theta \, d\theta$$

$$- 4 \int \sin^5 \theta \cos \theta \, d\theta$$

$$= 4 \int \frac{\sin \theta (\sin^2 \theta)}{\cos \theta} d\theta - 4 \int \sin^3 \theta \cos \theta \, d\theta$$

$$- 4 \int \sin^5 \theta \cos \theta \, d\theta$$

$$= 4 \int \frac{\sin \theta (1 - \cos^2 \theta)}{\cos \theta} d\theta - 4 \int \sin^3 \theta \cos \theta \, d\theta$$

$$- 4 \int \sin^5 \theta \cos \theta \, d\theta$$

$$= 4 \int \frac{\sin \theta - \sin \theta \cos^2 \theta}{\cos \theta} d\theta - 4 \int \sin^3 \theta \cos \theta \, d\theta$$

$$- 4 \int \sin^5 \theta \cos \theta \, d\theta.$$

Separating the first integral into the sum of two separate integrals, we obtain,

$$4 \int \frac{\sin \theta}{\cos \theta} d\theta - 4 \int \frac{\sin \theta \cos^2 \theta}{\cos \theta} d\theta - 4 \int \sin^3 \theta \cos \theta \, d\theta$$

$$- 4 \int \sin^5 \theta \cos \theta \, d\theta.$$

Now we can proceed to integrate. The first term can be integrated using the formula: $\int \frac{du}{u} = \ln u$, and the last three terms can be integrated using the formula: $\int u^n \, du = \frac{u^{n+1}}{n+1}$. Integrating, we obtain:

$$-4 \int \frac{(-\sin \theta)}{\cos \theta} d\theta - 4 \int \sin \theta \cos \theta \, d\theta$$

$$- 4 \int \sin^3 \theta \cos \theta \, d\theta$$

$$- 4 \int \sin^5 \theta \cos \theta \, d\theta$$

$$= -4 \ln(\cos \theta) - 4 \frac{\sin^2 \theta}{2} - 4 \frac{\sin^4 \theta}{4} - 4 \frac{\sin^6 \theta}{6} + C.$$

We now substitute back for θ in terms of t.
We previously let $\sin^2 \theta = \sqrt{t}$, and $\sin^4 \theta = t$. We can obtain $\sin^6 \theta$ from $(\sin^2 \theta)(\sin^4 \theta)$, therefore $\sin^6 \theta = (\sqrt{t})(t)$. Cos θ can be found, using the identity $\sin^2 \theta + \cos^2 \theta = 1$, as follows:

$$\sin^2 \theta = \sqrt{t}, \quad \text{and}$$

$$\sin^2 \theta + \cos^2 \theta = 1.$$

$$\sqrt{t} + \cos^2 \theta = 1.$$

$$\cos^2 \theta = 1 - \sqrt{t}.$$

$$\cos \theta = (1 - \sqrt{t})^{1/2}.$$

Substituting these values into the integrated function, we obtain:

$$-4 \ln(\cos \theta) - 4 \frac{\sin^2 \theta}{2} - 4 \frac{\sin^4 \theta}{4} - 4 \frac{\sin^6 \theta}{6} + C$$

$$= -4 \ln(1 - \sqrt{t})^{1/2} - 2\sqrt{t} - t - \frac{2}{3}t\sqrt{t} + C$$

$$= -4(\tfrac{1}{2}) \ln(1 - \sqrt{t}) - 2\sqrt{t} - t - \frac{2}{3}t\sqrt{t} + C$$

$$= -2 \ln(1 - \sqrt{t}) - 2\sqrt{t} - t - \frac{2}{3}t\sqrt{t} + C.$$

CHAPTER 27

INTEGRATION BY PARTS

This method of integration should be tried when encountering functions having exponential and trigonometric factors. The method often allows the factors to be separated into terms that are readily integrated. The selection of the terms to be represented by u and v is generally a matter of trial and error, and is best based on experience.

In applying this theorem of integration, it is possible that the original integral is again obtained. In that event the integral can be solved for as a function of the remaining terms in the equation.

• PROBLEM 751

Integrate the expression:

$$\int \ln x \, dx.$$

Solution: Here we use integration by parts. Then,

$$\int u dv = uv - \int v du.$$

Let $u = \ln x$, $dv = dx$, $du = \frac{1}{x} dx$, and

$$v = \int dv = \int dx = x.$$

Substituting into the above equation, we obtain:

$$\int \ln x \, dx = (\ln x)x - \int \frac{x}{x} dx$$

$$= x \ln x - x + c.$$

• PROBLEM 752

Integrate:

$$\int x e^x dx.$$

Solution: To integrate this expression, we use

integration by parts. We write:

$$\int udv = uv - \int vdu.$$

Let $u = x$, and $dv = e^x dx$. Then $du = dx$, and

$v = \int e^x dx = e^x$. Substituting into the equation,

we have:
$$\int xe^x dx = xe^x - \int e^x dx.$$

To integrate the last term we use the formula,

$$\int e^u du = e^u + C,$$

obtaining:

$$\int xe^x dx = xe^x - e^x + C,$$

or, $\int xe^x dx = e^x (x - 1) + C.$

• PROBLEM 753

Integrate:

$$\int x \ln x \, dx.$$

Solution: To integrate, we use the formula for integration by parts. We have:

$$\int udv = uv - \int vdu.$$

Let $u = \ln x$, and $dv = x\,dx$. Then,

$du = \frac{1}{x} dx$, and $v = \int x\,dx = \frac{x^2}{2}$, by use of the

formula: $\int u^n du = \frac{u^{n+1}}{n + 1}$.

Substituting into the equation, we have:

$$\int x \ln x \, dx = \frac{x^2}{2} \ln x - \int \frac{x^2}{2} \cdot \frac{1}{x} dx.$$

To integrate the last expression, we again use the formula:

$$\int u^n du = \frac{u^{n+1}}{n + 1},$$

obtaining:

$$\int x \ln x \, dx = \frac{x^2}{2} \ln x - \frac{x^2}{4} + C.$$

• PROBLEM 754

Integrate by parts the expression:

$$\frac{dy}{dx} = x^2 \ln x.$$

Solution:

$$dy = x^2 \ln x \cdot dx.$$

$$y = \int x^2 \ln x \, dx.$$

To integrate by parts we use the equation:

$$\int udv = uv - \int vdu .$$

Now, let $u = \ln \cdot x$.

Then, $$du = \frac{1}{x} \cdot dx.$$

Let $$dv = x^2 \cdot dx.$$

Then, $$v = \int dv = \int x^2 \, dx = \frac{x^3}{3} ,$$

by use of the formula for $\int u^n \, du$. Substituting into the above equation, we have:

$$y = \int \overset{u}{\ln x} \cdot \overset{dv}{x^2 \cdot dx} = \frac{x^3}{3} \ln x - \int \frac{x^3}{3} \cdot \frac{1}{x} \cdot dx.$$

We can now integrate

$$\int \frac{x^3}{3} \cdot \frac{1}{x} \cdot dx,$$

by using the formula for $\int u^n \, du$,

with $u = x$, $du = dx$, and $n = 2$. Doing this, we obtain:

$$y = \frac{x^3}{3} \ln x - \frac{x^3}{9} + C.$$

• PROBLEM 755

Integrate: $\int x \cdot \cos x \cdot dx$.

Solution: In this case we use integration by parts, the rule for which states: $\int udv = uv - \int vdu$. Let $u = x$ and $dv = \cos x \ dx$. Then $du = dx$ and $v = \int \cos x \cdot dx = \sin x$. $\int u \cdot dv = uv - \int v \cdot du$ becomes $\int x \cdot \cos x \cdot dx = x \cdot \sin x - \int \sin x \cdot dx$. To integrate $\int \sin x dx$ we use the formula, $\int \sin u du = -\cos u + C$. This gives:

$$\int x \cdot \cos x \cdot dx = x \sin x - (-\cos x) + C$$

$$= x \sin x + \cos x + C.$$

• PROBLEM 756

Integrate the expression:

$$\int \arctan x \, dx.$$

Solution: Performing integration by parts, we use the formula:

$$\int udv = uv - \int vdu.$$

Let $u = \arctan x$, $dv = dx$, $du = \frac{1}{1 + x^2} dx$,

and $v = \int dv = \int dx = x.$

Substituting into the above equation, we obtain:

$$\int \arctan x \, dx = (\arctan x)x - \int \frac{x}{1 + x^2} dx$$

We can now integrate, using the formula:

$$\int \frac{du}{u} = \ln u,$$

letting $u = (1 + x^2)$ and $du = 2x \, dx$. We obtain:

$$x \arctan x - \frac{1}{2} \ln(1 + x^2) + c.$$

• PROBLEM 757

Integrate:

$$\int \frac{x^5}{\sqrt{1 - 2x^3}} dx.$$

Solution: Here we use integration by parts. Then,

$$\int udv = uv - \int vdu.$$

Let $u = x^3$, $dv = \frac{x^2}{\sqrt{1 - 2x^3}} dx$, $du = 3x^3 \, dx$,

and

$$v = \int (1 - 2x^3)^{-1/2} x^2 \, dx = - \frac{1}{6} 2(1 - 2x^3)^{1/2}$$

$$= - \frac{1}{3} (1 - 2x^3)^{1/2}$$

by use of the formula:

$$\int u^n \, du = \frac{u^{n+1}}{n + 1}, \quad \text{letting}$$

$u = (1 - 2x^3)$, $du = - 6x \, dx$ and $n = - \frac{1}{2}$. Substituting into the above equation, we obtain:

$$\int \frac{x^5}{\sqrt{1 - 2x^3}} dx = x^3 \left[- \frac{1}{3}(1 - 2x^3)^{1/2}\right]$$

$$+ \frac{1}{3} \int (1 - 2x^3)^{1/2} 3x^2 \, dx.$$

We now integrate, again using the formula for $\int u^n \, du$, with $u = (1 - 2x^3)$, $du = -6x^2 dx$, and $n = \frac{1}{2}$, obtaining:

$$-\frac{x^3}{3}\sqrt{1 - 2x^3} - \frac{1}{9}\left(1 - 2x^3\right)^{3/2} + c.$$

• **PROBLEM 758**

Integrate the expression:

$$\int x \sqrt{1 - 3x} \, dx.$$

Solution: Here we use integration by parts. Then,

$$\int udv = uv - \int vdu.$$

$x = u$, $\sqrt{1 - 3x}\, dx = dv$, $dx = du$, and

$$v = \int \sqrt{1 - 3x}\, dx = -\frac{1}{3} \cdot \frac{2}{3}\left(1 - 3x\right)^{3/2}$$

$$= -\frac{2}{9}\left(1 - 3x\right)^{3/2}.$$ This is done by use of the integration formula:

$$\int u^n \, du = \frac{u^{n+1}}{n + 1},$$ letting $u = (1 - 3x)$, $du = -3dx$, and $n = \frac{1}{2}$. Substituting into the above equation, we obtain:

$$\int x \sqrt{1 - 3x}\, dx = -\frac{2x}{9}\left(1 - 3x\right)^{3/2} + \frac{2}{9}\int \left(1 - 3x\right)^{3/2} dx.$$

We now integrate, again using:

$$\int u^n \, du = \frac{u^{n+1}}{n + 1},$$ with $u = (1 - 3x)$, $du = -3\, dx$, and $n = \frac{3}{2}$, obtaining:

$$-\frac{2x}{9}\left(1 - 3x\right)^{3/2} - \frac{4}{135}\left(1 - 3x\right)^{5/2} + c.$$

• **PROBLEM 759**

Integrate the expression: $\int \sin (\ln x)\, dx.$

Solution: To integrate we must use integration by parts. We write:

$$\int udv = uv - \int vdu.$$

Let $u = \sin(\ln x)$ and $dv = dx$. Then

$du = \cos(\ln x)(1/x)dx$, and $v = \int dx = x$.

Substituting into the above equation we obtain:

$$\int \sin(\ln x)dx = x \sin(\ln x) - \int x \cdot \frac{1}{x} \cos(\ln x)dx$$

$$= x \sin(\ln x) - \int \cos(\ln x)dx$$

In the last integral, we use integration by parts once more, letting $u = \cos(\ln x)$, $dv = dx$,

$du = -\sin(\ln x) \cdot \frac{1}{x} dx$, $v = \int dx = x$.

Substituting we obtain:

$$\int \cos(\ln x)dx = x \cos(\ln x) + \int \sin(\ln x)\, dx.$$

Now, substituting for $\int \cos(\ln x)\, dx$, we obtain:

$$\int \sin(\ln x)dx = \frac{1}{2} x \sin(\ln x) - \frac{1}{2} x \cos(\ln x) + C$$

and substituting for: $\int \sin(\ln x)\, dx$, we obtain:

$$\int \cos(\ln x)dx = \frac{1}{2} x \sin(\ln x) + \frac{1}{2} x \cos(\ln x) + C.$$

• **PROBLEM 760**

$\frac{dy}{dx} = \cos^2 x$. Integrate by parts.

Solution: $dy = \cos^2 x \cdot dx$; $y = \int \cos^2 x \cdot dx$.

Using the equation for integration by parts, we have:

$$\int u\, dv = uv - \int v \cdot du .$$

Let $u = \cos x$

Then, $du = -\sin x \cdot dx$.

Let $dv = \cos x \cdot dx$.

Then, $v = \int dv = \int \cos x\, dx = \sin x$,

by use of the integration formula,

$$\int \cos u \, du = \sin u.$$

Substituting into the equation, we obtain:

$$y = \int \overset{u}{\cos x} \cdot \overset{dv}{\cos x \cdot dx} = \sin x \cdot \cos x - \int - \sin^2 x \cdot dx.$$

We now make use of the identities:

$$\sin x \cos x = \frac{\sin 2x}{2}, \text{ and } \sin^2 x = 1 - \cos^2 x.$$

We have: $$\int \cos^2 x \, dx = \frac{\sin 2x}{2} + \int (1 - \cos^2 x) dx.$$

$$\int \cos^2 x \cdot dx = \frac{\sin 2x}{2} + \int dx - \int \cos^2 x \cdot dx.$$

Adding the last term to the left side of the equation gives:

$$2 \int \cos^2 x \, dx = \frac{\sin 2x}{2} + x.$$

Division by 2 yields:

$$y = \int \cos^2 x \cdot dx = \frac{\sin 2x}{4} + \frac{x}{2} + C.$$

• PROBLEM 761

Integrate the expression:

$$\int x^2 \cos x \, dx.$$

Solution: Here we use integration by parts. We have:

$$\int udv = uv - \int vdu.$$

Let $u = x^2$, and $dv = \cos x \, dx$. Then, $du = 2x \, dx$, and

$$v = \int \cos x \, dx = \sin x,$$

so we have, by substituting into the equation:

$$\int x^2 \cos x \, dx = x^2 \sin x - 2 \int x \sin x \, dx + C.$$

The last integral can be evaluated by applying integration by parts once more.

Let $u = x$, $dv = \sin x \, dx$. Then, $du = dx$, and

$v = \int dv = \int \sin x \, dx = - \cos x$. Substituting in the above equation, we have:

$$\int x \sin x \, dx = (x)(-\cos x) - \int -\cos x \, dx$$

$$= -x \cos x + \int \cos x \, dx.$$

We now integrate, using the formula:

$$\int \cos u \, du = \sin u,$$

obtaining:

$$\int x \sin x \, dx = -x \cos x + \sin x + C.$$

Substituting, we obtain:

$$\int x^2 \cos x \, dx = x^2 \sin x - 2(-x \cos x + \sin x)$$

$$\int x^2 \cos x \, dx = x^2 \sin x + 2x \cos x - 2 \sin x + C.$$

• PROBLEM 762

$\frac{dy}{dx} = x^2 \sin x$. Integrate by parts.

Solution:

$$dy = x^2 \sin x \cdot dx \,.\; y = \int x^2 \sin x \, dx$$

Using the equation for integration by parts:

$$\int u \cdot dv = u \cdot v - \int v \cdot du.$$

Let $u = x^2$.

Then $du = 2x \cdot dx$.

Let $dv = \sin x \cdot dx$.

Then $v = \int dv = \int \sin x \, dx = -\cos x,$

by use of the integration formula:

$$\int \sin u \, du = -\cos u.$$

Substituting into the above equation, we obtain:

$$y = \int x^2 \sin x \cdot dx = -x^2 \cos x + \int 2x \cdot \cos x \cdot dx.$$

To integrate $\int x \cdot \cos x \cdot dx$, we use integration by parts again.

Let $u = x$

Then $du = dx$

Let $dv = \cos x \cdot dx$

Then $v = \int dv = \int \cos x\, dx = \sin x,$

by use of the integration formula:

$$\int \cos u\, du = \sin u.$$

By parts,

$$uv - \int v \cdot du = x \sin x - \int \sin x \cdot dx$$
$$= x \sin x + \cos x.$$

Substituting the value found for

$$\int x \cos dx,$$

we have:

$$y = \int x^2 \sin x\, dx = - x^2 \cos x + 2(x \sin x + \cos x)$$
$$= 2x \sin x + 2 \cos x - x^2 \cos x.$$

• PROBLEM 763

Integrate the expression: $\int e^x \sin x\, dx.$

Solution: We use integration by parts, applying the equation

$$\int udv = uv - \int vdu.$$

Let $u = e^x$, $dv = \sin x\, dx$. Then $du = e^x\, dx$,

$$v = \int dv = \int \sin x\, dx = - \cos x.$$

Substituting into the equation, we have:

$$\int e^x \sin x\, dx = - e^x \cos x + \int e^x \cos x\, dx + C.$$

The integral $\int e^x \cos x\, dx$ is treated the same way. We let $u = e^x$, $dv = \cos x\, dx$, $du = e^x\, dx$,

$v = \int dv = \int \cos x \, dx = \sin x$, obtaining:

$$\int e^x \cos x \, dx = e^x \sin x - \int e^x \sin x \, dx + C.$$

We now substitute the value for $\int e^x \cos x \, dx$ and obtain:

$$\int e^x \sin x \, dx = - e^x \cos x + e^x \sin x - \int e^x \sin x \, dx.$$

Adding $\int e^x \sin x \, dx$ to the left side of the equation, we have:

$$2 \int e^x \sin x \, dx = e^x(- \cos x + \sin x).$$

Dividing by 2, we obtain:

$$\int e^x \sin x \, dx = \frac{e^x}{2}(\sin x - \cos x) + C.$$

• PROBLEM 764

Integrate: $\int \varepsilon^{ax} \cdot \sin 3x \cdot dx.$

Solution: To integrate the given function we use integration by parts, employing the rule: $\int udv = uv - \int vdu$. Let

$$u = \varepsilon^{ax} \quad \text{and} \quad dv = \sin 3x \cdot dx$$

Then

$$du = a\varepsilon^{ax} \cdot dx \quad \text{and} \quad v = \int \sin 3x \cdot dx = -\frac{1}{3} \cos 3x,$$

by use of the formula $\int \sin u \, du = -\cos u$. Placing the values into the formula:

$$\int u \cdot dv = uv - \int v \cdot du,$$

we have:

$$\int \varepsilon^{ax} \cdot \sin 3x \cdot dx = \varepsilon^{ax} \cdot - \frac{1}{3} \cos 3x - \int - \frac{1}{3} \cos 3x \cdot a\varepsilon^{ax} \cdot dx \quad (1)$$

$$= -\frac{1}{3}\varepsilon^{ax} \cdot \cos 3x$$

$$+ \frac{a}{3} \int \varepsilon^{ax} \cdot \cos 3x \cdot dx$$

Now we apply integration by parts again, this time to $\int \varepsilon^{ax} \cdot \cos 3x \cdot dx$. Let $u = \varepsilon^{ax}$ and $dv = \cos 3x \cdot dx$.

Then $du = a\varepsilon^{ax} \cdot dx$ and $v = \int \cos 3x \cdot dx = \frac{1}{3} \sin 3x$.
Again: $\int u \cdot dv = uv - \int v \cdot du$

Therefore,

$$\int \varepsilon^{ax} \cdot \cos 3x \cdot dx = \varepsilon^{ax} \cdot \frac{1}{3} \sin 3x - \int \frac{1}{3} \sin 3x \cdot a\varepsilon^{ax} \cdot dx$$

$$= \frac{1}{3}\varepsilon^{ax} \cdot \sin 3x - \frac{a}{3} \int \varepsilon^{ax} \cdot \sin 3x \cdot dx.$$

Substitute the value found for $\int \varepsilon^{ax} \cos 3xdx$ into equation (1). Doing this we obtain:

$$\int \varepsilon^{ax} \cdot \sin 3xdx = -\frac{1}{3}\varepsilon^{ax} \cdot \cos 3x + \frac{a}{3}\left(\frac{1}{3}\varepsilon^{ax} \cdot \sin 3x - \frac{a}{3}\int \varepsilon^{ax} \cdot \sin 3x \cdot dx\right)$$

$$= -\frac{1}{3}\varepsilon^{ax} \cdot \cos 3x + \frac{a}{9} \cdot \varepsilon^{ax} \cdot \sin 3x - \frac{a^2}{9}\int \varepsilon^{ax} \cdot \sin 3x \cdot dx.$$

We can now add the two terms containing the integral, obtaining:

$$\int \varepsilon^{ax} \sin 3xdx + \frac{a^2}{9}\int \varepsilon^{ax} \sin 3xdx = -\frac{1}{3}\varepsilon^{ax} \cos 3x + \frac{a}{9}\varepsilon^{ax} \sin 3x,$$

and, factoring out the integral itself ($\int \varepsilon^{ax} \sin 3xdx$), we have:

$$\left(1 + \frac{a^2}{9}\right)\left(\int \varepsilon^{ax} \cdot \sin 3x \cdot dx\right) = -\frac{1}{3}\varepsilon^{ax} \cdot \cos 3x + \frac{a}{9}\varepsilon^{ax} \cdot \sin 3x.$$

Now divide by $\left(1 + \frac{a^2}{9}\right)$ and simplify, by factoring ε^{ax} from both terms in the numerator, and multiplying the fraction by $\frac{9}{9}$. Doing this we have:

$$\int \varepsilon^{ax} \cdot \sin 3x \cdot dx = \frac{\varepsilon^{ax}(a \sin 3x - 3 \cos 3x)}{9 + a^2} + C.$$

The method of integration by parts sometimes fails because it leads back to the original integral.

Solution: For example, try to integrate

$$\int x^{-1}\, dx \text{ by parts.}$$

We use the equation:

$$\int udv = uv - \int vdu.$$

Let $u = x^{-1}$ and $dv = dx$. Then $du = -x^{-2}\, dx$ and $v = \int dx = x$. Substituting, we obtain:

$$\int x^{-1}\, dx = x \cdot x^{-1} - \int x \cdot (-x^{-2})\, dx$$

$$\int x^{-1}\, dx = 1 + \int x^{-1}\, dx,$$

and we are back where we started. When integration by parts fails, we must consider another approach. Let us look at a more direct approach to this problem. We have,

$$\int x^{-1}\, dx,$$

which can be written as

$$\int \frac{dx}{x}.$$

To this we can apply the formula,

$$\int \frac{du}{u} = \ln u + C,$$

obtaining:

$$\int x^{-1}\, dx = \ln x + C.$$

CHAPTER 28

IMPROPER INTEGRALS

When an integral is improper, the integrand has one or more points of discontinuity, or at least one of the limits is infinite. Hence we cannot simply integrate and substitute in the limits as we would do with proper integrals. What we can do, however, is work at first with some finite number, such as u, perform the integration as we would for any finite number, and then, as the last step, allow u to go to ∞. In this way, the rules for integration are not violated. It may turn out, despite the function going to infinity, that the integral is finite. If the result is indeterminate, we can apply L'Hospital's rule, and if it turns out to be infinite, we can conclude that it does not have a limit.

• PROBLEM 766

Evaluate the integral: $\int_1^2 \frac{dx}{\sqrt{x-1}}$ if it converges.

Solution: Since $\frac{1}{\sqrt{x-1}}$ has an infinite discontinuity at x = 1, we have, by definition:

$$\int_1^2 \frac{dx}{\sqrt{x-1}} = \lim_{\varepsilon \to 0^+} \int_{1+\varepsilon}^2 \frac{dx}{\sqrt{x-1}} = \lim_{\varepsilon \to 0^+} \int_{1+\varepsilon}^2 (x-1)^{-1/2} dx.$$

We can now integrate, using the formula:

$$\int u^n du = \frac{u^{n+1}}{n+1},$$

letting u = (x-1), du = dx and $n = -\frac{1}{2}$. Applying the formula, we obtain:

$$\lim_{\varepsilon \to 0^+} \left[2(x-1)^{1/2} \right]_{1+\varepsilon}^2$$

$$= 2(2-1)^{1/2} - \lim_{\varepsilon \to 0^+} \left[2(1+\varepsilon-1)^{1/2} \right] = 2 - \lim_{\varepsilon \to 0^+} (2\sqrt{\varepsilon}).$$

As $\varepsilon \to 0^+$, the limit approaches 0. Therefore, the integral converges to the value 2.

• PROBLEM 767

Evaluate the integral: $\int_0^1 x \ln x\, dx$ if it is convergent.

Solution: The integrand has a discontinuity at the lower limit. By definition:

$$\int_0^1 x \ln x dx = \lim_{\varepsilon \to 0^+} \int_\varepsilon^1 x \ln x dx.$$

To integrate this function we use integration by parts, applying the formula, $\int u dv = uv - \int v du$. We let $u = \ln x$, $dv = xdx$. Then $du = \frac{1}{x}dx$ and $v = \int dv = \int xdx = \frac{x^2}{2}$, by the formula for $\int u^n du$. Substituting these values into the formula, we obtain:

$$\int x \ln x dx = \ln x \cdot \frac{x^2}{2} - \int \frac{x^2}{2} \cdot \frac{1}{x} dx = \frac{x^2}{2} \ln x - \frac{1}{2}\int xdx.$$

Integrate, again using the formula for $\int u^n du$, obtaining:

$$\lim_{\varepsilon \to 0^+} \left[\frac{1}{2}x^2 \ln x - \frac{1}{4}x^2\right]_\varepsilon^1$$

$$= \lim_{\varepsilon \to 0^+} \left[\frac{1}{2} \ln 1 - \frac{1}{4} - \frac{1}{2}\epsilon^2 \ln \epsilon - \frac{1}{4}\varepsilon^2\right].$$

We must recall that $\ln 1 = 0$, and $\lim_{\varepsilon \to 0^+} \left(-\frac{1}{4}\varepsilon^2\right) = 0$.

We now have:

$$\int_0^1 x \ln x dx = 0 - \frac{1}{4} - \frac{1}{2} \lim_{\varepsilon \to 0^+} \varepsilon^2 \ln \varepsilon - 0.$$

To evaluate,

$$\lim_{\varepsilon \to 0^+} \varepsilon^2 \ln \varepsilon = \lim_{\varepsilon \to 0^+} \frac{\ln \varepsilon}{\frac{1}{\varepsilon^2}},$$

we apply L'Hôpital's rule because $\lim_{\varepsilon \to 0^+} \ln \varepsilon = -\infty$ and $\lim_{\varepsilon \to 0^+} \frac{1}{\varepsilon^2} = +\infty$. We have $\frac{-\infty}{\infty}$, which is an indeterminate form. The rule states that, if $\frac{f_1}{f_2}$ is indeterminate, we can evaluate $\frac{f_1'}{f_2'}$. Here,

$$\frac{d}{dx}(\ln u) = \frac{1}{u}\frac{du}{dx}$$

in the numerator, and

$$\frac{d}{dx}\left(\frac{1}{u^2}\right) = \frac{d}{dx}\left(u^{-2}\right) = -\frac{2}{u^3}\frac{d}{dx}(u)$$

in the denominator. Replacing u by ε, we obtain:

$$\lim_{\varepsilon \to 0^+} \frac{\ln \varepsilon}{\frac{1}{\varepsilon^2}} = \lim_{\varepsilon \to 0^+} \frac{\frac{1}{\varepsilon}}{-\frac{2}{\varepsilon^3}} = \lim_{\varepsilon \to 0^+}\left[-\frac{\varepsilon^2}{2}\right] = 0.$$

Therefore, by substitution, we have:

$$\int_0^1 x \ln x dx = 0 - \frac{1}{4} - \frac{1}{2}(0) - 0 = -\frac{1}{4}.$$

• **PROBLEM** 768

Evaluate the integral: $\int_{-1}^{1} \frac{dx}{x^2}$.

Solution: The integrand becomes infinite at $x = 0$, a point between the upper and lower limits of the integral. We therefore substitute for the given integral, $\int_{-1}^{1} \frac{dx}{x^2}$, two integrals defined as:

$$\lim_{t \to 0}\int_{-1}^{-t} \frac{dx}{x^2} + \lim_{t' \to 0}\int_{t'}^{1} \frac{dx}{x^2}$$

$$= \lim_{t \to 0}\int_{-1}^{-t} x^{-2}dx + \lim_{t' \to 0}\int_{t'}^{1} x^{-2}dx.$$

Applying the formula for $\int u^n du$ we obtain:

$$\lim_{t \to 0}\left[-\frac{1}{x}\right]_{-1}^{-t} + \lim_{t' \to 0}\left[-\frac{1}{x}\right]_{t'}^{1}$$

$$= \lim_{t \to 0}\left[\frac{1}{t} - \left(-\frac{1}{1}\right)\right] + \lim_{t' \to 0}\left[-\frac{1}{1} + \frac{1}{t'}\right]$$

$$= \lim_{t \to 0}\left(1 + \frac{1}{t}\right) + \lim_{t' \to 0}\left(-1 + \frac{1}{t'}\right).$$

We see that, as t and t' approach 0, both quantities become infinite, so that neither limit exists. Therefore, the given integral diverges, hence it cannot be evaluated. Failure to recognize this fact would lead to substitution of the limits and an entirely erroneous result.

• **PROBLEM** 769

Evaluate the integral: $\int_0^2 \frac{dx}{(x-1)^2}$ if it is convergent.

Solution: The integrand has an infinite discontinuity at x = 1. By definition:

$$\int_0^2 \frac{dx}{(x-1)^2} = \lim_{\varepsilon\to 0^+}\int_0^{1-\varepsilon} \frac{dx}{(x-1)^2} + \lim_{\delta\to 0^+}\int_{1+\delta}^2 \frac{dx}{(x-1)^2}$$

$$= \lim_{\varepsilon\to 0^+}\int_0^{1-\varepsilon} (x-1)^{-2}dx + \lim_{\delta\to 0^+}\int_{1+\delta}^2 (x-1)^{-2}dx.$$

To integrate we apply the formula for $\int u^n du$, letting $u = (x-1)$, $du = dx$ and $n = -2$. We obtain:

$$\lim_{\varepsilon\to 0^+}\left[-\frac{1}{x-1}\right]_0^{1-\varepsilon} + \lim_{\delta\to 0^+}\left[-\frac{1}{x-1}\right]_{1+\delta}^2$$

$$= \lim_{\varepsilon\to 0^+}\left[-\frac{1}{(1-\varepsilon)-1} - \left(-\frac{1}{0-1}\right)\right] + \lim_{\delta\to 0^+}\left[-\frac{1}{2-1} - \left(-\frac{1}{1+\delta-1}\right)\right]$$

$$= \lim_{\varepsilon\to 0^+}\left[\frac{1}{\varepsilon} - 1\right] + \lim_{\delta\to 0^+}\left[-1 + \frac{1}{\delta}\right].$$

As both ε and δ approach 0^+, the limits approach ∞. Because neither of these limits exist, the integral is divergent and cannot be evaluated.

• PROBLEM 770

Evaluate the integral: $\int_{-1}^2 \frac{dx}{(x-1)^2}$, if possible.

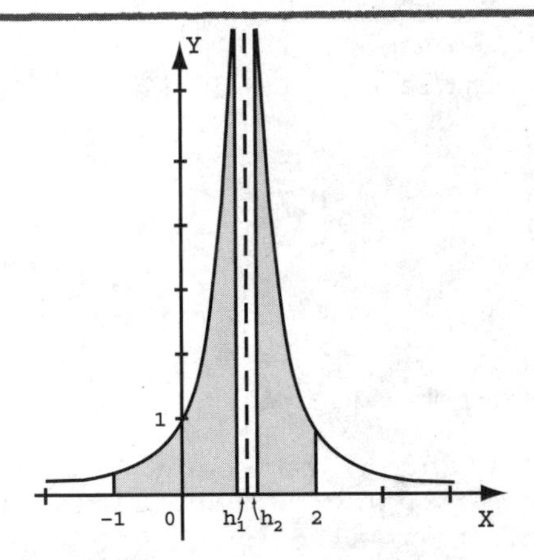

Solution: The graph of the curve represented by $y = \frac{1}{(x-1)^2}$ has a vertical asymptote, x = 1. Hence the integrand is discontinuous at x = 1, and by definition:

$$\int_{-1}^2 \frac{dx}{(x-1)^2} = \lim_{h_1\to 0}\int_{-1}^{1-h_1} \frac{dx}{(x-1)^2}$$

$$+ \lim_{h_2 \to 0} \int_{1+h_2}^{2} \frac{dx}{(x-1)^2}, \quad h_1, h_2 > 0.$$

We now integrate, using the formula:

$$\int u^n du = \frac{u^{n+1}}{n+1},$$

letting $u = (x-1)$, $du = dx$ and $n = -2$. We obtain:

$$\lim_{h_1 \to 0} \left[-\frac{1}{x-1} \right]_{-1}^{1-h_1} + \lim_{h_2 \to 0} \left[-\frac{1}{x-1} \right]_{1+h_2}^{2}$$

$$= \lim_{h_1 \to 0} \left(\frac{1}{h_1} - \frac{1}{2} \right) + \lim_{h_2 \to 0} \left(-1 + \frac{1}{h_2} \right) = \infty,$$

since $\frac{1}{h_1}$ and $\frac{1}{h_2} \to +\infty$ as h_1 and $h_2 \to 0$. Therefore the integral is divergent, and in this case the area under the curve has no meaning.

• PROBLEM 771

Evaluate the integral:

$$\int_1^{+\infty} \frac{dx}{x\sqrt{x^2 - 1}}$$

if it is convergent.

__Solution:__ For this integral, there is an infinite upper limit and there is also an infinite discontinuity of the integrand at the lower limit.
The number 2 is arbitrarily chosen to make two integrals out of one, so that each integral only approaches one limit.

$$\int_1^{+\infty} \frac{dx}{x\sqrt{x^2 - 1}} = \lim_{a \to 1^+} \int_a^2 \frac{dx}{x\sqrt{x^2 - 1}} + \lim_{b \to +\infty} \int_2^b \frac{dx}{x\sqrt{x^2 - 1}}.$$

To integrate, we apply the formula, $\int \frac{du}{u\sqrt{u^2 - a^2}} = \frac{1}{a} \sec^{-1} \frac{u}{a}$, letting $u = x$, $du = dx$ and $a = 1$. We obtain:

$$\lim_{a \to 1^+} \left[\sec^{-1} x \right]_a^2 + \lim_{b \to +\infty} \left[\sec^{-1} x \right]_2^b$$

$$= \lim_{a \to 1^+} \left(\sec^{-1} 2 - \sec^{-1} a \right) + \lim_{b \to +\infty} \left(\sec^{-1} b - \sec^{-1} 2 \right).$$

We recall that $\sec^{-1} 2 = \frac{\pi}{3}$, $\sec^{-1} 1 = 0$ and $\sec^{-1} \infty$

$= \frac{\pi}{2}$. Substituting, we obtain:

$$\frac{\pi}{3} - \lim_{a \to 1^+} (\sec^{-1} a) + \lim_{b \to \infty} \sec^{-1} b - \frac{\pi}{3}$$
$$= -\lim_{a \to 1^+} \sec^{-1} a + \lim_{b \to +\infty} \sec^{-1} b.$$

As $a \to 1$, $(\sec^{-1} a)$ approaches 0, and as $b \to \varphi$ $(\sec^{-1} b)$ approaches $\frac{\pi}{2}$. We therefore have the integral equal to

$$0 + \frac{1}{2}\pi = \frac{1}{2}\pi.$$

Hence the integral converges to the value $\frac{\pi}{2}$.

• PROBLEM 772

Evaluate the integral: $\int_1^\infty \frac{dx}{x}$.

Solution: The integral is defined as:

$$\lim_{t \to \infty} \int_1^t \frac{dx}{x} = \lim_{t \to \infty} \left[\ln x\right]_1^t = \ln t - \ln 1$$
$$= \ln t - 0 = \ln t.$$

Therefore, $\int_1^t \frac{dx}{x} = \ln t$, and since $\ln t \to +\infty$ as $t \to +\infty$, the given integral is divergent, since a limit of ∞ indicates that the limit does not exist. The integral thus cannot be evaluated.

• PROBLEM 773

Evaluate the integral: $\int_1^\infty \frac{1}{\sqrt{x}} dx$, if it converges.

Solution: By definition:

$$\int_1^\infty \frac{1}{\sqrt{x}} dx = \lim_{A \to \infty} \int_1^A \frac{1}{\sqrt{x}} dx = \int_1^A (x)^{-1/2} dx.$$

This can be integrated by using the formula for $\int u^n du$, letting $u = x$, $du = dx$, $n = -\frac{1}{2}$. We obtain:

$$\lim_{A \to \infty} \left[\frac{x^{1/2}}{\frac{1}{2}}\right]_1^A = \lim_{A \to \infty} \left[2\sqrt{x}\right]_1^A = \lim_{A \to \infty} (2\sqrt{A} - 2).$$

As $A \to \infty$, $(2\sqrt{A} - 2)$ also approaches infinity. Therefore the limit does not exist, and the integral does not converge.

• PROBLEM 774

Evaluate the improper integral: $\int_0^\infty \frac{dx}{1 + x^2}$.

<u>Solution:</u> The integral is defined as:

$$\lim_{t\to\infty} \int_0^t \frac{dx}{1 + x^2}.$$

We notice that $(1 + x^2)$ is the sum of two squares, hence we use the formula

$$\int \frac{du}{a^2 + u^2} = \frac{1}{a} \text{arc tan } \frac{u}{a},$$

with $a = 1$, $u = x$, $du = dx$. We obtain:

$$\lim_{t\to\infty} \int_0^t \frac{dx}{1 + x^2} = \lim_{t\to\infty} \Big[\text{arc tan } x\Big]_0^t$$

$$= \lim_{t\to\infty} \text{arc tan } t - \text{arc tan } 0$$

$$= \lim_{t\to\infty} \text{arc tan } t - 0$$

$$= \lim_{t\to\infty} \text{arc tan } t.$$

Since arc tan $t \to \frac{1}{2}\pi$ as $t \to +\infty$, the given integral converges to the value $\frac{1}{2}\pi$.

• PROBLEM 775

Evaluate the improper integral: $\int_0^\infty xe^{-x^2} dx.$

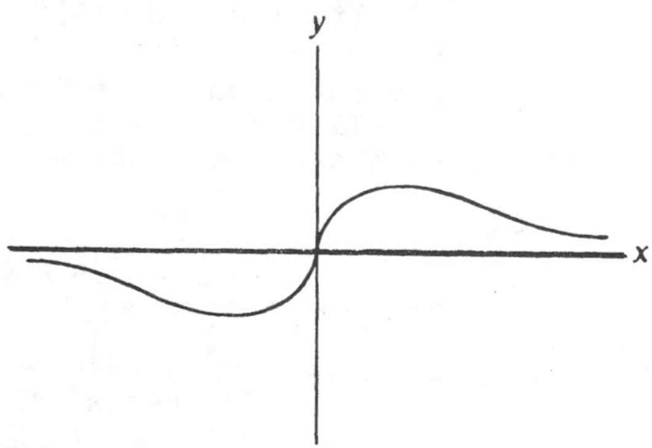

<u>Solution:</u> By definition,

$$\int_0^\infty xe^{-x^2} dx = \lim_{t\to\infty} \int_0^t xe^{-x^2} dx.$$

Using the formula $\int e^u du = e^u + C$ we obtain:

$$\int_0^\infty xe^{-x^2} dx = \lim_{t\to\infty} -\frac{1}{2}\Big[e^{-x^2}\Big]_0^t$$

$$= -\frac{1}{2}\Big[e^{-t^2} - e^0\Big]$$

$$= -\frac{1}{2}\left[\frac{1}{e^{t^2}} - 1\right]$$

$$= \frac{1}{2}\left[1 - \frac{1}{e^{t^2}}\right].$$

We see that as $t \to \infty$, $e^{t^2} \to \infty$, $\frac{1}{e^{t^2}} \to 0$, and the limit approaches $\frac{1}{2}$. Therefore,

$$\int_0^{\infty} xe^{-x^2} dx$$

converges and is equal to $\frac{1}{2}$.

• PROBLEM 776

Determine whether the integral:

$$\int_0^{+\infty} \sin x dx$$

is convergent or divergent.

Solution: By definition:

$$\int_0^{+\infty} \sin x dx = \lim_{b \to +\infty} \int_0^b \sin x dx.$$

We now apply the integration formula: $\int \sin u du = -\cos u$, letting $u = x$ and $du = dx$. We obtain:

$$\lim_{b \to +\infty}\Big[-\cos x\Big]_0^b = \lim_{b \to +\infty}\left[-\cos b - (-\cos 0)\right]$$
$$= \lim_{b \to +\infty}(-\cos b + 1).$$

For any integer n, as b takes on all values from $n\pi$ to $2n\pi$, cos b takes on all values from -1 to 1. Hence, $\lim_{b \to +\infty} \cos b$ does not exist. Therefore, the integral is divergent.

• PROBLEM 777

Evaluate the infinite integral: $\int_{-\infty}^{\infty} \frac{dx}{x^2 + 2x + 2}$.

Solution: The integral, being continuous, is defined as:

$$\int_{-\infty}^{+\infty} \frac{dx}{x^2+2x+2} = \lim_{t \to \infty} \int_0^t \frac{dx}{x^2+2x+2} + \lim_{t' \to -\infty} \int_{t'}^0 \frac{dx}{x^2+2x+2}.$$

We write the denominator, x^2+2x+2, as the sum of two squares, so as to use the formula. $\int \frac{du}{a^2+u^2} = \frac{1}{a} \arctan \frac{u}{a}$. Completing the square, we obtain: $x^2+2x+2 = (x^2+2x+2-1) + 1 = (x+1)^2 + (1)^2$. Now let $a = 1$, $u = (x+1)$, $du = dx$, and apply the formula. We have:

$$\lim_{t\to\infty}\Big[\text{arc tan}(x+1)\Big]_0^t + \lim_{t'\to-\infty}\Big[\text{arc tan}(x+1)\Big]_{t'}^0$$

$$= \Big[\text{arc tan}(t+1) - \text{arc tan}(0+1)\Big]$$

$$+ \Big[\text{arc tan}(0+1) - \text{arc tan}(t'+1)\Big]$$

$$= \Big[\text{arc tan}(t+1) - \frac{\pi}{4}\Big] + \Big[\frac{\pi}{4} - \text{arc tan}(t'+1)\Big].$$

Now, as $t \to \infty$, arc tan $(t+1) \to \frac{\pi}{2}$. As $t' \to -\infty$, arc $\tan(t'+1) \to -\frac{\pi}{2}$. We can substitute the values found for arc $\tan(t+1)$ and arc $\tan(t'+1)$, keeping in mind that t and t' approach these values, but will never actually reach them. We have:

$$\Big[\text{arc tan}(t+1) - \frac{1}{4}\pi\Big] + \Big[\frac{1}{4}\pi - \text{arc tan}(t'+1)\Big]$$

$$= \frac{1}{2}\pi - \frac{1}{4}\pi + \frac{1}{4}\pi - \left(-\frac{1}{2}\pi\right) = \frac{1}{2}\pi + \frac{1}{2}\pi = \pi.$$

The given integral is therefore convergent with the value π.

• PROBLEM 778

Find the value of the integral: $\int_{-\infty}^{\infty}\frac{x}{x^2+1}dx$, provided that it converges.

Solution: By definition,

$$\int_{-\infty}^{\infty}\frac{x}{x^2+1}dx = \lim_{t\to\infty}\int_a^t\frac{x}{x^2+1}dx + \lim_{t'\to-\infty}\int_{t'}^a\frac{x}{x^2+1}dx,$$

with the value a to be chosen. In this example we shall use a = 0. Then

$$\int_{-\infty}^{\infty}\frac{x}{x^2+1} = \lim_{t\to\infty}\int_0^t\frac{xdx}{x^2+1} + \lim_{t'\to-\infty}\int_{t'}^0\frac{xdx}{x^2+1}.$$

We can now integrate, using the formula, $\int\frac{du}{u} = \ln|u| + C$, and we have:

$$\lim_{t\to\infty}\frac{1}{2}\Big[\ln(x^2+1)\Big]_0^t + \lim_{t'\to-\infty}\frac{1}{2}\Big[\ln(x^2+1)\Big]_{t'}^0$$

$$= \lim_{t\to\infty}\frac{1}{2}\Big[\ln(t^2+1) - \ln(0+1)\Big]$$

$$+ \lim_{t'\to-\infty}\frac{1}{2}\Big[\ln(0+1) - \ln(t'^2+1)\Big]$$

Using the fact that ln 1 = 0, we **obtain**:

$$\lim_{t\to\infty}\frac{1}{2}\Big[\ln(t^2+1)\Big] + \lim_{t'\to-\infty}\frac{1}{2}\Big[-\ln(t'^2+1)\Big].$$

We can now observe that as $t \to \infty$ and $t' \to -\infty$, $\ln(t^2+1) \to \infty$ and $\ln(t'^2+1) \to \infty$. This means that the limit does not exist. Since this limit does not exist,

$$\int_0^{\infty} \frac{x}{x^2+1}dx + \int_{-\infty}^{0} \frac{x}{x^2+1}dx$$

diverges, and consequently

$$\int_{-\infty}^{\infty} \frac{x}{x^2+1}dx$$

diverges.

• PROBLEM 779

Evaluate the integral: $\displaystyle\int_{-\infty}^{+\infty} \frac{dx}{x^2 + 6x + 12}$ if it converges.

Solution: By definition we can rewrite the given integral.

$$\int_{-\infty}^{+\infty} \frac{dx}{x^2+6x+12} = \lim_{a\to-\infty} \int_a^0 \frac{dx}{(x+3)^2+3} + \lim_{b\to+\infty} \int_0^b \frac{dx}{(x+3)^2+3},$$

completing the square in the denominator so as to obtain two integrals in the form $\displaystyle\int \frac{du}{a^2+u^2}$. Applying the formula: $\displaystyle\int \frac{du}{a^2+u^2} = \frac{1}{a}\tan^{-1}\frac{u}{a}$, with $u = (x+3)$, $du = dx$, and $a = \sqrt{3}$, we obtain:

$$\lim_{a\to-\infty}\left[\frac{1}{\sqrt{3}}\tan^{-1}\frac{x+3}{\sqrt{3}}\right]_a^0 + \lim_{b\to+\infty}\left[\frac{1}{\sqrt{3}}\tan^{-1}\frac{x+3}{\sqrt{3}}\right]_0^b.$$

Evaluating at the limits, we have:

$$\lim_{a\to-\infty}\left(\frac{1}{\sqrt{3}}\tan^{-1}\sqrt{3} - \frac{1}{\sqrt{3}}\tan^{-1}\frac{a+3}{\sqrt{3}}\right)$$

$$+ \lim_{b\to+\infty}\left(\frac{1}{\sqrt{3}}\tan^{-1}\frac{b+3}{\sqrt{3}} - \frac{1}{\sqrt{3}}\tan^{-1}\sqrt{3}\right)$$

$$= \frac{1}{\sqrt{3}}\left[\lim_{a\to-\infty}\left(\tan^{-1}\sqrt{3} - \tan^{-1}\frac{a+3}{\sqrt{3}}\right)\right.$$

$$\left.+ \lim_{b\to\infty}\left(\tan^{-1}\frac{b+3}{\sqrt{3}} - \tan^{-1}\sqrt{3}\right)\right].$$

We now make use of the fact that $\tan^{-1}\sqrt{3} = \frac{\pi}{3}$ and $\tan^{-1}\infty = \frac{\pi}{2}$. We obtain:

$$\frac{1}{\sqrt{3}}\left\{\left[\frac{\pi}{3} - \left(-\frac{\pi}{2}\right)\right] + \left[\frac{\pi}{2} - \frac{\pi}{3}\right]\right\} = \frac{1}{\sqrt{3}}\left[\frac{\pi}{3} + \frac{\pi}{2} + \frac{\pi}{2} - \frac{\pi}{3}\right] =$$

$$\frac{1}{\sqrt{3}} \cdot \pi = \frac{\pi}{\sqrt{3}}.$$

• PROBLEM 780

Evaluate the improper integral: $\int_{-\infty}^{+\infty} \frac{dx}{(1 + x^2)}$.

Solution: Let $t_1 < 0$ and $t_2 > 0$. By definition,

$$\int_{-\infty}^{+\infty} \frac{dx}{1+x^2} = \lim_{t_2 \to +\infty} \int_0^{t_2} \frac{dx}{1+x^2} + \lim_{t_1 \to -\infty} \int_{t_1}^{0} \frac{dx}{1+x^2}.$$

To integrate, we use the formula:

$$\int \frac{du}{a^2+u^2} = \frac{1}{a} \text{ arc tan } \frac{u}{a},$$

letting $u = x$, $du = dx$ and $a = 1$. Applying the formula, we obtain:

$$\int_{-\infty}^{+\infty} \frac{dx}{1+x^2} = \lim_{t \to +\infty} \Big[\text{arc tan } x\Big]_0^{t_2} + \lim_{t' \to -\infty} \Big[\text{arc tan } x\Big]_{t_1}^{0}$$

$$= \lim_{t_2 \to +\infty} \left(\text{arc tan } t_2\right) - \text{arc tan } 0$$

$$+ \text{arc tan } 0 - \lim_{t_1 \to -\infty} \left(\text{arc tan } t_1\right).$$

We recall that arc tan $0 = 0$, arc tan $\infty = \frac{\pi}{2}$, and arc tan $-\infty = -\frac{\pi}{2}$. Using these facts, we have:

$$\lim_{t_2 \to +\infty} \left(\text{arc tan } t_2\right) - 0 + 0 - \lim_{t_1 \to -\infty} \left(\text{arc tan } t_1\right)$$

$$= \frac{\pi}{2} - 0 + 0 - \left(-\frac{\pi}{2}\right) = \pi.$$

Hence $\int_{-\infty}^{+\infty} \frac{dx}{(1+x^2)}$ converges and $\int_{-\infty}^{+\infty} \frac{dx}{(1+x^2)} = \pi$.

• PROBLEM 781

Evaluate the integral: $\int_{-\infty}^{\infty} xe^{-x^2} dx$, if it converges.

Solution: For convenience, we break this infinite range into two parts. We can write:

$$\int_{-\infty}^{\infty} xe^{-x^2} dx = \int_{-\infty}^{0} xe^{-x^2} dx + \int_0^{\infty} xe^{-x^2} dx,$$

which, by definition,

$$= \lim_{A' \to -\infty} \int_{A'}^{0} xe^{-x^2} dx + \lim_{A \to \infty} \int_0^{A} xe^{-x^2} dx.$$

We can now integrate, using the formula: $\int e^u du = e^u$, letting $u = -x^2$ and $du = -2xdx$. We obtain:

$$\lim_{A'\to-\infty}\left[\frac{e^{-x^2}}{2}\right]_{A'}^{0} + \lim_{A\to\infty}\left[-\frac{e^{-x^2}}{2}\right]_{0}^{A}$$

$$= \lim_{A'\to-\infty}\left(\frac{e^{-A'^2}}{2} - \frac{1}{2}\right) + \lim_{A\to\infty}\left(\frac{1}{2} - \frac{e^{-A^2}}{2}\right).$$

Now, as $A' \to -\infty$ and $A \to \infty$, $\frac{e^{-A'^2}}{2}$ and $\frac{e^{-A^2}}{2}$ approach 0, so that we obtain:

$$-\frac{1}{2} + \frac{1}{2} = 0.$$

The integral therefore converges to 0.

CHAPTER 29

ARC LENGTH

If we consider a curve in a plane, we can cut it into small sections. One of these sections may be designated with length Δs, where $(\Delta s)^2 = (\Delta x)^2 + (\Delta y)^2$ by the Pythagorean theorem. Allowing the section Δs to approach zero as a limit, yields the expression:

$$(ds)^2 = (dx)^2 + (dy)^2 .$$

Taking the square root, and bringing the dx factor outside, we have:

$$ds = \sqrt{1 + \left(\frac{dy}{dx}\right)^2} \cdot dx \ .$$

Hence, if we can obtain the derivative of y with respect to x, we have an expression for ds. That expression can then be integrated to obtain s,

$$s = \int_a^b \sqrt{1 + \left(\frac{dy}{dx}\right)^2} \cdot dx \ ,$$

where a and b are the values of x at the end points of the curve. The quantity s is the length of the arc.

Similar relations have been derived for polar coordinates.

• PROBLEM 782

Find the length of the curve $y^2 = -4x$ from (-4,4) to (0,0).

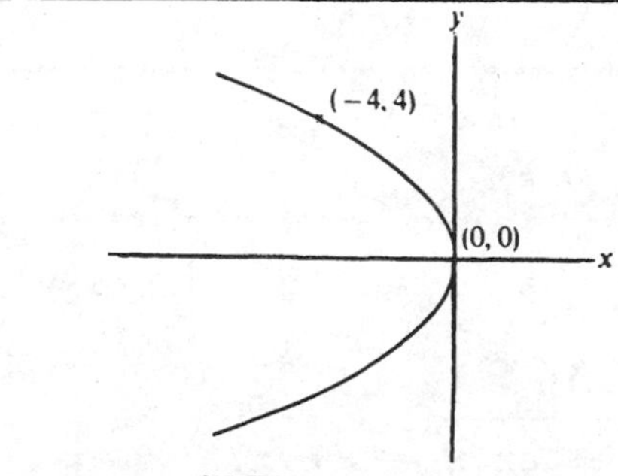

Solution:

$$s = \int_a^b \sqrt{1 + \left(\frac{dx}{dy}\right)^2} \ dy.$$

Since $y^2 = -4x$, $x = -\frac{y^2}{4}$, $\frac{dx}{dy} = -\frac{y}{2}$.

$$s = \int_0^4 \sqrt{1 + \frac{y^2}{4}}\, dy = \frac{1}{2} \int_0^4 \sqrt{4+y^2}\, dy$$

$\int_0^4 \sqrt{4+y^2}\, dy$ is of the form: $\int \sqrt{u^2+a^2}\, du$, **which is equal to:**

$$\frac{1}{2} u\sqrt{u^2+a^2} + \frac{1}{2} a^2 \ln\left(u + \sqrt{u^2+a^2}\right).$$

$$s = \frac{1}{2}\left[\frac{y}{2}\sqrt{4+y^2} + \frac{4}{2} \ln(y+\sqrt{4+y^2})\right]_0^4$$

$$= \frac{1}{2}\left[2\sqrt{20} + 2 \ln(4+\sqrt{20}) - 2 \ln 2\right]$$

$$= 2\sqrt{5} + \ln(2 + \sqrt{5}).$$

• PROBLEM 783

Find the length of the curve: $y = x^{3/2}$, from $x = 0$ to $x = 4$.

Solution: The formula for arc length with respect to the x-axis is:

$$s = \int_a^b \sqrt{1 + \left(\frac{dy}{dx}\right)^2}\, dx.$$

Since $y = x^{3/2}$, $y' = \frac{3}{2} x^{1/2}$.

$$s = \int_0^4 \sqrt{1 + \frac{9}{4} x}\, dx$$

$$= \frac{1}{2} \int_0^4 \sqrt{4+9x}\, dx = \left(\frac{1}{9}\right)\frac{1}{2}\left(\frac{2}{3}\right) (4+9x)^{3/2}\Big]_0^4$$

$$= \frac{1}{27}\left(40^{3/2} - 4^{3/2}\right) = \frac{8}{27}\left(10^{3/2} - 1\right).$$

• PROBLEM 784

Find the length of arc of the curve $y = \ln \cos x$ from $x = 0$ to $x = \frac{1}{4}\pi$.

Solution: We use the formula: $s = \int_a^b \sqrt{1+y'^2}\, dx$.

$$y = \ln \cos x. \quad y' = -\tan x.$$

$$\sqrt{1+y'^2} = \sqrt{1+\tan^2 x} = \sec x.$$

Therefore,

$$s = \int_0^{\pi/4} \sec x\, dx = \ln(\sec x + \tan x)\Big|_0^{\pi/4}$$

$$= \ln(\sqrt{2} + 1).$$

Find the length of the curve: $y = x^2$, from $x = 2$ to $x = 5$.

Solution: **Applying the expression:**

$$s = \int_a^b \sqrt{1 + \left(\frac{dy}{dx}\right)^2}\, dx,$$

we have $\quad y = x^2. \quad y' = 2x.$

Therefore, $\quad s = \int_2^5 \sqrt{1 + 4x^2}\, dx.$

Integrating by parts and recalling

$$\int udv = uv - \int vdu,$$

let $\quad u = (1 + 4x^2)^{1/2}, \quad du = 4x(1 + 4x^2)^{-1/2},$

$dv = dx, \quad v = x.$

$$\int_2^5 (1+4x^2)^{1/2} dx$$

$$= (1+4x^2)^{1/2} x - \int_2^5 \frac{4x^2}{(1+4x^2)^{1/2}} dx. \qquad (1)$$

But $\quad \int_2^5 (1+4x^2)^{1/2} dx = \int_2^5 \frac{(1+4x^2)}{(1+4x^2)^{1/2}} dx.$

After expanding the right side, we have:

$$\int_2^5 (1+4x^2)^{1/2} dx$$

$$= \int_2^5 \frac{dx}{(1+4x^2)^{1/2}} + \int_2^5 \frac{4x^2 dx}{(1+4x^2)^{1/2}}. \qquad (2)$$

By adding equations (1) and (2), the term:

$$\int \frac{4x^2 dx}{(1+4x^2)^{1/2}},$$

is eliminated.

At this point we obtain:

$$2\int_2^5 (1+4x^2)^{1/2} dx$$

$$= x(1+4x^2)^{1/2} + \int_2^5 \frac{dx}{(1+4x^2)^{1/2}},$$

or, $$\int_2^5 (1+4x^2)^{1/2}dx$$

$$= \frac{x}{2}(1+4x^2)^{1/2} + \int_2^5 \frac{dx}{\left(\frac{1}{4}+x^2\right)^{1/2}} \cdot$$

Note that
$$\int \frac{dx}{\left(\frac{1}{4}+x^2\right)^{1/2}}$$

is of the form:
$$\int \frac{dx}{(a^2+x^2)^{1/2}} = \ln\left(x+\sqrt{a^2-x^2}\right)$$

Thus, $$\int \frac{dx}{\left(\frac{1}{4}+x^2\right)^{1/2}} = \ln\left(x+\sqrt{\frac{1}{4}-x^2}\right).$$

$$\int_2^5 (1+4x^2)^{1/2}dx$$

reduces to: $$\left[\frac{x}{2}(1+4x^2)^{1/2} + \ln\left(x+\sqrt{\frac{1}{4}+x^2}\right)\right]_2^5 .$$

This expression, evaluated at $x = 5$, is equal to 27.43. At $x = 2$, it is equal to 5.52.

$$\int_2^5 (1+4x^2)^{1/2}dx = 27.43 - 5.52$$
$$= 21.91 \text{ units of arc}$$

• **PROBLEM 786**

Find the length of the arc of the curve: $y = \frac{2}{3}x^{3/2}$, from $x = 3$ to $x = 8$.

Solution: The arc length is given by:

$$s = \int_a^b \sqrt{1+y'^2}\,dx.$$

$$y' = \frac{2}{3}\cdot\frac{3}{2}x^{1/2} = \sqrt{x}. \quad 1+y'^2 = 1+x,$$

and therefore:
$$s = \int_3^8 \sqrt{1+x}\,dx$$

$$= \int_3^8 (1+x)^{1/2}dx = \frac{2}{3}\left(1+x\right)^{3/2}\Big|_3^8$$

$$= \frac{2}{3}(9^{3/2}-4^{3/2}) = \frac{2}{3}(27-8) = \frac{38}{3}.$$

• **PROBLEM 787**

Find the total length of the circumference of a

circle of radius r.

Solution: Let $x^2 + y^2 = r^2$.

Then $$x\,dx + y\,dy = 0,$$

or, $$\frac{dy}{dx} = y' = -\frac{x}{y}.$$

$$y'^2 = \frac{x^2}{y^2}.$$

$$1 + y'^2 = 1 + \frac{x^2}{y^2} = \frac{x^2+y^2}{y^2} = \frac{r^2}{y^2}.$$

$$\sqrt{1 + y'^2} = \frac{r}{y} = \frac{r}{\sqrt{r^2-x^2}}.$$

The circumference, C, is therefore given by:

$$C = 4\int_0^r \frac{r\,dx}{\sqrt{r^2-x^2}}$$

$$= 4r\left[\sin^{-1}\frac{x}{r}\right]_0^r$$

$$= 2\pi r, \text{ a well-known result.}$$

In polar coordinates, the differential of arc length is:

$$ds = \sqrt{d\rho^2 + \rho^2\,d\theta^2}$$

$$= \sqrt{\rho^2 + \left(\frac{d\rho}{d\theta}\right)^2}\,d\theta$$

$$= \sqrt{1 + \rho^2\left(\frac{d\theta}{d\rho}\right)^2}\,d\rho.$$

Length of curve is, therefore,

$$s = \int ds = \int_{\theta_1}^{\theta_2} \sqrt{\rho^2 + \rho'^2}\,d\theta, \text{ or, alternatively}$$

$$s = \int_{\rho_1}^{\rho_2} \sqrt{1 + \rho^2\theta'^2}\,d\rho.$$

The equation of a circle in polar coordinates is $\rho = r$, a constant, therefore $\frac{d\rho}{d\theta} = 0$.

$$s = \int_0^{2\pi} \sqrt{r^2}\,d\theta = r\theta\Big|_0^{2\pi} = 2\pi r$$

On the other hand, $\frac{d\theta}{d\rho}$ does not exist, and the other formula cannot be applied.

• PROBLEM 788

What is the length of the arc of $\rho = \varepsilon^{2\theta}$ between $\theta = 0$ and $\theta = 2$ radians?

Solution: If $\rho = \varepsilon^{2\theta}$,

then $\frac{d\rho}{d\theta} = 2\varepsilon^{2\theta} = 2\rho.$

$$s = \int \left[\rho^2 + \left(\frac{d\rho}{d\theta}\right)^2\right]^{\frac{1}{2}} \cdot d\theta$$

$$= \int_0^2 \left[\rho^2 + (2\rho)^2\right]^{\frac{1}{2}} \cdot d\theta = \int_0^2 \left(5\rho^2\right)^{\frac{1}{2}} \cdot d\theta$$

$$= \sqrt{5} \int_0^2 \rho \cdot d\theta = \sqrt{5} \int_0^2 \varepsilon^{2\theta} \cdot d\theta$$

$$= \frac{\sqrt{5}}{2} \int_0^2 \varepsilon^{2\theta} \cdot 2\, d\theta = \frac{\sqrt{5}}{2} \left[\varepsilon^{2\theta}\right]_0^2$$

$$= \frac{2.236}{2} (\varepsilon^4 - \varepsilon^0) = 1.12(54.598 - 1)$$

$$= 1.12 \cdot 53.598.$$

Hence, $s = 60.03$ in.

• PROBLEM 789

Find the total length, s, of the curve: $r = \cos\theta$.

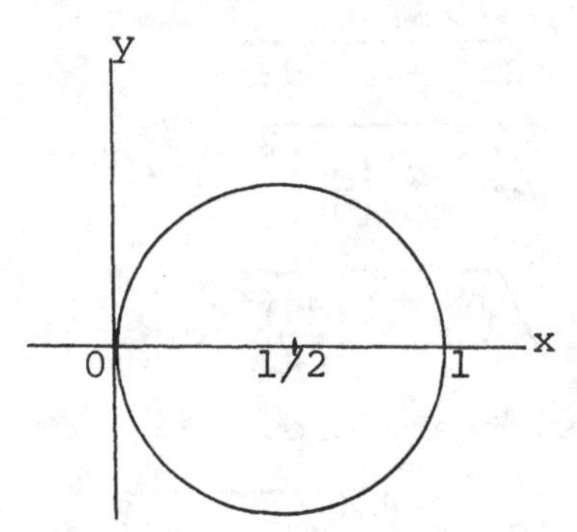

Solution: $dr/d\theta = -\sin\theta$. Hence,

$$ds = \sqrt{\left(\frac{dr}{d\theta}\right)^2 + r^2}\ (d\theta)$$

$$= \sqrt{\sin^2\theta + \cos^2\theta}\ (d\theta)$$

$$= d\theta, \quad (\text{since } \sin^2\theta + \cos^2\theta = 1).$$

$$s = \int_{-\pi/2}^{\pi/2} d\theta = 2\int_0^{\pi/2} d\theta = 2\theta\Big|_0^{\pi/2} = \pi.$$

Since the curve is a circle of radius 1/2, we can check the result by the formula: circumference = $2\pi r = 2\pi(1/2) = \pi$

• PROBLEM 790

Determine the length of the curve: $\rho = (1 - \cos\theta)$, from $\theta = 0$ to $\theta = \pi$.

Solution:

$$s = \int_0^{\pi}\left[\rho^2 + \left(\frac{d\rho}{d\theta}\right)^2\right]^{\frac{1}{2}} \cdot d\theta$$

If $\qquad \rho = (1 - \cos\theta)$,

then, $\qquad \dfrac{d\rho}{d\theta} = \sin\theta.$

$$s = \int_0^{\pi}[(1 - \cos\theta)^2 + \sin^2\theta]^{\frac{1}{2}} \cdot d\theta$$

$$= \int_0^{\pi}(1 - 2\cos\theta + \cos^2\theta + \sin^2\theta)^{\frac{1}{2}} \cdot d\theta$$

$$= \int_0^{\pi}(2-2\cos\theta)^{\frac{1}{2}} \cdot d\theta, \text{ (since } \cos^2\theta + \sin^2\theta = 1).$$

$$S = \int_0^{\pi} \sqrt{2}(1 - \cos\theta)^{\frac{1}{2}} \cdot \frac{\sqrt{2}}{\sqrt{2}} \cdot d\theta$$

$$= \int_0^{\pi} \sqrt{2}\,\sqrt{2}\left(\frac{1-\cos\theta}{2}\right)^{\frac{1}{2}} \cdot d\theta$$

$$= 2\int_0^{\pi}\sin\frac{\theta}{2} \cdot d\theta \cdot \frac{2}{2}.$$ This step is possible

because $\sin\dfrac{\theta}{2} = \left(\dfrac{1-\cos\theta}{2}\right)^{\frac{1}{2}}$.

$$s = 4\int_0^{\pi}\sin\frac{\theta}{2} \cdot \frac{d\theta}{2} = 4\left[-\cos\frac{\theta}{2}\right]_0^{\pi}$$

$$= \left[-4\cos\frac{\theta}{2}\right]_0^{\pi} = -4\cos\frac{\pi}{2} + 4\cos\frac{0}{2}$$

$$= -0 + 4 = 4.$$

• PROBLEM 791

Find the length of the cardioid $\rho = a(1 - \cos\theta)$.

Solution: The cardioid is traced as θ goes from 0 to 2π. The length s is equal to $\int_{\theta_1}^{\theta_2}\sqrt{\rho^2 + (\rho')^2}\, d\theta$.

$$\rho' = a \sin \theta .$$

$$\rho'^2 = a^2 \sin^2 \theta .$$

$$\rho^2 + \rho'^2 = a^2 (1 - \cos \theta)^2 + a^2 \sin^2 \theta$$

$$= 2a^2 (1 - \cos \theta).$$

$$s = \int_0^{2\pi} \sqrt{2a^2 (1 - \cos \theta)}\, d\theta$$

$$= 2a \int_0^{2\pi} \sqrt{\frac{1 - \cos \theta}{2}}\, d\theta$$

$$= 2a \int_0^{2\pi} \sin \frac{1}{2} \theta \, d\theta$$

$$= 2a \left[-2 \cos \frac{1}{2} \theta \right]_0^{2\pi}$$

$$= 8a.$$

• PROBLEM 792

> Find the length of the arc of the curve $x = t^2$, $y = t^3$, between the points for which $t = 0$ and $t = 2$.

Solution: The formula for the arc length is:

$$s = \int_a^b \sqrt{1 + y'^2}\, dx .$$

$$dx = 2t\, dt, \ dy = 3t^2 dt, \ \frac{dy}{dx} = y' = \frac{3}{2} t.$$

$$1 + y'^2 = 1 + \frac{9}{4} t^2 = \frac{1}{4}(4 + 9t^2).$$

Then

$$s = \int_0^2 \frac{1}{2} \sqrt{4 + 9t^2}\, (2t\, dt)$$

This can be written as:

$$s = \frac{1}{2} \int_0^2 (4 + 9t^2)^{\frac{1}{2}} \cdot 2t\, dt$$

$$= \frac{2}{3} \cdot \frac{1}{18} \left[4 + 9t^2 \right]^{3/2} \Bigg|_0^2$$

$$= \frac{1}{27} \left[40^{3/2} - 4^{3/2} \right]$$

$$= \frac{8}{27} \left[10^{3/2} - 1 \right] .$$

Find the length of arc of one arch of the cycloid

$$x = a(\theta - \sin\theta), \quad y = a(1 - \cos\theta).$$

Solution: We employ the formula:

$$s = \int_a^b ds = \int_a^b \sqrt{\left(\frac{dx}{d\theta}\right)^2 + \left(\frac{dy}{d\theta}\right)^2}.$$

$$dx = a(1 - \cos\theta)d\theta, \quad dy = a\sin\theta\, d\theta.$$

$$(ds)^2 = (dx)^2 + (dy)^2$$

$$= \left[a^2(1 - \cos\theta)^2 + a^2\sin^2\theta\right](d\theta)^2$$

$$= 2a^2(1 - \cos\theta)(d\theta)^2 = 4a^2\sin^2\frac{1}{2}\theta(d\theta)^2.$$

$$ds = 2a\sin\frac{1}{2}\theta\, d\theta.$$

Hence, $s = \int_0^{2\pi} 2a\sin\frac{1}{2}\theta\, d\theta = -4a\cos\frac{1}{2}\theta\Big|_0^{2\pi} = 8a$

Find the length of the curve given by the parametric equations:

$$\left.\begin{array}{l} x = 5\sin t \\ y = 5\cos t \end{array}\right\} \text{ from } t = -\frac{\pi}{3} \text{ to } t = \frac{\pi}{2}.$$

Solution: We employ the formula:

$$s = \int_a^b \sqrt{\left(\frac{dx}{dt}\right)^2 + \left(\frac{dy}{dt}\right)^2}.$$

$$\frac{dx}{dt} = 5\cos t, \quad \frac{dy}{dt} = -5\sin t,$$

$$s = \int_{-\pi/3}^{\pi/2} \sqrt{25\cos^2 t + 25\sin^2 t}\, dt$$

$$= \int_{-\pi/3}^{\pi/2} \sqrt{25(\cos^2 t + \sin^2 t)}\,(dt)$$

$$= \int_{-\pi/3}^{\pi/2} 5\, dt, \text{ (since } \cos^2 t + \sin^2 t = 1)$$

$$S = \frac{25}{6}\pi$$

CHAPTER 30

PLANE AREAS

The area of a square or rectangle is defined as the product of the lengths of two adjacent sides. The most complex or irregular-shaped surface can be sub-divided into small squares or rectangles, and the total area can be obtained from the sum of the areas of the individual squares or rectangles.

In calculus, if we subdivide the given surface into thin rectangles having sides y and dx, for example, then a summation (integration) of these rectangles over the surface results in the value of the area of the surface.

This same principle can be extended to polar coordinates where the elemental area is a triangle instead of a rectangle as used in Cartesian coordinates. The elemental triangle has sides r and ds, from which the area becomes $\frac{1}{2} r\, ds = \frac{1}{2} r \cdot r d\theta = \frac{1}{2} r^2 d\theta$.

• PROBLEM 795

Determine the area under the curve: $y = f(x) = x^2$ between $x = 2$ and $x = 3$.

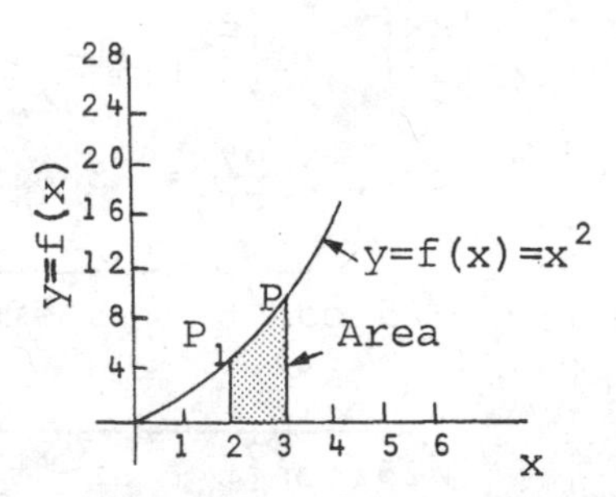

__Solution:__ It is given that the area to be evaluated is between $x = 2$ and $x = 3$, therefore, these are the limits of the integral which give us the required area. Area is equal to the integral of the upper function minus the lower function. From the diagram it is seen that the required area is between $y = x^2$ as the upper function and $y = 0$ (the x-axis) as the lower function. Therefore, we can write:

$$A = \int_2^3 (x^2 - 0)\,dx$$

$$= \int_2^3 x^2\,dx$$

$$= \left.\frac{x^3}{3}\right]_2^3$$

$$A = \frac{3^3}{3} - \frac{2^3}{3} = \frac{19}{3}.$$

• PROBLEM 796

Determine the area of the isosceles triangle formed by the equation y = x, with one of its sides equal to a.

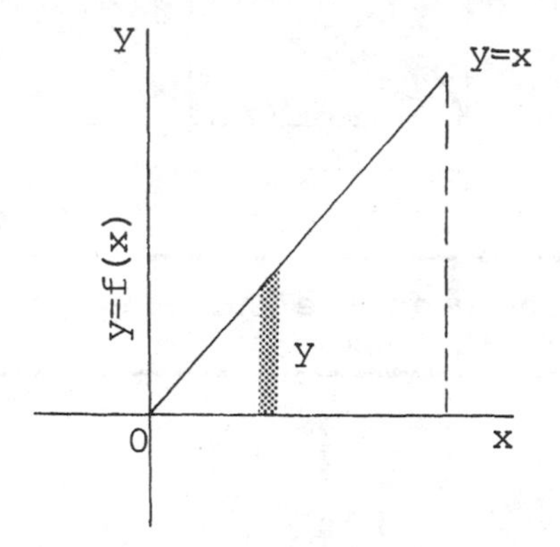

Solution: Looking at the diagram, we can see that the required area is to be evaluated between x = 0 and x = a. These values are the limits of the integral which give us the required area. The area is equal to the integral of the upper function minus the lower function. In this problem, the area is between y = x as the upper function and y = 0 (the x-axis) as the lower function. Therefore,

$$A = \int_0^a (x) - (0)dx = \int_0^a x\,dx$$

$$A = \left.\frac{x^2}{2}\right]_0^a = \frac{a^2}{2},$$

a result easily checked from elementary geometry.

• PROBLEM 797

Find the area formed by the curve: $y^2 = x$, the x axis, and the ordinates at x = 0 and x = a.

Solution: The limits of the integral which give

the required area are the boundary lines $x = 0$ and $x = a$. Area is equal to the integral of the upper function minus the lower function. In this problem the upper function is $y = \sqrt{x}$ and the lower function is $y = 0$ (the x-axis). Therefore, we can write:

$$A = \int_0^a \sqrt{x} - 0 \, dx = \int_0^a (x)^{1/2} dx.$$

$$= \frac{2}{3} x^{3/2} \Big|_0^a$$

$$= \frac{2}{3} a^{3/2}$$

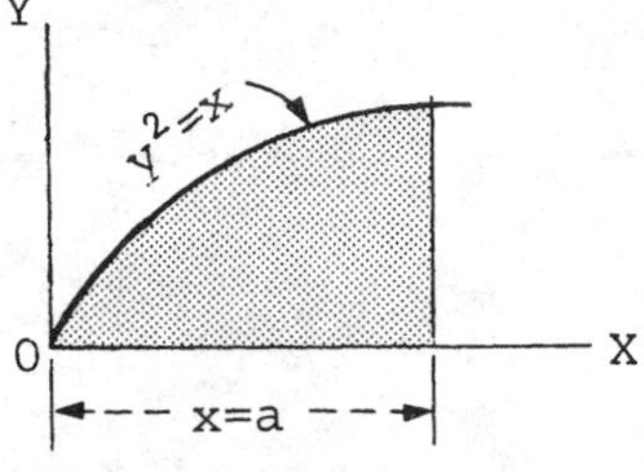

• **PROBLEM 798**

Find the area between the curve: $y = x^3$, and the x-axis, from $x = -2$ to $x = 3$.

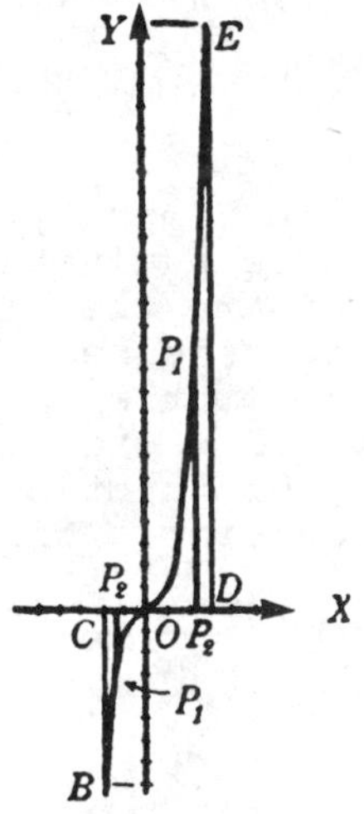

Solution: It is generally advantageous to sketch the curve, since parts of the curve may have to be considered separately, particularly when positive and negative limits are given. The desired area is composed of the two parts: BOC and ODE. To find the total area, we can evaluate each area separately and then add. The area is the integral of the upper function minus the lower function. In the first quadrant, the upper function is the curve $y = x^3$, the lower function is $y = 0$, (the x-axis) and the limits are $x = 0$ and $x = 3$. In the third quadrant, the upper function is $y = 0$, the lower function is the curve $y = x^3$, and the limits are $x = -2$ and $x = 0$. Hence, we can write,

$$A_{total} = \int_0^3 (x^3 - 0)\,dx + \int_{-2}^0 (0 - x^3)\,dx$$

$$= \int_0^3 x^3 dx + \int_{-2}^0 -x^3 dx$$

$$= \left[\frac{x^4}{4}\right]_0^3 + \left[-\frac{x^4}{4}\right]_{-2}^0$$

$$= \frac{81}{4} + \frac{16}{4}$$

$$= 24\tfrac{1}{4} \text{ sq. units.}$$

Note that refusal to consider this problem in two parts does not give area, but gives "net area" with one area considered positive and the other negative.

• PROBLEM 799

Find the area enclosed by the curve: $f(x) = x^2 - x - 2$, the x-axis, and the lines $x = 0$, $x = 3$.

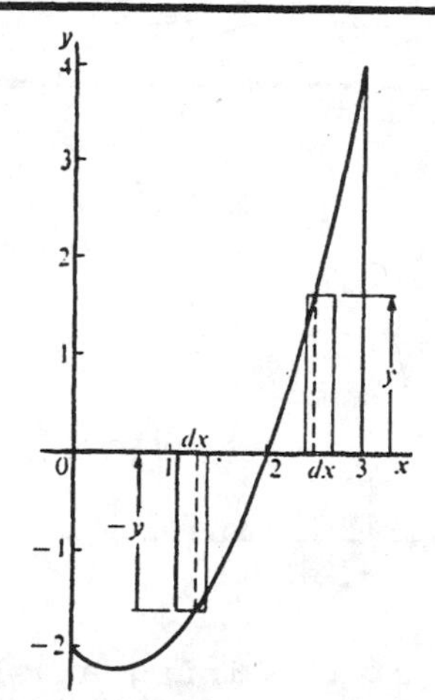

Solution: The required area is split into two parts, one above the x-axis and one below. To find the total area we can evaluate each of the two smaller areas separately, and then add the values we obtain for each. The area is defined as the integral of the upper function minus the lower function. For the area below the x-axis, we can see from the diagram that the limits are $x = 0$ and $x = 2$. The x-axis or $y = 0$ is above the curve $f(x) = x^2 - x - 2$. Therefore the integral for the area below the x-axis can be written as:

$$\int_0^2 \left[(0) - (x^2 - x - 2)\right]dx.$$

For the area above the x-axis, we can see from the diagram that the limits are: x = 2 and x = 3. In this case, however, the curve $f(x) = x^2 - x - 2$ is above y = 0 (the x-axis). Therefore, the integral for the area above the x-axis can be written as:

$$\int_2^3 (x^2 - x - 2) - (0)dx.$$

Therefore the total area is:

$$A_{total} = \int_0^2 \left[-(x^2 - x - 2)\right] dx + \int_2^3 (x^2 - x - 2)dx$$

$$= \left[-\left(\frac{x^3}{3} - \frac{x^2}{2} - 2x\right)\right]_0^2 + \left[\frac{x^3}{3} - \frac{x^2}{2} - 2x\right]_2^3$$

$$= \frac{10}{3} + \frac{11}{6} = \frac{31}{6}.$$

Note that refusal to consider this problem in two parts does not give the area, but gives "net area" with one area considered positive and the other negative.

• PROBLEM 800

Find the area of the region bounded by the x-axis, the curve: $y = 6x - x^2$, and the vertical lines: x = 1 and x = 4.

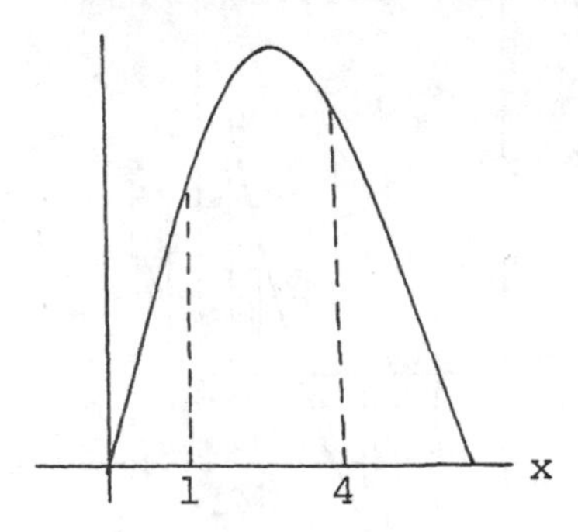

Solution: The limits of the integral which give the required area are x = 1 and x = 4. The function: $y = 6x - x^2$ is above the function y = 0 (the x-axis), therefore the area can be found by taking the integral of the upper function minus the lower function, or, $y = 6x - x^2$ minus y = 0, from x = 1 to x = 4. Therefore, we obtain:

$$A = \int_1^4 (6x - x^2) - 0\,dx = \left[3x^2 - \frac{x^3}{3}\right]_1^4$$

$$= \left(\frac{80}{3}\right) - \left(\frac{8}{3}\right) = 24.$$

• **PROBLEM 801**

Find the area under the curve: $y = 3x^{-2}$, from $x = 1$ to $x = \infty$.

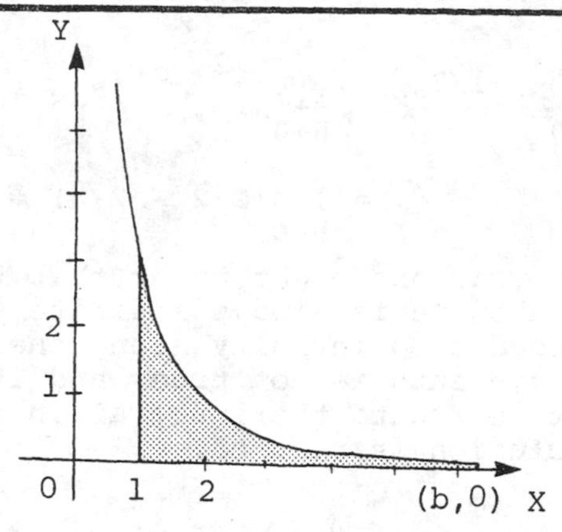

Solution: A figure is very important in all such problems as it shows the places where the area is not bounded. The curve: $y = 3x^{-2}$, has $y = 0$ as a horizontal asymptote. It is then necessary to find the value of the shaded area in the diagram and evaluate the limit of that area as $b \to \infty$, <u>if possible</u>. The area is evaluated as the integral of the upper function, $3x^{-2}$, minus the lower function, $y = 0$ (the x-axis). The limits are given as $x = 1$ and $x = \infty$. We can evaluate such an integral (an integral involving infinity as one of the limits), by the method of improper integrals. Thus,

$$A = \int_1^\infty (3x^{-2} - 0)\,dx = \int_1^\infty 3x^{-2}dx = \lim_{b\to\infty} \int_1^b 3x^{-2}dx$$

$$= 3 \lim_{b\to\infty} \left(-\frac{1}{x} \Big]_1^b \right)$$

$$= 3 \lim_{b\to\infty} \left(-\frac{1}{b} + 1 \right) = 3,$$

since $\frac{1}{b} \to 0$ as $b \to \infty$. Hence this area exists and equals 3 square units.

• **PROBLEM 802**

Find the area under the curve: $y = x^{-1/2}$ from $x = 0$ to $x = 2$.

Solution: The curve: $y = x^{-1/2}$ has $x = 0$ as a vertical asymptote. It is then necessary to find the value of the shaded area in the diagram and evaluate the limit of that area as $h \to 0$, if possible. The area is evaluated as the integral of the upper function which is $y = x^{-\frac{1}{2}}$, minus the lower function, $y = 0$ (the x-axis). The limits are given as $x = 0$ and $x = 2$.

We evaluate such an integral (one involving zero as one of the limits) by the method of improper integrals. Thus,

$$A = \int_0^2 (x^{-1/2} - (0))dx$$

$$= \int_0^2 x^{-1/2}dx = \lim_{h \to 0} \int_h^2 x^{-1/2}dx = \lim_{h \to 0}\left(2x^{1/2}\Big]_h^2\right)$$

$$= \lim_{h \to 0}(2\sqrt{2} - 2\sqrt{h}) = 2\sqrt{2} \text{ sq. units.}$$

Without analysis one might think that the area under such a curve is always infinite since its boundary recedes to infinity along the y-axis. But for this curve such is not true, and it is therefore advisable to evaluate the integral in each case before a conclusion is reached.

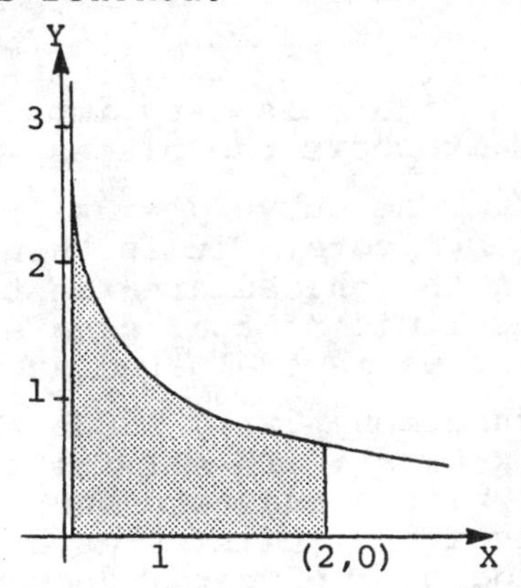

• PROBLEM 803

Determine, if possible, the area between $y = \frac{1}{x}$, $y = 0$, and $x = 1$.

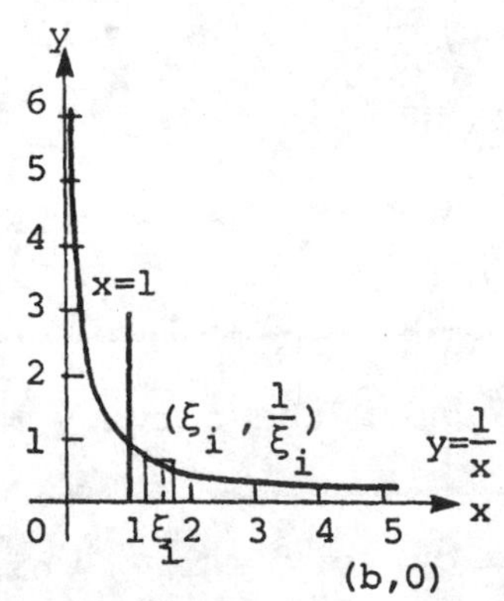

Solution: The curve $y = \frac{1}{x}$ has a horizontal asymptote on the x-axis, or $y = 0$. It is then necessary to find the value of the shaded area in the diagram in terms of a quantity b and then evaluate the limit of that area as b approaches infinity, if possible.

The area can be evaluated as the integral of the upper function, $y = \frac{1}{x}$, minus the lower function, $y = 0$ (the x-axis). The limits are from the boundary line $x = 1$ to $x = \infty$. Such an integral (one involving infinity as a limit) is evaluated by the method of improper integrals. Hence,

$$A = \int_1^\infty \left[\frac{1}{x} - (0)\right] dx = \int_1^\infty \left(\frac{1}{x}\right) dx$$

$$= \lim_{b \to +\infty} \int_1^b \frac{1}{x} dx$$

$$= \lim_{b \to +\infty} [\ln b - \ln 1]$$

$$= +\infty.$$

Therefore, it is not possible to assign a finite number to represent the measure of the area of the region.

• PROBLEM 804

Find the area bounded by the curve: $xy = 4$, the x-axis, $x = e$, and $x = 2e$, and draw a figure.

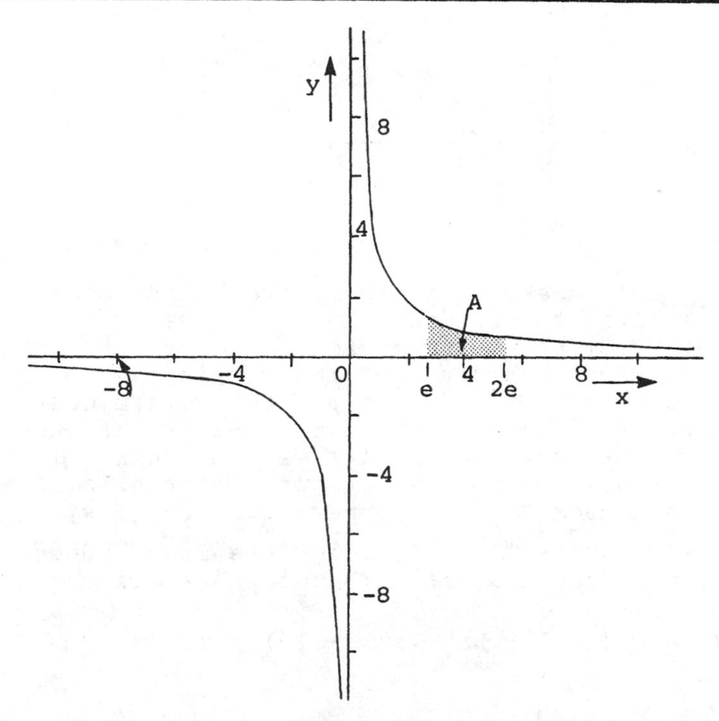

Solution: We write $xy = 4$ as $y = \frac{4}{x}$. The desired area is between $y = \frac{4}{x}$ and the x-axis or, $y = 0$. To obtain the area we evaluate $\int\left(\frac{4}{x} - 0\right)dx$ with 2e as an upper bound and e as a lower bound. Therefore,

$$A = \int_e^{2e} \frac{4}{x}\, dx = 4 \int_e^{2e} \frac{dx}{x}.$$

Applying the formula $\int \frac{dx}{x} = \ln x$, we obtain,

$$A = \Big[4 \ln x \Big]_e^{2e} = 4(\ln 2e - \ln e).$$

But $\ln 2e = \ln 2 + \ln e$. Hence,

$$A = 4(\ln 2 + \ln e - \ln e)$$

$$= 4 \ln 2.$$

Therefore, the area bounded by the curve: $xy = 4$, the x-axis, $x = e$, and $x = 2e$ is $4 \ln 2$, and is shown in the accompanying diagram.

• PROBLEM 805

> Find the finite area above the x-axis and under the curve: $y = x^3 - 9x^2 + 23x - 15$.

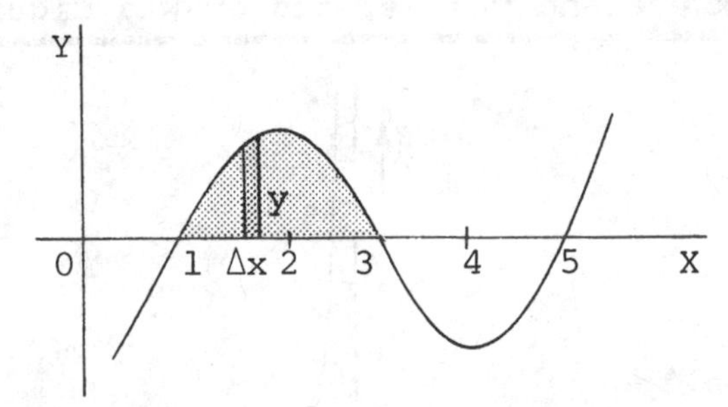

Solution: A sketch of the curve helps to determine the limits of the integral which give us the required area. From the diagram, the limits are the intersections of the curve and the x-axis, giving the limits: $x = 1$, $x = 3$, $x = 5$. But we are only concerned with the area above the x-axis and under the curve, therefore the limits we use are: $x = 1$ and $x = 3$. Now set up the integral. The area is equal to the integral of the upper function of x, $y = x^3 - 9x^2 + 23x - 15$, minus the lower function of x, $y = 0$ (the x-axis). Therefore we write:

$$A = \int_1^3 (x^3 - 9x^2 + 23x - 15) - (0)dx$$

$$= \int_1^3 (x^3 - 9x^2 + 23x - 15)dx$$

$$= \frac{x^4}{4} - 3x^3 + \frac{23x^2}{2} - 15x \Big]_1^3$$

$$= \left(\frac{81}{4} - 81 + \frac{207}{2} - 45\right) - \left(\frac{1}{4} - 3 + \frac{23}{2} - 15\right)$$

$$= 4.$$

• PROBLEM 806

Determine the area between the curve: $y = 4x - x^2$ and the x-axis.

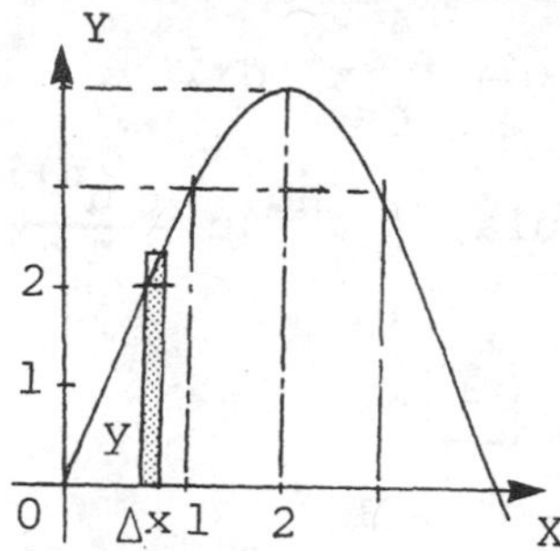

Solution: It is necessary to sketch this curve to see how to proceed and to determine the limits of integration. It is easiest to proceed to set up an integral with respect to x. The limits of the integral which give us the required area are the points of intersection of the two functions: $y = 4x - x^2$ and $y = 0$ (the x-axis). The points of intersection can be found by setting the two functions equal and solving for x.

$$4x - x^2 = 0$$

$$x(4 - x) = 0$$

$$x = 0 \quad x = 4.$$

The area is equal to the integral of the upper function minus the lower function. In this problem, the upper function is $y = 4x - x^2$ and the lower function is $y = 0$. Therefore we can write:

$$A = \int_0^4 (4x - x^2) - (0)dx = \int_0^4 (4x - x^2)dx$$

$$= \left[2x^2 - \frac{x^3}{3}\right]_0^4 = 32 - \frac{64}{3} = \frac{32}{3} \text{ sq. units.}$$

• PROBLEM 807

Find the area bounded by the parabola: $y = x^2 - 3x$, and the line: $y = x$.

Solution: We draw a figure to show the area and the upper and lower bounds of the integral. Setting $y = x^2 - 3x = x$ to find the points of intersection, we have $x^2 = 4x$, $x = 0, 4$. From the diagram we see that the desired area is between $y = x$ and

$y = x^2 - 3x$, with an upper bound of 4 and a lower bound of 0. Therefore,

$$A = \int_0^4 \left[x - (x^2 - 3x)\right]dx$$

$$= \int_0^4 [4x - x^2]dx$$

$$= 4\int_0^4 x\,dx - \int_0^4 x^2\,dx.$$

Applying the formula, $\int u^n\,du = \frac{u^{n+1}}{n+1}$, we obtain:

$$A = 4\left[\frac{x^2}{2}\right]_0^4 - \left[\frac{x^3}{3}\right]_0^4$$

$$= 4\left[\frac{16}{2} - 0\right] - \left[\frac{4^3}{3} - 0\right]$$

$$= 4(8) - \frac{64}{3}$$

$$= 32 - \frac{64}{3} = \frac{96}{3} - \frac{64}{3} = \frac{32}{3}.$$

Hence, the area bounded by the parabola $y = x^2 - 3x$ and the line $y = x$ is $\frac{32}{3}$.

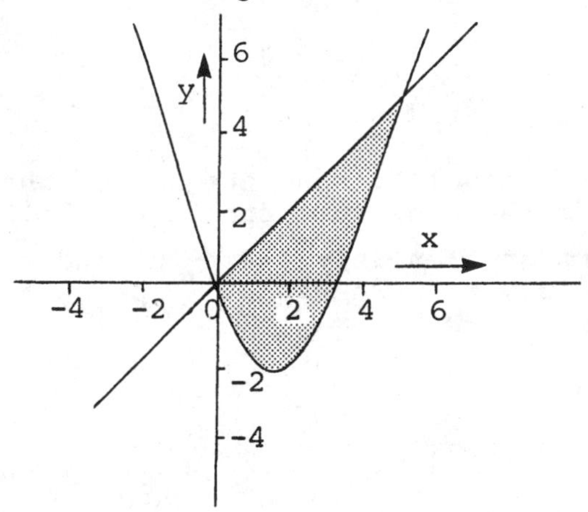

• PROBLEM 808

Find the area between the parabola: $y = x^2 - 6x + 5$, and the line: $y = x - 5$.

Solution: The limits of the integral which give the required area are found by solving for the points of intersection of the two functions. To find the points of intersection, we set

$y = x^2 - 6x + 5$ equal to $y = x - 5$, and solve for x.

$$x^2 - 6x + 5 = x - 5$$

$$x^2 - 7x + 10 = 0$$

$$(x - 2)(x - 5) = 0$$

$$x = 2 \quad x = 5.$$

The line: $y = x - 5$, is above the curve: $y = x^2 - 6x + 5$. Therefore, the area is found by taking the integral of the upper function minus the lower function, or $y = x - 5$ minus $y = x^2 - 6x + 5$, between $x = 2$ and $x = 5$. Hence,

$$A = \int_2^5 \left[(x - 5) - (x^2 - 6x + 5)\right]dx$$

$$= \int_2^5 (-x^2 + 7x - 10)dx$$

$$= -\frac{x^3}{3} + \frac{7x^2}{2} - 10x \Big|_2^5$$

$$= \left(-\frac{125}{3} + \frac{175}{2} - 50\right) - \left(-\frac{8}{3} + 14 - 20\right)$$

$$= \frac{27}{6}$$

$$= 4\frac{1}{2}.$$

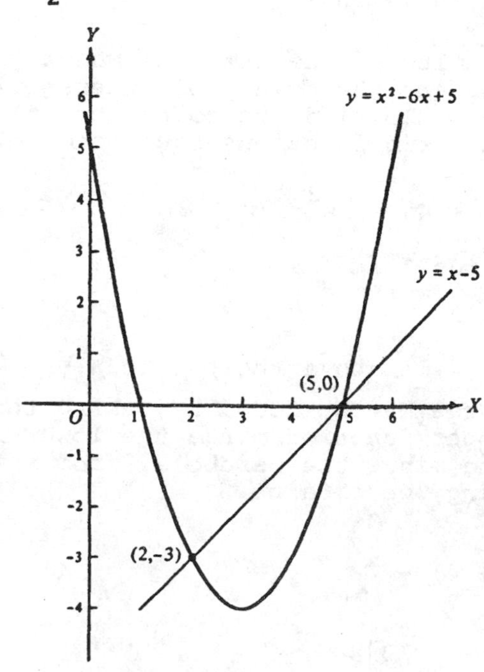

Find the area bounded by the parabola: $2y = x^2$, and the line: $y = x + 4$.

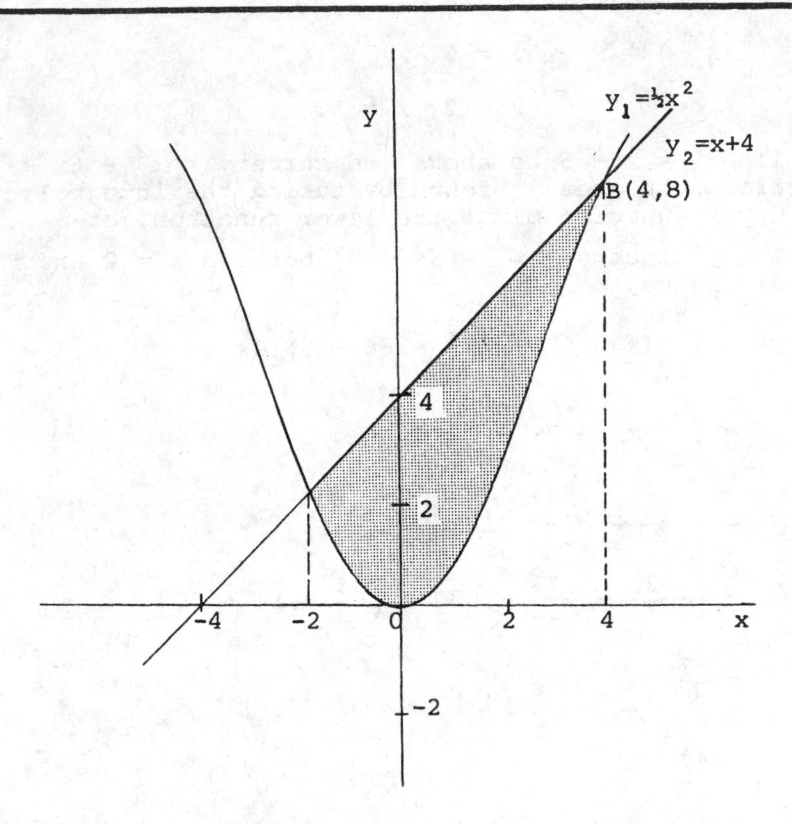

Solution: The limits of the integral which give the required area are the points of intersection of the two functions. To find the points of intersection, we set the two functions equal and solve for x.

$$\frac{x^2}{2} = x + 4. \quad x^2 = 2x + 8. \quad x^2 - 2x - 8 = 0.$$

$$(x - 4)(x + 2) = 0$$

$$x = 4 \quad x = -2.$$

The line: $y = x + 4$, is above the parabola: $y = \frac{x^2}{2}$, Therefore, the area can be found by taking the integral of the upper function minus the lower function, or the line minus the parabola, from $x = -2$ to $x = 4$. Solving, we obtain:

$$A = \int_{-2}^{4} \left[(x + 4) - \frac{x^2}{2}\right] dx$$

$$= \left[\frac{x^2}{2} + 4x - \frac{x^3}{6}\right]_{-2}^{4}$$

$$= \frac{40}{3} + \frac{14}{3}$$

$$= 18.$$

• PROBLEM 810

Determine the area A of the region to the right of the curve: $x = y^2$, and to the left of the line: $y = x - 2$.

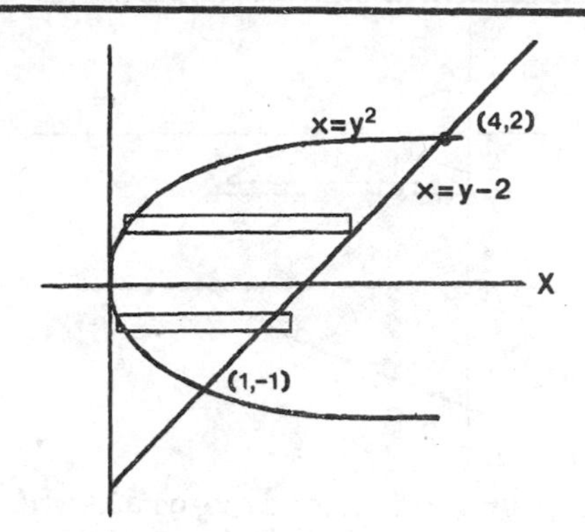

Solution: From the diagram we can see that the best way to proceed is to set up two integrals with respect to y, one for the area above the x-axis and one for the area below the x-axis. The limits for the two integrals are the points of intersection and y = 0 (the x-axis). The points of intersection can be found by setting the two functions equal and solving for y.

$$y^2 = y + 2. \qquad y^2 - y - 2 = 0. \quad (y+1)(y-2) = 0$$

$$y = -1 \quad y = 2.$$

Now we can set up the integrals. The area is equal to the integral of the upper function of y minus the lower function of y. For the area above the x-axis the integral is:

$$\int_{-1}^{0} \left[(y + 2) - y^2\right] dy;$$

and the integral for the area below the x-axis is:

$$\int_{0}^{2} \left[(y + 2) - y^2\right] dy.$$

Therefore the total area is;

$$A_{total} = \int_{-1}^{0} \left[(y + 2) - y^2\right] dy + \int_{0}^{2} \left[(y + 2) - y^2\right] dy$$

$$= \left[\frac{y^2}{2} + 2y - \frac{y^3}{3}\right]_{-1}^{0} + \left[\frac{y^2}{2} + 2y - \frac{y^3}{3}\right]_{0}^{2}$$

$$= \frac{9}{2}.$$

Note: Since we have chosen to do this problem in terms of y rather than x, the split is not really required (all area being to the right of the y-axix) but the result is the same either way.

• PROBLEM 811

Find the area bounded by the curve: $x = 8 + 2y - y^2$, the y-axis and the line: $y = 3$, and below $y = 3$.

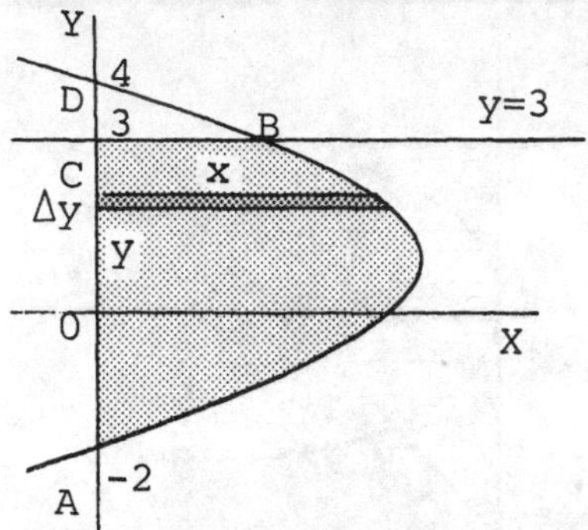

Solution: The limits of the integral which give the required area can be found by solving for the points of intersection. Since this curve is a function of y, the limits of integration are in terms of y. The required area is to be below $y = 3$, therefore the upper limit is $y = 3$. The lower limit is the point of intersection of $x = 8 + 2y - y^2$ and the y-axis, which can be expressed as $x = 0$. To find the points where the curves intersect, we set them equal and solve for y. Hence,

$$8 + 2y - y^2 = 0. \quad y^2 - 2y - 8 = 0.$$

$$(y + 2)(y - 4) = 0.$$

$$y = -2 \quad \text{or} \quad y = 4.$$

We do not employ the upper intersection, but use the upper limit of $y = 3$. The lower limit is $y = -2$. The area integral is with respect to y. We subtract the lower function of y from the higher function of y. We subtract $x = 0$ from $x = 8 + 2y - y$. Hence,

$$A = \int_{-2}^{3} \left[(8 + 2y - y^2) - 0\right] dy$$

$$= \int_{-2}^{3} (8 + 2y - y^2) dy$$

$$= \left[8y + y^2 - \frac{y^3}{3}\right]_{-2}^{3}$$

$$= (24) - \left(\frac{-28}{3}\right)$$

$$= \frac{100}{3}.$$

• PROBLEM 812

Find the area bounded by the curve: $y^2 = x^3$, the y-axis, and the line $y = 8$.

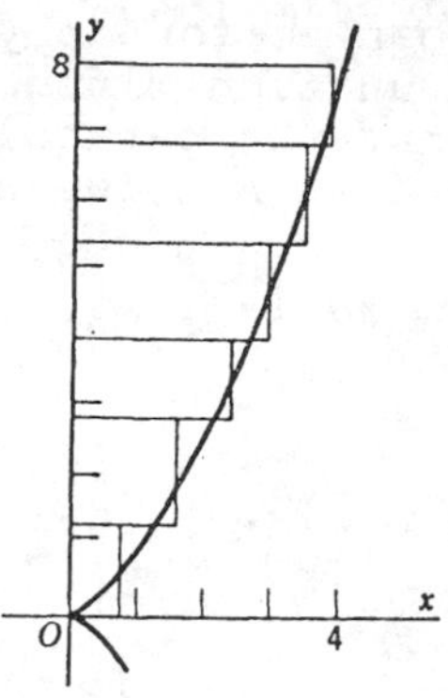

Solution: A sketch of the curve shows the best approach. In this case it seems best to set up an integral with respect to y. The upper limit is 8, and we can see from the diagram that the lower limit is zero. The area is the integral of the upper function of y minus the lower function of y. In this problem, the upper function of y is $y^2 = x^3$, or $x = \sqrt[3]{y^2}$, and the lower function of y is the y-axis, or $x = 0$. Hence,

$$A = \int_0^8 (\sqrt[3]{y^2} - 0)dy$$

$$= \int_0^8 \sqrt[3]{y^2}\, dy$$

$$A = \left[\frac{3}{5}y^{5/3}\right]_0^8$$

$$= \frac{96}{5}$$

$$= 19\tfrac{1}{5}.$$

• PROBLEM 813

Find the area between the parabola $y = x^2 - 4x + 4$ and the line $y = x$.

Solution: The limits of the integral which give the required area are the points of intersection of the two functions. So we find the points of intersection by setting the two functions equal and solving for x.

$$x^2 - 4x + 4 = x$$

$$x^2 - 5x + 4 = 0$$

$$(x-1)(x-4) = 0$$

$$x = 1, \; x = 4.$$

The line: $y = x$, is above the parabola: $y = x^2 - 4x + 4$. Therefore, the area can be found by taking the integral of the upper function minus the lower function, or the line minus the parabola, from $x = 1$ to $x = 4$. Thus, the area A between the two curves is:

$$A = \int_1^4 \left[x - (x^2 - 4x + 4)\right] dx$$

$$= \int_1^4 (-x^2 + 5x - 4)dx$$

$$= -\frac{x^3}{3} + \frac{5x^2}{2} - 4x \Big|_1^4$$

$$= \left(-\frac{64}{3} + 40 - 16\right) - \left(-\frac{1}{3} + \frac{5}{2} - 4\right)$$

$$= \frac{9}{2}.$$

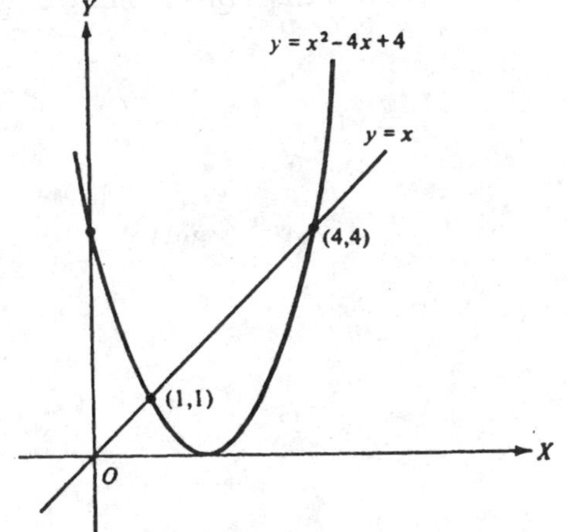

• PROBLEM 814

Compute the area between the graphs of f and g over the interval [-1,2] if $f(x) = x$ and $g(x) = \frac{x^3}{4}$.

Solution: Since the required area is split into two areas by the intersection of the two curves at the origin, we evaluate each area separately, using integration, and then add the two results to give total area.

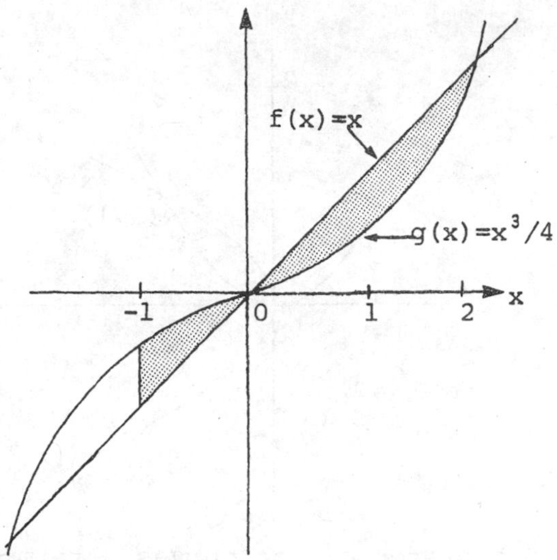

From the diagram we can see that the area in the third quadrant goes from $x = -1$ to $x = 0$, and the area in the first quadrant goes from $x = 0$ to $x = 2$. These are the limits of the integrals for the area. The area is equal to the integral of the upper function minus the lower function. In the third quadrant, the upper function is $g(x) = \frac{x^3}{4}$ and the lower function is $f(x) = x$. In the first quadrant, the upper function is $f(x) = x$ and the lower function is $g(x) = \frac{x^3}{4}$. We can now write the total area as:

$$A_{total} = \int_{-1}^{0} \left[g(x) - f(x)\right] dx + \int_{0}^{2} \left[f(x) - g(x)\right] dx$$

$$= \int_{-1}^{0} \left(\frac{x^3}{4} - x\right) dx + \int_{0}^{2} \left(x - \frac{x^3}{4}\right) dx$$

$$= \frac{x^4}{4.4} - \frac{x^2}{2} \Bigg|_{-1}^{0} + \frac{x^2}{2} - \frac{x^4}{4.4} \Bigg|_{0}^{2}$$

$$= -\frac{1}{4}\frac{(-1)^4}{4} + \frac{(-1)^2}{2} + \frac{2^2}{2} - \frac{1}{4}\frac{2^4}{4} = \frac{23}{16}.$$

Note that refusal to consider this problem in two parts does not give the area, but gives "net area" with one area considered positive and the other negative.

• PROBLEM 815

Find the area bounded by the two curves: $y = x^2$, $y^2 = x$.

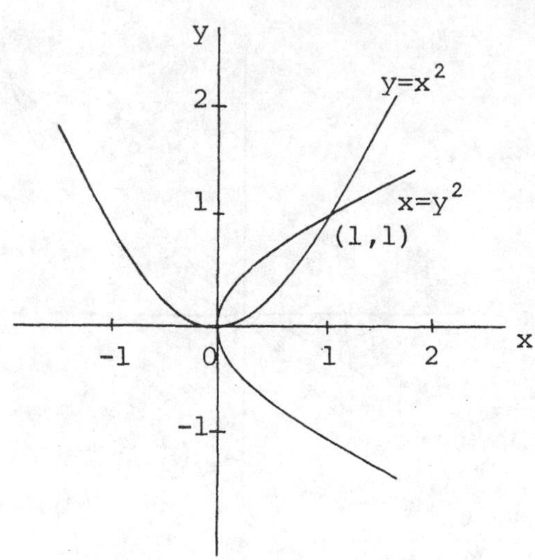

Solution: A sketch of the curves determines how best to proceed, which, in this case, is to set up an integral with respect to x. The limits of the integral which give us the required area are the points of intersection of the two curves. They are found by equating the two functions and solving for x.

$$x^2 = \sqrt{x} \ . \quad x^4 = x.$$

$$x^4 - x = 0$$

$$x(x^3 - 1) = 0$$

$$x = 0 \quad x = 1.$$

The area is the integral of the upper function of x, $y = \sqrt{x}$, minus the lower function of x, $y = x^2$. We write:

$$A = \int_0^1 (\sqrt{x} - x^2)dx$$

$$= \int_0^1 \sqrt{x}\, dx - \int_0^1 x^2\, dx$$

$$= \left[\frac{2}{3}x^{3/2} - \frac{x^3}{3}\right]_0^1$$

$$= \frac{1}{3} \text{ sq. unit.}$$

• PROBLEM 816

Find the area between the curves: $y = x^3$ and $y = x$.

Solution: Since the required area is split, by the intersection of the two curves at the origin, into two areas, we must evaluate each area separately using integration, and then add the two results to give us the total area. First we need to find the points of intersection, which give us the limits of integration. To find the points of intersection, we set the two curves equal and solve for x:

$$x^3 = x$$

$$x^3 - x = 0$$

$$x(x^2 - 1) = 0$$

$$x = 0 \quad x = \pm 1.$$

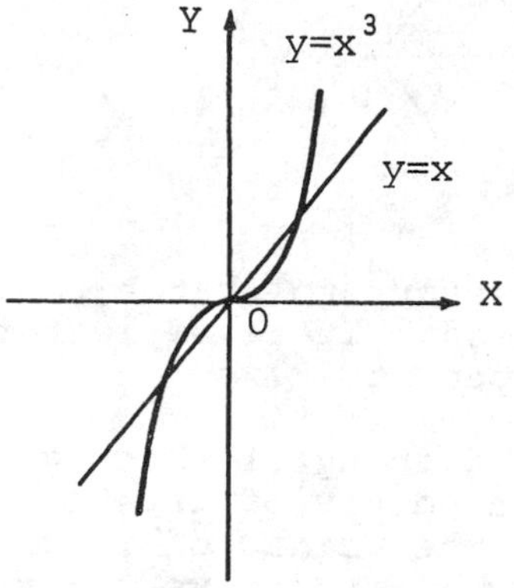

Y y=x³ y=x X 0

The area in the first quadrant is contained between the line $y = x$ above and the curve $y = x^3$ below, and is between $x = 0$ and $x = 1$, which we found by solving for the points of intersection. But the area in the third quadrant is between $y = x^3$ above and $y = x$ below, and is between $x = -1$ and $x = 0$. Thus, the area is found by adding the integral for the area in the third quadrant and the integral for the area in the first quadrant. (The integral for area is written as the integral of the upper function of x minus the lower function of x.) We obtain:

$$A = \int_{-1}^{0} (x^3 - x)\,dx + \int_{0}^{1} (x - x^3)\,dx$$

$$= \left.\frac{x^4}{4} - \frac{x^2}{2}\right|_{-1}^{0} + \left.\frac{x^2}{2} - \frac{x^4}{4}\right|_{0}^{1}$$

$$= \left[0 - \left(\frac{1}{4} - \frac{1}{2}\right)\right] + \left[\left(\frac{1}{2} - \frac{1}{4}\right) - 0\right]$$

$$= \frac{1}{4} + \frac{1}{4}$$

$$= \frac{1}{2}.$$

Note that refusal to consider this problem in two parts does not give the area, but gives "net area" with one area considered positive and the other negative.

• **PROBLEM 817**

Find the area bounded by the two curves: $y = x^3 - x^2$ and $y = x^2$.

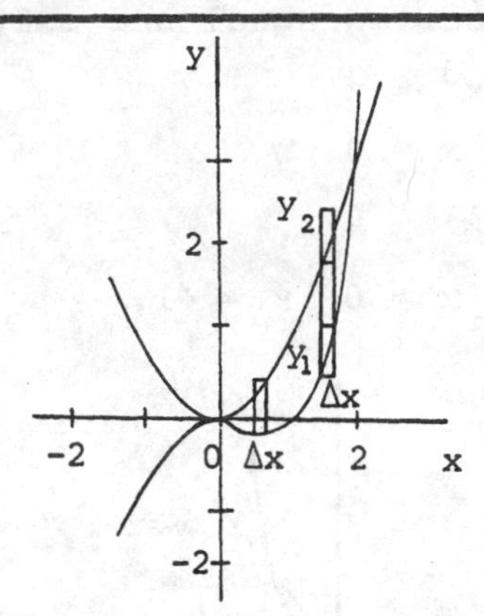

Solution: It is necessary to sketch these curves to see how best to proceed. It is easiest to set up an integral with respect to x.

The limits of the integral which give us the required area are the points of intersection of the two curves. To find the points of intersection, we set the two curves equal and solve for x.

$$x^3 - x^2 = x^2$$

$$x^3 - 2x^2 = 0$$

$$x^2(x - 2) = 0$$

$$x = 0 \quad x = 2.$$

The area is the integral of the upper function minus the lower function. In this problem, $y = x^2$ is above $y = x^3 - x^2$, therefore, we can write:

$$A = \int_0^2 \left[x^2 - (x^3 - x^2)\right] dx = \int_0^2 (2x^2 - x^3) dx$$

$$= \left(\frac{2x^3}{3} - \frac{x^4}{4}\right)\Bigg]_0^2 = \frac{16}{3} - \frac{16}{4} = \frac{4}{3} \text{ square units.}$$

• **PROBLEM 818**

Find the area enclosed by the curves: $f(x) = x + 1$, and $g(x) = (x - 1)^2$.

Solution: The limits of the integral which give the required area are the points of intersection of the two curves. To find the points of intersection we set the two functions equal and solve for x. Therefore,

$$x + 1 = (x - 1)^2$$

$$x + 1 = x^2 - 2x + 1$$

$$0 = x^2 - 3x$$

$$x = 0 \quad x = 3.$$

Since these are functions with respect to x, we find the area by taking the integral of the upper function of x minus the lower function of x from x = 0 to x = 3. We write,

$$A = \int_0^3 \left[(x + 1) - (x - 1)^2\right] dx$$

$$= \int_0^3 [3x - x^2] dx$$

$$= \left[\frac{3x^2}{2} - \frac{x^3}{3}\right]_0^3 = \frac{9}{2}.$$

• **PROBLEM 819**

Find the area enclosed by the curves: $x = f(y) = (y - 1)^2$, and $x = g(y) = y + 1$.

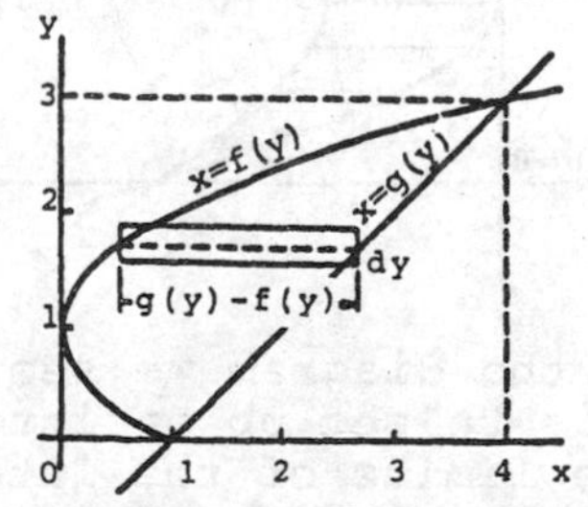

Solution: The limits of the integral which give us the required area are the points of intersection of the two curves. To find the points of intersection we set the two curves equal and solve for y. Hence,

$$(y - 1)^2 = y + 1$$

$$y^2 - 2y + 1 = y + 1$$

$$y^2 - 3y = 0$$

$$y = 0 \quad y = 3.$$

Since these are functions of y, we write the integral with respect to y and subtract the lower function of y from the higher function of y. Therefore, we take the integral of y + 1 minus $(y - 1)^2$ from y = 0 to y = 3. We write:

$$A = \int_0^3 (y + 1) - (y - 1)^2 dy$$

$$A = \int_0^3 (y + 1 - y^2 + 2y - 1)dy$$

$$A = \left[\frac{y^2}{2} + y - \frac{y^3}{3} + y^2 - y\right]_0^3$$

$$A = \left(\frac{9}{2} + 3 - \frac{27}{3} + 9 - 3\right)$$

$$A = \frac{9}{2}.$$

• PROBLEM 820

Find the area bounded by the curves: $2(y - 1)^2 = x$ and $(y - 1)^2 = x - 1$.

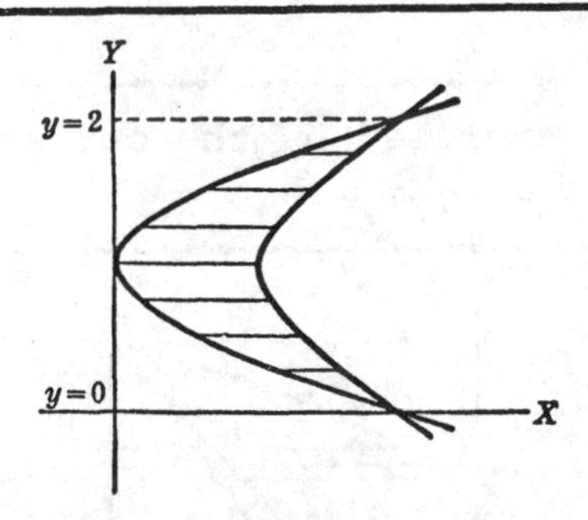

Solution: From the diagram we see that the best way to proceed is to set up an integral with respect to y. The limits of the integral which give us the required area are the points of intersection of the two curves, which can be found by setting the curves equal and solving for y.

$$x = 2(y - 1)^2 = (y - 1)^2 + 1$$

$$2y^2 - 4y + 2 = y^2 - 2y + 1 + 1$$

$$y^2 - 2y = 0$$

$$y = 0 \quad y = 2.$$

The area is the integral of the upper function of y minus the lower function of y. In this problem, the upper function of y is $x = (y - 1)^2 + 1$, and the lower function of y is $x = 2(y - 1)^2$. We can now write;

$$A = \int_0^2 \left[(y - 1)^2 + 1 - 2(y - 1)^2\right]dy$$

$$= \int_0^2 \left[1 - (y - 1)^2\right]dy$$

$$= \left[y - \frac{1}{3}(y - 1)^3\right]_0^2$$

$$= \left(2 - \frac{1}{3}\right) - \left(\frac{1}{3}\right)$$

$$= \frac{4}{3} \text{ sq. units.}$$

• PROBLEM 821

The figure shows sketches of the graphs of $y = 2^x$ and $y = 2^{-x}$ on the same set of axes. Find the area of the region bounded by these two graphs and the line $x = 2$.

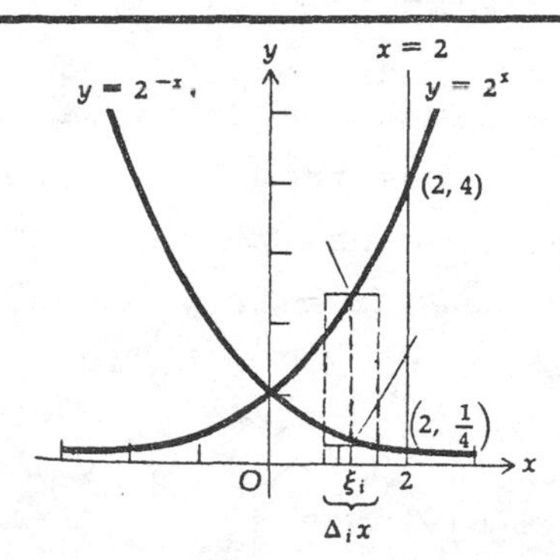

Solution: One limit of the integral which gives the required area is the boundary line: $x = 2$. The other limit is the point of intersection of the two curves. From the diagram we can see that the two curves intersect at $x = 0$. The area is equal to the integral of the upper function of x, $y = 2^x$, minus the lower function of x, $y = 2^{-x}$, Therefore, we can write:

$$A = \int_0^2 (2^x - 2^{-x})dx.$$ We use the formula:

$$\int a^u du = \frac{\ln u}{a}.$$

$$A = \frac{2^x}{\ln 2} + \frac{2^{-x}}{\ln 2}\Bigg]_0^2$$

$$= \frac{4}{\ln 2} + \frac{\frac{1}{4}}{\ln 2} - \frac{1}{\ln 2} - \frac{1}{\ln 2}$$

$$= \frac{9}{4 \ln 2}$$

$$\approx 3.25.$$

• PROBLEM 822

Find the area of an ellipse with semi-axes a and b.

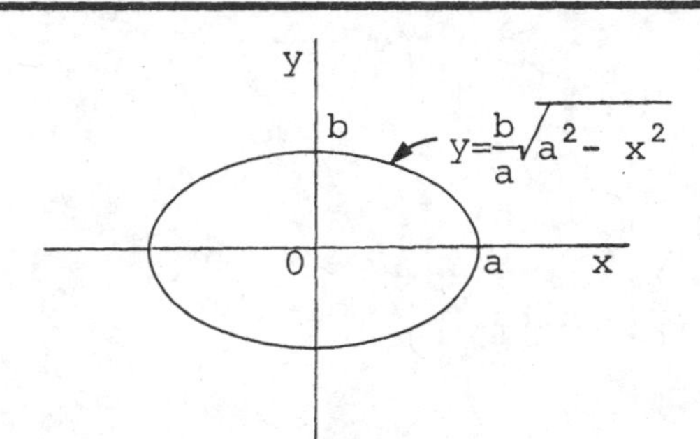

<u>Solution:</u> It is helpful to make a sketch of the curve to determine the limits of integration, and how best to proceed. It is easiest to consider the entire area of ellipse as four times the area in the first quadrant. Taking: $\frac{x^2}{a^2} + \frac{y^2}{b^2} = 1$ as the equation of the ellipse, and considering the portion of the ellipse in the first quadrant, we have $y = \frac{b}{a}\sqrt{a^2 - x^2}$, where x goes from 0 to a, which are the limits. The area is equal to the integral of the upper function of x, which is $y = \frac{b}{a}\sqrt{a^2 - x^2}$, minus the lower function of x, which is $y = 0$ (the x-axis). The total area is then

$$A_{total} = 4 \int_0^a \frac{b}{a}\sqrt{a^2-x^2} - (0)\,dx = 4\frac{b}{a} \int_0^a \sqrt{a^2-x^2}\,dx.$$

The formula for $\int\sqrt{a^2 - x^2}\,dx$ is:

$$\int\sqrt{a^2-x^2}\,dx = \frac{1}{2}\left(x\sqrt{a^2-x^2} + a^2 \sin^{-1} \frac{x}{a}\right).$$

Therefore,

$$A_{total} = 4\frac{b}{a}\left[\frac{1}{2}(x\sqrt{a^2-x^2} + a^2 \sin^{-1} \frac{x}{a}\right]_0^a$$

$$= 4\frac{b}{a}\left(\frac{1}{2} a^2 \sin^{-1} 1\right).$$

The $\sin^{-1} 1 = \frac{\pi}{2}$, therefore $A_{total} = \pi ab$.

• PROBLEM 823

Evaluate the integral: $\int_{-2}^{3} |x+1|dx$.

Solution: The expression as it stands is not integrable. But we can rewrite $f(x) = |x+1|$ as:

$$f(x) = \begin{cases} x + 1 \ , & x \geq -1, \\ - x - 1, & x < -1. \end{cases}$$

Then,

$$\int_{-2}^{3} |x+1|dx = \int_{-2}^{-1} (-x-1)dx + \int_{-1}^{3} (x+1)dx,$$

where the point of division is taken as the point at which the formula for f(x) is changed. The two integrals on the right are evaluated as follows:

$$\int_{-2}^{-1} (-x-1)dx = \int_{-2}^{-1} [-xdx - dx].$$

We make use of the formula for $\int u^n du$ on the first integral, obtaining:

$$\left[\frac{-x^2}{2} - x\right]_{-2}^{-1} = \frac{1}{2},$$

$$\int_{-1}^{3} (x+1)dx = \left[\frac{x^2}{2} + x\right]_{-1}^{3} = 8.$$

Consequently,

$$\int_{-2}^{3} |x+1|dx = 8 + \frac{1}{2} = \frac{17}{2}.$$

• PROBLEM 824

Find the area bounded by the sine curve between $x = 0$, $x = 4\pi$.

Solution: The area of the sine curve bounded by the sine curve and the x-axis is divided into 4 equal parts, 2 above and 2 below the x-axis. We can find the total area by multiplying the area between $x = 0$ and $x = \pi$ by four. The area is given by the integral of the function $y = \sin x$, between $x = 0$ to $x = \pi$. Hence,

$$A_{total} = 4 \int_{0}^{\pi} \sin x$$

$$= 4[-\cos x]_{0}^{\pi}$$

$$= 4\left[-(-1)-(-1)\right]$$
$$= 8 \quad .$$

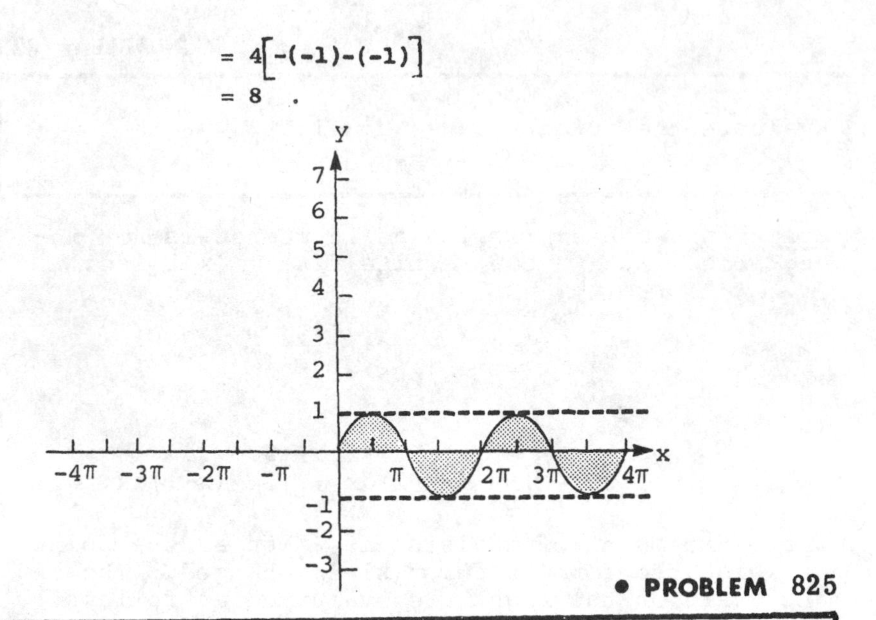

• PROBLEM 825

Find the area of the region in the first quadrant bounded by the curves y = sin x, y = cos x, and the y-axis.

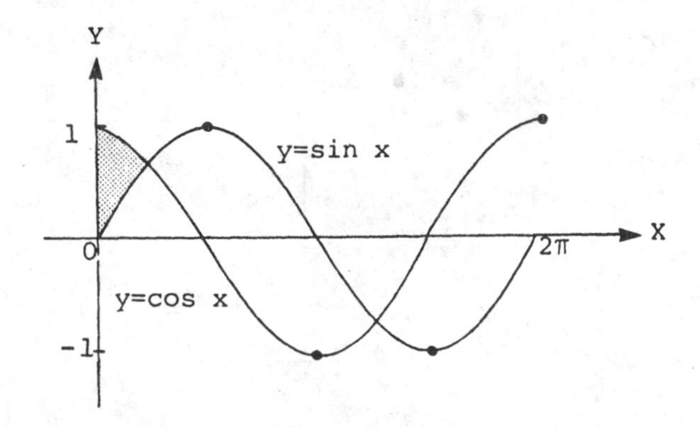

Solution: The limits of the integral which give the required area are the points of intersection of the two curves. To find the points of intersection, set y = sin x equal to y = cos x and solve for x. Solving, we obtain,

$$\sin x = \cos x$$

$$\frac{\sin x}{\cos x} = 1$$

$$\tan x = 1$$

$$x = \frac{\pi}{4}, \quad x = \frac{5\pi}{4}, \ \ldots$$

Since we are asked to find the region in the first quadrant bounded by the y-axis and the two curves, we only need the point of intersection where

$x = \frac{\pi}{4}$. For the required area, the curve: $y = \cos x$ is above the curve: $y = \sin x$, therefore the area is found by taking the integral from $x = 0$ to $x = \frac{\pi}{4}$ of the upper function minus the lower function, or the cosine curve minus the sine curve.

$$A = \int_0^{\pi/4} (\cos x - \sin x)dx$$

$$= (\sin x + \cos x)\Big]_0^{\pi/4}$$

$$= \left(\frac{\sqrt{2}}{2} + \frac{\sqrt{2}}{2}\right) - (0 + 1)$$

$$= \sqrt{2} - 1$$

• PROBLEM 826

Find the area bounded by $y = \cos(x) + 1$, $y = \frac{3}{2}$, $x = 0$, and $x = \pi$.

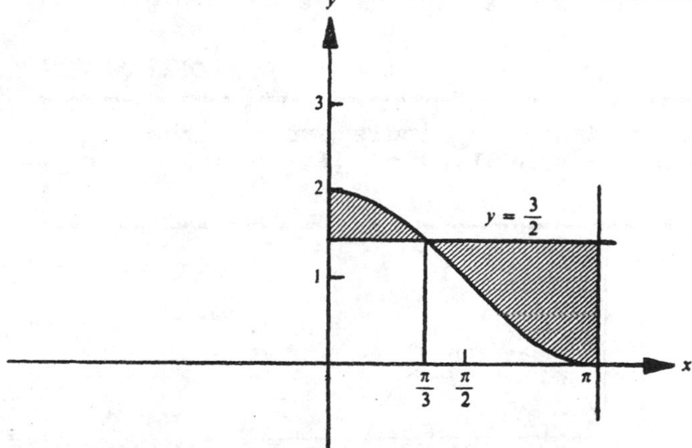

Solution: The total area required is split into two separate unequal areas by the intersection of the two curves. We can then evaluate this total area as the sum of the two areas. The area is the integral of the upper function minus the lower function. For the area to the left of the point of intersection of the two curves, (see the accompanying diagram), the upper function is $y = \cos(x) + 1$ and the lower function is $y = \frac{3}{2}$. The limits of the integral are $x = 0$ and the x-coordinate of the point of intersection. The point of intersection can be found by setting the two curves equal and solving for x. Hence,

$$\cos(x) + 1 = \frac{3}{2}$$

$$x = \frac{\pi}{3}, \text{ and}$$

$$y = \frac{3}{2}.$$

Therefore, the upper limit is $x = \frac{\pi}{3}$.

For the area to the right of the point of intersection, the upper function is $y = \frac{3}{2}$ and the lower function is $y = \cos(x) + 1$. The limits of this integral are $x = \frac{\pi}{3}$ and $x = \pi$. We can now write:

$$A_{total} = \int_0^{\frac{\pi}{3}} \left[(\cos(x) + 1) - \left(\frac{3}{2}\right)\right] dx$$

$$+ \int_{\frac{\pi}{3}}^{\pi} \left[\left(\frac{3}{2}\right) - (\cos(x) + 1)\right] dx$$

$$= \int_0^{\frac{\pi}{3}} \left(\cos x - \frac{1}{2}\right) dx + \int_{\pi/3}^{\pi} \left(\frac{1}{2} - \cos x\right) dx$$

$$= \sin x - \frac{x}{2}\Big]_0^{\pi/3} + \left(\frac{x}{2} - \sin x\right)\Big]_{\pi/3}^{\pi}$$

$$= \sin \frac{\pi}{3} - \frac{\pi}{6} + \left(\frac{\pi}{2} - \left[\frac{\pi}{6} - \sin \frac{\pi}{3}\right]\right)$$

$$= \frac{\sqrt{3}}{2} - \frac{\pi}{6} + \left(\frac{\pi}{2} - \frac{\pi}{6} + \frac{\sqrt{3}}{2}\right) = \sqrt{3} + \frac{\pi}{6}.$$

• PROBLEM 827

Find the area bounded by a single arch of the cycloid: $x = a(\theta - \sin \theta)$, $y = a(1 - \cos \theta)$; and the x-axis.

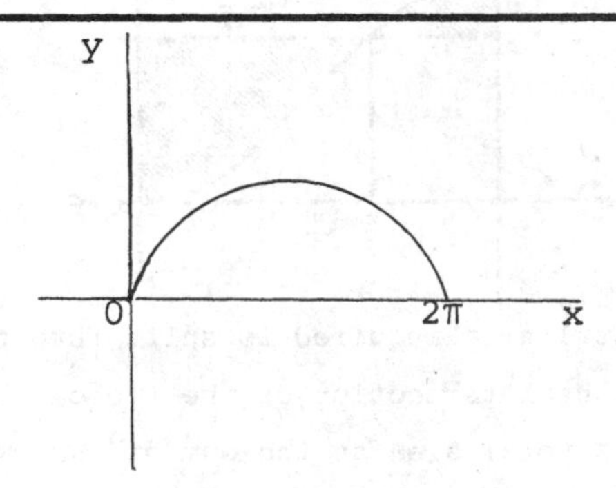

Solution: The area is defined by $A = \int y\,dx$. We can write, by definition: $dx = \left(\frac{dx}{d\theta}\right) d\theta =$ $\frac{d}{d\theta}\left[a(\theta - \sin \theta)\right] d\theta = (a - a \cos \theta) d\theta$.
$y = a(1 - \cos \theta)$, as given. One arch of the cycloid goes from $\theta = 0$ to $\theta = 2\pi$, as shown in the diagram. Hence, we can now write:

$$A = \int_0^{2\pi} y \frac{dx}{d\theta} d\theta$$

$$= \int_0^{2\pi} a(1 - \cos \theta) \cdot a(1 - \cos \theta)(d\theta)$$

$$= a^2 \int_0^{2\pi} (1 - \cos\theta)^2 d\theta$$

$$= a^2 \int_0^{2\pi} (1 - 2\cos\theta + \cos^2\theta) d\theta.$$

Substituting $\cos^2\theta = \frac{1 + \cos 2\theta}{2}$ and integrating, we obtain:

$$A = a^2 \left[\theta - 2\sin\theta + \frac{\theta}{2} + \frac{1}{4}\sin 2\theta\right]_0^{2\pi}.$$

$\therefore$ Area $= a^2(2\pi - 0 + \pi + 0) = 3\pi a^2$ sqaure units.

• PROBLEM 828

Find the area bounded by the catenary: $y = \frac{1}{2}\left(e^{x/2} + e^{-x/2}\right)$ the x-axis, and the ordinates: $x = -1$ and $x = 1$.

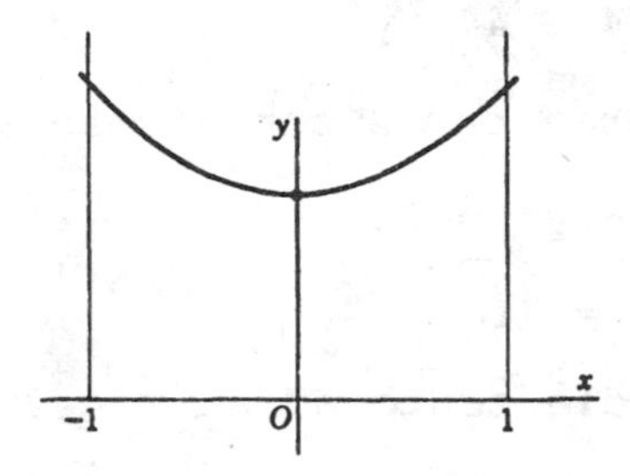

Solution: The catenary is the curve naturally assumed by a flexible, heavy cable having no load but its own weight, hanging freely between two supports. Because of the symmetry of the catenary about the y-axis, the required area is two times the area in the first quadrant. Area is the integral of the upper function minus the lower function. The limits of the integral are the y-axis, or $x = 0$, and the boundary line: $x = 1$. The upper function is $y = \frac{1}{2}\left(e^{x/2} + e^{-x/2}\right)$ and the lower function is $y = 0$ (which is the x-axis). We can now write:

$$A_{total} = 2 \int_0^1 \left[\frac{1}{2}\left(e^{x/2} + e^{-x/2}\right) - 0\right] dx$$

$$= 2 \int_0^1 \frac{1}{2}\left(e^{x/2} + e^{-x/2}\right) dx$$

$$= 2\left[e^{x/2} - e^{-x/2}\right]_0^1 = 2\left(e^{1/2} - e^{-1/2}\right)$$

$$= 2(1.649 - .607) = 2.084.$$

Find the smallest area bounded by the circles:

$$x^2 + y^2 = 4, \quad \text{and} \quad x^2 + y^2 = 4x.$$

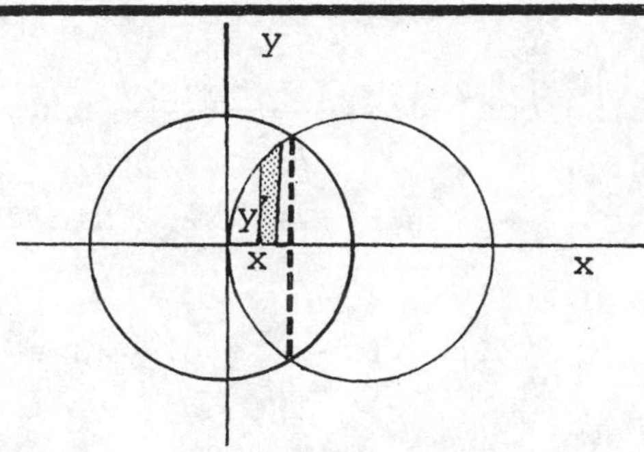

Solution: The two circles lie in the x-y plane. The first circle is centered at the origin with a radius of 2 units. We plot the second one by completing the square. The expression:

$$x^2 - 4x + y^2 = 0,$$

can be written as:

$$x^2 - 2(2x) + 4 - 4 + y^2 = 0,$$

where the first three terms constitute a square.

$$(x - 2)^2 - 4 + y^2 = 0,$$

or,

$$(x - 2)^2 + y^2 = 4.$$

We have a circle centered at $x = 2$ and a radius of 2 units.

Considering the symmetry about the point (1,0), we calculate one-fourth of the required area

$$A' = \int_0^1 y\, dx.$$

We note that the curve y, as shown by the shaded portion of the figure, belongs to the circle centered at (2,0). I.e.,

$$y = \sqrt{4 - (x - 2)^2}.$$

Hence,

$$A' = \int_{x=0}^{x=1} \left[4 - (x - 2)^2\right]^{\frac{1}{2}} dx.$$

Letting $x - 2 = t$, using integration by parts, and then substitution,

$$A' = \int_{t=-2}^{t=-1} \sqrt{4 - t^2}\, dt$$

$$= \frac{1}{2} \left[t \sqrt{4 - t^2} + 4 \sin^{-1} \frac{t}{2} \right|_{-2}^{-1}$$

$$= \frac{1}{2} \left[- 4 \sin^{-1} \frac{1}{2} + 4 \sin^{-1} 1 - \sqrt{3} \right].$$

Now, the required area is 4A', which is:

$$A = 2 \left[- \frac{2\pi}{3} - \sqrt{3} + 2\pi \right]$$

$$= 4\pi - \frac{4\pi}{3} - 2 \sqrt{3} = \frac{8\pi}{3} - 2 \sqrt{3}.$$

• PROBLEM 830

Find the area in the first quadrant under the curve $(x^3 + 8)y = 8$.

Solution: The curve is asymptotic to the x-axis and the line $x = -2$. In the first quadrant, $y = 1$ for $x = 0$, and then y decreases as x increases. We wish to find the area in the form:

$$A = \int_0^\infty y \, dx,$$

and hence,

$$A = \int_0^\infty \frac{8}{x^3 + 8} dx = 8 \int_0^\infty \frac{dx}{x^3 + 2^3}.$$

To integrate, we factor the denominator and use partial fractions. $x^3 + 2^3 = (x + 2)(x^2 - 2x + 4)$. Hence,

$$\frac{1}{x^3 + 2^3} = \frac{1}{(x + 2)(x^2 - 2x + 4)}$$

$$= \frac{A}{x + 2} + \frac{B(x - 2) + C}{x^2 - 2x + 4},$$

or,

$$\frac{1}{x^3 + 2^3}$$

$$= \frac{Ax^2 - 2Ax + 4A + Bx^2 - 4B + Cx + 2C}{(x + 2)(x^2 - 2x + 4)}.$$

Canceling denominators and equating like coefficients, we obtain:

for x^2 : $A + B = 0.$

x : $- 2A + C = 0.$

1 : $4A - 4B + 2C = 1.$

$$A = \frac{1}{12}, \quad B = - \frac{1}{12} \quad \text{and} \quad C = \frac{1}{6}.$$

Then,

$$A = 8\int_0^{\infty} \frac{dx}{x^3 + 2^3} = 8\int_0^{\infty} \frac{dx}{12(x + 2)}$$

$$- 8\int_0^{\infty} \frac{-\frac{1}{12}(x - 2) + \frac{1}{6}}{x^2 - 2x + 4} dx$$

$$= \frac{2}{3}\int_0^{\infty} \frac{dx}{x + 2} - \frac{2}{3}\int_0^{\infty} \frac{4 - x}{x^2 - 2x - 4}$$

$$= \frac{2}{3} \ln(x + 2) + \frac{1}{3}\int \frac{(2x - 2)dx}{x^2 - 2x - 4}$$

$$- \frac{1}{3}\int \frac{6\, dx}{x^2 - 2x - 4}$$

$$= \frac{2}{3} \ln(x + 2) + \frac{1}{3} \ln(x^2 - 2x - 4)$$

$$- 2\int \frac{dx}{(x - 1)^2 + 3}$$

$$= \frac{2}{3} \ln(x + 2) + \frac{1}{3} \ln(x^2 - 2x + 4)$$

$$- \frac{2}{\sqrt{3}} \text{arc tan} \frac{x - 1}{\sqrt{3}} .$$

• **PROBLEM 831**

Find the total area bounded by the curve:

$$x^2y^2 + 4x^2 - 4y^2 = 0$$

and its asymptotes.

Solution: The curve is symmetric about the origin and asymptotic to x = 2 and x = - 2 in the region of $-2 \le x \le 2$. We calculate the area enclosed by the positive y-axis and the given curve:

$$y = \frac{\pm\ 2x}{\sqrt{4 - x^2}} .$$

The improper integral:

$$\int_0^{\infty} x\, dy = \int_0^{\infty} \frac{\pm\ 2y}{\sqrt{y^2 + 4}}\, dy,$$

does not exist. Therefore we perform the integration with respect to x in the first quadrant:

$$A = 4\int_0^2 y\,dx = 4\int_0^2 \frac{2x}{\sqrt{4 - x^2}}\,dx$$

$$= 8\left(-\sqrt{4 - x^2}\right|_0^2$$

$$= 16.$$

• PROBLEM 832

What is the area bounded by the curve: $\rho = \sin\theta$, as θ varies from 0 to π?

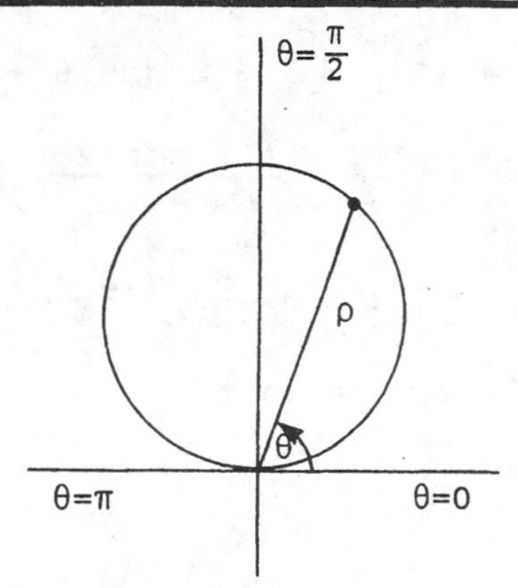

<u>Solution:</u> In polar coordinates, the area is given by:

$$A = \int_\alpha^\beta \frac{1}{2}\rho^2 d\theta.$$

Here, the limits are 0 and π. Hence we can write:

$$A = \int_0^\pi \frac{1}{2}(\sin\theta)^2 d\theta$$

$$= \frac{1}{2}\int_0^\pi \sin^2\theta\,d\theta.$$

Substituting: $\sin^2\theta = \frac{1 - \cos 2\theta}{2}$, and integrating, we obtain:

$$A = \frac{1}{2}\left[\frac{\theta}{2} - \frac{\sin 2\theta}{2}\right]_0^\pi$$

$$= \frac{\pi}{4} \text{ square units.}$$

This is a circle of radius $\frac{1}{2}$ centered at $\left(\frac{1}{2}, \frac{\pi}{2}\right)$. Its area can be checked by $\pi r^2 = \frac{\pi}{4}$.

• PROBLEM 833

Find the area of the cardioid: $r = a(1 + \cos\theta)$.

<u>Solution:</u> From the diagram we can see that the area

is split into two equal parts, therefore the total area can be written as 2 times the area above the polar line A. The formula for area in polar coordinates is:

$$A = \int_{\alpha}^{\beta} \frac{1}{2}r^2 d\theta.$$

For the area above the polar line A, θ goes from 0 to π which are the limits of the integral for area. Therefore we can write:

$$A = 2\int_0^{\pi} \frac{1}{2}\Big(a(1+\cos\theta)\Big)^2 d\theta$$

$$= \int_0^{\pi} a^2(1+\cos\theta)^2 d\theta$$

$$= a^2\int_0^{\pi}\Big(1 + 2\cos\theta + \cos^2\theta\Big)d\theta.$$

Now, substituting: $\cos^2\theta = \frac{1+\cos 2\theta}{2}$, and integrating, we obtain:

$$A = a^2\left(\theta + 2\sin\theta + \frac{\theta}{2} + \frac{\sin 2\theta}{4}\right)\Bigg|_0^{\pi}$$

$$= 3\frac{\pi a^2}{2} \text{ square units.}$$

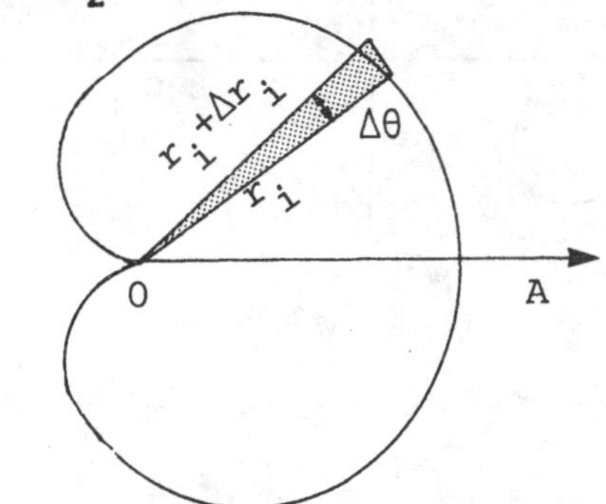

• PROBLEM 834

Find the total area enclosed by the curve: $r^2 = \cos\theta$.

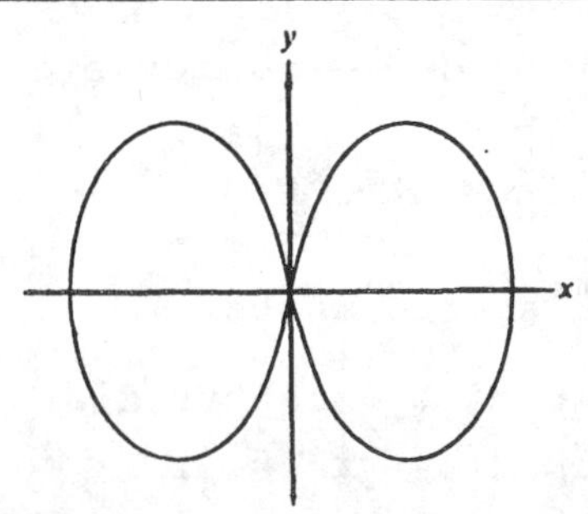

Solution: The total area of this curve can be expressed as 4 times the area of the curve in the first quadrant. In the first quadrant, θ goes from zero to $\frac{\pi}{2}$, which are the limits of the integral that gives the required area. The formula for the area of a curve in polar coordinates is:

$$A = \int_{\alpha}^{\beta} \frac{1}{2} r^2 d\theta.$$

The given curve is: $r^2 = \cos\theta$. The integral can therefore be written as follows:

$$A = 4\int_0^{\pi/2} \frac{1}{2}\cos\theta\, d\theta = 2\Big[\sin\theta\Big]_0^{\pi/2} = 2,$$

for the entire area.

• PROBLEM 835

What is the area enclosed by the curve:

$$\rho = a(1 - \cos\theta)^{1/2}?$$

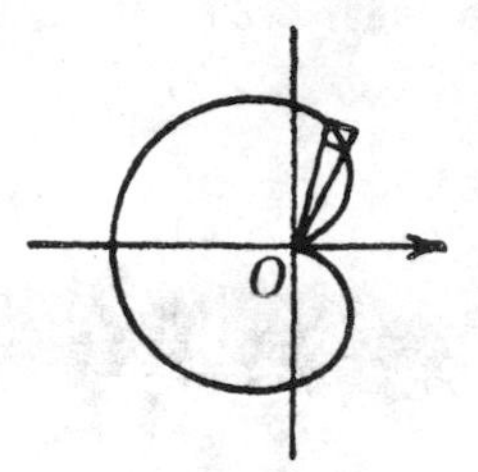

Solution: The total area of this curve covers each of the four quadrants. Therefore θ goes from 0 to 2π, which are the limits of the integral that gives the required area. The formula for area in polar coordinates is:

$$A = \int_{\alpha}^{\beta} \frac{1}{2}\rho^2 d\theta.$$

Therefore we can write:

$$A = \int_0^{2\pi} \frac{1}{2}\left(a(1 - \cos\theta)^{1/2}\right)^2 d\theta$$

$$= \frac{a^2}{2}\int_0^{2\pi} (1 - \cos\theta)d\theta = \frac{a^2}{2}\Big[\theta - \sin\theta\Big]_0^{2\pi}$$

$$= \frac{a^2}{2}(2\pi - 0 - 0 + 0) = \frac{a^2 \cdot 2\pi}{2} = \pi a^2.$$

• PROBLEM 836

The expression for a four-leaved rose is: $\rho = a \sin 2\theta$. Find its area.

Solution: The entire area of the four-leaved rose is divided evenly into each of the four quadrants. Therefore the total area can be expressed as four times the area in the first quadrant. The formula for area expressed in polar coordinates is:

$$A = \int_{\alpha}^{\beta} \frac{1}{2}\rho^2 d\theta.$$

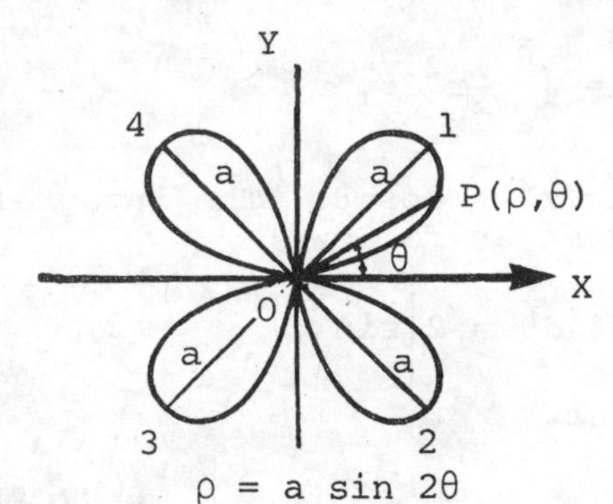

$\rho = a \sin 2\theta$

The integral for the area in the first quadrant is expressed as θ goes from 0 to $\frac{\pi}{2}$, which are the limits. We can then express the total area as:

$$A = 4 \int_0^{\pi/2} \frac{1}{2}(a \sin 2\theta)^2 d\theta.$$

$$= 2a^2 \int_0^{\pi/2} \sin^2 2\theta \, d\theta.$$

But $\sin^2 2\theta = \frac{1}{2} - \frac{1}{2} \cos 4\theta$. Substituting, we therefore obtain:

$$A = 2a^2 \int_0^{\pi/2} \left(\frac{1}{2} - \frac{1}{2} \cos 4\theta\right) d\theta = a^2 \int_0^{\pi/2} (1 - \cos 4\theta) d\theta$$

$$= a^2 \left[\theta - \frac{1}{4} \sin 4\theta\right]_0^{\pi/2} = a^2 \left[\frac{\pi}{2} - 0 - 0 + 0\right] = \frac{\pi a^2}{2},$$

the area of the 4 leaves.

• PROBLEM 837

Find the area inside the curve: $r = \cos \theta$, and outside the curve: $r = 1 - \cos \theta$.

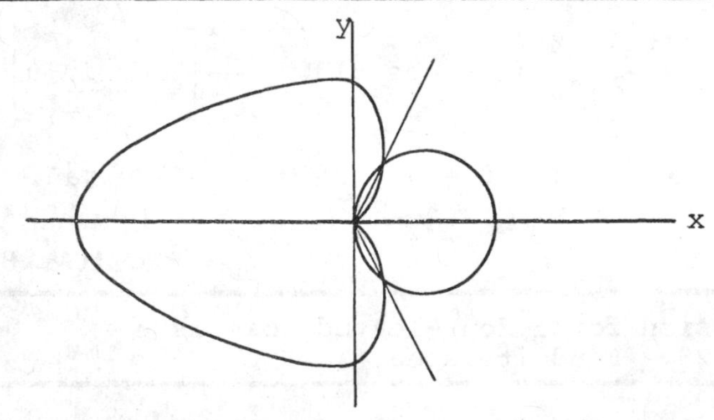

<u>Solution:</u> The limits of the integral which give us the required area are the points of intersection of the two curves. To find the points of intersection we set the two curves equal and solve for θ.

$$\cos \theta = 1 - \cos \theta$$
$$2 \cos \theta = 1$$

$$\cos\theta = \frac{1}{2}$$

$$\theta = \frac{\pi}{3}, -\frac{\pi}{3}.$$

Since the area required is divided evenly above and below the x-axis, we can multiply the value we obtain for the area above the x-axis by 2. The limits for the integral which give us the area above the x-axis are $x = 0$ and $x = \frac{\pi}{3}$. (The value of θ which is $-\frac{\pi}{3}$ would be in the fourth quadrant). Now we can set up the integral. The formula for the area in polar coordinates is:

$$A = \int_{\alpha}^{\beta} \frac{1}{2} r^2 d\theta.$$

The total area required in this problem is two times the difference between the areas of the two curves, above the x-axis. Hence,

$$A = 2\int_0^{\pi/3} \frac{1}{2}(\cos\theta)^2 d\theta - 2\int_0^{\pi/3} \frac{1}{2}(1 - \cos\theta)^2 d\theta$$

$$= \int_0^{\pi/3} \left[\cos^2\theta - (1 - \cos\theta)^2\right] d\theta$$

$$= \int_0^{\pi/3} \left(\cos^2\theta - 1 + 2\cos\theta - \cos^2\theta\right) d\theta$$

$$= \int_0^{\pi/3} (2\cos\theta - 1) d\theta$$

$$= \left[2\sin\theta - \theta\right]_0^{\pi/3} = 2\sin\frac{\pi}{3} - \frac{\pi}{3} = \sqrt{3} - \frac{\pi}{3}.$$

• PROBLEM 838

Find the area outside the circle: $r = 2a\cos\theta$ and inside the cardioid: $r = a(1 + \cos\theta)$.

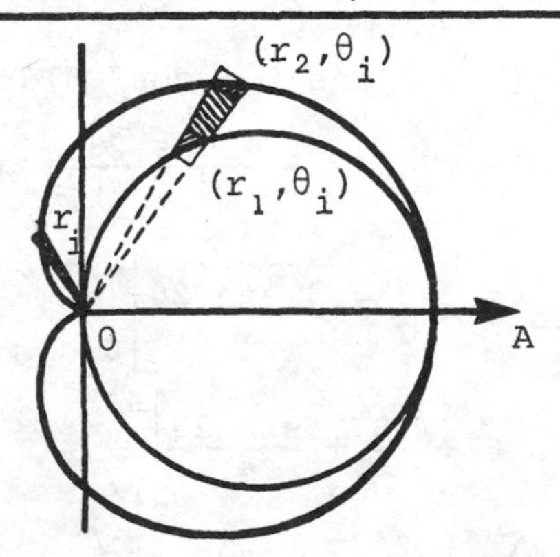

<u>Solution:</u> The required area is split evenly into two equal parts, above and below the x-axis. Therefore, to obtain the total area we can multiply the value for the area above the x-axis by 2. The

area above the x-axis can, in turn, be split into two parts and expressed as the sum of the area in the first quadrant and the area in the second quadrant. The area in the first quadrant can be expressed as the area of the cardioid minus the area of the circle, where θ goes from 0 to $\frac{\pi}{2}$. These are the limits of the integral for the area. The area in the second quadrant can be found by using the expression for the cardioid alone, and where θ goes from $\frac{\pi}{2}$ to π. These are the limits of the integral which give the required area.

The formula for area expressed in polar coordinates is:

$$A = \int_{\alpha}^{\beta} \frac{1}{2} r^2 d\theta.$$

Hence we have,

$$A_{total} = 2 \int_{0}^{\pi/2} \frac{1}{2}\Big(a(1 + \cos\theta)\Big)^2 d\theta - 2 \int_{0}^{\pi/2} \frac{1}{2}(2a\cos\theta)^2 d\theta + 2 \int_{\pi/2}^{\pi} \frac{1}{2}\Big[a(1 + \cos\theta)\Big]^2 d\theta$$

$$A_{total} = a^2 \int_{0}^{\pi/2} (1 + 2\cos\theta + \cos^2\theta)d\theta - a^2 \int_{0}^{\pi/2} 4\cos^2\theta\, d\theta + a^2 \int_{\pi/2}^{\pi} \cdot(1 + 2\cos\theta + \cos^2\theta)d\theta$$

$$A_{total} = a^2 \int_{0}^{\pi/2} (1 + 2\cos\theta - 3\cos^2\theta)d\theta + a^2 \int_{\pi/2}^{\pi} (1 + 2\cos\theta + \cos^2\theta)d\theta.$$

Now, substituting: $\cos^2\theta = \frac{1 + \cos 2\theta}{2}$, and integrating, we obtain,

$$A_{total} = a^2 \left[2\sin\theta - \frac{\theta}{2} - \frac{3\sin 2\theta}{4}\right]_{0}^{\pi/2} + a^2 \left[2\sin\theta + \frac{3\theta}{2} + \frac{\sin 2\theta}{4}\right]_{\pi/2}^{\pi}$$

$$A_{total} = a^2\left(2 - \frac{\pi}{4}\right) + a^2\left(\frac{3\pi}{4} - 2\right) = \frac{\pi a^2}{2}.$$

• PROBLEM 839

Derive the equation for the area of a circle.

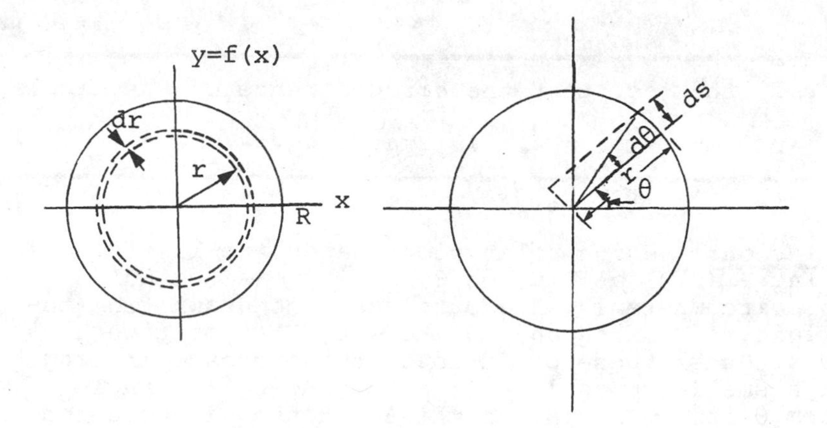

Solution: Looking at the diagram, we can say that the ring formed by the two dotted lines is a rectangle curved to close on itself. The rectangle is an infinitesimal part of the entire circle. It has a thickness dr and its length is $2\pi r$. The area of the rectangle is then $2\pi r dr$. The area of the entire circle is then the sum of all these infinitesimal areas, which can be written as the integral of $2\pi r\, dr$ from $r = 0$ to $r = R$. Hence,

$$A = \int_0^R 2\pi r\, dr$$

$$A = 2\pi \int_0^R r\, dr$$

$$= 2\pi \left[\frac{r^2}{2}\right]_0^R = 2\pi\left(\frac{R^2}{2} - \frac{0^2}{2}\right) = \pi R^2.$$

A different procedure can also be used to solve this problem. A formula can be derived for finding area using polar coordinates. The area of the infinitesimal rectangle shown is length r times width ds, or rds. Therefore the area of the triangle is $\frac{1}{2}$ rds. But, $ds = rd\theta$. The area of the entire circle is the sum of all the infinitesimal triangles, or the integral of $\frac{1}{2}r(rd\theta)$ which is:

$$\int_\alpha^\beta \frac{1}{2} r^2 d\theta,$$

where α and β are the limits. For the entire circle, θ varies from 0 to 2π which are the limits α and β.

$$A = \int_0^{2\pi} \frac{1}{2} r^2 d\theta.$$

But r is a constant. Therefore,

$$A = \frac{r^2}{2} \int_0^{2\pi} d\theta = \frac{r^2}{2}[2\pi - 0]$$

$$= \pi r^2.$$

Verify the fact that the circumference of the circle

$$x^2 + y^2 = a^2 \text{ is } 2\pi a.$$

Solution: Using polar coordinates, we locate a point P (x,y) as P(a cos Θ, a sin Θ) situated on the circumference. The straight line drawn from the origin to P makes an angle Θ with the positive x-axis. Under these conditions, the increase in length of a small arc as a result of changing the angle from Θ to Θ + ΔΘ is: Δs = a ΔΘ. Hence, integration over the whole range of Θ, after passing to the limit, yields the circumference:

$$S = \int_0^{2\pi} a d\Theta = 4 \int_0^{\pi/2} a d\Theta = 2\pi a.$$

To find the area, since it is known that the circumference of a circular loop of f radius r is $2\pi r$, and the thickness of a ring-like element of area is dr, it has an area $2\pi r\, dr$. Integration over $0 \leq r \leq a$ gives:

$$A = 2\pi \int_0^a r\, dr = 2\pi \left| \frac{1}{2} r^2 \right|_0^a = \pi a^2.$$

CHAPTER 31

SOLIDS: VOLUMES AND AREAS

Solids of revolution may be generated when a plane curve is revolved around an axis, thus generating the solid. If we consider just one point revolving, then it generates a circle. If, instead of a point, we now take a length, Δx or Δy, and revolve it around an axis, say, for example, the X-axis or the Y-axis, we have a small cylinder or disk. The volume of this cylinder is: $\pi x^2 dy$, when revolving around the Y-axis or, $\pi y^2 dx$, when revolving around the X-axis.

The volume of a solid of revolution is, then, the sum of an infinite number of such disks, which in calculus terms, is the integral of the infinitesimal volumes given above.

In some cases, the volume of revolution may be generated by two curves rather than by only one. In that case, we employ the coordinates of both curves. For example, if revolving around the Y-axis, and given that $x_1 = f(y)$ and $x_2 = g(y)$, we obtain a "washer" rather than a solid disk. The volume of such a washer or disk with hollow center is given by:

$$\pi\left(x_1^2 - x_2^2\right) dy,$$

where all x's are to be expressed in terms of y.

The "shell method" may also be used to find volumes of solids of revolution. In this method, the volume is subdivided into infinitesimal cylindrical shells. By summing the infinitesimal volumes of these shells by integration, the total volume of the solid may be found.

Depending on the character of the equation of the plane curve to be revolved, either the disk or the shell method may be applied best. Experience and trial-and-error must again be relied on to select the best method for a given problem.

AREA OF SURFACE OF REVOLUTION

If a given curve is revolved around an axis, it generates a surface of revolution. The area of the surface of revolution may be found from a summation of small (infinitesimal) cylinders. Consider that the curve is expressed as $y = f(x)$, and that it is revolved around the y-axis. Then the cylinder has a radius x, and hence a circumference of $2\pi x$. The height ds, of the cylinder is defined from

$$ds = \sqrt{dy^2 + dx^2} = \sqrt{1 + \left(\frac{dy}{dx}\right)^2}\, dx .$$

The surface of revolution, therefore, has an area equal to $\int 2\pi x\, ds$, or

$$2\pi\int_a^b x\sqrt{1 + \left(\frac{dy}{dx}\right)^2}\, dx,$$

where the limits a and b are the values of x at the beginning and end of the portion of the curve to be considered.

A corresponding formula can be derived for rotation around the x-axis.

• PROBLEM 841

Find the volume of: $y = f(x) = 2x$, when rotated about the x axis and bounded by $x = 2$.

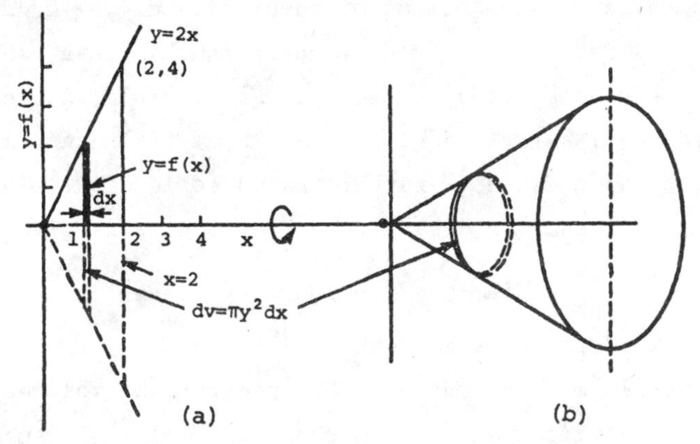

Solution: The shaded strip, as shown in figure (a), when rotated about the x-axis, sweeps a volume expressed by the disk as in (b). The radius of this disk is y. Hence its volume is $\pi y^2\, dx$, where dx is the thickness of the disk. The sum of the volumes of all such disks for $0 \le x \le 2$ and passing to the limit gives:

$$V = \int_0^2 (\pi y^2)\, dx = \pi \int_0^2 (2x)^2\, dx$$

$$= 4\pi \int_0^2 x^2\, dx = 4\pi \left.\frac{x^3}{3}\right|_0^2 = \frac{32\pi}{3} .$$

Since an increment of volume is a disk, this is called the disk method.

Solution #2: Shell Method

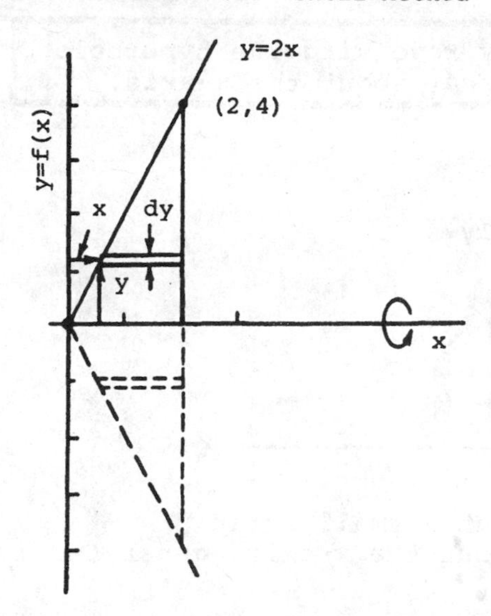

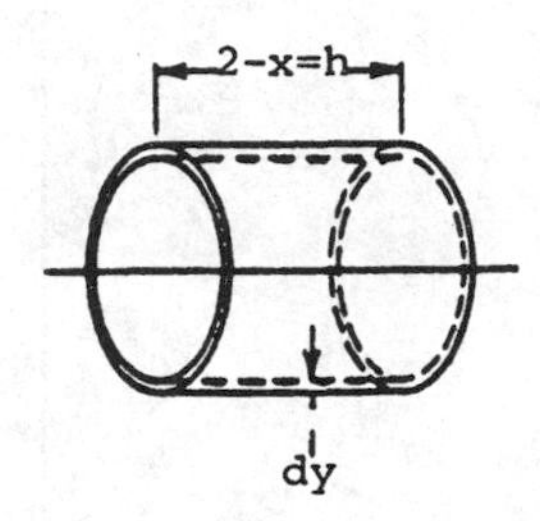

Element of volume is shell

This volume might be conceived of as a constantly expanding cylindrical shell, the radius y of which increases from 0 to 4. The height (x value) varies from 2 down to 0. The element of volume is obtained by multiplying circumference by height and by differential of thickness:

$$dV = 2\pi r\ h\ dy$$

$$= 2\pi y(2 - x)dy$$

One can observe that the height of the element is really the length on the coordinate axis between the outer boundary and the equation of the line y - 2x. Since x is the distance to the function, 2 - x equals the height of the shell. All the unknowns in the dV expression must be in terms of y, since the differential term is dy. Therefore, we substitute y/2 for x (from the original equation):

$$dV = 2\pi y \left(2 - \frac{y}{2}\right) dy$$

$$\int dV = 2\pi \int_0^4 \left(2y - \frac{y^2}{2}\right) dy$$

$$V = 2\pi \left(y^2 - \frac{y^3}{6}\right)\Bigg|_0^4 = \frac{32\pi}{3}.$$

This method is more work than the disk method. It is however, a convenient method to use when integration by another method becomes too complex.

● **PROBLEM 842**

Find the volume formed by revolving the hyperbola: $xy = 6$, from $x = 2$ to $x = 4$, about the x-axis.

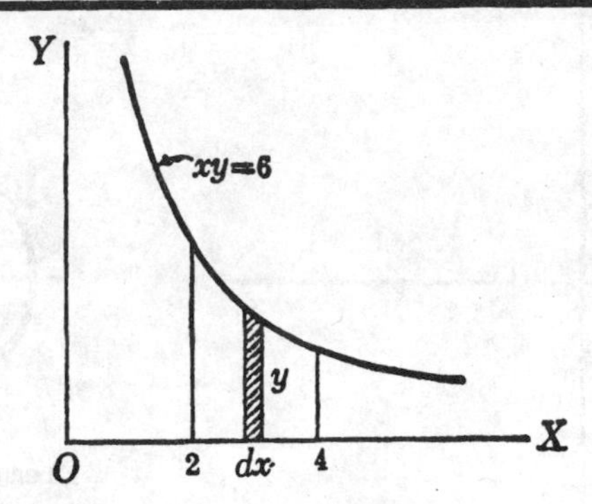

<u>Solution:</u> We assume that a small strip, y by dx, when rotated about the x-axis, generates an element of volume,

$$dV = \pi y^2 \cdot dx,$$

the differential volume of the cylindrical element. Since

$$xy = 6, \quad y = \frac{6}{x}.$$

Then

$$V = \int_2^4 \pi \cdot \left(\frac{6}{x}\right)^2 \cdot dx = 36\pi \int_2^4 x^{-2} \cdot dx$$

$$= 36\pi \left[\frac{x^{-2+1}}{-2+1}\right]_2^4 = -36\pi \left[\frac{1}{x}\right]_2^4$$

$$= -36\pi \left[\frac{1}{4} - \frac{1}{2}\right] = -36\pi \cdot -\frac{1}{4}$$

$$= 9\pi.$$

● **PROBLEM 843**

Find the volume generated by revolving about the x-axis the area bounded by the hyperbola: $xy = 9$, the x-axis, and the abscissae $x = 3$ and $x = 9$.

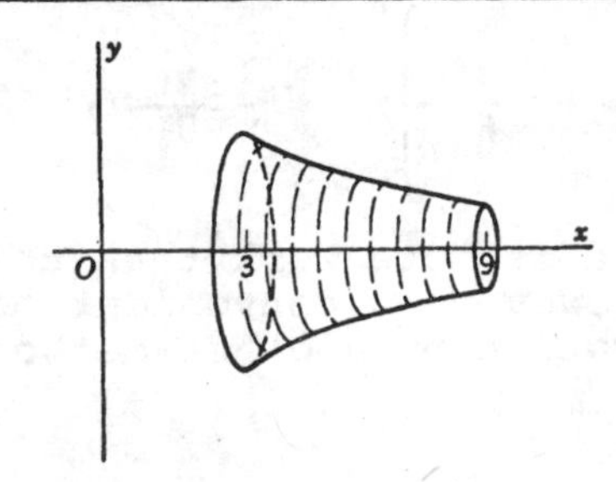

Solution: An element of volume is the disk. Its volume is:

$$dV = \pi y^2 \, dx. \quad y = \frac{9}{x}.$$

$$V = \pi \int_3^9 \left(\frac{9}{x}\right)^2 dx = 81\pi \int_3^9 \frac{1}{x^2} dx$$

$$= 81\pi \left[-\frac{1}{x}\right]_3^9 = 81\pi \left(-\frac{1}{9} + \frac{1}{3}\right)$$

$$= 18\pi.$$

• PROBLEM 844

Find the volume of the solid generated by revolving about the X-axis the region bounded by the parabola: $y^2 = 4x$, the X-axis and the lines: $x = 0$ and $x = 4$.

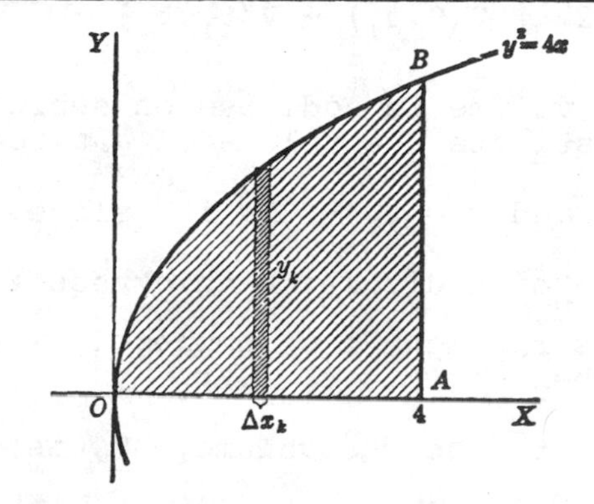

Solution: Here an elemental strip has the dimensions $y_k \, \Delta x_k$. When rotated, it generates a cylindrical disk with the volume:

$$\Delta V = \pi {y_k}^2 \, \Delta x_k.$$

In the limit, this becomes:

$$dV = \pi y^2 \, dx.$$

Hence,

$$V = \pi \int_0^4 y^2 \, dx = \pi \int_0^4 4\,x \, dx$$

$$= \pi \cdot \left.\frac{4x^2}{2}\right|_0^4 = 32\pi.$$

• PROBLEM 845

Compute the volume V of the solid formed by rotating about the x axis the region bounded by

$y = x^2$, the x axis, and the line $x = 2$.

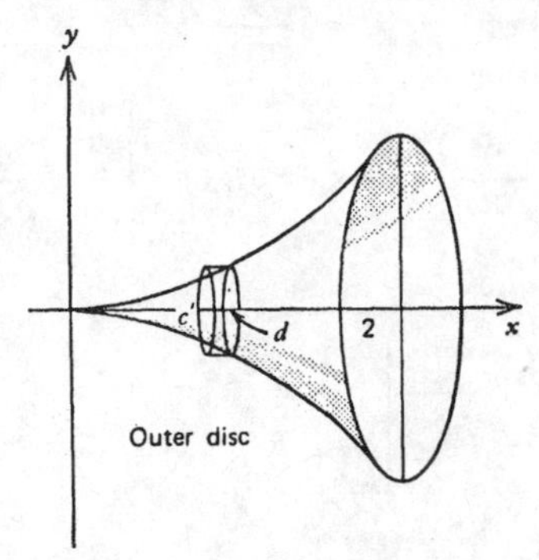

Solution: In this case, the desired interval is $(0,2)$. Let us denote the volume by $V(0,2)$. If we partition $(0,2)$ by points $t, \ldots, t_n$ such that $0 < t_1 < t_2 < \ldots < t_n < 2$, then $V(0,2)$ is the sum of the volumes of the respective strips. That is,

$$V(0,2) = V(0,t_1) + V(t_1,t_2) + \ldots + V(t_n,2).$$

Thus volume is additive on subintervals. Next we estimate the volume of a slice between $x = t_{i-1}$ and $x = t_i$ - the ith slice. Let v be the volume of a disc with thickness: $t_i - t_{i-1} = \Delta x$, and radius r, which lies completely within the region. Then, at $x = t_{i-1}$, $r = y$, $t_{i-1} \leq x \leq t_i$, and the volume, ΔV, is:

$$\pi r^2 \Delta x = \pi y^2 \Delta x.$$

We use the given equation of the curve, $y = x^2$, hence $\Delta V = \pi x^4 \Delta x$. If we let $r = y + \Delta y$ at $x = t_i$, then we obtain the inequality:

$$\pi y^2 \Delta x \leq \Delta V(t_{i-1},y) \leq \pi(y + \Delta y)^2 \Delta x.$$

Furthermore, if we make t_{i-1} so close to $x = t_i$ that the above inequality is not significant, the iterative sum of all such discs leads to an integral, in the region $0 \leq t_i \leq 2$, for $n \to \infty$.

Hence,

$$V = \pi \int_0^2 x^4 \, dx = \pi \frac{x^5}{5}\bigg|_0^2 = \frac{32\pi}{5}.$$

• PROBLEM 846

Find the volume of the solid obtained by revolving the area enclosed by $y = x^2$, $x = 1$, and $y = 0$, about the x-axis.

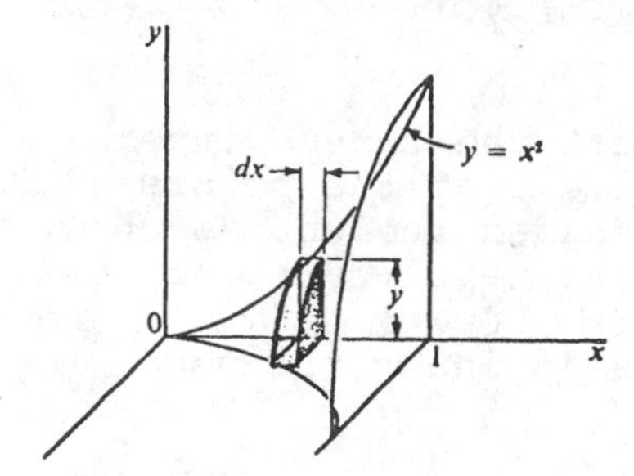

Solution: In the diagram, only one quarter of the volume is shown, in order to keep the figure as simple as possible. The volume can be partitioned into "slices", as indicated in the figure. The volume, dV, of a typical slice, which is a cylindrical disk, is:

$$dV = \pi r^2 \, dx,$$

where r is the radius of the disk. Evidently, the radius r is equivalent to the y-coordinate of the curve in this case, so that:

$$dV = \pi y^2 \, dx,$$

and

$$V = \pi \int_0^1 x^4 \, dx$$

$$= \frac{\pi}{5}.$$

• PROBLEM 847

Find the volume of the solid generated by revolving about the X-axis the region bounded by the curves: $y = x^2$, and $y = 2 - x^2$.

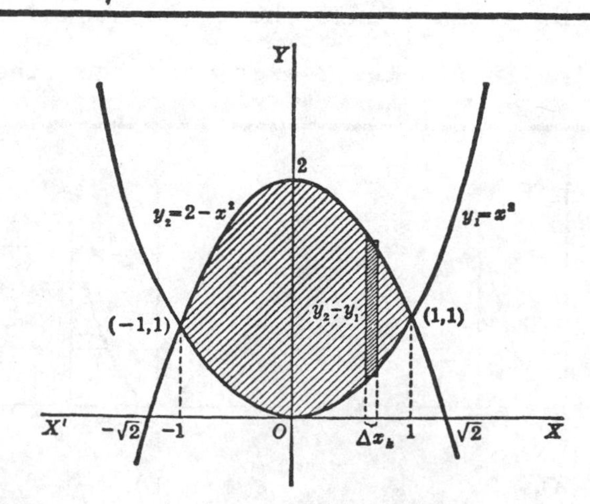

Solution: To find the points of intersection of the two curves (which we use as the boundary values for the region in question), we find the values of x, at which both curves have the same y-coordinates. Then, by substituting the values of x obtained into one of the given equations, we obtain the corresponding value of y.

$$y = x^2 = 2 - x^2,$$

which shows that the curves intersect at $x = 1$ and $x = -1$, i.e., at the points (1,1) and (- 1,1). The shaded region, as shown in the figure, is to be revolved about the X-axis, and it generates a hollow disk, or we can call it a washer. It has an inner radius:

$$y_1 = x_k^2,$$

outer radius: $y_2 = 2 - x_k^2$, and thickness Δx_k.

Its volume is:

$$V_k = \pi y_2^2 \Delta x_k - \pi y_1^2 \Delta x_k = \pi\left(y_2^2 - y_1^2\right) \Delta x_k.$$

Therefore, the required volume is:

$$V = \lim_{\Delta x \to 0} \Sigma V_k,$$

or,

$$V = \pi \int_{-1}^{1} \left(y_2^2 - y_1^2\right) dx$$

$$= \pi \int_{-1}^{1} \left[(2 - x^2)^2 - (x^2)^2\right] dx$$

$$= \pi \int_{-1}^{1} \left(4 - 4x^2\right) dx = \frac{16}{3} \pi.$$

• PROBLEM 848

Find the volume of the solid generated by revolving about the Y axis the region bounded by the parabola: $y = x^2$, the Y axis, and the line: $x = 2$.

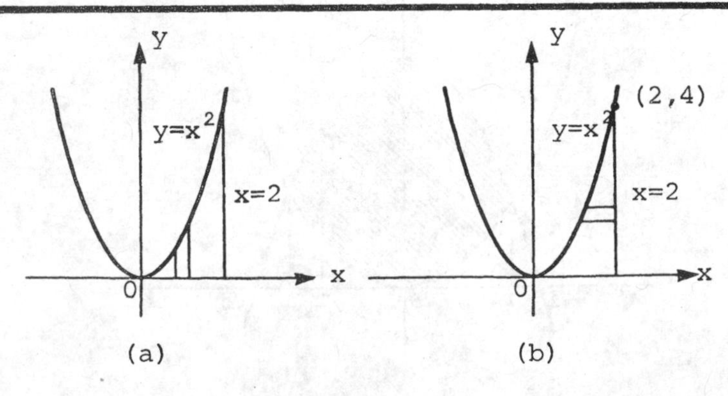

Solution: Method 1. We employ the method of cylindrical shells, figure (a). This method is based on the assumption that the volume can be considered as if constructed from many concentric shells of different lengths. Taking one sample shell of height y, x units away from the y-axis, with thickness Δx, we obtain its volume as:

$$\Delta V = 2\pi ry \; \Delta x,$$

where $2\pi ry$ is the surface area of the shell. The radius r, in this case, is x. Using the given equation of the curve, we substitute

x^2 for y. Since the sum of the volumes of all shells gives the approximate volume of the solid in question, we can pass to the limits - taking Δx infinitesimally small - and express the desired volume as an integral:

$$V = \int_0^2 2\pi ry \; dx,$$

or,

$$V = 2\pi \int_0^2 x \cdot x^2 \; dx = 2\pi \int_0^2 x^3 \; dx$$

$$= 2\pi \cdot \left.\frac{x^4}{4}\right|_0^4$$

$$= 2\pi \; (4 - 0) = 8\pi.$$

Method 2. We can also apply the method of cylindrical disks, figure (b). We assume that the volume is made up of many concentric washers, where the inner radius of each washer is dependent on its location. For a washer y units high, the radius is x. We can call it f(y). The outer radius is constant and equal to 2 for every washer. The thickness is Δy, y varies from 0 to 4. Within this region,

$$V = V_1 - V_2 = \pi \int_0^4 2^2 \; dy - \pi \int_0^4 f(y)^2 \; dy,$$

or,

$$V = \pi \int_0^4 \left(2^2 - f(y)^2\right) dy$$

$$= \pi \int_0^4 (4 - y) \; dy$$

$$= \pi \left(4y - \frac{y}{2} \right) \Bigg|_0^4$$

$$= \pi(16 - 8) = 8\pi.$$

• PROBLEM 849

Find the volume of the solid generated by revolving about the Y axis the region bounded by the parabola: $y = -x^2 + 6x - 8$, and the X-axis.

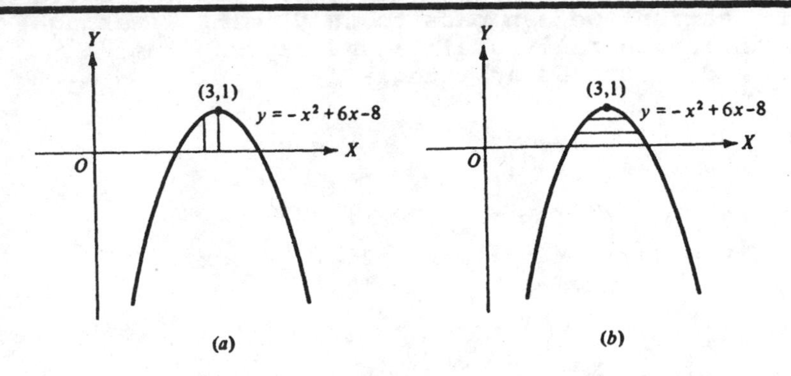

Solution: Method 1. We use the method of cylindrical shells. The curve:

$$y = -x^2 + 6x - 8,$$

cuts the X-axis at $x = 2$ and $x = 4$.

The cylindrical shells are generated by the strip formed by the two lines parallel to the Y-axis, at distances x and $x + \Delta x$ from the Y-axis, $2 \le x \le 4$, as shown in figure (a). When this strip is revolved about the Y-axis, it generates a cylindrical shell of average height y^*, $y \le y^* \le y + \Delta y$, thickness Δx, and average radius x^*, $x \le x^* \le x + \Delta x$. The volume of this element is:

$$\Delta V = 2\pi x^* y^* \Delta x,$$

where $2\pi x^* y^*$ is the surface area. Expressing y in terms of x and passing to the limits, the sum of the volumes of all such cylindrical shells is the integral:

$$V = 2\pi \int_2^4 x(-x^2 + 6x - 8)\, dx$$

$$= 2\pi \int_2^4 (-x^3 + 6x^2 - 8x)\, dx$$

$$= 2\pi \left(-\frac{x^4}{4} + 2x^3 - 4x^2 \right) \Bigg|_2^4$$

$$= 2\pi \left((-64 + 128 - 64) - (-4 + 16 - 16) \right)$$

$$= 8\pi.$$

Method 2. This can also be thought of as the volume comprising a series of concentric washers with variable outer and inner radii, as sectionally shown in fig.(b). The variable radii are as follows: Since

$$y = -x^2 + 6x - 8,$$

we solve for x.

To complete the square, we require a 9, so that

$$x^2 - 6x + 9$$

constitutes a perfect square. Rewriting the equation,

$$x^2 - 6x + 9 - 9 + 8 = -y.$$

$$x^2 - 6x + 9 = 1 - y.$$

$$(x - 3)^2 = 1 - y.$$

Therefore,

$$x = 3 \pm \sqrt{1 - y}$$

which shows the washers, y units from the X-axis, have an

inner radius: $x_{in} = 3 - \sqrt{1 - y}$, and an

outer radius: $x_o = 3 + \sqrt{1 - y}$.

(The particular one on the X-axis has $x_{in} = 2$ and $x_o = 4$.)

The volume of this washer with thickness dy is:

$$dV = \pi\left(x_o^2 - x_{in}^2\right) dy,$$

or $$dV = \pi\left[\left(x_o + x_{in}\right)\left(x_o - x_{in}\right)\right] dy.$$

Substituting the values for x_o and x_{in},

$$dV = \pi\left(\left[\left(3 + \sqrt{1 - y}\right) + \left(3 - \sqrt{1 - y}\right)\right]\right.$$

$$\left.\cdot\left[\left(3 + \sqrt{1 - y}\right) - \left(3 - \sqrt{1 - y}\right)\right]\right) dy$$

$$= \pi\left(12\sqrt{1 - y}\right) dy.$$

Since y varies from 0 to 1, the desired volume is:

$$V = 12\pi \int_0^1 (1 - y)^{\frac{1}{2}} dy$$

$$= 12\pi \left[-\frac{2}{3} (1 - y)^{\frac{3}{2}} \right]_0^1 = 8\pi.$$

• PROBLEM 850

Find the volume of the solid generated by revolving about the Y-axis the region bounded by the parabola: $y^2 = 4x$, the Y-axis and the line $y = 2$.

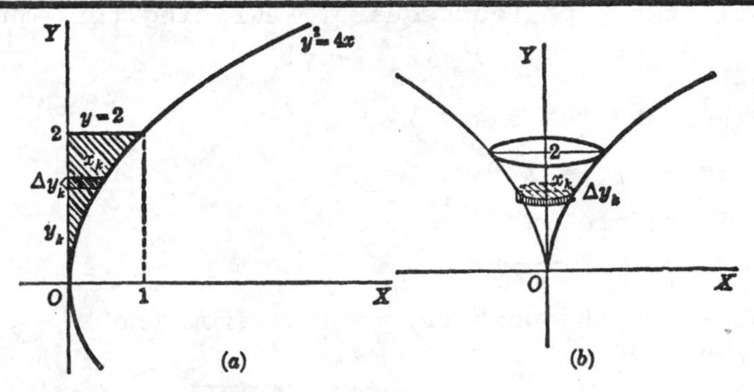

Solution: An element of volume is the disk generated by rotating a strip: x by dy, about the Y-axis. The volume of the disk is:

$$dV = \pi x^2 \, dy.$$

Hence,

$$V = \pi \int_0^2 x^2 \, dy$$

$$= \pi \int_0^2 \frac{1}{16} y^4 \, dy = \frac{2}{5} \pi.$$

• PROBLEM 851

Find the volume generated by revolving about the Y-axis the region bounded by the parabola:

$y = 1 - x^2$, and the coordinate axes.

Solution: The elemental strip has the dimensions: $y \, \Delta x$. The volume generated by rotating this strip about the Y-axis is a cylindrical shell of volume

$$\Delta V = 2\pi xy \, \Delta x,$$

which, in the limit, becomes:

$$dV = 2\pi xy\, dx.$$

$$V = 2\pi \int_0^1 xy\, dx$$

$$= 2\pi \int_0^1 x(1 - x^2)\, dx = \frac{1}{2}\pi.$$

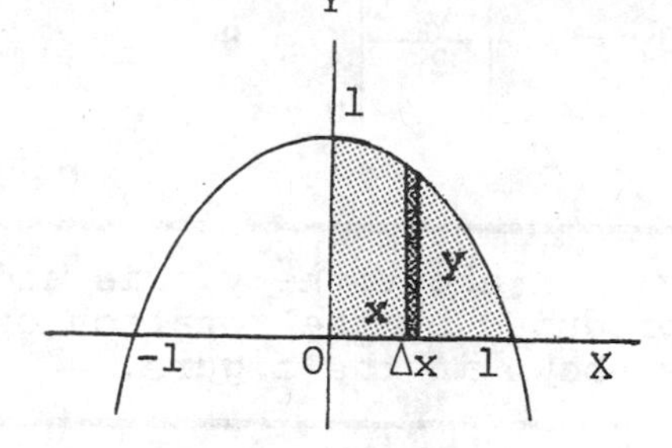

• PROBLEM 852

> Find the volume generated by revolving about the y-axis the area bounded by the curve $y = x^2$, the y-axis, and the line $y = 4$.

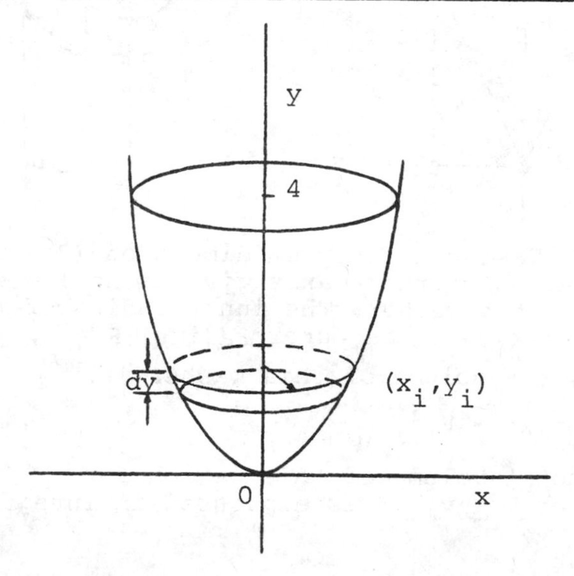

Solution: The cylindrical slices, by which this volume is approximated, have the radii x_i, instead of y_i, where $x_i^2 = y_i$, or $x_i = \sqrt{y_i}$. Then, the volume of a slice is:

$$dV_i = \pi x_i^2\, dy_i.$$

Substituting y_i for x_i^2, we obtain the volume element:

$$dV_i = \pi y_i \, dy_i.$$

Thus,

$$V = \lim_{n\to\infty} \sum_{i=1}^{n} \pi y_i \, dy_i,$$ which, in the limit, can be expressed as an integral. Since y is varying from 0 to 4, we have:

$$V = \pi \int_0^4 y \, dy = \pi \left[\frac{y^2}{2}\right]_0^4 = 8\pi.$$

• PROBLEM 853

Rotate the line: y = 2x, about the y axis and determine the volume produced by the rotation of the shaded portion as shown in the figure.

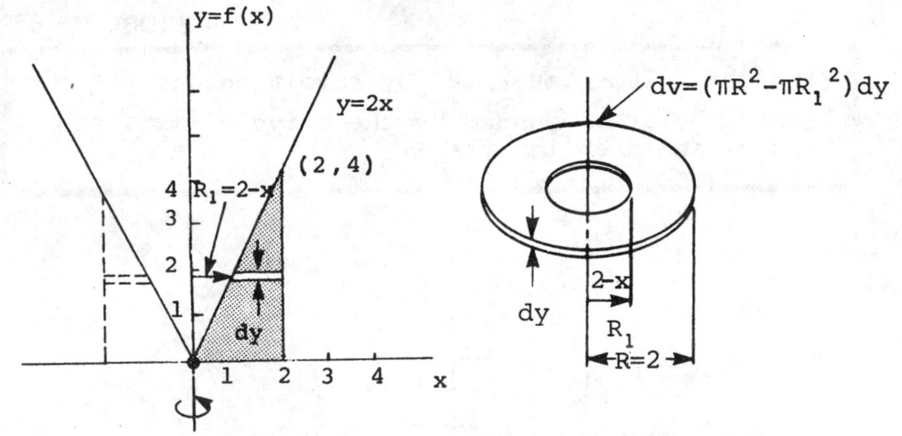

Solution: Taking a strip of dimension (2 - x)dy, when rotated about the y-axis gives a hollow disk. We can call it a washer. The inner radius of this washer, $R_1 = x$, and the outer radius, $R = 2$.

Thus, the volume of this washer is:

$$dV = \pi\left(R^2 - R_1^{\,2}\right).$$

The total volume of this region for $0 \le x \le 2$ or $0 \le y \le 4$ is expressed in integral form:

$$V = \int_0^4 \left(\pi R^2 - \pi R_1^{\,2}\right) dy$$

$$= \int_0^4 \left(\pi 2^2 - \pi(2 - x)^2\right) dy$$

$$= \pi \int_0^4 \left(4 - 4 - 4x + x^2\right) dy$$

Substituting the value for x in terms of y from the original equation,

$$V = \pi \int_0^4 \left(4\,\frac{y}{2} - \frac{y^2}{4}\right) dy$$

$$= \pi \left(y^2 - \frac{y^3}{4 \cdot 3}\right)\Bigg|_0^4 = \pi\left(16 - \frac{16}{3}\right)$$

$$= \frac{32\pi}{3}.$$

Since the increment of this volume is a "washer", this is called the washer method.

• PROBLEM 854

Find the volume of the solid generated by revolving the area enclosed by $y = x^2$, $x = 0$, $y = 0$, about the line $x = 1$.

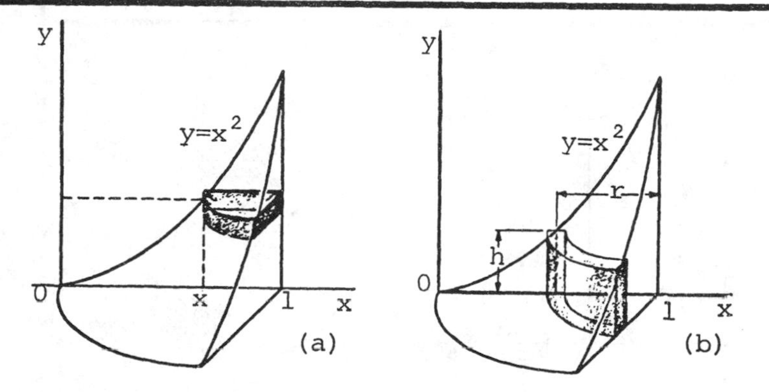

Solution: The typical element of volume, dV, is taken parallel to the plane of the base of the solid as in (a), so that:

$$dV = \pi r^2\, dy.$$

The radius of the element is the length $1 - x$. Since $x = \sqrt{y}$,

$$dV = \pi\left(1 - \sqrt{y}\right)^2 dy.$$

Thus,

$$V = \pi \int_0^1 \left(1 - \sqrt{y}\right)^2 dy = \frac{\pi}{6}.$$

The method used above can be called the disk method, because we have been dealing with the volume of an element shaped like a disk. The following method is the cylindrical shell method, since, as shown in the diagram (b) we work with elemental shells.

The base radius of the shell is approximately $(1 - x)$ and the circumference swept is $2\pi(1 - x)$ with surface area $2\pi(1 - x)h$, where $h = y$, and the thickness Δx. The volume of this shell is:

$$\Delta V = 2\pi(1 - x)y\ \Delta x,$$

or, $$dV = 2\pi(1 - x)x^2\ dx.$$

x varies from 0 to 1.

Thus,

$$V = 2\pi \int_0^1 (1 - x)x^2\ dx$$

$$= 2\pi \left[\frac{4x^3 - 3x^4}{12}\right]_0^1 = \frac{\pi}{6}.$$

• **PROBLEM 855**

Rotate the curve $y = 2x$ about the line $x = 4$ and find the volume produced by the rotation of the shaded portion.

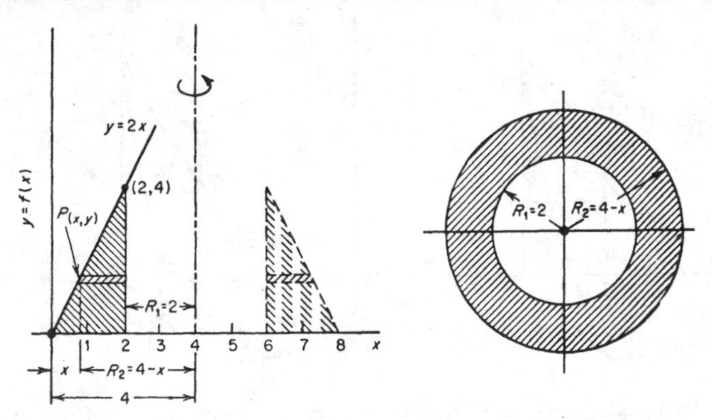

Solution: Area of washer $= \pi\left(R_2^{\,2} - R_1^{\,2}\right)$

Increment of volume, $dV = \pi\left(R_2^{\,2} - R_1^{\,2}\right) dy$.

$$V = \int_a^b \pi\left(R_2^{\,2} - R_1^{\,2}\right) dy$$, where a and b are limits of y.

$$V = \int_0^4 \pi\left((4 - x)^2 - 2^2\right) dy$$

$$= \int_0^4 \pi\left(16 - 8x + x^2 - 4\right) dy$$

$$= \pi \int_0^4 \left(12 - 8x + x^2\right) dy.$$

But the x terms must be in terms of y because of dy. We have: $x=y/2$, from the equation of the curve. Substituting,

$$V = \pi \int_0^4 \left(12 - 4y + \frac{y^2}{4}\right) dy$$

$$= \pi \left(12y - \frac{4y^2}{2} + \frac{y^3}{12}\right)\Bigg|_0^4$$

$$= \pi \left(48 - 32 + \frac{16}{3}\right) = \frac{64\pi}{3}.$$

This is the washer method. The important point to remember is to obtain the radii from the center line of rotation.

• PROBLEM 856

Find the volume of the solid generated by revolving about the line $y = 2$ the smaller region bounded by the curve: $y^2 = 4x$, and the lines: $y = 2$ and $x = 4$.

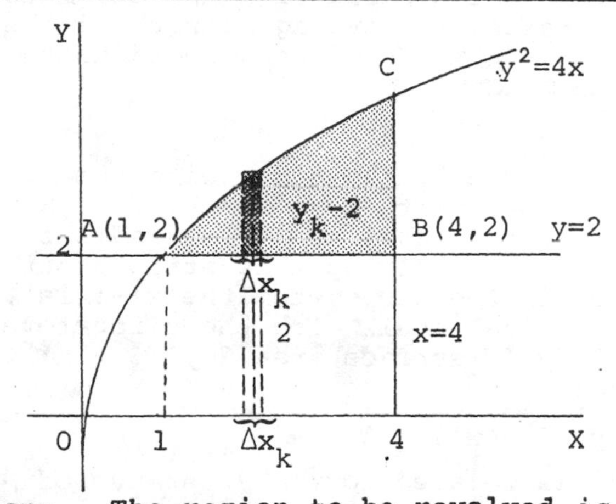

<u>Solution:</u> The region to be revolved is the area ABC, shaded in the diagram. We divide the segment AB into n parts Δx_k, and on each such part construct a rectangle of altitude $y_k - 2$ and base Δx_k, where y_k is the ordinate of the curve for some point in the sub-interval Δx_k. When revolved about the line $y = 2$, each such rectangle generates a cylindrical disk of radius $y_k - 2$ and thickness Δx_k. Then the required volume is:

$$V = \lim_{n \to \infty} \sum_{k=1}^{\infty} \pi (y_k - 2)^2 \Delta x_k$$

$$= \pi \int_1^4 (y - 2)^2 \, dx$$

$$= \pi \int_1^4 (y^2 - 4y + 4) \, dx$$

$$= \pi \int_1^4 (4x - 8\sqrt{x} + 4) \, dx = \frac{14}{3} \pi .$$

This can also be worked out using the method of cylindrical shells. For convenience, we shift the origin 0 to the point A(1,2), hence in the new x'- y'-axes we have:

$$x' = x - 1, \quad \text{or}, \quad x = x' + 1$$

$$y' = y - 2, \quad \text{or}, \quad y = y' + 2.$$

Then, drawing two parallel lines of width Δy along the x-axis, the average length of the strip formed is 4 - x. In the new coordinate system, **the length is:**

$$4 - (x' + 1) = 3 - x'.$$

Now, the length of the strip with one of its corners at the point (x,y) to the line x = 4 in the new coordinate axes is 3 - x' and its width is $\Delta y = \Delta y'$. Revolving this strip about the line y = 2, (or in the new system the x'-axis), we obtain a cylindrical shell with the circumference of its base $2\pi(y')$, surface area $2\pi y'(3 - x')$. Its volume is:

$$\Delta V = 2\pi y'(3 - x') \, \Delta y'.$$

Since x' is related to y' by the use of $y^2 = 4x$, we obtain:

$$(y' + 2)^2 = 4(x' + 1).$$

Solving for x', we obtain:

$$x' = \frac{(y' + 2)^2}{4} - 1,$$

and thus:

$$\Delta V = 2\pi y' \left[3 - \left(\frac{(y' + 2)^2}{4} - 1 \right) \right] \Delta y'.$$

Passing to the limit, from the definition of integrals, we obtain the required volume in integral form:

$$V = 2\pi \int y' \left[3 - \frac{(y')^2 + 4y' + 4}{4} + 1\right] dy'.$$

Taking the new limits into account, as $2 \le y \le 4$, $0 \le y' \le 2$, we have:

$$V = 2\pi \int_0^2 \left[3y' - \frac{(y')^3}{4} - (y')^2 - y'\right] dy'$$

$$= 2\pi \left[\frac{3(y')^2}{2} - \frac{(y')^4}{16} - \frac{(y')^3}{3}\right]_0^2$$

$$= 2\pi\left[6 - 1 - \frac{8}{3}\right]$$

$$\doteq \frac{14\pi}{3}.$$

• PROBLEM 857

Determine the volume of the solid obtained by rotating the region below the graph of: $y = x$, and above the graph: $y = x^2$, about the x-axis.

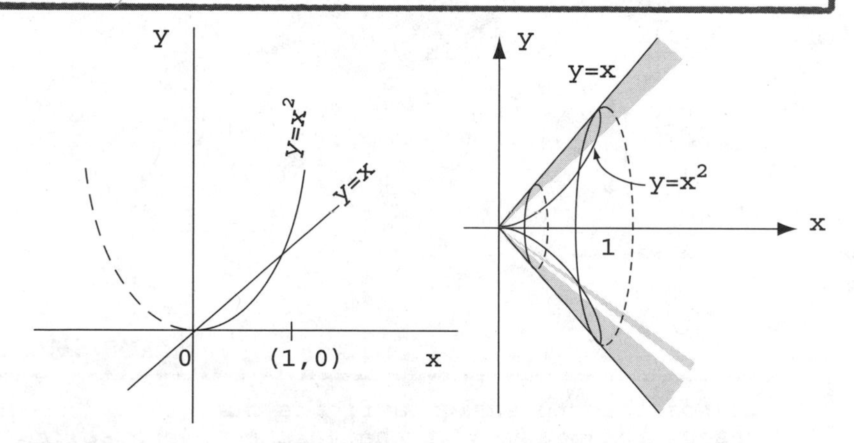

Solution: First we find the interval in which the given region is located. In accordance with the figure, we find the points at which the two given equations, $y = x$ and $y = x^2$, intersect. To do this, we equate the y-values.

$$y = x = x^2, \qquad \text{or}, x(1 - x) = 0.$$

Therefore, the two curves intersect at $x = 0$ with corresponding y-value $y_o = 0$: and at $x = 1$, $y_1 = 1$, as shown. Thus, the region is bounded by the curve $y = x^2$, the line $y = x$, and the x- and y-axes.

$$0 \le x \le 1 \qquad \text{and} \qquad 0 \le y \le 1.$$

To solve the problem, we first suppose that the interval $0 \leq x \leq 1$ is divided into n intervals, such that each interval is $\Delta x = \frac{1}{n}$ and expressed as: $0 = x_o \leq x_1 \leq x_2 \leq \ldots \leq x_i \leq \ldots \leq x_n = 1$.

The strip taken parallel to and x units from the y-axis, with width:

$$\Delta x = \frac{1}{n} = x_i - x_{i-1}$$

units, when rotated about the x-axis, appears as a hollow disk (washer) of inner radius, $r_{in} = x^2$, outer radius $r_o = x$, and thickness Δx. Then, its volume is:

$$\Delta V = \pi\left(r_o{}^2 - r_{in}{}^2\right)\Delta x.$$

Hence, the sum of the volumes ΔV of all the n discs gives the required volume of the solid. To be accurate, we take the number n big enough so that Δx is almost zero. Thereby the disks look like very fine washers with slanted edges. We let n approach infinity, $\Delta x \rightarrow 0$. The summation leads to the integration:

$$V = \int_0^1 \pi\left(r_o{}^2 - r_{in}{}^2\right) dx$$

$$= \pi \int_0^1 \left(x^2 - x^4\right) dx$$

$$= \pi \left(\frac{1}{3} x^3 - \frac{1}{5} x^5\right|_0^1 = \frac{2}{15}\pi.$$

• PROBLEM 858

Is it possible to assign a finite number to represent the measure of the volume of the solid formed by revolving the region bounded by

$$y = \frac{1}{x}, \; y = 0, \; x = 1?$$

Solution: Consider a small region, the shaded strip as shown in the figure, for example. If this strip is rotated about the x-axis, the volume swept by this strip is a disk having thickness $\Delta_i x$ and base radius of y_i. y_i is the value corresponding to $\frac{1}{\xi_i}$. Let L be the number we wish to assign to the measure of the volume of a finite bounded region formed by $y = \frac{1}{x}$, $y = 0$, $1 \le x \le b$ and the number $b > \xi_i$. The volume of the element we are considering is:

$$\Delta V_i = \pi y_i^2 \Delta_i x = \pi \left(\frac{1}{\xi_i} \right)^2 \Delta_i x,$$

for any $i = 1, 2, 3, \ldots, n$, no matter how large, and $\xi_n = b$. We obtain:

$$L = \lim_{|\Delta_i x| \to 0} \sum_{i=1}^{n} \pi \left(\frac{1}{\xi_i} \right)^2 \Delta_i x.$$

This leads to the integral:

$$L = \pi \int_1^b \left(\frac{1}{x} \right)^2 dx.$$

If $V = \lim_{b \to \infty} L$, the required volume, if the limit exists, we obtain:

$$V = \lim_{b \to \infty} \pi \int_1^b \frac{1}{x^2} dx = \pi \lim_{b \to \infty} \left(-\frac{1}{x} \right) \Bigg|_1^b$$

$$= \pi \lim_{b \to \infty} \left(-\frac{1}{b} + 1 \right) = \pi(0 + 1) = \pi.$$

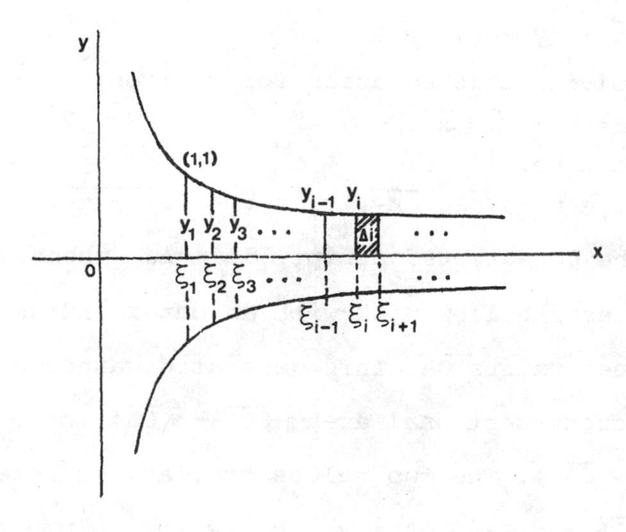

Therefore, we can assign a finite number during calculations, later finding the limit by extending the domain to infinity. We thus arrive at a finite value, provided the limit exists.

• PROBLEM 859

Find the volume of the solid generated by revolving a circle of radius a about an axis in its plane at a distance b from its center, when b > a. (This solid is called a torus.)

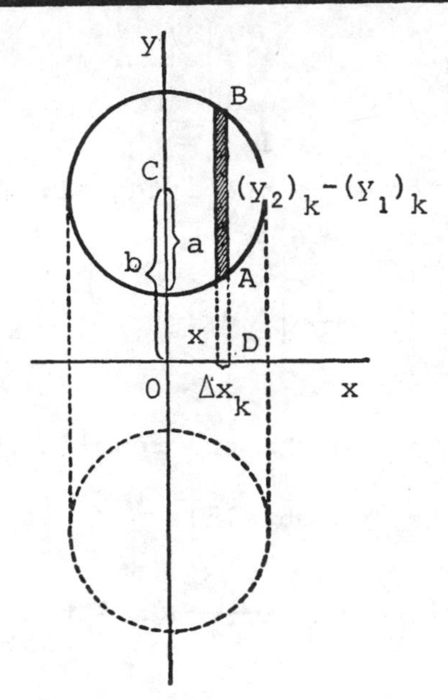

Solution: As shown in the figure, the circle with center C(0,b) and radius a is to be revolved about the X-axis, to generate the torus. The equation of this circle is:

$$x^2 + (y - b)^2 = a^2.$$

Solving this equation for y gives:

$$y = b \pm \sqrt{a^2 - x^2}.$$

We take:

$$y_1 = b - \sqrt{a^2 - x^2},\ y_2 = b + \sqrt{a^2 - x^2}.$$

The constructed strip, $\overline{AB}$ by Δx_k, then generates a hollow disk with an outer radius $\overline{OB}$ and inner radius $\overline{OA}$. This generated washer has a solid cross-sectional area: $\pi\,\overline{OB}^2 - \pi\,\overline{OA}^2$, or $\pi(\overline{OB}^2 - \overline{OA}^2)$. The two values of y are of interest, since $\overline{OB} = y_2$ and $\overline{OA} = y_1$. Hence the volume of the disk with thickness Δx_k is:

$$\Delta V_k = \pi \left(y_2^2 - y_1^2\right) \Delta x_k$$

$$= \pi \left(y_2 + y_1\right)\left(y_2 - y_1\right)\Delta x_k$$

Substituting the values for y_1 and y_2, we obtain:

$$\Delta V_k = \pi \left[\left(b + \sqrt{a^2 - x^2}\right) + \left(b - \sqrt{a^2 - x^2}\right)\right]$$

$$\cdot \left[\left(b + \sqrt{a^2 - x^2}\right) - \left(b - \sqrt{a^2 - b^2}\right)\right]$$

$$= \pi \left(4b \sqrt{a^2 - x^2}\right).$$

The required volume,

$$V \approx \sum_{k=1}^{n} \Delta V_k,$$

is a Riemann sum, which can be expressed as an integral:

$$V = \pi \int_{-a}^{a} 4b \sqrt{a^2 - x^2}\, dx.$$

By symmetry,

$$V = 2(4\pi)b \int_0^a \sqrt{a^2 - x^2}\, dx$$

$$= 8\pi b \left[\frac{1}{2}\left(x \sqrt{a^2 - x^2} + a^2 \sin^{-1} \frac{x}{a}\right)\right]_0^a$$

$$= 8\pi b \left(\frac{1a^2}{2} \cdot \frac{\pi}{2}\right)$$

$$= 2\pi^2 a^2 b.$$

• PROBLEM 860

Find the volume V of the segment of a sphere of radius r, intercepted by a plane at distance a from the center.

<u>Solution:</u> This is a solid of revolution generated by an arc of a circle. The equation of the circle of radius r is:

$$x^2 + y^2 = r^2.$$

By constructing a small strip, y_i by Δx_i units, parallel to the y-axis, we find the volume of the disk generated by the strip rotating on the x-axis. Noting that its radius is y_i and its thickness is Δx_i, its cross-sectional area is πy_i^2, and its volume is: $\Delta V = \pi y_i^2 \, \Delta x_i$,

where $a = x_1 < x_2 < \ldots < x_i < \ldots < x_n = r$.

By the use of the equation of a circle, y can be expressed in terms of x, and substituting this value for y_i^2 in the above equation, we have:

$$\Delta V = \pi \left(r^2 - x_i^2\right) \Delta x_i .$$

Hence,

$$V = \pi \lim_{n \to \infty} \sum_{i=1}^{\infty} \left(r^2 - x_i^2\right) \Delta x_i ,$$

which, on passing to the limit, becomes:

$$V = \pi \int_a^r \left(r^2 - x^2\right) dx.$$

Hence,

$$V = \pi \int_a^r r^2 \, dx - \pi \int_a^r x^2 \, dx$$

$$= \pi r^2 (r - a) - \frac{\pi}{3} \left(r^3 - a^3\right)$$

$$= \frac{\pi}{3} (r - a)\left(2r^2 - ar - a^2\right)$$

$$= \frac{\pi}{3} (r - a)^2 (2r + a).$$

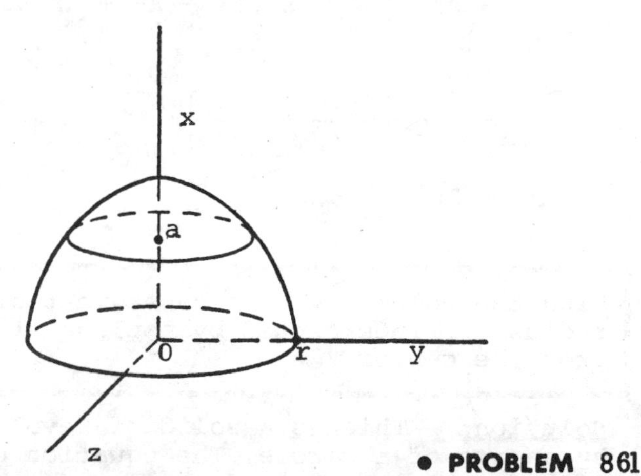

• **PROBLEM 861**

Find the volume of the solid generated by revolving about their common chord the region common to the circles:

$$x^2 + y^2 = 16, \quad \text{and} \quad x^2 + y^2 = 8x.$$

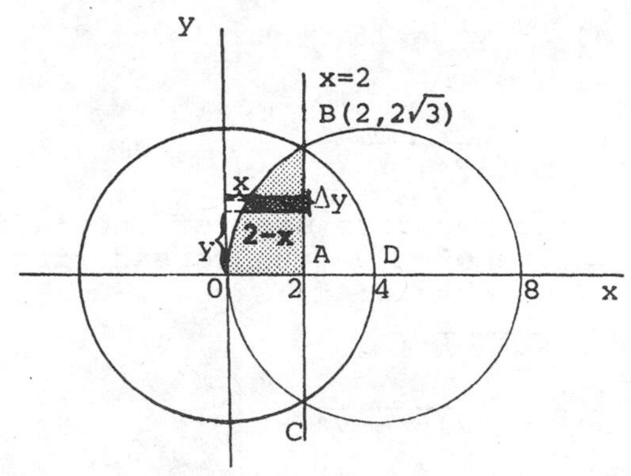

<u>Solution:</u> We first find the equation of the axis of rotation. To find the intersection points of the given circles, we determine x-coordinates at which the two circles have the same y-values. We let:

$$y^2 = 16 - x^2 = 8x - x^2.$$

This yields: $x = 2$. This is the equation of the axis. Substituting the value of x in either equation, more easily in: $x^2 + y^2 = 16$, we obtain

$$4 + y^2 = 16, \quad \text{or } y = \pm\sqrt{12} = \pm 2\sqrt{3}.$$

The intersection points are then $B\left(2, 2\sqrt{3}\right)$ and $C\left(2, -2\sqrt{3}\right)$ as shown in the figure. Now, we wish to find the volume of the elliptical solid constructed as the curves BOC and BDC rotate about the line $x = 2$.

As can be seen from the figure, the region OBDCO is symmetric about the line $x = 2$ and the x-axis. Thus, the shaded area AOBA generates half the desired volume. Using this area and the cylindrical disk method, we see that the radius of the disk, formed by rotating the strip of dimension $(2 - x)$ by Δy about the line: $x = 2$, is $(2 - x)$. The cross-sectional area is:

$$\pi(2 - x)^2.$$

With the thickness Δy, the disk has a volume:

$$\Delta V = \pi(2 - x)^2 \Delta y.$$

The required volume is thus:

$$V = 2\pi \int_0^{2\sqrt{3}} (2 - x)^2 \, dy,$$

where x is to be evaluated from the equation:

$$x^2 + y^2 = 8x.$$

To solve for x in the above equation, we first complete the square:

$$x^2 - 8x + y^2 + 16 = 16,$$

or, $(x - 4)^2 + y^2 = 16.$

Then,

$$x - 4 = \pm \sqrt{16 - y^2}.$$

Taking the negative sign of the radical, since the values of x we are considering are all less than 4, we obtain:

$$x = -\sqrt{16 - y^2} + 4,$$

or, $(2 - x)^2 = \left(\sqrt{16 - y^2} - 2\right)^2$

$$= 16 - y^2 - 4\sqrt{16 - y^2} + 4$$

$$= 20 - y^2 - 4\sqrt{16 - y^2}.$$

Thus

$$V = 2\pi \int_0^{2\sqrt{3}} \left(20 - y^2 - 4\sqrt{16 - y^2}\right) dy$$

$$= 2\pi \left[20y - \frac{1}{3} y^3 - 2y\sqrt{16 - y^2} - 32 \sin^{-1} \frac{y}{4}\right]_0^{2\sqrt{3}}$$

$$= 2\pi \left[40\sqrt{3} - \frac{8 \cdot 3\sqrt{3}}{3} - 4\sqrt{3} \times 2 - 32 \sin^{-1} \frac{\sqrt{3}}{2}\right]$$

$$= 48\pi\sqrt{3} - \frac{64}{3}\pi^2.$$

• PROBLEM 862

What is the volume generated by revolving the sine curve from 0 to π about the x axis?

Solution:

$$V = \pi \int_0^{\pi} y^2 \cdot dx = \pi \int_0^{\pi} \sin^2 x \cdot dx$$

$$= \pi \int_0^{\pi} \left(\frac{1}{2} - \frac{1}{2}\cos 2x\right) \cdot dx$$

$$= \pi \left[\frac{x}{2} - \frac{1}{4}\sin 2x\right]_0^{\pi}$$

$$= \pi \left(\frac{\pi}{2} - 0 - 0 + 0 \right) = \frac{\pi^2}{2},$$

the number of cubic units in the volume, provided that the unit on the x axis to represent one radian is the length of the unit cube.

• PROBLEM 863

Find the volume generated by revolving about the X-axis the region under the first arch of the curve y = sin x; also find the volume when this region is revolved about the Y-axis.

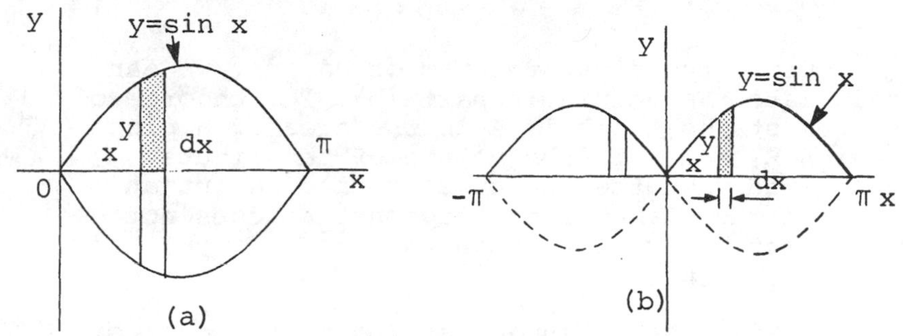

Solution: (a) To find the volume generated when the given region is revolved about the X-axis, the simplest method is the cylindrical disk method.

The shaded strip as shown in figure (a), with approximate dimensions y by dx, constitutes a disk when rotated about the X-axis, with approximate radius = y, thickness dx, and hence a volume:

$$dV = \pi y^2 \, dx.$$

Then the volume generated over the region $0 \le x \le \pi$ is obtained by the Riemann integral:

$$V = \int_0^{\pi} \pi y^2 \, dx = \pi \int_0^{\pi} y^2 \, dx$$

$$= \pi \int_0^{\pi} \sin^2 x \, dx = \pi \int_0^{\pi} \left(\frac{1}{2} - \frac{1}{2} \cos 2x \right) dx$$

$$= \pi \left(\frac{x}{2} - \frac{1}{4} \sin 2x \right) \Bigg|_0^{\pi} = \frac{\pi^2}{2}.$$

(b) To find the volume when the given region is revolved about the Y-axis, the simplest method is the cylindrical shell method. In this case, we can use the same strip with the base radius r = x, thickness dx and height $h = y$. Since here the circumference is approximately $2\pi x$, $2\pi x(y)$ gives the surface area, and hence $2\pi xy(dx)$ is the volume of the cylindrical shell generated by the strip $y\,dx$ about the Y-axis. Then, the volume generated about the Y-axis by the sine curve, $0 \leq x \leq \pi$, is:

$$V = 2\pi \int_0^\pi xy\,dx = 2\pi \int_0^\pi x \sin x\,dx.$$

Using integration by parts,

$$V = 2\pi \left(-x \cos x + \sin x \right\Big|_0^\pi = 2\pi^2.$$

Using the first theorem of Pappus, we can calculate the volume in part (b). The theorem of Pappus states that, if A is the area of a plane region R, and V is the volume of the figure generated when the region is rotated about an axis in the plane of the region that does not intersect the region, then

$$V = AL,$$

where L is the distance traveled by the centroid of R. Thus, in (b) we can see that, due to the symmetry of the sine curve, $0 < x \leq \pi$ (which, in this theorem, is the region R) about the line

$x = \frac{\pi}{2}$ (mid-way), $\bar{x} = \frac{\pi}{2}$. The distance $\bar{x}$ travels, i.e., L, is the circumference of the circle of radius x,

$$2\pi\bar{x} = 2\pi\left(\frac{\pi}{2}\right) = \pi^2;$$ and the area is:

$$\int_0^\pi \sin x\,dx = 2.$$

Thus,

$$V = 2\left(\pi^2\right) = 2\pi^2.$$

• PROBLEM 864

Find the volume of the solid generated by revolving about the X-axis the region in the first quadrant bounded by the hypocycloid:

$$x = a \cos^3 \theta, y = a \sin^3 \theta,$$

and the coordinate axes.

Solution: A hypocycloid is the locus of any fixed point on the circumference of a smaller circle, radius b, that rolls internally, without slipping, on a fixed larger circle.

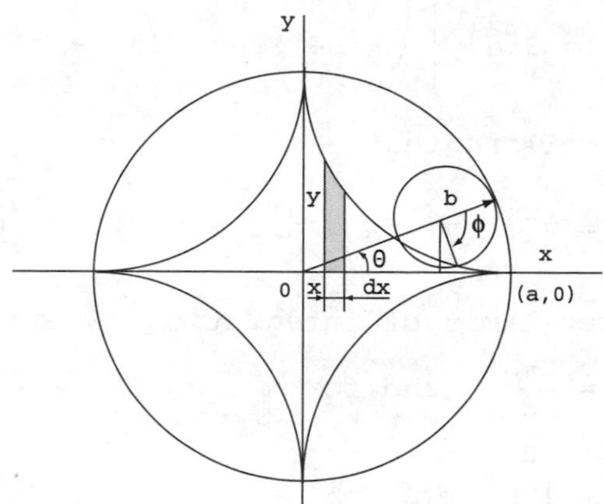

If $\frac{a}{b} = n$, then $a = nb$, where n is any integer, we can obtain the x and y expressions in parametric form - in terms of the angle Θ - and obtain the given equations:

$$x = a \cos^3 \Theta, \quad y = a \sin^3 \Theta.$$

Using the method of cylindrical disks, the volume of the shaded strip, as shown above, as it rotates about the X-axis, is:

$$dV = \pi y^2 \, dx.$$

The sum of the volumes of all such disks in the first quadrant leads to the integral:

$$V = \pi \int y^2 \, dx,$$

but $y^2 = a^2 \sin^6 \Theta$ and $dx = - 3 a \cos^2 \Theta \sin \Theta \, d\Theta$. Therefore,

$$V = \pi \int \left(a^2 \sin^6 \Theta\right)\left(- 3 a \cos^2 \Theta \sin \Theta \, d\Theta\right).$$

For simplicity, using the identity:

$$\sin^2 \Theta + \cos^2 \Theta = 1,$$

we obtain that

$$\sin^6 \Theta = \left(\sin^2 \Theta\right)^3 = \left(1 - \cos^2 \Theta\right)^3,$$

and $\sin \Theta \, d\Theta = - d (\cos \Theta)$.

Then the integral takes the form:

$$V = 3\pi a^3 \int \left(1 - \cos^2 \Theta\right)^3 \cos^2 \Theta \, d(\cos \Theta),$$

which is very similar to:

$$V_t = 3\pi a^3 \int \left(1 + t^2\right)^3 t^2 \, dt,$$

where t is any temporary variable of integration, or dummy variable. The integral becomes:

$$V_t = 3\pi a^3 \int \left(t^2 - 3t^4 + 3t^6 - t^8\right) dt.$$

Recalling that

$$V = \pi \int_0^a y^2\, dx, \quad \text{and } x = a \cos^3 \Theta,$$

the lower limit of integration, where x = 0, occurs when $\Theta = \frac{\pi}{2}$, and for x = a, Θ = 0. Thus,

$$V = \int_{\pi/2}^{0} f(\Theta)\, d\Theta.$$

Furthermore, we assumed $\cos \Theta \equiv t$. Now we infer that for $\Theta = \frac{\pi}{2}$, as a lower limit, t = 0, and for Θ = 0 we have t = 1. Thus $0 \leq t \leq 1$, and

$$V = 3\pi a^3 \int_0^1 \left(t^2 - 3t^4 + 3t^6 - t^8\right) dt$$

$$= 3\pi a^3 \left[\frac{1}{3} t^3 - \frac{3}{5} t^5 + \frac{3}{7} t^7 - \frac{1}{9} t^9 \right|_{t=0}^{t=1}$$

$$= 3\pi a^3 \frac{105 - 189 + 135 - 35}{315}$$

$$= \frac{16}{105} \pi a^3.$$

• PROBLEM 865

The area bounded by the cycloid: x = a (Θ - sin Θ), y = a (1 - cos Θ), and the x-axis from x = 0 to x = 2πa, is revolved about the y-axis. Find the volume formed.

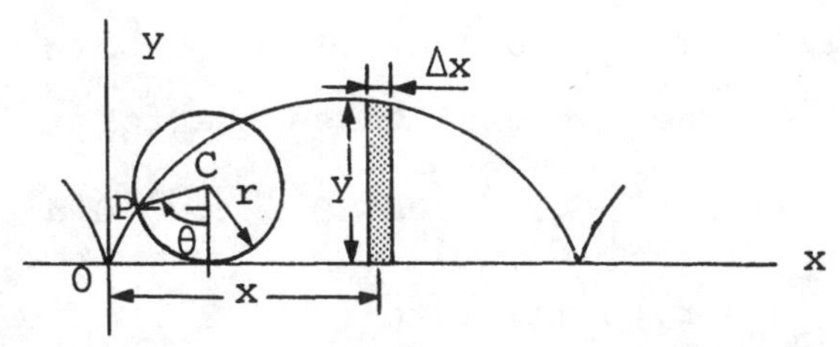

<u>Solution:</u> A cycloid is the locus of any fixed point on the circumference of a circle that rolls, without slipping, on a fixed straight line, namely the x-axis. The limiting points are, for x = 0, by substitution,

$$\Theta = \sin \Theta,$$

which holds only for $\Theta = 0$. For $x = 2\pi a$, substituting and cancelling out a yields

$$2\pi = \Theta + \sin\Theta.$$

This holds only if $\Theta = 2\pi$, so that the contribution by $\sin\Theta$ is zero. Substituting 0 and 2π for Θ in the $y = f(\Theta)$ expression yields $y(0) = y(2\pi) = 0$. Thus 2π is the period.

Returning to the problem, the strip of dimensions y by Δx, the shaded rectangle in the figure, when rotated about the y-axis, generates a hollow cylinder or a shell with base radius $r = |x|$, height $h = y$ and thickness Δx. The surface area, $s = 2\pi rh = 2\pi xy$, and the volume is $\Delta V = 2\pi xy\Delta x$. Thus we can generalize that the region R, $0 \leq x \leq 2\pi a$, $0 \leq \Theta \leq 2\pi$ generates a volume V which is equal to the sum of the volumes of all concentric shells. Going to the limits, the volume can be expressed in an integral form:

$$V = 2\pi \int_0^{2\pi a} xy\, dx.$$

Using the parametric expressions:

$$x = a(\Theta - \sin\Theta),\ y = a(1 - \cos\Theta),$$

$$0 \leq x \leq 2\pi.$$

Hence,

$$V = 2\pi \int_0^{2\pi a} xy\, dx$$

$$= 2\pi a^3 \int_0^{2\pi} (\Theta - \sin\Theta)(1 - \cos\Theta)d(\Theta - \sin\Theta)$$

$$= 2\pi a^3 \left[\int_0^{2\pi a} (\Theta - \sin\Theta)d(\Theta - \sin\Theta) - \int_0^{2\pi} \cos\Theta\ (\Theta - \sin\Theta)d(\Theta - \sin\Theta) \right].$$

The first integral is the same as $\int_0^{2\pi} t\, dt$, where $t = \Theta - \sin\Theta$. Hence,

$$\int_0^{2\pi} (\Theta - \sin\Theta)d(\Theta - \sin\Theta)$$

$$= \frac{1}{2}(\Theta - \sin\Theta)^2 \Bigg|_{\Theta - \sin\Theta = 0}^{\Theta - \sin\Theta = 2\pi}$$

$$= \frac{1}{2} (\Theta - \sin \Theta)^2 \Big|_0^{2\pi}$$

$$= 2\pi^2.$$

For the second integral, expansion is required. Noting that $d(\Theta - \sin \Theta) = (1 - \cos \Theta)d\Theta$

$$\int_0^{2\pi} \cos \Theta \, (\Theta - \sin \Theta) d(\Theta - \sin \Theta)$$

$$= \int_0^{2\pi} \cos \Theta \, (\Theta - \sin \Theta)(1 - \cos \Theta) d\Theta$$

$$= \int_0^{2\pi} (\Theta \cos \Theta - \sin \Theta \cos \Theta - \Theta \cos^2 \Theta + \sin \Theta \cos^2 \Theta) d\Theta,$$

where the two terms of the form:

$$\int_0^{2\pi} \sin \Theta \, f(\cos \Theta) d\Theta,$$

can be integrated as

$$- \int_1^1 f(\cos \Theta) d(\cos \Theta) = 0.$$

Using integration by parts on the remaining terms,

$$\int_0^{2\pi} \Theta \cos \Theta \, d\Theta = (\cos \Theta + \Theta \sin \Theta) \Big|_0^{2\pi} = 0,$$

and

$$\int_0^{2\pi} \Theta \cos^2 \Theta \, d\Theta$$

$$= \frac{1}{4} \left\{ \cos \Theta \left(\frac{\cos \Theta}{2} + 2 \sin \Theta \right) + \Theta^2 - \frac{\sin^2 \Theta}{2} \right\} \Bigg|_0^{2\pi} = \pi^2$$

Combining all the results obtained, and substituting for the volume expression,

$$V = 2\pi a^3 (2\pi^2 - \pi^2) = 2\pi^3 a^3.$$

As a second approach, we use the theorem of Papus: since the moment about the y-axis is expressed as

$$\int_0^{2\pi a} xy\,dx,$$

and the integral just calculated,

$$2\pi \int_0^{2\pi a} xy\,dx = 2\pi^3 a^3,$$

then $$\int_0^{2\pi a} xy\,dx = \pi^2 a^3.$$

This is the moment.

The area,

$$A = \int_0^{2\pi a} y\,dx.$$

$$A = \int_0^{2\pi a} a^2\,(1 - \cos\Theta)d(\Theta - \sin\Theta)$$

$$= a^2\,\frac{3}{2}\,\Theta\Big|_0^{2\pi} = 3\pi a^2.$$

If we divide the moment by the area,

$$\overline{x} = \frac{\pi^2 a^3}{A},$$

and the circumference swept by this line: $x = x$, **by making one revolution about the y-axis, is:**

$$L = 2\pi\overline{x} = 2\pi\left(\frac{\pi^2 a^3}{A}\right).$$

By Papus' theorem,

$V = LA.$

Thus

$$V = 2\pi\left(\frac{\pi^2 a^3}{A}\right) A = 2\pi^3 a^3.$$

• PROBLEM 866

The ellipse: $b^2x^2 + a^2y^2 = a^2b^2$, is revolved about a tangent line at one end of the major axis. Find the volume generated.

Solution: The equation of the ellipse is also given by:

$$\frac{x^2}{a^2} + \frac{y^2}{b^2} = 1, \quad y = \frac{b}{a}\sqrt{a^2 - x^2}.$$

Using the shell method - by dividing the x-axis, $-a \le x \le a$, into n equal subdivisions - we construct a series of parallel strips about the y-axis with respective dimensions of y by Δx, where $\Delta x = \frac{2a}{n}$. When the whole region is revolved about a tangent line at one end of the major axis, namely $|x| = a$ (for $a > b$), these strips generate concentric shells with corresponding base radii expressed as $r = |a - x|$, height $h = y$ and thickness of Δx. Thus, for the shell located x units away from the origin, the surface area is:

$$\Delta s = 2\pi rh = 2\pi|a - x|y.$$

Its volume is:

$$\Delta V = 2\pi|a - x|y\ \Delta x, \text{ above the x-axis.}$$

For brevity, suppose the line of generation to be at $x = -a$, then $r = |a - x| = a + x$. Then $\Delta V = 2\pi(a + x)y\ \Delta x$. The total volume generated above the x-axis is the sum of volumes of all such shells. Passing to the limits, as we make n sufficiently large, or as Δx shrinks to zero, the sum is:

$$V = \int_{-a}^{a} 2\pi\ (a + x)\ y\ dx.$$

Expressing y in terms of x,

$$V = \int_{-a}^{a} \frac{2\pi b}{a}\ (x + a)\ \sqrt{a^2-x^2}\ dx.$$

As an alternative approach, we can shift the axis a units behind the y-axis, so that the new equation of the ellipse is:

$$\frac{(x - a)^2}{a^2} + \frac{y^2}{b^2} = 1.$$

Using the same method of construction, as above, $r = x$, $h = y$ and thickness is Δx. Hence

$$V = \int_{0}^{2a} 2\pi xy\ dx\ .$$

Here, $y = \frac{b}{a}\sqrt{a^2 - (x - a)^2}$, and thus:

$$V = \frac{2\pi b}{a}\int_{0}^{2a} x\ \sqrt{a^2 - (x - a)^2}\ dx.$$

Seeing that the expressions for the volume in both systems are the same, we proceed with the integration:

$$V = \frac{2\pi b}{a}\left[\int_{-a}^{a} a\ \sqrt{a^2 - x^2}\ dx\right.$$

$$+ \int_{-a}^{a} x \sqrt{a^2 - x^2}\, dx \Bigg] ,$$

The first integration is carried out by recalling the derivative of a product. If u and V are variables, then

$$d(uV) = udV + Vdu,$$

or, $udV = d(uV) - Vdu.$

We can integrate the entire equation and write it as:

$$\int udV = \int d(uV) - \int Vdu = uV - \int Vdu,$$

where the integrand udV is assumed to be difficult to integrate. This technique (integration by parts), is quite important to reduce a **difficult** integration to a simpler one. Let

$u = \sqrt{a^2 - x^2}$, and $dV = dx$. Then

$$\int \sqrt{a^2 - x^2}\, dx = x \sqrt{a^2 - x^2} + \int \frac{x^2}{\sqrt{a^2 - x^2}}\, dx.$$

The last integral can be obtained by setting $x = a \sin \Theta$ and remembering the trigonometric identities: $\sin^2 \Theta + \cos^2 \Theta = 1$, $\cos 2\Theta = 1 - 2 \sin^2 \Theta$, and $\sin 2\Theta = 2 \sin \Theta \cos \Theta$. We obtain:

$$\int \frac{x^2}{\sqrt{a^2 - x^2}}\, dx = \frac{a^2}{2} \sin^{-1} \frac{x}{a} - \frac{x}{2} \sqrt{a^2 - x^2}.$$

$$\int \sqrt{a^2 - x^2}\, dx$$

$$= \frac{1}{2}\left(x \sqrt{a^2 - x^2} + a^2 \sin^{-1} \frac{x}{a} \right) .$$

The second integral for V, by setting $u = a^2 - x^2$, yields:

$$\int x \sqrt{a^2 - x^2}\, dx = - \frac{1}{3} \left(a^2 - x^2\right)^{\frac{3}{2}}.$$

Consequently,

$$V = \frac{2\pi b}{a} \int_{-a}^{a} (a + x) \sqrt{a^2 - x^2}\, dx$$

$$= \frac{2\pi b}{a}\left(\frac{a}{2}\left(x\sqrt{a^2 - x^2} + a^2 \sin^{-1}\frac{x}{a}\right) - \frac{1}{3}(a^2 - x^2)^{\frac{3}{2}}\right|_{-a}^{a}$$

$$= \frac{2\pi b}{a}\left(a^3 \frac{\pi}{2}\right)$$

$$= \pi^2 a^2 b.$$

Remember that this is half of the volume required, since the height of the arbitrary shell was taken as y only (above the x-axis), considering symmetry about the x-axis. The total height is 2y, thus the required volume generated by the ellipse $= 2\pi^2 a^2 b$.

As another approach, we use the theorem of Papus, which states: The volume generated by a region R revolving about a line is equal to the product of the cross-sectional area of the region multiplied by the distance which the centroid has traveled for one revolution about this line. Using this approach we know that the ellipse is symmetric about the origin. Hence, its center of mass lies on the origin, and the area $A = \pi ab$. The distance traveled by the centroid $0(0,0)$, a units away from the line of revolution, $L = 2\pi a$. Thus, using Papus theorem, the required volume,

$$V = AL = \pi ab(2\pi a) = 2\pi^2 a^2 b.$$

• PROBLEM 867

Find the volume of a right pyramid having a height of h units and a square base a units on a side.

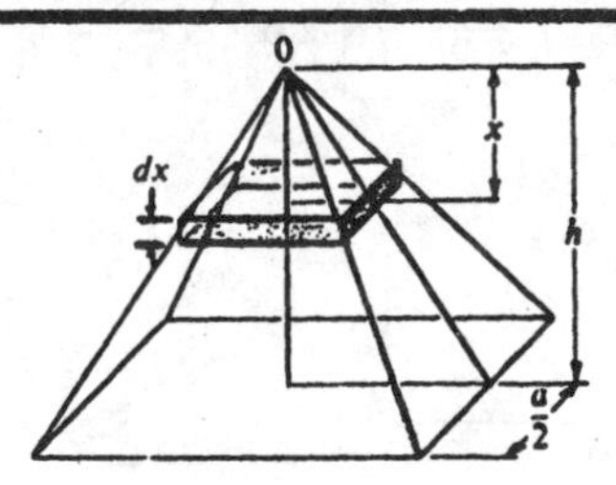

Solution: The volume can be approximated by means of square slabs as indicated in the figure. If the origin is taken at the vertex 0, and the x-axis is along the altitude of the pyramid, then the element of volume dV of a typical slab of side s is given by:

$$dV = s^2\, dx.$$

The volume of the pyramid is therefore

$$V = \int_0^h s^2 dx.$$

In order to evaluate the integral, we must be able to express s in terms of x. By considering similar triangles as indicated in the diagram, we write:

$$\frac{s/2}{x} = \frac{a/2}{h},$$

or,

$$s = \frac{a}{h} x.$$

Hence,

$$V = \int_0^h \frac{a^2}{h^2} x^2 \, dx$$

$$= \left[\frac{a^2 x^3}{3h^2} \right]_0^h = \frac{1}{3} a^2 h.$$

• PROBLEM 868

> A circle of varying size moves always parallel to the y-z-plane and with the ends of a diameter on the curves $y = 1 - x^2$, $z = 0$, and $x + y = 1$, $z = 0$. Find the portion in the first octant of the volume generated.

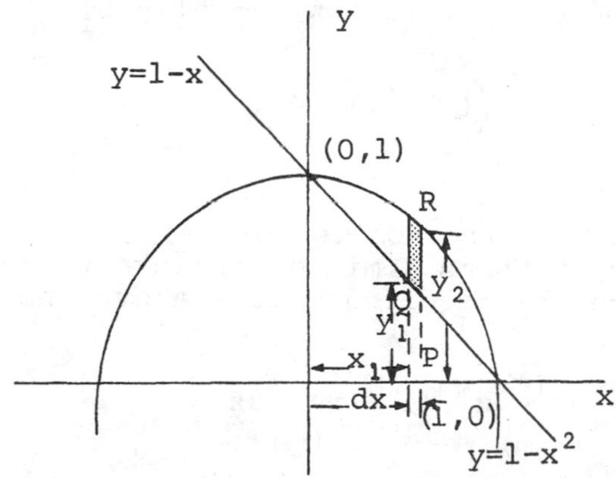

Solution: A line $x = p(x_1, 0)$, drawn parallel to the y-axis, intersects the curves at points $Q(x_1, y_1)$ and $R(x_1, y_2)$ as shown in the figure. The distance $\overline{QR} = y_2 - y_1$ is then the diameter of the circle. The equation of the circle centered at a point $\left(x_1, \frac{y_2 - y_1}{2}, 0\right)$ is:

$$\left(y - \left(y_1 + \frac{y_2 - y_1}{2}\right)\right)^2 + z^2 = \left(\frac{y_2 - y_1}{2}\right)^2.$$

Similarly, by constructing another line: $x = x_1 + dx$, we generate another circle with a diameter approximately $y_2 - y_1$. (The accuracy for the new diameter is improved if dx is chosen sufficiently small.) Under these conditions, the volume generated by the shaded strip, as shown above, is a disk with radius

$r = \frac{y_2 - y_1}{2}$, and thickness dx. It has a volume:

$$dV = \pi r^2 \, dx = \frac{\pi}{4} (y_2 - y_1)^2 \, dx,$$

where y_1 is the value obtained from the straight line: $y = 1 - x$, and y_2 is from: $y = 1 - x^2$.

In a similar fashion, the region bounded by $y = 1 - x^2$ and $y = 1 - x$ can be represented by a series of vertical strips. Each strip generates a disk given by:

$$dV_i = \frac{\pi}{4} \left(y_{2i} - y_{1i}\right)^2 dx_i$$

$$= \frac{\pi}{4} \left[\left(1 - x_i^2\right) - \left(1 - x_i\right)\right]^2 dx_i$$

$$= \frac{\pi}{4} \left[\left(1 - x_i\right)\left(1 + x_i\right) - \left(1 - x_i\right)\right]^2 \; dx;$$

$$= \frac{\pi}{4} \; x_i^2 \left(1 - x_i\right)^2 dx.$$

The sum of the volumes of all such elementary disks, as limit dx shrinks to zero, gives twice the required volume above the xy-plane (for $z \geq 0$).

$$V = \frac{\pi}{4} \int_0^1 x^2 \; (1 - x)^2 \; dx$$

$$= \frac{\pi}{4} \int_0^1 x^2 \; (1 - 2x + x^2) dx$$

$$= \frac{\pi}{4} \int_0^1 (x^4 - 2x^3 + x^2) dx$$

$$= \frac{\pi}{4} \left[\frac{1}{5} x^5 - \frac{1}{2} x^4 + \frac{1}{3} x^3\right]_0^1$$

$$= \frac{\pi}{120} \, .$$

The desired volume in the first octant is

$$\frac{\pi}{240}.$$

• PROBLEM 869

> The base of the solid, shown in the diagram, is a circle of radius 5 inches, and each section of the solid by a plane perpendicular to a fixed diameter AB is an isosceles triangle with altitude 6 inches. Find the volume of the solid.

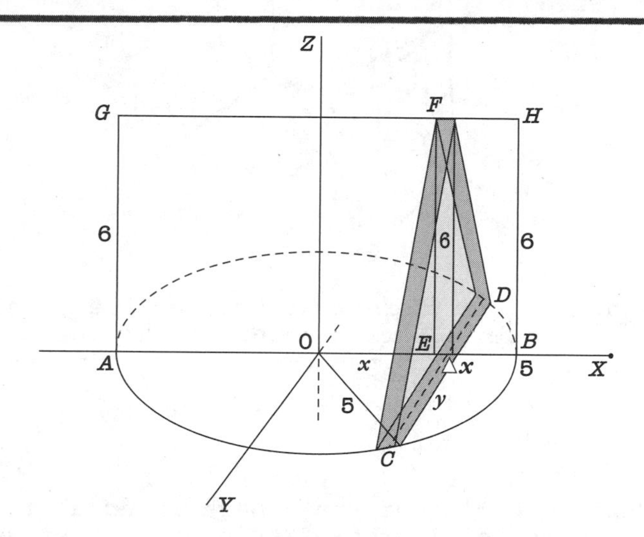

<u>Solution:</u> Let CDF be a section of the solid perpendicular to AB at a distance x from the center 0. Using the general equation for a circle in the x-y plane,

$$x^2 + y^2 = 25,$$

we obtain $CE = y = \sqrt{25 - x^2}$. The cross-sectional area CDF is $A(x) = 6\sqrt{25 - x^2}$. Then the volume of the right-hand half of the solid is:

$$V = \int_0^5 A(x)\, dx = 6 \int_0^5 \sqrt{25 - x^2}\, dx,$$

or,

$$V = 6\left(\frac{1}{2}\left[x\sqrt{25 - x^2} + 25 \sin^{-1}\frac{x}{5}\right]\right|_0^5$$

$$= \frac{75}{2}\pi.$$

The required volume of the entire solid is therefore 75π.

A solid mass has square cross-sections. The side of a square at any cross-section is equal to the square of the distance between the cross-section and one end of the solid mass. Compute the volume of the mass for 8 inches from the end.

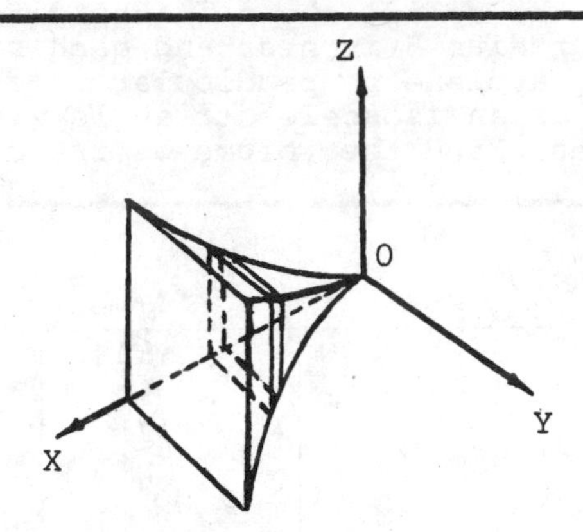

Solution: The differential volume, i.e., the volume of a thin square element of thickness dx and x units away from the y-z plane, is given by:

$$dV = Adx.$$

Since one side of the square is equal to x^2, the cross-sectional area of this element at any cross-section is:

$$A = (x^2)^2 = x^4,$$

thus

$$dV = x^4 \, dx.$$

The volume of this figure at a distance x units from the origin to the base cross-section is:

$$V_x = \int_0^x t^4 \, dt,$$

where t is a dummy variable of integration.

$$V_x = \frac{1}{5} t^5 \Big|_0^x = \frac{1}{5} x^5.$$

For a distance 8 inches from the end, x = 8 in., we obtain

$$V_8 = \frac{1}{5}(8)^5$$
$$= 6553\frac{3}{5}\ (\text{in})^3.$$

• PROBLEM 871

The arc of the curve: $y = \ln x$, lying in the fourth quadrant is revolved about the y-axis. Find the area of the surface generated.

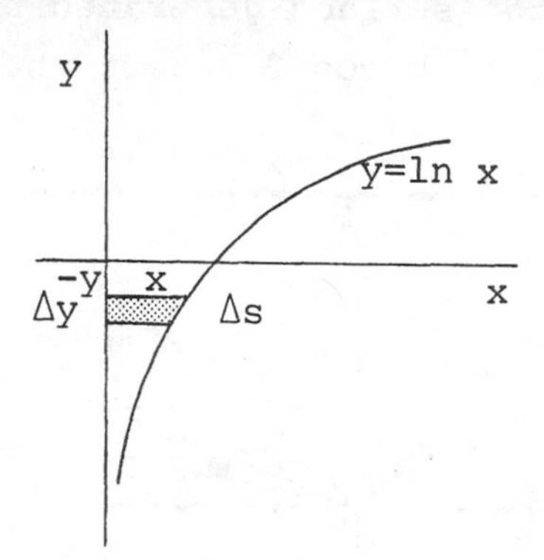

Solution: The shaded strip in the figure (x by Δy), when rotated about the y-axis, sweeps a volume approximately equal to that of a truncated cone, base radius x and thickness Δy, with a slanted edge Δs. If we were interested in finding the volume we would merely use Δy. However, for the surface area, Δs is involved, which converges to:

$$(ds)^2 = (dx)^2 + (dy)^2,$$

or,

$$ds = \left[1 + \left(\frac{dx}{dy}\right)^2\right]^{\frac{1}{2}} dy.$$

The surface area of an elementary disk, under the limit, is:

$$ds = 2\pi x\, ds$$
$$= 2\pi x\left[1 + \left(\frac{dx}{dy}\right)^2\right]^{\frac{1}{2}} dy.$$

$$y = \ln x\ .\ \frac{dy}{dx} = \frac{1}{x}\ \cdot\ \frac{dx}{dy} = x.$$

$$dy = \frac{1}{x}\,dx.$$

Upon substitution,

$$ds = 2\pi x\ (1 + x^2)^{\frac{1}{2}}\ \frac{1}{x}\ dx.$$

Integration from $x = 0$ to $x = 1$ yields the required area in the region bounded by the x- and y-axes.

$$s = 2\pi\int_0^1 \sqrt{1 + x^2}\ dx$$

$$= \pi \left\{ x\sqrt{1+x^2} + \ln\left(x + \sqrt{1+x^2}\right) \right|_0^1$$

$$= \pi \left[\sqrt{2} + \ln\left(\sqrt{2}+1\right) \right].$$

(Integration was carried out by the use of integration by parts and then substitution.)

• PROBLEM 872

> Find the area of the surface generated by revolving the lemniscate: $r^2 = a^2 \cos 2\Theta$, about the line: $\Theta = \frac{\pi}{2}$.

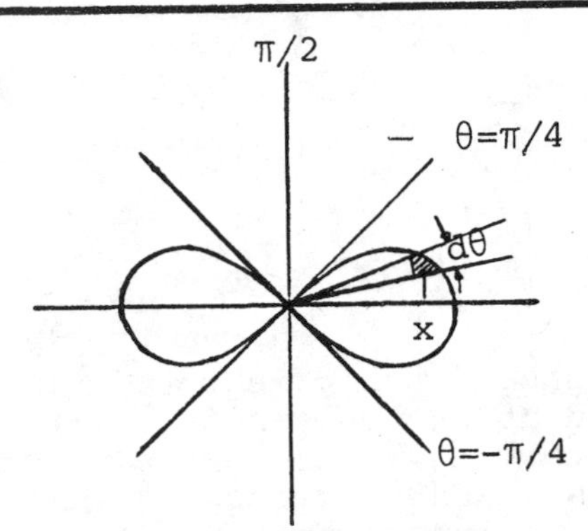

Solution: As shown in the figure, the area generated by the shaded region, when revolved about the $\pi/2$-axis, is:

$$2\pi x ds,$$

where x is the radius of generation, and ds is the length of the arc subtended by $d\Theta$. Since, at any point P on the edge of the lemniscate,

$$x = r\cos\Theta, \quad r d\Theta = ds,$$

and due to symmetry,

$$A = 4\pi \int_0^{\pi/4} r\cos\Theta (r d\Theta)$$

$$= 4\pi a^2 \int_0^{\pi/4} \cos\Theta \cos 2\Theta \, d\Theta = 4\pi a^2 \int_0^{\pi/4} \cos\theta(1-2\sin^2\theta)d\theta$$

$$= 4\pi a^2 \left[\sin\Theta - \frac{2\sin^3\Theta}{3} \right]_0^{\pi/4}$$

$$= 4\pi a^2 \left(\frac{1}{\sqrt{2}} - \frac{2}{3}\,\frac{1}{2\sqrt{2}} \right)$$

$$= 4\pi a^2\, \frac{2}{3} \left(\frac{1}{\sqrt{2}} \right) = 4\frac{\pi a^2\sqrt{2}}{3}$$

Note, that multiplying both sides of the equation

by r^2, we obtain:

$$r^4 = r^2a^2 \cos 2\Theta = a^2 \left(r^2 \cos^2 \Theta - r^2 \sin^2 \Theta\right),$$
$$= a^2 \quad x^2 - y^2 ,$$

On the other hand,

$$r^2 = x^2 + y^2,$$

Thus,

$$\left(x^2 + y^2\right)^2 = a^2 \left(x^2 - y^2\right).$$

Therefore, insertion of $x = r \cos \theta$ does not affect the calculation.

• **PROBLEM** 873

Find the area of the surface formed by revolving about the x-axis the parabola: $y^2 = 2x - 1$, from $y = 0$ to $y = 1$.

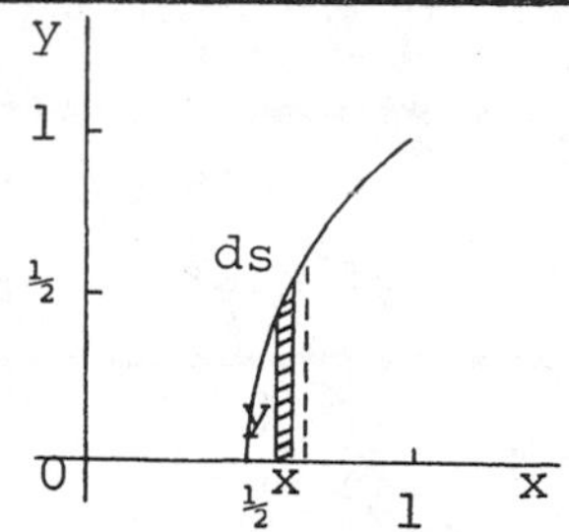

Solution: We consider the shaded strip as shown in the figure. Its length is y units and width Δx units. If this strip is rotated about the x-axis, it sweeps a volume that looks like a disk with radius y units, and thickness of Δx units. The circumference $2\pi r = 2\pi y$. Hence its surface area is

$$\Delta S = 2\pi y \ \Delta s.$$

Noting that the parabola is made up of such disks for

$$\frac{1}{2} \le x \le 1,$$

the required surface area is the sum of all surface areas of all the disks which compose the entire given solid. Mathematically, if S is the surface area required, then

$$S \simeq \sum_{i=1}^{n} 2\pi y_i \ \Delta s_i,$$

or

$$S = \lim_{n\to\infty} \sum_{i=1}^{n} 2\pi y_i \ \Delta s_i, \quad \frac{1}{2} \le x_i \le 1$$

Since $2y \, dy = 2 \, dx$, and $\frac{dx}{dy} = y$, the differential of arc length is given by:

$$(ds)^2 = (dx)^2 + (dy)^2$$

$$= \left[1 + \left(\frac{dx}{dy}\right)^2\right](dy)^2 \cdot$$

$$ds = \sqrt{1 + \left(\frac{dx}{dy}\right)^2} \cdot dy.$$

Furthermore, $y \geqq 0$, so that $|y| = y$, and hence:

$$S = 2\pi \int_0^1 y\sqrt{1 + y^2}\, dy$$

$$= \left[\frac{2\pi}{3}(1 + y^2)^{\frac{3}{2}}\right]_0^1$$

$$= \frac{2\pi}{3}(2\sqrt{2} - 1).$$

• PROBLEM 874

Find the area of the surface generated by revolving the hypocycloid: $x^{\frac{2}{3}} + y^{\frac{2}{3}} = a^{\frac{2}{3}}$, about the x-axis.

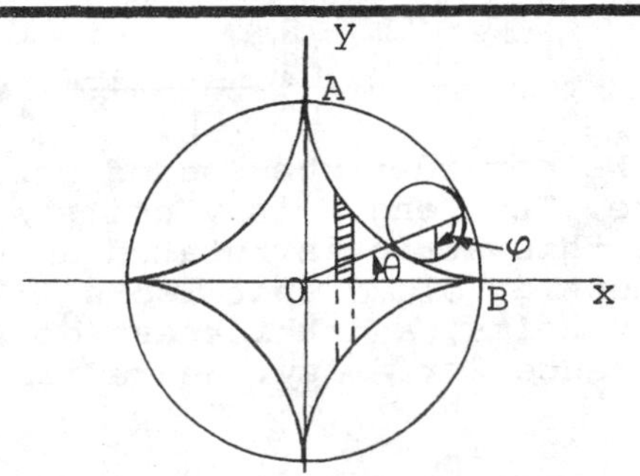

Solution: A hypocycloid is the locus of any fixed point on the circumference of a circle (radius = b) that rolls internally, without slipping, on a fixed circle of radius a, where b divides a.

Using the disk method and noting the symmetry of the region about the x-axis, the surface area is given by:

$$s = 2\int_A^B 2\pi y\, ds = 4\pi \int_0^a y\, ds.$$

ds is the arc length, given by:

$$(ds)^2 = (dy)^2 + (dx)^2,$$

Multiplying and dividing the right-hand side of the equation by $(dy)^2$, we have:

$$(ds)^2 = \left[1 + \left(\frac{dx}{dy}\right)^2\right](dy)^2.$$

Taking the square root, we obtain:

$$ds = \left[1 + \left(\frac{dx}{dy}\right)^2\right]^{\frac{1}{2}} dy.$$

From the given equation: $x^{\frac{2}{3}} + y^{\frac{2}{3}} = a^{\frac{2}{3}}$, we differentiate implicitly:

$$\frac{2}{3} x^{-\frac{1}{3}} dx + \frac{2}{3} y^{-\frac{1}{3}} dy = 0,$$

or,

$$\frac{dx}{dy} = -\frac{x^{\frac{1}{3}}}{y^{\frac{1}{3}}}.$$

Substitution in the ds expression yields:

$$ds = \left[1 + \left(\frac{x^{\frac{1}{3}}}{y^{\frac{1}{3}}}\right)^2\right]^{\frac{1}{2}} dy = \left(\frac{y^{\frac{2}{3}} + x^{\frac{2}{3}}}{y^{\frac{2}{3}}}\right)^{\frac{1}{2}} dy$$

$$= \left(\frac{a^{\frac{2}{3}}}{y^{\frac{2}{3}}}\right)^{\frac{1}{2}} dy = \frac{a^{\frac{1}{3}}}{y^{\frac{1}{3}}} dy.$$

Thus,

$$s = 4\pi \int_0^a y \frac{a^{\frac{1}{3}}}{y^{\frac{1}{3}}} dy = 4\pi a^{\frac{1}{3}} \int_0^a y^{\frac{2}{3}} dy$$

$$= 4\pi a^{\frac{1}{3}} \left(\frac{3}{5} y^{\frac{5}{3}}\right|_0^a = \frac{12\pi a^{\frac{1}{3}}}{5} a^{\frac{5}{3}}$$

$$= \frac{12\pi a^2}{5}.$$

CHAPTER 32

CENTROIDS

The centroid or center of gravity of a mass is the point at which all the mass can be considered concentrated. If, for example, the mass were to be suspended at its centroid, the mass would be in complete balance. This balanced relationship may be expressed by the condition that if we take moments about any reference axis, the algebraic sum of moments of all individual elements of the mass about the reference axis is equal to the entire mass multiplied by the distance from the centroid to the reference axis. Expressing this mathematically, and taking moments about the y-axis, we have:

$$M\bar{x} = m_1x_1 + m_2x_2 + m_3x_3 \ldots ,$$

where $\bar{x}$ is the distance from the centroid to the y-axis, M is the total mass, and mx are the moments of the individual elements. Care must be taken to use the proper algebraic sign for the moments taken.

An element of mass be computed from the product of its volume and density. When dealing with plane areas or plate members of uniform density, the computation for the centroid does not involve the density.

CENTROIDS OF ARCS AND SURFACES OF REVOLUTION

To find the centroid of an arc, we subdivide the arc, s, into elements, ds, and obtain the moments of ds with respect to the x- and y-axes. Now

$$ds = \sqrt{dx^2 + dy^2} = \sqrt{1 + \left(\frac{dy}{dx}\right)^2}\, dx.$$

If we are taking moments about the y-axis, then an element of moment,

$$dM_y = x\sqrt{1 + \left(\frac{dy}{dx}\right)^2}\, dx,$$

and

$$M_y = \int x\sqrt{1 + \left(\frac{dy}{dx}\right)^2}\, dx .$$

The length of the arc is:

$$s = \int ds = \int \sqrt{1 + \left(\frac{dy}{dx}\right)^2}\, dx .$$

The x coordinate of the centroid, $\bar{x}$, is equal to: $\frac{M_y}{s}$.

A similar expression exists for $\bar{y} = \frac{M_x}{s}$.

For surfaces of revolution, about the y-axis, an element of area is:

$$dA = 2\pi x \, ds,$$

and the moment arm is x. Therefore,

$$dM_y = 2\pi x^2 \, ds$$

and

$$M_y = 2\pi \int x^2 \sqrt{1 + \left(\frac{dy}{dx}\right)^2} \, dx .$$

The total area is:

$$A = 2\pi \int x \sqrt{1 + \left(\frac{dy}{dx}\right)^2} \, dx .$$

The x coordinate of the centroid, $\bar{x}$, is equal to: $\frac{M_y}{A}$, while

$$\bar{y} = \frac{M_x}{A} .$$

• PROBLEM 875

Find the centroid of the solid generated by revolving about tye Y-axis the area in the first quadrant bounded by the curve: $y = 4 - x^2$, and the coordinate axes.

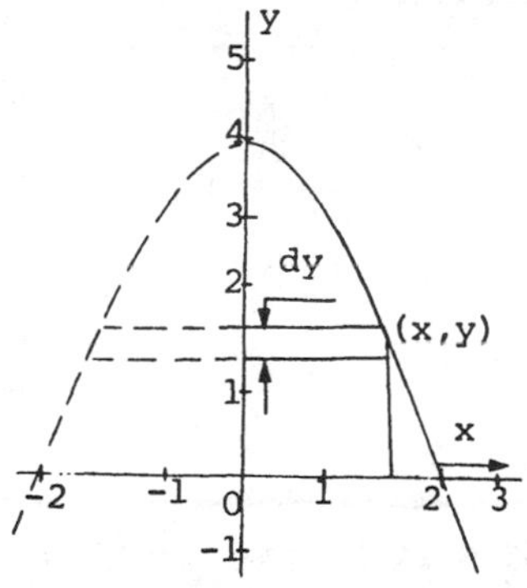

__Solution:__ The moment md of an object is the turning effort with respect to a point, a line, or a plane. d is the distance from the object to the point or is a perpendicular distance to the line or the plane. Under a Cartesian three-coordinate system, the moment with respect to the x-y plane is zm, with respect to the y-z plane we have xm, and ym for the z-x plane, where m is the mass, which may be expressed in terms of density and volume.

In this problem, we are interested in finding the volume generated by revolving about the y-axis, and hence calculate the moment with respect to the three orthogonal planes and the centroid. From the figure it appears that this paraboloid is symmetric on the y-z and x-y planes. Thus the centroid lies at the point $(0, \bar{y}, 0)$, where $\bar{y}$ is to be determined.

The small rectangle, as shown, with the dimensions (x) and (dy), generates a disc about the y-axis with radius x, thickness dy, and element volume:

$$dv = \pi x^2 \, dy \, .$$

Thus the volume of the paraboloid under the given boundaries is:

$$v = \int_0^4 \pi x^2 \, dy = \pi \int_0^4 (4 - y)dy = \pi\left(4y - \frac{1}{2} y^2\right)\Bigg|_0^4$$

$$= 8\pi \, .$$

The moment with respect to the z-x plane is:

$$M_{zx} = \int_0^4 y\pi x^2 dy = \pi\int_0^4 y(4 - y)dy = \pi\int_0^4 \left(4y - y^2\right)dy$$

$$= \frac{32}{3}\pi \, .$$

Remembering that the density, P, as long as it is a constant, is taken outside the integral sign and cancelled, we have:

$$\bar{y} = \frac{PM_{zx}}{m} = \frac{\rho\pi\int_0^4 yx^2 dy}{\rho\pi\int_0^4 x^2 dy} = \frac{\frac{32\pi}{3}}{8\pi} \, ,$$

from which we obtain:

$$\bar{y} = \frac{32}{3}/8 = 4/3 \, .$$

The centroid is at the point $(0, 4/3, 0)$.

• PROBLEM 876

Find the centroid of a right circular cone formed by rotating the line formed by $2x + y = 4$ between the x and y axes about the y axis.

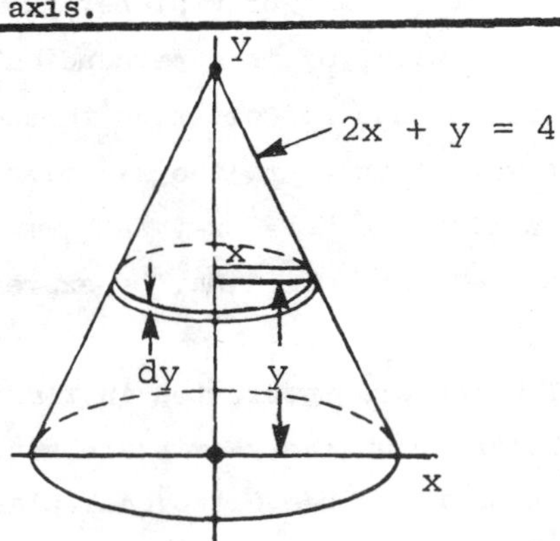

Solution: The centroid is on the y axis. Hence $\bar{x} = \bar{z} = 0$.

The area of an element $= \pi r^2 = \pi x^2 = \pi\left(2 - \frac{y}{2}\right)^2$.

$$\bar{y} = \frac{\int_a^b \pi y\left(x_2 - x_1\right)^2 dy}{\int_a^b \pi\left(x_2 - x_1\right)^2 dy}.$$

$x_1 = 0$ (the axis). $x_2 = 2 - \frac{y}{2}$.

$$\bar{y} = \frac{\int_0^4 y\pi\left(2 - y/2\right)^2 dy}{\int_0^4 \pi\left(2 - y/2\right)^2 dy} = \frac{\int_0^4 \left(4y - 2y^2 + \frac{y^3}{4}\right) dy}{\int_0^4 \left(4 - 2y + \frac{y^2}{4}\right) dy}$$

Since π is a multiplying factor in both the numerator and denominator, it may be canceled.

$$\bar{y} = \frac{4y^2/2 - 2y^3/3 + y^4/16]_0^4}{4y - y^2 + y^3/12]_0^4} = 1 .$$

An analysis of this result indicates that the centroid of a right circular cone is 1/4 the distance from its base, since the height of the cone is 4.

• PROBLEM 877

Find the coordinates of the centroid of the first quadrant arc of the circle $x^2 + y^2 = a^2$.

Solution: The arc of the circle can be considered the sum of infinitesimal arcs ds. The arc length ds is given by the expression:

$$ds = \sqrt{1 + \left(\frac{dy}{dx}\right)^2} .$$

$x^2 + y^2 = a^2$. By implicit differentiation, $2x + 2yy' = 0$.

$$y' = \frac{dy}{dx} = -\frac{x}{y},$$

$$1 + y'^2 = 1 + \frac{x^2}{y^2} = \frac{x^2 + y^2}{y^2} = \frac{a^2}{y^2} .$$

$$ds = \sqrt{1 + y'^2}\, dx = \frac{a}{y}\, dx.$$

Taking moments about the y-axis,

$$M_y = \int_{z-0}^{x-a} x\, ds = a\int_0^a \frac{x}{y}\, dx = a\int_0^a \frac{x\, dx}{\sqrt{a^2 - x^2}} = -a\sqrt{a^2 - x^2}\Big|_0^a = a^2 .$$

Therefore,

$$\bar{x} = \frac{M_y}{s} = \frac{a^2}{\frac{2\pi a}{4}} = \frac{a^2}{\frac{1}{2}\pi a} = \frac{2a}{\pi} .$$

By symmetry, $\bar{y} = \bar{x} = \frac{2a}{\pi}$. As an alternative, we could take moments about the x-axis,

$$M_x = \int_{z=0}^{x=a} y \, ds = \int_0^a y \cdot \frac{a}{y} dx = a\int_0^a dx = a^2 ,$$

$$\bar{y} = \frac{M_x}{s} = \frac{a^2}{\frac{1}{2} \pi a} = \frac{2a}{\pi} .$$

• PROBLEM 878

Find the centroid of the curve: $y^2 = -4x$, from (-4,4) to (0,0).

Solution: We know that the moment with respect to the x-axis is:

$$M_x \simeq \sum_{i=1}^{n} y_i \, \Delta s_i ,$$

where Δs_i is the infinitesimal length of curve in the neighborhood of the point $p\left(x_i, y_i\right)$. Similarly, the moment with respect to the y-axis is:

$$M_y \simeq \sum_{i=1}^{n} x_i \, \Delta s_i .$$

Passing to the limits, the first moments of this element are defined by the integrals

$$M_x = \int y \, ds \quad \text{and} \quad M_y = \int x \, ds .$$

If we construct parallel lines of dimensions Δx and Δy, we can approximate the curve within the rectangle formed by the intersections by a straight line and call it Δs. Using the Pythagorean theorem for this triangle, we obtain:

$$(\Delta s)^2 = (\Delta x)^2 + (\Delta y)^2 .$$

Passing to limits, making Δx and Δy sufficiently small:

$$(ds)^2 = (dx)^2 + (dy)^2 ,$$

or

$$(ds)^2 = \left[\left(\frac{dx}{dy}\right)^2 + 1\right](dy)^2 . \; ds = \sqrt{1 + \left(\frac{dx}{dy}\right)^2} \, dy .$$

From the given equation of the curve,

$$\frac{dx}{dy} = -\tfrac{1}{2}y,$$

and thus

$$ds = \left((-\tfrac{1}{2}y)^2 + 1\right)^{\frac{1}{2}} \cdot dy,$$

or

$$ds = \sqrt{1 + y^2/4} \; dy .$$

Hence

$$M_x = \int_0^4 \sqrt{1 + \frac{y^2}{4}}\, y\, dy = \tfrac{1}{2}\int_0^4 \sqrt{4 + y^2}\, y\, dy = \frac{1}{6}\left(4 + y^2\right)^{3/2}\Bigg]_0^4$$

$$= \frac{4}{3}\left[5^{3/2} - 1\right].$$

$$M_y = \tfrac{1}{2}\int_0^4 \sqrt{4 + y^2}\, x\, dy = \tfrac{1}{2}\int_0^4 \sqrt{4 + y^2}\left(- \frac{y^2}{4}\right)dy = - \frac{1}{8}\int_0^4 \sqrt{4 + y^2}\; y^2 dy$$

$$= - \frac{1}{8}\left[\frac{y}{8}\left(2y^2 + 4\right)\sqrt{4 + y^2} - \frac{16}{8}\ln\left(y + \sqrt{4 + y^2}\right)\right]_0^4$$

$$= - \frac{1}{8}\left[18\sqrt{20} + 2\ln\frac{2}{4 + \sqrt{20}}\right].$$

After finding the length of the above curve between the given points as

$$s = \int_4^0 \sqrt{1 + y^2/4}\; dy$$

$$= 2\sqrt{5} + \ln\left(2 + \sqrt{5}\right),$$

the point $(\bar{x},\bar{y})$, at which the center of mass lies, may be obtained by dividing the respective moments by s.

• PROBLEM 879

Find the centroid of the area bounded by the parabola: $y = x^2$, and the line: $y = 4$, by using a) horizontal and b) vertical strips.

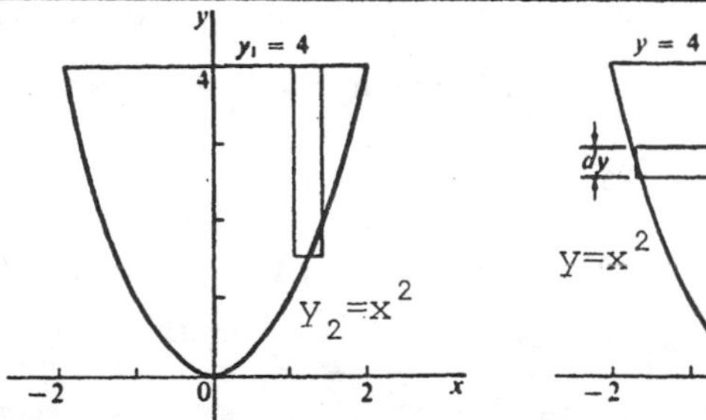

Solution:

a) taking rectangular strips parallel to the x-axis as shown in the figure, the moment of the strip with respect to the x-axis is ydA.

Since the parabola is symmetric with respect to the y-axis, we can make our calculations in the first quadrant and multiply each sum by 2. Obviously, due to this symmetry, we can conclude that $\bar{x} = 0$. For $\bar{y}$, we wish to calculate $A\bar{y}$ and divide it by A, the area of the bounded curve.

$$A\bar{y} = \int_0^4 y(2x dy) = \int_0^4 2xy dy .$$

Since $x = y^{\frac{1}{2}}$,

$$A\bar{y} = \int_0^4 2y^{3/2}\, dy$$

$$= \left[\frac{4}{5} y^{5/2}\right]_0^4$$

$$= \frac{128}{5} .$$

We now determine the area.

$$A = 2\int_0^4 x dy = 2\int_0^4 y^{\frac{1}{2}} dy = 2\left[\frac{2}{3} y^{3/2}\right]_0^4 = \frac{32}{3} ,$$

so that

$$\bar{y} = \left(\frac{128}{5}\right)\left(\frac{3}{32}\right) = \frac{12}{5} .$$

Therefore, the centroid of the parabola $y = x^2$ cut by the line $y = 4$ is

$$\left(0, \frac{12}{5}\right) .$$

b) Using vertical strips, as shown, the y-coordinate of the centroid of the rectangular strip is $\frac{1}{2}\left(y_1 + y_2\right)$. The midpoint of the line segment has the coordinate $\left(y_1 + y_2\right)/2$. Then the vertical strip of width Δx has a moment about the x-axis approximately equal to:

$$\frac{y_1 + y_2}{2} \Delta A = \frac{y_1 + y_2}{2} \left(y_1 - y_2\right) \Delta x ,$$

where $\left(y_1 - y_2\right)\Delta x$ is the area of the strip.

Taking the sum of such strips, and noting that the parabola is symmetric about the y-axis, we obtain the following result:

$$A\bar{y} = 2\int_0^2 \frac{1}{2}\left(y_1^2 - y_2^2\right) dx ,$$

after taking the limit $\Delta x \to 0$ in the usual way.

Since $y_1 = 4$ and $y_2 = x^2$, the expression for $A\bar{y}$ becomes:

$$A\bar{y} = \int_0^2 \left(16 - x^4\right) dx = \frac{128}{5} .$$

It follows that

$$\bar{y} = \left(\frac{128}{5}\right)\left(\frac{3}{32}\right) = \frac{12}{5} .$$

• PROBLEM 880

Find the centroid of the circle:

$$\left.\begin{matrix} x = 5 \sin t \\ y = 5 \cos t \end{matrix}\right\} \text{ from } t = -\frac{\pi}{3} \text{ to } t = \frac{\pi}{2} .$$

Solution: We divide the given bounded section into n segments. The length of the ith segment is $r\Delta t_i$, where the radius in this problem is 5 units. Since the density of this arc is constant, the mass of the ith element is proportional to $5\Delta t_i$. Assuming that the mass of this segment is concentrated at the single point (ξ_i, ζ_i), where $x_{i-1} \le \xi_i \le x_i$ and $y_{i-1} \le \zeta_i \le y_i$, the moment of this segment about the x-axis is between $5y_{i-1}\Delta t_i$ and $5y_i\Delta t_i$, and about the y-axis the moment is between $5x_{i-1}\Delta t_i$ and $5x_i\Delta t_i$. If the moments of the sector bounded by the points $A(5\cos\pi/3, -5\sin\pi/3)$ and $B(5\cos\pi/2, 5\sin\pi/2)$ about the x- and y-axes are M_x and M_y, then the Riemann sums at the average values for $i = 1,2,3,\ldots,n$ are:

$$M_x \approx \sum_{i=1}^{n} \zeta_i\left(5\Delta t_i\right), \text{ and } M_y \approx \sum_{i=1}^{n} \xi_i\left(5\Delta t_i\right),$$

which reduce to:

$$M_x = 5\int_{t=-\pi/3}^{t=\pi/2} y\,dt, \text{ and } M_y = 5\int_{t=-\pi/3}^{t=\pi/3} x\,dt .$$

Using the x and y relationships with t, we substitute $5\cos t$ and $5\sin t$ for x and y, respectively. Thus we obtain:

$$M_x = \int_{-\pi/3}^{\pi/2} 5\cdot 5\cos t\,dt = 25\sin t\Big]_{-\pi/3}^{\pi/2} = \frac{25}{2}\left(2 + \sqrt{3}\right),$$

$$M_y' = \int_{-\pi/3}^{\pi/2} 5\cdot 5\sin t\,dt = -25\cos t\Big]_{-\pi/3}^{\pi/2} = \frac{25}{2}.$$

The length of the arc is $\left(\frac{\theta_1-\theta_2}{2\pi}\right)2\pi r = r\left(\theta_1-\theta_2\right) = 5\left(\frac{\pi}{2}+\frac{\pi}{3}\right) = 25\frac{\pi}{6}$. Dividing M_x and M_y by this factor, we have

$$\bar{x} = \frac{3}{\pi},$$

$$\bar{y} = \frac{6+3\sqrt{3}}{\pi} .$$

• PROBLEM 881

Find the center of gravity of a cone of radius r and altitude h. (Measure from the vertex.)

__Solution:__ We are looking for a point, such that, if the mass of the cone were concentrated at that point, say $(\bar{x}, \bar{y}, \bar{z})$, the turning effect of this mass with respect to a given point is the same as that obtained in the following expression:

$$m\bar{x} = M_{yz}, \quad m\bar{y} = M_{zx}, \quad m\bar{z} = M_{xy},$$

where M_{yz} is the moment (turning effect) of the cone with respect to the plane formed by the y- and z-axes. The rest of the moments have the same definition with respect to their corresponding planes.

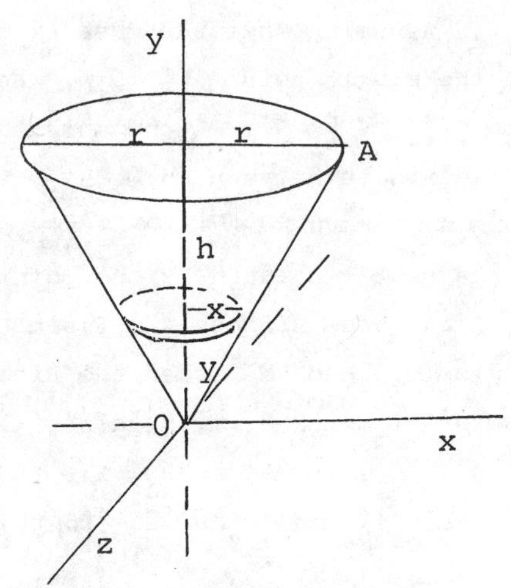

The whole cone can be thought of as made up of n small particles such that $m = \sum_{i=1}^{n} m_i$, and hence,

$$M_{yz} \simeq \sum_{i=1}^{n} x_i m_i \ , \ M_{zx} \simeq \sum_{i=1}^{n} y_i m_i \ \text{ and } \ M_{xy} \simeq \sum_{i=1}^{n} z_i m_i \ .$$

We can furthermore assume that these particles are so small that the summation becomes integration. We express the masses in terms of their corresponding infinitesimal volumes and densities, ρ. Thus, we obtain:

$$M_{yz} = \int_v \rho x dv \ , \ M_{zx} = \int_v \rho y dv \ \text{ and } \ M_{xy} = \int_v \rho z dv$$

where $\int_v f(x,y,z)dv$ indicates that integration performed over the entire volume.

If the density of each element is the same, we say that the object, in this case, the cone, is homogeneous, and we can take ρ out of the integral sign. Assuming constant density,

$$m\bar{x} = \rho v\bar{x} = \rho\int xdv,$$

or,

$$v\bar{x} = \int xdv,$$

and

$$\bar{x} = \frac{\int xdv}{v} = \frac{\int xdv}{\int dv} \ .$$

Similarly,

$$\bar{y} = \frac{\int ydv}{\int dv} \ , \quad \bar{z} = \frac{\int zdv}{\int dv} \ .$$

As can be seen from the figure, the cone is symmetric about the y-axis. Therefore we can conclude that there are equal distributions of mass. The net moments of these particles on the y-z and x-y planes are zero -- suggesting $\left(0,\bar{y},0\right)$ is the point in question.

Now we try to express the equation of the cone in mathematical form to find the value of $\bar{y}$.

We can find the equation of the line OA in the diagram in terms of the radius r and height h. The equation of a straight line is: $y = mx + b$, where m is the tangent of the angle made by the line with the positive x-axis and b is the intercept. In this case,

$$y = \frac{hx}{r}.$$

If we assume this line rotating about the y-axis, the generated volume is the desired cone. The small strip of rectangle with width dy drawn from the y-axis horizontally, one of its corners to touch a point (x,y) on the line, generates a disc with approximate volume

$$\Delta v = \pi x^2 \Delta y ,$$

since x is its radius and it has a thickness of Δy. If the cone is composed of such discs, we can find the entire volume in integral form:

$$v = \int \pi x^2 dy.$$

Returning to the previous equations, we write:

$$\bar{y} = \frac{\int y dv}{\int dv}.$$

We obtain: $$\bar{y} = \int yx^2 dy \Big/ \int x^2 dy,$$

and, using the expression: $y = \frac{hx}{r}$, we substitute $\frac{r^2}{h^2} y^2$ for x^2 in the above expression.

$$\bar{y} = \int_0^h y^3 dy \Big/ \int_0^h y^2 dy$$

$$= 3/4\ h.$$

• PROBLEM 882

Find the volume of the torus generated by revolving the circle $(x - b)^2 + y^2 = a^2$ $(b > a)$ about the y-axis.

Solution: We use the theorem of Pappus which says: Let A denote the area of a plane region R, and let V be the volume of the figure generated when the region is rotated about an axis in the plane of the region that does not intersect the region. Then

$$V = AL$$

where L is the distance traveled by the centroid of R.

Since the torus is the volume generated by the circle:

$$(x - b)^2 + y^2 = a^2,$$

centered at (b, 0) and with radius $r = a < b$ about the y-axis, we have the cross-sectional area πa^2, the centroid at (b,0), and $L = 2\pi b$. Thus, the volume of the torus,

$$V = \pi a^2 \ (2\pi b)$$

$$= 2\pi^2 \ a^2 b.$$

• **PROBLEM** 883

Determine the center of gravity of the area between the parabola: $y^2 = 2px$, and the line: $y = b$ and $x = a$.

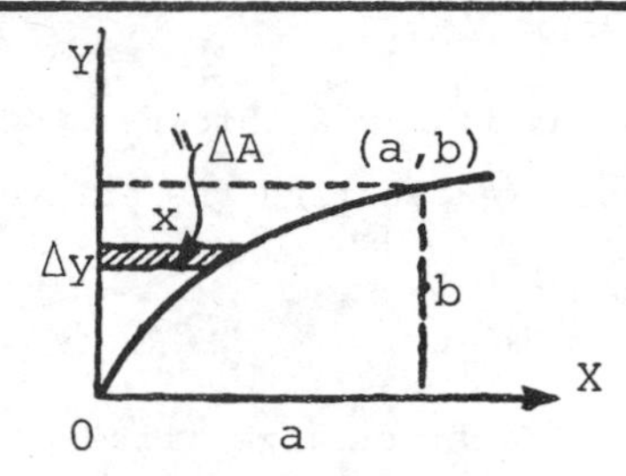

Solution: $\Delta A = x \cdot \Delta y$, $y = \sqrt{2px}$. Passing to the limit,

$$A = \int_0^b x \cdot dy.$$

Since $x = \frac{y^2}{2p}$,

$$A = \frac{1}{2p} \int_0^b y^2 \cdot dy = \left[\frac{1}{2p} \cdot \frac{y^3}{3}\right]_0^b = \frac{b^3}{6p}.$$

Since y varies from 0 to b, two constant values, the average value of y is then just y. But x varies from 0 to $y^2/2p$, with a variable ending, hence the average value for x is $x/2$. Under these conditions, we have

$$M_x = \Delta A \cdot y \quad \text{and} \quad M_y = \Delta A \cdot \frac{x}{2}.$$

Passing to the limit,

$$M_x = \int_0^b y \cdot x \cdot dy = \int_0^b \frac{y^2}{2p} \cdot y \cdot dy = \frac{1}{2p} \int_0^b y^3 \cdot dy = \frac{1}{2p} \left[\frac{y^4}{4}\right]_0^b$$

$$= \frac{b^4}{8p}.$$

$$M_y = \int_0^b \frac{x}{2} \cdot x \cdot dy = \frac{1}{2} \int_0^b \frac{y^4}{4p^2} \cdot dy = \frac{1}{8p^2} \int_0^b y^4 \cdot dy = \frac{1}{8p^2} \left[\frac{y^5}{5}\right]_0^b$$

$$= \frac{1}{8p^2} \cdot \frac{b^5}{5} = \frac{b^5}{40p^2}.$$

Then

$$\bar{x} = \frac{M_y}{A} = \frac{b^5}{40p^2} \cdot \frac{6p}{b^3} = \frac{3b^2}{20p}$$

$$\bar{y} = \frac{M_x}{A} = \frac{b^4}{8p} \cdot \frac{6p}{b^3} = \frac{3b}{4} .$$

But $x = a$, $y = b$, and $y^2 = 2px$; therefore, $b^2 = 2pa$.

$$\bar{x} = \frac{3b^2}{20p} = \frac{3 \cdot 2pa}{20p} = \frac{3a}{10}$$

$$\bar{y} = \frac{3b}{4} .$$

• PROBLEM 884

Determine what are the coordinates of the center of gravity of the area under one arch of the sine curve : $y = \sin x$.

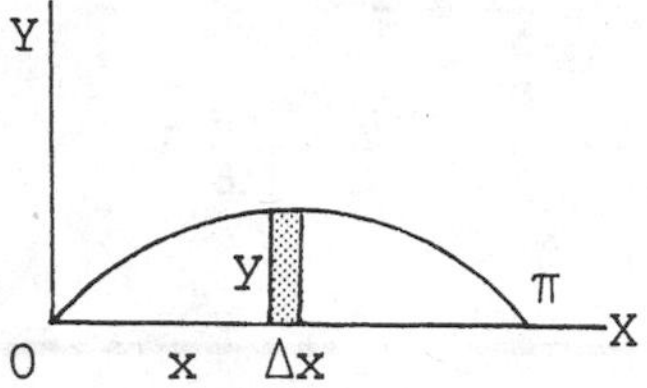

Solution: $\Delta A = y\Delta x = \sin x \cdot \Delta x$, as shown.

x varies from 0 to π, both constants, and y varies from zero to sin x. Thus the average values of x and y are, respectively, x and y/2. Therefore the moment of ΔA about its x-axis is

$$\Delta M_x = \frac{y}{2} \cdot y\Delta x, \quad \text{or} \quad \Delta M_x = \frac{y^2}{2} \cdot \Delta x = \tfrac{1}{2} \sin^2 x \cdot \Delta x \quad .$$

The moment of ΔA about the y-axis is:

$$\Delta M_y = x \cdot y\Delta x = x \sin x \cdot \Delta x \quad .$$

The limits are: $x = 0$ and $x = \pi$. Passing to the limit,

$$M_y = \int_0^{\pi} x \cdot \sin x \cdot dx \; .$$

To integrate this, we remember that, from the derivative of a product of two variables, say u and v, we have:

$$d(uv) = udv + vdu,$$

or

$$udv = d(uv) - vdu \; .$$

Integration of the whole differential equation with respect to the corresponding variables gives:

$$\int udv = uv - \int vdu,$$

which is applicable whenever $\int udv$ is difficult to integrate. This is known as integration by parts. Now, we let $u = x$ and $dv = \sin x \cdot dx$; then $du = dx$ and $v = \int \sin x \cdot dx = -\cos x$. Then $\int x \cdot \sin x \cdot dx = uv - \int v \cdot du = -x \cos x - \int -\cos x \cdot dx$. Now,

$$M_y = [-x \cos x + \sin x]_0^{\pi} = \left[-\pi \cdot (-1) + 0 + 0 \cdot 1 - 0\right] = \pi \; .$$

$$M_x = \frac{1}{2}\int_0^{\pi} \sin^2 x \cdot dx .$$

But $\sin^2 x = \frac{1}{2} - \frac{1}{2}\cos 2x$. Then,

$$M_x = \frac{1}{4}\int dx - \frac{1}{4}\int \cos 2x \cdot dx = \left[\frac{x}{4} - \frac{1}{8}\sin 2x\right]_0^{\pi}$$

$$= \left(\frac{\pi}{4} - \frac{1}{8}\sin 2\pi - 0 + \frac{1}{8}\sin 0\right) = \frac{\pi}{4} .$$

Now $A = \text{area} = \int_0^{\pi} \sin x \cdot dx = [-\cos x]_0^{\pi} = -\cos \pi + \cos 0 = 1 + 1 = 2.$

Therefore,

$$\bar{x} = \frac{M_y}{A} = \frac{\pi}{2}$$

$$\bar{y} = \frac{M_x}{A} = \frac{\pi/4}{2} = \frac{\pi}{8}$$

• PROBLEM 885

Find the centroid of the area in the first quadrant bounded by the hypocycloid

$$x^{\frac{2}{3}} + y^{\frac{2}{3}} = a^{\frac{2}{3}}.$$

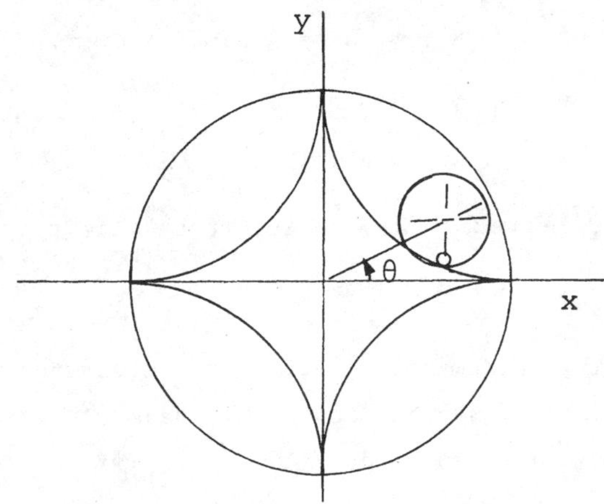

Solution: As can be seen from the figure, a hypocycloid of equation:

$$x^{\frac{2}{3}} + y^{\frac{2}{3}} = a^{\frac{2}{3}},$$

is a locus of any fixed point on the circumference of a smaller circle that internally rolls, without slipping, on a fixed larger circle.

Realizing that:

$$a \cos 3\theta = a(4\cos^3\theta - 3\cos\theta),$$

$$a\cos^3\theta = a\left[\frac{\cos 3\theta + 3\cos\theta}{4}\right]$$

$$= \frac{1}{4}\ a\cos 3\theta + \frac{3}{4}\ a\cos\theta,$$

the equation for the hypocycloid can be expressed in a parametric form as:

$$x = \frac{3a}{4} \cos \Theta + \frac{a}{4} \cos 3\Theta, \text{ and}$$

$$y = \frac{3a}{4} \sin \Theta - \frac{a}{4} \sin 3\Theta,$$

from which the quantities (area and moments):

$$A = \int_0^a y\, dx, \qquad M_x = \int_0^a y^2\, dx,$$

and

$$M_y = \int_0^a xy\, dx,$$

can be determined.

$$A = \int_0^a dx$$

is most easily integrable if we use

$$x = a \cos^3 \Theta, \qquad y = a \sin^3 \Theta,$$

$$dx = -\, 3a \cos^2 \Theta \sin \Theta\, d\Theta.$$

$$A = \int_{\frac{\pi}{2}}^{0} (a \sin^3 \Theta)(-\, 3a \cos^2 \Theta \sin \Theta) d\Theta$$

$$= \int_{\frac{\pi}{2}}^{0} -\, 3a^2 \cos^2 \Theta \sin^4 \Theta\, d\Theta.$$

Recognizing that $\cos^2 \Theta = \frac{1 + \cos 2\Theta}{2}$,

and $\sin^2 \Theta = \frac{1 - \cos 2\Theta}{2}$,

and reversing the limits to eliminate the minus sign, we have:

$$A = 3a^2 \int_0^{\frac{\pi}{2}} \frac{1 + \cos 2\Theta}{2} \left[\frac{1 - 2 \cos 2\Theta + \cos^2 2\Theta}{4}\right] d\Theta$$

$$= \frac{3a^2}{8} \int_0^{\frac{\pi}{2}} (1 + \cos 2\Theta) \left[1 - 2 \cos 2\Theta + \frac{1 + \cos 4\Theta}{2}\right] d\Theta$$

$$= \frac{3a^2}{8} \int_0^{\frac{\pi}{2}} \left(1 - 2\cos 2\Theta + \frac{1}{2} + \frac{\cos 4\Theta}{2}\right.$$

$$+ \cos 2\Theta - 2\cos^2 2\Theta + \frac{\cos 2\Theta}{2}$$

$$\left. + \frac{\cos 2\Theta \cos 4\Theta}{2}\right) d\Theta$$

$$= \frac{3a^2}{8} \int_0^{\frac{\pi}{2}} \left(\frac{3}{2} - \frac{\cos 2\Theta}{2} + \frac{\cos 4\Theta}{2} - 2\cos^2 2\Theta\right.$$

$$\left. + \frac{\cos 6\Theta}{4} + \frac{\cos 2\Theta}{4}\right) d\Theta$$

$$= \frac{3a^2}{8} \int_0^{\frac{\pi}{2}} \left(\frac{3}{2} - \frac{\cos 2\Theta}{2} + \frac{\cos 4\Theta}{2} - 2\right.$$

$$\left. \left(\frac{1 + \cos 4\Theta}{2}\right) + \frac{\cos 6\Theta}{4} + \frac{\cos 2\Theta}{4}\right) d\Theta$$

$$= \frac{3a^2}{8} \int_0^{\frac{\pi}{2}} \left(\frac{1}{2} - \frac{\cos 2\Theta}{4} - \frac{\cos 4\Theta}{2}\right.$$

$$\left. + \frac{\cos 6\Theta}{4}\right) d\Theta$$

$$= \frac{3a^2}{8} \left(\frac{\Theta}{2} - \frac{\sin 2\Theta}{8} - \frac{\sin 4\Theta}{8}\right.$$

$$\left. + \frac{\sin 6\Theta}{12}\right|_0^{\frac{\pi}{2}}$$

$$= \frac{3a^2}{8} \left(\frac{\pi}{4}\right) = \frac{3a^2 \pi}{32}.$$

$$M_x = \int_0^a y^2 \, dx$$

$$= \int_{\frac{\pi}{2}}^0 a^2 \sin^6 \Theta \left(-3a\cos^2 \Theta \sin \Theta\right) d\Theta$$

$$= 3a^3 \int_0^{\frac{\pi}{2}} \cos^2 \Theta \sin^7 \Theta \, d\Theta$$

$$= 3a^3 \int_0^{\frac{\pi}{2}} \cos^2 \Theta \sin^6 \Theta \sin \Theta \, d\Theta$$

$$= 3a^3 \int_0^{\frac{\pi}{2}} \cos^2 \Theta \, (\sin^2)^3 \sin \Theta \, d\Theta$$

$$= 3a^3 \int_0^{\frac{\pi}{2}} \cos^2 \Theta \, (1 - \cos^2 \Theta)^3 \sin \Theta \, d\Theta$$

$$= 3a^3 \int_0^{\frac{\pi}{2}} \Big(\cos^2 \Theta (1 + 3 \cos^4 \Theta - 3 \cos^2 \Theta - \cos^6 \Theta) \cdot \sin \Theta\Big) d\Theta$$

$$= 3a^3 \int_0^{\frac{\pi}{2}} (\cos^2 \Theta + 3 \cos^6 \Theta - 3 \cos^4 \Theta - \cos^8 \Theta) \sin \Theta \, d\Theta$$

$$= 3a^3 \left(- \frac{\cos^3 \Theta}{3} - \frac{3 \cos^7 \Theta}{7} + \frac{3 \cos^5 \Theta}{5} + \frac{\cos^9 \Theta}{9} \right|_0^{\frac{\pi}{2}}$$

$$= - 3a^3 \left[- \frac{1}{3} - \frac{3}{7} + \frac{3}{5} + \frac{1}{9} \right]$$

$$= - 3a^3 \left(\frac{- 105 - 135 + 189 + 35}{315} \right)$$

$$= \frac{- 3a^3 (- 16)}{315} = \frac{48a^3}{315} .$$

$$\bar{x} = \bar{y} = \frac{M_x}{A} = \frac{48a^3}{315} \cdot \frac{32}{3a^2\pi} = \frac{16 \cdot 32a}{315\pi}$$

$$= \frac{512a}{315 \pi} = a(.517).$$

CHAPTER 33

MOMENTS OF INERTIA

Computations for the Moment of Inertia are an extension of the computations used for centroids. In computations for moment of inertia, the moment arm is raised to the second power. The moment of inertia is otherwise computed with respect to reference axes, similar to that in centroids. The moment of inertia is involved in problems dealing with dynamics, and the mechanics of materials, for example.

• PROBLEM 886

A homogeneous straight bar has a constant linear density of ρ slugs/ft. Find the moment of inertia of the bar about an axis perpendicular to the bar and passing through one end.

Solution: Let the bar be of length a ft, and let it extend along the x-axis from the origin. We find its moment of inertia about the y-axis. Divide the bar into n segments; the length of the ith segment is $\Delta_i x$ ft. The mass of the ith segment is then $\rho\Delta_i x$ slugs. Assume that the mass of the ith segment is concentrated at a single point ξ_i, where $x_{i-1} \le \xi_i \le x_i$. The moment of inertia of the ith segment about the y-axis lies between $\rho x_{i-1}^2 \Delta_i x$ slug-ft^2 and $\rho x_i^2 \Delta_i x$ slug-ft^2 and is approximated by $\rho\xi_i^2 \Delta_i x$ slug-ft^2, where $x_{i-1} \le \xi_i \le x_i$. If the moment of inertia of the bar about the y-axis is I_y slug-ft^2, then

$$I_y = \lim_{\Delta_i x \to 0} \sum_{i=1}^{n} \rho\xi_i^2 \Delta_i x = \int_0^a \rho x^2\, dx = \frac{1}{3}\rho a^3 .$$

Therefore, the moment of inertia is: $\frac{1}{3}\rho a^3$ slug-ft^2.

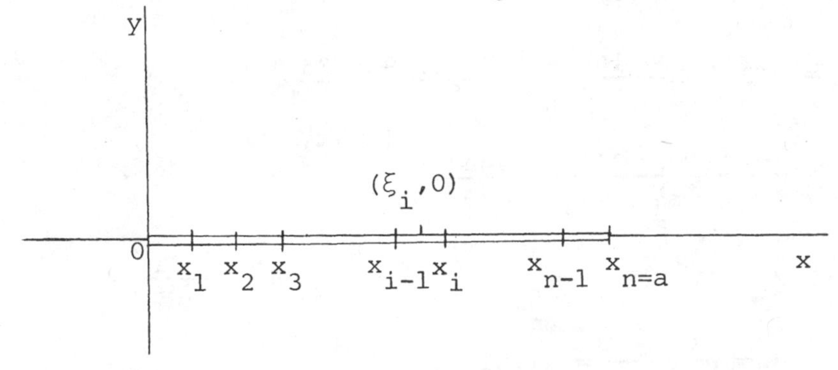

• PROBLEM 887

Find the moment of inertia about the x-axis of the curve:

$$4x = 2y^2 - \ln y, \text{ from } y = 2 \text{ to } y = 4.$$

Solution: If an arc length Δs, is between x and $x + \Delta x$, its moment of inertia with respect to the x-axis is

$$\Delta I_x = y^2 \Delta s ,$$

where Δs is a one-dimensional arc. Therefore it is not necessary to find its area or volume. The sum of all such infinitesimal curves, each multiplied by the square of its ordinate, gives the required moment of inertia. From different calculus, we have:

$$ds = \sqrt{dx^2 + dy^2} ,$$

where $ds = \Delta s$ when Δs decreases to an infinitesimal dimension. The expression for ds can also be expressed as:

$$ds = \left[\left(\frac{dx}{dy}\right)^2 + 1\right]^{\frac{1}{2}} dy .$$

From the given equation,

$$x = \frac{2y^2}{4} - \frac{1}{4} \ln y, \text{ and}$$

$$\frac{dx}{dy} = y - \frac{1}{4y}$$

$$ds = \left[(y - \frac{1}{4y})^2 + 1\right]^{\frac{1}{2}} dy = \left(y^2 - \frac{1}{2} + \frac{1}{16y^2} + 1\right)^{\frac{1}{2}} dy$$

$$= \left(y + \frac{1}{4y}\right) dy.$$

$$I_x = \int \Delta I_x = \int_2^4 \left(y + \frac{1}{4y}\right) y^2 dy = \int_2^4 \left(y^3 + \frac{y}{4}\right) dy = \left.\frac{y^4}{4} + \frac{y^2}{8}\right]_2^4 = \left[64+2-4-\frac{1}{2}\right]$$

$$= \frac{123}{2}.$$

• PROBLEM 888

Find the moment of inertia with respect to the y-axis of the curve: $3y = (2 + x^2)^{3/2}$ from $x = 0$ to $x = 1$.

Solution: The moments of inertia with respect to the coordinate axes of an arc Δs of a curve (a piece of homogeneous fine wire, for example) are given, respectively, by

$$I_x = y^2 \Delta s \quad \text{and} \quad I_y = x^2 \Delta s ,$$

The sum of all such infinitesimal curves, each multiplied by the square of the respective distances to either axis, gives the required moment of inertia. With curve limits A and B,

$$I_x = \int_A^B y^2 ds , \quad I_y = \int_A^B x^2 ds .$$

From differential calculus, ds can be expressed in terms of either dx or dy by the relationship:

$$ds^2 = dx^2 + dy^2.$$

Since x is the primary variable in this problem, we wish to express ds in terms of dx so that:

$$ds = \sqrt{1 + (dy/dx)^2} \cdot dx \; ,.$$

But $y = \frac{1}{3}(2 + x^2)^{3/2}$

$$\frac{dy}{dx} = \frac{1}{3} \cdot \frac{3}{2}(2 + x^2)^{\frac{1}{2}} \cdot 2x = x\sqrt{2 + x^2} \; .$$

Hence $ds = \left[1 + x^2(2 + x^2)\right]^{\frac{1}{2}} dx = (x^2 + 1)dx.$

$$I_y = \int_0^1 x^2(x^2 + 1)dx = \int_0^1 (x^4 + x^2)dx = \left.\frac{x^5}{5} + \frac{x^3}{3}\right]_0^1 = \frac{1}{5} + \frac{1}{3} = \frac{8}{15} \; .$$

• PROBLEM 889

Find the moments of inertia, I_x and I_y, and the corresponding radii of gyration, k_x and k_y, for the area bounded by the semi-cubical parabola: $y^2 = x^3$, and the straight line: $y = x$.

Solution: The y limits are found by solving simultaneously the two curves: $y = x$, and $y = \sqrt{x^3}$, to obtain the point of intersection.

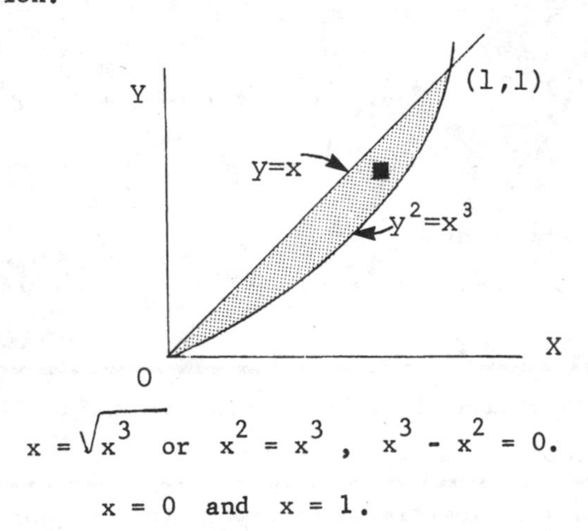

$$x = \sqrt{x^3} \text{ or } x^2 = x^3 , \; x^3 - x^2 = 0.$$

$$x = 0 \text{ and } x = 1.$$

We substitute in $y = \sqrt{x^3}$ to find $y = 0$ and $y = 1$.

Then,

$$I_x = \int_A k_x^2 da = \int_0^1 \int_y^{y^{2/3}} y^2 \cdot dx \cdot dy = \int_0^1 y^2 \Big[x\Big]_y^{y^{2/3}} \cdot dy$$

$$= \int_0^1 y^2(y^{2/3} - y)dy = \int_0^1 (y^{8/3} - y^3)dy$$

$$= \left(\frac{3y^{11/3}}{11} - \frac{y^4}{4}\right)_0^1 = \left(\frac{3}{11} - \frac{1}{4}\right) = \frac{1}{44} \; .$$

and

$$I_y = \int_A k_y^2 da = \int_0^1 \int_y^{y^{2/3}} x^2 \cdot dx \cdot dy = \int_0^1 \left[\frac{x^3}{3}\right]_y^{y^{2/3}} \cdot dy = \int_0^1 \left(\frac{y^2}{3} - \frac{y^3}{3}\right) dy$$

$$= \left[\frac{y^3}{9} - \frac{y^4}{12}\right]_0^1 = \left(\frac{1}{9} - \frac{1}{12}\right) = \frac{1}{36} .$$

The area,

$$A = \int_0^1 \int_y^{y^{2/3}} dx \cdot dy = \int_0^1 \left[x\right]_y^{y^{2/3}} \cdot dy = \int_0^1 (y^{2/3} - y) dy$$

$$= \left[\frac{3}{5} y^{5/3} - \frac{y^2}{2}\right]_0^1 = \frac{3}{5} - \frac{1}{2} = \frac{1}{10} .$$

Since $I_x = Ak_x^2$,

$$k_x = \sqrt{\frac{I_x}{A}} = \sqrt{\frac{\frac{1}{44}}{\frac{1}{10}}} = \sqrt{\frac{5}{22}} = 0.477$$

Since $I_y = Ak_y^2$,

$$k_y = \sqrt{\frac{I_y}{A}} = \sqrt{\frac{\frac{1}{36}}{\frac{1}{10}}} = \sqrt{\frac{5}{18}} = 0.527$$

• PROBLEM 890

Find the moment of inertia of a homogeneous right circular cylinder with respect to its axis.

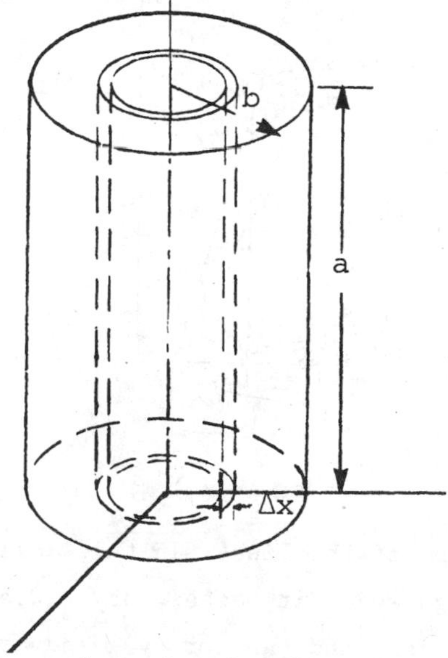

Solution: Let a and b be the altitude and radius of base of the cylinder, and δ its density. Consider the cylinder as built up of hollow cylinderical shells. Using a typical shell as indicated in the diagram, we obtain its surface area, $A = 2\pi xa$. Its infinitesimal volume is $dv = 2\pi xadx$, and its mass in terms of its density, $dm = 2\pi x\delta adx$. Since all the points in any one shell are roughly the same distance x from the y-axis, we have, by the definition of the moment of inertia, that:

$$I_y = \int r^2\, dm\ ,$$

$$I_y = 2\pi\delta a \int_0^b x^3 dx = 2\pi\delta a \left[\frac{x^4}{4}\right]_0^b = \frac{\pi\delta a b^4}{2}\ .$$

Since the mass of the cylinder is $m = \pi\delta a b^2$, we have

$$I_y = \tfrac{1}{2}\, m b^2\ .$$

• PROBLEM 891

Find the moment of inertia of a homogenous right circular cone with respect to its axis and its radius of gyration.

Solution: Let a and b be, respectively, the altitude and radius of base of the cone and δ its density. We place the cone as in the diagram, with its axis of symmetry along the y-axis and the center of its base at the origin. From the familiar expression for a straight line:

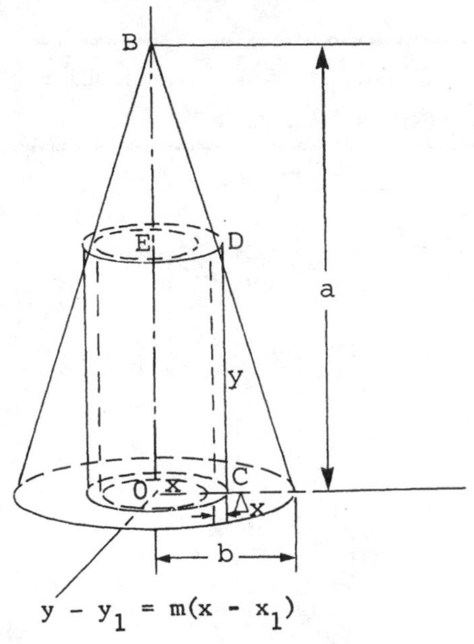

$$y - y_1 = m(x - x_1)$$

where m is the slope of the line. In this case, $m = \pm\, a/b$. We use a point through which it passes, say $(0,a)$. After substitution of 0 for x_1 and a for y_1, and using $m = -\, a/b$ for the line AB, we obtain the equation of the line AB:

$$y = \frac{a}{b}(b - x).$$

(This relation between x and y could also be obtained from the figure by use of similar triangles. Considering triangles BED and BOA, where point $D = D(x,y)$, we have:

$$\frac{a - y}{x} = \frac{a}{b},$$

or,

$$y = a/b(b - x).)$$

Now, dv is an element of volume in the shape of a cyclindrical shell. Hence, $dv = 2\pi xydx$. Moment of inertia,

$$I_y = \int x^2\delta dv = \int x^2 \cdot 2\pi\delta xydx = 2\pi\delta \cdot \frac{a}{b}\int_0^b x^3(b - x)dx = \frac{1}{10}\pi\delta ab^4 .$$

Since the mass is $m = \delta \cdot \frac{1}{3}\pi b^2 a$, we may write I_y in the form

$$I_y = \frac{3}{10} mb^2 .$$

To find the radius of gyration of the cone with respect to its axis, we use the definition $I = mR^2$. We have $\frac{3}{10} mb^2 = mR_y^2$, from which we find $R_y = \frac{1}{10}\sqrt{30}\, b$.

• PROBLEM 892

Find the moment of inertia of the homogeneous solid bounded by the paraboloid of revolution: $y^2 + z^2 = 4x$, and the plane: $x = 4$, with respect to the x-axis.

Solution: We consider the solid as built up of thin disks as indicated in the diagram. The limit volume of this disk is $dv = \pi r^2 dx$ and its mass in terms of its density, $dm = \delta\pi r^2 dx$. Since the solid is homogeneous, its density δ is constant. r is the radius which, in terms of x, y, and z, is expressed by the relation:

$$r^2 = y^2 + z^2 = 4x.$$

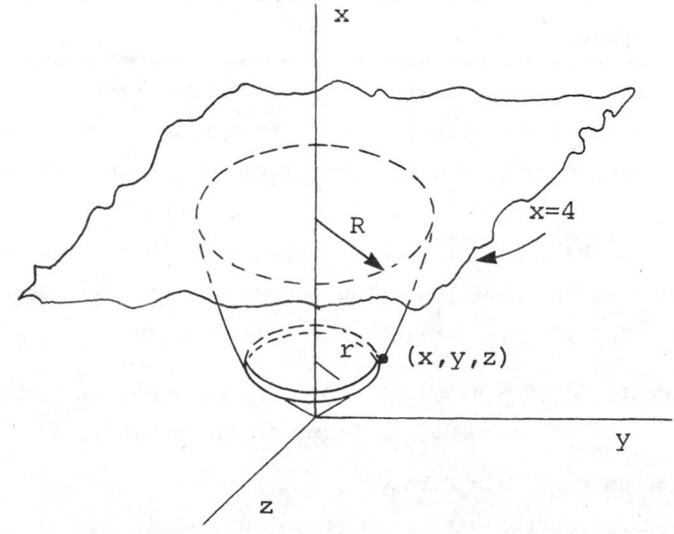

Since the moment of inertia of a disk along a perpendicular axis through its center is $I_y = \frac{1}{2} R^2 m$, the moment of inertia of the particular disk is $\frac{1}{2} r^2 dm$. Substituting $\delta\pi r^2 dx$ for dm and

$4x$ for r^2, we obtain:

$$I_x = \int_0^4 \tfrac{1}{2}(4x)\delta\pi(4x)dx,$$

or,

$$I_x = 8\pi\delta \int_0^4 x^2 dx = \frac{512}{3} \pi\delta .$$

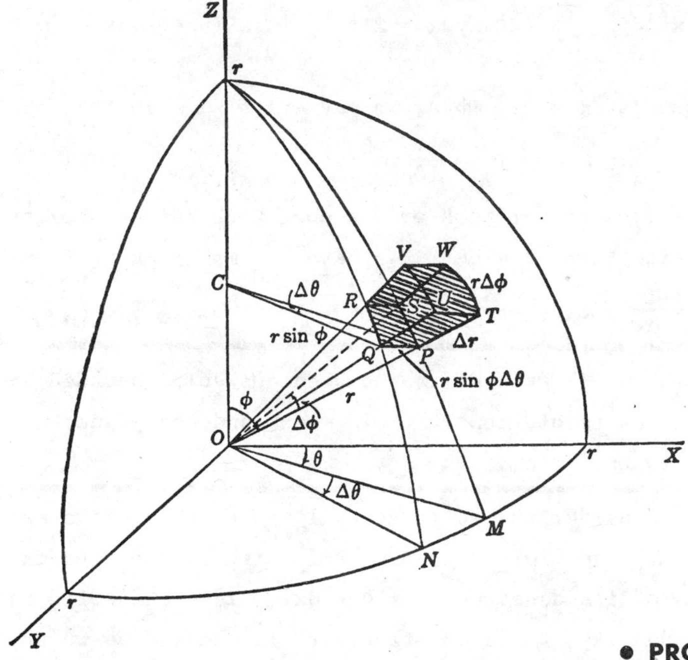

• PROBLEM 893

Find the moment of inertia of a sphere of radius a with respect to a diameter, if the density varies as the distance from the center of the sphere.

Solution: We take the moment with respect to the Z-axis; then in a spherical co-ordinate system $p(r,\theta,\varphi)$ is a point in the element of mass dm, the distance of this point from the Z-axis is $r \sin \varphi$ as shown in the figure. The volume of this element is the product of the lines $\overline{PQ}$, $\overline{PS}$ and $\overline{PT}$; where $PQ = r\Delta\theta$, $PS = r \sin \varphi\Delta\varphi$ and $PT = \Delta r$. Going to the limit, we find the volume of this element

$$dv = (rd\theta)(r \sin \varphi d\varphi)(dr) = r^2 \sin \varphi dr d\varphi d\theta .$$

Since the density is $\delta = kr$, where k is a constant of proportionality, the mass of the element, in terms of its density, is:

(a) $$dm = kr^3 \sin \varphi dr d\varphi d\theta ,$$

and its moment of inertia with respect to the Z-axis is:

$$I'_z = (r \sin \varphi)^2 dm.$$

Using equation (a) for dm and integrating,

$$I_z = k\int_0^{2\pi}\int_0^{\pi}\int_0^{a}(r^2 \sin^2\varphi)r(r^2\sin \varphi drd\varphi d\theta)$$

$$= k\int_0^{2\pi}\int_0^{\pi}\int_0^{a} r^5\sin^3\varphi drd\varphi d\theta = \frac{4}{9}\pi ka^6.$$

• PROBLEM 894

Find the moment of inertia about its axis of an ellipsoid of revolution.

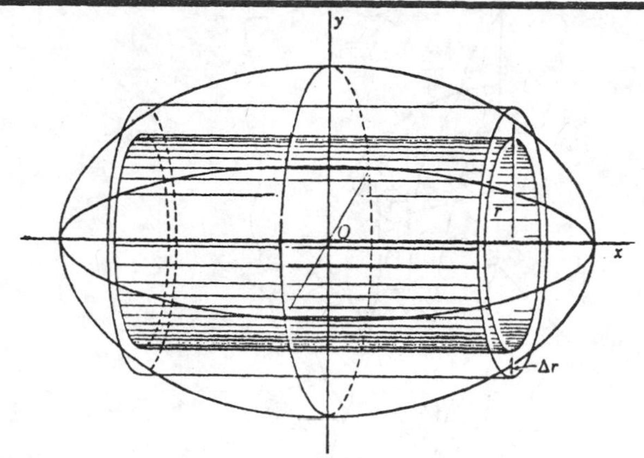

Solution: Let the equation of the ellipse be:

$$x^2/a^2 + y^2/b^2 = 1,$$

and let it be revolved about the x-axis. Consider a cylindrical shell of radius r, thickness Δr, and length $2x$.(as shown in the figure). Its volume is approximately:

$$2\pi r \cdot \Delta r \cdot 2x = 4\pi xr \cdot \Delta r ,$$

and its moment of inertia about the x-axis is approximately:

$$4\pi xr \cdot \Delta r \cdot r^2 = 4\pi xr^3 \cdot \Delta r.$$

We conclude that the moment of inertia of the whole solid about the x-axis is:

$$I = \int_0^b 4\pi xr^3 dr ,$$

since r assumes its maximum value when $x = 0$, then $r = b$. Here x and r are related by the equation of the ellipse, since the radius of a shell of length $2x$ is the ordinate of the point on the ellipse having the abscissa x , i.e., $r = y$. Hence,

$$\frac{x^2}{a^2} + \frac{r^2}{b^2} = 1, \quad \text{and} \quad x = \frac{a}{b}\sqrt{b^2-r^2} .$$

$$I = \int_0^b 4\pi \frac{a}{b}\sqrt{b^2-r^2}\, r^3 dr$$

$$= \frac{8}{15}\pi ab^4 .$$

Find the moment of inertia with respect to the z-axis of the mass in the first octant bounded by the cylinder: $x^2 + z^2 = 4$, and the planes $y = x$, $y = 0$, and $z = 0$, if the density is $\delta = kz$.

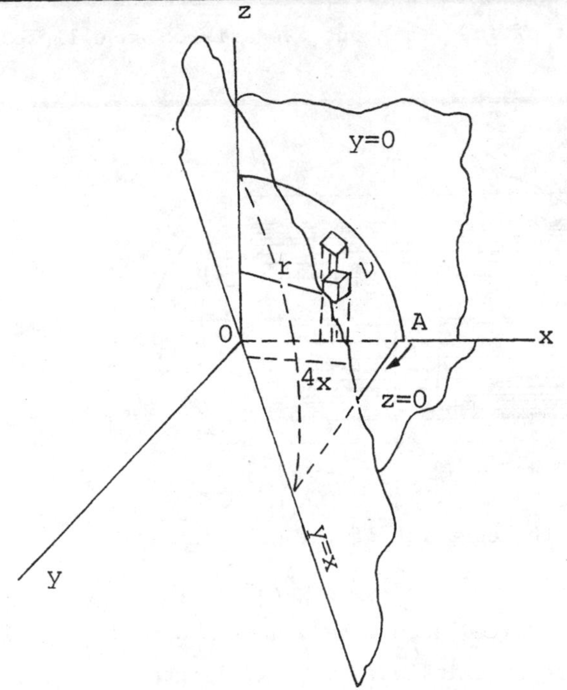

Solution: Let ν indicate the volume of the element cube as shown in the figure. Thus the total volume of the segment bounded by the planes can be obtained by a triple integral where x varies from zero to the point A(2,y,0); y varies from zero to a point on the plane $y = x$; and z varies from zero to the surface of the cylinder $\sqrt{4-x^2}$. The mass of this element volume in terms of its density is: $dm = \delta\nu = kzdzdydx$. Its moment of inertia with respect to the z-axis is $I_z = r^2dm$, where, by Pythagorean Theorem, $r^2 = x^2 + y^2$. Hence, the moment of inertia of the given mass with respect to the z-axis is the triple integral, first with respect to z and then with respect to y:

$$I_z = \int_v (x^2 + y^2)dm = \int_0^2 \int_0^x \int_0^{\sqrt{4-x^2}} (x^2 + y^2)kzdzdydx$$

$$= \frac{k}{2}\int_0^2 \int_0^x (4 - x^2)(x^2 + y^2)dydx$$

$$= k\int_0^2 \left(\frac{8}{3}x^3 - \frac{2}{3}x^5\right) dx$$

$$= k\left[\frac{8x^4}{12} - \frac{2x^6}{18}\right]_0^2$$

$$= k \left[\frac{8 \cdot 16}{12} - \frac{2(64)}{18} \right]$$

$$= k \left[\frac{32}{3} - \frac{64}{9} \right]$$

$$\frac{32k}{9}.$$

• PROBLEM 896

> Find the second moment, or moment of inertia, with respect to the z-axis for the volume above the xy-plane, below the cone: $x^2 + y^2 = z^2$, and inside the cylinder: $x^2 + y^2 - 2y = 0$ (see figure).

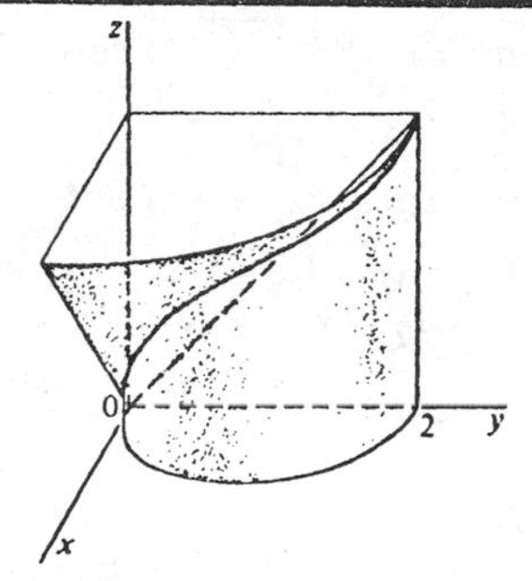

Solution: Using a spherical coordinate system, for convenience, we let φ be the angle made by a line ρ (extending from the origin to the point $p(\rho,\varphi,\theta)$) from the positive z-axis, and θ be the angle made by the projection of ρ on the x-y plane from the positive x-axis. Hence

$$\text{(a)} \qquad \begin{aligned} x &= \rho \sin \varphi \cos \theta \\ y &= \rho \sin \varphi \sin \theta \\ z &= \rho \cos \varphi . \end{aligned}$$

The volume of an elemental cube is:

$$\text{(b)} \qquad \begin{aligned} \delta v &= (\rho \sin \varphi \delta\theta)(\rho \delta\varphi)(\delta\rho) \\ &= \rho^2 \sin \varphi \, \delta\rho\delta\varphi\delta\theta . \end{aligned}$$

To find the equation of the cone in the spherical coordinate system, we substitute for the variables. Since $x^2 + y^2 = z^2$, we have:

$$\rho^2 \cos^2\varphi = \rho^2 \sin^2\varphi \cos^2\theta + \rho^2 \sin^2\varphi \sin^2\theta .$$

Factoring out $(\rho^2 \sin^2\varphi)$ and using the trigonometric identity:

$$\cos^2\theta + \sin^2\theta = 1, \text{ we obtain:}$$

$$\rho^2 \cos^2\varphi = \rho^2 \sin^2\varphi, \text{ or } \sin \varphi = \cos \varphi ,$$

which holds only if $\varphi = \pi(\frac{1}{4} + 2n)$, $n = 1,2,3,\ldots$. Since we are dealing with the first octant, we have:

$$\varphi = \pi/4,$$

as the equation of the cone.

Similarly we find the equation of the cylinder, by substituting in: $x^2 + y^2 - 2y = 0$.

$$\rho^2 \sin^2\varphi \cos^2\theta + \rho^2 \sin^2\varphi \sin^2\theta - 2\rho \sin\varphi \sin\theta = 0,$$

or,

$$\rho^2 \sin^2\varphi - 2\rho \sin\varphi \sin\theta = 0$$

which gives:

$$\rho = 2\,\frac{\sin\theta}{\sin\varphi} = 2 \sin\theta \csc\phi.$$

Noting that ρ varies from the origin to the surface of the cylinder, φ from $\pi/4$ to the x-y plane $(\pi/2)$, and since we can infer from the symmetry of the volume and of the distance that the moment of inertia of the entire volume is twice that of the volume in the first octant, we can let θ vary from 0 to $\pi/2$. Since

$$r^2 = x^6 + y^2 = \rho^2 \sin^2\varphi,$$

after multiplying by r^2 we integrate equation (b) over the given volume to find the moment of inertia with respect to the Z-axis.

Hence,

$$I_z = 2\int_0^{\pi/2}\int_{\pi/4}^{\pi/2}\int_0^{2\sin\theta\csc\varphi} \rho^4 \sin^3\varphi \, d\rho \, d\varphi \, d\theta$$

$$= \frac{64}{5}\int_0^{\pi/2}\int_{\pi/4}^{\pi/2} \sin^5\theta \csc^2\varphi \, d\varphi \, d\theta = \frac{512}{75}.$$

$$= \frac{64}{5}\int_0^{\pi/2} \sin^5\theta \left(-\cot\phi\right)\Big|_{\frac{\pi}{4}}^{\frac{\pi}{2}} d\theta$$

$$= \frac{64}{5}\int_0^{\pi/2} \sin^5\theta \, d\theta$$

$$= \frac{64}{5}\int_0^{\pi/2} \sin^4\theta \sin\theta \, d\theta$$

$$= \frac{64}{5}\int_0^{\pi/2} (1-\cos^2\theta)^2 \sin\theta \, d\theta$$

$$= \frac{64}{5}\int_0^{\pi/2} (1-2\cos^2\theta+\cos^4\theta) \sin\theta \, d\theta$$

$$= \frac{64}{5}\left[-\cos\theta + 2\,\frac{\cos^3\theta}{3} + \frac{\cos^5\theta}{5}\right]_0^{\frac{\pi}{2}}$$

$$= \frac{64}{5}\left[1 - \frac{2}{3} + \frac{1}{5}\right]$$

$$= \frac{64}{5}\left[\frac{8}{15}\right] = \frac{512}{75}.$$

A thin,flexible chain hangs in the form of a catenary: $y = \cosh x = \frac{1}{2}\left(e^{x} + e^{-x}\right)$. Find the moment of inertia of the catenary from $x = -1$ to $x = 1$, with respect to the x- and y-axes.

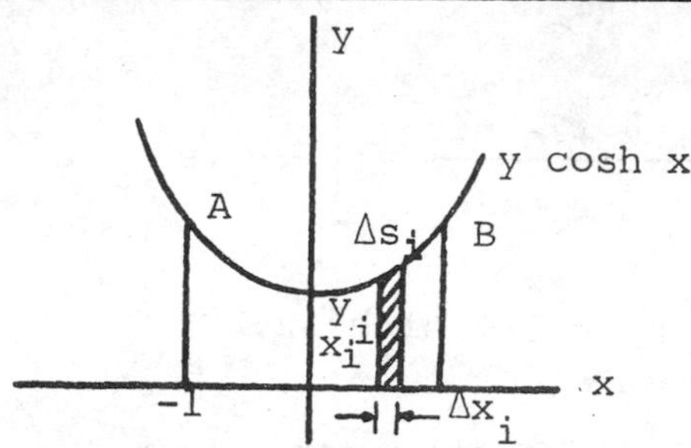

Solution: Assuming flexibility and uniformity, let the length density of the chain be ρ. Then the mass of the ith element of the chain, with a length Δs_i, is:

$$\Delta m_i = \rho\ \Delta s_i ,$$

where $i = 1, 2, 3, \ldots, n;\ -1 \le x \le 1.$

The moment of inertia of this element about, say, the x-axis, is the second moment of Δs_i, which, in a mathematical notation, is:

$$\Delta I_{x_i} = y_i^{\,2}\ \Delta m_i = \rho y_i^{\,2}\ \Delta s_i .$$

On the other hand, the approximation of Δs_i as a straight line, the hypotenuse of a triangle constructed of Δx_i and Δy_i, leads to the use of the Pythagorean theorem, to give:

$$(\Delta s_i)^2 = (\Delta x_i)^2 + (\Delta y_i)^2 .$$

$$(ds_i)^2 = \lim_{\substack{\Delta x_i \to 0 \\ \Delta y_i \to 0}} \left[(\Delta x_i)^2 + (\Delta y_i)^2\right]$$
$$= (dx_i)^2 + (dy_i)^2 .$$

Dividing and multiplying the right-hand side by $(dx_i)^2$, and taking the square-root, yields:

$$ds_i = \left[1 + \left(\frac{dy}{dx}\right)_i^2\right]^{\frac{1}{2}} dx_i .$$

But, since $y = \frac{1}{2}(e^x + e^{-x}) = \cosh x$,

$$\frac{dy}{dx} = \frac{1}{2}(e^x - e^{-x}) = \sinh x$$

$$1 + \left(\frac{dy}{dx}\right)^2 = 1 + \frac{1}{4}(e^x - e^{-x})^2$$

$$= \frac{e^{2x} + 2 + e^{-2x}}{4} = \left(\frac{1}{2}(e^x + e^{-x})\right)^2$$

$$= \cosh^2 x = 1 + \sinh^2 x.$$

Hence, $ds_i = \cosh x_i \, dx_i$.

Thus,

$$dI_{x_i} = \rho y^2 \cosh x_i \, dx_i = \rho \cosh^3 x_i \, dx_i,$$

and the total moment of inertia of the chain from point A to point B is the sum of all contributions of ds_i $(i = 1, 2, 3, \ldots, n)$,

$$I_x = \lim_{n \to \infty} \sum_{i=1}^{n} \rho \cosh^3 x_i dx_i, \quad -1 \le x_i \le 1.$$

With this limit, the Riemann integral follows:

$$I_x = \rho \int_{-1}^{1} \cosh^3 x \, dx.$$

It was derived earlier that

$1 + \sin h^2 x = \cos h^2 x$. Hence,

$$I_x = \rho \int_{-1}^{1} (1 + \sinh^2 x) \cosh x \, dx$$

$$= \rho \int_{s=A}^{s=B} (1 + \sinh^2 x) \, d(\sinh x).$$

The integral is now simplified. Upon introducing a dummy variable t in place of sinh x, we obtain

$$I_x = \rho \left(\sin h \, x + \frac{1}{3} \sin h^3 x\right|_{x=-1}^{x=1}$$

$$= 2\rho\left(\sinh 1 + \frac{1}{3}\sinh^3 1\right).$$

Since the last two terms are of degree 1 and 3, simplification of this result in terms of exponential expressions is irrelevant.

In a similar argument, the moment of inertia of an elementary mass dm and length ds about the y-axis with moment arm x is

$$dI_y = \rho\, x^2\, ds = \rho\, x^2 \cosh x\, dx.$$

Integration over $-1 \leq x \leq 1$ gives:

$$I_y = \rho \int_{-1}^{1} x^2 \cosh x\, dx.$$

Using integration by parts,

$$Iy = \rho\left(x^2 \sinh x \Big|_{-1}^{1} - 2\int x \sinh x\, dx\right)$$

$$= \rho\left(2 \sinh 1 - 2\ (x \cosh x - \sinh x\Big|_{-1}^{1}\right.$$

$$= \rho\ (2 \sinh 1 - 4 \cos h\ 1 + 4 \sinh 1)$$

$$= \rho\ (3e - 3e^{-1} - 2e - 2e^{-1})$$

$$= \rho\ (e - 5e^{-1}).$$

CHAPTER 34

DOUBLE/ITERATED INTEGRALS

In many problems, such as those involving areas, volumes, moments, there arises the double integral. This is written in the form: $\iint_R F(x,y)\,dA$ where R is the region of integration. A double integral can be evaluated by expressing it in terms of two single integrals called iterated integrals. That is the double integral can be written as,

$$\int_a^b \int_{g_1(x)}^{g_2(x)} f(x,y)\,dy\,dx ,$$

where the variable of the first differential, in this case, y, is integrated first. While this is done, x is considered a constant. After integration with respect to y, the two limits $g_2(x)$, and $g_1(x)$, are substituted. This results in an equation in x and dx alone, which is then integrated, and the limits a and b are substituted for x. It is possible to reverse the roles of the variables, in which case the integral can be written as:

$$\int_c^d \int_{h_2(y)}^{h_1(y)} f(x,y)\,dx\,dy .$$

Here, integration takes place with respect to x first, and $h_1(y)$ and $h_2(y)$ are the limits. The constants c and d are the limits of y. If the work is done correctly, the result is the same, and the choice of the sequence depends on the complexity of the integration for each possibility.

In polar coordinates, an element of area is: $r\,dr\,d\theta$. The total area, A, may then be found from one of the following:

$$A = \int_{\theta_2}^{\theta_1} \int_{f_2(\theta)}^{f_1(\theta)} f(r,\theta)\,dr\,d\theta . \qquad A = \int_{r_2}^{r_1} \int_{g_2(r)}^{g_1(r)} f(r,\theta)\,d\theta\,dr$$

If we wish to take the moment of an elementary area about the y-axis, then the element of moment is,

$$dM = x\,dy\,dx ,$$

and the total moment is,

$$M_y = \int_a^b \int_{f_2(x)}^{f_1(x)} x\,dy\,dx ,$$

where $f_1(x)$ and $f_2(x)$ are the limits of y, with respect to which the first integration is carried out, and a and b are the limits of x. It is, of course, possible to arrive at the same result by integrating first with respect to x, as discussed in the previous section. Corresponding integrations can be carried out for M_x, the moment about the x-axis.

The area, A, obtained by double integration, is

$$A = \int_a^b \int_{f_2(x)}^{f_1(x)} dy\, dx .$$

The distances to the centroid are:

$$\bar{x} = \frac{M_y}{A}, \text{ and } \bar{y} = \frac{M_x}{A} .$$

• PROBLEM 898

Find $u = \iint \left(x^3 + y^3\right) dy \cdot dx$.

Solution: We integrate first with respect to y, treating x as a constant; and then with respect to x, treating y as a constant. We write:

$$u = \int dx \int \left(x^3 + y^3\right) dy$$

$$= \int dx \left[x^3 y + \frac{y^4}{4} + \psi(x)\right]$$ where $\psi(x)$ is an arbitrary function of x.

$$u = y\int x^3 dx + \frac{y^4}{4}\int dx + \int \psi(x)dx$$

$$= \frac{x^4 y}{4} + \frac{y^4 x}{4} + \varphi_1(x) + \varphi_2(y),$$

where $\varphi_1(x)$ and $\varphi_2(y)$ are arbitrary functions.

• PROBLEM 899

Calculate: $\int_0^1 \int_y^1 ye^{-x^3}\, dx\, dy$.

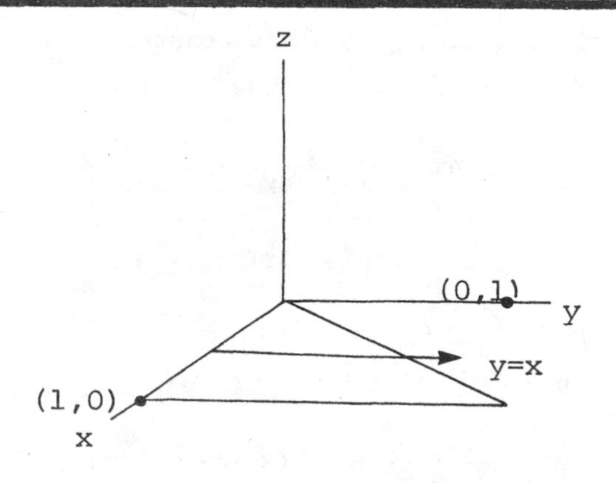

Solution: Here it is difficult to carry out the first or inner integration. However, we can tell from the limits of integration that this iterated integral represents the volume of the solid in the first octant bounded by the planes, $y = 0$, $y = 1$, $x = 1$ and $x = y$ and the surface $z = ye^{-x^3}$ (see the accompanying figure). We can therefore reverse the order of integration and integrate first with respect to y and then with respect to x. For the integral with respect to x the limits are $x = 0$ and $x = 1$. For the integral with respect to y, the limits of integration can be found by drawing a line parallel to the y axis. The line first meets $y = 0$, the x-axis and lower limit, and then meets the line $y = x$, the upper limit. We can now write the new integral as:

$$\int_0^1 \int_0^x ye^{-x^3}\, dy\, dx.$$

This is more easily evaluated.

$$\int_0^1 \int_0^x ye^{-x^3}\, dy\, dx = \int_0^1 dx \int_0^x ye^{-x^3}\, dy$$

$$= \int_0^1 dx \left[\frac{y^2}{2} e^{-x^3}\right]_0^x$$

$$= \int_0^1 \frac{x^2}{2} e^{-x^3}\, dx$$

$$= -\frac{1}{6}\int_0^1 e^{-x^3}\left(-3x^2\right)dx$$

$$= -\frac{1}{6}\left[e^{-x^3}\right]_0^1 = -\frac{1}{6}\left(\frac{1}{e} - 1\right) = \frac{e-1}{6e}.$$

• **PROBLEM** 900

If $u = \iint e^{4x} \cdot y^3 \cdot dy \cdot dx$, find u.

Solution: We can say that this double integral is a product of two single integrals and we can solve it by first integrating with respect to y, treating x as a constant, and then integrating with respect to x, treating y as a constant. We can write:

$$u = \int e^{4x}\, dx \int y^3\, dy$$

$$= \int e^{4x}\, dx \left[\frac{y^4}{4} + \varphi(x)\right],$$ where $\varphi(x)$ is an arbitrary function of x.

$$u = \frac{y^4}{4}\int e^{4x}\, dx + \int \varphi(x)dx.$$

$$u = \frac{e^{4x}}{4} \cdot \frac{y^4}{4} + \int \varphi(x)\cdot dx + F(y)$$

And, since $\varphi(x)$ was arbitrary, its integral, $\int\varphi(x)\cdot dx$ is arbitary and equal to $\psi_1(x)$, so that

$$u = \frac{4x_y{}^4}{16} + \psi_1(x) + \psi_2(y),$$

where both $\psi_1(x)$ and $\psi_2(y)$ are arbitrary.

• PROBLEM 901

> Find the area between the parabolas: $y^2 = x$, and $y = x^2$.

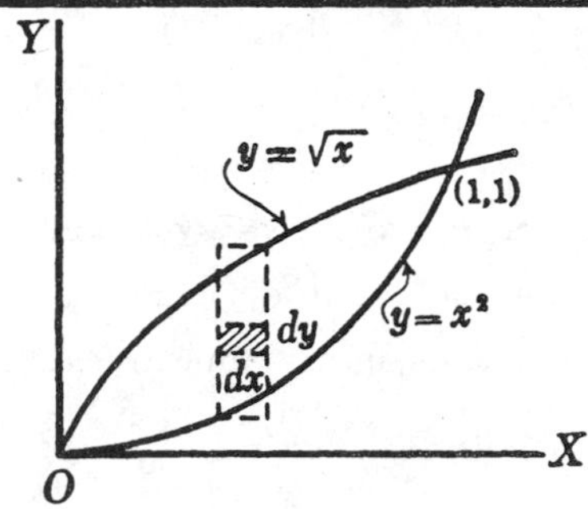

Solution: The formula for area, using double integrals is:

$$A = \iint dy\,dx = \int_a^b dx \int_{f(x)}^{F(x)} dy.$$

To find the limits, a and b, of the integral with respect to x, we set the two functions equal and solve for x to give the points of intersection. These values are the limits. Hence, it is given that $y = \sqrt{x}$, $y = x^2$, $\sqrt{x} = x^2$, $x = x^4$, $x\left(1 - x^3\right) = 0$, or $x = 0$, $x = 1$.

The limits of the integral with respect to y are the two functions. The lower limit is the lower parabola, $y = x^2$, and the upper limit is the upper parabola, $y = \sqrt{x}$. Therefore,

$$A = \int_0^1 dx \int_{x^2}^{\sqrt{x}} dy$$

$$= \int_0^1 dx\,[y]_{x^2}^{\sqrt{x}}$$

$$= \int_0^1 \left(\sqrt{x} - x^2\right)dx$$

$$= \int_0^1 x^{1/2}\cdot dx - \int_0^1 x^2\cdot dx$$

$$= \left[\frac{2}{3}x^{3/2} - \frac{x^3}{3}\right]_0^1$$

$$= \frac{2}{3} - \frac{1}{3}$$

$$= \frac{1}{3} \text{ square units.}$$

Use double integration to find the total area enclosed by the two curves: $y = 2x$, and $y^3 = 2x$.

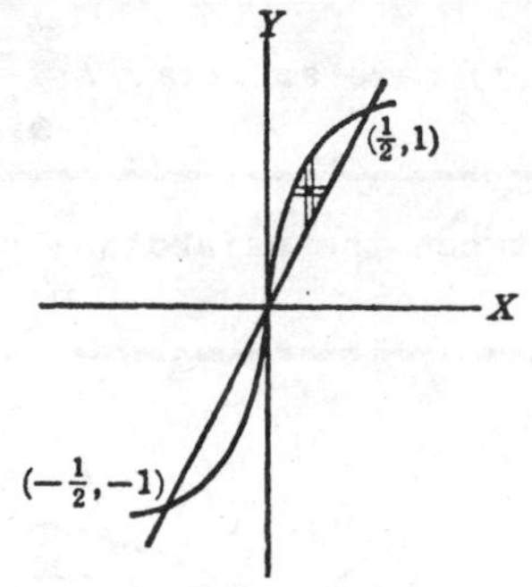

Solution: Since the curves have symmetry, the total area is twice that lying in the first quadrant. The formula for area in Cartesian coordinates, using double integrals, is:

$$A = \iint_R dy\,dx = \int_a^b dx \int_{f(x)}^{F(x)} dy.$$

The limits a and b of the integral with respect to x are the x-coordinates of the points of intersection of the two curves. To find the points of intersection, we set $y = 2x$ equal to $y^3 = 2x$ and solve for x.

$$2x = \sqrt[3]{2x}\ ,\ 8x^3 = 2x,\ x\left(4x^2-1\right) = 0.$$

$$x = 0,\ 4x^2 = 1,\ x = \pm\frac{1}{2}.$$

In the first quadrant, $x = 0$ is the lower limit of the integral with respect to x and $x = +\frac{1}{2}$ is the upper limit. The limits of the integral with respect to y are the two functions. The lower function, $y = 2x$, is the lower limit and the upper function, $y = (2x)^{1/3}$, is the upper limit. Hence,

$$A_{\text{Total}} = 2\int_0^{\frac{1}{2}} dx \int_{2x}^{(2x)^{1/3}} dy$$

$$= 2\int_0^{\frac{1}{2}} dx\, [y]_{2x}^{(2x)^{1/3}}$$

$$= 2\int_0^{\frac{1}{2}} \left[(2x)^{1/3} - 2x\right] dx$$

$$= 2\left[\frac{3}{4}(2x)^{4/3}\cdot\frac{1}{2}\right]_0^{\frac{1}{2}} - 4\left[\frac{x^2}{2}\right]_0^{\frac{1}{2}}$$

$$= \frac{3}{4} - \frac{1}{2} = \frac{1}{4}\ \text{sq. unit.}$$

• PROBLEM 903

Use double integration to find the area enclosed by $y = x^2$ and $x + y - 2 = 0$.

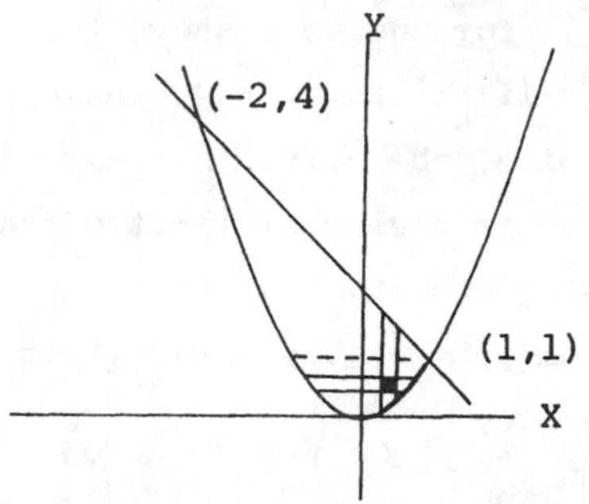

Solution: The formula for area in Cartesian coordinates, using double integrals, is:

$$A = \iint dy\, dx = \int_a^b dx \int_{f(x)}^{F(x)} dy.$$

The limits a and b of the integral with respect to x are the x-coordinates of the points of intersection of the two curves. To find the points of intersection, we set $y = x^2$ equal to $y = 2-x$ and solve for x.

$$x^2 = 2 - x \qquad (x + 2)(x - 1) = 0$$

$$x = -2 \quad x = 1$$

The limits of the integral with respect to y are the two functions. The lower limit is the lower function, the parabola $y = x^2$. The upper limit is the upper function, the line $y = 2 - x$. Therefore,

$$A = \int_{-2}^{1} dx \int_{x^2}^{2-x} dy$$

$$= \int_{-2}^{1} dx \Big[y \Big]_{x^2}^{2-x}$$

$$= \int_{-2}^{1} \left[2 - x - x^2 \right] dx = \left[2x - \frac{x^2}{2} - \frac{x^3}{3} \right]_{-2}^{1}$$

$$= \frac{7}{6} + \frac{10}{3} = \frac{27}{6}$$

$$= \frac{9}{2} \text{ sq. units.}$$

We can also find the area by reversing the order of integration. Instead of integrating first with respect to y and then x we employ:

$$A = \iint dx\, dy = \int_a^b dy \int_{f(y)}^{F(y)} dx.$$

However, from the diagram we can see that the total area must be considered as the sum of two areas, one above the dotted line in the diagram, and one below. The reason for this is that the limits for the integrals are different for each area. For the area above the dotted line, the limits of the integral with respect to x are $x = -\sqrt{y}$, the lower limit, and $x = 2 - y$, the upper limit. For the integral with respect to y for the area above the dotted line, y goes from $y = 1$, the lower limit, to $y = 4$, the upper limit. For the area below the dotted line, the curve $x = -\sqrt{y}$ is the lower limit and the curve $x = \sqrt{y}$ is the upper limit of the integral with respect to x. For the integral with respect to y, for the area below the dotted line, x goes from $y = 0$, the lower limit, to $y = 1$, the upper limit. We can now write,

$$A_{Total} = \int_1^4 dy \int_{-\sqrt{y}}^{2-y} dx + \int_0^1 dy \int_{-\sqrt{y}}^{\sqrt{y}} dx$$

$$= \int_1^4 dy \left[x\right]_{-\sqrt{y}}^{2-y} + \int_0^1 dy \left[x\right]_{-\sqrt{y}}^{\sqrt{y}}$$

$$= \int_1^4 2-y+\sqrt{y}\, dy + \int_0^1 \sqrt{y}+\sqrt{y}\, dy$$

$$= \left[2y - \frac{y^2}{2} + \frac{2y^{3/2}}{3}\right]_1^4 + \left[\frac{4y^{3/2}}{3}\right]_0^1$$

$$= \frac{9}{2} \text{ sq. units.}$$

Clearly, the first method is preferable here.

• PROBLEM 904

Determine the region in the XY-plane over which the iterated integral $\int_0^1 dx \int_{2x}^{2\sqrt{x}} dy$ extends, and then reverse the order of integration.

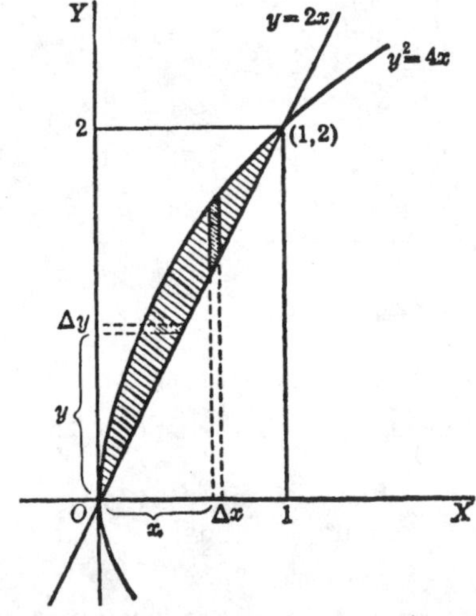

Solution: The limits of the integral with respect to x are obtained from the x coordinates of the points of intersection in the diagram. The upper limit of the integral with respect to y is the upper function $y = 2\sqrt{x}$. The lower limit of the integral with respect to y is the lower function, $y = 2x$. By double integration we find:

$$\int_0^1 dx \int_{2x}^{2\sqrt{x}} dy = \int_0^1 dx \left[y\right]_{2x}^{2\sqrt{x}}$$

$$= \int_0^1 (2\sqrt{x} - 2x)dx$$

$$= \left[\frac{4}{3} x^{3/2} = x^2\right]_0^1$$

$$= \frac{1}{3} \; .$$

When we reverse the order of integration we integrate first with respect to x and then with respect to y. The limits of integration for the integral with respect to y are the y coordinates of the points of intersection in the diagram, $y = 0$ and $y = 2$. We can find the limits of the integral with respect to x by drawing a line parallel to the x-axis. The line first meets the function $x = y^2/4$, the lower limit, and then the function $x = y/2$, the upper limit. By double integration we find:

$$\int_0^2 dy \int_{y^2/4}^{y/2} dx = \int_0^2 dy \left[x\right]_{y^2/4}^{y/2}$$

$$= \int_0^2 \left(\frac{y}{2} - \frac{y^2}{4}\right) dy$$

$$= \left[\frac{y^2}{4} - \frac{y^3}{12}\right]_0^2$$

$$= \frac{1}{3} \; .$$

The choice of the order of integration is of no consequence, except that it affects the amount of work involved.

• PROBLEM 905

Sketch the volume represented by the iterated integral:

$$\int_0^1 \int_{y^2}^{\sqrt{y}} (2 - x - y)\, dx\, dy,$$

and compute its value.

Solution:

$$V = \int_0^1 \int_{y^2}^{\sqrt{y}} (2 - x - y) dx\, dy$$

$$= \int_0^1 \left(2x - \frac{1}{2} x^2 - yx \Big|_{y^2}^{\sqrt{y}}\right) dy$$

$$= \int_0^1 \left(2\sqrt{y} - \frac{1}{2} y - y\sqrt{y} - 2y^2 + \frac{1}{2} y^4 + y^3\right) dy$$

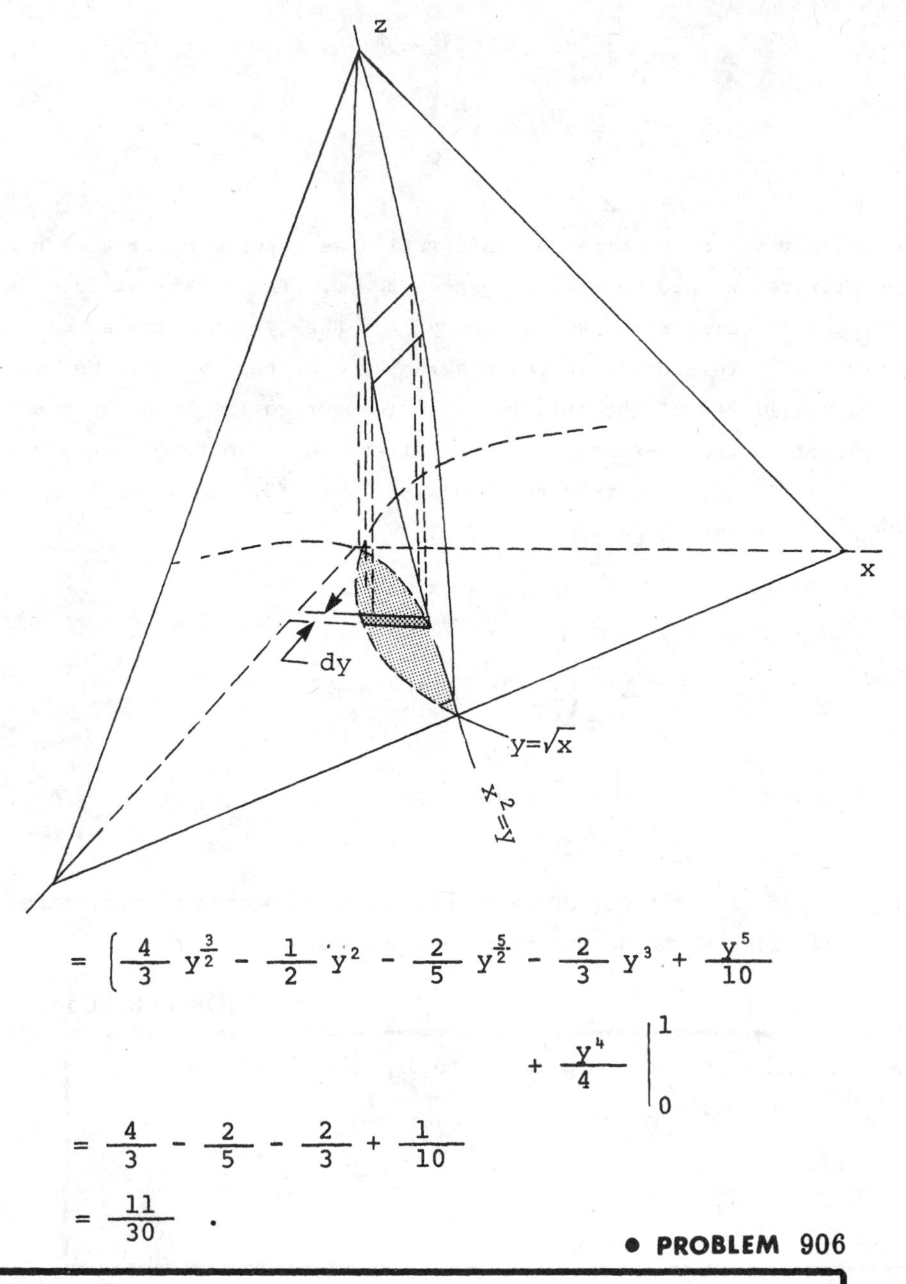

$$= \left(\frac{4}{3} y^{\frac{3}{2}} - \frac{1}{2} y^2 - \frac{2}{5} y^{\frac{5}{2}} - \frac{2}{3} y^3 + \frac{y^5}{10} + \frac{y^4}{4} \right|_0^1$$

$$= \frac{4}{3} - \frac{2}{5} - \frac{2}{3} + \frac{1}{10}$$

$$= \frac{11}{30} .$$

• PROBLEM 906

Sketch the volume represented by the iterated integral:

$$\int_0^1 \int_0^{\sqrt{1-y^2}} 4y \, dx \, dy,$$

and compute its volume.

Solution: The volume of any solid can be represented as the sum of the volumes of all the approximate parallelepipeds, the dimensions of which are: the height z and the elementary area

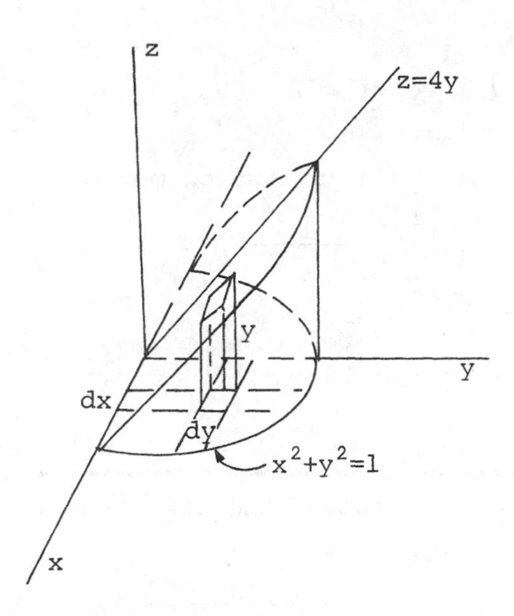

on the xy-plane (dx by dy). This volume of the parallelepepiped is:

$$dV = z\ dx\ dy$$

Integration over the region R (the area of the surface projected on the xy-plane) gives the required volume:

$$V = \int_R z\ dA.$$

Comparing to the given expression, we see that $z = 4y$ represents the equation of the surface. It is a plane. Since, from the limits of the integration, x varies from 0 to $\sqrt{1 - y^2}$, the variation of x in terms of y is:

$$x^2 + y^2 = 1,$$

which is a circle of unity radius on the xy-plane, namely the region R. $0 \leq y \leq 1$ indicates the positive y-axis, and the fact that x starts from zero indicates the positive x-axis. Thus, the volume to be calculated is the volume in the first octant.

According to the process of evaluating multiple integrals:

$$V = \int_0^1 4y \left[\int_0^{\sqrt{1-y^2}} dx \right] dy$$

$$= \int_0^1 4y \left(x \Big|_0^{\sqrt{1-y^2}} \right) dy$$

$$= 4 \int_0^1 y\sqrt{1-y^2}\, dy.$$

Using the substitution method,

$$V = 4 \int_0^1 y\sqrt{1-y^2}\, dy$$

$$= 4\left(-\frac{1}{3}(1-y^2)^3\right)\Bigg|_0^1 = \frac{4}{3}.$$

• PROBLEM 907

Compute the value of the iterated integral:

$$\int_0^1 \int_0^{\sqrt{1+x^2}} \frac{dy\, dx}{x^2+y^2+1}.$$

<u>Solution:</u> Let the integral represent a volume V. Then,

$$V = \int_0^1 \int_0^{\sqrt{1+x^2}} \frac{dy\, dx}{x^2+y^2+1}$$

shows the volume of the region bounded by the x- and y-axes, the line x = 1, the curve: $y^2 = 1+x^2$ (a hyperbola with index at $y = \pm 1$, asymptotic to the lines $y = \pm x$) and the surface:

$$z = \frac{1}{x^2+y^2+1}.$$

To carry out the integration, we consider $1 + x^2$ in the integrand as a constant, c^2. By substituting $y = \sqrt{1+x^2}\tan\Theta$ we obtain:

$$V = \int_0^1 \int_0^{\sqrt{1+x^2}} \frac{dy\, dx}{(x^2+1)+y^2}$$

$$= \int_0^1 \left(\frac{1}{\sqrt{1+x^2}} \tan^{-1} \frac{y}{\sqrt{1+x^2}}\Bigg|_0^{\sqrt{1+x^2}}\right) dx$$

$$= \int_0^1 \frac{1}{\sqrt{1+x^2}} (\tan^{-1} 1 - \tan^{-1} 0)\, dx$$

$$= \frac{\pi}{4} \int_0^1 \frac{dx}{\sqrt{1 + x^2}}.$$

With a similar argument we set

$$V = \frac{\pi}{4} \int_0^1 \frac{dx}{\sqrt{1 + x^2}} = \frac{\pi}{4} \ln \left(x + \sqrt{1 + x^2}\right) \Big|_0^1$$

$$= \frac{\pi}{4} \ln \left(1 + \sqrt{2}\right).$$

• PROBLEM 908

Evaluate the double integral:

$$\iint e^{-(x^2+y^2)} dA,$$

in the first quadrant and bounded by the circle: $x^2 + y^2 = a^2$ and the coordinate axes.

Solution: This double integral is to be evaluated over a region which is a quarter of a circle with center at the origin and radius a, therefore the circle intersects the x-axis at (a,0). For the first quadrant, θ varies from 0 to $\frac{\pi}{2}$. The point of intersection on the x-axis and the variation of θ give the limits for the double integral. The equation of a circle can be written as: $r^2 = x^2 + y^2$, and we can express an element of area, dA, as dxdy. However, by using the polar coordinate form it is easier to evaluate this double integral. Substituting r^2 for $x^2 + y^2$, and $rdrd\theta$ for dA, we can write:

$$\iint_R e^{-(x^2+y^2)} dA = \iint_R e^{-r^2} rdrd\theta.$$

This can be split into two single integrals, and evaluated. Hence

$$\iint_R e^{-r^2} rdrd\theta = \int_0^{\pi/2} d\theta \int_0^a e^{-r^2} rdr.$$

(The limits of θ are 0 and $\frac{\pi}{2}$, and the limits of r are 0 and a.) Solving, we find:

$$\int_0^{\pi/2} d\theta \int_0^a e^{-r^2} rdr = \int_0^{\pi/2} d\theta \left[\left(-\frac{1}{2}\right) e^{-r^2}\right]_0^a$$

$$= -\frac{1}{2} \int_0^{\pi/2} (e^{-a^2} - 1) d\theta = -\frac{1}{2}\left[\theta(e^{-a^2} - 1)\right]_0^{\pi/2}$$

$$= -\frac{\pi}{4}\left(e^{-a^2} - 1\right) = \frac{\pi}{4}\left(1 - e^{-a^2}\right).$$

Determine the area of one leaf of a rose, given by the equation: $r = \sin 3\theta$, by double integration.

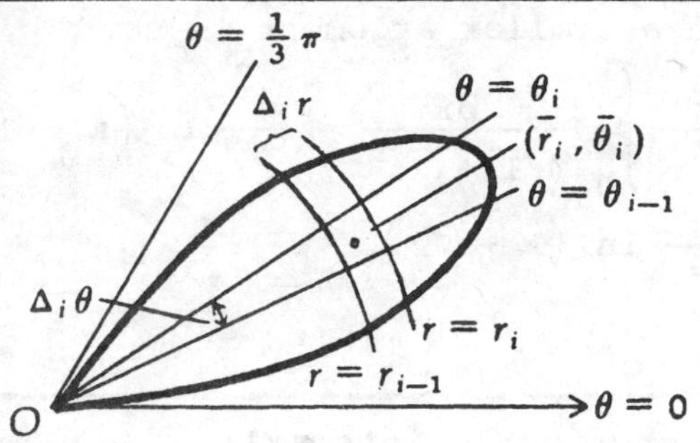

Solution: The formula for area using double integrals in polar coordinates is: $A = \iint r\,dr\,d\theta$. The lower limit on r is 0 and the upper limit is the one leaf of the rose $r = \sin 3\theta$. The limits on θ are the points where $r = 0$. Therefore,

$$r = \sin 3\theta = 0 \cdot 3\theta = 0, \pi.$$

$$\theta = 0, \ \theta = \frac{1}{3}\pi.$$

We can now write:

$$= \int_0^{\pi/3} d\theta \int_0^{\sin 3\theta} r\,dr$$

$$= \int_0^{\pi/3} d\theta \left[\frac{r^2}{2}\right]_0^{\sin 3\theta}$$

$$= \frac{1}{2}\int_0^{\pi/3} \sin^2 3\theta \, d\theta.$$

Now, using the formula:

$$\int \sin^2 x \, dx = \frac{1}{2}x - \frac{1}{4}\sin 2x,$$

we obtain:

$$A = \frac{1}{4}\theta - \frac{1}{24}\sin 6\theta \Big]_0^{\pi/3}$$

$$= \frac{1}{12}\pi.$$

Find the area inside the cardioid: $r = a(1 + \sin\theta)$, and outside the circle: $r = a$.

Solution: The area is the difference between the two areas, one bounded by the cardioid, and the other bounded by the circle. We can use the formula for area in polar coordinates, which is:

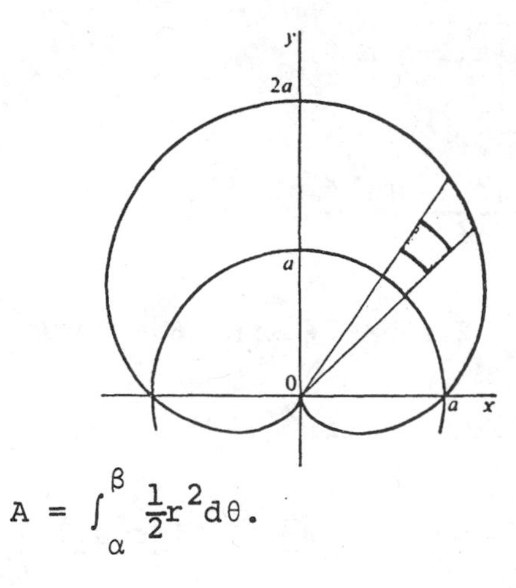

$$A = \int_{\alpha}^{\beta} \frac{1}{2} r^2 d\theta.$$

The limits are the points of intersection of the two curves, which are at $\theta = 0$ and at $\theta = \pi$. Hence, we can write:

$$A = \int_0^{\pi} \frac{1}{2}\left(a(1 + \sin\theta)\right)^2 d\theta - \int_0^{\pi} \frac{1}{2} a^2 \, d\theta$$

$$= \frac{a^2}{2} \int_0^{\pi} (1 + 2\sin\theta + \sin^2\theta) d\theta - \frac{a^2}{2} \int_0^{\pi} d\theta.$$

Now, substituting: $\sin^2\theta = \frac{1 - \cos 2\theta}{2}$, and combining the two integrals, we obtain:

$$A = \frac{a^2}{2} \int_0^{\pi} \left(2\sin\theta + \frac{1 - \cos 2\theta}{2}\right) d\theta$$

$$= \frac{a^2}{2} \left[\frac{\theta}{2} - \sin 2\theta - 2\cos\theta\right]_0^{\pi}$$

$$= \frac{a^2}{4}(8 + \pi).$$

We can also obtain this value by using double integrals. The formula for area using double integrals in polar coordinates is:

$$A = \iint r dr \, d\theta.$$

It is convenient in this problem to integrate first with respect to r. The lower limit on r is the circle: $r = a$ and the upper limit is the cardioid: $r = a(1 + \sin\theta)$. The limits on θ are the θ-coordinates of the points of intersection of the two curves, which in this example are on the rays $\theta = 0$ and $\theta = \pi$. Therefore, the desired area is given by

$$A = \int_0^{\pi} d\theta \int_a^{a(1+\sin\theta)} r dr$$

$$= \int_0^{\pi} d\theta \left[\frac{r^2}{2}\right]_a^{a(1+\sin\theta)}$$

$$= \int_0^{\pi} \left[\frac{(a(1+\sin\theta))^2}{2} - \frac{a^2}{2}\right] d\theta$$

$$= \int_0^{\pi} \frac{a^2}{2}(1 + 2\sin\theta + \sin^2\theta - 1)d\theta$$

$$= \int_0^{\pi} \frac{a^2}{2}(2\sin\theta + \sin^2\theta)d\theta$$

$$= \frac{a^2}{2}(-2\cos\theta)\Big|_0^{\pi} + \frac{a^2}{2}\int_0^{\pi}\left(\frac{1-\cos 2\theta}{2}\right) d\theta$$

$$= -a^2\cos\theta + \frac{a^2}{2}\left(\frac{\theta}{2} + \frac{\sin 2\theta}{4}\right)\Big|_0^{\pi}$$

$$= \frac{a^2}{4}\left[(1+1)4 + \pi + 0\right]$$

$$= \frac{a^2}{4}(8 + \pi).$$

• PROBLEM 911

Using double integration, find the area of the upper half of the cardioid $\rho = 2a(1 - \cos\theta)$.

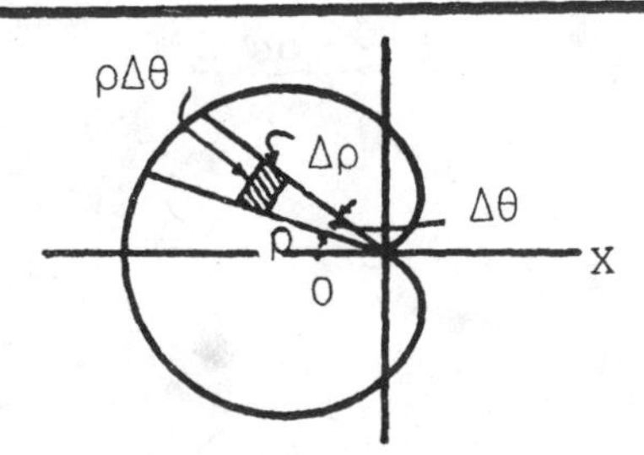

Solution: The formula for area in polar coordinates using double integrals is $A = \iint \rho\, d\rho\, d\theta = \int_{\theta_1}^{\theta_2} d\theta \int_{F(\theta)}^{f(\theta)} \rho\, d\rho$. Since the upper half of the cardioid covers the first and second quadrants, and since the cardioid intersects the axes at $\theta = 0$ and $\theta = \pi$, the integral with respect to θ goes from $\theta = 0$ to $\theta = \pi$. Since the upper half of the cardioid extends from the x-axis and up, the limits are from $\rho = 0$ to $\rho = 2a(1 - \cos\theta)$. Therefore,

$$A = \int_0^{\pi} d\theta \int_0^{2a(1-\cos\theta)} \rho\, d\rho$$

$$= \int_0^{\pi} d\theta \left[\frac{\rho^2}{2}\right]_0^{2a(1-\cos\theta)}$$

$$= \int_0^{\pi} \frac{(2a(1-\cos\theta))^2}{2}\, d\theta$$

$$= 2a^2 \int_0^{\pi} (1 - \cos\theta)^2 \cdot d\theta$$

$$= 2a^2 \int_0^{\pi} (1 - 2\cos\theta + \cos^2\theta) d\theta$$

$$= 2a^2 \int_0^{\pi} \left(1 - 2\cos\theta + \frac{1+\cos 2\theta}{2}\right) d\theta$$

$$= 2a^2 \left[\theta - 2\sin\theta + \frac{\theta}{2} + \frac{\sin 2\theta}{4}\right]_0^{\pi}$$

$$= 2a^2 \left(\pi - 0 + \frac{\pi}{2} + 0\right) = 2a^2 \cdot \frac{3\pi}{2} = 3\pi a^2$$

• PROBLEM 912

Use double integration to determine the area inside the circle $p = a\cos\theta$ and outside the cardioid $p = a(1 - \cos\theta)$.

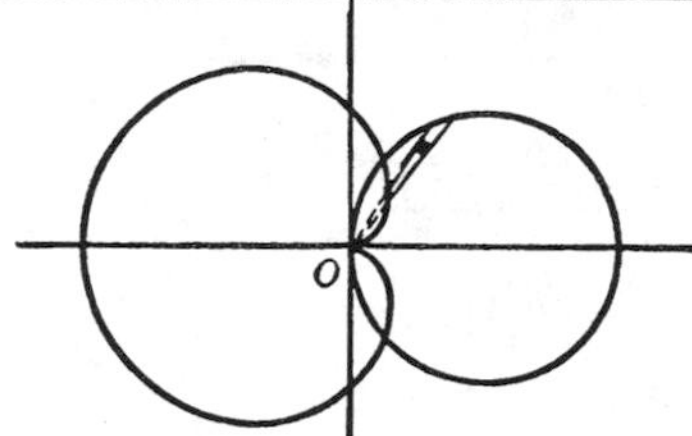

Solution: The formula for area in polar coordinates using double integrals is

$$\iint p\, dp\, d\theta = \int_{\theta_1}^{\theta_2} d\theta \int_{f(\theta)}^{F(\theta)} p\, dp.$$

The total area required can be expressed as twice the area above the polar line, since the area is divided equally above and below the line. For the area above the polar line, the limits of the integral with respect to θ are $\theta = 0$ and the point of intersection of the two curves. The point of intersection is found by setting the two curves equal and solving for θ,

$$a(1 - \cos\theta) = a\cos\theta,$$
$$\frac{1}{2} = \cos\theta, \quad \theta = \frac{\pi}{3} .$$

The limits of the integral with respect to p can be found by drawing a line from the pole intersecting the two curves and passing through the required area. The line first intersects the cardioid, $p = a(1 - \cos\theta)$, which is the lower limit. The line then meets the circle, $p = a\cos\theta$, which is the upper limit. Therefore,

$$A = 2\int_0^{\frac{\pi}{3}} d\theta \int_{a(1-\cos\theta)}^{a\cos\theta} p\, dp$$

$$= \int_0^{\frac{\pi}{3}} d\theta \left[\rho^2 \right]_{a(1-\cos\theta)}^{a\cos\theta}$$

$$= \int_0^{\frac{\pi}{3}} [a^2\cos^2\theta - a^2(1 - \cos\theta)^2]d\theta$$

$$= a^2 \int_0^{\frac{\pi}{3}} (2\cos\theta - 1)d\theta$$

$$= a^2 \left[2\sin\theta - \theta \right]_0^{\frac{\pi}{3}}$$

$$= \frac{a^2}{3}\left(3\sqrt{3} - \pi\right) \text{sq. units.}$$

APPLIED PROBLEMS USING DOUBLE INTEGRATION

• PROBLEM 913

A homogeneous rectangular lamina has constant area density of ρ slugs/ft^2. Find the moment of inertia of the lamina about one corner.

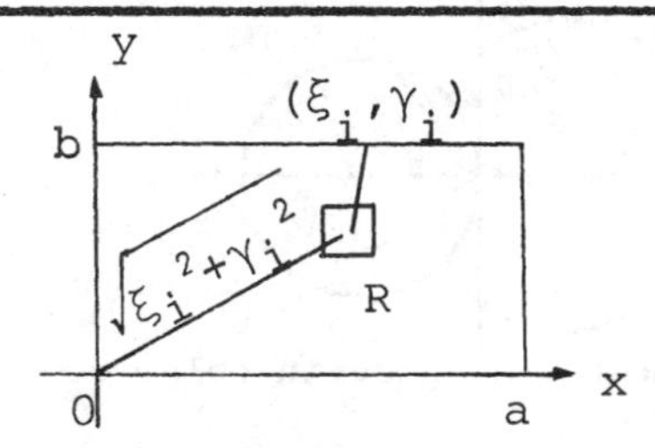

Solution: Let the lamina be bounded by the lines $x = a$, $y = b$, the x-axis, and the y-axis. Assume that the lamina is divided into n rectangles. If the ith rectangle with mass $\rho\Delta_i A$ is considered, with center (ξ_i, γ_i), and distance from the origin is
$$r = \sqrt{\xi_i^2 + \gamma_i^2},$$
then its moment of inertia with respect to the origin is:
$$r^2\rho\Delta_i A = \rho(\xi_i^2 + \gamma_i^2)\Delta_i A .$$
The sum of the moments of inertia of all rectangles in the whole region R about the orgin, in slug-ft^2, is:

$$\lim_{\Delta_i A \to 0} \sum_{i=1}^{n} \rho r^2 \Delta_i A = \lim_{\Delta_i A \to 0} \sum_{i=1}^{n} \rho(\xi_i^2 + \gamma_i^2)\Delta_i A$$

$$= \iint_R \rho(x^2 + y^2)dA$$

$$= \rho\int_0^b \int_0^a (x^2 + y^2)dx\, dy$$

$$= \rho\int_0^b \left[\frac{1}{3}x^3 + xy^2\right]_0^a dy$$

$$= \rho \int_0^b \left(\frac{1}{3} a^3 + ay^2\right) dy$$

$$= \frac{1}{3} \rho ab(a^2 + b^2) .$$

Therefore, the moment of inertia is: $\frac{1}{3} \rho ab(a^2 + b^2)$ slug-ft^2.

• PROBLEM 914

> An area consists of a rectangle of base a and altitude b surmounted by a semicircle of diameter a. Find the distance of the centroid of this area from the lower horizontal base of the rectangle.

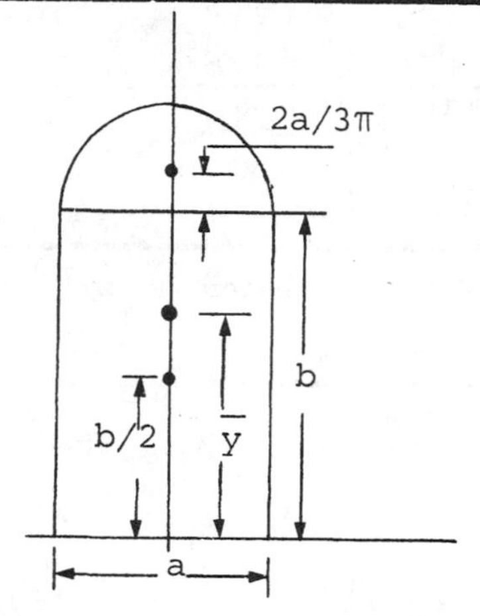

Solution: The centroid of the semicircle is found first. An element of area in polar coordinates is $rd\Theta dr$. The moment arm is $y = r \sin \Theta$. The moment,

$$M_{x_c} = \int_0^{\pi} \int_0^{a/2} r^2 \sin \Theta dr d\Theta = \int_0^{\pi} \sin \theta \, d\theta \left[\frac{r^3}{3} \right|_0^{a/2}$$

$$= \frac{1}{3} \frac{a^3}{8} \int_0^{\pi} \sin \Theta d\Theta = \frac{-a^3}{24} \cos \Theta \Big|_0^{\pi} = \frac{a^3}{12}$$

The area is $\pi a^2/8$. Hence,

$$\bar{y}_c = \frac{a^3}{12} \Big/ \frac{\pi a^2}{8} = \frac{2a}{3\pi} .$$

The centroid of the rectangular part is:

$$\bar{y}_r = b/2.$$

Now we find a point $\bar{y}$, such that the moments of the semicircle plus the rectangle are balanced by a moment of: (total area) $\bar{y}$. In other words, using the dimensions in the diagram,

$$\frac{\pi a^2}{8}\left(\frac{2a}{3\pi}+b\right)+(ab)\frac{b}{2}=\bar{y}\left(\frac{\pi a^2}{8}+ab\right).$$

Hence,

$$\bar{y}=\frac{\frac{a^3}{12}+\frac{\pi a^2 b}{8}+\frac{ab^2}{2}}{\pi\frac{a^2}{8}+ab}$$

$$=\frac{2a^2+3\pi ab+12b^2}{3\pi a+24b}=\frac{2a+3\pi ab+12b^2}{3(8b+\pi a)}.$$

• PROBLEM 915

Find the centroid of the region R with vertices A,0, and B as shown in the diagram.

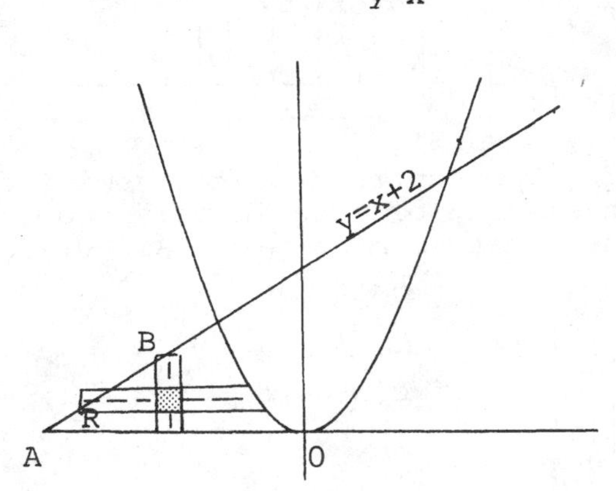

Solution: We use double integrals.

Considering the region enclosed by the functions given in the problem, we determine the point of intersection by setting $x + 2 = x^2$, $x = -1$, $y = +1$. Hence,

$$\int_0^1 \int_{y-2}^{-\sqrt{y}} f(x,y)dx\,dy \text{ ,}$$

is the double integral satisfying all variations of x and y within the region in question. The factor f(x,y) is either y or x, to obtain M_x and M_y, respectively. First, we find the area.

$$A=\int_0^1 \int_{y-2}^{-\sqrt{y}} dx\,dy=\int_0^1 [x]_{y-2}^{-\sqrt{y}} dy$$

$$= \int_0^1 \left[-\sqrt{y} - (y-2)\right] dy = \left[-\frac{2}{3} y^{3/2} - \frac{1}{2} y^2 + 2y\right]_0^1 = 5/6.$$

Now we take moments about the x-axis:

$$M_x = \int_0^1 \int_{y-2}^{-\sqrt{y}} y dx\, dy = \int_0^1 y[x]_{y-2}^{-\sqrt{y}} dy$$

$$= \int_0^1 y\left[-\sqrt{y} - (y-2)\right] dy = \left[-\frac{2}{5} y^{5/2} - \frac{1}{3} y^3 + y^2\right]_0^1$$

$$= \frac{4}{15}.$$

and similarly,

$$M_y = \int_0^1 \int_{y-2}^{-\sqrt{y}} x dx\, dy = \frac{1}{2} \int_0^1 \left[x^2\right]_{y-2}^{-\sqrt{y}} dy = \frac{1}{2} \int \left(y - (y-2)^2\right) dy$$

$$= \frac{1}{2} \int (y - y^2 + 4y - 4) dy = \frac{1}{2}\left(\frac{y^2}{2} - \frac{y^3}{3} + 2y^2 - 4y\right)_0^1$$

$$= \frac{1}{2}\left(\frac{1}{2} - \frac{1}{3} + 2 - 4\right)$$

$$= -\frac{11}{12}$$

Then the centroid is given by:

$$\bar{x} = \frac{4}{15} \div \frac{5}{6} = \frac{24}{75} = \frac{8}{25}$$

$$\bar{y} = -\frac{11}{12} \div \frac{5}{6} = -\frac{66}{60} = -\frac{11}{10}.$$

• PROBLEM 916

Find the centroid of the area of the region bounded by the line: $y = x$, and the parabola: $y = 4x - x^2$.

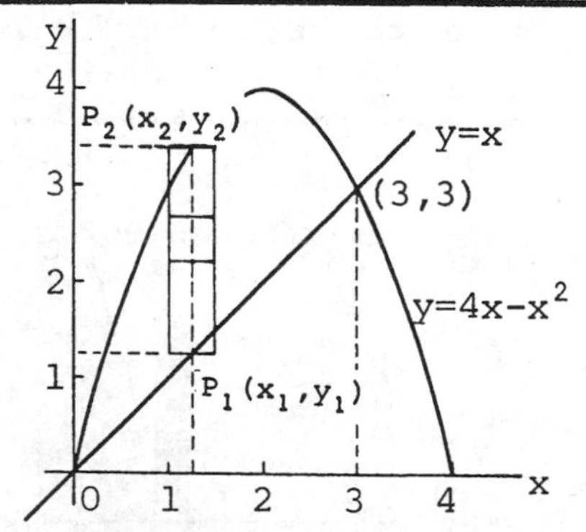

Solution: The points of intersection of the curves are (0,0) and (3,3) as shown in the diagram. It is apparent from the figure that if we integrate first with respect to x, then we must divide the region into two parts and consider them separately (see the note at the end of this example). This is not necessary if we integrate first with respect to y. Since the lower boundary is given by $y = x$, the upper boundary by $y = 4x - x^2$, and the extreme values of x are 0 and 3, we write, in order to obtain area A:

$$A = \int_0^3 \int_x^{4x-x^2} dydx = \int_0^3 (4x - x^2 - x)dx = \frac{9}{2} .$$

For the first moment with respect to the y-axis, we have:

$$A\bar{x} = \int_0^3 \int_x^{4x-x^2} x \, dy \, dx = \int_0^3 (3x^2 - x^3)dx = \frac{27}{4} ,$$

Similarly, for the first moment with respect to the x-axis, we have:

$$A\bar{y} = \int_0^3 \int_x^{4x-x^2} y \, dy \, dx$$

$$= \int_0^3 \tfrac{1}{2}\left[y^2\right]_x^{4x-x^2} dx$$

$$= \tfrac{1}{2}\int_0^3 (15x^2 - 8x^3 + x^4)dx = \frac{54}{5} .$$

Thus,

$$\bar{x} = \frac{A\bar{x}}{A} = \left(\frac{27}{4}\right)\left(\frac{2}{9}\right) = \frac{3}{2}, \text{ and } \bar{y} = \frac{A\bar{y}}{A} = \left(\frac{54}{5}\right)\left(\frac{2}{9}\right) = \frac{12}{5}.$$

Note: If, instead, we insisted on integrating first with respect to x we would have to evaluate two iterated integrals, one taken over each of the subregions obtained by drawing a horizontal line through the point (3,3). Thus,

$$A = \int_0^3 \int_{2-\sqrt{4-y}}^{y} dx \, dy + \int_3^4 \int_{2-\sqrt{4-y}}^{2+\sqrt{4-y}} dx \, dy.$$

• PROBLEM 917

> Find $\bar{x}$ and $\bar{y}$ of the region bounded by the curve $y = x^2$ and $y = x + 2$.

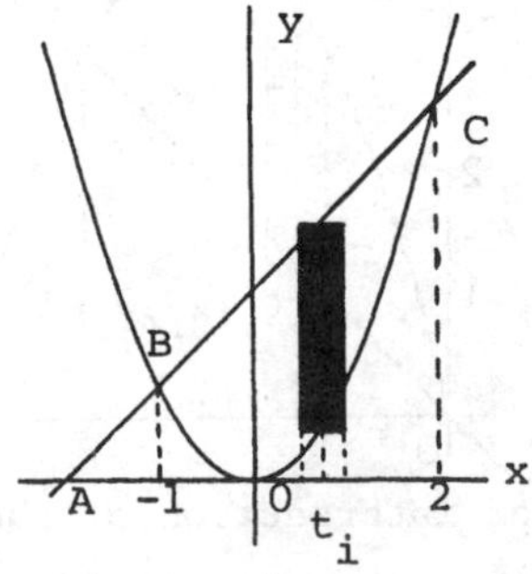

Solution: By equating $y = x^2 = x + 2$ and solving for x, we can obtain the points of intersection as B(-1,1) and C(2,4).

We consider a sequence of subdivisions of $\left[-1,2\right]$, each a rectangle and with a width converging to zero. The geometric center of the rectangle with base $[x_i, x_{i+1}]$ and altitude $t_i + 2 - t_i^2$ is on the horizontal line $y = (t_i^2 + t_i + 2)/2$. Hence

$$M_x = \lim\left(\sum_i (t_i + 2 - t_i^2)(x_{i+1} - x_i)\frac{t_i^2 + t_i + 2}{2}\right)$$

$$= \int_{-1}^{2} \frac{(x+2)^2 - x^4}{2}\,dx = \int_{-1}^{2} \frac{x^2 + 4x + 4 - x^4}{2}\,dx$$

$$= \tfrac{1}{2}\left[\frac{1}{3}x^3 + 2x^2 + 4x - \frac{1}{5}x^5\right]_1^2 = \frac{36}{5}.$$

In this problem, the area is found as follows:

$$A = \int_{-1}^{2}\int_{x^2}^{x+2} dy\,dx = \int_{-1}^{2} (x + 2 - x^2)dx = \frac{9}{2},$$

Since the area of the region is $\frac{9}{2}$, $\bar{y} = \frac{36}{5} \div \frac{9}{2} = \frac{8}{5}$

$$M_y = \lim\left(\sum_i (t_i + 2 - t_i^2)(x_{i+1} - x_i)t_i\right)$$

$$= \int_{-1}^{2} (x + 2 - x^2)x\,dx = \int_{-1}^{2} (x^2 + 2x - x^3)dx = \frac{9}{4}.$$

Then $\bar{x} = \frac{9}{4} \div \frac{9}{2} = \frac{1}{2}$.

• PROBLEM 918

Find the center of mass of the four particles having masses 2,6,4, and 1 slugs located at the points (5,-2), (-2,1), (0,3), and (4,-1), respectively.

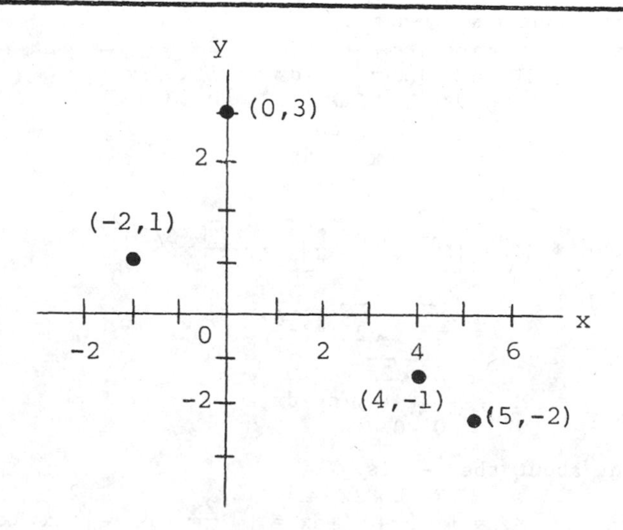

Solution: Taking moments about the y-axis, we have:

$$M_y = \sum_{i=1}^{4} m_i x_i = 2(5) + 6(-2) + 4(0) + 1(4) = 2 .$$

Taking moments about the x-axis:

$$M_x = \sum_{i=1}^{4} m_i y_i = 2(-2) + 6(1) + 4(3) + 1(-1) = 13 .$$

Total mass,

$$M = \sum_{i=1}^{4} m_i = 2 + 6 + 4 + 1 = 13 .$$

Therefore,

$$\bar{x} = \frac{M_y}{M} = \frac{2}{13}, \text{ and } \bar{y} = \frac{M_x}{M} = \frac{13}{13} = 1 .$$

The center of mass is at $\left(\frac{2}{13}, 1\right)$.

• PROBLEM 919

Determine the center of gravity of the area bounded by $y^2 = 2x$, $x = 2$, and $y = 0$.

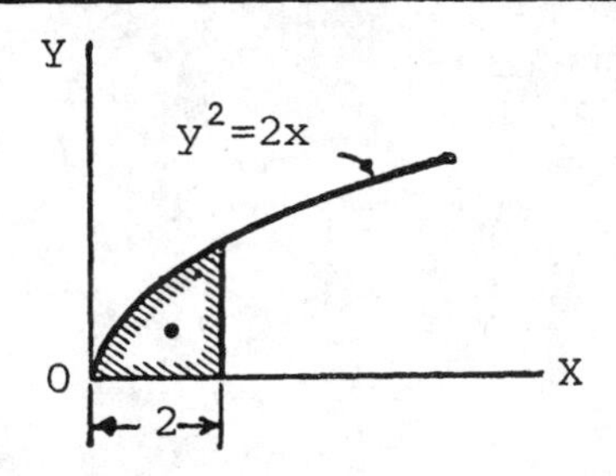

Solution: $\Delta M_y = x\Delta A$ and $\Delta M_x = y\Delta A$. Using a double integral over the given region,

$$M_y = \int_0^2 \int_0^{\sqrt{2x}} x \cdot dy \cdot dx,$$

the moment about the y-axis.

$$M_y = \int_0^2 \Big[x \cdot y\Big]_0^{\sqrt{2x}} \cdot dx = \int_0^2 x\sqrt{2x} \cdot dx$$

$$= \sqrt{2}\int_0^2 x^{3/2} \cdot dx .$$

Finally,

$$M_y = \left[\sqrt{2} \cdot x^{5/2} \cdot \frac{2}{5}\right]_0^2 = \frac{2\sqrt{2}}{5}\left[x^{5/2}\right]_0^2$$

$$= \frac{2\sqrt{2}}{5} \cdot \sqrt{2^5} = \frac{16}{5} .$$

Also,

$$M_x = \int_0^2 \int_0^{\sqrt{2x}} y \cdot dy \cdot dx,$$

the moment about the x-axis.

$$M_x = \int_0^2 \left[\frac{y^2}{2}\right]_0^{\sqrt{2x}} \cdot dx = \int_0^2 \frac{2x}{2} \cdot dx = \int_0^2 x \cdot dx = \left[\frac{x^2}{2}\right]_0^2 .$$

Finally, $M_x = \frac{4}{2} = 2 .$

Now $A = \text{area} = \int_0^2 \int_0^{\sqrt{2x}} dy \cdot dx = \int_0^2 \Big[y\Big]_0^{\sqrt{2x}} \cdot dx$.

$$A = \int_0^2 \sqrt{2x} \cdot dx = \sqrt{2} \int_0^2 x^{\frac{1}{2}} \cdot dx \; .$$

Thus, $\quad A = \sqrt{2}\left[x^{3/2} \cdot \frac{2}{3}\right]_0^2 = \frac{2\sqrt{2}}{3}\left[x^{3/2}\right]_0^2 = \frac{2\sqrt{2}}{3} \cdot \sqrt{2^3} = \frac{2}{3} \cdot 4 = \frac{8}{3}$,

and hence
$$\bar{x} = \frac{M_y}{A} = \frac{16}{5} \cdot \frac{3}{8} = \frac{6}{5}$$
$$\bar{y} = \frac{M_x}{A} = 2 \cdot \frac{3}{8} = \frac{3}{4} \; .$$

Therefore, the c.g. is at $\bar{x} = \frac{6}{5}$, $\bar{y} = \frac{3}{4}$.

• PROBLEM 920

Find the center of gravity of a homogeneous hemisphere of radius r.

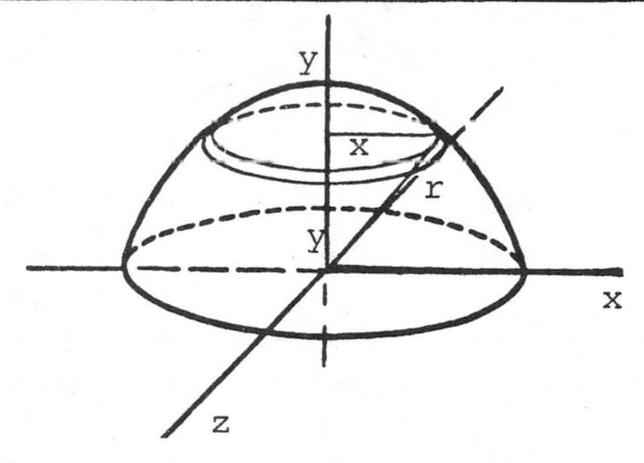

Solution: Since the hemisphere is symmetric on the y - z and x - y planes, the net moment with respect to these planes is zero, making $\bar{x} = \bar{z} = 0$. Then, the center of gravity will lie on the diameter perpendicular to the base at a distance $\bar{y}$ above the base.

To find the volume and moment in integral forms, the hemisphere can be thought of as a volume generated by a semicircle in the first quadrant about the y-axis. Hence, the small strip made by a line drawn from the y-axis to an arbitrary point p(x,y) parallel to the x-axis and another parallel line Δy units above it will look like a disc with radius x units, thickness Δy, and volume

$$\Delta v = \pi x^2 \; \Delta y \; .$$

Passing to the limits, we obtain the volume expression in an integral form:
$$v = \int_0^r \pi x^2 \; dy \; .$$

The turning effect of the mass of the hemisphere with respect to the z-x plane can be expressed as:

$$M_{zx} = \rho \int_v y du \text{ , where } \rho \text{ is the density,}$$

We wish to obtain an analogous expression, where the mass would be thought of as concentrated at a point some distance $\bar{y}$ above the origin, such that $\bar{y}m = M_{zx}$, or, under constant density,

$$\bar{y}V = M_{zx}/\rho = \int_v y \, dv.$$

$$\bar{y} = \frac{\int_v y \, du}{V} = \frac{\int_0^r y\pi x^2 \, dy}{\int dv} \quad . \quad x^2 = r^2 - y^2.$$

Hence,

$$\bar{y} = \frac{\int_0^r \pi y(r^2 - y^2)dy}{\frac{2}{3}\pi r^3}$$

$$= \frac{3}{2r^3}\left[\frac{r^2y^2}{2} - \frac{y^4}{4}\right]_0^r$$

$$= \frac{3}{8} r .$$

Therefore the center of gravity of the given hemisphere with radius r is at $(0, 3/8\ r, 0)$.

• PROBLEM 921

Find the radius of gyration of the area under one arch of the cycloid $x = a(\Theta - \sin \Theta)$, $y = a(1 - \cos \Theta)$, with respect to the x-axis.

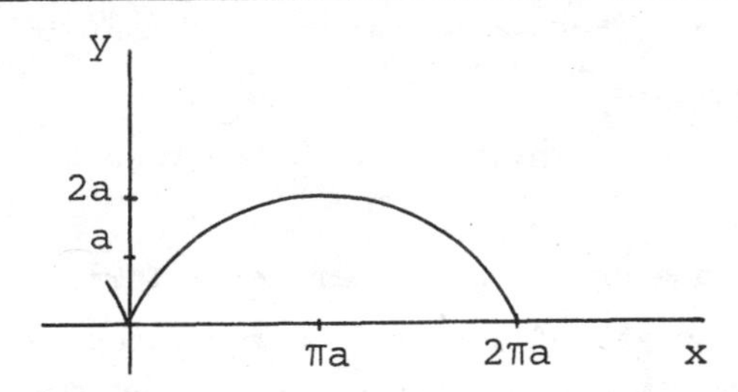

Solution: The moment of inertia,

$$I_x = \int y^2 \, dA,$$

where the integration over the region in the arch of the cycloid is given by:

$$A = \int\int dydx,$$

$$dx = a(d\Theta - \cos \Theta \, d\Theta) = a\ (1 - \cos \Theta)d\Theta .$$

Hence,

$$I_x = \int_{x=0}^{x=2\pi a} \int_0^{a(1-\cos\Theta)} y^2 \, dy \, d\big(a(\Theta - \sin\Theta)\big)$$

$$= \int_{\Theta=0}^{\Theta=2\pi} \frac{1}{3} a(1 - \cos\Theta) y^3 \Bigg|_{y=0}^{y=a(1-\cos\Theta)} d\Theta$$

$$= \frac{a^4}{3} \int_0^{2\pi} (1 - \cos\Theta)^4 \, d\Theta$$

$$= \frac{a^4}{3} \int_0^{2\pi} (1 - 4\cos\Theta + 6\cos^2\Theta - 4\cos^3\Theta + \cos^4\theta)\, d\theta.$$

Since integration of $\cos\Theta$ and $\cos^3\Theta$ does not contribute anything, as they are odd functions, we merely integrate the following:

$\int_0^{2\pi} \cos^2\Theta d\Theta = \pi$. Using integration by parts, we have:

$$\int_0^{2\pi} \cos^4\Theta d\Theta = \frac{1}{4}\cos^3\Theta \sin\Theta \Bigg|_0^{2\pi} + \frac{3}{4}\int_0^{2\pi} \cos^2\Theta d\Theta = \frac{3\pi}{4}$$

Thus,

$$I_x = \frac{a^4}{3}\left(2\pi + 6\pi + \frac{3\pi}{4}\right) = \frac{35a^4\pi}{12}.$$

For the area ,

$$A = \int_0^{2\pi} \int_0^{a(1-\cos\Theta)} dy \, dx$$

$$= \int_0^{2\pi} \int_0^{a(1-\cos\Theta)} a(1 - \cos\Theta) dy d\Theta$$

$$A = a^2 \int_0^{2\pi} (1 - \cos\Theta)^2 \, d\Theta$$

$$= a^2 \int_0^{2\pi} (1 - 2\cos\Theta + \cos^2\Theta)d\Theta = a^2(2\pi + \pi)$$

$$= 3a^2\pi.$$

Thus, the radius of gyration,

$$y_R = \frac{\int y^2\, dA}{\int dA} = \frac{I_x}{A},$$

or

$$y_R = \frac{35a^4\pi}{12} \div 3a^2\pi = \frac{35}{36}a^2.$$

• PROBLEM 922

Find the volume bounded by the cylinder $x^2 + y^2 = 4$ and the hyperboloid $x^2 + y^2 - z^2 = 1$.

Solution: On the xy-plane, the required volume is bounded by the circle $r = 1$ (obtained by setting $z = 0$ in the equation of the hyperboloid),and the base radius of the cylinder is 2. The volume is then obtained by integrating

$$z = (x^2 + y^2 - 1)^{\frac{1}{2}} = \sqrt{r^2 - 1}$$

over the region bounded by the two circles. Since z can assume a positive or negative sign, and due to symmetry over the plane $z = 0$,the height for the required volume is

$$2z = 2\sqrt{r^2 - 1}.$$

Hence,

$$V = \int_0^{2\pi}\int_1^2 2r\sqrt{r^2 - 1}\, dr d\Theta.$$

The integral:

$$\int_1^2 2r\sqrt{r^2 - 1}\, dr = \int_1^2 2r(r^2 - 1)^{\frac{1}{2}} dr$$

$$= \frac{2}{3}(r^2 - 1)^{\frac{3}{2}}\Big|_1^2 = 2\sqrt{3},$$

and

$$V = \int_0^{2\pi} 2\sqrt{3}\, d\Theta = 4\pi\sqrt{3}.$$

• PROBLEM 923

Find the volume of the solid in the first octant bounded by the cylinder $y = e^x$ and the planes $x = 0$, $x = 1$, $z = 0$, and $z = y$.

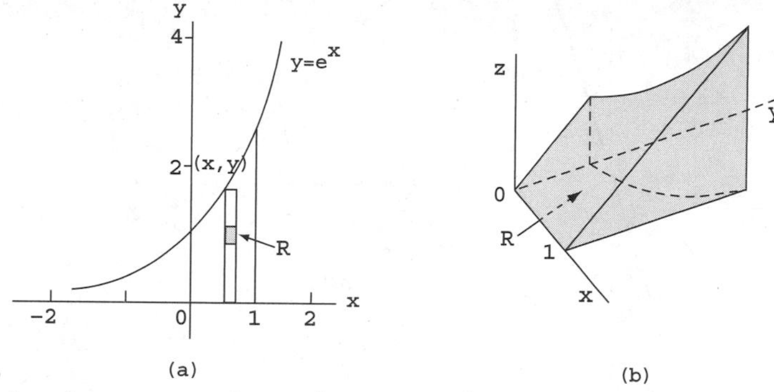

(a) (b)

Solution: Before integration, we investigate the projection of the bounded surface, as given by $y = e^x$, $0 \leq x \leq 1$, and $0 \leq z \leq y$, on the x - y plane. We call this region R as shown in figure (a) and also in three dimensional figure (b).

In (a), if the shaded portion of the strip has area dA, which is dx dy, then the area of the region R is given by double integration, where x varies from zero to 1, and y varies from zero to e^x. Integration of the equation of the surface $z = y$ over the region R gives the required volume. This means dA in R is the base of a column from the x - y plane to the surface of height $z = y$. This height, multiplied by the cross-sectional area, gives the volume.

$$V = \int_R z\, dA = \int_R y\, dA$$

$$= \int_0^1 \int_0^{e^x} y\, dy\, dx$$

$$= \int_0^1 \frac{1}{2} \left[y^2 \right]_0^{e^x} dx$$

$$= \frac{1}{2} \int_0^1 e^{2x}\, dx = \frac{1}{2} \left[\frac{1}{2} e^{2x} \right]_0^1$$

$$= \frac{1}{4} \left(e^2 - 1 \right).$$

Find the volume of the tetrahedron bounded by the plane $x + y + z = 1$ and the coordinate planes.

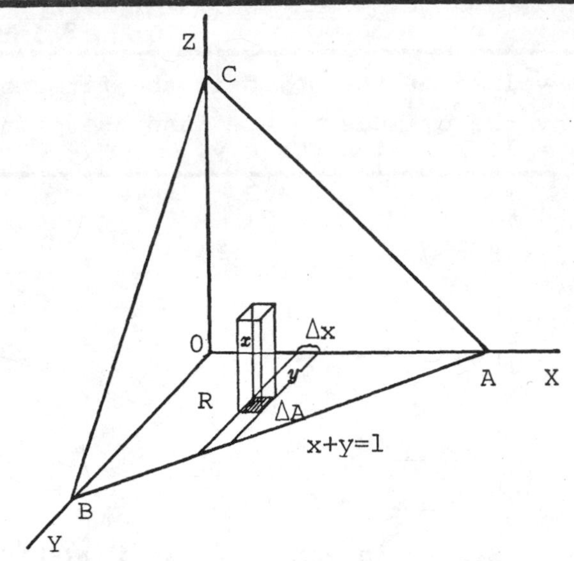

Solution: The equation of the upper surface of the tetrahedron is:

$$z = 1 - x - y,$$

so that the function $z = f(x,y)$ of the given formula is $z = 1 - x - y$, and the region R is the triangle in the XY-plane bounded by the line $x + y = 1$ or $y = 1 - x$ and the X- and Y-axes. We then evaluate the expression:

$$V = \iint_R (1 - x - y)\, dA.$$

Transforming this into an iterated integral, if we integrate first with respect to y, we see that y varies from $y = 0$ to $y = 1 - x$. In the second integration, x varies from 0 to 1. Then:

$$V = \int_0^1 \left[\int_0^{1-x} z\, dy \right] dx$$

$$= \int_0^1 \left[\int_0^{1-x} (1 - x - y)\, dy \right] dx$$

$$= \int_0^1 \left[\left((1 - x)\, y - \frac{y^2}{2} \right|_0^{1-x} \right] dx$$

$$= \int_0^1 \left[(1 - x)(1 - x) - \frac{(1 - x)^2}{2} \right] dx$$

$$= \int_0^1 \frac{(1-x)^2}{2}\,dx = -\frac{1}{2}\,\frac{(1-x)^3}{3}\Bigg|_0^1$$

$$= \frac{1}{6}\ .$$

• PROBLEM 925

Find the volume under the paraboloid: $x^2 + y^2 = az$, above the XY-plane and inside the cylinder: $x^2 + y^2 = 2ax$.

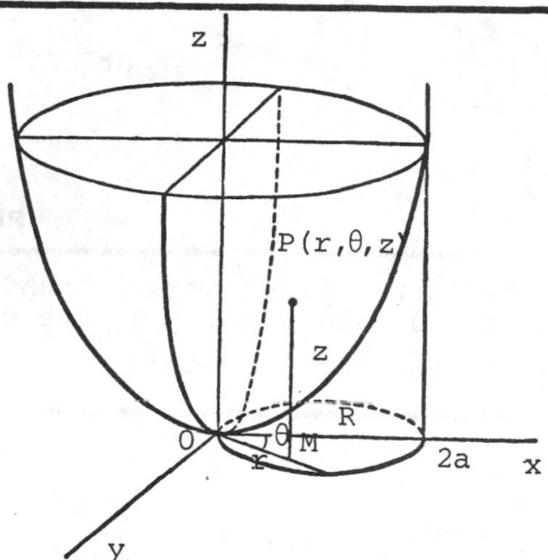

Solution: To calculate the volume required, we deal with the region R, formed as a projection on the x-y plane of the surface in question. This is

$$f(x,y) = z = (x^2 + y^2)/a.$$

Hence, the integral of this function, f(x,y) over the region gives the volume of the surface

$$V = \int_R f(x,y)\ dA.$$

This is an alternative form of the triple integral. For the sake of simplicity, we use the cylindrical coordinate system and express the integral in terms of the radius r and the angle Θ, as shown in the figure.

In the cylindrical coordinate system, $x^2 + y^2$ is replaced by r^2, and, as a result of the relationships:

$$x = r \cos \Theta,\ y = r \sin \Theta \text{ and } z = z,$$

the expression:

$$r^2 = az, \text{ or } z = \frac{r^2}{a}\ ,$$

is the equation of the paraboloid. Also,

$$r^2 = 2ar \cos \Theta, \text{ or } r = 2a \cos \Theta,$$

the equation of the cylinder. The function f(x,y)

is thus $\frac{r^2}{a}$. Due to symmetry about the x-axis we carry out the integration in the first octant and double it, letting Θ vary from 0 to π/2 rad., and r vary in the interior of the cylinder:

$$0 < r < 2a \cos \Theta.$$

The required volume is then given by:

$$V = 2 \int_0^{\frac{\pi}{2}} \int_0^{2a \cos \Theta} z(r\, dr\, d\Theta)$$

$$= 2 \int_0^{\frac{\pi}{2}} \int_0^{2a \cos \Theta} \frac{r^2}{a} r\, dr\, d\Theta$$

$$= \frac{3}{2} \pi a^3.$$

• PROBLEM 926

Find the volume in the first octant bounded by the planes: $x = 0$, $z = 0$, and $y = 2 - x$, the parabolic cylinder: $y = x^2$, and the surface: $z = x^2 + 2y^2$.

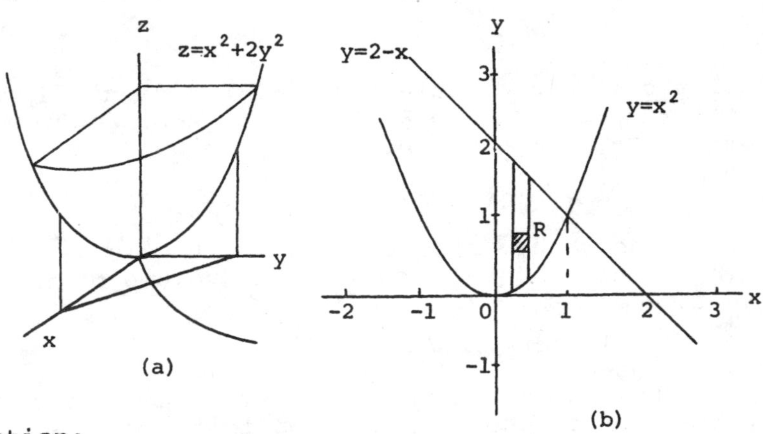

Solution:

The equation of the surface, $f(x,y) = z = x^2 + 2y^2$, shows that $z = 0$, if and only if $x = 0$ and $y = 0$ simultaneously. Here we understand that the region R is not the projection of the surface $f(x,y)$ on the x-y plane. We examine whether the expression:

$$V = \int_R f(x,y)\, dA ,$$

holds, to obtain the required volume.

Consider a set of rectangles, and in each rectangle there are points of the region R. Choose any such rectangle such as the shaded one shown in figure (b), and let it be located at point (x_i, y_i). We have the **inequality:**

$$m_i(n) \le f(x_i, y_i) \le M_i(n),$$

where $m_i(n)$ and $M_i(n)$ are, respectively, the greatest lower bound and the least upper bound of $f(x,y)$ on the intersection of R with the ith rectangle. We have:

$$s_n \le \sum_{i=1}^{n} m_i(n)\ \Delta A \le \sum_{i=1}^{n} f(x_i, y_i)\ \Delta A$$

$$\le \sum_{i=1}^{n} M_i(n)\ \Delta A = S_n.$$

If $f(x,y)$ is integrable on R, and the limit:

$$\lim_{n\to\infty} \sum_{i=1}^{n} m_i(n)\ \Delta A = \lim_{n\to\infty} \sum_{i=1}^{n} M_i(n)\ \Delta A = S,$$

then this proves that:

$$\lim_{n\to\infty} \sum_{i=1}^{n} f(x_i, y_i)\ \Delta A = \int_R f(x,y)\ dA = S$$

We can use such a sum in order to help set up double integrals. Noting that $f(x_i, y_i)$ is taken as the approximate mean height of the ith rectangle to the surface, we set

$$V = \int_R f(x,y)\ dA.$$

From figure (b), we see that y is within the region bounded by x^2 and the line $2 - x$, or

$$x^2 \le y \le 2 - x,$$

whereas x varies from zero to the intersection point:

$$y = x^2 = 2 - x.$$

$$(x + 2)(x - 1) = 0,$$

at $x = 1$ and $x = -2$.

Thus, $0 \le x \le 1$.

This leads to

$$V = \int_0^1 \int_{x^2}^{2-x} \left(x^2 + 2y^2\right)\ dy\ dx,$$

$$\int_x^{2-x} (x^2 + 2y^2)\, dy = x^2 y + \frac{2}{3} y^3 \Bigg|_x^{2-x}$$

$$= x^2 (2 - x) + \frac{2}{3} (2 - x)^3 - x^4 - \frac{2}{3} x^6 .$$

$$V = \int_0^1 \left[\frac{2}{3} (2 - x)^3 + 2x^2 - x^3 - x^4 - \frac{2}{3} x^6\right] dx$$

$$= - \frac{1}{6} (2 - x)^4 + \frac{2}{3} x^3 - \frac{1}{4} x^4 - \frac{1}{5} x^5 - \frac{1}{21} x^7 \Bigg|_0^1$$

$$= \frac{367}{140} .$$

• PROBLEM 927

Find the volume bounded by the cone: $r = z$, the cylinder: $r = a \cos \Theta$, and the plane $z = 0$.

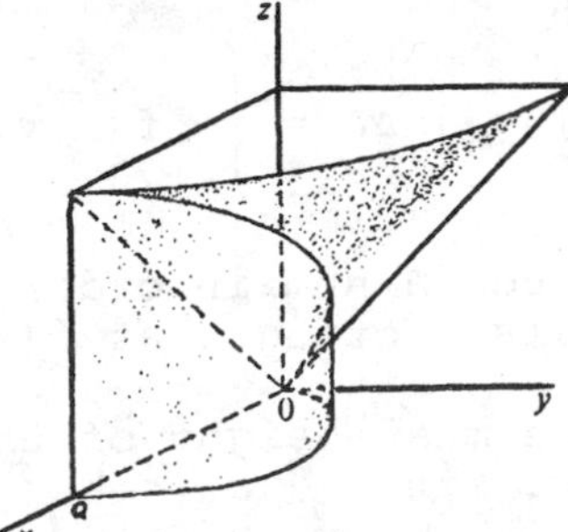

Solution: By taking small increments of Θ and r in the $x - y$ plane in the region R, bounded by $r = a \cos \Theta$, we obtain an elemental area with dimension $rd\Theta$ and dr. Hence its area is:

$$dA = r\, dr\, d\Theta.$$

To find the volume, we have two ways of setting up the integral equation: The first one is to integrate:

$$\int_R z\, dA ,$$

over the region R. This means the elemental volume is constructed so that the height is z, which extends from the $x - y$ plane to the surface. The second way is using the triple integral where, instead of using the whole column to represent the element volume, we assign an infinitesimal height, say dz, and hence the volume in the cylindrical coordinate system (as suggested in this problem) is:

$$dV = (dz)(rd\Theta)(dr);$$

the required volume is given by:

$$V = \int_V dV = \int_V r\, dr\, d\Theta\, dz.$$

Both ways of setting up the integral expressions are necessarily the same. The region R is symmetric relative to the xz-plane, so that the required volume is twice the volume of the region in the first octant. We use $r\, dr\, d\Theta$ for dA, and note that z varies from the x - y plane to the surface of the cone $r = z$, with fixed values of the angle Θ. r varies within the region R that describes the base cross-sectional region of the cylinder along a straight line that makes a constant angle Θ with the positive x-axis. Due to symmetry, the angle Θ sweeps over $\pi/2$ radians, i.e., $0 \leq \Theta \leq \pi/2$ in the integration. We obtain:

$$V = 2\int_0^{\frac{\pi}{2}}\int_0^{a\cos\Theta} r(r\, dr\, d\Theta),$$

$$= \frac{2}{3}\int_0^{\frac{\pi}{2}} (a\cos\Theta)^3\, d\Theta$$

$$= \frac{2}{3} a^3 \int_0^{\frac{\pi}{2}} \cos^3\Theta\, d\Theta$$

$$= \frac{4}{9} a^3$$

• PROBLEM 928

Find the area of a circular sector of radius a and central angle α.

Solution: To solve this problem, we use the polar coordinate system.

$x = a\cos\Theta,$

$y = a\sin\Theta,$ and $0 \leq \Theta \leq \alpha.$

The area is:

$$A = \int_0^{\alpha}\int_0^{a} r\, dr\, d\Theta$$

$$= \int_0^{\alpha} \frac{1}{2} r^2 \Big|_0^{a}$$

$$= \frac{1}{2} r^2 \alpha.$$

• PROBLEM 929

Find the area in the first quadrant bounded by

the ellipse:

$$x^2 + 4y^2 = 5,$$

and the hyperbola: $xy = 1$.

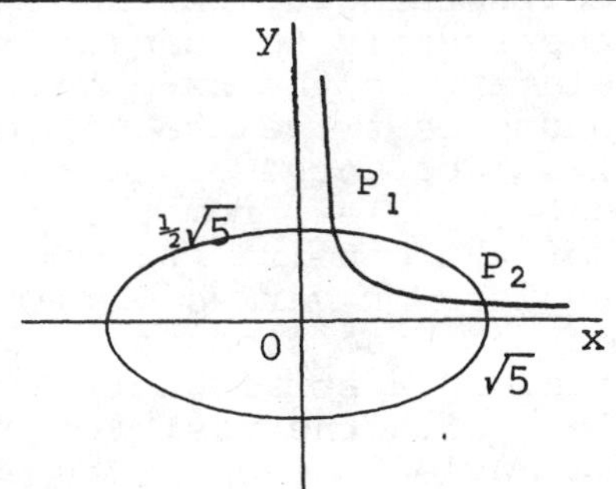

Solution: We first find the intersection points P_1 and P_2 as shown in the figure, by equating the y-values for the ellipse,

$$y = \frac{\sqrt{5\ x^2}}{2},$$

and that of the hyperbola,

$$y = \frac{1}{x}.$$

Or, we can equate the y^2 values:

$$\frac{1}{x^2} = \frac{1}{4}(5 - x^2), \text{ or } 5x^2 - x^4 = 4.$$

By inspection of the expression:

$$5x^2 - x^4 = 4,$$

the first root is $x = \pm 1$. Dividing the polynomial:

$$x^4 - 5x^2 + 4, \text{ by } (x - 1),$$

gives:

$$(x^3 + x^2 - 4x - 4)(x - 1) = x^4 - 5x^2 + 4 = 0.$$

Division by $(x + 1)$ gives:

$$x^3 + x^2 - 4x - 4 = (x^2 - 4)(x + 1).$$

Hence,

$$(x^2 - 4)(x + 1)(x - 1) = x^4 - 5x^2 + 4 = 0$$

the roots $x = -1$ and $x = -2$ are trivial. The desired values are $x = 1$ and $x = 2$.

Now, using double integration, the area of the required region is given by:

$$A = \int_1^2 \int_{1/x}^{1/2\sqrt{5-x^2}} dy\, dx$$

$$= \frac{1}{2}\int_1^2 \sqrt{5 - x^2}\, dx - \int_1^2 \frac{1}{x}\, dx$$

$$= \left[\frac{1}{4} x \sqrt{5 - x^2} + \frac{5}{4} \sin^{-1} \frac{x}{\sqrt{5}} - \ln x \right|_1^2$$

$$= \frac{1}{4} \cdot 2 + \frac{5}{4} \sin^{-1} \frac{2}{\sqrt{5}} - \ln 2 - \frac{1}{4} \cdot 2 - \frac{5}{4} \sin^{-1} \frac{1}{\sqrt{5}}$$

$$= \frac{5}{4} \left(\sin^{-1} \frac{2}{\sqrt{5}} - \sin^{-1} \frac{1}{\sqrt{5}} \right) - \ln 2.$$

• PROBLEM 930

Find the area of the portion of the cylinder: $x^2 + z^2 = a^2$, lying inside the cylinder: $x^2 + y^2 = a^2$.

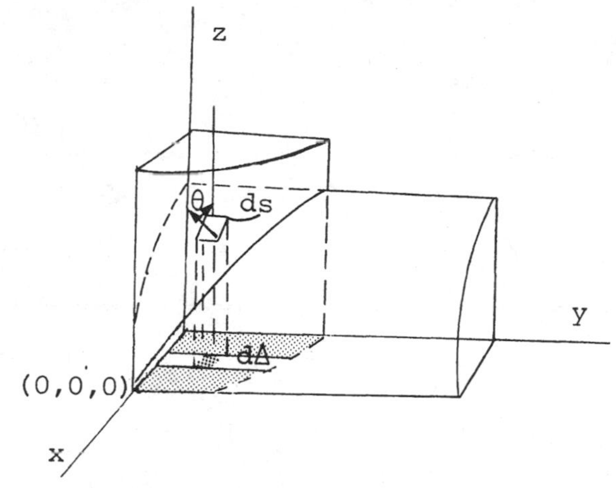

Solution: One-eighth of the required area is shown in the diagram. Its projection on the xy-plane is a quadrant of the circle:

$$x^2 + y^2 = a^2.$$

Now, take any rectangular area on the xy-plane in the region of the circle: $x^2 + y^2 = a^2$. Let this elemental area be denoted by dA. Then the corresponding area on the surface of the cylinder: $x^2 + z^2 = a^2$, is ds, which can be related to dA as:

$$dA = ds \cos \Theta,$$

where Θ is the angle that the plane tangent to ds makes with the xy-plane. Furthermore, the angle which a line normal to the plane tangent to ds makes with the positive z-axis is also Θ. To find the angle Θ in the expression above, we wish to find the vector normal to ds at a point somewhere within ds, which, in the limit reduces to a point, through which the tangential plane passes.

$$\cos \Theta = \frac{z}{a}$$

$$ds = a\,\frac{dA}{z} = a\,\frac{dA}{\sqrt{a^2 - x^2}}$$

where dA is in the region: $(x^2 + y^2 = a^2)$ for

$-a \le x \le a$, and $-(a^2 - x^2)^{\frac{1}{2}} \le y \le (a^2 - x^2)^{\frac{1}{2}}$.

Noting the symmetry about the origin, the required surface area is:

$$s = 8\int_0^a \int_0^{\sqrt{a^2-x^2}} a\,\frac{dy\,dx}{\sqrt{a^2 - x^2}}$$

$$= 8a\int_0^a \left(y\,\Big|_0^{\sqrt{a^2-x^2}} \cdot \frac{1}{\sqrt{a^2 - x^2}} \right) dx$$

$$= 8a\int_0^a dx = 8a^2.$$

• PROBLEM 931

Find the area of the first-octant portion of the cylinder $x^2 + z^2 = a^2$ that is included between the planes $y = 0$ and $y = x$.

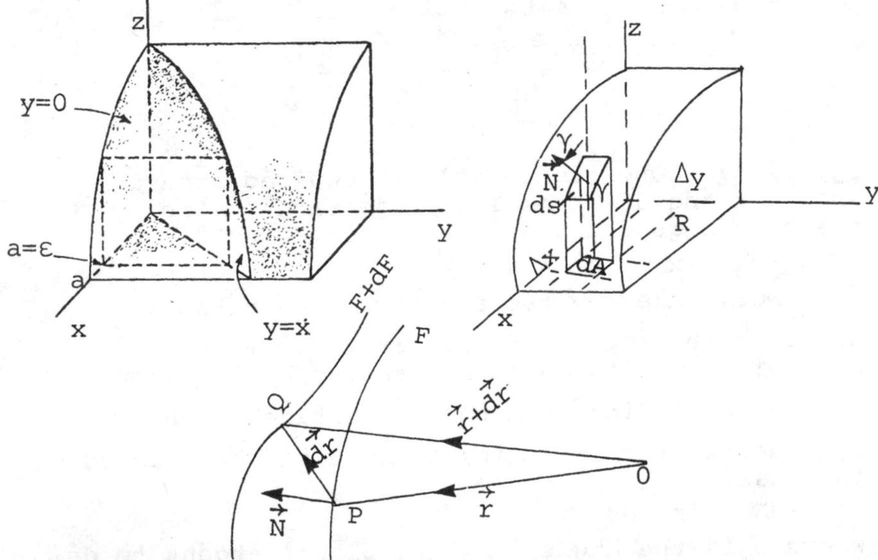

Solution: Let ΔA be the area of a rectangle in the region R on the x-y plane, and let the dimensions of this rectangle be Δx by Δy. By drawing lines parallel to the z-axis from the element area to the surface, we can obtain a corresponding surface area Δs on the surface. ΔA can be taken as the projected area of Δs on the x-y plane. If Δs can be thought of as part of a plane which makes angle γ with the x-y plane, then the approximate relationship between Δs and ΔA is:

$$\Delta s \cos \gamma = \Delta A, \quad \text{or}$$

$$\Delta s = \Delta A \sec \gamma.$$

We now find the angle γ. Let $F(x, y, z)$ be the equation of the surface of the cylinder. We call this a scalar function of position, and

$$F(x, y, z) = x^2 + z^2 - a^2.$$

The differential dF of $F(x, y, z)$ is given by:

$$dF = \frac{\partial F}{\partial x} dx + \frac{\partial F}{\partial y} dy + \frac{\partial F}{\partial z} dz,$$

where the operator $\left(\frac{\partial}{\partial x}\right)$ represents the partial differential with respect to x, considering the rest variables as constants. The above partial-differential equation may be regarded as the scalar (or dot) product of two vectors, the components of which are $\frac{\partial F}{\partial x}$, $\frac{\partial F}{\partial y}$, $\frac{\partial F}{\partial z}$, and dx, dy, dz, respectively. Consequently, we have:

$$dF = \left(\frac{\partial F}{\partial x}\hat{\imath} + \frac{\partial F}{\partial y}\hat{\jmath} + \frac{\partial F}{\partial z}\hat{k}\right) \cdot (dx\hat{\imath} + dy\hat{\jmath} + dz\hat{k}).$$

If $\vec{r} = x\hat{\imath} + y\hat{\jmath} + z\hat{k}$, the second vector evidently represents $d\vec{r}$; and the first vector can be denoted by the operator ∇ (called "del") such that:

$$\nabla F \equiv \frac{\partial F}{\partial x}\hat{\imath} + \frac{\partial F}{\partial y}\hat{\jmath} + \frac{\partial F}{\partial z}\hat{k}.$$

Thus,

$$dF = \nabla F \cdot d\vec{r}.$$

Furthermore, let $\vec{r} = x\hat{\imath} + y\hat{\jmath} + z\hat{k}$ be the vector from the origin 0 to the point $P(x, y, z)$ on the surface $F(x, y, z)$, and let $\vec{N}$ be a unit normal to the surface at P in the direction of increasing F. If ds denotes the magnitude of the differential vector $d\vec{r} = dx\hat{\imath} + dy\hat{\jmath} + dz\hat{k}$ from P to a point Q on a neighboring surface, and dN the projection of ds onto $\vec{N}$ as shown below, we have:

$$dN = \vec{N} \cdot d\vec{r},$$

so that the differential dF of the function $F(x, y, z)$ is:

$$dF = \frac{dF}{dN} dN = \frac{dF}{dN} \vec{N} \cdot d\vec{r},$$

From the derivation above, $dF = \nabla F \cdot d\vec{r}$. Combining both equations and canceling $d\vec{r}$ gives:

$$\nabla F = \frac{dF}{dN} \vec{N}$$

This guarantees that the gradient vector ∇F is in the direction of the vector normal to the surface.

Here, the angle $\vec{N}$ makes with the z-axis is also γ. Hence,

$$\vec{N} \cdot \hat{k} = \cos \gamma,$$

$\vec{N}$ is the unit normal vector.

From the equation above, the magnitude of the gradient vector,

$$|\nabla F| = \frac{dF}{dN},$$

which in this problem is the magnitude of:

$$\nabla F = \frac{\partial}{\partial x} (x^2 + z^2 - a^2)\, \hat{\imath}$$

$$+ \frac{\partial}{\partial y} (x^2 + z^2 - a^2)\, \hat{\jmath}$$

$$+ \frac{\partial}{\partial z} (x^2 + z^2 - a^2)\, \hat{k}$$

$$= 2x\hat{\imath} + 2z\hat{k}.$$ We have:

$$|\nabla F| = \sqrt{(2x)^2 + (2z)^2} = 2\sqrt{x^2 + z^2} = 2a.$$

Thus,

$$\vec{N} \cdot \hat{k} = \cos \gamma = \frac{\nabla F}{|\nabla F|} \cdot \hat{k} = \frac{(2x\hat{\imath} + 2z\hat{k}) \cdot \hat{k}}{2a}$$

$$= \frac{2z}{2a} = \frac{z}{a}.$$

Or, by the use of the equation of the cylinder,

$$z = \sqrt{a^2 - x^2},$$

$$\cos \gamma = \frac{\sqrt{a^2 - x^2}}{a},$$

from which,

$$ds = \frac{dA}{\cos \gamma} = \frac{a}{\sqrt{a^2 - x^2}} dA.$$

Since the required surface projects into the triangle in the xy-plane bounded by $y = 0$, $y = x$, $x = a$, we have:

$$S = \int_R \frac{a}{\sqrt{a^2 - x^2}} \, dA = \int_0^a \int_0^x \frac{a \, dy \, dx}{\sqrt{a^2 - x^2}}$$

$$= \int_0^a \frac{ax}{\sqrt{a^2 - x^2}} \, dx = a^2.$$

• PROBLEM 932

(a) Find a formula for the approximate error in the area of a circular sector, due to errors in measuring the radius and central angle.

(b) If the radius is 8 $\pm$ 0.1 in. and the central angle is $30^\circ \pm 10'$, find the maximum possible error in the area.

Solution: The area of a sector of radius r with central angle Θ can be obtained as:

$$A = \int_0^\Theta \int_0^r r \, dr \, d\Theta = \frac{1}{2} r^2 \Theta.$$

The effect of the errors in measuring the radius and the angle on the area can be expressed as:

$$\Delta A \simeq \frac{\partial A}{\partial r} \Delta r + \frac{\partial A}{\partial \Theta} \Delta\Theta,$$

and this is the approximate error in the area. For the sector,

$$\frac{\partial A}{\partial r} = r\Theta, \quad \frac{\partial A}{\partial \Theta} = \frac{r^2}{2}.$$

$$\Delta A \simeq r\Theta \, \Delta r + \frac{1}{2} r^2 \, \Delta\Theta.$$

(b) $r = 8$ in. $\qquad \Theta = 30^\circ = \frac{\pi}{6}$ radian.

$|\Delta r| = 0.1$ in., and

$$|\Delta\Theta| = 10' = \left(\frac{10}{60}\right)^0 = \frac{1}{6} \frac{\pi}{180}.$$

Hence,

$$\Delta A \simeq 8 \left(\frac{\pi}{6}\right)(0.1) + \frac{1}{2} (8)^2 \left(\frac{1}{6} \frac{\pi}{180}\right)$$

$$= \pi \left(\frac{0.8}{6} + \frac{32}{1080}\right) = \pi \; (0.133 + 0.0296)$$

$$= 0.51 \text{ in.}^2.$$

CHAPTER 35

TRIPLE INTEGRALS

In the determination of such quantities as volumes and their centroids, a triple integral may arise. This takes the form:

$$V = \int_a^b \int_{h_2(z)}^{h_1(z)} \int_{g_2(y,z)}^{g_1(y,z)} f(x,y,z)dx\ dy\ dz\ ,$$

where $g_1(y,z)$ and $g_2(y,z)$ are the limits of x, $h_1(z)$ and $h_2(z)$ are the limits of y, and a and b are the limits of z. However, the order, in which the integrations are performed, can be changed as desired. In each case, as integration is performed with respect to one variable, the others are treated as constants. After completion of each integration, the limits are substituted before proceeding to the next integration. The order of integration can be chosen on the basis of the apparent comparative simplicity of the functions to be integrated. Some trial-and-error may be necessary before the best sequence of integration is found.

Triple integration is an essential method when the function such as volume of a highly irregular body is to be evaluated.

• PROBLEM 933

Evaluate the integral:

$$\int_0^1 \int_x^{x^2} \int_{xy}^{x^2y^3} xy\ dz\ dy\ dx.$$

Solution: In evaluating the integral, we insert grouping symbols to indicate clearly the order of integration:

$$\int_0^1 \int_x^{x^2} \int_{xy}^{x^2y^3} xy\ dz\ dy\ dx$$

$$= \int_0^1 \left[x \int_x^{x^2} \left[y \int_{xy}^{x^2y^3} dz \right] dy \right] dx$$

$$= \int_0^1 \left[x \int_x^{x^2} (yz) \Big|_{z=xy}^{z=x^2y^3} dy \right] dx$$

$$= \int_0^1 \left[x \int_x^{x^2} y \; (x^2 \; y^3 \; - \; xy) \; dy \right] dx$$

$$= \int_0^1 \left[x \int_x^{x^2} (x^2 \; y^4 \; - \; xy^2) \; dy \right] dx$$

$$= \int_0^1 \left[x \left[x^2 \int_x^{x^2} y^4 \; dy \; - \; x \int_x^{x^2} y^2 \; dy \right] \right] dx$$

$$= \int_0^1 \left[x \left[\frac{1}{5} \; x^2 \; y^5 \; - \; \frac{1}{3} \; xy^3 \Big|_{y=x}^{y=x^2} \right] dx$$

$$= \int_0^1 \left(\frac{1}{5} \; x^{13} \; - \; \frac{1}{3} \; x^8 \; - \; \frac{1}{5} \; x^8 \; + \; \frac{1}{3} \; x^5 \right) dx$$

$$= \left[\frac{1}{70} \; x^{14} \; - \; \frac{8}{135} \; x^9 \; + \; \frac{1}{18} \; x^6 \right]_0^1$$

$$= \frac{1}{70} - \frac{8}{135} + \frac{1}{18} = \frac{2}{189} \; .$$

• PROBLEM 934

Integrate:

$$\int_0^a \int_0^{a-z} \int_0^{a-y-z} yz \; dx \; dy \; dz.$$

Solution: If we let

$$M = \int_0^a \int_0^{a-z} \int_0^{a-y-z} yz\, dx\, dy\, dz,$$

then

$$M = \int_0^a \int_0^{a-z} xyz \Big|_{x=0}^{x=a-y-z} dy\, dz$$

$$= \int_0^a \int_0^{a-z} yz(a - y - z)dy\, dz$$

$$= \int_0^a \frac{1}{2} y^2 za - \frac{1}{3} y^3 z - \frac{1}{2} y^2 z^2 \Big|_{y=0}^{y=a-z} dz$$

$$= \int_0^a (a - z)^2 \left[\frac{1}{2} za - \frac{1}{3}(a - z)z - \frac{1}{2} z^2\right] dz$$

$$= \frac{1}{6} \int_0^a z\, (a - z)^3\, dz$$

$$= \frac{z^2}{6} \left[\frac{1}{2} a^3 - a^2 z + \frac{3}{4} az^2 - \frac{1}{5} z^3\right] \Big|_0^a$$

$$= \frac{a^5}{120}.$$

• PROBLEM 935

Find the limits of the iterated integral for

$$\iiint_V f(x,y,z)\, dV$$

when the region V is that bounded by the paraboloid $x^2 + 4y^2 = 16 - z$ and the XY-plane, and in the first octant.

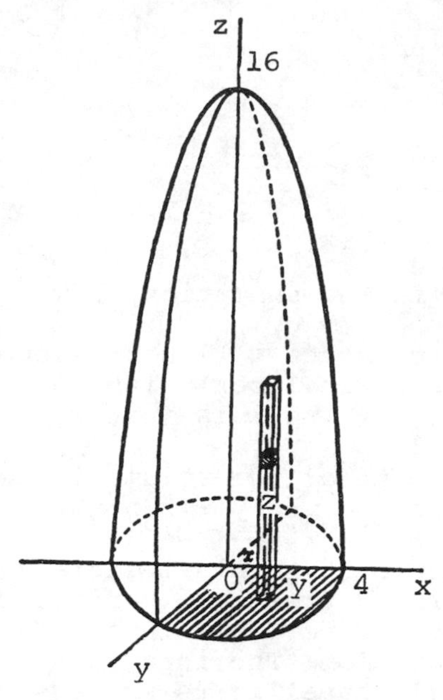

Solution: We integrate first with respect to z. The integration extends from the XY-plane to the surface of the paraboloid: $z_1 = 0$, and $z_2 = 16 - x^2 - 4y^2$. Next, we integrate across the quarter ellipse in the XY-plane, with respect to y, from y = 0 to $y = \frac{1}{2}\sqrt{16 - x^2}$, since the equation of the ellipse in the XY-plane is $x^2 + 4y^2 = 16$. Finally, we integrate with respect to x from x = 0 to x = 4. The required iterated integral is:

$$\int_0^4 \int_0^{\frac{1}{2}\sqrt{16-x^2}} \int_0^{(16-x^2-4y^2)} f(x,y,z)\ dz\ dy\ dx.$$

• **PROBLEM 936**

Evaluate the iterated triple integral:

$$\int_0^1 \int_{x^2}^1 \int_0^{1-y} x\ dz\ dy\ dx.$$

Solution: In general, the triple integral over a volume enclosed by surface Q is defined to be:

$$\int_Q f(x, y, z)\ dV = \lim_{n\to\infty} \sum_{i=1}^{n^3} f(x_i, y_i, z_i)\ \Delta_n V,$$

which is evaluated in terms of a thrice-iterated integral, or:

$$\int_Q f(x,\ y,\ z)\ dV$$

$$= \int_a^b \int_{g_1(x)}^{g_2(x)} \int_{f_1(x,y)}^{f_2(x,y)} f(x,y,z)\ dz\ dy\ dx, \qquad (a)$$

where the point in the summation, $f(x_i, y_i, z_i)$, is assumed to be within the ith box (a cube, cylindrical or spherical box) with volume ΔV. In this problem, $f(x_i, y_i, z_i)$ is just x_i, which might mean that the moment of the volume is being calculated with respect to the y - z plane.

To integrate such triple integrals, we follow the order and treat all variables as constants except the variable involved in that particular integration. The integrated variable should necessarily have limits expressed either as constants, or in terms of the other variables, which are treated as constants. Thus the interdependence of all the variables is preserved in the evaluation.

We apply the general statement about triple integrals to the integration of the given expression. The given integral is:

$$\int_0^1 \int_{x^2}^1 \int_0^{1-y} x\ dz\ dy\ dx,$$

where, according to (a),

$$f(x,y,z) = x,\ f_1(x,y) = 0,\ f_2(x,y) = 1 - y$$

$$g_1(x) = x^2, \qquad g_2(x) = 1,$$

and the constant boundaries are $a = 0$ and $b = 1$. Hence, the integral can be expressed as:

$$\int_0^1 \int_{x^2}^1 \int_0^{1-y} x\ dz\ dy\ dx$$

$$= \int_0^1 x \left[\int_{x^2}^1 \left[\int_0^{1-y} dz \right] dy \right] dx$$

First,

$$\int_0^{1-y} dz = z \Big|_{z=0}^{z=1-y} = 1 - y - 0 = 1 - y.$$

The given integral reduces to:

$$\int_0^1 x \left[\int_{x^2}^1 (1 - y)\, dy \right] dx.$$

$$\int_{x^2}^1 (1 - y)\, dy = \left(y - \frac{1}{2} y^2 \right) \Big|_{y=x^2}^{y=1}$$

$$= 1 - \frac{1}{2} - \left(x^2 - \frac{1}{2} x^4 \right)$$

$$= \frac{1}{2} - x^2 + \frac{1}{2} x^4.$$

By substituting this in the integral, we obtain:

$$\int_0^1 x \left(\frac{1}{2} - x^2 + \frac{1}{2} x^4 \right) dx$$

$$= \int_0^1 \left(\frac{1}{2} x - x^3 + \frac{1}{2} x^5 \right) dx$$

$$= \frac{1}{4} x^2 - \frac{1}{4} x^4 + \frac{1}{12} x^6 \Big|_0^1$$

$$= \frac{1}{12}.$$

• PROBLEM 937

Find the volume bounded by the surfaces:

$$x^2 + y^2 = z, \quad \text{and } x^2 + y^2 = 4$$

and **the XY-plane.**

Solution: The surface: $x^2 + y^2 = z$, is a paraboloid passing through the origin and symmetric about the Z-axis, while the surface: $x^2 + y^2 = 4$, is a cylinder about the Z-axis, as shown. Using the general definition of triple integral for volumes, we have:

$$V = \int_S dV = \int_a^b \int_{g_1(x)}^{g_2(x)} \int_{f_1(x,y)}^{f_2(x,y)} dz\, dy\, dx,$$

where the symbol $\int_S dV$ means integration over the given surface S. We can also express this in the following form

$$V = \int_R f(x,y)\, dA,$$

which means integration over the region R projected on the x-y plane. This is an alternative form of the triple integral. Now, $f(x,y) = z = x^2 + y^2$, y varies in the interior of a circle $x^2 + y^2 = 4$, or $-\sqrt{4 - x^2} \leq y \leq \sqrt{4 - x^2}$. $x = -2$ is the lower limit, and $x = 2$ is the upper limit.

By symmetry, the required volume is 4 times the volume in the first octant. Hence,

$$V = 4 \int_0^2 \left[\int_0^{\sqrt{4-x^2}} (x^2 + y^2)\, dy \right] dx$$

$$= 4 \int_0^2 \left[x^2\sqrt{4-x^2} + \frac{(4-x^2)^{3/2}}{3} \right] dx$$

$$= 4 \left[4 \text{ arc sin } \frac{x}{2} + \sqrt{4-x^2}\ (x) \left(\frac{x^2}{6} + \frac{1}{3} \right) \right|_0^2$$

$$= 4 \left(\frac{4\pi}{2} + 0 - 0 - 0 \right)$$

$$= 8\pi.$$

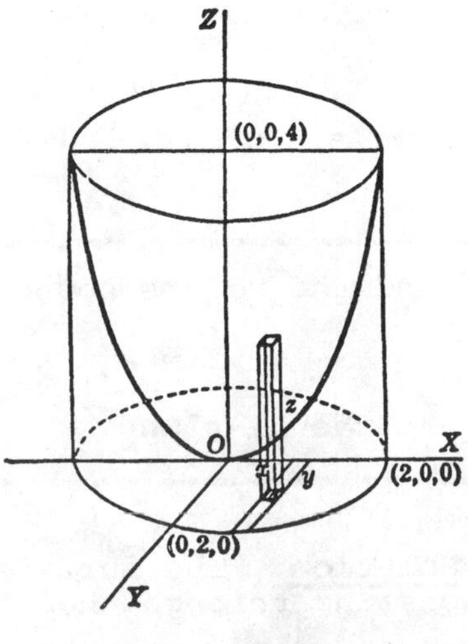

Find the volume common to the two cylinders

$$x^2 + y^2 = a^2, \quad y^2 + z^2 = a^2.$$

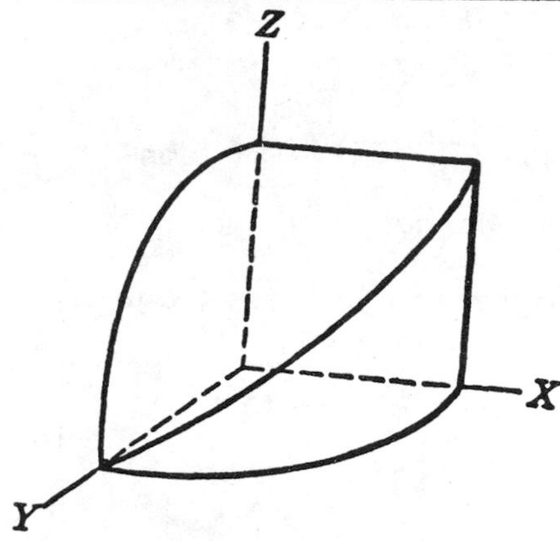

Solution: For reasons of symmetry, we work with the part of the volume lying in the first octant. Since the curve of intersection lies on the cylinders, it projects into

$$x^2 + y^2 = a^2$$

in the XY-plane.

$$V = 8\int_0^a \int_0^{\sqrt{a^2-x^2}} \int_0^{\sqrt{a^2-y^2}} dz\, dy\, dx$$

$$= 8\int_0^a \int_0^{\sqrt{a^2-x^2}} \sqrt{a^2 - y^2}\, dy\, dx.$$

Using integration by parts,

$$V = 8\int_0^a \frac{1}{2}\left(y\sqrt{a^2 - y^2} + a^2 \sin^{-1}\frac{y}{a}\right\vert_{y=0}^{y=\sqrt{a^2-x^2}} dx$$

$$= 4\int_0^a \left(x\sqrt{a^2 - x^2} + a^2 \sin^{-1}\frac{\sqrt{a^2 - x^2}}{a}\right) dx.$$

Since $\sin^{-1}\dfrac{\sqrt{a^2 - x^2}}{a}$ represents a certain angle, we can construct a right-angled triangle with dimensions such that, for the angle Θ, $\sin\Theta = \dfrac{\sqrt{a^2 - x^2}}{a}$. Thus, the side opposite to the angle Θ is $\sqrt{a^2 - x^2}$, the hypotenuse is a units, and by the Pythagorean theorem, the adjacent side is x. Thus, Θ can also be represented as

$$\Theta = \cos^{-1} \frac{x}{a} .$$

The integral then becomes:

$$V = 4 \int_0^a \left(x \sqrt{a^2 - x^2} + a^2 \cos^{-1} \frac{x}{a} \right) dx.$$

Using integration by parts again,

$$\int_0^a x \sqrt{a^2 - x^2}\, dx = - \frac{1}{3} (a^2 - x^2)^{\frac{3}{2}} \Big|_0^a$$

$$= \frac{1}{3} a^3 .$$

For

$$a^2 \int_0^a \cos^{-1} \frac{x}{a}\, dx, \text{ we let } \frac{x}{a} = u.$$

To find the lower and upper limits, for x = 0, u = 0, and for x = a, u = 1. Hence,

$$a^2 \int_0^a \cos^{-1} \frac{x}{a}\, dx$$

$$= a^2 \int_0^1 \cos^{-1} u \; (a\, du),$$

or

$$a^3 \int_0^1 \cos^{-1} u\, du$$

$$= a^3 \left(u \cos^{-1} u - \sqrt{1 - u^2} \right|_{u=0}^{u=1} = a^3 .$$

Thus,

$$V = 4 \int_0^a \left(x \sqrt{a^2 - x^2} + a^2 \sin^{-1} \frac{\sqrt{a^2 - x^2}}{a} \right) dx$$

$$= 4\left(\frac{1}{3}a^3 + a^3\right) = \frac{16}{3}a^3.$$

The calculation can be simplified if we interchange the order of integration in such a way that the intergrand can be treated as a constant. Recalling

$$V = 8\int_0^a\int_0^{\sqrt{a^2-x^2}} \sqrt{a^2 - y^2}\, dy\, dx,$$

and keeping in mind that the variation of x can be expressed in terms of y, we introduce the new limits

$$0 \le x \le \sqrt{a^2 - y^2},\ 0 \le y \le 1$$

Hence,

$$V = 8\int_0^a\int_0^{\sqrt{a^2-x^2}} \sqrt{a^2 - y^2}\, dx\, dy$$

$$= 8\int_0^a \left[x\sqrt{a^2 - y^2}\right]^{\sqrt{a^2-y^2}} dy$$

$$= 8\int_0^a (a^2 - y^2)\, dy$$

$$= \frac{16}{3}a^3.$$

• PROBLEM 939

Find the volume cut from the cone: $x^2 + y^2 - z^2 = 0$, by the sphere: $x^2 + y^2 + (z - 2)^2 = 4$.

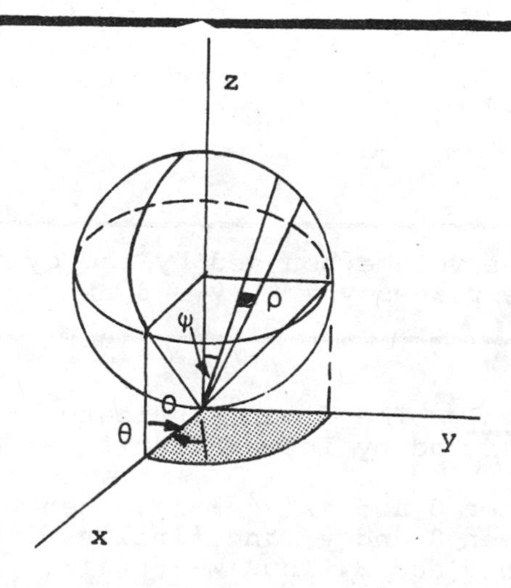

Solution: The calculation can be straightforward if we reformulate the equations of the cone and the sphere in the spherical coordinate system, **based on** the transformation

$$x = \rho \sin \phi \cos \Theta$$

$$y = \rho \sin \phi \sin \Theta$$

$$z = \rho \cos \phi.$$

The equation of the sphere is:
$\rho = 4 \cos \phi,$

and that of the cone is:

$\phi = 1/4\ \pi.$

Due to symmetry, we find the volume in the first octant, and the required volume (over the xy-plane) is four times this. Hence

$$V = 4 \int_0^{\pi/2} \int_0^{\pi/4} \int_0^{4 \cos \phi} \rho^2 \sin \phi \, d\rho \, d\phi \, d\Theta$$

$$= \frac{256}{3} \int_0^{\pi/2} \int_0^{\pi/4} \cos^3 \phi \sin \phi \, d\phi \, d\Theta$$

$$= \frac{16}{3} \int_0^{\pi/2} d\theta$$

$$= \frac{8\pi}{3} .$$

• PROBLEM 940

Find the volume bounded by the cylinder $z = 4/(y^2+1)$ and the planes $y = x$, $y = 3$, $x = 0$, and $z = 0$.

Solution: The figure indicates that the volume may be found by integrating first with respect to

z between 0 and $4/(y^2 + 1)$, then with respect to x between 0 and y, and finally with respect to y, between 0 and 3. Thus we **obtain:**

$$V = \int_0^3 \int_0^y \int_0^{4/(y^2+1)} dz\, dx\, dy$$

$$= \int_0^3 \int_0^y \frac{4}{y^2 + 1} dx\, dy$$

$$= \int_0^3 \frac{4y}{y^2 + 1} dy = 2 \left[\ln (y^2 + 1) \right|_0^3$$

$$= 2 (\ln 10 - \ln 1) = 2 \ln 10.$$

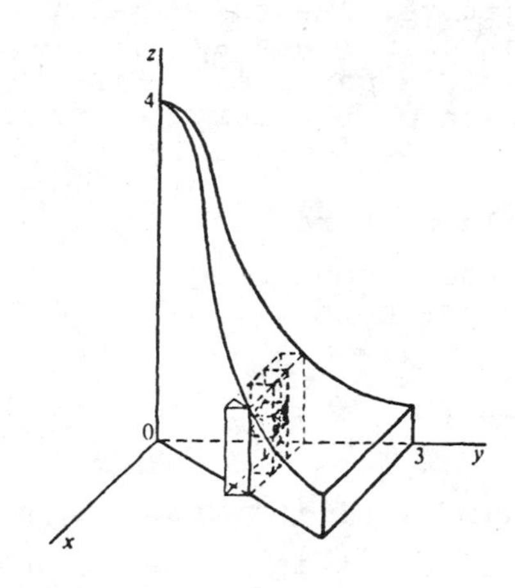

• PROBLEM 941

Find the volume bounded by the sphere: $x^2 + y^2 + z^2 = 9$ and the paraboloid: $x^2 + y^2 = 8x$.

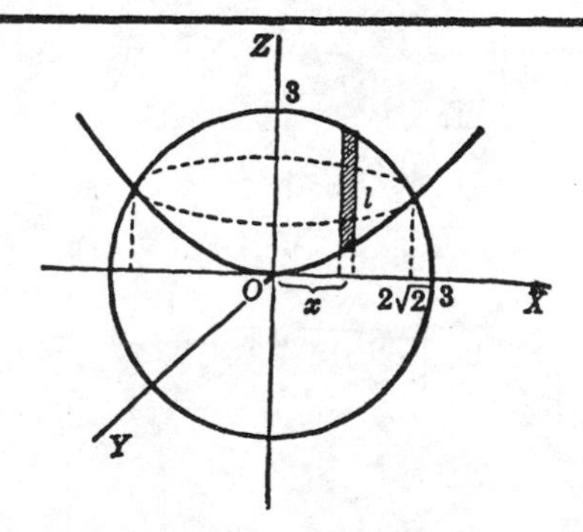

Solution: Using cylindrical coordinates (r, Θ, z), the equations of the sphere and the paraboloid become: $r^2 + z^2 = 9$, and $r^2 = 8z$. Their respective surface equations are:

$$f_s(r, \Theta, z) = r^2 + z^2 - 9,$$

and

$$f_\rho(r,\Theta,z) = r^2 - 8z.$$

To determine the limits of integration, we find the surface at which both solids intersect. This is carried out by **equating the surfaces:**

$$f_s(r,\Theta,z) = f_\rho(r,\Theta,z), \quad r^2 + z^2 - 9$$

$$= r^2 - 8z,$$

or, $(z - 1)(z + 9) = 0.$

The points of intersection are at $z = 1$, $z = -9$. The latter is meaningless because r^2 can never be negative. Due to symmetry of both solids over the positive z-axis, the plane of intersection, $z = 1$, is parallel to the x-y plane, and the equation of the curve of intersection is:

$r^2 = 8$, a circle with radius $r = \sqrt{8} = 2\sqrt{2}$. This is obtained from the equation of the paraboloid. Under these conditions, z is within the surfaces,

$$\frac{r^2}{8} \le z \le \sqrt{9 - r^2}.$$

r is inside the cylinder of intersection, varying from zero to $2\sqrt{2}$. Since it is symmetric for positive z, the volume required is four times the volume calculated in the first octant, thus making $0 \le \Theta \le \pi/2$. Hence, integration yields:

$$V = 4\int_0^{\frac{\pi}{2}}\int_0^{2\sqrt{2}}\int_{\frac{r^2}{8}}^{\sqrt{9-r^2}} r\,dz\,dr\,d\Theta$$

$$= 4\int_0^{\frac{\pi}{2}}\int_0^{2\sqrt{2}} r\left(\sqrt{9 - r^2} - \frac{1}{8}r^2\right) dr\,d\Theta$$

$$= 4\int_0^{\frac{\pi}{2}} \left(-\frac{1}{3}(9 - r^2)^{\frac{3}{2}} - \frac{1}{32}r^4\right|_0^{2\sqrt{2}} d\Theta$$

$$= 4\int_0^{\frac{\pi}{2}} \frac{20}{3}\,d\Theta$$

$$= \frac{40}{3} \pi.$$

Alternatively, using the figure, the length of a rectangular element extending between the parabola and the circle is:

$$l = \sqrt{9 - x^2} - \frac{1}{8} x^2,$$

so that:

$$V = 2 \pi \int_0^{2\sqrt{2}} x \left(\sqrt{9 - x^2} - \frac{1}{8} x^2 \right) dx$$

$$= \frac{40}{3} \pi.$$

• PROBLEM 942

Find the volume cut from the elliptic paraboloid $z = x^2 + 4y^2$ by the plane $z = 1$.

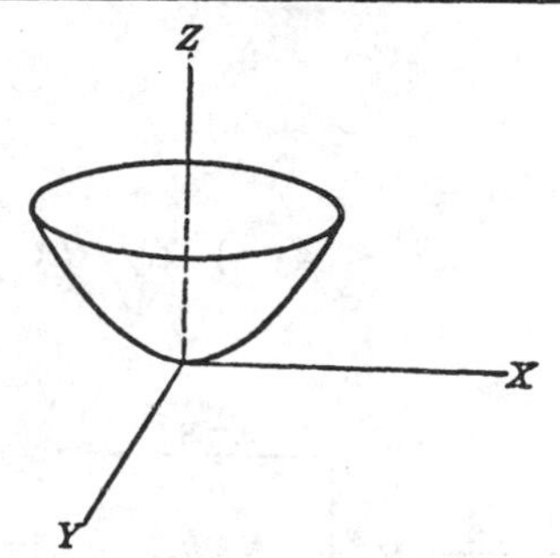

Solution: We wish to express the volume in terms of a generalized triple integral:

$$V = \int_V dV = \int_a^b \int_{y_1(x)}^{y_2(x)} \int_{z_1(x,y)}^{z_2(x,y)} dz\, dy\, dx.$$

According to the above expression, the integration starts with z, where the integration extends from the surface of the elliptic paraboloid to the plane:

$z = 1$. $f(x,y) \le z \le 1$. **Hence,** $x^2 + 4y^2 \le z \le 1$,

and thus,

$$z_1(x,y) = x^2 + 4y^2, \text{ and } z_2(x,y) = 1.$$

We next ingetrate across the ellipse

$x^2 + 4y^2 = 1$ for $z = 1$ on the x-y plane with respect to y. Thus,

$$-\frac{\sqrt{1 - x^2}}{2} \le y \le \frac{\sqrt{1 - x^2}}{2} .$$

Due to symmetry about the x-axis (inferred from $- y_1(x) \le y \le + y_1(x)$ we let y start from

zero to the point on the ellipse and multiply the integral by two. Thus we have

$$y_1(x) = 0 \quad \text{and} \quad y_2(x) = \frac{1}{2}\sqrt{1 - x^2}.$$

Finally, we integrate over the major axis, where $- 1 \le x \le 1$. For reasons of symmetry, we let $0 \le x \le 1$ and multiply by two again. In the integral, $a = 0$, $b = 1$. Hence,

$$V = 4 \int_0^1 \int_0^{\frac{\sqrt{1-x^2}}{2}} \int_{x^2+4y^2}^{1} dz\, dy\, dx$$

$$= 4 \int_0^1 \int_0^{\frac{\sqrt{1-x^2}}{2}} (1 - x^2 - 4y^2)\, dy\, dx$$

$$= 4 \int_0^1 \left(y(1 - x^2) - \frac{4}{3} y^3 \right)\Bigg|_0^{\frac{\sqrt{1-x^2}}{2}} dx$$

$$= 4 \int_0^1 \frac{1}{3} \left(1 - x^2\right)^{\frac{3}{2}} dx$$

$$= \frac{1}{3} \left(x\left(1 - x^2\right)^{\frac{3}{2}} + \frac{3}{2} x \left(1 - x^2\right)^{\frac{1}{2}} \right.$$

$$\left. + \frac{3}{2} \sin^{-1} x \right)\Bigg|_0^1$$

$$= \frac{\pi}{4} \text{ cu. units.}$$

Find the volume of a sphere of radius a.

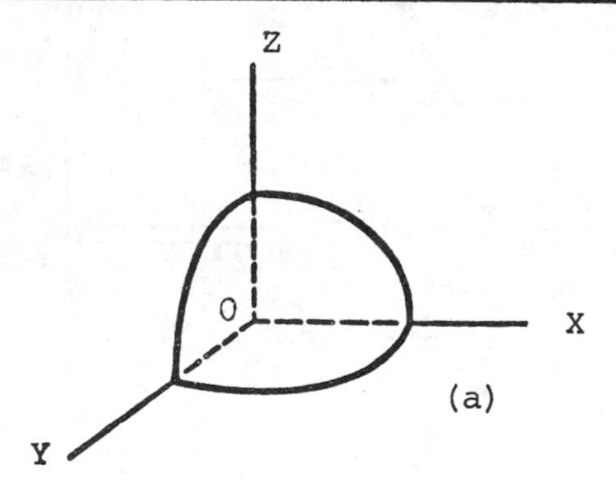

(a)

Solution: I. Rectangular coordinates: Assuming a small cube constructed by lines parallel to the x-, y- and z-axes, with dimensions: Δx, Δy and Δz, we obtain its volume:

$$\Delta V = \Delta x \ \Delta y \ \Delta z.$$

Passing to the limits, $dV = dz\ dy\ dx$. Integration, using the proper limits, gives an integral of the form:

$$V = \int_a^b \int_{g_1(x)}^{g_2(x)} \int_{f_1(x,y)}^{f_2(x,y)} dz\ dy\ dx.$$

This is the volume of an object of any shape. a and b are constants. In particular, for reasons of symmetry for the given sphere, the equation of the sphere is

$$x^2 + y^2 + z^2 = a^2.$$

We work only with the part lying in the first octant.

$$V = 8 \int_0^a \int_0^{\sqrt{a^2-x^2}} \int_0^{\sqrt{a^2-x^2-y^2}} dz\ dy\ dx.$$

In the first integration, z sweeps from surface $(z = 0)$ to surface

$z = \sqrt{a^2 - x^2 - y^2}$. In the second integration

y varies from curve $(y = 0)$ to curve $y = \sqrt{a^2 - x^2}$. This last is obtained by putting $z = 0$ in the equation of the surface, since this gives the curve of intersection of the surface with the XY-plane. In the final integral x goes from 0 to a, the extent of the figure in the X-direction.

$$V = 8 \int_0^a \int_0^{\sqrt{a^2-x^2}} \sqrt{a^2 - x^2 - y^2}\, dy\, dx$$

$$= 8 \int_0^a \left[\frac{y}{2} \sqrt{a^2 - x^2 - y^2} + \frac{a^2 - x^2}{2} \sin^{-1} \frac{y}{\sqrt{a^2 - x^2}} \right]_0^{\sqrt{a^2-x^2}} dx$$

$$= 2\pi \int_0^a (a^2 - x^2)\, dx$$

$$= 2\pi \left[a^2 x - \frac{x^3}{3} \right]_0^a$$

$$= \frac{4}{3} \pi a^3 \text{ cu. units.}$$

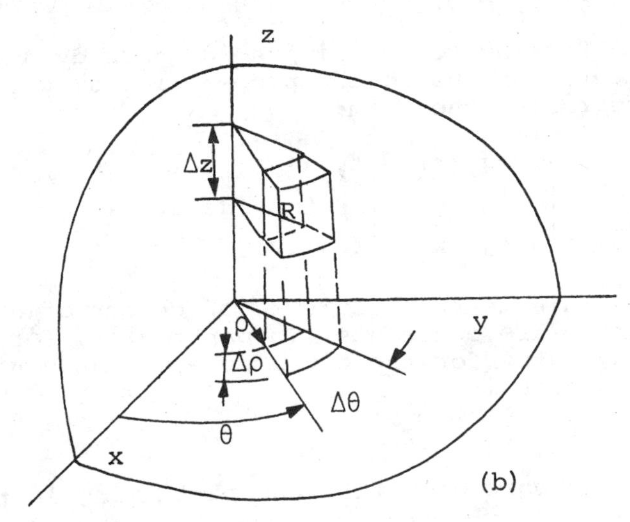

II. Cylindrical coordinates. Recalling the equation of the sphere,

$$x^2 + y^2 + z^2 = a^2,$$

we substitute ρ^2 for $x^2 + y^2$ (the equation of a cylinder of base radius ρ) in the Cartesian equation of the sphere. Hence, the corresponding cylindrical equation of the sphere is:

$$\rho^2 + z^2 = a^2.$$

We also explicitly express x and y in terms of (ρ, Θ, z), as seen in figure (b) which shows the relative size and location of ρ on a plane parallel to, and z-units away from the x-y plane:

$$x = \rho \cos \Theta,$$

$$y = \rho \sin \Theta,$$

and $z = z.$

Considering the right circular cylindrical solid in the figure,

$$R = \left\{ (\rho, \Theta, z) : -a \le \rho \le a,\ 0 \le \Theta \le 2\pi, -\sqrt{a^2 - \rho^2} \le z \le \sqrt{a^2 - \rho^2} \right\}.$$

Its volume is given by

$$dV = (\rho\, d\Theta)(d\rho)(dz)$$

Hence, the volume of the sphere,

$$V = \int_0^{2\pi} \left[\int_{-a}^{a} \left(\int_{-\sqrt{a^2-\rho^2}}^{\sqrt{a^2-\rho^2}} \rho\, dz \right) d\rho \right] d\Theta.$$

Due to symmetry about the axes, we can obtain the volume in the first octant and multiply it by 8. The integral then becomes:

$$V = 8 \int_0^{\frac{\pi}{2}} \left[\int_0^{a} \rho \left(\int_0^{\sqrt{a^2-\rho^2}} dz \right) d\rho \right] d\Theta$$

$$= 8 \int_0^{\frac{\pi}{2}} \left[\int_0^{a} \rho \sqrt{a^2 - \rho^2}\, d\rho \right] d\Theta$$

$$= 8 \int_0^{\frac{\pi}{2}} \left(-\frac{1}{3}(a^2 - \rho^2)^{\frac{3}{2}} \right) \Bigg|_{\rho=0}^{\rho=a} d\Theta$$

$$= \frac{8}{3} \int_0^{\frac{\pi}{2}} a^3\, d\Theta = \frac{8}{3} a^3 \frac{\pi}{2}$$

$$= \frac{4}{3} \pi a^3 \text{ cu. units.}$$

As in Cartesian coordinates, the integral was evaluated in the following manner:

$$V = \int_a^b \int_{g_1(\Theta)}^{g_2(\Theta)} \int_{f_1(\rho,\Theta)}^{f_2(\rho,\Theta)} \rho \, dz \, d\rho \, d\Theta,$$

where a,b, $g_1(\Theta)$ and $g_2(\Theta)$ are constants, and $f_1(\rho,\Theta)$ and $f_2(\rho,\Theta)$ are respectively expressed as $-\sqrt{a^2 - \rho^2}$ and $\sqrt{a^2 - \rho^2}$.

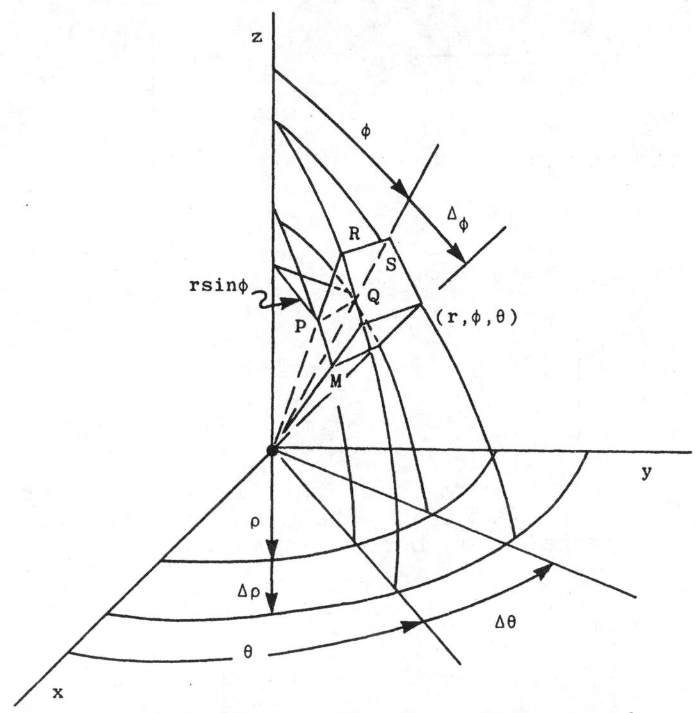

III. Spherical coordinates. The equation of the sphere is r = a, as shown in figure (c).

From the rotation r = a, and the expression:

$$r^2 = x^2 + y^2 + z^2,$$

we infer that the three component variables determine the relative size and position of the vector r to locate a point which, in the Cartesian coordinate system, is (x, y, z). We introduce angles ϕ and Θ, such that (r, ϕ, Θ) locates the same point. In this case,we express the following relationships:

$$\rho = r \sin \phi, \quad x = \rho \cos \Theta = r \sin \phi \cos \Theta$$

$$z = r \cos \phi, \quad y = \rho \sin \Theta = r \sin \phi \sin \Theta$$

$$\Theta = \Theta \quad , \quad z = z \quad = r \cos \phi .$$

Hence,

$$(x, y, z) = (r \sin \phi \cos \Theta,\ r \sin \phi \sin \Theta,\ r \cos \phi).$$

Considering the diagram, we compute that a small area of region S, with dimensions

$$\overline{PR} = \Delta r, \quad \overline{PQ} = r \sin \phi\ \Delta\Theta \quad \text{and } \overline{PM} = r\ \Delta\phi,$$

has a volume given approximately by:

$$\Delta V \simeq (\Delta r)(r \sin \phi\ \Delta\Theta)(r\ \Delta\phi).$$

Passing to the limits,

$$dV = r^2 \sin \phi\ dr\ d\phi\ d\Theta.$$

In terms of triple integrals,

$$V = \int_S dV,$$

that is:

$$V = 8 \int_0^{\frac{\pi}{2}} \int_0^{\frac{\pi}{2}} \int_0^a r^2 \sin \phi\ dr\ d\Theta\ d\phi$$

$$= \frac{8}{3} a^3 \int_0^{\frac{\pi}{2}} \int_0^{\frac{\pi}{2}} \sin \phi\ d\Theta\ d\phi$$

$$= \frac{4}{3} \pi a^3 \int_0^{\frac{\pi}{2}} \sin \phi\ d\phi$$

$$= \frac{4}{3} \pi a^3 \text{ cu. units.}$$

IV. Volume Generation. Recognizing the symmetry of a figure with respect to an axis, the volume generated by revolving this figure about the axis of symmetry gives the volume of the solid required. For example, a circle, radius a, with center located at (b,0) where $b > a$, when revolved about the y-axis gives a torus. A rectangle, r by h, bounded by the x- and y-axes and the lines $x = r$ and $y = h$ generates a cylinder of base radius r and height h when rotated about the y-axis. Similarly, we can think of the sphere, radius a, as generated by rotating a circle of radius a, and centered at the origin, about either axis. But in this case, to find the volume, we use the following two methods:

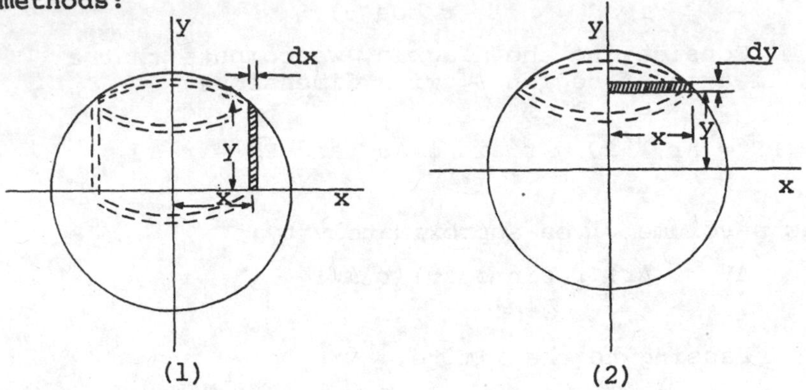

1. The shell method. If a strip of length y and width dx, parallel and x units away from the y-axis, is revolved about the y-axis, it generates a cylindrical shell of base radius x, height y and thickness dx, the surface area of which is: $2\pi x(y)$. The volume, fig. (d), is:

$$dV = 2\pi xy(dx).$$

Now, we can think the hemisphere above the x-axis as comprising concentric shells, where the sum of the volumes of all such shells, multiplied by two, gives the volume of the sphere. The sum leads to the integration:

$$V = 2\int_0^a 2\,\pi\,xy\,dx$$

$$= 4\pi\int_0^a x\sqrt{a^2 - x^2}\;dx,$$

since x and y are related to each other by the equation:

$$x^2 + y^2 = a^2$$

for the circle,

$$V = 4\pi\left\{-\frac{1}{3}(a^2 - x^2)^{\frac{3}{2}}\right|_0^a$$

$$= \frac{4}{3}\pi a^3.$$

2. The disk method. Similarly, using the shaded strip in fig. (e), rotation of the strip about the y-axis gives the disk with cross-sectional area πx^2 or a volume:

$$dV = \pi x^2 \ dy.$$

Then, with similar reasoning as for the shell method,

$$V = 2 \int_0^a \pi x^2 \ dy = 2\pi \int_0^a (a^2 - y^2) \ dy$$

$$= 2\pi \left(a^2 y - \frac{1}{3} y^3 \right|_0^a$$

$$= \frac{4}{3} \pi a^3.$$

• PROBLEM 944

Find the moment of inertia of a cube of edge a with respect to one of three given forces, if the density varies as the sum of the distances from three adjacent faces.

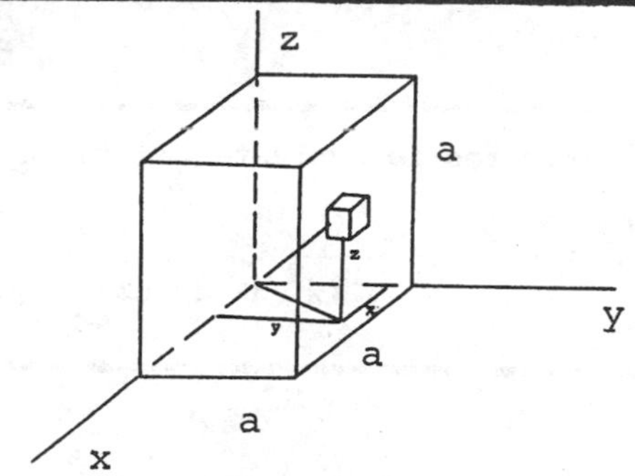

<u>Solution:</u> The density of the element cube at P(x, y, z)

$$\rho = k \ (x + y + z)$$

where k is the constant of proportionality. The moment of inertia of this element of mass,

$$dm = k \ (x + y + z) dv,$$

about the xy-plane is given by:

$$\Delta I_{xy} = z^2 \ k \ (x + y + z) \ dv.$$

Integration over the volume of the cube yields:

$$I_{xy} = k \int_0^a \int_0^a \int_0^a z^2 \ (x + y + z) \ dz \ dy \ dx$$

$$= k \int_0^a \int_0^a \frac{1}{3} z^3 x + \frac{1}{3} z^3 y$$

$$+ \frac{1}{4} z^4 \Bigg|_{z=0}^{z=a} dy\, dx$$

$$= k \int_0^a \int_0^a \left(\frac{1}{3} a^3 (x + y) + \frac{1}{4} a^4\right) dy\, dx$$

$$= k \int_0^a \frac{1}{3} a^3 \left(xy + \frac{1}{2} y^2\right) + \frac{1}{4} a^4 y \Bigg|_{y=0}^{y=a} dx$$

$$= k \int_0^a \left(\frac{1}{3} a^4 \left(x + \frac{1}{2} a\right) + \frac{1}{4} a^5\right) dx$$

$$= k \left[\frac{1}{3} a^4 \left(\frac{1}{2} x^2 + \frac{1}{2} ax\right) + \frac{1}{4} a^5 x\right]_0^a$$

$$= \frac{7}{12} a^6 k.$$

• PROBLEM 945

Use spherical coordinates to evaluate the integral:

$$\int_0^a \int_0^{\sqrt{a^2-z^2}} \int_0^{\sqrt{a^2-y^2-z^2}} \sqrt{x^2+y^2+z^2}\, dx\, dy\, dz.$$

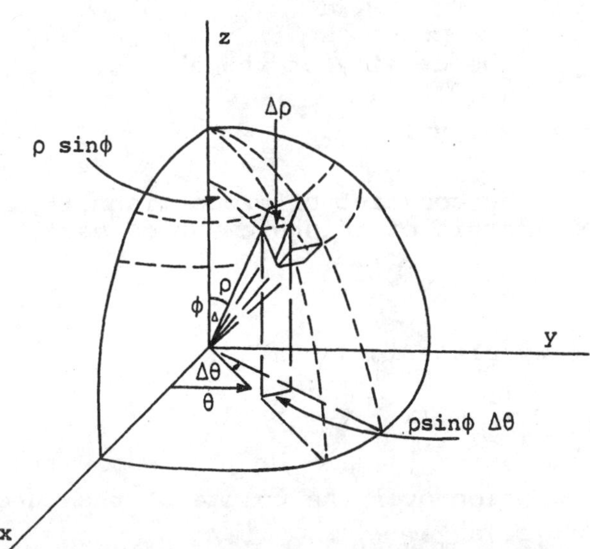

Solution: From the limits on the integral, it is seen that the region of integration is the portion of the first octant that is bounded by the coordinate planes and the sphere $x^2+y^2+z^2 = a^2$, as shown in the diagram. For spherical coordinates,

$$x^2 + y^2 + z^2 = \rho^2.$$

$$x = \rho \cos \varphi \cos \theta.$$

$$y = \rho \cos \varphi \sin \theta.$$

$$z = \rho \sin \varphi .$$

The given integral obtains volume by summing infinitesimal cubes dxdydz. The sides of these cubes in polar coordinates are as follows:

one side is $d\rho$,

" " " $\rho\, d\theta$,

" " " $\rho \sin \varphi\, d\varphi$.

The cube, therefore, is $\rho^2 \sin \varphi\, d\rho\, d\varphi\, d\theta$. The quantity $\sqrt{x^2+y^2+z^2} = \rho$. This gives the integrand: $\rho^3 \sin \varphi\, d\rho\, d\varphi\, d\theta$. The limits are obtained as follows: When

$$x = \sqrt{a^2-y^2-z^2}\ , \quad \rho = a .$$

$$y = \sqrt{a^2-z^2}\ , \quad \theta = \pi/2 .$$

$$z = a, \quad \varphi = \pi/2 .$$

Therefore, the given integral is equivalent to:

$$\int_0^{\pi/2} \int_0^{\pi/2} \int_0^{a} \rho^3 \sin \varphi\, d\rho\, d\varphi\, d\theta = \int_0^{\pi/2} \int_0^{\pi/2} \frac{\rho^4}{4} \sin \varphi\, d\varphi\, d\theta \Big]_0^a$$

$$= \int_0^{\pi/2} \frac{a^4}{4} \left(-\cos \varphi\right)\Big]_0^{\pi/2} d\theta = \frac{a^4}{4}\ \theta\Big]_0^{\pi/2}$$

$$= \frac{\pi}{2}\ \frac{a^4}{4} = \frac{\pi\ a^4}{8} .$$

CHAPTER 36

MASSES OF VARIABLE DENSITY

With the basic expression:

$m = Dv$, where D is the density, v is the volume, and m is the mass, we can deal with problems where the density is variable. In such problems we cannot find the total mass by simply multiplying total volume by the density. However, we can write the expression:

$dm = D\,dv$, where dm is an element of mass, and dv is an element of volume. In a Cartesian coordinate system,

$$dv = dx\,dy\,dz\ .$$

The density, D, has to be expressed as some function of the variables: x, y, and z.

We can now write the expression for the mass as:

$$m = \iiint D(x,y,z)dx\,dy\,dz\ .$$

The triple integration is performed in the usual manner by integrating with respect to each of the variables, left to right, in turn, each time considering the other variables as constants.

• PROBLEM 946

Find the mass of a sphere of radius a if the density at every point is proportional to the distance from the center.

Solution: Let dV denote an elemental volume in the sphere at a distance r from the center. Then, using spherical coordinates, its dimensions are: $r \sin\phi\, d\Theta$, $rd\phi$ and dr, where Θ and ϕ are the angles made by the line r with the positive x- and z-axes, respectively. Hence, dV is approximated by:

$$dV = r^2 \sin\phi\, dr\, d\phi\, d\Theta.$$

Since the density of this volume is proportional to its distance from the center, and introducing a constant of proportionality k, we obtain the density:

$$\rho = kr.$$

The mass is given by:

$$dm = \rho\, dV = krdV.$$

Therefore, the mass of the sphere is:

$$m = k \int_V r dV,$$

or,

$$m = k \int_0^{2\pi} \int_0^{\pi} \int_0^{a} r \cdot r^2 \sin \phi \, dr \, d\phi \, d\Theta$$

$$= k \, \pi a^4.$$

• PROBLEM 947

Find the mass of the body in the form of a tetrahedron cut from the first octant by the plane $x + y + z = 1$, if the density varies as the product of the distances from the three coordinate planes.

Solution: Let ΔV be the volume element, and let (x, y, z) be the coordinates of any point in this element. Then the density of the element is $kxyz$ (k constant), the mass of the element is $kxyz \, \Delta V$, and the required mass of the body is given by

$$m = \iiint_V kxyz \, dV$$

$$= \int_0^1 \int_0^{1-x} \int_0^{1-x-y} kxyz \, dz \, dy \, dx$$

$$= k \int_0^1 \int_0^{1-x} xy \left(\frac{z^2}{2} \right|_0^{1-x-y} dy \, dx$$

$$= k \int_0^1 \int_0^{1-x} \frac{1}{2} \left(1-x-y-x+x^2+xy-y+xy+y^2 \right) \quad xy \, dy \, dx$$

$$= k \int_0^1 x \int_0^{1-x} \frac{1}{2} \left(y-2xy-2y^2+2xy^2+x^2y+y^3 \right) dy \, dx$$

$$= k \int_0^1 x \left(\frac{y^2}{4} - \frac{xy^2}{2} - \frac{y^3}{3} + \frac{xy^3}{3} \right.$$

$$\left. + \frac{x^2y^2}{4} + \frac{y^4}{8} \right|_0^{1-x} dx$$

$$= \frac{k}{24} \int_0^1 x \, (1 - x)^2 \Big[6 - 12x - 8 \, (1 - x)$$

$$+ 8x \, (1 - x) + 6x^2 + 3 \, (1 - x)^2 \Big) \, dx$$

$$= k \int_0^1 x \, \frac{(1 - x)^2}{24} \, (6 - 12x - 8 + 8x + 8x$$

$$- 8x^2 + 6x^2 + 3 - 6x + 3x^2)\, dx$$

$$= k \int_0^1 \frac{(1 - x)^2\, x}{24} (1 - 2x + x^2)\, dx$$

$$= \frac{k}{24} \int_0^1 (1 - 4x + 6x^2 - 4x^3 + x^4)\, x\, dx$$

$$= \frac{k}{24} \int_0^1 (x - 4x^2 + 6x^3 - 4x^4 + x^5)\, dx$$

$$= \frac{k}{24} \left(\frac{x^2}{2} - \frac{4x^3}{3} + \frac{6x^4}{4} - \frac{4x^5}{5} + \frac{x^6}{6} \right) \Bigg|_0^1$$

$$= \frac{k}{24} \left(\frac{1}{2} - \frac{4}{3} + \frac{3}{2} - \frac{4}{5} + \frac{1}{6} \right)$$

$$= \frac{k}{24} \left(\frac{15 - 40 + 45 - 24 + 5}{30} \right) = \frac{k}{720} .$$

• PROBLEM 948

Find the center of mass of one octant of an ellipsoid of revolution with major axis 2a and minor axis 2b, and the axis of revolution forming one bounding edge, if the density varies as the distance from the axis of revolution.

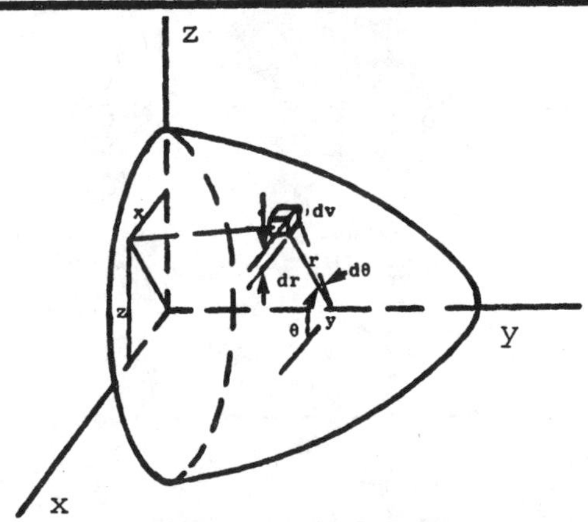

Solution: We assume that the ellipse is rotated about the y-axis. Then the bounding edge is the plane xz. Since the density is proportional to the distance from the y-axis, an element volume, r units away from the y-axis, has a density:

$$\rho\ (x,y,z) = cr,$$

where c is a proportionality constant.

Here, $r = \sqrt{x^2 + z^2}$,

as shown projected onto the x-z plane. We can see that, as a result of revolution,

$$\frac{x^2}{b^2} + \frac{y^2}{a^2} + \frac{z^2}{b^2} = 1,$$

the equation of the ellipsoid. In polar coordinates, as indicated in the diagram, the equation of the ellipsoid is:

$$\frac{r^2}{b^2} + \frac{y^2}{a^2} = 1,$$

where the area of one face of the element cube on the xz-plane is $dr(rd\Theta)$, and its volume is:

$$dV = r\ dr\ d\Theta\ dy.$$

For the first octant, $0 \le \Theta \le \frac{\pi}{2}$,

$0 \le r \le \frac{b}{a}\sqrt{a^2 - y^2}$ (the surface), and $0 \le y \le a$.

Now,

$$M\bar{y} = \int_V y\ \rho\ dV.$$

First, we find the mass. Since

$$dm = \rho\ dV = cr\ dV,$$

or, $$dm = cr^2\ dr\ d\Theta\ dy,$$

$$M = c\int_0^a \int_0^{\frac{\pi}{2}} \int_0^{\frac{b}{a}\sqrt{a^2-y^2}} r^2\ dr\ d\Theta\ dy$$

$$= c\int_0^a \int_0^{\frac{\pi}{2}} \frac{1}{3}\left(\frac{b}{a}\sqrt{a^2 - y^2}\right)^3 d\Theta\ dy$$

$$= c\ \frac{\pi}{6}\ \frac{b^3}{a^3}\int_0^a \left(a^2 - y^2\right)^{\frac{3}{2}} dy.$$

Since Θ does not depend on any other variable, we can just multiply the integrand by $\pi/2$. Furthermore, since the constant of proportionality c cancels out in the division for the center of mass, there is no need of using it in every step. Using integration by parts and rearranging factors:

$$\int_0^a \left(a^2-y^2\right)^{\frac{3}{2}} dy = \frac{1}{4}\left\{y\left(a^2 - y^2\right)^{\frac{3}{2}} + \frac{3a^2y}{2}\sqrt{a^2 - y^2} + \frac{3a^4}{2}\sin^{-1}\frac{y}{a}\right|_0^a$$

$$= \frac{3}{16}\pi a^4.$$

Thus the mass is proportional to $\frac{\pi^2}{32}$ ab^3.

$$M\bar{y} = \frac{\pi}{2}\int_0^a \int_0^{\frac{b}{a}\sqrt{a^2-y^2}} y\, r^2\, dr\, dy$$

$$= \frac{\pi}{2}\int_0^a \frac{1}{3}\, y \left(\frac{b}{a}\sqrt{a^2 - y^2}\right)^3 dy$$

$$= \frac{\pi}{6}\,\frac{b^3}{a^3}\int_0^a y\,(a^2 - y^2)^{\frac{3}{2}}\, dy,$$

where, upon substitution:

$$M\bar{y} = \frac{\pi}{6}\,\frac{b^3}{a^3}\left(-\frac{1}{5}\,(a^2 - y^2)^{\frac{5}{2}}\right|_0^a$$

$$= \frac{\pi}{30}\, a^2\, b^3.$$

Thus,

$$\bar{y} = \frac{\int_V y\rho\, dV}{\int_V \rho\, dV} = \frac{\frac{\pi}{30}\, a^2\, b^3}{\frac{\pi^2}{32}\, a\, b^3}$$

$$= \frac{16}{15}\,\frac{a}{\pi}.$$

• PROBLEM 949

Find the mass of a plate in the form of a cardioid, $r = a(1 + \cos\theta)$, if the density varies as the distance from the cusp.

Solution: A cusp is a double point at which a curve has a single tangent. Inspection of the shape of the cardioid shows that the cusp occurs at the origin. The density at a point p (x,y) in the given cardioid varies as the distance from the origin to the point. This can be written in a mathematical form as the density.

$$p = cr,$$

where c is a constant and r is the distance expressed as:

$$r = \sqrt{x^2 + y^2}.$$

The required mass is expressed by the integral

$$M = \int_0^{2\pi}\int_0^{a(1+\cos\theta)} (cr)\, r\,dr\,d\theta$$

$$= c\int_0^{2\pi} \frac{1}{3}\, r^3 \Bigg|_0^{a(1+\cos\theta)} d\theta$$

$$= \frac{c}{3} a^3 \int_0^{2\pi} (1 + 3 \cos \Theta + 3 \cos^2 \Theta + \cos^3 \Theta) d\Theta$$

$$= \frac{c}{3} a^3 \int_0^{2\pi} \left[1 + 3 \cos \Theta + \frac{3}{2} (1 + \cos 2\Theta) + \frac{1}{4} (\cos 3\Theta + 3 \cos \Theta) \right] d\Theta,$$

employing trigonometric identities.

The integral:

$$\int_0^{2\pi} \cos n\Theta d\Theta = \int_{-\pi}^{\pi} \cos n\Theta d\Theta$$

$$= \int_c^{c+2\pi} \cos n\Theta d\Theta = 0.$$

Hence, all the cosine terms vanish, and the integral becomes:

$$\frac{5}{2} \int_0^{2\pi} d\Theta.$$

$$M = \frac{c}{3} a^3 \frac{5}{2} \left(2\pi \right) = \frac{5}{3} \pi a^3 c.$$

• PROBLEM 950

Find the mass of a plate in the form of a right triangle with legs a and b, if the density is proportional to the square of the distance from the vertex of the right angle.

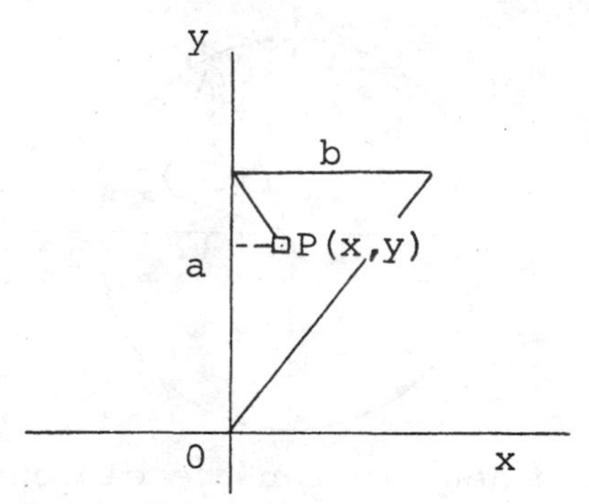

Solution: The density,

$$\rho = c\left(x^2 + (a - y)^2\right),$$

where c is a constant.

The total mass of the plate is given by:

$$M = c \int_0^b \int_{\frac{ax}{b}}^a \left(x^2 + (a - y)^2\right) dy\, dx,$$

where $y = \frac{ax}{b}$ is the equation of the diagonal line.

$$M = c \int_0^b \left. x^2y + a^2y - ay^2 + \frac{1}{3} y^3 \right|_{\frac{ax}{b}}^{a} dx$$

$$= c \int_0^b \left[x^2 \left(a - \frac{a}{b} x\right) + a^2 \left(a - \frac{a}{b} x\right) - a \left(a^2 - \left(\frac{a}{b} x\right)^2\right) + \frac{1}{3} \left(a^3 - \left(\frac{a}{b} x\right)^3\right)\right] dx$$

$$= c \left| \frac{1}{3} ax^3 - \frac{1}{4} \frac{a}{b} x^4 + a^3x - \frac{a^3}{2b} x^2 - a^3x + \frac{a^3}{3b^2} x^3 + \frac{1}{3} a^3x - \frac{1}{12} \left(\frac{a}{b}\right)^3 x^4 \right|_0^b$$

$$= \frac{abc}{12} \left(b^2 + a^2\right).$$

• PROBLEM 951

Find the mass of a circular plate of a radius a, if the density varies as the square of the distance from a point on the circumference.

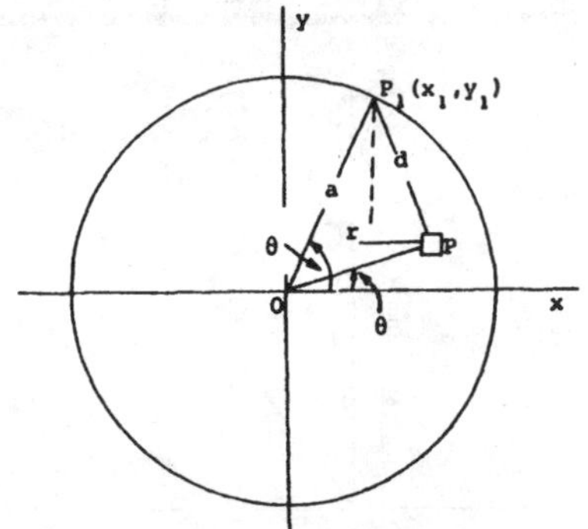

Solution: The density for the element at P(x,y) is

$$P = c\, d^2$$

where c is a constant, and

$$d^2 = (x - x_1)^2 + (y_1 - y)^2.$$

Using polar coordinates,

$$d^2 = (r \cos \Theta - a \cos \Theta_1)^2 + (a \sin \Theta_1 - r \sin \Theta)^2.$$

Expansion and regrouping yields:

$$d^2 = r^2 + a^2 - 2ar \cos (\Theta_1 - \Theta),$$

which corresponds to the law of cosines.

Therefore, the mass of the plate

$$M = c \int_0^{2\pi} \int_0^a \left(a^2 + r^2 - 2ar \cos (\Theta_1 - \Theta)\right) r \, dr \, d\Theta,$$

where $r \, dr d\Theta$ is an elemental area.

$$M = c \int_0^{2\pi} \left| \frac{1}{2} r^2 a^2 + \frac{1}{4} r^4 - \frac{2}{3} ar^3 \cos (\Theta_1 - \Theta) \right|_{r=0}^{a} d\Theta$$

$$= a^4 c \int_0^{2\pi} \left\{ \frac{3}{4} - \frac{2}{3} \cos (\Theta_1 - \Theta) \right\} d\Theta.$$

Now, let $\Theta_1 - \Theta = \alpha$. Then $d\Theta = - d\alpha$. For $0 \leq \Theta \leq 2\pi$, we have $\Theta_1 \leq \alpha \leq \Theta_1 - 2\pi$. Since $f(\Theta_1 + 2\pi) = f(\Theta_1)$ for $\cos \Theta$, and due to symmetry about the x - axis, the integration over a period 2π is zero. Hence,

$$M = - a^4 c \int_{\Theta_1}^{\Theta_1 - 2\pi} \frac{3}{4} \, d\alpha = \frac{3\pi a^4 c}{2}.$$

• PROBLEM 952

A plate has as its edges one arch of the curve $y = \sin x$ and the x-axis. If the density varies as the square of the distance from one corner, find the mass of the plate.

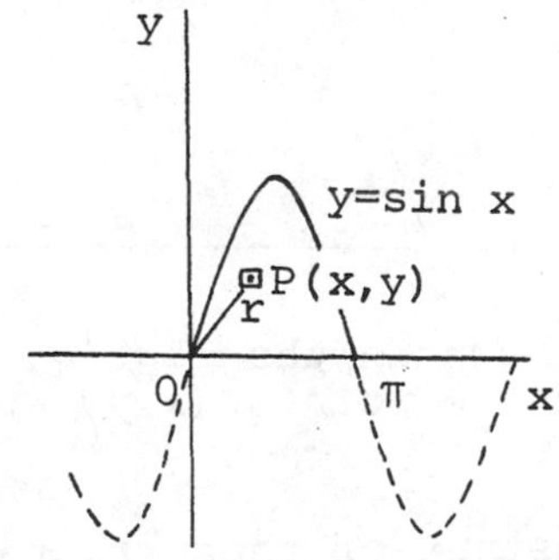

Solution: The density $\rho = cr^2$, for a point P(x,y)

$$\rho = c(x^2 + y^2).$$

Hence, the mass is given by:

$$M = c\int_0^{\pi}\int_0^{\sin x}(x^2 + y^2)dydx$$

$$= c\int_0^{\pi} x^2y + \frac{1}{3}y^3\Bigg|_{y=0}^{y=\sin x} dx$$

$$= c\int_0^{\pi}\left(x^2 \sin x + \frac{1}{3}\sin^3 x\right)dx.$$

Using integration by parts,

$$M = c\left|2x \sin x - (x^2 - 2)\cos x - \frac{1}{9}\cos x\,(\sin^2 x + 2)\right|_0^{\pi}$$

$$= c\left(\pi^2 - 2 + \frac{2}{9} - 2 + \frac{2}{9}\right)$$

$$= \frac{1}{9}c(9\pi^2 - 32).$$

• PROBLEM 953

Find the mass of a plate in the form of a right triangle with legs a and b, if a < b and the density varies as a square of the distance from the vertex of the larger acute angle.

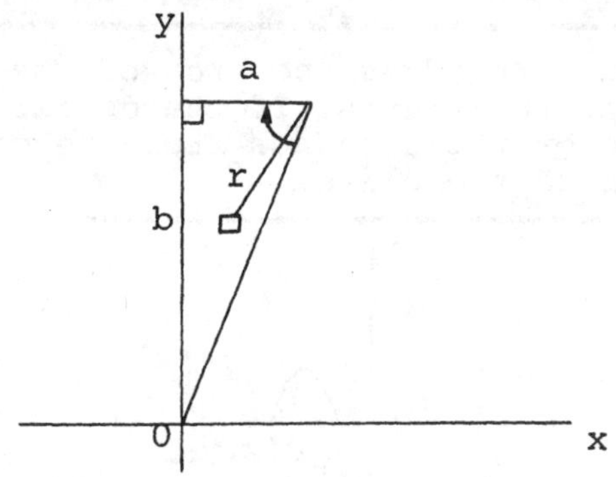

Solution: As shown in the figure, the density of an element at the point P(x,y) is given by:

$$\rho = cr^2,$$

where c is a constant of proportionality, and r is expressed as:

$$r^2 = (a - x)^2 + (b - y)^2,$$

and the equation of the right-hand edge of the

plate is $y = \frac{b}{a}x$. Thus, the mass of the plate,

$$M = c\int_0^a\int_{b/a\,x}^b \left((a - x)^2 + (b - y)^2\right)dy\,dx,$$

or, for simplicity, changing the order of integration,

$$M = c\int_0^b\int_0^{a/b\,y} ((a - x)^2 + (b - y)^2)\,dx\,dy$$

$$= c\int_0^b \left| a^2x - ax^2 + \frac{1}{3}x^3 + b^2x - 2byx + y^2x \right|_{x=0}^{x=\frac{a}{b}y} dy$$

$$= c\int_0^b \left(\frac{a^3}{b}y - \frac{a^3}{b^2}y^2 + \frac{1}{3}\frac{a^3}{b^3}y^3 + aby - 2ay^2 + \frac{a}{b}y^3\right)dy$$

$$= c\left|\frac{1}{2}\frac{a^3}{b}y^2 - \frac{1}{3}\frac{a^3}{b^2}y^3 + \frac{1}{12}\frac{a^3}{b^3}y^4 + \frac{ab}{2}y^2 - \frac{2}{3}ay^3 + \frac{1}{4}\frac{a}{b}y^4\right|_0^b$$

$$= c\left(\frac{1}{2}a^3b - \frac{1}{3}a^3b + \frac{1}{12}a^3b + \frac{1}{2}ab^3 - \frac{2}{3}ab^3 + \frac{1}{4}ab^3\right)$$

$$= \frac{1}{12}c\,ab\,(3a^2 + b^2).$$

• PROBLEM 954

Find the mass of a right circular cylinder of altitude h and base radius a, if the density varies as the distance from the base.

Solution: Since the disk method can be used to evaluate the volume of the cone, for an elementary disk, y units above the x axis, we have

$$dv = 2\pi x dy dx$$

This was done on the basis that, in the neighborhood of a point P(x,y), a small rectangle is constructed with dimensions dy by dx. When revolved about the y-axis it generates a volume in the shape of a washer with an average radius $x + \frac{dx}{2} \approx x$ and thickness dy. Since the density of this point is a function of its altitude above the x-axis, we have

$$\rho = cy,$$

where c is a constant of proportionality.

Since density,

$$\rho = \frac{m}{v},$$

we have $\quad dm = 2\pi c\ xy\ dy\ dx.$

Integration yields the mass of the cylinder (radius a and altitude h)

$$M = 2\pi c \int_0^h \int_0^a xy\ dx\ dy$$

$$= 2\pi c \int_0^h \frac{1}{2} a^2\ y\ dy$$

$$= \pi c a^2 \frac{1}{2} y^2 \Big|_0^h$$

$$= \frac{\pi c a^2 h^2}{2}.$$

• PROBLEM 955

Find the mass of an ellipsoid of revolution with major axis 2a and minor axes 2b, if the density varies as the distance from the axis of revolution.

Solution: Taking the xy plane as the region, we construct an element rectangle of dy by dx within the region of the ellipse:

$$\frac{x^2}{a^2} + \frac{y^2}{b^2} = 1, \qquad x > 0.$$

The volume generated by revolving the rectangle about the y-axis is given by

$$dV = 2\pi x\ dx\ dy.$$

Since its density is proportional to the distance from the axis of revolution, and the volume generated by half of the ellipse about the y-axis is an ellipsoid;

$$\frac{x^2}{a^2} + \frac{y^2}{b^2} + \frac{z^2}{a^2} = 1,$$

we calculate the total mass. We use the relationship:

$$dm = (2\pi x dx dy)(cx),$$

where c is the constant of proportionality.

$$M = \int_0^a \int_{-\frac{b}{a}\sqrt{a^2-x^2}}^{\frac{b}{a}\sqrt{a^2-x^2}} 2c\pi x^2 \, dy \, dx$$

$$= 2c\pi \int_0^a \left| x^2 y \right|_{y=-\frac{b}{a}\sqrt{a^2-x^2}}^{y=\frac{b}{a}\sqrt{a^2-x^2}} dx$$

$$= 4 \frac{b}{a} c\pi \int_0^a x^2 \sqrt{a^2 - x^2} \, dx$$

$$= 4 \frac{b}{a} c\pi \left| -\frac{x}{4} (a^2 - x^2)^{\frac{3}{2}} + \frac{a^2}{8} \left(x\sqrt{a^2 - x^2} + a^2 \sin^{-1} \frac{x}{a} \right) \right|_0^a$$

$$= 4 \frac{b}{a} c\pi \cdot \frac{a^4}{8} \cdot \frac{\pi}{2}$$

$$= \frac{c}{4} \pi^2 a^3 b.$$

• PROBLEM 956

A plate has as its edges the hyperbola $xy = 1$, the lines $x = 1$ and $x = 2$, and the x-axis. If the density varies as the square of the distance from the origin, find the center of mass of the plate with respect to the x-axis.

<u>Solution:</u> The density,

$$\rho = c\,(x^2 + y^2)$$

where c is the constant of proportionality, $x^2 + y^2$ is the square of the distance from the origin to the point $P(x,y)$. The total mass within the given boundary is:

$$M = \int_1^2 \int_0^{\frac{1}{x}} c(x^2 + y^2)dy\ dx = c \int_1^2 x^2y + \frac{1}{3}y^3 \Bigg|_0^{\frac{1}{x}} dx$$

$$= c \int_1^2 \left(x + \frac{1}{3}x^{-3}\right)dx = c \left|\frac{1}{2}x^2 - \frac{1}{6}x^{-2}\right|_1^2$$

$$= \frac{13}{8}c.$$

$$M_x = c \int_1^2 \int_0^{\frac{1}{x}} y(x^2 + y^2)dydx$$

$$= c \int_1^2 \left|\frac{1}{2}x^2y^2 + \frac{1}{4}y^4\right|_0^{\frac{1}{x}} dx$$

$$= c \int_1^2 \left(\frac{1}{2} + \frac{1}{4}x^{-4}\right)dx$$

$$= c \left|\frac{1}{2}x - \frac{1}{12}x^{-3}\right|_1^2$$

$$= \frac{55}{96}c.$$

$$M_y = c \int_1^2 \int_0^{\frac{1}{x}} x(x^2 + y^2)dydx$$

$$= c \int_1^2 x^3y + \frac{xy^3}{3}\Bigg|_0^{\frac{1}{x}} dx$$

$$= \left(x^2 + \frac{1}{3}x^{-2}\right)dx = \frac{x^3}{3} - \frac{x^{-1}}{3}\Bigg|_1^2$$

$$= \frac{5}{2}\ c.$$

Thus, the center of gravity is given by

$$(\bar{x},\bar{y}) = \left(\frac{M_y}{M}\right), \left(\frac{M_x}{M}\right) = \left(\frac{20}{13}\right), \left(\frac{55}{156}\right).$$

• PROBLEM 957

If the density of a square plate of side a is at any point proportional to the square of the distance from one corner, find the center of mass of the plate with respect to an edge through the corner from which the density is measured.

Solution: The density,

$$\rho = c(x^2 + y^2),$$

where c is a constant of proportionality. The area of the element is dxdy, and the mass of the element is:

$$dm = c(x^2 + y^2)dxdy.$$

$$M = \int_0^a \int_0^a (x^2 + y^2)\ dxdy$$

$$= \int_0^a \left(\frac{x^3}{3} + y^2 x\right|_0^a dy$$

$$= \int_0^a \left(\frac{a^3}{3} + ay^2\right) dy = \frac{a^3 y}{3} + \frac{ay^3}{3}\Bigg|_0^a$$

$$= \frac{a^4}{3} + \frac{a^4}{3} = \frac{2a^4}{3}.$$

$$M_y = \int_0^a \int_0^a (x^2 + y^2)\ x\ dxdy$$

$$= \int_0^a \left(\frac{x^4}{4} + \frac{x^2 y^2}{2}\right|_0^a dy$$

$$= \int_0^a \left(\frac{a^4}{4} + \frac{a^2 y^2}{2}\right) dy$$

$$= \left(\frac{a^4 y}{4} + \frac{a^2 y^3}{6}\right|_0^a$$

$$= \frac{a^5}{4} + \frac{a^5}{6} = \frac{10a^5}{24} = \frac{5a^5}{12}$$

$$\bar{x} = \frac{M_y}{M} = \frac{5a^5}{12} \cdot \frac{3}{2a^4}$$

$$= \frac{a \cdot 15}{24}$$

$$= \frac{5a}{8} .$$

CHAPTER 37

SERIES

Since series are widely used in many branches of mathematics, an understanding of convergence of series is necessary. A series is a group of terms, often infinite, where the terms are formed according to some general rule. In order for the series to be of practical use, they must be convergent, i.e., as we add more and more terms, the sum of these terms must approach a limiting value. The sum must not increase indefinitely. Hence, before we employ a series we usually test for the convergence of that series. When we speak of absolute convergence, the series is convergent regardless of the value we assign to the variable. However, a series can also be convergent for only a range of values of the variable. In any case, we must know for what values the series converges, and we are then justified in using it only within that range.

The most generally used test for convergence is the ratio test, where we first formulate the general term for the series, and then show that the (n+1) term is smaller than the (n)th term.

An alternating series is a series in which the terms are alternately positive and negative. Thus, if

$u_1 - u_2 + u_3 - u_4 + \dots$ is an alternating

series, and if $\lim_{n \to \infty} u_n = 0$, the series is convergent.

POSITIVE SERIES

• PROBLEM 958

Establish the convergence or divergence of the series:

$$\frac{1}{1+\sqrt{1}} + \frac{1}{1+\sqrt{2}} + \frac{1}{1+\sqrt{3}} + \frac{1}{1+\sqrt{4}} + \dots .$$

<u>Solution:</u> To establish the convergence or divergence of the given series we first determine the nth term of the series. By studying the law of formation of the terms of the series we find the nth term to be

$\frac{1}{1 + \sqrt{n}}$. To determine whether this series is convergent or divergent we use the comparison test. We choose $\frac{1}{n}$, which is a known divergent series since it is a p-series, $\frac{1}{n^p}$, with p = 1. If we can show $\frac{1}{1 + \sqrt{n}} > \frac{1}{n}$, then $\frac{1}{1 + \sqrt{n}}$ is divergent. But we can see this is true, since $1 + \sqrt{n} < n$ for $n > 1$. Therefore the given series is divergent.

• PROBLEM 959

Establish the convergence or divergence of the series:

$$\sin \frac{\pi}{2} + \frac{1}{4} \sin \frac{\pi}{4} + \frac{1}{9} \sin \frac{\pi}{6} + \frac{1}{16} \sin \frac{\pi}{8} + \ldots .$$

__Solution:__ To establish the convergence or divergence of the given series, we first determine the nth term of the series. By studying the law of formation of the terms of the series, we find the nth term to be: $\frac{1}{n^2} \sin \frac{\pi}{2n}$. To determine whether this series is convergent or divergent, we use the comparison test. We choose $\frac{1}{n^2}$, which is a known convergent series, since it is a p-series, $\frac{1}{n^p}$, with p = 2. If we can show $\frac{1}{n^2} \sin \frac{\pi}{2n} < \frac{1}{n^2}$, then $\frac{1}{n^2} \sin \frac{\pi}{2n}$ is convergent. But we can see this is true since $\sin \frac{\pi}{2n}$ is less than 1 for $n > 1$. Therefore, the given series is convergent.

• PROBLEM 960

Test the series:

$$1 + \frac{2!}{2^2} + \frac{3!}{3^3} + \frac{4!}{4^4} + \ldots$$

by means of the ratio test. If this test fails, use another test.

__Solution:__ To make use of the ratio test, we find the nth term of the given series, and the (n+1)th term. If we let the first term, $1 = u_1$, then $\frac{2!}{2^2} = u_2$, $\frac{3!}{3^3} = u_3$, etc., up to $u_n + u_{n+1}$. We examine the terms of the series to find the law of formation, from which we conclude:

$$u_n = \frac{n!}{n^n} \quad \text{and,} \quad u_{n+1} = \frac{(n+1)!}{(n+1)^{n+1}}.$$

Forming the ratio $\frac{u_{n+1}}{u_n}$ we obtain:

$$\frac{(n+1)!}{(n+1)^{n+1}} \times \frac{n^n}{n!}$$

$$\frac{(n+1)(n!)}{(n+1)^n(n+1)} \times \frac{n^n}{n!} = \frac{n^n}{(n+1)^n}.$$

Now, we find $\lim_{n\to\infty}\left|\frac{n^n}{(n+1)^n}\right|$. This can be rewritten as:

$$\lim_{n\to\infty} \frac{n^n}{\left[n\left(1+\frac{1}{n}\right)\right]^n} = \lim_{n\to\infty} \frac{n^n}{n^n \cdot \left(1+\frac{1}{n}\right)^n} = \lim_{n\to\infty} \frac{1}{\left(1+\frac{1}{n}\right)^n}.$$

We now use the definition:

$$e = \lim_{x\to 0}(1 + x)^{\frac{1}{x}}.$$

If we let $x = \frac{1}{n}$ in this definition, we have:

$\lim_{\frac{1}{n}\to 0}\left(1 + \frac{1}{n}\right)^{\frac{1}{\frac{1}{n}}} = \lim_{n\to\infty}\left(1 + \frac{1}{n}\right)^n$, which is what we have above. Therefore, $\lim_{n\to\infty} \frac{1}{\left(1 + \frac{1}{n}\right)^n} = \frac{1}{e}$. Since $e \simeq 2.7$, $\frac{1}{e} \simeq \frac{1}{2.7}$ which is less than 1.

Hence, by the ratio test, the given series in convergent.

• PROBLEM 961

Test the series:

$$\frac{1 + \sqrt{2}}{2} + \frac{1 + \sqrt{3}}{4} + \frac{1 + \sqrt{4}}{8} + \frac{1 + \sqrt{5}}{16} + \ldots$$

by means of the ratio test. If this test fails, use another test.

Solution: To make use of the ratio test, we find the nth term of the given series, and the (n+1)th term. If we let the 1st term, $\frac{1 + \sqrt{2}}{2} = u_1$, then $\frac{1 + \sqrt{3}}{4} = u_2$, $\frac{1 + \sqrt{4}}{8} = u_3$, etc. up to $u_n + u_{n+1}$. We examine the terms of the series to find the law of formation, from which we find:

$$u_n = \frac{1 + \sqrt{n+1}}{2^n}.$$

Therefore,

$$u_{n-1} = \frac{1 + \sqrt{n}}{2^{n-1}}.$$

Forming the ratio $\frac{u_n}{u_{n-1}}$, we obtain:

$$\frac{1+\sqrt{n+1}}{2^n} \cdot \frac{2^{n-1}}{1+\sqrt{n}} = \left(1+\sqrt{n+1}\right)\frac{2^{n-1-n}}{1+\sqrt{n}}$$

$$= \left(1+\sqrt{n+1}\right)\frac{1}{2\left(1+\sqrt{n}\right)}$$

$$= \frac{1+\sqrt{n+1}}{2+2\sqrt{n}}.$$

We find:

$$\lim_{n\to\infty}\left|\frac{1+\sqrt{n+1}}{2+2\sqrt{n}}\right|.$$

Upon evaluating this limit, we obtain $\frac{\infty}{\infty}$, which is an indeterminate form. Applying L'Hospital's Rule we have:

$$\lim_{n\to\infty}\left|\frac{\frac{1}{2}(n+1)^{-1/2}}{2\cdot\frac{1}{2}n^{-1/2}}\right| = \lim_{n\to\infty}\left|\frac{\frac{1}{2}(n+1)^{-1/2}}{n^{-1/2}}\right| = \lim_{n\to\infty}\left|\frac{1}{2}\ \frac{\sqrt{n}}{\sqrt{n+1}}\right|$$

$$= \lim_{n\to\infty}\left|\frac{1}{2}\cdot\sqrt{\frac{n}{n+1}}\right|.$$

Now we divide both numerator and denominator of the expression under the radical by n, obtaining:

$$\lim_{n\to\infty}\left|\frac{1}{2}\sqrt{\frac{1}{1+\frac{1}{n}}}\right| = \left|\frac{1}{2}\sqrt{1}\right| = \frac{1}{2}.$$

Since $\lim_{n\to\infty}\left|\frac{u_n}{u_{n-1}}\right| = \frac{1}{2}$ and $\frac{1}{2} < 1$, the given series converges.

• PROBLEM 962

Find all values of x for which the series:

$$1 + \frac{1}{2}x + \frac{1\cdot 3}{2\cdot 4}x^2 + \frac{1\cdot 3\cdot 5}{2\cdot 4\cdot 6}x^3 + \ldots$$

converges.

<u>Solution:</u> To find the values for which the given series converges we use the ratio test. Upon examining the terms of the series we find the nth term to be:

$$\frac{1\cdot 3\cdot 5\cdot \ldots(2n-3)}{2\cdot 4\cdot 6\cdot \ldots(2n-2)}x^{n-1}.$$

Then the (n+1)th term is:

$$\frac{1 \cdot 3 \cdot 5 \cdot \ldots (2n-3)(2n-1)}{2 \cdot 4 \cdot 6 \cdot \ldots (2n-2)(2n)} x^n .$$

Forming the ratio $\frac{u_{n+1}}{u_n}$, we have:

$$\frac{1 \cdot 3 \cdot 5 \cdot \ldots (2n-3)(2n-1)x^n}{2 \cdot 4 \cdot 6 \cdot \ldots (2n-2)(2n)} \times \frac{2 \cdot 4 \cdot 6 \cdot \ldots (2n-2)}{1 \cdot 3 \cdot 5 \cdot \ldots (2n-3)x^{n-1}}$$

$$= \frac{(2n-1)x^n}{(2n)x^{n-1}} = \frac{(2n-1)x}{2n} = \frac{2nx - x}{2n}.$$

We now find:

$$\lim_{n\to\infty} \left| \frac{2nx - x}{2n} \right| .$$

Dividing both numerator and denominator by n we obtain:

$$\lim_{n\to\infty} \left| \frac{2x - \frac{x}{n}}{2} \right| = \left| \frac{2x}{2} \right| = |x| .$$

The given series is convergent when $|x| < 1$, and divergent when $|x| > 1$. Therefore, we conclude that the series converges when $1 > x > -1$. But we must still test the endpoints of the interval since the ratio test gives no information when $|x| = 1$. We find that both endpoints are not included in the interval of convergence since, when $x = \pm 1$, $\lim_{n\to\infty}$ of the nth term is not 0. We find:

$$\lim_{n\to\infty} \frac{1 \cdot 3 \cdot 5 \cdot \ldots (2n - 3)}{2 \cdot 4 \cdot 6 \cdot \ldots (2n - 2)} = \frac{\infty}{\infty},$$

an indeterminate form. Using L'Hospital's Rule: $\lim_{n\to\infty} \frac{2}{2} = 1 \neq 0$. Therefore the given series converges for $1 > x > -1$.

SERIES WITH NEGATIVE TERMS

• PROBLEM 963

Test the series:

$$1 - \frac{3^2}{2^2} + \frac{3^4}{2^2 \cdot 4^2} + - \frac{3^6}{2^2 \cdot 4^2 \cdot 6^2} + \ldots$$

by means of the ratio test. If this test fails, use another test.

Solution: To make use of the ratio test, we find the nth term of the given series, and the (n+1)th term. If we let the first term, $1 = u_1$, then $\frac{3^2}{2^2} = u_2$, $\frac{3^4}{2^2 \cdot 4^2} = u_3$, etc. up to $u_n \pm u_{n+1}$. We examine the terms of the series to find the law of formation, from which we conclude:

$$u_n = \frac{3^{2n-2}}{2^2 \cdot 4^2 \cdot \ldots (2n-2)^2},$$

and

$$u_{n+1} = \frac{3^{2(n+1)-2}}{2^2 \cdot 4^2 \cdot \ldots (2n-2)^2 [2(n+1)-2]^2}$$

$$= \frac{3^{2n}}{2^2 \cdot 4^2 \cdot \ldots (2n-2)^2 (2n)^2}.$$

Forming the ratio $\frac{u_{n+1}}{u_n}$, we obtain:

$$\frac{3^{2n}}{2^2 \cdot 4^2 \cdot \ldots (2n-2)^2 (2n)^2} \times \frac{2^2 \cdot 4^2 \cdot \ldots (2n-2)^2}{3^{2n-2}}$$

$$= \frac{3^{2n}}{(2n)^2 \times 3^{2n-2}} = \frac{3^{2n-(2n-2)}}{(2n)^2} = \frac{3^2}{4n^2}.$$

Now, we find:

$$\lim_{n\to\infty} \left| \frac{3^2}{4n^2} \right| = 0.$$

Since $\lim_{n\to\infty} \left| \frac{u_{n+1}}{u_n} \right| = 0$ and $0 < 1$, the given series converges.

• PROBLEM 964

Test the alternating series:

$$\frac{1+\sqrt{2}}{2} - \frac{1+\sqrt{3}}{4} + \frac{1+\sqrt{4}}{6} - \frac{1+\sqrt{5}}{8} + \ldots$$

for convergence.

Solution: An alternating series is oonvergent if (a) the terms, after a certain nth term, decrease numerically, i.e., $u_{n+1} < u_n$, and (b) the general term approaches 0 as n becomes infinite. Therefore, we determine the nth term of the given alternating series. By discovering the law of formation, we find that the general term is $\pm \frac{1+\sqrt{n+1}}{2n}$. Therefore, the preceding term is $\mp \frac{1+\sqrt{n}}{2(n-1)}$. To satisfy condition (a) stated above, we must show that:

$$\frac{1+\sqrt{n+1}}{2n} < \frac{1+\sqrt{n}}{2(n-1)}.$$

Obtaining a common denominator for both these terms,

$$\frac{1}{2} \cdot \frac{\left(1 + \sqrt{n+1}\right)(n-1)}{n(n-1)} < \frac{\left(1 + \sqrt{n}\right)(n)}{n(n-1)} \cdot \frac{1}{2}.$$

Since the denominators are the same, to prove condition (a) we must show,

$$1 + \sqrt{n+1}(n-1) < \left(1 + \sqrt{n}\right)(n),$$

which is obvious, since subtracting 1 from n has a greater effect than adding 1 to $\sqrt{n}$. Since $u_{n+1} < u_n$, we have the first condition for convergence.

Now we must show that

$$\lim_{n \to \infty} \frac{1 + \sqrt{n}}{2n - 2} = 0.$$

We find that $\lim\limits_{n \to \infty} \frac{1 + \sqrt{n}}{2n - 2} = \frac{\infty}{\infty}$, which is an indeterminate form. We therefore apply L'Hospital's Rule, obtaining:

$$\lim_{n \to \infty} \frac{\frac{1}{2}n^{-1/2}}{2} = \lim_{n \to \infty} \frac{1}{4\sqrt{n}} = 0.$$

Since both conditions hold, the given alternating series is convergent.

• PROBLEM 965

Find all values of x for which the series:

$$1 + \frac{1}{3}x - \frac{1 \cdot 2}{3 \cdot 6}x^2 + \frac{1 \cdot 2 \cdot 5}{3 \cdot 6 \cdot 9}x^3 - \frac{1 \cdot 2 \cdot 5 \cdot 8}{3 \cdot 6 \cdot 9 \cdot 12}x^4 + \ldots$$

converges.

Solution: To find the values for which the given series converges we use the ratio test. Upon examining the terms of the series we find the nth term to be:

$$\pm \frac{1 \cdot 2 \cdot 5 \cdot \ldots (3n-7)}{3 \cdot 6 \cdot 9 \cdot \ldots (3n-3)} x^{n-1}.$$

Then the (n+1)th. term is:

$$\pm \frac{1 \cdot 2 \cdot 5 \cdot \ldots (3n-7)(3n-4)}{3 \cdot 6 \cdot 9 \cdot \ldots (3n-3)(3n)} x^n.$$

Forming the ratio $\frac{u_{n+1}}{u_n}$, we have:

$$\frac{1 \cdot 2 \cdot 5 \cdot \ldots (3n-7)(3n-4)x^n}{3 \cdot 6 \cdot 9 \cdot \ldots (3n-3)(3n)} \times \frac{3 \cdot 6 \cdot 9 \cdot \ldots (3n-3)}{1 \cdot 2 \cdot 5 \cdot \ldots (3n-7)x^{n-1}}$$

$$= \frac{(3n-4)x^n}{(3n)x^{n-1}} = \frac{(3n-4)x}{3n} = \frac{3nx - 4x}{3n}.$$

We now find,

$$\lim_{n\to\infty}\left|\frac{3nx - 4x}{3n}\right|.$$

Dividing both numerator and denominator by n, we obtain:

$$\lim_{n\to\infty}\left|\frac{3x - \frac{4x}{n}}{3}\right| = \left|\frac{3x}{3}\right| = |x|.$$

The given series is convergent when $|x| < 1$, and divergent when $|x| > 1$. Therefore, we conclude that the series converges when $1 > x > -1$. But we must still test the endpoints of the interval, since the ratio test gives no information when $x = 1$. We find that both endpoints are not included in the interval of convergence since, when $x = \pm 1$, $\lim_{n\to\infty}$ of the nth term is not 0. We find:

$$\lim_{n\to\infty} \frac{1 \cdot 2 \cdot 5 \cdot \ldots (3n-7)}{3 \cdot 6 \cdot 9 \cdot \ldots (3n-3)} = \frac{\infty}{\infty}.$$

Using Gauss's test (an advanced test) we find that the given series converges for

$$1 \geq x \geq -1.$$

COMPUTATIONS WITH SERIES

Series are very useful for numerical computations of such constants as e, π, etc., and for the computations of terms such as log x, sin x, etc.

Hence, if, by one of a number of possible procedures, we can find a series for a function, then that series can be used for computational purposes, but only IF IT CONVERGES. I.e., the only kind of series that can be used for computation is a convergent series. The reason for this is that, if a series were not convergent but divergent, as we would add more and more terms, a continually different result would be obtained. With a convergent series, however, the series approaches a limit, and gives an ever more accurate result, the more terms we add. Adding more terms does not change the result substantially after the first several terms, it only makes the result more accurate.

Assuming, then, that a series converges, it can be differentiated, integrated, added to other series, subtracted from other series, and multiplied by a constant, for example. A series, therefore, has a great deal of utility. If, for example, we know the series for sin x, we can find the series for cos x by differentiating that series term by term, since we know that $\frac{d}{dx}(\sin x) = \cos x$.

In a computation, once convergence is established, the number of terms to be used in the computation depends only on the accuracy desired.

• PROBLEM 966

Determine the general term of the sequence:

$$\frac{1}{2}, \frac{1}{12}, \frac{1}{30}, \frac{1}{56}, \frac{1}{90}, \ldots$$

Solution: To determine the general term, it is necessary to find how the adjacent terms differ. In this example, it is sufficient to consider the denominator because the numerator is the same for all the terms. The difference between the first two terms is 10. For the second and third terms, the difference is 18. By continuing this process, the results are tabulated as:

$$10, 18, 26, 34, \ldots .$$

Note that each difference is larger by 8 than for the preceding term.

Now we try to write an expression that generates the series. By inspection, each term is the product of 2 successive integers, for example:

$$\frac{1}{2} = \frac{1}{1} \cdot \frac{1}{2}, \quad \frac{1}{12} = \frac{1}{3} \cdot \frac{1}{4},$$

$$\frac{1}{30} = \frac{1}{5} \cdot \frac{1}{6}, \quad \frac{1}{56} = \frac{1}{7} \cdot \frac{1}{8}$$

This fact can be expressed as

$$\frac{1}{(2n - 1)(2n)}$$

and this is the desired answer.

• PROBLEM 967

Determine the general term of the sequence:

$$\frac{1}{5^3},\ \frac{3}{5^5},\ \frac{5}{5^7},\ \frac{7}{5^9},\ \frac{9}{5^{11}},\ \dots$$

Solution: The numerators of the terms in the series are consecutive odd numbers beginning with 1. An odd number can be represented by $2n - 1$.

In the denominators, the base is always 5, and the power is a consecutive odd integer beginning with 3.

The general term can therefore be expressed by

$$\frac{2n - 1}{5^{2n+1}},$$

and the series is generated by replacing n with $n = 1, 2, 3, 4, \dots$.

• PROBLEM 968

From the series for $(1+x)^{-1}$, obtain the series for $\ln(1+x)$.

Solution: First, we determine the series for $(1+x)^{-1}$. To do this we find $f(x)$, $f(0)$, $f'(x)$, $f'(0)$, $f''(x)$, $f''(0)$, etc. We find:

$$f(x) = (1 + x)^{-1} \qquad f(0) = 1$$

$$f'(x) = -(1+x)^{-2} \qquad f'(0) = -1$$

$$f''(x) = 2(1+x)^{-3} \qquad f''(0) = 2$$

$$f'''(x) = -6(1+x)^{-4} \qquad f'''(0) = -6$$

$$f^4(x) = 24(1+x)^{-5} \qquad f^4(0) = 24$$

$$f^5(x) = -120(1+x)^{-6} \qquad f^5(0) = -120.$$

We develop the series as follows:

$$f(x) = f(0) + f'(0)x + \frac{f''(0)}{2!}x^2 + \frac{f''(0)}{3!}x^3 + \frac{f^4(0)}{4!}x^4 + \frac{f^5(0)}{5!}x^5 + \dots .$$

By substitution:

$$(1+x)^{-1} = 1 - x + \frac{2}{2!}x^2 - \frac{6}{3!}x^3 + \frac{24}{4!}x^4 - \frac{120}{5!}x^5 + \ldots$$

$$= 1 - x + x^2 - x^3 + x^4 - x^5 + \ldots \quad \pm x^n \pm \ldots .$$

To obtain the series for ln(1 + x), we find

$$\int_0^x \frac{dx}{1+x}.$$

$$\int_0^x \frac{dx}{1+x} = \int_0^x \left(1 - x + x^2 - x^3 + \ldots \quad \pm x^n + \ldots\right)dx$$

$$= \Big[\ln(1+x)\Big]_0^x$$

$$= x - \frac{x^2}{2} + \frac{x^3}{3} - \frac{x^4}{4} + \frac{x^5}{5} + \ldots \pm \frac{x^n}{n} \pm \frac{x^{n+1}}{n+1} \quad \ldots .$$

Therefore,

$$\ln(1+x) = x - \frac{x^2}{2} + \frac{x^3}{3} - \frac{x^4}{4} + \ldots + \frac{(-1)^{n+1} x^n}{n} + \ldots .$$

Mac LAURIN'S AND TAYLOR'S FORMULAS

• PROBLEM 969

Find the Maclaurin series for the function: f(x) = sin x, and the interval of convergence.

Solution: To find the Maclaurin series for the given function, we must determine f(0), f'(x), f'(0), f''(x), f''(0), etc. We find:

$$f(x) = \sin x \qquad f(0) = 0$$

$$f'(x) = \cos x \qquad f'(0) = 1$$

$$f''(x) = -\sin x \qquad f''(0) = 0$$

$$f'''(x) = -\cos x \qquad f'''(0) = -1$$

$$f^4(x) = \sin x \qquad f^4(0) = 0$$

$$f^5(x) = \cos x \qquad f^5(0) = 1$$

$$f^6(x) = -\sin x \qquad f^6(0) = 0$$

$$f^7(x) = -\cos x \qquad f^7(0) = -1.$$

Now, we develop the series as follows:

$$f(x) = f(0) + f'(0)x + \frac{f''(0)}{2!}x^2 + \frac{f'''(0)}{3!}x^3$$

$$+ \frac{f^4(0)}{4!}x^4 + \frac{f^5(0)}{5!}x^5 + \frac{f^6(0)}{6!}x^6$$

$$+ \frac{f^7(0)}{7!}x^7 + \ldots .$$

By substitution:

$$\sin x = 0 + x + 0 - \frac{x^3}{3!} + 0 + \frac{x^5}{5!} + 0 - \frac{x^7}{7!} + \ldots$$

$$= x - \frac{x^3}{3!} + \frac{x^5}{5!} - \frac{x^7}{7!} + \ldots .$$

We examine the terms of this series to determine the law of formation. We find the nth term of the series to be:

$$\frac{x^{2n-1}}{(2n-1)!}.$$

Then, the (n+1)th term is:

$$\frac{x^{2n+1}}{(2n+1)!}.$$

Therefore, the Maclaurin series is:

$$\sin x = x - \frac{x^3}{3!} + \frac{x^5}{5!} - \frac{x^7}{7!} + \ldots \pm \frac{x^{2n-1}}{(2n-1)!}$$

$$\pm \frac{x^{2n+1}}{(2n+1)!} + \ldots .$$

To find the interval of convergence, we use the ratio test. We set up the ratio:

$$\frac{u_{n+1}}{u_n},$$

obtaining:

$$\frac{x^{2n+1}}{(2n+1)!} \times \frac{(2n-1)!}{x^{2n-1}} = \frac{x^{2n+1}}{(2n+1)(2n-1)!} \times \frac{(2n-1)!}{x^{2n-1}}$$

$$= \frac{x^2}{2n+1}.$$

Now, we find $\lim_{n\to\infty}\left|\frac{x^2}{2n+1}\right| = |0| = 0$. By the ratio test we know that, if

$$\lim_{n\to\infty}\left|\frac{u_{n+1}}{u_n}\right| < 1,$$

the series converges. Since 0 is always less than 1, the series converges for all values of x.

Find the Maclaurin series and the interval of convergence for the function $f(x) = \cos x$.

Solution: To find the Maclaurin series for the given function, we determine $f(0)$, $f'(x)$, $f'(0)$, $f''(x)$, $f''(0)$, etc. We find:

$$\begin{array}{rlrl} f(x) &= \cos x & f(0) &= 1 \\ f'(x) &= -\sin x & f'(0) &= 0 \\ f''(x) &= -\cos x & f''(0) &= -1 \\ f'''(x) &= \sin x & f'''(0) &= 0 \\ f^4(x) &= \cos x & f^4(0) &= 1 \\ f^5(x) &= -\sin x & f^5(0) &= 0 \\ f^6(x) &= -\cos x & f^6(0) &= -1. \end{array}$$

We develop the series as follows:

$$f(x) = f(0) + f'(0)x + \frac{f''(0)}{2!}x^2 + \frac{f'''(0)}{3!}x^3 + \frac{f^4(0)}{4!}x^4 + \frac{f^5(0)}{5!}x^5 + \frac{f^6(0)}{6!}x^6 + \ldots$$

By substitution:

$$\cos x = 1 + 0 - \frac{x^2}{2!} + 0 + \frac{x^4}{4!} + 0 - \frac{x^6}{6!} + \ldots$$

$$= 1 - \frac{x^2}{2!} + \frac{x^4}{4!} - \frac{x^6}{6!} + \ldots .$$

We examine the terms of this series to determine the law of formation. We find the nth term of the series to be $\frac{x^{2n-2}}{(2n-2)!}$. Then the (n+1)th term is $\frac{x^{2n}}{(2n)!}$. Therefore, the Maclaurin series is:

$$\cos x = 1 - \frac{x^2}{2!} + \frac{x^4}{4!} - \frac{x^6}{6!} + \ldots \pm \frac{x^{2n-2}}{(2n-2)!} \pm \frac{x^{2n}}{(2n)!} \ldots .$$

To find the interval of convergence we use the ratio test. We set up the ratio $\frac{u_{n+1}}{u_n}$, obtaining:

$$\frac{x^{2n}}{(2n)!} \times \frac{(2n-2)!}{x^{2n-2}} = \frac{x^{2n}}{(2n)(2n-2)!} \times \frac{(2n-2)!}{x^{2n-2}} = \frac{x^2}{2n}.$$

Now, we find $\lim_{n\to\infty}\left|\frac{x^2}{2n}\right| = |0| = 0$. By the ratio test we know that if $\lim_{n\to\infty}\left|\frac{u_{n+1}}{u_n}\right| < 1$ the series converges Since 0 is always less than 1, the series converges for all values of x.

• PROBLEM 971

Find the Maclaurin series for the function: $\frac{1}{2}\left(e^x + e^{-x}\right)$, and the interval of convergence.

Solution: Let $f(x) = \frac{1}{2}\left(e^x + e^{-x}\right)$. To find the Maclaurin series, we determine $f(0)$, $f'(x)$, $f'(0)$, $f''(x)$, $f''(0)$ etc. We find:

$$f(x) = \frac{1}{2}\left(e^x + e^{-x}\right) \qquad f(0) = 1$$

$$f'(x) = \frac{1}{2}\left(e^x - e^{-x}\right) \qquad f'(0) = 0$$

$$f''(x) = \frac{1}{2}\left(e^x + e^{-x}\right) \qquad f''(0) = 1$$

$$f'''(x) = \frac{1}{2}\left(e^x - e^{-x}\right) \qquad f'''(0) = 0$$

$$f^4(x) = \frac{1}{2}\left(e^x + e^{-x}\right) \qquad f^4(0) = 1, \text{ etc.}$$

Now, we develop the series as follows:

$$f(x) = f(0) + f'(0)x + \frac{f''(0)}{2!}x^2 + \frac{f'''(0)}{3!}x^3 + \dots .$$

By substitution:

$$\frac{1}{2}\left(e^x + e^{-x}\right) = 1 + 0 + \frac{x^2}{2!} + 0 + \frac{x^4}{4!} + \dots$$

$$= 1 + \frac{x^2}{2!} + \frac{x^4}{4!} + \dots .$$

We examine the terms of this series to determine the law of formation. We find the nth term of the series to be:

$$\frac{x^{2n-2}}{(2n-2)!};$$

Then, the (n+1)th term is:

$$\frac{x^{2n}}{(2n)!}.$$

Therefore the Maclaurin series is:

$$\frac{1}{2}(e^x + e^{-x}) = 1 + \frac{x^2}{2!} + \frac{x^4}{4!} + \ldots + \frac{x^{2n-2}}{(2n-2)!} + \frac{x^{2n}}{(2n)!}.$$

To find the interval of convergence we use the ratio test. We set up the ratio $\frac{u_{n+1}}{u_n}$ obtaining:

$$\frac{x^{2n}}{(2n)!} \cdot \frac{(2n-2)!}{x^{2n-2}} = \frac{x^{2n}}{(2n)(2n-2)!} \cdot \frac{(2n-2)!}{x^{2n-2}} = \frac{x^2}{2n}.$$

We find:
$$\lim_{n\to\infty}\left|\frac{x^2}{2n}\right| = |0| = 0.$$

By the ratio test we know that, if

$$\lim_{n\to\infty}\left|\frac{u_{n+1}}{u_n}\right| < 1,$$

the series converges. Since 0 is always less than 1, the series converges for all values of x.

• PROBLEM 972

Find three non-vanishing terms of the Maclaurin series for the function $e^{\cos x}$.

Solution: Let $f(x) = e^{\cos x}$. To find the terms of the Macluarin series, we determine $f(0)$, $f'(x)$, $f'(0)$, $f''(x)$, $f''(0)$, etc. We find:

$$f(x) = e^{\cos x}; \qquad f(0) = e.$$

$$f'(x) = -\sin x\, e^{\cos x}; \qquad f'(0) = 0.$$

$$f''(x) = \left[(-\sin x)e^{\cos x}(-\sin x)\right] + \left[e^{\cos x} - \cos x\right]$$

$$= \sin^2 x\, e^{\cos x} - \cos x\, e^{\cos x} \qquad f''(0) = -e.$$

$$f'''(x) = \left[\sin^2 x\, e^{\cos x}(-\sin x) + e^{\cos x}(2 \sin x \cos x)\right] + \left[-\cos x\, e^{\cos x}(-\sin x) + e^{\cos x} \sin x\right]$$

$$= -\sin^3 x\, e^{\cos x} + 2 \sin x \cos x\, e^{\cos x}$$

$$+ \sin x \cos x\, e^{\cos x}$$

$$+ \sin x\, e^{\cos x}.$$

By use of the formulas:

$2 \sin x \cos x = \sin 2x$, $\sin x \cos x = \frac{1}{2} \sin 2x$,
and, substituting, we obtain:

$$f'''(x) = -\sin^3 x\, e^{\cos x} + \sin 2x\, e^{\cos x}$$

$$+ \frac{1}{2} \sin 2x\, e^{\cos x} + \sin x\, e^{\cos x};\ f'''(0) = 0.$$

$$f^4(x) = \left[\left(-\sin^3 x\right)e^{\cos x}(-\sin x)\right.$$

$$\left. + e^{\cos x}\left(-3 \sin^2 x \cos x\right)\right]$$

$$+ \left[\sin 2x\, e^{\cos x}(-\sin x)\right.$$

$$\left. + e^{\cos x} \cos 2x \cdot 2\right]$$

$$+ \left[\frac{1}{2} \sin 2x\, e^{\cos x}(-\sin x)\right.$$

$$\left. + e^{\cos x}\, \frac{1}{2} \cos 2x \cdot 2\right]$$

$$+ \left[\sin x\, e^{\cos x}(-\sin x)\right.$$

$$\left. + e^{\cos x} \cos x\right]$$

$$= -\sin^4 x\, e^{\cos x} - 3 \sin^2 x \cos x\, e^{\cos x}$$

$$- \sin x \sin 2x\, e^{\cos x} + 2 \cos 2x\, e^{\cos x}$$

$$- \frac{1}{2}\sin x \sin 2x\, e^{\cos x} + \cos 2x\, e^{\cos x}$$

$$- \sin^2 x\, e^{\cos x} + \cos x\, e^{\cos x}.$$

In determining $f^4(0)$, all the terms containing $\sin x$ equal 0; therefore, $f^4(0) = 2e + e + e = 4e$. We develop the series as follows:

$$f(x) = f'(0) + f'(0)x + \frac{f''(0)}{2!}x^2 + \frac{f'''(0)}{3!}x^3 + \frac{f^4(0)}{4!}x^4 + \ldots$$

We need only establish the first five terms of the series to obtain three non-vanishing terms. Therefore, by substitution:

$$e^{\cos x} = e + 0 + \frac{-e}{2!}x^2 + 0 + \frac{4e}{4!}x^4 + \ldots$$

$$= e - \frac{e}{2!}x^2 + \frac{4e}{4!}x^4 + \ldots$$

$$= e\left(1 - \frac{x^2}{2!} + \frac{4x^4}{4!} + \ldots\right).$$

• PROBLEM 973

> Expand the function: cos x, in powers of x - a, where $a = -\frac{\pi}{4}$, and determine the interval of convergence.

Solution. Expanding the given function in powers of x-a is equivalent to finding the Taylor Series for the function. To find the Taylor Series we determine f(x), f(a), f'(x), f'(a), f''(x), f''(a), etc. We find:

$$f(x) = \cos x; \qquad f(a) = f\left(-\frac{\pi}{4}\right) = \frac{\sqrt{2}}{2}.$$

$$f'(x) = -\sin x; \quad f'(a) = f'\left(-\frac{\pi}{4}\right) = \frac{\sqrt{2}}{2}.$$

This is a positive value because the value of sin x in the 2nd. quadrant is negative, therefore -sin x is positive.

$$f''(x) = \cos x; \qquad f''(a) = f''\left(-\frac{\pi}{4}\right) = \frac{\sqrt{2}}{2}$$

$$f'''(x) = -\sin x; \quad f'''(a) = f'''\left(-\frac{\pi}{4}\right) = \frac{\sqrt{2}}{2}.$$

We develop the series as follows:

$$f(x) = f(a) + f'(a)[x-a] + \frac{f''(a)}{2!}[x-a]^2$$

$$+ \frac{f'''(a)}{3!}[x-a]^3 + \ldots .$$

By substitution:

$$\cos x = \frac{\sqrt{2}}{2} + \frac{\sqrt{2}}{2}\left(x + \frac{\pi}{4}\right) + \frac{\frac{\sqrt{2}}{2}}{2!}\left(x + \frac{\pi}{4}\right)^2$$

$$+ \frac{\frac{\sqrt{2}}{2}}{3!}\left(x + \frac{\pi}{4}\right)^3 + \ldots .$$

We examine the terms of this series to determine the law of formation. We find the nth term of the series to be:

$$\frac{\frac{\sqrt{2}}{2}}{(n-1)!}\left(x + \frac{\pi}{4}\right)^{n-1};$$

Then, the (n+1)th term is:

$$\frac{\frac{\sqrt{2}}{2}}{n!}\left(x + \frac{\pi}{4}\right)^n.$$

Therefore the Taylor series is:

$$\cos x = \frac{\sqrt{2}}{2} + \frac{\sqrt{2}}{2}\left(x+\frac{\pi}{4}\right) + \frac{\frac{\sqrt{2}}{2}}{2!}\left(x+\frac{\pi}{4}\right)^2 + \frac{\frac{\sqrt{2}}{2}}{3!}\left(x+\frac{\pi}{4}\right)^3 + \ldots$$

$$+ \frac{\frac{\sqrt{2}}{2}}{(n-1)!}\left(x+\frac{\pi}{4}\right)^{n-1} + \frac{\frac{\sqrt{2}}{2}}{n!}\left(x+\frac{\pi}{4}\right)^n + \ldots$$

To find the interval of convergence we use the Ratio Test. We set up the ratio $\frac{U_{n+1}}{U_n}$, obtaining:

$$\frac{\sqrt{2}\,(x+\frac{\pi}{4})^n}{2(n!)} \times \frac{2(n-1)!}{\sqrt{2}(x+\frac{\pi}{4})^{n-1}} =$$

$$\frac{(x+\frac{\pi}{4})^n}{n(n-1)!} \times \frac{(n-1)!}{(x+\frac{\pi}{4})^{n-1}} = \frac{x+\frac{\pi}{4}}{n} .$$

Now, we find $\lim_{n\to\infty}\left|\frac{x+\frac{\pi}{4}}{n}\right| = |0| = 0$. By the ratio test we know that if $\lim_{n\to\infty}\left|\frac{U_{n+1}}{U_n}\right| < 1$ the series converges. Since $0 < 1$, the series converges for all values of x.

FOURIER SERIES

• PROBLEM 974

Expand $f(x) = e^x$, $-\pi < x < \pi$, in a Fourier series.

Solution: The familiar expression for the Fourier series is:

$$f(x) = e^x = a_0 + \sum_{n=1}^{\infty} a_n \cos nx + b_n \sin nx \quad \text{for} \quad -\pi < x < \pi.$$

By understanding the behavior of the trigonometric functions under integration in a closed interval of length 2π, say $c \le x \le c+2\pi$ (which in this case is $-\pi < x < \pi$), we have arrived at the following values:

$$a_0 = \frac{1}{2\pi}\int_c^{c+2\pi} f(x)dx$$

$$a_n = \frac{1}{\pi}\int_c^{c+2\pi} f(x)\cos nx\,dx, \text{ and}$$

$$b_n = \frac{1}{\pi}\int_c^{c+2\pi} f(x)\sin nx\,dx .$$

In the given problem,

$$a_0 = \frac{1}{2\pi}\int_{-\pi}^{\pi} e^x\,dx = \frac{1}{\pi}\,\frac{e^{\pi}-e^{-\pi}}{2} = \frac{1}{\pi}\sinh\pi.$$

$$a_n = \frac{1}{\pi}\int_{-\pi}^{\pi} e^x \cos nx\,dx = \frac{1}{\pi}\left[e^x\frac{(\cos nx + n\sin nx)}{n^2+1}\right]_{-\pi}^{\pi} = (-1)^n\,\frac{e^{\pi}-e^{-\pi}}{(n^2+1)\pi}$$

$$= (-1)^n \frac{2 \sinh \pi}{(n^2+1)\pi} .$$

$$b_n = \frac{1}{\pi} \int_{-\pi}^{\pi} e^x \sin nx \, dx = \frac{1}{\pi}\left[e^x \frac{(\sin nx - n\cos nx)}{n^2+1} \right|_{-\pi}^{\pi} = (-1)^{n+1} \frac{e^{\pi} - e^{-\pi}}{(n^2+1)\pi} n$$

$$= - \frac{(-1)^n \, 2n \sinh \pi}{(n^2+1)\pi} .$$

Substituting these values back into the Fourier expression,

$$f(x) = a_0 + \sum_{n=1}^{\infty} a_n \cos nx + b_n \sin nx$$

$$= \frac{\sinh \pi}{\pi} + \sum_{n=1}^{\infty} \frac{(-1)^n 2 \sinh \pi}{(n^2+1)\pi} \cos nx - \frac{(-1)^n 2n \sinh \pi}{(n^2+1)\pi} \sin nx$$

$$= \frac{\sinh \pi}{\pi} \left[1 + 2 \sum_{n=1}^{\infty} \frac{(-1)^n}{n^2+1} (\cos nx - n \sin nx) \right] .$$

From the given function $f(0) = e^0 = 1$, and from the Fourier expansion,

$$f(0) = \frac{\sinh \pi}{\pi} \left[1 + 2 \sum_{n=1}^{\infty} \frac{(-1)^n}{n^2+1} \right] = 1,$$

Hence,

$$\frac{\pi}{\sinh \pi} = 1 - \frac{2}{2} + \sum_{n=2}^{\infty} \frac{(-1)^n}{n^2+1} ,$$

or,

$$\frac{\pi}{\sinh \pi} = \sum_{n=2}^{\infty} \frac{(-1)^n}{n^2+1} .$$

• PROBLEM 975

Expand the function:

$$f(x) = \begin{cases} x, & 0 < x < \pi \\ 0, & \pi < x < 2\pi \end{cases}$$

as a Fourier series.

Solution: The general form of Fourier series to represent a function is given by:

$$f(x) = \sum_{n=0}^{\infty} a_n \cos nx + b_n \sin nx,$$

or, removing the first term from the summation:

$$f(x) = a_0 + \sum_{n=1}^{\infty} a_n \cos nx + b_n \sin nx.$$

Integration over $[0,2\pi]$ of the entire equation term by term leads to:

$$\int_0^{2\pi} f(x) \, dx = \int_0^{2\pi} a_0 \, dx + \sum_{n=1}^{\infty} \int_0^{2\pi} a_n \cos nx \, dx + \int_0^{2\pi} b_n \sin nx \, dx ,$$

from which:

$$a_0 = \frac{1}{2\pi} \int_0^{2\pi} f(x) \, dx .$$

Since

$$\int_0^{2\pi} a_n \cos nx\, dx = \int_0^{2\pi} b_n \sin nx\, dx = 0,$$

and the integration $\int_0^{2\pi} f(x)dx$ can be split to a sum of finite integrals characterized by the different expressions for $f(x)$ corresponding to their intervals, we have:

$$a_0 = \frac{1}{2\pi}\int_0^{\pi} x\, dx + \frac{1}{2\pi}\int_{\pi}^{2\pi} 0\, dx = \pi/4 .$$

Noting that multiplication of the equation for $f(x)$ by either $\cos nx$ or $\sin nx$ does not affect the convergence of the series, since $|\cos nx| \le 1$ and $|\sin nx| \le 1$, we first multiply by $\cos nx$ and integrate over the domain $0 \le x \le 2\pi$.

$$\int_0^{2\pi} f(x)\cos nx\, dx = \int_0^{2\pi} a_0 \cos nx\, dx + \sum_{n=1}^{\infty} \int_0^{2\pi} a_n \cos^2 nx\, dx + \int_0^{2\pi} b_n \sin nx \cos nx dx.$$

This gives:

$$\int_0^{2\pi} b_n \sin nx \cos nx\, dx = b_n/2 \int_0^{2\pi} \sin 2nx\, dx = 0, \int_0^{2\pi} a_0 \cos nx\, dx = 0$$

and $\int_0^{2\pi} a_n \cos^2 nx\, dx = a_n \pi$ thus

$$\int_0^{2\pi} f(x)\cos nx\, dx = a_n\pi \text{ , or } a_n = \frac{1}{\pi}\int_0^{2\pi} f(x)\cos nx\, dx = \frac{1}{\pi}\int_0^{\pi} x \cos nx\, dx$$

$$= \frac{1}{\pi}\left[\frac{x}{n}\sin nx + \frac{\cos nx}{n^2}\right|_0^{\pi}$$

$$= \frac{1}{n^2\pi}(\cos n\pi - 1).$$

For $n = 2,4,6,\ldots$ $\cos n\pi = 1$ and hence $a_n = 0$, and for $n = 1,3,5,\ldots$ $\cos n\pi = -1$ and $a_n = \frac{-2}{n^2\pi}$, n = odd. Similarly, to evaluate b_n, we multiply the entire equation by $\sin x$ and integrate:

$$\int_0^{2\pi} f(x)\sin nx\, dx = \int_0^{2\pi} a_0 \sin nx\, dx + \sum_{n=1}^{\infty} \int_0^{2\pi} a_n \cos nx \sin nx\, dx + \int_0^{2\pi} b_n \sin^2 nx\, dx$$

Since the first two integrals are zero,

$$\int_0^{2\pi} f(x)\sin nx\, dx = \int_0^{2\pi} b_n \sin^2 nx\, dx = b_n\pi \text{ , for a particular } n.$$

Hence,

$$b_n = \frac{1}{\pi}\int_0^{2\pi} f(x)\sin nx\, dx = \frac{1}{\pi}\int_0^{\pi} x \sin x\, dx = \frac{1}{\pi}\left[\frac{\sin nx}{n^2} - \frac{x \cos nx}{n}\right|_0^{\pi}$$

$$= \frac{-\pi \cos n\pi}{\pi n} \text{ for all } n = 1,2,3,\ldots$$

$$= \frac{-\pi(-1)^n}{\pi n} = \frac{(-1)^{n+1}}{n} .$$

So far, we have obtained the following:

$$a_0 = \pi/4$$

$$a_n = \frac{-2}{n^2\pi}, \quad n = 1,3,5\ldots$$

$$b_n = \frac{(-1)^{n+1}}{n}, \quad n = 1,2,3,\ldots .$$

To make a_n valid for all n, we ensure that n is odd by replacing n by $(2n-1)$. Finally, we obtain:

$$a_n = \frac{-2}{(2n-1)^2\pi}, \quad n = 1,2,3,\ldots .$$

Now, referring back to the Fourier expression,

$$f(x) = a_0 + \sum_{n=1}^{\infty} a_n \cos nx + b_n \sin nx$$

and substituting back their respective values,

$$f(x) = \pi/4 = \sum_{n=1}^{\infty} \frac{-2}{(2n-1)^2\pi} \cos(2n-1)x + \frac{(-1)^{n+1}}{n} \sin nx$$

or

$$f(x) \simeq \pi/4 - 2/\pi\left(\cos x + \frac{\cos 3x}{3^2} + \frac{\cos 5x}{5^2} + \ldots\right) + \sin x - \frac{\sin 2x}{2} + \frac{\sin 3x}{3} - \ldots$$

Moreover, $\lim_{h\to 0^-} f(\pi-h) = \pi$

and $\lim_{h\to 0^+} f(\pi+h) = 0$

the mean value is then $f(\pi) = \pi/2$. Using the Fourier expansion and assuming its convergence,

$$f(\pi) = \pi/2 = \pi/4 + 2/\pi \sum_{n=1}^{\infty} \frac{1}{(2n-1)^2}, \text{ from which } \frac{\pi^2}{8} = 1 + \frac{1}{3^2} + \frac{1}{5^2} + \ldots .$$

• PROBLEM 976

Find a half-range sine expansion for the function:

$$f(x) = \begin{cases} x, & 0 < x < 2 \\ 6 - 2x, & 2 < x < 3. \end{cases}$$

By setting $x = 2$, obtain a series for π^2.

Solution: From the basic theory of Fourier series for an interval L, where $L \neq \pi$ or 2π,

$$a_n = \frac{1}{\pi} \int_{-\pi}^{\pi} f\left(\frac{Lz}{\pi}\right) \cos nz \, dz$$

$$= \frac{1}{L} \int_{-L}^{L} f(x) \cos \frac{n\pi x}{L} \, dx,$$

and

$$b_n = \frac{1}{\pi} \int_{-\pi}^{\pi} f\left(\frac{Lz}{\pi}\right) \sin nz \, dz$$

$$= \frac{1}{L} \int_{-L}^{L} f(x) \sin \frac{n\pi x}{L} \, dx,$$

where the transformation is $z = \frac{\pi x}{L}$. In the same fashion, we can also derive the half-range series for the interval $0 < x < L$: For the cosine series,

$$f(x) = \frac{a_o}{2} \sum_{n=1}^{\infty} a_n \cos nx,$$

we have:

$$a_n = \frac{2}{L} \int_0^L f(x) \cos \frac{n\pi x}{L} \, dx.$$

Similarly,

$$b_n = \frac{2}{L} \int_0^L f(x) \sin \frac{n\pi x}{L} \, dx$$

for the half-range sine series.

In the given problem, $L = 3$ and, noting that the integration is a Riemannian sum under a limit, we can split integrations, so that, for the half-range sine expansion,

$$b_n = \frac{2}{3} \int_0^3 f(x) \sin \frac{n\pi x}{L} \, dx$$

$$= \frac{2}{3} \int_0^2 x \sin \frac{n\pi x}{3} \, dx + \frac{2}{3} \int_2^3 (6 - 2x)$$

$$\sin \frac{n\pi x}{3}\, dx$$

$$b_n = \frac{2}{3}\left\{\left(-\frac{3x}{n\pi}\cos\frac{n\pi x}{3} + \left(\frac{3}{n\pi}\right)^2 \sin\frac{n\pi x}{3}\right|_0^2\right.$$

$$+ \left(-6\,\frac{3}{n\pi}\cos\frac{n\pi x}{3} + 2\,\frac{3x}{n\pi}\cos\frac{n\pi x}{3}\right.$$

$$\left.\left. - 2\left(\frac{3}{n\pi}\right)^2 \sin\frac{n\pi x}{3}\right|_2^3\right\}$$

$$= \frac{2}{3}\left\{\frac{3}{n\pi} + \left(\frac{3}{n\pi}\right)^2\frac{\sqrt{3}}{2} - (-1)^n\left(\frac{3}{n\pi}\right)6\right.$$

$$+ (-1)^n\left(\frac{9}{n\pi}\right)2 - 3\left(\frac{3}{n\pi}\right) + \left(\frac{6}{n\pi}\right)$$

$$\left. + \left(\frac{3}{n\pi}\right)^2\sqrt{3}\right\}$$

$$= \frac{2}{3}\left(\frac{3}{n\pi}\right)^2 \cdot \frac{3}{2}\sqrt{3} = \frac{9\sqrt{3}}{\pi^2} \cdot \frac{1}{n^2}\,.$$

Hence

$$f(x) = \frac{9\sqrt{3}}{\pi^2}\sum_{\substack{n=1\\ n\neq 3m}}^{\infty} \frac{\sin\frac{n\pi x}{3}}{n^2}\,,$$

$n, m = 1, 2, 3, \ldots$.

By the definition of convergence of Fourier series,

$$\lim_{x\to a} f(x) = \frac{f(a^-) + f(a^+)}{2}$$

and

$$\lim_{x\to 2} f(x) = \frac{f(2^-) + f(2^+)}{2}$$

$$= \frac{2 + (6 - 4)}{2} = 2.$$

To evaluate for $\sin \frac{2}{3} \pi n$ (where n is not a multiple of 3), we have

$$\sin \frac{2}{3} \pi n = \frac{\sqrt{3}}{2}$$

for n = 1, 4, 7, 10, ..., (3n - 2), ...
and

$$\sin \frac{2}{3} \pi n = - \frac{\sqrt{3}}{2}$$

for n = 2, 5, 8, 11, ..., (3n - 1), ...

Hence, by

$$f(2) = \frac{9\sqrt{3}}{\pi^2} \sum_{n=1}^{\infty} \frac{\sin \frac{2}{3} \pi n}{n^2}$$

we can write:

$$f(2) = \frac{9\sqrt{3}}{\pi^2} \sum_{n=1}^{\infty} \frac{\sqrt{3}}{2} \left[\frac{1}{(3n - 2)^2}\right] - \frac{\sqrt{3}}{2} \left[\frac{1}{(3n - 1)^2}\right]$$

$$= \frac{27}{2\pi^2} \sum_{n=1}^{\infty} \frac{1}{(3n - 2)^2} - \frac{1}{(3n - 1)^2} .$$

And from the given function

$$f(2) = \lim_{x \to 2} f(x) = 2,$$

hence,

$$2 = \frac{27}{2\pi^2} \sum_{n=1}^{\infty} \frac{1}{(3n-2)^2} - \frac{1}{(3n-1)^2}$$

or

$$\pi^2 = \frac{27}{4} \sum_{n=1}^{\infty} \frac{1}{(3n-2)^2} - \frac{1}{(3n-1)^2}$$

$$= \frac{27}{4} \left[1 - \frac{1}{2^2} + \frac{1}{4^2} - \frac{1}{5^2} + \frac{1}{7^2} - \ldots \right].$$

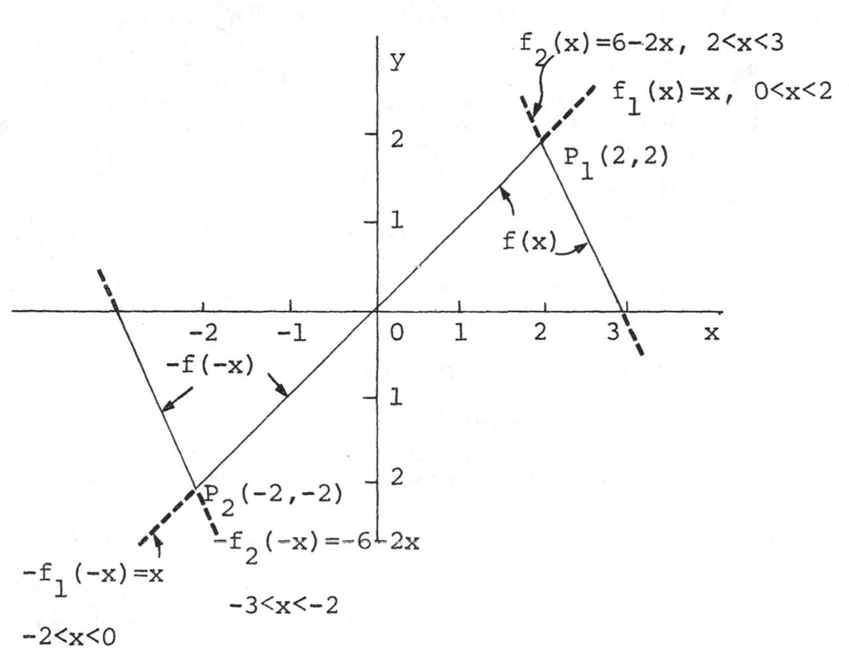

• PROBLEM 977

Find the sine and cosine half-range series for the function $f(x) = x^2$, $0 < x < \pi$.

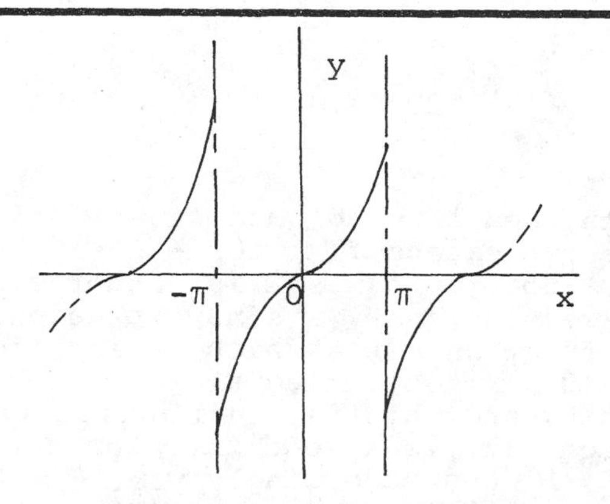

Solution: To discuss the concept of half-range series, we first explain the parity of functions: A function f,(x) is said to be odd if the identity:

$$f(-x) = -f(x)$$

holds.

For example, sin x is an odd function, since sin (- x) = - sin x. Similarly, tan x, cosec x and polynomials with only odd exponents of x (like x^3, x^5, ...) are odd functions. On the other hand, an even function leads to the relationship $f(-x) = f(x)$, such as x^2, cos x, sec x, etc. Under these conditions, we have the following facts:

$$\int_{-a}^{a} f(x)\, dx = 0$$

when f(x) is an odd function, and

$$\int_{-a}^{a} f(x)\, dx = 2\int_{0}^{a} f(x)\, dx$$

when f(x) is an even function.

In the Fourier series expansion of f(x) in the interval (- L, L) all coefficients of sin nx (b_n) must vanish for an even function while the coefficients a_n are equivalent to those obtained by integrating from zero to L only and multiplying by two. Similarly, for an odd f(x), $a_n = 0$ (n = 1, 2, 3, ...), and hence.

$$a_n = \frac{1}{\pi}\int_{-L}^{L} f(x) \cos nx\, dx$$

$$= \frac{2}{\pi}\int_{0}^{L} f(x) \cos nx\, dx.$$

With this idea in mind, and given f(x) for $0 < x < L$, we can extend f(x) for $-L < x < 0$, such that f(x) for $-L < x < L$ is either even or odd. For, if we wish, say, a sine expansion, we may create a function F(x), which is identical with f(x) for $0 < x < L$ but equal to $-f(-x)$ for $-L < x < 0$, and this is a half-range sine expansion. Then F(x) is an odd function for the interval (- L, L) and accordingly yields a Fourier series involving only sine terms. Likewisely, for the cosine expansion, we can write:

$$F(x) = f(x) \qquad 0 < x < L,$$

$$F(x) = f(-x) \qquad -L < x < 0.$$

Hence F(x) is even, and therefore can be expressed in a cosine series.

In the problem, $f(x) = x^2$, $0 < x < \pi$, its sine expansion can be written as:

$$F(x) = f(x) = x^2 \qquad 0 < x < \pi$$

$$F(x) = -f(-x) = -x^2 \qquad -\pi < x < 0.$$

Hence,

$$b_n = \frac{2}{\pi}\int_0^\pi x^2 \sin nx\, dx$$

$$= \frac{2}{\pi}\left[-\left(\frac{x^2}{n} - \frac{2}{n^3}\right)\cos nx + \frac{2x}{n^2}\sin nx\right|_0^\pi,$$

obtained by use of integration by parts.

$$b_n = \frac{2}{\pi}\left[\left(\frac{\pi^2}{n} - \frac{2}{n^3}\right)(-1)^{n+1} - \frac{2}{n^3}\right]$$

$$= \frac{2\pi}{n}(-1)^{n+1} + \frac{4}{n^3\pi}\left[(-1)^n - 1\right].$$

For every even n, the last term vanishes. Hence,

$\left[(-1)^n - 1\right]$ always $= -2$, when n is odd.

$$b_n = \frac{2}{\pi}\left[(-1)^{n+1}\frac{\pi^2}{n} - \frac{4}{n^3}\right] \quad \text{for n odd.}$$

To make n odd, we need to replace n by 2n - 1, so that we have:

$$b_n = \frac{2}{\pi}\left[(-1)^{2n}\frac{\pi^2}{2n-1} - \frac{4}{(2n-1)^3}\right]$$

hence the Fourier sine series reads:

$$f(x)_s = \frac{2}{\pi}\sum_{n=1}^{\infty}\left[(-1)^{2n}\frac{\pi^2}{(2n-1)} - \frac{4}{(2n-1)^3}\right]$$

$$\sin nx.$$

For the cosine half-range series, x^2 is always even, and the nature of Fourier series results in the answer:

$$f(x)_c = \frac{\pi^2}{3} + \sum_{n=1}^{\infty} (-1)^n \frac{4}{n^2} \cos nx.$$

• PROBLEM 978

From the cosine series for $f(x) = x$, $0 < x < \pi$, derive the relation:

$$\frac{\pi^4}{96} = 1 + \frac{1}{3^4} + \frac{1}{5^4} + \frac{1}{7^4} + \dots .$$

<u>Solution:</u> The half-range cosine series for $f(x) = x$, $0 < x < \pi$, suggests that $f(x) = |x|$, $-\pi < x < \pi$, and, due to the symmetry about the y-axis, we have:

$$\int_{-\pi}^{\pi} f(x)dx = 2\int_0^{\pi} f(x)dx .$$

Since the mean square value for the interval $(-\pi,\pi)$ is

$$\bar{y}^2 = \frac{1}{2\pi}\int_{-\pi}^{\pi}[f(x)]^2 dx = \frac{1}{\pi}\int_0^{\pi}[f(x)]^2 dx = \frac{1}{\pi}\int_0^{\pi} x^2\, dx = \frac{\pi^2}{3},$$

and the desired coefficients are:

$$a_n = \frac{2}{\pi}\int_0^{\pi} x \cos nx\, dx = \frac{2}{\pi}\left[\frac{x \sin nx}{n} + \frac{\cos nx}{n^2}\right|_0^{\pi}$$

$$= \frac{2}{\pi n^2}\left[(-1)^n - 1\right]$$ (n = 1,2,3,...), and, for n = 0, we have

$a_0 = \pi$. $b_n = 0$.

This gives $a_n = -\frac{4}{\pi n^2}$ (n = odd), and $a_n = 0$ (n = even). We can write $a_{2n-1} = -\frac{4}{\pi(2n-1)^2}$ (n = 1,2,3,...). This leads to the Fourier cosine expansion

$$x_c = \frac{\pi}{2} - \frac{4}{\pi} \sum_{n=1}^{\infty} \frac{\cos(2n-1)x}{(2n-1)^2} .$$

Using the relation: $\bar{y}^2 \equiv [f(x)]^2 = \frac{a_0^2}{4} + \frac{1}{2} \sum_{n=1}^{\infty} (a_n^2 + b_n^2)$. For

$f(x) = x$, we have:

$$\bar{y}^2 = \frac{\pi^2}{4} + \frac{8}{\pi^2} \sum_{n=1}^{\infty} \frac{1}{(2n-1)^4}, \text{ since } a_{2n-1}^2 = \frac{16}{\pi^2} \cdot \frac{1}{(2n-1)^4} .$$

Equating both mean square values,

$$\frac{\pi^2}{3} = \frac{\pi^2}{4} + \frac{8}{\pi^2} \sum_{n=1}^{\infty} \frac{1}{(2n-1)^4} ,$$

or

$$\frac{\pi^2}{12} = \frac{8}{\pi^2} \sum_{n=1}^{\infty} \frac{1}{(2n-1)^4} ,$$

which yields:

$$\frac{\pi^4}{96} = 1 + \frac{1}{3^4} + \frac{1}{5^4} + \frac{1}{7^4} + \dots .$$

• PROBLEM 979

Using the sine series for $f(x) = x$, $0 < x < \pi$, and the mean square value, obtain the relation:

$$\frac{\pi^2}{6} = 1 + \frac{1}{2^2} + \frac{1}{3^2} + \frac{1}{4^2} + \dots .$$

Solution: The root-mean-square value of the function $y = f(x)$ over an interval from $x = a$ to $x = b$ is defined as:

$$\bar{y} = \left[\frac{\int_a^b y^2 dx}{b-a} \right]^{1/2} .$$

If, in particular, the interval is of length 2π, say $s < x < s+2\pi$, where s is any number, the above equation gives:

$$\bar{y}^2 = \frac{1}{2\pi} \int_s^{s+2\pi} y^2 \, dx .$$

On the other hand, for $f(x)$ expressed in its Fourier series as

$$y = f(x) = \frac{a_0}{2} + \sum_{n=1}^{\infty} a_n \cos nx + b_n \sin nx ,$$

squaring this expression gives terms of the following type:

$$y^2 = \frac{a_0^2}{4} + \sum_{n=1}^{\infty} \left(a_n^2 \cos^2 nx + b_n^2 \sin^2 nx\right) + \frac{a_0}{2} \sum_{n=1}^{\infty} \left(a_n \cos nx + b_n \sin nx\right)$$

$$+ \sum_{n=1}^{\infty} \sum_{m=n+1}^{\infty} \left[a_n \cos nx(a_m \cos mx + b_{m-1} \sin(m-1)x) + b_n \sin nx(b_m \sin mx + a_{m-1} \cos(m-1)x) \right]$$

Integrations from s to $s+2\pi$ of the last two terms vanish: the third term vanishes due to its symmetry about the x-axis, and the last one does so due to the orthogonality property, m = n + 1.

$$\int_s^{s+2\pi} \sin nx \cos nx \, dx = 0.$$ Hence we have:

$$\bar{y}^2 = \frac{1}{2\pi}\int_s^{s+2\pi} y^2 dx = \frac{1}{2\pi}\int_s^{s+2\pi} \frac{a_0^2}{4} dx + \frac{1}{2\pi}\sum_{n=1}^{\infty}\int_s^{s+2\pi}\left(a_n^2\cos^2 nx + b_n^2 \sin^2 nx\right)dx$$

or

$$\bar{y}^2 = \frac{a_0^2}{4} + \frac{1}{2\pi}\sum_{n=1}^{\infty} a_n^2\pi + b_n^2\pi = \frac{a_0^2}{4} + \frac{1}{2}\sum_{n=1}^{\infty}\left(a_n^2 + b_n^2\right).$$

For the sine expansion for $f(x) = x$, $0 < x < \pi$, we have:

$$b_n = \frac{2}{\pi}\int_0^{\pi} x \sin nx \, dx = \frac{2}{\pi}\left[-\frac{x\cos nx}{n} + \frac{\sin nx}{n^2}\right|_0^{\pi}$$

$$= (-1)^{n+1}\frac{2}{n}.$$

$a_n = 0$, $n = 0,1,2,\ldots$; hence

$$x = 2\sum_{n=1}^{\infty}(-1)^{n+1}\frac{\sin nx}{n},$$

since $b_n^2 = \frac{4}{n^2}$, we have:

$$\bar{y}^2 = 2\sum_{n=1}^{\infty}\frac{1}{n^2}.$$

On the other hand,

$$\bar{y}^{2\prime} = \frac{1}{2\pi}\int_{-\pi}^{\pi} x^2 dx = \frac{1}{\pi}\int_0^{\pi} x^2 dx = \frac{\pi^2}{3}.$$

Equating the two results, we finally obtain the relation:

$$\frac{\pi^2}{3} = 2\sum_{n=1}^{\infty}\frac{1}{n^2}, \quad \text{or,}$$

$$\frac{\pi^2}{6} = 1 + \frac{1}{2^2} + \frac{1}{3^2} + \frac{1}{4^2} + \ldots .$$

• PROBLEM 980

Find an expansion for the function $f(x) = 2x-1$ in the interval $-1 < x < 1$.

Solution: Many expansions of functions in Fourier series deal with trigonometric series with a period 2π. However, in most engineering and physics applications we require an expansion of a given function over an interval different from π or 2π. Then we stretch or compress this interval $(-\pi,\pi)$ by means of a transformation of variable to suit the circumstances.

Suppose, in general, we are given a function defined over the interval $(-L,L)$, where L is any positive number. Depending on the value of L compared to that of π (elongation or compression) the interval from $-\pi$ to π can be expressed by the ratio L/π. Thus, if we denote by z the variable referring to the new interval, we must have $x/z = L/\pi$, or $z = \pi x/L$. Now $f(x) = f(Lz/\pi)$, regarded as a function of z. Finally the Fourier series representation is:

$$f\left(\frac{Lz}{\pi}\right) = \frac{a_0}{2} + \sum_{n=1}^{\infty} a_n \cos nz + b_n \sin nz,$$

valid for $-\pi < z < \pi$, which, under transformation, becomes:

$$f(x) = \frac{a_0}{2} + \sum_{n=1}^{\infty} a_n \cos \frac{n\pi x}{L} + b_n \sin \frac{n\pi x}{L},$$

valid for $-L < x < L$. From this, we have:

$$a_n = \frac{1}{\pi}\int_{-\pi}^{\pi} f\left(\frac{Lz}{\pi}\right) \cos nz\, dz = \frac{1}{L}\int_{-L}^{L} f(x) \cos \frac{n\pi x}{L}\, dx,$$

$$b_n = \frac{1}{\pi}\int_{-\pi}^{\pi} f\left(\frac{Lz}{\pi}\right) \sin nz\, dz = \frac{1}{L}\int_{-L}^{L} f(x) \sin \frac{n\pi x}{L}\, dx.$$

In this problem, $L = 1$, hence we have shrinkage of the periodicity by the factor $1/\pi$. For its Fourier representation,

$$a_0 = \frac{1}{2}\int_{-1}^{1} (2x-1)dx = -1$$

$$a_n = \int_{-1}^{1} (2x-1)\cos n\pi x\, dx = 2\int_{-1}^{1} x \cos n\pi x\, dx - \int_{-1}^{1} \cos n\pi x\, dx.$$

The first integration gives:

$$2\left[\frac{x}{n\pi} \sin n\pi x + \frac{1}{n^2\pi^2} \cos n\pi x\right|_{-1}^{1} = 0.$$

The second integration is obviously zero, hence $a_n = 0$.

$$b_n = \int_{-1}^{1} (2x-1)\sin n\pi x\, dx = 2\int_{-1}^{1} x \sin n\pi x\, dx - \int_{-1}^{1} \sin n\pi x\, dx$$

$$= 2\left(-\frac{x}{n\pi} \cos n\pi x + \frac{1}{n^2\pi^2} \sin n\pi x\right|_{-1}^{1} = -\frac{4}{n\pi} \cos n\pi = (-1)^{n+1} \frac{4}{n\pi}$$

Thus

$$f(x) = 2x-1 = -1 + \frac{4}{\pi}\sum_{n=1}^{\infty} \frac{(-1)^{n+1}}{n} \sin n\pi x.$$

• PROBLEM 981

Expand $f(x) = x^2$ in Fourier series for $-\pi \le x \le \pi$.

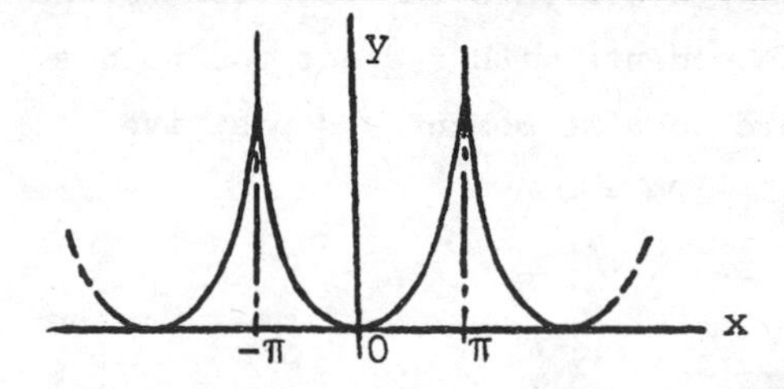

Solution: The Fourier representation is given by:

$$f(x) = \frac{a_0}{2} + \sum_{n=1}^{\infty} a_n \cos nx + b_n \sin nx .$$

$$\frac{a_0}{2} = \frac{1}{2\pi}\int_{-\pi}^{\pi} f(x)dx = \frac{1}{2\pi}\int_{-\pi}^{\pi} x^2 dx = \frac{1}{6\pi} x^3 \Big|_{-\pi}^{\pi} = \frac{\pi^2}{3} .$$

$$a_n = \frac{1}{\pi}\int_{-\pi}^{\pi} f(x)\cos nx\, dx = \frac{1}{\pi}\int_{-\pi}^{\pi} x^2 \cos nx\, dx = \frac{1}{\pi}\left[\left(\frac{x^2}{n} - \frac{2}{n^3}\right)\sin nx + \frac{2x}{n^2}\cos nx \right|$$

obtained through integration by parts.

$a_n = 4/n^2 \cos n\pi$, where the sign of $\cos n\pi$ depends on n and is $(-1)^n$. Therefore,

$$a_n = \frac{(-1)^n 4}{n^2} .$$

Similarly,

$$b_n = \frac{1}{\pi}\int_{-\pi}^{\pi} f(x)\sin nx dx = \frac{1}{\pi}\int_{-\pi}^{\pi} x^2 \sin nx dx = \frac{1}{\pi}\left[-\left(\frac{x^2}{n} - \frac{2}{n^3}\right)\cos nx + \frac{2x}{n^2}\sin nx\right|_{-\pi}^{\pi}$$

$$= 0 .$$

Thus, the Fourier expression is:

$$f(x) = x^2 = \frac{\pi^2}{3} + \sum_{n=1}^{\infty} (-1)^n \left(\frac{4}{n^2}\right) \cos nx .$$

The periodic result is shown in the diagram.

• PROBLEM 982

A horizontal beam L ft. long is simply supported at its ends. Using trigonometric series find the equation of the elastic curve of the beam if it carries a load of P lb., c ft. from one end. (Neglect the weight of the beam).

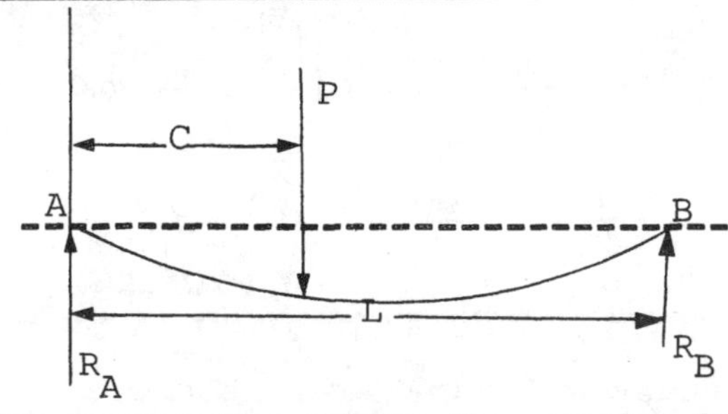

Solution: First we wish to evaluate the reaction forces R_A and R_B by finding the moments with respect to, points A and B. Taking counterclockwise moments as positive,

$$R_B L - Pc = 0.$$

$$P(L-c) - R_A L = 0.$$

Hence,

$$R_A = \frac{P}{L}(L-c),$$

and

$$R_B = \frac{Pc}{L}\ .$$

If we take an arbitrary point, x ft. away from A, then the bending moment of the constraint forces with respect to this point are given by:

$$M_1 = -R_A x = -\frac{P}{L}(L-c)x, \quad 0 < x < c$$

$$M_2 = -R_B(L-x) = -\frac{Pc}{L}(L-x), \quad c < x < L.$$

Now, using the Fourier half-range sine series, the bending moment can be expressed, in the interval (0,L), as

$$M = \sum_{n=1}^{\infty} b_n \sin nx,$$

where

$$b_n = \frac{2}{L}\int_0^L \left(M_1 + M_2\right)\sin\frac{n\pi x}{L}\,dx = \frac{2}{L}\int_0^c -\frac{P}{L}(L-c)x\,\sin\frac{n\pi x}{L}\,dx$$

$$+\frac{2}{L}\int_c^L -\frac{Pc}{L}(L-x)\sin\frac{n\pi x=}{L}dx.$$

$$b_n = \frac{2}{L}\left\{-\frac{P(L-c)}{L}\int_0^c x\sin\frac{n\pi x}{L}\,dx - \frac{Pc}{L}\int_c^L (L-x)\sin\frac{n\pi x}{L}\,dx\right\}$$

$$= \frac{2}{L}\left\{\frac{P(c-L)}{L}\left[-\frac{Lx}{n\pi}\cos\frac{n\pi x}{L} + \frac{L^2}{n^2\pi^2}\sin\frac{n\pi x}{L}\right|_0^c - \frac{Pc}{L}\left[-\frac{L^2}{n\pi}\cos\frac{n\pi x}{L} + \frac{Lx}{n\pi}\cos\frac{n\pi x}{L}\right.\right.$$

$$\left.\left. - \frac{L^2}{n^2\pi^2}\sin\frac{n\pi x}{L}\right|_c^L\right\}$$

$$= \frac{2}{L}\left[-\frac{Pc^2}{n\pi}\cos\frac{n\pi c}{L} + \frac{PLc}{n^2\pi^2}\sin\frac{n\pi c}{L} + \frac{PLc}{n\pi}\cos\frac{n\pi c}{L} - \frac{PL^2}{n^2\pi^2}\sin\frac{n\pi c}{L} + (-1)^n\frac{PLc}{n\pi}\right.$$

$$\left. + (-1)^{n+1}\frac{PLc}{n\pi} - \frac{PLc}{n\pi}\cos\frac{n\pi c}{L} + \frac{Pc^2}{n\pi}\cos\frac{n\pi c}{L} - \frac{PLc}{n^2\pi^2}\sin\frac{n\pi c}{L}\right].$$

$$b_n = -\frac{2PL}{n^2\pi^2}\sin\frac{n\pi c}{L}.$$

Hence,

$$M = -\frac{2PL}{\pi^2}\sum_{n=1}^{\infty}\frac{\sin\frac{n\pi c}{L}}{n^2}\sin\frac{n\pi x}{L}.$$

Now, **we recall** the differential equation of the elastic curve of the beam:

$$EI\frac{d^2y}{dx^2} = M,$$

where $E(\text{lb./ft}^2)$ is Young's modulus, I is the moment of inertia (in. ft.^4) of the cross-sectional area of the beam with respect to the neutral axis, and M is the bending moment.

Hence, combining the two equations,

$$EI\frac{d^2y}{dx^2} = -\frac{2PL}{\pi^2}\sum_{n=1}^{\infty}\frac{\sin\frac{n\pi c}{L}}{n^2}\sin\frac{n\pi x}{L},$$

or,

$$\frac{d^2y}{dx^2} = -\frac{2PL}{EI\pi^2} \sum_{n=1}^{\infty} \frac{\sin\frac{n\pi c}{L}}{n^2} \sin\frac{n\pi x}{L} .$$

Since $dy/dx \simeq 0$ and $y = 0$ at $x = 0$, double successive integration yields the desired equation of the bend:

$$y = 2\frac{PL}{EI\pi^2} \sum_{n=1}^{\infty} \frac{\sin\frac{n\pi c}{L}}{n^2} \left(-\frac{L}{n\pi}\right)\left(\frac{L}{n\pi}\right) \sin\frac{n\pi x}{L}$$

$$= \frac{2PL^3}{EI\pi^4} \sum_{n=1}^{\infty} \frac{\sin\frac{n\pi c}{L}}{n^4} \sin\frac{n\pi x}{L} .$$

As an alternative derivation, we consider the following facts; The work done in bending a piece of beam, and hence the potential energy stored in the piece, is given by:

$$\Delta w = \frac{1}{2} M \Delta s/R,$$

where Δs is the length of this element, and R is the radius of curvature of the bend, which can be approximated as $R \simeq 1/d^2y/dx^2$.

$$\frac{ds}{R^2} \simeq \frac{d^2y}{dx^2} .$$

Thus,

$$W = \frac{EI}{2} \int_0^L \left(\frac{d^2y}{dx^2}\right)^2 dx.$$

Again the equation of the elastic curve is desired to be represented as:

$$y = \sum_{n=1} b_n \sin\frac{n\pi x}{L}.$$

Hence,

$$\frac{d^2y}{dx^2} = -\frac{\pi^2}{L^2} \sum_{n=1}^{\infty} b_n n^2 \sin\frac{n\pi x}{L} .$$

$$\left(\frac{d^2y}{dx^2}\right)^2 = \frac{\pi^4}{L^4} \sum_{n=1}^{\infty} b_n^2 n^4 \sin^2\frac{n\pi x}{L} + \frac{\pi^4}{L^4} \sum_{n=1}^{\infty} \sum_{m=n-1}^{\infty} b_n n^2 \sin\frac{n\pi x}{L}\left(b_m m^2 \sin\frac{n\pi x}{L}\right) ,$$

where upon integration from zero to L the second **sum** vanishes $(n \neq m)$. We have:

$$\int_0^L \left(\frac{d^2y}{dx^2}\right)^2 dx = \frac{\pi^4}{L^4} \sum_{n=1}^{\infty} b_n^2 n^4 \int_0^L \sin^2\frac{n\pi x}{L} dx ,$$

where

$$\int_0^L \sin^2\frac{n\pi x}{L} dx = \frac{L}{2} .$$

Hence,

$$W = \frac{EI\pi^4}{4L^3} \sum_{n=1}^{\infty} b_n^2 n^4 .$$

We know that the potential energy, w, is imparted due to the weight p at $x = c$, producing a deflection y_c at that point, where

$$y_c = \sum_{n=1}^{\infty} b_n \sin\frac{n\pi c}{L}.$$

Now, considering the load p acting through an infinitesimal displacement dy_c, it does an amount of work pdy_c, equal to the infinitesimal increase in dw, which, according to the relation obtained, is a function of b_n only. Thus,

$$\frac{EI\pi^4}{4L^3} \sum_{n=1}^{\infty} 2n^4 \, b_n \, db_n = p \sum_{n=1}^{\infty} \sin \frac{n\pi c}{L} \, db_n \; .$$

This holds if and only if $b_n \neq 0$, which leads to:

$$\frac{EI\pi^4}{4L^3} n^4 b_n = p \sin \frac{n\pi c}{L} \; .$$

That is,

$$b_h = \frac{2L^3 p}{EI\pi^4} \; \frac{\sin \frac{n\pi c}{L}}{n^4} \; .$$

Hence the Fourier half-range expansion, with values of b_n above, becomes:

$$y = \frac{2pL^3}{EI\pi^4} \sum_{n=1}^{\infty} \frac{\sin \frac{n\pi c}{L}}{n^4} \sin \frac{n\pi x}{L} ,$$

as obtained before.

CHAPTER 38

THE LAW OF THE MEAN

The Mean Value Theorem can be stated as follows:

Let f be continuous on the interval [a,b] and be differentiable at each interior point. Then there exists a point x_0, $a < x_0 < b$, such that

$$f'(x_0) = \frac{f(b) - f(a)}{b - a} .$$

The relationships involved are best illustrated by the diagram below:

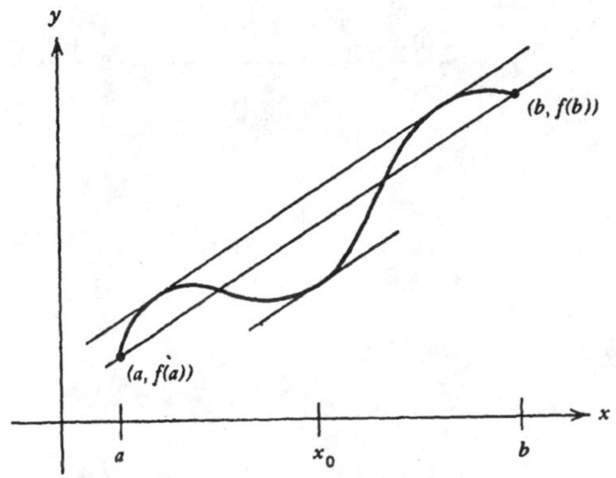

Slope of tangent line $= f'(x_0) =$ slope of chord $= \frac{f(b) - f(a)}{b - a}$.

• PROBLEM 983

If $f(x) = 3x^2 - x + 1$, find the point x_o at which $f'(x)$ assumes its mean value in the interval [2,4].

Solution: Recall the mean value theorem. Given a function f(x) which is continuous in [a,b] and differentiable in (a,b), there exists a point x_o where $a < x_o < b$ such that:

$$\frac{f(b) - f(a)}{b - a} = f'(x_o),$$

where x_o is the mean point in the interval.

In our problem, $3x^2 - x + 1$ is continuous, and the derivative exists in the interval (2,4). We have:

$$\frac{f(4) - f(2)}{4 - 2} = \frac{\left[3(4)^2 - 4 + 1\right] - \left[3(2)^2 - 2 + 1\right]}{4 - 2}$$

$$= f'(x_o),$$

or

$$\frac{45 - 11}{2} = 17 = f'(x_o) = 6x_o - 1 .$$

$$6x_o = 18$$

$$x_o = 3.$$

$x_o = 3$ is the point where $f'(x)$ assumes its mean value.

• PROBLEM 984

If $y = \sin x$, find the point x_o where y' assumes its mean value in the interval $[0, \Pi/2]$.

Solution: Applying the mean value theorem, we have:

$$\frac{\sin(\Pi/2) - \sin(0)}{\Pi/2 - 0} = f'(x_o) = \left[\cos(x_o)\right],$$

where the interval is $[0, \Pi/2]$.

Therefore,

$$\frac{1 - 0}{\Pi/2} = \cos(x_o)$$

$$\frac{2}{\Pi} = \cos(x_o) \quad \text{or} \quad x_o = \cos^{-1}\left(\frac{2}{\Pi}\right) = \text{arc} \cos\left[\frac{2}{\pi}\right].$$

• PROBLEM 985

Show that the following inequality holds:

$$4 + \frac{2}{17} < \sqrt{17} < 4 + \frac{1}{8} .$$

Solution: To solve the problem, we will use the mean value theorem. Recall that the theorem states that for a function that is continuous in the closed interval and continuously differentiable in the open interval, we have:

$$\frac{f(b) - f(a)}{b - a} = f'(x_o),$$

where our interval is $[a, b]$ $(b \neq a)$ and $a < x_o < b$. For this problem,

$$f(x) = \sqrt{x} \quad \text{and} \quad a = 16, \ b = 17$$

Divide the problem into two cases.

First case. For the right hand side of the inequality

$$\frac{f(17) - f(16)}{17 - 16} = \frac{1}{2}\frac{1}{\sqrt{x_o}} \quad \left[= f'(x_o)\right] \qquad (1)$$

therefore,

$$\frac{\sqrt{17} - 4}{1} = \frac{1}{2}\frac{1}{\sqrt{x_o}} \cdot \sqrt{17}$$

$$= 4 + \frac{1}{2\sqrt{x_o}} \qquad (2)$$

Since $x_o > 16$, $\sqrt{x_o} > \sqrt{16}$, $\sqrt{x_o} > 4$.

Dividing both sides of the inequality by x_o (and noting $x_o > 0$)

$$\frac{1}{\sqrt{x_o}} > \frac{4}{\sqrt{x_o}} \qquad (2)$$

Since $x_o < 17$, $\frac{1}{x_o} > \frac{1}{17}$

$$\frac{4}{x_o} > \frac{4}{17}$$

Therefore, using equations (1) and (2)

$$\sqrt{17} = 4 + \frac{1}{2}\frac{1}{\sqrt{x_o}} > 4 + \frac{1}{2}\frac{4}{17} > 4 + \frac{2}{17}$$

Second case. For the left hand side of the inequality, we know from the previous case that

$\sqrt{x_o} > 4$, hence $\frac{1}{\sqrt{x_o}} < \frac{1}{4}$.

Using this result and equation (2)

$$\sqrt{17} = 4 + \frac{1}{2}\frac{1}{\sqrt{x_o}} < 4 + \frac{1}{2}\frac{1}{4} < 4 + \frac{1}{8}$$

Combining the results from each case, we have the desired inequality.

• PROBLEM 986

Show: $e^x \geq 1 + x$ for all real numbers x.

Solution: Divide the problem into 3 cases; $x = 0$, $x > 0$, $x < 0$.

Case 1. $x = 0$

For $x = 0$, we have $e^0 \geq 1 + 0$, or $1 = 1$.

Case 2. $x > 0$

For this case, we apply the mean value theorem. We let

$$f(x) = e^x$$

and the interval will be [0, x]. Applying the theorem, we have

$$\frac{f(x) - f(0)}{x - 0} = \frac{e^x - e^o}{x - 0} = f'(x_o) = e^{x_o},$$

where $0 < x_o < x$.

Simplifying, we have:

$$e^x = e^{x_o} \cdot x + e^o = xe^{x_o} + 1$$

since $x_o > 0$, $e^{x_o} > 1$. Therefore,

$$e^x > x + 1$$

Case 3. $x < 0$

Solution is similar to Case 2 and will be left to the reader as an exercise.

Combining the three results, we have the desired inequality.

The mean value theorem for the integral has a very simple geometric interpretation.

The mean value theorem says that for a continuous function on the closed interval [a,b], there exists a point x_o, where $a < x_o < b$, such that:

$$f(x_o) = \frac{1}{b - a} \int_a^b f(x)dx .$$

If we multiply both sides by (b - a) we have

$$(b - a)\ f(x_o) = \int_a^b f(x)dx,$$

which states that the integral from a to b is equal to the area of a rectangle of length (b - a) and height $f(x_o)$. In the diagram, it means that

the area in region 1 can be put in region 2 , thus forming a rectangle.

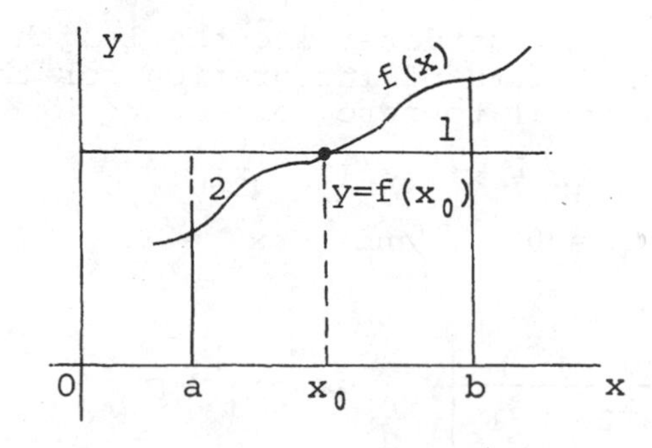

• **PROBLEM** 987

Find the mean value of $y = \sin x$ with respect to x from $x = 0$ to $x = 1/2\ \pi$.

Solution: Recalling the definition for the mean value theorem of the integral, we have:

$$f(x_o) = \frac{1}{b-a}\int_a^b f(x)\,dx,$$

where $a < x_o < b$, and a,b are the endpoints of the closed interval [a,b].

Therefore,

$$\sin(x_o) = \frac{1}{\frac{\pi}{2} - 0}\int_0^{\pi/2} \sin x\,dx$$

$$= \frac{2}{\pi}\left[-\cos(x)\right]\Big|_0^{\pi/2}$$

$$= \frac{2}{\pi}\left[-\cos(\pi/2) - \left[-\cos(0)\right]\right]$$

$$= \frac{2}{\pi}[0+1] = \frac{2}{\pi}.$$

Hence,

$$\sin(x_o) = \frac{2}{\pi}.$$

• **PROBLEM** 988

What is the mean value or mean ordinate of the positive part of the curve $y = 2x - x^2$?

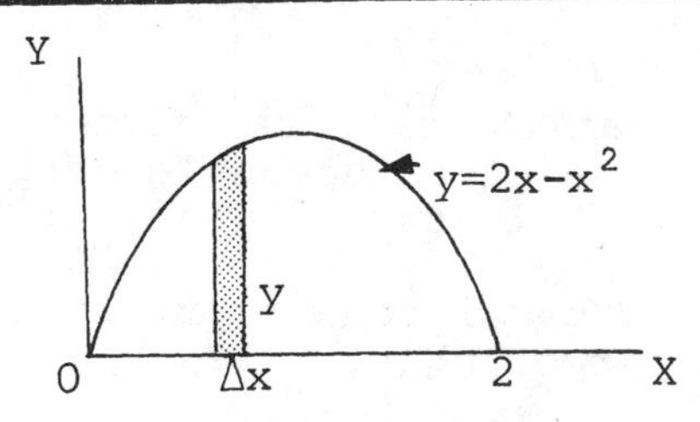

Solution: First determine the length of the base to fix the limits of integration for the area by setting y equal to zero, or:

$$0 = 2x - x^2 = x(2 - x).$$

Then, $x_1 = 0$, and $x_2 = 2$.

Now,

$$\bar{y}_x = \frac{1}{x_2 - x_1}\int y \cdot dx$$

$$= \frac{1}{2-0}\int_0^2 (2x - x^2)\,dx$$

$$= \int_0^2 \left(x - \frac{x^2}{2}\right) dx$$

$$= \left| \frac{x^2}{2} - \frac{x^3}{6} \right|_0^2 = \frac{4}{2} - \frac{8}{6} = \frac{2}{3},$$

the mean ordinate.

• PROBLEM 989

Find the mean value of the ordinates of the circle

$$x^2 + y^2 = a^2$$

in the first quadrant.
(a) With respect to the radius along the x-axis
(b) With respect to the arc-length.

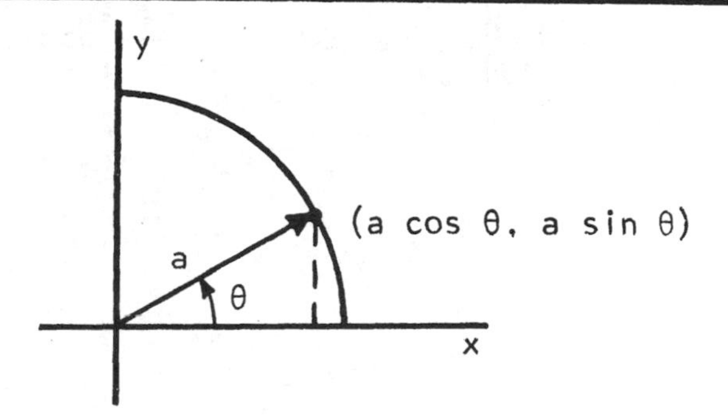

Solution:

(a) Noting that the mean value is defined by:

$$f(x_o) = \frac{1}{b-a}\int_a^b f(x)dx, \text{ where } a < x_o < b,$$

we have:

$$f(x_o) = \frac{1}{a-o}\int_0^a \sqrt{a^2 - x^2}\,dx,$$

where $f(x) = y = \pm\sqrt{a^2 - x^2}$. We take $y = \sqrt{a^2 - x^2}$, because we are in the first quadrant. Also, noting that:

$$\int \sqrt{a^2 - x^2}\,dx = \frac{1}{2} x \sqrt{a^2 - x^2}$$

$$+ \frac{1}{2} a^2 \text{ arc sin } \frac{x}{a} + C,$$

we have:

$$f(x_o) = \frac{1}{4}\Pi a .$$

(b) The coordinates of any point on the circle can be expressed in terms of Θ by the following method. For the point in question, drop a perpendicular line to the x-axis. We have a right triangle with hypothenuse of length a.

$$\cos(\Theta) = \frac{x}{a},$$

or

$$x = a \cos(\Theta). \qquad \sin(\Theta) = \frac{y}{a},$$

or $y = a \sin(\Theta)$.

Now an element of the arc length is $a\Delta\Theta$. (Recall that the length of an arc of a circle is equal to $S = R\Theta$ where R is the radius and Θ the angle in radians.) Therefore the length of the arc of the circle in the first quadrant is $\frac{1}{2}\Pi a$.

By taking the limit, $a\Delta\Theta$ becomes $ad\Theta$.

$$f(\Theta) = \frac{1}{\frac{1}{2}\Pi a}\int_0^{\Pi/2} a \sin \Theta (ad\Theta)$$

$$= \frac{2a}{\Pi}\int_0^{\Pi/2} \sin \Theta d\Theta = \frac{2a}{\Pi}\Big[\cos \Theta\Big]_0^{\frac{\pi}{2}}$$

$$= \frac{2a}{\Pi}(1 - 0) = \frac{2a}{\Pi} .$$

CHAPTER 39

MOTION: RECTILINEAR AND CURVILINEAR

When dealing with motion problems the following relationships are usually involved:

$$v = \frac{ds}{dt} \; ; \quad a = \frac{dv}{dt} = \frac{d^2s}{dt^2} \; ; \quad a_n = \frac{v^2}{r} = w^2 r \; ; \quad v = wr \; ;$$

$$s = \sqrt{s_x^2 + s_y^2} \; ; \quad v = \sqrt{v_x^2 + v_y^2} \; ; \quad a = \sqrt{a_x^2 + a_y^2} \; ;$$

$$v_x = \frac{dx}{dt} \; ; \quad v_y = \frac{dy}{dt} \; ; \quad a_x = \frac{d^2x}{dt^2} \; ; \quad a_y = \frac{d^2y}{dt^2} \; .$$

Where s = distance; v = velocity; and a = acceleration. All of these quantities are vector quantities which require to be fully specified by both magnitude and direction. The components of these vectors as, for example, v_x and v_y specify fully the magnitude and direction,

of the vector v. $\left(\theta = \tan^{-1} \frac{v_y}{v_x}\right)$,

For rectilinear motion (along a straight-line path), all relationships apply except the normal acceleration a_n. The latter is also applicable where the motion is along a curved path. Note that all differentiations above, are carried out with respect to time. If the path $y = f(x)$ is to be found from velocity components, for example, the chain rule may be applied as follows:

$$\frac{dy}{dx} = \frac{dy}{dt} \div \frac{dx}{dt} \; .$$

When the motion is due to gravitational effects, $g = 32.2$ ft./sec^2. is generally substituted for acceleration, a.

• PROBLEM 990

A particle moves in a straight line according to the law of motion:

$$s = t^3 - 4t^2 - 3t.$$

When the velocity of the particle is zero, what is its acceleration?

Solution: The velocity, v, can be found by differentiating this equation of motion with respect to t. Further differentiation gives the acceleration.

Hence, the velocity, v, and acceleration, a, are:

$$v = \frac{ds}{dt} = 3t^2 - 8t - 3,$$

$$a = \frac{dv}{dt} = 6t - 8.$$

The velocity is zero when

$$3t^2 - 8t - 3 = (3t + 1)(t - 3) = 0,$$

from which

$$t = -\frac{1}{3} \quad \text{or } t = 3.$$

The corresponding values of the acceleration are

$$a = -10 \text{ for } t = -\frac{1}{3}, \text{ and}$$

$$a = +10 \text{ for } t = 3.$$

• PROBLEM 991

Discuss the circular motion, the equation of which is $\Theta = \frac{t}{t+1}$, Θ being measured in radians and t in seconds. The radius of the circle, r = 2 feet.

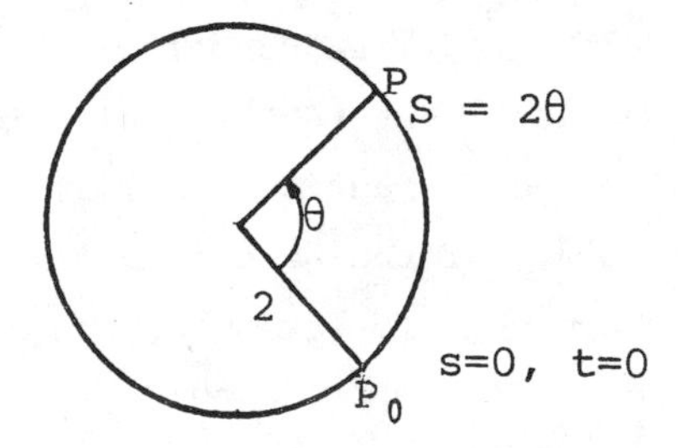

Solution: The angular displacement is:

$$\Theta = \frac{t}{t+1}.$$

The time-rate of change of the angle Θ is the angular velocity and is denoted by ω (rad./sec.).

$$\omega = \frac{d\Theta}{dt} = \frac{1}{(t+1)^2}.$$

The time-rate of change of angular velocity is called the angular acceleration, α.

$$\alpha = \frac{d\omega}{dt} = \frac{d^2\Theta}{dt^2} = -\frac{2}{(t+1)^3}.$$

For linear correspondence, if s (ft.) denotes the displacement measured algebraically along the arc of the circle from P_o to P_1 we have:

$$s = r\,\Theta, \quad \text{where } r, \text{ in this case, is } \alpha.$$

The velocity v (ft./sec.) and acceleration a (ft./sec^2.) along the circle are:

$$v = \frac{ds}{dt} = 2\,\frac{d\Theta}{dt} = 2\omega,$$

and

$$a = \frac{dv}{dt} = 2\,\frac{d\omega}{dt} = 2\alpha.$$

Now, following the diagram, let the particle be situated at point P_o for t = 0. Hence the particle starts its motion from P_o with initial velocity and acceleration as follows:

$$\omega_o = 1 \text{ (rad./sec.)} \quad \text{or} \quad v_o = 2 \text{ (ft./sec.)}.$$

$$\alpha_o = -\,2 \text{ (rad./sec}^2\text{.)} \text{ or } a_o = -\,4 \text{ (ft./sec}^2\text{.)}.$$

The negative sign in the acceleration indicates that the particle is undergoing deceleration, thereby losing velocity with time. For $t >> 1$, we can approximate:

$$\omega \simeq \frac{1}{t^2},$$

which shows that the velocity varies inversely proportionally to the square of time. Neglecting any resisting medium, the particle stops only when $t = \infty$. In practice, however, this does not happen.

• PROBLEM 992

The equations:

$$x = 2t - 1, \quad y = 4t^2 - 1,$$

give the coordinates (x,y) of a moving particle in feet when time t is measured in seconds. Find the tangential and normal components of acceleration.

<u>Solution:</u> Acceleration is given by:

$$\vec{A} = \frac{d^2s}{dt^2}\,\hat{T} + K\left(\frac{ds}{dt}\right)^2\hat{N},$$

where d^2s/dt^2 is the scalar acceleration of the particle along the curve s and is the tangential component of the acceleration vector $\vec{A}$. K is the radius of curvature, expressed, in parametric forms, as:

$$K = \frac{\frac{dx}{dt}\frac{d^2y}{dt^2} - \frac{dy}{dt}\frac{d^2x}{dt^2}}{\left[\left(\frac{dx}{dt}\right)^2 + \left(\frac{dy}{dt}\right)^2\right]^{\frac{3}{2}}} .$$

$$\frac{ds}{dt} = \left[\left(\frac{dx}{dt}\right)^2 + \left(\frac{dy}{dt}\right)^2\right]^{\frac{1}{2}}$$

$$= (4 + 64t^2)^{\frac{1}{2}} = 2(1 + 16t^2)^{\frac{1}{2}}.$$

The tangential component of the acceleration is:

$$\frac{d^2s}{dt^2} = \frac{32t}{\sqrt{1 + 16t^2}} .$$

Using the above relation for K, we have:

$$K = \frac{4}{\left(1 + 16t^2\right)^{\frac{3}{2}}}$$

and

$$\left(\frac{ds}{dt}\right)^2 = 4(1 + 16t^2) \cdot$$

Thus, the normal acceleration is:

$$K\left(\frac{ds}{dt}\right)^2 = \frac{16}{\sqrt{1 + 16t^2}} .$$

$\vec{R} = x\hat{i} + y\hat{j} = (2t + 1)\,\hat{i} + (4t^2 - 1)\hat{j}$, hence the tangential unit vector is obtained as;

$$\hat{T} = \frac{d\vec{R}}{ds} = \frac{d\vec{R}}{dt}\,\frac{dt}{ds} ,$$

where

$$\frac{d\vec{R}}{dt} = 2\,\hat{i} + 8t\,\hat{j},$$

and

$$\frac{dt}{ds} = \frac{1}{2\sqrt{1 + 16t^2}} .$$

Hence,

$$\hat{T} = \frac{(\hat{\imath} + 4t\,\hat{\jmath})}{\sqrt{1 + 16t^2}} .$$

By use of the relation: $\hat{T} \cdot \hat{N} = 0$, the normal vector is:

$$\hat{N} = \frac{-4t\,\hat{\imath} + \hat{\jmath}}{\sqrt{1 + 16t^2}} .$$

• PROBLEM 993

Find the tangential and normal components of acceleration at any time t for the motion determined by the parametric equations:

$$x = t^2 + 1, \ y = 2t.$$

Solution:

$$v_x = \frac{dx}{dt} = 2t, \quad v_y = \frac{dy}{dt} = 2,$$

Therefore,

$$v^2 = v_x{}^2 + v_y{}^2 = (2t)^2 + 2^2$$

$$= 4\left(t^2 + 1\right).$$

To find the tangential acceleration - acceleration along the path - we solve for v from the above expression and differentiate:

$$v^2 = 4\left(t^2 + 1\right)$$

$$v = 2\sqrt{t^2 + 1}$$

$$a_T = \frac{dv}{dt} = 2 \cdot \frac{1}{2} \cdot 2t \cdot \frac{1}{\sqrt{t^2 + 1}}$$

$$a_T = \frac{2t}{\sqrt{t^2 + 1}} .$$

Furthermore, from the law of centripetal forces,

$$F = \frac{mv^2}{R} = ma_N ,$$

where m is the mass of the particle, R is the radius of curvature, and a_N is the normal component of the acceleration.

Thus

$$a_N = v^2/R.$$

The radius of curvature can be obtained from the formula:

$$R = \left[\left(\frac{dx}{dt}\right)^2 + \left(\frac{dy}{dt}\right)^2\right]^{\frac{3}{2}} \div \left|\frac{dx}{dt}\frac{d^2y}{dt^2} - \frac{dy}{dt}\frac{d^2x}{dt^2}\right| ,$$

or, in this case,

$$R = \frac{v^3}{v_x a_y - v_y a_x}$$

$$= \frac{\left[2\sqrt{t^2 + 1}\right]^3}{(2t)(0) - (2)(2)} = \frac{8(t^2 + 1)^{\frac{3}{2}}}{-4}$$

$$= 2(t^2 + 1)^{\frac{3}{2}},$$

but, since R is always a positive quantity,

$$R = 2(t^2 + 1)^{\frac{3}{2}} .$$

Recalling that

$$a_N = \frac{v^2}{R} = \frac{4(t^2 + 1)}{2(t^2 + 1)^{3/2}},$$

$$= 2(t^2 + 1)^{1-\frac{3}{2}} = 2(t^2 + 1)^{-\frac{1}{2}}$$

$$= \frac{2}{\sqrt{t^2 + 1}} .$$

• PROBLEM 994

A particle moves along the curve: $y^2 = x^3$. If its horizontal velocity is constant and equal to

2 ft/sec., find (a) the angular velocity (ω) and angular acceleration (α) and (b) ω and α at the point (1,1).

Solution: The angular velocity and angular acceleration can be found when the motion of the particle is expressed in terms of its angular displacement Θ. Θ is the angle made by a straight line from the origin to the momentary position of the particle (called the position vector) on the positive x-axis. This gives

$$\Theta = \tan^{-1} (y/x).$$

Since the particle's trajectory is specified by the relationship

$$y^2 = x^3, \qquad \text{or} \qquad y = x^{3/2}$$

we obtain

$$\Theta = \tan^{-1}\left(\frac{x^{3/2}}{x}\right) = \tan^{-1} x^{1/2}.$$

(a) Now, the angular velocity is the rate of change of the angle Θ, and the angular acceleration is the rate of change of the angle swept in a unit of time. In mathematical form:

$$\omega = \frac{d\Theta}{dt} = \frac{1}{1 + x}\left(\frac{1}{2} x^{-\frac{1}{2}} \frac{dx}{dt}\right).$$

(This is obtained using the chain rule, letting $x^{1/2} = u$, then $\frac{d\Theta}{dt} = \frac{d\Theta}{du}\frac{du}{dt}$.)

$$\omega = \frac{1}{\sqrt{x}\,(1 + x)}, \text{ using } \frac{dx}{dt} = 2.$$

Furthermore,

$$\alpha = \frac{d^2\Theta}{dt^2} = -\frac{\frac{d}{dt}\left[\sqrt{x}\,(1 + x)\right]}{x(1 + x)^2}$$

$$= \frac{-\frac{1}{2}\left(\frac{1 + x}{\sqrt{x}}\right) + \sqrt{x}}{x(1 + x)^2} \cdot \frac{dx}{dt}$$

$$= \frac{-\left(\frac{1}{2} x^{-\frac{1}{2}} + \frac{3}{2} x^{\frac{1}{2}}\right)}{x(1 + x)^2} \frac{dx}{dt}$$

$$= -\frac{1 + 3x}{x^{\frac{3}{2}}(1 + x)^2}$$

(b) $\omega|_{(1,1)} = \frac{1}{2}$ rad./sec.

$\alpha|_{(1,1)} = -1$ rad./sec^2.

• PROBLEM 995

Find the vector acceleration (magnitude and direction) of a body traveling on the path:

$$x^2 = 4y,$$

from left to right with a constant speed of 8, at the point (2, 1).

Solution: From the equation of the path, we obtain, by differentiation, that

$$2x \frac{dx}{dt} = \frac{4dy}{dt}, \text{ or,}$$

$$2x \cdot v_x = 4v_y. \qquad \text{(a)}$$

For $x = 2$, we have $4v_x = 4v_y$, or,

$$v_x = v_y. \qquad \text{(b)}$$

Since $v = 8$ at all times,

$$v_x{}^2 + v_y{}^2 = v^2 = 64. \qquad \text{(c)}$$

Solving equations (b) and (c) for v_x and v_y, we find:

$$v_x = 4\sqrt{2} = v_y.$$

To determine the vertical and horizontal components of acceleration, differentiation of equations (a) and (c) with respect to time gives:

$$\frac{d}{dt}\left(2x\, v_x\right) = \frac{d}{dt}\left(4v_y\right),$$

$$2x \frac{dv_x}{dt} + 2\frac{dx}{dt} \cdot v_x = 4\frac{dv_y}{dt}, \text{ or,}$$

$$2x\, a_x + 2 v_x{}^2 = 4 a_y \qquad \text{(d)}$$

$$\frac{d}{dt}\left(v_x{}^2\right) + \frac{d}{dt}\left(v_y{}^2\right) = \frac{d}{dt}\left(v^2\right)$$

$$2\,v_x \cdot a_x + 2\,v_y \cdot a_y = 0. \qquad \text{(e)}$$

Substituting $x = 2$, $v_x = 4\sqrt{2}$, $v_y = 4\sqrt{2}$

in (d) and (e) and solving the resulting equations for a_x and a_y, we find $a_x = -8$, $a_y = 8$. Then,

$$a = \sqrt{a_x^2 + a_y^2} = 8\sqrt{2},$$

$$\tan\phi = \frac{8}{-8} = -1,$$

$$\phi = 135°,$$

since a_x is negative and a_y is positive.

• PROBLEM 996

Find the vector acceleration (magnitude and direction) of a body traveling on the path:

$$y = \frac{x^3}{3},$$

with $v_y = 8$, at the point $\left(2, \frac{8}{3}\right)$.

Solution: Differentiating both sides of the equation of the path with respect to t, we obtain:

$$\frac{dy}{dt} = x^2 \frac{dx}{dt}, \text{ or,}$$

$$v_y = x^2 \cdot v_x. \qquad \text{(a)}$$

Differentiating once more gives the x- and y-components of the acceleration.

$$\frac{dv_y}{dt} = a_y = \frac{d}{dt}\left(x^2 v_x\right)$$

$$= 2x \frac{dx}{dt}\left(v_x\right) + x^2 \frac{dv_x}{dt}.$$

$$a_y = x^2 a_x + 2x\,v_x^2. \qquad \text{(b)}$$

But $v_y = 8$ for all values of t, so that

$a_y = \frac{dv_y}{dt} = 0$ at all times. Substitution of

$x = 2$ and $v_y = 8$ in (a) gives: $8 = 4\,v_x$, or $v_x = 2$.

Substitution of $x = 2$, $v_x = 2$ and $a_y = 0$ in (b)

gives: $0 = 4\,a_x + 16$, or $a_x = -4$. Then, from

$a^2 = a_x{}^2 + a_y{}^2$, we find $a = 4$. Also, since $a_y = 0$,

it follows that the acceleration vector is parallel to the X-axis.

• PROBLEM 997

A particle moves in a plane according to the parametric equations of motion:

$$x = -t^2, \quad y = t^3.$$

Find the magnitude and direction of the acceleration when

$$t = \frac{2}{3}.$$

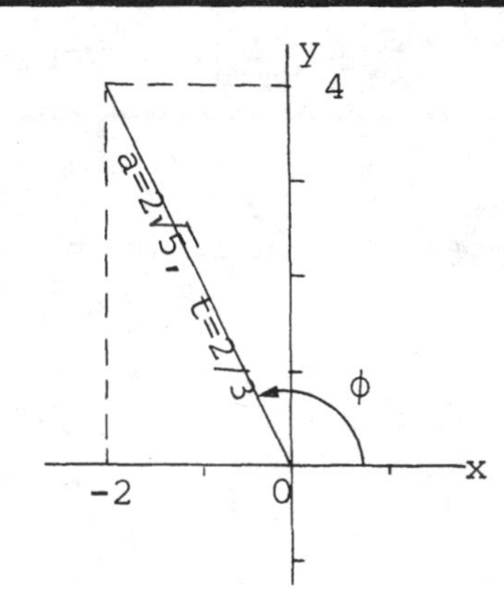

Solution:

$$v_x = \frac{dx}{dt} = -2t, \quad v_y = \frac{dy}{dt} = 3\,t^2.$$

Then,

$$a_x = \frac{dv_x}{dt} = -2, \quad a_y = \frac{dv_y}{dt} = 6\,t.$$

At $t = \frac{2}{3}$, we obtain: $a_x = -2$, $a_y = 4$.

Then,

$$a^2 = a_x{}^2 + a_y{}^2$$

$$= (-2)^2 + (4)^2$$

$$= 20.$$

$$a = 2\sqrt{5}.$$

$$\tan\phi = \frac{4}{-2} = -2, \ \phi = 116° \ 34',$$

since $\cos\phi = -\dfrac{2}{2\sqrt{5}}$ is negative, and

$\sin\phi = \dfrac{4}{2\sqrt{5}}$ is positive.

• PROBLEM 998

A particle moves up and to the right along the parabola $y^2 = 8x$ in the first quadrant. If it passes through the point (2, 4) with a speed of 2, how fast is it rising vertically at that point?

Solution: In order to obtain the velocities of the particle along the x- and y-axes, we differentiate the equation of the curve with respect to time:

$$2y \frac{dy}{dt} = 8 \frac{dx}{dt},$$

or

$$2 y v_y = 8 v_x.$$

At the point (2, 4) this becomes

$8 v_y = 8 v_x$, or,

$$v_y = v_x. \quad \text{(a)}$$

Since the speed, $v = 2$, we also have:

$$v_x^2 + v_y^2 = v^2 = 4. \quad \text{(b)}$$

Solving equations (a) and (b) for v_y, we find:

$$v_y = \sqrt{2}.$$

Alternatively, we find the slope of the curve at the point (2, 4) and resolve the given speed into its horizontal and vertical components:

$$y^2 = 8x, \quad \text{or} \quad y = 2\sqrt{2x}.$$

Then,

$$\frac{dy}{dx} = 2\sqrt{2} \cdot \frac{1}{2} \cdot \frac{1}{\sqrt{x}} = \frac{\sqrt{2}}{\sqrt{x}}.$$

At $x = 2$, we have $\frac{dy}{dx} = 1$.

Introducing dt into the above differential equation, and keeping the chain rule in mind:

$$\frac{dy}{dx} = \frac{dy}{dt} \cdot \frac{dt}{dx} = 1,$$

or,

$$\frac{dy}{dt} = \frac{dx}{dt}.$$

Thus,

$$\left(\frac{dy}{dt}\right)^2 + \left(\frac{dx}{dt}\right)^2 = v^2 = 4, \text{ or,}$$

$$2\left(\frac{dy}{dt}\right)^2 = 4.$$

Hence,

$$\frac{dy}{dt} = \sqrt{2}.$$

• PROBLEM 999

A particle moves along the right-hand part of the curve:

$$4y^3 = x^2, \text{ with a velocity } v_y = \frac{dy}{dt},$$

constant at 2. Find the speed and direction of motion at the point where $y = 4$.

Solution: Differentiating the equation of the path with respect to t, we have:

$$12y^2 \frac{dy}{dt} = 2x \frac{dx}{dt}, \text{ or,}$$

$$6y^2 \cdot v_y = x \cdot v_x. \tag{a}$$

From the equation of the path, we find that $x = 16$ when $y = 4$. Substituting these values of x and y and the value: $v_y = 2$, in equation (a), and solving for v_x, we obtain $v_x = 12$. Then,

$$v^2 = v_x^{\,2} + v_y^{\,2} = 144 + 4 = 148.$$

$$v = 2\sqrt{37}.$$

$$\tan \alpha = \frac{v_y}{v_x} = \frac{1}{6}, \quad \alpha = 9°28',$$

since $\cos \alpha = \dfrac{6}{2\sqrt{37}}$ and $\sin \alpha = \dfrac{2}{2\sqrt{37}}$

are both positive.

• PROBLEM 1000

> The equations of motion of a particle moving in a plane are: $x = t^2$, $y = 3t - 1$, where t is the time and x, y are rectangular coordinates. Find the path of the particle, and find the speed and direction of motion at the instant when t = 2.

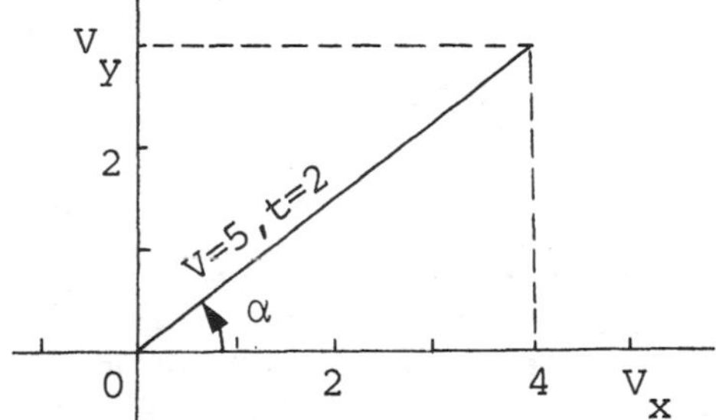

Solution: From $x = t^2$, $y = 3t - 1$, we find that the speed of the particle along the x-axis is:

$$v_x = \frac{dx}{dt} = 2t,$$

and the speed along the y-axis is:

$$v_y = \frac{dy}{dt} = 3.$$

When t = 2, we have $v_x = 4$, $v_y = 3$. Then,

$$v^2 = v_x^{\,2} + v_y^{\,2} = 4^2 + 3^2 = 25,$$

$$v = 5. \ \tan \alpha = \frac{3}{4}, \ \alpha = 36°52',$$

since $\cos \alpha = \frac{4}{5}$ and $\sin \alpha = \frac{3}{5}$ are both positive. Furthermore, eliminating t from the given equations of motion, we have the following:

solving $x = t^2$ for t, we have $t = \sqrt{x}$.

On the other hand, from $y = 3t - 1$,

$$t = \frac{1}{3}(y + 1).$$

Equating the expressions for t,

$$\frac{1}{3}(y + 1) = \sqrt{x}, \text{ or,}$$

$$\frac{1}{9}(y + 1)^2 = x.$$

Hence, the rectangular equation of the path is:

$$9x = (y + 1)^2,$$

which represents a parabola.

• PROBLEM 1001

A body moves along a straight line in such a way that its distance s (in feet) from a fixed point of the line, at time t seconds, is given by the equation:

$$s = t^3 - \frac{15}{2}t^2 + 12t + 10.$$

(a) When and where is its velocity momentarily 0? (b) During what periods and over what distances is it moving forward? Backward? (c) During what periods is its acceleration positive? Negative? (d) Sketch its motion.

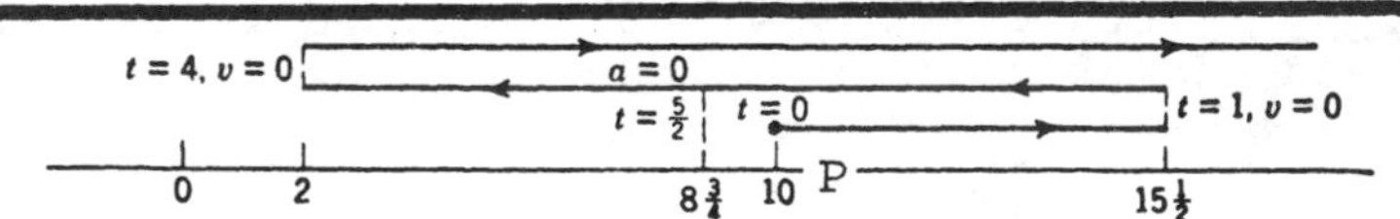

Solution: (a) Differentiating s with respect to t,

$$\frac{ds}{dt} = D_t s = v = 3t^2 - 15t + 12$$

$$= 3(t - 1)(t - 4).$$

Setting $v = 0$ to find when the body is momentarily stationary, we obtain:

$$t = 1, \quad \text{and} \quad t = 4.$$

To find the value of s at $t = 1$, we substitute $t = 1$ in the given expression for $s = f(t)$. Thus we find:

$$s = \frac{31}{2}, \text{ at } t = 1.$$

Similarly, for t = 4, s = 2. Hence the body is momentarily at rest at the end of 1 sec. when it is $\frac{31}{2}$ ft, positively, from the fixed starting point P, and again at the end of 4 sec., when it is 2 ft, positively, from P. The distance of starting point P from the origin, found by substituting t = 0 in the given equation, is 10.

(b) By examination of $v = 3(t - 1)(t - 4)$, we see that, for $t < 1$, v is positive, since for $t < 1$ both components $[(t - 1)$ and $(t - 4)]$ are negative, hence their product is positive. Similarly, by analyzing in this manner, for t between 1 and 4, v is negative, and for $t > 4$, v is positive.

(c) Differentiating v with respect to t, we have

$$\frac{d^2s}{dt^2} = D_t^{\,2}s = D_t v = a = 6t - 15 = 3(2t - 5).$$

It can be seen that a is positive for $t > \frac{5}{2}$ and negative for $t < \frac{5}{2}$. At $t = \frac{5}{2}$, the acceleration is 0. At this time the body is $8\frac{3}{4}$ ft. positively from 0. Thus, the body starts at t = 0, s = 10 with a positive velocity which is retarded by the negative acceleration until it stops at t = 1, $s = \frac{31}{2}$. The acceleration is still negative and produces a negative velocity which increases (negatively) until $t = \frac{5}{2}$, $s = 8\frac{3}{4}$, when the acceleration becomes momentarily 0. Thereafter, the acceleration is positive. This has the effect of slowing down the negative velocity (increasing it algebraically) until it becomes 0 at t = 4, s = 2. The acceleration is still positive, producing a positive velocity which increases continually after $t = \frac{5}{2}$.

The motion is shown diagrammatically in Fig. 1.

(d) If s is graphed against t on a rectangular coordinate system we obtain the graph in Fig. 2, and graphs for a and v are given in Fig. 3.

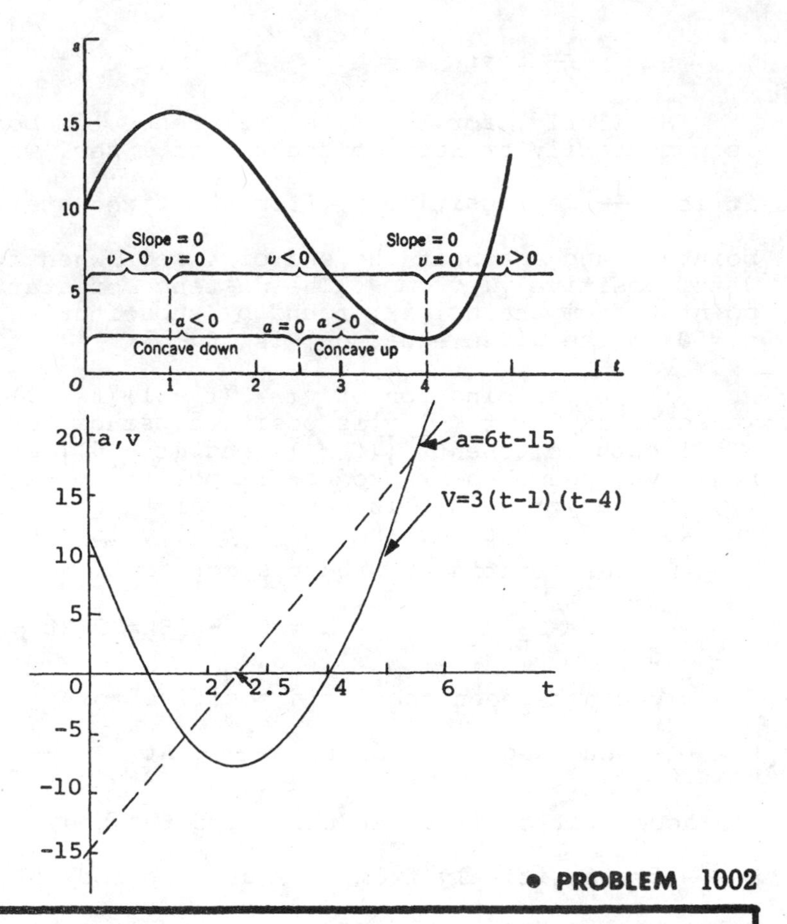

• **PROBLEM 1002**

A particle starts at the origin and travels along the curve: $y = x^2$. As it passes through the point (3,9), its velocity is such that the x-component, $V_x = D_t x$, is 2 ft./sec. Given that $D_x y = \frac{D_t y}{D_t x}$, find the y-component: $V_y = D_t y$, and the resultant speed $V = \sqrt{V_x^2 + V_y^2}$. Use the Δ-method.

Solution: We can find $V_y = D_t y$, using the formula: $D_x y = \frac{D_t y}{D_t x}$. We are given $V_x = D_t x = 2$. We find $D_x y$ of the curve $y = x^2$ by using the Δ method.

$$D_x y = \lim_{\Delta x \to 0} \left(\frac{\Delta y}{\Delta x}\right).$$

$$\frac{\Delta y}{\Delta x} = \frac{f(x + \Delta x) - f(x)}{\Delta x}$$

$$\frac{\Delta y}{\Delta x} = \frac{(x + \Delta x)^2 - x^2}{\Delta x}$$

$$= \frac{x^2 + 2x\Delta x + (\Delta x)^2 - x^2}{\Delta x}$$

$$= 2x + \Delta x$$

$$\lim_{\Delta x \to 0} \frac{\Delta y}{\Delta x} = \lim_{\Delta x \to 0} 2x + \Delta x = 2x.$$

$$D_x y = 2x.$$

It is given that $D_t x = 2$, $D_t y = (D_x y)(D_t x) = (2x)(2) = 4x$. At the point $(3,a)$, $D_t y = 4x = 4(3) = 12$. We can also calculate the speed V, using $V_x = D_t x = 2$, $V_y = D_t y = 12$, and the formula:

$$V = \sqrt{V_x^2 + V_y^2}.$$

$$V = \sqrt{(2)^2 + (12)^2}$$

$$= \sqrt{148}$$

$$= 12.25 \frac{\text{ft.}}{\text{sec.}}$$

• PROBLEM 1003

A particle moves along the parabola: $y^2 = 4x$, with a constant horizontal component of velocity of 2 ft./sec. Find the vertical components of velocity and acceleration at the point (1,2).

Solution: To solve this problem we use the fact that the horizontal component of velocity $= \frac{dx}{dt} =$ 2ft./sec. The vertical component of velocity $= \frac{dy}{dt}$. The horizontal component of acceleration $= \frac{d^2x}{dt^2} = 0$, and the vertical component of acceleration $= \frac{d^2y}{dt^2}$. Differentiating the given equation, we obtain:

$$y^2 = 4x.$$

$$2y\frac{dy}{dt} = 4\frac{dx}{dt}.$$

To find the vertical component of velocity, we solve for $\frac{dy}{dt}$, obtaining:

$$\frac{dy}{dt} = \frac{4}{2y}\frac{dx}{dt}.$$

We substitute $\frac{dx}{dt} = 2$ and $y = 2$, since we are concerned with the velocity at the point (1,2). Therefore, $\frac{dy}{dt} = \frac{4}{2(2)}(2) = 2$ft./sec., and the vertical component of velocity at the given point is 2 ft./sec.

To find the vertical component of acceleration, we determine $\frac{d^2y}{dt^2}$ by differentiating $\frac{dy}{dt}$. We find:

$$\frac{d^2y}{dt^2} = \left(\frac{4}{2y}\right)\frac{d^2x}{dt^2} + \left(\frac{dx}{dt}\right)\left[\frac{-4(2)}{(2y)^2}\frac{dy}{dt}\right].$$

We substitute $\frac{dx}{dt} = 2$, $\frac{d^2x}{dt^2} = 0$, $y = 2$, $\frac{dy}{dt} = 2$, and obtain:

$$\frac{d^2y}{dt^2} = \frac{4}{2(2)}(0) + (2)\frac{-8}{(2 \cdot 2)^2} \cdot 2$$

$$= 0 + 2\left(-\frac{16}{16}\right)$$

$$= -2.$$

Therefore, the vertical component of acceleration at the point is -2ft./sec.^{2}.

• PROBLEM 1004

A particle travels along the parabola:

$$y = \frac{1}{3}x^2,$$

keeping its vertical velocity component constant at the value 8. Find the magnitude of the resultant velocity when the particle is at the point $\left(2, \frac{4}{3}\right)$.

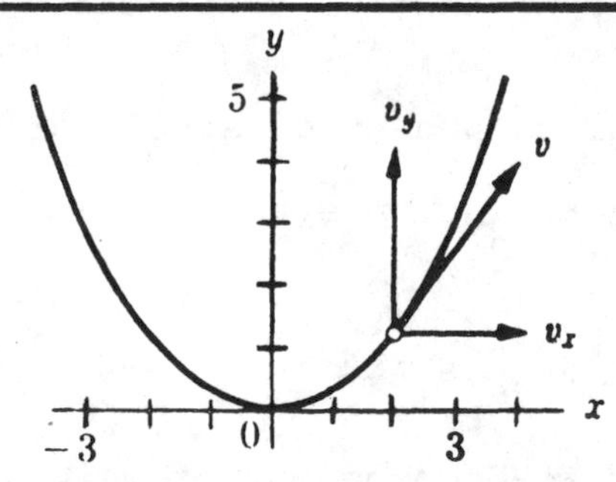

Solution: Since both y and x change with time, both can be considered as functions of time. Thus we can take derivatives of

$$y = \frac{1}{3}x^2$$

with respect to time. We obtain the following result:

$$\frac{dy}{dt} = \frac{2}{3}x\frac{dx}{dt},$$

dy/dt and dx/dt are, respectively, the vertical and horizontal components of the particle's velocity, and dy/dt = 8 is given.

At the point $\left(2, \frac{4}{3}\right)$ we have $8 = \frac{2}{3}(2)\frac{dx}{dt}$.

Thus, $\frac{dx}{dt} = 6$ at this point. Therefore, the resultant velocity,

$$v = \sqrt{\left(\frac{dx}{dt}\right)^2 + \left(\frac{dy}{dt}\right)^2}, \text{ or}$$

$$v = \sqrt{64 + 36} = 10.$$

• PROBLEM 1005

A vehicle moves along a trajectory having coordinates given by:

$$x = t^3, \quad \text{and:} \quad y = 1 - t^2.$$

The acceleration of the vehicle at any point on the trajectory is a vector, having magnitude and direction. Find the acceleration when t = 2.

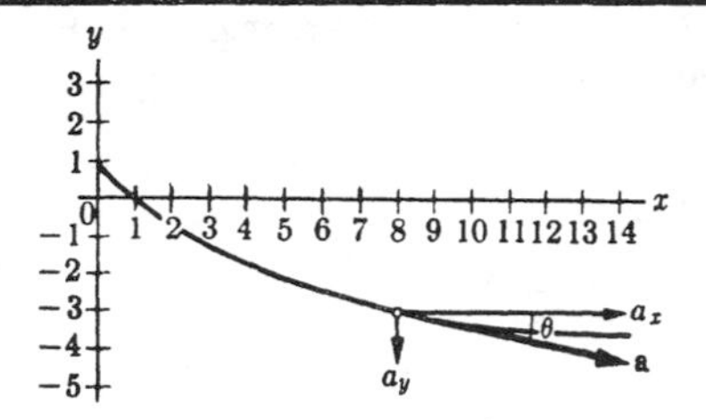

Solution: We first evaluate the velocity and acceleration components. Then we combine vectorially the acceleration components to find the resultant magnitude and direction of the acceleration.

$$v_x = \frac{dx}{dt} = 3t^2,$$

$$a_x = \frac{dv_x}{dt} = \frac{d^2x}{dt^2} = 6t, \qquad a_x|_{t=2} = 12,$$

$$v_y = \frac{dy}{dt} = -2t,$$

$$a_y = \frac{dv_y}{dt} = \frac{d^2y}{dt^2} = -2, \qquad a_y|_{t=2} = -2,$$

$$a|_{t=2} = \sqrt{144 + 4} = \sqrt{148} = 12.2,$$

$$\tan\Theta = \frac{a_y}{a_x} = -\frac{2}{12} = -\frac{1}{6}, \quad \Theta = -9.5°.$$

A particle moves on the circle $r = 2\cos\Theta$ in a counterclockwise direction with a constant speed of 4 units per second. If the particle is at the origin when $t = 0$, determine a relation describing the motion as a function of time.

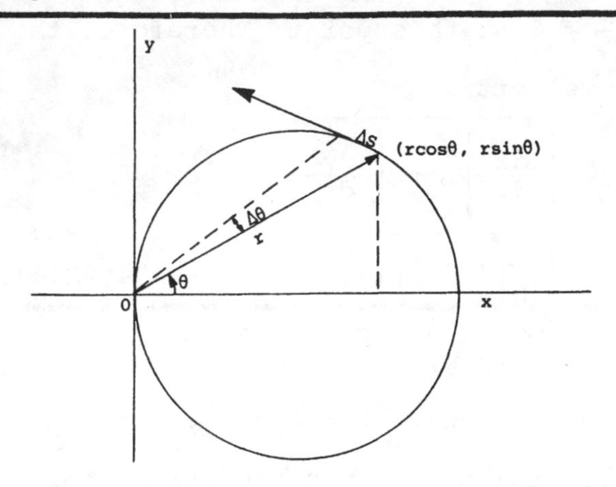

Solution: Using the fact that the tangential velocity of the particle, 4 units per second, can be expressed in terms of its components along the x- and y-axes, we have

$$v^2 = 16 = \left(\frac{dx}{dt}\right)^2 + \left(\frac{dy}{dt}\right)^2.$$

The x- and y- components of the position vector, r can be obtained as

$$x = r\cos\Theta \qquad \text{and} \qquad y = r\sin\Theta,$$

Hence, differentiation gives:

$$\frac{dx}{dt} = \frac{dr}{dt}\cos\Theta - r\sin\Theta\,\frac{d\Theta}{dt}\,.$$

$$\frac{dy}{dt} = \frac{dr}{dt}\sin\Theta + r\cos\Theta\,\frac{d\Theta}{dt}\,.$$

Thus,

$$v^2 = 16 = \left(\frac{dr}{dt}\right)^2\cos^2\Theta - 2r\sin\Theta\cos\Theta\,\frac{dr}{dt}\,\frac{d\Theta}{dt}$$

$$+ r^2\sin^2\Theta\left(\frac{d\Theta}{dt}\right)^2 + \left(\frac{dr}{dt}\right)^2\sin^2\Theta$$

$$+ 2r\sin\Theta\cos\Theta\,\frac{dr}{dt}\,\frac{d\Theta}{dt}$$

$$+ r^2\cos^2\Theta\left(\frac{d\Theta}{dt}\right)^2$$

$$= \left(\frac{dr}{dt}\right)^2 (\cos^2 \Theta + \sin^2 \Theta)$$

$$+ r^2 \left(\frac{d\Theta}{dt}\right)^2 (\sin^2 \Theta + \cos^2 \Theta)$$

$$= \left(\frac{dr}{dt}\right)^2 + r^2 \left(\frac{d\Theta}{dt}\right)^2 ,$$

where the **identity**: $\sin^2 \Theta + \cos^2 \Theta = 1$, is used.

But, since

$r = 2 \cos \Theta$,

$$\frac{dr}{dt} = - 2 \sin \Theta \frac{d\Theta}{dt} .$$

Substituting this gives:

$$v^2 = 16 = 4 \sin^2 \Theta \left(\frac{d\Theta}{dt}\right)^2 + 4 \cos^2 \Theta \left(\frac{d\Theta}{dt}\right)^2$$

$$= 4 \left(\frac{d\Theta}{dt}\right)^2 (\sin^2 \Theta + \cos^2 \Theta).$$

Using the trigonometric identity again, we obtain:

$$4 = \left(\frac{d\Theta}{dt}\right)^2$$

or $$\frac{d\Theta}{dt} = \pm 2.$$

Since the object is moving counterclockwise, Θ increases with time, thus we choose the positive sign for the angular velocity. Now, integration gives:

$$\Theta = 2t + k.$$

The particle is at the origin when $t = 0$.

$$r(t) = 2 \cos \Theta(t_o) = 0$$

$\Theta(t=0) = \pi/2 + n\pi$, where $n = 1, 2, 3, \ldots$.

We choose $\frac{\pi}{2}$, since Θ is a periodic

function, hence the choice is immaterial. Therefore,

$$\Theta = 2t + \frac{\pi}{2}.$$

Now,

$$x = r \cos \Theta = 2 \cos \Theta(\cos \Theta) = 2 \cos^2 \Theta.$$

$$= 2\left(\frac{1 + \cos 2\Theta}{2}\right) = 1 + \cos 2\Theta$$

$$= 1 + \cos 2\left(2t + \frac{\pi}{2}\right)$$

$$= 1 + \cos (4t + \pi)$$

$$y = r \sin \Theta = 2 \cos \Theta \sin \Theta = \sin 2\Theta$$

$$= \sin 2\left(2t + \frac{\pi}{2}\right) = \sin(4t + \pi).$$

• PROBLEM 1007

Discuss the rectilinear motion, the equation of which is:

$$s = e^{-t} \sin t,$$

s being measured in feet and t in seconds.

Solution: The equation of motion indicates that the motion is a harmonic oscillation with fixed period but subject to a resistance force proportional to the velocity. In most cases it is called the damped harmonic oscillation. Since

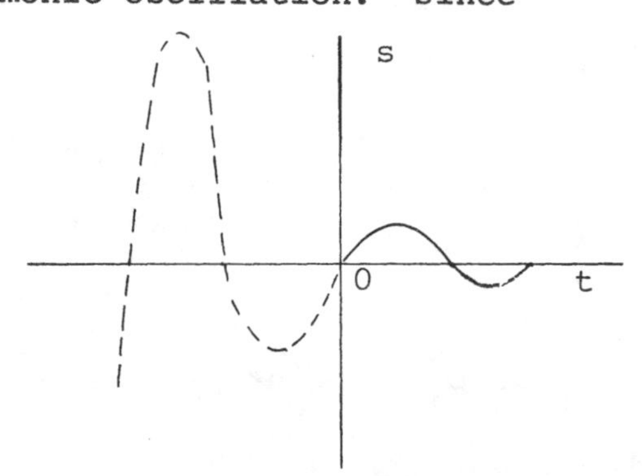

$$V = \frac{ds}{dt} = e^{-t} (\cos t - \sin t),$$

the particle comes to rest when time increases infinitely. Its velocity also goes through zero at

$$t = \frac{(2n - 1)\pi}{2}, \quad n = \pm 1, 2, 3, \ldots\ldots.$$

The angular frequency is unity.

• PROBLEM 1008

A particle has a trajectory given parametrically by $x = a\Theta - b \sin \Theta$ and $y = a - b \cos \Theta$. (a) Express the trajectory in terms of Cartesian coordinates alone. (b) Derive an expression for the velocity of the particle along the trajectory. (c) Determine the point at which the trajectory has the greatest inclination (angle of elevation).

Solution: From the given expressions:

$$x = a\Theta - b \sin \Theta, \text{ and}$$
$$y = a - b \cos \Theta,$$

we can eliminate Θ.

$$\cos \Theta = \frac{a - y}{b}$$

or

$$\Theta = \cos^{-1} \frac{a - y}{b}.$$

The trigonometric identity gives:

$$\sin^2 \Theta + \cos^2 \Theta = 1.$$

We wish to express $\sin \Theta$ in terms of $\cos \Theta$. Hence,

$$\sin \Theta = \sqrt{1 - \cos^2 \Theta} = \left[1 - \left(\frac{a - y}{b}\right)^2\right]^{\frac{1}{2}}$$

$$= \frac{\sqrt{b^2 - (a - y)^2}}{b}.$$

Thus,

$$x = a \cos^{-1}\left(\frac{a - y}{b}\right) - \sqrt{b^2 - (a - y)^2},$$

the required equation.

(b) Since x, y and Θ must be functions of t,

$$v_x = \frac{dx}{dt} = a \frac{d\Theta}{dt} - b \cos \Theta \frac{d\Theta}{dt}$$

$$= (a - b \cos \Theta) \frac{d\Theta}{dt}$$

$$v_y = \frac{dy}{dt} = b \sin \Theta \frac{d\Theta}{dt}.$$

The velocity of the particle along its trajectory comprises these two components and, using the Pythagorean theorem, we have:

$$v^2 = u_x{}^2 + u_y{}^2$$

$$= (a - b\cos\Theta)^2 \left(\frac{d\Theta}{dt}\right)^2 + b^2 \sin^2\Theta \left(\frac{d\Theta}{dt}\right)^2$$

$$= (a^2 - 2ab\cos\Theta + b^2\cos^2\Theta)\left(\frac{d\Theta}{dt}\right)^2$$

$$+ b^2 \sin^2\Theta \left(\frac{d\Theta}{dt}\right)^2$$

$$= \left[a^2 + b^2(\cos^2\Theta + \sin^2\Theta) - 2ab\cos\Theta\right]\left(\frac{d\Theta}{dt}\right)^2$$

$$= (a^2 + b^2 - 2ab\cos\Theta)\left(\frac{d\Theta}{dt}\right)^2 .$$

$$v = (a^2 + b^2 - 2ab\cos\Theta)^{1/2}\frac{d\Theta}{dt} .$$

(c) The inclination is greatest where the slope reaches a maximum.

$$y' = \frac{dy}{dx} = \frac{\frac{dy}{dt}}{\frac{dx}{dt}} = \frac{Vy}{Vx} = \frac{b\sin\Theta}{a - b\cos\Theta} .$$

$$\frac{dy'}{d\Theta} = \frac{b\cos\Theta\,(a - b\cos\Theta) - b^2\sin^2\Theta}{(a - b\cos\Theta)^2}$$

$$= \frac{ab\cos\Theta - b^2\cos^2\Theta - b^2\sin^2\Theta}{(a - b\cos\Theta)^2}$$

$$= \frac{ab\cos\Theta - b^2(\cos^2\Theta + \sin^2\Theta)}{(a - b\cos\Theta)^2}$$

$= \dfrac{b(a\cos\Theta - b)}{(a - b\cos\Theta)^2}$. For a maximum, we set

$\dfrac{dy'}{d\Theta} = 0$, occurring when $\cos\Theta = b/a$, or

$$\Theta = \cos^{-1}\left(\frac{b}{a}\right) .$$

This is meaningful only for $b \leq a$. For $b > a$, the expression: $(a - b \cos \Theta)$ goes to 0 when $\cos \Theta = \frac{a}{b}$, giving $y' = \infty$, a vertical slope.

If $b < a$, $= \cos^{-1}(b/a)$ gives the maximum slope.

If $a < b$, direction of motion is vertical when $\Theta = \cos^{-1}(a/b)$.

• PROBLEM 1009

The acceleration of a free-falling body is g. Find the distance-time equation.

Solution: We shall assume that other forces, such as air resistance, are absent and that the force of gravity is constant. Results obtained with these assumptions are generally fairly close approximations to the facts as long as the motion takes place near the surface of the earth at relatively low speeds. Experiments have shown that near the surface of the earth the acceleration produced by the force of gravity is approximately 32 ft per second per second in the downward vertical direction. Since acceleration is the rate of change of velocity, it is expressed in units of velocity per unit time. Thus, if velocity is expressed in feet per second, 32 ft per sec. per sec. means the rate of change of 32 units of velocity per unit time. For brevity, this is also shown as

32 ft per sec^2. or 32 ft/sec^2.

Since acceleration is equal to the derivative of velocity, we see that $a = g$ results from the variable velocity, $v = gt + C$. This is a result of using the relationship of acceleration and velocity in a differential form:

$$a = \frac{dv}{dt} = g, \quad \text{or} \quad dv = gdt.$$

The first integration gives

$$v = gt + C, \qquad \text{(a)}$$

where C is any integration constant, the value of which is determined under the boundary conditions: when $t = 0$, the free-falling body has an initial velocity, say v_i. Then,

putting $t = 0$ in equation (a), we find $v_i = C$.

Since v, the velocity, is equal to $\frac{ds}{dt}$, the derivative of s, the distance, with respect to time, i.e,

$$v = \frac{ds}{dt}, \text{ or, conversely,}$$

$$s = \int v \, dt .$$

$$v = gt + C .$$

Therefore,

$$s = \int v \, dt = \int (gt + C) \, dt = \int gt \, dt + \int C \, dt$$

$$= \frac{1}{2} gt^2 + Ct + C_1$$

$$= \frac{1}{2} gt^2 + v_i t + C_1 .$$

When $t = 0$, the falling body is at a height s_i, which is the distance of the body from the reference point. This is usually called the initial distance.

Thus,

$$s(\text{at } t = 0) = s_i = C_1 ,$$

and we have the complete distance-time equation:

$$s = \frac{1}{2} gt^2 + v_i t + s_i .$$

The above expressions for s, v and a are based on the assumption that the downward direction is positive, since the force of gravity produces a downward acceleration, which has been given the positive sign. If we wish to consider the upward direction, the sign of the equation should be changed to:

$$a = - g$$

$$v = - gt + v_i$$

$$s = - \frac{1}{2} gt^2 + v_i t + s_i ,$$

where v_i and s_i are, once more, respectively, the initial velocity and distance.

• PROBLEM 1010

A body falls under the influence of gravity ($g \approx 32$ ft./sec^2.) so that its speed is $v = 32t$. Determine the distance it falls in 3 sec. Let x = distance.

Solution:

$$v = f(t) = 32t$$

The velocity is dependent on time because of the following general relationship:

$$v = gt + v_i$$

where v increases indefinitely as time goes on -

neglecting air resistance and some other factors. The initial velocity v_i is zero in this case because the body starts from rest.

Assuming the distance covered is dx in time t, we can represent the velocity in a differential form:

$$\frac{dx}{dt} = v = 32t.$$

Integrating to find the relationship between x and t yields:

$$\int dx = \int 32t\ dt.$$

$$x = \frac{32t^2}{2} + C.$$

x = 0 when t = 0. Therefore, $0 = 16(0)^2 + C$, $C = 0$.

$$x = 16t^2.$$

The distance the body falls from the reference point,

$$x = 16t^2 = 16(3)^2 = 144 \text{ ft.}$$

• PROBLEM 1011

A projectile is fired with a velocity of 1800 ft./sec. at an angle of 27° with the horizontal. Find the equations that describe its motion. Ignore air resistance.

Solution: Assuming the projectile to be moving in a vertical plane, we write the vector expression of the physical law - that the acceleration of gravity is constant:

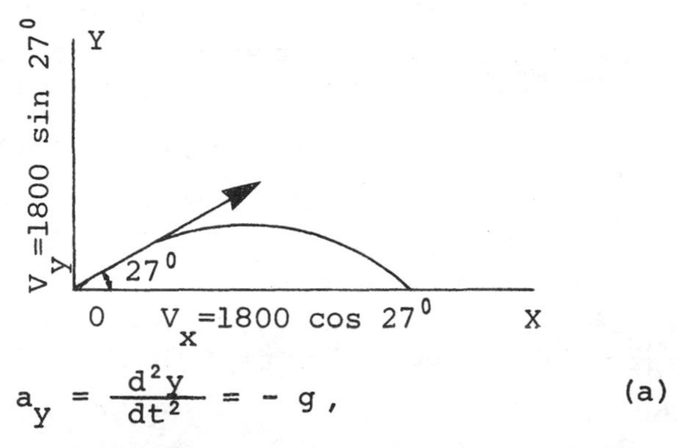

$$a_y = \frac{d^2y}{dt^2} = -g, \tag{a}$$

where g is the constant value of the acceleration of gravity. Direct measurements show that

$$g = 32.2 \text{ ft./sec}^2$$

(approximately), and this constant is known as the gravitational constant. Furthermore, since there are no other forces or accelerations on the body, the horizontal acceleration is obviously zero. Hence,

$$a_x = \frac{d^2x}{dt^2} = 0. \qquad (b)$$

Integrating equation (a) with respect to t, we have:

$$v_y = \int - g\ dt = \int - 32.2\ dt$$

$$v_y = - 32.2\ t + C_1,$$

where C_1 is a constant of integration.

To find the equation of motion of the particle along the y-axis, we integrate further:

$$y = \int v_y d_y = - 16.1t^2 + C_1 t + C_2 \qquad (c)$$

where C_2 is another constant of integration.

In a similar manner, we wish to find the velocity and the equation of motion along the x-axis:

$$v_x = \int a_x dt = 0 + C,$$

and

$$x = \int v_x dt = \int Cdt .$$

$$x = Ct + C_3, \qquad (d)$$

where C_2 and C_3 are constants of integration, the values of which are determined under the actual boundary conditions: when t = 0, x = 0 and y = 0.

Substitution of t = 0 in equations (c) and (d) gives:

$$C_2 = 0 \text{ and } C_3 = 0.$$

Also, when t = 0,

$$\frac{dx}{dt} = C \quad \text{and} \quad \frac{dy}{dt} = C_1 .$$

But

$$\frac{dy}{dt} = v_y = \text{the vertical component of velocity}$$

$$= 1,800 \sin 27°$$

$$v_y = 1,800 \cdot 0.454 = 817.2 = C_1,$$

and

$$\frac{dx}{dt} = v_x = \text{the horizontal component of}$$

$$\text{velocity} = 1{,}800 \cos 27°.$$

$$v_x = 1{,}800 \cdot 0.891 = 1{,}603.8 = C.$$

Substituting back in equations (c) and (d) gives:

$$\left.\begin{aligned} x &= 1{,}603.8t \\ y &= 817.2t - 16.1t^2 \end{aligned}\right\} \text{equations of motion of projectile.}$$

• **PROBLEM 1012**

A body is dropped from the top of a building 120 ft high. When will it reach the ground, and with what velocity?

Solution: Since the body is merely dropped, its initial velocity is 0. Keeping in mind the expressions for s, v and a, i.e.:

$$a = g \tag{1}$$

$$v = gt + v_o \tag{2}$$

$$s = \frac{1}{2} gt^2 + v_o t + s_o, \tag{3}$$

we can solve problems of this sort generally, problems involving rectilinear motion of bodies moving vertically under the influence of gravity. The positive or negative sign of the gravitational acceleration (g) depends on our choice of taking downward and upward directions as positive. Now, if we think of the distances as measured from the ground, and the upward direction is positive, the initial distance is 120. Putting:

$$v_o = 0, \quad s_o = 120,$$

and recalling that

$$g = -32 \text{ ft/sec}^2.,$$

in equations (2) and (3), we have:

$$v = -32t, \tag{4}$$

$$s = -16t^2 + 120. \tag{5}$$

By choosing the upward direction positive, we mean that our measurements are performed on the surface of the earth, so that the initial distance of the object was 120 ft. high. Thus, to find when the body reaches the ground, we set s equal to 0. From (4,5), we obtain:

$$-16t^2 + 120 = 0,$$

$$16t^2 = 120, \qquad t^2 = 15/2,$$

and $t = \sqrt{15/2} = 2.7$ approximately.

Substituting $t = \sqrt{15/2}$ in (4), we find:

$v = -32\sqrt{15/2} = -86.4$, approximately.

Hence, the body strikes the ground in approximately 2.7 sec., with a velocity of approximately 86.4 ft. per sec. downward.

• PROBLEM 1013

The path of a projectile is described by the equation:

$$y = x \tan\alpha - \frac{gx^2}{2v^2\cos^2\alpha} .$$

The initial velocity is 1000 ft./sec. at an angle of 30° with the horizontal. What is the maximum height and what is the range (horizontal distance)?

Solution: $$y = x\tan\alpha - \frac{g}{2v^2\cdot\cos^2\alpha}\cdot x^2$$

We find the slope of the curve and set that slope equal to 0 to find the maximum.

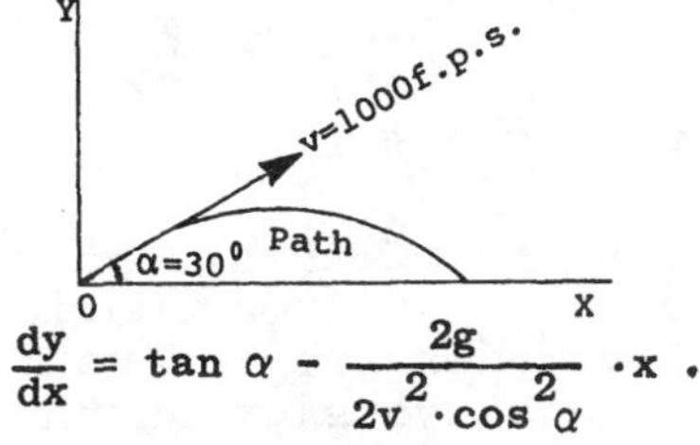

$$\frac{dy}{dx} = \tan\alpha - \frac{2g}{2v^2\cdot\cos^2\alpha}\cdot x .$$

For a maximum, $$\tan\alpha - \frac{g}{v^2\cos^2\alpha}\cdot x = 0 .$$

$$x = \frac{\tan\alpha\cdot v^2\cdot\cos^2\alpha}{g} = \frac{\sin\alpha}{\cos\alpha}\cdot\frac{v^2\cos^2\alpha}{g} .$$

$x = \frac{v^2}{g}\cdot\sin\alpha\cdot\cos\alpha$, the point where the maximum height occurs, and $2x = \frac{2v^2}{g}\cdot\sin\alpha\cdot\cos\alpha = \frac{v^2}{g}\cdot\sin 2\alpha$, the range (horizontal distance).

We substitute the value of x in the expression for y above, getting

$$y = \frac{\sin\alpha}{\cos\alpha}\cdot\sin\alpha\cdot\cos\alpha\,\frac{v^2}{g} - \frac{g}{2v^2\cos^2\alpha}\cdot\frac{v^4}{g^2}\sin^2\alpha\cos^2\alpha.$$

Finally, $y = \frac{v^2}{g}\sin^2\alpha - \frac{v^2}{2g}\sin^2\alpha = \frac{v^2}{2g}\sin^2\alpha$, the maximum height.

For $v = 1{,}000$ ft. per sec., $\alpha = 30^\circ$, $g = 32.2$ ft./sec.^{2},

$\sin\alpha = 0.5$, $\sin 2\alpha = \sin 60^\circ = 0.866$.

Then $y = \frac{(1{,}000^2)}{2\cdot 32.2}\cdot(\tfrac{1}{2})^2 = \frac{1{,}000{,}000}{8\cdot 32.22} = 3{,}882$ ft., the maximum height,

and $2x = \frac{1{,}000{,}000}{32.2}\cdot 0.866 = 26{,}894$ ft., the range.

• **PROBLEM 1014**

A ball is thrown vertically upward, and its distance from the ground is given by:

$$s = 104t - 16t^2$$

(a) Find the height to which the ball will rise, if s is expressed in feet when t is in seconds.
(b) Compute the acceleration.

Solution: (a) The velocity is given by:

$$v = \frac{ds}{dt} = (104 - 32t)\text{ft/sec.},$$

The ball rises until s is a maximum, or the velocity is zero, that is:

$$v = \frac{ds}{dt} = 104 - 32t = 0. \qquad (a)$$

Solving for t in the above expression gives the time to travel to the maximum height.

Thus, from equation (a),

$$t = \frac{104}{32} = 3\frac{1}{4} \text{ sec.}$$

The height to which it rises is the value of s for $t = 3\frac{1}{4}$ sec., or,

$$s = 104\left(3\frac{1}{4}\right) - 16\left(3\frac{1}{4}\right)^2$$

$$s = 169 \text{ ft.}$$

(b) The acceleration of the particle is the derivative of the expression for velocity:

$$a = \frac{dv}{dt} = -32 \text{ ft/sec}^2.$$

The acceleration is a negative quantity in this case. This implies that the velocity has been reduced with time, due to the gravitational force. This kind of negative acceleration is also called deceleration.

• **PROBLEM 1015**

A ball is thrown upward from a building 91 ft. high and strikes the ground 6.5 sec. later.
(a) What is the initial velocity? (b) How high does the ball rise? (c) With what velocity does it strike the ground?

Solution: To solve this problem - since it involves rectilinear vertical motion of the ball under the influence of gravity - we use the familiar s, v, and a expressions, namely:

$$a = g \qquad (1)$$

$$v = gt + v_o \qquad (2)$$

$$s = \frac{1}{2} gt^2 + v_o t + s_o . \qquad (3)$$

(a) As suggested by the problem, we assume that measurements are performed on the ground, so that the initial height (s_o) is 91 ft. $t = 6.5$ is the time elapsed to bring the ball to the ground, thus $s = 0$ at $t = 6.5$. With these data, and using equation (3), we obtain:

$$0 = -16(6.5)^2 + v_o(6.5) + 91.$$

Solving this equation for v_o, we obtain $v_o = 90$. Thus, the initial velocity is 90 ft. per sec.

(b) With $v_o = 90$ ft/sec. and $s_o = 91$ ft. we once more use equation (3) to obtain the highest point the ball reaches. We first evaluate how long the ball takes to reach this highest point. Using (2),

$$v = -32t + 90, \qquad (4)$$

and since the highest point is reached when the upward or positive velocity becomes 0, immediately after this time the velocity is negative and the body falls. Setting $v = 0$ in equation (4), we have:

$$0 = -32t + 90,$$

whence

$$t = \frac{45}{16} = 2\frac{13}{16} .$$

The height of the ball at this time is found by substituting

$t = \frac{45}{16}$ in equation (3).

$$s = -16t^2 + 90t + 91,$$

and hence,

$$s = -16\left(\frac{45}{16}\right)^2 + 90\left(\frac{45}{16}\right) + 91 = 217\frac{9}{16} .$$

Therefore, the ball reaches a maximum height of $217\frac{9}{16}$ ft., $2\frac{13}{16}$ sec. after it is thrown.

To find its velocity when it strikes the ground we put $t = 6.5$ in. (4).

$$v = -32(6.5) + 90 = -118.$$

Hence, the ball strikes the ground with a velocity of 118 ft. per sec. downward.

A ball is thrown horizontally from a roof 100 ft. high with the initial velocity 60 ft. per sec. Find (a) the formula for the horizontal distance moved in t sec; (b) the formula for the height after t sec; (c) the Cartesian equation of the path; and (d) when, where, and with what velocity the ball strikes the ground.

Solution: Since there is no horizontal acceleration, the horizontal velocity must be constant at 60 ft. per sec. That is,

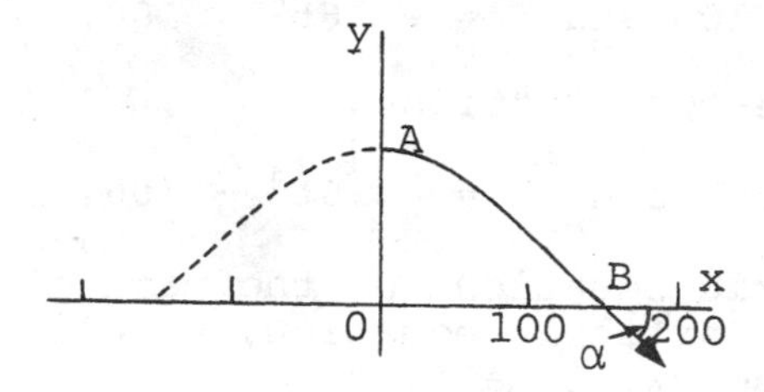

$$v_x = 60 \quad \text{and} \quad X = 60t + C_1.$$

If the measurements are made on the roof, at a point where the ball is thrown, and recalling the equation of rectilinear motion with constant velocity, where the gravitational force does not have any influence on the motion, we have:

$$X = 60t + C_1$$

$X = 0$ when $t = 0$, and therefore $C_1 = 0$. Hence,

$X = 60\ t.$

Now, if we transform the measurements to a point on the ground directly under the point from which the ball is thrown, the equation still is the same - x (on the ground) = X (on the roof), since the movement is perpendicular - which does not involve any horizontal displacement. Thus, on the ground,

$$X = 60t,$$

which is the answer to (a).

(b) The acceleration in the vertical direction is that due to gravity, assumed to be - 32 ft. per sec. per sec. Hence,

$$a_y = \frac{dv_y}{dt} = -32$$

and, consequently,

$$v_y = -32t + C_2.$$

Since the ball is thrown horizontally, the initial velocity in the vertical direction is zero, making $C_2 = 0$, and

$$v_y = -32t.$$

Moreover, since $v_y = \frac{dY}{dt}$, and

$\frac{dY}{dt} = -32t$, integration yields:

$$Y = -16t^2 + C_3$$

But the ball is at height 100 when $t = 0$; hence, $C_3 = 100$ and $Y = -16t^2 + 100$.

(c) The two equations:

$$X = 60t, \quad \text{and} \quad Y = -16t^2 + 100,$$

are parametric equations of the path of the ball. To find the Cartesian equation, we eliminate t between these two equations.

Squaring $X = 60t$, substituting

$$t^2 = \frac{X^2}{(60)^2} \quad \text{in} \quad Y = -16t^2 + 100,$$

and performing some numerical calculations, we obtain:

$$Y = -\frac{X^2}{225} + 100$$

This is the equation of the parabola shown **in the figure.** Only the solid part of the curve represents the path of the ball, which we imagine thrown to the right from A and reaching the ground at B. (d) The ball strikes the ground when t has the positive value that makes $Y = 0$ in the equation

$Y = -16t^2 + 100$. To find this value we solve the

equation: $-16t^2 + 100 = 0$, obtaining $t = 10/4$, or

$2\frac{1}{2}$. The ball strikes the ground at B, to the

right of A, the distance OB being the horizontal

distance it moves in $2\frac{1}{2}$ sec. Putting $t = 2\frac{1}{2}$

in the equation $X = 60t$, we find the distance to be 150. The ball therefore strikes the ground

150 ft. away from 0, in $2\frac{1}{2}$ sec.

When the ball strikes the ground, the horizontal component of velocity is still 60 ft. per sec. The vertical component is found by putting $2\frac{1}{2}$ in the equation $v_y = -32t$. Hence, at this time,

$$v_y = -32 \cdot \left(2\frac{1}{2}\right) = -80.$$

The magnitude of the vector v is

$$|v| = \sqrt{(60)^2 + (-80)^2} = 100.$$

The direction of v is such that

$\tan\alpha = 80/60 = 1.3333$, and $\alpha = 53^\circ$, approximately.

• PROBLEM 1017

A body thrown upward at an angle of 45°, with an initial speed of 100 ft/sec., travels in the parabolic path:

$$y = \frac{-gx^2}{10{,}000} + x,$$

where x and y are, respectively, the horizontal and vertical distances from the starting point, g is the gravitational constant = 32.2 ft/sec.^{2} and the horizontal speed has a constant value $100/\sqrt{2}$. (a) Find the vertical speed at any time t, and (b) find a point where it is zero. Neglect air resistance.

<u>Solution:</u> (a) From the given equation, we can find the time rate of change of y with respect to x, or dy/dx, by differentiating:

$$\frac{dy}{dx} = \frac{-gx}{5{,}000} + 1.$$

Since dy/dx can be expressed in terms of dx/dt and dy/dt, the horizontal and vertical speeds as follows:

$$\frac{dy}{dx} = \frac{dy}{dt} \div \frac{dx}{dt},$$

we obtain:

$$\frac{dy}{dt} = \left(\frac{-gx}{5{,}000} + 1\right)\left(\frac{dx}{dt}\right).$$

Substituting $\frac{100}{\sqrt{2}}$ for the horizontal speed, $\frac{dx}{dt}$,

$$\frac{dy}{dt} = \frac{-gx}{50\sqrt{2}} + \frac{100}{\sqrt{2}}.$$

(b) This vertical speed is zero where

$$\frac{-gx}{50\sqrt{2}} + \frac{100}{\sqrt{2}} = 0,$$

from which

$$x = \frac{5,000}{g} = 155.3 \text{ ft.},$$

corresponding to

$$y = -\frac{2,500}{g} + 155.3 = 77.7 \text{ ft.}$$

• PROBLEM 1018

A ball is rolled up a plane making an angle Θ with the horizontal. If the initial velocity is V_o (ft./sec.), the distance s (ft.) traveled in time t (sec.) is given by:

$$s = V_o t - (g \sin \Theta)t^2/2,$$

where $g = 32.2$ ft./sec.^{2}. Find the distance traveled by the ball before it comes to rest.

Solution:

$$s = V_o t - \frac{g \sin \Theta\, t^2}{2}.$$

We first calculate the time taken for the ball to move a certain distance and stop. This happens when

$$V = \frac{ds}{dt} = 0.$$

Thus,

$$\frac{ds}{dt} = V_o - g \sin \Theta\, t = 0.$$

Solving for t,

$$t = \frac{V_o}{g \sin \Theta}.$$

The distance covered within this time is:

$$s = V_o\left(\frac{V_o}{g \sin \Theta}\right) - \frac{g \sin \Theta}{2}\left(\frac{V_o}{g \sin \Theta}\right)^2$$

$$= \frac{1}{2}\frac{V_o^2}{g \sin \Theta}.$$

• PROBLEM 1019

When a stone is dropped into a pond, ripples are formed as concentric circles of increasing radius. If the radius is increasing at a rate of 0.5 ft./sec. at what rate is the area of one of these circles increasing when the radius is 4 ft?

Solution: Here we desire the relationship between the change in area and the change in radius, or dA/dt and dr/dt, where A = area, and r = radius. To obtain this, we note that A is related to r by the formula

$$A = \pi r^2$$

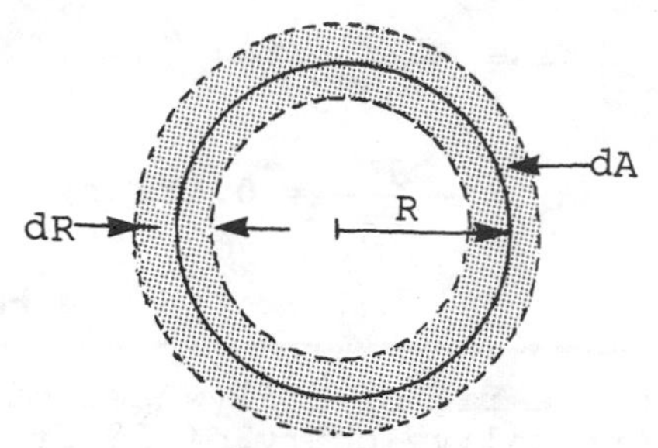

Since A and r are implied functions of time, we may differentiate both sides with respect to t to obtain:

$$\frac{d}{dt}(A) = \frac{d}{dt}(\pi r^2)$$

$$\frac{dA}{dt} = 2\pi r \frac{dr}{dt},$$

Substituting r = 4 and dr/dt = 0.5 into this equation, we obtain:

$$\frac{dA}{dt} = 2\pi(4)(0.5) = 12.56 \text{ ft}^2/\text{sec}.$$

• PROBLEM 1020

A winch on a truck is used to move a load. At what rate is the angle Θ changing when there are 10 ft of cable out, if the winch winds in the cable at the rate of 2ft/sec., and if the truck is 5 ft. high?

Solution: Let s represent the length of the cable between load and winch.

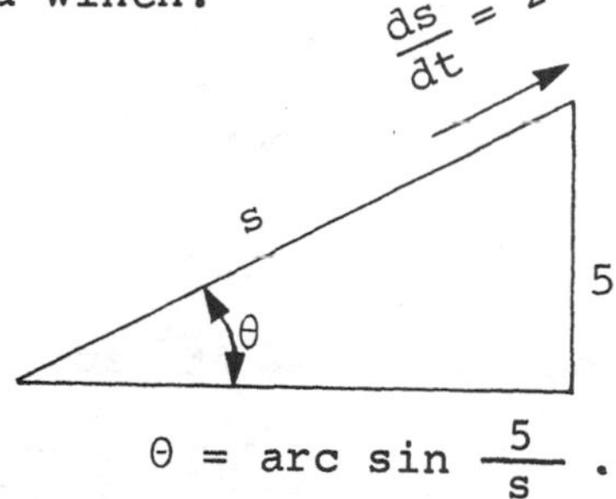

$$\Theta = \arc \sin \frac{5}{s}.$$

The rate at which Θ is changing is then found by differentiating this function with respect to t.

$$\frac{d\Theta}{dt} = \frac{1}{\sqrt{1 - (25/s^2)}} \cdot \frac{d}{dt}\left(\frac{5}{s}\right)$$

$$= \frac{1}{\sqrt{(s^2 - 25)/s^2}}\left(-\frac{5}{s^2}\right)\left(\frac{ds}{dt}\right)$$

$$= \frac{-5\, ds/dt}{s\sqrt{s^2 - 25}}$$

The cable is wound in at a rate: ds/dt = - 2 ft/sec. Therefore, when s = 10 ft,

$$\frac{d\Theta}{dt} = \frac{-5(-2)}{10\sqrt{75}}$$

$$= \frac{1}{\sqrt{75}} = 0.115 \text{ rad/sec.}$$

• **PROBLEM 1021**

A point on the rim of a flywheel of radius 5 ft. has a vertical velocity of 50 ft./sec. at a point P, 4 ft. above the x-axis (see diagram). What is the angular velocity of the wheel?

Solution: The point P has coordinates

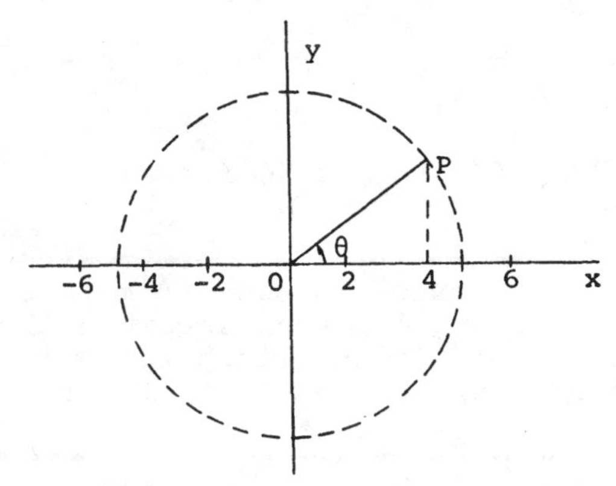

$x = 5 \cos \Theta$, $y = 5 \sin \Theta$,

the equations of a circle.

v_y, the vertical velocity, is:

$$v_y = \frac{dy}{dt} = 5 \cos \Theta \frac{d\Theta}{dt} = 50 \text{ ft./sec.}$$

When the point P is 4 ft. above the x-axis,

$$y = 5 \sin \Theta = 4 \quad \text{or} \quad \sin \Theta = \frac{4}{5}.$$

By using the identity: $\sin^2 \Theta + \cos^2 \Theta = 1$,

$$\left(\frac{4}{5}\right)^2 + \cos^2 \Theta = 1$$

$$\cos \Theta = \frac{3}{5}.$$

Using this in the expression for vertical velocity,

$$v_y = 5 \cos \Theta \frac{d\Theta}{dt} = 50,$$

$$\frac{d\Theta}{dt} = \frac{50}{5 \cos \Theta}$$

$$= \frac{50}{5\ (3/5)} = 16\ \frac{2}{3} \text{ radians per second.}$$

A ray of light travels through air with velocity V_1 ft./sec. from a point A at a distance of a ft. above water level, and enters the water at a point P such that the angle between AP and the vertical is α. It then proceeds with velocity V_2 ft./sec. at an angle β with the vertical to a point B at a distance b ft. below water level, the horizontal distance between A and B being c ft. It is found by experiment that

$$\frac{\sin \alpha}{\sin \beta} = \frac{V_1}{V_2} .$$

This relation is known as Snell's law. Show that Snell's law is obeyed when the time of travel from A to B is minimum.

Solution: Let the total length of the ray be l. We wish to find the location of P, such that the time of travel of the light be a minimum.

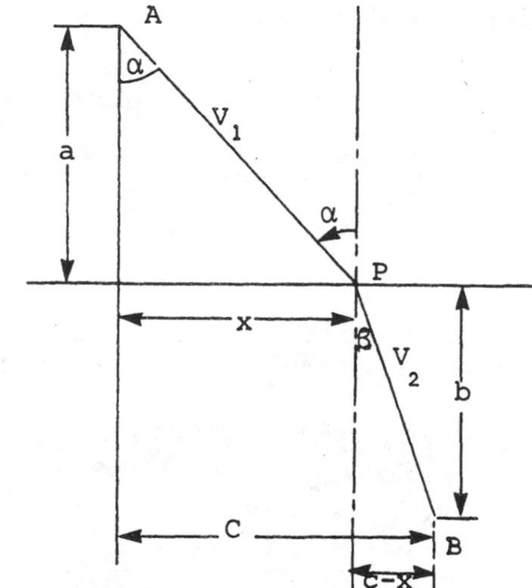

The time t is given by:

$$t = \frac{l_1}{V_1} + \frac{l_2}{V_2}$$

where the geometrical path length of the ray is

$$l = l_1 + l_2 .$$

Using the relation: $n = c/V$, the expression for time can be written as:

$$t = \frac{n_1 l_1}{c} + \frac{n_2 l_2}{c} = \frac{l}{c}$$

where the value: $n_1 l_1 + n_2 l_2$, is the optical length of the ray.

For the minimum time required, the differential of l with respect to x must be zero.

$$l(x) = n_1 l_1 + n_2 l_2 = n_1\sqrt{a^2 + x^2} + n_2\sqrt{b^2 + (c-x)^2}.$$

Differentiation yields:

$$\frac{dl}{dx} = n_1\left(\frac{1}{2}\right)(a^2 + x^2)^{-\frac{1}{2}}(2x)$$

$$+ n_2\left(\frac{1}{2}\right)\left(b^2 + (c - x)^2\right)^{-\frac{1}{2}}(2)(c - x)(-1)$$

$$= 0.$$

This can be rewritten as

$$n_1 \frac{x}{\sqrt{a^2 + x^2}} = n_2 \frac{c - x}{\sqrt{b^2 + (c - x)^2}}$$

Comparison with the figure gives:

$$n_1 \sin \alpha = n_2 \sin \beta,$$

or

$$\frac{\sin \alpha}{\sin \beta} = \frac{n_2}{n_1}.$$

But the constants (indeces of refraction), according to the above relation, are defined as:

$n_1 = \dfrac{c}{V_1}$ and $n_2 = \dfrac{c}{V_2}$. Hence,

$$\frac{\sin \alpha}{\sin \beta} = \frac{V_1}{V_2}.$$

Substitution of values either side of the critical value shows the condition to be a minimum.

• PROBLEM 1023

When a simple pendulum of length L ft. is displaced from the vertical by a small angle Θ_o rad. and then released, its angular displacement Θ rad. at any subsequent time t sec. may be shown to be given approximately by

$$\Theta = \Theta_o \cos\sqrt{\frac{g}{L}}\, t,$$

where g = 32.2 ft./sec.^{2}. If L = 3 ft. and $\Theta_o = 5^o$, find the displacement and velocity of the bob in its circular orbit after 2 sec.

Solution: The displacement is defined as the length of curve the pendulum moves in t sec., measured from its initial position, namely when

$$\Theta = \Theta_o = 5^\circ.$$

Since the pendulum moves in a circular arc, $L\Delta\Theta(t)$ is the length of arc, where $\Theta(t)$ is given by:

$$\Theta(t) = \Theta_o \cos\sqrt{\frac{g}{L}}\, t.$$

For time t = 2 sec., we find whether the pendulum is within its first oscillation. To do this, we calculate the period T. For this period we write, neglecting any resisting medium,

$$\Theta(T) = 4\,\Theta_o,$$

since, to complete a full cycle, the pendulum must travel four times its original displacement.

$$\Theta\left(\frac{T}{4}\right) = \Theta_o = \Theta_o \cos\sqrt{\frac{g}{L}}\,\frac{T}{4}\,.$$

For the above equation to hold

$$\cos\sqrt{\frac{g}{L}}\,\frac{T}{4} = 1, \text{ or } \sqrt{\frac{g}{L}}\,\frac{T}{4} = \frac{\pi}{2}\,,$$

and, solving for the period T,

$$T = 2\pi\sqrt{\frac{L}{g}} = 2\pi\sqrt{\frac{3}{32.2}}$$

$$= 1.92 \text{ sec.}$$

Thus, for 2 sec. the pendulum is in its second oscillation, (2 - 1.92) = 0.08 sec. after the first oscillation is completed. The part of the second cycle completed is

$$.08\sqrt{\frac{32.2}{3}} = 0.8\ (3.27) = .262 \text{ radian}$$

equivalent to 15°, $\cos 15^\circ = .9689$.

The length traveled for t = 0.08 sec. is

$$s_1 = L\Delta\Theta \quad \text{where}$$

$$\Delta\Theta = \Theta_o - \Theta(0.08) = \Theta_o\ (1 - 0.9689)$$

$$= 0.0311\ \Theta_o.$$

For the full oscillation,

$$s_2 = L\ 4\Theta_o\,,$$

hence the total length is:

$$s = s_1 + s_2 = L\Theta_o\ (4.0311)$$

$$= 3 \times 0.0873 \times 4.0311 = 1.044 \text{ ft.}$$

$\Theta_o = 5^\circ = 0.0873$ rad.

The linear velocity of the pendulum is:

$$L \frac{d\Theta}{dt} = V,$$

and $$\frac{d\Theta}{dt} = -\Theta_o \sqrt{\frac{g}{L}} \sin \sqrt{\frac{g}{L}}\, t.$$

Thus,

$$V = -\Theta_o L \sqrt{\frac{g}{L}} \sin \sqrt{\frac{g}{L}}\, t$$

for t = 2 sec., or t = 0.08 sec. in the second oscillation, we have:

$$V(2) = -\Theta_o L \sqrt{\frac{g}{L}} \sin \sqrt{\frac{g}{L}}\,(0.08)$$

$$= -0.0873 \times 3 \times 3.27 \times 0.2474$$

$$= -0.2088 \text{ ft./sec.}$$

$$= -2.5 \text{ in./sec.}$$

The negative sign shows $\Theta(t)$ decreases with time, which guarantees the bob is in the initial region.

• PROBLEM 1024

A moving circle of varying size always has its plane parallel to the yz-plane, passes through the line z = a, y = 0, and has a chord in common with the circle $x^2 + y^2 = a^2$, z = 0. Find the volume generated.

Solution: To meet the given conditions at x = 0, the moving circle must have a radius a and its center at the origin. To meet these conditions at $x = \pm a$, the moving circle has a radius $\frac{a}{2}$ and its center at $z = \frac{a}{2}$. From this it can be seen that the equation of the moving circle is of the form:

$$y^2 + (z - p)^2 = r^2,$$

where r is the radius and p is the displacement of the center and both p and r are functions of x.

Expanding,

$$y^2 + z^2 - 2pz + p^2 = r^2.$$

When x = any value, say x_1, y, in the stationary cirle, is equal to

$$y_1 = \sqrt{a^2 - x_1^2},$$

the length of half a chord. This is also a point on the moving circle, namely a point having $z = 0$ and the same y_1. Imposing this condition on the moving circle, we have:

$$y_1^2 + p^2 = r^2$$

Another condition to be imposed on the moving circle is that it must pass through $z = a$. This occurs at $y = 0$, the highest point, due to symmetry.

Hence, for $y = 0$, $z = a$. Substituting this in the equation, we have:

$$a^2 - 2ap + p^2 = r^2.$$

Solving simultaneously:

$$y_1^2 + p^2 = r^2,$$

and $a^2 - 2ap + p^2 = r^2$,

gives: $y_1^2 = a^2 - 2ap$, from which,

$$p = \frac{a^2 - y_1^2}{2a} = \frac{a^2 - (a^2 - x_1^2)}{2a} = \frac{x_1^2}{2a}.$$

$$r^2 = y_1^2 + p^2$$

$$= a^2 - x_1^2 + \frac{x_1^4}{4a^2}.$$

We consider a disk located at x_1. This has a thickness Δx, the radius r, and hence the volume:

$$\pi r^2 \Delta x = \pi \left(a^2 - x_1^2 + \frac{x_1^4}{4a^2} \right) \Delta x.$$

Total volume is the summation of such disks, with the limits + a and - a, i.e., in the limit:

$$V = \int_{-a}^{a} \pi \left(a^2 - x^2 + \frac{x^4}{4a^2} \right) dx$$

$$= \pi \left[a^2x - \frac{x^3}{3} + \frac{x^5}{20a^2} \right|_{-a}^{a}$$

$$= 2\pi \left(a^3 - \frac{a^3}{3} + \frac{a^5}{20a^2} \right)$$

$$= 2\pi a^3 \left(\frac{60 - 20 + 3}{60} \right)$$

$$= \frac{2\pi a^3 (43)}{60} = \pi a^3 \left(\frac{43}{30} \right) .$$

• PROBLEM 1025

A block of weight W is to be moved along a flat table by a force inclined at an angle Θ with the line of motion, where $0 \leq \Theta \leq \frac{1}{2}\pi$, as shown in the Figure. Assume the motion is resisted by a frictional force which is proportional to the normal force with which the block presses perpendicularly against the surface of the table. Find the angle Θ for which the propelling force needed to overcome friction will be as small as possible.

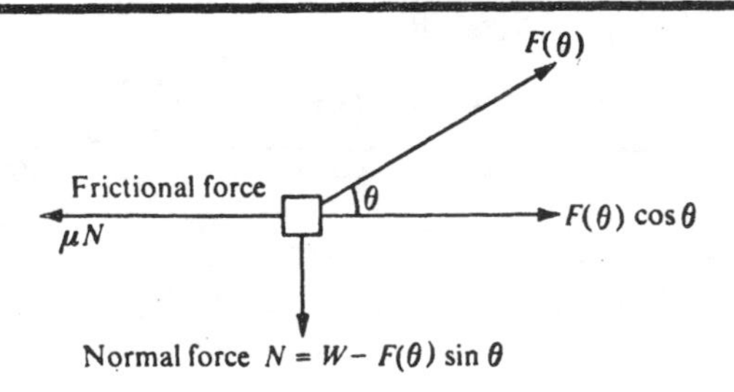

Solution: Let $F(\Theta)$ denote the propelling force. It has an upward vertical component $F(\Theta) \sin \Theta$, so the net normal force pressing against the table is $N = W - F(\Theta) \sin \Theta$.

The frictional force is proportional to the normal force. We introduce μ as a constant of proportionality (called the coefficient of friction). The friction force,

$$f_{fr} = \mu N = \mu\left(W - F(\Theta)\sin \Theta\right).$$

Since motion is prevented by this friction force, the block is in a state of equilibrium, and there is no sliding. This means that the horizontal and vertical components of force cancel each other, and as a result,

$$f_{fr} = \mu\left(W - F(\Theta)\sin \Theta\right) = F(\Theta)\cos \Theta,$$

where $F(\Theta)\cos\Theta$ is the horizontal component of the propelling force $F(\Theta)$. From this we find that:

$$\mu W - F(\Theta)\mu \sin\Theta - F(\Theta)\cos\Theta = 0$$

or

$$F(\Theta) = \frac{\mu W}{\cos\Theta + \mu \sin\Theta} .$$

To find the angle Θ for which $F(\Theta)$ is a minimum, we find the derivative of the propelling force $F(\Theta)$ with respect to the angle Θ and set it to zero, from which we solve for the angle.

$$\frac{dF(\Theta)}{d\Theta}$$

$$= \frac{(\cos\Theta+\mu\sin\Theta)d(\mu W)-\mu W\, d(\cos\Theta+\mu\sin\Theta)}{(\cos\Theta+\mu\sin\Theta)^2} = 0.$$

Since uW is a constant, d(uW) = 0, and we have:

$$\frac{-\mu W(-\sin\Theta + \mu\cos\Theta)}{(\cos\Theta + \mu\sin\Theta)^2} = 0,$$

which implies:

$$-\sin\Theta + \mu\cos\Theta = 0,$$

or, $\mu\cos\Theta = \sin\Theta$,

since the weight W and μ are never zero.

To solve for the angle, we proceed by squaring both sides of the equation, and using the trigonometric identity:

$$\sin^2\Theta + \cos^2\Theta = 1.$$

We obtain:

$$\mu^2\cos^2\Theta = \sin^2\Theta = 1 - \cos^2\Theta.$$

Combining like terms and solving for $\cos\Theta$,

$$\cos\Theta = \frac{1}{\sqrt{1+\mu^2}} .$$

Since $\dfrac{d^2 F(\Theta)}{d\Theta^2}$ is positive for $0 \le \Theta \le \pi/2$, this guarantees that F(Θ) is a minimum at this angle . $\mu \cos^{-1} \dfrac{1}{\sqrt{1 + \mu^2}} = \alpha$.

Sin α can be expressed in terms of μ by constructing a right-angled triangle, the sides of which, using Θ, as one of its angles, satisfy the expression:

$$\cos \alpha = \frac{1}{\sqrt{1 + \mu^2}} \cdot$$

Using the Pythagorean theorem, the opposite side is μ, the adjacent side is unity, and the **hypotenuse is**

$\sqrt{1 + \mu^2}$. From this,

$$\sin \alpha = \frac{\mu}{1 + \mu^2} \cdot$$

Thus, the minimum force required is

$$F(\Theta_1) = \frac{\mu W}{\dfrac{1}{\sqrt{1 + \mu^2}} + \mu\left(\dfrac{\mu}{\sqrt{1 + \mu^2}}\right)}$$

$$= \frac{\mu \ W}{1 + \mu^2} \sqrt{1 + \mu^2} \ .$$

As a simpler alternative, we use the following approach:

Recalling,

$$F(\Theta) = \frac{\mu \ W}{\cos \Theta + \ \sin \Theta} ,$$

to minimize F(Θ), we attempt to maximize the denominator: $g(\Theta) = \cos \Theta + \mu \sin \Theta$ in the interval $0 \leq \Theta \leq \frac{1}{2}\pi$. At the endpoints, we have $g(0) = 1$ and $g\left(\frac{1}{2}\pi\right) = \mu$. In the interior of the interval, we have:

$$g'(\Theta) = - \sin \Theta + \mu \cos \Theta.$$

Hence, g has a critical point at $\Theta = \alpha$, where $\sin \alpha = \mu \cos \alpha$. This gives $g(\alpha) = \cos \alpha + \mu^2 \cos \alpha = (1 + \mu^2)\cos \alpha$. We can express cos α in terms of μ. Since $\mu^2 \cos^2 \alpha = \sin^2 \alpha = 1 - \cos^2 \alpha$, we find $(1 + \mu^2)\cos^2 \alpha = 1$, so $\cos \alpha = 1/\sqrt{1 + \mu^2}$. Thus $g(\alpha) = \sqrt{1 + \mu^2}$.

Since g(α) exceeds g(0) and $g\left(\frac{1}{2}\pi\right)$, the maximum of g occurs at the critical point. Hence the minimum force required is

$$F(\alpha) = \frac{\mu W}{g(\alpha)} = \frac{\mu W}{\sqrt{1 + \mu^2}} .$$

• PROBLEM 1026

If m(x) is the slope of the tangent line to the curve $y = x^3 - 2x^2 + x$ at the point (x,y), find the instantaneous rate of change of m per unit change in x at the point (2,2).

Solution: $m = D_x y = \frac{dy}{dx} = 3x^2 - 4x + 1.$

The instantaneous rate of change of m per unit change in x is given by $D_x m$.

$$D_x m = D_x{}^2 y = \frac{d^2 y}{dx^2} = 6x - 4.$$

At the point (2,2), $D_x{}^2 y = 8$. Hence, at the point (2,2) the change in m(x) is 8 times the change in x.

• PROBLEM 1027

A piston is actuated by a rod attached to a revolving crankshaft one foot from the center. What are the expressions for velocity and acceleration of the piston at any time t? What are the values of t for maximum and minimum velocities of the piston if the crankshaft rotates at a constant angular speed of $\frac{1}{2\pi}$ revolutions per second?

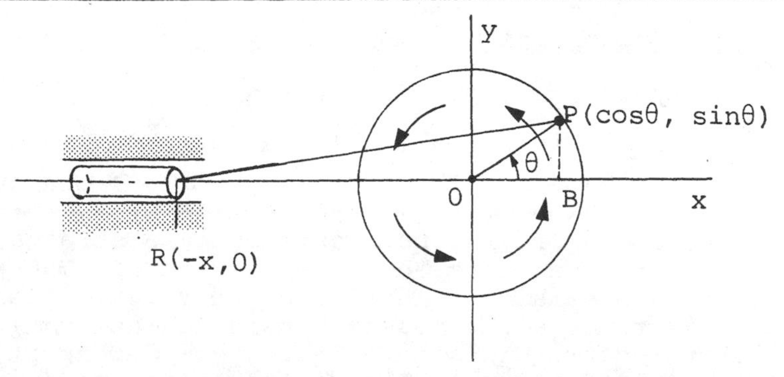

Solution: Let P be the point of attachment, L be the length of the rod and θ be the angle made with the horizontal axis by the radius from the origin to P. Hence P has the coordinates $x = \cos\theta$, $y = \sin\theta$ at time t as illustrated in the figure. Let the right edge of the piston be situated at R(-x,0).

We wish to express R(-x,0) as a function of θ in the form of $R\left[x(\theta),0\right]$, for any values of θ.

$$\left(L^2 - \sin^2\theta\right)^{1/2} = \overline{RB}.$$

and $$\cos\theta = \overline{OB}$$

thus, $$\left(L^2 - \sin^2\theta\right)^{1/2} - \cos\theta = \overline{RB} - \overline{OB} = -x$$

$$x = \cos\theta - \left(L^2 - \sin^2\theta\right)^{1/2}$$

Hence the right edge of the piston at any time t is located at

$$R\left[\cos\theta - \left(L^2 - \sin^2\theta\right)^{1/2}, 0\right].$$

The rate of change of this position at time t is the velocity of the piston, which is

$$\frac{dx}{dt} = \left(-\sin\theta + \frac{2}{2}\frac{\sin\theta\ \cos\theta}{\sqrt{L^2 - \sin^2\theta}}\right)\frac{d\theta}{dt}$$

If L is large compared to 1, then the velocity of the piston is approximately equal to $-\sin\theta\frac{d\theta}{dt}$. But the angular speed is given as $\frac{1}{2\pi}$ revolutions/sec. Therefore, $\frac{d\theta}{dt} = 1$ radian/sec.

$$\frac{dx}{dt} = -\sin\theta$$

$$\frac{d^2x}{dt^2} = -\cos\theta$$

The velocity is a maximum at $\theta = \frac{\pi}{2}, -\frac{\pi}{2}$

The acceleration is a maximum at $\theta = 0, \pi$

The velocity is a minimum (0) at $\theta = 0, \pi$

The acceleration is a minimum (0) at $\theta = \frac{\pi}{2}, \frac{-\pi}{2}$

• **PROBLEM** 1028

A helicopter flying horizontally in a straight line at 60 miles an hour and an elevation of 1760 feet crosses a straight, level road at right angles. At the same time, a car passes underneath at 30 miles per hour. One minute later, how far apart are the two, and at what rate do they separate?

Solution: In the diagram let C represent the position of the helicopter at the instant when the car is vertically below it at the point O. Then CP is the direction of the helicopter and OB that of the car. If the arrows indicate the motion, then after time t (minutes) they occupy the positions

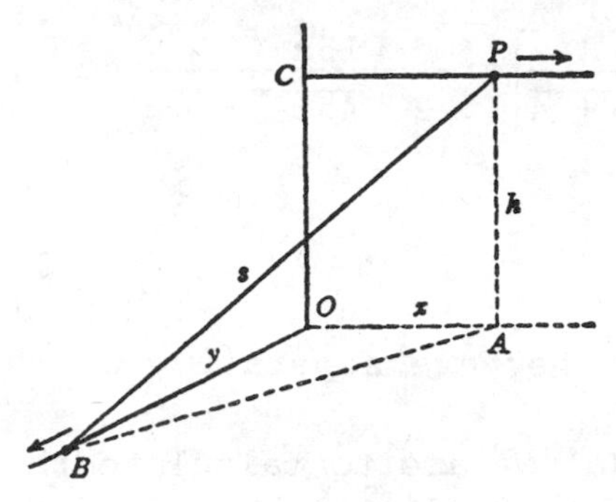

P and B, and the straight line distance between them is the diagonal PB = s. We are to find an expression for s at any time t and also its rate of change, $\frac{ds}{dt}$.

In the diagram, OA is parallel to CP. AB is a vertical line under P and x, y, h are as shown. Then OAB is a right triangle with legs OA and OB and hypotenuse AB, and PAB is a right triangle with legs AP and AB and hypotenuse PB = s. Therefore,

$$s^2 = \overline{AB}^2 + h^2, \text{ and } \overline{AB}^2 = x^2 + y^2. \quad \text{Hence,}$$

$$s^2 = x^2 + y^2 + h^2. \qquad \text{(a)}$$

Since the helicopter is traveling at a rate: dx/dt = 60 miles per hour = 1 mi./min., and the car at the rate: dy/dt = 30 miles per hour $= \frac{1}{2}$ mi./min., then, at the end of t minutes, they are at the distances from the crossing point (in miles) of :

$$CP = x = t, \; OB = y = \frac{1}{2}t, \; OC = h = \frac{1760}{5280} = \frac{1}{3} \qquad \text{(b)}$$

Using these values of x, y, h in (a) we have

$$s^2 = t^2 + \left(\frac{1}{2}t\right)^2 + \left(\frac{1}{3}\right)^2 = \frac{5t^2}{4} + \frac{1}{9} = \frac{45t^2 + 4}{36}.$$

$$s = \frac{1}{6}\sqrt{45t^2 + 4} \qquad \text{(c)}$$

is the distance between the helicopter and the car at any time t minutes after the crossing, and the rate at which they are separating is ds/dt. From (c)

$$ds = \frac{1}{6} \cdot d(\sqrt{45t^2 + 4}) = \frac{1}{6}\left(\frac{d(45t^2 + 4)}{2\sqrt{45t^2 + 4}}\right)$$

$$= \frac{1}{6}\left(\frac{90t\ dt}{2\sqrt{45t^2+4}}\right) = \frac{15t}{2\sqrt{45t^2+4}}\ dt.$$

$$\frac{ds}{dt} = \frac{15t}{2\sqrt{45t^2+4}} \qquad \text{(d)}$$

is the rate at which they are separating at the time t.

Using (c) and (d) we are to calculate the distance s, and the rate ds/dt at the end of one minute. Thus t = 1, and, by use of (c),

$$s = \frac{1}{6}\sqrt{45+4} = \frac{7}{6}\ \text{miles}.$$

By use of (d),

$$\frac{ds}{dt} = \frac{15}{2\sqrt{45+4}} = \frac{15\ \text{mi.}}{14\ \text{min.}} = 64\ \frac{2}{7}\ \text{mi./hr.}$$

• PROBLEM 1029

The crank and connecting rod of a pump are three and ten feet long respectively, and the crank revolves at a uniform rate of 120 r.p.m. At what rate is the crosshead moving when the crank makes an angle of 45 degrees with the dead-center line?

Solution: Let OC represent the dead-center line and the circle the path of the crank pin P. C represents the crosshead, CP the connecting rod and OP the crank. As P moves steadily about the circle in the direction shown, C moves back and forth at different rates along OC. If OC = x and angle POC = Θ, then we are to find dx/dt when

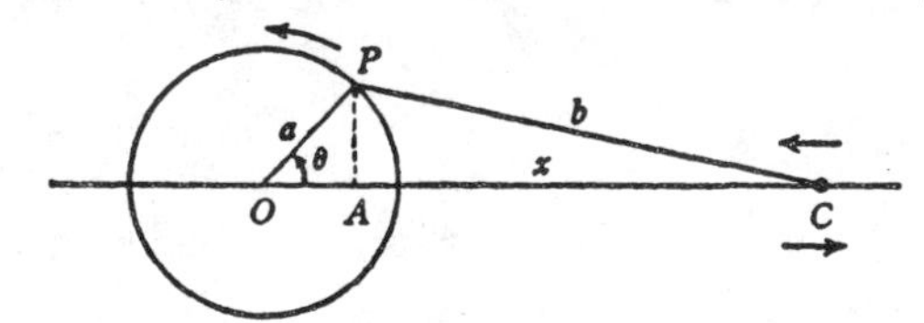

$$\frac{d\Theta}{dt} = 120\ \text{r.p.m.} = 2\ \frac{\text{rev}}{\text{sec}}$$, Or, since one revolution sweeps 360° or 2 π radians, we have

$$\frac{d\Theta}{dt} = 4\pi\ \text{rad/sec.}$$

We are to determine

$$\frac{dx}{dt} = \frac{dx}{d\Theta} \cdot \frac{d\Theta}{dt}.$$

In the figure, let a = crank length = 3 feet, and b = length of connecting rod = 10 feet. PA is perpendicular to OC. Then, for varying positions of P, A and C have different positions, but always:

$$x = \overline{OA} + \overline{AC}. \qquad \text{(a)}$$

In the right triangle PAC, the hypotenuse formula gives:

$$\overline{AC} = \sqrt{b^2 - \overline{AP}^2}, \qquad \text{(b)}$$

and in the right triangle POA,

$$\overline{OA} = a \cos \Theta, \ \overline{AP} = a \sin \Theta. \qquad \text{(c)}$$

Substituting this value of $\overline{AP}$ in (b), we have:

$$\overline{AC} = \sqrt{b^2 - a^2 \sin^2 \Theta},$$

and this value of $\overline{AC}$, together with the value of $\overline{OA}$ in the first of equations (c), when used in equation (a) gives:

$$x = a \cos \Theta + \sqrt{b^2 - a^2 \sin^2 \Theta}, \qquad \text{(d)}$$

which expresses x as a function of the angle Θ. In order to find the rate dx/dt, this equation must be differentiated to obtain dx/dΘ.

Differentiating equation (d) and carrying out the transformations and simplifications, this gives:

$$dx = d(a \cos \Theta) + d\left(\sqrt{b^2 - a^2 \sin^2 \Theta}\right)$$

$$= a \cdot d(\cos \Theta) + \frac{d(b^2 - a^2 \sin^2 \Theta)}{2\sqrt{b^2 - a^2 \sin^2 \Theta}}$$

$$= a(- \sin \Theta \, d\Theta) + \frac{- d(a^2 \sin^2 \Theta)}{2\sqrt{b^2 - a^2 \sin^2 \Theta}}$$

$$= - a \sin \Theta \, d\Theta - \frac{a^2 (2 \sin^{2-1} \Theta \cdot \cos \Theta \ d\Theta)}{2\sqrt{b^2 - a^2 \sin^2 \Theta}}$$

$$= - a \sin \Theta \, d\Theta - \frac{a^2 \sin \Theta \cos \Theta \ d\Theta}{\sqrt{b^2 - a^2 \sin^2 \Theta}}$$

$$= - a \sin \Theta \, d\Theta - \frac{a^2 \sin \Theta \cos \Theta \, d\Theta}{a \sqrt{\frac{b^2}{a^2} - \sin^2 \Theta}}$$

$$dx = - a \sin \Theta \left(1 + \frac{\cos \Theta}{\sqrt{\left(\frac{b}{a}\right)^2 - \sin^2 \Theta}} \right) d\Theta.$$

$$\frac{dx}{dt} = - a \sin \Theta \left(1 + \frac{\cos \Theta}{\sqrt{\left(\frac{b}{a}\right)^2 - \sin^2 \Theta}} \right) \frac{d\Theta}{dt}.$$

Now $a = 3, \; b = 10, \; \left(\frac{b}{a}\right)^2 = 11.1, \; \frac{d\Theta}{dt} = 4\pi.$
Substitution gives:

$$\frac{dx}{dt} = -\,12\pi\left[1 + \frac{\cos\Theta}{\sqrt{11.1 - \sin^2\Theta}}\right]\sin\Theta. \qquad (e)$$

When $\Theta = 45°$, $\sin\Theta = \cos\Theta = .707$,
and hence

$$\frac{dx}{dt} = -\,12\pi\left[1 + \frac{.707}{\sqrt{11.1 - (.707)^2}}\right]\left(.707\right)$$

$$= -\,32.44 \text{ ft./sec.}$$

The sign indicates that C is approaching 0.

• PROBLEM 1030

A ship is sailing due north at 20 miles per hour. At a certain time another ship crosses its route 40 miles north sailing due east at 15 miles per hour. (a) At what rate are the ships approaching or separating after one hour? (b) After two hours? (c) After how long are they momentarily neither approaching nor separating? (d) At that time, how far apart are they?

Solution: We express the distance between the ships as a function of the time elapsed after the second crossed the path of the first. The rate of change of this distance is then their speed of approach or separation. Let P represent the position of the first ship when the second crosses its path at 0, 40 miles due north. After a certain time t, the ship sailing east will have reached a point A, and the ship sailing north will have reached a point B. The distance between them is then AB. This distance is to be expressed as a function of the time t that has elapsed since A passed O and B left P.

With O as a reference point, we let OA = x, OB = y, AB = s. Then OP = 40, and

$$s = \sqrt{x^2 + y^2}. \qquad (1)$$

The rate of ship B is $dy/dt = 20$, and that of ship A, $dx/dt = 15$ miles per hour. Therefore, after a time t hours has passed, B has covered the distance $\overline{PB} = 20 \cdot t$, and the ship A the distance $\overline{OA} = 15 \cdot t$. Then, $\overline{OB} = \overline{OP} - \overline{PB} = 40 - 20 \cdot t$. Therefore,

$$x = 15t, \qquad y = 40 - 25t. \qquad (2)$$

Using these values of x, y in equation (1) gives:

$$s = \sqrt{(15t)^2 + (40 - 20t)^2} = \sqrt{625t^2 - 1600t + 1600}$$

$$s = 5\sqrt{25t^2 - 64t + 64}. \tag{3}$$

This is the desired relation between the distance between the ships, s, and the time t after the crossing at 0. If at any time the rate ds/dt is positive, the distance is increasing, that is, the ships are separating. If at any time it is negative, they are approaching. To find the rate ds/dt, we must differentiate equation (3). Using the square root formula and taking account of the constant multiplier 5, we obtain:

$$ds = 5\left(\frac{d(25t^2 - 64t + 64)}{2\sqrt{25t^2 - 64t + 64}}\right).$$

Differentiating the expression in the numerator of this fraction by the sum rule and each of the individual terms by the appropriate formula, this becomes:

$$ds = \frac{5}{2}\,\frac{d(25t^2) - d(64t) + d(64)}{\sqrt{25t^2 - 64t + 64}}$$

$$= \frac{5}{2}\,\frac{50t\,dt - 64\,dt}{\sqrt{25t^2 - 64t + 64}}$$

$$= \frac{5(25t - 32)\,dt}{\sqrt{25t^2 - 64t + 64}}$$

$$\frac{ds}{dt} = \frac{5(25t - 32)}{\sqrt{25t^2 - 64t + 64}}. \tag{4}$$

We calculate the required results (a) to (d).

(a) After 1 hour, t = 1, and

$$ds/dt = 5(25 - 32)/\sqrt{25 - 64 + 64}.$$

ds/dt = - 7 mi./hr. and the ships are approaching.

(b) After 2 hours, t = 2, and

$$ds/dt = 5(50 - 32)/\sqrt{100 - 128 + 64}.$$

ds/dt = + 15 mi./hr. and they are separating.

(c) At the instant when they just cease to approach and begin to separate they are at their nearest positions and momentarily are neither approaching nor separating, therefore ds/dt = 0.

$$\frac{5(25t - 32)}{\sqrt{25t^2 - 64t + 64}} = 0.$$

For this fraction to equal zero, the numerator must be zero.

$$5(25t - 32) = 0, \quad \text{or } 25t - 32 = 0$$

$$t = 1\frac{7}{25} \text{ hours} = 1.28 \text{ hr.}$$

(d) After a time $t = 1\frac{7}{25}$ hours, equations (2) give:

$$x = 96/5 \text{ mi.}, \qquad y = 72/5 \text{ mi.}$$

The distance between the ships, according to equation (1), is:

$$s = \sqrt{\left(\frac{96}{5}\right)^2 + \left(\frac{72}{5}\right)^2} = 24 \text{ mi.}$$

Or, directly by equation (3),

$$s = \sqrt{25\left(1\frac{7}{25}\right)^2 - 64\left(1\frac{7}{25}\right) + 64} = 24 \text{ mi.}$$

• PROBLEM 1031

An elliptical cam rotates about its focus F, causing a roller at P move up and down parallel to the y-axis. If the diameters of the cam are six and ten inches and it rotates at the rate of 240 r.p.m., how fast is the roller moving at the moment when the long axis of the cam makes an angle of 60° with the line of motion of the roller?

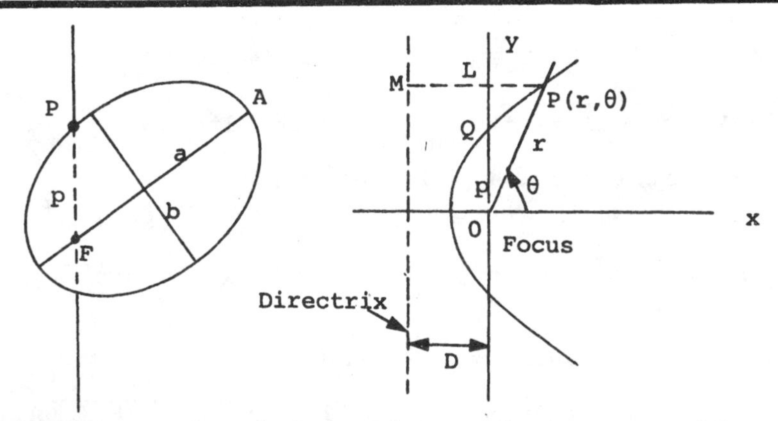

Solution: In solving this problem, we consider some general ideas about conics:

If a point P (r,Θ) moves so that its distance from a fixed point (called the focus), divided by its distance from a fixed line (called the directrix), is a constant e (called the eccentricity), then the curve described by P is called a conic (so-called because such curves can be obtained by cutting a cone at different angles).

If the focus is chosen at the origin 0, the equation of a conic in polar coordinates (r,Θ) is, for $0Q = P$ and $LM = D$,

$$r = \frac{eD}{1 - e\cos\theta}.$$

Which one of the four possible conics we **have depends on e, as follows:**

1. $e = 0$ for a circle,
2. $e < 1$ an ellipse,
3. $e = 1$ a parabola, and
4. $e > 1$ a hyperbola.

For the ellipse in question, the relation between P and Θ is expressed as:

$$P = \frac{b^2}{a(1 - e\cos\Theta)} \qquad \text{(a)}$$

where

$$e = \frac{\sqrt{a^2 - b^2}}{a}$$

In the above equation, b and e are constants. We find dp by writing equation (a) as

$$p = \frac{b^2}{a} \cdot \left(\frac{1}{1 - e\cos\Theta}\right).$$

$$dp = \frac{b^2}{a} \cdot d\left(\frac{1}{1 - e\cos\Theta}\right),$$

and by the formula for the reciprocal,

$$dp = \frac{b^2}{a}\left[-\frac{d(1 - e\cos\Theta)}{(1 - e\cos\Theta)^2}\right]$$

$$= \frac{b^2}{a}\left[\frac{d(e\cos\Theta)}{(1 - e\cos\Theta)^2}\right]$$

$$= \frac{b^2}{a}\left[\frac{-e\sin\Theta\, d\Theta}{(1 - e\cos\Theta)^2}\right].$$

$$\frac{dp}{dt} = -\frac{b^2 e\sin\Theta}{a(1 - e\cos\Theta)^2}\frac{d\Theta}{dt} \qquad \text{(b)}$$

Since a, half the long diameter, = 5, and b, half the short diameter, = 3, we obtain

$$e = \frac{\sqrt{(5)^2 - (3)^2}}{5} = \frac{4}{5}.$$

Also, $b^2 = 9$, and $d\Theta/dt = 240$ r.p.m. = 4 rev./sec.

One full revolution sweeps 2Π radians, hence $d\Theta/dt = 8\Pi$ rad./sec. Using these values, equation (b) for any angle Θ becomes:

$$\frac{dp}{dt} = - \frac{9\left(\frac{4}{5}\right) \quad \sin \Theta}{5\left(1 - \frac{4}{5} \cos \Theta\right)^2} \cdot 8\Pi$$

$$= - \frac{57.6 \ \Pi \quad \sin \Theta}{5\left(1 - \frac{4}{5} \cos \Theta\right)^2} .$$

For $0 \leq \Theta \leq \Pi$, p is decreasing, and therefore, the roller is moving down. At $\Theta = \Pi$ the velocity is zero. $p = 2a - \overline{FA}$, which is the lowest position of the roller.

Now, we wish to find the instantaneous rate of motion of the roller at $\Theta = 60° = \Pi/3$ rad. At this moment,

$$\sin \Theta = \sqrt{3/2}, \ \cos \Theta = 1/2 , \text{ and hence,}$$

$$\frac{dp}{dt} = - \frac{57.6 \ \Pi \ (\sqrt{3/2})}{5\left[1 - 4/3(1/2)\right]^2} = - 16 \ \Pi \ \sqrt{3}$$

$$= - 87.1 \text{ in/sec.}$$

Therefore, the roller is moving downward at a rate of 87.1 in/sec. or 7.25 ft/sec.

• PROBLEM 1032

> If the path of a particle is a curve with an inflection point, show that the normal component of acceleration vanishes at such a point. Illustrate with the curve:
>
> $$x = t, \quad y = t^3.$$

Solution: Let $\vec{R} = x\hat{\imath} + y\hat{\jmath}$ be the position vector of the particle in a plane. Then its velocity,

$$\vec{V} = \frac{d\vec{R}}{dt} = \frac{d\vec{R}}{ds} \frac{ds}{dt} .$$

ds/dt is the scalar velocity of the particle along the curve, s. But, since

$$\vec{R} = x\hat{\imath} + y\hat{\jmath},$$

$$d\vec{R} = dx\hat{\imath} + dy\hat{\jmath},$$

and, since

$$ds = (dx^2 + dy^2)^{\frac{1}{2}},$$

$$\frac{d\vec{R}}{ds} = \frac{dx\hat{\imath} + dy\hat{\jmath}}{\left[(dx)^2 + (dy)^2\right]^{\frac{1}{2}}} .$$

We can call this a unit vector $\hat{T}$.

$$d\vec{R} = \lim_{\Delta R \to 0} \left(\vec{R}_2 - \vec{R}_1\right),$$

where $\vec{R}_1$ and $\vec{R}_2$ are the initial and final locations of the particle in time Δt. Hence, $\Delta\vec{R}$ is the third side of a triangle constructed by the vectors $\vec{R}_1$ and $\vec{R}_2$ and, passing to the limit as $\vec{R}_2$ approaches $\vec{R}_1$, $\Delta\vec{R}$ diminishes in magnitude but still is in the direction of $\vec{R}_2$. The shrinkage of $\Delta\vec{R}$ per unit length of the curve under a limit becomes a unit vector pointing tangentially in the direction of increasing $\vec{R}$. Thus $\hat{T}$ is a unit tangent vector. The velocity is given by:

$$\vec{V} = V\,\hat{T},$$

where

$$V = \frac{ds}{dt},$$

and $$\left|\hat{T}\right| = \left|\frac{d\vec{R}}{ds}\right| = 1.$$

Furthermore, the acceleration,

$$\vec{A} = \frac{d\vec{V}}{dt} = \frac{d}{dt}\,(V\,\hat{T}),$$

in other words,

$$\vec{A} = \frac{dV}{dt}\,\hat{T} + V\,\frac{d\hat{T}}{dt}.$$

To express $\frac{d\hat{T}}{dt}$ in a physically obtainable form, there is no change in substituting:

$$\frac{d\hat{T}}{dt} = \frac{d\hat{T}}{d\Theta}\,\frac{d\Theta}{ds}\,\frac{ds}{dt},$$

where Θ is the angle which the vector $\hat{T}$ makes with the positive x-axis. In doing so,

$$\frac{d\hat{T}}{d\Theta}$$

is a unit vector in the direction of increasing Θ, in other words,

$$\hat{N} = \frac{d\hat{T}}{d\Theta}$$

is normal to the unit tangential vector T from the concave side of the curve. $d\Theta/ds$ is the radius of curvature K and ds/dt is the velocity V. Hence we can rewrite the acceleration as:

$$\vec{A} = \frac{dV}{dt}\hat{T} + K V^2 \hat{N}.$$

$$K V^2 = a_N$$

is called the normal acceleration, and

$$\frac{dV}{dt} = a_T$$

is the tangential acceleration. From the differential calculus,

$$K = \frac{d^2y/dx^2}{\left[1 + (dy/dx)^2\right]^{\frac{3}{2}}}.$$

Thus

$$a_N = \frac{d^2y/dx^2 \; V^2}{\left[1 + (dy/dx)^2\right]^{\frac{3}{2}}}.$$

But at the point of inflection $d^2y/dx^2 = 0$,

hence $K = 0$. This leads to $\vec{A} = \frac{dV}{dt}\hat{T}$, since $K = 0$

makes the normal acceleration $a_N = 0$.

We can also parametrize the equation of the curve by setting $x = t$ and $y = f(t)$, hence

$$\frac{dx}{dt} = 1, \quad \frac{d^2x}{dt^2} = 0; \quad \frac{dy}{dt} = \frac{dy}{dx}, \quad \frac{d^2y}{dt^2} = \frac{d^2y}{dx^2}.$$

This leads to

$$K = \frac{\frac{dx}{dt}\frac{d^2y}{dt^2} - \frac{dy}{dt}\frac{d^2x}{dt^2}}{\left[\left(\frac{dx}{dt}\right)^2 + \left(\frac{dy}{dt}\right)^2\right]^{\frac{3}{2}}}.$$

Now, given that $x = t$, $y = t^3$, we have $\frac{dx}{dt} = 1$, $\frac{dy}{dt}$ the position vector as

$= 3t^2$. We can write down the position vector as:

$$\vec{R}(t) = t\hat{\imath} + t^3\hat{\jmath}.$$

$$\vec{V}(t) = \frac{d\vec{R}}{dt} = \hat{\imath} + 3t^2\hat{\jmath}. \quad \frac{ds}{dt} = \sqrt{\left(\frac{dx}{dt}\right)^2 + \left(\frac{dy}{dt}\right)^2} = \sqrt{1+9t^4}.$$

$$\vec{T}(t) = \frac{d\vec{R}}{dt}\frac{dt}{ds} = \frac{\hat{\imath}}{\sqrt{1 + 9t^4}} + \frac{3t^2\hat{\jmath}}{\sqrt{1 + 9t^4}}.$$

From this we have:

$$\left|\vec{V}(t)\right| = \frac{ds}{dt} = \sqrt{1 + 9t^4},$$

which proves the relationship

$$\vec{V}(t) = \frac{ds}{dt}\ \hat{T}$$

quantitatively.

For the acceleration,

$$\vec{A}(t) = \frac{d\vec{V}}{dt} = 6t\ \hat{j},$$

but in terms of the normal and tangential components,

$$\frac{d^2s}{dt^2} = \frac{18t^3}{\sqrt{1 + 9t^4}}, \quad \left(\frac{ds}{dt}\right)^2 = 1 + 9t^4.$$

$$K = \frac{6t}{(1 + 9t^4)^{\frac{3}{2}}},$$

Since $\hat{T} \cdot \hat{N} = 0$, we have

$$\hat{N} = \frac{-\ 3t^2\hat{i}}{\sqrt{1 + 9t^4}} + \frac{\hat{j}}{\sqrt{1 + 9t^4}}.$$

Hence, for any time t,

$$\vec{A}(t) = \frac{d^2s}{dt^2}\ \hat{T} + K\left(\frac{ds}{dt}\right)^2\ \hat{N}$$

$$= \frac{18t^3\ (\hat{i} + 3t^2\hat{j}) + 6t(-\ 3t^2\hat{i} + \hat{j})}{1 + 9t^4}$$

$$= \frac{(18t^3 - 18t^3)\hat{i} + 6t(1 + 9t^4)\ \hat{j}}{1 + 9t^4}$$

$$= 6t\ \hat{j}$$

which is exactly equal to what was obtained from

$$\vec{A}(t) = \frac{d\vec{V}}{dt} = \frac{d^2\vec{R}}{dt^2}.$$

This concludes the proof.

CHAPTER 40

ADVANCED INTEGRATION METHODS

• PROBLEM 1033

Express the value of the integral

$$\int_0^{\pi/2} \frac{dy}{\sqrt{\cos y}}$$

in terms of gamma functions. Hence show that

$$K\left(\frac{\sqrt{2}}{2}\right) = \frac{\sqrt{2\pi}}{2} \frac{\Gamma\left(\frac{1}{4}\right)}{\Gamma\left(\frac{3}{4}\right)} .$$

Solution: From the expressions obtained in another problem,

$$\int_0^{\pi/2} \cos^n y \, dy = \frac{\Gamma\left(\frac{n+1}{2}\right)}{\Gamma\left(\frac{n}{2}+1\right)} \frac{\sqrt{\pi}}{2} .$$

We then have:

$$\int_0^{\pi/2} \cos^{-\frac{1}{2}} y \, dy = \frac{\Gamma\left(\frac{1}{4}\right)}{\Gamma\left(\frac{3}{4}\right)} \cdot \frac{\sqrt{\pi}}{2} .$$

Let $\cos y = \cos^2 \Theta$, then

$$dy = \frac{2 \sin \Theta \cos \Theta \, d\Theta}{\sin y} = \frac{2 \sin \Theta \cos \Theta}{\sqrt{1 - \cos^4 \Theta}} \, d\Theta.$$

Hence,

$$\int_0^{\pi/2} \frac{dy}{\sqrt{\cos y}} = 2 \int_0^{\pi/2} \frac{\sin \Theta \cos \Theta \, d\Theta}{\cos \Theta \sqrt{1 - \cos^4 \Theta}}$$

$$= 2 \int_0^{\pi/2} \frac{(1 - \cos^2 \Theta)^{\frac{1}{2}} d\Theta}{(1 + \cos^2 \Theta)^{\frac{1}{2}}(1 - \cos^2 \Theta)^{\frac{1}{2}}}$$

$$= 2 \int_0^{\pi/2} \frac{d\Theta}{(1 + 1 - \sin^2 \Theta)^{\frac{1}{2}}}$$

$$= \sqrt{2} \int_0^{\pi/2} \frac{d\Theta}{\sqrt{1 - \frac{1}{2} \sin^2 \Theta}},$$

where the last integral is called the elliptic integral of the first kind and is denoted by

$K\left(\frac{1}{\sqrt{2}}\right)$, since the general expression is of the form:

$$\int_0^{\pi/2} \frac{d\Theta}{\sqrt{1 - k^2 \sin^2 \Theta}} \equiv K(k).$$

Hence,

$$\int_0^{\pi/2} \frac{dy}{\sqrt{\cos y}} = \sqrt{2}\, K\left(\frac{\sqrt{2}}{2}\right),$$

and, using the beta function for the given integral, we obtain:

$$K\left(\frac{\sqrt{2}}{2}\right) = \frac{\sqrt{2\pi}}{2} \frac{\Gamma\left(\frac{1}{4}\right)}{\Gamma\left(\frac{3}{4}\right)}.$$

• **PROBLEM 1034**

Evaluate :

$$\int_0^{\infty} \frac{dx}{x^2 \sqrt{x^2 - 9}}, \text{ if it exists.}$$

Solution: Let

$$\Gamma = \int_3^x \frac{dt}{t^2\sqrt{t^2 - 9}} .$$

Then, by substituting $t = 3 \sec \Theta$, and, by the use of $t = \frac{3}{\cos \Theta}$,

$$dt = \frac{\cos \Theta \, d(3) - 3d(\cos \Theta)}{\cos^2 \Theta} = \frac{3 \sin \Theta}{\cos^2 \Theta} d\Theta$$

$$= 3 \sec \Theta \tan \Theta \, d\Theta.$$

Hence,

$$\int \frac{dt}{t^2 \sqrt{t^2 - 9}} = \int \frac{3 \sec \Theta \tan \Theta \, d\Theta}{9 \sec^2 \Theta \sqrt{9 \sec^2 \Theta - 9}} .$$

By using the trigonometric identity:

$$\sec^2 \Theta = \tan^2 \Theta + 1, \sqrt{9 \sec^2 \Theta - 9} = 3 \tan \Theta.$$

Substitution in the integrand and cancelation results in:

$$\int \frac{dt}{t^2 \sqrt{t^2 - 9}} = \int \frac{3 \sec \Theta \tan \Theta \, d\Theta}{9 \sec^2 \Theta \cdot 3 \tan \Theta}$$

$$= \frac{1}{9} \int \cos \Theta \, d\Theta = \frac{\sin \Theta}{9} .$$

But

$$\sin^2 \Theta + \cos^2 \Theta = 1, \text{ and } \frac{1}{\cos \Theta} = \frac{t}{3} .$$

$$\cos \Theta = \frac{3}{t} .$$

Hence,

$$\sin \Theta = \sqrt{1 - \cos^2 \Theta} = \sqrt{1 - 9/t^2} = \frac{1}{t} \sqrt{t^2 - 9}.$$

Therefore,

$$\Gamma = \int_3^x \frac{dt}{t^2\sqrt{t^2 - 9}} = \frac{1}{9} \left(\frac{1}{t} \sqrt{t^2 - 9} \right|_3^x$$

$$= \frac{1}{9x} \sqrt{x^2 - 9}.$$

$$\int_0^{\infty} \frac{dx}{x^2 \sqrt{x^2 - 9}} = \lim_{x \to \infty} \Gamma = \lim_{x \to \infty} \frac{1}{9x} \sqrt{x^2 - 9}$$

$$= \lim_{x \to \infty} \frac{1}{9} \sqrt{1 - 9/x^2} = \frac{1}{9}.$$

• PROBLEM 1035

Evaluate $\int_0^{\infty} e^{-x} \cos x \, dx$, if it exists.

Solution: Let Γ be the area enclosed in the region for $0 \leq t \leq x$. Then,

$$\Gamma = \int_0^{x} e^{-t} \cos t \, dt.$$

The required value of the integral:

$$\int_0^{\infty} e^{-x} \cos x \, dx,$$

is obtained in terms of Γ, where

$$\int_0^{\infty} e^{-x} \cos x \, dx = \lim_{x \to \infty} \Gamma.$$

To integrate the expression, we use the method of integration by parts.

$$\int e^{-t} \cos t \, dt = e^{-t} \int \cos t \, dt - \int \left(\int \cos t \, dt \right) d(e^{-t})$$

$$= e^{-t} \sin t + \int e^{-t} \sin t \, dt,$$

where e^{-t} is taken as u and cos t = dV. Again,

$$\int e^{-t} \sin t \, dt = e^{-t} \int \sin t \, dt$$

$$- \int \left(\int \sin t \, dt \right) d(e^{-t})$$

$$= - e^{-t} \cos t - \int e^{-t} \cos t \, dt.$$

Thus,

$$\int e^{-t} \cos t \, dt = e^{-t} \sin t - e^{-t} \cos t - \int e^{-t} \cos t \, dt.$$

Combining both integrands, we obtain:

$$\Gamma = \int_{t=0}^{t=x} e^{-t} \cos t \, dt$$

$$= \frac{1}{2} \left[e^{-t} (\sin t - \cos t) \right]_{t=0}^{t=x}$$

$$= \frac{e^{-x} (\sin x - \cos x) - (-1)}{2}.$$

$$\int_0^\infty e^{-x} \cos x \, dx = \lim_{x\to\infty} \Gamma$$

$$= \lim_{x\to\infty} \frac{1}{2} \left(e^{-x} (\sin x - \cos x) + 1 \right)$$

$$= \frac{1}{2}.$$

• PROBLEM 1036

Show that a) $\int_0^\infty e^{-\alpha^2 x^2} dx = \frac{1}{2} \frac{\sqrt{\pi}}{\alpha}$; and hence, b) for n any positive integer,

$$\int_0^\infty x^{2n} e^{-\alpha^2 x^2} dx = \frac{\sqrt{\pi}}{2} \cdot \frac{1 \cdot 3 \cdot 5 \ldots (2n-1)}{2^n \alpha^{2n+1}}.$$

<u>**Solution:**</u> a) Let $\alpha^2 x^2 = u$, then $dx = \frac{du}{2\alpha^2 x} = \frac{du}{2\alpha^2 \left(\frac{\sqrt{u}}{\alpha}\right)} = \frac{du}{2\alpha\sqrt{u}}$.

Substitution in the integrand gives

$$\int_0^\infty e^{-\alpha^2 x^2} dx = \frac{1}{2\alpha} \int_0^\infty u^{-1/2} e^{-u} du.$$

But from the theory of gamma functions we have, for any positive number n,

$$\int_0^\infty x^{n-1} e^{-x} dx = \Gamma(n), \text{ and,}$$

for an integer n we have $\Gamma(n) = (n-1)!$

To apply the integral, we rewrite the integrand as

$$\frac{1}{2\alpha}\int_0^\infty u^{-1/2} e^{-u} du = \frac{1}{2\alpha}\int_0^\infty u^{1/2-1} e^{-u} du = \frac{1}{2\alpha}\Gamma(\tfrac{1}{2})$$

$$= \frac{1}{2\alpha}\sqrt{\pi},$$

since $\Gamma(\frac{1}{2}) = \sqrt{\pi}$.

b) Using the same procedure, since $x^2 = \frac{u}{\alpha^2}$, $x^{2n} = \frac{u^n}{\alpha^{2n}}$.

Hence,

$$\int_0^\infty x^{2n} e^{-\alpha^2 x^2} dx = \frac{1}{2\alpha}\int_0^\infty \frac{u^n}{\alpha^{2n}} \cdot u^{-1/2} e^{-u} du = \frac{1}{2\alpha^{2n+1}}\int_0^\infty u^{n-1/2} e^{-u} du$$

$$= \frac{1}{2\alpha^{2n+1}}\int_0^\infty u^{(n-\frac{1}{2}+1)-1} e^{-u} du$$

$$= \frac{1}{2\alpha^{2n+1}}\Gamma(n-\tfrac{1}{2}+1)$$

$$= \frac{1}{2\alpha^{2n+1}}\Gamma(\tfrac{2n+1}{2}).$$

By definition, $\Gamma(n+1) = n!$; hence $\Gamma\left(\frac{2n+1}{2}\right) = \left(\frac{2n-1}{2}\right)! = \frac{2n-1}{2} \cdot \frac{2n-2}{2} \cdot \frac{2n-3}{2} \cdot$ $\ldots \frac{5}{2} \cdot \frac{3}{2} \cdot \frac{1}{2}\Gamma\left(\frac{1}{2}\right)$. Since the 2's in the denominators exist n times, we can factor them out and obtain $\frac{1}{2^n}(2n-1)(2n-2)(2n-3)\ldots 3\cdot 1\cdot\Gamma(\frac{1}{2})$

In this case, $\Gamma(\frac{2n+1}{2}) = \frac{1}{2^n} 1\cdot 3\cdot 5 \cdots (2n-1)\Gamma(\frac{1}{2})$,

which yields the final result. By substitution,

$$\int_0^\infty x^{2n} e^{-\alpha^2 x^2} dx = \frac{1}{2\alpha^{2n+1}}\int_0^\infty u^{n-1/2} e^{-u} du = \frac{1}{2\alpha^{2n+1}}\Gamma(\tfrac{2n+1}{2})$$

$$= \frac{1}{2\alpha^{2n+1}} \cdot \frac{1}{2^n} \cdot 1\cdot 3\cdot 5 \ldots (2n-1)\sqrt{\pi},$$

where, again, $\Gamma(\frac{1}{2}) = \sqrt{\pi}$. Furthermore,

$$\int_0^\infty e^{-\alpha^2 x^2} dx = \lim_{n\to 0}\int_0^\infty x^{2n} e^{-\alpha^2 x^2} dx = \lim_{n\to 0}\frac{(2n-1)!\sqrt{\pi}}{2\alpha\, \alpha^{2n} 2^n} = \frac{\sqrt{\pi}}{2\alpha}$$

as obtained above.

• PROBLEM 1037

Find the expression, in terms of n, for

$$\int_0^1 \frac{dx}{\sqrt{1-x^n}}.$$

Evaluate the result for $n = 6$.

Solution: Let $1 - x^n = \cos^2\theta$. Then, $dx = \dfrac{2\sin\theta\,\cos\theta\,d\theta}{nx^{n-1}} =$

$\dfrac{2x^{1-n}\sin\theta\,\cos\theta}{n}\,d\theta$. Since $x^n = 1-\cos^2\theta = \sin^2\theta$, $x = \sin^{2/n}\theta$, and

$x^{1-n} = \sin^{\frac{2(1-n)}{n}}\theta$. Hence $dx = \dfrac{2}{n}\sin^{\frac{2(1-n)}{n}}\theta\,\sin\theta\,\cos\theta\,d\theta =$

$\dfrac{2}{n}\sin^{2-n/n}\theta\,\cos\theta\,d\theta$. For $x = 0$, $\cos\theta = 1$, $\theta = 0$ and for $x = 1$,

$\theta = \pi/2$. Thus

$$\int_0^1 \frac{dx}{\sqrt{1-x^n}} = 2\int_0^{\pi/2} \frac{\sin^{\frac{2-n}{n}}\theta\,\cos\theta\,d\theta}{n\cos\theta}$$

$$= \frac{2}{n}\int_0^{\pi/2} \sin^{\frac{2-n}{n}}\theta\,d\theta\,.$$

From the Beta functions, we have:

$$\int_0^{\pi/2} \sin^n\theta\,d\theta = \int_0^{\pi/2} \cos^n\theta\,d\theta = \frac{\Gamma\left(\frac{n+1}{2}\right)}{\Gamma\left(\frac{n}{2}+1\right)} \cdot \frac{\sqrt{\pi}}{2}\,.$$

Hence,

$$\int_0^{\pi/2} \sin^{2-n/n}\theta\,d\theta = \frac{\Gamma(1/n)}{\Gamma\left(\frac{2+n}{2n}\right)} \cdot \frac{\sqrt{\pi}}{2} = \frac{\Gamma(1/n)}{\Gamma\left(\frac{1}{n}+\frac{1}{2}\right)}\,\frac{\sqrt{\pi}}{2}\,.$$

Therefore,

$$\int_0^1 \frac{dx}{\sqrt{1-x^n}} = \frac{2}{n}\int_0^{\pi/2} \sin^{2-n/n}\theta\,d\theta = \frac{\Gamma(1/n)}{n\Gamma\left(\frac{1}{n}+\frac{1}{2}\right)}\sqrt{\pi}\,.$$

For $n = 6$, we have

$$\int_0^1 \frac{dx}{\sqrt{1-x^6}} = \frac{1}{3}\int_0^{\pi/2} \sin^{-2/3}\theta\,d\theta = \frac{\Gamma(1/6)}{6\Gamma(2/3)}\sqrt{\pi}$$

$$= \frac{\sqrt{\pi}}{6}\,\frac{\Gamma(0.166)}{\Gamma(0.666)}\,.$$

Using the relationship: $\dfrac{\Gamma(n+1)}{n} = \Gamma(n)$, we have:

$$\Gamma(1/6) = \frac{\Gamma(1/6+1)}{1/6} = 6\Gamma(1.166)$$

$$\Gamma(2/3) = \frac{\Gamma(2/3+1)}{2/3} = \frac{3}{2}\Gamma(1.666).$$

From tables, $\Gamma(1.166) = 0.93$ and $\Gamma(1.666) = 0.902$. Hence,

$$\int_0^1 \frac{dx}{\sqrt{1-x^6}} = \frac{\sqrt{\pi}}{6} \cdot \frac{6\Gamma(1.166)}{\frac{3}{2}\Gamma(1.666)} = \frac{2}{3}\sqrt{\pi}\,\frac{0.93}{0.902}$$

$$= 1.216.$$

• PROBLEM 1038

Evaluate the integral: $F = \displaystyle\int_0^1 \frac{x-1}{\ln x}\,dx$.

Solution: Consider, for example, the area of a circle. The well-known formula, $A = \pi r^2$, is obtained by the integration: $\int_R r dr\, d\theta$, over a region R occupied by it. This region is dependent on the value of the radius r, such that $R = R(r)$, and this leads to the fact that the expression for the area of a circle is nothing more than a relationship showing the dependence of the area A on the radius r. The area of a circle is proportional to the square of its radius, expressed by $A(r) = \pi r^2$, where the constant of proportionality, π, is the ratio of the circumference to its diameter. This concept says that, for a given r (say $r = a$), the corresponding A-value is πa^2, and this holds for any complete circle occupying a certain region expressed by $r = a$. Under these conditions, r is a parameter, and we have summed the complete circle by letting the angle $\theta = 2\pi$ radians. This simple problem introduces the concept of parametrization.

Likewise, a function in general, parametrized by α, may be defined as:

$$F(\alpha) = \int_a^b f(x,\alpha)dx$$

where a and b are fixed constants. Since F is a function of α, there **exists, in general, a derivative of F with respect** to α,

$$\frac{dF}{d\alpha} = \frac{d}{d\alpha}\int_a^b f(x,\alpha)dx,$$

and the above equation suggests that the integration $f(x,\alpha)$ is to be performed first, and the resulting function of α is then differentiated. Our goal here is to simplify functions **that are difficult to integrate into intergrable functions by employing parameters.** The given integral is difficult to perform as it stands:

$$F = \int_0^1 \frac{x - 1}{\ln x} dx .$$

Hence we introduce a parameter α, such that

$$F(\alpha) = \int_0^1 \frac{x^\alpha - 1}{\ln x} dx,$$

where the case of

$$F(1) = \int_0^1 \frac{x^1 - 1}{\ln x} dx$$

reduces the integral to the given form. We now have:

$$\frac{dF(\alpha)}{d\alpha} = \frac{d}{d\alpha}\int_0^1 \frac{x^\alpha - 1}{\ln x} dx = \int_0^1 \frac{\partial}{\partial\alpha}\left(\frac{x^\alpha - 1}{\ln x}\right)dx ,$$

under the assumption that the order of the integration and differentiation is inconsequential. To differentiate $\frac{x^\alpha}{\ln x}$ with respect to α, we let $y = x^\alpha$, from which we are interested in obtaining $\frac{dy}{d\alpha}$. Taking the logarithm, $\ln y = \ln x^{\alpha} = \alpha \ln x$, and, differentiating implicitly:

$$\frac{1}{y}\,dy = \ln x\,d\alpha, \quad \text{or} \quad \frac{dy}{d\alpha} = y\ln x = x^{\alpha}\ln x.$$

And

$$\frac{\partial}{\partial\alpha}\left(\frac{1}{\ln x}\right) = 0 .$$

Hence,

$$\frac{dF(\alpha)}{d\alpha} = \int_0^1 \frac{x^{\alpha}\ln x}{\ln x}\,dx = \int_0^1 x^{\alpha}dx = \left.\frac{x^{\alpha+1}}{\alpha+1}\right|_{x=0}^{x=1} = \frac{1}{\alpha+1}$$

This leads to:

$$F(\alpha) = \int \frac{d\alpha}{\alpha+1} + c = \ln(\alpha+1) + c.$$

Now, to determine c, when $\alpha = 0$, $F(\alpha) = \int_0^1 0\cdot dx = 0$. Hence

$F(0) = \ln 1 + c$, which implies that $c = 0$. The required value of the integral now becomes the limiting value of the parametrized integral:

$$\int_0^1 \frac{x-1}{\ln x}dx = \lim_{\alpha\to 1}\int_0^1 \frac{x^{\alpha}-1}{\ln x}\,dx = F(1) = \ln 2 = 0.693.$$

• PROBLEM 1039

Evaluate the integral: $\int_0^{\infty} \frac{e^{-x}\sin x}{x}\,dx$.

Solution: We parametrize the given integrand as:

$$F(\alpha) = \int_0^{\infty} \frac{e^{-\alpha x}\sin x}{x}\,dx,$$

and we attempt to find the value of $F(1)$, which is the value of the given integral. We write:

$$\frac{dF}{d\alpha} = \frac{d}{d\alpha}\int_0^{\infty}\frac{e^{-\alpha x}\sin x}{x}\,dx = \int_0^{\infty}\frac{\partial}{\partial\alpha}\,\frac{e^{-\alpha x}\sin x}{x}\,dx$$

$$= -\int_0^{\infty} e^{-\alpha x}\sin x\,dx .$$

Using integration by parts, i.e., $\int u\,dv = uv - \int v\,du$, we obtain:

$$\frac{dF}{d\alpha} = -\left.\frac{e^{-\alpha x}}{\alpha^2+1}(-\alpha\sin x - \cos x)\right|_0^{\infty}$$

$$= -\frac{1}{\alpha^2+1} .$$

Then

$$F(\alpha) = -\int \frac{d\alpha}{\alpha^2+1} + c$$

$$= -\tan^{-1}\alpha + c .$$

When $\alpha = \infty$, then $F(\alpha) = F(\infty) = 0$. From the integration just performed, $F(\infty) = -\tan^{-1}\infty + c$. $-\tan^{-1}\infty + c = 0$ implies that

$c = \tan^{-1}\infty = \pi/2$. Finally, $F(\alpha) = \pi/2 - \tan^{-1}\alpha$ and

$$F = F(1) = \lim_{\alpha \to 1} \int_0^\infty \frac{e^{-\alpha x}\sin x}{x}\,dx = \int_0^\infty \frac{e^{-x}\sin x}{x}\,dx$$

$$= \pi/2 - \tan^{-1} 1 = \pi/4 .$$

• PROBLEM 1040

Evaluate: $\int_0^\infty \frac{dx}{x^2+\alpha^2}$, $\alpha > 0$, and hence show that

$$\int_0^\infty \frac{dx}{(x^2+\alpha^2)^{n+1}} = \frac{\pi}{2} \cdot \frac{1\cdot 3\cdot 5 \ldots (2n-1)}{2\cdot 4\cdot 6\cdot \ldots 2n\alpha^{2n+1}} .$$

Solution:

$$\int_0^\infty \frac{dx}{x^2+\alpha^2} = \frac{1}{\alpha}\tan^{-1}\frac{x}{\alpha}\Big|_0^\infty = \frac{\pi}{2\alpha} .$$

Now, let

$$F_1(\alpha) = \int_0^\infty \frac{dx}{x^2+\alpha^2} = \int_0^\infty (x^2+\alpha^2)^{-1}dx = \frac{\pi}{2\alpha} .$$

Then, using the prime operator, $F'(\alpha) = dF/d\alpha$, and noting that

$$\frac{dF}{d\alpha} = \frac{d}{d\alpha}\int_0^\infty f(x,\alpha)dx = \int_0^\infty \frac{\partial}{\partial\alpha}f(x,\alpha)dx ,$$

we have:

$$F_1'(\alpha) = (-1)(2\alpha)\int_0^\infty (x^2+\alpha^2)^{-2}dx = (-1)\frac{\pi}{2\alpha^2},$$

or,

$$\int_0^\infty (x^2+\alpha^2)^{-2}dx = \frac{\pi}{2\cdot 2\alpha^3} \equiv F_2(\alpha).$$

Again,

$$F_2'(\alpha) = (-2)(2\alpha)\int_0^\infty (x^2+\alpha^2)^{-3}dx = \frac{(-3)\pi}{2\cdot 2\alpha^4},$$

or,

$$\int_0^\infty (x^2+\alpha^2)^{-3}dx = \frac{3\pi}{2.2\cdot 4\alpha^5} \equiv F_3(\alpha).$$

$$F_3'(\alpha) = (-3)(2\alpha)\int_0^\infty (x^2+\alpha^2)^{-4}dx = \frac{3\cdot 5\ \pi}{2\cdot 2\cdot 4\alpha^6},$$

or,

$$\int_0^\infty (x^2+\alpha^2)^{-4}dx = \frac{3\cdot 5\ \pi}{2\cdot 2\cdot 4\cdot 6\cdot \alpha^7} .$$

By mathematical induction, for any intermediate value $1 < k < n$, we have:

$$\int_0^\infty (x^2+\alpha^2)^{-k}dx = \frac{3\cdot 5\cdot 7 \ldots (2k-3)\pi}{2\cdot 2\cdot 4\cdot 6. \ldots (2k-2)\alpha^{2k-1}} \equiv F_k(\alpha) .$$

Again, by the theorem of mathematical induction, if the above relationship holds for k, then, it should also hold for $k+1$. Therefore we can replace k by $k+1$:

$$\int_0^\infty (x^2+\alpha^2)^{-(k+1)}dx = \frac{3\cdot5\cdot7\cdot \ldots \left[2(k+1)-3\right]\pi}{2\cdot2\cdot4\cdot6\cdot \ldots \left[2(k+1)-2\right]\alpha^{2(k+1)-1}}$$

$$= \frac{3\cdot5\cdot7\cdot \ldots (2k-1)\pi}{2\cdot2\cdot4\cdot6\cdot \ldots (2k)\alpha^{2k+1}} .$$

On the other hand, we can differentiate $F_k(\alpha)$, to obtain:

$$(-k)(2\alpha)\int_0^\infty (x^2+\alpha^2)^{-(k+1)}dx = -\frac{3\cdot5\cdot7\cdot \ldots (2k-3)(2k-1)\pi}{2\cdot2\cdot4\cdot6\cdot \ldots (2k-2)\alpha^{2k-1+1}},$$

or,

$$F_{k+1}(\alpha) = \int_0^\infty (x^2+\alpha^2)^{-(k+1)}dx = \frac{3\cdot5\cdot7\cdot \ldots (2k-3)(2k-1)\pi}{2\cdot2\cdot4\cdot6\cdot \ldots (2k-2)(2k)\alpha^{2k+1}} .$$

Hence, for any integer $n \geq k$, we generally have:

$$F_{n+1}(\alpha) = \int_0^\infty \frac{dx}{(x^2+\alpha^2)^{n+1}} = \frac{3\cdot5\cdot7\cdot \ldots (2n-1)\pi}{2\cdot2\cdot4\cdot6\cdot \ldots 2n\ \alpha^{2n+1}}.$$

$$= \frac{\pi}{2}\ \frac{1\cdot3\cdot5\cdot7\cdots\cdots(2n-1)}{2\cdot4\cdot6\cdots\cdots 2n\alpha^{2n+1}}.$$

• PROBLEM 1041

Express the integral: $\int_0^\varphi \frac{\sin^2\varphi d\varphi}{\sqrt{1-k^2\sin^2\varphi}}$ $(0 < k < 1)$

in terms of elliptic integrals. What is the result when the upper limit is $\pi/2$?

Solution: Rewriting the given integral, we attempt to achieve some kind of similarity between the numerator and denominator in the integral. Since, by definition, an integration is nothing more than a summation, multiplication of any individual term by a constant is the same as multiplying the integral by the same constant. Thus,

$$\int_0^\varphi \frac{\sin^2\varphi d\varphi}{\sqrt{1-k^2\sin^2\varphi}} = \frac{1}{k^2}\int_0^\varphi \frac{k^2\sin^2\varphi d\varphi}{\sqrt{1-k^2\sin^2\varphi}} .$$

Also,

$$\frac{1}{k^2}\int_0^\varphi \frac{k^2\sin^2\varphi d\varphi}{\sqrt{1-k^2\sin^2\varphi}} = \frac{1}{k^2}\int_0^\varphi \frac{1-1+k^2\sin^2\varphi}{\sqrt{1-k^2\sin^2\varphi}}\,d\varphi .$$

In this case, we can write:

$$\int_0^\varphi \frac{\sin^2\varphi d\varphi}{\sqrt{1-k^2\sin^2\varphi}} = \frac{1}{k^2}\int_0^\varphi \frac{1-(1-k^2\sin^2\varphi)}{\sqrt{1-k^2\sin^2\varphi}}\,d\varphi$$

$$= \frac{1}{k^2}\int_0^\varphi \frac{d\varphi}{\sqrt{1-k^2\sin^2\varphi}} - \frac{1}{k^2}\int_0^\varphi \frac{1-k^2\sin^2\varphi}{\sqrt{1-k^2\sin^2\varphi}}\,d\varphi .$$

If we let functions denote the integrals:

$$F(k,\varphi) \equiv \int_0^{\varphi} \frac{d\varphi}{\sqrt{1-k^2\sin^2\varphi}} \quad \text{and} \quad E(k,\varphi) \equiv \int_0^{\varphi} \sqrt{1-k^2\sin^2\varphi}\, d\varphi,$$

which is the last integral, we obtain:

$$\int_0^{\varphi} \frac{\sin^2\varphi d\varphi}{\sqrt{1-k^2\sin^2\varphi}} = \frac{1}{k^2}\left[F(k,\varphi) - E(k,\varphi)\right].$$

The new functions introduced are called elliptic integrals: $F(k,\varphi)$ is of the first kind and $E(k,\varphi)$ is of the second kind for $0 < k < 1$. The number k is called the modulus of the elliptic integral (in conics, k is called the eccentricity); and the upper limit φ is called the amplitude of the elliptic integral. If the amplitude is $\pi/2$, then, by the known notation, we can write:

$$F(k,\pi/2) \equiv F(k) \text{ or just } F, \text{ and}$$
$$E(k,\pi/2) \equiv E(k) \equiv E.$$

Thus, for $\varphi = \pi/2$,

$$\int_0^{\pi/2} \frac{\sin^2\varphi d\varphi}{\sqrt{1-k^2\sin^2\varphi}} \equiv \frac{1}{k^2}\left[F(k) - E(k)\right].$$

Furthermore, if we are given:

$$\int_0^{\varphi} \frac{\cos^2\varphi d\varphi}{\sqrt{1-k^2\sin^2\varphi}},$$

we can use the identity $\cos^2\varphi = 1-\sin^2\varphi$, and hence,

$$\int_0^{\varphi} \frac{\cos^2\varphi d\varphi}{\sqrt{1-k^2\sin^2\varphi}} = \int_0^{\varphi} \frac{1-\sin^2\varphi}{\sqrt{1-k^2\sin^2\varphi}} = F(k,\varphi) - \frac{1}{k^2}\left[F(k,\varphi) - E(k,\varphi)\right]$$

$$= \frac{\left(k^2-1\right)F(k,\varphi) + E(k,\varphi)}{k^2}.$$

• PROBLEM 1042

The integrals: $F(k,\varphi) = \int_0^{\varphi} \frac{d\varphi}{\sqrt{1-k^2\sin^2\varphi}}$, and $E(k,\varphi) = \int_0^{\varphi} \sqrt{1-k^2\sin^2\varphi}\, d\varphi$,

are defined as elliptic integrals of the first and second kind, respectively. If $k > 1$, these integrals may also be expressed in terms of elliptic integrals. Show that

a) $\int_0^{\varphi} \frac{d\varphi}{\sqrt{1-k^2\sin^2\varphi}} = \frac{1}{k} F\left(\frac{1}{k}, x\right)$, and b) $\int_0^{\varphi} \sqrt{1-k^2\sin^2\varphi}\, d\varphi$

$$= \left(\frac{1}{k} - k\right)F\left(\frac{1}{k}, x\right) + kE\left(\frac{1}{k}, x\right),$$

(for $k > 1$), where the amplitude x is expressed in terms of the upper limit φ by $x = \sin^{-1}(k \sin\varphi)$, and the integrals have real values when $k \sin\varphi \le 1$.

Solution: Let $k \sin\varphi = \sin x$ Then $d\varphi = \frac{\cos x}{k\cos\varphi}dx$, $1-k^2\sin^2\varphi = 1-\sin^2 x = \cos^2 x$ and $x = \sin^{-1}(k \sin\varphi)$. Hence, by substitution,

a)
$$\int \frac{d\varphi}{\sqrt{1-k^2\sin^2\varphi}} = \int \frac{dx}{k\cos\varphi} = \frac{1}{k}\int \frac{dx}{\cos\varphi} .$$

By virtue of the fact that $k \sin\varphi = \sin x$ and the trigonometric identity $\cos^2\varphi + \sin^2\varphi = 1$, we have:

$$\cos\varphi = \sqrt{1-1/k^2 \sin^2 x} .$$

Hence

$$\int_0^\varphi \frac{d\varphi}{\sqrt{1-k^2\sin^2\varphi}} = \frac{1}{k}\int_0^x \frac{dx}{\sqrt{1-1/k^2 \sin^2 x}} .$$

$F(k,\varphi)$ is the elliptic integral:

$$\int_0^\varphi \frac{d\varphi}{\sqrt{1-k^2\sin^2\varphi}} \quad (0 \le k \le 1).$$

Then we can write:

$$F(1/k,x) \equiv \int_0^x \frac{dx}{\sqrt{1-1/k^2\sin^2 x}} \quad (k > 1 \text{ or } 0 < 1/k \le 1).$$

Thus, by this definition,

$$\int_0^\varphi \frac{d\varphi}{\sqrt{1-k^2\sin^2\varphi}} = \frac{1}{k}\int_0^x \frac{dx}{\sqrt{1-1/k^2\sin^2 x}} \equiv \frac{1}{k}F(1/k,x) \quad \text{for } k > 1.$$

b) By the use of the above argument and another problem, we have

$$\int\sqrt{1-k^2\sin^2\varphi}\, d\varphi = \int \cos x\left(\frac{\cos x}{k\cos\varphi}dx\right) = \frac{1}{k}\int \frac{\cos^2 x}{\cos\varphi}dx ,$$

or,

$$\int_0^\varphi \sqrt{1-k^2\sin^2\varphi}\, d\varphi = \frac{1}{k}\int_0^x \frac{\cos^2 x dx}{\sqrt{1-1/k^2\sin^2 x}} .$$

If we replace $1/k$ by m in the radical, the relationship proved in another problem shows that:

$$\int_0^x \frac{\cos^2 x dx}{\sqrt{1-m^2\sin^2 x}} = \frac{(m^2-1)F(m,x)+E(m,x)}{m^2} \quad \text{for } 0 < m < 1 \text{ or } k > 1.$$

Substituting the value of m back in the above expression yields:

$$\frac{1}{k}\int \frac{\cos^2 x dx}{\sqrt{1-m^2\sin^2 x}} = \frac{1}{k}\cdot\frac{(1/k^2-1)F(1/k,x)+E(1/k,x)}{(1/k)^2}$$

$$= (1/k - k)F(1/k,x)+kE(1/k,x)$$

$$\equiv \int_0^\varphi \sqrt{1-k^2\sin^2\varphi}\, d\varphi .$$

Note that, for $k > 1$,

$$F(1/k,x) \equiv \int_0^x \frac{dx}{\sqrt{1-1/k^2 \sin^2 x}} ,$$ the elliptic integral of

the first kind.

$$E(1/k,x) \equiv \int_0^x \sqrt{1-1/k^2 \sin^2 x}\, dx,$$ the elliptic integral of

the second kind. Both integrals have the modulus 1/k and amplitude x.

• PROBLEM 1043

Calculate $\int_0^{\frac{1}{2}} e^{\cos x}\, dx$.

Solution: Let $v = e^u = 1 + u + \frac{u^2}{2!} + \frac{u^3}{3!} + \ldots + \frac{u^{n-1}}{(n-1)!} + \ldots$.

We know that:

$$\cos x - 1 = -\frac{x^2}{2!} + \frac{x^4}{4!} - \ldots + (-1)^{n-1} \frac{x^{2n-2}}{(2n-2)!},$$

where v and cos x are convergent for every u and x respectively. We substitute for u in the first series its value in terms of x from the second series so as to obtain v as a power series in x, convergent for $|x|$ sufficiently small. Considering only three terms from the second series, we have:

$$v(x) = \left\{1+\left(-\frac{x^2}{2!} + \frac{x^4}{4!} - \frac{x^6}{6!}\right) + \frac{1}{2!}\left(\frac{x^4}{(2!)^2} - \frac{x^6}{2!4!}\right) + \frac{1}{3!}\left(-\frac{x^6}{(2!)^3}\right) + \ldots\right\} e.$$

The multiplying factor e appears in the series as a result of the relationship:

$$e^{\cos x} = e \cdot e^{\cos x-1} .$$

Then, letting u = cos x-1, we proceeded. Thus

$$\int_0^{\frac{1}{2}} e^{\cos x}\, dx = e \int_0^{\frac{1}{2}} \left(1-x^2/2 + x^4/6 - 31x^6/720 + \ldots\right)$$

$$\cong e(0.48)$$

$$\cong 1.305.$$

• PROBLEM 1044

Find $\int_0^x \frac{\ln(1-x)}{x}\, dx$ in series form.

Solution: The given function can be expanded into a series form term by term. Since the validity of dividing a series by a series, and term by term differentiation or integration are confirmed, we expand ln(1-x) in Maclaurin's series. Before doing so, we have an expression for:

$$\frac{1}{1-x} = 1 + x + x^2 + \ldots + x^n .$$

The relationship between these two functions is:

$$\ln(1-x) = -\int \frac{dx}{1-x} .$$

Hence,

$$\ln(1-x) = -x - \frac{x^2}{2} - \frac{x^3}{3} - \ldots - \frac{x^{n+1}}{n+1} - \ldots,$$

and

$$\frac{\ln(1-x)}{x} = -1 - \frac{x}{2} - \frac{x^2}{3} - \ldots - \frac{x^n}{n+1} - \ldots .$$

In this case,

$$\int_0^x \frac{\ln(1-x)}{x}\,dx = -\int_0^x \left\{1 + \frac{x}{2} + \frac{x^2}{3} + \ldots + \frac{x^n}{n-1} + \ldots\right\} dx ,$$

$$= -\left[x + \frac{x^2}{4} + \frac{x^3}{9} + \ldots + \frac{x^{n+1}}{n^2}\right].$$

Both series are convergent for $-1 \le x < 1$.

• PROBLEM 1045

Evaluate: $\displaystyle\int_0^{0.3} \frac{\cos x}{\sqrt{1-x}}\,dx$.

Solution: To find the value of the given integral, we express the integrand in series form. Two series are known:

$$\cos x = 1 - \frac{x^2}{2!} + \frac{x^4}{4!} - \ldots (-1)^{n-1}\frac{x^{2n-2}}{(2n-2)!} , \text{ and}$$

$$\frac{1}{\sqrt{1-x}} = 1 + \frac{x}{2} + \frac{3x^2}{8} + \frac{5}{16}x^3 + \ldots .$$

For example, from the theory of magnetic and electrostatic potentials we are familiar with $u(r) \equiv k/r$, where k, a constant, is the product of a constant of proportionality and either the masses or charges. $u(r)$ represents the potential at a point $P(x,y)$, on a plane, due to a certain mass or charge at, say, $P_1(x_1,y_1)$. Hence $r = \left[(x-x_1)^2 + (y-y_1)^2\right]^{1/2}$. Now, if r approaches zero, the theoretical expression for the potential **increases** to infinity. To overcome this, r has to be expressed in terms of the relative values of the x and y components. If, for instance, the x component is taken as constant and P_1 is at the origin, we can approximate the potential in a binomial form as either:

$$u(r) = kx\left(1+y^2/x^2\right)^{-\frac{1}{2}} \simeq kx \sum_{t=0}^{n} \binom{-\frac{1}{2}}{t}\left(\frac{y^2}{x^2}\right)^t \simeq k\left(x - \frac{1}{2}\frac{y^2}{x} + \frac{3}{8}\frac{y^4}{x^3} - \frac{5}{16}\frac{y^6}{x^5} + \ldots\right)$$

for $y < x$; or:

$$u(r) = ky\left(\frac{x^2}{y^2} + 1\right)^{-\frac{1}{2}} \simeq k\left(y - \frac{1}{2}\frac{x^2}{y} + \frac{3}{8}\frac{x^4}{y^3} - \frac{5}{16}\frac{x^6}{y^5} + \ldots\right) \text{ for } x < y .$$

The binomial coefficient $\displaystyle\binom{a}{b} = \frac{a!}{(a-b)!\,b!}$.

In elliptical integrals, since $k < 1$ and $|\sin \varphi| \le 1$, we can approximate the integrand in terms of a series. For the first kind,

$$\int_0^{\varphi}\left(1-k^2\sin^2\varphi\right)^{-\frac{1}{2}}d\varphi = \int_0^{\varphi}\left(1+\frac{k^2}{2}\sin^2\varphi + \frac{3k^4}{8}\sin^4\varphi + \ldots\right)d\varphi.$$

The approximate series for the given integrand:

$$\frac{\cos x}{\sqrt{1-x}} \simeq \left(1 - \frac{x^2}{2!} + \frac{x^4}{4!} - \ldots\right)\left(1 + \frac{x}{2} + \frac{3x^2}{8} + \frac{5}{16}x^3 + \ldots\right)$$

$$\simeq 1 + \frac{x}{2} - \frac{1}{8}x^2 + \frac{1}{16}x^3 - \ldots .$$

Integration within the given limits gives:

$$x + \frac{x^2}{4} - \frac{x^3}{24} + \frac{x^4}{64} - \ldots\Big|_0^{0.3} \quad . \quad \simeq 0.323.$$

Thus

$$\int_0^{0.3} \frac{\cos x}{\sqrt{1-x}}\,dx \simeq 0.323.$$

• PROBLEM 1046

Evaluate: $\int_0^1 e^{-x^2}\,dx$.

Solution: We use the theorem: If the series $f(x)$ converges for $|x| < r$, the integral of $f(x)$ may be found by integrating the series term by term, and the integral series converges for $|x| < r$. We first attempt to obtain the region of convergence of the given integrand.

The function $f(x) = e^{-x^2}$ can be expanded using Maclaurin's formula:

$$f(x) = \sum_{n=0}^{\infty} \frac{f^{(n)}(0)}{n!}x^n ,$$

where $f^{(n)}(0)$ is the nth derivative of the function evaluated at $x = 0$. However, to minimize the effort, since e^u is well-known, (for example, from the Euler formula) we have:

$$e^u = 1 + u + \frac{u^2}{2!} + \ldots + \frac{u^{n-1}}{(n-1)!} + \ldots ,$$

substituting $-x^2$ for u ,

$$e^{-x^2} = 1 - x^2 + \frac{x^4}{4!} - \frac{x^6}{3!} + \ldots + \frac{(-1)^{n-1}x^{2(n-1)}}{(n-1)!} + \ldots .$$

This kind of series is called an alternating series, since the real constant terms are alternately positive and negative. This is a power series, the convergence of which can be determined, using one of the following interrelated methods:

i) if $\lim_{n\to\infty}\left|\frac{a_{n-1}}{a_n}\right| = r$, ($a_n$ is the coefficient of x to the nth degree), then the series is absolutely convergent for $|x| < r$ and divergent for $|x| > r$;

ii) Cauchy's ratio test, states that where the ratio:

$$\frac{u_{n+1}}{u_n} \left(u_n \text{ is the } n^{th} \text{ sequence of the function}\right),$$

exists as n becomes infinite, and is less than unity, the given series converges absolutely. If this limit does not exist, or if it is greater than unity, the series diverges. The behavior of the series cannot be determined for $x = r$, in the first case, and the limit equals unity in the second one. The series is thus absolutely convergent for all values of x, $|x| < \infty$. Integration, under the given limit yields

$$\int_0^1 e^{-x^2} dx = \int_0^1 \left\{1 - x^2 + \frac{x^4}{4!} - \frac{x^6}{6!} + \dots\right\} dx$$

$$\simeq 0.747.$$

Note that the integrated terms are also absolutely convergent for all x.

PROBLEMS SOLVED BY ADVANCED INTEGRATION METHODS

• PROBLEM 1047

If, for a radioactive substance, $A = A_0 e^{-kt}$, what is the relationship between the half-life, t_h, of a substance and the constant k?

Solution: The definition of half-life, t_h, of a radioactive substance is the time necessary for a given amount of the specific substance to decay to one-half of the original amount. Therefore when $t = t_h$, $A = \frac{1}{2}A_0$. Thus,

$$\tfrac{1}{2}A_0 = A_0 e^{-kt_h}.$$

$$\tfrac{1}{2} = e^{-kt_h},$$

$$\ln(\tfrac{1}{2}) = -kt_h.$$

$$-\ln 2 = -kt_h.$$

$$kt_h = \ln 2,$$

or,

$$t_h = \frac{\ln 2}{k},$$

or,

$$k = \frac{\ln 2}{t_h}.$$

• PROBLEM 1048

The half-life of radium is 1,590 years. Find the equation describing the amount of radium present as a function of time. Then determine how long it would take a given sample of radium to decay to $\frac{1}{6}$ of its present amount.

Solution: It has been shown that the relationship between the constant k and t_h is $k = \frac{\ln 2}{t_h}$. Therefore, when $t_h = 1,590$,

$$k = \frac{\ln 2}{1,590}.$$

Thus,

$$A = A_0 e^{\frac{-\ln 2}{1,590} t}.$$

To determine the time necessary for A to equal $\frac{A_0}{6}$,

$$\frac{A_0}{6} = A_0 e^{\frac{-\ln 2}{1,590} t}.$$

$$\frac{1}{6} = e^{\left(\frac{-\ln 2}{1,590}\right) t}.$$

$$\ln \frac{1}{6} = \frac{-\ln 2}{1,590} t.$$

$$-\ln 6 = \frac{-\ln 2}{1,590} t.$$

$$t = \frac{1,590(\ln 6)}{\ln 2} = \frac{1,590(1.792)}{1.693}$$

$$= 4,110 \text{ years.}$$

• PROBLEM 1049

Since the half-life of C^{14} is 5,570 years, how old is an organic object if it has $\frac{1}{10}$ of the normal amount of C^{14} present?

Solution: Since C^{14} is a radioactive substance, it obeys the radioactive decay principle described by the equation

$$A = A_0 e^{-kt},$$

where

$$k = \frac{\ln 2}{t_h} = \frac{\ln 2}{5,570}.$$

The problem is to find t such that

$$A = \frac{1}{10} A_0.$$

$$\frac{A_0}{10} = A_0 e^{\frac{-\ln 2}{5,570} t}.$$

$$\frac{1}{10} = e^{\frac{-\ln 2}{5,570} t}.$$

$$\ln \frac{1}{10} = \frac{-\ln 2}{5,570} t.$$

$$-\ln(10) = \frac{-\ln 2}{5,570} t.$$

$$t = \frac{5,570 \ln 10}{\ln 2} = \frac{5,570(2.303)}{0.693}$$

$$= 18,510 \text{ years.}$$

• PROBLEM 1050

The revenue in dollars from a certain source is known to be given by:

$$f(t) = (t + 1)^{-3/2}(3,000),$$

where t is given in years. What is the total revenue expected from this source over the next 18 months?

Solution: The revenue is given by:

$$\int_0^{3/2} 3,000(t + 1)^{-3/2}dt = 3,000(-2)(t + 1)^{-\frac{1}{2}}\Big]_0^{3/2}$$

$$= -6,000\left[\left(\frac{3}{2} + 1\right)^{-\frac{1}{2}} - (1)^{-\frac{1}{2}}\right]$$

$$= -6,000\left[\left(\frac{5}{2}\right)^{-\frac{1}{2}} - 1\right]$$

$$= -6,000\left[\left(\frac{2}{5}\right)^{+\frac{1}{2}} - 1\right]$$

$$= 6,000\left[1 - \sqrt{\frac{2}{5}}\right] \text{ dollars.}$$

This type of problem leads in turn to a problem of a more complex nature. If a businessman knows that he will need a certain sum of money at some later point in time, he might ask how much he would have to invest today at a fixed interest rate in order to have this required amount at the later date. We need not consider the actual amount, but ask,instead,how much he would have to invest now to have \$1 at the later point in time. Or else, we can assume that he invests \$1 at a rate r per year. We need only consider the \$1 figure, as any other dollar amount can be found by multiplying the \$1 figure by the actual amount. For example, if \$1 yields \$1.05 then \$7.300 yields \$7.300(1.05)=\$7.525.

At the end of one year, he would have $(1 + r)$ dollars. In two years he would have

$$(1 + r) + (1 + r)r = (1 + r)(1 + r) = (1 + r)^2$$

dollars, and so forth. In general he would have $(1 + r)^t$ dollars if his interest were compounded yearly.

If the amount is compounded n times per year, the one dollar becomes:

$$\left(1 + \frac{r}{n}\right)^{tn} \text{ dollars} = \left(1 + \frac{r}{n}\right)^{(n/r)(tr)} = \left(\left[1 + \frac{r}{n}\right]^{n/r}\right)^{tr}$$

Now consider what happens as n tends to infinity, i.e., the number of times interest is compounded becomes large.

$$\lim_{n \to \infty} \left(1 + \frac{r}{n}\right)^{n/r} = \lim_{x \to 0} (1 + x)^{1/x} = e,$$

where $\frac{r}{n} = x$. (This limit is sometimes used as a definition of e.)

Thus,

$$\lim_{n \to \infty} \left(\left[1 + \frac{r}{n}\right]^{n/r} \right)^{tr} = e^{tr} .$$

One dollar today becomes e^{tr} dollars in t years at an interest rate r when interest is compounded continuously. However, the original problem was: What amount, a, becomes 1 dollar in t years. Using the result for one dollar,

$$ae^{tr} = 1,$$

or

$$a = e^{-tr} .$$

In other words, the present value of 1 dollar, t years in the future is e^{-tr} dollars.

• PROBLEM 1051

Consider a constant income of \$100 per month over a period covering the next 5 years. What is the capital value of this income if the interest rate is 5%?

Solution: The capital value is given by

$$\int_0^{t'} e^{-rt} f(t)dt,$$

where t' = time in years, f(t) = amount of money in one year, and r = interest rate in decimal notation. Therefore, in this example,

$$t' = 5, \; f(t) = \$1,200. \text{ (\$100 a month = \$1200 per year.)}$$
$$r = 0.05.$$

Therefore, we have:

$$\int_0^5 e^{-0.05t}\, 1,200dt = 1,200 \int_0^5 e^{-0.05t}dt$$

$$= 1,200 \left(\frac{e^{-0.05t}}{-0.05}\right)\Bigg|_0^5$$

$$= \frac{-1,200}{0.05}\left[e^{-0.25} - e^0\right]$$

$$= 24,000\left[1 - \frac{1}{e^{0.25}}\right]$$

$$= 24,000\left[1 - \frac{1}{1.2840}\right]$$

$$= 24,000[1 - 0.7788]$$

$$= 24,000[0.2212]$$

$$= 5,308.80.$$

• PROBLEM 1052

How much would have to be invested at 6% interest in order to have 5,000 dollars in 30 months?

Solution: t=5/2 years. r 0.06 It has been shown that the equation which determines what amount, a, becomes 1 dollar in t years at an interest rate r is:

$$a = e^{-tr}$$

Since we are interested in 5,000 dollars, e^{-tr} will have a coefficient of 5,000. Therefore,

$$5{,}000e^{-\frac{5}{2}(0.06)} = 5{,}000e^{-0.15}$$

$$= \frac{5{,}000}{e^{0.15}}.$$

Using a table or slide rule to evaluate $e^{0.15} = 1.1618$,

$$\frac{5000}{e^{0.15}} = \frac{5000}{1.1618} = 4{,}303.67.$$

The ideas of present value and future revenue can be combined. That is, we can formulate an expression for the total present value or worth of future revenue. Consider the actual amount of a future revenue. Assuming that the rate of revenue is given by a function f(t), the actual amount of revenue over a short period of time, Δt, is $f(t)\Delta t$. The present value of this is $e^{-rt}f(t)\Delta t$ if the interest rate is r.

The total present value of this revenue over an interval of time from 0 to t' is given by:

$$\int_0^{t'} e^{-rt} f(t)dt.$$

This integral is called the capital value of the future income.

• PROBLEM 1053

The results of a certain experiment correspond to real-number values between 0 and π, and the probability density function for the results is known to be $P(x) = \frac{1}{2}\sin x$. What is the probability that when the experiment is performed an outcome between $\frac{\pi}{4}$ and $\frac{\pi}{2}$ will occur?

Solution: The area is the shaded portion in the figure. The probability corresponds to the area bounded by the x axis and the probability density function. Specifically, the probability of an outcome between $\frac{\pi}{4}$ and $\frac{\pi}{2}$ is:

$$\int_{\pi/4}^{\pi/2} \frac{1}{2} \sin x \, dx = \frac{1}{2}(-\cos x)\Big]_{\pi/4}^{\pi/2}$$

$$= \frac{1}{2}\left[-\cos \frac{\pi}{2} - \left(-\cos \frac{\pi}{4}\right)\right]$$

$$= \frac{1}{2}\left[\cos \frac{\pi}{4}\right]$$

$$= \frac{1}{2} \cdot \frac{\sqrt{2}}{2} = \frac{\sqrt{2}}{4} \approx \frac{1.414}{4}$$

$$= 0.3535.$$

Thus, approximately 35.35% of the time an experimental result between $\frac{\pi}{4}$ and $\frac{\pi}{2}$ would be expected for this experiment. Note that it is often the case that the experimental values in question involve infinite limits. This requires the use of improper integrals.

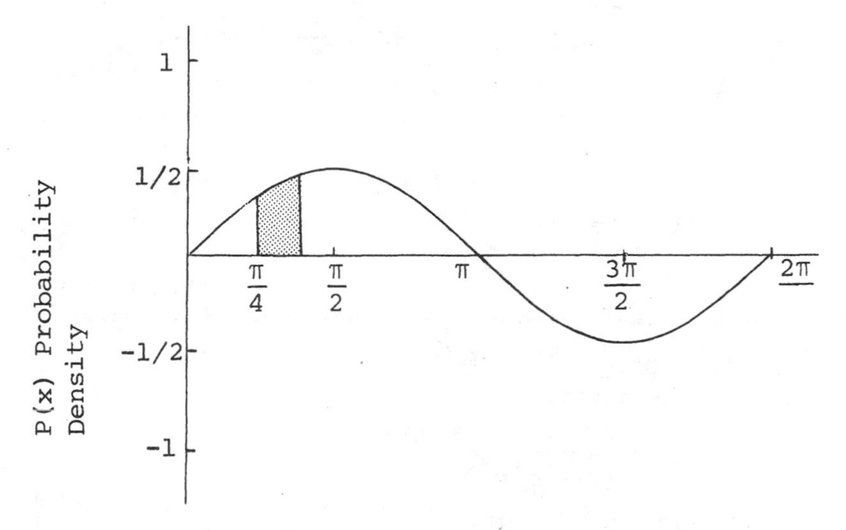

• PROBLEM 1054

The results of a certain experiment correspond to the positive real numbers, and the applicable probability density function is

$$P(x) = \frac{1}{(x+1)^2} \quad \text{where} \quad 0 \le x < \infty .$$

What is the probability of an experimental outcome greater than or equal to 3?

<u>Solution:</u> First we should check that the total probability equals 1. Therefore

$$\int_0^\infty \frac{1}{(x+1)^2} dx = 1,$$

an improper integral that should approach a finite limit if it is to be useful.

$$\int_0^{\infty} \frac{1}{(x+1)^2}\,dx = \lim_{b\to\infty} \int_0^{b} \frac{1}{(x+1)^2}\,dx$$

$$= \lim_{b\to\infty} \left[\frac{-1}{(x+1)}\right]_0^b$$

$$= \lim_{b\to\infty} \left[\frac{-1}{b+1} + \frac{1}{1}\right] = 1.$$

Then the probability of an experimental outcome greater than or equal to 3 is given by:

$$\int_3^{\infty} \frac{1}{(x+1)^2}\,dx = \lim_{b\to\infty} \int_3^{b} \frac{1}{(x+1)^2}\,dx$$

$$= \lim_{b\to\infty} \left[\frac{-1}{(x+1)}\right]_3^b$$

$$= \lim_{b\to\infty} \left[\frac{-1}{b+1} + \frac{1}{3+1}\right]$$

$$= \frac{1}{4}.$$

Thus, for this experiment, $\frac{1}{4}$ of the time an outcome greater than or equal to 3 can be expected.

• PROBLEM 1055

The mean square error μ, arising in the theory of probability, is defined by the relation:

$$\mu^2 = \frac{h}{\sqrt{\pi}} \int_{-\infty}^{\infty} x^2 e^{-h^2x^2}\,dx.$$

Derive a formula for μ in terms of h.

Solution: First, we consider the expression:

$$F(h) = \int_{-\infty}^{\infty} e^{-h^2x^2}\,dx,$$

where the integrand $f(x,h)$ is an even function, so that

$$f(x,h) = f(-x,h) = e^{-h^2x^2}.$$

This means that $f(x,h)$ is symmetric about the y-axis, and we can therefore deduce that:

$$\int_{-\infty}^{\infty} e^{-h^2x^2}\,dx = 2\int_0^{\infty} e^{-h^2x^2}\,dx = \frac{1}{h}\int_0^{\infty} u^{1/2-1} e^{-u}\,du$$

$$= \frac{1}{h}\Gamma\left(\frac{1}{2}\right) = \frac{\sqrt{\pi}}{h},$$

which is obtained by substituting u for h^2x^2, and, hence, $\frac{du}{2h\sqrt{u}}$ for dx.

Now, if we let

$$F(h) = \int_{-\infty}^{\infty} e^{-h^2x^2}\,dx = \frac{\sqrt{\pi}}{h}, \text{ then, } \frac{dF(h)}{dh} = \sqrt{\pi}\left(\frac{-1}{h^2}\right) = \frac{-\sqrt{\pi}}{h^2}.$$

Also,

$$\frac{dF(h)}{dh} = \frac{d}{dh}\int_{-\infty}^{\infty} e^{-h^2x^2}dx = \int_{-\infty}^{\infty} \frac{\partial}{\partial h} e^{-h^2x^2}dx = \int_{-\infty}^{\infty} -2hx^2e^{-h^2x^2}dx.$$

But, above it has been found that

$$\frac{dF(h)}{dh} = -\frac{\sqrt{\pi}}{h^2} .$$

Hence,

$$-2h\int_{-\infty}^{\infty} x^2e^{-h^2x^2}dx = -\frac{\sqrt{\pi}}{h^2} ,$$

or,

$$\mu^2 = \frac{\sqrt{\pi}}{h^2} \cdot \frac{1}{2\sqrt{\pi}} .$$

Thus,

$$\mu = \frac{1}{h\sqrt{2}} .$$

• PROBLEM 1056

Consider the experiment, the results of which range from $-\infty$ to $+\infty$, and having a probability density function given by:

$$P(x) = \frac{1}{\pi} \frac{1}{x^2+1} .$$

What is the probability that a result corresponding to values between $x = 1$ and $x = \sqrt{3}$ will occur?

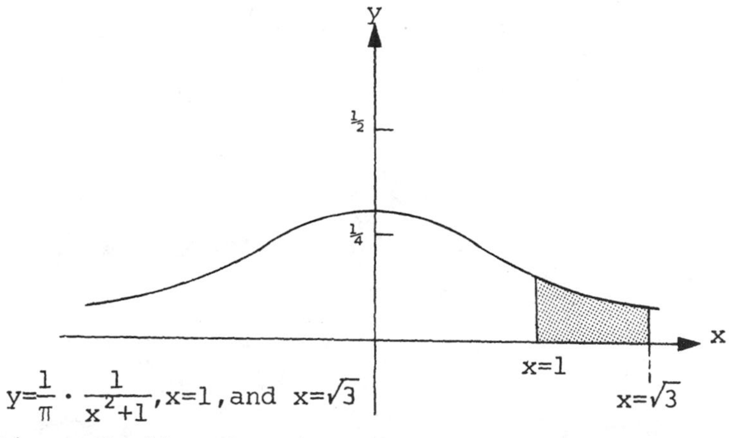

Solution: Checking that this function meets the requirements of a probability density function,

$$\int_0^{\infty} \frac{1}{\pi} \frac{1}{x^2+1} dx = \frac{1}{\pi} \lim_{b\to\infty} \int_{-b}^{b} \frac{1}{x^2+1} dx$$

$$= \frac{1}{\pi} \lim_{b\to\infty} \tan^{-1} x \Big]_{-b}^{b}$$

$$= \frac{1}{\pi}\left[\lim_{b\to\infty} \{\tan^{-1}(b) - \tan^{-1}(-b)\}\right]$$

$$= \frac{1}{\pi}\left[\frac{\pi}{2} - \left(-\frac{\pi}{2}\right)\right] = 1$$

Then the probability of an outcome between 1 and $\sqrt{3}$ is given by:

$$\frac{1}{\pi}\int_1^{\sqrt{3}} \frac{1}{x^2 + 1}\, dx ,$$

and is shown in the figure.

$$\frac{1}{\pi}\int_1^{\sqrt{3}} \frac{1}{x^2 + 1}\, dx = \frac{1}{\pi} \tan^{-1} x \Big]_1^{\sqrt{3}}$$

$$= \frac{1}{\pi}\left[\tan^{-1} \sqrt{3} - \tan^{-1} 1\right]$$

$$= \frac{1}{\pi}\left[\frac{\pi}{3} - \frac{\pi}{4}\right] = \frac{1}{12} .$$

Thus, $\frac{1}{12}$ of the time the experiment is expected to have an outcome between 1 and $\sqrt{3}$.

• PROBLEM 1057

Using the formula:

$$\varphi(D)(e^{mx}u) = e^{mx}\, \varphi(D+m)u$$

where $\varphi(j)$ is a polynomial of degree n in j, carry out the operation: $(D^2-D+1)(e^x \sin x)$.

Solution: i) Expanding the given formula, we have:

$$(a_0D^n + a_1D^{n-1} + a_2D^{n-2} + \ldots + a_{n-1}D + a_n)(e^{mx}u)$$

$$= e^{mx}\left[a_0(D+m)^n + a_1(D+m)^{n-1} + \ldots + a_n\right]u$$. This

implies that:

$$D^n(e^{mx}u) = e^{mx}(D+m)^n\, u.$$

In this case, D^2-D+1 is a polynomial of degree 2 with $a_0 = a_2 = 1$ and $a_1 = -1$ and $m = 1$. Thus,

$$(D^2-D+1)(e^x \sin x) = e^x\left[(D+1)^2 - (D+1) +1\right](\sin x)$$

$$= e^x(D^2+D+1)(\sin x) ,$$

where

$$D(\sin x) = \cos x \quad \text{and} \quad D^2(\sin x) = D(\cos x) = -\sin x.$$

Hence,

$$(D^2-D+1)(e^x \sin x) = e^x(-\sin x + \cos x + \sin x)$$
$$= e^x \cos x.$$

ii) $(D^2-D+1)(e^x \sin x) = D^2(e^x\sin x) - D(e^x\sin x) + e^x\sin x$

and

$$D(e^x\sin x) = e^x\sin x + e^x\cos x = e^x(\sin x + \cos x)$$

$$D^2(e^x\sin x) = D\left[e^x(\sin x + \cos x)\right] = e^x(\sin x + \cos x)$$
$$+ e^x\cos x - e^x\sin x = e^x(2\cos x).$$

$$(D^2-D+1)(e^x \sin x) = e^x(2\cos x) - e^x(\sin x + \cos x) + e^x \sin x$$

$$= e^x(2\cos x - \sin x - \cos x + \sin x) = e^x \cos x.$$

• PROBLEM 1058

Given the ellipse: $x^2 + 2y^2 = 8$, find a) its entire length; b) the length of the arc from $x = 1$ to $x = 2$.

Solution: The equation of the ellipse can be rewritten as:

$$\frac{x^2}{8} + \frac{y^2}{4} = 1, \quad \text{or,} \quad \frac{x^2}{a^2} + \frac{y^2}{b^2} = 1,$$

from which we infer that the semi-major axis, $a = \sqrt{8} = 2\sqrt{2}$, and the semi-minor axis, $b = \sqrt{2}$. Consequently, the eccentricity,

$$k = \frac{\sqrt{a^2-b^2}}{a} = \frac{1}{\sqrt{2}}.$$

Using the equation of the ellipse in the polar coordinate system, we have

$$x = a \sin \varphi, \quad \text{and} \quad y = \frac{b}{a}\sqrt{a^2-x^2} = b \cos \varphi,$$

where φ is the angle which a line from the origin to the point $P(x,y)$ makes with the positive x-axis. **Using the relationship:**

$$ds = \sqrt{(dx)^2+(dy)^2},$$

and the above parametric expression, the length of the ellipse is expressed by

$$S = \int \sqrt{dx^2+dy^2} = \int_0^{2\pi} \sqrt{a^2\cos^2\varphi + b^2\sin^2\varphi}\, d\varphi = 4\int_0^{\pi/2} \sqrt{a^2-(a^2-b^2)\sin^2\varphi}\, d\varphi$$

$$= 4a\int_0^{\pi/2} \sqrt{1 - \frac{a^2-b^2}{a^2}\sin^2\varphi}\, d\varphi.$$

$$S = 4a\int_0^{\pi/2} \sqrt{1 - k^2\sin^2\varphi}\, d\varphi,$$

since, as shown above,

$$k = \frac{\sqrt{a^2-b^2}}{a} = \frac{1}{\sqrt{2}}.$$

For a portion of this, corresponding to a point $P(x,y)$ from the positive x-axis,

$$s = a\int_0^{\varphi} \sqrt{1 - k^2\sin^2\varphi}\, d\varphi.$$

This integral appearing here is the so-called elliptic integral. It cannot be evaluated in finite form in terms of elementary functions. Instead, the values of the integrals for their corresponding k and φ are given in tabular form.* In most cases, these values are given cor-

*For example, see "A Short Table of Integrals" by B.O. Peirce, Ginn and Company, 1929, pages 121, 122 and 123.

responding to values of $\sin^{-1}k$ in steps of, perhaps, 1°, rather than the values of k itself.

From tables of trigonometric functions, $\sin^{-1}\frac{1}{\sqrt{2}} = 45^\circ$.

a) To find the entire length of the ellipse, we have:

$$S = 4a\int_0^{\pi/2}\sqrt{1 - k^2\sin^2\varphi}\, d\varphi.$$

For k corresponding to $\sin 45^\circ$, the integral value is 1.3506, and

$$S = 4(2\sqrt{2}) \times 1.3506$$
$$\simeq 15.28.$$

b) For the length of the arc from $x = 1$ to $x = 2$, we have

$$s\Big|_{x=1}^{x=2} = s\Big|_{x=0}^{x=2} - s\Big|_{x=0}^{x=1}, \text{ and since } x = a\sin\varphi = 2\sqrt{2}\sin\varphi,$$

we can alternatively write:

$$s\Big|_{\varphi=20^\circ 45'}^{\varphi=45^\circ} = s\Big|_{\varphi=0}^{\varphi=45^\circ} - s\Big|_{\varphi=0}^{\varphi=20^\circ 45'}.$$

This gives the required length.

$$s = a\int_0^{45^\circ}\sqrt{1 - k^2\sin^2\varphi}\, d\varphi - a\int_0^{20^\circ 45'}\sqrt{1 - k^2\sin^2\varphi}\, d\varphi.$$

By the use of interpolation we find the elliptic integral corresponding to $\varphi' = 20^\circ 45'$ under $k = \sin 45^\circ$. From the table, we see that, for $\varphi_1 = 20^\circ$, and, the next step, $\varphi_2 = 25^\circ$, both under $\alpha = 45^\circ$ for k, we have: $E_1 = 0.3456$ and $E_2 = 0.4296$, respectively, and hence, by interpolation, for $\varphi' = 20^\circ 45'$ we have:

$$E' = \frac{E_2 - E_1}{\varphi_2 - \varphi_1}(\varphi' - \varphi_1) + E_1 = \frac{0.0420}{(5\times 60)'} 45' + 0.3456$$

$$= 0.3582.$$

The value of the integral for $\varphi = 45^\circ$ under $k = \sin 45^\circ$ is given as 0.7482, and hence the length of the arc for $1 \le x \le 2$ is

$$s = a(0.7482) - a(0.3582) = 0.7482(2\sqrt{2}) - 0.3582(2\sqrt{2}),$$

or,

$$s \simeq 1.10.$$

• PROBLEM 1059

By substituting $y = e^{-x}$ in the expression:

$$\Gamma(n) = \int_0^\infty x^{n-1} e^{-x}\, dx \quad (n > 0).$$

Obtain another form of $\Gamma(n)$.

<u>Solution:</u> We understand that the integration (by parts) of:

$$\int_0^\infty x^n e^{-x}\, dx$$

leads to $\left. -x^n e^{-x} \right|_0^\infty + n \int_0^\infty x^{n-1} e^{-x} dx$, (n = integer or real number),

where, at $x = 0$, $x^n e^{-x} = 0$. But $\lim_{x\to\infty} x^n e^{-x} = \lim_{x\to\infty} \frac{x^n}{e^x}$, which

is of the indeterminate form ∞/∞, for $n > 0$. To use L'Hospital's rule, we let $x^n = f(x)$ and $e^x = g(x)$, and, by definition:

$$\lim_{x\to\infty} \frac{x^n}{e^x} = \lim_{x\to\infty} \frac{f(x)}{g(x)} \equiv \lim_{x\to\infty} \frac{f^{(p)}(x)}{g^{(p)}(x)}$$

where the operation $f^{(p)}(x) \equiv \frac{d^p f(x)}{dx^p}$ and $p \geq n$ is an integer.

Hence we have:

$$\lim_{x\to\infty} \frac{x^n}{e^x} = \lim_{x\to\infty} \frac{n(n-1)(n-2)\ldots(n-p+1)}{x^{p-n} e^x} = 0.$$

Therefore, under these conditions,

$$\int_0^\infty x^n e^{-x} dx = n \int_0^\infty x^{n-1} e^{-x} dx = n(n-1)(n-2)\ldots(n-k+1) \int_0^\infty x^{n-k} \cdot e^{-x} dx$$

where $k < n$ is another integer. If we make $k = n$, then

$$\int_0^\infty x^n e^{-x} dx = n(n-1)(n-2)\ldots 3\cdot 2\cdot 1 = n!,$$

which means that integration has been carried out n times, and

$$\int_0^\infty e^{-x} dx = - \left. \frac{1}{e^x} \right|_0^\infty = 1 \quad \text{for } k = n.$$

This process leads to the concept of a generalization of the factorial called the gamma function represented by the capital Greek letter Γ (gamma). Thus, the expression for the gamma function is

$$\int_0^\infty x^n e^{-x} dx = \Gamma(n+1) = n\Gamma(n) = n(n-1)! \;.$$

We now set $y = e^{-x}$. Then $dy = -e^{-x} dx$, or $dx = -e^x dy$. By taking the logarithm, $x = -\ln y = \ln \frac{1}{y}$, and $x^n = \left(\ln \frac{1}{y}\right)^n$, with corresponding limits $1 < y < 0$ for $0 < x < \infty$. Substitution yields:

$$\int_0^\infty x^n e^{-x} dx = - \int_1^0 \left(\ln \frac{1}{y}\right)^n dy = \int_0^1 \ln\left(\frac{1}{y}\right)^n dy = \Gamma(n+1).$$

By replacing n by n-1, we obtain:

$$\int_0^\infty x^{n-1} e^{-x} dx = \Gamma(n) = \int_0^1 \left(\ln \frac{1}{y}\right)^{n-1} dy.$$

This final integration is convenient for tabulating $\Gamma(t)$, where t is between 1 and 2. The values of $\Gamma(n)$ for $1 > n > 2$ can be obtained by use of the relationships:

$$\Gamma(n) = \frac{\Gamma(n+1)}{n} \quad \text{for } n < 1$$

$$\Gamma(n+1) = n\Gamma(n) \quad \text{for } n \geqslant 2 \;.$$

Find the moment of inertia, with respect to the x-axis, of the area bounded by one loop of the curve $\rho^3 = \sin^2\theta$.

Solution: The moment of inertia with respect to the x-axis is given by

$$I_x = \int_R y^2 \, dm,$$

where the elementary mass can be represented by the corresponding area in the plane, dxdy, or, in polar coordinates, $\rho d\rho d\theta$. $y = \rho\sin\theta$. Hence,

$$I_x = \int_0^{\pi}\int_0^{\sin^{2/3}\theta} \rho^3 \sin^2\theta \, d\rho \, d\theta .$$

Due to symmetry about the y-axis, we can rewrite the above expression after performing the integral with respect to ρ. This gives a factor of 2.

$$I_x = 2\int_0^{\pi/2} \frac{\left(\sin^{2/3}\theta\right)^4}{4} \sin^2\theta \, d\theta = \frac{1}{2}\int_0^{\pi/2} \sin^{8/3}\theta \sin^2\theta \, d\theta$$

$$= \frac{1}{2}\int_0^{\pi/2} \sin^{14/3}\theta \, d\theta.$$

To carry out the integration, we use the Gamma function.

$$\Gamma(n) = \int_0^{\infty} x^{n-1} e^{-x} \, dx .$$

Using this definition, we can write:

$$\Gamma(m) \cdot \Gamma(n) = \int_0^{\infty} s^{m-1} e^{-s} \, ds \cdot \int_0^{\infty} t^{n-1} e^{-t} \, dt.$$

Letting $s = x^2$, $ds = 2x \, dx$; $t = y^2$, $dt = 2y \, dy$, we have:

$$\Gamma(m)\,\Gamma(n) = \int_0^{\infty} 2x^{2m-1} e^{-x^2} dx \cdot \int_0^{\infty} 2y^{2n-1} e^{-y^2} dy$$

$$= 4\int_0^{\infty}\int_0^{\infty} x^{2m-1} y^{2n-1} e^{-\left(x^2+y^2\right)} dy \, dx.$$

Changing to polar coordinates, we have: $x = \rho\cos\theta$, $dx = -\rho\sin\theta \, d\theta$, $y = \rho\sin\theta$, $dy = \rho\cos\theta \, d\theta$, and $x^2+y^2 = \rho^2$. Furthermore, since $dydx = \rho d\rho d\theta$, the area bounded by the curves over the entire region in the first quadrant gives rise to the variations of ρ, from 0 to ∞, and θ, from 0 to $\pi/2$. Hence,

$$\Gamma(m)\,\Gamma(n) = 4\int_0^{\pi/2}\int_0^{\infty} (\rho\cos\theta)^{2m-1} (\rho\sin\theta)^{2n-1} e^{-\rho^2} \rho d\rho d\theta$$

$$= 4\int_0^{\pi/2} \cos^{2m-1}\theta \sin^{2n-1}\theta d\theta \int_0^{\infty} \rho^{2m+2n-2} e^{-\rho^2} \rho d\rho.$$

The second integral can be evaluated by setting $\rho^2 = z$. Hence,

$$\int_0^{\infty} \rho^{2(m+n-1)} e^{-\rho^2} \rho d\rho = \frac{1}{2}\int_0^{\infty} z^{(m+n)-1} e^{-z} dz = \frac{1}{2}\Gamma(m+n).$$

Then

$$\Gamma(m)\,\Gamma(n) = 2\Gamma(m+n)\int_0^{\pi/2} \cos^{2m-1}\theta \, \sin^{2n-1}\theta \, d\theta,$$

which holds if and only if

$$\int_0^{\pi/2} \cos^{2m-1}\theta \, \sin^{2n-1}\theta \, d\theta = \frac{1}{2}\frac{\Gamma(m)\Gamma(n)}{\Gamma(m+n)} \equiv B(m,n), \quad (m > 0,\ n > 0).$$

By the use of the Beta function, the integral: $\frac{1}{2}\int_0^{\pi/2} \sin^{14/3}\theta \, d\theta$,

can be evaluated by setting $m = 1/2$ and $n = 17/6$. Hence,

$$\frac{1}{2}\int_0^{\pi/2} \sin^{14/3}\theta \, d\theta \equiv \frac{1}{2} B(1/2, 17/6) = \frac{1}{4}\frac{\Gamma(1/2)\Gamma(17/6)}{\Gamma(10/3)}.$$

Now, using the relationship: $\Gamma(n+1) = n\Gamma(n)$, we have:

$$\Gamma(17/6) = \Gamma(11/6 + 1) = 11/6\,\Gamma(11/6) = 11/6\,\Gamma(1.8333)$$

$$\Gamma(10/3) = \Gamma\left[(1+4/3)+1\right] = (1+4/3)\Gamma(4/3+1)$$

$$= \frac{7}{3}\cdot\frac{4}{3}\Gamma(4/3) = \frac{28}{9}\Gamma(1.333).$$

For $\Gamma(1/2)$, we use $\Gamma(n) = \frac{\Gamma(n+1)}{n}$ $\quad \Gamma(1/2) = \frac{\Gamma(1/2+1)}{1/2} = 2\Gamma(1.5)$.

Using a table* for $\Gamma(t)$, where $1 < t \le 2$, we have:

$$\Gamma(1.83) = 0.94$$

$$\Gamma(1.33) = 0.893$$

$$\Gamma(1.50) = 0.886$$

Finally,

$$I_x = \frac{1}{2}\int_0^{\pi/2} \sin^{14/3}\theta \, d\theta = \frac{1}{4}\frac{\Gamma(1/2)\Gamma(17/6)}{\Gamma(10/3)} = \frac{1}{4}\,\frac{2(0.886)\,\frac{11}{6}(0.94)}{\frac{28}{9}\,(0.893)}$$

$$\simeq 0.275.$$

• PROBLEM 1061

Show that the sum of the squares of the intercepts on the coordinate axes of the tangent plane to the surface, $x^{\frac{2}{3}} + y^{\frac{2}{3}} + z^{\frac{2}{3}} = a^{\frac{2}{3}}$, is constant.

Solution:

$$F(x,y,z) = x^{\frac{2}{3}} + y^{\frac{2}{3}} + z^{\frac{2}{3}} - a^{\frac{2}{3}} = 0$$

The tangent plane can be expressed as

$$\nabla F \cdot (\vec{r} - \vec{r}_1) = 0,$$

where $(\vec{r} - \vec{r}_1)$ is a vector tangent to the surface extending from (x_1, y_1, z_1) to some neighboring point (x,y,z) on the surface. ∇F is the gradient vector of the function F normal to the surface.

*For example, see A Short Table of Integrals by B.O. Peirce, p. 120.

Rewriting the above equation explicitly, we obtain

$$\frac{\partial F}{\partial x}\left(x - x_1\right) + \frac{\partial F}{\partial y}\left(y - y_1\right) + \frac{\partial F}{\partial z}\left(z - z_1\right) = 0$$

At the point $\left(x_1, y_1, z_1\right)$,

$$\frac{\partial F}{\partial x} = \frac{2}{3} x_1^{-\frac{1}{3}}, \qquad \frac{\partial F}{\partial y} = \frac{2}{3} y_1^{-\frac{1}{3}}, \quad \text{and}$$

$$\frac{\partial F}{\partial z} = \frac{2}{3} z_1^{-\frac{1}{3}}.$$

Combining this, and canceling out $\frac{2}{3}$, we have:

$$x_1^{-\frac{1}{3}}\left(x - x_1\right) + y_1^{-\frac{1}{3}}\left(y - y_1\right) + z_1^{-\frac{1}{3}}\left(z - z_1\right) = 0.$$

This is the equation of the plane tangent to the surface at $\left(x_1, y_1, z_1\right)$. We find the intercepts on the coordinate axes. (For example, for the x-intercept, we let $y = z = 0$, and solve for x.)

$$x_1^{-\frac{1}{3}} x - x_1^{\frac{2}{3}} - y_1^{\frac{2}{3}} - z_1^{\frac{2}{3}} = 0.$$

Thus the x-intercept is at

$$x_{in} = x_1^{\frac{1}{3}}\left(x_1^{\frac{2}{3}} + y_1^{\frac{2}{3}} + z_1^{\frac{2}{3}}\right) = c_1.$$

Similarly, for the y and z intercepts,

$$y_{in} = y_1^{\frac{1}{3}}\left(x_1^{\frac{2}{3}} + y_1^{\frac{2}{3}} + z_1^{\frac{2}{3}}\right) = c_2, \quad \text{and}$$

$$z_{in} = z_1^{\frac{1}{3}}\left(x_1^{\frac{2}{3}} + y_1^{\frac{2}{3}} + z_1^{\frac{2}{3}}\right) = c_3.$$

Thus

$$x_{in}^2 + y_{in}^2 + z_{in}^2 = \left(x_1^{\frac{2}{3}} + y_1^{\frac{2}{3}} + z_1^{\frac{2}{3}}\right)^3 \text{ is a constant.}$$

• **PROBLEM 1062**

Using Leibniz's formula for the rth derivative of a product,

$$D^r(uv) = vD^r u + r(Dv)\left(D^{r-1}u\right) + \frac{r(r-1)}{2!}\left(D^2 v\right)\left(D^{r-2}u\right) + \ldots$$
$$+ r\left(D^{r-1}v\right)(Du) + u\,D^r v,$$

where the coefficients are the binomial coefficients, derive the expression, for a polynomial φ of degree n,

$$\varphi(D)\left(e^{mx}u\right) = e^{mx}\varphi(D+m)u.$$

Solution: We express Leibniz's formula as:

$$D^r(uv) = \sum_{t=0}^{r} \binom{r}{t} \frac{D^{r-t}uD^tv}{t!},$$

where $\binom{r}{t} = \frac{r!}{(r-t)!t!}$, and $r! = r(r-1)(r-2)\ldots\left(r-[r-i]\right)\ldots(3)(2)(1)$.

$1 \le i \le r$, and $0! = 1$.

Now, if $v = e^{mx}$, then

$$D^0e^{mx} = e^{mx},\ De^{mx} = me^{mx},\ D^2e^{mx} = m^2e^{mx},\ \ldots,\ D^re^{mx} = m^re^{mx}.$$

Hence the above formula for $v = e^{mx}$ can be expressed as:

$$D^r\left(ue^{mx}\right) = \sum_{t=0}^{r} \binom{r}{t} \frac{D^{r-t}u\,D^tv}{t!} = e^{mx}\sum_{t=0}^{r} \binom{r}{t} \frac{m^tD^{r-t}u}{t!}.$$

From the binomial expansion,

$$(a+b)^r = \sum_{t=0}^{r} \binom{r}{t} \frac{a^t\,b^{r-t}}{t!},$$

we can assert that:

$$\sum_{t=0}^{r} \binom{r}{t} \frac{m^tD^{r-t}u}{t!} = (D+m)^r_u.$$

Thus

$$D^r\left(ue^{mx}\right) = e^{mx}(D+m)^r_u.$$

Furthermore, if $\varphi(D)$ is a polynomial of degree n in D, or

$$\varphi(D) = a_0D^n + a_1D^{n-1} + a_2D^{n-2} + \ldots + a_{n-1}D + a_n$$

$$\varphi(D)(e^{mx}u) = a_0D^n(e^{mx}u) + a_1D^{n-1}(e^{mx}u) + \ldots + a_{n-1}D(e^{mx}u) + a_n.$$

Recalling that:

$$D^r(e^{mx}u) = e^{mx}(D+m)^r_u,\quad 0 \le r \le n$$

$$\varphi(D)(e^{mx}u) = e^{mx}\left[a_0(D+m)^n + a_1(D+m)^{n-1} + a_2(D+m)^{n-2} + \ldots + a_{n-1}(D+m) + a_n\right]u.$$

But the expression in parentheses is a polynomial of degree n in $(D+m)$, which can be written as

$$\varphi(D+m) = a_0(D+m)^n + a_1(D+m)^{n-1} + \ldots + a_n.$$

Therefore,

$$\varphi(D)(e^{mx}u) = e^{mx}\varphi(D+m)u,$$

which completes the proof.

From this, we can also assert that:

$$\varphi(D-m)(e^{mx}u) = e^{mx}\varphi(D)u.$$

• PROBLEM 1063

According to Planck's radiation law, the radiation density ψ is:

$$\psi = \frac{8\pi h}{c^3}\int_0^\infty \frac{\nu^3}{e^{h\nu/KT}-1}\,d\nu,$$

where ν = frequency (sec^{-1}), T = temperature (°K.), h = Planck's constant = 6.554×10^{-27} ergsec., c = velocity of light = 2.998×10^{10}cm/sec., and K = Boltzmann's constant = 1.372×10^{-16} erg/deg. Show that the above formula reduces to the Stefan-Boltzmann law, $\psi = \sigma T^4$.

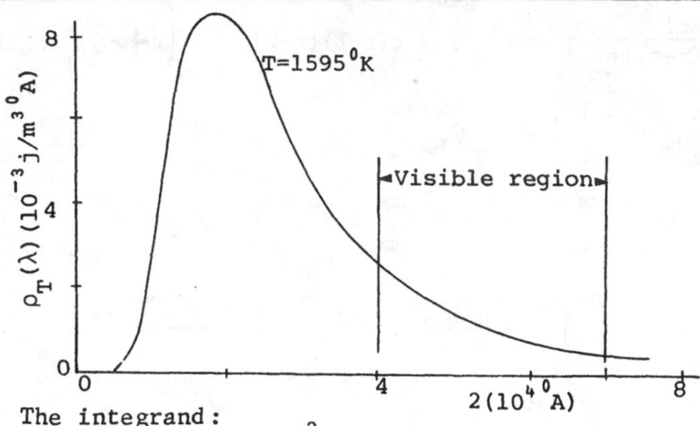

Solution: The integrand:

$$\frac{\nu^3}{e^{h\nu/KT}-1},$$

can be expanded in two ways: as ν and T are positive quantities, $h\nu/KT > 0$. Hence, by using: $e^x = 1 + x + x^2/2! + \dots$, we have: $e^{h\nu/KT} = 1 + h\nu/KT + \frac{1}{2!}(h\nu/KT)^2 + \frac{1}{3!}(h\nu/KT)^3 + \dots$ Or else, by using: $1/(1-x) = 1 + x + x^2 + \dots$, we obtain:

$$\frac{1}{e^{h\nu/KT}-1} = e^{-h\nu/KT}\left[1 + h\nu/KT + (h\nu/KT)^2 + \dots\right].$$

Taking the first term of either result, the integration reduces to

$$\psi \simeq \frac{8\pi h}{c^3}\int_0^\infty \nu^3 e^{-h\nu/KT}\, d\nu .$$

By letting $h\nu/KT = x$ we have:

$$\psi \simeq \frac{8\pi K^4 T^4}{c^3 h^3}\int_0^\infty x^3 e^{-x}\, dx = \frac{48K^4\pi}{c^3h^3}T^4 .$$

A second method is the use of integral tables, where we find:

$$\int_0^\infty \frac{x^3 dx}{e^x-1} = \frac{\pi^4}{15} .$$

By again letting $h\nu/KT = x$, $\nu^3 = (KT/h)^3x^3$, $d\nu = KT/h\, dx$, we have:

$$\psi = \frac{8\pi h}{c^3}(KT/h)^4\int_0^\infty \frac{x^3}{e^x-1}dx = \frac{8\pi h}{c^3}(KT/h)^4(\pi^4/15)$$

$$= \frac{8\pi^5K^4}{15c^3h^3}T^4$$

which results in $\psi = \sigma T^4$ where $\sigma = \frac{8\pi^5K^4}{15c^3h^3} = 5.67 \times 10^{-8}\ \frac{\text{watt}}{m^2(°\text{abs.})^4}$

in the MKSA system.

The expression for ψ, the spectral radiancy (energy emitted per unit time per unit area), has great importance in physics. It was based on the thermal radiation of bodies corresponding to their temperatures, for example, the self-luminosity of a body at high temperatures.

Historically, at about the time Einstein was carrying out his work on the behavior of bodies of very large size or moving with very high velocities, classical physics had reached its limits in attempting to explain the behaviors of natural phenomena. At the other end of the scale, there came aid from an unexpected quarter when Planck proposed his quantum theory in 1900, as a result of his analysis of certain properties of thermal radiation. Subsequent developments went far beyond the scope of Planck's original theory and are known by different names such as the "wave mechanics" and "quantum mechanics".

The problem studied by Planck was the distribution of energy among different wave lengths in the thermal radiation emitted by hot bodies as shown in the figure.

Based on the assertion that surfaces, which are good emitters of radiation, are also good absorbers, emission with relatively higher efficiency results from black bodies. For this purpose, a small cavity in a body can provide the desired efficiency, and very small cavities have been used for the study of radiation. Under these conditions, we can imagine the energy as being associated with the radiation in the act of passing between the walls of the cavity. The origin of radiation must lie in the oscillations or vibrations of atoms (any oscillating dipole acts as a source of electromagnetic radiation). The fact that the amount of energy in the enclosure corresponding to a given wave length does not depend on the walls suggests that the form of the distribution is in itself a fundamental fact of physics. A number of physicists advanced theories based on classical physics which, however, had only limited success. On the basis of Planck's radical hypotheses, he was able to modify the existing theory so as to account exactly for the form of the curves in the energy distribution of thermal radiation. His hypotheses were: An oscillator cannot have a continuum of energy, but only such discrete energies as are given by : $E = nh\nu$, where ν is the oscillator frequency, h is Planck's constant and n is an integral quantum number. Thus, the above equation asserts that the oscillator energy is of integral value (quantized). The second hypothesis concerned the stationary and excited states (quantized states). This suggested that radiation could be emitted in <u>pulses</u>, a concept which was out of harmony with the existing idea that light and other electromagnetic radiation were transmitted as a continuous train of waves.

Aside from the quantization theory in cavity radiation (black body radiation) there are numerous advantages and uses of the black body as a means of measurment. From experimental results, it had been understood that

$$\lambda_{max} T = \text{constant}$$

with the empirical value of the constant (Wien's constant) being 2.898×10^{-3}(m)($^\circ$K). Assuming that stellar surfaces behave like black bodies, we can get a good estimate of their temperatures. For example, for the sun λ_{max} = s 100 Å (Å, Angstrom unit = 10^{-8}cm.), whereas for the North Star λ_{max} = 3500 Å . By virtue of:

$$T = \frac{\text{Wien's constant}}{\lambda_{max}},$$

the surface temperatures of the sun and the North Star can be obtained as 5700° K. and 8300°K., respectively. At 5700°K the sun's surface is near the temperature at which the greatest part of its radiation lies within the visible region of the spectrum. This suggests that over ages of human evolution our eyes adopted to the sun to become most sensitive to those wave lengths that it radiates most intensively. Using other methods, it is generally accepted today that the temperature at the center of the sun must be of the order of 12 million$^\circ$c. This may seem high by our standards, but is very modest for celestial temperatures. This temperature is important for maintaining the generation of energy by the sun and sun-like stars (fusion reaction) for the transmutation of hydrogen into helium. Furthermore, the Stefan-Boltzmann's law (derived above) and the temperatures obtained give the power radiated from 1 cm^2 of stellar surface. For the sun,

$$\psi = \sigma T^4 = 6000 \text{ watts/cm}^2,$$

and for the North Star,

$$\psi = 27{,}000 \text{ watts/cm}^2.$$

CHAPTER 41

BASIC DIFFERENTIAL EQUATIONS

A differential equation is of the order of the highest derivative involved. For example, a second-order equation involves only the first and second derivatives. Differential equations can have constant coefficients, or it can have coefficients that are functions of the variables. A differential equation can also be homogeneous, or it can be non-homogeneous.

The solution of a differential equation can present a difficult task, and for this reason numerous methods have been developed in the attempt to systematize their solutions. The simplest differential equations can be handled by the separation of variables, in which each variable is on only one side of the equation. This leads to direct integration. For the linear equation, a sum of solutions known as the complementary function and the particular function may be used. For homogeneous equations, substitution is often useful. Sometimes, integrating factors can be employed.

Differential equations are applicable to a wide variety of complex problems in the physical and social sciences.

• PROBLEM 1064

Solve the equation:

$$(2x-y)dx + (2y-x)dy = 0.$$

Solution: By regrouping terms, we have

$$(2xdx + 2ydy) - (ydx + xdy) = 0$$

or

$$d\left(x^2+y^2\right) - d(xy) = 0.$$

Integrating, we get

$$x^2 + y^2 - xy = C.$$

Some of the simpler integrable combinations are the following:

(a) $xdy + ydx = d(xy)$;

(b) $2xdx + 2ydy = d\left(x^2+y^2\right)$;

(c) $\frac{xdy-ydx}{x^2} = d\left(\frac{y}{x}\right)$; $\frac{ydx-xdy}{y^2} = d\left(\frac{x}{y}\right)$;

(d) $\frac{xdy-ydx}{x^2+y^2} = d\left(\text{Arc tan } \frac{y}{x}\right)$.

• PROBLEM 1065

Find the general solution of: $\frac{dy}{dx} = x^2y^3$.

Solution: Separating the variables, the given differential equation may be written as:

$$\frac{dy}{y^3} = x^2dx .$$

The general solution may be obtained by integrating. We obtain:

$$\int \frac{dy}{y^3} = \int x^2dx + c,$$

or,

$$-\frac{1}{2y^2} = \frac{x^3}{3} + c .$$

This can be rearranged to yield:

$$2y^2x^3 + cy^2 + 3 = 0.$$

This equation satisfies the given differential equation, for differentiation leads to:

$$\frac{dy}{dx} = x^2y^3 .$$

• PROBLEM 1066

Solve the differential equation:

$$y' = \frac{x}{y^2\sqrt{1+x^2}} .$$

Solution: This differential equation can be solved by separating the variables.

Setting $y' = \frac{dy}{dx}$,

then multiplying the given differential equation by y^2, we obtain:

$$y^2 \frac{dy}{dx} = \frac{x}{\sqrt{1+x^2}}$$

$$y^2dy = \frac{x}{\sqrt{1+x^2}}dx.$$

Integrating both sides with respect to their respective variables,

$$\int y^2 dy = \int \frac{xdx}{\sqrt{1+x^2}} + C$$

or

$$\frac{y^3}{3} = \sqrt{1+x^2} + C.$$

where the constant C is a suitable constant of integration, the value of which can be determined under given boundary conditions, imposed on the differential equation.

• PROBLEM 1067

Solve the differential equation $dy/dx = 2xy$ to obtain a general solution. Also find the particular solution if $y=5$ when $x=0$.

Solution: From $dy = 2xy\ dx$, by dividing by y, we obtain

$$\frac{dy}{y} = 2x\ dx,$$

in which the variables are separated. Integrating, we have

$$\ln y = x^2 + \ln C,$$

or

$$\ln y - \ln C = x^2, \quad \ln(y/C) = x^2.$$

Therefore, $y/C = e^{x^2}$ or $y = Ce^{x^2}$.

This is the general solution of the differential equation.

Substituting the initial conditions: $x=0$, $y=5$ in the general solution $y=Ce^{x^2}$, we get $5=C$, so that the required particular solution is $y=5e^{x^2}$.

• PROBLEM 1068

Solve: $x\ dy + \frac{x^2}{y} dy = y\ dx.$

Solution: Transposing ydx to the left-hand side and dividing the entire equation by x^2, we get

$$\frac{xdy-ydx}{x^2} + \frac{dy}{y} = 0.$$

But

$$\frac{xdy-ydx}{x^2} = d\left(\frac{y}{x}\right) \quad \text{and} \quad \frac{dy}{y} = d(\ln y),$$

thus $d\left(\frac{y}{x}\right) + d(\ln y) = 0.$

Integration yields:

$$\frac{y}{x} + \ln y = \ln C,$$

or,

$$\frac{y}{x} + \ln \frac{y}{C} = 0.$$

• PROBLEM 1069

Solve the differential equation:

$$\frac{d^2y}{dx^2} = \cos 2x.$$

Solution: $\frac{dy}{dx} = \frac{1}{2} \sin 2x + a.$ Hence

$y = -\frac{1}{4} \cos 2x + ax + b$ is the general solution.

This solution is obtained by setting $p = \frac{dy}{dx}$, and hence

$$\frac{dp}{dx} = \frac{d^2y}{dx}.$$

If the given equation had been expressed as (y'', y', y), we would (by the chain rule) have used

$$\frac{d^2y}{dx^2} = \frac{dp}{dy}\frac{dy}{dx} = p\frac{dp}{dy}.$$

Substituting $\frac{d^2y}{dx^2} = \frac{dp}{dx}$ in the given differential equation, and integrating yields an expression from which we solve for p. Substituting $\frac{dy}{dx}$ back for p, we obtain the first-order differential equation. Further integration gives the desired general solution. Once we are used to this technique, we can automatically write down the first-order and the general equations.

• PROBLEM 1070

Solve the differential equation:

$$(xy^2-x)\,dx + (x^2y+y)\,dy = 0.$$

Solution: Factoring the coefficients of dx and dy, we have

$$x(y^2-1)\,dx + y(x^2+1)\,dy = 0.$$

Dividing each term by $(x^2+1)(y^2-1)$, we separate variables and obtain

$$\frac{xdx}{x^2+1} + \frac{ydy}{y^2-1} = 0.$$

Using the substitution method to integrate the left-hand side of the above differential equation, we let

$$v = x^2+1 \qquad\qquad w = y^2-1$$

$$dv = 2xdx \qquad\qquad dw = 2ydy.$$

Therefore, the given equation becomes:

$$\frac{1}{2}\frac{dv}{v} + \frac{1}{2}\frac{dw}{w} = 0.$$

Integrating, we have:

$$\frac{1}{2}\ln v + \frac{1}{2}\ln w = k$$

or

$$\frac{1}{2}\ln\left(x^2+1\right) + \frac{1}{2}\ln\left(y^2-1\right) = k.$$

Putting $k = \frac{1}{2}\ln C$, we may write this:

$$\ln\left[\left(x^2+1\right)\left(y^2-1\right)\right] = \ln C,$$

from which we have

$$\left(x^2+1\right)\left(y^2-1\right) = C.$$

This expression is the required general solution of the given differential equation.

• PROBLEM 1071

Obtain the general solution for: $\frac{dy}{dx} = e^x + y$.

Solution: The given differential equation is of the general form:

$$\frac{dy}{dx} + py = Q,$$

where p and Q are constants and functions of x only. In this case, $p = -1$ and $Q = e^x$. This differential equation is of degree one and is called a linear differential equation. To find the general solution, we first let Q be identically zero. Hence,

$$\frac{dy}{dx} + py = 0 ,$$

from which, upon separating the variables and integrating, we obtain:

$$\ln y + \int pdx = \ln c,$$

or

$$\ln\left(\frac{c}{y}\right) = e^{\int pdx} .$$

$$c = y\, e^{\int pdx} .$$

Now, we can write:

$$\left(\frac{dy}{dx} + py\right)e^{\int pdx} = \frac{d}{dx}\left(y\, e^{\int pdx}\right) = Q\, e^{\int pdx} .$$

Integration of the last equation yields:

$$y\, e^{\int p dx} = \int Q\, e^{\int p dx}\, dx + c\ ,$$

from which y is obtained by division.

To solve the given problem, $p = -1$ and $Q = e^x$. Hence,

$$y = e^{-\int p dx} \int Q\, e^{\int p dx}\, dx + c\, e^{-\int p dx}$$

$$= e^{\int - dx} \int e^x\, e^{\int - dx}\, dx + c\, e^{-\int - dx}$$

$$= e^x x + c\, e^x .$$

$$y = e^x (x+c)$$

is the general solution.

• PROBLEM 1072

Solve the differential equation:

$$y' + y = x.$$

Solution: This is a differential equation of the general form:

$$\frac{dy}{dx} + P(x)y = Q(x). \qquad (1)$$

We wish to solve the above general differential equation assuming $P(x)$ and $Q(x)$ are either constants or functions of x alone.

Setting $\quad M = e^{\int P(x) dx} \qquad (2)$

and observing that, by using the chain rule,

$$\frac{dM}{dx} = e^{\int P(x) dx} \cdot P(x) = P(x)M. \qquad (3)$$

If we multiply both sides of (1) by M we have

$$M\frac{dy}{dx} + MP(x)y = MQ(x)$$

Using (3), and the formula for the derivative of a product, this can be written as:

$$\frac{d}{dx}\left(My\right) = MQ(x).$$

Since $M(x)$ is a function of x alone this gives

$$My = MQ(x)dx + C \qquad (4)$$

as a general solution of (1). Dividing both sides of (4) by M and using (2), the solution can be put in the form

$$y = e^{-\int P(x)dx}\left(\int Q(x)e^{\int P(x)dx}dx + C\right). \tag{5}$$

For convenience, the expression $e^{\int P(x)dx}$ is called the integrating factor.

Since, in our case, P(x) = 1 (constant), the integrating factor becomes

$$e^{\int (1)dx} = e^x.$$

Thus, the general solution of the given differential equation takes the form

$$y = e^{-x}\left(\int xe^x dx + C\right) \tag{6}$$

Integrating the right-hand side by parts, we obtain

$$\begin{aligned} y &= e^{-x}\left(xe^x - e^x + C\right) \\ &= (x-1) + Ce^{-x} \end{aligned} \tag{7}$$

The constant C may be determined by imposing an initial condition on the solution y. If y(0) = 1, for example, then substituting x = 0, y = 1 in (7) we obtain 1 = -1 + C, or C = 2. Thus the solution to the initial value problem y' + y = x, y(0) = 1 is given by

$$y = (x-1) + 2e^{-x}.$$

The equation

$$\frac{dy}{dx} + P(x)y = Q(x)$$

is called a first order differential equation, since it involves only the first **derivative** and linear because the general solution is of the form y = f(x) + g(x). The f(x) part of the solution is the socalled particular solution of the above differential equation and g(x) is a non-zero solution of the homogeneous equation:

$$\frac{dy}{dx} + P(x)y = 0.$$

• PROBLEM 1073

Solve: $2\,xy\,dy = \left(y^2 - x^2\right)dx.$

Solution: This equation has homogeneous coefficients of the second degree. Homogeneous equations are of the form

$$f(xt, yt) = t^n f(x, y)$$

for all $t > 0$. Quantity n is called the degree. For example, if we let $f(x,y) = 2xy$ and $g(x,y) = y^2-x^2$, then

$$f(xt,yt) = 2xtyt = t^2(2xy) = t^2 f(x,y) \quad \text{and}$$

$$g(xt,yt) = (yt)^2-(xt)^2 = t^2\left(y^2-x^2\right) = t^2 g(x,y).$$

Separation of variables appears to be possible if we set $y=vx$ in this kind of equation. If we put $y=vx$, then $dy = v\,dx + x\,dv$. The equation becomes

$$2x\cdot vx\left(v\,dx + x\,dv\right) = \left(v^2x^2-x^2\right)dx,$$

or

$$\left(v^2+1\right)dx + 2xv\,dv = 0.$$

Separating variables, we get

$$\frac{dx}{x} + \frac{2v\,dv}{v^2+1} = 0.$$

Integrating, we have

$$\ln x + \ln\left(v^2+1\right) = \ln C,$$

$$\ln\left[x\left(v^2+1\right)\right] = \ln C,$$

$$\therefore\ x\left[\frac{y^2}{x^2} + 1\right] = C, \quad \text{and} \quad y^2 + x^2 = Cx.$$

Transposing Cx to the left-hand side of the general solution and completing the square we obtain

$$y^2 + \left(x-\frac{C}{2}\right)^2 = \frac{C^2}{4},$$

which represents a family of circles centered at $(C/2,0)$ with radii of $C/2$.

• PROBLEM 1074

Find the general solution for: $\dfrac{dy}{dx} = \dfrac{x - 2y}{2x - y}$.

Solution: The above differential equation can be rewritten as:

$$(2x - y)dy = (x - 2y)dx,$$

where, after expansion, we can group terms as follows:

$$ydy + xdx = 2(xdy + ydx).$$

Now, $\int(ydy + xdx) = 2\int(xdy + ydx) + c$, But the integral,

$$\int(xdy + ydx) = \int d(xy) = xy.$$

Hence,

$$\tfrac{1}{2}y^2 + \tfrac{1}{2}x^2 = 2xy + c_1,$$

or

$$x^2 + y^2 - 4xy + c = 0,$$

where

$$c = -2c_1.$$

• PROBLEM 1075

> Solve the differential equation: $y'' - 2y' - y = 0$.

Solution: This kind of differential equation is called a second-order, linear, homogeneous equation with constant coefficients.

Setting $y = e^{mx}$ to see if it is a solution of the given equation, and since $y' = me^{mx}$ and $y'' = m^2e^{mx}$, substitution gives:

$$\text{(a)} \quad y'' - 2y' - y = e^{mx}\left(m^2-2m-1\right) = 0.$$

e^{mx} is different from zero in a finite range of x unless m is trivially chosen. Thus, the only way for (a) to hold is for

$$m^2 - 2m - 1 = 0.$$

The roots are $m = 1 \pm\sqrt{2}$.

Thus, either $y = c_1e^{\left(1+\sqrt{2}\right)x}$ or $y = c_2e^{\left(1-\sqrt{2}\right)x}$ solves the equation. The general solution is the sum of the two terms.

$$y = c_1e^{(1+\sqrt{2})x} + c_2e^{(1-\sqrt{2})\,x}$$

• PROBLEM 1076

> Solve the differential equation: $2y\,\frac{d^2y}{dx^2} = \left(\frac{dy}{dx}\right)^2$.

Solution: Let $\frac{dy}{dx} = p$. Then

$$\frac{d^2y}{dx} = \frac{dp}{dx} = \frac{dp}{dy}\frac{dy}{dx} = \frac{dp}{dy}\,p\,.$$

Thus, the given differential equation can be written in terms of p as a dependent variable, and y, as follows:

$$2y\,\frac{d^2y}{dx^2} - \left(\frac{dy}{dx}\right)^2 = 2y\left(p\,\frac{dp}{dy}\right) - p^2 = 0,$$

or,

$$\frac{dp}{dy} - \frac{1}{2y}\,p = 0\,.$$

The last expression is a homogeneous and linear differential equation in p. Separating variables,

$$\frac{dp}{p} = \frac{dy}{2y}\,.$$

Integration yields:

$$\ln p = \tfrac{1}{2}\ln y + \ln c\,.$$

Solving for p,

$$\frac{dy}{dx} = p = c_1\sqrt{y} .$$

$$\frac{dy}{\sqrt{y}} = c_1 \, dx ,$$

and the general solution is:

$$2\sqrt{y} = c_1x + c_2 ,$$

where the factor 2 can be absorbed into the arbitrary constants c_1 and c_2.

$$y = \left(c_1'x + c_2'\right)^2$$

is the general solution.

• PROBLEM 1077

Solve the following differential equations:

a) $$\frac{dy}{dx} = \frac{x^4+2y}{x}$$

b) $$\frac{dy}{dx} = \frac{2xy}{y^2-x^2} .$$

Solution: a) Rewriting the first equation,

$$\frac{dy}{dx} - \frac{2}{x}y = x^3, \text{ and}$$

we note that it is a nonhomogeneous and linear differential equation. Under the general expression:

$$\frac{dy}{dx} + py = Q,$$

where $p = -2/x$ and $Q = x^3$, we obtain the general solution as:

$$y = e^{-\int pdx}\int Q\, e^{\int pdx}dx + c\, e^{-\int pdx}.$$

the integrating factor, $e^{\int pdx}$, in this case is: $e^{\int -2/x\, dx} = e^{-2\ln x} = 1/x^2$.

Hence,

$$y = x^2\int \frac{1}{x^2}x^3\, dx + x^2c$$

$$= \tfrac{1}{2}x^4 + x^2c .$$

b) $$\frac{dy}{dx} = \frac{2xy}{y^2-x^2} .$$

This is a homogeneous equation of degree zero, where, by substitution of xt for x and yt for y, we have:

$$f(xt,yt) = \frac{2(xt)(yt)}{(yt)^2-(xt)^2} = \frac{2xy}{y^2-x^2} = f(x,y) = t^0f(x,y),$$

The function: $dy/dx = f(x,y)$ is not affected. Under this condition, f(x,y) can be expressed as a function of a single argument: $f(x,y)=f(1,y/x)$ or $f(x/y,1)$ Now, setting $y=vx$,

$$v + x\frac{dv}{dx} = \frac{2vx^2}{v^2x^2-x^2} = \frac{2v}{v^2-1}.$$

Separating the variables, the differential equation can be written as

$$\frac{-3(v^2-1)}{3v-v^3}dv = \frac{3dx}{x}.$$

Hence,

$$\ln(3v-v^3) = -3\ln x + \ln c,$$

or,

$$3x^2y - y^3 = c.$$

• PROBLEM 1078

Solve the differential equation: $\frac{dy}{dx} = \frac{y^3-2x^3}{xy^2}$.

Solution: To solve this problem, we try to investigate whether the given differential equation is homogeneous or not. By homogeneous equations we mean that, if $f(x,y)$ is a given function, then it is called a homogeneous function of degree n if and only if the following identity, involving a quantity t, holds for the function:

$$f(tx,ty) = t^n f(x,y).$$

If substitution of tx and ty for every x and y, respectively, does not affect the function, then $t^n = 1$ or $n = 0$ which indicates that the function is homogeneous of degree zero. Under this condition, if the homogeneity requirement is met, the given function can be expressed as a function of a single argument: either $f(1,y/x)$ or $f(x/y,1)$.

This is a very important procedure in solving homogeneous differential equations, for we can substitute vx or vy for y or x, respectively.

The given differential equation is homogeneous of degree zero, since

$$f(tx,ty) = \frac{(ty)^3-2(tx)^3}{(tx)(ty)^2} = \frac{t^3y^3-2t^3x^3}{t^3xy^2} = \frac{y^3-2x^3}{xy^2} = f(x,y) = t^0f(x,y).$$

Thus, letting $y = vx$, we obtain: $dy/dx = v + x\,dv/dx$, hence,

$$\frac{dy}{dx} = v + x\frac{dv}{dx} = \frac{(vx)^3-2x^3}{x(vx)^2} = \frac{v^3-2}{v^2}.$$

$$x\frac{dv}{dx} = \frac{v^3-2}{v^2} - v = -\frac{2}{v^2}.$$

Separating variables,

$$-\frac{v^2}{2}\,dv = \frac{dx}{x}.$$

Integration yields:

$$-\frac{v^3}{6} = \ln x + c.$$

Returning to the expression in terms of y,

$$-\frac{y^3}{6x^3} = \ln x + c,$$

or,

$$y = -6x^3(\ln x + c).$$

• PROBLEM 1079

Find the general solution of: $\frac{dy}{dx} = \frac{x^2+y^2+y}{x}$.

Solution: Expressing the equation in an implicit form,

$$x^2dx + y^2dx + ydx = xdy,$$

or,

$$x^2dx + y^2dx = xdy - ydx.$$

Dividing the whole equation by $x^2 + y^2$,

$$dx = \frac{xdy - ydx}{x^2 + y^2}.$$

To integrate the right-hand side, we let $\tan\theta = y/x$. Then, differentiation yields:

$$\sec^2\theta\ d\theta = \frac{xdy - ydx}{x^2},$$

where $1 + \tan^2\theta = \sec^2\theta$ and $\tan^2\theta$, in terms of x and y, is y^2/x^2. Thus,

$$1 + \frac{y^2}{x^2}\,d\theta = \frac{xdy - ydx}{x^2}.$$

By using a common denominator, and canceling out x^2,

$$d\theta = \frac{xdy - ydx}{x^2 + y^2}.$$

Integrating,

$$\int dx = \int \frac{xdy - ydx}{x^2 + y^2} + c = \int d\theta + c$$

or

$$x = \theta + c = \tan^{-1}(y/x) + c.$$

Thus the general equation with an arbitrary constant c is:

$$y = x\tan(x - c).$$

• PROBLEM 1080

Find the general solution of: $\frac{dy}{dx} = \frac{y}{x}\ln x$.

Solution: Separating variables, we obtain:

$$\frac{dy}{y} = \frac{\ln x}{x}\,dx$$

Integration yields $\displaystyle\int \frac{dy}{y} = \int \frac{\ln x}{x}\,dx$

To integrate $\displaystyle\int \frac{\ln x}{x}\,dx$, we set $\ln x = u$. Then $\frac{1}{x}\,dx = du$.

Hence the integral is in the form

$$\int u\,du = \frac{u^2}{2} + C_1 = \frac{(\ln x)^2}{2} + C_1$$

and $\displaystyle\int \frac{dy}{y} = \ln y = \frac{(\ln x)^2}{2} + C_1$

Since C_1 is just an arbitrary constant, we can set $C_1 = \ln C$.

Therefore, $\displaystyle\ln y = \frac{(\ln x)^2}{2} + \ln C$

from which $\displaystyle\ln y - \ln C = \frac{(\ln x)^2}{2}$

or $\displaystyle\ln \frac{y}{c} = \frac{(\ln x)^2}{2}$

Taking antilogarithms

$$\frac{y}{c} = x^{\frac{1}{2}\ln x}$$

and

$$y = Cx^{\frac{1}{2}\ln x}$$

• PROBLEM 1081

Solve the equation:

$$y'' - 2y' - 3y = 4e^{3x}.$$

Solution: This is a second-order non-homogeneous linear differential equation with constant coefficients.

Setting $\displaystyle\frac{d^2y}{dx^2} - 2\frac{dy}{dx} - 3y = 0$

First we obtain its general solution, called the complementary function. This equation may be written: $(D^2-2D-3)y = 0$, where $D = \frac{d}{dx}$. The auxiliary equation is: $m^2-2m-3 = 0$, which has the roots $m = -1$ and $m = 3$. The general solution for the homogeneous part of the differential equation is then

$$y_c = c_1e^{3x} + c_2e^{-x}.$$

To find the particular solution for the given differential equation we use the method of undetermined coefficients: that there is some solution which is of an exponential form as in the right-hand side of the given equation. But this is an exceptional case, since e^{3x} occurs in y_c. Hence we do not assume an e^{3x} term, but employ, instead, an xe^{3x} term. We therefore assume

$$y_p = xAe^{3x}$$

we find

$$Dy_p = x\ 3Ae^{3x} + Ae^{3x},$$

$$D^2y_p = x\ 9Ae^{3x} + 3Ae^{3x} + 3Ae^{3x} = Axe^{3x} + 6Ae^{3x}.$$

Substituting these in the given differential equation, we get

$$9Axe^{3x} + 6Ae^{3x} - 2(3Axe^{3x}+Ae^{3x}) - 3Axe^{3x} = 4e^{3x},$$

the terms involving xe^{3x} cancel, and we have

$$6Ae^{3x} - 2Ae^{3x} = 4e^{3x},$$

from which we find $A = 1$. We recall the theorem: If $y = u(x)$ is a solution of the given differential equation, and $y = v(x)$ is a solution of the homogeneous part of that equation, then $y = u(x) + v(x)$ is the general solution for the given equation. By this theorem, we merely add y_c and y_p to obtain

$$y = c_1e^{3x} + c_2e^{-x} + xe^{3x},$$

the required general solution.

• PROBLEM 1082

Solve the equation:

$$\frac{d^2y}{dx^2} + \frac{dy}{dx} - 2y = 4x^2 - 10x + 1.$$

Solution: The equation,

$$\frac{d^2y}{dx^2} + \frac{dy}{dx} - 2y = 4x^2 - 10x + 1,$$

is called a second-order, linear non-homogeneous equation with constant coefficients.

Setting $\frac{d^2y}{dx^2} + \frac{dy}{dx} - 2y = 0.$

First we obtain its general solution, called the complementary function. This equation may be written: $(D^2+D-2)y = 0$, where $D = \frac{d}{dx}$. The auxilary equation is: $m^2+m-2 = 0$, which has the roots $m = 1$, $m = -2$. The general solution for this homogeneous differential equation is

$$y_c = c_1e^x + c_2e^{-2x}.$$

To find the particular solution of the given differential equation we use the method of undetermined coefficients: that there is some solution which is a polynomial in x. Since Q(x) is a second-degree polynomial, the solution y_p will also be a second-degree polynomial. But it is instructive to find what happens if we guess that y_p is a third-degree polynomial. Let

$$y_p = Ax^3 + Bx^2 + cx + D,$$

where the constants A, B, C and D are to be determined.

On differentiating and substituting in the left side of the given equation we have

$$(6Ax+2B) + (3Ax^2+2B+C) - 2(Ax^3+Bx^2+Cx+D) = 4x^2-10x+1,$$

or

$$-2Ax^3 + (3A-2B)x^2 + 2(3A+B-C)x + 2B+C-2D = 4x^2-10x+1.$$

This last will be an identity if and only if the coefficients of like powers of x are the same on both sides. Equating like coefficients gives:

For x^3: $-2A = 0$

For x^2: $3A - 2B = 4$

For x : $3A + B - C = -5$

For $\chi^\circ = 1$: $2B + C - 2D = 1.$

Solving this set, we obtain A = 0, B = -2, C = 3, D = -1. Then

$$y_p = -2x^2 + 3x - 1$$

is the particular solution and is a second-degree polynomial in x as predicted.

We recall a theorem: If $y = u(x)$ is a solution of a given differential equation, and $y = v(x)$ is a solution of the homogeneous part of that equation, then $y = u(x) + v(x)$ is the general solution of the given equation.

By this theorem, we merely add y_c and y_p to obtain

$$y = c_1e^x + c_2e^{-2x} - 2x^2 + 3x - 1,$$

the required general solution.

• PROBLEM 1083

Solve the equation:

$$\frac{d^2y}{dx^2} + 2\frac{dy}{dx} + y = 0.$$

Solution: Assume $y = ce^{mx}$ is the solution for the differential equation. Then, since $\frac{dy}{dx} = mce^{mx}$ and $\frac{d^2y}{dx^2} = m^2ce^{mx}$, substitution for the given equation gives

$$\frac{d^2y}{dx^2} + 2\frac{dy}{dx} + y = m^2ce^{mx} + 2mce^{mx} + ce^{mx} = 0$$

or $$ce^{mx}\left(m^2+2m+1\right) = 0.$$

Assuming e^{mx} to be different from zero, and considering $c = 0$ leads to a trivial solution. The only non-trivial solution occurs when

$$m^2 + 2m + 1 = 0,$$

or

$$(m+1)^2 = 0$$

with a repeated root: $m = -1$.

Since a multiple root occurs here, in order to obtain a complete solution, we seek another linearly independent expression that satisfies the obtained auxiliary equation besides ce^{mx}. To get this, we take $y = c_2xe^{mx}$ and hence:

$$xc_2e^{mx}\left(m^2+2m+1\right) + c_1e^{mx}\left(m^2+2m+1\right) = 0.$$

Thus
$$y = \left(c_1+c_2x\right)e^{mx}$$

is the desired general solution.

Using $m = -1$ as a root, the general solution of the differential equation is

$$y = \left(c_1+c_2x\right)e^{-x}.$$

• PROBLEM 1084

Solve the equation: $\frac{d^2y}{dx^2} + 16y = 0$, subject to the initial conditions: $y = 0$ and $\frac{dy}{dx} = 5$ when $x = 0$.

Solution: This is a second-order, homogeneous differential equation of the form,

$$\frac{d^2y}{dx^2} + a_1\frac{dy}{dx} + a_2y = 0,$$

with $a_1 = 0$ and $a_2 = 16$.

Assuming $y = ke^{mx}$ is the general solution, and, since $\frac{d^2y}{dx^2} = km^2e^{mx}$, we obtain:

$$\frac{d^2y}{dx^2} + 16y = ke^{mx}\left(m^2+16\right) = 0.$$

Taking $k = 0$ leads to trivial solution. A non-trivial solution is found by the auxiliary equation:

$$m^2 + 16 = 0$$

with roots $m = \pm 4i$.

The solution $y_1 = k_1e^{iax}$ or $y_2 = k_2e^{-4x}$ is adequate, but as we are looking for the general solution, the sum of all possible solutions is of importance. Therefore

$$y = y_1 + y_2$$

or
$$y = k_1e^{i4x} + k_2e^{-i4x}$$

is the desired general solution.

Recalling Euler's formula for complex components

$$e^{\pm ix} = \cos x \pm i \sin x,$$

and expressing our general solution in terms of the circular functions, after some algebraic manipulation, we obtain

$$y = c_1 \cos 4x + c_2 \sin 4x$$

where $c_1 = k_1 + k_2$ and $c_2 = i(k_1 - k_2)$.

Substituting $y = 0$ and $x = 0$ gives $c_1 = 0$.

$$y = c_2 \sin 4x.$$

Differentiating this, we have

$$\frac{dy}{dx} = 4c_2 \cos 4x;$$

substituting $\frac{dy}{dx} = 5$ and $x = 0$, we get $5 = 4c_2$ or $c_2 = \frac{5}{4}$. The required particular solution is then $y = \frac{5}{4} \sin 4x$.

It is frequently convenient to write the solution

$$y = c_1 \cos bx + ic_2 \sin bx,$$

for the case of complex roots, in another form. By trigonometry,

$$c_1 \cos bx + ic_2 \sin bx = c \sin(bx+\alpha) = c \cos(bx+\beta),$$

where $c = \sqrt{c_1^2 + c_2^2}$, $\tan \alpha = c_1/c_2$, $\tan \beta = -c_2/c_1$. The general solution then becomes:

$$y = ce^{ax} \sin(bx+\alpha) = ce^{ax} \cos(bx+\beta).$$

In these forms, c and α, or c and β are the real arbitrary constants.

• PROBLEM 1085

Solve the equation: $\frac{d^2y}{dx^2} = -y$.

Solution: Let $p = \frac{dy}{dx}$, then $\frac{d^2y}{dx^2} = \frac{dp}{dx} = \frac{dp}{dy}\frac{dy}{dx} = p\frac{dp}{dy}$.

Substitution gives

$$p\frac{dp}{dy} = -y.$$

Separating variables and integrating, we obtain:

$$p^2 = -y^2 + a^2 \text{ or } p = \frac{dy}{dx} = \pm\sqrt{a^2-y^2},$$

where a^2 is a constant of integration.

Separating variables and integrating once more, we have

$$\sin^{-1}(y/a) = \pm x + c,$$

or

$$y = a \sin(\pm x+c).$$

Another approach is the following: Since this equation is a homogeneous, linear and second-order differential equation with constant coefficients we have, in a general form:

$$\frac{d^2y}{dx^2} + a_1 \frac{dy}{dx} + a_2 y = 0$$

where $a_1 = 0$ and $a_2 = 1$.

The auxiliary equation is then

$$m^2+1 = 0$$

with roots $m = \pm i$.

Thus, the general solution is $y = c_1 e^{ix} + c_2 e^{-ix}$.

From Euler's formula about complex exponents

$$e^{\pm ix} = \cos x \pm i \sin x.$$

Setting $c_1 + c_2 = k_1$ and $c_1 - c_2 = k_2$, and factoring out the cosine and sine parts, the general solution we obtain will be, in a harmonic form,

$$y = k_1 \cos x + ik_2 \sin x.$$

• PROBLEM 1086

Solve the differential equation:

$$\frac{d^2y}{dx^2} - \frac{dy}{dx} - 6y = 0.$$

Solution: This is a second-order, linear, homogeneous differential equation with constant coefficients.

Let $y = e^{mx}$ to see if it is a solution of the given equation. Since $\frac{dy}{dx} = me^{mx}$ and $\frac{d^2y}{dx^2} = m^2e^{mx}$, substitution gives

$$\frac{d^2y}{dx^2} - \frac{dy}{dx} - by = e^{mx}\left(m^2-m-6\right) = 0.$$

e^{mx} is obviously different from zero, and hence

$$m^2 - m - 6 = 0$$

is what is known as the auxiliary equation, with roots $m = 3$ and $m = -2$. Therefore, either $y = c_1e^{3x}$ or $y = c_2e^{-2x}$ can solve the equation. However, since we are looking for the general solution, the sum of all possible solutions is of importance. Thus, the general solution of the differential equation is:

$$y = c_1e^{3x} + c_2e^{-2x}.$$

Here, c_1 and c_2 are arbitrary constants, but under given initial conditions for x, y and $\frac{dy}{dx}$ we can substitute the values and solve for these constants. Once the values of c_1 and c_2 are known, our solution is a particular solution from among the family of curves obtained in the general solution.

• PROBLEM 1087

Find the particular solution of the differential equation $y'' + y' - 6y = 0$, subject to the initial conditions: $y = 4$ and $y' = 3$ when $x = 0$.

Solution: Consider equations of the type:

$$\text{(a)} \qquad \frac{d^2y}{dx^2} + a_1\frac{dy}{dx} + a_2y = Q(x).$$

This type is called the second-order non-homogeneous differential equation with constant coefficients. Q(x) may be a constant or function of x alone.

Assuming $Q(x) = 0$, a homogeneous equation, we wish to find the general solution. Let us try the function $y = e^{mx}$ to see if it is a solution of (a). Since $y' = me^{mx}$ and $y'' = m^2e^{mx}$, substitution in (a) gives

(b) $$\frac{d^2y}{dx^2} + a_1 \frac{dy}{dx} + a_2 y = m^2 e^{mx} + a_1 m e^{mx} + a_2 e^{mx} = 0$$

$e^{mx} \neq 0$, therefore:

(c) $$m^2 + a_1 m + a_2 = 0.$$

If (b) has two distinct roots r_1 and r_2, then $y = e^{r_1 x}$ and $y = e^{r_2 x}$ are the two solutions. The general solution is thus

$$y = c_1 e^{r_1 x} + c_2 e^{r_2 x}.$$

This way, our differential equation is ready to be solved: the roots of the auxiliary equation $m^2 + m - 6 = 0$ and $m = 2$ and $m = -3$, and the general solution of the differential equation is

(d) $$y = c_1 e^{2x} + c_2 e^{-3x}.$$

Substituting $x = 0$ and $y = 4$ in this equation, we have $c_1 + c_2 = 4$. Differentiating (d), we have

$$y' = 2c_1 e^{2x} - 3c_2 e^{-3x}.$$

Substituting $x = 0$ and $y' = 3$ in this equation, we have $2c_1 - 3c_2 = 3$. Solving these two equations for c_1 and c_2, we find $c_1 = 3$, $c_2 = 1$. The required particular solution is therefore $y = 3e^{2x} + e^{-3x}$.

• **PROBLEM** 1088

Solve the differential equation:

$$y\frac{d^2y}{dx^2} + \left(\frac{dy}{dx}\right)^2 = 1.$$

Solution: In this kind of differential equation, of the form

$$f(y'', y', y),$$

we substitute $y' = \frac{dy}{dx} = p$ which, in turn, gives an expression for y'', namely $y'' = \frac{dp}{dx}$.

But, if we set $\frac{dy}{dx} = p$, using the chain rule, we can also write, instead, that:

$$y'' = \frac{d^2y}{dx} = \frac{dp}{dy}\frac{dy}{dx} = p\frac{dp}{dy},$$

a more useful formulation here, since it does not involve χ. This gives:

$$yp\frac{dp}{dy} + p^2 = 1.$$

Separating variables:

$$\frac{dy}{y} = \frac{pdp}{1-p^2}.$$

Integration yields:

$$\ln y = -\frac{1}{2}\ln\left(1-p^2\right) + \ln c,$$

$$2\ln y + \ln\left(1-p^2\right) = 2\ln c$$

or

$$\ln\left|y^2\left(1-p^2\right)\right| = \ln c^2.$$

Letting $\ln c^2 = c_1$, we have

$$y^2\left(1=p^2\right) = c_1, \text{or } \frac{dy}{dx} = \frac{\pm\sqrt{y^2-c_1}}{y}.$$

Separating variables again,

$$\pm\frac{ydy}{\sqrt{y^2-c_1}} = dx,$$

and, integrating: $\sqrt{y^2-c_1} = \pm x + c_2$,

or

$$y^2 = \left(\pm x + c_2\right)^2 + c_1.$$

• **PROBLEM** 1089

Solve the equation:

$$y' = \frac{2xy}{y^2-x^2}.$$

Solution: The above differential equation is called homogeneous since it fulfills the following property:

$$f(tx,ty) = t^n f(x,y)$$

for all $t > 0$ and all (x,y) in a given region. In our case, the degree $n = 0$

$$f(tx,ty) = \frac{2txty}{t^2y^2-t^2x^2} = \frac{2xy}{y^2-x^2} = f(x,y).$$

For such differential equations, the substitution $y = vx$ will transform the differential equation into one in v and x in which the variables are separable. The substitution $y = vx$ yields

$$\frac{d(vx)}{dx} = \frac{2x^2v}{x^2(v^2-1)} = \frac{2v}{v^2-1}.$$

Using the formula for the derivative of a product,

$$x\frac{dv}{dx} + v = \frac{2v}{v^2-1}.$$

Therefore,

$$x\frac{dv}{dx} = \frac{2v}{v^2-1} - v = \frac{3v-v^3}{v^2-1}.$$

Separating the variables;

$$\frac{dv}{\frac{3v-v^3}{v^2-1}} = \frac{dx}{x}$$

$$\int \frac{v^2-1}{3v-v^3}dv = \int \frac{dx}{x}.$$

Using the substitution method to integrate the left-hand side, we set

$$3v - v^3 = M$$

and

$$\left(3-3v^2\right)dv = dM \quad \text{or } dv = dM/3\left(1-v^2\right)$$

thus

$$\int \frac{v^2-1}{3v-v^2}dv = \int \frac{v^2-1}{M} \frac{dM}{3\left(1-v^2\right)}$$

$$= -\frac{1}{3}\int \frac{dM}{M} = -\frac{1}{3}\ln|M| = -\frac{1}{3}\ln|3v-v^3|.$$

Then we have

$$-\frac{1}{3}\ln|3v-v^3| = \ln|x| + C_1$$

Multiplying the whole equation by 3 and combining the ln terms, we have

$$\ln \left(3v-v^3\right)x^3 = C_1$$

Since C_1 is an arbitrary constant of integration, we can choose C_1 in such a way that it is the logarithm of another constant, C. Here we prefer to set $C_1 = \ln C$, so that we can take the antilogarithm.

Hence, $$\left(3v-v^3\right)x^3 = C.$$

Substituting $v = y/x$ yields

$$\left(\frac{3y}{x} - \frac{y^3}{x^3}\right)x^3 = C$$

or $$3x^3y-y^3 = C.$$

The desired solution $y = f(x)$ to the differential equation is defined implicitly by the above equation.

• PROBLEM 1090

If the operator $D \equiv \frac{d}{dx}$, perform the operation:

$$\left(D^2 + 5D-1\right)\left(\tan 2x - 3/x\right).$$

Solution: Multiplying out the expression,

$$\left(D^2+5D-1\right)(\tan 2x-3/x) = D^2(\tan 2x)-D^2(3/x)+5D(\tan 2x)-5D(3/x)-(\tan 2x-3/x)$$

$$D(3/x) = -3/x^2 \quad , \quad -5D(3/x) = 15/x^2$$

$$D^2(3/x) = 6/x^3 \quad , \quad -D^2(3/x) = -\ 6/x^3 \ ;$$

$$D(\tan 2x) = 2\sec^2 2x \quad , \quad 5D(\tan 2x) = 10\sec^2 2x$$

$$D^2(\tan 2x) = D(2\sec^2 2x) = 8\sec^2 2x \tan 2x.$$

Thus,

$$\left(D^2+5D-1\right)(\tan 2x-3/x) = 9\sec^2 2x+\tan 2x-6/x^3+10\sec^2 2x+15/x^2-\tan 2x+3/x \ .$$

• PROBLEM 1091

Find the general solution of $\left(D^2-1\right)y = 2x + e^{2x}$.

Solution: Here the auxiliary equation is: $m^2-1 = 0$ with roots $m = 1$ and $m = -1$. Hence, the complementary solution is:

$$y_c = Ae^x + Be^{-x} \ .$$

Letting u be the particular solution, $u = \left(D^2-1\right)y$, and we have: $\left(D^2-1\right)u = 2x + e^{2x}$. We choose $u = c_1x + c_2e^{2x}$. Hence,

$$\left(D^2-1\right)u = 4c_2e^{2x} - c_1x - c_2e^{2x} = 2x + e^{2x} \ .$$

Equating like coefficients,

$$3c_2 = 1 \qquad \text{for } e^{2x}$$

$$-c_1 = 2 \qquad \text{for } x \ ,$$

from which

$$c_2 = 1/3 \text{ and } c_1 = -2.$$

Thus,

$$u = -2x + \frac{1}{3}e^{2x}.$$

The general solution for the given differential equation is $y_c + u$.

$$y = Ae^x + Be^{-x} - 2x + \frac{1}{3}e^{2x}.$$

• PROBLEM 1092

Solve the differential equation:

$$\frac{d^2y}{dx^2} + 4\frac{dy}{dx} = 10 \sin 2x.$$

Solution: First we set

$$\text{(a)} \qquad \frac{d^2y}{dx^2} + 4\frac{dy}{dx} = 0,$$

and write (a) in the form:

$$\left(D^2+4D\right)y = 0$$

where the D's represent derivatives, i.e. $D = \frac{d}{dx}$, $D^2 = \frac{d}{dx^2}$, etc. We obtain the roots of the auxiliary equation, $m^2 + 4m = 0$. They are $m = 0, -4$, hence the complementary function is

$$y_c = c_1 + c_2e^{-4x}.$$

Since the right-hand side of the given nonhomogeneous differential equation is a periodic circular function with $W = 2$ we assume, by the method of undetermined coefficients, that the solution is of the same form, namely,

$$y_p = A \sin 2x + B \cos 2x.$$

Then

$$Dy_p = 2A \cos 2x - 2B \sin 2x, \; D^2y_p = -4A \sin 2x - 4B \cos 2x.$$

Substitution of these values in the differential equation gives

$$-4A \sin 2x - 4B \cos 2x + 8A \cos 2x - 8B \sin 2x = 10 \sin 2x.$$

Equating coefficients of sin 2x and cos 2x separately on both sides, we get

$$-4A - 8B = 10, \; -4B + 8A = 0.$$

Solution of these simultaneous equations gives

$A = -\frac{1}{2}$, $B = -1$. Using the theorem: if $y = u(x)$ is a solution of the homogeneous part of the given differential equation, and $y = v(x)$ is the particular solution of the whole given equation, then $y = u(x) + v(x)$ is the complete solution for this given equation. By this theorem, we merely add y_c and y_p to obtain

$$y = c_1 + c_2 e^{-4x} - \frac{1}{2}\sin 2x - \cos 2x,$$

the required general solution.

• PROBLEM 1093

Resolve the given operator $\varphi(D)$ into linear factors, and arrange these factors in all possible orders. Operate upon the given function u with $\varphi(D)$, and verify the fact that the results obtained are the same.

$$\varphi(D) = 2D^2 + D - 6;\ u = \sin x + 3e^x .$$

Solution: i) $(2D^2 + D - 6)(\sin x + 3e^x) = 2D^2(\sin x + 3e^x) + D(\sin x + 3e^x) - 6(\sin x + 3e^x)$

$$D(\sin x + 3e^x) = \cos x + 3e^x$$

$$2D^2(\sin x + 3e^x) = 2D(\cos x + 3e^x) = -2\sin x + 6e^x.$$

Therefore,

$$(2D^2 + D - 6)(\sin x + 3e^x) = -2\sin x + 6e^x + \cos x + 3e^x - 6\sin x - $$
$$= -8\sin x + \cos x - 9e^x.$$

ii) $2D^2 + D - 6 = (2D - 3)(D + 2)$

$$\begin{aligned}(2D^2 + D - 6)(\sin x + 3e^x) &= (2D - 3)(D + 2)(\sin x + 3e^x)\\ &= (2D - 3)\left[D(\sin x + 3e^x) + 2(\sin x + 3e^x)\right]\\ &= (2D - 3)\left[(\cos x + 3e^x) + 2(\sin x + 3e^x)\right]\\ &= 2D\left[(\cos x + 3e^x) + 2(\sin x + 3e^x)\right]\\ &\quad - 3\left[(\cos x + 3e^x) + 2(\sin x + 3e^x)\right]\\ &= 2(-\sin x + 3e^x) + 4(\cos x + 3e^x)\\ &\quad - 3(\cos x + 3e^x) - 6(\sin x + 3e^x)\\ &= -8\sin x + \cos x - 9e^x .\end{aligned}$$

iii)
$$\begin{aligned}(2D^2 + D - 6)(\sin x + 3e^x) &= (D + 2)(2D - 3)(\sin x + 3e^x)\\ &= (D + 2)\left[2D(\sin x + 3e^x) - 3(\sin x + 3e^x)\right]\\ &= (D + 2)\left[(2\cos x + 6e^x) - 3(\sin x + 3e^x)\right]\\ &= D\left[(2\cos x + 6e^x) - 3(\sin x + 3e^x)\right]\\ &\quad + 2\left[(2\cos x + 6e^x) - 3(\sin x + 3e^x)\right]\\ &= -2\sin x + 6e^x - 3\cos x - 9e^x\\ &\quad + 2(2\cos x + 6e^x) - 6(\sin x + 3e^x)\\ &= -8\sin x + \cos x - 9e^x.\end{aligned}$$

CHAPTER 42

ADVANCED DIFFERENTIAL EQUATIONS

• PROBLEM 1094

Using Picard's method of successive approximations, obtain a solution of the equation: $dy/dx = y+x$, such that $y = 1$ when $x = 0$. Carry out the work through the fourth approximation, and check your result by finding the exact particular solution.

Solution: Picard's method involves a successive approximation of the y-values on the basis of the given differential equation. Consider the following differential equation, to which we are attempting to find the general solution under a given boundary condition, say $y = b$ when $x = a$,

$$dy/dx = f(x,y).$$

Integration yields:

$$\int_b^y dy = y - b = \int_a^x f(x,y)dx$$

or

$$y = b + \int_a^x f(x,y)dx.$$

If we substitute the initial value of y, which is b, in the integrand, the first approximation for y is given by:

$$y_1 = b + \int_a^x f(x,b)dx.$$

So far, we have the value:

$$\frac{dy}{dx} = f\left(x,y_1\right), \text{ where } y_1 = f(x) \text{ as shown above.}$$

Under these conditions, we can write, as before:

$$y = b + \int_a^x f\left(x,y_1\right)dx.$$

But y_1 is not the correct expression for the function. In other words, it does not converge to every point on the curve, and hence suggests more accurate values of y. Therefore, the above expression is not sufficient and can be called the second approximation. This way, we can approximate, step by step, to the nth value given as:

$$y_{n+1} = b + \int_a^x f\left(x,y_n\right)dx.$$

The limit, as n tends to infinity, under the restriction that $y_{n+1} = b$ for $x = a$, gives the desired solution.

On this basis, we try to find the solution for the given equation

$$dy/dx = y + x , \quad y = 1 \text{ when } x = 0 .$$

The first approximation is:

$$y_1 = 1 + \int_0^x (1+x)\,dx$$

$$= 1 + x + \frac{x^2}{2!} .$$

Likewise, the second approximation is:

$$y_2 = 1 + \int_0^x (y_1+x) = 1 + \int_0^x \left(1+x+ \frac{x^2}{2!} + x\right)dx$$

$$= 1 + x + \frac{x^2}{2!} + \frac{x^3}{3!} + \frac{x^2}{2!} .$$

$$y_3 = 1 + \int_0^x (y_2+x) = 1 + \int_0^x \left(1+x+ \frac{x^2}{2!} + \frac{x^2}{3!} + \frac{x^2}{2!} + x\right)dx$$

$$= 1 + x + \frac{x^2}{2!} + \frac{x^3}{3!} + \frac{x^4}{4!} + \frac{x^3}{3!} + \frac{x^2}{2!} .$$

Algebraic manipulations were not made because the study of the behavior of the series, as it is, is of more interest.
From what we know so far, one can write down the fourth approximation without performing the integration. It would be:

$$y_4 = 1 + x + \frac{x^2}{2!} + \frac{x^3}{3!} + \frac{x^4}{4!} + \frac{x^5}{5!} + \frac{x^4}{4!} + \frac{x^3}{3!} + \frac{x^2}{2!} .$$

Adding $1 + x$ to the whole equation, we can write:

$$y_4+1+x = 1+x + \frac{x^2}{2!} + \frac{x^3}{3!} + \frac{x^4}{4!} + \frac{x^5}{5!} + \frac{x^4}{4!} + \frac{x^3}{3!} + \frac{x^2}{2!} + x + 1,$$ or, in a shorter form,

$$y_4 + 1 + x = 2\left(1 + x + \frac{x^2}{2!} + \frac{x^3}{3!} + \frac{x^4}{4!}\right) + \frac{x^5}{5!} .$$

By induction, we can write the kth approximation for y

$$y_k + 1 + x = 2\left(1 + x + \frac{x^2}{2!} +\ldots+ \frac{x^k}{k!}\right) + \frac{x^{k+1}}{(k+1)!}$$

or

$$y_k + 1 + x = 2 \sum_{i=0}^{i=k} \frac{x^i}{(i)!} + \frac{x^{k+1}}{(k+1)!} .$$

The mathematical induction can be proved to hold by adding $2\frac{x^{k+1}}{(k+1)!} + \frac{x^{k+2}}{(k+2)}$ in the last term and checking with the (k+1)th approximation. Taking the limit,

$$\lim_{n\to\infty}(y_n+1+x) = y+1+x = 2 \lim_{n\to\infty} \left[\sum_{n=0}^{n} \frac{x^n}{n!} + \frac{x^{n+1}}{(n+1)!}\right] = 2 \sum_{n=0}^{\infty} \frac{x^n}{n!} .$$

But
$$\sum_{n=0}^{\infty} \frac{x^n}{n!} = e^x .$$

Thus
$$y + 1 + x = e^x,$$

or, $y = 2e^x - 1 - x$, the desired solution by Picard's method.

Rewriting the given differential equation to obtain the exact solution, $dy/dx - y = x$, its integrating factor is $e^{\int(-1)dx} = e^{-x}$. The general solution is:

$$y = e^x \int xe^{-x}\, dx + ce^x$$
$$= e^x(-x-1)e^{-x} + ce^x$$
$$= -x-1 + ce^x .$$

$y = 1$ for $x = 0$, which implies that $c = 2$ hence, $y = 2e^x - x - 1$, as before.

• PROBLEM 1095

> Find the general solution of $4x\frac{d^2y}{dx^2} + 2\frac{dy}{dx} - y = 0$, using the method of Frobenius.

Solution: In the Frobenius method,

$$y = \sum_{n=0}^{n} a_n x^{n+c}$$

is assumed to be the general solution. The constants $a_n (n = 0,1,2,\ldots)$ and c are to be determined. Assuming the expression for y to hold,

$$\frac{dy}{dx} = \sum_{n=0}^{n} a_n(n+c)x^{n+c-1} . \quad 2\frac{dy}{dx} = 2\sum_{n=0}^{n} a_n(n+c)x^{n+c-1} .$$

$$\frac{d^2y}{dx^2} = \sum_{n=0}^{n} a_n(n+c)(n+c-1)x^{n+c-2} . \quad 4x\frac{d^2y}{dx^2} = 4\sum_{n=0}^{n} a_n(n+c)(n+c-1)x^{n+c-1} .$$

Hence,

$$4x\frac{d^2y}{dx^2} + 2\frac{dy}{dx} - y \equiv 4\sum_{n=0}^{n} a_n(n+c)(n+c-1)x^{n+c-1} + 2\sum_{n=0}^{n} a_n(n+c)x^{n+c-1} - \sum_{n=0}^{n} a_n x^{n+c} = 0.$$

Combining the first two terms, we obtain:

$$2\sum_{n=0}^{n} a_n(n+c)(2n+2c-1)x^{n+c-1} - \sum_{n=0}^{n} a_n x^{n+c} = 0 .$$

Now, we wish to combine the two summations. To do this, we wish to make the powers of x the same by removing the first, $n = 0$, term, from the left-hand side of the equation.

$$2\, a_0(0+c)(0+2c-1)x^{c-1} + 2\sum_{n=0}^{n} a_n(n+c)(2n+2c-1)x^{n+c-1} - \sum_{n=0}^{n} a_n x^{n+c} = 0.$$

Both summations start from x^c and continue from there, and, in order to combine these, we let $n = n+1$ for any factor in the first summation. This gives:

$$2\, a_0 c(2c-1)x^{c-1} + 2\sum_{n=0}^{n} a_{n+1}(n+1+c)(2n+2+2c-1)x^{n+c} - \sum_{n=0}^{n} a_n x^{n+c} .$$

Combination under the same sign yields:

$$2\, a_0 c(2c-1)x^{c-1} + \sum_{n=0}^{n} [2a_{n+1}(n+1+c)(2n+2+2c-1)-a_n]x^{n+c} = 0 .$$

Since all the coefficients of x have to equal zero(they must be independent) $2\, a_0 c(2c-1) = 0$ (the coefficients of the lowest power of x). If we choose $a_0 \neq 0$, then $c = 0$ or $c = \frac{1}{2}$. From the other side,we have the recursive formula:

$$a_{n+1} = \frac{a_n}{2(n+1+c)(2n+2+2c-1)} .$$

For $c = 0$,

$$a_{n+1} = \frac{a_n}{2(n+1)(2n+1)} , \text{ or } a_n = \frac{a_{n-1}}{2n(2n-1)} .$$

Now, in order to obtain significant recurrence in terms of a_0, we try to replace n by $n-1$ successively, and observe:

$$a_{n-1} = \frac{a_{n-2}}{(2n-2)(2n-3)} ,$$

and

$$a_{n-2} = \frac{a_{n-3}}{(2n-4)(2n-5)} .$$

So far, we have

$$a_n = \frac{a_{n-3}}{(2n)(2n-1)(2n-2)(2n-3)(2n-4)(2n-5)} .$$

Proceeding this way, we ultimately find:

$$a_n = \frac{a_{n-n}}{(2n)(2n-1)(2n-2)\ldots\left[2n-(2n-1)\right]} = \frac{a_0}{(2n)!} .$$

Substituting for $a_1, a_2, \ldots$ their values as given above, we obtain the series solution

$$y = a_0 \sum_{n=0}^{\infty} \frac{x^n}{(2n)!}$$

where a_0 in the first term is arbitrary. By inspection, the series is merely the expansion of the function $\cosh\sqrt{x}$, so that a solution of the given differential equation, when $c = 0$, is:

$$y = a_0 \cosh\sqrt{x} .$$

For $c = \frac{1}{2}$, we use, the previously obtained recursion formula:

$$a_n = \frac{a_{n-1}}{2n(2n+1)}$$

by substitution of $n-1$ for n, we have:

$$a_n = \frac{a_{n-3}}{2^3 n(2n+1)(n-1)(2n-1)(n-2)(2n-3)} .$$

Hence,

$$a_n = \frac{a_{n-n}}{2^n n!\ 3\cdot5\cdot7\cdot \ldots (2n+1)} = \frac{a_0\ 2\cdot4\cdot6\cdot \ldots (2n)}{2^n n!\ (2n+1)!} = \frac{a_0\ 2^n n!}{2^n n! (2n+1)!}$$

$$= \frac{a_0}{(2n+1)!} = \frac{a_0}{(2n+1)(2n)!} \approx \frac{a_0}{(2n)!}$$

which leads to the solution: $y = a_0\sqrt{x} \cosh \sqrt{x}$.

• PROBLEM 1096

Using the method of Frobenius, find the general solution of the differential equation

$$2x(1-x)\frac{d^2y}{dx^2} + (1+x)\frac{dy}{dx} - y = 0.$$

Solution: By the use of Frobenius' method, we let the general solution be:

$$y = \sum_{n=0}^{\infty} a_n x^{n+c} .$$

Then,

$$\frac{dy}{dx} = \sum_{n=0}^{\infty} a_n (n+c)x^{n+c-1} \quad \text{and} \quad \frac{d^2y}{dx^2} = \sum_{n=0}^{\infty} a_n (n+c)(n+c-1)x^{n+c-2} .$$

Substitution in the given differential equation yields:

$$2x(1-x)\frac{d^2y}{dx^2} + (1+x)\frac{dy}{dx} - y$$

$$= 2\sum_{n=0}^{\infty} a_n (n+c)(n+c-1)x^{n+c-1} - 2\sum_{n=0}^{\infty} a_n (n+c)(n+c-1)x^{n+c} + \sum_{n=0}^{\infty} a_n (n+c)x^{n+c-1}$$

$$+ \sum_{n=0}^{\infty} a_n (n+c-1)x^{n+c} = 0 .$$

Combining like coefficients,

$$\sum_{n=0}^{\infty} a_n (n+c)(2n+2c-1)x^{n+c-1} - \sum_{n=0}^{\infty} a_n (n+c-1)(2n+2c-1)x^{n+c} = 0 .$$

Removing the first, $n = 0$, term,

$$a_0 c(2c-1)x^{c-1} + \sum_{n=0}^{\infty} a_n (n+c)(2n+2c-1)x^{n+c-1} - \sum_{n=0}^{\infty} a_n (n+c-1)(2n+2c-1)x^{n+c} = 0.$$

In both expressions under the summation signs, the exponents of x start from c and go on to $n+c$ for $n = 1,2,\ldots$. Thus, it is simple to combine them by substituting $n+1$ for n in the left-hand side and start from $n = 0$. In doing so, we have:

$$a_0 c(2c-1)x^{c-1} + \sum_{n=0}^{\infty} a_{n+1} (n+c+1)(2n+2c+1)x^{n+c} - \sum_{n=0}^{\infty} a_n (n+c-1)(2n+2c-1)x^{n+c} = 0,$$

or

$$a_0c(2c-1)x^{c-1} + \sum_{n=0}^{\infty}\left[a_{n+1}(n+c+1)(2n+2c+1) - a_n(n+c-1)(2n+2c-1)\right]x^{n+c} = 0.$$

Noting that all the like coefficients of x (coefficients of x under a certain power) must independently be equal to zero, we automatically have:

$$a_0c(2c-1) = 0.$$

If a_0 is chosen to be different from zero, c has to be either zero or 1/2. Furthermore, by setting

$$a_{n+1}(n+c+1)(2n+2c+1) - a_n(n+c-1)(2n+2c-1) = 0,$$

we finally obtain the recursion formula:

$$a_{n+1} = \frac{(n+c-1)(2n+2c-1)}{(n+c+1)(2n+2c+1)} a_n .$$

Or, by replacing $n-1$ for n, we have:

$$a_n = \frac{(n+c-2)(2n+2c-3)}{(n+c)(2n+2c-1)} a_{n-1} .$$

For the case $c = 0$,

$$a_n = \frac{(n-2)(2n-3)}{n(2n-1)} a_{n-1} .$$

We wish to express a_n in terms of a_0, which is a constant to be obtained under a given boundary condition. Starting from a_1 and using the above equation,

$$a_1 = a_0,\ a_2 = a_3 = \ldots = a_n = 0$$

hence, we have

$$y = a_0(1+x) \quad \text{as a general solution.}$$

For $c = \frac{1}{2}$, we recall:

$$a_n = \frac{(n+c-2)(2n+2c-3)}{(n+c)(2n+2c-1)} a_{n-1} ,$$

and obtain:

$$a_n = \frac{(2n-3)(n-1)}{n(2n+1)} a_{n-1} .$$

This gives $a_0' =$ constant, but $a_1 = a_2 = a_3 = \ldots = a_n = 0$.
Hence using:

$$y = \sum_{n=0}^{\infty} a_n x^{n+c} ,$$

for $c = \frac{1}{2}$, $a_0' =$ constant and the rest are zero. We have the second general solution:

$$y = a_0' \sqrt{x} .$$

Since the sum of all general solutions of a given differential equation is also a solution, we have:

$$y = a_0(1+x) + a_0' \sqrt{x} ,$$

as a general solution.

• PROBLEM 1097

Using the method of Frobenius, find the particular solution for

$$x^2 \frac{d^2y}{dx^2} + x\frac{dy}{dx} + (x^2-\nu^2)y = 0 .$$

Solution: Assuming the solution to be in the form:

$$y = \sum_{n=0}^{\infty} a_n x^{n+c} ,$$

the given differential equation (of order two) in terms of the assumed solution, after combining like coefficients, becomes:

$$\sum_{n=0}^{\infty}\left[a_n(c+n)(c+n-1) + a_n(c+n)-a_n\nu^2\right]x^{n+c} + \sum_{n=0}^{\infty} a_n x^{n+c+2} = 0 ,$$

or

$$\sum_{n=0}^{\infty} a_n\left[(c+n)^2-\nu^2\right]x^{n+c} + \sum_{n=0}^{\infty} a_n x^{n+c+2} = 0 .$$

Removing the first terms in the first sum, we have:

$$a_0(c^2-\nu^2)x^c + a_1\left[(c+1)^2-\nu^2\right]x^{c+1} + \sum_{n=2}^{\infty} a_n\left[(c+n)^2-\nu^2\right]x^{n+c} + \sum_{n=0}^{\infty} a_n x^{n+c+2} = 0.$$

We replace n by $n+2$ in the first sum, and combine the two sums.

$$a_0(c^2-\nu^2)x^c+a_1\left[(c+1)^2-\nu^2\right]x^{c+1}+ \sum_{n=0}^{\infty}\left\{a_{n+2}\left[(c+n+2)^2-\nu^2\right] + a_n\right\}x^{n+c+2} = 0 .$$

The indicial equation is:

$$a_0\left(c^2-\nu^2\right)= 0 ,$$

and for $a_0 \neq 0$, arbitrary a_0, $c = \nu$.

From the second lowest power,

$$a_1\left[(c+1)^2-\nu^2\right] = 0$$

yields $c+1 = \nu$ under arbitrary a_1. In order to keep the correct relations in the first equation, the second equation suggests that a_1 must be zero. Hence, we can replace c by ν and obtain the recursion formula from:

$$a_{n+2}\left[(\nu+n+2)^2-\nu^2\right] + a_n = 0$$

$$a_{n+2} = \frac{-a_n}{(\nu+n+2)^2-\nu^2} .$$

Hence,

$$a_2 = \frac{-a_0}{2(2\nu+2)} ;$$

$$a_3= \frac{a_1}{(\nu+3)^2-\nu^2} = 0 ;$$

$$a_4 = \frac{-a_2}{4(2+4)} = \frac{a_0}{2\cdot 4(2\nu+2)(2\nu+4)} ;$$

$$a_5 = 0.$$

In general, for any number $k \le n$ in the subscript we have, $a_{2k-1} = 0$ but $a_{2k} \neq 0$. Thus,

$$a_{2k} = (-1)^k \frac{a_0}{2.4.6. \ldots . 2k(2\nu+2)(2\nu+4)(2\nu+6)\ldots(2\nu+2k)}$$

$$= (-1)^k \frac{a_0}{2^{2k}k!(\nu+1)(\nu+2)\ldots(\nu+k)}$$

The general solution is, therefore:

$$y = \sum_{n=0}^{\infty} (-1)^n \frac{a_0 x^{\nu+2n}}{2^{2n}n!(\nu+1)(\nu+2)\ldots(\nu+n)} .$$

We use the operational notation: $\Gamma(n+1) = n!$, and let

$$a_0 = \frac{1}{2^{\nu}\Gamma(\nu+1)} .$$

By substitution for the general solution, we have:

$$y = \sum_{n=0}^{\infty} (-1)^n \frac{x^{\nu+2n}}{2^{\nu+2n}n!\,\Gamma(\nu+n+1)} .$$

For a particular value of ν, we have the particular solution known as the Bessel function $J_{\nu}(x)$:

$$J_{\nu}(x) = \sum_{n=0}^{\infty} (-1)^n \frac{x^{\nu+2n}}{2^{\nu+2n}n!\Gamma(\nu+n+1)} \qquad (\nu \ge 0) .$$

This is the definition of the Bessel function of order ν, where ν is a positive, real number or zero. From this, the expression for $\nu = \frac{1}{2}$ can be obtained as:

$$J_{\frac{1}{2}}(x) = \sum_{n=0}^{\infty} (-1)^n \frac{x^{2n+\frac{1}{2}}}{2^{2n+\frac{1}{2}}n!(\frac{1}{2}+n)!}$$

$$= \sqrt{2/x} \sum_{n=0}^{\infty} (-1)^n \frac{x^{2n+1}}{2^{2n+1}n!\,\frac{2n-1}{2}\cdot\frac{2n-1}{2}\cdot\frac{2n-3}{2}\cdot}$$

Using the definition of Γ,

$$\Gamma(\tfrac{1}{2}+n+1) = (\tfrac{1}{2}+n)! = \frac{2n+1}{2}\cdot\frac{2n-1}{2}\cdot \ldots \tfrac{1}{2}\Gamma(\tfrac{1}{2})$$

where $\Gamma(\frac{1}{2}) = \sqrt{\pi}$, we have

$$J_{\frac{1}{2}}(x) = \sqrt{2/\pi x} \sum_{n=0}^{\infty} (-1)^n \frac{x^{2n+1}}{2^n n!(2n+1)(2n-1)\ldots 1} .$$

Multiplying 2^n into $n!$, in other words, as a result of multiplying each factor of $n!$ by 2, we have $2n(2n-2)(2n-4) \ldots 2$. Hence, when 2^n is combined with $(2n+1)(2n-1)(2n-3)(2n-7)\ldots 1$, it yields:

$(2n+1)(2n)(2n-1)(2n-2)(2n-3)\ldots 1 = (2n+1)!$ Hence,

$$J_{\frac{1}{2}}(x) = \sqrt{2/\pi x} \sum_{n=0}^{\infty} (-1)^n \frac{x^{2n+1}}{(2n+1)!},$$

where the summation shows the familiar series for $\sin x$. Finally

$$J_{\frac{1}{2}}(x) = \sqrt{2/\pi x} \quad \sin x .$$

• PROBLEM 1098

Find the solution of the differential equation:

$$(1-x^2) + 2x \frac{dy}{dx} - \lambda y = 0$$

where λ is a real number.

Solution: The Frobenius series: $y = \sum_{n=0}^{\infty} a_n x^{n+c}$, leads to

$$\sum_{n=0}^{\infty} a_n(n+c)(n+c-1)x^{n+c-2} - \sum_{n=0}^{\infty} \left[a_n(n+c)(n+c-1) + 2a_n(n+c) - \lambda\right]x^{n+c} = 0.$$

In order to relate the two sums, we remove the first two terms from the first sum and let n start from 2.

$$a_0c(c-1)x^{c-2} + a_1c(1+c)x^{c-1} + \sum_{n=2}^{\infty} a_n(n+c)(n+c-1)x^{n+c-2} - \sum_{n=0}^{\infty} a_n\left[(n+c)(n+c+1)-\lambda\right]x^{n+c}$$
$$= 0.$$

Standardizing the notation, by replacing n by $n+2$ in the first summation, we have:

$$a_0c(c-1)x^{c-2}+a_1c(c+1)x^{c-1}+ \sum_{n=0}^{\infty} \{a_{n+2}(n+c+2)(n+c+1)-a_n\left[(n+c)(n+c+1)-\lambda\right]\}x^{n+c}$$
$$= 0.$$

The indicial equation $a_0c(c-1) = 0$ gives rise to the roots $c_1 = 0$, $c_2 = 1$, $a_0 \neq 0$. The equation for a_1 (power x^{c-1}) is: $a_1c(c+1) = 0$, and yields $a_1 = 0$ if $c_2 = 1$. a_1 is arbitrary for $c_1 = 0$. The recursion formula is:

$$a_{n+2} = \frac{\left[(n+c)(n+c+1)-\lambda\right]}{(n+c+2)(n+c+1)} a_n ,$$

or, by replacing n by $n-2$ throughout,

$$a_n = \frac{(n+c-2)(n+c-1)-\lambda}{(n+c)(n+c-1)} a_{n-2}.$$

The root $c = 0$ yields:

$$a_n = \frac{(n-2)(n-1)-\lambda}{n(n-1)} a_{n-2} ,$$

giving rise to the general solution of the form:

$$y_1 = \sum_{n=\text{even}}^{\infty} f(a_0)x^n + \sum_{n=\text{odd}}^{\infty} f(a_1)x^n$$

where $a_n = f(a_0)$ and $a_n = f(a_1)$ for even and odd n respectively. The root $c = 1$ yields the odd part of this solution (since for $c \neq 0$, $a_1 = 0$ as shown). Hence, we have:

$$y_2 = \sum_{n=\text{even}} f(a_0')x^n .$$

Considering the convergence by the ratio test:

$$\left| \frac{a_n'x^n}{a_{n-2}'x^{n-2}} \right| = \left| \frac{(n-2)(n-1)-\lambda}{n(n-1)} \right| \cdot x^2 .$$

Both series converge for $|x| < 1$, whatever the value of λ may be, and by the integral test, using the dummy variable t to represent n, we have

$$\int^M \frac{(t-2)(t-1)-\lambda}{t(t-1)} dt = \int^M \frac{(t-2)}{t} dt - \lambda\int^M \frac{dt}{t(t-1)} .$$

Since

$$\int^M \frac{t-2}{t} dt \to 0 \quad \text{as} \quad M \to \infty .$$

Both series diverge for $|x| = 1$. There is, however, an exception: if λ happens to be of the form $\lambda = \ell(\ell+1)$ where ℓ is a non-negative integer, then one of the series terminates, giving rise to a polynomial which is valid for all values of x. This kind of differential equation is called the Legendre differential equation and is one of the most important ones in physics. In particular, physical conditions usually provide a $\lambda = \ell(\ell+1)$, which assures the convergence at $x = 1$. Hence,

$$y = a_0 \sum_{n=\text{even}}^{\infty} a_n x^n + a_1 \sum_{n=\text{odd}}^{\infty} a_n' x^n$$

by setting $a_1 = 0$ (if ℓ is even) or $a_0 = 0$ (if ℓ is odd).

Furthermore, it is customary to standardize the solutions by the following choice of coefficients with the lowest powers of x:

$$a_0 a_0' = (-1)^{\ell/2} \frac{\ell!}{2^{\ell}[(\ell/2)!]^2} , \quad (\ell = \text{even})$$

$$a_1 a_1' = (-1)^{(\ell-1)/2} \frac{\ell!}{2^{\ell-1}\left[(\frac{\ell-1}{2})!\right]^2} , \quad (\ell = \text{odd}).$$

These standard solutions are denoted by $P_\ell(x)$, and are called Legendre Polynomials (of the first kind). The lowest ones are

$P_0(x) = 1$

$P_1(x) = x$

$P_2(x) = \frac{1}{2}(3x^2-1)$

$P_3(x) = \frac{1}{2}(5x^3-3x)$

$P_4(x) = \frac{1}{8}\left(35x^4-30x^2+3\right)$

etc.

Remark: So far, we have been dealing with non-homogeneous differential equations of the second order to find their solutions by the use of Frobenius method, namely, assuming the solutions are of the form

$$y = \sum_{n=0}^{\infty} a_n x^{n+c}$$

where the a_n's and c are to be determined for $n = 0,1,2,\ldots,$ and we have seen how long this process is. Hence, for simplification and efficiency, differential equations of the following forms have been worked out by mathematicians, who have compiled various tables of their solutions and corresponding generating functions. Some of these are Hermite's differential equations of the form:

$$\frac{d^2y}{dx^2} - 2x\frac{dy}{dx} + 2ny = 0,$$

with a generating function of the solution by Rodrigue's formula:

$$H_n(x) = (-1)^n e^{x^2} \frac{d^n}{dx^n}(e^{-x^2});$$

the Chebyshev's differential equation

$$(1-x^2)\frac{d^2y}{dx^2} - x\frac{dy}{dx} + n^2y = 0$$

and the solution (Chebyschev polynomial)

$$\Gamma_n(x) = \cos(n\cos^{-1}x) = x^n - \binom{n}{2}x^{n-2}(1-x^2) + \binom{n}{4}x^{n-4}(1-x^2)^2 - \ldots,$$

where in both cases $n = 0,1,2,\ldots$ and the notation

$$\binom{n}{t} = \frac{n!}{t!(n-t)!}, \quad t = 0,1,2,\ldots$$

• PROBLEM 1099

Find the solution for: $x^2\frac{d^2y}{dx^2} + 2x\frac{dy}{dx} + (x^2-2)y = 0.$

Solution: The solution is assumed to be in the form:

$$y = \sum_{n=0}^{\infty} a_n x^{n+c}.$$

Then,

$$\frac{dy}{dx} = \sum_{n=0}^{\infty} a_n(n+c)x^{n+c-1}, \quad \frac{d^2y}{dx^2} = \sum_{n=0}^{\infty} a_n(n+c)(n+c-1)x^{n+c-2}.$$

Substitution in the given differential equation gives:

$$\sum_{n=0}^{\infty} a_n(n+c)(n+c-1)x^{n+c} + 2\sum_{n=0}^{\infty} a_n(n+c)x^{n+c} + \sum_{n=0}^{\infty} a_n x^{n+c+2} - 2\sum_{n=0}^{\infty} a_n x^{n+c} = 0.$$

To compare the coefficients of various powers of x it is convenient to standardize the notation, so that the powers of x have the form x^{n+c} in the sum involving x^{n+c+2}. The differential equation, after combining like terms and replacing n by $n-2$ in the third sum, assumes the form:

$$\sum_{n=0}^{\infty} a_n\left[(c+n)(c+n-1) + 2(n+c)-2\right]x^{n+c} + \sum_{n=2}^{\infty} a_{n-2}x^{n+c} = 0.$$

After factoring out and recombining, we have:

$$\sum_{n=0}^{\infty} a_n\left[(c+n)(c+n+1)-2\right]x^{n+c} + \sum_{n=2}^{\infty} a_{n-2}x^{n+c} = 0 .$$

We write separately the powers of x which appear in the first but not in the second sum, namely, the terms with $n = 0$.

$$a_0\left[c(c+1)-2\right]x^c + a_1\left[(c+1)(c+2)-2\right]x^{c+1} + \sum_{n=2}^{\infty} \{a_n\left[(c+n)(c+n+1)-2\right] + a_{n-2}\}x^{n+c} =$$

It is necessary, first of all, that

$$a_0\left[c(c+1)-2\right] = 0.$$

Since $a_0 \neq 0$ (x^c is assumed to be the lowest power of x appearing in the series), it follows that:

$$c(c+1)-2 = 0$$

such an equation arising from the lowest power of x on the left-hand side of the differential equation is called the <u>indicial equation</u>: it determines possible values of c. In this particular case, $c = 1$ and $c = -2$. The second condition to be satisfied is:

$$a_1\left[(c+1)(c+2)-2\right] = 0.$$

Whether $c = 1$ or $c = -2$, it follows from this equation that a_1 must be zero. The next condition, valid for the values of $n \geq 2$, reads:

$$a_n\left[(c+n)(c+n+1)-2\right] + a_{n-2} = 0,$$

or

$$a_n = \frac{1}{2 - (c+n)(c+n+1)}\, a_{n-2} \qquad (n \geq 2).$$

This recursion formula determines all other coefficients. (Note that the denominator must not vanish if $n \geq 2$).
For $c = 1$, we have:

$$a_2 = \frac{1}{2 - 3.4}\, a_0 = -\frac{1}{10}\, a_0$$

$$a_4 = \frac{1}{2 - 5.6}\, a_2 = \frac{1}{280}\, a_0 \text{ , etc.}$$

Evidently, by virtue of $a_1 = 0$, $a_3 = a_5 = a_7 = \ldots = 0$).

This yields the series

$$y_1 = a_0\left(x - \frac{x^3}{10} + \frac{x^5}{280} - \ldots\right),$$

which resulted from

$$y = \sum_{n=0}^{\infty} a_n x^{n+c} = a_0 x^c + a_1 x^{c+1} + \ldots ; \; c = 1, \; a_1 = a_3 = \ldots = 0, \; a_n = f(a_0).$$

The series is also absolutely convergent (by ratio test) for all values of x. Note, if we set $a_0 = 1/3$, then the function becomes:

$$y_1 = \frac{\sin x}{x^2} - \frac{\cos x}{x},$$

which is called the spherical Bessel function of order one. Now, for $c = -2$, we have

$$a'_2 = \frac{1}{2-0} a'_0 = \tfrac{1}{2} a'_0$$

$$a'_4 = \frac{1}{2-2.3} a'_2 = -\frac{1}{8} a'_0, \text{ etc.}$$

Again $a'_3 = a'_5 = \ldots = 0$. The series is

$$y_2 = a'_0 \left(\frac{1}{x^2} + \frac{1}{2} - \frac{x^2}{8} + \ldots \right)$$

and is absolutely convergent for $|x| > 0$. It is called a Laurent series. If we set $a'_0 = -1$, then

$$y_2 = -\frac{\cos x}{x^2} - \frac{\sin x}{x} \quad \text{(spherical Neumann function of order one)}$$

The general solution is thus

$$y = y_1 + y_2 = a_0 \left(x - \frac{x^3}{10} + \frac{x^5}{280} - \ldots \right) + a'_0 \left(\frac{1}{x^2} + \frac{1}{2} - \frac{x^2}{8} + \ldots \right), \text{ or if}$$

$a_0 = 1/3$ and $a'_0 = -1$, we have

$$y = \frac{\sin x}{x^2} - \frac{\cos x}{x} - \frac{\cos x}{x^2} - \frac{\sin x}{x} = \frac{1}{x^2}\left[(1-x)\sin x - (1+x)\cos x\right]$$

• PROBLEM 1100

Using the method of Frobenius, find the general solution for:

$$x^2 \frac{d^2 y}{dx^2} + x \frac{dy}{dx} + (x^2 - \tfrac{1}{4})y = 0 .$$

Solution: Assuming the solution to be expressed as:

$$y = \sum_{n=0}^{\infty} a_n x^{n+c},$$

and substitution in the given differential equation yields:

$$\sum_{n=0}^{\infty} a_n (n+c)(n+c-1)x^{n+c} + \sum_{n=0}^{\infty} a_n (n+c)x^{n+c} + \sum_{n=0}^{\infty} a_n x^{n+c+2} - \sum_{n=0}^{\infty} \tfrac{1}{4} a_n x^{n+c} = 0.$$

Combining like terms and factoring,

$$\sum_{n=0}^{\infty} a_n \left[(n+c)^2 - \tfrac{1}{4}\right] x^{n+c} + \sum_{n=0}^{\infty} a_n x^{n+c+2} = 0.$$

In the second sum, we can replace n by $n-2$, and hence:

$$\sum_{n=0}^{\infty} a_n x^{n+c+2} = \sum_{n=2}^{\infty} a_{n-2} x^{n+c} .$$

Taking the first two terms (for $n = 0$ and $n = 1$) out from the first sum, we have:

$$\sum_{n=0}^{\infty} a_n\left[(n+c)^2-\tfrac{1}{4}\right]x^{n+c} = a_0(c^2-\tfrac{1}{4})x^c + a_1\left[(1+c)^2-\tfrac{1}{4}\right]x^{1+c} + \sum_{n=2}^{\infty} a_n\left[(n+c)^2-\tfrac{1}{4}\right]x^{n+c}.$$

Thus, the general statement appears as:

$$a_0(c^2-\tfrac{1}{4})x^c + a_1\left[(1+c)^2-\tfrac{1}{4}\right]x^{1+c} + \sum_{n=2}^{\infty} a_n\left[(n+c)^2-\tfrac{1}{4}\right]x^{n+c} + \sum_{n=2}^{\infty} a_{n-2}x^{n+c} = 0,$$

or

$$a_0(c^2-\tfrac{1}{4})x^c + a_1\left[(1+c)^2-\tfrac{1}{4}\right]x^{1+c} + \sum_{n=2}^{\infty} \{a_n\left[(n+c)^2-\tfrac{1}{4}\right] + a_{n-2}\}x^{n+c} = 0.$$

The indicial equation is:

$a_0(c^2-\tfrac{1}{4}) = 0$, yielding the roots $c_1 = \tfrac{1}{2}$ and $c_2 = -\tfrac{1}{2}$ for $a_0 \neq 0$. Considering $c_1 = \tfrac{1}{2}$, the second term:

$a_1\left[(1+c)^2-\tfrac{1}{4}\right] = 0$ yields $a_1(g/4-\tfrac{1}{4}) = 0$, so that $a_1 = 0$.

For any $n \geq 2$, we have:

$$\sum_{n=2}^{\infty} \{a_n\left[(n+\tfrac{1}{2})^2-\tfrac{1}{4}\right] + a_{n-2}\}x^{n+c} = 0.$$

Thus,

$$a_n = \frac{1}{\tfrac{1}{4} - (n+\tfrac{1}{2})^2}\, a_{n-2}\, .$$

Expanding, $\tfrac{1}{4} - (n+\tfrac{1}{2})^2 = -n(n+1)$, we have:

$$a_n = -\frac{1}{n(n+1)}\, a_{n-2}.$$

Since we know that $a_1 = 0$, the above recursion formula holds for even numbers of n, where

$$a_n = -\frac{1}{2\cdot 3}\, a_0\ ,\ a_4 = \frac{1}{2\cdot 3\cdot 4\cdot 5}\, a_0,\ldots,a_{2n} = \frac{(-1)^n}{(n+1)!}\, a_0\ , \text{ and}$$

$a_1 = a_3 = a_5 = \ldots = 0$. We can replace n by $2n$, such that

$a_{2n} = -\dfrac{1}{2n(2n+1)}\, a_{2(n-1)}$, and the general solution is

$$y_1 = a_0x^{\frac{1}{2}} \sum_{n=0}^{\infty} \frac{(-1)^n}{(2n+1)!}\, x^{2n}\ .$$

Furthermore, for $c_2 = -\tfrac{1}{2}$, the second root from the indicial equation, the recursion formula is just:

$$a'_n = \frac{1}{\tfrac{1}{4} - (n-\tfrac{1}{2})^2}\, a'_{n-2} = -\frac{1}{n(n-1)}\, a'_{n-2}\ ..$$

We recall that the x^{1+c} coefficient, i.e., that $a_1\left[(1+c)^2-\tfrac{1}{4}\right] = 0$, and $(1+c)^2-\tfrac{1}{4} = 0$. For $c = -\tfrac{1}{2}$, a'_1 need not be zero. It is thus no longer true that all odd coefficients vanish. The series contains both even and odd powers of x, and hence by:

$$a'_n = -\frac{1}{n(n-1)}\, a'_{n-2}\ ,$$

we obtain the following terms:

$$a_2' = -\frac{1}{2\cdot1}a_0'$$

$$a_4' = -\frac{1}{4\cdot3}a_2' = \frac{1}{4\cdot3\cdot2\cdot1}a_0'.$$

For the nth term,

$$a_n' = \frac{(-1)^n}{n!}a_0', \quad (n = \text{even}).$$

Or, we can write:

$$a_{2n}' = \frac{(-1)^n}{(2n)!}a_0', \quad n = 0,1,2,\ldots$$

and hence, the corresponding series is:

$$y_{2_1} = a_0'x^{-\frac{1}{2}}\sum_{n=0}^{\infty}\frac{(-1)^n}{(2n)!}x^{2n},$$

or,

$$y_{2_1} = a_0'x^{-\frac{1}{2}}\cos x.$$

$$a_3' = -\frac{1}{3\cdot2}a_1'$$

$$a_5' = -\frac{1}{5\cdot4}a_3' = \frac{1}{5\cdot4\cdot3\cdot2}a_1'.$$

For the nth term,

$$a_n' = \frac{(-1)^n}{n!}a_1', \quad (n = \text{odd}).$$

In other words,

$$a_{2n+1}' = \frac{(-1)^n}{(2n+1)!}a_1', \quad n = 0,1,2,\ldots$$

yielding the series:

$$y_{2_2} = a_1'x^{-\frac{1}{2}}\sum_{n=0}^{\infty}\frac{(-1)^n}{(2n+1)!}x^{2n+1}.$$

This gives

$$y_{2_2} = a_1'x^{-\frac{1}{2}}\sin x$$

Finally, we have the second solution

$$y_2 = x^{-\frac{1}{2}}(a_0'\cos x + a_1'\sin x).$$

Recalling the first solution,

$$y_1 = a_0x^{\frac{1}{2}}\sum_{n=0}^{\infty}\frac{(-1)^n}{(2n+1)!}x^{2n},$$

then, the general solution is given compactly by:

$$y = x^{-\frac{1}{2}}\left[(a_0x)\sum_{n=0}^{\infty}\frac{(-1)^n}{(2n+1)!}x^{2n} + a_0'\cos x + a_1'\sin x\right].$$

Further: the functions:

$$J_{-\frac{1}{2}}(x) = \sqrt{2/\pi}\,x^{-\frac{1}{2}}\cos x, \quad \text{and}$$

$$J_{\frac{1}{2}}(x) = \sqrt{2/\pi}\,x^{\frac{1}{2}}\sin x,$$

are known as the Bessel functions (of the first kind) of order of $-\frac{1}{2}$ and $\frac{1}{2}$, respectively, and the solution can be written as:

$$y = a_0\sum_{n=0}^{\infty}\frac{(-1)^n x^{(4n+1)/2}}{(2n+1)!} + AJ_{\frac{1}{2}}(x) + BJ_{-\frac{1}{2}}(x).$$

Remark: The differential equation $x^2\frac{d^2y}{dx^2} + x\frac{dy}{dx} + (x^2-\nu^2)y = 0$ (ν = any real number or even complex) is known as the Bessel differential equation of order ν. Its solutions, known as the cylinderical functions (or Bessel function of various kinds), occur in many physical problems.

• PROBLEM 1101

Show that the Bernoulli's equation, $\frac{dy}{dx} + py = Qy^n$, where p and Q are functions of x only, and n is a constant other than zero or unity, may be transformed into a linear equation by the substitution $t = y^{1-n}$. Hence solve the equation

$$\frac{dy}{dx} = \frac{x^2y^2+2y}{x}.$$

If n is zero or unity, how may the equation be solved?

Solution: Setting $t = y^{1-n}$, we obtain: $y = t^{1/(1-n)}$, from which

$$\frac{dy}{dx} = \frac{1}{1-n} t^{\frac{1}{1-n}-1} \frac{dt}{dx} = \frac{1}{1-n} t^{\frac{n}{1-n}} \frac{dt}{dx}.$$

Substituting $1/(1-n)t^{n/(1-n)}\frac{dt}{dx}$ for $\frac{dy}{dx}$ and $t^{1/(1-n)}$ for y in Bernoulli's equation yields:

$$\frac{1}{1-n} t^{\frac{n}{1-n}} \frac{dt}{dx} + pt^{\frac{1}{1-n}} = Qt^{\frac{n}{1-n}},$$

where, upon division the whole equation by $1/(1-n)t^{n/(1-n)}$, we obtain:

$$\frac{dt}{dx} + (1-n)p\ t^{[1/(1-n) - n/(1-n)]} = (1-n)Q\ t^{[n/(1-n) - n/(1-n)]}$$

or

$$\frac{dt}{dx} + (1-n)p\ t = (1-n)Q.$$

The above equation is thus a linear non-homogeneous differential equation, the integrating factor for which can be calculated as:

$$\text{I.f.} = e^{(1-n)/pdx}.$$

By the familiar expression for the general solution of the equation of the type:

$$\frac{dy}{dx} + py = Q,$$

we have that:

$$y = e^{-\int pdx} \int Q\, e^{\int pdx} + ce^{-\int pdx}.$$

Thus the corresponding solution for t is:

$$t = y^{1/(1-n)} = \exp\left[(n-1)\int pdx\right]\int Q\cdot\exp\left[(1-n)\int pdx\right]dx + c\ \exp\left[(n-1)\int pdx\right] \ldots (1)$$

To solve the equation:

$$\frac{dy}{dx} = \frac{x^2y^2+2y}{x},$$

we rewrite

$$\frac{dy}{dx} - \frac{2}{x}y = xy^2,$$

a Bernoulli's equation. Setting $t = y^{1-2} = 1/y$ and, using the same argument as above, the differential equation reduces to:

$$\frac{dt}{dx} + \frac{2}{x}t = -x.$$

For the solution for t, we use the integrating factor:

$$e^{\int pdx} = e^{\int \frac{2}{x}\, dx} = e^{2\ln x}.$$

Since $e^{2\ln x} = x^2$, $e^{\int pdx} = x^2$. Now,

$$t = x^{-2} \int -x \cdot x^2 dx + cx^{-2}$$

$$= \frac{c}{x^2} - \tfrac{1}{4} x^2$$

$$= \frac{4c - x^4}{4x^2}.$$

Using the relationship: $t = 1/y$, and substituting for t,

$$y = \frac{4x^2}{4c - x^4},$$

or,

$$4x^2 + x^4 y = c_1 y \quad \text{(as a general solution)},$$

where $c_1 = 4c$. For the case of $n = 0$, we simply substitute $n = 0$ in the Bernoulli equation (the general solution for n, different from unity), and we obtain :

$$t = y^{1/(1-0)} = \exp\left(-\int pdx\right)\int Q \cdot \exp\left(\int pdx\right)dx + c\, \exp\left(-\int pdx\right)$$

or

$$y = e^{-\int pdx} \int Q\, e^{\int pdx}\, dx + ce^{-\int pdx}$$

which is a general solution of the form

$$\frac{dy}{dx} + py = Q = Qy^0.$$ This

suggests that Bernoulli's general solution is valid for $n = 0$. For $n = 1$, the solution does not hold, and hence we study the differential equation under this condition. Bernoulli's equation for $n = 1$ is:

$$\frac{dy}{dx} + py = Qy,$$

and this can be factored as:

$$\frac{dy}{dx} + (p - Q)y = 0,$$

which is now reduced to a homogeneous differential equation. Since p and Q are constants or functions of x only, $p - Q$ can be combined into one variable, say r, and hence:

$$\frac{dy}{dx} + ry = 0,$$

resulting in

$$\ln y + \int rdx = \ln c,$$

or

$$y = c\, e^{-\int rdx}$$

CHAPTER 43

APPLIED PROBLEMS IN DIFFERENTIAL EQUATIONS

• PROBLEM 1102

If the population of the earth was found to be 3.5 billion in 1970, and is increasing at a rate of 2% per year, when will a population of 50 billion be reached?

Solution: If P is the population, then

$$\frac{dP}{dt} = kP = 0.02P$$

$$\frac{dP}{P} = 0.02dt .$$

$$\int \frac{dP}{P} = \int 0.02dt .$$

$$\ln P = 0.02t + C .$$

$$P = e^{0.02t+C} = e^C e^{0.02t} .$$

$$P = P_0 e^{0.2t} . (P_0 \text{ is the population when } t = 0 .)$$

We are measuring time from 1970, when $P_0 = 3.5$ (in billions). Hence,

$$P = 3.5e^{0.02t} .$$

Setting $P = 50$ billion,

$$50 = 3.5e^{0.02t} .$$

$$e^{0.02t} = \frac{50}{3.5} .$$

$$0.02t = \ln\frac{50}{3.5} .$$

$$t = \frac{\ln \frac{50}{3.5}}{0.02}$$

$$= \frac{\ln 50 - \ln(3.5)}{0.02}$$

$$= \frac{3.91 - 1.25}{0.02}$$

$$= 133 \text{ years.}$$

Hence the population will reach 50 billion in 1970 + 133 years or 2103 A.D.

• PROBLEM 1103

If it is assumed that the earth cannot support a population greater than 20 billion persons, and that the rate of population growth is proportional to the difference between how close the world population is to this limiting value, what is the mathematical expression describing the world population as a function of time?

Solution: If P is the world population, then, according to the described model, $\frac{dP}{dt} = k(P - 20)$, where k is negative to make $\frac{dP}{dt}$ positive, since P must be less than 20 billion.

Then,

$$\frac{dP}{P - 20} = kdt .$$

$$\int \frac{dP}{P - 20} = \int kdt .$$

$$\ln|P - 20| = kt + C .$$

$$|P - 20| = e^{kt+C} = e^C e^{kt} .$$

Let

$$e^C = B, \text{ then}$$

$$|P - 20| = Be^{kt} .$$

However, $|P - 20| = -(P - 20)$, since P is assumed less than 20 billion. Thus,

$$-(P - 20) = Be^{kt} .$$

$$P - 20 = -Be^{kt} .$$

$$P = 20 - Be^{kt},$$

where B is positive and k is negative.

What does B represent? Let us assume that $P = 3.5$ billion in 1970. If we call $t = 0$ in 1970,

$$3.5 = 20 - Be^{k \cdot 0} .$$

$$3.5 = 20 - B .$$

$$B = 16.5 .$$

B is the difference between the world population at $t = 0$ and the 20-billion limit.

$$P = 20 - 16.5e^{kt} .$$

• PROBLEM 1104

Radioactive substances are those elements that naturally break down into other elements, releasing energy as they do. The rate at which such a substance decays is proportional to the mass of the material present. If A is the amount present, then $\frac{dA}{dt} = -kA$, where k is positive and constant.

The problem is to find A, the amount present, as a function of the time t.

Solution: Since $\frac{dA}{dt} = -kA$, by separation of variables,

$$\frac{dA}{A} = -kdt.$$

$$\int \frac{dA}{A} = \int -kdt.$$

$$\ln A = -kt + C.$$

$$A = e^{-kt+C} = e^{C}e^{-kt} .$$

Letting

$$e^{C} = A_0 ,$$

$$A = A_0 e^{-kt} .$$

Notice that when $t = 0$, $A = A_0$, hence A_0 is the amount of substance present when $t = 0$.

• PROBLEM 1105

If the marginal cost of producing a certain item is

$$\frac{dy}{dx} = y' = 3 + x + \frac{e^{-x}}{4} ,$$

what is the cost of producing 1 item if there is a fixed cost of $4?

Solution:

$$y = \int y' \, dx = \int \left(3 + x + \frac{e^{-x}}{4}\right) dx$$

$$= 3x + \frac{x^2}{2} - \frac{e^{-x}}{4} + C.$$

Solving for the constant C, we know that the fixed cost is 4 when $x = 0$. Therefore, $y = 4$. Hence,

$$y = 4 = 0 + \frac{0^2}{2} - \frac{1}{4} + C.$$

Thus,

$$C = 4 + \frac{1}{4} = \frac{17}{4} ,$$

and

$$y = 3x + \frac{x^2}{2} - \frac{e^{-x}}{4} + \frac{17}{4} .$$

when $x = 1$,

$$y = 3 + \frac{1}{2} - \frac{1}{4e} + \frac{17}{4}$$

$$= \frac{12 + 2 + 17}{4} - \frac{1}{4e}$$

$$= \frac{31}{4} - \frac{1}{4e}$$

$$= \frac{31e - 1}{4e} .$$

• PROBLEM 1106

Let the marginal revenue (how much more revenue is obtained by selling one more item) relative to the sales of a certain object be given by

$$\frac{dR}{dx} = 4x^2 - 3x + 2.$$

What is the total revenue produced by the sale of 5 of these items?

Solution: Since $\frac{dR}{dx} = 4x^2 - 3x + 2$,

$$dR = (4x^2 - 3x + 2)dx.$$

$$R = \int(4x^2 - 3x + 2)dx$$

$$= \frac{4x^3}{3} - \frac{3x^2}{2} + 2x + C.$$

To solve for the constant C, we note that $R = 0$ if $x = 0$. Thus

$$0 = \frac{4}{3}(0)^3 - \frac{3}{2}(0)^2 + 2(0) + C, \text{ and}$$

$$C = 0.$$

Then,

$$R = \frac{4x^3}{3} - \frac{3x^2}{2} + 2x.$$

If $x = 5$,

$$R = \frac{4}{3}(125) - \frac{3}{2}(25) + 2(5)$$

$$= \frac{500}{3} - \frac{75}{2} + 10$$

$$= \frac{1,000 - 225 + 60}{6}$$

$$= \frac{835}{6}.$$

• PROBLEM 1107

Find the equation in polar coordinates of the curve through the point $(a,\pi/4)$, from which is derived the relation $dr/d\theta = (a^2/r) \cos 2\theta$.

Solution: Separating the variables r and θ, the given relation becomes

$$rdr = a^2\cos 2\theta d\theta.$$

Integrating

$$\frac{r^2}{2} = \frac{a^2}{2} \sin 2\theta + C.$$

Here, C is an arbitrary constant of integration. The expression thus represents a family of curves, any one of which can give rise to the original given equation. However, only one of these curves passes through the given point $(a,\pi/4)$.

Therefore, substitution of $(a,\pi/4)$ in the equation gives:

$$\frac{a^2}{2} = \frac{a^2}{2} \sin \frac{\pi}{2} + C,$$

from which $C = 0$.

Therefore the desired equation is

$$r^2 = a^2\sin 2\theta.$$

• **PROBLEM 1108**

The slope of a certain curve at any point is the square of the reciprocal of the abscissa of the point. If the curve passes through (2, 4), find its equation.

Solution: Writing the conditions expressed in the problem in mathematical form,

$$(1) \qquad \frac{dy}{dx} = \frac{1}{x^2} .$$

Rewriting in differential notation,

$$(2) \qquad dy = \frac{dx}{x^2} .$$

Integrating each member of this equation, we have

$$(3) \qquad y = -\frac{1}{x} + C,$$

where C represents both constants of integration, one of which might otherwise have been written on each side of the equation.

The equation (3) represents a family of curves such that the slope of each satisfies the condition (1). To find the particular member of this family that passes through (2, 4) we impose the condition that (2, 4) satisfy (3) and thereby find that $C = 9/2$. Therefore, the desired equation is

$$y = -\frac{1}{x} + \frac{9}{2} , \quad \text{or} \quad 2xy - 9x + 2 = 0,$$

which is a *rectangular hyperbola.*

• **PROBLEM 1109**

Find the orthogonal trajectories of the system of circles which pass through the origin and have their centers on the X-axis.

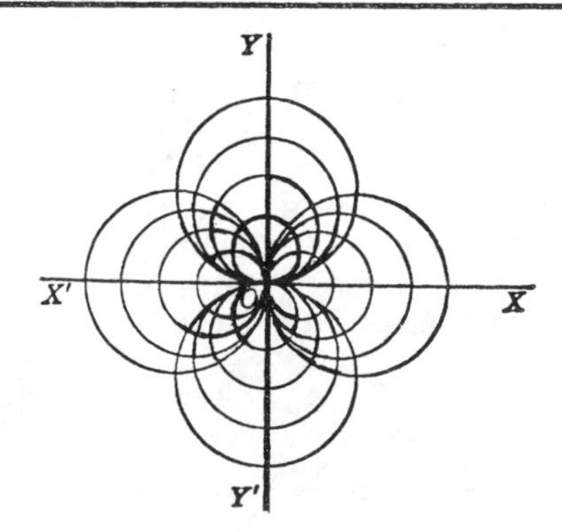

Solution: The equation of the system of circles given in the figure is

$$x^2 + y^2 = 2ax,$$

from which, upon transposing and completing the

square, we obtain

$$y^2 + (x-a)^2 = a^2.$$

This is the desired family of circles centered on the X-axis, namely at points (a,0). The parameter of the system, a, can assume positive or negative values. By differentiation,

$$x + yy' = a.$$

To eliminate the parameter a, we substitute this value of a in the given equation:

$$x^2 + y^2 = 2x(x+yy') \text{ or } y' = \frac{y^2-x^2}{2xy}.$$

The slope of the circles is thus

$$\frac{dy}{dx} = y' = \frac{a-x}{y} = \frac{y^2-x^2}{2xy}.$$

The negative of the reciprocal of this slope is:

$$\frac{-2xy}{y^2-x^2} \text{ or } \frac{2xy}{x^2-y^2}.$$

This is the slope of all curves having perpendicular intersections with the given circles. These are the orthogonal trajectories, the equations of which were to be found. For these,

$$\frac{dy}{dx} = \frac{2xy}{x^2-y^2}, \text{ or } (x^2-y^2)dy - 2xydx = 0.$$

This equation has homogeneous coefficients of the second degree, and the relation: $f(xt,yt) = t^2f(x,y)$ is satisfied. Thus we set $x = vy$. Then $dx = ydv + vdy$. The equation becomes:

$$(v^2y^2-y^2)dy - 2yvy(ydv+vdy) = 0,$$

canceling y^2 and arranging like terms results in:

$$(1+v^2)dy + 2vydv = 0.$$

Separation of variables and integration gives

$$\ln y + \ln(1+v^2) = \ln c_1$$

or

$$\ln\left[y(1+v^2)\right] = \ln c_1.$$

Let $\ln c_1 = 2c$. Then

$$y(1+v^2) = 2c.$$

Substituting back for v, and after some algebraic manipulation, we get

$$\frac{1}{y}\left(y^2+x^2\right) = 2c$$

or

$$x^2 + y^2 = 2cy.$$

This is a family of circles centered on the Y-axis at points (0,c) and passing through the origin. The two families of curves always intersect orthogonally.

• PROBLEM 1110

Find an equation of a curve, the slope of which is equal to 2x at any point on the curve.

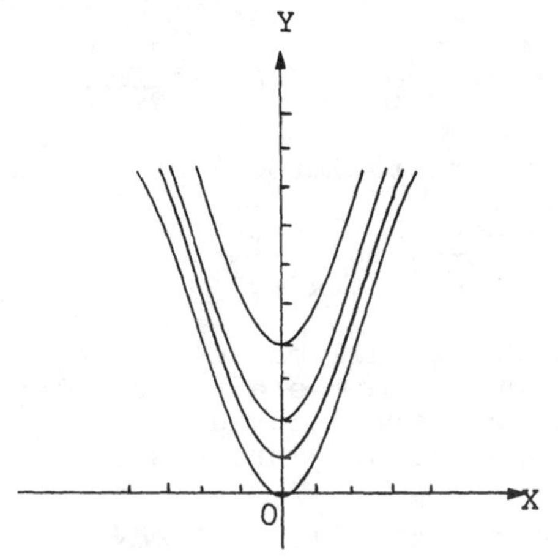

Solution: We know that the slope m is equal to $y' = \frac{dy}{dx}$. Thus $y' = 2x$. Integrating, we get

$$y = x^2 + C.$$

If we choose C = 0, we get the curve

$$y = x^2.$$

If we choose C = 1, we get the curve

$$y = x^2 + 1,$$

which has the same slope as the curve $y = x^2$ but has an ordinate which is always one greater than the ordinate for the corresponding x values of the curve $y = x^2$. $y = x^2 + C$ is called the general solution of the differential equation $\frac{dy}{dx} = 2x$, the constant C can have any desired value for a family of curves, all of which have a slope of 2x.

Find the curves that cut the circles $x^2 + y^2 = c$ at angles of 45°.

Solution: If we let m_1 be the slope of a curve c_1 at a point, and m_2 the slope of another curve perpendicular to c_1 at the same point, then the relationship of the slopes of the two curves is expressed by

$$m_1 = dy/dx = \tan\theta,$$

and

$$m_2 = \tan(90+\theta) = -\cot\theta = -\frac{1}{\tan\theta}$$

or

$$m_2 = -\frac{1}{m_1} = -\frac{1}{dy/dx}.$$

Similarly for intersections by 45° we find m_3, where

$$m_3 = \tan(45^\circ+\theta) = \frac{\tan45^\circ+\tan\theta}{1-\tan45^\circ\tan\theta},$$

or

$$m_3 = \frac{1+m_1}{1-m_1} = \frac{1+dy/dx}{1-dy/dx}.$$

For the given family of circles, $dy/dx = -x/y$ and, for the family of curves to be found: $m_3 = \frac{1-x/y}{1+x/y} = \frac{y-x}{y+x}$.

To find the equation of these curves for the same x- and y-axes

$$\frac{dy}{dx} = \frac{y-x}{y+x}.$$

The above equation is homogeneous of degree zero. Hence we set $y = vx$, $dy/dx = v + x\frac{dv}{dx}$. Substitution gives:

$$v+x\frac{dv}{dx} = \frac{vx-x}{vx+x} = \frac{v-1}{v+1}$$

which can be rewritten as:

$$\frac{dx}{x} = -\left(\frac{1+v}{1+v^2}\right)dv.$$

Integrating the entire equation and using substitution,

$$\ln x = -\tan^{-1}v - \tfrac{1}{2}\ln(v^2+1) + \ln c'.$$

Substituting the value of v,

$$\ln x = -\tan^{-1}y/x - \tfrac{1}{2}\ln[(y/x)^2+1] + \ln c'$$

$$= -\tan^{-1}y/x - \ln\left[\frac{y^2+x^2}{x^2}\right]^{\frac{1}{2}} + \ln c'.$$

Gathering like terms, we obtain:

$$\ln\frac{1}{c'}\sqrt{x^2+y^2} + \tan^{-1}y/x = 0.$$

This is the equation of a family of curves, with a parameter c', which intersects the circles at 45°.

A point moves on a curve in the x-y plane in such a way that the angle made by the tangent to the curve with the x-axis is three times the angle between the radius vector and the x-axis. Find the Cartesian equation of the family of curves satisfying this condition.

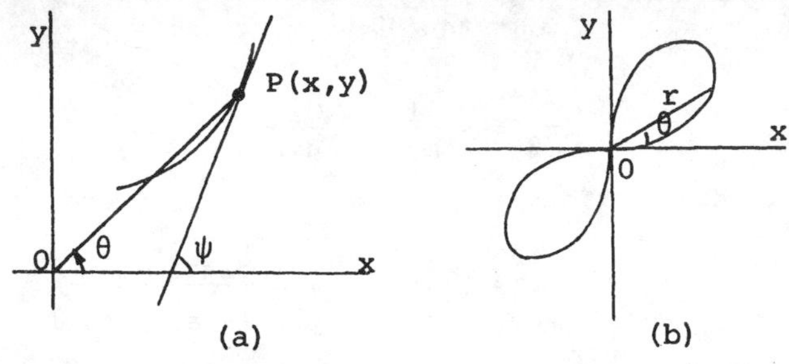

Solution: In the notation of the figure, $\psi = 3\theta$, so that

$$\frac{dy}{dx} = \tan\psi$$

$$= \tan 3\theta.$$

Using the trignometric identity:

$$\tan 3\theta = \tan(2\theta+\theta) = \frac{\tan 2\theta + \tan\theta}{1-\tan 2\theta \tan\theta},$$

and, furthermore, using the identity:

$$\tan 2\theta = \frac{2\tan\theta}{1-\tan^2\theta},$$

we have:

$$\tan 3\theta = \frac{2\frac{\tan\theta}{1-\tan^2\theta} + \tan\theta}{1-\frac{2\tan\theta}{1-\tan^2\theta}\tan\theta}$$

$$= \frac{2\tan\theta + \tan \quad - \tan^3\theta}{1-\tan^2\theta - 2\tan^2\theta}$$

$$= \tan\theta \; \frac{3-\tan^2}{1-3\tan^2\theta}$$

thus,

$$\frac{dy}{dx} = \tan \; \frac{3-\tan^2\theta}{1-3\tan^2\theta}.$$

Remembering that $\tan\theta = \frac{y}{x}$, we substitute $\frac{y}{x}$ for $\tan\theta$ in the above differential equation, and, after some algebraic manipulation, we obtain:

$$\frac{dy}{dx} = \frac{y}{x}\,\frac{3x^2 - y^2}{x^2 - 3y^2}\,.$$

This is a homogeneous differential equation of degree zero. (A function f(x,y) is said to be **homogeneous of degree n if**

$$f(tx,\ ty) = t^n f(x,y)$$

for all $t > 0$.)

Under these conditions, we wish to introduce a new variable v, in order that the differential equation has a form suitable for separating the variables.

Setting $y = vx$

$$\frac{d(vx)}{dx} = v + x\frac{dv}{dx}.$$

The differential equation becomes:

$$v + x\frac{dv}{dx} = \frac{vx}{x} \cdot \frac{3x^2 - (vx)^2}{x^2 - 3(vx)^2}$$

or

$$v + x\frac{dv}{dx} = v\frac{3-v^2}{1-3v^2}$$

and

$$x\frac{dv}{dx} = \frac{2v\left(1+v^2\right)}{1-3v^2}.$$

Now, we can easily use the method of separation of variables and, after some algebraic manipulation, we obtain

$$2\frac{dx}{x} = \frac{1-3v^2}{v\left(1+v^2\right)}dv.$$

The left-hand side can be integrated to 2 ln x or $\ln x^2$. The right-hand side is integrated by the use of partial fractions. Put

$$\frac{1-3v^2}{v\left(1+v^2\right)} = \frac{A}{v} + \frac{B+Cv}{1+v^2}.$$

Then $1-3v^2 = A\left(1+v^2\right) + (B+Cv)v$, for a range of v. Hence, equating like coefficients, we have

$$A = 1,\ B = 0,\ C = -4,$$

and $$2\frac{dx}{x} = \frac{dv}{v} - \frac{4v\,dv}{1+v^2}.$$

Integrating, we have

$$\ln x^2 = \ln v - 2 \ln(1+v^2) + \ln \lambda,$$

$\ln \lambda$ can be employed instead of an ordinary number c for subsequent use since it is an arbitrary constant. Combining the ln-terms on the right-hand side of the above equation into one, canceling, and taking the antilogarithm, we obtain:

$$x^2 = \frac{\lambda v}{(1+v^2)^2}.$$

Substituting for $v = \frac{y}{x}$ in the above equation,

$$x^2 = \frac{\lambda x^3 y}{(x^2+y^2)^2},$$

and the required family of curves is

$$(x^2+y^2)^2 = \lambda xy.$$

The form of these curves is best seen by writing this in polar coordinates using:

$$r^2 = x^2+y^2, \ x = r \cos \theta \text{ and } y = r \sin \theta.$$

We now have

$$r^2 = \frac{1}{2}\lambda \sin 2\theta.$$

A typical member of the family is shown in part (b) of the figure. The curves are Bernoulli's lemniscates.

• PROBLEM 1113

Find the equation of the family of curves having the slope of the tangent always equal to $3x^2 - 2x + 1$. Find the particular curve of the family that passes through (1,3).

Solution: In mathematical form, we write

$$\frac{dy}{dx} = 3x^2 - 2x + 1$$

or $dy = (3x^2 - 2x+1)\, dx$, (in differential form).

Integration yields:

$$y = x^3 - x^2 + x + c.$$

This is the required equation of the family. To find the particular curve that passes through (1,3), we suppose that the equation is satisfied by $x = 1$, $y = 3$. Thus

$$3 = 1^3 - 1^2 + 1 + c, \text{ and } c = 2.$$

Hence,

$$y = x^3 - x^2 + x + 2$$

is the equation of the particular curve required. The curves of the family corresponding to c = -4,0,2, and 5 are shown in the diagram. The slopes of all these curves at any value of x are the same.

In a similar way, we can find a curve for which the slope is given by any formula of the form $\frac{dy}{dx} = Dy = f(x)$, provided that we can carry out the inverse differentiation $D^{-1}f(x)$, or, in other words, integration. After performing this operation, a value of the arbitrary constant can be found so that the curve will pass through any specified point.

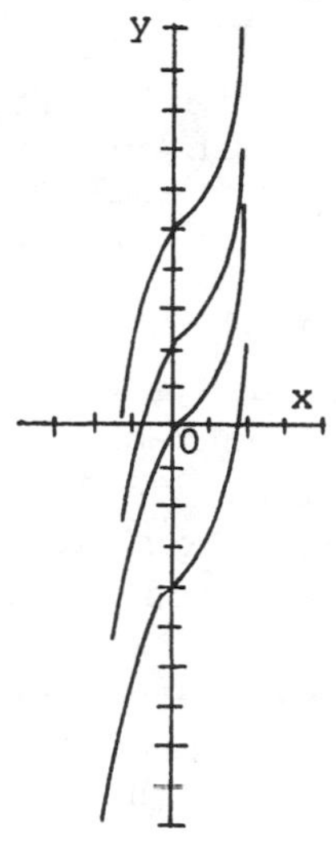

• PROBLEM 1114

Find a curve having its slope always equal to half the abscissa, and passing through (0,-3).

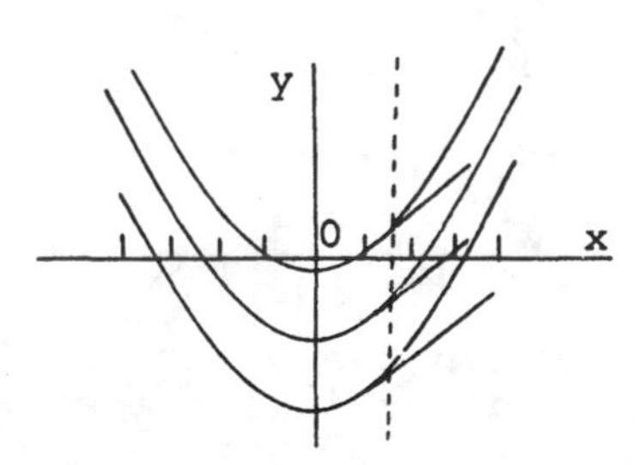

Solution: In mathematical form, we write

$$\frac{dy}{dx} = \frac{1}{2}x,$$

or

$$dy = \frac{1}{2}xdx.$$

Integration gives:

$$y = \frac{1}{4}x^2 + c.$$

There can be infinitely many curves which satisfy the condition of the problem, corresponding to the infinitely many values which c can assume, which we call the family of curves. Several of these are shown in the Figure. If we require a particular curve which passes through the point (0,-3), we must substitute 0 for x and -3 for y to determine the value of the parameter c. Upon doing so we obtain:

$$-3 = \frac{1}{4}0^2 + c, \text{ or } c = -3.$$

The equation is then

$$y = \frac{1}{4}x^2 - 3.$$

It is obvious geometrically from this example that the process of inverting differentiation ought to lead to infinitely many results, for, if one of the parabolas $y = \frac{1}{4}x^2 + c$ has its slope always equal to $\frac{1}{2}x$, so does any other in the family. The whole system may be thought of as generated by moving any one of them parallel to the Y-axis. At points of intersection with any line parallel to the Y-axis, all the slopes are equal. For example, wherever one of these curves meets the line x = 2, the slope of the tangent line is 1.

• PROBLEM 1115

Find the equation of the curve, the slope of which is 4 - 2x, and which passes through the point (2,6).

Solution: Since

$$\frac{dy}{dx} = 4 - 2x,$$

we integrate both sides.

$$\int dy = \int (4-2x)dx$$

or

$$y = 4x - x^2 + C.$$

The curve passes through the point (2,6), and if we substitute these values into the equation, we get

$$6 = 4(2) - (2)^2 + C$$

$$6 = 4 + C$$

$$2 = C$$

Therefore, $y = 4x - x^2 + 2$ is the equation of the desired curve.

• **PROBLEM 1116**

Find a first-order differential equation satisfied by all circles with center at the origin.

Solution: A circle with center at the origin and radius C satisfies the equation $x^2 + y^2 = C^2$. As C varies over all positive numbers, we obtain every circle with center at the origin. To find a first-order differential equation having these circles as integral curves, we simply differentiate the Cartesian equation to obtain

$$2xdx + 2ydy = 0$$

or $$x + yy' = 0 \text{ where } y' = \frac{dy}{dx}.$$

Thus, the entire family of circles having different values of the parameter C, satisfies the differential equation

$$y' + \frac{x}{y} = 0.$$

• **PROBLEM 1117**

Find a first-order differential equation for the family of all circles passing through the origin and having their centers on the x-axis.

Solution: If the center of a **circle** is at (C,0) and if it passes through the origin, the theorem of Pythagoras tells us that each point (x,y) on the circle satisfies the Cartesian equation $(x-C)^2 + y^2 = C^2$, which can be written as

(a) $$x^2 + y^2 - 2Cx = 0.$$

To find a differential equation having these circles as integral curves, we differentiate (a) to obtain $2x + 2yy' - 2C = 0$, or

(b) $$x + yy' = C.$$

Since this equation contains C, it is satisfied only by that circle in (a) corresponding to the same C. To obtain one differential equation satisfied by all the curves in (a), we must eliminate C. We can obtain a first-order equation by eliminating C algebraically from (a) and (b). Substituting $x + yy'$ for C in (a), we obtain $x^2 + y^2 - 2x(x+yy')$, a first order equation which can be solved for y' and written as $y' = (y^2-x^2)/(2xy)$.

• **PROBLEM 1118**

Find the equation of the curve which is perpendicular to the line joining any point on the curve to the point (3,4), if the curve also passes through the origin.

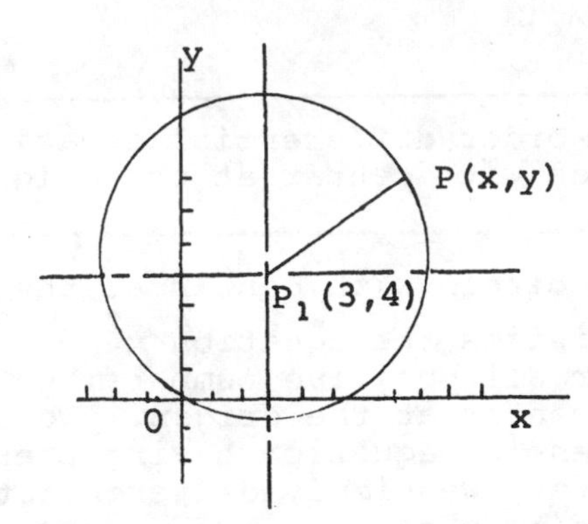

Solution: Let P_1 represent the point (3,4) and P(x,y) any point of the curve. The slope of the curve at P is the negative reciprocal of that of the line P_1P. The slope of

$$P_1P = \frac{y-4}{x-3}.$$

Therefore,
$$\frac{dy}{dx} = -\frac{x-3}{y-4}.$$

Separating the variables,

$$(y-4)dy = (3-x)dx.$$

Integration gives the family of curves

$$\frac{y^2}{2} - 4y = 3x - \frac{x^2}{2} + C.$$

Since the desired curve passes through the origin, we substitute x = 0, y = 0 in this equation and find that C = 0. Clearing and transposing, we obtain:

$$x^2 + y^2 - 6x - 8y = 0,$$

which is a circle with (3,4) as its center.

• PROBLEM 1119

If the temperature is constant, the rate of change of the atmospheric pressure at any height is proportional to the pressure at that height:

$$\frac{dp}{dh} = -kp,$$

where p = pressure, h = height.

The minus sign is used since the pressure decreases as the height increases. Express the relationship between p and h.

Solution: Since $\frac{dp}{dh} = -kp$, we wish to find the general solution of the given differential equation. We find, by separating the variables, that:

$$\frac{dp}{p} = -kdh,$$

Integrating,

$$\ln p = -kh + \ln c.$$

Here, c is an arbitrary constant of integration. Hence

$$\frac{p}{c} = e^{-kh},$$

or,

$$p = ce^{-kh}.$$

The pressure at zero elevation (h = 0) is designated p_0, and thus $p_0 = p$ (at $h = 0$) $= ce^{-k(0)} = c$. Therefore, $p = p_0 e^{-kh}$. Hence as the height increases, the pressure decreases exponentially. The maximum pressure occurs at zero elevation.

• PROBLEM 1120

For a certain curve, the length of the radius of curvature at any point p is numerically equal to the length of the normal drawn from p to the x-axis. a) Show that the differential equation has one of the forms

$$1 + y'^2 = \pm y'' y .$$

b) Explain the significance of the positive and negative signs, and show that one choice of sign leads to a family of catenaries whereas the other choice leads to a family of circles.

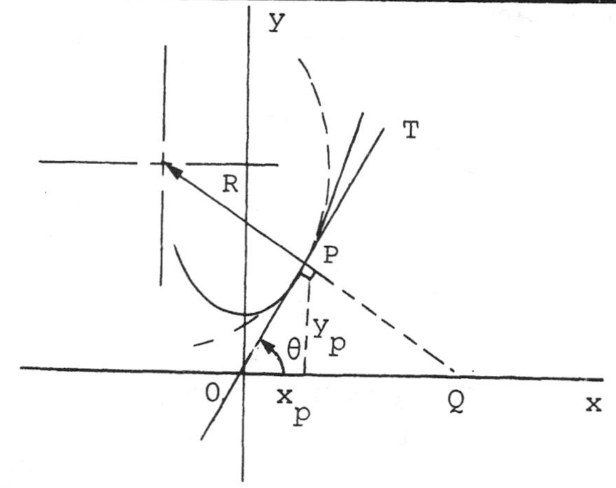

Solution: This problem states that $|PQ| = |R|$, as shown in the figure. The value of R as expressed in differential calculus at the point $P(x_p, y_p)$ is

$$|R| = \frac{\left(1+y'^2\right)^{3/2}}{|y''|}.$$

For $|PQ|$, we first find the equation of the line. Since its slope is $-\cot\theta = -1/y'$, its general equation is:

$$\left(y-y_p\right) = -1/y' \left(x-x_p\right).$$

When $y = 0$,

$$|OQ| = x_Q = y_p y' + x_p .$$

Hence,

$$|PQ| = \left[y_p^2 + \left(x_Q - x_p\right)^2\right]^{1/2} = y_p\sqrt{1+y'^2} .$$

From the relationship for $|R|$ and $|PQ|$,

$$y_p\sqrt{1+y'^2} = \frac{\left(1+y'^2\right)\sqrt{1+y'^2}}{|y''|} ,$$

canceling out like terms,

$$y_p|y''| = 1 + y'^2$$

Since the above expression holds for curves concave upward or downward, y'' can assume a positive or a negative sign. Therefore, we are not restricted to its magnitude only. As a result, the differential equation can appear, in general (for any P) as:

$$1 + y'^2 = \pm yy'' .$$

i) For $y'' > 0$, by setting $y' = dy/dx = p$, $y'' = p\frac{dp}{dy}$. Upon substitution we have:

$$yp\frac{dp}{dy} = 1 + p^2 .$$

Separation of variables and integration results in:

$$\frac{1}{2}\ln\left(1+p^2\right) = \ln y + \ln c.$$

Taking the antilogarithm,

$$\left(1+p^2\right)^{1/2} = yc, \text{ or } 1 + p^2 = y^2c^2.$$

Solving for p,

$$p = \frac{dy}{dx} = \sqrt{y^2c^2-1} .$$

By separating the variables and integrating once more, (by setting $yc = \cosh\theta$) we obtain:

$$\frac{1}{c}\cosh^{-1}y_c = x + k_1 ,$$

or,

$$y = \cosh(cx+k), \text{ where } k = k_1c.$$

This is a family of catenaries.

ii) For $y'' < 0$, we use the same procedure as above.

$$-yp\frac{dp}{dy} = 1 + p^2.$$

Integrating, separating the variables, and solving for p, we obtain:

$$p = \frac{dy}{dx} = \frac{1}{y}\sqrt{k^2-y^2} .$$

Integrating again:

$$\int \frac{y}{\sqrt{k^2-y^2}}\,dy = \int dx + c$$

or

$$-\sqrt{k^2-y^2} = x + c.$$

After rearrangement, we have:

$$(x+c)^2 + y^2 = k^2$$

a family of circles centered along the x-axis.

• PROBLEM 1121

A tank initially holds 100 gallons of brine containing 150 lbs. of salt dissolved in solution. Additional solution containing 1 lb. of salt per gallon enters the tank at the rate of 2 gal./min., and the brine, which is kept uniform by stirring, flows out at the same rate. Find the amount of salt in the tank at the end of one hour.

Solution: Let Q be the number of pounds of salt in the tank at the end of t minutes. Then dQ/dt is the rate of change in the amount of salt at time t. Since incoming salt water contains 1 pound of salt per gallon and the incoming solution flows at 2 gallons per minute, there is a gain of 2 pounds of salt a minute. Similarly, there is Q/100 pound of salt per gallon in the tank at **any time**, and, since the solution flows out at 2 gallons per minute, there is a loss of $\frac{2Q}{100}$ or 0.02Q pounds of salt a minute. The difference between the gain and the loss is the resultant rate of change of quantity of salt:

$$\text{(a)} \qquad \frac{dQ}{dt} = 2 - 0.02Q.$$

The initial condition is: Q = 150 when t = 0.

Separating variables in (a), we have

$$\frac{dQ}{0.02Q - 2} = -dt;$$

integrating,

$$\text{(b)} \qquad \ln(0.02Q - 2) = -0.02t + C.$$

Substituting Q = 150 and t = 0, we find C = ln 1 = 0. Then

$$0.02Q - 2 = e^{-0.02t},$$

or

$$\text{(c)} \qquad Q = 100 + 50e^{-0.02t}.$$

Now set t = 60 (minutes), and we find

$$Q = 100 + 50e^{-1.2} = 115.06.$$

The required amount of salt at the end of one hour is then 115.06 **pounds** . In fact, equation (a) can also be more briefly solved by the method for linear differential equations.

The general form of (a) is:

$$\frac{dy}{dx} + p(x)y = Q(x),$$

where the integrating factor is $e^{\int p(x)dx}$, with a general solution:

$$y = e^{-\int p(x)dx} \int \left[Q(x)e^{\int p(x)dx}dx + C\right].$$

Here, p(x) and Q(x) are either constants or functions of x alone.

In our case, p(x) = 0.02 and Q(x) = 2. Hence,

$$Q = e^{-\,0.02dt} \int \left(2e^{\,0.02dt}dt + C\right)$$

or

$$Q = e^{-0.02t}\left(\frac{2}{0.02}e^{0.02t} + C\right)$$

$$= 100 + Ce^{-0.02t}.$$

Thus, using the given boundary condition, i.e., the initial value, we have C = 50.

$$Q = 50e^{-0.02t} + 100,$$

as before. As t approaches infinity, Q approaches 100. Hence the concentration of the outgoing solution approaches 1 pound of salt per gallon which is equal to the concentration of the incoming solution.

• PROBLEM 1122

A substance in solution, for example, cane sugar, is decomposed in a chemical reaction into other substances through the presence of acids, and the rate at which the reaction takes place is proportional to the mass of sugar still unchanged. We then have

$$\frac{dx}{dt} = k(a-x),$$

where x is the amount of sugar converted in time t and a is the original amount of sugar. Find the dependence of the sugar converted on time t.

Solution: In the equation:

$$\frac{dx}{dt} = k(a-x),$$

we wish to find the solution by the use of separation of variables.

$$\frac{dx}{a-x} = kdt, \text{ or } \frac{dx}{x-a} = -kdt.$$

Integration yields

$$\ln(x-a) = -kt + \ln C,$$

or

$$\ln\frac{x-a}{C} = -kt,$$

then

$$x-a = Ce^{-kt},$$

is the general solution. Before the **decomposition** starts the amount of sugar in solution is a, therefore $x = 0$. Substituting the condition: $x = 0$, $t = 0$, into the above equation, $x = Ce^{-kt} + a$,

$$Ce^{-k(0)} + a = 0,$$

hence

$$C = -a.$$

The particular solution is then:

$$x-a = -ae^{-kt}$$

or

$$x = a\left(1-e^{-kt}\right).$$

All of the sugar will be decomposed into other substances as time approaches infinity, and therefore x approaches a.

• PROBLEM 1123

According to Newton's law of cooling, the rate at which a body loses heat, and therefore the change in temperature, is proportional to the difference in temperature between the body and the surrounding medium:

$$\frac{dT}{dt} = -k\left(T - T_0\right),$$

where T is the temperature of the body, T_0 is the temperature of the surrounding medium, and t is the time. Show that $T - T_0 = \left(T_1 - T_0\right)e^{-kt}$, where T_1 is the value of T when $t = 0$.

Solution: Dividing the given equation, $\frac{T}{dt} = -k\left(T-T_0\right)$, by $\left(T-T_0\right)$ and multiplying by dt, thereby separating the variables, we obtain

$$\frac{dT}{T-T_0} = -kdt.$$

Integrating, we have

$$\ln\left(T-T_0\right) = -kt + \ln C$$

$$\ln\left(T-T_0\right) - \ln C = -kt$$

$$\ln\frac{T-T_0}{C} = -kt$$

therefore, $$\frac{T-T_0}{C} = e^{-kt}, \text{ or } T-T_0 = Ce^{-kt}.$$

We now apply the initial conditions. At t = 0, the temperature of the body is T_1, from which we obtain the particular solution with a definite value of C. Remember that under arbitrary conditions, the integration constant C can assume many parametric values. Thus we get a family of curves, all of which have the property determined by the given equation. According to the initial condition,

$$T_1 = T(\text{at } t = 0) = Ce^{-k(0)} + T_0.$$

Therefore $$T_1 = C + T_0 \text{ or } C = T_1 - T_0,$$

hence $$T - T_0 = \left(T_1 - T_0\right)e^{-kt}.$$

As t approaches ∞, T goes to T_0.

• PROBLEM 1124

When light radiation enters a medium, its rate of absorption with respect to the depth of penetration t is proportional to the amount of light that is incident on a unit area at that depth. Find the law relating L, the quantity of light, and t, if the incident light is L_o and the emerging light after passing through a thickness t_1 is L_1.

Solution: The incident light is L at any penetration t. Then the rate of change of L (being absorbed) is proportional to the quantity of light L. Writing this in a mathematical form,

$$\frac{dL}{dt} = -kL$$

where the negative sign arises from the fact that the light L decreases with increasing penetration, and k is a constant of proportionality.

Separating the variables, we obtain:

$$\frac{dL}{L} = -kdt,$$

and, integrating this, we have:

$$\ln L = -kt + \ln C.$$

Rewriting $$L = Ce^{-kt}.$$

Considering the medium to be of sufficient thickness, as the penetration t approaches infinity, L will go to zero. Since at t = 0, $L = L_0$, then $C = L_0$. Also, when $t = t_1$, $L = L_0$, and therefore $L_1 = L_0 e^{-kt_1}$. Regrouping and raising the whole equation to the $\left(\frac{1}{t_1}\right)$th power we obtain

$$e^{-k} = \left(\frac{L_1}{L_0}\right)^{1/t_1},$$

from which, multiplying the exponents by t, and substituting L for $L_0 e^{-kt}$, we have

$$L = L_0\left(\frac{L_1}{L_0}\right)^{t/t_1}, \text{ for } t_1 \neq 0.$$

• PROBLEM 1125

A chain has its ends fastened at the same level to two posts 40 ft. apart. (a) Find the dip in the chain if it is 50 ft. long. (b) How long is the chain if the dip is 6 ft? (c) Find the tension at the lowest point, and at a point of suspension, of the chain of part (b) if the weight is 2 lb./ft.

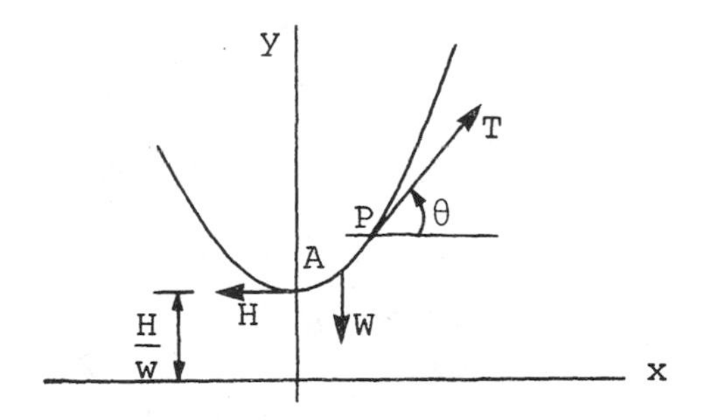

Solution: Taking a point P(x,y) on the suspended chain, it is evident that the direction of the tension force is tangent to this point. The horizontal and vertical forces are:

$$T \sin \Theta = W,$$

and

$$T \cos \Theta = H,$$

Here W is the weight of the portion of the cable from a fixed reference, H is the horizontal tensile force, and Θ is the angle which the chain makes with a horizontal line, as shown in the figure.

Division gives:

$$\tan \Theta = \frac{W}{H} ,$$

where $\tan \Theta$ is the slope of the chain. Considering the component forces at the lowest point (where the slope is zero), the weight contributes nothing ($W = 0$) but its contribution increases as the angle Θ increases. Thus, this point is the reference, from which the length of the chain is measured. We substitute $W = ws$ for the weight, where w is the weight per unit length. Hence,

$$\tan \Theta = \frac{dy}{dx} = \frac{w}{H} \; s.$$

This is the differential equation of the cable, but it contains three variables: x, y, and s. In order to eliminate s, we differentiate with respect to x and replace ds/dx by

$$\left[1 + \left(\frac{dy}{dx}\right)^2\right]^{\frac{1}{2}}.$$

Then

$$\frac{d^2y}{dx^2} = \frac{w}{H} \sqrt{1 + \left(\frac{dy}{dx}\right)^2}.$$

To solve this differential equation, we write:

$$\frac{dy}{dx} = p, \qquad \frac{d^2y}{dx^2} = \frac{dp}{dx} ,$$

and separate the variables:

$$\frac{dp}{\sqrt{1 + p^2}} = \frac{w}{H} \; dx.$$

Integrating, we have:

$$\sin h^{-1} \; p = \frac{w}{H} x + c_1.$$

We set the y-axis at the point where

$$\frac{dy}{dx} = p = 0,$$

from which s is measured. Thus, $p = 0$ for $x = 0$ implies that $c_1 = 0$. Then,

$$\frac{dy}{dx} \; \sinh^{-1} \frac{w}{H} \; x,$$

and integration yields

$$y = \frac{H}{w} \cosh \frac{w}{H} x + c_2.$$

We choose the x-axis at a distance $\frac{H}{w}$ below the lowest point on the chain, so that, when $y = \frac{H}{w}$, $x = 0$ and $c_2 = 0$. Thus, the equation of deflection of the chain is expressed by:

$$y = \frac{H}{w} \cosh \frac{w}{H} x.$$

(a) If we let $a = \frac{H}{w}$, then

$$y = a \cosh \frac{x}{a}, \quad \text{and} \quad \frac{dy}{dx} = \sinh \frac{x}{a}.$$

$$ds = \left[1 + \left(\frac{dy}{dx}\right)^2\right]^{\frac{1}{2}} dx = \left[1 + \sinh^2 \frac{x}{a}\right]^{\frac{1}{2}} dx$$

$$= \cosh \frac{x}{a} dx,$$

for the length of the chain. Observing the symmetry due to the uniform distribution of weight,

$$s = 2 \int_0^{20} \cosh \frac{x}{a} dx = 50 \text{ ft}$$

Integration gives:

$$s = 2a \sinh \frac{20}{a} = 50,$$

or $\frac{25}{a} = \sinh \frac{20}{a}$.

By plotting a graph for both $\frac{25}{a}$ and $\sinh \frac{20}{a}$ as functions of a on the same axes, we obtain the intersection point at approximately $a = 20.6$.

(a) As shown in the figure, the dip is expressed by

$$d = a\left(\cosh\frac{20}{a} - \cosh 0\right)$$

$$= 20.6\left(\cosh\frac{20}{20.6} - 1\right)$$

$$= 20.6\ (1.509 - 1)$$

$$= 10.2.$$

(b) d = 6ft., from which we determine a.

$$6 = a\left(\cosh\frac{20}{a} - 1\right),$$

and solving for a, we have a = 34.3. Hence,

$$s = 2\int_0^{20} \cosh\frac{x}{a}\,dx = 2a\ \sinh\frac{20}{a}$$

$$= 2 \times 34.3\ \sinh\frac{20}{34.3}$$

$$= 2 \times 34.3 \times 0.613$$

$$\simeq 42 \text{ ft.}$$

Using the relationship:

$$y = a\ \cosh\frac{x}{a},$$

to find the tension at x = 0,

$$y = a = \frac{H}{w} \quad \text{at } x = 0.$$

Hence, at x = 0, $H = a\,w$

$$= 34.3 \times 2$$

$$= 68.6 \text{ lb.}$$

At the point of suspension,

$$y = a \cosh \frac{20}{a} \simeq 34.3 \cosh 0.612$$

$$\simeq 40.5 \text{ ft.}$$

Hence,

$$T = wy = 81 \text{ lb.}$$

The above result is obtained by using the relationship:

$$T \cos \Theta = H.$$

Since H is the same everywhere, we can rewrite

$$T = H \sec \Theta = H \sqrt{1 + \left(\frac{dy}{dx}\right)^2}$$

$$= H \cosh \frac{x}{a}$$

$$= wy.$$

• PROBLEM 1126

Find the deflection of the free end of a uniform beam of weight w lb./in. carrying a load Q lb. at the free end.

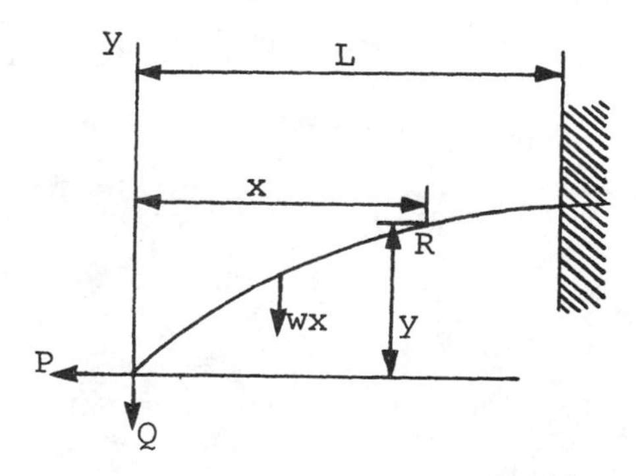

Solution: From the study of the deflection of beams, the moment of all acting forces with respect to a point in a cross-section is given by:

$$M = \frac{EI}{R},$$

where E is Young's modulus, I is the moment of inertia of the cross-sectional area of the beam with respect to a neutral axis, and R is the radius of curvature at this point.

From differential calculus, for a small bend, R can be approximated as

$$R = 1/y''.$$

Hence,

$$M = EIy''.$$

Neglecting the couple within the supporting medium at the other end, the moment at point R (considering clockwise moment positive) is:

$$M_p = Py - \frac{1}{2}x(wx) - Qx = EIy'',$$

where P is the horizontal tensile force, and

$$\frac{d^2y}{dx^2} = \frac{Py}{EI} - \frac{w}{2EI}x^2 - \frac{Q}{EI}x.$$

Setting $a^2 = P/EI$ and using the differential operator $D \equiv \frac{d}{dx}$

$$\left(D^2 - a^2\right) = -\frac{Qa^2}{p}x - \frac{wa^2}{2p}x^2.$$

The auxiliary equation of this is: $m^2 - a^2 = 0$, with roots $m = \pm a$, and its complementary solution is:

$$y_c = Ae^{ax} + Be^{-ax}.$$

The particular solution appears as:

$$y_p = c_1x^2 + c_2x + c_3.$$

$$\frac{dy_p}{dx} = 2c_1x + c_2, \quad \frac{d^2y_p}{dx^2} = 2c_1.$$

Substitution into the main differential equation yields:

$$\left(D^2 - a^2\right)y_p = 2c_1 - a^2c_1x^2 - a^2c_2x - a^2c_3 = -\frac{Qa^2}{p}x - \frac{wa^2}{2p}x^2.$$

Equating like coefficients,

$$x^2: \quad a^2c_1 = \frac{wa^2}{2p}, \quad c_1 = w/2p$$

$$x: \quad a^2c_2 = \frac{Qa^2}{p}, \quad c_2 = Q/p$$

$$1: \quad 2c_1 - a^2c_3 = 0, \quad c_3 = \frac{2c_1}{a^2} = w/pa^2.$$

Hence, the general solution is:

$$y = Ae^{ax} + Be^{-ax} + \frac{w}{2p}x^2 + \frac{Q}{p}x + \frac{w}{pa^2}.$$

But, since

$$Ae^{ax} = A(\cosh ax + \sinh ax),$$

$$Be^{-ax} = B(\cosh ax - \sinh ax),$$

$$Ae^{ax} + Be^{-ax} = (A+B)\cosh ax + (A-B)\sinh ax.$$

Calling $A+B = k_1$ and $A-B = k_2$

$$y = k_1 \cosh ax + k_2 \sinh ax + \frac{w}{2p}x^2 + \frac{Q}{p}x + \frac{w}{pa^2}.$$

When $x = 0$, $y = 0$. Hence,

$$k_1 + w/pa^2 = 0, \quad k_1 = -w/pa^2.$$

Again, when $x = L$, $dy/dx = 0$. That is:

$$ak_1 \sinh ax + ak_2 \cosh ax + \frac{w}{p}x + \frac{Q}{p}\Big|_{x=L} = 0.$$

Substitution for k_1 and solving for k_2 yields:

$$k_1 = -\frac{w}{pa^2}, \quad k_2 = \frac{w}{pa^2}\tanh aL - \frac{wL}{pa}\operatorname{sech} aL - \frac{Q}{pa}\operatorname{sech} aL.$$

Thus, substituting back the value of $a = \sqrt{p/EI}$, and calling $aL = \theta_1$ the general solution is:

$$y = \left[\frac{w}{p^2}EI \tanh\theta_1 - \frac{1}{p\sqrt{p/EI}}(wL+Q)\operatorname{sech}\theta_1\right]\sinh\theta - \frac{w}{p^2}EI\cosh\theta$$

$$+ \frac{w}{2p}x^2 + \frac{Q}{p}x + \frac{w}{p^2}EI.$$

• PROBLEM 1127

A rope is wound on a rough circular cylinder of radius a (ft.). Let the coefficient of friction be μ, let T (lb.) be the tension in the rope at any point p, and let θ be the angle with the horizontal axis. Neglecting the weight of the rope, show that, when the rope is on the point of slipping,

$$\frac{dT}{d\theta} = \mu T,$$

whence

$$T = T_0 e^{\mu\theta},$$

where T_0 is the tension at A.

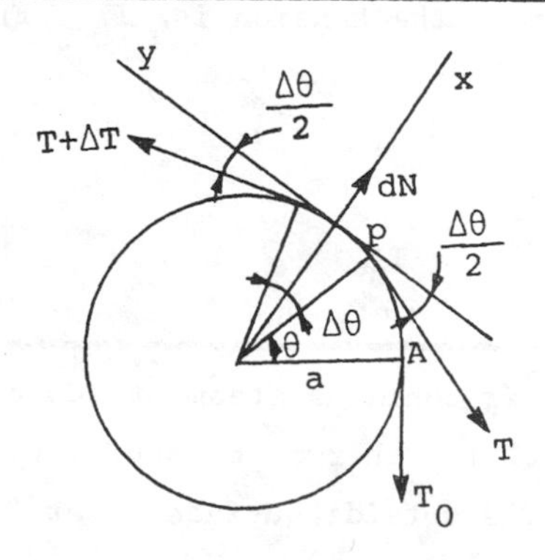

Solution: Taking the origin at point p as shown, we compute the x- and y-components of the forces acting at p. This is accurate when, in the limit, $\Delta\theta$ approaches zero from both directions.

For equilibrium conditions,

$$(T+\Delta T)\cos\frac{\Delta\theta}{2} - T\cos\frac{\Delta\theta}{2} - u\, dN = 0$$

and

$$-(T+\Delta T)\sin\frac{\Delta\theta}{2} - T\sin\frac{\Delta\theta}{2} + dN = 0 .$$

To find the limit, as $\Delta\theta$ approaches zero, we use the facts that:

$$\lim_{\Delta\theta\to 0}\cos\frac{\Delta\theta}{2} = 1,\ \lim_{\Delta\theta\to 0}\sin\frac{\Delta\theta}{2} = \frac{d\theta}{2},\ \text{and}\ \lim_{\Delta\theta\to 0}\Delta T = dT .$$

Thus, for the y-components, we have:

$$(T+dT) - T - \mu\, DN = 0,$$

or

$$dT = \mu\, dN.$$

For the x-component,

$$-(T+dT)\frac{d\theta}{2} - T\frac{d\theta}{2} + dN = 0,$$

or,

$$-T\, d\theta - dT\frac{d\theta}{2} + dN = 0.$$

Neglecting $dT\, d\theta/2$, we obtain:

$$dN = T\, d\theta .$$

Using the y-component to eliminate dN in the above expression, we finally arrive at:

$$\frac{dT}{\mu} = T\, d\theta\ ,\ \text{or}\ \frac{dT}{d\theta} = \mu T.$$

Separating variables, we wish to find the general solution of:

$$\frac{dT}{T} = \mu\, d\theta .$$

Integration yields:

$$\ln T = \mu\theta + \ln c,$$

where c is an arbitrary constant. This, in exponential form, is:

$$T = ce^{\mu\theta} .$$

At point A, where $\theta = 0$, the tension is T_0. From the above expression for $\theta = 0$,

$$T = c.$$

Thus $c = T_0$ and hence

$$T = T_0 e^{\mu\theta} .$$

• PROBLEM 1128

A pipe 10 cm. in diameter contains steam at 100°c. It is covered with asbestos 5cm. thick, the thermal conductivity of which, k = 0.00060 cal./cm$^\circ$sec. The outside surface is at 30°c. Find the heat loss per hour from a meter length of pipe.

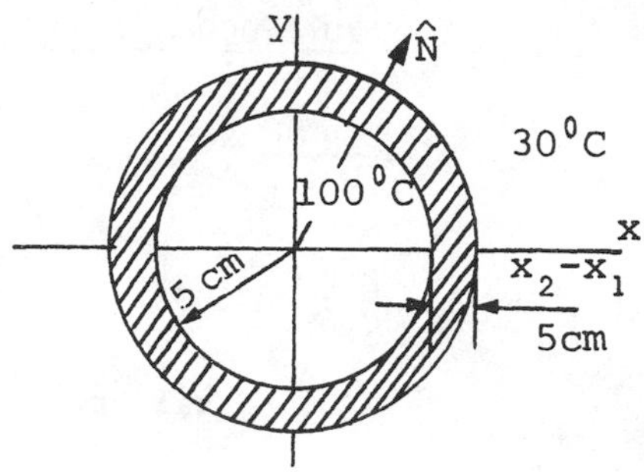

Solution: We first obtain the heat equation, based on experimental studies on a body. We base the analysis on the following empirical laws:

a) The quantity of heat in a body is proportional to its mass and to its temperature.

b) Heat flows from a higher to lower temperature.

c) The rate of flow across an area is proportional to the area and the rate of change of temperature with respect to the distance normal to the area.

Let q (cal/sec.) be the constant quantity of heat flowing through an area $A(cm^2)$, perpendicular to the direction of flow in each second, let $u(^\circ c)$ be the temperature at any point p of the body, and let x(cm) be the distance from the origin to the point p. Then, by (c), q is proportional to $A\frac{du}{dx}$, and by b) the temperature decreases in the direction of flow. Hence

$$q = -KA\frac{du}{dx},$$

where K is a constant of proportionality (cal./cm. deg.sec.), the thermal conductivity as suggested in the problem.

For the pipe in question, sectionally drawn in the figure, and under steady-state conditions, heat flows radially. The area A perpendicular to the direction of flow is the lateral surface of the cylinder of radius x, where $x_1 \le x \le x_2$. For the length L (cm) of the pipe we have $A = 2\pi xL$, and consequently the differential equation here takes the form:

$$q = 2\pi KLx\frac{du}{dx}.$$

Separating variables and integrating between limits, we obtain:

$$q\int_5^{10}\frac{dx}{x} = -2\pi KL\int_{100}^{30} du$$

This yields:

$$q \ln x\Big|_5^{10} = q \ln\frac{10}{5} = q \ln 2 = 2\pi KL(100-30),$$

for which K = 0.00060 cal./cm. deg.sec., L = 100 cm. Hence,

$$q(\text{cal./sec.}) = \frac{2\pi(0.00060)(100)(70)}{\ln 2}$$

$$= \frac{84\pi}{0.692},$$

or,

$$q(\text{cal./hr.}) = \frac{84\pi}{0.692}\, 3600$$

$$\cong 140{,}000 \text{ cal/hr.}\ .$$

• PROBLEM 1129

What is the velocity of a projectile at an altitude of 8,000 feet after it was fired directly upward from the ground with a muzzle velocity of 1,000 feet per second?

Solution: Neglecting air resistance, the acceleration $\cong -32$ ft./sec.$^2 = \alpha$.

Then $$\alpha = \frac{dv}{dt} = -32$$

and $$dv = -32\ dt$$

$$\int dv = -32\int dt$$

or $$v = -32t + C_1.$$

When $t = 0$, $v = 1{,}000$. Therefore,

$$C_1 = 1{,}000.$$

$$v = \frac{ds}{dt} = -32t + 1{,}000$$

or, $$ds = -32t \cdot dt + 1{,}000\ dt$$

$$\int ds = -32\int t \cdot dt + 1{,}000\int dt.$$

Then $$s = -32\frac{t^2}{2} + 1{,}000t + C_2$$

$$s = 0 \text{ when } t = 0.$$

Therefore, $C_2 = 0$.

And $$s = 16t^2 + 1{,}000t.$$

We calculate the time t for an elevation of 8,000 feet.

$$8{,}000 = -16t^2 + 1{,}000t$$

$$16t^2 - 1{,}000t + 8{,}000 = 0$$

$$4t^2 - 250t + 2{,}000 = 0$$

$t = \frac{-b \pm \sqrt{b^2-4ac}}{2a}$ (from the formula for the root of a quadratic equation).

Where $a = 4,$

$b = -250,$

$c = 2{,}000.$

$$t = \frac{+250 \pm \sqrt{62{,}5000 - 32{,}000}}{8}$$

or $$t = \frac{250 \pm \sqrt{30{,}500}}{8} = \frac{250 \pm 174.6}{8}$$

Finally, t = 53.1 or 9.4 sec. to reach 8,000 ft.

Substituting t = 9.4 sec. in: $v = -32t + 1{,}000.$

$v = -32 \cdot 9.4 + 1{,}000 = -300.8 + 1{,}000 \approx 700$ ft./sec.

The value of t = 53.1 gives a negative velocity, corresponding to the velocity when the elevation of 8,000 ft. is reached while descending, and is not used.

• PROBLEM 1130

A body falls with an initial velocity of 1000 ft./sec. and is subject to the acceleration of gravity ($g \approx 32$ ft/sec^2). What distance does it fall in 3 sec.?

Solution: Let v = velocity in ft./sec.

$$v = \frac{ds}{dt} = gt + v_0 = 32t + v_0$$

or $ds = (32t + v_0)dt$ (differential-equation form).

Integrating, $s = \int 32t \cdot dt + v_0 \int dt + C$

$$= 32 \cdot \frac{t^2}{2} + v_0 t + C = 16t^2 + v_0 t + C.$$

Now s = 0 when t = 0, which we substitute.

$$0 = 0 + 0 + C \text{ and } C = 0.$$

Therefore, $s = 16t^2 + v_0 t.$

When t = 3 and $v_0 = 1{,}000$, as given, then

$s = 16 \cdot (3)^2 + 1{,}000 \cdot 3 = 3{,}144$-ft. drop in 3 sec.

A projectile is fired straight upwards with an initial velocity of 1600 ft./sec. What is its velocity at 40,000 ft.? Assume $g = 32\text{ft./sec.}^2$

Solution: The projectile decelerates at the rate of $g = 32\ \text{ft/sec}^2$. Therefore,

$$\frac{dv}{dt} = -32, \quad dv = -32 \cdot dt$$

$$v = -32\int dt = -32t + C_1.$$

Now v = 1,600 when t = 0, or

$$1{,}600 = -0 + C_1, \quad \text{and} \quad C_1 = 1{,}600.$$

$$\text{Hence,} \quad v = -32t + 1{,}600. \qquad (1)$$

$$\text{Also, } v = \frac{ds}{dt} = -32t + 1{,}600$$

$$\text{or} \quad ds = -32t \cdot dt + 1{,}600dt$$

$$\text{and} \quad s = -32 \int t \cdot dt + 1{,}600 \int dt = -32\frac{t^2}{2} + 1{,}600t$$

$$= -16t^2 + 1{,}600t + C_2.$$

Now s = 0 when t = 0, or

$$0 = -0 + 0 + C_2$$

$$C_2 = 0$$

$$s = -16t^2 + 1{,}600t.$$

Now when s = 40,000,

$$40{,}000 = -16t^2 + 1{,}600t$$

$$\text{and} \quad (4t - 200)^2 = 0$$

$$\text{whence} \quad 4t = 200 \quad \text{or} \quad t = 50.$$

Since the time t has only one value, we can automatically understand that v = 0. Usually, we obtain two values of t, one is the time to reach a certain elevation ascending (t_1) and the time to reach the same elevation in descending, (t_2). Also $t_1 < t_2$.

Substitute t = 50 in Eq. (1) to get

$$v = -32 \cdot (50) + 1{,}600 = 0 \text{ ft./sec.}$$

• **PROBLEM 1132**

A body falls from a height of 300 ft. What distance has it traveled after 4 sec. if subject to g, the earth's acceleration?

Solution: The velocity of a falling body for any time t sec. after it starts to fall is 32.2t ft./sec. Let H = the height in feet of the body above earth at any time. H decreases as t increases.

Then $\frac{dH}{dt} = -32.2t$ = the velocity of the body,

or $dH = -32.2t \cdot dt$, the differential-equation form.

Now sum up all the differentials of height for every differential of time elapsed to get the total height or distance above the surface of the earth.

Then $$\int dH = -32.2\, t\int \cdot dt$$

or $$H = -32.2 \cdot \frac{t^2}{2} = -16.1t^2 + C.$$

To determine C, we know H = 300 when t = 0. Substitute these values.

$$300 = -16.1 \cdot 0 + C \quad \text{or} \quad C = 300.$$

Then at any time t sec. after the body starts to fall

$$H = -16.1t^2 + 300.$$

Now, when t = 4,

$$H = -16.1 \cdot 16 + 300 = 42.4 \text{ ft.}$$

At the end of 4 sec, the body is 42.4 ft above the earth's surface. It has traveled 257.6 ft.

• **PROBLEM 1133**

A body falls from rest under the action of gravity. The fall takes place in a viscous medium offering resistance proportional to the velocity. Find expressions for its velocity and distance fallen at any time t.

Solution: Take the origin at the starting point, and let y denote the distance of the body from this origin measured as positive downward. We know that the downward force of the body, with no external force, is its weight mg, where m is the mass and g is the gravitational acceleration. This force is opposed by an upward force which tends to retard any downward motion of the body. This is the resistance force (friction) which, according to the given data, is proportional to the velocity. Thus

$$F_r = kvm \quad \text{or resistance,} \quad R = kv$$

where k is the constant of proportionality.

The net downward force is then

$$m\frac{d^2y}{dt^2} = m(g-kv)$$

or

$$\text{(a)} \qquad \frac{d^2y}{dt^2} = \frac{dv}{dt} = g - kv,$$

the differential equation of motion.

Separating variables, we obtain:

$$\frac{dv}{g-kv} = dt.$$

Integration gives

$$\text{(b)} \quad -\frac{1}{k}\ln(g-kv) = t + c_1.$$

Since the body falls from rest, we have the initial condition v=0 when t=0; substituting these values in (b), we find $c_1 = -1/k \ln g$. Then (b) becomes

$$-\frac{1}{k}\ln(g-kv) = t - \frac{1}{k}\ln g, \quad \text{or} \quad \ln(g-kv) - \ln g = -kt,$$

or

$$\ln\frac{g-kv}{g} = -kt, \quad \text{hence} \quad \frac{g-kv}{g} = e^{-kt}$$

from which we have

$$\text{(c)} \qquad v = \frac{g}{k}\left(1-e^{-kt}\right).$$

From this we see that $v \to g/k$ when $t \to \infty$. We therefore call the constant g/k the limiting velocity; from (a) we find the acceleration dv/dt = 0 when v = g/k.

Replacing v by $\frac{dy}{dt}$ and integrating again, we get

$$y = \frac{g}{k}t + \frac{g}{k^2}e^{-kt} + c_2.$$

Using the initial condition y=0 when t=0, we find $c_2 = -g/k^2$. Then

$$\text{(d)} \qquad y = \frac{g}{k}t - \frac{g}{k^2}\left(1-e^{-kt}\right).$$

• PROBLEM 1134

A body falls from rest subject to gravity in a medium offering resistance proportional to the square of the velocity. Find expressions for velocity and distance.

Solution: Take the origin at the starting point and let y denote the distance of the body from this origin measured as positive downward. Treating this problem in terms of forces, the downward force of the body with no external force is obviously its weight mg. But this force is opposed by an upward force, which tends to retard any movement of the body. This is the resistance force (friction) which, according to the given data, is proportional to the square of the velocity. Thus, we have

$$F_r = kv^2m, \quad \text{or resistance} \quad R = kv^2$$

where k is the constant of proportionality and m is the mass of the object. For brevity and future use, we let $k = g/a^2$. Thus, the net downward force is f_r+mg (due to gravity). Therefore,

$$m\frac{d^2y}{dt^2} = mg\left(1 - \frac{v^2}{a^2}\right),$$

or

$$\text{(a)} \qquad \frac{d^2y}{dt^2} = \frac{dv}{dt} = g\left(1 - \frac{v^2}{a^2}\right).$$

Separating variables and integrating, we obtain:

$$\int \frac{dv}{1 - \frac{v^2}{a^2}} = \int g\ dt, \quad \text{or} \quad a \tanh^{-1}\frac{v}{a} = gt + c_1.$$

Since v=0 when t=0, we find

$$a \tanh^{-1}\left(\frac{0}{a}\right) = g(0) + c_1.$$

If $\tanh^{-1}0=\alpha$, then $\tanh \alpha = 0$

or

$$\frac{\sinh \alpha}{\cosh \alpha} = 0.$$

We know that $\cosh \alpha$ is different from zero, always positive and, recalling the behavior of $\sinh \alpha$, the only possible value for $\sinh \alpha = 0$ is $\alpha=0$. Thus we find that $c_1=0$. Then

$$\text{(b)} \qquad v = a \tanh \frac{gt}{a}.$$

From this we see that $v \to a$ when $t \to \infty$, since $\tanh x \to 1$ as $x \to +\infty$. We therefore call the constant a the *limiting velocity*, from (a) we find the acceleration dv/dt=0 when v=a. Replacing v in (b) by dy/dt and integrating, we have

$$\text{(c)} \qquad y = a\int \tanh\frac{gt}{a}\,dt = \frac{a^2}{g}\ln\cosh\frac{gt}{a} + c_2.$$

Since our origin is at the starting point, y=0 when t=0, hence we have $c_2=0$. Therefore, the distance of the body from the starting point at any time t is given by

$$\text{(d)} \qquad y = \frac{a^2}{g}\ln\cosh\frac{gt}{a},$$

where a is the limiting velocity.

A body falls in a medium offering resistance proportional to the square of the velocity. If the limiting velocity is numerically equal to $g/2 = 16.1$ ft./sec., find a) the velocity at the end of 1 sec.; b) the distance fallen at the end of 1 sec.; c) the distance fallen when the velocity equals 1/2 the limiting velocity; d) the time required to fall 100 ft.

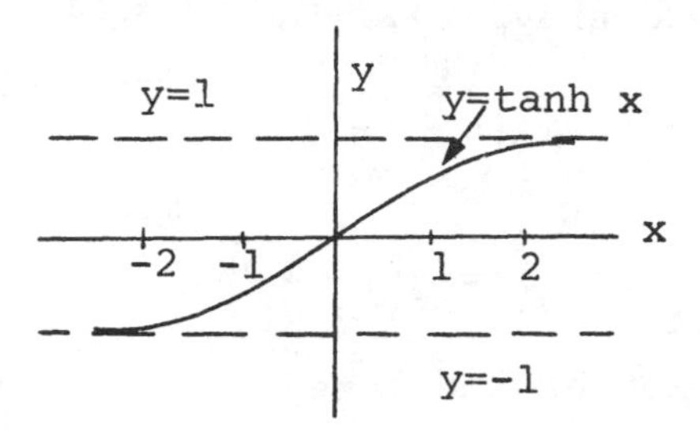

Solution: Taking the frame of reference at a point where the body starts to fall and assuming the downward direction as positive, this particle is acted upon by its weight (positive) and the resisting force (proportional to the square of the velocity and negative with respect to the selected origin). The difference between these forces is characteristic of its motion, and since this quantity equals mass × acceleration, we write:

$$m \frac{d^2y}{dt^2} = mg - k\left(\frac{dy}{dt}\right)^2$$

where k is a constant of proportionality. Noting that the limiting velocity occurs when the acceleration goes to zero,

$$mg = k\, v_{lim}^2$$

or,

$$k = \frac{mg}{v_{lim}^2}$$

where $v_{lim} = dy/dt$ is the maximum velocity and is constant. Hence

$$m \frac{dv}{dt} = mg - \frac{mg}{v_{lim}^2} v^2 .$$

Separating variables and using $v_{lim} = g/2 = 16.1$ ft./sec., we have:

$$\frac{dv}{1 - \frac{v^2}{v_{lim}^2}} = g\, dt.$$

Integrating both sides of the equation with g/2 substituted for v_{lim},

$$\int \frac{dv}{\frac{1-4v^2}{g^2}} = g \int dt + c_1 .$$

Setting $v = g/2 \tanh x$, $dv = g/2 \operatorname{sech}^2 x\, dx$, and noting that $1-\tanh^2 x = \operatorname{sech}^2 x$ we have:

$$\int \frac{dv}{\frac{1-4v^2}{g^2}} = g/2 \int \frac{\operatorname{sech}^2 x\, dx}{1-\tanh^2 x} = g/2 \int dx = g/2[x] = g/2 \tanh^{-1} \frac{2v}{g}$$

Thus

$$g/2 \tanh^{-1} \frac{2v}{g} = gt + c_1 .$$

Solving for v,

$$v = g/2 \tanh\left(2t+c_2\right).$$

Since the body was dropped initially, $v = 0$ for $t = 0$, making $c_2 = 0$. (See the diagram for evaluation.) Furthermore,

$$v = \frac{dy}{dt} = g/2 \tanh 2t.$$

This yields, as a result of integration:

$$y = g/4 \ln(\cosh 2t) + c_3,$$

where, for $t = 0$, $y = 0$, implying that $c_3 = 0$. The equation of motion is therefore:

$$y = g/4 \ln(\cosh 2t).$$

a) To find the velocity at the end of 1 sec.:

$$v_1 = g/2 \tanh(1) = g/2(0.9640) = 15.5 \text{ ft./sec.}$$

b) The corresponding distance covered is:

$$y_1 = g/4 \ln\left[\cosh 2(1)\right] = g/4 \ln(3.762) = g/4(1.32495) = 10.7 \text{ ft.}$$

c) When the speed reaches $1/2\, v_{lim}$, then $v = g/4$ ft./sec., and from the expressions:

$$v = g/4 = g/2 \tanh 2t,$$

and

$$y = g/4 \ln(\cosh 2t),$$

we eliminate t and solve for y in terms of v. Since

$$v = g/2 \tanh 2t = g/2 \frac{\sinh^2 t}{\cosh 2t} = g/2 \frac{(\cosh^2 2t-1)^{1/2}}{\cosh 2t},$$

and, solving for $\cosh 2t$, we obtain:

$$\cosh 2t = g/\sqrt{g^2-4v^2} .$$

Hence
$$y = g/4 \ln(\cosh 2t) = g/4 \ln\left[\frac{g}{\sqrt{g^2-4v^2}}\right] .$$

Using $v = g/4$

$$y = g/4 \ln \frac{2}{\sqrt{3}} = 1.13 \text{ ft.}$$

d) To calculate the time required for a fall of 100 ft., we use

$$y = g/4 \ln(\cosh 2t).$$

Solving for t,

$$t = 1/2 \cosh^{-1}\left(e^{4y/g}\right) = 1/2 \cosh^{-1}\left(e^{12.4}\right), \text{ for } y = 100.$$

If we neglect the factor $e^{-12.4}$ in comparison to $e^{12.4}$, we can approximate the above equation. Since

$$\cosh 2t = e^{12.4},$$

we let $\cosh 2t = 1/2\, e^{2t}$. Thus,

$$1/2\, e^{2t} = e^{12.4}.$$

Taking the antilogarithm,

$$2t = \ln 2 + 12.4,$$

or,

$$t = 1/2 \ln 2 + 6.2 = 1/2(0.69) + 6.2 = 6.545 \text{ sec.}$$

• PROBLEM 1136

A particle accelerates from t=0 in accordance with the law that the acceleration, a=6-3t.

a) Compute the distance that the particle travels while the velocity is increasing starting from zero,

b) find the distance traveled by the particle during the interval from t=1 to t=5.

Solution:

a) The velocity increases during the interval in which the acceleration is positive. For this example, this interval begins at t=0 and continues until t=2. (This is obtained from examining the equation for the acceleration.) Hence the distance moved during the interval from t=0 to t=2 is desired. The acceleration is

$$a = \frac{d^2s}{dt^2} = \frac{dv}{dt} = 6 - 3t.$$

Then

$$dv = (6-3t)dt,$$

$$v = 6t - \frac{3}{2}t^2 + c.$$

Since v=0 when t=0, the constant of integration c is zero.

Before evaluating s over the interval from t=0 to t=2, we must see if the motion is in the same direction during the interval. Since the direction of motion may change when v=0, we see from

$$v = 6t - \frac{3}{2}t^2 = \frac{3}{2}(4-t)t = 0$$

that the direction of motion can change only at t=0 or t=4. Hence it does not change between t=0 and t=2. Now, writing s as a definite integral we have

$$s = \int_0^2 \left(6t - \frac{3}{2}t^2\right)dt$$

$$= 3t^2 - \frac{1}{2}t^3 \Big]_0^2 = 12 - 4 = 8 \text{ units.}$$

b) The velocity is positive for $0 < t < 4$, negative for $t > 0$ and zero for $t=0$ and 4. For $t > 4$ the motion of the particle is opposite to the direction in which positive s is measured. To find the total (or absolute) distance traveled we must compute s separately for the intervals $t=1$ to $t=4$ and $t=4$ to $t=5$, and add the two results numerically. Thus

$$s = \int_1^4 \frac{3}{2}(4-t)tdt + \int_4^5 \frac{3}{2}(4-t)tdt = s_1 + s_2.$$

$$s = \int_1^4 \frac{3}{2}(4-t)tdt + \int_4^5 \frac{3}{2}(4-t)tdt = \int_1^5 \left(6t - \frac{3t^2}{2}\right)dt$$

or

$$s = 3t^2 - \frac{t^3}{2}\Bigg|_1^5 = 75 - \frac{125}{2} - 3 + \frac{1}{2} = 10 \text{ units},$$

the total distance traveled.

• PROBLEM 1137

A flywheel runs at $\omega = 0.1t$, where ω is the angular speed in radians per second and t is the time in seconds. How long will this flywheel require to complete 1 revolution (2π radians)?

Solution: Let θ = the number of radians the flywheel has turned up to some time t.

Then $\omega = \frac{d}{dt} = 0.1t$

or $d\theta = 0.1t \cdot dt$

and $\theta = 0.1 \int t \cdot dt = 0.1 \cdot \frac{t^2}{2} + C.$

Now $\theta = 0$ when $t = 0$, or $0 = 0 + C$. Therefore, $C = 0$.

Then $\theta = 0.1\frac{t^2}{2}$ or $t = \sqrt{20.\theta}$

Now when $\theta = 2\pi$ (one revolution), then $t = \sqrt{20 \cdot \theta}$. $t = \sqrt{20 \cdot 2\pi} = 11.21$ sec., the time for the first revolution.

• PROBLEM 1138

Find the "escape velocity", i.e., the speed with which a particle would have to be projected from the surface of the earth in order never to return, by determining the speed with which a particle would strike the earth's surface if it started from rest at a very great (supposedly infinite) distance and traveled subject only to the earth's gravitation.

Solution: From the law of gravitation,

$$F = -k_1 \frac{mM}{r^2},$$

where k_1 is the gravitational constant, m and M are the masses of the particle and the earth, respectively, and r^2 is the distance separating them.

We can call $(k_1M) = k_2$, and

$$F = -\frac{k_2 m}{r^2}.$$

On the surface of the earth, with the radius of the earth $R = r = 3959$ mi., the force is just the weight of the particle, that is:

$$F = -\frac{k_2 m}{R^2} = -mg,$$

and hence, for $r \neq R$,

$$F = -\frac{R^2 gm}{r^2}.$$

Calling $R^2 g = k^2$, we have:

$$F = -\frac{k^2}{r^2} = \frac{d^2r}{dt^2}.$$

To integrate, we set $p = dr/dt$. Then $\frac{d^2r}{dt^2} = \frac{dp}{dt} = \frac{dp}{dr}\frac{dr}{dt} = p\frac{dp}{dr}$

$$p\frac{dp}{dr} + \frac{k^2}{r^2} = 0.$$

Separation of variables and integration yields:

$$\frac{p^2}{2} = \frac{k^2}{r} + c.$$

To find c, we let $r \to \infty$, where the motion starts, and $p = \frac{dr}{dt} = 0$. This makes $c = 0$. Furthermore, solving for $p = dr/dt$,

$$\frac{dr}{dt} = k\sqrt{2/r}.$$

Using $k = R\sqrt{g}$, we finally obtain:

$$v = R\sqrt{2g/r}.$$

Since we are interested in finding the velocity of this particle on the earth's surface,

$$v_0 = r\sqrt{2g/R} = \sqrt{2gR} = \left(\frac{2 \times 32.2 \times 3959}{5280}\right)^{1/2}$$

$$= 7 \text{ miles/second.}$$

This value is independent of the mass of the particle.

• PROBLEM 1139

The attraction of a spherical mass on a particle within the mass is directed toward the center of the sphere and is proportional to the distance from the center. Suppose a straight tube were bored through the center of the earth and a particle of mass m lb. were dropped into the tube. If the radius of the earth is 3960 mi., find how long it will take a) to pass through the tube; b) to drop halfway to the center. Neglect resistance.

Solution: Force = Mass × acceleration

$$F = m\frac{d^2x}{dt^2} = -cx ,$$

where c is a constant of proportionality. We can write:

$$\frac{d^2x}{dt^2} = - k^2x ,$$

where $k^2 = c/m$, m is constant. To find the solution, we use the operator $D \equiv d/dt$,

$$(D^2 + k^2)x = 0.$$

The auxiliary equation is: $m^2 + k^2 = 0$, with roots $m = ik$ and $m = -ik$, and the solution appears as:

$$x = A \cos kt + B \sin kt.$$

Using the boundary condition: at time $t = 0$ the initial velocity is zero,

$$\left.\frac{dx}{dt}\right|_{t=0} = -Ak \sin kt + Bk \cos kt = 0,$$

Since $\sin 0 = 0$ and $\cos 0 = 1$, this suggests that B has to be zero. Hence

$$x = A \cos kt$$

is the required solution with the constants A and k to be determined. At time $t = 0$, the initial position of the particle is on the surface of the earth which is at $x_0 = -3960$ mi. from the center of the earth (with respect to the reference origin). Then, so far, we have:

$$x = -3960 \cos kt.$$

To find the value of k, we investigate the following condition: when the particle is on the surface of the earth, say at $t = 0$, $ma = mg$ or $a = g$. In other words,

$$\frac{d^2x}{dt^2} = -k^2x_0 = g.$$

But at $t = 0$, $x_0 = 3960$. Hence,

$$k^2(3960) = g$$

or

$$k = \sqrt{\frac{g}{3960}} .$$

Thus,

$$x = -3960 \cos\sqrt{\frac{g}{3960}}\ t$$

is the particular solution for the motion of the particle.

a) For the particle to pass through the origin, it has to cover a distance 2(3960)mi. in time t_1 and its momentary location is at $x = 3960$ mi. From this, we wish to calculate for t_1:

$$3960 = -3960 \cos \sqrt{\frac{g}{3960}}\, t_1 ,$$

or

$$\cos\sqrt{\frac{g}{3960}}\, t_1 = -1 ,$$

from which

$$\left(\frac{g}{3960}\right)^{\frac{1}{2}} t_1 = \pi .$$

Quantity g in this expression is expressed in miles/sec.^{2} $g = \dfrac{32.2}{5280} \dfrac{\text{mi.}}{\text{sec}^2}$.

Finally, $$t_1 = \pi\left(\frac{3960 \times 5280}{32.2}\right)^{\frac{1}{2}} \Big/ 60 \text{ (min.)}$$

$$= 42.2 \text{ min.}$$

b) For the particle to be halfway to the center, $x = \dfrac{-3960}{2}$. To calculate for the time t_2, we set:

$$\frac{-3960}{2} = -3960 \cos\left(\frac{g}{3960}\right)^{\frac{1}{2}} t_2 \cdot \cos\left[\left(\frac{g}{3960}\right)^{\frac{1}{2}} t_2\right] = \tfrac{1}{2} ,$$

or

$$\left(\frac{g}{3960}\right)^{\frac{1}{2}} t_2 = \pi/3.$$

Therefore,

$$t = \pi/3\left(\frac{3960 \times 5280}{32.2}\right)^{\frac{1}{2}} \Big/ 60 \text{ (min.)}$$

$$= 14.1 \text{ min.}$$

• PROBLEM 1140

A particle slides freely in a tube which rotates in a vertical plane about its midpoint with constant angular velocity w. If x is the distance of the particle from the midpoint of the tube at time t, and if the tube is horizontal with $t = 0$, show that the motion of the particle along the tube is given by

$$\frac{d^2x}{dt^2} - w^2x = -g \sin wt.$$

Solve this equation if $x = x_0$, $dx/dt = v_0$ when $t = 0$. For what values of x_0 and v_0 is the motion simple harmonic?

Solution: In order to determine the x- and y-components of motion, we choose the frame of reference at the center of mass of the particle. The particle is subjected to horizontal and vertical forces. Restricting our attention to the horizontal components, the two unbalanced forces acting on the particle are a portion of its weight, (proportional to $\sin\theta$), and the centripetal force, the difference between which gives an acceleration along the x-axis. Mathematically,

$$mg \sin\theta + \frac{mv^2}{r} = -m \frac{d^2x}{dt^2} ,$$

where $v = wr$, $r = -x$, and $\theta = wt$. Then

$$\frac{d^2x}{dt^2} = w^2x - g \sin wt.$$

To find the solution of this non-homogeneous, second order differential equation with constant coefficients, we use the D operator, where $D = d/dt$, hence the differential equation can be rewritten as:

$$\left(D^2 - w^2\right)x = -g \sin wt.$$

The auxiliary equation is $m^2 - w^2 = 0$, $m = w$, or $m = -w$, and the complementary solution is:

$$x_c = Ae^{wt} + Be^{-wt} .$$

Furthermore, the particular solution appears as:

$$x_p = C \sin wt + D \cos wt,$$

where A, B, C and D are constants to be determined. Since the general solution is

$$x = Ae^{wt} + Be^{-wt} + C \sin wt + D \cos wt,$$

$$\frac{dx}{dt} = Awe^{wt} - Bwe^{-wt} + Cw \cos wt - Dw \sin wt,$$

and

$$\frac{d^2x}{dt^2} = w^2\left(Ae^{wt} + Be^{-wt} - C \sin wt - D \cos wt\right).$$

Substitution in the main differential equation yields:

$$\left(D^2 - w^2\right)x = -2Cw^2 \sin wt - 2Dw^2 \cos wt = -g \sin wt,$$

which implies that:

$$-2Cw^2 = -g, \text{ or } C = \frac{g}{2w^2} ,$$

and $D = 0$. Thus,

$$x = Ae^{wt} + Be^{-wt} + \frac{g}{2w^2} \sin wt .$$

To find the values of A and B, we use the boundary conditions for $t = 0$, $dx/dt = v_0$ and $x = x_0$. And since

$$\frac{dx}{dt} = Awe^{wt} - Bwe^{-wt} + \frac{g}{2w} \cos wt,$$

$$\left.\frac{dx}{dt}\right|_{t=0} = Aw - Bw + \frac{g}{2w} = v_0,$$

and

$$x\Big|_{t=0} = A + B = x_0 .$$

These are simultaneous equations with A and B unknown. Substitution of $B = x_0 - A$ in the expression for v_0 yields:

$$A = \frac{v_0 - g/2w + x_0 w}{2w} , \text{ and } B = \frac{x_0 w - v_0 + g/2w}{2w} .$$

Hence,

$$x = \frac{v_0 - g/2w + x_0 w}{2w} e^{wt} + \frac{x_0 w - v_0 + g/2w}{2w} e^{-wt} + \frac{g}{2w^2} \sin wt .$$

After some algebraic manipulation,

$$x = \tfrac{1}{2}x_0(e^{wt} + e^{-wt}) + (v_0/2w - g/4w^2)(e^{wt} - e^{-wt}) + (g/2w^2)\sin wt.$$

For the motion to be of the simple harmonic type, we use the expression:

$$\frac{d^2x}{dt^2} = w^2x - g \sin wt.$$

Hence, we must have the relationship:

$$|w^2x| \gg |g \sin wt|, \text{ or } \left|\frac{w^2x}{g}\right| \gg 1 .$$

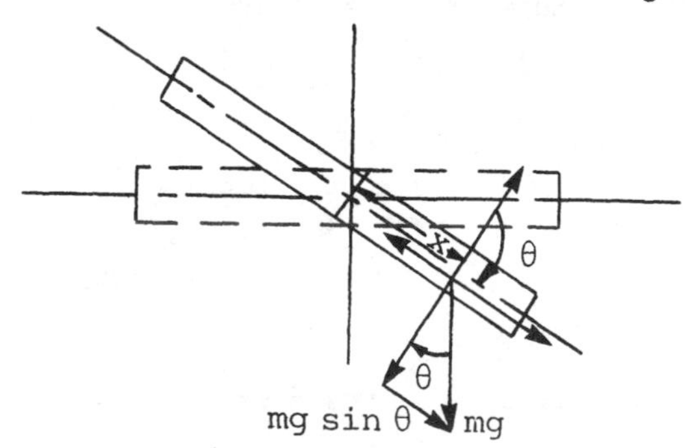

• PROBLEM 1141

A spring is stretched 3 inches by a 6-pound weight. A 12-pound weight is attached to the spring and pulled down 6 inches below its equilibrium point. If the weight is then given an initial upward velocity of 2 feet per second, describe the resulting motion of the weight. (Neglect resistive forces.)

Solution: The spring constant, k, is determined from the amount of stretch produced by a given weight. From Hooke's Law, $F = kx$, and for $F = 6$ pounds, $x = 3$ in $= 1/4$ ft.

Therefore,

$$6 = k(1/4)$$

or

$$k = 24 \text{ pounds per foot.}$$

Neglecting all external forces other than the tension of the spring, $F = kx$, we wish to obtain the net downward force equation. Under equilibrium conditions the weight is perfectly supported by the spring and thus

$$F = mg$$

where m is the mass of the body, x is the length of the deformation.

Assume that the spring is pulled down dx units below its equilibrium position and released. Thus the new elongation is $x + dx$ and the new equation for the forces acting on the body is

$$k(x+dx) = mg + f,$$

where f is the force applied. But at the time the weight is released, $t=0$, $f=0$, $kdx=0$ or $kx = -\theta$.

Now, θ is balanced by the force resulting from acceleration, namely ma or md^2x/dt^2. Inserting numerical values for m and k and transposing θ to the left-hand side of the equation, we obtain the differential equation of the motion

$$\frac{12}{32}\frac{d^2x}{dt^2} + 24x = 0,$$

or

$$\frac{d^2x}{dt^2} + 64x = 0.$$

The initial conditions state that when $t=0$, then

$$x = \frac{1}{2} \text{ foot, and } \frac{dx}{dt} = -2 \text{ feet per second.}$$

The velocity is given a negative sign, since it is in the direction of decreasing x. Note also that all dimensions must be consistent, preferably in feet and seconds, since the value of g is taken as 32 feet per second per second.

The auxiliary equation of our differential equation is $m^2+64=0$ which has the complex roots $m = \pm 8i$. Thus we obtain $y=c_1e^{i8x}+c_2e^{-i8x}$ as a general solution, which reduces, by the use of Euler's formula of complex exponents, to

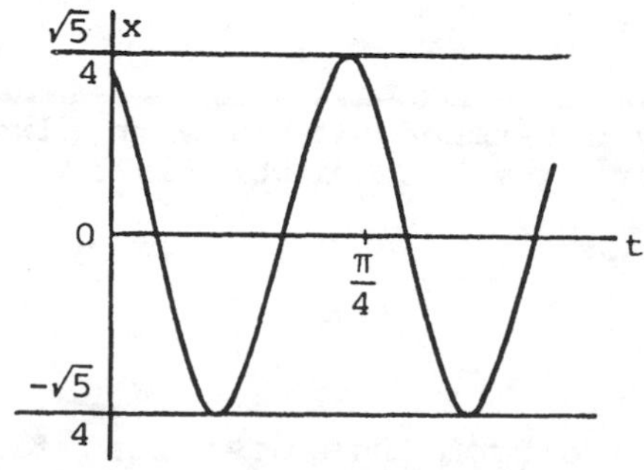

$$x = c_1 \cos 8t + c_2 \sin 8t.$$

Then

$$\frac{dx}{dt} = -8c_1 \sin 8t + 8c_2 \cos 8t,$$

so that putting t=0 and using the initial values of x and dx/dt, we get

$$\frac{1}{2} = c_1,$$

and

$$-2 = +8c_2 \;, \quad \text{or} \quad c_2 = -\frac{1}{4}.$$

Therefore, the position of the weight at time t is given by

$$x = \frac{1}{2} \cos 8t - \frac{1}{4} \sin 8t.$$

By rewriting this equation in another form, we can get a better insight into the motion of the weight. If the right side of the equation is multiplied and divided by

$$\sqrt{(1/2)^2 + (1/4)^2} = \frac{1}{4}\sqrt{5}.$$

then the formula for x becomes

$$x = \frac{1}{4}\sqrt{5}\left(\frac{4}{\sqrt{5}} \cdot \frac{1}{2} \cos 8t - \frac{4}{\sqrt{5}} \cdot \frac{1}{4} \sin 8t\right)$$

or, rationalizing denominators and cancelling, we get

$$x = \frac{1}{4}\sqrt{5}\left(\frac{2}{5}\sqrt{5} \cos 8t - \frac{1}{5}\sqrt{5} \sin 8t\right).$$

Let $\frac{2}{5}\sqrt{5} = \cos\alpha$ and $\frac{1}{5}\sqrt{5} = \sin\alpha$ where $\tan\alpha = \frac{1}{2}$ and substitute in the above equation. Then we have

$$x = \frac{1}{4}\sqrt{5}(\cos\alpha \cos 8t - \sin\alpha \sin 8t)$$

$$= \frac{1}{4}\sqrt{5} \cos(8t+\alpha).$$

This equation reveals that the weight oscillates with an amplitude of $\sqrt{5}/4$ feet. We find the period $T: 8T+\alpha=2\pi$, or $T=\pi/4-\alpha$. Since α is constant we have a period of $\pi/4$ seconds. The motion may be represented graphically, as shown in the figure.

• PROBLEM 1142

A tightly stretched string with fixed end points x = 0 and x = L is initially in a position given by

$$y(x,0) = y_o \sin^3 \frac{a\pi x}{L}.$$

If it is released from this position, find the displacement y at any distance x from one end and

at any time t.

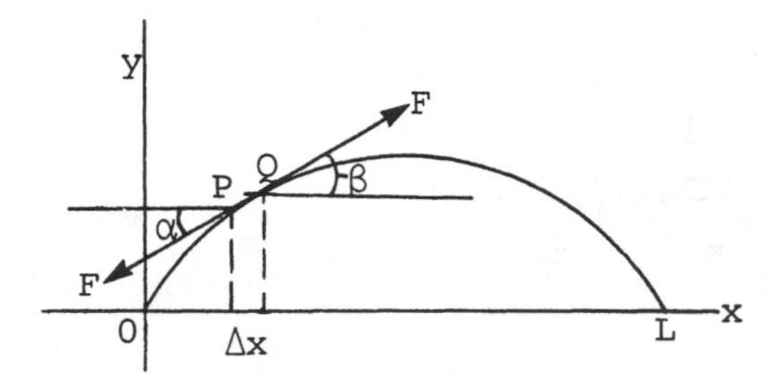

<u>Solution:</u> The displacement y is a function of two variables, namely, the distance x along the string and the time t. The vibration is caused by the forces due to the internal structure of the string (its elasticity) attempting to regain its original length L. These forces can be expressed by Newton's second law of motion. The weight of a small segment PQ, originally of length Δx, is $w\Delta x$ where w is the weight of the string per unit length. Since PQ moves only up and down, it experiences acceleration given by

d^2y/dt^2. Hence, we have the force:

$$F_R = \frac{w\Delta x}{g} \frac{\partial^2 y}{\partial t^2},$$

which must be the resultant force of the y-components of the tensions (F) at the two ends. If α and β represent the angles made by the tangential forces F with the horizontal, then the resultant force is:

$$F_R = F(\sin\beta - \sin\alpha).$$

The horizontal component forces give:

$$F_h = F(\cos\beta - \cos\alpha) = 0$$

Hence, for the resultant force,

$$\frac{w\Delta x}{g} \frac{\partial^2 y}{\partial t^2} = F(\sin\beta - \sin\alpha)$$

Next, we express $\sin\alpha$ and $\sin\beta$ in terms of the slopes of the string at points $P(x, f(x))$ and $Q(x + \Delta x, f(x + \Delta x))$, respectively. Since, under the relationship: $y = f(x)$ independent of time t,

$$\frac{df(x)}{dx} = \tan\alpha, \quad \frac{df(x + \Delta x)}{dx} = \tan\beta,$$

and $$\sin\alpha = \frac{\sin\alpha}{\cos\alpha}\cos\alpha = \tan\alpha \cdot \frac{1}{\frac{1}{\cos\alpha}}$$

$$= \frac{\tan \alpha}{\sec \alpha} = \frac{\tan \alpha}{\sqrt{1 + \tan^2 \alpha}} \simeq \tan \alpha \left(1 - \frac{1}{2} \tan^2 \alpha\right)$$

$$\sin \alpha = \frac{\frac{df(x)}{dx}}{\sqrt{1 + \left(\frac{df(x)}{dx}\right)^2}}$$

$$\simeq \frac{df(x)}{dx} \left[1 - \frac{1}{2} \left(\frac{df(x)}{dx}\right)^2\right] .$$

Similarly,

$$\sin \beta = \frac{\frac{df(x + \Delta x)}{dx}}{\sqrt{1 + \left(\frac{df(x + \Delta x)}{dx}\right)^2}}$$

$$\simeq \frac{df(x + \Delta x)}{dx} \left[1 - \frac{1}{2} \left(\frac{df(x + \Delta x)}{dx}\right)^2\right].$$

But for small vibrations assuming tight stretching, dy/dx can be considered small, and its square may be neglected with respect to unity. Hence, approximately,

$$\sin \alpha = \frac{df(x)}{dx} \quad \text{and} \quad \sin \beta = \frac{df(x + \Delta x)}{dx} .$$

For the resultant forces:

$$\frac{w\Delta x}{g} \frac{\partial^2 y}{\partial t^2} = F \frac{d}{dx} (f(x + \Delta x) - f(x))$$

or, since $\frac{d}{dx}$ is just an operator and Δx is a number, we have

$$\frac{\partial^2 y}{\partial t^2} = \frac{Fg}{w} \frac{d}{dx} \left[\frac{f(x + \Delta x) - f(x)}{\Delta x}\right] .$$

Letting $\frac{Fg}{w} = a^2$, and passing to the limit:

$$\frac{\partial^2 y}{\partial t^2} = a^2 \frac{d}{dx}\left(\frac{df(x)}{dx}\right) = a^2 \frac{d^2 f(x)}{dx^2}.$$

which, in consideration of the fact that y is dependent on time, is:

$$\frac{\partial^2 y}{\partial t^2} = a^2 \frac{\partial^2 y}{\partial x^2}$$

This is the partial differential equation of a vibrating string, the solutions of which represent the displacement y at any distance x from the origin and at any time t.

The solution of the above equation can be assumed to be expressed in the form of a product of a function of x alone and a function of t alone, where y is a time dependent function:

$$y = y(x,t) = X(x) \cdot T(t).$$

Thus, the above partial differential equation can be rewritten as

$$X(x) \frac{d^2 T(t)}{dt^2} = a^2 T(t) \frac{d^2 X(x)}{dx^2}$$

Separating variables of like functions and using the prime notation for corresponding differentiation, we have:

$$a^2 \frac{X''}{X} = \frac{T''}{T}.$$

Now, let t vary while x remains fixed. Then the left-hand side of the above equation is a constant. Therefore the right member is a constant and vice versa. Consequently, the common value of the two members must be a constant. We call it K.

Thus,

$$a^2 \frac{X''}{X} = K,$$

or $$X'' - \frac{K}{a^2} X = 0,$$

which is the familiar homogeneous second-order differential equation. Its auxiliary equation is:

$$m^2 - \frac{K}{a^2} = 0,$$

with roots $r_1 = \frac{\sqrt{K}}{a}$, and $r_2 = -\frac{\sqrt{K}}{a}$.

Hence,

$$X = c_1 e^{\frac{\sqrt{K}}{a}x} + c_2 e^{-\frac{\sqrt{K}}{a}x}, \quad (K > 0)$$

$$X = c_3 e^{i\frac{\sqrt{-K}}{a}x} + c_4 e^{\frac{-i\sqrt{-K}}{a}x}$$

$$\equiv c_3 \cos \frac{\sqrt{-K}\,x}{a} + c_4 \sin \frac{\sqrt{-K}\,x}{a}, \quad (K < 0)$$

$$X = c_5 + c_6 x, \qquad (K = 0)$$

are the possible solutions for the function dependent on x alone.

Similarly, for the time dependent function, $T'' - K T = 0$, and we have:

$$T(t) = \begin{cases} c_1' e^{\sqrt{K}t} + c_2' e^{-\sqrt{K}t}, & (K > 0) \\ c_3' \cos \sqrt{-K}\,t + c_4' \sin \sqrt{-K}\,t, & (K < 0) \\ c_5' + c_6' t, & (K = 0). \end{cases}$$

and

Since we are dealing with motion periodic in time (vibration of the string), we choose expressions that involve $K < 0$. Thus the required solution is:

$$y(x,t) = X(x) \cdot T(t) = \left(c_3 \cos \frac{\sqrt{-K}\,x}{a} + c_4 \sin \frac{\sqrt{-K}\,x}{a}\right)\left(c_3' \cos \sqrt{-K}\,t + c_4' \sin \sqrt{-K}\,t\right),$$

where all the constants are to be determined in such a way that the solution is in agreement with the physical requirements.

These requirements are called the boundary conditions, where

$$y(0,t) = y(L,t) = 0, \qquad t \geq 0$$

$$y(x,0) = y_o \sin^3 \frac{a\pi x}{L} \quad \text{(which is given)}.$$

At $t = 0$, the string is in its equilibrium condition under the initial force.

$$y_o \sin^3 \frac{a\pi x}{L}.$$

Hence,

$$\left.\frac{\partial y}{\partial t}\right|_{t=0} = 0$$

must be satisfied.

When $x = 0$ and $t = t$,

$$y(0,t) = c_3 \left(c_3' \cos \sqrt{-K}\, t + c_4' \sin \sqrt{-K}\, t\right) = 0$$

which is easily met by taking $c_3 = 0$.

For the second boundary condition,

$$y(L,t) = c_4 \sin \frac{\sqrt{-K}\, L}{a} \left(c_3' \cos \sqrt{-K}\, t + c_4' \sin \sqrt{-K}\, t\right) = 0,$$

making $c_4 = 0$ leads to a trivial solution, suggesting that the string is always in its equilibrium position. Likewise, making $c_3' = c_4' = 0$ leads to the same condition. The expression within the brackets yields particular values of t, when set to zero, constituting another trivial solution. Thus, the only way to achieve a non-trivial solution is for

$\sin \dfrac{\sqrt{-K}\, L}{a}$ to equal zero, leading to

$$\frac{\sqrt{-K}\, L}{a} = n\pi, \text{ or } \sqrt{-K} = \frac{an\pi}{L},$$

where n is any integer. This yields:

$$y(x,t) = c_4 \sin \frac{an\pi x}{L} \left[c_3' \cos \frac{n\pi t}{L} + c_4' \sin \frac{n\pi t}{L} \right].$$

Furthermore, using the boundary condition:

$$\left. \frac{\partial y}{\partial t} \right|_{t=0} = 0,$$

$$\left. \frac{\partial y(x,t)}{\partial t} \right|_{t=0} = c_4 \sin \frac{an\pi x}{L}$$

$$\left. \left(- \frac{c_3' n\pi}{L} \sin \frac{n\pi t}{L} + \frac{c_4' n\pi}{L} \cos \frac{n\pi t}{L} \right) \right|_{t=0}$$

$$= c_4 \sin \frac{an\pi x}{L} \left(\frac{c_4' n\pi}{L} \right) = 0,$$

which suggests $c_4' = 0$. Thus,

$$y(x,t) = c_4 \sin \frac{an\pi x}{L} c_3' \cos \frac{n\pi t}{L}.$$

Taking the imposed force on the string into account at $t = 0$, we have

$$y(x,0) = c_3' c_4 \sin \frac{an\pi x}{L} = y_o \sin^3 \frac{a\pi x}{L}.$$

We use the theorem which states: If y_1, y_2, ..., y_n are n solutions of a linear homogeneous partial differential equation, then

$$a_1 y_1 + a_2 y_2 + \ldots + a_n y_n,$$

where the a's are any constants, is also a solution of the given homogeneous equation. We use this theorem and the identity:

$$y_o \sin^3 \frac{a\pi x}{L} = \frac{y_o}{4}\left[3 \sin \frac{a\pi x}{L} - \sin \frac{3a\pi x}{L}\right] .$$

Choosing n = 1 and n = - 3, we have:

$$c_3' \, c_4 \left[\sin \frac{a\pi x}{L} - \sin \frac{3a\pi x}{L}\right]$$

$$= \frac{y_o}{4}\left[3 \sin \frac{a\pi x}{L} - \sin \frac{3a\pi x}{L}\right] .$$

Hence, the desired relation takes the form:

$$y(x,t) = \frac{y_o}{4}\left[3 \sin \frac{a\pi x}{L} \cos \frac{\pi t}{L} - \sin \frac{3a\pi t}{L} \cos \frac{3\pi t}{L}\right] .$$

• PROBLEM 1143

(a) A tightly stretched string with fixed end points x = 0 and x = L is initially at rest in its equilibrium position. If it is set vibrating by giving each of its points a velocity

$$3(Lx - x^2),$$

find the displacement of any point on the string at any time t. (b) If the string is 2 ft. long, weighs 0.1 lb., and is subjected to a constant tension of 6 lb., find the maximum displacement when t = 0.01 sec.

Solution: The partial differential equation of a stretched string is given by

$$\frac{\partial^2 y}{\partial t^2} = a^2 \frac{\partial^2 y}{\partial x^2}$$

where

$$a^2 = \frac{Fg}{w} ,$$

F is the tension lb., g is the gravitational acceleration 32.2 ft./sec.^{2}, and w is the weight of the string per unit length in lb./ft.

The given boundary conditions are

$$y(0,t) = y(L,t) = 0 \qquad t \geq 0$$

$$y(x,0) = 0 \qquad x \geq 0$$

and

$$\left.\frac{\partial y}{\partial t}\right|_{t=0} = 3\left(Lx - x^2\right).$$

To seek the solution for the partial differential equation, we assume:

$$y(x,t) = X(x) \cdot T(t)$$

i.e., a solution in product form. Hence

$$\frac{\partial^2 y}{\partial t^2} = X(x)\ T''(t), \text{ and}$$

$$\frac{\partial^2 y}{\partial x^2} = T(t)X''(x).$$

Substitution yields:

$$X(x)\ T''(t) = a^2\ T(t)\ X''(x).$$

Separating like variables,

$$\frac{X''(x)}{X(x)} = \frac{T''(t)}{a^2\ T(t)},$$

where the position of the constant a^2 does not make any difference in the calculation.

Now, we let the left-hand and right-hand sides of the above equation be equal to some constant $\left(-k^2\right)$, on the basis that the equality of equations in two different variables holds only if both functions equal the same constant number. Hence,

$$\frac{X''}{X} = -k^2$$

or $\quad X'' + k^2x = 0,$

the auxiliary equation of which is: $m^2 + k^2 = 0$. Its roots are: $r_1 = ki$, and $r_2 = -ki$; $(i = \sqrt{-1})$, yielding two distinct particular solutions $X(x) = \cos kx$, $X(x) = \sin kx$. Hence, the general solution is given by:

$$X(x) = c_1 \cos kx + c_2 \sin kx.$$

Similarly, for

$$\frac{T''(t)}{a^2\, T(t)} = -k^2,$$

the solution is given by

$$T(t) = c_3 \cos akt + c_4 \sin akt.$$

Consequently

$$y(x,t) = X(x) \cdot T(t) = \left(c_1 \cos kx + c_2 \sin kx\right)$$

$$\left(c_3 \cos akt + c_4 \sin akt\right).$$

The first boundary condition gives:

$$y(0,t) = 0 = c_1 \left(c_3 \cos akt + c_4 \sin akt\right)$$

which suggests $c_1 = 0$.

Secondly,

$$y(L,t) = 0 = c_2 \sin kL \left(c_3 \cos akt + c_4 \sin akt\right),$$

giving $\sin kL = 0$. If we have $kL = n\pi$ with n an integer, we have $\sin kL = \sin n\pi = 0$. Hence, by $kL = n\pi$, $k = \frac{n\pi}{L}$. Substitution for k in the remaining terms gives:

$$y(x,t) = c_2 \sin \frac{n\pi}{L} x \left(c_3 \cos \frac{n\pi a}{L} t + c_4 \sin \frac{n\pi a}{L} t \right).$$

Thirdly, the string is initially at rest, which means that:

$$y(x,0) = 0 = c_2 \sin \frac{n\pi}{L} x \left(c_3 \right).$$

Since c_2 cannot be equal to zero - which would show that the string is at rest in its equilibrium position (a trivial result) - c_3 must be equal to zero. Under these conditions, for any x and t:

$$y(x,t) = c_2 \sin \frac{n\pi}{L} x \; c_4 \sin \frac{n\pi a}{L} t.$$

Since the string is set into motion under a driving force which gives every point of the string the velocity $3\left(Lx - x^2\right)$, we have

$$\left. \frac{\partial y(x,t)}{\partial t} \right|_{t=0} = 3\left(Lx - x^2\right)$$

$$= c_2 \sin \frac{n\pi}{L} x \left(c_4 \frac{n\pi a}{L} \right),$$

or, $$Lx - x^2 = \frac{c_2 \, c_4}{3} \frac{n\pi a}{L} \sin \frac{n\pi}{L} x.$$

This relationship appears impossible since a sine curve cannot be made to coincide with a parabolic arc over the entire x-range. But this difficulty suggests that the equality can be satisfied if the parabolic arc is approximated by many sine expressions of period 2L, leading to the Fourier sine expansion of

$\left(Lx - x^2\right)$ over $0 < x < L$.

Mathematically,

$$Lx - x^2 = \sum_{n=1}^{\infty} b_n \sin \frac{n\pi x}{L}, \qquad 0 < x < L$$

Here, from the theory of Fourier series,

$$b_n = \frac{2}{L} \int_0^L \left(Lx - x^2\right) \sin \frac{n\pi x}{L} \, dx$$

$(n = 1, 2, 3, \ldots; \ 0 < x < L)$.

Expanding the integrand and using integration by parts,

$$b_n = \frac{2}{L} \left[\left(\frac{L^3}{n^2\pi^2} \sin \frac{n\pi x}{L} - \frac{xL^2}{n\pi} \cos \frac{n\pi x}{L} \right|_0^L - \left(- \frac{x^2 L}{n\pi} \cos \frac{n\pi x}{L} + \frac{2xL^2}{n^2\pi^2} \sin \frac{n\pi x}{L} + \frac{2L^3}{n^3\pi^3} \cos \frac{n\pi x}{L} \right|_0^L \right]$$

$$= \frac{4\ L^2}{n^3\pi^3} \left(1 - (-1)^n\right) .$$

This shows that, for every even n, $b_n = 0$, and $b_n \neq 0$ for odd n. The odd n can be assured by replacing n by 2n - 1. Hence we have:

$$b_{2n-1} = \frac{8\ L^2}{(2n - 1)^3 \pi^3} .$$

Thus, the Fourier half-range expansion reads:

$$Lx - x^2 = \frac{8L^2}{\pi^3} \sum_{n=1}^{\infty} \frac{\sin (2n - 1)\pi x/L}{(2n - 1)^3} .$$

For the initial velocity of the string,

$$Lx - x^2 = \frac{8L^2}{\pi^3} \sum_{n=1}^{\infty} \frac{\sin \frac{(2n-1)\pi x}{L}}{(2n-1)^3}$$

$$= \frac{c_2 c_4}{3} \frac{n\pi a}{L} \sin \frac{n\pi x}{L} .$$

We recall the theorem that says: If y_1, y_2, ..., y_n are n solutions of a linear homogeneous partial differential equation, $c_1y_1 + c_2y_2 + \ldots + c_ny_n$ is also a solution, where the c's are non-zero constants. Under these conditions, the above equation suggests the solution in a summation form, where the constants c_2c_4 are no longer arbitrary fixed constants but are dependent on n. We can call them $(c_2c_4)_{2n-1}$ where thev are zero for all even n's. Hence we have:

$$\frac{24L^3}{\pi^4 a} \sum_{n=1}^{\infty} \frac{\sin \frac{(2n-1)\pi x}{L}}{(2n-1)^4}$$

$$= \sum_{n=1}^{\infty} (c_2c_4)_{2n-1} \sin \frac{(2n-1)\pi x}{L} ,$$

and $(c_2c_4)_{2n-1} = \frac{24L^3}{\pi^4 a} \cdot \frac{1}{(2n-1)^4}$.

Once the constants are determined,

$$y(x,t) = c_2 \; c_4 \sin \frac{n\pi x}{L} \sin \frac{n\pi a}{L} t$$

becomes

$$y(x,t) = \frac{24L^3}{\pi^4 a} \sum_{n=1}^{\infty} \frac{1}{(2n-1)^4} \sin \frac{(2n-1)\pi}{L} x$$

$$\sin \frac{(2n-1)\pi a}{L} t,$$

which is the general solution. We replaced n by 2n - 1 in the time function because, as we know, the product of zero with a number is zero.

(b) For maximum displacement we have to find the value of x at which

$$\frac{\partial y}{\partial x} = 0:$$

$$\frac{\partial y}{\partial x} = -\frac{24L^2}{\pi^3 a} \sum_{n=1}^{\infty} \frac{1}{(2n-1)^3} \cos \frac{(2n-1)\pi}{L} x$$

$$\sin \frac{(2n-1)\pi a}{L} t = 0,$$

which means that for any n,

$$\cos \frac{(2n-1)\pi}{L} x = 0$$

But we know that

$$\cos \frac{(2n-1)\pi}{2} = 0,$$

that is,

$$\cos \frac{\pi}{2} = \cos \frac{3\pi}{2} = \ldots = 0.$$

Thus, the only possibility for x is the value

$$x = \frac{L}{2}.$$

Hence, for maximum displacement at any time t:

$$y\left(\frac{L}{2}, t\right) = \frac{24L^3}{\pi^4 a} \sum_{n=1}^{\infty} \frac{1}{(2n-1)^4}$$

$$\sin \frac{(2n-1)\pi}{2} \sin \frac{(2n-1)\pi a}{L} t$$

$$= \frac{24L^3}{\pi^4 a} \sum_{n=1}^{\infty} \frac{(-1)^{n+1}}{(2n-1)^4} \sin \frac{(2n-1)\pi a}{L} t.$$

Recalling the value of a,

$$a = \sqrt{\frac{Fg}{w}},$$

and since the weight of the string wL = 0.1 lb., the tension F = 6 lb., the length of the string L = 2ft. and at time t = 0.01 sec., we have:

$$y\left(\frac{L}{2}, 0.01\right) = \frac{24(2)^3}{\pi^4\sqrt{\frac{6x32}{0.05}}} \sum_{n=1}^{\infty} \frac{(-1)^{n+1}}{(2n-1)^4}$$

$$\sin \frac{0.01\ (2n-1)\pi\sqrt{\frac{6x32}{0.05}}}{2}$$

$$\simeq \frac{1}{26}\left[\sin(0.31\pi) - \frac{\sin(0.93\pi)}{3^4} + \frac{\sin(1.55\pi)}{5^4} - \ldots\right]$$

$$\simeq \frac{1}{26}\left[0.8 - \frac{0.07}{3^4} + \ldots\right]$$

$$\simeq 0.29.$$

• PROBLEM 1144

A vibrating string, fixed at x = 0 and x = L(ft.), is subjected to a damping force which is proportional to the speed at each point. If it starts from rest with an initial displacement f(x), show that the displacement function y(x,t) is given by

$$y(x,t) = e^{-bt} \sum_{n=1}^{\infty} A_n \left(\cos B_n t + \frac{b}{B_n} \sin B_n t\right) \sin \frac{n\pi x}{L},$$

where $B_n = \frac{\sqrt{n^2\pi^2 a^2 - b^2L^2}}{L}$

and $A_n = \frac{2}{L}\int_0^L f(x) \sin \frac{n\pi x}{L} dx.$

Solution: Consider an elementary deformed length of the string Δs which was, under steady conditions, of length Δx on the x-axis. The deformation due to some external force, which resulted in displacing every point of the string to a curve expressed by $f(x)$, was balanced by the tension forces F which can be thought constant throughout, on the condition that the string is assumed to be perfectly flexible.

Starting with the time the force was removed, $t = 0$, we use Newton's second law of motion: that the resultant tension force on the string expressed by

$$F\left[\frac{\partial}{dx}\, y(x+\Delta x,0) - \frac{\partial}{dx}\, y(x,0)\right]$$

must be equal to the product of the mass of Δs and its acceleration. To find its mass, we let w be the weight of the undeformed string per unit length, hence $\frac{w}{g}\Delta x$ is the mass of the segment Δs. Thus, by dividing the entire equation by Δx and passing to the limit, we have:

$$\frac{w}{g}\frac{\partial^2 y}{\partial t^2} = F\frac{\partial^2 y}{\partial x^2},$$

or

$$\frac{\partial^2 y}{\partial t^2} = a^2\frac{\partial^2 y}{\partial x^2}, \quad (t \geq 0) \quad a^2 = \frac{Fg}{w}.$$

For purposes of simplicity, the velocity at $t \geq 0$ is considered negligible, but not its acceleration. At some other time t the elementary length ds is not only acted upon by the resultant tension of the string but also by a resistance force proportional to its speed : $2b\,\partial y/\partial t$, where $b\ (\text{sec.}^{-1})$ is a constant. Hence,

$$\frac{\partial^2 y}{\partial t^2} = a^2\frac{\partial^2 y}{\partial x^2} - 2b\frac{\partial y}{\partial t}.$$

Assuming the solution of the above differential equation to be expressed in a product form:

$$y(x,t) = X(x)\,T(t),$$

where

$$\frac{\partial^2 y}{\partial t^2} = X(x)\, T''(t),$$

$$\frac{\partial^2 y}{\partial x^2} = T(t)X''(x) \quad \text{and} \quad \frac{\partial y}{\partial t} = X(x)T'(t).$$

Substitution yields:

$$XT'' = a^2\, T\, X'' - 2b\, XT'$$

Separation of the variables, and assigning a constant $-k^2$, we have:

$$\frac{T'' + 2b\, T'}{a^2\, T} = \frac{X''}{X} = -k^2.$$

Here, k is chosen negative since the motion of points on the string is periodic vibration. Hence, the solution for the function of x only is given by:

$$X(x) = c_1 \cos kx + c_2 \sin kx.$$

For the other differential equation, involving time, we have:

$$T'' + 2b\, T' + (ka)^2\, T = 0.$$

with the auxiliary equation:

$$m^2 + 2bm + (ka)^2 = 0,$$

and roots $r = -b \pm \sqrt{b^2 - (ka)^2}$. The particular solution is:

$$T(t) = e^{-bt}\left(c_3'\, e^{\sqrt{b^2-(ka)^2}} + c_4'\, e^{-\sqrt{b^2-(ka)^2}}\right).$$

Physically, for harmonic oscillations to appear, $ka > b$, so that the string is not over-damped. Hence,

$$\sqrt{b^2 - (ka)^2} = \sqrt{-\left((ka)^2 - b^2\right)} = i\sqrt{(ka)^2 - b^2}$$

$$T(t) = e^{-bt}\left[c_3 \cos \sqrt{(ka)^2 - b^2}\, t + c_4 \sin \sqrt{(ka)^2 - b^2}\, t\right].$$

Letting $\sqrt{(ka)^2 - b^2} = B$, we have:

$$y(x,t) = \left(c_1 \cos kx + c_2 \sin kx\right)e^{-bt}\left(c_3 \cos Bt + c_4 \sin Bt\right).$$

Here, the condition: $y(0,t) = 0$, rules out the function $\cos kx$, and the condition $y(L,t) = 0$ gives rise to $\sin kL = 0$, yielding $k = \frac{n\pi}{L}$.

Since k depends on n as shown above, we can replace B by B_n and k by k_n, which is called the eigenvalue. Replacing the value of k in the expression for B_n,

$$B_n = \left(\frac{n^2\pi^2a^2}{L^2} - b^2\right)^{\frac{1}{2}} = \frac{\sqrt{n^2\pi^2a^2 - b^2L^2}}{L}.$$

The displacement function now is:

$$y(x,t) = e^{-bt}\, c_2 \left(c_3 \cos B_n t + c_4 \sin B_n t\right) \sin \frac{n\pi}{L} x.$$

Using the condition that the string is initially at rest, we have:

$$\left.\frac{\partial}{\partial t} y(x,t)\right|_{t=0} = 0$$

$$= -be^{-bt}\, c_2 \left(c_3 \cos B_n t + c_4 \sin B_n t\right) + \left. e^{-bt}\, c_2 B_n \left(-c_3 \sin B_n t + \cos B_n t\right)\right|_{t=0}$$

$$= - bc_3 + c_4 B_n .$$

$$c_4 = \frac{b}{B_n} c_3 .$$

Thus, so far:

$$y(x,t) = c_2 c_3 e^{-bt} \left(\cos B_n t + \frac{b}{B_n} \sin B_n t \right) \sin \frac{n\pi x}{L} x,$$

with the final constants $c_2 c_3$ to be determined.

It is given that the initial displacement, in its Fourier expansion, is:

$$y(x,0) = c_2 c_3 \sin \frac{n\pi x}{L} = f(x)$$

$$= \sum_{n=1}^{\infty} b_n \sin \frac{n\pi x}{L} ,$$

which suggests the fact that the linear sum of distinct solutions of a partial differential equation is also a solution. Hence

$$c_2 c_3 = (c_2 \cdot c_3)_n = b_n ,$$

where b_n is the coefficient of the sine expansion of $f(x)$ in its Fourier series:

$$b_n = \frac{2}{L} \int_0^L f(x) \sin \frac{n\pi x}{L} dx.$$

Thus,

$$y(x,t) = e^{-bt} \sum_{n=1}^{\infty} b_n \left(\cos B_n t + \frac{b}{B_n} \sin B_n t \right) \sin \frac{n\pi x}{L} .$$

As an alternative approach, we use the following method. Since $y(x,t)$ is a smooth function with respect to x and vanishes at the end points, namely $x = 0$ and $x = L$, then it can be expressed in a Fourier sine series appropriate to this interval.

$$y(x,t) = \sum_{n=1}^{\infty} c_n(t) \sin \frac{n\pi x}{L} .$$

Differentiating this,

$$\frac{\partial y}{\partial t} = \sum_{n=1}^{\infty} \frac{d}{dt} c_n(t) \sin \frac{n\pi x}{L} ,$$

$$\frac{\partial^2 y}{\partial t^2} = \sum_{n=1}^{\infty} \frac{d^2}{dt^2} c_n(t) \sin \frac{n\pi x}{L}$$

and

$$\frac{\partial^2 y}{\partial x^2} = - \sum_{n=1}^{\infty} \frac{n^2 \pi^2}{L^2} c_n(t) \sin \frac{n\pi x}{L} .$$

Substitution in the known expression for a stretched string:

$$\frac{\partial^2 y}{\partial t^2} = a^2 \frac{\partial^2 y}{\partial x^2} - 2b \frac{\partial y}{\partial t} ,$$

yields:

$$\sum_{n=0}^{\infty} \left(\frac{d^2}{dt^2} c_n(t) + 2b \frac{d}{dt} c_n(t) + \frac{n^2 \pi^2 a^2}{L^2} c_n(t) \right) \sin \frac{n\pi x}{L} = 0$$

This is satisfied if:

$$\frac{d^2}{dt^2} c_n(t) + 2b \frac{d}{dt} c_n(t) + \frac{n^2 \pi^2 a^2}{L^2} c_n(t)$$

$$= 0.$$

The auxiliary equation is:

$$m^2 + 2bm + \frac{n^2 \pi^2}{L^2},$$

with roots

$$r = -b \pm i \frac{\sqrt{n^2 \pi^2 a^2 - b^2 L^2}}{L}, \text{ where } b < n\pi a.$$

The solution is of the form:

$$c_n(t) = e^{-bt} \left(d_n \cos B_n t + e_n \sin B_n t \right),$$

where

$$B_n = \frac{\sqrt{n^2 \pi^2 a^2 - b^2 L^2}}{L}.$$

Substitution in the sine series yields:

$$y(x,t) = e^{-bt} \sum_{n=1}^{\infty} \left(d_n \cos B_n t + e_n \sin B_n t \right)$$

As before,

$$\sin \frac{n\pi x}{L}.$$

$$\left. \frac{\partial y}{\partial t} \right|_{t=0}, \text{ yielding: } e_n = \frac{b}{B_n} d_n.$$

Hence,

$$y(x,t) = e^{-bt} \sum_{n=1}^{\infty} d_n \left[\cos B_n t + \frac{b}{B_n} \sin B_n t \right]$$

$$\sin \frac{n\pi x}{L}.$$

The second equation is:

$$y(x,0) = f(x) = \sum_{n=1}^{\infty} d_n \sin \frac{n\pi x}{L} .$$

But f(x) can be expanded into a Fourier sine series. Hence,

$$y(x,0) = f(x) = \sum_{n=1}^{\infty} c_n \sin \frac{n\pi x}{L}$$

$$= \sum_{n=1}^{\infty} d_n \sin \frac{n\pi x}{L} ,$$

where

$$c_n = d_n = \frac{2}{L} \int_0^L f(x) \sin \frac{n\pi x}{L} dx.$$

CHAPTER 44

FLUID PRESSURES/FORCES

When a fluid exerts a pressure p on an area A, the total force, F, is found by the expression:

$$F = pA.$$

In terms of differentials, $dF = pdA$.

The pressure exerted by a fluid is proportional to the height of fluid. The proportionality constant is the density, w, of the fluid (water). Consequently,

$$p = wh,$$

where h is the height of the fluid column above the point at which the pressure, p, is being considered.

By expressing p and A in terms of the parameters given in the problem, and integrating dF over the entire area, the total force against the area may be found.

• PROBLEM 1145

A vertical floodgate in the form of a rectangle 6 feet long and 4 feet deep has its upper edge 2 feet below the surface of the water, as shown in the diagram. Find the force which it must withstand.

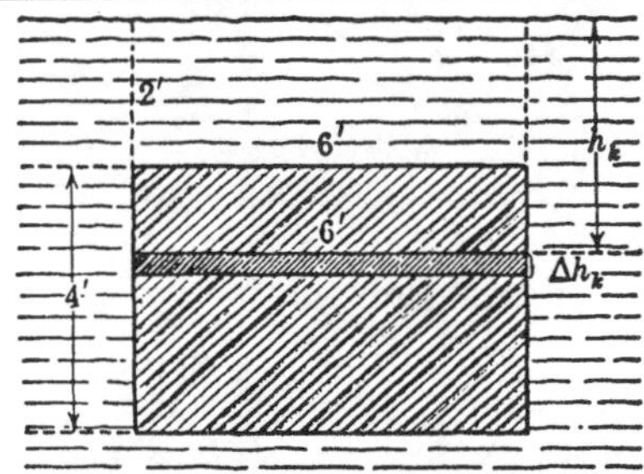

Solution: The pressure at a point on the floodgate is a function of its depth. Thus $p = wh$, where p denotes the pressure, w the density of the fluid, and h the depth of the point. The total force on the floodgate is a sum of the infinitesimal forces on rectangular strips of the gate with lengths parallel to the upper edge. $dF = pdA$, where dF denotes the differential force on an infinitesimal area dA of the floodgate.

A typical infinitesimal rectangular strip of the gate surface has length $l = 6$ and infinitesimal height dh, where h varies from

2 to 6. Hence, $dF = p \cdot 6dh = 6wh\ dh$. The density of water, w, is 62.5 pounds/cu.ft. Then the total force on the gate is:

$$F = \int_2^6 6wh\ dh = \frac{6wh^2}{2}\Bigg|_2^6 = 3w(36-4)$$

$$= 96w = 96(62.5) = 6000 \text{ pounds.}$$

• PROBLEM 1146

A floodgate in the shape of an equilateral triangle of side 2a is submerged vertically in water until the upper edge is at the surface of the water. Find the total force on one side of such a plate.

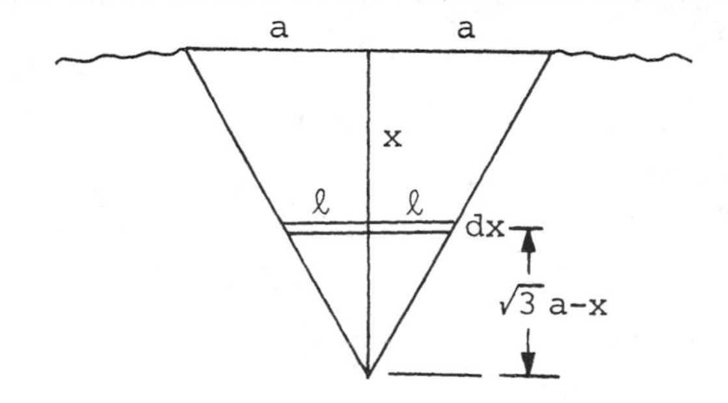

Solution: If 2ℓ is the length of a differential element of the plate submerged to a depth x, then, from the laws of similar triangles in plane geometry,

$$\frac{\ell}{\sqrt{3}a - x} = \frac{a}{\sqrt{3}a},$$

since the perpendicular distance from a corner of the equilateral triangle to the opposite side, i.e., the altitude, is:

$$\sqrt{(2a)^2 - a^2} = \sqrt{3}a,$$

and $\sqrt{3}a - x$ is the distance from the vertex of the triangle to the rectangular element,

$$\ell = a - \frac{\sqrt{3}}{3}x.$$

Area of the element, $dA = 2\ell\ dx$. Pressure of the water on the element, $p = wx$, where w is the density of water and x is the depth. The force exerted by the water on the element is then $dF = pdA = wx \cdot 2\ell\ dx$. x varies from 0 to $\sqrt{3}a$. The total force is then given by:

$$F = 2\int_0^{\sqrt{3}a} wx\left(a - \frac{\sqrt{3}}{3}x\right)dx$$

$$= 2w\left[\frac{ax^2}{2} - \frac{\sqrt{3}\,x^3}{9}\right]_0^{\sqrt{3}a}$$

$$= wa^3 \text{ lbs.}$$

• PROBLEM 1147

Compute the total force on a gate closing a circular pipe 4 feet in diameter when the pipe is half full.

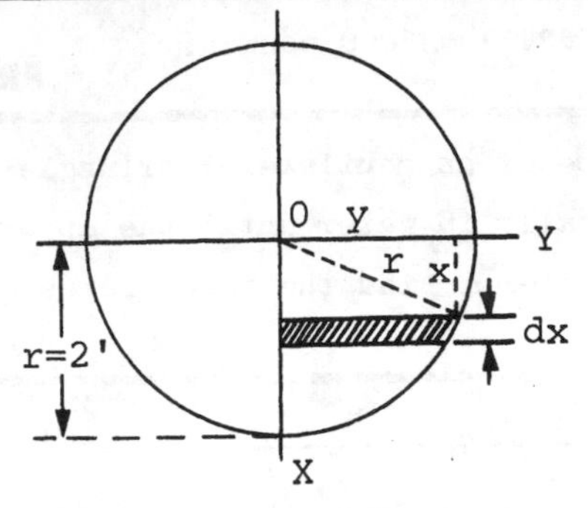

Solution: $y \cdot dx$ = a differential of area

xw = fluid pressure

= depth X weight of 1 cu. ft. of fluid, or density, w.

Therefore, $xw \cdot y \cdot dx$ = force on the differential area.

Summing up the differential fluid forces, F = total force = $w\int y \cdot x \cdot dx$.

For water, w = 62.5 lb./cu.ft. But

$$y = \sqrt{r^2 - x^2} = \sqrt{4 - x^2} .$$

Hence, $F = 62.5 \int_0^2 \sqrt{4 - x^2} \cdot x \cdot dx$ = total force on one-half of the water area.

To integrate, let $u = 4 - x^2$. For $x = 0$, $u = 4$. For $x = 2$, $u = 0$. This yields the new limits. Then,

$$du = -2x \cdot dx ,$$

and
$$F = \frac{2}{2} \cdot 62.5 \int_0^2 \sqrt{4 - x^2} \cdot x \cdot dx$$

$$= -\tfrac{1}{2} \cdot 62.5 \int_4^0 \sqrt{4 - x^2}(-2x)dx .$$

$$F = -\frac{62.5}{2} \int_4^0 u^{\frac{1}{2}} \cdot du$$

$$= -\frac{62.5}{2} \cdot \frac{u^{3/2}}{3/2}\Big]_4^0$$

$$= \frac{62.5}{3}\sqrt{4^3} = \frac{62.5}{3} \cdot 8 = \frac{500}{3} = 167 \text{ lb. for one-}$$

half of the water area. Hence, the force exerted by the water on the gate = 2·167 = 334 lb.

• PROBLEM 1148

The center of a circular floodgate of radius 2' in a reservoir is at a depth of 6' below the water surface. Find the total force on the gate.

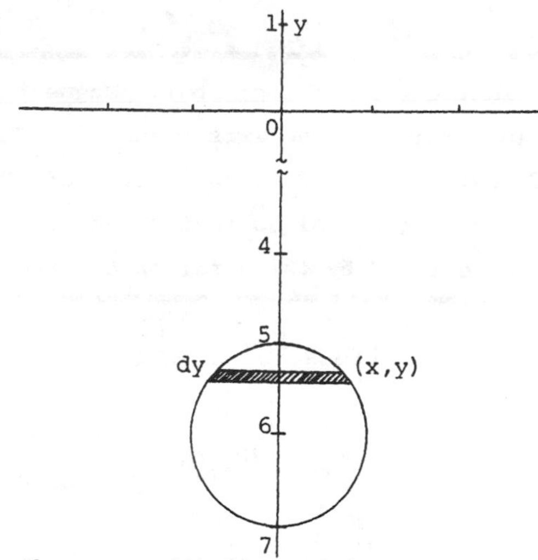

Solution: Choose the axes with the origin at the center of the circle as shown in the figure. The equation of the circle is: $x^2 + y^2 = 4$, hence $x = \sqrt{4 - y^2}$. The area of the rectangular strip shown is: $dA = 2x\, dy$. y varies from $-2'$ to $2'$. Pressure on the strip, $p = wh$, where w is the density of water, and $h = 6 - y$ is the depth of the strip below the water surface. The force on the strip: $dF = pdA = wh \cdot 2x\, dy$. Therefore, the total force on the gate is:

$$F = \int_{-2}^{2} w(6 - y) \cdot 2x\, dy$$

$$= 2w\int_{-2}^{2} (6 - y)\sqrt{4 - y^2}\,dy$$

$$= 2w\int_{-2}^{2} 6\sqrt{4 - y^2}\,dy - 2w\int_{-2}^{2} y\sqrt{4 - y^2}\,dy .$$

The second integral has an integrand with an odd power of y and hence vanishes in the integration from -2 to 2. The first integral has an integrand of even power in y, and so

$$\int_{-2}^{2} \sqrt{4 - y^2}\,dy = 2\int_{0}^{2} \sqrt{4 - y^2}\,dy .$$

Therefore, $F = 24w\int_{0}^{2} \sqrt{4 - y^2}\,dy.$

Put $y = 2 \sin \theta$, then $\sqrt{4 - y^2} = 2 \cos \theta$ and $dy = 2 \cos \theta\, d\theta$. For $y = 0$, $\theta = 0$; for $y = 2$, $\theta = \pi/2$. Therefore,

$$F = 24w\int_{0}^{\pi/2} 2 \cos \theta \cdot 2 \cos \theta\, d\theta$$

$$= 48w\int_{0}^{\pi/2} (1 + \cos 2\theta)d\theta$$

$$= 48w\left[\theta + \frac{\sin 2\theta}{2}\right]_{0}^{\pi/2}$$

$$= 48w\left[\frac{\pi}{2}\right] = 24\,\pi w \text{ lbs.}$$

A floodgate in the form of a parabolic segment is submerged vertically in water with the axis vertical. The base of the plate is 12 feet across and is at a depth of 10 feet, while the vertex is at a depth of 30 feet as shown in the figure. Find the force exerted by the water on one face of the plate.

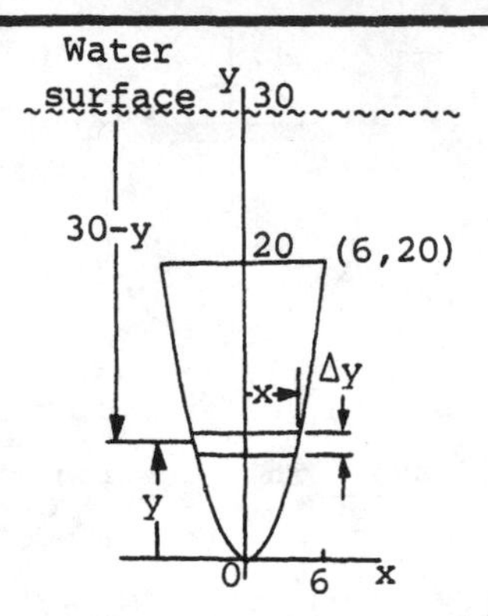

Solution: If the origin and axes are chosen as in the figure, then a typical element of area ΔA, on the face of the plate has dimensions $2x$ by Δy. The pressure on the rectangular element is $p = wh$ where $w = 62.5$ lb./sq.ft. is the density of water; $h = 30 - y$ is the depth of the rectangular element below the surface of water. An approximation to the force on this element is given by:

$$\Delta F = p\Delta A = 62.5(30 - y)2x\,\Delta y.$$

In order to express x in terms of y, we need to know the equation of the parabola. For the chosen axes, the equation must be of the form: $x^2 = ay$, and the curve passes through the point $(6,20)$. Therefore we find $a = 9/5$ and $x = \left(3/\sqrt{5}\right)y^{1/2}$.

It follows that $$\Delta F = 62.5(30 - y)\frac{6}{\sqrt{5}}\,y^{1/2}\,\Delta y,$$

where y varies from 0 to 20, and the total force on the face of the plate is given by:

$$F = \int_0^{20} \frac{6}{\sqrt{5}}(62.5)(30 - y)y^{1/2}\,dy$$

$$= \frac{375}{\sqrt{5}}\int_0^{20}\left(30y^{1/2} - y^{3/2}\right)dy$$

$$= \frac{375}{\sqrt{5}}\left[\frac{30y^{3/2}}{3/2} - \frac{y^{5/2}}{5/2}\right]_0^{20}$$

$$= \frac{375}{\sqrt{5}}\left[1788 - 715\right]$$

$$= 180{,}000 \text{ pounds.}$$

The ends of a trough filled with water are parabolic segments, as shown. If the width of the trough is 4 feet at the top and the depth is 4 feet, find the force on one end.

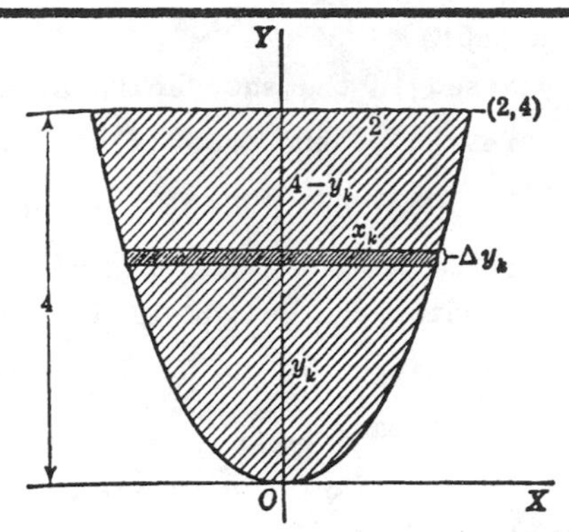

Solution: If we take a set of rectangular axes as shown in the figure, with the origin at the vertex of the parabola, then the equation of the parabola is: $y = x^2$. This can be checked at the points of intersection of the parabola with the top of the segment (2,4) and (-2,4). Taking a typical rectangular element of area ΔA_k with ordinate y_k, width Δy_k and length $2x_k$, where (x_k, y_k) denotes a point of the rectangular strip on the curve, and the depth of the element below the surface of the water is $4 - y_k$. To cover the parabolic segment, y must vary from 0 to 4. $2x_k = 2\sqrt{y_k}$ from the equation of the parabola. The pressure at a point on the rectangular strip is $p_k = wh_k$, where w is the density of water, $h_k = 4 - y_k$ is the depth of the rectangular strip. The increment of the force on the rectangular strip is $\Delta F_k = p_k \Delta A_k$, where $\Delta A_k = 2x_k \Delta y_k$. In the limit, $dF = pdA = w(4 - y)2\sqrt{y}\, dy$. Hence, the required force is:

$$F = w\int_0^4 (4 - y)2\sqrt{y}\, dy$$

$$= \frac{256}{15} w = 1066\,\frac{2}{3} \text{ pounds.}$$

CHAPTER 45

WORK/ENERGY

Work is expressed in the same units as energy. Work or energy may be expressed as the product of a force and the distance over which the force moves or is displaced. Work or energy may also be expressed as the product of a weight and the height through which the weight is displaceable, or in the form of kinetic energy, $\left(\frac{1}{2} mv^2\right)$, where m is mass and v is the velocity of the mass.

If the work involves a force varying over the distance moved, we write the relationship of the force, f, acting over an infinitesimal distance, ds, to obtain the total work done from

$$W = \int_a^b f \cdot ds ,$$

where f is a function of s, and the limits a and b are the starting and end points of the path, over which the force acts.

It should be remembered that f is the force that acts <u>along</u> the path ds. If the force does not coincide entirely with the tangent along the path, then the component along the path must be used. This component is equal to $F \cos \theta$, where F is the full force, and θ is the angle between the force and the tangent to the path.

Whereas F and s are both vector quantities specified by both magnitude and direction, W is not a vector quantity. W is a scalar quantity specified by magnitude only.

• PROBLEM 1151

Find the instantaneous rate of change of force if the equation for work is 1,000 = Fx.

Work	=	force	X distance.
(joules)		(newtons)	(meters)

Solution: In this problem,

$$F = \frac{\text{work}}{\text{distance}} = \frac{1,000}{x}$$

Find dF/dx by the Δ method:

$$F + \Delta F = \frac{1,000}{x + \Delta x} .$$

Subtracting F from both sides,

$$\Delta F = \frac{1,000}{x + \Delta x} - F.$$

Substituting the value of F,

$$\Delta F = \frac{1,000}{x + \Delta x} - \frac{1,000}{x} .$$

Using the common denominator,

$$\Delta F = - \frac{1,000(\Delta x)}{(x + \Delta x)x}$$

Dividing the equation by Δx,

$$\frac{\Delta F}{\Delta x} = - \frac{1,000}{(x + \Delta x)x} = - \frac{1,000}{x^2 + x\Delta x}$$

$$\lim_{\Delta x \to 0} \frac{\Delta F}{\Delta x} = \frac{dF}{dx} = - \frac{1,000}{x^2} \text{ newtons/m}^2 .$$

• PROBLEM 1152

A 20-lb. weight is being raised by a 100-ft. rope weighing $\frac{1}{2}$ lb/ft. Determine the work needed to raise the weight 50 ft.

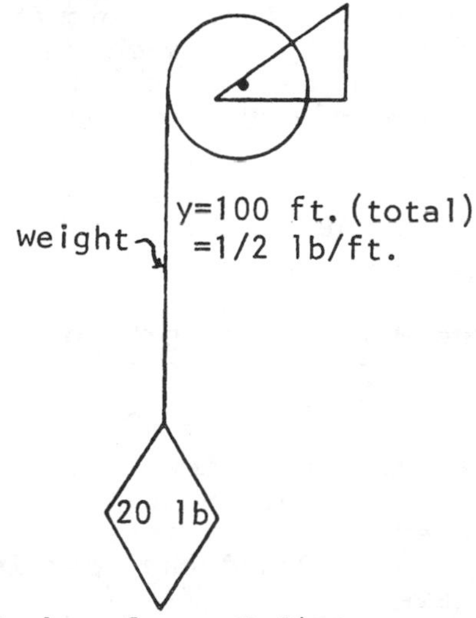

Solution: Work = force X distance.

$$F = 20 + (100 - y) \frac{1}{2}$$

Let y be the distance moved at any time, hence, the remaining length of the rope is (100 - y)-ft., weighing

$$(100 - y)\text{ft.} \cdot \frac{1}{2} \text{ lb./ft.} = (100 - y)\frac{1}{2} \text{ lb.}$$

Since work is the product of the force applied and the distance moved, a change of work can be expressed in terms of the corresponding change in distance. Thus, for an infininitesimal work we have the following differential equation:

$$dW = Fdy$$

$$dW = \left[20 + (100 - y)\frac{1}{2}\right] dy$$

$$= \left[70 - \frac{y}{2}\right] dy.$$

Integrating both sides,

$$\int dW = \int \left(70 - \frac{y}{2}\right) dy.$$

The weight has to be raised 50 ft., i.e. $0 \le y \le 50$, therefore,

$$W = \int_0^{50} \left[70 - \frac{y}{2}\right] dy$$

$$= 70d - \frac{y^2}{4}\Bigg|_0^{50} = 2,875 \text{ ft.-lb.}$$

• PROBLEM 1153

A ship's anchor weighs a ton (2000 lbs.) and the anchor chain weighs 50 lbs./linear ft. What is the work done in pulling up the anchor if 100 ft. of chain are out, assuming that the lift is vertical?

Solution: Let x be the number of feet of anchor chain out at any time. The portion of the chain that is out weighs 50x pounds.

After some time, say Δt, the anchor has moved up Δx ft. The work done within time Δt, noting that the work is the product of the weight and perpendicular distance moved, is:

$$\Delta w = (50x + 2000)\Delta x.$$

Δx can be made infinitesimally small, so that the total work done in pulling up the anchor 100 ft. can be evaluated by integration. Since x varies from 100 to 0 and the work is a positive quantity, we have:

$$w = \int_0^{100} (50x + 2000)dx$$

$$= \int_0^{100} 50x\, dx + \int_0^{100} 2000\, dx$$

$$= 25\, x^2\Big|_0^{100} + 2000\, x\Big|_0^{100}$$

$$= 250,000 + 200,000$$

$$= 450,000 \text{ ft.lbs.}$$

• PROBLEM 1154

A particle free to move on the X-axis is attracted toward the origin by a force of magnitude equal to kx^2 (k constant). Find the work done when the particle moves from the position x = 4 to the position x = 2.

Solution: Since the attraction is in the negative direction on the X-axis, we take

$$F(x) = -kx^2.$$

Then,

$$W = \int_4^2 F(x)\,dx = -\int_4^2 kx^2 dx$$

$$= -\frac{k}{3}\left|x^3\right|_4^2 = \frac{k}{3}(64 - 8) = \frac{56}{3}k.$$

• PROBLEM 1155

By Hooke's law, any force producing an elongation of a helical spring is proportional to the elongation produced, the constant of proportionality being called the spring constant. A spring, the constant of which is 2 lb./in., has a natural length of 10 in. Find the work done in stretching the spring from a length of 12 in. to a length of 15 in.

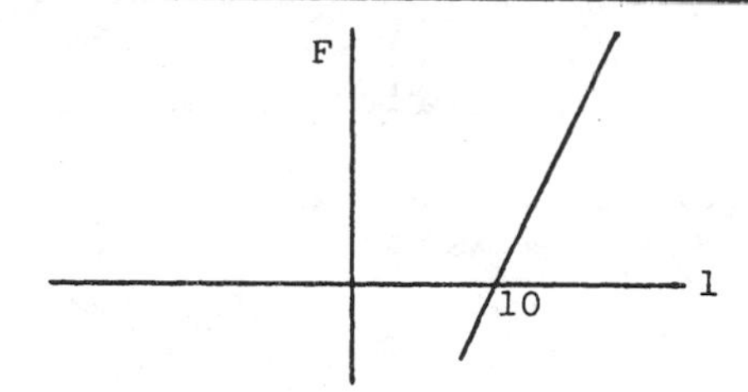

Solution: Under the given conditions, the length with no external force is $L_o = 10$. Then, for every force applied, the length increases numerically by one-half of the magnitude of the force. Assuming linearity, the dependence in differential form is:

$$dF = 2\,dL.$$

Under the given boundary conditions,

$$F = 2L - 20. \qquad (1)$$

At some length L_i, the corresponding force is $F_i = 2L_i - 20$. For a small increment of length to l_{i+1}, we have $F_{i+1} = 2L_{i+1} - 20$. If the increment in length $\overline{L_i L_{i+1}}$ is sufficiently small, then we can regard $F_i \simeq F_{i+1}$. The work performed in this elongation, taking F_i as the constant force for this length, and assuming the force is applied along the axis of the spring, is $F_i \cdot dL_i$, where $L_{i+1} - L_i = dL_i$. Hence,

$$dw_i = F_i dL_i = (2L_i - 20)\, dL_i;\ i = 1,2,3,\ldots,n$$

The work done in stretching the spring from L_1 = 12 in. to L_n = 15 in. is the sum of all contributions of dw_i's. Passing to the limit - as n approaches infinity - the sum can be expessed as:

$$w = \int_{12}^{15} (2L - 20)\, dL = L^2 - 20L = L(L - 20)\Big|_{20}^{15}$$

$$= 15\ (15 - 20) - 12\ (12 - 20)$$

$$= 15\ (-5) - 12\ (-8)$$

$$= 96 - 75 = 21 \text{ lb.in.}$$

• PROBLEM 1156

Hooke's law states that an elastic body such as a spring stretches an amount proportional to the force applied. If a spring is stretched 1/2 inch when a 2 pound force is applied to it, how much work is done in stretching the spring an additional 1/2 inch?

Solution: If the work,

$$W = \int_a^b F(x)\, dx,$$

the problem is to determine the force function F and the interval [a,b]. By Hooke's law, if x denotes the distance the spring has been stretched, then the force function,

$$F(x) = kx.$$

But when $x = \frac{1}{2}$, $F(x) = 2$. Hence $k/2 = 2$, or $k = 4$, and it follows that $F(x) = 4x$. The work done in stretching the spring an additional 1/2 inch is:

$$W = \int_{1/2}^{1} 4x\,dx = 4\ \frac{x^2}{2}\Big|_{1/2}^{1} = \frac{3}{2}\ .$$

If the force is measured in pounds and the distance in feet (inches), then the unit of measurement for work is the foot-pound (inch-pound). Hence for this problem, where the dimensions are in inches,

$$W = \frac{3}{2} \text{ inch-pounds.}$$

• PROBLEM 1157

How much work is done in stretching a spring 4 in. if it takes 500 lb. of force to stretch it that amount?

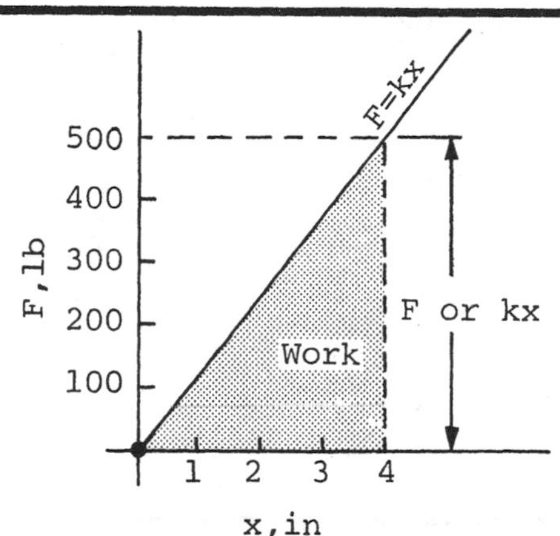

Force-Distance

Solution: The constant k is the spring constant and represents the number of pounds required either to stretch or to compress the spring 1 in., provided the material of the spring remains within its elastic range, i.e.,

$$F = kx$$

$$k = \frac{\text{total load on spring}}{\text{change in length}}$$

$$= \frac{500 \text{ lb}}{4 \text{ in.}} = 125 \text{ lb/in.}$$

Work is defined as the product of the force and distance when the force is in the direction of motion or displacement x: $W = Fx$, or:

$$W = \int F\,dx.$$

x varies from 0 to 4.

Thus, the required work in stretching the spring 4 in. is:

$$W = \int_0^4 F\,dx = \int_0^4 kx\,dx$$

$$= k \left| \frac{x^2}{2} \right|_0^4 \qquad k = 125 \text{ lb/in.}$$

$$\text{Work} = 125\ \frac{(4)^2}{2} = 1{,}000 \text{ in.-lb.}$$

Alternatively, since $\int_0^4 F\,dx$ represents the area of the triangle in the figure, one can evaluate the work by finding the shaded area:

$$A = W = \int_0^4 F\,dx.$$

$$\text{But} \quad A = \frac{1}{2}\,bh$$

$$= \frac{1}{2}\,(4)(4k)$$

$$= \frac{1}{2}\,(4)^2\,125 = 1{,}000 \text{ in.-lb.}$$

• PROBLEM 1158

The force required to stretch a spring is proportional to the elongation. If a force of one pound stretches a certain spring half an inch, what is the work done in stretching the spring 2 inches?

Solution: Call L the natural length of the spring and x the elongation. Then

$$F = kx,\ 1 = k \cdot \frac{1}{2},\ k = 2.$$

Hence $F = 2x$, and the work is:

$$W = \int_0^2 2x\,dx$$

$$= x^2 \Big|_0^2$$

$$= 4 \text{ in.-lbs.}$$

• PROBLEM 1159

A spring is 6 in. long and is stretched 1 in. by a 5 lb. weight. What is the work done in stretching this spring from 8 to 11 inches?

Solution: The force varies with the stretch or elongation. Therefore, $F = ks$. 5 lb. = k · 1 in. (A 5 lb. force stretches the spring 1 in.)

Therefore,

$k = 5$, and $F = 5s$. $dW = F \cdot ds$.

A differential of work = force x differential of stretch.

$$dW = 5s \cdot ds$$

and

$$W = 5 \int_{2 \text{ in.}}^{5 \text{ in.}} s \cdot ds .$$

When the spring is 8 in. long, the stretch is 2 in.; when 11 in. long, the stretch is 5 in. Therefore, the limits of integration are 2 and 5; not, as might be assumed without due consideration, 8 and 11.

$$W = \left| 5 \cdot \frac{s^2}{2} \right|_2^5 = \left(5 \cdot \frac{25}{2} - 5 \cdot \frac{4}{2} \right)$$

$$= 52.5 \text{ in.-lb.}$$

• PROBLEM 1160

Hooke's law states that, if a spring is stretched x in. beyond its unstretched length, it is pulled back with a force equal to kx lb, where k is a constant depending on the material used.

A spring has an unstretched length of 14 in. If a force of 5 lb is required to keep the spring stretched 2 in., how much work is done in stretching the spring from its unstretched length to a length of 18 in.?

Solution: Place the spring along the x axis as shown, with the origin at the point where the stretching starts.

Let x = the number of inches the spring is stretched;

$f(x)$ = the number of pounds in the force acting on the spring x in. beyond its unstretched length.

Then, by Hooke's law, $f(x) = kx$. Because $f(2) = 5$, we have:

$$5 = k \cdot 2$$

$$k = \frac{5}{2} .$$

Therefore, $f(x) = \frac{5}{2} x .$

If W = the number of inch-pounds of work done in stretching the spring from its unstretched length of 14 in. to a length of 18 in., then the infinitesimal work done in stretching the spring from x to x + dx, is

$$dW = f(x)\, dx.$$

x varies from 0 to 4.

We have:

$$W = \int_0^4 f(x)\, dx$$

$$= \int_0^4 \frac{5}{2} x\, dx$$

$$= \frac{5}{4} x^2 \Big|_0^4$$

$$= 20.$$

Therefore, the work done in stretching the spring is 20 in.-lb.

• PROBLEM 1161

A spring of unstretched length 15 inches requires a force of 8 pounds to stretch it 1 inch. Find the work necessary to stretch the spring from a length of 16 inches to a length of 18 inches.

Solution: By Hooke's law, the force F(x) required to stretch the spring x inches is F(x) = kx. Since F(x) = 8 when x = 1, we have k = 8, so that in this case F(x) = 8x. The infinitesimal work done in stretching the spring an infinitesimal distance dx is

$$dW = F(x)dx. \text{ Here,}$$

x varies from 16 - 15 = 1" to 18 - 15 = 3". The required work is then:

$$W = 8\int_1^3 x\, dx = 8\left|\frac{x^2}{2}\right|_1^3 = 8\left(\frac{9}{2} - \frac{1}{2}\right)$$

$$= 32 \text{ inch-pounds} = 2\,\frac{2}{3} \text{ foot-pounds.}$$

• PROBLEM 1162

A cylindrical tank has a diameter of 10 ft., is 15 ft. high, and contains water to a depth of 10 ft. What is the work done in pumping the water to a height of 15 ft. above the top of the tank?

Solution: A differential of work = dW = F · ds, where F = the weight of water in the differential of height ds, equal to $62.5\pi r^2 \cdot ds$, and s is the total height this element of the weight is raised. The density of water is 62.5 lbs./cu.ft. Then

$$dW = 62.5\pi r^2 \cdot ds \cdot s$$

The total work,

$$W = 62.5\pi r^2 \int_{20}^{30} s \cdot ds$$

$$= 62.5\pi r^2 \left| \frac{s^2}{2} \right|_{20}^{30}$$

The limits are 20 ft. and 30 ft. down from the line to which the water is to be raised.

$$W = 31.25\pi r^2 \left| s^2 \right|_{20}^{30} = 31.25\pi r^2 (900 - 400)$$

$$= 31.25 \cdot 3.1416 \cdot 25 \cdot 500$$

$$= 1{,}227{,}187.5 \text{ ft.-lb.} = \text{work.}$$

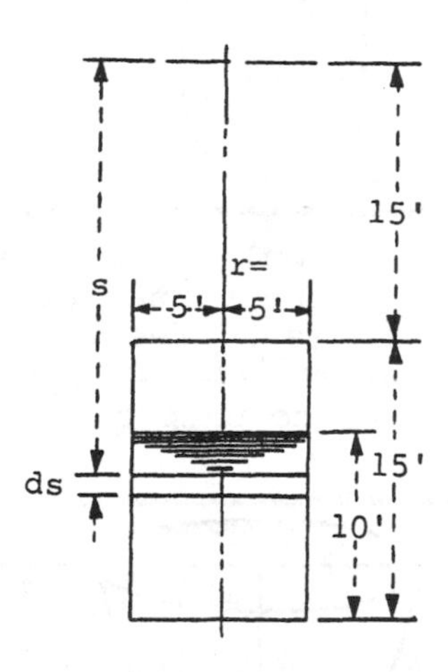

• PROBLEM 1163

A vertical cylindrical tank of radius γ ft. and height h ft. is filled with water. By pumping the water out over the rim at the top, the tank is emptied. Find the work done in emptying it.

Solution: Consider an element of water (a disk) at depth x from the top. The volume of this disk is $\pi\gamma^2 dx$. Its weight is therefore $w\pi\gamma^2 dx$, where the weight of 1 cu. ft. of the liquid is w (density); Here, w = 62.5 lbs., per cu. ft., since the liquid is water. In general, w depends upon the liquid considered.

Work done = Force x distance.

The work done in lifting this weight (force x distance) is, therefore:

$$dW = 62.5\ \pi\gamma^2 x\ dx,$$

x varies from 0 to h, and the total work is:

$$W = \int_0^h 62.5\ \pi\gamma^2 x\ dx$$

$$= \frac{62.5\ \pi\gamma^2 h^2}{2}\ \text{ft.-lbs.},$$

in the case of water, or

$$\frac{w}{2}\ \pi\gamma^2\ h^2$$

in the case of a liquid of density w.

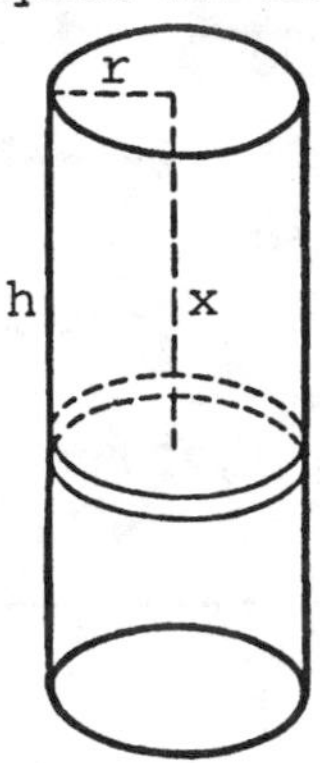

• **PROBLEM 1164**

A conical tank with circular cross-section has its axis vertical and the vertex at the bottom, The radius of the top is 10 feet, the altitude is 30 feet, and water in the tank is 12 feet deep. Find the work required to pump the water out over the top.

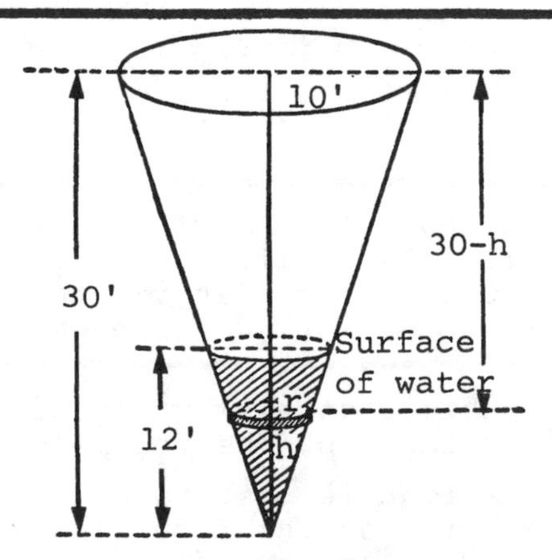

Solution: Let us consider the water to be divided into n layers by horizontal planes, and let us approximate these layers by thin cylindrical disks (as in the usual method for finding volumes). If such a typical disk is at a distance h_k above the vertex and is of thickness Δh_k and of radius r_k, then the weight of this disk is $w \cdot \pi r_k^2 \Delta h_k$, where $w = 62.5$ is the approximate weight of 1 cubic foot of water, and r_k and h_k are expressed in feet. From similar triangles in the figure, we have:

$$\frac{r_k}{h_k} = \frac{10}{30} \quad \text{or} \quad r_k = \frac{1}{3} h_k.$$

Then, the weight of the disk is:

$$\frac{1}{9} \pi w h_k{}^2 \Delta h_k.$$

In pumping out the water over the top, this typical layer (disk) is raised a distance $30 - h_k$ feet. Work done in lifting the disk of water to the top is:

$$dW = \text{weight of disk(force)} \times \text{distance lifted}$$

$$= \frac{1}{9} \pi w h^2 dh \cdot (30 - h).$$

Hence, the required work is:

$$W = \frac{1}{9} \pi w \int_0^{12} h^2 (30 - h) dh = 1344 \pi w$$

$$= 84{,}000\pi \text{ foot-pounds.}$$

• PROBLEM 1165

Determine the amount of work required to pump the water out through the top of a 3-ft radius sphere which is half-filled. Let the density of water be 62 lb/ft^3.

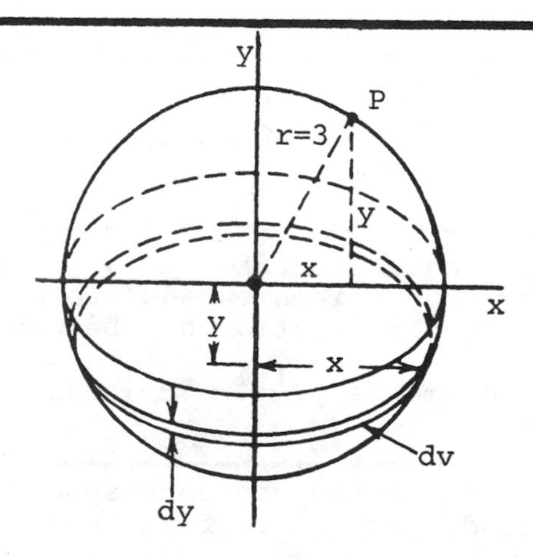

Solution: The equation of circle is:

$$x^2 + y^2 = r^2, \text{ hence,}$$

in this case,

$$x^2 + y^2 = 9.$$

Solving for x as a function of y,

$$9 - y^2 = x^2.$$

The volume of a slice (cylinder with height dy),

$$dV = \pi x^2 dy.$$

The weight of a slice, $dF = 62\pi x^2 dy$.

Work = force X distance.

While pumping, the water level inside the sphere is lowered from 3 ft to 6 ft. below the top of the sphere.

At any one instant, the water level is 3 + y ft. below the top of the sphere.

The work required to bring a circular slice of height dy from 3 + y ft. depth to the top of the sphere is

$$W = F(3 + y).$$

$$dW = (3 + y)dF = (3 + y)62\pi x^2 dy$$

Therefore, the total work required in emptying the sphere is:

$$\int dW = 62\pi \int_0^3 x^2(3 + y)dy$$

We now substitute to express the integral in terms of y alone.

(see step 2):

$$W = 62\pi \int_0^3 (9 - y^2)(3 + y)dy$$

Multiply, so as to obtain the terms for integration:

$$W = 62\pi \int_0^3 (27 - 3y^2 + 9y - y^3)dy$$

$$= 62\pi \left| 27y - y^3 + \frac{9y^2}{2} - \frac{y^4}{4} \right|_0^3$$

$$= 14,462 \text{ ft/lb.}$$

• PROBLEM 1166

A volume of gas equal to 2ft.^{3} at a pressure of 100 pounds per square inch expands to a volume of 3 cubic feet. If the relationship between the pressure and volume is $pv^{1.4} = c$ (constant), find the work done.

Solution: Substituting the corresponding values, p = 14400 (pounds per square foot) and v = 2 (cubic feet) in the formula $pv^{1.4} = c$, we find c = 38002. Then $p = 38002\ v^{-1.4}$. The work done by an expanding gas depends only on the volume and not on the shape of the container. We can therefore suppose that the gas is in a cylinder and expands against a piston head of cross-sectional area A. If the gas expands an amount Δv, the piston moves a distance $\Delta v/A$(feet). The force on the piston head is pA. By remembering that work = Force x distance, we have:

$$W = \int_{v_1}^{v_2} (pA) \cdot \frac{dv}{A} = \int_{v_1}^{v_2} p\, dv,$$

where v_1 and v_2 are the limiting volumes. Note that the cancellation of the area confirms the assumption that the shape of the container is immaterial. In the present problem,

$$W = \int_2^3 p\, dv = 38002 \int_2^3 v^{-1.4} dv$$

$$= 10{,}781 \text{ foot-pounds.}$$

• PROBLEM 1167

If a gas expands isothermally, its pressure p lb./in.^{2} and volume V in.^{3} are connected by the relation $pV = k$, where k is a constant. (a) Find the work done in compressing the gas in a cylinder to half its original volume. (b) Find the work done if the gas expands adiabatically in accordance with the relation $pV^{1.4} = k$.

Solution: (a) Let the cross-sectional area of the cylinder be $A(\text{in.})^2$. Then the force applied by the gas to the cylinder is given by:

$$F = pA.$$

The work done in moving the piston against this force for a length dh is:

$$dw = F\, dh = pA dh.$$

The small change in volume is:

$dV = A dh$. Hence, $dw = pdV$.

Since $p = \frac{k}{V}$, the total work done in compressing the volume from $V = V_o$ to $V = \frac{1}{2} V_o$ (where V_o is the original volume of the gas) is:

$$W = \int_{V_o}^{1/2\, V_o} \frac{k}{V} dV = k\left[\ln \frac{V_o}{2} - \ln V_o\right] = k \ln \tfrac{1}{2} \text{lb. - in.}$$

(b) If

$$pV^{1.4} = k, \text{ then } p \frac{k}{V^{1.4}}$$

$$W = \int_{V_o}^{1/2\ V_o} \frac{k}{V^{1.4}}\, dV = - \frac{1}{0.4} \frac{k}{V^{0.4}} \Bigg|_{V_o}^{1/2\ V_o}$$

$$= -2.5k \left(\frac{1}{\left(\frac{V_o}{2}\right)^{0.4}} - \frac{1}{\left(V_o\right)^{0.4}} \right)$$

If work done in compression is positive, work done in expansion is negative.

• PROBLEM 1168

Air expands against a movable piston in a closed cylinder from a volume of 6 cubic feet to a volume of 10 cubic feet. If the expansion takes place acording to the well-known law,

$$pV^{1.41} = k,$$

and the pressure is 60 pounds per square inch at the 6 cubic ft. volume, compute the work done.

Solution: Let A = area of piston, V = volume, s = piston movement.

Force F = pA = unit pressure x area.

But $pV^{1.41} = k$.

$$p = \frac{k}{V^{1.41}}, \text{ the unit pressure.}$$

$$F = \frac{kA}{V^{1.41}}.$$

dw = F · ds (differential of work = force x differential of expansion),

or
$$dw = \frac{kA}{V^{1.41}} \cdot ds = A \cdot s. \qquad (1)$$

Therefore,

$$s = \frac{V}{A},$$

and, differentiating this, we obtain:

$$ds = \frac{dV}{A}$$

Substituting in eq. (1), we have:

$$dw = \frac{kA}{V^{1.41}} \cdot \frac{dV}{A} = \frac{k}{V^{1.41}} \cdot dV,$$

which is to be integrated. Then

$$w = k \int_6^{10} V^{-1.41} \cdot dV$$

$$= \left| -\frac{k}{0.41} \cdot V^{-0.41} \right|_6^{10},$$

or, $$w = -\frac{k}{0.41}\left(10^{-0.41} - 6^{-0.41}\right). \qquad (2)$$

Now we find k from $pV^{1.41} = k$. When $p = 60$ lb./sq.in. $= 8,640$ lb./sq.ft., then

$$k = 8,640 \cdot 6^{1.41}.$$

Substituting this in eq. (2) yields:

$$w = -\frac{8,640 \cdot 6^{1.41}}{0.41}\left(10^{-0.41} - 6^{-0.41}\right)$$

$$= -\frac{8,640 \cdot 6^{1.41}}{0.41}(0.3890 - 0.4797)$$

$$= \frac{8,640 \cdot 6^{1.41} \cdot 0.0907}{0.41}$$

$= 23,907$ ft.-lb. of work.

• PROBLEM 1169

Under certain conditions that are usually approximated in a gas engine, the product of the number of units of pressure per square unit of area and the number of units of volume of gas are related by the équation:

$$pv^{1.4} = k = \text{a constant}.$$

If the gas is confined in a cylinder with a piston at one end, the pressure of the gas causes the piston to move so as to increase the volume, thus lowering the pressure. Suppose the diameter of the cylinder is 12 in. and the pressure is 100 lbs. per square in. when the piston is 10 in. from the inner end of the cylinder. Find the work done by the force due to the pressure of the gas in moving the piston 12 in., i.e., from 10 in. to 22 in. from the inner end.

Solution: Since $pv^{1.4} = k$, we first find k from the fact that $p = 100$ when $v =$ the volume of the cylinder 12 in. in diameter and 10 in. long. The original volume of the cylinder is:

$$v = \pi r^2 h = \pi(6)^2 10 = (36\pi)10.$$

To simplify the expressions, we let $A = 36\pi$. We have $p = 100$ when $v = 10A$, so that

$k = 100(10A)^{1.4}$

Letting x represent the variable length of the cylindrical chamber, we have $v = Ax$, and, hence,

$$p = \frac{k}{(Ax)^{1.4}} = \frac{100(10A)^{1.4}}{(Ax)^{1.4}} = \frac{100(10)^{1.4}}{x^{1.4}}$$

Since p denotes the pressure per square inch, the force acting on the piston is the product of the pressure and the cross-sectional area A. Thus

$$F = pA = \frac{100(10)^{1.4}A}{x^{1.4}}$$

Since the piston is moving from 10 in. to 22 in., the work done to move the piston is:

$$W = \int F\,ds = \int_{10}^{22} \frac{100(10)^{1.4}A}{x^{1.4}}\,dx$$

$$= 100\ (10)^{1.4}A \int_{10}^{22} x^{-1.4}\,dx$$

$$= 100\ (10)^{1.4}A \left| \frac{x^{-1.4+1}}{-\ 1.4+1} \right|_{10}^{22}$$

$$= 100(10)^{1.4}A \left| \frac{x^{-.4}}{-\ .4} \right|_{10}^{22}$$

$$= -\ 250(10)^{1.4}\ A \left(\frac{1}{(22)^{.4}} - \frac{1}{(10)^{.4}} \right)$$

By use of logarithms,

$10^{1.4} = 25.12$ and $1/(10)^{.4} - 1/(22)^{.4} = .1077$.

Hence,

$W = 250(25.12)(.1077)A = 676.3A$.

Inserting $A = 36\pi$ in this result,

$W = 76,490$ in.-lb. or 6374.2 ft.-lb.

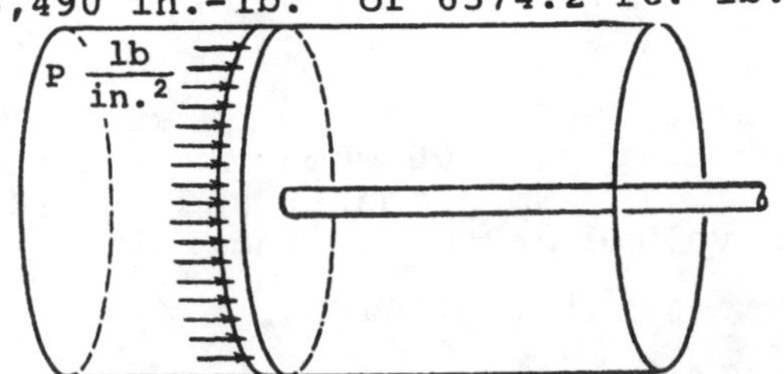

By Coulomb's law, the force of attraction between two unlike charges varies inversely as the square of the distance between them. If the force is A (dynes) when the charges are originally a distance a (cm.) apart, find the work done in moving one charge an additional distance a (cm.) from the other.

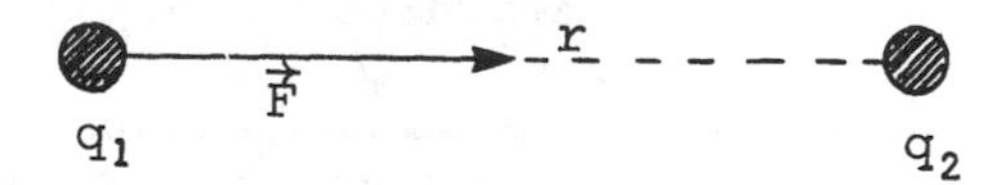

Solution: Coulomb's law, in mathematical form, is:

$$F = \frac{k\, q_1\, q_2}{r^2},$$

where k is a constant of proportionality, q_1 and q_2 are the respective charges, and r is the distance between these unlike charges. With no external influence, the force is directed along the line joining the charges, as shown in the diagram.

From the definition of work, the work done in moving an object from one place to another along a prescribed path is the product of the force projected on the line of displacement and the distance covered. This, in a mathematical expression, is:

$$w = F \cdot s \cos \Theta,$$

where Θ is the angle between the direction of the force and the displacement. Assuming that the displacement occurs along the line joining the charges, $\Theta = 0$, $\cos 0 = 1$ and $s = r$. In differential form, the work is:

$$dw = F\, dr, \qquad a \leq r \leq 2a.$$

q_1 and q_2 are constants..

$$w = \int_a^{2a} F\, dr = k\, q_1 q_2 \int_a^{2a} \frac{1}{r^2}\, dr$$

$$= kq_1q_2 \left(-\frac{1}{r}\right|_a^{2a} = k\, q_1 q_2 \left(\frac{1}{a} - \frac{1}{2a}\right) = k\, q_1 q_2 \left[\frac{1}{2a}\right]$$

But $A = \dfrac{kq_1q_2}{a^2}$. Hence,

$$w = k\, q_1 q_2 \left(\frac{1}{2a}\right) = \frac{Aa}{2}.$$

If, instead of 2a, the charge were moved to an infinite distance, then the work would be:

$$w = k\, q_1 q_2 \left(\frac{1}{a} - \frac{1}{\infty}\right) = k\, \frac{q_1\ q_2}{a} \quad = Aa.$$

This is called the potential energy.

• PROBLEM 1171

> The work done by a particle in an electric field is given by $W = s^3$, where s is the displacement. What is the work done in moving the object from 1.99 to 2.01 meters. Work is measured in joules if displacement is given in meters.

Solution: The instantaneous work can be obtained by differentiating the given expression,

$w = s^3$:

$$\frac{dw}{ds} = 3s^2, \quad \text{or } dw = 3s^2\, ds.$$

The above differential form is easier to manipulate by using:

$$\Delta w = w_f - w_i = s_f^3 - s_i^3$$

where w_f and w_i are, respectively, the work corresponding to the particle's final position $(s_f = 2.01)$ and initial position $(s_i = 1.99)$.

Remember that dw is only an approximation to Δw and, therefore, we assign s the most convenient value - the average location of the particle, namely, s = 2.00 meters. Letting $ds = s_f - s_i = 0.02$ meters, and substituting in the equation, we have:

$$dw = 3(2.00)^2(0.02) = 0.24 \text{ joules}.$$

• PROBLEM 1172

> An exploratory positive charge of electricity of strength P_1 is moved from a distance a to a greater distance b from another positive charge of strength P_2. What is the work done?

Solution: Coulomb's Law states: Like charges of electricity repel each other with a force that is directly proportional to the product of the charges and inversely proportional to the square of the distance between them, or:

$$F = \text{force} = k \cdot \frac{P_1 \cdot P_2}{s^2}.$$

Differential of work = force X differential of distance moved, or

Substituting $dW = F \cdot ds$. for F in terms of s,

$$dW = k \cdot \frac{P_1 \cdot P_2}{s^2} \cdot ds,$$

and, since P_1 moves from a to b,

$$W = kP_1 \cdot P_2 \int_a^b \frac{ds}{s^2}$$

$$= kP_1 \cdot P_2 \int_a^b s^{-2} \cdot ds$$

$$= kP_1 \cdot P_2 \left| -s^{-1} \right|_a^b ,$$

or, $$W = kP_1 \cdot P_2 \left| -\frac{1}{s} \right|_a^b .$$

Finally,

$$W = kP_1 \cdot P_2 \left(\frac{1}{a} - \frac{1}{b} \right),$$ the

work in dyne-centimeters if s is in centimeters and F in dynes.

The unit strength of a charge is the repulsion force of 1 dyne at 1 cm. distance. Then two charges are equal and of unit strength.

• PROBLEM 1173

In an electric circuit, the power P (measured in watts) is defined as the rate of change of work (measured in joules) with respect to time t (measured in seconds). Thus, the instantaneous power P is represented by

$$P = \lim_{\Delta t \to 0} \frac{\Delta W}{\Delta t} .$$

If the work done by a certain current is given by

the equation $W = 8t^2 - 2t$ joules, find the power in the circuit at any time t.

(b) Find the power when t = 2 sec.

Solution: We have:

$$W = 8t^2 - 2t .$$

$$\Delta W = 8(t + \Delta t)^2 - 2(t + \Delta t) - \left(8t^2 - 2t\right)$$

$$= 8t^2 + 16t\,\Delta t + 8\Delta t^2 - 2t - 2\Delta t - 8t^2 + 2t$$

$$= 16t\,\Delta t + 8\Delta t^2 - 2\Delta t .$$

$$\frac{\Delta W}{\Delta t} = 16t + 8\Delta t - 2 .$$

$$P = \lim_{\Delta t \to 0} \frac{\Delta W}{\Delta t} = \lim_{\Delta t \to 0} (16t + 8\Delta t - 2)$$

$$= 16t - 2 \text{ watts.}$$

When $t = 2$ sec., we have:

$$P = 16(2) - 2 = 30 \text{ watts.}$$

CHAPTER 46

ELECTRICITY

Differential equations involving electrical circuits are generally derived from Ohm's law, Kirchof's law, and the definition relating current, I, and charge, q, by $I = \frac{dq}{dt}$. Even when applied to only relatively simple circuits and networks, the differential equations can be unwieldy. Advanced techniques, beyond the scope treated here, are essential when dealing with networks having, for example, several circuit loops.

• PROBLEM 1174

> Find the force of attraction of a thin straight wire of length L cm. upon a unit particle in the line of the wire at a distance h cm. from the nearer end.

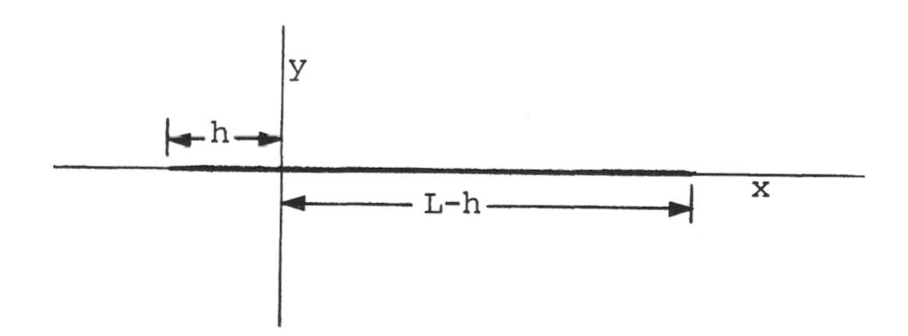

Solution: Let the density of the wire per unit length be ρ. Then an elemental length dx with a corresponding mass ρ dx, interacting with the unit mass situated at the origin, experiences a force:

$$dF = k\,\frac{\rho dx}{x^2}\,.$$

Taking the direction of the force along the negative x-axis, the net attraction is:

$$F = \lim_{\varepsilon\to 0}\int_{\varepsilon}^{-h}\frac{k\rho\ dx}{x^2} - \lim_{\varepsilon\to 0}\int_{\varepsilon}^{L-h}\frac{k\rho\ dx}{x^2}$$

$$= -k\rho \lim_{\varepsilon\to 0}\left(\frac{1}{x}\Bigg|_{\varepsilon}^{-h} - \frac{1}{x}\Bigg|_{\varepsilon}^{L-h}\right)$$

$$= k\rho\left(\frac{1}{h} + \frac{1}{L-h}\right)$$

$$= \frac{k\rho L}{h(L-h)} .$$

• PROBLEM 1175

Find the attraction of a thin circular disc of radius a cm. and surface density ρ gm./cm.^{2} upon a unit particle at a distance h cm. above the center of the disc.

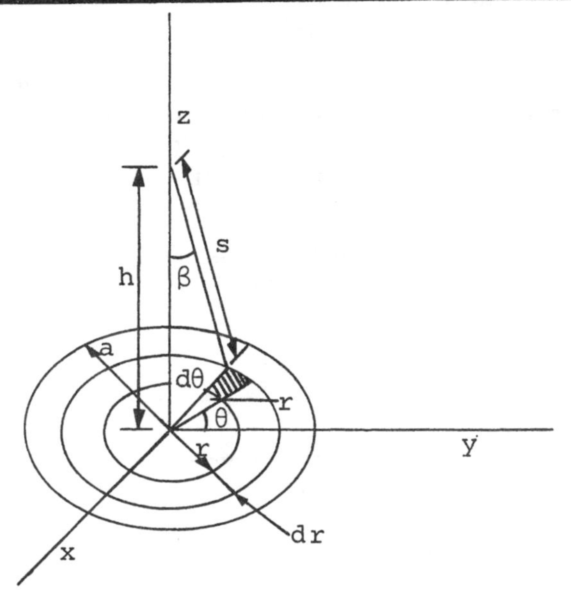

Solution: Consider the shaded element of the surface ds constructed by r, r + dr, and Θ, Θ + dΘ. The area of this elemental surface is ds = (rdΘ) (dr). Its mass, therefore, is:

$$dm = \rho\ r\ dr\ d\Theta.$$

Using the equation of force:

$$F = k\,\frac{m_1\ m_2}{s^2},$$

where K is the constant of proportionality, m_1 and m_2 are the respective masses, and s is the distance between them. In this case, one of the masses is unity and the second mass is expressed in terms of its density and surface area. The force of attraction between ds and the unit particle is;

$$dF = \frac{k\ \rho r\ dr\ d\Theta}{s^2} .$$

Due to symmetry, assuming homogeneity, the net component of this force along the xy-plane is zero, so that only the component perpendicular to this plane (along the z-axis) has effect. This component is

$$dF_z = dF \cos \beta = dF \frac{h}{s} .$$

$$dF_z = k\rho h \frac{r\ dr\ d\Theta}{s^3} = k\rho h \frac{r\ dr\ d\Theta}{(h^2 + r^2)^{\frac{3}{2}}} ,$$

where $s^2 = h^2 + r^2$.

Therefore the total vertical component is the integral over the whole surface region of the circle.

$$F_z = k\rho h \int_0^{2\pi} \int_0^a \frac{r\ dr\ d\Theta}{(h^2 + r^2)^{\frac{3}{2}}} .$$

The integral,

$$\int_0^a \frac{r\ dr}{(h^2 + r^2)^{\frac{3}{2}}} = - \left. \frac{1}{\sqrt{h^2 + r^2}} \right|_0^a$$

$$= \frac{1}{h} - \frac{1}{\sqrt{h^2 + a^2}} = \left(1 - \frac{h}{\sqrt{h^2 + a^2}}\right) \cdot \frac{1}{h} ,$$

where the integration is performed by the substitution method. Since there is no dependence between r and the angle Θ, integration with respect to Θ yields just the multiplication of 2π. Thus, the result is:

$$F_z = k\rho h\ (2\pi) \left(1 - \frac{h}{\sqrt{h^2 + a^2}}\right) \cdot \frac{1}{h} ,$$

or, $$F_z = 2\pi k\rho \left(1 - \frac{h}{\sqrt{h^2 + a^2}}\right) .$$

Since

$$F_{xy} = 0,$$

$$F = \left(F_{xy}{}^2 + F_z{}^2\right)^{\frac{1}{2}} = 2\pi k\rho \left(1 - \frac{h}{\sqrt{h^2 + a^2}}\right).$$

• PROBLEM 1176

Find the attraction of a right circular cone of

altitude h cm., density ρ gm./cm.^{3}, and generating angle α rad., on a unit particle at its vertex.

Solution: We assume that the cone is formed by a straight line, of height h and making an angle α with the vertical, being rotated about the vertical. The volume is generated by an elemental rectangle within the region in question, namely the x- and y-axes and the line $y = x \cot \alpha + h$. Since this elemental rectangle generates a washer-like volume when revolved, by specifying its location with respect to the origin, we can calculate the volume. Let the center of this rectangle be at point P (x,y). Since the distance x becomes the base radius of the disk of width dx and thickness dy when rotated about the y-axis, its volume is:

$$dV = 2\pi x \, dxdy$$

with the density ρ gm/cm^3, we have:

$$dm = 2\pi\rho \quad x \, dxdy.$$

The force is expressed as:

$$F = k \frac{m_1 m_2}{r^2}$$

where, in this case, the mass, with which dm interacts at the vertex, is unity. Hence,

$$dF = \frac{k\rho 2\pi \; x \; dxdy}{r^2}.$$

r is the distance between these two elements.

$$r = \left[x^2 + (h - y)^2\right]^{\frac{1}{2}}.$$

Furthermore, due to symmetry about the y-axis, the net horizontal force is zero and the remaining net value is the force along the y-axis, which is obtained by:

$$F_y = F \cos \Theta = F \frac{h - y}{\sqrt{x^2 + (h - y)^2}}.$$

Hence,

$$dF = k\rho 2\pi \frac{x \; dy \; dx}{\left[x^2 + (h - y)^2\right]^{\frac{3}{2}}}$$

Integration yields:

$$F = k\rho 2\pi \int_0^h \int_0^{-(h-y)\tan \alpha} \frac{x(h - y)}{\left[x^2 + (h - y)^2\right]^{\frac{3}{2}}} dxdy$$

$$= k\rho 2\pi \int_0^h (h - y) \left| - \frac{1}{\sqrt{x^2 + (h - y)^2}} \right|_{x=0}^{x=-(h-y)\tan\alpha} dy$$

$$= k\rho 2\pi \int_0^h \left(\frac{- h - y}{\left[(h - y)^2 \tan^2 \alpha + (h - y)^2\right]^{\frac{1}{2}}} + 1 \right) dy$$

$$= k\rho 2\pi \int_0^h (1 - \cos \alpha) dy$$

$$= 2\pi k\rho h\ (1 - \cos \alpha).$$

• PROBLEM 1177

Current is defined as the time rate of charge flow at any instant: $i = dq/dt$. Find the charge in coulombs transmitted per second if $i = t^2 - t$ amperes and the time is $t = 1$ to $t = 3$ sec.

Solution: In differential equation form,

$$dq = i\ dt.$$

Integrating both sides:

$$\int dq = \int i\ dt .$$

Substituting for i in terms of t:

$$\int dq = \int_1^3 (t^2 - t) dt.$$

$$q = \left. \frac{t^3}{3} - \frac{t^2}{2} \right]_1^3$$

$$= \left[\frac{(3)^3}{3} - \frac{(3)^2}{2} \right] - \left[\frac{(1)^3}{3} - \frac{(1)^2}{2} \right] = \frac{14}{3} \text{ coulomb}$$

• PROBLEM 1178

(a) The charge transferred past a point in an electric circuit is given by the equation

$q = 4t^2 + 3t + 0.002$ coulombs. What is the average current between $t = 0.01$ sec and $t = 0.03$ sec?

(b) What is the instantaneous current in the circuit at any time t?

(c) What is the current at $t = 0.04$ sec?

Solution:

(a) We have

$$q = 4t^2 + 3t + 0.002.$$

$$q + \Delta q = 4(t+\Delta t)^2 + 3(t+\Delta t) + 0.002.$$

Hence,
$$\Delta q = 4(t + \Delta t)^2 + 3(t + \Delta t) + 0.002 - (4t^2 + 3t + 0.002)$$

$$= 4t^2 + 8t\Delta t + 4(\Delta t)^2 + 3t + 3\Delta t + 0.002 - 4t^2 - 3t - 0.002$$

$$= 8t\Delta t + 4(\Delta t)^2 + 3\Delta t.$$

$$I_{av} = \frac{\Delta q}{\Delta t} = 8t + 4\Delta t + 3.$$

We wish to find the average current between $t = 0.01$ and $t = 0.03$. Thus, we start with a value of $t = 0.01$ and $\Delta t = 0.03 - 0.01 = 0.02$

$$\frac{\Delta q}{\Delta t} = 8(0.01) + 4(0.02) + 3 = 3.16 \text{ amperes.}$$

(b) The instantaneous current is: $I = \lim_{\Delta t \to 0} \frac{\Delta q}{\Delta t}$.

$$I = \lim_{\Delta t \to 0} \frac{\Delta q}{\Delta t} = \lim_{\Delta t \to 0} (8t + 4\Delta t + 3)$$

$$= 8t + 3 \text{ amperes.}$$

(c) At $t = 0.04$:

$$I = 8(0.04) + 3 = 3.32 \text{ amperes.}$$

• **PROBLEM 1179**

The relation between the electric resistance R of a wire at the temperature T°C. and the resistance R_0 of the same wire at 0°C. is found to be:

$$\frac{R}{R_0} = (1 + aT + b\sqrt[3]{T}),$$

where a and b are constants. What is the rate of variation of R with temperature?

Solution:

$$R = R_0(1 + aT + b\sqrt[3]{T})$$

$$= R_0 + R_0 aT + R_0 bT^{1/3}.$$

R_0 is a constant.

$$\frac{dR}{dT} = 0 + R_0 a + R_0 b \frac{1}{3} T^{1/3-1}$$

$$= aR_0 + \frac{bR_0}{3} T^{-2/3}$$

$$= R_0\left(a + \frac{b}{3T^{2/3}}\right)$$

$$= R_0\left(a + \frac{b}{3\sqrt[3]{T^2}}\right)$$

• PROBLEM 1180

In a capacitor with constant capacitance C (measured in farads) the charge q can be expressed in terms of the capacitor voltage by the formula:

$$q = CV \text{ coulombs}.$$

$$\frac{\Delta q}{\Delta t} = C\frac{\Delta V}{\Delta t}.$$

$I_{av} = \frac{\Delta q}{\Delta t}$, and the instantaneous current,

$I = \lim_{\Delta t \to 0} \frac{\Delta V}{\Delta t} = \frac{dV}{dt}$. Or, using a different notation

$$I = D_t V.$$

Replacing q in terms of V,

$$I_{av} = \frac{\Delta(CV)}{\Delta t}.$$

But the capacitance C is a constant. Therefore, it can be taken out of the brackets, giving:

$$I_{av} = C\frac{\Delta V}{\Delta t},$$

or, $\quad I = CD_t V.$

(a) If a voltage of $V = 100t^2$ volts is applied to a 0.5-microfarad capacitor (1 microfarad = 0.000001 farad), what is the average current in the capacitor from t = 2 to t = 5 sec?
(b) What is the current at any time t?
(c) What is the current at t = 2 sec?

Solution:

(a) We have:

$$V = 100t^2, \quad V + \Delta V = 100(t + \Delta t)^2$$

$$\Delta V = 100(t + \Delta t)^2 - 100\,t^2$$

$$= 100t^2 + 200t\Delta t + 100(\Delta t)^2 - 100t^2$$

$$= 200t\Delta t + 100(\Delta t)^2.$$

$$\frac{\Delta V}{\Delta t} = 200t + 100\Delta t.$$

We wish to find the average current from t = 2 to t = 5. Thus, we start with a value of t = 2 and $\Delta t = 3$.

$$\frac{\Delta V}{\Delta t} = 200(2) + 100(3)$$

$$= 700.$$

$$I_{av} = C(700)$$

$$= 0.0000005(700) = 0.00035 \text{ ampere.}$$

(b) The instantaneous current, $I = CD_tV$.

$$D_tV = \lim_{\Delta t \to 0} \frac{\Delta V}{\Delta t} = \lim_{\Delta t \to 0} (200t + 100\Delta t) = 200t$$

$$I = CD_tV$$

$$= 0.0000005(200t)$$

$$= 0.0001t \text{ ampere .}$$

(c) At t = 2,

$$I = 0.0001(2)$$

$$= 0.0002 \text{ ampere.}$$

• **PROBLEM** 1181

The charge of a capacitor q in coulombs is directly proportional to its capacitance C in farads and its impressed voltage ν_c in volts. If a 1-μf capacitor is connected to a circuit, in which the current $i = t^3$ amperes, determine the voltage across the capacitor 0.1 sec after the connection is made.

Solution: $q = C\nu_c$

$$i = \frac{dq}{dt}$$

$$dq = i\ dt$$

$$q = \int i\ dt = C\nu_c .$$

Divide both sides by C:

$$\nu_c = \frac{1}{C}\int i\ dt$$

Substitute the conditions:

$$\nu_c = \frac{1}{10^{-6}}\int_0^{0.1} t^3 dt,$$

since 1-μf = 10^{-6}f, and t, the time elapsed after

the instantaneous connection at t=0 is 0.1 sec.

Thus,

$$\nu_c = 10^6 \left[\frac{t^4}{4}\right]_0^{0.1}$$

$$= 10^6 \left(\frac{1}{4}\right)(0.1)^4 - 0$$

$$= 10^6(0.25)(10^{-4}) - 0 = 25 \text{ volts.}$$

• PROBLEM 1182

The general equation expressing the relation between electromotive force and current in an inductive electric circuit is:

$$E = Ri + L\frac{di}{dt},$$

where E is the electromotive force, R is the resistance, L is the inductance, i is the current, and t is the time. Let the electromotive force be removed, and, at the instant of removal when t = 0, let the current be i_0. Then the general equation becomes: $0 = Ri + L(di/dt)$, from which we obtain

$$\frac{di}{dt} = -\frac{Ri}{L}.$$

Prove that the solution of this equation is $i = i_0 e^{-Rt/L}$.

Solution: By the use of separation of variables, the equation takes the form

$$\frac{di}{i} = -\frac{R}{L}dt.$$

Integration of both sides of the above differential equation yields:

$$\ln i = -\frac{R}{L}t + \ln k,$$

where k is a constant,

or,

$$\ln(i/k) = -\frac{R}{L}t.$$

Expressing the above equation in an exponential form,

$$i/k = e^{-Rt/L},$$

or,

$$i = ke^{-Rt/L}.$$

But at t = 0, $i = i_0$, making $k = i_0$.

Thus,

$$i = i_0 e^{-Rt/L}.$$

When a simple electric circuit contains resistance R and inductance L in series, and an electromotive force E is impressed on it, the differential equation for the current i at any time t is:

$$\text{(a)} \qquad L\frac{di}{dt} + Ri = E.$$

Solve this equation when: (a) $E = E_0$ (constant), and (b) $E = E_0 \sin \omega t$, where L, R E_0 and ω are constant.

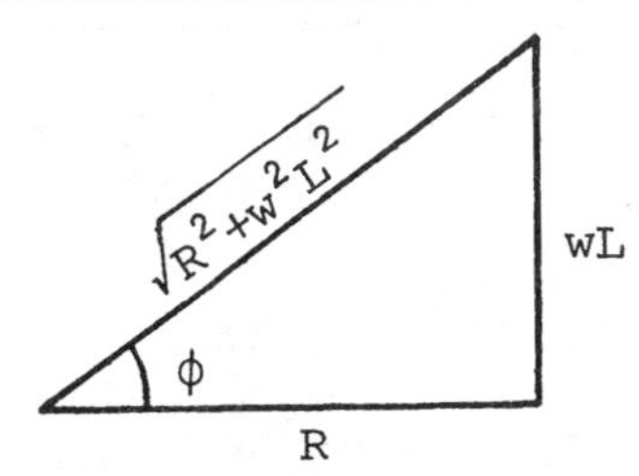

Solution: The above differential equation is called a first-order linear differential equation of the general form of:

$$\text{(b)} \qquad \frac{dy}{dx} + p(x)y = q(x),$$

where p and q are either constants or functions of x alone. Setting

$$\text{(c)} \qquad M = e^{\int p(x)dx},$$

and observing that, by using the chain rule,

$$\text{(d)} \qquad \frac{dM}{dx} = e^{\int p(x)dx} \cdot p(x) = p(x) \cdot M.$$

If we multiply both sides of equation (b) by M, we have:

$$M\frac{dy}{dx} + Mp(x)y = Mq(x).$$

Using (c), this can be written in the following form, using the derivative of a product:

$$\frac{d}{dx}(My) = Mq(x).$$

Since Mq(x) is a function of x alone this gives

$$\text{(e)} \qquad My = \int Mq(x)dx + c,$$

a general solution of (b). Dividing both sides of (e) by M and using (c), the solution can be put in the form

$$\text{(f)} \qquad y = e^{-\int p(x)dx}\left[\int q(x)e^{\int p(x)dx} + c\right].$$

For convenience, the expression: $e^{\int p(x)dx}$, is called the integrating factor. In case (a), where $E = E_0$, equation (a) becomes

$$\frac{di}{dt} + \frac{R}{L}\cdot i = \frac{E_0}{L},$$

a linear differential equation, which may be solved by the formula (f). Since $p(x) = R/L$ (constant), the integrating factor becomes $e^{\int R/L\, dt} = e^{\frac{R}{L}t}$, and the solution is given by

$$i\, e^{\frac{R}{L}t} = \int \frac{E_0}{L} e^{\frac{R}{L}t} dt + C = \frac{E_0}{R} e^{\frac{R}{L}t} + C,$$

whence, dividing the whole equation by $e^{\frac{R}{L}t}$,

$$\text{(g)} \qquad i = \frac{E_0}{R} + Ce^{-\frac{R}{L}t}.$$

If $i = 0$ when $t = 0$, we find $C = -E_0/R$, and then

$$\text{(h)} \qquad i = \frac{E_0}{R}(1 - e^{-\frac{R}{L}t}).$$

This shows that the current gradually builds up toward the limiting value E_0/R, for t approaching infinity, which is the current that would flow if no inductance were present.

In case (b), where $E = E_0 \sin \omega t$, equation (a) becomes

$$\text{(i)} \qquad \frac{di}{dt} + \frac{R}{L}\cdot i = \frac{E_0}{L} \sin \omega t,$$

which again is a linear differential equation. By formula (f), its solution is given by

$$i\cdot e^{\frac{R}{L}t} = \frac{E_0}{L}\int e^{\frac{R}{L}t} \sin \omega t\, dt + C.$$

Evaluating this integral by parts, $\int u dv = uv - \int v du$, we get

$$\text{(j)} \qquad i\cdot e^{\frac{R}{L}t} = \frac{E_0}{L}\cdot\frac{e^{\frac{R}{L}t}}{\frac{R^2}{L^2} + \omega^2}\left(\frac{R}{L} \sin \omega t - \omega \cos \omega t\right) + C.$$

Multiplying both sides of the equation by $e^{-\frac{R}{L}t}$, and some algebraic manipulation, yield:

$$\text{(k)} \qquad i = \frac{E_0}{R^2 + L^2\omega^2}(R \sin \omega t - \omega L \cos \omega t) + Ce^{-\frac{R}{L}t}.$$

By trigonometry, this may also be written:

$$i = \frac{E_0}{\sqrt{R^2+L^2\omega^2}}\left[\frac{R}{\sqrt{R^2+L^2\omega^2}}\sin\omega t - \frac{\omega L}{\sqrt{R^2+L^2\omega^2}}\cos\omega t\right] + Ce^{-\frac{R}{L}t}$$

$$= \frac{E_0}{\sqrt{R^2+L^2\omega^2}}(\cos\phi\sin\omega t - \sin\phi\cos\omega t)\cdot$$

$$(1)\qquad i = \frac{E_0}{\sqrt{R^2+L^2\omega^2}}\sin(\omega t - \phi) + Ce^{-\frac{R}{L}t}$$

where $\tan\phi = \omega L/R$.

The term $Ce^{-Rt/L}$ in formula (k) or (1) is called the transient term, because it usually becomes negligibly small after a short lapse of time. The first term, called the steady-state term, varies periodically, with the same period, $2\pi/\omega$, as the impressed electromotive force, but lags behind the latter, as is indicated by the term ϕ.

The triangle in the figure shows the relationship between the quantities

• PROBLEM 1184

An electric circuit has an impressed electromotive force $e = E_0 \sin 100t$, a resistance R of 10 ohms, a capacitance c of 0.002 farad, and an inductance L of 0.1 henry. Determine the current in the circuit at time t, if $i = 0$ and $q = 0$ when $t = 0$.

Solution: The differential equation for the circuit is:

$$L\frac{d^2q}{dt^2} + R\frac{dq}{dt} + \frac{q}{c} = e.$$

Substituting:

$$0.1\frac{d^2q}{dt^2} + 10\frac{dq}{dt} + 500q = E_0 \sin 100t,$$

or

$$\frac{d^2q}{dt^2} + 100\frac{dq}{dt} + 5000q = 10E_0 \sin 100t.$$

This is a second order, linear, non-homogeneous differential equation with constant coefficients. To find the general solution, we set the auxiliary equation:

$$m^2 + 100m + 5000 = 0.$$

Solving for m yields:

$$m = -50 \pm 50j$$

i.e., the roots are $m_1 = -50 + 50j$ and $m_2 = -50 - 50j$, where $j = \sqrt{-1}$ has been written in order to avoid confusion with i as the current. Thus, the complementary solution of the differential equation is

$$q_c = c_1 e^{(-50+50j)t} + c_2 e^{(-50-50j)t} ,$$

or

$$q_c = e^{-50t}(c_1 \sin 50t + c_2 \cos 50t).$$

To find the particular solution, we use the method of undetermined coefficients.

In connection with the angular frequency of the electromotive force given by $E_0 \sin 100t$, where $\omega = 100$, we assume the particular solution of the form

$$q_p = A \cos 100t + B \sin 100t.$$

Upon substituting q_p in the differential equation, we find:

$$-10^4 A \cos 100t - 10^4 B \sin 100t$$
$$+ 100(-100A \sin 100t + 100B \cos 100t)$$
$$+ 5 \times 10^3 (A \cos 100t + B \sin 100t)$$
$$= 10E_0 \sin 100t.$$

Equating coefficients of corresponding terms gives:

$$-10^4 B - 10^4 A + 5 \times 10^3 B = 10E_0,$$
$$10^4 B - 10^4 A + 5 \times 10^3 A = 0.$$

Hence, from the above simultaneous equations,

$$B = \frac{1}{2} A,$$

and, in terms of E_0,

$$A = -\frac{E_0}{1250}, \qquad B = -\frac{E_0}{2500}.$$

Therefore,

$$q_p = -\frac{E_0}{2500} (2 \cos 100t + \sin 100t).$$

At time t = 0 (or before applying the voltage to the circuit), the capacitor is not yet charged. Using this as an initial condition to determine the constants c_1 and c_2, we set q = 0 at t = 0 into

$$q = q_c + q_p$$
$$= e^{-50t}(c_1 \sin 50t + c_2 \cos 50t)$$
$$- \frac{E_0}{2500} (2 \cos 100t + \sin 100t).$$

We obtain

$$0 = c_2 - \frac{E_0}{1250}$$

Moreover, since we know that current is the time rate of charge flow at any instant, we can use $i = dq/dt = 0$ at $t = 0$ to obtain the other constant in the expression.

$$i = \frac{dq}{dt} = e^{-50t}(-50c_1 \sin 50t - 50c_2 \cos 50t$$

$$+ 50c_1 \cos 50t - 50c_2 \sin 50t)$$

$$- \frac{E_0}{2500}(-200 \sin 100t + 100 \cos 100t).$$

We find:

$$0 = -50c_2 + 50c_1 - \frac{E_0}{25}.$$

These equations give

$$c_2 = \frac{E_0}{1250}, \text{ and } \quad c_1 = \frac{E_0}{625}.$$

Hence,

$$i = \frac{E_0}{25}(2 \sin 100t - \cos 100t$$

$$+ e^{-50t} \cos 50t - 3e^{-50t} \sin 50t).$$

For large t, the terms having e^{-50t} as a factor are negligible, so that

$$i \approx \frac{E_0}{25}(2 \sin 100t - \cos 100t).$$

That is, some time after $t = 0$ the current flow settles to a "steady state." Hence the particular solution of the differential equation is called the steady-state solution. The complementary part of the solution is, for obvious reasons, called the transient solution.

• PROBLEM 1185

Derive the three-dimensional heat conduction equation, starting with Fourier's Law of heat transfer. Also write the special cases of the conduction equation and its transformation into cylindrical and spherical coordinates.

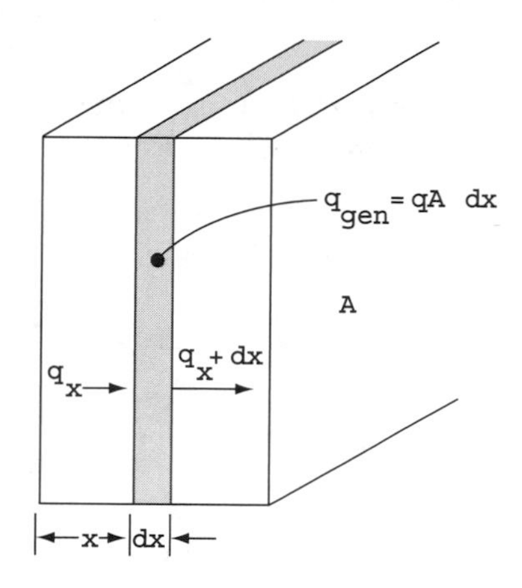

Fig. 1 Elemental volume for one-dimensional heat conduction.

Solution: Fourier's Law is

$$q = -kA \frac{\partial T}{\partial x} \tag{1}$$

Consider the one-dimensional system shown in Fig. 1. If the system is in a steady state, i.e., if the temperature does not change with time, then the problem is a simple one and need only integrate eq.(1) and substitute the appropriate values to solve for the desired quantity. However, if the temperature of the solid is changing with time, or if there are heat sources or sinks within the solid, the situation is more complex. We consider the general case where the temperature may be changing with time and heat sources may be present within the body. For the element of thickness dx the following energy balance may be made:

Energy conducted in left face + heat generated within element = change in internal energy + energy conducted out right face

These energy quantities are given as follows:

$$\text{Energy in left face} = q_x = -kA \frac{\partial T}{\partial x}$$

$$\text{Energy generated within element} = \dot{q}A\, dx$$

$$\text{Change in internal energy} = \rho c A \frac{\partial T}{\partial \tau} dx$$

$$\text{Energy out right face} = q_{x+dx} = -kA \left.\frac{\partial T}{\partial x}\right|_{x+dx}$$

$$= -A\left[k \frac{\partial T}{\partial x} + \frac{\partial}{\partial x}\left(k \frac{\partial T}{\partial x}\right) dx\right]$$

where $\dot{q}$ = energy generated per unit volume
c = specific heat of material
ρ = density

Combining the relations above,

$$-kA \frac{\partial T}{\partial x} + \dot{q}A\, dx = \rho c A \frac{\partial T}{\partial \tau} dx - A\left[k \frac{\partial T}{\partial x} + \frac{\partial}{\partial x}\left(k \frac{\partial T}{\partial x}\right) dx\right] \tag{2}$$

or

$$\frac{\partial}{\partial x}\left(k \frac{\partial T}{\partial x}\right) + \dot{q} = \rho c \frac{\partial T}{\partial \tau}$$

This is the one-dimensional heat-conduction equation. To treat more than one-dimensional heat flow, only consider the heat conducted in and out of a unit volume in all three coordinate directions, as shown in Fig. 2a. The energy balance yields

$$q_x + q_y + q_z + q_{gen} = q_{x+dx} + q_{y+dy} + q_{z+dz} + \frac{dE}{d\tau}$$

and the energy quantities are given by

$$q_x = -k\, dy\, dz \frac{\partial T}{\partial x}$$

$$q_{x+dx} = -\left[k \frac{\partial T}{\partial x} + \frac{\partial}{\partial x} k \left(\frac{\partial T}{\partial x}\right) dx\right] dy\, dz$$

$$q_y = -k\, dx\, dz \frac{\partial T}{\partial y}$$

$$q_{y+dy} = -\left[k \frac{\partial T}{\partial y} + \frac{\partial}{\partial y}\left(k \frac{\partial T}{\partial y}\right) dy\right] dx\, dz$$

$$q_z = -k\, dx\, dy \frac{\partial T}{\partial z}$$

$$q_{z+dz} = -\left[k \frac{\partial T}{\partial z} + \frac{\partial}{\partial z}\left(k \frac{\partial T}{\partial z}\right) dz\right] dx\, dy$$

$$\frac{\partial^2 T}{\partial x^2} + \frac{\partial^2 T}{\partial y^2} + \frac{\partial^2 T}{\partial z^2} = \frac{1}{\alpha}\frac{\partial T}{\partial t}$$

2. Poisson equation (steady state with internal energy conversion)

$$\frac{\partial^2 T}{\partial x^2} + \frac{\partial^2 T}{\partial y^2} + \frac{\partial^2 T}{\partial z^2} + \frac{q'''}{k} = 0$$

3. Laplace equation (steady state and no internal energy conversion)

$$\frac{\partial^2 T}{\partial x^2} + \frac{\partial^2 T}{\partial y^2} + \frac{\partial^2 T}{\partial z^2} = 0$$

Equation (4), the general conduction equation in cylindrical and spherical coordinates is given by:

Cylindrical coordinates (Fig. 2b)

$$\frac{\partial^2 T}{\partial r^2} + \frac{1}{r}\frac{\partial T}{\partial r} + \frac{1}{r^2}\frac{\partial^2 T}{\partial \phi^2} + \frac{\partial^2 T}{\partial z^2} + \frac{\dot{q}}{k} = \frac{1}{\alpha}\frac{\partial T}{\partial \tau}$$

Spherical coordinates (Fig. 2c)

$$\frac{1}{r}\frac{\partial^2}{\partial r^2}(rT) + \frac{1}{r^2 \sin\theta}\frac{\partial}{\partial \theta}\left(\sin\theta \frac{\partial T}{\partial \theta}\right) + \frac{1}{r^2 \sin^2\theta}\frac{\partial^2 T}{\partial \phi^2} + \frac{\dot{q}}{k} = \frac{1}{\alpha}\frac{\partial T}{\partial \tau}$$

• PROBLEM 1186

The inside temperature of a furnace is 1500°F and its walls are lined with a 4 in. thick refractory brick. The ambient temperature is 100°F. Compute the rate of heat flow through the walls and the temperature profile across the wall, if the variation of thermal conductivity of the refractory is given by

$$k = 0.10 + 5 \times 10^{-5} T$$

Solution: The rate of heat transfer is expressed by the Fourier equation

$$q = -k \frac{dT}{dx}$$

For steady state

$$q\int_0^L dx = -\int_{T_1}^{T_2} k \, dT$$

or

$$q\int_0^{1/3 \text{ ft}} dx = -\int_{1,500}^{100} (0.10 + 5 \times 10^{-5} T)dT$$

$$q = 3[(0.1)(1,500 - 100) + (5 \times 10^{-5})(1.12 \times 10^{6})]$$

$$q = 588 \text{ Btu/hr ft}^2$$

For obtaining the profile

$$q\int_0^x dx = -\int_{1,500}^{T} (0.10 + 5 \times 10^{-5}T)dT$$

$$588\,x = 0.1(1,500 - T) + \frac{5 \times 10^{-5}}{2}(2.25 \times 10^{6} - T^2)$$

$$x = 0.352 - 1.7 \times 10^{-4}T - 4.28 \times 10^{-8}T^2$$

The above equation is a quadratic one and shows the non-linear characteristic of temperature variation across the furnace wall.

• PROBLEM 1187

Determine the temperature distribution in a semi-infinite two-dimensional flat plate shown in the figure, if the base temperature is F(x) and the ambient temperature is T_∞.

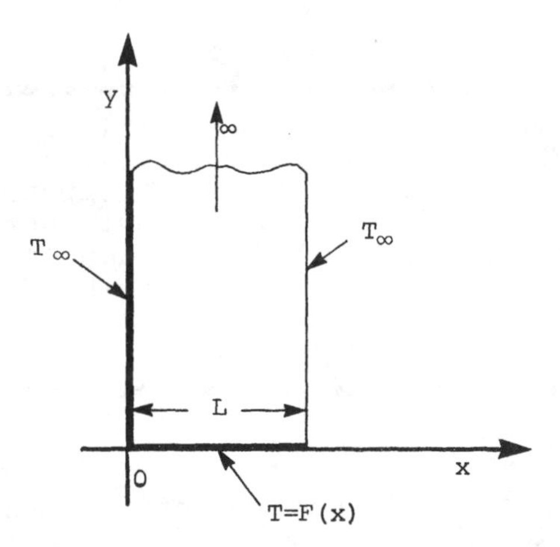

Solution: The controlling differential equation is

$$\frac{\partial^2 T}{\partial x^2} + \frac{\partial^2 T}{\partial y^2} = 0$$

and the given boundary conditions are

$$T(x,0) = F(x); \quad T(L,y) = T_\infty$$

$$T(0,y) = T_\infty \quad ; \quad T(x,\infty) = T_\infty$$

Since the boundary conditions are not homogeneous, the differential equation cannot be solved by the method of separation of variables. It requires the differential equation and three of the boundary conditions to be homogeneous. This can be achieved by a simple transformation

$$\theta = T - T_\infty$$

Writing the differential equation and boundary conditions in terms of θ,

$$\frac{\partial^2\theta}{\partial x^2} + \frac{\partial^2\theta}{\partial y^2} = 0 \quad (1)$$

$$\theta(x,0) = F(x) - T_\infty = f(x) \quad (2)$$

$$\theta(0,y) = 0 \quad (3)$$

$$\theta(L,y) = 0 \quad (4)$$

$$\theta(x,\infty) = 0 \quad (5)$$

The solution to the equation (1) by the method of separation of variables is

$$\theta(x,y) = (A\cos\lambda x + B\sin\lambda x)(Ce^{\lambda y} + De^{-\lambda y})$$

where

A, B, C, D and λ are constants.

Using the boundary condition (3)

$$\theta(0,y) = (A+0)(Ce^{\lambda y} + De^{-\lambda y}) = 0$$

Therefore $A = 0$

so,

$$\theta(x,y) = B\sin\lambda x(Ce^{\lambda y} + De^{-\lambda y})$$

The constant B can be absorbed into the constants C and D.

Therefore,

$$\theta(x,y) = \sin\lambda x(Ce^{\lambda y} + De^{-\lambda y})$$

Using the boundary condition (5)

$$\theta(x,\infty) = \sin\lambda x(Ce^{\infty} + De^{-\infty}) = 0$$

From the above equation $C = 0$. Therefore the differential equation becomes

$$\theta(x,y) = \sin\lambda x(De^{-\lambda y})$$

Now using the boundary condition (4)

$$\theta(L,y) = De^{-\lambda y} \sin\lambda L = 0$$

Since $D \neq 0$, $\sin\lambda L = 0$

Therefore $\lambda L = n\pi \qquad n = 1,\ldots\infty$

or

$$\lambda = \frac{n\pi}{L} \tag{6}$$

The solution $\theta(x,y) = De^{-\lambda y} \sin\lambda x$, is a particular solution of the equation (1).

The general solution is obtained by summing all the particular solutions.

Therefore $\theta(x,y) = \sum_{n=L}^{\infty} D_n e^{-\lambda_n y} \sin\lambda_n x$ is the general solution.

Using the non-homogeneous boundary condition (2)

$$f(x) = \sum_{n=L}^{\infty} D_n \sin\lambda_n x$$

This is a Fourier series. The value of D_n is

$$D_n = \frac{\int_o^L f(x)\sin\lambda_n x \, dx}{\int_o^L \sin^2\lambda_n x \, dx}$$

Simplifying, this reduces to

$$D_n = \frac{2}{L}\int_o^L f(x)\sin\lambda_n x \, dx$$

Substituting the value of D_n into the general solution,

$$\theta(x,y) = \frac{2}{L} \sum_{n=1}^{\infty} \left[\int_o^L f(\eta)\sin\lambda_n \eta \, d\eta\right] e^{-\lambda_n y} \sin\lambda_n x$$

where η is a dummy variable.

Equation (7) is the required temperature distribution.

Find the solution for a region, with steady state heat conduction, having boundaries at x=0, x=a, y=0 and y=f(x) and boundary conditions as shown in the figure. Use the Galerkin method of partial integration, with respect to the y variable. Take the heat generation rate to be constant.

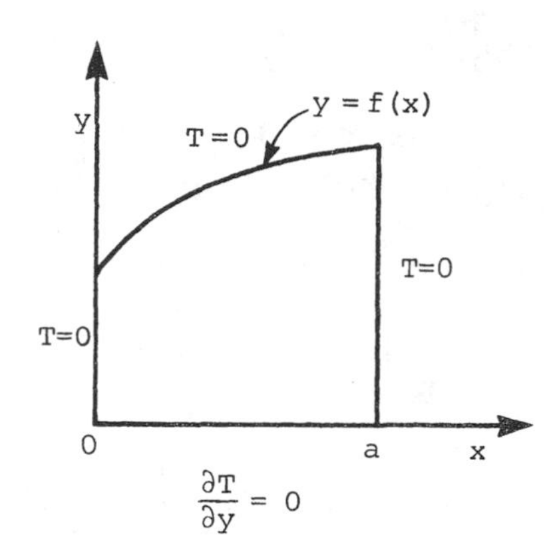

Solution: The appropriate equations and boundary conditions for this problem are

$$\frac{\partial^2 T}{\partial x^2} + \frac{\partial^2 T}{\partial y^2} + \frac{g}{k} = 0 \quad \text{in } 0 < x < a,\ 0 < y < f(x) \tag{1a}$$

$$T = 0 \qquad \text{at } x=0, x=a, \text{ and } y=f(x) \tag{1b}$$

$$\frac{\partial T}{\partial y} = 0 \qquad \text{at } y=0 \tag{1c}$$

Using the Galerkin method, partial integration of equation (1a) with respect to y, yields

$$\int_{y=0}^{f(x)} \left[\frac{\partial^2 T}{\partial x^2} + \frac{\partial^2 T}{\partial y^2} + \frac{g}{k}\right] \phi_i(y)dy = 0 \tag{2}$$

Now, taking a one-term trial solution

$$T_1(x,y) = X(x) \cdot \phi_1(y) \tag{3a}$$

where

$$\phi_1(y) = [y^2 - f^2(x)] \tag{3b}$$

The boundary conditions y=0 and y=f(x) are satisfied by this solution, but the function X(x) is not known. Substituting the trial solution into equation (2) and doing a partial integration with respect to the y variable, gives an ordinary differential equation which can be used to determine the function X(x),

$$2/5\ f^2X'' + 2ff'X' + (ff''+f'^2 - 1)X = -\frac{g}{2k} \text{ in } 0<x<a \quad (4a)$$

$$\text{for} \quad X = 0 \quad \text{at} \quad x = 0 \quad \text{and} \quad x = a \quad (4b)$$

The function X(x) can be determined, once the function f(x), defining the form of the boundary, is specified. There are two special cases:

1. $y = f(x) = b$: The region is rectangular and the equation (4a) becomes

$$X'' - \frac{5}{2b^2} X = \frac{-5g}{4b^2 k} \quad \text{in } 0 < x < a \quad (5a)$$

$$X = 0 \quad \text{at } x = 0 \text{ and } x = a, \quad (5b)$$

and the one term approximate solution becomes

$$T_1(x,y) = (y^2 - b^2)X(x) \quad (6)$$

where

$$X(x) = \frac{g}{2k}\left[1 - \frac{\cos h\left(\sqrt{2.5}\ \frac{x}{b}\right)}{\cos h\left(\sqrt{2.5}\ \frac{a}{b}\right)}\right]$$

2. $y = f(x) = \beta x$: For this case eq.(4a) becomes

$$x^2X'' + 5xX' + \frac{5(\beta^2-1)}{2\beta^2} X = \frac{5g}{4\beta^2 k} \quad \text{in } 0 < x < a \quad (7a)$$

$$x = 0 \quad \text{at} \quad x = 0 \quad \text{and} \quad x = a \quad (7b)$$

Equation (7a) is an Euler type equation that can be solved by finding a solution for X(x) in the form x^n. Substitution of $X = x^n$ into the homogenous part of eq.(7a), gives the expression

$$n^2 + 4n + \frac{5(\beta^2-1)}{2\beta^2} = 0 \quad (8a)$$

Hence

$$n_1,\ n_2 = -2 \pm \sqrt{4 - B} \quad (8b)$$

$$\text{where} \quad B \equiv \frac{5(\beta^2-1)}{2\beta^2} \quad (8c)$$

Hence the complete solution for X(x) can be written as

$$X(x) = c_1 x^{n_1} + c_2 x^{n_2} - P(x) \quad (8d)$$

where P(x) is the particular solution of equation (7a)

$$P(x) = g/2k(\beta^2-1)$$

and the coefficients c_1 and c_2 are determined by the application of the boundary conditions given in equation (7b). Since the solution of X(x) is now known, the one term approximate solution is

$$T_1(x,y) = (y^2-\beta^2x^2)\ X(x)$$

Suppose there is a plane area on which charges are distributed uniformly with a surface density σ coul/m². Calculate the field at a point P a distance a from the plane as shown in Fig. 1. Assume that the dimensions of the plane are much greater than a.

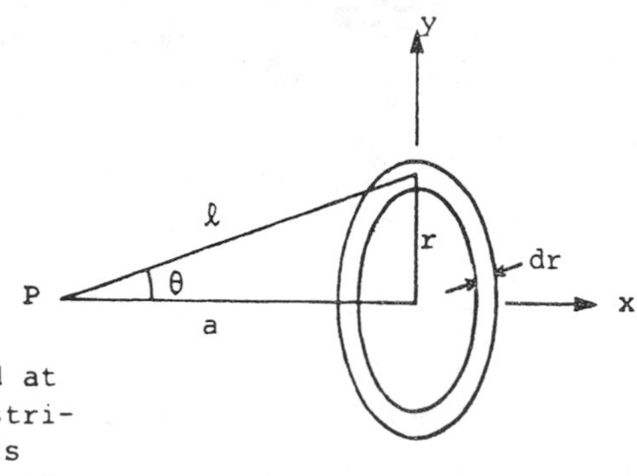

The electric field at P from a plane distribution of charge is obtained by integrating the contributions from concentric ring.

Fig. 1

Solution: Use

$$E = \frac{1}{4\pi\varepsilon_0} \int \frac{dq}{r^2} \hat{r}$$

to add up the vector contributions of all charges at the point P. Begin by calculating the contribution from the ring of charge of radius r and width dr and then integrate for all such rings that make up the total plane charge distribution. The calculation of the contribution of charges on the ring is really a two-dimensional integration problem, but because of the symmetry it can be reduced to a simple summation. The contribution of an elementary charge on the ring to the field at P makes an angle θ with the x axis as shown on the figure. Because of symmetry, however, components of $\bar{E}$ perpendicular to the x axis cancel. Thus one needs to consider only the x components of $\bar{E}$ at P. Then the field at P due to the ring of charge is

$$dE = \frac{1}{4\pi\varepsilon_0} \frac{\sigma 2\pi r dr}{l^2} \cos\theta$$

In order to reduce this to a single variable, substitute a tan θ = r and a/cos θ = a sec θ = l. By differentiation dr = a sec² θ dθ. Making these substitutions yields,

$$dE = \frac{\sigma}{2\varepsilon_0} \frac{\tan\theta \sec^2\theta \cos\theta}{\sec^2\theta} d\theta = \frac{\sigma}{2\varepsilon_0} \sin\theta \, d\theta$$

This is the contribution to the field at P from the ring chosen. To obtain the total field at P from all rings making up the plane charge distribution, integrate this expression over the entire plane. The limits of integration are from θ = 0 to θ = π/2. Thus,

$$E = \frac{\sigma}{2\varepsilon_0}\int_0^{\pi/2} \sin\theta \, d\theta = -\frac{\sigma}{2\varepsilon_0}[\cos\theta]_0^{\pi/2}$$

$$= \frac{\sigma}{2\varepsilon_0} \text{ newtons/coul}$$

The resultant field is in the x direction and is independent of the distance from the plane as long as the plane is very large compared with a.

• PROBLEM 1190

Find the electric field intensity about the finite line charge of uniform ρ_ℓ distribution along the z axis, as shown in Fig. 1. Use this result to find the electric field intensity about an infinite line charge.

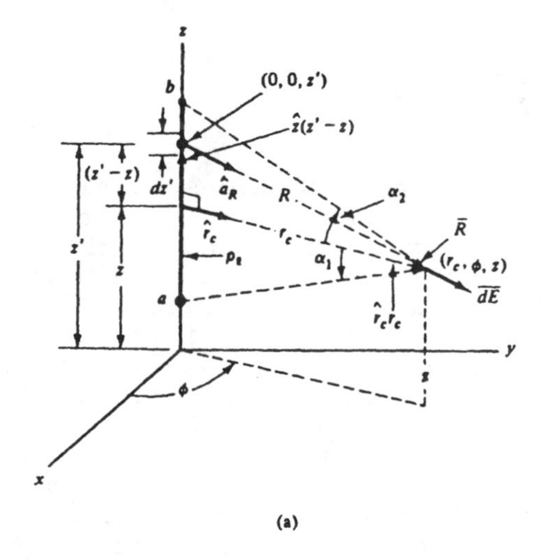

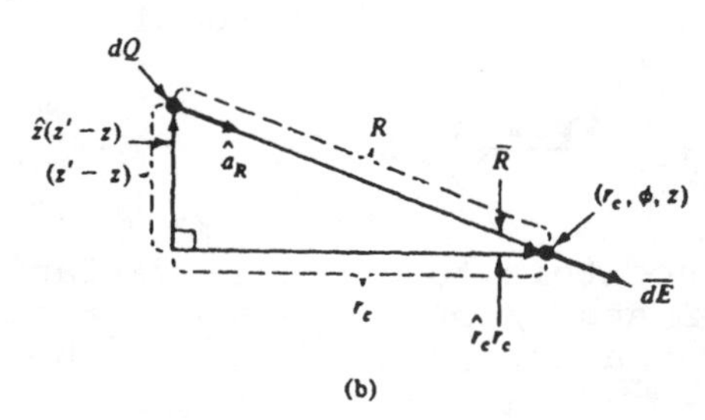

(a) Graphical construction to evaluate $\overline{dE}$ and thus $\bar{E}$ about a finite length of line charge of uniform ρ_ℓ. (b) Isolated view for evaluating $\hat{a}_R$.

Fig. 1

Solution: The differential electric field intensity $\overline{dE}$ can be found through

$$\overline{dE} = \frac{\rho_\ell d\ell}{4\pi\varepsilon_0 R^2}\,\hat{a}_R$$

Use the primed variables to locate points on the line of charge and the unprimed variables to locate the electric field point. Thus, point charge dQ is located at $P_{cyl}(r_c', \phi', z')$. With the aid of the graphical construction in Fig. 1, the following is obtained:

$$R = (r_c^2 + (z' - z)^2)^{1/2}$$

$$\bar{R} = \hat{r}_c r_c - \hat{z}(z' - z) = \hat{a}_R R$$

$$\hat{a}_R = \frac{\bar{R}}{R} = \frac{\hat{r}_c r_c - \hat{z}(z' - z)}{(r_c^2 + (z' - z)^2)^{1/2}}$$

$$d\ell = dz'$$

Substituting into the $\overline{dE}$ expression,

$$\overline{dE} = \frac{\rho_\ell dz'}{4\pi\varepsilon_0 (r_c^2 + (z' - z)^2)}\left[\frac{\hat{r}_c r_c - \hat{z}(z' - z)}{(r_c^2 + (z' - z)^2)^{1/2}}\right]$$

Now,

$$\bar{E} = \int_a^b \overline{dE} = \frac{\rho_\ell}{4\pi\varepsilon_0}\int_a^b \frac{(\hat{r}_c r_c - \hat{z}(z' - z))}{(r_c^2 + (z' - z)^2)^{3/2}}\,dz'$$

Over the range of the above integral, the only variable is z', while $\hat{r}_c$ is a constant, since the point of $\bar{E}$ is fixed for the integration over the line from a to b. The $\bar{E}$ integral can be rewritten as

$$\bar{E} = \frac{\rho_\ell}{4\pi\varepsilon_0}\left\{\hat{r}_c r_c \int_a^b \frac{dz'}{(r_c^2 + (z' - z)^2)^{3/2}} - \hat{z}\int_a^b \frac{(z' - z)dz'}{(r_c^2 + (z' - z)^2)^{3/2}}\right\}$$

The integrals found in the above expression are of the forms

$$\int \frac{dx}{(c^2 + x^2)^{3/2}} = \frac{x}{c^2(c^2 + x^2)^{1/2}} \tag{1}$$

$$\int \frac{xdx}{(c^2 + x^2)^{3/2}} = \frac{-1}{(c^2 + x^2)^{1/2}} \tag{2}$$

Using (1) and (2), $\bar{E}$ becomes

$$\bar{E} = \frac{\rho_\ell}{4\pi\varepsilon_0}\left\{\hat{r}_c r_c \frac{(z' - z)}{r_c^2(r_c^2 + (z' - z)^2)^{1/2}}\Bigg|_a^b + \hat{z}\frac{1}{(r_c^2 + (z' - z)^2)^{1/2}}\Bigg|_a^b\right\}$$

$$= \frac{\rho_\ell}{4\pi\varepsilon_0}\left\{\frac{\hat{r}_c}{r_c}\left[\frac{(b - z)}{(r_c^2 + (b - z)^2)^{1/2}} - \frac{(a - z)}{(r_c^2 + (a - z)^2)^{1/2}}\right] + \hat{z}\left[\frac{1}{(r_c^2 + (b - z)^2)^{1/2}} - \frac{1}{(r_c^2 + (a - z)^2)^{1/2}}\right]\right\} \text{ (Vm}^{-1}) \tag{3}$$

In terms of α_1 and α_2, see Fig. 1a, eq. (3) becomes

$$\bar{E} = \frac{\rho_\ell}{4\pi\varepsilon_0}\left\{\frac{\hat{r}_c}{r_c}(\sin\alpha_2 + \sin\alpha_1) + \frac{\hat{z}}{r_c}(\cos\alpha_2 - \cos\alpha_1)\right\} \text{ (Vm}^{-1}) \tag{4}$$

The electric field intensity $\bar{E}$ about an infinite line of charge of uniform ρ_ℓ distributed along the z axis, can then be found.

In (3), let $b = +\infty$, $a = -\infty$ to obtain

$$\bar{E} = \frac{\rho_\ell}{2\pi\varepsilon_0 r_c}\hat{r}_c \text{ (Vm}^{-1})$$

Also, let $\alpha_1 = \pi/2$ and $\alpha_2 = \pi/2$ in (4) to obtain

$$\bar{E} = \frac{\rho_\ell}{2\pi\varepsilon_0 r_c} \hat{r}_c \ (\text{Vm}^{-1}) \tag{5}$$

It should be noted that only the radial component exists for the infinite line of charge of uniform ρ_ℓ.

• **PROBLEM 1191**

Find the current density $\bar{J}$ between two concentric spherical electrodes, then find the resistance from the previous result.

The figure illustrates two concentric spheres of respective radii a and b. The inner sphere is kept at the potential V_o relative to the outer sphere. The medium between the spheres is assumed to be homogeneous of conductivity σ.

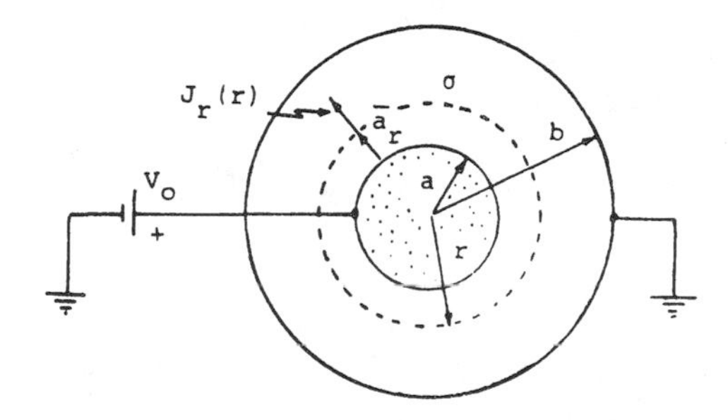

Solution: The configuration has perfect spherical symmetry, and it may be assumed that the current density $\bar{J}$ is in the radial direction and is a function of r only. Let the total current supplied by the center electrode be I; thus I is the total source strength. Now apply Gauss' law to the spherical surface of radius r, to get

$$\oint_S \bar{J} \cdot d\bar{S} = I$$

Because of the symmetry, the integration can be performed to get

$$\oint_S \bar{J} \cdot d\bar{S} = J_r(r) \oint \bar{a}_r \cdot d\bar{S} = 4\pi r^2 J_r(r) = I$$

The existence of symmetry is a crucial requirement in order to assume that the radial current density J_r (r) is constant everywhere on the spherical surface of radius r and to permit the performing of integration.

From the above result,

$$J_r(r) = \frac{1}{4\pi r^2}$$

Now writing in spherical coordinates

$$\bar{J}_r(r) = \sigma\bar{E} = -\sigma \frac{\partial\phi}{\partial r} \bar{a}_r$$

and so

$$\frac{\partial\phi}{\partial r} = \frac{-1}{4\pi\sigma r^2}$$

from which, by integration,

$$\phi = \frac{1}{4\pi\sigma r} + C$$

From the figure the boundary condition is: r = b, $\phi = 0$; from this $C = -I/4\pi\sigma b$. Further, for ϕ to equal V_o at r = a, the current I must then have the value

$$I = \frac{4\pi\sigma ab}{b-a} V_o$$

and the current density is

$$J_r(r) = \frac{\sigma abV_o}{(b-a)r^2}$$

The ratio of V_o to the total current I is the total or effective lumped resistance to current flow.

$$R = \frac{V_o}{I} = \frac{b-a}{4\pi\sigma ab}$$

GRAPHS COMMONLY USED IN CALCULUS

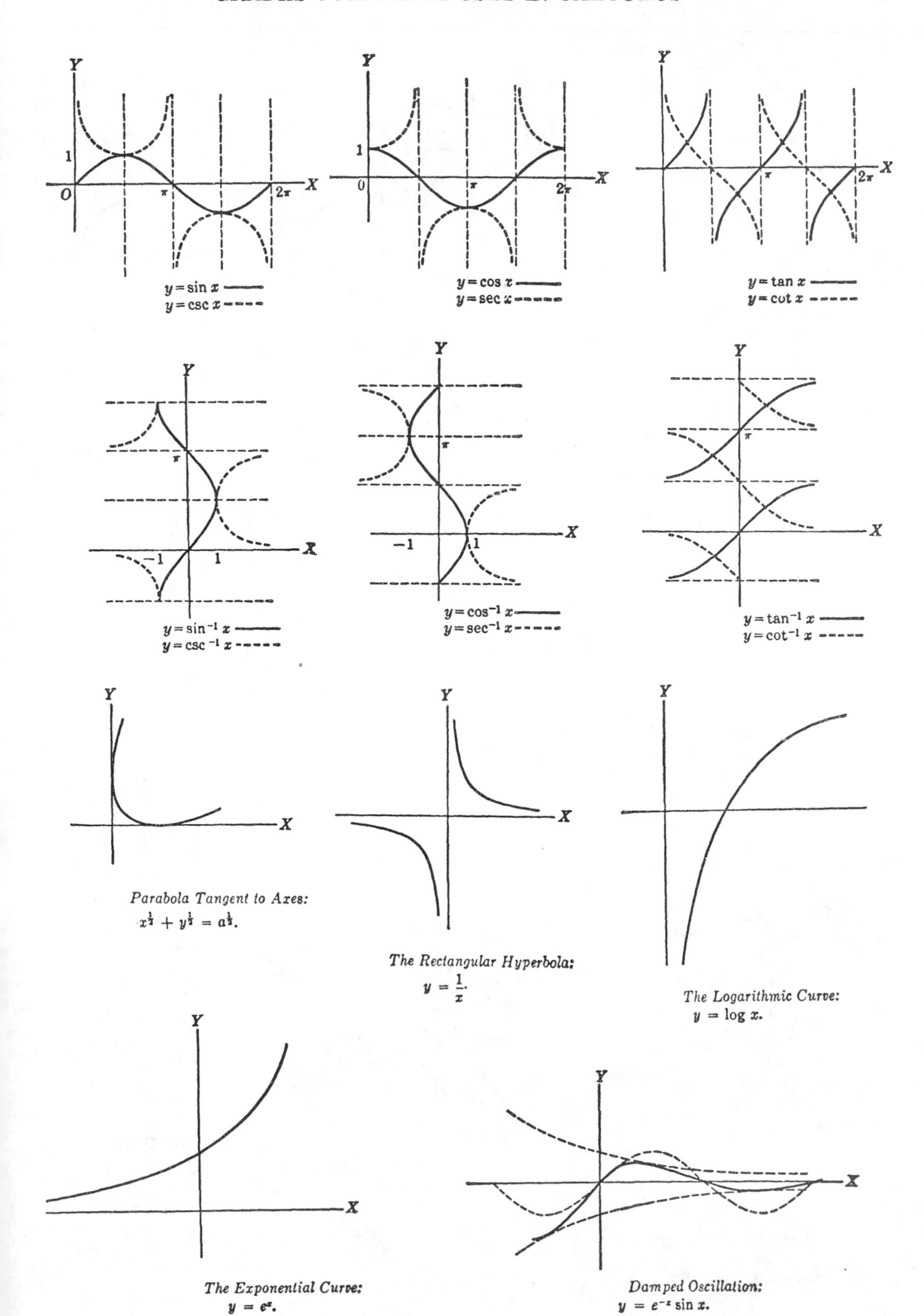

Parabola Tangent to Axes:
$x^{\frac{1}{2}} + y^{\frac{1}{2}} = a^{\frac{1}{2}}$.

The Rectangular Hyperbola:
$y = \frac{1}{x}$.

The Logarithmic Curve:
$y = \log x$.

The Exponential Curve:
$y = e^x$.

Damped Oscillation:
$y = e^{-x} \sin x$.

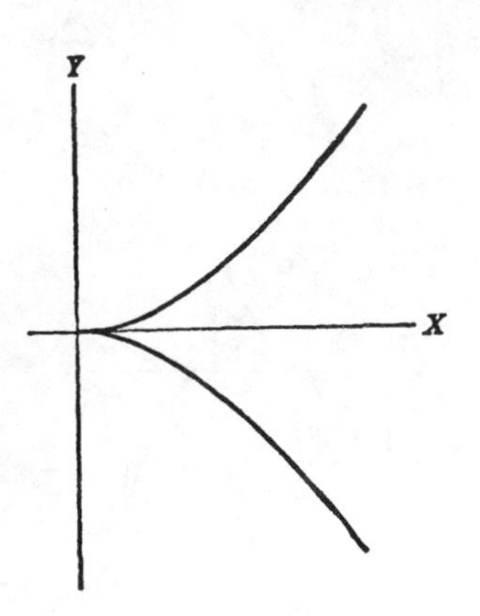

Semicubical Parabola:
$y^2 = ax^3.$

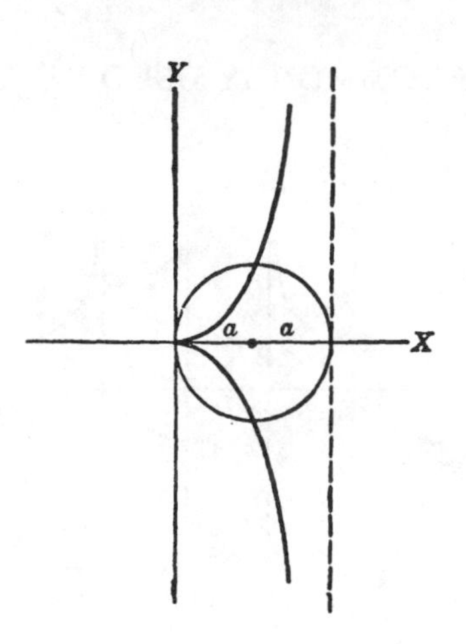

The Cissoid of Diocles:
$$y^2 = \frac{x^3}{2a - x}.$$

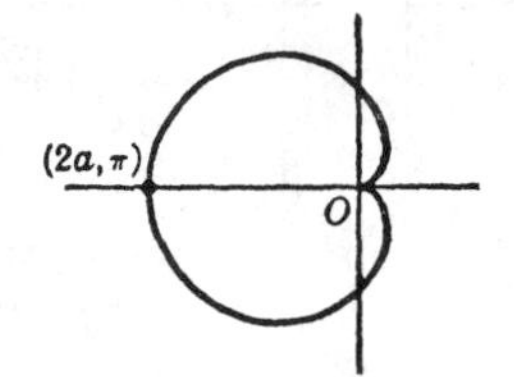

The Cardioid:
$\rho = a(1 - \cos\theta).$

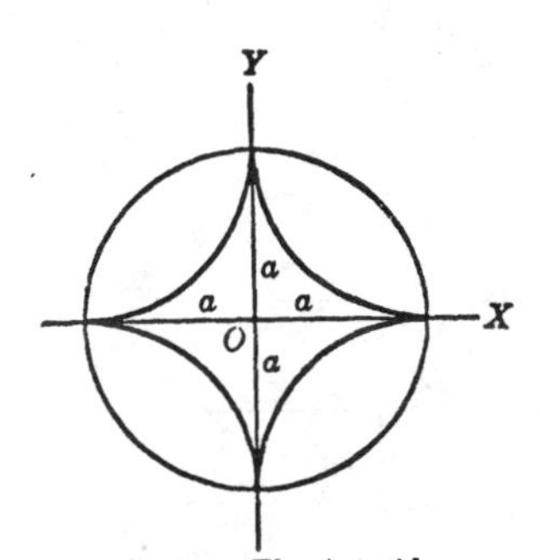

The Astroid:
(Hypocycloid of four cusps)
$x^{\frac{2}{3}} + y^{\frac{2}{3}} = a^{\frac{2}{3}}.$

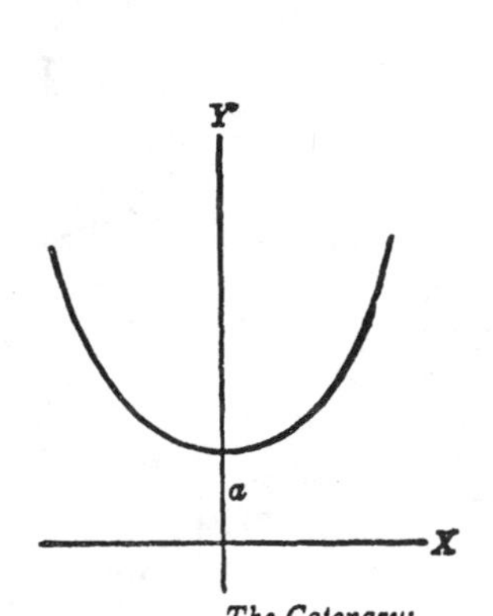

The Catenary:
$$y = \frac{a}{2}\left(e^{\frac{x}{a}} + e^{-\frac{x}{a}}\right).$$

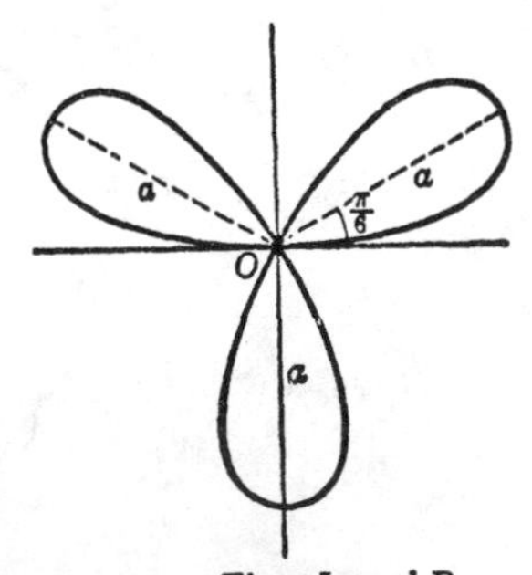

Three-Leaved Rose:
$\rho = a \sin 3\theta.$

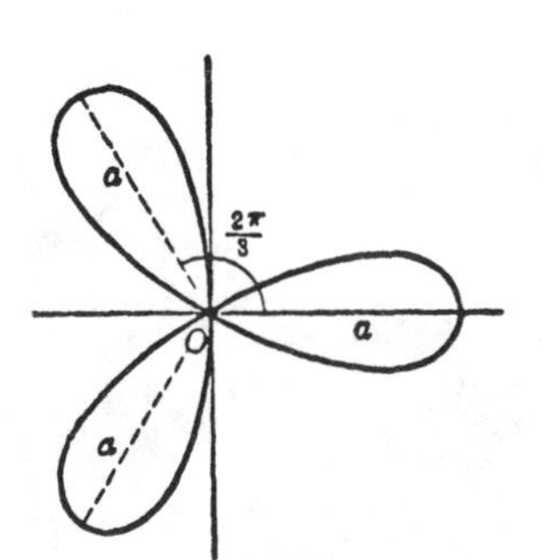

Three-Leaved Rose:
$\rho = a \cos 3\theta.$

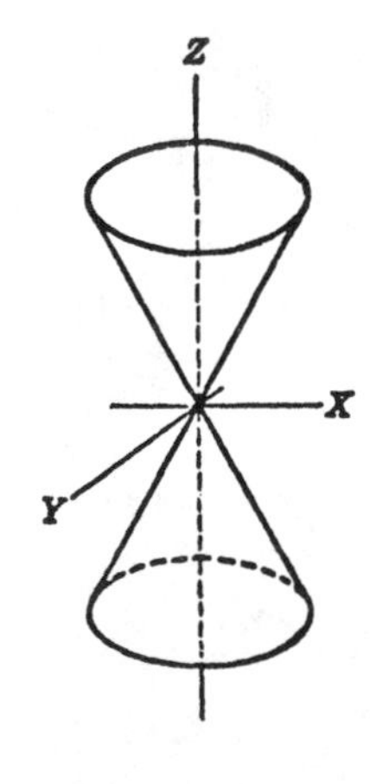

Elliptic Cone:
$$\frac{x^2}{a^2} + \frac{y^2}{b^2} - \frac{z^2}{c^2} = 0$$

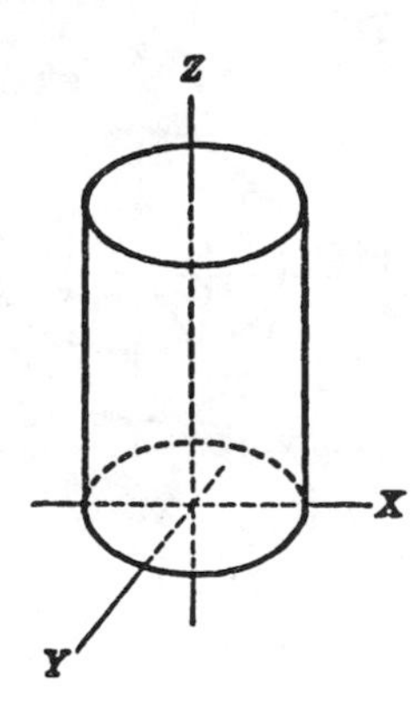

ELLIPTIC CYLINDER:

$$\frac{x^2}{a^2} + \frac{y^2}{b^2} = 1$$

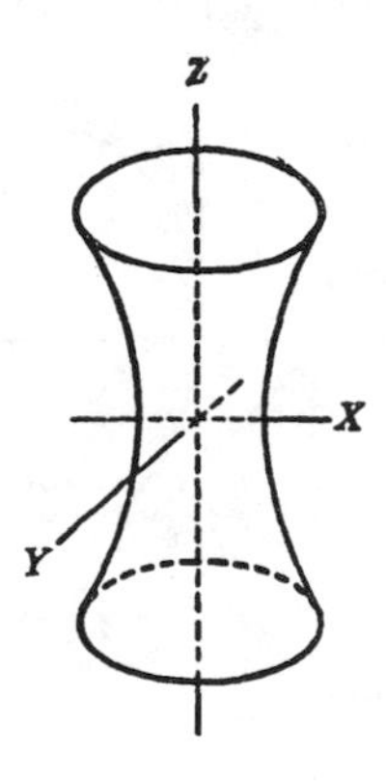

HYPERBOLOID OF ONE SHEET:

$$\frac{x^2}{a^2} + \frac{y^2}{b^2} - \frac{z^2}{c^2} = 1$$

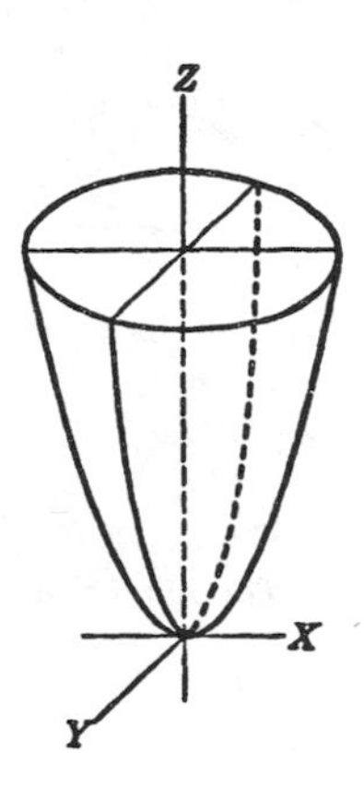

Elliptic Paraboloid:

$$\frac{x^2}{a^2} + \frac{y^2}{b^2} = cz$$

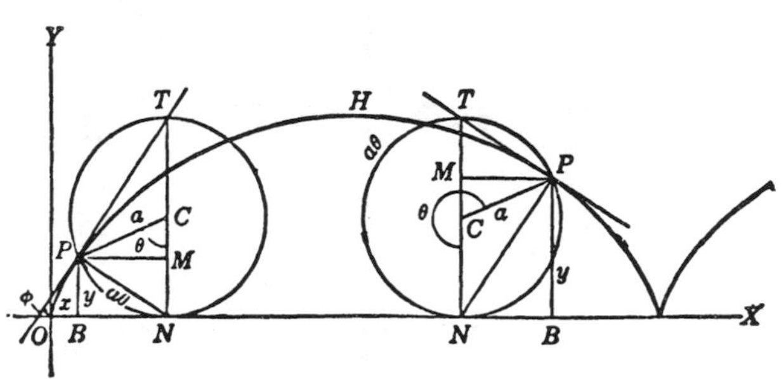

The Cycloid:
$x = a(\theta - \sin\theta), \quad y = a(1 - \cos\theta).$

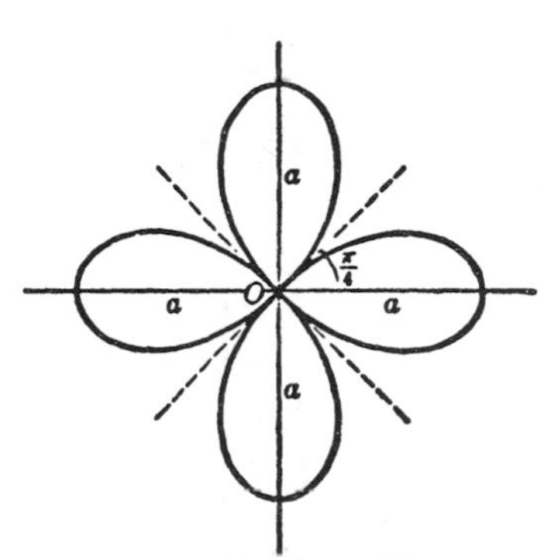

Four-Leaved Rose:
$\rho = a\cos 2\theta.$

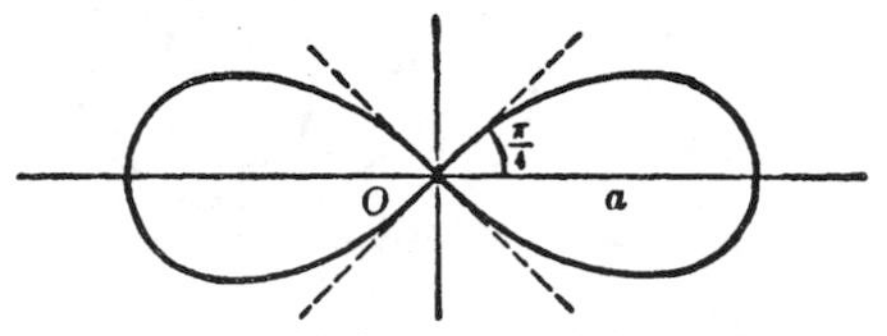

The Lemniscate of Bernoulli:
$\rho^2 = a^2\cos 2\theta;$
$x = a\cos\theta\sqrt{\cos 2\theta}, \quad y = a\sin\theta\sqrt{\cos 2\theta}.$

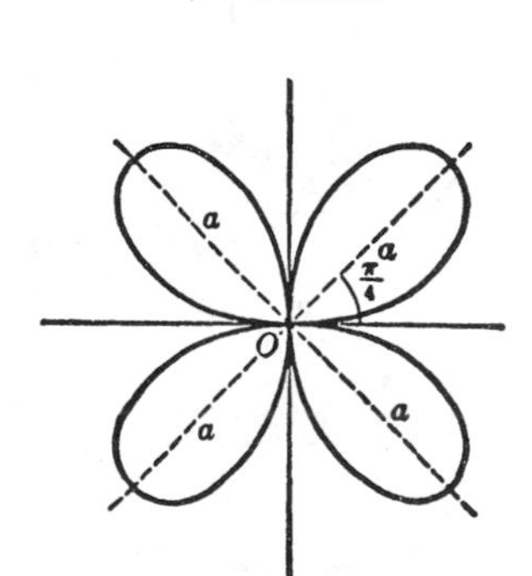

Four-Leaved Rose:
$\rho = a\sin 2\theta.$

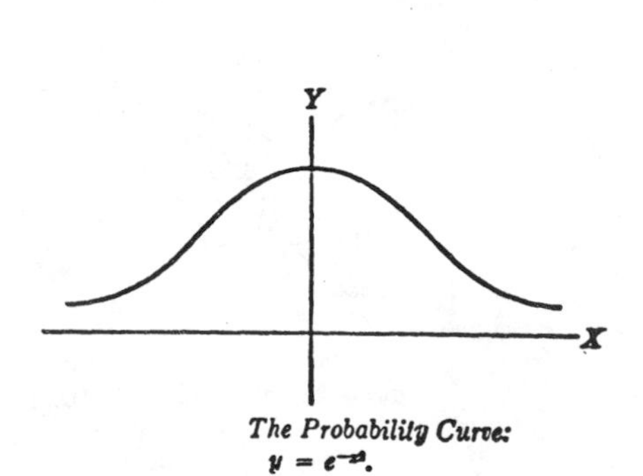

The Probability Curve:
$y = e^{-x^2}.$

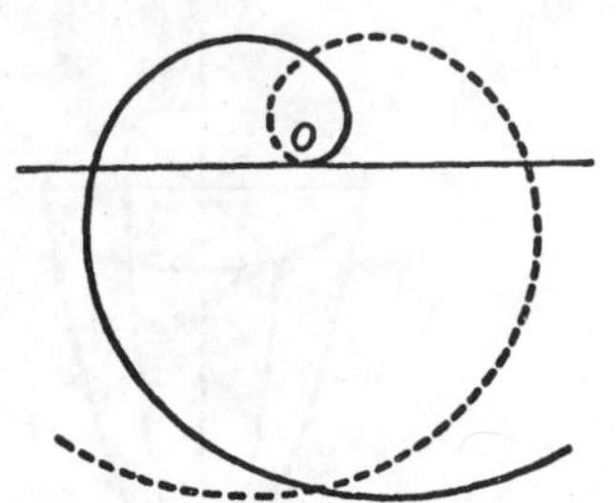

The Spiral of Archimedes: $\rho = a\theta$.

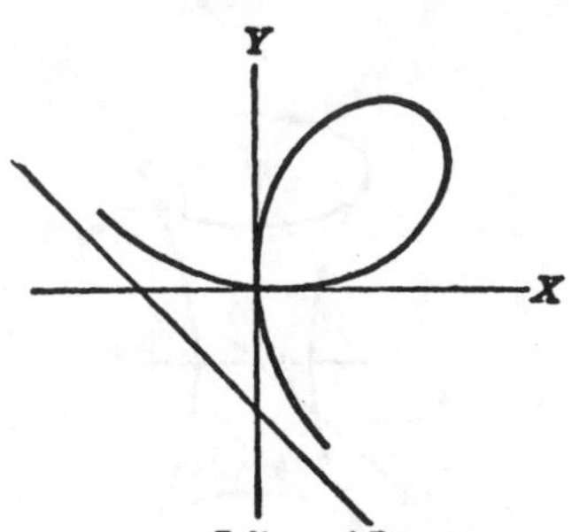

Folium of Descartes: $x^3 + y^3 - 3axy = 0$.

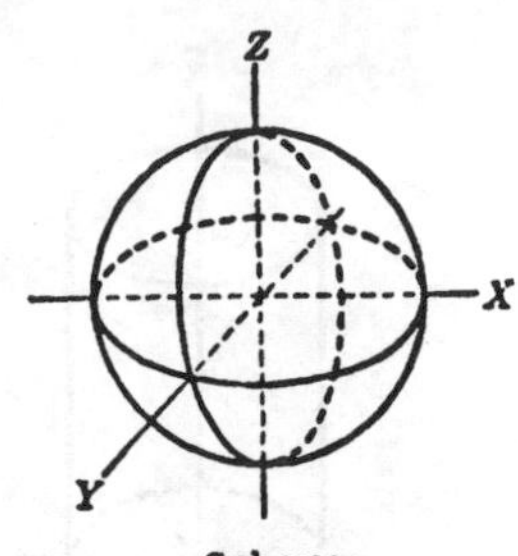

Sphere: $x^2 + y^2 + z^2 = r^2$

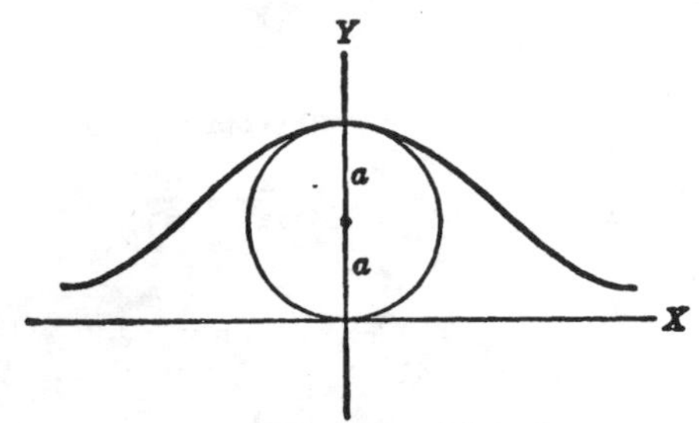

The Witch of Agnesi:

$$y = \frac{8a^3}{4a^2 + x^2}.$$

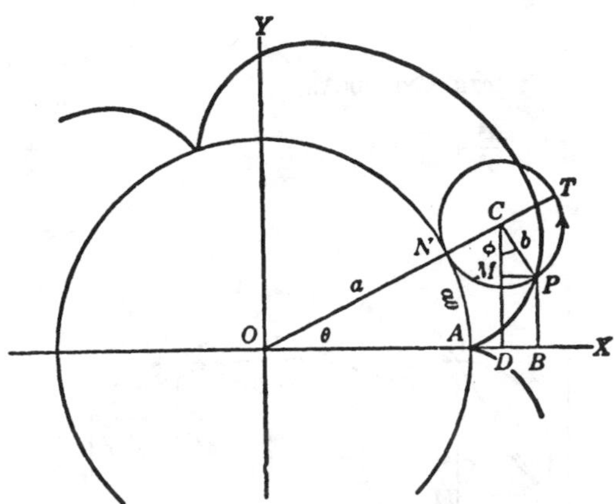

The Epicycloid:

$$x = (a+b)\cos\theta - b\cos\frac{a+b}{b}\theta,$$

$$y = (a+b)\sin\theta - b\sin\frac{a+b}{b}\theta.$$

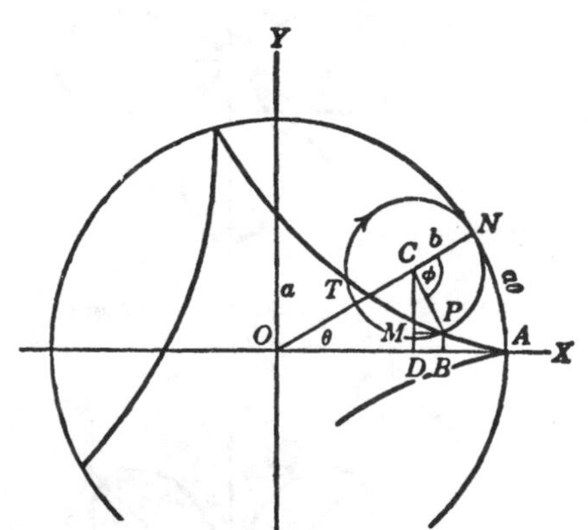

The Hypocycloid:

$$x = (a-b)\cos\theta + b\cos\frac{a-b}{b}\theta,$$

$$y = (a-b)\sin\theta - b\sin\frac{a-b}{b}\theta.$$

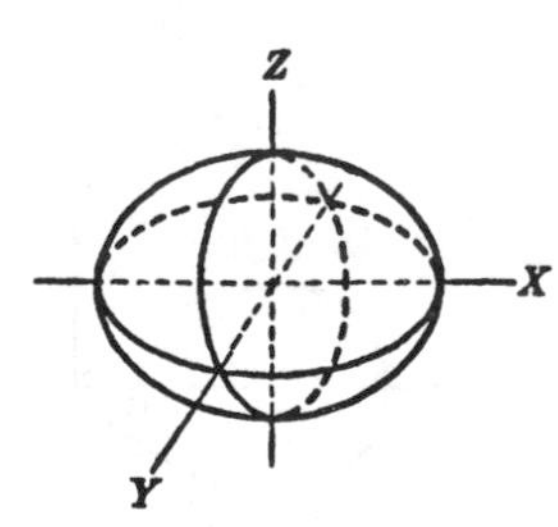

Ellipsoid:

$$\frac{x^2}{a^2} + \frac{y^2}{b^2} + \frac{z^2}{c^2} = 1$$

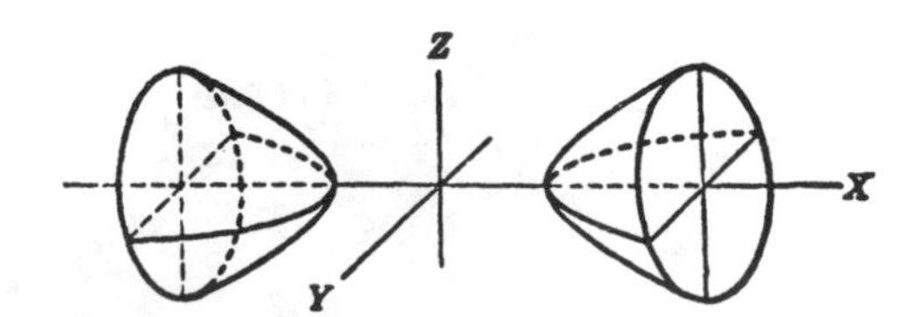

HYPERBOLOID OF TWO SHEETS:

$$\frac{x^2}{a^2} - \frac{y^2}{b^2} - \frac{z^2}{c^2} = 1$$

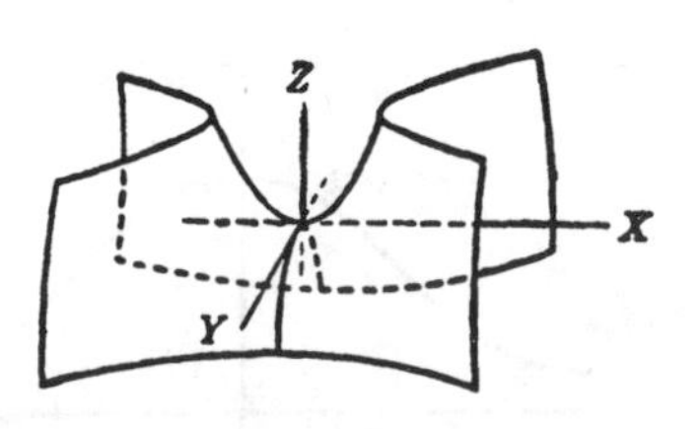

Hyperbolic Paraboloid:

$$\frac{x^2}{a^2} - \frac{y^2}{b^2} = cz$$

INDEX

Numbers on this page refer to PROBLEM NUMBERS, not page numbers

Numbers on this page refer to PROBLEM NUMBERS, not page numbers

Numbers on this page refer to PROBLEM NUMBERS, not page numbers

Numbers on this page refer to PROBLEM NUMBERS, not page numbers

Numbers on this page refer to PROBLEM NUMBERS, not page numbers

Numbers on this page refer to **PROBLEM NUMBERS**, not page numbers

Numbers on this page refer to PROBLEM NUMBERS, not page numbers

Research & Education Association
61 Ethel Road W., Piscataway, NJ 08854
Phone: (732) 819-8880